S0-EJV-372

HEAD AND NECK CANCER

Emerging Perspectives

HEAD AND NECK CANCER

Emerging Perspectives

Edited by

JOHN F. ENSLEY

Department of Medicine, Oncology, and
Otolaryngology—Head and Neck Surgery
Barbara Ann Karmanos Cancer Institute
Wayne State University
Detroit, Michigan

J. SILVIO GUTKIND

Oral and Pharyngeal Cancer Branch
National Institute of Dental and Craniofacial Research
National Institutes of Health
Bethesda, Maryland

JOHN R. JACOBS

Department of Otolaryngology—Head and Neck Surgery
Barbara Ann Karmanos Cancer Institute
Wayne State University
Detroit, Michigan

SCOTT M. LIPPMAN

Department of Clinical Cancer Prevention
The University of Texas M.D. Anderson Cancer Center
Houston, Texas

Amsterdam Boston London New York Oxford Paris
San Diego San Francisco Singapore Sydney Tokyo

This book is printed on acid-free paper. 

Academic Press
An imprint of Elsevier Science.
525 B Street, Suite 1900, San Diego, California 92101-4495, USA
http://www.academicpress.com

Academic Press
84 Theobald's Road, London WC1X 8RR, UK
http://www.academicpress.com

Library of Congress Catalog Card Number: 2002114102

International Standard Book Number: 0-12-239990-0

PRINTED IN THE UNITED STATES OF AMERICA
02 03 04 05 06 07 8 7 6 5 4 3 2 1

Contents

Contributors xiii
Foreword xvii

PART
I
INTRODUCTION

1. Clinical Perspectives in Head and Neck Cancer
JOHN F. ENSLEY

I. Clinical Patterns in Head and Neck Cancer 3
II. Human Cancer Surrogates and Models 5
III. Translational or Correlative Cancer Research 6
IV. Conclusion 6
References 7

2. Head and Neck Cancers with Unusual Natural Histories
HAROLD KIM AND JOHN F. ENSLEY

a. Unknown Primary Carcinoma Metastatic to Cervical Nodes

I. Introduction 9
II. Clinical Presentation 9
III. Evaluation 10
IV. Treatment 12
V. Treatment Results 13
References 15

b. Adenoid Cystic Carcinoma

I. Introduction 16
II. Incidence 16
III. Natural History 17
IV. Pathology 17
V. Evaluation 18
VI. Treatment 19
VII. Results 19
References 21

3. Head and Neck Cancer Imaging
VAL J. LOWE, BRENDAN C. STACK, JR., AND ROBERT E. WATSON, JR.

I. Head and Neck Cancer Imaging 23
II. Anatomic Cross-Sectional Imaging Techniques for the Head and Neck 23
III. Positron Emission Tomography (PET) Imaging Techniques for the Head and Neck 24
IV. Staging Head and Neck Cancer 25
V. Evaluation of Therapy of Head and Neck Cancer 28
VI. Assessment of Recurrent Head and Neck Cancer 31
VII. Conclusion 32
References 32

4. Preneoplastic Lesions of the Upper Aerodigestive Tract
F. LONARDO AND W. SAKR

I. Introduction 35
II. Normal Anatomy 35
III. Pathological Features 37
IV. Clinical Features 45

V. Correlation between Clinical and Histological Findings 47
VI. Malignant Progression 48
References 50

PART

II

BASIC MOLECULAR MECHANISMS

5. Animal Models in Head and Neck Cancer

D. T. WONG, R. TODD, G. SHKLAR, AND A. RUSTGI

I. Introduction 57
II. History of Experimental Head and Neck Cancer Research 57
III. Genetic Models for Head and Neck Cancer Research 59
IV. Integrative Human-Based Discoveries and Animal Model Testing for Head and Neck Cancer Research 61
References 61

6. Adhesion Receptors in Oral Cancer Invasion

BARRY L. ZIOBER AND RANDALL H. KRAMER

I. Introduction 65
II. Cell–Substrate Adhesion 66
III. Extracellular Matrix–Integrins Interactions and Signal Transduction 70
IV. Cadherins and Intracellular Adhesion 71
V. Conclusions 74
References 74

7. Angiogenesis, Basic Mechanisms, and Role in Head and Neck Squamous Cell Carcinoma

ELENA TASSI AND ANTON WELLSTEIN

I. Epidemiology, Pathology, and Biology of Head and Neck Squamous Cell Carcinoma (HNSCC) 81
II. Basic Mechanisms of Angiogenesis 82
III. Angiogenesis in HNSCC 83
IV. Antiangiogenic Therapies in HNSCC 91
V. Conclusions 92
References 92

8. Cell Cycle Regulatory Mechanisms in Head and Neck Squamous Cell Carcinoma

W. ANDREW YEUDALL AND KATHARINE H. WRIGHTON

I. Introduction 101
II. Progression through the Mammalian Cell Cycle 101
III. Cyclin-Dependent Kinase Inhibitors (CKIs) 104
IV. Alterations in Cell Cycle Regulators 108
References 110

9. Oncogenes and Tumor Suppressor Genes in Oral or Head and Neck Squamous Cell Carcinoma

CRISPIAN SCULLY, J. K. FIELD, AND HIDEKI TANZAWA

I. Introduction 117
II. Cancer 118
III. Cell Regulation, Oncogenes, and Tumor Suppressor Genes 118
IV. Tumor Suppressor Genes 120
V. Detection of Individuals at Risk 125
VI. Molecular Diagnosis 125
VII. Staging 125
VIII. Prognostication 125
References 126

10. Proteolysis in Carcinogenesis

THOMAS H. BUGGE

I. Tumor-Associated Protease Systems 137
II. Molecular Targets for Proteases in Tumor Progression 140
III. Extracellular Proteolysis in Head and Neck Cancer Progression 143
References 144

11. Papillomaviruses in Head and Neck Squamous Cell Carcinoma

S. KIM AND E. J. SHILLITOE

I. Introduction 151
II. Human Papillomaviruses (HPVs) 152
III. Interaction between Cellular Host Factors and E6 of High-Risk HPVs 153
IV. Interaction between Cellular Host Factors and E7 of High-Risk HPVs 154
V. Methods of Detection of HPVs 155
VI. HPV and Normal Mucosa of the Upper Aerodigestive Tract 155
VII. HPV and Benign Lesions of the Head and Neck 156

VIII. HPVs and Premalignant Lesions of the Head and Neck 157
IX. HPV and SCCA of Head and Neck 159
X. Conclusions 161
References 162

12. Clinical Correlations of DNA Content Parameters, DNA Ploidy, and S-Phase Fraction in Head and Neck Cancer

Z. MACIOROWSKI AND JOHN F. ENSLEY

I. Introduction 167
II. Technical Considerations 168
III. DNA Content Parameters for Squamous Cell Carcinomas of the Head and Neck 169
IV. DNA Content Parameters and Nonsquamous Cell Carcinomas of the Head and Neck 176
V. Multiparameter Studies 177
VI. Conclusions 178
References 179

PART

III

PREVENTION AND DETECTION

13. Smoking Behavior and Smoking Cessation among Head and Neck Cancer Patients

ROBERT A. SCHNOLL AND CARYN LERMAN

I. Introduction 185
II. Adverse Effects of Smoking for Head and Neck Cancer Patients 186
III. Smoking Rates among Head and Neck Cancer Patients 187
IV. Why Head and Neck Cancer Patients May Have Trouble Quitting Smoking 188
V. Stage of the Science in Smoking Cessation Interventions 192
VI. Directions for Future Research 195
References 196

14. Nutrients, Phytochemicals, and Squamous Cell Carcinoma of the Head and Neck

OMER KUCUK AND ANANDA PRASAD

I. Introduction 201
II. Tobacco and Nutrients 202
III. Alcohol and Nutrients 203
IV. Cancer Treatment and Nutrients 203
V. Nutritional Consequences of Radiation and Chemotherapy 203
VI. Effects of Selected Micronutrients on Radiation and Chemotherapy Toxicity 204
VII. Nutrients and Antitumor Effects of Radiation/Chemotherapy 205
VIII. Nutrients and Pharmaceuticals in the Prevention of Radiation and Chemotherapy Toxicity 206
IX. Conclusions 208
References 208

15. Molecular Epidemiology of Head and Neck Cancer

QINGYI WEI, PETER SHIELDS, ERICH M. STURGIS, AND MARGARET R. SPITZ

I. Introduction 213
II. Epidemiology of Head and Neck Cancer 213
III. Molecular Epidemiology of Head and Neck Cancer 215
IV. Xenobiotic Metabolism of Carcinogens in Head and Neck Cancer 215
V. DNA Repair Phenotype and Risk of Head and Neck Cancer 216
VI. Polymorphisms in DNA Repair Genes 219
VII. Polymorphisms in Cell Cycle Genes 220
References 221

16. Head and Neck Field Carcinogenesis

WALTER N. HITTELMAN

I. Introduction 227
II. Clinical Evidence for Head and Neck Field Cancerization 229
III. Histopathologic Evidence of Field Cancerization 230
IV. Genetic Evidence for Field Cancerization 231
V. Phenotypic Changes Associated with Field Carcinogenesis 236
VI. Clinical Implication of Field Cancerization 236
References 238

17. Molecular Markers of Oral Premalignant Lesion Risk

MIRIAM P. ROSIN, LEWEI ZHANG, AND CATHERINE POH

I. Introduction 245
II. Oral Premalignant Lesions (OPLs): Traditional and Evolving Definitions 246

III. Lack of a Reliable System for Predicting Progression Risk of OPLs 247
IV. Lack of a Consensus on Treatment for OPLs 247
V. Use of Microsatellite Analysis to Identify Oral Mucosal Regions at Risk for Progression 248
VI. Use of Molecular Markers to Manage OPLs 251
VII. Exfoliated Cell Sampling: A Noninvasive Method for Identification, Risk Prediction, and Management of OPLs 253
VIII. The Need for Longitudinal Studies 255
References 256

18. Chemoprevention of Oral Premalignant Lesions

SUSAN T. MAYNE, BRENDA CARTMEL, AND DOUGLAS E. MORSE

I. Oral Premalignant Lesions: Definition and Significance 261
II. Chemoprevention Trials: General Design and Outcome Assessment 262
III. Review of Trials 263
IV. Validity of Oral Precancerous Lesions in Predicting Efficacy of Agents for Oral Cancer Prevention 266
V. Summary and Conclusions 267
References 268

19. Aerodigestive Tract Chemoprevention Trials and Prevention of Second Primary Tumors

EDWARD S. KIM AND FADLO R. KHURI

I. Introduction 271
II. Epidemiology 272
III. Natural History 272
IV. Risk Factors 273
V. Biology of Head and Neck Cancer 275
VI. Chemoprevention 276
VII. Chemoprevention Trials 276
VIII. Biochemoprevention 281
IX. Summary: Future Directions 281
References 281

20. Statistical Methods for Biomarker Analysis for Head and Neck Carcinogenesis and Prevention

J. JACK LEE

I. Introduction 287
II. Type of Biomarker Measures 288
III. Standard Statistical Methods for Analyzing Biomarkers 289
IV. Distribution of Continuous Biomarkers 290
V. Choosing Cut Points for Continuous Biomarkers 291
VI. Analysis of Surrogate End Point Biomarkers 292
VII. Predicting Cancer Development Using Clinical, Histological, Epidemiologic, and Multiple Biomarker Information 295
VIII. Summary and Design Considerations for Biomarker-Integrated Translational Studies 301
References 302

21. Molecular Detection of Head and Neck Cancer

WAYNE M. KOCH AND DAVID SIDRANSKY

I. Introduction 305
II. Detection of Head and Neck Squamous Cell Carcinoma in Exfoliated Cell Samples 306
III. Detection of Circulating Tumor Markers 310
IV. Microarrays 311
V. Spectroscopic Analysis 311
VI. Diagnostic or Prognostic Biomarkers in Biopsy Material 311
VII. Conclusion 312
References 312

PART IV

CURRENT APPROACHES

Surgical Considerations

22. Management of the Neck in Squamous Cell Carcinomas of the Head and Neck

JESUS E. MEDINA

I. Introduction 317
II. Incidence of Cervical Metastases 317
III. Patterns of Lymphatic Spread 319
IV. Prognostic Implications of Neck Node Metastases 320
V. Current Management of Lymph Nodes of the Neck 321
VI. Types of Neck Dissection 324
References 325

23. Management of the Carotid Artery in Advanced Head and Neck Cancer

JOSE E. OTERO-GARCIA, GEORGE H. YOO, AND JOHN R. JACOBS

I. Management of the Carotid Artery in Advanced Head and Neck Cancer 329
II. Patent Presentation: Clinical and Radiological Diagnosis 329
III. Clinical Outcome 330
IV. Nonsurgical Treatment Options 331
V. Surgical Treatment Options 332
VI. Preoperative Evaluation 332
VII. Surgical Procedure 333
VIII. Postoperative Complications 335
IX. Conclusion 335
References 336

24. Skull Base Surgery

TERRY Y. SHIBUYA, WILLIAM B. ARMSTRONG, AND JACK SHOHET

I. Introduction 339
II. Evaluation of Patient 340
III. Anesthetic Considerations 341
IV. Physiological Monitoring 342
V. Pathology 342
VI. Surgery of Skull Base 346
VII. Complications 355
VIII. Conclusion 356
References 356

25. Management of Laryngeal Cancer

R. KIM DAVIS

I. Introduction 359
II. Treatment of Carcinoma *in Situ* and Minimally Invasive T_1 Glottic Cancer 360
III. Treatment of T_1 Glottic Cancer 361
IV. Treatment of T_2 Glottic Cancer 363
V. Treatment of T_3 and T_4 Glottic Cancer 366
VI. Treatment of Supraglottic Cancer 369
VII. Summary 371
References 372

26. Partial Laryngeal Procedures

GREGORY S. WEINSTEIN

I. Introduction 375
II. Basic Concepts: Beyond Conservation Laryngeal Surgery to Organ Preservation Surgery of the Larynx 376
III. Principle One: Nonsurgical Organ Preservation Strategies versus Organ Preservation Surgery 377
IV. Principle Two: The Cricoarytenoid Unit 380
V. Principle Three: The Spectrum Concept of Laryngeal Carcinoma and Laryngeal Cancer Surgery 381
VI. Principle Four: Resection of Normal Tissue to Maintain Postoperative Function 382
VII. Principle Five: The Importance of Local Control 383
VIII. Organ Preservation Surgical Techniques 383
IX. Organ Preservation Surgery for Carcinomas Arising at the Glottic Level 384
X. Organ Preservation Surgery for Carcinomas Arising at the Supraglottic Level 386
XI. Summary 388
References 388

27. High-Dose Intraarterial Chemotherapy/Radiation for Advanced Head and Neck Cancer

K. THOMAS ROBBINS

I. Introduction 393
II. RADPLAT Drug Delivery Technique 394
III. RADPLAT Results 395
IV. RADPLAT Experience at Other Centers 398
V. New Directions Using the RADPLAT Concept 399
VI. Conclusions 402
References 403

28. Thyroid Cancer

STEPHEN Y. LAI AND RANDAL S. WEBER

I. Introduction 405
II. Molecular Basis for Thyroid Cancer 406
III. Risk Factors and Etiology 406
IV. Tumor Staging/Classification 407
V. Evaluation of a Thyroid Nodule 408
VI. Review of Thyroid Cancers 412
VII. Surgical Management and Technique 419
VIII. Postoperative Management and Special Considerations 426
IX. Conclusions 427
References 427

29. Assessment of Outcomes in Head and Neck Cancer

URJEET PATEL AND JAY PICCIRILLO

I. Background 433
II. Survival 434

III. Determinant Survival 434
IV. Actuarial Survival 435
V. Kaplan–Meier Survival 436
VI. Morbidity 436
VII. Health Status 436
VIII. Quality of Life 438
IX. Cost of Care 441
X. Summary 441
References 442

Nonsurgical Considerations

30. Combination of Chemotherapy with Radiation for Head and Neck Cancer

NAOMI R. SCHECHTER AND K. KIAN ANG

I. Introduction 445
II. Radiotherapy Fractionation Regimes 445
III. Goals of Combining Chemotherapy and Radiation 447
IV. Randomized Studies of Induction Chemotherapy Followed by Locoregional Therapy 447
V. Randomized Trials Addressing Concurrent Chemotherapy with Radiation 449
VI. Meta-Analyses Addressing the Role of Chemotherapy 454
VII. Summary of Available Trial Results 456
VIII. Future Research Directions 457
References 459

31. Clinical Trials in Advanced Unresectable Head and Neck Cancer

DAVID J. ADELSTEIN

I. Altered Fractionation Radiation 462
II. Induction Chemotherapy 463
III. Alternating Chemotherapy and Radiation 466
IV. Concurrent Chemotherapy and Radiation 466
V. Future Directions 469
References 470

32. Organ Preservation in Head and Neck Cancer

ARLENE A. FORASTIERE AND MAURA GILLISON

I. Introduction 475
II. Laryngeal Preservation: Cancers of the Larynx and Hypopharynx 475
III. Role of Chemotherapy in Preservation of the Oropharynx 478
References 489

33. Nasopharyngeal Cancer

P. G. SHANKIR GIRI AND MUHYI AL SARRAF

I. Introduction 491
II. Epstein–Barr Virus 491
III. Genetics 492
IV. Environmental Factors 492
V. Anatomy 492
VI. Pathology 492
VII. Natural History 493
VIII. Workup 493
IX. Pretreatment Evaluation 493
X. Staging 493
XI. Radiation Therapy 494
XII. Volume Treated 494
XIII. Boost Volume 494
XIV. Dose Response 494
XV. Reirradiation 495
XVI. Survival 496
XVII. Chemotherapy 496
XVIII. Chemotherapy for Recurrent/Metastatic Disease 497
XIX. Chemotherapy for Locally Advanced and Previously Untreated Nasopharyngeal Cancer 498
XX. Induction Chemotherapy 498
XXI. Concurrent Chemoradiotherapy 499
XXII. Adjuvant Chemotherapy 500
XXIII. Randomized Trials 500
XXIV. Conclusion 501
References 501

PART

NOVEL APPROACHES

34. Squamous Carcinomas of the Head and Neck: Novel Genomic Approaches

YYOMESH PATEL, CHIDCHANOK LEETHANAKUL, PANOMWAT AMORNPHIMOLTHAM, AND J. SILVIO GUTKIND

I. Introduction 509
II. Genetic Alterations in Head and Neck Squamous Cell Carcinomas (HNSCC) 510
III. Techniques Used for Molecular Profiling of HNSCC 510
IV. Use of Gene Assays to Evaluate Gene Expression Profiles in HNSCC 514

V. Gene Discovery Efforts in HNSCC: The Cancer Genome Anatomy Project 514
VI. Proteomics 518
VII. Conclusion 519
References 519

35. Molecular Assessment of Surgical Margins

WAYNE M. KOCH

I. Introduction 523
II. Insight from the Molecular Biology Revolution 524
III. The Margin Dilemma 524
IV. Technical Factors Influencing Margin Assessment 525
V. Is Margin Status Reflective of Biologic Aggressiveness? 527
VI. Theoretical Basis for Molecular Margin Analysis 528
VII. Published Molecular Margin Studies 528
VIII. Clinical Response to Positive Margins 530
IX. Concluding Remarks 531
References 532

36. Novel Agents and Modalities for the Treatment of Squamous Carcinoma of the Head and Neck

ADRIAN M. SENDEROWICZ, CARTER VAN WAES, JANET DANCEY, AND BARBARA CONLEY

I. Novel Targets and Agents for the Treatment of Advanced Head and Neck Squamous Cell Carcinoma (HNSCC) 535
II. Conclusions 547
References 547

37. Gene Therapy for Patients with Head and Neck Cancer

GEORGE H. YOO AND GARY CLAYMAN

I. Background 556
II. Approaches to Gene Therapy for Cancer 557
III. Genes Used in Combination Therapy 557
IV. Vectors 558
V. Gene Transfer Delivery Sites 559
VI. Gene Therapy Trials in Head and Neck Squamous Cell Carcinoma 560
VII. Future of Gene Therapy for Cancer 565
References 566

38. Immunology and Immunotherapy of Head and Neck Cancer

TERRY Y. SHIBUYA, LAWRENCE G. LUM, THOMASZ PAWLOWSKI, AND THERESA L. WHITESIDE

I. Introduction 569
II. Principles of Immunotherapy 570
III. Immune Therapeutic Strategies 571
IV. Immunology of Head and Neck Squamous Cell Carcinoma (HNSCC) 574
V. Immunotherapy of Head and Neck Cancer 576
VI. Conclusion 581
References 581

Index 593

Contributors

Numbers in parentheses indicate the pages on which the authors' contributions begin.

David J. Adelstein (461) Department of Hematology and Medical Oncology, Cleveland Clinic Foundation, Cleveland, Ohio 44195

Muhyi Al Sarraf (491) Department of Radiation Oncology, Eastern Virginia Medical School, Norfolk, Virginia 23507

Panomwat Amornphimoltham (509) Oral and Pharyngeal Cancer Branch, National Institute of Dental and Craniofacial Research, National Institutes of Health, Bethesda, Maryland 20892

K. Kian Ang (445) Department of Radiation Oncology, The University of Texas M.D. Anderson Cancer Center, Houston, Texas 77030

William B. Armstrong (339) Department of Otolaryngology—Head and Neck Surgery, University of California, Irvine, College of Medicine and Craniofacial/Skull Base Surgery Center, University of California, Irvine, Medical Center, Orange, California 92868

Thomas H. Bugge (137) Oral and Pharyngeal Cancer Branch, National Institute of Dental and Craniofacial Research, National Institutes of Health, Bethesda, Maryland 20892

Brenda Cartmel (261) Department of Epidemiology and Public Health,Yale University School of Medicine, New Haven, Connecticut 06520

Gary Clayman (555) Department of Surgery, The University of Texas M.D. Anderson Cancer Center, Houston, Texas 77030

Barbara Conley (535) Cancer Treatment Evaluation Program, National Cancer Institute, National Institutes of Health, Bethesda, Maryland 20892

Janet Dancy (535) Cancer Treatment Evaluation Program, National Cancer Institute, National Institutes of Health, Bethesda, Maryland 20892

R. Kim Davis (359) Division of Otolaryngology—Head and Neck Surgery, University of Utah School of Medicine, Salt Lake City, Utah 84132

John F. Ensley (3, 9, 167) Department of Medicine, Oncology, and Otolaryngology—Head and Neck Surgery, Karmanos Cancer Institute, Wayne State University, Detroit, Michigan 48201

J. K. Field (117) Molecular Genetics and Oncology Group, Clinical Dental Sciences, University of Liverpool and Roy Castle International Centre for Lung Cancer Research, Liverpool L69 3BX, United Kingdom

Arlene A. Forstiere (475) Department of Oncology, Johns Hopkins University School of Medicine, Baltimore, Maryland 21231

Maura Gillison (475) Department of Oncology, Johns Hopkins University School of Medicine, Baltimore, Maryland 21231

P. G. Shankir Giri (491) Department of Radiation Oncology, Eastern Virginia Medical School, Norfolk, Virginia 23507

J. Silvio Gutkind (509) Oral and Pharyngeal Cancer Branch, National Institute of Dental and Craniofacial Research, National Institutes of Health, Bethesda, Maryland 20892

Walter N. Hittelman (227) Department of Experimental Therapeutics, The University of Texas M.D. Anderson Cancer Center, Houston, Texas 77030

John R. Jacobs (329) Department of Otolaryngology—Head and Neck Surgery, Karmanos Cancer Institute, Wayne State University, Detroit, Michigan 48201

Fadlo R. Khuri (271) Department of Thoracic/Head and Neck Medical Oncology, The University of Texas M.D. Anderson Cancer Center, Houston, Texas 77030

Edward S. Kim (271) Department of Thoracic/Head and Neck Medical Oncology, The University of Texas M.D. Anderson Cancer Center, Houston, Texas 77030

Harold Kim (9) Department of Radiation Oncology, Karmanos Cancer Institute, Wayne State University, Detroit, Michigan 48201

S. Kim (151) Department of Otolaryngology and Communication Sciences, SUNY Upstate Medical University, Syracuse, New York 13210

Wayne M. Koch (305, 523) Department of Otolaryngology—Head and Neck Surgery, Johns Hopkins University School of Medicine, Baltimore, Maryland 21286

Randall H. Kramer (65) Oral Cancer Center—Department of Stomatology, School of Dentistry, University of California, San Francisco, San Francisco, California 94143

Omer Kucuk (201) Karmanos Cancer Institute, Wayne State University, Detroit, Michigan 48201

Stephen Y. Lai (405) Department of Otolaryngology—Head and Neck Surgery, University of Pennsylvania Medical Center, Philadelphia, Pennsylvania 19104

J. Jack Lee (287) Department of Biostatistics, The University of Texas M.D. Anderson Cancer Center, Houston, Tesax 77030

Chidchanok Leethanakul (509) Oral and Pharyngeal Cancer Branch, National Institute of Dental and Craniofacial Research, National Institutes of Health, Bethesda, Maryland 20892

Caryn Lerman (185) University of Pennsylvania Health Sciences, Philadelphia, Pennsylvania 19104

F. Lonardo (35) Department of Pathology, Wayne State University, School of Medicine, Detroit, Michigan 48201

Val J. Lowe (23) Department of Radiology, Mayo Clinic and Foundation, Rochester, Minnesota 55905

Lawrence G. Lum (569) Roger Williams Cancer Center, Providence, Rhode Island 02908

Z. Maciorowski (167) Institut Curie, Paris 75005, France

Susan T. Mayne (261) Department of Epidemiology and Public Health,Yale University School of Medicine, New Haven, Connecticut 06520

Jesus E. Medina (317) Department of Otorhinolaryngology, The University of Oklahoma Health Sciences Center, Oklahoma City, Oklahoma 43190

Douglas E. Morse (261) New York University College of Dentistry, New York, New York 10010

Jose E. Otero-Garcia (329) Department of Otolaryngology—Head and Neck Surgery, Karmanos Cancer Institute, Wayne State University, Detroit, Michigan 48201

Urjeet Patel (433) Department of Otolaryngology, Washington University, St. Louis, Missouri 63110

Vyomesh Patel (509) Oral and Pharyngeal Cancer Branch, National Institute of Dental and Craniofacial Research, National Institutes of Health, Bethesda, Maryland 20892

Tomasz Pawlowski (569) Department of Pathology, University of California, Irvine, College of Medicine, Orange, California 92868

Jay Piccirillo (433) Department of Otolaryngology, Washington University, St. Louis, Missouri 63110

Catherine Poh (245) School of Kinesiology, Simon Fraser University, Burnaby, British Columbia, Canada V5A 1S6 and Faculty of Dentistry, University of British Columbia, Vancouver, British Columbia, Canada V6T 1Z3

Ananda Prasad (201) Karmanos Cancer Institute, Wayne State University, Detroit, Michigan 48201

K. Thomas Robbins (393) Department of Otolaryngology—Head and Neck Surgery, University of Florida, Gainesville, Florida

Miriam R. Rosin (245) British Columbia Cancer Agency/Cancer Research Centre, Vancouver, British Columbia, Canada V5Z 4E6 and School of Kinesiology, Simon Fraser University, Burnaby, British Columbia, Canada V5A 1S6

A. Rustgi (57) Laboratory of Molecular Pathology, Harvard University, School of Dental Medicine, Boston, Massachusetts 02115

W. Sakr (35) Department of Pathology, Wayne State University, School of Medicine, Detroit, Michigan 48201

Naomi R. Schechter (445) Department of Radiation Oncology, The University of Texas M.D. Anderson Cancer Center, Houston, Texas 77030

Robert A. Schnoll (185) Fox Chase Cancer Center, Cheltenham, Pennsylvania 19012

Crispian Scully (117) Eastman Dental Institute for Oral Health Care Sciences, University College London, University of London, London WC1X 8LD, United Kingdom

Adrian M. Senderowicz (535) Molecular Therapeutics Unit, Oral and Pharyngeal Cancer Branch, National Institute of Dental and Craniofacial Research, National Institutes of Health, Bethesda, Maryland 20892

Terry Y. Shibuya (339, 569) Department of Otolaryngology—Head and Neck Surgery and the Chao Family Comprehensive Cancer Center, University of California, Irvine, College of Medicine and Craniofacial/Skull Base Surgery Center, University of California, Irvine, Medical Center, Orange, California 92868

Peter Shields (213) Cancer Genetics and Epidemiology Program, Lombardi Cancer Center, Georgetown University Medical Center, Washington, DC 2007

E. J. Shillitoe (151) Department of Microbiology and Immunology, SUNY Upstate Medical University, Syracuse, New York 13210

G. Shklar (57) Laboratory of Molecular Pathology, Harvard University, School of Dental Medicine, Boston, Massachusetts 02115

Jack Shohet (339) Department of Otolaryngology—Head and Neck Surgery, University of California, Irvine, College of Medicine and Craniofacial/Skull Base Surgery Center, University of California, Irvine, Medical Center, Orange, California 92868

David Sidransky (305) Department of Otolaryngology—Head and Neck Surgery, Johns Hopkins University School of Medicine, Baltimore, Maryland 21286

Margaret R. Spitz (213) Department of Epidemiology, The University of Texas M.D. Anderson Cancer Center, Houston, Texas 77030

Brendan C. Stack, Jr. (23) Department of Otolaryngology—Head and Neck Cancer, Pennsylvania State University, College of Medicine, Hershey, Pennsylvania

Erich M. Sturgis (213) Department of Epidemiology, The University of Texas M.D. Anderson Cancer Center, Houston, Texas 77030

Hideki Tanzawa (117) Department of Oral Surgery, Chiba University, Chiba 260-8670, Japan

Elena Tassi (81) Lombardi Cancer Center, Georgetown University, Washington, DC 20007

R. Todd (57) Laboratory of Molecular Pathology, Harvard University, School of Dental Medicine, Boston, Massachusetts 02115

Carter Van Waes (535) National Institute of Deafness and Communication Disorders, National Institutes of Health, Bethesda, Maryland 20892

Robert E. Watson, Jr. (23) Department of Radiology, Mayo Clinic and Foundation, Rochester, Minnesota 55905

Randal S. Weber (405) Department of Otolaryngology—Head and Neck Surgery, University of Pennsylvania Medical Center, Philadelphia, Pennsylvania 19104

Qingyi Wei (213) Department of Epidemiology, The University of Texas M.D. Anderson Cancer Center, Houston, Texas 77030

Gregory S. Weinstein (375) Department of Otolaryngology—Head and Neck Surgery, The Center of Head and Neck Cancer, The University of Pennsylvania Medical Center, Philadelphia, Pennsylvania 19104

Anton Wellstein (81) Lombardi Cancer Center, Georgetown University, Washington, DC 20007

Theresa L. Whiteside (569) Departments of Otolaryngology—Head and Neck Surgery and Pathology, Pittsburgh Cancer Institute, Pittsburgh, Pennsylvania 15213

D. T. Wong (57) Laboratory of Molecular Pathology, Harvard University, School of Dental Medicine, Boston, Massachusetts 02115

Katharine H. Wrighton (101) Molecular Carcinogenesis Group, Head and Neck Cancer Program, Guy's King's and St. Thomas' Schools of Medicine and Dentistry, King's College London, London SE1 9RT, United Kingdom

W. Andrew Yeudall (101) Molecular Carcinogenesis Group, Head and Neck Cancer Program, Guy's King's and St. Thomas' Schools of Medicine and Dentistry, King's College London, London SE1 9RT, United Kingdom

George H. Yoo (329, 555) Department of Otolaryngology—Head and Neck Surgery, Karmanos Cancer Institute, Wayne State University, Detroit, Michigan 48201

Lewei Zhang (245) Faculty of Dentistry, University of British Columbia, Vancouver, British Columbia, Canada V6T 1Z3

Barry L. Ziober (65) Department of Otorhinolaryngology—Head and Neck Surgery, University of Pennsylvania, Philadelphia, Pennsylvania 19104

Foreword

You hold in your hands an important new volume detailing the most advanced, cutting-edge research that is changing the way we diagnose, treat, and prevent head and neck cancer. These cancers are among the leading cause of death from cancer worldwide, and the survival rate has stubbornly refused to change to any significant degree. However, new approaches are finally beginning to offer realistic hope for improvement in these patients' survival rates and in their quality of life.

You will find here a highly comprehensive review of the state-of-the-art clinical, basic, and translational research in this area, with contributions from some of today's most esteemed head and neck cancer scientists. I recommend that it is worthwhile to purchase it, read it, and keep it on your bookshelf for frequent reference. It's a guide to the most effective, cutting-edge treatments available today, and a map to the future of head and neck cancer—to the prevention, early detection, diagnosis, treatment, and survival of our patients.

Waun Ki Hong, M.D.
Head, Division of Cancer Medicine
Chair, Department of Thoracic/Head and
Neck Medical Oncology
M. D. Anderson Cancer Center

PART I

INTRODUCTION

C H A P T E R

1

Clinical Perspectives in Head and Neck Cancer

JOHN F. ENSLEY

Department of Medicine,
Oncology, and Otolaryngology—Head and Neck Surgery
Karmanos Cancer Institute
Wayne State University
Detroit, Michigan 48201

I. Clinical Patterns in Head and Neck Cancer 3
A. TNM Differences 3
B. Histopathology Differences 4
C. Treatment Outcome Differences 4
D. Correlative Parameters and Outcome 5
II. Human Cancer Surrogates and Models 5
III. Translational or Correlative Cancer Research 6
IV. Conclusion 6
References 7

At first glance, head and neck cancers appear to be a heterogeneous group of tumors consisting of multiple histopathologies, primary sites, TNM stages, natural histories, and treatment outcomes [1]. However, as one gains clinical experience with these cancers, predictable and stable groupings, with recognizable patterns of tumor/host behavior, emerge. For example, approximately 70% of patients with stage T2N0 squamous cell carcinomas (SCC) of the endolarynx are cured when treated with full course, conventionally administered, radiation therapy and 30% are not. Such observations suggest that there are fundamental differences in the mechanisms that control the phenotypic expression of these tumor subgroups and that these differences are not evident at the time the disease is diagnosed. Because these variations in the natural history and treatment outcome certainly result from differences in underlying pathophysiological mechanisms, they provide a unique and identifiable opportunity for experimental investigation. Unfortunately, such clinical observations rarely become the basis for experimental study, and when they do, they are rarely performed with the rigor and thoroughness required to shed light on the underlying molecular genetic mechanisms involved. The potential value of these clinical observations for translational and basic cancer research is underappreciated and clearly underutilized. In the past, justified limitations on experimental research in humans may have precluded such a direct study of these diseases [2]. However, with today's sophisticated, relatively noninvasive sampling procedures and a patient population that is usually willing to participate, cancer research can and should move closer to the direct study of these diseases rather than using nonrepresentative surrogates and nonvalidated models. The following are a few illustrative examples, many of which will be addressed more thoroughly in the chapters in this text.

I. CLINICAL PATTERNS IN HEAD AND NECK CANCER

A. TNM Differences

What factors determine the TNM stage of a patient's tumor when they present with their head and neck cancers? Certainly time can be implicated in accounting for such differences. Patient delay in seeking treatment and their differences in tolerance thresholds for noxious symptoms are clearly important in this respect. Moreover, certain primary sites, such as the endolarynx, tend to be detected earlier due to their anatomy and the subsequent early disruption of function, whereas tumors such as pyriform sinus, supraglottic larynx, and base of tongue tend to present at a later stage. Time alone, however, does not adequately account for the spectrum of TNM stages observed at presentation. The inexperienced clinician is often skeptical when confronted with a patient who claims that their N3 neck mass was not present a few weeks ago, but such skepticism dissipates rapidly when an N1 neck node is actually observed to progress to the N3 stage in a week following diagnosis. The degree and rate at which a tumor progresses are determined mainly by molecular–genetic factors that control the (1) rate of local growth and invasion, which involves the induction of angiogenesis and adequate stromal substructure formation; (2) ability of the cancer cells to migrate to and access local vasculature (motility and diapedesis); and (3) ability of cancer cells to survive in and exit the vasculature and implant at distant sites, which also involves the successful induction of angiogenesis and adequate stromal substructure formation at the distant site.

During this process, to be successful, tumors must deploy or develop mechanisms to suppress the local and systemic host immune systems. At one end of this spectrum are the unresectable, *massive* "T4" cancers that invade through multiple local tissue and bony barriers, including complex vasculature, *but do not, or cannot,* develop regional or distant metastatic disease [1]. At the other end are patients with SCC of unknown primary. In these instances, cells, which, by definition, do not or cannot propagate or invade sufficiently to be detected at their site of origin, can and do access local vasculature, disseminate, implant, and often progress vigorously at regional and distant sites [3]. Implications for the inherent biological differences between these two well-recognized tumor groups, and the potential they provide for studying cancer mechanisms, are extraordinary.

Other examples, illustrative of observations that raise similar opportunities for investigation, include the difference in the rate of distant metastases, which are a feature at presentation or recurrence in SCC of the nasopharynx, as opposed to other squamous cell carcinomas of the head and neck (SCCHN) [4], the relatively high distant metastatic rate for paranasal sinus carcinomas [5], and, most recently, the site-specific difference in the rate and type of the secondary upper aerodigestive tract, which can be associated with the site of the original primary tumor [6]. Finally, the clinical implications, particularly the risk of distant metastases, for local/regional lymph node involvement in SCCHN are substantially different when compared with other carcinomas, such as breast, lung, prostate, melanoma, and colon. Even with advanced N2C and N3 lymph node status, a substantial proportion of these patients can still be cured with local modalities alone [7]. Are these lymph nodes in this region more efficient as "traps" or is it that the biology of SCCHN itself is different? Does this explain the relatively good outcome for patients with SCCHN of unknown primary as compared to other types of unknown primaries, such as adenocarcinomas or nonhead and neck SCC of unknown primary [8]?

B. Histopathology Differences

The microscopic appearance of the tumor, as described by the conventional morphological grade, is not particularly predictive of the natural or treatment outcome for patients with SCCHN [9]. However, local tumor–host features, such as tumor-stromal borders, pattern of invasion, degree of stromal or inflammatory response, access to microvasculature, or vascularity (angiogenesis) and extracapsular lymph node involvement [10,11], have been shown to have a marked effect on the ability to achieve and determine negative margins at surgery and the subsequent rates of local relapse and cure. The subsequent local recurrence rates are two to three times higher for patients in surgical series who present with histological "high-risk" histological phenotypes, and the survival is only one-third to one-half as good as those that do not have such features [12,13]. However, conventional grading is extremely significant for salivary gland tumors, such as adenocarcinomas and mucoepidermoid carcinomas, where high-grade or poorly differentiated lesions commonly recur very often with distant metastases [14–16]. Perhaps the most intriguing tumor among all human cancers is the subset of salivary gland tumors with adenoidcystic histopathology [17]. It is *common* for these tumors to either present with or later evolve distant lung, bone, and liver metastases. Distant metastases may remain undetected clinically for years or even decades. Occasionally, these metastases, once documented, may persist for decades, with minimal morbidity to the host. How these cancers and their host remain in such a symbiosis for decades is unclear and unique in cancer. The unique biological aspects of these tumors and their host–tumor relationships have not been the subjects of adequate scientific investigation.

C. Treatment Outcome Differences

Why do subsets of patients with tumors in seemingly homogeneous clinical subgroups behave so differently following treatment? This is true regardless of which modality is chosen as the initial therapy. The ability to achieve microscopically negative surgical margins is possible in nearly 100% of early staged SCCHN and the majority of patients with advanced, resectable tumors [7]. Although microscopically negative surgical margins remain the hallmark of successful surgery, when achieved, substantial percentages of patients still fail locally with increasing frequency as the TNM stage increases [1,7]. Although this observation has been made repeatedly for decades, it is only recently that studies have demonstrated that negative histological margins are not negative "molecular margins" and that histologically normal-appearing tissue may be molecularly/genetically fated to become malignant or may be malignantly transformed already [18].

Similar examples can be provided for patients treated initially with radiation therapy. Clearly, tumor oxygenation and whether there is an intact mechanism for programmed cell death following lethal radiation are keys to the success or failure when radiotherapy is employed [19,20]. However, these two factors alone cannot account for the broad differences in treatment outcome seen in early staged, T1 and T2 N0 tumors.

The emergence of chemotherapy as the primary treatment for patients with SCCHN, first in recurrent tumors [21], then in advanced, unresectable disease [22], and now in organ preservation strategies [23,24], has also produced outcome subgroups that are not evident before treatment. Even with the best regimens, 10–20% of previously untreated SCCHN patients will not respond or actually progress on treatment, 50% will not achieve a complete clinical remission, and no more than 25–30% have histologically negative cytotoxic responses. Each of these subgroups, defined by treatment outcome, has dramatically different survival characteristics [25,26]. It is also clear from organ preservation strategies that substantial differences in clinical complete response rates are specific for the primary tumor site [24,25].

As the current treatment for patients with advanced SCCHN is often multimodality, it is becoming more difficult to sort out these patterns of failure, and some of these subgroups tend to blur or disappear. For example, organ preservation approaches have been developed on the principle that patients failing to achieve at least a partial clinical response with induction chemotherapy can only be salvaged with surgery [27,28]. However, the use of sequential concurrent chemo/radiation following induction therapy has changed this algorithm [29]. Finally, and most importantly, regardless of which modality is used initially, it is not clear that the *same* subpopulation of patients are cured with each modality. If indeed they are different populations, as some correlative studies have suggested [30,31], how are they different and to what extent? This has important implications for individualizing patient therapy, increasing the overall cure rates for patients with SCCHN, and understanding the underlying mechanisms that regulate the behavior of these tumors.

D. Correlative Parameters and Outcome

Few, if any, correlative parameters have been tested thoroughly and adequately enough in SCCHN clinical trials to be considered validated and prospectively useful [30,31]. Most correlative studies are single institution trials, which, more often than not, produce conflicting results. In SCCHN, the most commonly studied correlative laboratory parameter has been DNA content (DNA ploidy and %S phase fraction), most often determined by flow cytometry. These parameters have undergone large numbers of single institutional investigations over a period of decades and, most recently, validation at the cooperative and intergroup phase III level [30,31]. Other experimental parameters, including p53, epidermal growth factor (EGF), chromosomal polysomy, loss of heterozygosity at 3p or 9p, C-erbB2 expression, Ki67 expression, tumor angiogenesis, cell cycle (cyclin pathway) regulatory aberrations, and integrity of apoptotic pathways, have only been studied anecdotally in small, preliminary trials [32–35]. In the few instances where newer parameters, such as p53 mutation or functional status, have been evaluated more extensively, conflicting results have been published [36–38]. These conflicting results raise another concern about validation of another sort; i.e., tumor preparative and experimental technique methodology validation, which is required before routine research with new parameters and technologies is performed [39–41]. This is rarely if ever done and accounts for the considerable "noise" in the literature concerning the clinical value of such correlative research.

II. HUMAN CANCER SURROGATES AND MODELS

Experimental models are employed in scientific research as substitutes for the subject or object of the research when such subjects are not available or cannot be used. To the extent that such models do not represent the subjects or phenomenon under investigation, the research conclusions suffer proportionately. This is true regardless of how carefully the work is done or how sophisticated the technology utilized. Laboratory models employed to investigate human cancers are of three major types.

1. *In vitro* and *in vivo*, nonhuman (usually murine) systems. These models exist as either *in vivo* nonhuman cell lines or tissue models [42–45]. These tumors are almost always induced in syngeneic animals rather than spontaneously occurring and, other than microscopic appearance, bare little clinical resemblance to their human counterparts. The use of syngeneic animals, inbred for decades, further distances these models from the genetic heterogeneity and complexity with which human tumors arise.
2. More recently, cell lines derived from human cancers have become models for human cancer [46,47]. They suffer from the same limitations as any cell line in that most human cancers, except for leukemia, are tissues with all the inherent additional complexities inherent in tissues, organs, and organisms. Often these lines are passaged for decades (MCF 7, HeLa cells). More recently, three-dimensional human tumor models (organotypic or raft cultures) derived from human tumors have been developed [48,49]. Although they are derived from human cancers and often have similar appearances, they may not represent human tumors any better than other models, particularly because the important tumor–host interaction is still missing.
3. Often human tumor cell lines are transplanted into immune-suppressed murine systems in an attempt to simulate an *in vivo* system [50,51], but again, these models are orders of magnitude removed from the spontaneously occurring cancers arising in the genetically heterogeneous and immunocompetent human population. However, because they are derived from human cancers, grow as tissues, have

three dimensions, and interact with an intact organism, it is at least possible to compare some features with human tumors *in vivo* for model validation such as histopathology, local regional invasion, and metastatic potential.

A potential, but unused, resource for experimental and clinical cancer research exists in spontaneously occurring cancers in household or domesticated animals [52]. Indeed, in this population, which numbers in the millions, spontaneous cancers occur at approximately the same rate as in the humans. Unfortunately, the more common cancers, with the exception of canine breast cancer, occur with much less frequency than in humans (lung, prostate, and gastrointestinal). This is an untapped potential resource for studying spontaneously arising cancers in more genetically diverse and less inbred animals.

This leads to the most important aspect of model use in experimental research. As stated previously, model use should be restricted to those instances in which the actual subject of the experiment is not available or cannot be used. Given the current state of human correlative research, it is becoming increasingly difficult to argue that such subjects are not available. When and if it is proposed that models must substitute for the subject under investigation, it is imperative that the models be validated before their use. In laboratory cancer research, cancer models are rarely validated and, more importantly, attempts to validate them are never undertaken; i.e., attempts must be made to determine with what fidelity these models represent the spectrum of natural history and biology seen in a particular human cancer. Humbling reminders of these differences have come from experiences such as during the "stem cell assay" era [53] where investigators often found it impossible to grow a human cancer line *in vitro* while, simultaneously, finding it impossible to stop its growth in the human host from which it was derived!

III. TRANSLATIONAL OR CORRELATIVE CANCER RESEARCH

The terms translational or correlative research are meant to represent that area of cancer research that brings together clinical observations, laboratory research, and the application of the product of this interaction for the treatment of patients with cancer [54]. This process also implies the potential to individualize treatment strategies that are tailored to the patient's tumor biology [55]. However, where is the optimum point, if any, to initiate this process? While important and valid molecular–genetic observations are often made at cellular and subcellular levels in the laboratory, how such observations become properly integrated into the extraordinary complex panorama of human cancer is often problematic. Guidelines for research that have evolved over the centuries have culminated in an agreed upon format generally referred to as the scientific method [56–58]. Its components include (1) an observation, (2) the generation of a hypothesis to be tested based on the observation, (3) the development and validation of an experimental design and methodology to test the hypothesis, (4) the conduct of the experiment, (5) data gathering and analysis, and (6) conclusions derived from data regarding the hypothesis, which is relevant to the observation.

This is basically a unidirectional system and one that works best when not begun in the middle or run backward! The observational step of the scientific method is, and has usually been, the critical starting point for most clinically relevant medical research. It is *always* the anvil on which any and all conclusions must ultimately be tested [56]. Without this component, data derived from laboratory experiments "dangle" without meaningful context. Unfortunately, much, if not most, experimental cancer research currently falls into this category. It seems intuitive that when the goal of an investigation is the study of *human* cancer, human cancer must be included somewhere in the process [2]. While it is becoming fairly common for clinical cancer scientists to spend substantial portions of their training time in gaining experimental laboratory experience, it is still quite rare for basic cancer scientists to spend comparable periods during their training learning about human cancer in the clinic. Clinical observations provide the scaffolding on which laboratory data can be organized in a meaningful manner. As the eminent cancer researcher Van Rensselaer Potter opined over a decade ago as he anticipated the emerging human molecular–genetic revolution, "...the assumption that the total sequencing of the normal human genome will tell us how life is organized can only lead to a technologic blitzkrieg that will produce more descriptive data than the computerized human mind can organize into a blueprint for life—unless some cybernetics system...is used as a scaffolding" [59,60]. Therefore, the "closing of the loop" of the scientific method, if you will, can and should be practiced as religiously as possible in experimental cancer research.

IV. CONCLUSION

It is highly unlikely that models that faithfully represent the broad spectrum of biology seen in human cancers, in its entirety, will ever be created. This will certainly be unlikely for as long as the observational background required for their development, which can be provided only by the clinical knowledge of the spectrum and diversity of these diseases, is lacking in the formulation and validation of such models. It is therefore imperative that basic researchers and clinicians involved in human cancer research work closely in multidisciplinary teams. Cross training and continual learning for both clinical and basic scientific specialties in a multimodality environment will be critical for translational strategies to reach their full potential. It is also critical that natural experiments in human cancer, detected and catalogued by clinical observation, become the focused subject of experimental research whenever and to whatever extent possible.

References

1. Al-Sarraf, M., Kish, J. A., and Ensley, J. F. (1991). The Wayne State University experience with adjuvant chemotherapy of head and neck cancer. *Hem/Oncol. Clin. North Am.* **5**(4), 687–700.
2. Ensley, J. F. (1993). Molecular Medicine, the forests, trees, and leaves. *Arch. Otolaryngol. Head Neck Surg.* **119**, 1173–1177.
3. de Braud, F., Heilbrun, L. K., Ahmed, K., Sakr, W., Ensley, J., Kish, J., Tapazoglou, E., and Al-Sarraf, M. (1989). Squamous cell carcinoma of unknown primary metastatic to the neck: Advantages of an aggressive treatment. *Cancer* **64**, 510–515.
4. Ensley, J. F., Kim, H., and Yoo, G. (2001). Locally advanced nasopharyngeal carcinoma. *Curr. Treat. Opin. Oncol.* **2**, 15–23.
5. LoRusso, P., Tapazoglou, E., Kish, J. A., Ensley, J. F., Cummings, G., Kelly, J., and Al-Sarraf, M. (1988). Chemotherapy for paranasal sinus carcinoma: A ten-year experience at Wayne State University. *Cancer* **62**, 1–5.
6. Kim, E. S., Khuri, F. R., Lee, J. J., Winn, R, Cooper, J. M., Fu, K, Lippman, S. M., Vokes E. E., Pajak, T. F., Goepfert, H., and Hong W. K. (2000). Second primary tumor incidence related primary index tumor and smoking status in a randomized chemoprevention study of head and neck squamous cell cacner. *Proc. ASCO* **19**, 1642 [Abstract].
7. Looser, K. G., Shah, J. P., Strong, E. W., Salesiotis, A. N., and Cullen, K. J. (1978). The significance of "positive" margins in surgically resected epidermoid carcinomas. *Head Neck Surg.* **1**, 107–111.
8. Spiro, R. H., DeRose, G., and Strong, E. W. (1983). Cervical node metastasis of occult origin. *Am. J. Surg.* **146**, 441–446.
9. Ensley, J., Crissman, J., Kish, J., Jacobs, J., Weaver, A., Kinzie, J., Cummings, G., and Al-Sarraf, M. (1986). The impact of morphological analysis on response rates and survival in patients with squamous cell cancer of the head and neck. *Cancer* **57**, 711–717.
10. Ensley, J., Crissman, J., Kish, J., Weaver, A., Jacobs, J., Cummings, G., and Al-Sarraf, M. (1989). The correlation of individual parameters of tumor differentiation with response rate and survival in patients with advanced head and neck cancer treated with induction therapy. *Cancer* **63**, 1487–1492.
11. Sakr, W., Hassan, M., Zarbo, R., Ensley, J., and Crissman, J. D. (1989). DNA quantitation and histologic characteristics of squamous cell carcinomas of the upper aerodigestive tract. *Arch. Pathol. Lab. Med.* **113**, 1009–1014.
12. Laramore, G. E., Scott, C. B., Al-Sarraf, M., Haselow, R. E., Ervin, T. J., Wheeler, R., Jacobs, J.R, Schuller, D. E., Gahbauer, R. A., and Schwade, J. G. (1992). Adjuvant chemotherapy for resectable squamous cell carcinomas of the head and neck: Report on intergroup study 0034. *Int. J. Radiat. Oncol. Biol. Phys.* **23**, 705–713.
13. Looser, K. G., Shah, J. P., and Strong, E. W. (1978). The significance of positive margins in surgically resected epidermoid carcinomas. *Head Neck Surg.* **1**, 107–111.
14. Hoffman, H. T., Karnel, L. H., Robinson, R. A., Pinkston J. A., and Menck, H. R. (1999). National Cancer Data Base report on cancer of the head and neck: Acinic cell carcinoma. *Head Neck* **21**, 297–309.
15. Hicks, M. J., el-Naggar, A. K., Flaitz, C. M., Luna, M. A., and Batsakis, J. G. (1995) Histocytologic grading of mucoepidermoid carcinoma of major salivary glands in prognosis and survival: A clinicopathologic and flow cytometric investigation. *Head Neck* **17**, 89–95.
16. Spiro, R. H., Huvos, A. G., Berk, R, Strong, E. W. (1978). Mucoepidermoid carcinoma of salivary gland origin. A clinicopathologic study of 367 cases. *Am. J. Surg.* **136**, 461–468.
17. Spiro, R. H., Huvos, A. G., and Strong, E. W. (1979). Adenoid cystic carcinoma: Factors influencing survival. *Am. J. Surg.* **138**, 579–583.
18. Koch, W. M., Brennan, J. A., Zahurak, M., Goodman, S. N., Westra, W. H., Schwab, D., Yoo, G. H., Lee, D. J., Forastiere, A. A., and Sidransky, D. (1996). p53 mutations and locoregional treatment failure in head and neck squamous cancer. *J. Natl. Cancer Inst.* **88**, 1580–1586.
19. Vanselow, B., Eble, M. J., Ruda, V., Wollensack, P., Conradt, C., and Dietz, A. (2000). Oxygenation of advanced head and neck cancer: prognostic marker for the response to primary radiochemotherapy. *Otolaryngol. Head Neck Surg.* **122**, 856–862.
20. Shintani, S., Mihara, M., Nakahara, Y., Terakado, N., Yoshihama,Y., Kiyota, A., Ueyama. Y., and Matsumura, T. (2000). Apoptosis and p53 are associated with effect of preoperative radiation in oral squamous cell carcinomas. *Cancer Lett.* **154**, 71–77.
21. Amer, M.H, Al-Sarraf, M., and Vaitkevicius, V. K. (1979). Factors that affect response to chemotherapy and survival of patients with advanced head and neck cancer. Cancer **43**, 2202–2206.
22. Amer, M. H., Izbick, R. M., Vaitkevicius, V. K., and Al-Sarraf, M. (1980) Combination chemotherapy with cis-diamminedichloroplatinum, oncovin, and bleomycin (COB) in advanced head and neck cancer: Phase II. *Cancer* **45**, 217–223.
23. (1991). Induction chemotherapy plus radiation compared with surgery plus radiation in patients with advanced laryngeal cancers. *N. Engl. J. Med.* **324**, 1685–1690.
24. Lefebvre, J. L., Chevalier, D., Luboinski, B., Kirkpatrick, A., Collette, L., and Sahmoud, T. (1996). Larynx preservation in pyriform sinus cancer: Preliminary results of a European Organization for Research and Treatment of Cancer phase III trial. EORTC Head and Neck Cancer Cooperative Group. *J. Natl. Cancer Inst.* **88**, 890–899.
25. Rooney, M., Kish, J., Jacobs, J., Kinzie, J., Weaver, A., Crissman, J., and Al-Sarraf, M. (1985). Improved complete response rate and survival in advanced head and neck cancer after three-course induction therapy with 120–hour 5-FU infusion and cisplatin. *Cancer* **55**, 1123–1128.
26. Al-Kourainy, K., Kish, J. A., Ensley, J. F., Tapazoglou, E., Jacobs, J., Weaver, A., Crissman, J., Cummings, G., and Al-Sarraf, M. (1987). Achievement of superior survival for histologically negative vs histologically positive clinically complete responders to cisplatinum combination in patients with locally advanced head and neck cancer. *Cancer* **59**, 233–238.
27. Ensley, J., Jacobs, J., Weaver, A., Kinzie, J., and Al-Sarraf, M. (1984). The correlation between response to cisplatinum combination chemotherapy and subsequent radiotherapy in previously untreated patients with advanced squamous cell cancers of the head and neck. *Cancer* **54**, 811–814.
28. Ensley, J. F., Kish, J. A., Jacobs, J., Weaver, A., Crissman, J., and Al-Sarraf, M. (1984). Incremental improvements in median survival associated with degree of response to adjuvant chemotherapy in patients with advanced squamous cell cancer of the head and neck. *In* "Adjuvant Therapy of Cancer" (S. E. Jones and S. E. Salmon, eds.), Vol. IV, pp. 117–126. Grune & Stratton, Orlando, FL.
29. Ensley, J., Ahmad, K., Kish, J., Tapazoglou, E., and Al-Sarraf. M. (1990). Salvage of patients with advanced squamous cell cancers of the head and neck (SCCHN) following induction chemotherapy failure using radiation and concurrent cisplatinum (CACP). *In* "Adjuvant Therapy of Cancer" (S. E. Salmon, ed.), Vol. VI, pp. 92–100. Saunders, New York.
30. Ensley, J., and Maciorowski M. (1994). Clinical application of DNA content parameters in patients with squamous cell carcinomas of the head and neck. *Semin. Oncol.* **21**, 330–339.

31. Ensley, J. F. (1996). The clinical application of DNA content and kinetic parameters in the treatment of patients with squamous cell carcinomas of the head and neck. *Cancer Metast. Res.* **15**, 133–141.
32. Salesiotis, A. N., and Cullen, K. J. (2000). Molecular markers predictive of response and prognosis in the patient with advanced squamous cell carcinoma of the head and neck: Evolution of a model beyond TNM staging. *Curr. Opin. Oncol.* **12**, 229–239.
33. Chiesa, F., Mauri, S., Tradati, N., Calabrese, L., Giugliano, G., Ansarin, M., Andrle, J. Zurrida, S., Orecchia, R., and Scully, C. (1999). Surfing prognostic factors in head and neck cancer at the millennium. *Oral Oncol.* **35**, 590–6.
34. Liu, M., Weynand, B., Delos, M., and Marbaix, E. (2000). Prognostic factors in squamous cell carcinomas of the head and neck. *Cancer J. Sci. Am.* **6**, 2–10.
35. Tralongo, V., Rodolico, V., Luciani, A., Marra, G., and Daniele, E. (1999). Prognostic factors in oral squamous cell carcinoma. A review of the literature. *Anticancer Res.* **19**, 3503–3510.
36. Chomchai, J. S., Du, W., Sarkar F. H., Li, Y. W., Jacobs, J. R., Ensley, J.F, Sakr, W., and Yoo, G. H. (1999). Prognostic significance of p53 gene mutations in laryngeal cancer. *Laryngoscope* **109**, 455–459.
37. Narayana, A., Vaughan, A. T., Kathuria, S., Fisher, S. G., Walter, S. A., and Reddy, S. P. (2000). P53 overexpression is associated with bulky tumor and poor local control in T1 glottic cancer. *Int. J. Radiat. Oncol. Biol. Phys.* **46**, 21–26.
38. Bradford, C. R., Wolf, G. T., Carey, T. E., Zhu, S., Beals, T. F., Truelson, J. M., McClatchey, K. D., and Fisher, S. G. (1999). Predictive markers for response to chemotherapy, organ preservation, and survival in patients with advanced laryngeal carcinoma. *Otolaryngol. Head Neck Surg.* **121**, 534–538.
39. Hitchcock, C., Zalupski, M., and Ensley, J. (1995). Solid tumor preparative technology, *In* "Cytometry 2000. Flow Cytometric Analysis of Solid Tumors" (A. Nakeff, F. Valeriote, J. Crissman, S. Lerman, and J. Ensley, eds.), pp. 95–124. Kluwer Academic, Boston, MA.
40. Ensley, J. F., Maciorowski, Z., Pietraszkiewicz, H., Klemic, G., KuKuruga, M., Sapareto, S., Corbett, T., and Crissman, J. D. (1989). Solid tumor preparation for flow cytometry using a standard murine tumor model. *Cytometry* **8**, 479–487.
41. Ensley, J., Maciorowski, Z., Hassan, M., Pietraszkiewicz, H., Sakr, W., and Hielbrun, L. (1993).Variations in DNA aneuploid cell content during tumor dissociation in human colon and head and neck cancers analyzed by flow cytometry. *Cytometry* **14**, 550–558.
42. Kage, T., Mogi, M., Katsumata, Y., and Chino, T. (1987). Regional lymph node metastasis created by partial excision of carcinomas induced in hamster cheek pouch with 9,10-dimethyl-1,2-benzanthracene. *J. Dent. Res.* **66**, 1673–1679.
43. Hawkins, B. L., Heniford, B. W., Ackermann, D. M., Leonberger, M., Martinez, S. A., and Hendler, F. J. (1994). 4NQO carcinogenesis: A mouse model of oral cavity squamous cell carcinoma. *Head Neck* **16**, 424–432.
44. Hier, M.P., Black, M. J., Shenouda, G., Sadeghi, N., and Karp, S. E. (1995). A murine model for the immunotherapy of head and neck squamous cell carcinoma. *Laryngoscope* **10**, 1077–1080.
45. Sung, M. W., Lee, S. G., Yoon, S. J. Lee, H. J., Heo, D. S., Kim. K. H., Koh, T. Y., Choi, S. H., Park, S. W., Koo, J. W., and Kwon, T. Y. (2000). Cationic liposome-enhanced adenoviral gene transfer in a murine head and neck cancer model. *Anticancer Res.* **20**(3A), 1653.
46. Cardinalli, M., Pietraszkiewicz, H., Ensley, J. F., and Robbins, K. C. (1995). Tyrosine phosphorylation as a marker for aberrantly regulated growth promoting pathways in cells derived from head and neck malignancies. *Int. J. Cancer* **61**, 98–103.
47. Shridhar, R., Shridhar, V., Rivard, S., Siegfried, J. M., Pietraszkiewicz, H., Ensley, J. F., Pauley, R., Grignon, D., Sakr, W., Miller, O. J., and Smith, D. I. (1997). Mutations in the arginine-rich protein (ARP) gene in lung, breast prostate cancers and in squamous cell carcinomas of the head and neck. *Cancer Res.* **56**, 5576–5578.
48. Park, N. H., Gujuluva, C. N., Baek, J. H., Cherrick, H. M., Shin, K. H., and Min, B. M. (1995). Combined oral carcinogenicity of HPV-16 and benzo(a)pyrene: An in vitro multistep carcinogenesis model. *Oncogene* **10**, 2145–2153.
49. Eicher, S. A., Clayman, G. L., Liu, T. J., Shillitoe, E. J., Storthz, K. A., Roth, J. A., and Lotan, R. (1996). Evaluation of topical gene therapy for head and neck squamous cell carcinoma in an organotypic model. *Clin. Cancer. Res.* **10**, 1659–1666.
50. Yeudall, W. A., Jakus, J., Ensley, J., and Robbins, K. C. (1997). Functional characterization of p53 molecules expressed in human squamous cell carcinomas. *Mol. Carcinogen.* **18**, 89–96.
51. Cardinali, M., Kratochvil, J., Ensley, J. F., Robbins, K. C., and Yeudall, W. A. (1997). Functional characterization in vivo of mutant p53 molecules derived from squamous cell carcinomas of the head and neck. *Mol. Carcinogen.* **18**, 78–88.
52. Gardner, D. G. (1996). Spontaneous squamous cell carcinomas of the oral region in domestic animals: A review and consideration of their relevance to human research. *Oral Dis.* **2**,148–154.
53. Hamburger, A. W., and Salmon, S. E. (1977). Primary bioassay of human tumor stem cells. *Science* **197**, 461–463.
54. Chabner, B. A., Boral, A. L., and Multani, P. (1998). Translational research: walking the bridge between idea and cure—seventeenth Bruce F. Cain Memorial Award lecture. *Cancer Res.* **58**, 4211–4216.
55. Bartelink, H., Begg, A. C., Martin, J. C. van Dijk, M. Moonen, L., van't Veer, L. J., Van de Vaart, P., and Verheij, M. (2000). Translational research offers individually tailored treatments for cancer patients. *Cancer J. Sci. Am.* **6**, 2–10.
56. Krause, R. M. (1983). The beginning of health is to know the disease. *Public Health Rep.* **98**, 531–535.
57. Rhodes, P. (1983). Research. *Br. Med. J. (Clin. Res. Ed.)* **286**, 1271–1272.
58. Medawar, S. P. (1975). Scientific method in science and medicine. *Perspect. Biol. Med.* **18**, 345–52.
59. Potter, V. R. (1988). On the road to the blocked ontogeny theory. *Adv. Oncol.* **4**, 43–46.
60. Potter, V. R. (1987). Blocked ontogeny. *Science* **237**(4818), 964.

C H A P T E R

2

Head and Neck Cancers with Unusual Natural Histories

HAROLD KIM* and JOHN F. ENSLEY†

†*Department of Medicine, Oncology, and Otolaryngology—Head and Neck Surgery and*
**Department of Radiation Oncology, Karmanos Cancer Institute, Wayne State University, Detroit, Michigan 48201*

a. Unknown Primary Carcinoma Metastatic to Cervical Nodes

I. Introduction 9
II. Clinical Presentation 9
III. Evaluation 10
IV. Treatment 12
A. Extent of Optimal Neck Surgery 12
B. Extent of Radiation Field 12
C. Sequence of Radiation and Neck Dissection 13
D. Role of Adjuvant Chemotherapy 13
V. Treatment Results 13
References 15

I. INTRODUCTION

Cervical lymphadenopathy is a common presentation of malignant tumor in the head and neck area, but in 2–4% of patients [1–4] (Table 2a.1), the primary tumor cannot be identified following extensive diagnostic workup. It is assumed that the primary site has undergone spontaneous regression [5], which is observed in other malignancies, such as neuroblastoma, renal cell carcinoma, melanoma, and breast cancer. However, others postulate immune-modulated destruction of primary cancer [6], a faster proliferation rate of lymph node metastases [7], and removal of the primary site by sloughing of a necrotic tumor as possible causes [5]. The biological mechanisms underlying these phenomena are not yet understood. The optimal diagnostic evaluations and management of unknown primary carcinoma in the head and neck region are controversial.

II. CLINICAL PRESENTATION

Cervical adenopathy from an undetectable primary source accounts for about 2 to 4% of head and neck malignancies. The most common age at presentation is 50 to 60 years old, and the male-to-female ratio is 3:1 [8,9]. A painless and solitary neck mass is the most common presentation, but about 15% have multiple ipsilateral nodes and the remaining 10% have bilateral adenopathy [8,9]. The location of cervical adenopathy may suggest the possible primary tumor site based on known patterns of lymphatic spread (Table 2a.2) and should help guide diagnostic evaluation. Table 2a.3 shows distribution of metastatic lymph nodes from unknown primary tumors. The most commonly involved nodal group is jugulodiagatric and midjugular, which are involved in about 60% of patients, where squamous cell and undifferentiated carcinoma accounts for 70–80% of histology. However, in low neck and supraclaviclar nodes, the incidence of adenocarcinoma increases up to 40%, which suggests of possible primary in the lung, breast, and gastrointestinal tract [10]. The

TABLE 2a.1 Incidence of Metastasis to Cervical Lymph Nodes from Unknown Primary Tumors

Author	No. of patients (%)
de Braud *et al.* [1]	1577 (3%)
Fried *et al.* [2]	1900 (2.6)
Richard and Michaeu [3]	5137 (3.3)
Lefevre *et al.* [4]	8500 (2.2)

TABLE 2a.2 Probable Site of the Primary Tumor According to the Location of Cervical Metastases[a]

Location of nodes	Primary tumor site
Submental	Floor of mouth, lips, and anterior tongue
Submaxillary	Retromolar trigone and glossopalatine pilar
Jugulodigastric	Epipharynx, base of tongue, tonsil, nasopharynx, and larynx
Midjugular	Epipharynx, oropharynx, base of tongue, and larynx
Low jugular	Thyroid, epipharynx, and nasopharynx
Supraclavicular	Lung (40%), thyroid (20%), gastrointestinal (12%), and genitourinary (8%)
Posterior triangle	Nasopharynx

[a]From F. de Braud and M. Al-Sarraf (1993). Diagnosis and management of squamous cell carcinoma of unknown primary tumor site of the neck. *Semin. Oncol.* **20**, 273–278.

remainder of the chapter focuses on squamous cell carcinoma metastatic to cervical node from an unknown primary source.

III. EVALUATION

Table 2a.4 shows diagnostic workup for cervical lymph node metastasis of unknown primary. Figure 2a.1 shows a flowchart for workup of a suspicious neck node. A detailed physical examination, including a head and neck examination with indirect laryngoscope and careful attention to size, location, tenderness, consistency, and mobility of enlarged node, should be performed. A fine needle aspiration (FNA) of neck mass is preferred for tissue diagnosis over open biopsy, as it is not known to be associated with tumor seeding along the needle tract. FNA is safe, performed easily, and provides a diagnosis with a sensitivity of 90–99% and a specificity of 94–100% for squamous carcinoma [11,12]. However, the small amount of cytological specimen obtained by FNA may not be sufficient for the diagnosis of lymphoma or undifferentiated carcinoma. In patients for whom a nasopharyngeal carcinoma is high, probability includes an involvement of posterior chain nodes, a histology of lymphoepithelioma or undifferentiated carcinoma, and an ethnic background with a high incidence of nasopharyngeal carcinoma, such as Chinese. In these patients, an IgA titer against the viral capsid antigen of Epstein–Barr virus (EBV) may be helpful in ruling out a primary nasopharyngeal carcinoma. Using polymerase chain reaction (PCR), genome products of Epstein–Barr virus can be identified from a FNA specimen from neck nodes [13,14]. In a study of 41 FNA specimens, Feinmesser *et al.* [14] reported a presence of EBV in specimens from nine patients. Seven of these patients were found to have a nasopharyngeal primary; in the remaining two patients, a nasopharyngeal primary appeared within 1 year. Another study reported a high sensitivity of the *in situ* hybridization technique in detecting EBV in a FNA specimen of the neck node [15].

TABLE 2a.3 Distribution of Cervical Metastases from Unknown Primary Tumors[a]

Site	Incidence(%)
Jugulodigastric	71
Midjugular	22
Supraclavicular	18
Submaxillary	12
Low jugular	12
Submental	8

[a]From F. de Braud and M. Al-Sarraf (1993). Diagnosis and management of squamous cell carcinoma of unknown primary tumor site of the neck. *Semin. Oncol.* **20**, 273–278.

TABLE 2a.4 Diagnostic Workup for Cervical Lymph Node Metastases: Unknown Primary Tumor[a]

- General
  - History
  - Physical examination
    - Careful examination of the neck and supraclavicular regions
    - Examination of the oral cavity, pharynx, and larynx (indirect laryngoscopy)
- Radiographic studies
  - Chest roentgenogram
  - Computed tomography of the head and neck (special attention to nasopharynx, pharynx, and larynx)
  - Upper gastrointestinal series and barium enema (in patients with adenocarcinoma involving supraclavicular lymph nodes)
- Laboratory studies
  - Complete blood cell count
  - Blood chemistry profile
- Direct laryngoscopy and directed biopsies
  - Nasopharynx, both tonsils, base of tongue, both pyriform sinuses, and any suspicious or abnormal mucosal areas
  - Fine needle aspirate or core needle biopsy of the cervical node

[a]From W. M. Mendenhall, J. T. Parsons, A. A. Mancuso, S. P. Stringer, and N. J. Cassisi (1997). Head and neck: Management of the neck. In "Principles and Practice of Radiation Oncology," (C. A. Perez and L. W. Brady, eds.), 3rd Ed., pp. 1151–1154. Lippincott-Raven, Philadelphia.

If a physical examination is unrevealing, panendoscopy (nasopharyngoscopy, laryngoscopy, bronchoscopy, and esophagoscopy) should be performed under general anesthesia with a biopsy of all suspicious lesions. If no suspicious lesions are noted, a directed biopsy of potential primary sites (nasopharynx, tonsil, base of tongue, and pyriform sinus) should be performed. If repeated physical examination and computed tomography (CT) or magnetic resonance imaging (MRI) is negative, the yield of a directed biopsy is low. The base of tongue and tonsil are sites of highest positive biopsy yield. Because a superficial biopsy of the tonsil can

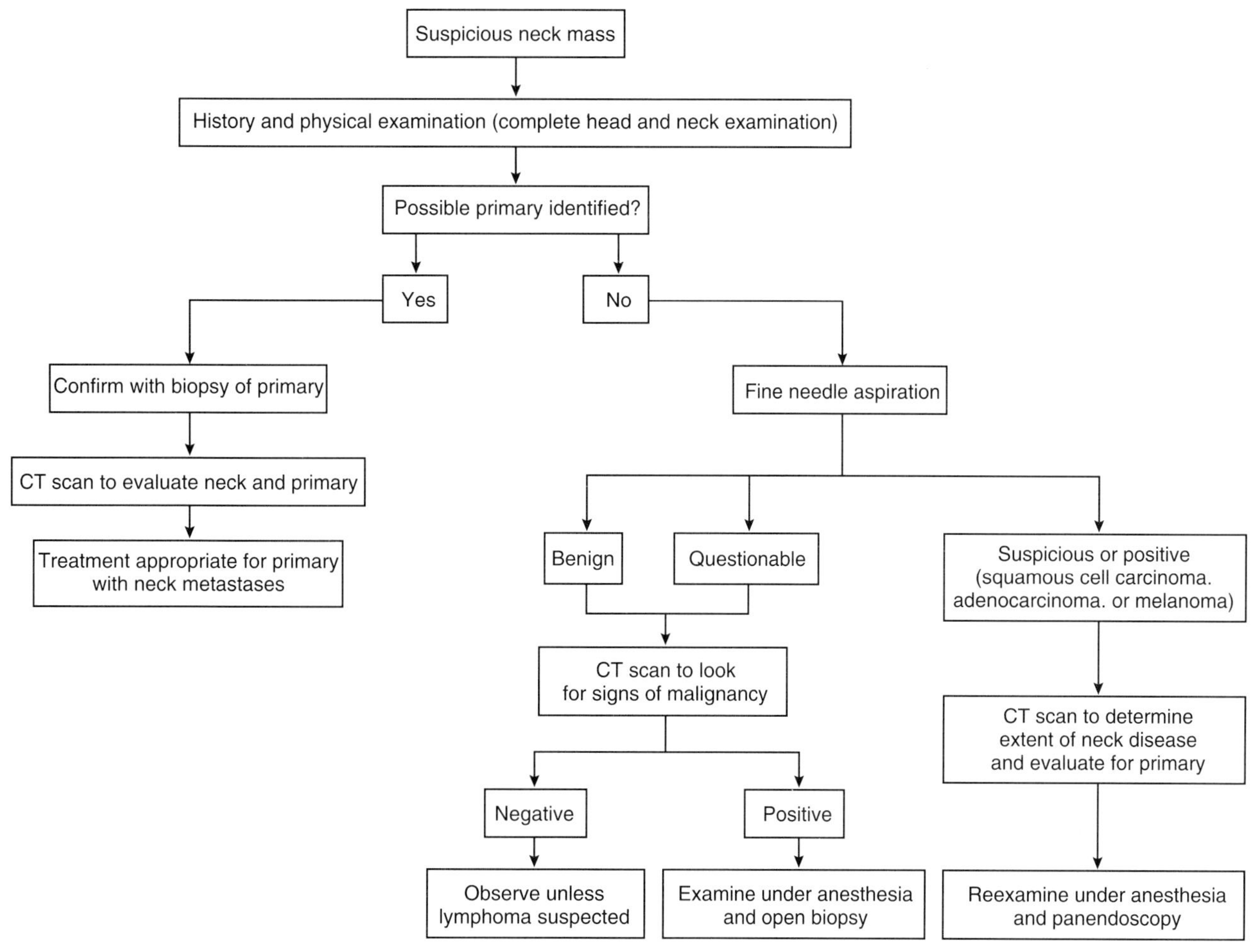

FIGURE 2a.1 Flowchart for workup of a suspicious neck node.

result in a high false-negative rate, tonsillectomy has been performed in the past. Several studies showed the benefit of tonsillectomy in. In a study by Randal *et al.* [16], 6 (18%) out of 34 patients with unknown primary were diagnosed as having primary tonsil carcinoma after having tonsillectomy. In a series of 87 patients with an unknown primary site, Lapeyre *et al.* [17] reported that subclinical disease had been found in the tonsil in 23 patients (26%).

Advances in molecular biology may be helpful in further identifying the primary site in these instances. Based on theories of tumor progression and field cancerization, Califano *et al.* [18] compared microsatellite analysis of tumors obtained in 18 patients with unknown squamous carcinoma cervical nodes with that of benign specimens obtained from directed biopsy. In 10 (55%) of the patients, at least one histopathologically benign mucosal specimen from defined anatomic sites demonstrated a pattern of genetic alterations identical to that present in cervical lymph node metastasis. These genetic changes include identical losses on multiple chromosomal arms or chromosomal breakpoints.

A chest radiograph needs to be obtained to rule out a pulmonary lesion or mediastinal adenopathy. A CT scan is obtained to evaluate the extent of neck disease, and the involvement of retropharyngeal nodes, as well as potential primary sites. It also provides information on the presence of necrosis and the involvement of extranodal tissues, soft tissue of the neck or the carotid sheath.

2[^{18}F]-Fluoro-2-deoxy-D-glucose positron emission tomography (18-FDG-PET) imaging of tumor metabolism may also be useful in the search for the primary site. This scan is based on metabolic differences between malignant and normal tissues, such as a greater number of glucose transporters, molecular changes of the hexokinases, and a reduced number of glucose 6-phosphate, which leads to trapping of FDG-6-PO_4 in tumor cells. In a study by Jungehulsing *et al.* [19] of 27 patients with unknown primary carcinoma metastatic to cervical nodes, a primary tumor was identified in 7 patients and additional metastases were detected on 18-FDG-PET. If PET shows an uptake in a particular location, it can render further diagnostic approaches more specific in certain patients.

Therefore, if PET is going to be obtained as a part of routine workup, it is advisable to do so prior to panendoscopy so that any areas of suspicious uptake can be examined with biopsy during panendoscopy.

IV. TREATMENT

All published data on the results of treatment of metastatic neck nodes from unknown primary site are from retrospective single institutional studies with nonuniform diagnostic workup and treatment. No prospective study has been conducted on this disease thus far. Therefore, definitive conclusions from these studies are difficult and remain controversial. However, some generalizations can be drawn. Most series used a combined modality treatment approach with neck dissection followed by postoperative radiation treatments. Combined modality treatment has superior control of neck disease, as well as a lower rate of subsequent emergence of primary tumor compared to surgical treatment of neck disease alone, especially for N2 and N3 disease.

The main controversies arising from these studies are (1) the extent of optimal neck surgery, whether neck dissection is needed or excisional biopsy is sufficient for early stage neck disease; (2) the extent of optimal radiation field, whether the mucosal/submucosal potential primary site should be irradiated given toxicities of radiation therapy; (3) the sequence of radiation and neck dissection; and (4) the role of adjuvant chemotherapy.

A. Extent of Optimal Neck Surgery

N1 disease can be managed with single modality, either radiation treatment or surgery. Some authors advocate neck dissection alone for a single node less than 3 cm in size without extranodal extension. However, only a few patients present with disease that is suitable for this approach of neck dissection alone. In a retrospective review of 117 patients with an unknown primary at the Mayo Clinic between 1965 and 1985, Coster *et al.* [20] reported on 24 patients (14 N1, 10 N2-3) who presented with unilateral adenopathy. They were treated with curative resection of all gross neck disease with neck dissection or excisional biopsy only without radiation therapy. Twenty-five percent of the patients developed recurrence in the dissected neck within a median of 3 months after surgery. This report also provides an excellent review of literature (Fig 2a.1) comparing different treatment results from several institutions. These patients, however, all had high risk factors, such as pathologic stage N2 or higher or the presence of extracapsular extension (ECE). The authors suggested that pathologic stage N1 patients without ECE could be managed with surgery alone. However, the proportion of patients suitable for this approach is small. N1 and N2a neck also could be managed with primary radiation therapy followed by close observation if a complete response is obtained [21]. For incomplete responses, neck dissection is indicated. The disadvantages of primary radiation therapy are, of course, that (1) the complete pathologic extent of neck disease is not known and (2) the radiation field and doses cannot be tailored accordingly. Several series reported equivalent neck control and survival rates with excisional biopsy only compared to neck dissection when they are followed by postoperative radiation treatment. Nguyen and colleagues [22] reported equivalent survival and neck control in 54 patients treated with either excisional biopsy (30 patients) or neck dissection (24 patients) followed by radiation therapy to a median dose of 60 Gy. In a retrospective study of 136 patients from the M.D. Anderson Cancer Center, Colletier *et al.* [23] reported no regional relapse in 39 patients treated with excisional biopsy followed by radiation therapy. Patients with N2b, N2c, and N3 disease are at a high risk of failure and should be approached with a combined modality treatment of neck dissection and radiation treatment. The relapse rate in the involved neck for N2b, N2c, and N3 is as high as 40–50% after surgery alone and postoperative radiation therapy is recommended.

B. Extent of Radiation Field

The optimal field of radiation therapy remains controversial with respect to inclusion of the potential primary site, and contralateral neck. The complete inclusion of the potential primary site, such as nasopharynx, oropharynx, hypopharynx, and supraglottic larynx, in the radiation field with 5000 to 6000 cGy will inevitably lead to more side effects, mostly mucositis and xerostomia. A radiation port encompassing the nasopharynx inevitably includes bilateral parotid glands and is a major cause of xerostomia. However, radiation to bilateral neck fields generally treats tonsil, base of tongue, larynx, and hypopharynx to a nearly therapeutic dose and yet can spare significant portions of parotid glands. Therefore, the nasopharynx may not be included in radiation field unless the patient presents with features suggestive of a nasopharyngeal primary, such as positive EBV serology, undifferentiated carcinoma or lymphoepithelioma, bilateral high neck nodes, or posterior cervical node involvement. This approach minimizes radiation-induced xerostomia.

However, some authors advocate withholding irradiation to the entire pharyngeal mucosa for the following reasons.

1. The probability of emergence of a primary site is approximately 10%. The probability of side effects, whether acute or chronic, from radiation treatment is 100%. This means that 90% of patients will develop side effects without a potential benefit from radiation.
2. The major site of failure after surgical treatment of the neck is ipsilateral neck and distant metastasis, not in the occult primary site.
3. No clear data exist for a proven survival benefit of radiation therapy to potential primary sites.
4. Almost 50% of tumors that develop after 5 years of initial neck treatment are most likely second primary tumors, which can be managed accordingly with less morbidity without previous radiation.

Others, however, continue to advocate irradiation of potential primary sites and bilateral neck. Their reasons include the following.

1. In patients treated with surgery alone, several retrospective studies reported up to 40% of development of primary tumor eventually and up to 15% of contralateral neck failure eventually. Radiation treatment to potential primary sites reduces the incidence of mucosal failure to about 10%.
2. Data on the effectiveness of salvage treatments of primary and contralateral neck relapse are lacking.
3. If only the involved neck was irradiated, further irradiation of the primary site is technically more difficult and compromised if disease progresses at the primary site. It may also be associated with a higher morbidity.
4. Any untreated microscopic cancer can be a potential source of distant metastasis and may compromise survival.
5. Advances in the radiation technique, such as intensity modulated radiation treatment (IMRT) planning, and pharmaceutical agents, such as Amifostine and Pilocarpine, may minimize toxicities from radiation treatment.

C. Sequence of Radiation and Neck Dissection

N1 and N2a neck diseases have up to an 80% chance of control with radiation alone with neck dissection reserved as the salvage procedure. This approach enables a significant portion of patients to avoid the morbidity of a neck dissection. However, a true pathologic assessment of neck disease is not possible, and radiation field and doses cannot be tailed based on pathologic information. Also, neck dissection after high-dose radiation may increase the risks of postoperative wound and soft tissue complications.

For stage N2b or above, generally, initial treatment is neck dissection followed by radiation treatment. Unless the neck is inoperable, in most centers, neck dissection is performed first followed by postoperative radiation treatment. However, it is still common, especially in European centers, to deliver radiation treatment first followed by planned neck dissection for an incomplete response. With recent advances in chemoradiation therapy for advanced head and neck disease, the role of planned neck dissection remains controversial. There are no clear data on pathologic residual disease after a complete response with chemoradiation therapy. It is also unclear whether planned neck dissection after a complete response improves neck control or survival. PET may prove to be a valuable guide in the metabolic status of nodal disease after chemoradiation therapy. Inoperable neck disease is treated with radiation followed by neck dissection if it is downstaged to operable.

D. Role of Adjuvant Chemotherapy

The presence of multiple node involvement or extracapsular extension is considered a high-risk factor for local failure. Limited data exist regarding the benefit of chemotherapy in patients with an unknown primary tumor. However, concurrent postoperative radiation therapy and chemotherapy showed benefits in terms of improved local and regional control in several studies with known primary tumors. A recently completed intergroup study comparing postoperative radiation therapy alone and concurrent postoperative radiation therapy and Cisplatin in high-risk patients with known primary tumors will indirectly provide valuable data on the role of adjuvant chemotherapy in patients with an unknown primary tumor but with high-risk factors for recurrence. Also, the issue of chemoprevention remains to be investigated.

V. TREATMENT RESULTS

Table 2a.5 provides a compilation of treatment results from several institutions with three different approaches. It is difficult to draw conclusions from these retrospective reviews, but it can be seen that the percentage of patients developing either primary tumor or delayed metastasis in the contralateral neck is lower in patients who received radiation therapy.

Most studies identified the extent of neck disease as the most significant prognostic factor. The N stage has been shown to have a direct correlation with neck control, disease-free survival, and overall survival [9,20,22,24]. Jesse *et al.* [24] from M.D. Anderson, reported a 3-year, disease-free survival of 79, 67, 45, and 38% for NX, N1, N2, and N3 stage diseases, respectively, in 104 patients treated with surgery alone. A direct correlation was also seen between neck stage and 3-year disease absolute survival in patients treated with radiation therapy alone (52 patients) or a combination of surgery and radiation therapy (28 patients). Weir *et al.* [9], from Princess Margaret Hospital, reported on 144 patients treated with radiation therapy. Local control rates were 75, 67, 36, and 22% and 3-year actuarial survival rates were 76, 55, 51, and 24% for stages NX, N1, N2, and N3, respectively. The number of involved nodes and the fixation of nodes were reported to be independent prognostic factors for neck control and survival [25].

Extracapsular extension is also an important prognostic factor for local control as well as survival. In 136 patients treated with surgery followed by postoperative radiotherapy at M.D. Anderson, Colletier and co-workers [23] reported 5-year actuarial rates of neck relapse in patients with and without extracapsular extension of 16 and 0%, respectively.

The reported overall incidence of primary tumor development in the head and neck ranges from 12 to 25% after

TABLE 2a.5 Literature Review of "Unknown Primary" Squamous Cell Carcinoma[a]

Series	Treatment	Patients (no.)	Primary tumor developed[b] (no.)	Primary tumor developed[b] (%)	Recurrence in dissected neck (no.)	Recurrence in dissected neck (%)	Delayed metastases in contralateral neck (no.)	Delayed metastases in contralateral neck (%)
Surgery alone								
Wang *et al.*		57	6	11	7	12	—	—
Jesse *et al.*		104	17	16	25	24	16	15
Coker *et al.*		26	3	12	2	8	1	4
Current series		24	1	4	6	25	2	8
Total		211	27	13	40	19	19/154	12
Radiation therapy to neck only, with or without surgery								
Marcial-Vega *et al.*		19	1	5	10	53	0	0
Fitzpatrick and Kotalik		95	8	8	—		—	
Fermont		129	4	3	14	11	—	
Lee *et al.*		11	0	0	—		—	
Carlson *et al.*		20	1	5	2	10	2	10
Glynne-Jones *et al.*		49	4	8	14	29	1	2
Total		323	18	6	40/217	18	3/88	3
Radiation therapy to both sides of neck and mucosa, with or without surgery								
Wang *et al.*	RT	56	3	5	16	29	—	—
	RT+S	44	5	11	10	23	—	—
Marcial-Vega *et al.*	RT±S	53	7	13	30	57	3	6
Carlson *et al.*	RT±S	73	2	3	2	3	0	0
Harper *et al.*	RT±S	65	5	8	14	22	0	0
de Braud *et al.*	RT	16	2	13	8	50	—	—
McCunniff and Raben	RT±S	19	1	5	7	37	—	—
Jesse *et al.*	RT	52	1	2	11	21	0	0
	RT+S	28	3	11	4	14	0	0
Silverman *et al.*	RT±S	83	13	16	28	34	0	0
Total		489	42	9	130	27	3/354	1

[a]Modified from Coster, J. R., Foote, R. L., Olsen, K. D., *et al.*: Cervical node metastasis of squamous cell carcinoma of unknown origin: Indications for withholding radiation therapy. *Int. J. Radiat. Oncol. Biol. Phys.* **23**, 743, 1992. With permission from Elsevier Science.

[b]Primary squamous cell carcinoma developed above the clavicles within the nasopharynx, oropharynx, hypopharynx, or larynx within 5 years of neck operation for nodal metastases.

Abbreviations: RT, radiation therapy; S, surgery.

surgery alone [24,26–28], but it is probably underrepresented due to death from neck recurrence, distant metastasis, and intercurrent disease. At the University of Florida, Harper *et al.* [21] compared the incidence of subsequent mucosal lesion in 393 patients with known primary head and neck tumors to 65 patients with unknown primary sites but received 5000 to 6000 cGy of radiation to neck and mucosa. The incidence of developing subsequent mucosal lesion was about the same, 24% compared to 25% at 11 years, suggesting that radiotherapy was highly effective in preventing the appearance of a primary lesion. From a review by Coster *et al.* [20] a relapse in contralateral neck averages 12, 3, and 1% after surgery alone, involved neck irradiation, and bilateral neck irradiation, respectively. In a retrospective study by Reddy and Marks [29], contralateral neck control was 86 and 56% in patients who received radiation therapy to bilateral neck and involved neck alone, respectively. However, no disease survival benefit was observed with bilateral neck irradiation.

The 5-year actuarial survival reported in the literature is in the range of 40 to 60%, which is favorable compared to a similar neck stage disease with a known primary site. It is difficult to evaluate survival benefits of radiation treatment to bilateral neck and potential primary sites, as it is based on data from retrospective studies with small heterogeneous patient groups from single institutions over long spans of time. Appearance of the primary lesion, however, has a direct impact on survival. Jesse *et al.* [24] reported a 3-year survival rate of 31% in patients when the primary appears, compared to 58% for patients who never developed a primary tumor. In summary, multicenter prospective randomized clinical trials are needed to investigate the critical issues involved in the management of patient with this intriguing disease.

References

1. de Braud, F., Ahmad, K., Ensley, J. F., Tapazogolou, E., Kish, J. A., and AL Sarraf, M. (1988). Chemotherapy as part of the treatment of patients with squamous cell carcinoma of unknown primary metastasic to the neck. *Proc. Am. Soc. Clin. Oncol.* **7**, 156 [Abstract].
2. Fried, M. P., Diehl, W. H., Browson, Jr., *et al.* (1975). Cervical metastasis from unknown primary. *Ann. Oncol.* **84**, 152.
3. Lefevre, J. L., Coche-Dequeant, B., Ton Van, J., *et al.* (1990). Cervical lymph nodes from unknown primary tumor in 190 patients. *Am. J. Surg.* **160**, 443–446.
4. Richard, J. M., and Michaeu, C. (1977). Malignant: Cervical adenopathies from carcinoma of unknown origin. *Tumori* **63**, 249–258.
5. Abbuzzese, J. L., Lenzi, R., Raber, M. N., *et al.* (1993). The biology of unknown primary tumors. *Semin. Oncol.* **20**, 238–243.
6. Frost, P., and Levin, B. (1992). Clinical implications of metastatic process. *Lancet* **339**, 1458–1461.
7. Frost, P. (1991). Unknown primary tumors: An example of accelerated (type 2) tumor progression. *Basic Life Sci.* **57**, 233–237.
8. Colletier, P. J., Garden, A. S., Morrison, W. H., Goepfert, H., Geara, F., and Ang, K. K. (1998). Postoperative radiation for squamous cell carcinoma metastatic to cervical lymph nodes from unknown primary site: Outcomes and patterns of failure. *Head Neck* **20**, 674–681.
9. Weir, L., Keane, Th., Cummings, B., Goodman, P. H., O Sullivan, B., Payne, D., and Warde, P. (1995). Radiation treatment of cervical lymph node metastases from an unknown primary: An analysis of outcome by treatment volume and other prognostic factors. *Radiother. Oncol.* **35**, 206–211.
10. Wang, R. C., Geopfert, H., Barber, A., and Wolf, P. F. (1986). Squamous cell carcinoma metastatic to the neck from an unknown primary site. *In* "Cancer in the Neck: Evaluation and Treatment" (D. L. Larson, A. J. Ballnytne, and O. M. Guillamondegui, eds.), pp. 183–192.
11. Frable, M. A., and Frable, W. J. (1982). Fine-needle aspiration biopsy revisited. *Laryngoscopy* **92**, 1414–1418.
12. Tsai, S. T., and Jin, Y. T. (1995). Detection of Epstein-Barr viruses in metastatic neck squamous cell carcinoma by polymerase chain reaction and EBER1 in situ hybridization for differentiating the primary site. *J. Otolaryngol. Soc. Roc.* **30**, 66–72.
13. Macdonald, M. R., Freeman, J. L., Hui, M. F., Cheung, R. K., Warde, P., Mclvor, N. P., Irish, J., and Dosch, H. M. (1995). Role of Epstein Barr virus in fine needle aspirates of metastatic neck nodes in the diagnosis of nasopharyngeal carcinoma. *Head Neck* **17**(6), 487–493.
14. Feinmesser, R., Miyazaki, I., Cheung, R., Freeman, J. L., Noyek, A. M., and Dosch, H. M. (1992). Diagnosis of nasopharyngeal carcinoma by DNA amplification of tissue obtained by fine needle aspiration. *N. Engl. J. Med.* **326**(1), 17–21.
15. Lee, W. Y., Hsiao, J. R., Jin, Y. T., and Tsai, S. T. (2000). Epstein-Barr virus detection in neck metastases by in-situ hybridization in fine needle aspiration cytologic studies: An aid for differentiating the primary site. *Head Neck* **22**, 336–340.
16. Randal, D. A., Johnstone, P. A., Foss, R. D., and Martin, P. J. (2000). Tonsillectomy in diagnosis of the unknown primary tumor of the head and neck. *Otolaryngol. Head Neck Surg.* **122**, 52–55.
17. Lapeyre, M., Malissard, L., Peiffert, D., Hoffstetter, S., Toussaint, B., Renier, S., Dolivet, G., Simon, C., and Bey, P. (1997). Cervical lymph node metastasis from an unknown primary: Is a tonsillectomy necessary? *Int. J. Radiat. Oncol. Biol. Phys.* **39**(2), 291–296.
18. Califano, J., Westra, W., Kock, W., Meininger, G., Reed, A., Yip, L., Boyle, J., Lonardo, F., and Sidransky, D. (1999). Unknown primary head and neck squamous cell carcinoma: Molecular identification of the site of origin. *J. Natl. Cancer Inst.* **91**, 599–604.
19. Jungehulsing, M., Scheidhauer, K., Damm, M., Pietrzyk, U., Eckel, H., Schicha, H., and Stennert, E. (2000). 2[^{18}F]- fluoro-2-deoxy-D-glucose positron emission tomography is a sensitive tool for the detection of occult primary cancer (carcinoma of unknown primary syndrome) with head and neck lymph node manifestation. *Otolaryngol Head Neck* **123**, 294–301.
20. Coster, J. R., Foote, R. L., Olsen, K. D., Jack, S. M., Schaid, D. J., and DeSanto, L. W. (1992). Cervical nodal metastasis of squamous cell carcinoma of unknown origin: Indications for withholding radiation therapy. *Int. J. Radiat. Oncol. Biol. Phys.* **23**(4), 743–749.
21. Harper, C. S., Mendenhall, W. M., Parsons, J. T., Stringer, S. P., Cassissi, N. J., and Million, R. R. (1990). Cancer in neck nodes with unknown primary site: Role of mucosal radiotherapy. *Head Neck* **12**(6), 463–469.
22. Nguyen, C., Shenouda, G., Black, M. J., Vuong, T., Donath, D., and Yassa, M. (1994). Metastatic squamous cell carcinoma to cervical lymph nodes from unknown primary mucosal sites. *Head Neck* **16**(1), 58–63.
23. Colletier, P. J., Garden, A. S., Morrison, W. H., Geopfert, H., Geara, F., and Ang, K. K. (1998). Postoperative radiation for squamous cell carcinoma metastatic to cervical lymph nodes from unknown primary site: Outcomes and patterns of failure. *Head Neck* **20**(8), 674–681.
24. Jesse, R. H., Perez, C. A., and Fletcher, G. H. (1973). Cervical lymph node metastases: Unknown primary cancer. *Cancer* **31**, 854–859.
25. Maulard, C., Housset, M., Brunel, P., Huart, J., Ucla, L., Rozec, C., Delanian, S., and Baillet, F. (1992). Postoperative radiation therapy for cervical lymph node metastases from occult squamous cell carcinoma. *Laryngoscope* **102**(8), 884–890.
26. Barrie, J. R., Knapper, W. H., and Strong, E. W. (1970). Cervical nodal metastases of unknown origin. *Am. Surg.*$\,$**120**, 467–470.
27. Coker, D.D., Cesterline, P. F., Chambers, R. G., and Jaques, D. A. (1977). Metastases to lymph nodes of the head and neck from an unknown primary site. *Am. J. Surg.* **134**, 517–522.
28. Wang, R. C., Geopfert, H., Bonker, A. E., and Wolf, P. (1990). Unknown primary squamous cell carcinoma metastatic to the neck. *Arch. Otolaryngol Head Neck Surg* **116**, 1388–1393.
29. Reddy, S. P., and Marks, J. E. (1997). Metastatic carcinoma in the cervical lymph nodes from an unknown primary site: Results of bilateral neck plus mucosal irradiation VS ipsilateral neck irradiation. *Int. J. Radiat. Oncol. Biol. Phys.* **37**(4), 797–802.

b. Adenoid Cystic Carcinoma

I. Introduction 16
II. Incidence 16
III. Natural History 17
IV. Pathology 17
V. Evaluation 18
VI. Treatment 19
A. Surgery 19
B. Radiation Therapy 19
C. Chemotherapy 19
VII. Results 19
References 21

I. INTRODUCTION

Since Bilroth first described adenoid cystic carcinoma (ACC) in 1854, with the term cylindroma [1], the unique biological behavior and natural history have been well documented. Despite aggressive local treatments, the local control rate is still low and long-term survival is often compromised by the development of metastasis. It has characteristically a protracted and insidious development of local recurrence and metastasis, which often occur after decades. However, considerable differences in the clinical course have been observed among patients with ACC. Currently the biologic mechanisms underlying this peculiar tumor and its widerange in clinical behavior remain unresolved. More effective treatments need to be investigated for this aggressive disease with its intriguing biology. The most common site of ACC is in salivary glands, especially in minor salivary glands. Although ACC is known to occur in other organs besides salivary glands, incidence is rare and published data are scarce. This chapter focuses on ACC occurring in salivary glands.

II. INCIDENCE

ACC comprises 2 to 5% of all tumors of the salivary glands. It is the most common malignancies of the minor salivary glands. As shown in Fig. 2b.1, ACC is reported to be 36 to 58% of all malignancies of the submandibular glands [2–4]. It is less common in the parotid gland, where only 7 to 18% are ACC. The most common age at presentation is 50 to 60 years old with no predilection for either gender. The most common presenting symptom is a painless mass, often enlarging rapidly, but presenting indolently for years prior to presentation. Pain is a relatively uncommon symptom and may indicate involvement of deep structures. Lymph node metastasis at presentation occurs in about 10 to 30% [5–7]. About 25% of patients present with facial palsy due to cranial nerve invasion, but pathologic involvement of perineural space (PNS) is reported to be as high as 80% [8,9]. Approximately, 40 to 50% of patients present with stage III and IV disease [10,11]. Although etiologic factors for salivary gland tumors are not well defined, an association with a lack of vitamins A and C in the diet and exposure to radiation has been reported. However, no such specific etiologic factors are linked to ACC.

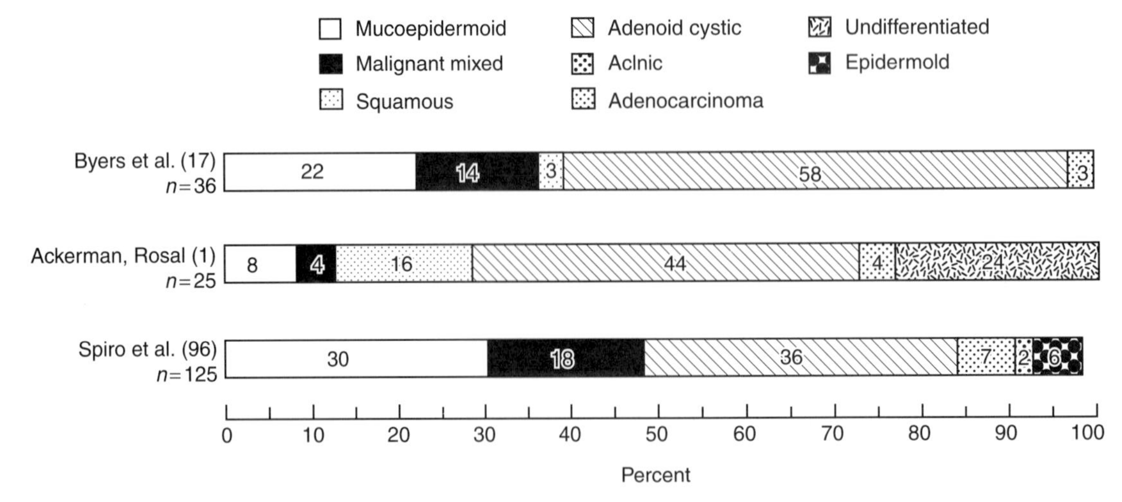

FIGURE 2b.1 Distribution of malignant tumors of the submandibular salivary gland.

III. NATURAL HISTORY

ACC has a high rate of local recurrence and distant metastasis despite aggressive initial local treatment. Because of frequent late development of metastasis, despite initial local control, it may take decades of follow-up to appreciate the insidious clinical course in some patients. In an analysis of 196 patients with ACC at Memorial Sloan Kettering between 1936 and 1986, Spiro *et al.* [12] noted that the disease-free interval ranged from 1 month to 19 years and exceeded 10 years in 8%. Prognostic factors of ACC include tumor site and size, nodal involvement, histology, perineural invasion, and DNA content. These factors are discussed in the following sections. The pattern of local invasion is characterized by diffuse infiltration into adjacent structures such as skin, soft tissue, bone, and nerve. Its unique feature, however, is perineural invasion, which has been reported to be as high as 80% histologically. In 198 patients with ACC of the head and neck treated with surgery and postoperative radiation therapy at M.D. Anderson between 1962 and 1991, Garden *et al.* [13] found 69% had perineural spread and 28% had invasion of a major cranial nerve. Crude failure rate was 18 and 12%, respectively, with and without major nerve invasion. Failure along the base of skull developed only in 2% in this study. However, failure of the path of cranial nerves and the base of skull is not uncommon in patients with extensive perineural invasion treated with surgery alone. Seaver *et al.* [14] reported perineural involvement in 11 out of 19 patients with recurrence. Vrielinck *et al.* [15] found perineural invasion in 53% of their patients and noted its inverse correlation with survival. However, Perzin and co-workers [16] found no such correlation. Most local recurrence occurs within a few years of initial treatment. In 37 patients with ACC at McGill, Haddad *et al.* [17] noted that average time for local failure was 3.5 years and 62% of local failure was seen within 2 years.

Distant metastasis is reported as high as 50% and is more common than nodal relapse. It is not unusual to occur long after initial local treatment. In Haddad's study, median time for appearance of distant metastasis was 8.1 years [17]. Metastasis is usually associated with local failure, but it is not infrequent as the only site of failure. Metastatic disease may progress in an indolent course, often over years, and survival can be protracted in some patients even without treatment. In Spiro's study, 74 patients (38%) developed distant metastasis, and in 23 patients (31%), metastasis occurred without locoregional failure [12]. Survival with distant metastasis was less than 3 years in 54 % and more than 10 years in 10% of patients. Fordice and co-workers [18] reported 22% distant metastasis as the only site of failure in 160 patients with ACC at M.D. Anderson between 1977 and 1996. Matsuba *et al.* [10] found that distant metastasis occurred in 51% of 71 patients, and distant metastasis occurred in 35%, despite local control. Lung is the most common site of metastasis (70%), followed by bone and liver. Isolated pulmonary metastasis can follow an indolent course and should not discourage treatment of locoregional disease. In Spiro's analysis of 74 patients with metastasis [12], the lung was the only site of metastasis in 50 out of 74 patients, and in 17 patients the lung was involved in addition to other sites. Matsuba *et al.* [10] reported that pulmonary metastasis occurred in 63% of all metastasis. The prognosis of patients with pulmonary was more favorable than that for those with other metastasis, especially osseous metastasis [19].

IV. PATHOLOGY

ACC is believed to originate from the reserve epithelial calls in the intercalated ducts, which can differentiate into epithelial and myoepithelial cell forms [20]. Two major types of cells in ACC are ductal and nonductal cells. Ductal cells express CEA and EMA, and nonductal cells express muscle-specific actin characteristic of myoepithelium on immunohistochemical staining [21].

The three histological classifications of ACC are tubular, cribriform, and solid pattern. However, most ACC are composed of a mixture of different patterns. The tubular pattern is characterized by an elongated tubular formation with a central lumen. In the cribriform pattern, cells are arranged in nests that are fenestrated by round and oval spaces. The solid pattern has a high cellularity with at least 30% composed of solid cellular arrangement with infrequent lumen or fenestrations. The degree of cellularity correlates with grade and prognosis. Low-grade ACC is generally a mixture of tubular and cribriform patterns, and the solid pattern is considered a highgrade. Perineural invasion, however, is seen frequently in both low- and high-grade tumors [22]. Several studies suggested that the best prognosis occurs with the tubular pattern, an intermediate prognosis with cribriform and the worst prognosis with a solid pattern [23–25]. Szanto *et al.* [26] divided ACC into grades I, II, and III based on the specimen that was a solid pattern: grade I had no solid component, grade II had less than 30% solid areas, and grade III had at least 30% solid pattern. With 15-year follow-up, cumulative survival correlated well with grade: 39, 26, and 5% [26]. Spiro [2], however, found no such correlation between grade and survival in ACC.

DNA content (DNA ploidy and % S phase) has been reported to have prognostic significance. In a study by Tytor *et al.* [27], aneuploid tumors had a higher S-phase value and recurrence rate compared with diploid tumors. Frazen *et al.* [28] reported a higher grade and stage with aneuploid tumor than diploid tumor.

p53 expression was present in 69% of ACC in a study by Karja *et al.* [29]. In this study, the highest prevalence for p53 expression was seen in ACC among 219 salivary gland tumors. However, no correlation was found between p53 expression and local recurrence, metastatic disease or survival of patients.

Estrogen (ER) and the progesterone receptor (PR) were studied with immunohistochemical staining in 27 cases of ACC by Dori *et al.* [30], but ER was not present in any tumor and PR was positive only in 2 cases. This result does not support the use of hormonal treatment for ACC.

Neural adhesion molecules (N-CAMs) were expressed in 14 out of 15 (93%) tumors with perineural invasion in a study by Gandour-Edwards *et al.* [31]. The biological mechanism and clinical significance of this finding remain to be investigated.

V. EVALUATION

It is essential to obtain a detailed history, particularly the duration of mass and change in the intensity of presenting symptoms, in order to assess clinical progression. A thorough and detailed examination of the head and neck area, including adjacent structures, cranial nerve function, and cervical lymph node status, should be performed. Standard radiological evaluation includes a chest X-ray, computed tomographic scan of the head and neck area, including the base of skull, and a bone scan. The CT scan is useful in evaluating the extent of the lesion, as well as lymph node metastasis. Magnetic resonance imaging provides a better image of tumor margins and internal architecture than a CT scan and is also useful for a better delineation of tumor from surrounding soft tissues, particularly blood vessels and nerves. ACC usually appears as a well-delineated to highly infiltrative homogeneous mass on CT and MRI. CT and MRI have about the same ability to detect perineural invasion. Nerve enlargement or fat displacement by the tumor around the nerve on a MRI scan and enlargement of bony foramina from nerve expansion on a CT scan are some of the indications of perineural invasion. A chest CT scan is useful for the detection of early metastasis not visualized on a chest X-ray.

ACC in major salivary glands are staged according to current American Joint Committee (AJC) TNM staging classification as shown in Table 2b.1 [32]. ACC arising from minor salivary glands are staged according to the AJCC system by the site of origin. For example, ACC of the hard palate is staged according to the oral cavity staging system. Dental evaluation should be obtained if radiation therapy is anticipated.

For a tissue diagnosis of salivary glad tumor, the commonly performed procedure is a fine needle aspiration. FNA is performed easily with a low complication rate and is accurate with a sensitivity and specificity of over 90% [33–37]. However, the rate of inadequate sampling ranges from 4 to 7% [38–41]. It is also difficult to accurately diagnose histological types of salivary neoplasm by FNA, which lacks histological architecture. Therefore, the main utility of FNA is in establishing the diagnosis of malignancy and is, therefore, helpful for planning appropriate surgery and extent of

TABLE 2b.1 Staging System for Major Salivary Gland Malignancy[a,b]

Primary tumor (T)

Tx	Primary tumor cannot be assessed
T0	No evidence of primary tumor
T1	Tumor 2 cm or less in greatest dimension
T2	Tumor more than 2 cm but not more than 4 cm in greatest dimension
T3	Tumor more than 4 cm but not more than 6 cm in greatest dimension
T4	Tumor more than 6 cm in greatest dimension

Regional lymph nodes (N)

Nx	Regional lymph nodes cannot be assessed
N0	No regional lymph node metastasis
N1	Metastasis in a single ipsilateral lymph node, 3 cm or less in greatest dimension
N2	Metastasis in a single ipsilateral lymph node, more than 3 cm but not more than 6 cm in greatest dimension; in multiple ipsilateral lymph nodes, none more than 6 cm in greatest dimension; or in bilateral or contralateral lymph nodes, none more than 6 cm in greatest dimension
N2a	Metastasis in a single lymph node more than 3 cm but not more than 6 cm in greatest dimension
N2b	Metastasis in multiple lymph nodes, none more than 6 cm in greatest dimension
N2c	Metastasis in bilateral or contralateral lymph nodes, none more than 6 cm in greatest dimension
N3	Metastasis in a lymph node more than 6 cm in greatest dimension

Distant metastasis (M)

Mx	Presence of distant metastasis cannot be assessed
M0	No distant metastasis
M1	distant metastasis

Stage grouping

Stage I	T1a	N0	M0
	T2a	N0	M0
Stage II	T1b	N0	M0
	T2b	N0	M0
	T3a	N0	M0
Stage III	T3b	N0	M0
	T4a	N0	M0
	Any T	N1	M0 (except T4b)
Stage IV	T4b	Any N	M0
	Any T	N2	M0
	Any T	N3	M0
Any T	Any N	M1	

[a]From American Joint Committee on Cancer (1997): "AJCC Staging Manual," 5th ed., p.57. Lippincott-Raven, Philadelphia.

[b]All categories are subdivided: (a) no local extension or (b) local extension. Local extension is clinical or macroscopic evidence of invasion of skin, soft tissues, bone, or nerve. Microscopic evidence alone is not local extension for classification purposes.

surgery. An incisional biopsy is generally not recommended due to the possibility of tumor implantation. Excisional biopsy for minor salivary gland tumor is a complete excision and for parotid tumors is a radical parotidectomy, total parotidectomy, or superficial parotidectomy with facial nerve preservation depending on the extent of the lesion.

VI. TREATMENT

A. Surgery

Primary treatment of ACC of the salivary gland is surgery. The surgical technique depends on location and extent of the primary lesion, on the status of regional lymph nodes, and is well described in surgical literature [42]. Because a positive surgical margin is associated with a poor outcome in patients with ACC [13,16,18], an attempt should be made to obtain negative margins. However, it may not be possible to obtain negative margins for lesions involving the paranasal sinus, nasopharynx, and nasal cavity without extensive disfiguring surgery. If the patient does not have signs of cranial nerve invasion, a nerve-sparing operation is generally done. However, it is not infrequent to find tumor adherence to major nerves requiring sacrifice of the nerve. In such situations, controversy still exits whether the nerve can be preserved with the delivery of postoperative radiation therapy. Based on retrospective studies demonstrating increased local recurrence with perineural invasion, some advocate the sacrifice of involved major nerves [43]. However, radiation therapy literature suggests that local control with postoperative radiation is equivalent to radical surgery for microscopically positive margins allowing for major nerve preservation [25]. A prospective trial is needed to evaluate the efficacy of postoperative radiation therapy in ACC.

Neck dissection is generally not performed electively for early stage, low-grade ACC, but should be performed for involved node, high-grade or advanced stage ACC. In analysis of 407 patients with major salivary gland malignancy by Armstrong *et al.* [44], elective neck node dissection revealed a positive node in only 2 out of 54 (4%) patients with ACC. However, first echelon lymph nodes are generally included in the surgical specimen of the salivary gland tumor. First echelon nodes are also readily accessible for frozen section examination; if the frozen section is positive, neck dissection can proceed.

B. Radiation Therapy

The main roles for radiation therapy in the management of ACC are (1) as an adjunct to surgery as postoperative treatment for patients with a high risk for local-regional recurrence and (2) for primary treatment of inoperable or unresectable lesions.

Indications of postoperative radiation therapy in ACC include (1) High-grade tumor, (2) advanced stage, (3) microscopic positive or close margins, (4) gross residual disease, (5) invasion of skin, soft tissue, or bone, (6) involvement of lymph node, and (7) recurrent disease.

Detailed techniques of radiation therapy for ACC involving various major salivary glands are well described in the radiation therapy literature [42,45,46]. For postoperative radiotherapy, careful attention should be given to include the entire pathologic extent of the lesion. Generally, the entire surgical bed with a 2-cm margin is included. Due to the high propensity of ACC for perineural invasion, cranial nerve pathways to the base of the skull need to be included if perineural invasion is present. Postoperative radiation therapy should be given to the ipsilateral neck, including the lower neck, for positive node neck dissections, as well as high grade or recurrent disease. The postoperative radiation dose ranges from 50 to 64 Gy over 5 to 6 weeks depending on the grade and pathologic extent of the disease. For gross residual disease, inoperable, or unresectable lesions, a dose of up to 70 Gy is required. Most centers use a combination of 4 to 6-MV photons and electrons, but a wedge pair photon beam technique is also used. Neutron beam therapy has been used in the treatment of ACC in a number of centers in Europe and the United States with the local control rate reported to be photon radiation [47–51]. However, neutron treatment facilities are limited to only a few centers in the world.

C. Chemotherapy

To date, no effective adjuvant chemotherapeutic agent has been reported for ACC. Data on efficacy chemotherapy for ACC come from the treatment of metastatic lesions or for the palliation of local-regional disease involving a relatively small number of patients, usually combined together with other histological types of salivary gland tumors. Among the different agents reported, cisplatin-based regimens showed moderate response rates of 30 to 40% [52–56]. A southwest oncology group phase II trial with mitoxantrone in ACC showed no significant antitumor activity [57]. Vermorken and colleagues [58] reported a local-regional response rate of 48% and a 15% response rate for metastatic disease based on pooled data from several studies. However, these responses have not been shown to translate into improved survival in ACC. Prospective clinical trials are needed to facilitate the development of more active systemic agents for metastatic disease, which is a major determinant of survival.

VII. RESULTS

Because the incidence of ACC is relatively low, most treatment results are from retrospective studies on salivary

TABLE 2b.2 Local/Regional Tumor Control Rates for Malignant Salivary Gland Tumors Using Low LET (Photon/Electron) Radiotherapy in Conventional Fractionation Schemas

Author/rate	Patient number
Fitzpatrick and Thoriault/12%(6/50)	50
Vikram/4%(2/49)	49
Borthne/23%(8/35)	35
Rafla/36%(9/25)	25
Fu/32%(6/19)	19
Stewart/47%(9/19)	19

gland tumors, which include other histology or involve a relatively small number of patients without sufficient follow-up. These studies, however, all reported improved local control with postoperative radiation therapy [59–64]. One of the largest series in these early studies is by Yu and Ma [61]. They analyzed 405 cases of salivary gland cancer and reported that patients with ACC had a higher survival rate after combined treatment with surgery followed by radiation compared to surgery alone. Other studies, however, failed to show a survival benefit for postoperative radiation [65].

Neutron radiation treatment has been investigated extensively for salivary gland cancers. Batterman *et al.* [66] investigated the relative biological effectiveness (RBE) of the fast neutron in metastatic lesions in lung from various malignancies. They reported the highest RBE of 8.0 in ACC compared to 3.0–3.5 for normal tissues. Various neutron facilities have conducted phase I/II trials of neutron treatment in inoperable or recurrent salivary gland tumors since then. Results of these trials are summarized in Table 2b.2 [67]. An overall local-regional control of 67% is significantly higher compared to 24% (Table 2b.3) with photon/electron treatment in inoperable or recurrent salivary gland tumor (Fig. 2.1). Based on these reports, the Radiation Therapy Oncology Group (RTOG) in the United States and the Medical Research Council in England conducted a randomized phase III trial comparing neutron treatment and photon treatment in inoperable salivary gland tumors. The 2-year local-regional control for neutron therapy was significantly higher than photon treatment: 67% vs 17% ($P<0.005$), respectively. Survival was initially reported to be higher in the neutron group than in the photon group: 62% vs 25%, respectively [68]. Because of this finding, the trial was closed early for ethical reasons. The latest report of this trial in 1993 continues to show significant superiority for neutron therapy in terms of local-regional control. However, the survival benefit of neutron was preserved due to the development of metastasis [69].

The two studies discussed here are the largest reported studies for ACC. Each involves over 150 patients treated in a single institution. The first study is postoperative treatment and uses photon therapy. The second study is primary treatment for

TABLE 2b.3 Local/Regional Tumor Control Rates for Malignant Salivary Gland Tumors Using High LET Fast Neutron Radiotherapy

Author/rate	Patient number	Control (%)
Saroja (71/113)	113	63
Catterall and Errington (50/65)	65	77
Buchholz (40/52)	52	77
Batterman and Mijnheer (21/32)	32	66
Duncan (12/22)	22	55

gross residual or recurrent disease and uses neutrons. Both studies demonstrate the unique natural history of ACC and address some of the key issues in the management of ACC.

Garden *et al.* [13] reviewed 30 years experience with postoperative photon radiation treatment in 198 patients at M.D. Anderson Hospital. Actuarial 5-, 10-, and 15-year freedom from relapse rate was 68, 52, and 45%, respectively. In this study, 83 out of 198 patients (42%) had positive margins and 55 patients (28%) had invasion of a major named nerve. Local recurrence developed in 23 patients (12%), and 5-, 10-, and 15-year actuarial local control was 95, 86, and 79%, respectively. Two adverse factors for local control reported in the analysis are positive margins and involvement of a major (named) nerve. The 10-year actuarial local control rate for patients with positive margins was 77% compared to 93% for patients with negative positive margins ($P=0.006$). The 10-year actuarial local control rate for patients with negative and close or uncertain margins was 93 and 92%, respectively. The 10-year actuarial survival rate with and without major (named) nerve involvement was 80 and 88%, respectively ($P=0.02$). The 10-year actuarial survival rate correlated with the number of adverse factors present: none, 93%; one, 83%; and two, 73%. However, the author does not mention whether the cranial nerve pathway was treated routinely in patients with a major (named) nerve invasion. The most common site of failure was distant metastasis, which developed in 74 patients (37%) in which 61 out of the 74 patients (82%) had local control. In contrast, in the study by Spiro *et al.* [12], 69% of the metastasis was associated with local-regional recurrence. This study demonstrates the efficacy of postoperative radiation treatment in ACC in preventing local recurrence. However, 37% still developed metastasis despite a high local control rate. Therefore, without the development of more effective treatment for systemic disease, survival is unlikely to improve significantly.

Douglas *et al.* [47] reported a 5-year actuarial local–regional control rate of 57% in 151 patients with gross residual disease or unresectable disease after neutron treatment. The 5-year actuarial overall survival and cause-specific survival were 72 and 77%, respectively. These results are encouraging considering a patient population with gross residual disease or recurrent disease. In multivariate analysis, the factors that had a significant negative impact on local-regional control were base of skull involvement and surgery consisting of a biopsy only. Local-regional control with and without the base of skull involvement was 68% vs. 23% ($P<0.01$), respectively. Patients who had a biopsy only had a recurrence rate of 71% compared to 43% in those who had at least surgical resection attempted ($P<0.03$). Patients without base of skull invasion who under went surgical resection had a 5-year local-regional control rate of 80%. It is interesting that all 8 patients with only microscopic disease had 100% 5-year actuarial local-regional control. On multivariate analysis, two factors influencing distant metastasis are involvement of the base of skull and regional lymph nodes. About 50% of the node-positive patients develop metastasis compared to 26% of node-negative patients at 5 years ($P<0.001$). With base of skull invasion, development of metastasis was seen in 56% of patients compared to 24% in patients with no invasion ($P<0.01$). In this study, perineural invasion did not influence either local-regional control or metastasis, and the authors suggest that neutron therapy is effective in eradicating microscopic disease along nerves, as the entire nerve pathway to the base of skull was treated routinely with 12 neutron Gy. The 5-year actuarial grade 3 and 4 complication rate was reported as 13.5%.

At Wayne State University Medical Center, treatment with a mixed beam of 30 Gy of photon and 10 to 15 Gy of neutron has been used in a postoperative setting, as well as for unresectable and inoperable lesions. Our experience also confirms a high rate of local control with a low complication rate.

In summary, the following conclusions can be made from available data on ACC.

1. Surgery is the primary treatment of resectable disease.
2. Postoperative radiation treatment improves local-regional control.
3. For unresectable and inoperable lesions, neutron radiation treatment offers superior local-regional control compared to photon radiation treatment.
4. Overall survival is not improved with postoperative radiation treatment due to distant metastasis, which occurs in 40–50% of patients.
5. Prospective clinical trials are needed to develop effective systemic agents for the improvement of overall survival.

References

1. Bilroth, T. (1856). Die Cylindergeschwulst (cylindroma) in Untersuchungen uber die Entwicklung der Blutgerfarse.nebst Beobachtungen aus der koniglichen chirurgischen Universats-Kilink zu Berlin. Berlin G. Reimer. 1855–1869.
2. Spiro, R. H. (1986). Salivary neoplasm: Overview of a 35 year experience with 2807 patients. *Head Neck Surg.* **8**, 177.
3. Byers, R. N., Jesse, R. H., Guillamondegui, O. M., *et al.* (1973). Malignant tumors of the submaxillary gland. *Am. J. Surg.* **126**, 458.
4. Ackerman, L. V., and Rosai, J. (1974). Major and minor salivary glands. *In* "Surgical Pathology" (L. V. Ackerman, and J. Rosai, eds.), 5th Ed., Vol. 19, p. 515. CV Mosby, St. Louis, MO.
5. Kumar, R. V., Kini, L., Bhargava, A. K., Mukherje, G., *et al.* (1989). Salivary duct carcinoma. *J. Surg. Oncol.* **12**, 316–319.
6. Spiro, R. H., Huvos, A. G., and Stang, E. W. (1974). Adenoid cystic carcinoma of salivary origin: A clinicopathologic study of 242 cases. *Am. J. Surg.* **128**, 512.
7. Jones, A. S., Hamilton, J. W., Rowley, H., Husband, D., and Helliwell, T. R. (1997). Adenoid cystic carcinoma of the head and neck. *Clin. Otolaryngol.* **22**(5), 434–443.
8. Matsuba, H. M., Spector, G. J., Thawley, S. E., *et al.* (1986). Adenoid cystic salivary gland carcinoma: A histopathologic review of treatment failure patterns. *Cancer* **57**, 519–524.
9. Conley, J., and Dingman, D. L. (1974). Adenoid cystic carcinoma in the head and neck (cylindroma). *Arch. Otolaryngol.* **100**, 81–90.
10. Matsuba, H. M., Spector, G. J., Thawley, S. E., Simpson, J. R., Mauney, M., and Pikul, F. J. (1986). Adenoid cystic salivary gland carcinoma. A histopathologic review of treatment failure patterns. *Cancer* **57**(3), 519–524.
11. Feldmann, H. J., Budach, V., Budach, W., Molls, M., and Sack, H. (1991). Postoperative radiotherapy of salivary gland tumors. Prognostic factors and treatment results. *Strahlenther Onkol.* **167**(5), 261–266.
12. Spiro, R. H., Huvos, A. G., and Strong, E. W. (1997). Adenoid cystic carcinoma: Factors influencing survival. *Am. J. Surg.* **138**(4), 579–583.
13. Garden, A. S., Weber, R. S., Morrison, W. H., Ang, K. K., and Peters, L. J. (1995). The influence of positive margins and invasion in adenoid cystic carcinoma of the head and neck treated with surgery and radiation. *Int. J. Radiat. Oncol. Biol. Phys.* **32**(3), 619–626.
14. Seaver, P. R., Jr., and Kuehn, P. G. (1979). Adenoid cystic carcinoma of the salivary glands. A study of ninety-three cases. *M. J. Surg. A* **137**(4), 449–455.
15. Vrielinck, L. J., Ostyn, F., Van Damme, B., *et al.* (1988). The significance of perineural spread in adenoid cystic carcinoma of the major and minor salivary glands. *Int. J. Oral. Maxillofac. Surg.* **17**, 190–193.
16. Perzin, K. H., Gullane, P., and Clairmont, A. C. (1978). Adenoid cystic carcinoma arising in salivary glands: A correlation of histologic features and clinical course. *Cancer* **4**, 265–282.
17. Haddad, A., Enepekides, D. J., Manolidis, S., and Black, M. (1995). Adenoid cystic carcinoma of the head and neck: A clinicopathologic study of 37 cases. *J. Otolaryngol.* **24**(3), 201–205.
18. Fordice, J., Kershaw, C., El Naggar, A., and Goepfert, H. (1999). Adenoid cystic carcinoma of the head and neck: Predictors of - morbidity and mortality. *Arch. Otolaryngol, Head Neck Surg.* **125**(2), 149–152.
19. Million, R. R., and Cassissi, J. N. (1994). Minor salivary gland tumors. *In* "Management of Head and Neck Cancer: A multidisciplinary Approach," (R. R. Million, and J. N. Cassissi eds). Lippincott, Philadelphia.
20. Batsakis, J. G., Regezi, J. A., Luna, M. A., *et al.* (1989). Histogenesis of salivary gland neoplasms: A postulate with prognostic implications. *J. Laryngol. Otolaryngol.* **103**, 939.
21. Azumi, N., and Battifora, H. (1987). The cellular composition of adenoid cystic carcinoma: An immunohistochemical study. *Cancer* **60**, 1589.
22. Goepfert, H., Luna, M. A., Lindberg, R. D., *et al.* (1983). Malignant salivary gland tumors of the paranasal sinuses and nasal cavity. *Arch. Otolaryngol.* **109**, 662–668.
23. Matsuba, H. M., Simpson, J. R., Mauney, M., *et al.* (1986). Adenoid cystic salivary gland carcinoma: A clinicopathologic correlation. *Arch. Otolaryngol Head Neck Surg.* **57**, 519.
24. Press, M. F., Pike, M. C., Hung, G., *et al.* (1994). Amplification and overexpression of HER–2/neu in carcinomas of the salivary gland: Correlation with poor prognosis. *Cancer Res.* **54**, 5675–5682.
25. Simpson, J. R., Thawley, S. E., and Matsuba, H. M. (1984). Adenoid cystic salivary gland carcinoma: Treatment with irradiation and surgery. *Radiology* **15**, 509.

26. Szanto, P. A., Luna, M. A., Tortoledo, E., *et al.* (1984). Histologic grading of adenoid cystic carcinoma of the salivary glands. *Cancer* **54**, 1062–1069.
27. Tytor, M., Genryd, I., Grenko, R., *et al.* (1995). Adenoid cystic carcinoma: Significance of DNA ploidy. *Head Neck* **17**, 319–327.
28. Franzen, G., Nordgard, S., Boysenn, M., *et al.* (1995). DNA content in adenoid cystic carcinomas. *Head Neck* **17**, 49–55.
29. Karja, V. J., Syrjanen, K. J., Kurvinen, A. K., and Syrjanen, S. M. (1997). Expression and mutation of p53 in salivary gland tumours. *J. Oral. Pathol. Med.* **26**(5), 217–223.
30. Dori, S., Trougouboff, P., David, R., and Buchner, A. (2000). Immunohistochemical evaluation of estrogen and progesterone receptors in adenoid cystic carcinoma of salivary gland origin. *Oral Oncol.* **36**(5), 450–453.
31. Gandour-Edwards, R., Kapadia, S. B., Barnes, L., Donald, P. J., and Janecka, I. P. (1997). Neural cell adhesion molecule in adenoid cystic carcinoma invading the skull base. *Otolaryngol. Head Neck Surg.* **117**(5), 453–458.
32. Staging of salivary gland tumors. *In* "Manual for Staging of Cancer," 5th Ed. Lippincott-Raven, Philadelphia.
33. Frable, M. A., and Frable, W. J. (1991). Fine needle aspiration biopsy of salivary glands. *Laryngoscope* **101**(3), 245–249.
34. Cristallini, E. G., Ascani, S., Farabi, R., Liberati, F., Maccio, T., Peciarolo, A., and Bolis, G. B. (1997). Fine needle aspiration biopsy of salivary gland, 1985–1995. *Acta Cytol.* **41**(5), 1421–1425.
35. Shaha, A. R., Webber, C., DiMaio, T., and Jaffe, B. M. (1990). Needle aspiration biopsy in salivary gland lesions. *Am. J. Surg.* **160**(4), 373–376.
36. Cajulis, R. S., Gokaslan, S. T., Yu, G. H., and Frias-Hidvegi, D. (1997). Fine needle aspiration of the salivary glands: A five-year experience with emphasis on diagnostic pitfalls. *Acta Cytol.* **41**(5), 1412–1420.
37. Kapadia, S. B., Dusenbery, D., and Dekker, A. (1997). Fine needle aspiration of pleomorphic adenoma and adenoid cystic carcinoma of salivary gland origin. *Acta Cytol.* **41**(2), 487–492.
38. Das, D. K., Gulati, A., Bhatt, N. C., Mandal, A. K., *et al.* (1993). Fine needle aspiration cytology of oral and pharyngeal lesions: A study of 45 cases. *Acta Cytol.* **37**, 333–342.
39. Hanna, D. C., Gaisford, J. C., Richardson, G. S., and Bindra, R. N. (1968). Tumors of the deep lobe of the parotid gland. *Am. J. Surg.* **116**, 524.
40. Macleod, C. B., and Frable, W. J. (1993). Fine needle aspiration biopsy of the salivary gland: Problem cases. *Diagn. Cytopathol.* **9**, 216–224.
41. Nigro, M. F., and Spiro, R. H. (1977). Deep lobe parotid tumors. *Am. J. Surg.* **134**, 523.
42. Luna, M. A. (1999). Pathology of tumors of the salivary glands. *In* "Comprehensive Management of Head and Neck Tumors," 2nd Ed., pp. 1106–1146. Saunders, Philadelphia.
43. Friedman, M., Rice, D. H., and Spiro, R. H. (1986). Difficult decisions in parotid surgery. *Otolaryngol. Clin. North Am.* **19**, 637–645.
44. Armstrong, J. G., Harrison, L. B., Thaler, H. T., *et al.* (1992). The indication of elective treatment of the neck in cancer of the major salivary glands. *Cancer* **69**, 615–619.
45. Million, R. R., and Cassisi, N. J. (1994). Minor salivary gland tumors. *In* "Management of Head and Neck Cancer: A Multidisciplinary Approach" (R. R. Million and N. J. Cassisi, eds.), 2nd Ed. Lippincott, Philadelphia.
46. Simpson, J. R., and Lee, H. K. (1997). Salivary glands. *In* "Principles and Practice of Radiation Oncology" (C. A. Perez and L. W. Brady, eds.), 3rd Ed. Lippincott-Raven, Philadelphia.
47. Douglas, J. G., Laramore, G. E., Austin-Seymour, M., Koh, W., Stelzer, K., and Griffin, T. W. (2000). Treatment of locally advanced adenoid cystic carcinoma of the head and neck with neutron radiotherapy. *Int. J. Radiat. Oncol. Biol. Phys.* **46**(3), 551–557.
48. Potter, R., Prott, F. J., Micke, O., Haverkamp, U., Wagner, W., and Willich, N. (1999). Results of fast neutron therapy of adenoid cystic carcinoma of the salivary glands. *Strahlenther Onkol.* **175**(Suppl. 2), 65–68.
49. Krull, A., Schwarz, R., Heyer, D., Brockmann, W. P., Junker, A., Schmidt, R., and Hubener, K. H. (1990). Results of fast neutron therapy of adenoid cystic carcinoma of the head and neck at the neutron facility Hamburg-Eppendorf. *Strahlenther Onkol.* **166**(1), 107–110.
50. Buchholz, T. A., Shimotakahara, S. G., Weymuller, E. A., Jr., Laramore, G. E., and Griffin, T. W. (1993). Neutron radiotherapy for adenoid cystic carcinoma of the head and neck. *Arch. Otolatyngol. Head Neck Surg.* **119**(7), 747–752.
51. Debus, J., Engenhart-Cabillic, R., Kraft, G., and Wannenmacher, M. (1999). The role of high-LET radiotherapy compared to conformal photon radiotherapy in adenoid cystic carcinoma. *Strahlenther Onkol.* **175**(Suppl. 2), 63–65.
52. Hill, M. E., Constenla, D. O., Hern, R. P., Henk, J. M., Rhys-Evans, P., Breach, N., Archer, D., and Gore, M. E. (1997). Cisplatin and 5-flourouracil for symptom control in a advanced salivary adenoid cystic carcinoma. *Oral Oncol.* **33**(4), 257–258.
53. Dreyfuss, A. I., Clark, J. R., Fallon, B. G., Posner, M. R., Norris, C. M., and Miller, D. (1987). Cyclophosphamide, doxorubicin, and cisplatin combination chemotherapy for advanced carcinomas of salivary gland origin. *Cancer* **15**, 2869–2872.
54. de Hann, L. D., De Mudler, P. H., Vermorken, J. B., Schornagel, J. H., Vermey, A. and Verweij, J. (1992). Cisplatin-based chemotherapy in advanced adenoid cystic carcinoma of the head and neck. *Head Neck* **14**(4), 273–277.
55. Tsukuda, M., Kokatsu, T., Ito, K., Mochimatsu, I., Kubota, A., and Sawaki, S. (1993). Chemotherapy for recurrent adeno-and adenoid cystic carcinoma in the head and neck. *J. Cancer Res. Clin. Oncol.* **119**(12), 756–758.
56. Sessions, R. B., Lehane, D. E., Smith, R. J., Bryan, R. N., and Suen, J. Y. (1982). Intra-arterial cisplatin treatment of adenoid cystic carcinoma. *Arch. Otolaryngol.* **108**(4), 221–224.
57. Mattox, D. E., Von Hoff, D. D., and Balcerzak, S. P. (1990). Southwest Oncology Group study of Mitoxantrone for treatment of patients with advanced adenoid cystic carcinoma of the head and neck. *Invest New Drugs* **8**(1), 105–107.
58. Vermorken, J. B., Verweij, J., deMulder, P. H., *et al.* (1994). Epirubicin in patients with advanced or recurrent adenoid cystic carcinoma of the head and neck: A phase II study of the EORTC Head and Neck Cancer Cooperative Group. *Ann. Oncol.* **4**, 785–788.
59. Shidnia, H., Hornback, N. B., Hamaker, R., *et al.* (1980). Carcinoma of the major salivary glands. *Cancer* **45**, 693–697.
60. Kagan, A. R., and Nussbaum, H. (1976). Recurrence from malignant parotid salivary gland tumors. *Cancer* **37**, 2600–2604.
61. Yu, G. Y., and Ma, D. Q. (1987). Carcinoma of the salivary gland: A clinicopathologic study of 405 cases. *Semin Surg. Oncol.* **3**, 240–244.
62. Fitzpatrick, P. J., and Theriault, C. (1986). Malignant salivary gland tumors. *Int. J. Radiat. Oncol. Biol. Phys.* **12**, 1743–1746.
63. Fletcher, G. H., and Jesse, R. H. (1977). The place of irradiation in the management of the primary lesion in head and neck cancers. *Cancer* **39**, 862–867.
64. Reddy, S. P., and Marks, J. E. (1988). Treatment of locally advanced high grade malignant tumors of major salivary glands. *Laryngoscope* **98**, 450–454.
65. Shingaki, S., Ohtake, K., Nomura, T., *et al.* (1992). The role of radiotherapy in the management of salivary gland carcinomas. *J. Craniomaxillofac. Surg.* **20**, 220–224.
66. Batterman, J. J., Breur, K., Hart, G. A. M., and van Peperzeel, H. A. (1981). *Eur. J. Cancer* **17**, 539–548.
67. Griffin, T. W., Pajak, T. F., Laramore, G. E., Duncan, W., *et al.* (1988). *Int. J. Radiat. Oncol. Biol. Phys.* **15**, 1085–1090.
68. Douglas, J. G., Laramore, G. E., Austin-Seymour, M., *et al.* (1996). Neutron radiotherapy for adenoid cystic carcinoma of minor salivary glands. *Int. J. Radiat. Oncol. Biol. Phys.* **36**, 87–93.

C H A P T E R

3

Head and Neck Cancer Imaging

VAL J. LOWE,* BRENDAN C. STACK, Jr.,† and ROBERT E. WATSON, Jr.*

*Department of Radiology, Mayo Clinic and Foundation, Rochester, Minnesota 55905

†Department of Otolaryngology—Head and Neck Surgery, Pennsylvania State University, College of Medicine, Hershey, Pennsylvania

I. Head and Neck Cancer Imaging 23
II. Anatomic Cross-Sectional Imaging Techniques for the Head and Neck 23
A. Computed Tomography Imaging 24
B. Magnetic Resonance Imaging 24
III. Positron Emission Tomography (PET) Imaging Techniques for the Head and Neck 24
IV. Staging Head and Neck Cancer 25
A. Clinical Considerations 25
B. PET Staging 27
V. Evaluation of Therapy of Head and Neck Cancer 28
A. Clinical Considerations 28
B. PET Imaging of Treated Head and Neck Cancer 28
VI. Assessment of Recurrent Head and Neck Cancer 31
A. Clinical Considerations 31
B. PET Detection of Recurrent Disease 31
VII. Conclusion 32
References 32

I. HEAD AND NECK CANCER IMAGING

Radiological assessment of head and neck cancer has largely depended on the demonstration of anatomical changes in the head and neck as an indication of tumor involvement. Computed tomography (CT) and magnetic resonance imaging (MRI) have both been used with similar success in evaluating head and neck cancer in this regard [1]. Distortion of normal anatomic spaces as seen on these images can imply the presence of tumor. Tumor size can be measured accurately. Destruction of bone or cartilage seen on CT can be an indication of tumor involvement. Nevertheless, CT only has marginal sensitivity to detect such involvement and it cannot be ruled out when tumors are in close proximity to bone [2]. Demonstration of enlarged nodes, indications of possible great vessel involvement, and cartilage or bone destruction can all help assess the overall stage and resectability of a tumor.

Anatomic data as obtained from CT and/or MRI, although extremely helpful, does have some limitations. Identification of small volumes of tumor can be difficult in areas such as the larynx or base of tongue. Also, postoperative scans can be made difficult by the anatomic changes caused by surgery and/or radiation. Assessing head and neck cancer using molecular imaging in conjunction with anatomic imaging has shown to be very helpful in the evaluation of head and neck cancer patients. Molecular imaging capitalizes on the biochemical differences between tumor and normal tissue rather than the anatomic abnormalities that occur. This difference in imaging can be an advantage in many occasions, but also has its own compliment of difficulties. After a short review of CT and MRI imaging techniques, this chapter focuses primarily on the indications for using molecular or functional imaging such a positron emission tomography (PET) in the evaluation of head and neck cancer patients.

II. ANATOMIC CROSS-SECTIONAL IMAGING TECHNIQUES FOR THE HEAD AND NECK

Cross-sectional imaging with CT and MRI are both useful and can be complementary for the evaluation of head and neck lesions. There are, however, situations when one imaging method may be preferred over the other. The longer scan acquisition times with MRI can be an impediment, particularly in sick patients, those short of breath, or those who are unable to be still during the image acquisition. In imaging the larynx, CT can be considered the imaging method of choice as normal motion may be more likely to hamper MRI. The multiplanar capabilities of MRI are especially helpful in the assessment of complex lesions, including those that involve the skull base, bone involvement, and those that may involve perineural tumor spread. MRI is also likely to be more sensitive in detecting bone involvement than CT.

A. Computed Tomography Imaging

For lesions involving the pharynx, imaging is generally performed with the patient in the supine position, and sections are obtained in plane with the infraorbital–meatal line. Scanning is performed from the skull base through the thoracic inlet following the administration of iodinated contrast especially to evaluate for nodal disease. The examination is performed during a breath hold to minimize motion. Generally, contiguous sections, 3–5 mm in thickness, are attained. Field of view is kept as small as possible. Appropriate angling of the gantry can be helpful in minimizing artifacts associated with dental hardware.

In addition to soft tissue windows, high-resolution bone reconstruction algorithms can be helpful in evaluating skull base lesions, as well as potential perineural tumor spread. Especially in patients with recurrent tumors, evaluation with lung windows at the lung apices is helpful in assessing metastatic disease. While most CT imaging is performed in the axial plane, reformatted images permitting visualization in the coronal and sagittal planes are also useful. In addition, direct coronal CT imaging is possible with lesions involving the face, skull base, and nasopharynx.

For CT evaluation of the larynx, in addition to a general survey of the neck, a focused helical examination, employing 2- to 3-mm sections, is performed between the hyoid bone and the inferior aspect of the cricoid cartilage. As any motion severely degrades image quality, the patient is admonished to stay as still as possible. Sagittal, axial, and coronal images are reconstructed using soft tissue algorithms. High-resolution bone algorithms can also be helpful in identifying invasion of laryngeal cartilage.

B. Magnetic Resonance Imaging

Because of the comparatively greater number of options open in protocoling MRI examinations, knowledge of patient history is especially important so that the examination can be tailored appropriately. Examinations of the oropharynx and nasopharynx generally can be performed with a head coil, whereas a dedicated neck coil is required for studies focused more inferiorly in the neck.

Whereas only postcontrasted images are necessary in CT studies, MRI examinations benefit from a combination of pregadolinium and postgadolinium sequences. As fat and proteinaceous products are high signal on T1-weighted images, it is important to be able to compare pre- and post-contrasted images. Noncontrasted T1-weighted images are especially useful for their fine anatomic detail, and because fat is high signal, any distortion of fat planes, including that of the parapharyngeal space, can be evaluated. In addition, replacement of fatty marrow by a lesion can be detected on noncontrasted T1-weighted images. Precontrast sagittal and axial T1-weighted sequences are routinely attained. T1 echo time should be minimized to reduce the magnetic-susceptibility artifact. Flow and other motion artifacts can be reduced with motion-compensation gradients and spatial-presaturation pulses.

Generally, lesions and abnormalities are more evident on T2-weighted sequences. Because fast spin echo techniques result in fat being of high signal intensity, the conspicuity of lesions and adenopathy can be increased on T2-weighted images by employing fat saturation. Most studies employ axial and coronal fat-saturated T2 fast spin echo sequences.

Fat-saturated T1-weighted images following gadolinium contrast administration are especially useful, especially in cases of skull base invasion and perineural tumor spread. Multiplanar imaging capabilities can be especially helpful in these circumstances.

Motion artifact can be troublesome with MRI head and neck examinations, and a tailored examination kept as brief as possible can be helpful in minimizing patient fatigue. Especially with examinations of the larynx, a continued development of faster MRI sequences will be helpful. Another troubling artifact in examinations of the head and neck is pulsation artifact imparted by large vessels. Depending on the lesion examined, alteration of the direction of the phase-encoding gradients can address this potential problem.

III. POSITRON EMISSION TOMOGRAPHY (PET) IMAGING TECHNIQUES FOR THE HEAD AND NECK

PET imaging of head and neck cancer depends on the increased metabolism and rapid cell proliferation of head and neck neoplasms. In the 1930s, malignant cells were shown to have increased glucose metabolism [3]. The largest PET experience with head and neck neoplasms has been with FDG, which capitalizes on the increased anaerobic metabolism of these malignancies. Greater than 90% of these tumors have a squamous cell pathology, which demonstrates high levels of glucose metabolism. Various other tumors, such as adenoid cystic tumors and adenocarcinomas, demonstrate elevated levels of glucose metabolism but at times not to the extent of squamous cell tumors. Other PET tracers, such as C-11 thymidine, have been use to evaluate head and neck neoplasms but to a lesser extent. These tracers depend on abnormalities in protein or nucleic acid synthesis in tumor cells.

FDG imaging is performed in the fasting state to minimize the competitive inhibition of FDG uptake by glucose. The effect of diabetes on the uptake of FDG is not fully elucidated, but elevated serum glucose levels may result in decreased FDG accumulation in cancer cells.

Because tumor uptake of FDG continues to increase even up to 2.5 h after injection, standardizing the delay from injection to imaging is advantageous. It is generally felt to be necessary to perform FDG emission scans at least 50 min after the intravenous administration of FDG [4]. A dose of 10.0 mCi of FDG can be used routinely but higher doses are commonly used and have the advantage of providing higher count rates. A range of 10–20 mCi would be considered adequate.

Images should be obtained to include at least regions from the maxillary sinuses to the aortic arch. These emission data should be corrected using measured attenuation correction when performing head and neck imaging. The acute curvature around the mouth, nose, mandible, and neck results in severe edge artifacts when PET studies are performed without attenuation correction. Neck lymph nodes can lie near the skin surface, and edge artifacts created without attenuation correction can hamper their identification. The anatomic relationships of airways and boney structures to soft tissue can also be assessed more reliably when an attenuation map is available for comparison. Utilization of semiquantitative analysis of data with standardized uptake ratio (SUR) calculation may also be of aid in assessing data and this can only be provided when attenuation correction is used. Imaging should also be performed to exclude distant metastatic disease in all but early stage tumors (T1 or T2). Images to include the liver are recommended, and nonattenuation corrected images may be sufficient outside of the head and neck region, although we recommend attenuation correction of all data when possible.

Imaging the head and neck region presents challenges for PET that are unique. The region has substantial normal variation in uptake that can present dilemmas in the identification of pathology (Fig. 3.1). Uptake in adenoidal, palatine, and lingual tonsil tissue is a normal variant that needs to be recognized. Uptake in the floor of mouth and laryngeal musculature is also common. Myelohyoid and vocalis muscles are usually seen as symmetric regions of uptake in "v-like" patterns that make them recognizable. On occasion, various other muscle uptake can be seen. Most commonly, sternocleidomastoid (SCM) muscle uptake may be seen in the neck. Scalene muscle uptake may also be seen and is seen more often in patients after neck dissection where the SCM is removed. The lower neck images may sometimes only demonstrate uptake at the insertion sites of the neck muscles. This can be confused easily with bilateral supraclavicular lymph node disease. The muscle uptake is, however, usually symmetric and palpation does not demonstrate nodes large enough to cause the amount of uptake seen on the scan. On rarer occasions, temporalis, ptyergoid, masseter, or other head and neck muscles will accumulate tracer depending on use by the patient, and obtaining a history may be helpful.

Taking steps to reduce or eliminate muscle uptake is crucial. Making sure that the patient is not chewing gum, reading, or talking during the uptake phase is important. Some have advocated the use of Valium to suppress muscle uptake. This appears to work well but deserves an extra note of caution in head and neck cancer patients, as existing airway compromise may be a variable to contend with. Technical attention to the position of the patient's head during the uptake phase is of prime importance. Experimenting with chair designs and pillow placement to ensure slight head flexion and complete head relaxation is essential to obtaining a scan without muscle uptake interference. Also, with new three-dimensional surface projections and careful examination of all three orthogonal views, most muscle uptake can be differentiated from disease by its anatomic pattern of distribution. This cannot, however, replace proper patient preparation.

IV. STAGING HEAD AND NECK CANCER

A. Clinical Considerations

The single most important factor in patient assessment, treatment planning, and survival prognostication is accurate staging [5,6]. The current staging guidelines in use for head and neck cancer are from the 1997 manual of the American Joint Committee on Cancer. Staging involves an accurate assessment of tumor at the primary site (T), regional lymphatic metastases (N), and distant metastases (M). Each primary tumor has a unique propensity for local and regional spread, which must be appreciated when evaluating head and neck cancer patients.

Evaluation of the patient begins at the first consultation with a physical examination. This includes direct visualization of the primary tumor either through the mouth or nose or by using office endoscopy. Next, the neck is palpated to determine the presence of cervical lymph node enlargement. Lymph nodes less than 1 cm in diameter are not reliable appreciated on physical examination. Body habitus such as obesity or short necks may make this assessment problematic. The next phase of the evaluation consists of a pathologic diagnosis (direct biopsy of the tumor or a needle biopsy of a neck mass) and/or anatomic imaging if the neoplastic process is not visualized, palpable, or accessible.

The most commonly used imaging modality for head and neck cancer CT is with an intravenous iodinated contrast. Lymph nodes greater than 1 cm in diameter (1.5 cm for

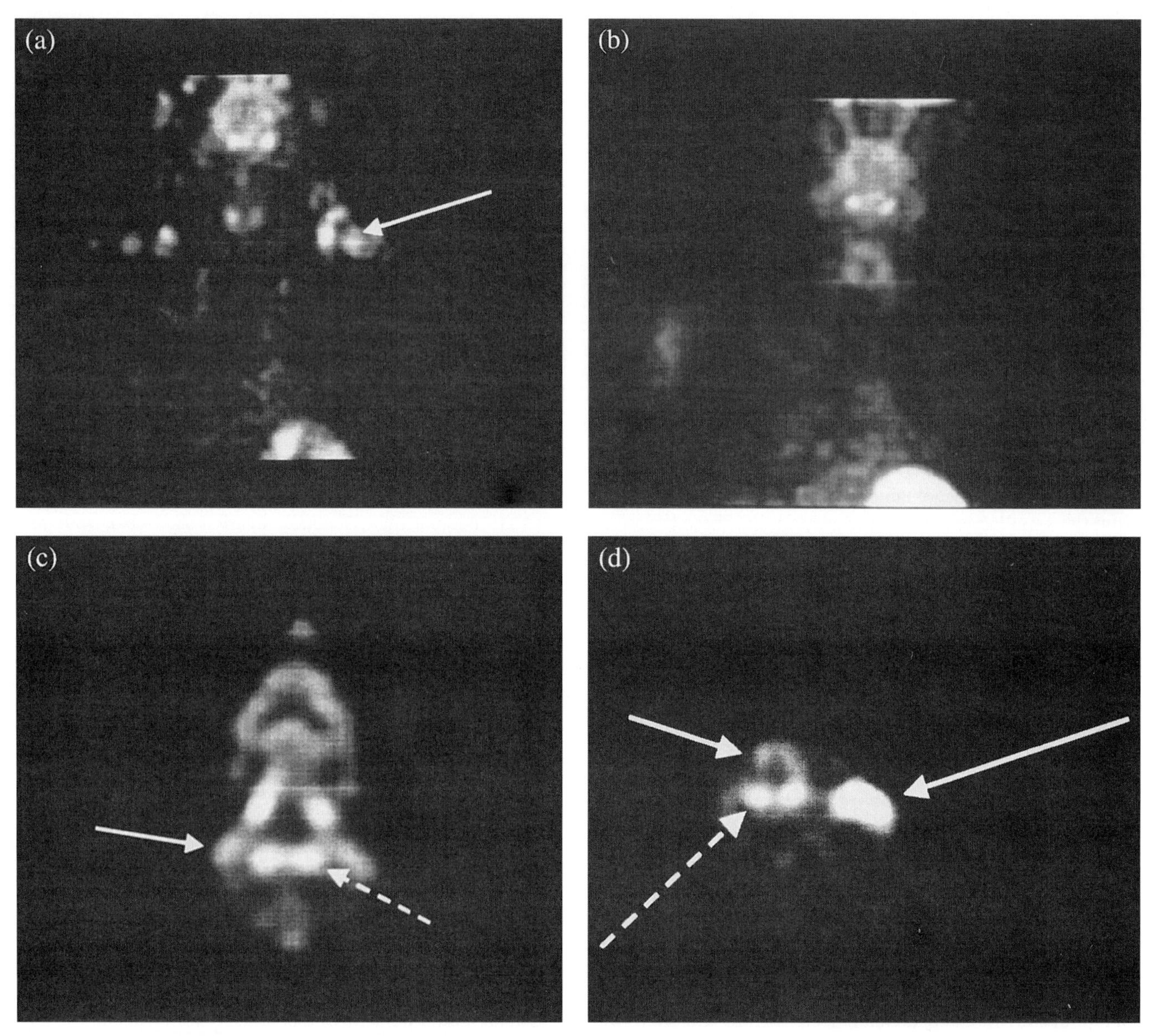

FIGURE 3.1 (a) PET scan coronal projection showing high metabolism in neck muscles, probably sternocleidomastoid muscles (arrow). (b) PET scan coronal projection of the same patient showing resolution of neck metabolism on a PET scan obtained 2 days later using Valium. (c) PET scan axial view showing high metabolism in salivary glands (arrow) and palatine tonsils (dashed arrow). (d) PET scan axial view showing high metabolism in vocalis muscles (short arrow), crico-arytenoid posterior muscle (dashed arrow), and a tumor-bearing lymph node (long arrow).

jugulodigastric nodes) or with central necrosis are considered abnormal and suspicious for metastasis. The obvious shortcoming to this approach is that the determination of metastasis is based on anatomic criteria alone and excludes the possibility of early nodal metastases, which have failed to enlarge the lymph node.

Standard assessment for distant metastases in head and neck cancer, which is uncommon for patients presenting with a new tumor, can include chest X-ray and liver function tests. CT of the chest or abdomen are used most commonly to evaluate abnormalities found on the preceding two studies.

Second primary disease (either synchronous or metachronous) is an occasional dilemma in head and neck cancer. These lesions are usually in the head and neck, lung, or esophagus. Synchronous lesions are defined as the discovery of a second primary within 6 months of the diagnosis of the first. Metachronous primaries are discovered at an interval greater than 6 months. The standard approach to head and neck cancer used to rule out a second primary has been operative endoscopy (laryngoscopy, bronchoscopy, esophagoscopy). Improved office endoscopy and CT (neck and chest) have resulted in a decrease in operative endoscopy. The precision of whole body PET may prove useful for synchronous second primary detection and for surveillance for metachronous lesions.

The standard treatment for advanced (stage III and IV) head and neck cancer is surgery followed by postoperative radiation therapy [7]. Which particular advanced stage head and neck tumors to treat with standard surgical resection and which to treat with chemotherapy and radiation can be a challenge. A greater impetus to treat with nonoperative therapy may result if distant metastatic disease is found by FDG PET.

B. PET Staging

The ability of PET to change the disease stage by finding undetected malignancy will have treatment implications. The evaluation of stage can be divided into contributions made in assessing the primary tumor, local nodal disease, and metastatic disease. Staging of the primary tumor (T stage) using PET has been described in several published articles [8,9]. In none did PET show an advantage over conventional techniques when the primary was seen by conventional techniques. Primary tumor staging with PET will likely contribute little over conventional staging in most patients with the possible exception of unknown primaries. Standard tumor staging using CT and physical examination with endoscopy will provide more anatomic information that is important to tumor staging than what can be provided by PET. In around 5% of cases, however, the primary may not be identified by standard techniques. Some of these primaries become obvious as the patient is followed over time. Others are thought to regress spontaneously, whereas most are never diagnosed by conventional means. Following these clinical evaluations, PET may identify the unknown primary in about 20–50% of cases as reported by several authors [10–15] (Fig. 3.2). There is some evidence that PET should only be performed after clinical assessment because routine panendoscopy and physical examination will identify some small lesions that may not be seen by PET [16].

Some early evidence suggests that PET may be able to predict radiocurabililty of patients with head and neck cancer. Treatment options for some head and neck tumors include surgery or radiation, and a decision between the two may be based on relative treatment morbidity. A higher level of metabolic activity seems to predict tumor radiocurabililty in a small group reported recently [17]. Larger trials will be needed to assess this finding.

Previous authors have described the high accuracy of FDG PET in local nodal staging of head and neck cancer [8,9,18–23] (Fig. 3.3). All studies have shown PET to be equivalent or superior to anatomic methods of nodal staging (Table 3.1). In a study by Adams *et al.* [18], about 1400 lymph nodes were sampled in 60 patients, and PET had a 10% advantage over CT, MRI, or US in sensitivity for local nodal disease. The specificity was also 10% higher for PET. The authors showed highly statistically significant differences in the performance of these modalities.

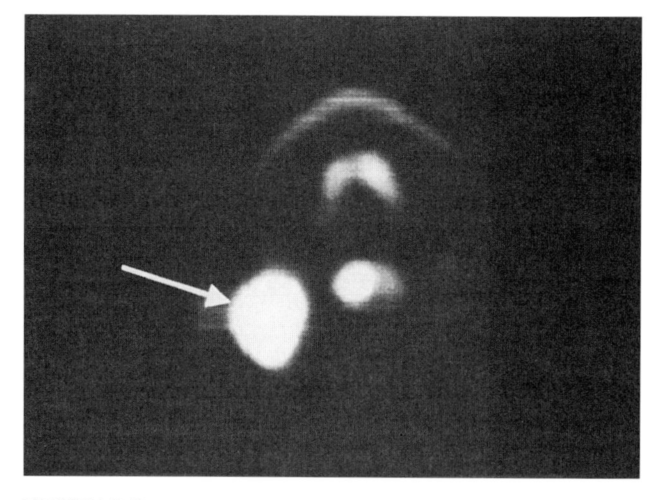

FIGURE 3.2 PET image of a patient with a right neck mass showing squamous cell cancer on biopsy (arrow) and an unknown primary even after review of the CT, physical examination, and negative panendoscopic, biopsies. Thereafter, PET showed a right base of tongue primary medial to the large right lymph node.

The decision to perform a neck dissection is the most challenging in lower stage primary tumors where there is no clinical evidence of nodal disease. Some of these patients have up to a 30% incidence of occult cervical metastasis. These N0 patients can be evaluated by PET, and the advantage over other methods for the detection of disease has also been demonstrated. Myers and Wax [24] showed that PET was more than twice as sensitive as CT in identifying nodal disease in patients with clinically N0 necks.

Metastatic disease to distant regions is not common with head and neck cancer. This may relate to an earlier detection of head and neck cancer due to obvious symptoms. It will likely be rare for PET to identify metastatic disease in initial staging that will impact a large proportion of patients due to the low incidence (probably $<5\%$). Imaging of the body with PET is still recommended at initial staging, as this data may be helpful as a baseline evaluation for comparison with later imaging. Subtle uptake from inflammatory lung lesions, for example, can be documented so as to not raise a concern of metastasis on future examinations. Such imaging would also address the issue of second primary disease and possibly reduce the need for other additional testing. There is no additional radiation exposure for the patient, and additional imaging can usually be completed in about 5 min.

In summary, if therapy is based in part on the more accurate staging by FDG PET, patients will have a chance to receive therapy that will be more appropriate. Patients with locoregional disease that is more extensive than what is identified

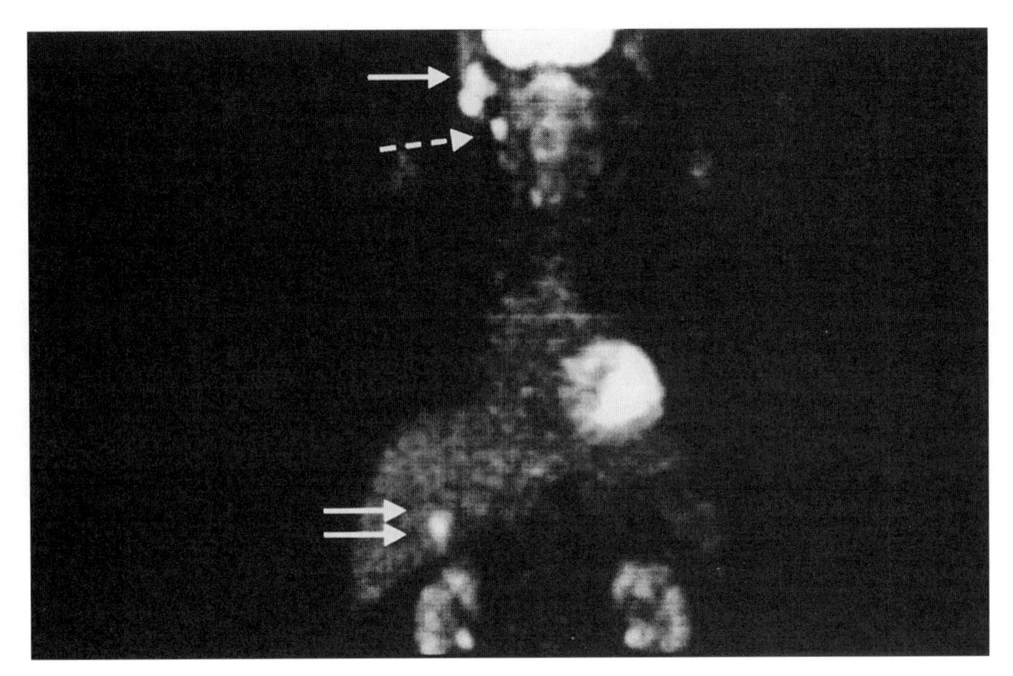

FIGURE 3.3 PET image demonstrating improved staging using PET of a patient with a right parotid squamous cell cancer (arrow) who had an MRI scan showing no definite tumor (inflammation vs infection) and no lymph node disease. The PET showed hypermetabolism in the parotid, right neck nodes (dashed arrow), and liver (double arrows), indicating metastasis proven to be disease by biopsy.

originally may have bilateral neck dissection rather than unilateral dissection. Patients with unknown primary disease of the head and neck may undergo radiation therapy with extensive radiation ports from the maxillary sinuses to the lower neck. Identification of an unknown primary malignancy by PET in such cases could prevent extensive morbidity from such treatment. Patients with distant metastasis that may otherwise go undetected may be prevented from receiving unnecessary surgery of the head and neck or offered conjoint head and neck or thoracic procedures.

V. EVALUATION OF THERAPY OF HEAD AND NECK CANCER

A. Clinical Considerations

The potential role of PET in evaluating tumor response to nonoperative therapies is promising. Many times, artifact produced by chemotherapy or radiotherapy (fibrosis, erythema, edema) may confound the practitioner's ability to evaluate tumor response to therapy either by physical examination or by anatomic imaging. Standard anatomic imaging using CT or MRI has a limited ability to evaluate the effect of radiation or chemotherapy on malignancy. This is most commonly true due to contrast enhancement and/or soft tissue distortion that is apparent in posttherapy regions seen on conventional imaging. Changes in tumor size may also lag behind metabolic effects of therapy. Because PET is a functional study, the persistence of tumor in the setting of no anatomic abnormality or the absence of tumor in the setting of persistent anatomic abnormality can be assessed.

B. PET Imaging of Treated Head and Neck Cancer

FDG PET can identify changes in tumor metabolism during therapeutic interventions. The usefulness of FDG PET to identify therapeutic effects may differ depending on whether radiation or chemotherapy is used. Data suggest that radiotherapy may induce early acute inflammatory hypermetabolism that can be confused with tumor hypermetabolism [25]. As well, some investigators have concluded that an early decrease in FDG uptake does not necessarily indicate a good prognosis.

A significantly increased FDG uptake can be seen in soft tissue regions that have been irradiated recently. Commonly, uptake will be seen in tissues that are exposed more intensely. Some normal deep structures may not show radiation-related changes in metabolism [26]. The duration following radiation therapy during which increased uptake occurs is of interest for study interpretation. Increased FDG accumulation in regions of radiation therapy can be statistically significant at least 12–16 months after treatment in some body regions [27]. The SUR of radiation related uptake is

TABLE 3.1 Studies Comparing PET and CT and/or MRI in Nodal Staging of Head and Neck Cancer

Author, yr	Patient #	Study type	Nodal status: Malignant/benign (numbers of)	Sensitivity PET	Specificity PET	Sensitivity (comp. test)	Specificity (comp. test)	Statistical significance PET vs TEST
Benchaou, 1996 [19]	48	Blinded, prospective, single	54/414 (nodes)	72%	99%	67% (CT)	97% (CT)	"P" = 0.25
Braams, 1995 [11]	12	Blinding not noted, single institution	22/177 (nodes)	91%	88%	36% (MRI)	94% (MRI)	Not done
Laubenbacher, 1995 [8]	22	Blinding not noted, prospective, single	83/438 (nodes)	90%	96%	78% (MRI)	71% (MRI)	"P" <0.05
McGuirt, 1995 [20]	49	Blinding not noted, prospective, single	23/22 (necks)	83%	82%	78% (CT)	86% (CT)	Not done
Myers, 1998 [21]	14 (NO necks by palpation)	Blinding not noted, prospective, single institution	9/15 (necks)	78%	100%	57% (CT)	90% (CT)	"P" = 0.11
Rege, 1994 [22]	34	Blinding not noted, prospective, single	16/18 (stations)	88%	89%	81% (MRI)	89% (MRI)	Not done
Wong, 1997 [9]	16	Blinding not noted, prospective, single	67%	100%	67%	25% (CT/MRI)	Not done (CT/MRI)	
Adams, 1998 [18]	60	Blinded, prospective, single	117/1284 (nodes)	90%	94%	82% (CT) 80 % (MRI) 72% (US)	85% (CT) 79% (MRI)	"P" < 0.017 "P" < 0.012 "P" < 0.0001
Kau, 1999 [23]	70	Blinded, prospective, single	98/1281 (nodes)	87%	94%	65% (CT) 88% (MRI)	47% (CT) 41% (MRI)	"P" < 0.001 "P" < 0.001

generally less than what is found in recurrent tumor. Nevertheless, FDG uptake from radiation effects can occasionally be in a range that is worrisome for malignancy and needs to be recognized.

Metabolic changes in tumor occurring during chemo-therapy may be somewhat more specific to the tumor response. Some tumors demonstrate a significant reduction in metabolism that is associated with a good pathologic response. Generally, a reduction of 80% in the SUR is predictive of a pathologic complete response. Hypermetabolism secondary to inflammation does not appear to be a significant problem in contrast to radiation-treated tissue. Tumor metabolism can be at baseline soft tissue levels as early as 1 week after therapy in responding patients [28]. PET has 90% sensitivity and specificity in detecting residual disease after chemotherapy in this setting. Papers by Lowe *et al.* [28] and Greven *et al.* [29] use pathologic standards to assess the use of PET in the posttherapy evaluation of residual disease. Lowe *et al.* [28] describe PET as being as sensitive as needle biopsy (90%) in detecting residual disease after therapy and as having an 83% specificity (Fig. 3.4). This study was a blinded, prospective, single institution study and included 27 patients. Greven and co-workers [29], in a non-blinded, prospective, single institution study, assessed 31 patients after radiation therapy and found an 80% sensitivity and an 81% specificity for the detection of residual or recurrent head and neck cancer.

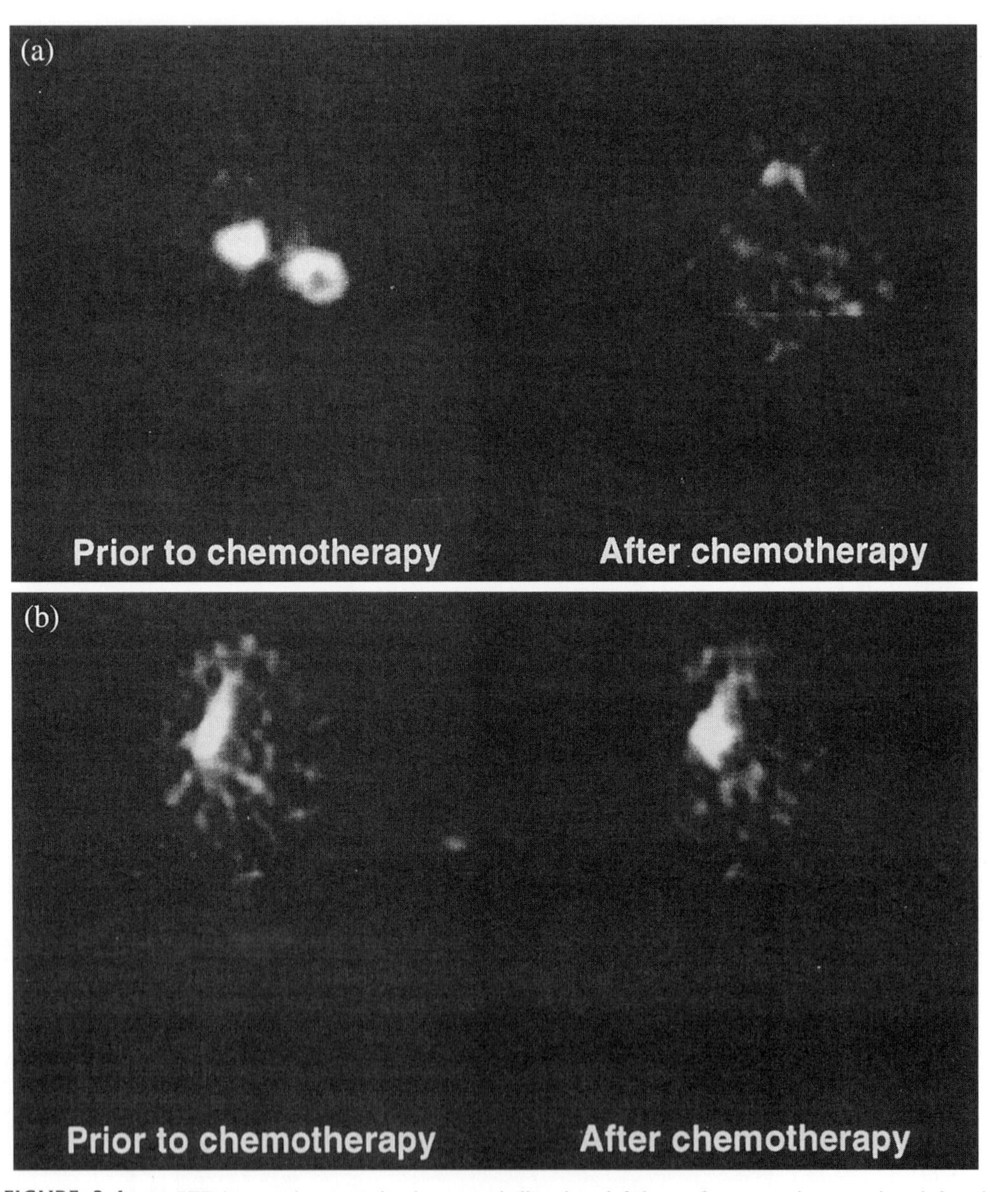

FIGURE 3.4 (a) PET images demonstrating hypermetabolism in a left base of tongue primary and neck lymph node prior to therapy and resolution of all abnormalities 2 weeks after neoadjuvant chemotherapy. (b) PET images demonstrating hypermetabolism in a floor of mouth cancer prior to therapy and persistent activity after neoadjuvant chemotherapy. Needle biopsy documented residual disease.

Several investigators are evaluating metabolic changes in tumor in the early course of chemotherapy. This work is promising because chemotherapeutic regimens could be altered earlier, or earlier salvage surgery could be performed, if the chemotherapeutic response could be assessed after the first treatment.

VI. ASSESSMENT OF RECURRENT HEAD AND NECK CANCER

A. Clinical Considerations

All head and neck cancer patients are at high risk for recurrence and a second primary disease. This must always be stressed in the treatment planning process. This is likely due to an underlying molecular or cellular abnormality of many of the cells lining the upper aerodigestive tract mucosal lining, which have been bathed in the previously mentioned carcinogens. Risks for recurrence arise from stage, treatment, and ongoing exposure to risk factors. The more advanced the stage of the head and neck cancer at presentation, the greater the risk for recurrence.

Most recurrences occur in the first 24 months following therapy for head and neck cancer. Later occurring lesions (lesions at different locations and with distinct histology) are probably second primaries. Local recurrences can present many challenges, but can often be reexcised when detected early. Reexcision will further compromise any preexisting dysfunction (speech, voice, swallowing, or airway) and will impact negatively on a patient's quality of life. This scenario may present significant surgical reconstruction challenges.

Regional recurrences can present in the nontreated neck, the operated neck, the radiated neck, or a neck that has had both treatments. Carotid artery involvement is a significant issue in this population and can result in stroke or death from acute arterial hemorrhage. Treatment for these recurrences can include reoperation, reirradiation (external beam or implant), chemotherapy (with palliative intent), or comfort measures and support.

Distant metastases are more common in patients that recur than at initial presentation and may be as high as 30% [30]. The lungs are the most common sites of distant recurrence. Prior to embarking on therapy of locally recurrent disease, distant disease recurrence should be excluded, as it would obviate the need for any attempt at curative resection of locally recurrent disease.

Head and neck cancer patients require long-term surveillance for recurrence and are only deemed "cured" after 5 years of being disease free. The traditional approach to this includes serial physical examinations, annual chest X-rays, and liver function tests. Other tests (CT or MR) are ordered when physical examination findings or patient complaints arouse a suspicion for recurrent or second primary neoplasm. Data suggest that PET may serve as a posttreatment surveillance tool for detecting new disease at a subclinical level or earlier than conventional means.

B. PET Detection of Recurrent Disease

Standard anatomic imaging can be especially problematic in evaluating recurrent disease due to contrast enhancement and/or soft tissue distortion after surgery or other therapy. Usually, changes from previous imaging are the most helpful method of indicating possible recurrent disease. A comparison of MRI and CT showed that both had a sensitivity of about 50% with substantial interobserver variability ($k = 0.563$) for identifying recurrent nasopharyngeal cancer [31].

PET is clearly superior to other modalities in identifying recurrence of head and neck cancer after therapy. The use of PET for deciding if head and neck cancer has recurred has been described by several authors [9,22,29,32–37] as shown in Table 3.2. Data show improved detection of

TABLE 3.2 Studies Comparing PET and CT and/or MRI in Detecting Recurrent Head and Neck Cancer

Author, yr	Patient #	Sensitivity PET	Specificity PET	Sensitivity (comp. test)	Specificity (comp. test)	Stat significance PET vs TEST
Anzai, 1996 [31]	12	88%	100%	25% (CT/MRI)	75% (CT/MRI)	"P" = 0.03
Greven, 1997 [29]	31	80%	81%	58% (CT)	100% (CT)	Not done
Lapela, 1995 [32]	22	88%	86%	92% (CT)	50% (CT)	Not done
Rege, 1994 [22]	17	90%	100%	60% (MRI)	57% (MRI)	Not done
Wong, 1997 [9]	12	100% (acc)		54% (accuracy) (CT/MRI)		NA
Fischbein, 1998 [36]	35	100% 93%	64% 1° 77% LN	NA		
Kao, 1998 [34]	36	100%	96%	72%	88%	
Farber, 1999 [33]	28	86%	93%	NA		
Lowe, 2000 [35]	30 Prospective	100%	93%	44% (PE) 38% (CT)	100% (PE) 85% (CT)	"P" = 0.013 "P" = 0.002

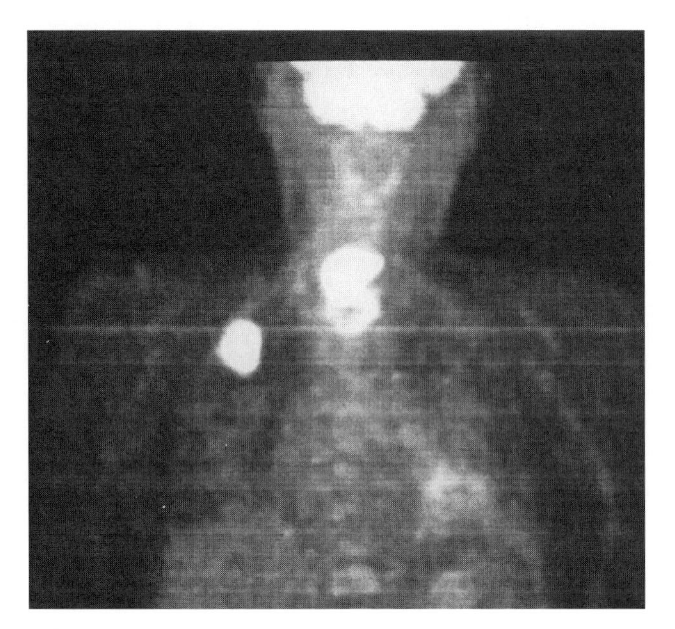

FIGURE 3.5 PET image of a patient with a history of treated larynx cancer who returned with local recurrence. PET showed metastatic disease in the right lung apex.

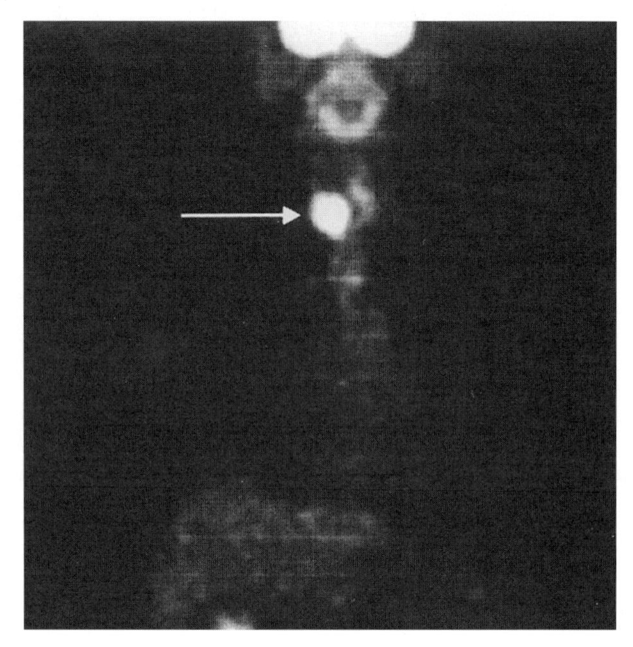

FIGURE 3.6 PET image of a patient who had previous larynx cancer treated by radiation who returns with what is thought to be edema that shows local recurrence (arrow). Biopsies were initially negative, and PET findings could not be confirmed for 4 months.

recurrence by PET with the exception of Lapela and colleagues [33], who showed a slightly higher sensitivity of CT (difference of 4%) for recurrence in their series (PET had a 36% higher specificity in this group). One paper that did a statistical analysis of their results showed a statistically significant advantage for PET over CT/MRI in detecting recurrent disease. When PET is used in the surveillance of patients who have completed treatment for head and neck cancer, PET can identify nearly twice as many cases of recurrent tumor as a regular physical examination or routine CT imaging [35]. Most of these early recurrences are understandably small and require biopsies for confirmation. If an initial biopsy is not positive where PET shows an abnormality, a repeat biopsy or close follow-up should be undertaken. Surgical exploration should also be considered if the PET results are impressive, as they are rarely incorrect (Figs. 3.5 and 3.6).

VII. CONCLUSION

A variety of imaging methods are available to assess head and neck malignancy. These methods can be complementary, although each has its strengths and weaknesses. Cross-sectional imaging with CT and MRI provides detailed anatomic information essential for treatment planning. New methods of metabolic imaging such as PET can make a significant contribution to the management of head and neck cancer as well. Improved detection of unknown primaries and local nodal disease by PET may alter initial therapeutic plans. The use of PET for assessing therapy is in its infancy but has the potential to allow physicians to have clearer information about the results of their treatment. Detecting early recurrence more accurately with PET may provide a means of improving the dismal survival from head and neck cancer recurrence.

References

1. Bootz, F., Lenz, M., Skalej, M., and Bongers, H. (1992). Computed tomography (CT) and magnetic resonance imaging (MRI) in T-stage evaluation of oral and oropharyngeal carcinomas. *Clin. Otolaryngol.* **17**, 421–429.
2. Bahadur, S. (1990). Mandibular involvement in oral cancer. *J. Laryngol. Otol.* **104**, 968–971.
3. Warburg, O. (1930). "The Metabolism of Tumors." Constable, London.
4. Lowe, V. J., Delong, D. M., Hoffman, J. M., Patz, E. F., and Coleman, R. E. (1995). Dynamic FDG-PET imaging of focal pulmonary abnormalities to identify optimum time for imaging. *J. Nuclear Med.* **36**, 883–887.
5. Bocca, E., Calearo, C., Marullo, T., DeVincentis, I., Motta, G., and Ottaviani, A. (1984). Occult metastases in cancer of the larynx and their relationship to clinical and histological aspects of the primary tumor: A four year multicentric research. *Laryngoscope* **94**, 1086–1090.
6. Shuller, D. E., McGuirt, W. F, McCabe, B. F., and Young, D. (1980). The prognostic significance of metastatic cervical lymph nodes. *Laryngoscope* **90**, 557–570.
7. Al-Sarraf, M., and Hussein, M. (1995). Head and neck cancer: Present status and future prospects of adjuvant chemotherapy. *Cancer Invest.* **13**, 41–53.
8. Laubenbacher, C., Saumweber, D., Wagner, M. C., *et al.* (1995). Comparison of fluorine-18-fluorodeoxyglucose PET, MRI and

endoscopy for staging head and neck squamous-cell carcinomas. *J. Nuclear Med.* **36**, 1747–1757.

9. Wong, W. L., Chevretton, E. B., McGurk, M., *et al.* (1997). A prospective study of PET-FDG imaging for the assessment of head and neck squamous cell carcinoma. *Clin. Otolaryngol. Appl. Sci.* **22**, 209–214.
10. Kole, A. C., Nieweg, O. E., Pruim, J., *et al.* (1998). Detection of unknown occult primary tumors using positron emission tomography. *Cancer* **82**, 1160–1166.
11. Braams, J. W., Pruim, J., Kole, A. C., *et al.* (1997). Detection of unknown primary head and neck tumors by positron emission tomography. *Inter. J. Oral. Maxillofacial. Surg.* **26**, 112–115.
12. Assar, O. S., Fischbein, N. J., Caputo, G. R., *et al.* (1992). Metastatic head and neck cancer: Role and usefulness of FDG PET in locating occult primary tumors. *Radiology* **210**, 177–181.
13. Lassen, U., Daugaard, G., Eigtved, A., Damgaard, K., and Friberg, L. (1999). 18F-FDG whole body positron emission tomography (PET) in patients with unknown primary tumours (UPT). *Eur. J. Cancer* **35**, 1076–1082.
14. Safa, A. A., Tran, L. M., Rege, S., *et al.* (1999). The role of positron emission tomography in occult primary head and neck cancers. *Cancer J. Sci. Am.* **5**, 214–218.
15. Jungehulsing, M., Scheidhauer, K., Damm, M., *et al.* (2000). 2[F]-fluoro-2-deoxy-D-glucose positron emission tomography is a sensitive tool for the detection of occult primary cancer (carcinoma of unknown primary syndrome) with head and neck lymph node manifestation. *Otolaryngol. Head Neck Surg.* **123**, 294–301.
16. Greven, K. M., Keyes, J. J., Williams, D. R., McGuirt, W. F., and Joyce W. R. (1999). Occult primary tumors of the head and neck: Lack of benefit from positron emission tomography imaging with 2-[F-18]fluoro-2-deoxy-D-glucose. *Cancer* **86**, 114–118.
17. Rege, S., Safa, A. A., Chaiken, L., Hoh, C., Juillard, G., and Withers, H. R. (2000). Positron emission tomography: An independent indicator of radiocurability in head and neck carcinomas. *Am. J. Clin. Oncol.* **23**, 164–169.
18. Adams, S., Baum, R. P., Stuckensen, T., Bitter, K., and Hor, G. (1998). Prospective comparison of 18F-FDG PET with conventional imaging modalities (CT, MRI, US) in lymph node staging of head and neck cancer. *Eur. J. Nuclear Med.* **25**, 1255–1260.
19. Benchaou, M., Lehmann, W., Slosman, D. O., *et al.* (1996). The role of FDG-PET in the preoperative assessment of N-staging in head and neck cancer. *Acta. Oto. Laryngol.* **116**, 332–335.
20. McGuirt, W. F, Williams, D., Keyes, J. J, *et al.* (1995). A comparative diagnostic study of head and neck nodal metastases using positron emission tomography. *Laryngoscope*.
21. Myers, L. L., Wax, M. K., Nabi, H., Simpson, G. T., and Lamonica, D. (1998). Positron emission tomography in the evaluation of the N0 neck. *Laryngoscope* **108**, 232–236.
22. Rege, S., Maass, A., Chaiken, L., *et al.* (1994). Use of positron emission tomography with fluorodeoxyglucose in patients with extracranial head and neck cancers. *Cancer* **73**, 3047–3058.
23. Kau, R. J., Alexiou, C., Laubenbacher, C., Werner, M., Schwaiger, M., and Arnold, W. (1999). Lymph node detection of head and neck squamous cell carcinomas by positron emission tomography with fluorodeoxyglucose F 18 in a routine clinical setting. *Arch. Otolaryngol. Head Neck Surg.* **125**, 1322–1328.
24. Myers, L. L., and Wax, M. K. (1998). Positron emission tomography in the evaluation of the negative neck in patients with oral cavity cancer. *J. Otolaryngol.* **27**, 342–347.
25. Hautzel, H., and Muller, G. H. (1997). Early changes in fluorine-18-FDG uptake during radiotherapy. *J. Nuclear Med.* **38**, 1384–1386.
26. Rege, S. D., Chaiken, L., Hoh, C. K., *et al.* (1993). Change induced by radiation therapy in FDG uptake in normal and malignant structures of the head and neck: Quantitation with PET. *Radiology* **189**, 807–812.
27. Lowe, V. J., Heber, M. E., Anscher, M. S., and Coleman, R. E. (1998). Chest wall FDG accumulation in serial FDG-PET images in patients being treated for bronchogenic carcinoma with radiation. *Clin. Positr. Imag.* **1**, 185–191.
28. Lowe, V., Dunphy, F., Varvares, M., *et al.* (1997). Evaluation of chemotherapy response in patients with advanced head and neck cancer using FDG-PET. *Head Neck* **19**, 666–674.
29. Greven, K. M., Williams, D., Keyes, J. J., McGuirt, W. F., Watson, N. J., and Case, L. D. (1997). Can positron emission tomography distinguish tumor recurrence from irradiation sequelae in patients treated for larynx cancer? *Cancer J. Sci. Am.* **3**, 353–357.
30. Budach, V. G., Haake, K., Stueben, G., Sack, H., Jahnke, K., Baumann, M., Herrmann, T., Baudach, W., Bamberg, M., Sauer, R., Hinkelbein, W., Frommhold, H., and Wernecke, K. (2001). Accelerated Chemoradiation to 70.6 Gy is more effective than accelerated radiation to 77.6 Gy alone: Two year's results of a German multicenter randomized trial (Aro95-6). *Proc. ASCO* 2001, 224a.
31. Chong. V. F. H, and Fan, Y.-F. Detection of recurrent nasopharyngeal carcinoma: MR imaging versus CT. *Radiology* **202**, 463–470.
32. Anzai, Y., Carroll, W. R., Quint, D.J., *et al.* (1996). Recurrence of head and neck cancer after surgery or irradiation: Prospective comparison of 2-deoxy-2-[F-18]fluoro-D-glucose PET and MR imaging diagnoses. *Radiology* **200**, 135–141.
33. Lapela, M., Grenman, R., Kurki, T., *et al.* (1995). Head and neck cancer: Detection of recurrence with PET and 2-[F-18]fluoro-2-deoxy-D-glucose. *Radiology* **197**, 205–211.
34. Farber, L. A., Benard, F., Machtay, M., *et al.* (1999). Detection of recurrent head and neck squamous cell carcinomas after radiation therapy with 2-18F-fluoro-2-deoxy-D-glucose positron emission tomography. *Laryngoscope* **109**, 970–975.
35. Kao, C. H., ChangLai, S. P., Chieng, P. U., Yen, R. F, and Yen, T. C. (1998). Detection of recurrent or persistent nasopharyngeal carcinomas after radiotherapy with 18-fluoro-2-deoxyglucose positron emission tomography and comparison with computed tomography. *J. Clin. Oncol.* **16**, 3550–3555.
36. Lowe, V. J., Boyd, J. H., Dunphy, F. R., *et al.* (2000). Surveillance for recurrent head and neck cancer using positron emission tomography. *J. Clin. Oncol.* **18**, 651–658.
37. Fischbein, N. J, Assar, O. S., Caputo, G. R., *et al.* (1998). Clinical utility of positron emission tomography with 18F-fluorodeoxyglucose in detecting residual/recurrent squamous cell carcinoma of the head and neck. *Am. J. Neuroradiol.* **19**, 1189–1196.

CHAPTER

4

Preneoplastic Lesions of the Upper Aerodigestive Tract

F. LONARDO and W. SAKR

Department of Pathology
Wayne State University, School of Medicine
Detroit, Michigan 48201

I. Introduction 35
II. Normal Anatomy 35
 A. Oral Cavity 35
 B. Pharynx 36
 C. Larynx 36
III. Pathological Features 37
 A. Definitions 37
 B. Histological Classification 40
 C. Molecular Markers of Dysplasia 44
IV. Clinical Features 45
 A. Oral Cavity 45
 B. Larynx 47
V. Correlation between Clinical and Histological Classifications 47
VI. Malignant Progression 48
 A. Oral Cavity 48
 B. Larynx 49
 References 50

I. INTRODUCTION

Several lines of evidence, including clinical, experimental, and morphological data, support the concept that squamous cell carcinoma of the upper aerodigestive tract (UADT) arises from noninvasive lesions of the squamous mucosa. These lesions encompass a histological continuum between the normal mucosa at one end and high-grade dysplasia/carcinoma *in situ* at the other, establishing a model of neoplastic progression. This continuum of preinvasive neoplasia is encountered in many other epithelia, including the lower respiratory tract, and the cervix uteri. Increasing genetic abnormalities occur in increasing histological grades in the UADT, as well as in other known examples of preneoplastic progression, supporting a model envisioning cancer progression as the phenotypical result of accumulation of genetic abnormalities.

The identification of preneoplastic lesions of the UADT, through clinical, morphological, and more recently, molecular means, lays the premises for early detection and treatment of head and neck squamous cell carcinoma (HNSCC). Furthermore, understanding and documenting the morphological and molecular abnormalities associated with this progression may shed light on the biology of SCC, while identifying markers of transformation, which may serve as a surrogate of clinical end points in chemoprevention clinical trials.

Our discussion focuses on the upper aerodigestive tract, which includes the oral cavity, the pharynx, and the larynx. These sites form a functional and anatomical unit; more importantly, they share exposure to the same etiopathogenetic factors and show the same spectrum of preinvasive changes associated with the development of SCC, which is their most common malignancy.

II. NORMAL ANATOMY

A. Oral Cavity

A nonkeratinized, stratified squamous epithelium lines most of the oral cavity, while in specialized regions the epithelium becomes keratinized. The nonkeratinized mucosa includes, from the basement membrane to the surface, a stratum basale, a stratum spinosum (prickly), and a superficial layer. In the keratinized epithelium, a stratum granulosum and a keratinized (corneum) layer are present above the prickly layer. From the stratum basale to the stratum corneum, a progressive decrease in the nucleus/cytoplasmic ratio occurs, along with intracytoplasmic accumulation of keratin filaments, denoting increasing differentiation and specialization. The stratum basale is constituted by cuboidal cells with a high N/C ratio, resting on the basement membrane (Fig. 4.1). Molecular data show that they are the only cell type that expresses proliferation-associated antigens [1] and the RNA component of telomerase [2] within the normal mucosa. By replacing committed cells, which undergo terminal differentiation in the more superficial layers of the epithelium, these basal cells assure the turnover of the epithelium and thus have the role of stem cells [3]. In keratinized epithelium, the superficial layer is constituted entirely by anucleated cells, showing accumulation of intermediate filaments, whereas in nonkeratinized epithelium, small nuclei are still retained. Furthermore, in keratinized epithelium, an intermediate layer may be present between the keratin layer and the prickly layer, similar to the epidermis, characterized by large intracellular granules, called stratum granulosum. Regional variations in the composition of the epithelium, including its degree of keratinization, reflect differences in the extent of mechanical stress during mastication, which in turn depends on the resiliency of the exposed areas. Thus, the squamous (masticatory) mucosa of the gingiva and hard palate, fastened to the underlying bone by heavy collagen bundles, not allowing it to stretch it, is keratinized. The thickness of the stratum granulosum is also more pronounced in the hard palate. Areas of the oral cavity, such as lips, soft palate, cheeks, and floor of mouth, characterized by higher resiliency and subject to lesser mechanical stress, are lined by nonkeratinized mucosa. Some individuals show an anatomic variant of this distribution, characterized by keratinization occuring in the malar mucosa, along a line starting from the labial commissure and running parallel to the occlusion line of premolars and molars. This line is visible clinically as a white line and is designated *linea alba*.

The structure of the interface with the underlying stroma also reflects the amount of mechanical stress to which the mucosa is subject. Thus, the buccal mucosa has prominent mucosal ridges, anchoring it to a heavily collagenized lamina propria. In contrast, areas protected from stress, such as the floor of mouth, possess thinner and more shallow rete ridges and a less collagenized lamina propria [4].

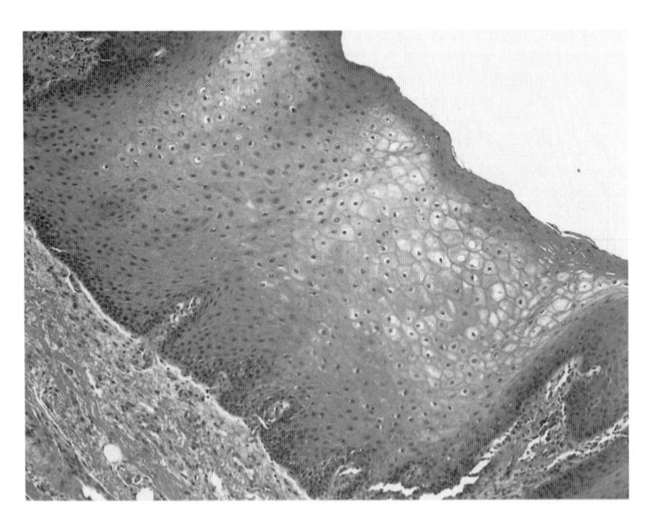

FIGURE 4.1 Normal squamous mucosa. Notice bland cytological features and progressive and orderly maturation from the basal to the superficial layer.

The mucosa of the tongue has peculiar histological features, pertaining to its function as a taste organ, which are of no relevance in this setting. In its posterior third, the lingual mucosa becomes enriched with lymphoid tissue, part of the mucosa-associated lymphoid tissue (MALT) of the UADT.

The degree of keratinization, thickness, presence of pigments, and degree of vascularization of the mucosa and its lamina propria all affect the color of the mucosa, an issue of relevance in correlating the clinical appearance of mucosal lesions with its microscopic composition, as discussed later.

The oral cavity hosts minor salivary glands within its submucosa. Notably, dysplasia and carcinoma *in situ* (CIS) can involve the acini as well as the excretory ducts of minor salivary glands, mimicking invasive carcinoma [5].

B. Pharynx

The pharynx is divided into three anatomical compartments: oropharynx, nasopharynx, and hypopharynx. The oropharynx and hypopharynx are lined by stratified, nonkeratinized squamous epithelium. The submucosa contains seromucinous glands and aggregates of lymphoid tissue. The nasopharynx is lined with approximately 60% stratified, nonkeratinized squamous epithelium and with approximately 40% ciliated, respiratory-type epithelium. The latter predominates in the posterior nares and in the roof of the posterior wall, whereas the remaining areas reveal an alternation of the two types of epithelia. Notably, at the transition between the two types, the mucosa assumes an "intermediate" or "transitional" appearance that may mimic dysplasia [6]. Similar features are observed at the transition betweeen squamous and respiratory epithelium in the larynx in normal conditions [6], and during the process of squamous metaplasia in the bronchial-ciliated epithelium, in response to irritants [7].

C. Larynx

The type of epithelial lining of the larynx changes according to the location and shows an alternation of ciliated, respiratory-type and squamous epithelium. The supraglottic compartment (extending from the tip of the epiglottis to the true vocal cord) shows respiratory-type epithelium, which merges into squamous epithelium in the posterior surface of the epiglottis superiorly and, inferiorly, at the glottis (composed of the true vocal cords and the anterior commissure). Thus, the false vocal cords are lined by respiratory epithelium, whereas the true vocal cords are lined by squamous epithelium. The squamous epithelium merges into respiratory mucosa at the lower border of the true vocal cord, covering the subglottic larynx (the portion of larynx comprised between the lower border of the true vocal cord and the first tracheal ring) and blends inferiorly in the respiratory epithelium of the trachea.

The respiratory epithelium is a ciliated, pseudostratified epithelium. Its basal layer is composed of basal cells, connected to the basement membrane by hemidesmosomes, which do not reach to the lumen. They have a high nucleus/cytoplasmic ratio and, like in squamous mucosa, represent the regenerative component of the epithelium. The differentiated cells warding the luminal surface are composed of ciliated, brush, and goblet cells, allowing mucociliary clearance. A minor component of the epithelium, detectable only by performing electron microscopy or special immunohistochemical stains, is constituted by small granular cells. These cells, which are morphologically similar to the basal cells by regular light microscopy, have neurosecretory granules and belong to the diffuse neuroendocrine system [8]. Areas between the squamous and the respiratory-type epithelium have a transitional appearance, characterized by progressive flattening of luminal cells, a progressively more elongated shape, and an arrangement parallel to the basement membrane. The mucosa in these areas may assume a relatively disorganized appearance, referred to also as "incomplete metaplasia," which may mimic true dysplasia. However, frank cytological atypia is absent and maturation is preserved [9]. In approximately half of smokers, patches of metaplastic squamous mucosa are present in the supraglottic larynx [9].

III. PATHOLOGICAL FEATURES

It should be stressed from the outset that the terms leukoplakia and erythroplakia are clinical and descriptive terms, referring to the appearance of the mucosa as a white (leuko) or red (erythro) patch. These terms, particularly leukoplakia, were widely used in the older reports, including seminal prospective studies establishing their potential for developing into overt cancer. However, even when defined clinically, these are diagnoses of exclusion, as they can be reproduced by inflammatory conditions. More importantly, in series controlling for histology, it has been shown that the degree of cytoarchitectural alterations present in the mucosa predicts the risk of progression. Thus, from a pathologist's standpoint, the terms leukoplakia and erythroplakia can be considered descriptive terms of the gross morphology of these lesions, whose identifying features rest on their microscopic appearance, namely on the presence and extent of dysplasia. Thus, histological features of UADT lesions, including their classification and their molecular features, are treated first. This is followed by a description of their clinical features, the correlation existing between histological and clinical classifications (i.e., gross and microscopic features), and their risk of progression.

A. Definitions

Discussion of the histological features of the preinvasive lesions of the upper aerodigestive tract is not possible without a definition of the morphogical nomenclature currently used. A short description of commonly encountered histological terms is provided, describing pathological changes affecting the UADT mucosa [10,11].

1. Squamous Metaplasia

Metaplasia describes the replacement of one type of specialized epithelium with another, i.e., replacement of the ciliated, respiratory-type epithelium of the false vocal cord by squamous epithelium. It is a reversible process that may progress into overt dysplasia or revert to normal and may also be seen in association with inflammatory conditions. It differs from dysplasia in that it lacks any cytological atypia.

2. Simple Hyperplasia

Hyperplasia is an increase in thickness of the epithelium, secondary to an increase in one or more of its component layers: the basal layer, the prickly layer (i.e., acanthosis), or the superficial layer (i.e., hyper/parakeratosis), usually a combination of the former with one or two of the latter, without perturbations in maturation and without any accompanying cytological atypia (Figs. 4.2–4.4). Notably, minimal cell crowding and cytological atypia may also occur in association with inflammation (Fig. 4.5).

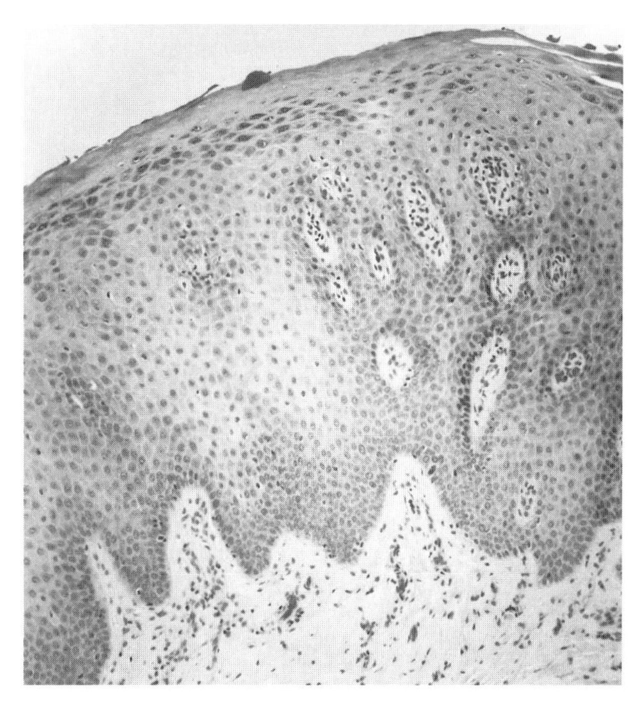

FIGURE 4.2 Hyperkeratosis. A granular layer is present below the superficial keratinized layer.

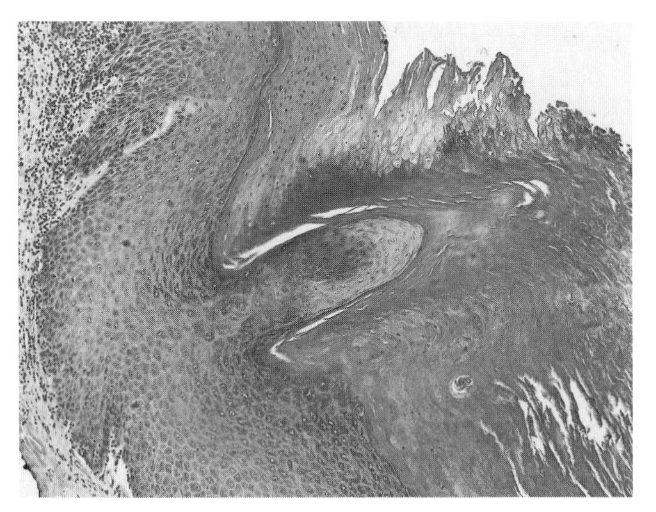

FIGURE 4.3 Hyperkeratosis. Extreme thickening of the mucosa is present, secondary to marked accumulation of superficial squames.

3. Dysplasia/Intraepithelial Neoplasia

This is a mucosal alteration characterized histologically by cellular atypia and loss of maturation, which has the biological potential of developing into overt cancer (see Table 4.1).

4. Mild Dysplasia

Cells with slightly abnormal cytological features are present but are limited to the lower third of the epithelium. Orderly maturation into prickly and squamous layers in the upper two-thirds of the epithelium is preserved. Mitoses may be present but are limited to the basal layer and are of normal configuration; keratosis may be present (Figs. 4.6–4.8).

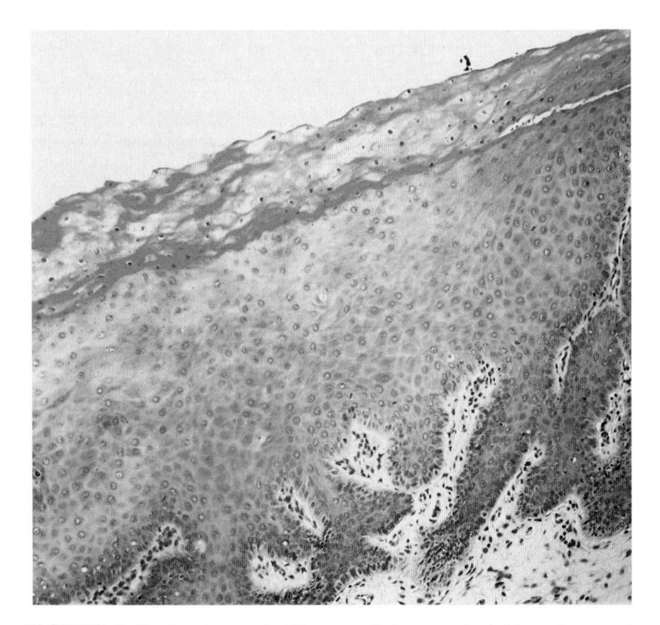

FIGURE 4.4 Parakeratosis. The superficial layer is thickened, secondary to the presence of multiple layers of parakeratotic squames.

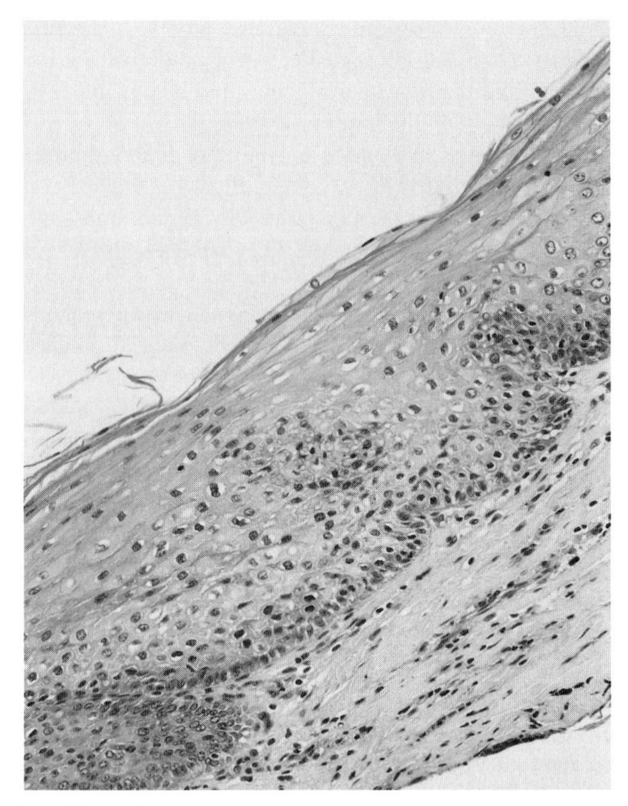

FIGURE 4.5 Inflammatory atypia. Minimal nuclear enlargement occurs in the basal and parabasal layers, associated with a mild lymphocytic infiltrate.

5. Moderate Dysplasia

Cells with atypical cytological features occupy the lower two-thirds of the mucosa. Cytological atypia is more pronounced than in mild dysplasia; nucleoli tend to be prominent. Maturation is preserved in the upper third of the mucosa; normal-appearing mitoses may be found in the parabasal and intermediate layers (Figs. 4.9–4.12).

TABLE 4.1 Histological Criteria Used in the Grading of Dysplasia

Cytological atypia
Increased N/C ratio, presence of nucleoli, hyperchromasia, pleomorphism
Mitotic activity
Increased number and level of mitoses within the mucosa; presence of abnormalities in the mitotic spindle, i.e., tripolar mitoses
Abnormal maturation and polarity
Decreased ratio between the differentiated component, composed of the prickly and squamous layers, and the undifferentiated component, composed of atypical cells. Also reflected by the occurrence of premature keratinization within the epithelium rather than at its luminal surface

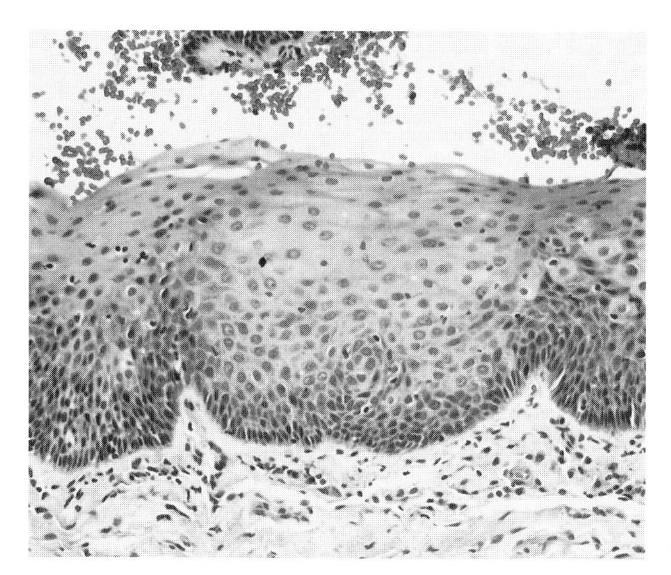

FIGURE 4.6 Low-grade dysplasia (mild). Slight enlargement and crowding of the basal layer, with progressive and orderly maturation in the upper two-thirds of the mucosa.

6. Severe Dysplasia

Atypical cells, showing marked nuclear abnormalities and prominent mitotic activity, occupy more than two-thirds of the epithelium. They are not as crowded and, most importantly, not as cytologically atypical as in CIS. Maturation is preserved, as evidenced by focal superficial squamous maturation and focal preservation of intercellular bridges. Mitoses, including atypical ones, may extend to the upper third of the epithelium. Associated keratosis may be present (Figs. 4.13–4.16).

7. High-Grade Keratinizing Dysplasia

Cells with high-grade cytological features, similar to those found in CIS, occupy the lower two-thirds of the mucosa; mitoses are frequent, including abnormal ones. Abnormal maturation is highlighted by the occurrence of single cell keratinization or keratin pearl formation within the epithelium rather than at its luminal surface. However, in contrast to "classic" CIS, the uppermost component of the epithelium shows a prominent keratinized layer (Figs. 4.17–4.22).

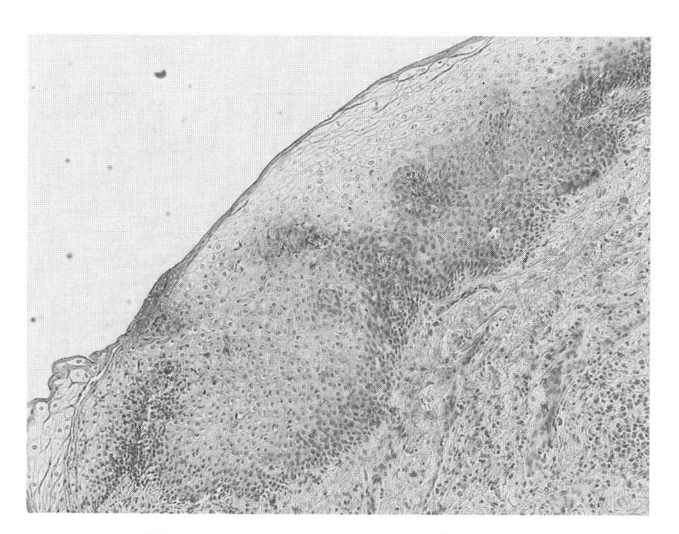

FIGURE 4.7 Low-grade dysplasia (mild to moderate).

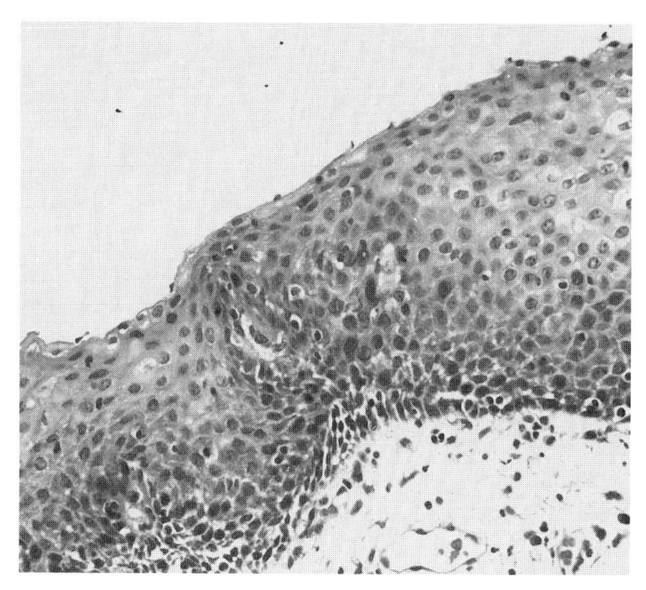

FIGURE 4.8 Low-grade dysplasia. Dysplastic cells are limited predominantly to the lower third of the epithelium. Progressive and orderly maturation occurs in thc parabasal and superficial layers.

8. Carcinoma *in Situ*

Cells with frankly malignant cytological features occupy the whole thickness of the epithelium; squamous differentiation is entirely absent. Abnormal cells have the cytological hallmarks of malignancy, including a high N/C ratio, prominent single or multiple nucleoli, nuclear hyperchromasia, and pleomorphism. Mitotic figures, including

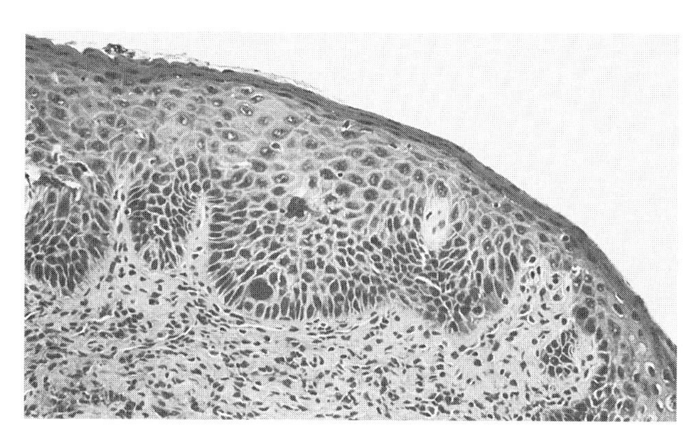

FIGURE 4.9 Moderate dysplasia. Dysplastic cells, including forms with giant and multinucleated nuclei, are limited to the lower two-thirds of the epithelium. Progressive and orderly maturation occurs in the superficial layers.

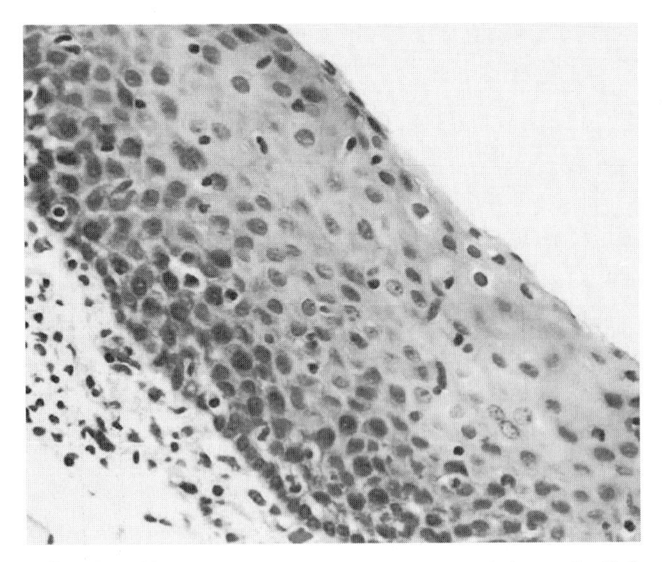

FIGURE 4.10 Moderate dysplasia. Accumulation of abnormal cells is limited to the lower two-thirds of the mucosal thickness.

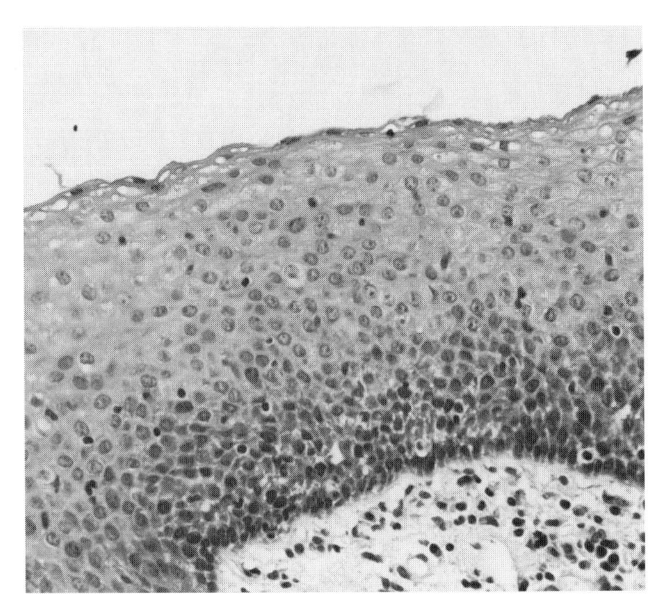

FIGURE 4.12 Low-grade dysplasia (mild to moderate). Abnormal cells occupy the lower two-thirds of the mucosa. Cytological atypia s minimal.

atypical ones, are frequent and extend throughout the entire mucosa, including its upper third. By definition, the changes are limited to the epithelium and stromal invasion is absent (Figs. 4.23–4.25).

B. Histological Classification

Microscopic grading of preneoplastic lesions of the UADT is of paramount importance biologically and clinically, as the probability of malignant progression into invasive cancer of these lesions is dictated by their grade. This correlation, which is discussed more extensively later, highlights the importance of obtaining a biopsy of all suspect lesions and the priority to be given to microscopic over clinical examination.

The grading system, as described by the WHO [10,11], follows criteria similar to those accepted for other organs, particularly the cervix uteri [12]. Cervical preneoplastic lesions have been studied extensively and have represented for years a standard histological model of squamous preneoplastic lesions. Thus the grading criteria applied to the cervix have been extended to all other squamous

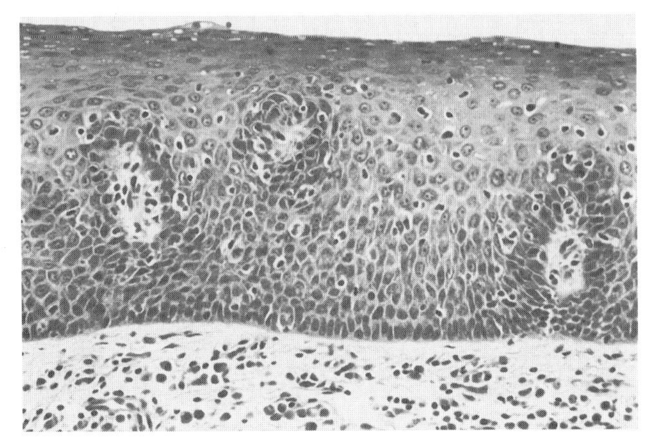

FIGURE 4.11 Moderate dysplasia. Superficial maturation is evident as a layer of spindle-shaped, parakeratotic squames, arranged parallel to the basement membrane.

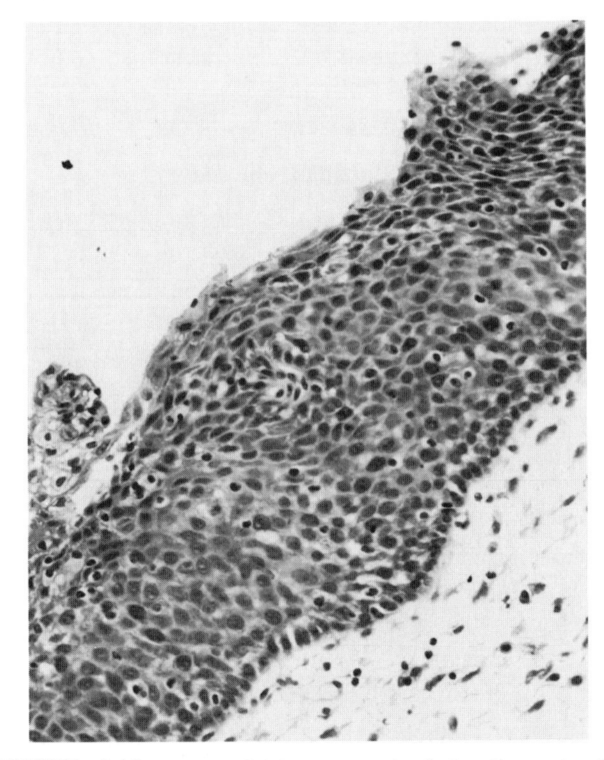

FIGURE 4.13 Nonkeratinizing severe dysplasia. Abnormal cells occupy the full thickness of the epithelium. Minimal differentiation is preserved as parakeratotic squames in the surface of the epithelium.

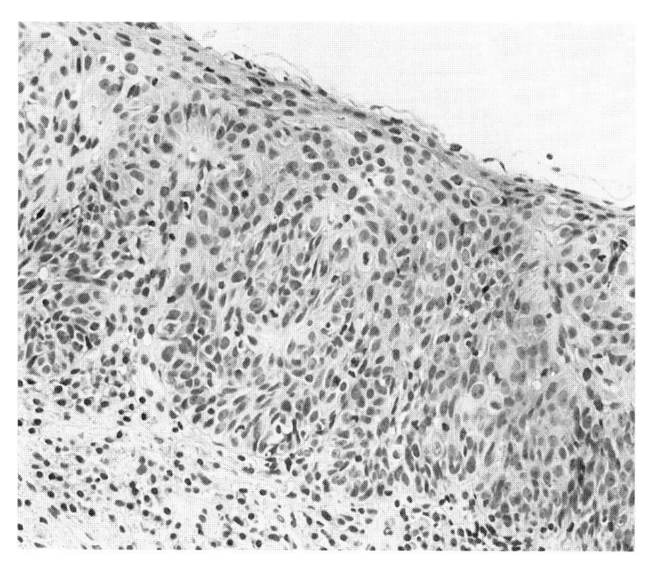

FIGURE 4.14 Nonkeratinizing severe dysplasia.

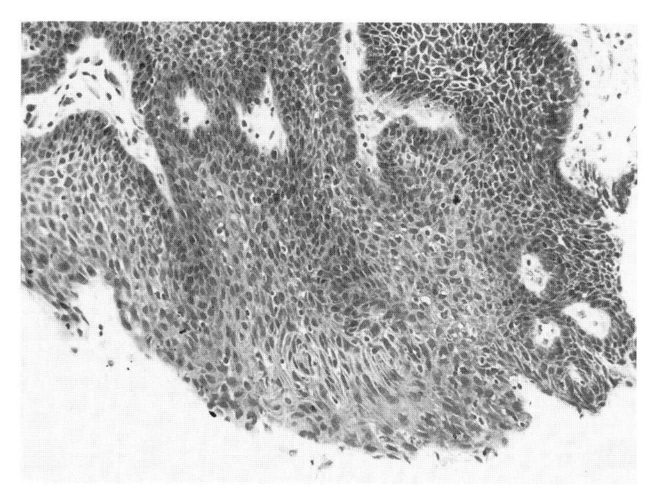

FIGURE 4.16 Nonkeratinizing severe dysplasia. Dysplastic cells occupy more than two-thirds of the thickness of the epithelium; however, keratinization is observed in the most superficial portions of the epithelium.

preneoplastic lesions, including those of the UADT. This grading system relies on the extent of distribution of the abnormal cells within the epithelium. These are limited to the lower third of the epithelium in mild dysplasia, extend to about two-thirds in moderate dysplasia, and to more than two-thirds of the epithelium in severe dysplasia. The difference between severe dysplasia and CIS is that abnormal cells in CIS have higher grade and frankly malignant cytological features. In addition, in CIS they involve the entire epithelium, including its most apical component and thus a complete lack of maturation/differentiation is present. Lesser grade cytological features and some maturation, however, characterize severe dysplasia [10,11]. Several authors have stressed that the etiopathogenetic factors associated with dysplasia and squamous cancer of the upper aerodigestive tract (i.e., alcohol and smoking) and cervix (HPV infection) are different [13]. Furthermore, it has been pointed out that UADT dysplastic lesions may show unique morphological features not seen in cervical dysplasia. Namely, superficial maturation may be seen in association with high-grade cytological features in the lower third or two-thirds of the mucosa. Thus, the mere application of criteria used for the cervix to the UADT preneoplastic lesions would result in an underassessment of their grade [14,15]. Two main classification schemes in alternative to the WHO system have been proposed [13,16–19]. They are both characterized by an emphasis on cytological features rather than the relative ratio

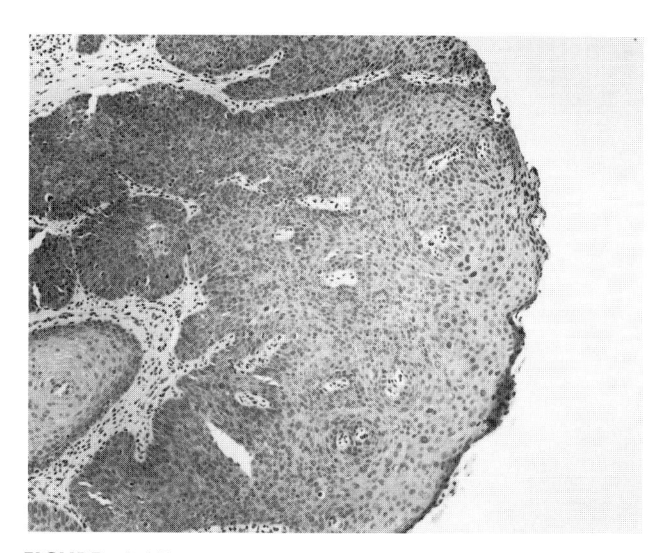

FIGURE 4.15 Nonkeratinizing severe dysplasia. Dysplastic cells occupy the full thickness of the epithelium. Differentiation is preserved focally as a thin layer of parakeratotic, strongly eosinophilic squames on the surface of the epithelium.

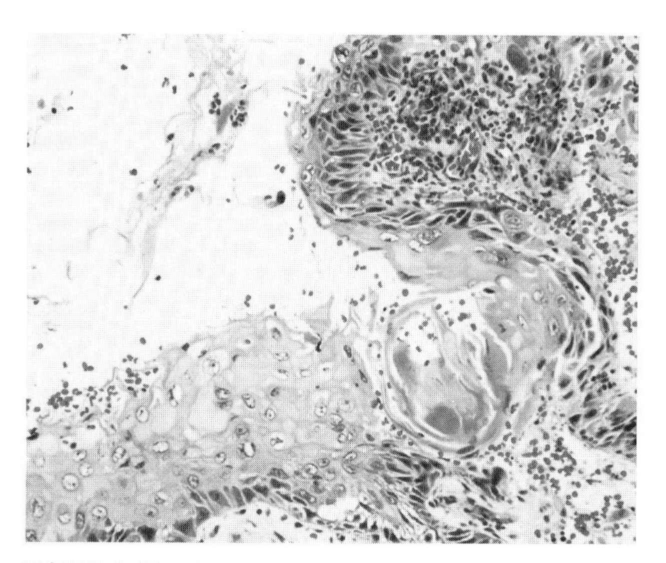

FIGURE 4.17 High-grade (severe) keratinizing dysplasia. Maturation is preserved, as keratinization occurs focally; however, this is associated with marked cytological atypia.

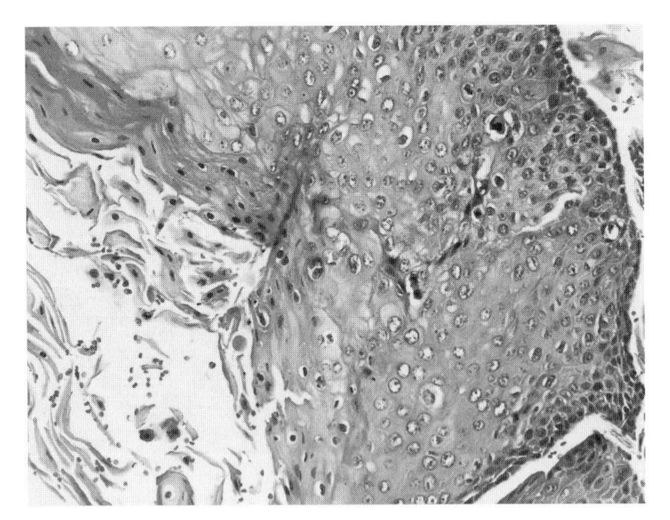

FIGURE 4.18 Dyskeratosis in high-grade keratinizing dysplasia. Single cell intraepithelial keratinization (dyskeratosis) is visible as eosinophilic, round cells, with a pyknotic nucleus.

of abnormal to differentiated cellular components within the epithelium. The system advocated by Crissman recognizes a category of "keratinizing dysplasia" to designate a lesion showing superficial keratinization in association with high-grade cytological features in the remaining mucosa. In these authors' experience, such lesions have a high incidence of local relapse, a high progression rate into invasive cancer [14,20], and a high content of aneuploidy [21]. Thus, they are included in a high-grade dysplasia group (SIN-LIN 3) [16,17]. These authors further stressed that abnormal differentiation was present in these lesions in the form of aberrant keratinization (dyskeratosis), represented by single cell keratinization and keratin pearls, occurring in the midst of the epithelium [16]. Underreporting of "keratinizing dysplasia"

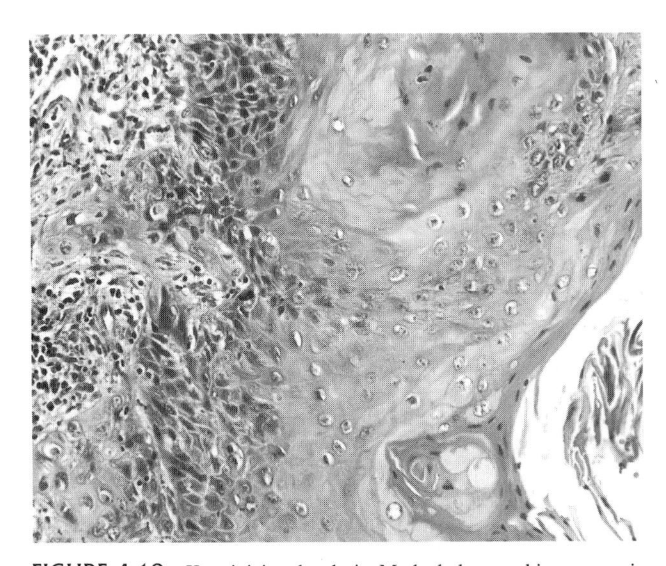

FIGURE 4.19 Keratinizing dysplasia. Marked pleomorphism occurs in the basal layer, associated with maturation in the keratinized superficial layer.

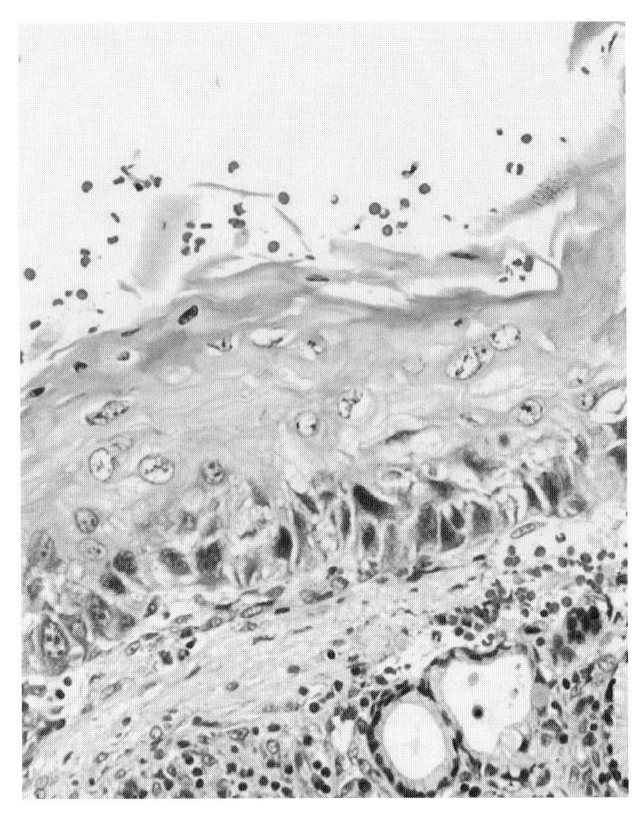

FIGURE 4.20 Severe keratinizing dysplasia.

and its difference from "classic CIS" were confirmed by others, on a systematic review of laryngeal biopsies [15]. The system advocated by Crissman and colleagues proposed the designation squamous intraepithelial neoplasia (SIN) or

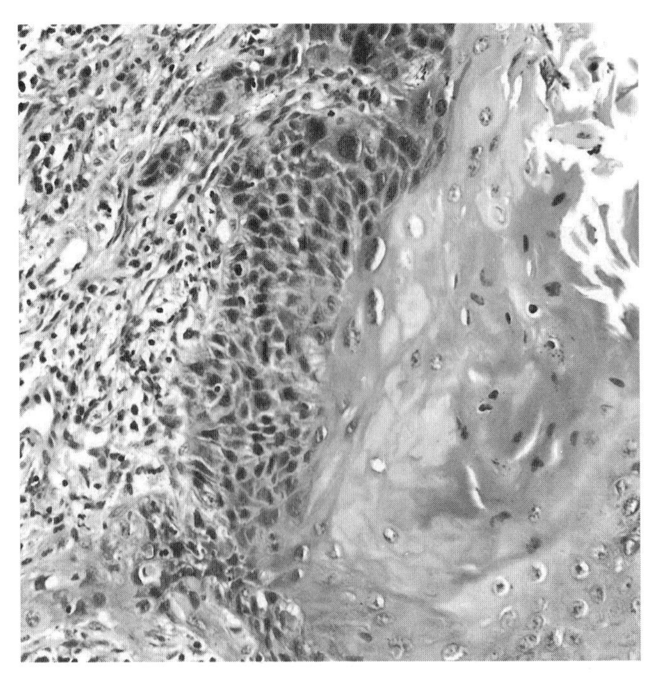

FIGURE 4.21 High-grade keratinizing dysplasia associated with microinvasive carcinoma.

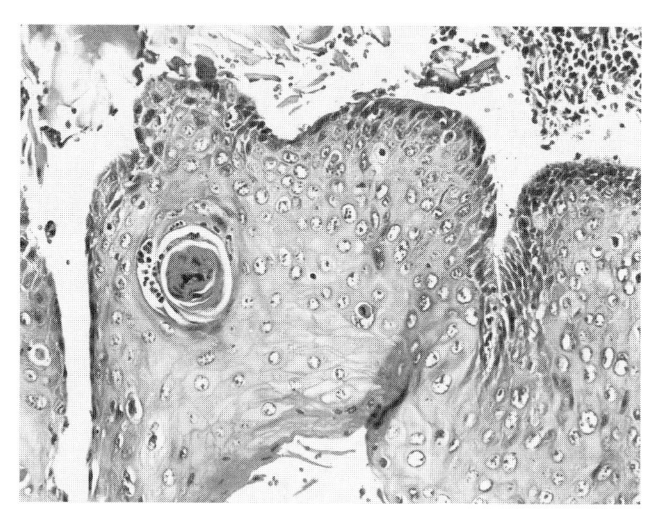

FIGURE 4.22 Dyskeratosis involving a single cell and a cell nest within the epithelium.

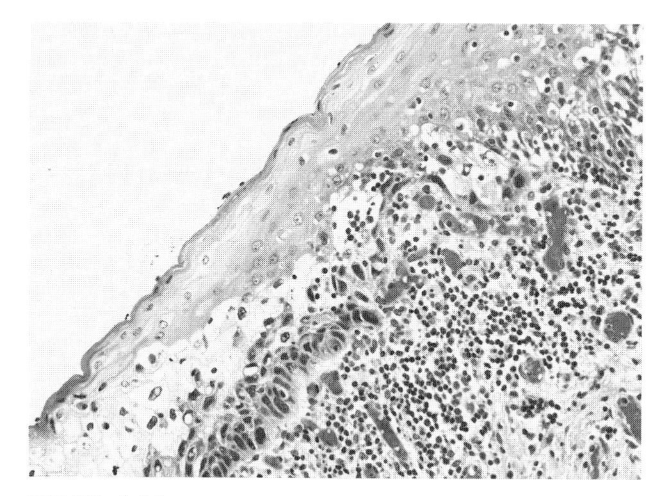

FIGURE 4.24 High-grade dysplasia. The basal layer is crowded by cells with high nucleus/cytoplasmic ratios, with enlarged pyknotic nuclei. Parabasal layers are occupied by cells with clear cytoplasm and enlarged nuclei. Contrast these features with the bland cytology of contiguous normal squamous epithelium, present in the upper portion of the figure.

laryngeal intraepithelial neoplasia (LIN) [16,17], as an alternative to dysplasia. The proposed advantages of the designation "intraepithelial neoplasia" are manifold. This definition matches the currently used designation for the cervix uteri [cervical intraepithelial neoplasia CIN], which has replaced the old designation of dysplasia and thus allows more standardized reporting of preneoplastic lesions across different anatomical sites. Furthermore, in the authors' intentions, this designation is more clinically oriented and broader than "dysplasia" and allows the inclusion of nonmorphological parameters, i.e., molecular markers, within the grading system. Although biologically sound, this suggestion, however, has not met acceptance in routine clinical practice.

Kambic and Gale proposed a distinction of dysplasia for the larynx into simple, abnormal, atypical hyerplasia, and CIS. In this classification, known as Ljubljana classification, the emphasis is on cytological features [13,19,22]. Simple hyperplasia shows an increase in epithelial thickness secondary to an increase in the stratum spinosum, with no cellular atypia in the basal and parabasal layers. In abnormal hyperplasia, basal or parabasal layers are increased, encompassing up to one-half of the mucosal thickness, but their nuclei, although enlarged, lack significant cytological atypia. The occurrence of significant nuclear atypia and dyskeratosis, associated with preservation of the overall epithelial architecture, characterizes atypical hyperplasia. In CIS, cells with the cytological features of malignancy occupy the majority of the epithelium, which has lost its regular stratification and shows very frequent mitoses [13,19,22]. The approximate correlation between the

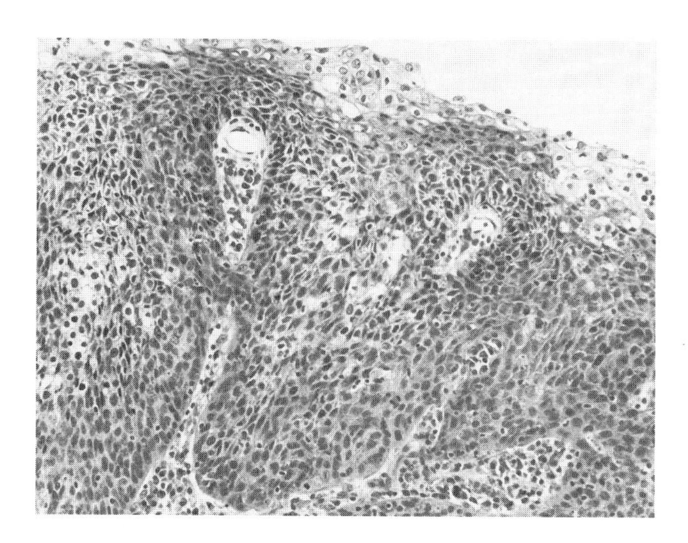

FIGURE 4.23 High-grade, nonkeratinizing dysplasia (CIS).

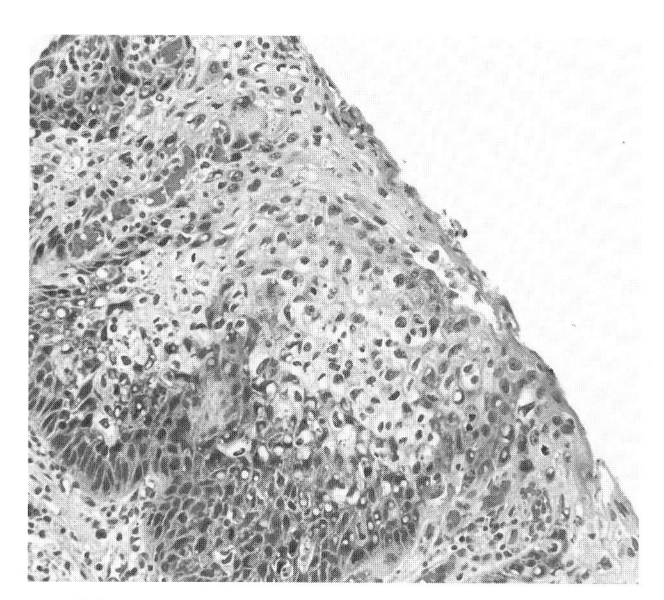

FIGURE 4.25 High-grade, nonkeratinizing dysplasia (CIS).

TABLE 4.2 Major Classification Schemes of Dysplasia and Approximate Correspondence of Definitions

WHO	Mild	Moderate	Severe dysplasia	CIS
Crissman	SIN-LINI	SIN-LINII	SIN-LINIII	CIS
Kambic	Simple	Abnormal	Atypical hyperplasia	CIS

different types of dysplasia is sketched in Table 4.2. However, the Ljubjana and the WHO systems are not easily reducible one to the other, as demonstrated by large variations in the classification of lesions between the two groups. Thus, of 13 cases diagnosed as mild dysplasia in the WHO system, 3 were reclassified as simple, 8 as abnormal, and 1 as atypical hyperplasia; of 20 cases of moderate dysplasia, 2 were reclassified as simple, 10 as abnormal, 6 as atypical hyperplasia, and 1 as carcinoma *in situ*; and of 14 cases of severe dysplasia, 1 was reclassified as simple, 8 as atypical hyperplasia, and 1 as carcinoma *in situ* [23]. Nine lesions were placed in a group of "large cell hyperplasia," a category not described in the original Ljubjana classification [23]. Like other multi-tiered grading systems, the WHO system, which is used most commonly in pathological practice, has high inter- and intraobserver variability [24]. The morphological distinction between severe dysplasia and CIS is particularly difficult and is not reproducible [10]; however, these lesions have a similar risk of progression into overt cancer [10]. These findings justify a distinction into two tiers: low-grade (including mild and moderate dysplasia) and high-grade (including severe dysplasia and CIS), dysplasia, provided that a mutual understanding of the terminology is instituted among pathologists, surgeons, and oncologists.

C. Molecular Markers of Dysplasia

Since the initial histological characterization of premalignant conditions of the UADT, one of the major advances in this field has been identification of the molecular alterations with which they are associated. While biologically justified and crucial in reducing the variability intrinsic in a morphology-based system [24], the inclusion of molecular markers has yet to find a role in the classification of dysplasia. However, understanding the molecular basis of progression for UADT preneoplastic lesions is invaluable in understanding the biology of SCC. Furthermore, it may be clinically relevant if molecular markers of dysplasia are used as screening tools for early detection or as intermediate markers in chemoprevention trials. The prototype model of molecular alterations associated with preneoplastic progression is the colon, where sequential molecular alterations have first been described to occur alongside the morphological progression from normal to cancer [25]. A similar accumulation of alterations has been described in the UADT in increasing morphological grades of dysplasia.

Molecular alterations occurring in preneoplastic lesions of the upper aerodigestive tract belong to two main groups, reflecting abnormalities in either cellular differentiation or cell cycle control. Changes of the first group include those affecting the profile of keratin expression and were the first to be reported. In the normal epithelium, low molecular weight keratins (LMW) are expressed in the basal layer and high molecular weight keratins (HMW) are expressed in the stratum spinosum. Abnormal cells express, regardless of their position, LMW keratins, whereas HMW keratins are expressed either in the uppermost, keratinized layer or not at all in dysplasia. Thus, suprabasal expression of LMK, such as CK19, has been proposed to constitute a marker of dysplasia [18,26], altough this finding cannot reliably distinguish hyperplasia from true dysplasia [1].

The foremost alteration in cell cycle regulation occurring in dysplasia is the occurrence of an increased proliferative rate in association with increasing morphological grades. This has been assessed traditionally by morphological evaluation, i.e., counting mitoses. However, more recently, the discovery of PCNA and Ki-67 antigens, expressed exclusively by proliferating cells, has allowed an objective evaluation of the proliferative rate [3,27–30]. Using both methods, a continuum of increasing proliferative rates is seen in increasing grades of dysplasia. In contrast, the only population to show infrequent positivity in normal mucosa is the basal cell, compatible with its role as a progenitor cell [3,29,30]. Interestingly, in simple mucosal hyperplasia, only the basal layer shows positive staining for proliferative antigens, setting it aside from true dysplasia, where the expression of this antigen extends to suprabasal cells [1].

Alterations in many molecules controlling the cell cycle are frequent in dysplasia and are likely responsible for its hyperproliferative state. A continuum of increasing positivity is observed in the rate of p53 positivity, as detected by immunohistochemistry in lesions of increasing histological grades [30–32]. While negative in normal epithelium, p53 is found in 9.4 to 32% of low-grade and 33 to 50% of high-grade dysplasia [30,32]. Suprabasal expression of CK 19 and PCNA is associated with positivity for p53 in the majority of cases of dysplasia [1], highlighting the link existing between abnormal differentiation and cell cycle alterations. Alterations in the distribution of the cyclin kinase inhibitor p21 also occur. Whereas only the intermediate layer of the normal epithelium expresses this marker, the entire dysplastic epithelium shows positivity [33].

Increasing percentages of EGFR positivity are also seen, seemingly associated with the dysplastic component [31]. Increasing percentages of aneuploid populations are also observed in increasing grades, progressing from 33% of SINI to 78% of SINII and 100% of SINII [21].

Apoptosis increases in parallel to the proliferative rate, as observed morphologically or by DNA *in situ*-labeling techniques [30,34]. The relevance of alterations in the apoptotic rate is highlighted by data showing a change in expression of the antiapoptotic protein bcl-2 (the target gene deregulated as a consequence of the 14; 18 translocation occurring in most follicular lymphomas) in dysplasia. While normally found only in 37% of normal squamous mucosa, where its expression is restricted to the basal layer, its expression raises to 71% of dysplastic lesions and 80% of invasive ones in the nasopharynx [35].

Genetic studies of preneoplastic lesions for loss of heterozygosity (LOH) have shown that increasing grades of dysplasia show accumulation of genetic deletions, compatible with the commonly accepted model envisioning cancer as the result of multiple genetic "hits." The earliest and most common changes occur at sites 3p and 9p. The 9p locus harbors genes for the kinase inhibitors p16 and p19, which are altered frequently in HNSCC, as well as in malignancies from other sites. The 3p14 and 3p21 sites harbor several candidate tumor suppressor genes, including the gene deleted in Von Hippel Lindau (VHL) disease, the DNA mismatch repair enzyme hMLH1, affected in hereditary nonpolyposis colon cancer, the retinoic acid receptor β, and the fragile histidine triad (FHIT) gene [36]. While the identification of the specific suppressor gene(s) located at 3p14 and 3p21, whose loss is responsible for the development of cancer, is still unresolved, the FHIT gene has been proposed to be a specific target of cigarette smoke carcinogens [37–39] and shown to behave as a tumor suppressor gene *in vitro* [40]. In a retrospective study correlating molecular alterations with progression to overt cancer, losses at 3p and 9p loci were the most common lesions in both nonprogressing and progressing premalignant lesions. However, they were virtually always present in progressing lesions, compatible with a model whereby they are necessary but not sufficient for malignant transformation. The occurrence of additional chromosomal losses was shown to confer a much higher risk of progression, comparable to losses at 3p and 9p alone [41]. These included chromosomal sites 4q, 8p, 11q, and, notably, 17p [41,42]. Because this latter site is the locus of p53, these data are compatible with histochemical data quoted previously, showing alterations in p53 as an important event in the progression of premalignant lesions. These data are overall compatible with a model envisioning loss of genetic material in chromosome 3p and 9p loci, as involved in initiation [41], and additional genetic losses, including p53, as involved in progression. Data are summarized in Table 4.3.

Changes in the basement membrane have also been described in dysplasia. Normal and hyperplastic mucosa is usually associated with a prominent and continuous basement membrane, as assessed by an immunohistochemical stain with collagen type IV and laminin. The basement membrane is usually prominent and continuous in mild to moderate dysplasia; in contrast, in severe dysplasia/carcinoma *in situ* it is often thinned and discontinuous. However, some invasive cancers retain a continuous pattern of basement membrane stain and thus this stain cannot be used to reliably discern noninvasive from invasive proliferations [43].

TABLE 4.3 Incidence of Molecular Alterations in Low- and High-Grade Dysplasia

	LGD	HGD	Ref.
EGFR	+	++	31, 93
Mib-1/PCNA	+	++	1,3,30
p21	+	+	33
p53	9–67%	33–85%	31,32
Aneuploidy	33%	100%	14
Apoptosis	+	++	34
bcl-2	+	++	35

Dysplastic lesions of the nasopharynx have been shown to harbor clonal integration of the Epstein–Barr virus (EBV), in concordance with the accepted role this virus plays in the development of nasopharyngeal carcinoma [44]. Expression of EBV antigens has also been shown to occur focally in the basal layer of normal squamous mucosa of the tongue and to be augmented in oral leukoplakia occurring in HIV patients [45].

IV. CLINICAL FEATURES

A. Oral Cavity

1. Definitions

a. Leukoplakia

The term leukoplakia (literally white patch) was coined by Schwimmer, a Hungarian pathologist, in the second half of the 19th century [46]. Several studies have, more recently, confirmed the preneoplastic nature of leukoplakia associated with dysplasia, defined its clinical features, and identified clinical and histological features associated with its progression to overt cancer.

The definition of leukoplakia, according to the WHO, is the following: "A predominantly white lesion of the oral mucosa that cannot be characterized as any other definable lesion" [11]. As such, leukoplakia is a descriptive, clinical term and a diagnosis of exclusion, not a pathological term. This appearance reflects changes in the thickness and/or composition of the epithelium (such as increased keratinization and parakeratosis) and/or in the lamina propria (such as fibrosis), altering its translucence. Lesions as heterogeneous as candidiasis, discoid lupus, carcinoma *in situ*, and invasive squamous cell carcinoma may all appear clinically as a white patch (Table 4.4). Thus, leukoplakia is a clinical diagnosis, whose true nature can only be unveiled by histological examination. The distinguishing feature of those leukoplakias that are truly preneoplastic, which is but a small subset of all clinically diagnosed leukoplakias, is the occurrence of *dysplasia* [11].

TABLE 4.4 Mimics of Leukoplakia

Lichen planus	Verruciform xanthoma
Candidiasis	Verruca
Discoid lupus	Granular cell tumor
White sponge nevus	Papillary hyperplasia 2° to ill-fitting dentures
Morsicatio buccorum	

b. Erythroplakia

Erythroplakia, a "fiery red patch that cannot be characterized clinically or pathologically as other definable lesion" [11], is included in older reports with LP, of the variegated or speckled type, but is now recognized by the WHO as an entity separate from LP, with distinctive clinical pathological features. This macroscopic appearance of the mucosa is due to the presence of a thin mucosa, associated with telangiectatic vessels and a chronic inflammatory infiltrate in the submucosa. This lesion may be associated with invasive cancer, harbors a high incidence of high-grade dysplasia and a high progression rate, and thus should be looked at with far greater suspicion clinically than leukoplakia [17,47–49]. The term erythroplasia was first used by Qeyrat in 1911 to indicate a lesion of the glans penis with the appearance of an erythematous area representing a premalignant process [48].

2. Etiology

When so defined, oral leukoplakia is associated in the Western world with the same etiological agents responsible for most head and neck squamous cell carcinomas: smoking and alcohol abuse [50]. In India, areca nut chewing produces a similar lesion, characterized histologically by oral submucosal fibrosis [11]. Other distinct clinical conditions have been shown to be associated with leukoplakia and an increased risk of oral cancer and some are discussed briefly.

Syderopenic dysphagia (Plummer Vinson syndrome), a condition characterized by iron deficient anemia, with frequent associated autoimmune diseases [51], affects the upper aerodigestive tract with atrophy of the oral mucosa and a predisposition to the development of multiple oral carcinomas, predominantly in the posterior oropharynx [11]. Esophageal dysphagia also typically occurs [51].

The premalignant potential of lichen planus (LP) [11] is not accepted by all authors [52,53] and is not supported by recent molecular data, which indicate that the percentage of LOH at several chromosomal sites in LP is lower in this lesion than in simple hyperplasia [54]. In contrast, the occurrence of a distinct type of dysplasia, which shares some histological features with LP, i.e., hyperkeratosis, band-like lymphocytic infiltrate, hypergranulosis, called lichenoid dysplasia, and distinct from true LP, is generally accepted [54–56].

Syphilis, now rarely seen in the industrialized world, may be associated with the development of leukoplakia, which may undergo malignant transformation [11]. Few cases of SCC of the lip arising in an atrophic epithelium in discoid lupus have been described [11]. Xeroderma pigmentosum, an autosomal recessive disease caused by defective DNA repair, is characterized by early onset SCC, arising in the setting of actinic keratoses, which may involve the lips. Epidermolysis bullosa, an inherited disease affecting the skin and the oral mucosa, has been described to be associated with SCC of the tongue [11].

The association of an increased risk of HNSCC with AIDS is controversial. A specific type of dysplasia, termed "hairy leukoplakia" can occur in homosexual AIDS patients [57]. This dysplasia is characterized by a ballooning, ground-glass appearance of the cells in the upper half of the epithelium and intranuclear inclusions and typically affects the lateral edges of the tongue [56]. It is associated with a high incidence of HPV and EBV infection [58]. Most authors report that AIDS is associated with an increased incidence of malignancy, including cervical and anogenital squamous cell cancer [59,60] although some contest this finding [61]. However, it is unclear whether AIDS confers an increased risk for HNSCC [62,63]. However, it is intriguing that the same institution reporting the occurrence of "hairy leukoplakia" in AIDS homosexual patients also reported at follow-up the largest incidence of transformation into overt cancer of leukoplakias (17.5%, with a follow-up of 7 years vs the 1–6% figure of all other reports, see Table 4.6) [57]. The presence of an association between HPV infection and squamous cell carcinoma has been documented in small subsets of patients who develop SCC in the absence of the usual predisposing factors of cigarette and alcohol use in the setting of a particular type of preneoplastic lesion, named proliferative verrucous hyperplasia [64].

An increased incidence of squamous cell carcinoma and associated premalignant lesions of the epidermis has been described in renal transplant recipient patients [65].

3. Clinical Aspects

a. Incidence

The peak age of incidence of oral leukoplakia is in the fifth to seventh decade [66]. Interestingly, the peak age of occurrence of dysplasia is the sixth decade, whereas the peak incidence of invasive cancer occurs a decade later. A similar time interval is present in the incidence of dysplasia vs invasive cancer in the cervix uteri [67] and in the bronchus [68]. The presence of this time interval indirectly supports the precursor role of dysplasia.

Leukoplakia affects predominantly males, with percentages ranging from 54 [66] to 78% [50].

b. Sites of Occurrence

Leukoplakia occurs most frequently at the buccal gingival gutter (maxillary and mandibular sulcus), followed by buccal mucosa, palate, and lips [66]. This distribution is different from that of dysplasia, which arises most commonly in the floor of mouth, tongue, and lips, which are also the sites of highest incidence of oral cancer [56,66]. In men, 54% of combined cases of dysplasia and invasive cancer are in the floor of the mouth. Thus, leukoplakias at different sites have different risks of harboring significant pathological alterations [50]: floor of mouth, tongue, and lips representing the sites most at risk for harboring dysplasia or invasive cancer [69].

This discrepancy further stresses the conceptual and biological difference existing between the term leukoplakia and the histologically defined terms of dyplasia and/or intraepithelial neoplasia.

c. Clinical Subtypes

Many descriptive, clinical variants of LP have been described. These can be reduced to two main types: those with a uniformly white appearance and those with a variegated appcarance [11], designated respectively as homogeneous and speckled by some authors [70]. Others have subdivided the first type in LP simplex and verrucosa and designated the latter LP erosiva [71]. Several descriptive clinical variants are recognized within the homogeneous and nonhomogeneous groups [11].

B. Larynx

1. Definition

a. Laryngeal Leukoplakia

While the term leukoplakia was coined by Schwimmer, laryngeal leukoplakia was first described by Durant in 1880 and the entity was studied further in the 1920s by Pierce [72] and Jackson [72,73]. The latter eloquently defined atypia in 1930 as the "mobilization of an army preparatory to invasion," recognizing its malignant potential. Remarkably, this conclusion, later endorsed by James Ewing, was made at a time when the concept of preinvasive neoplasia was not widely recognized.

The premalignant potential of areas appearing as whitish patches in the larynx, variously called hyperkeratosis, leukoplakia, and pachydermia laryngis, has long been recognized. Their laryngoscopic appearance is heterogeneous: they may be flat or raised, their surface rough or smooth, and they may be adjacent to normal or inflamed mucosa [74]. Similar to oral leukoplakia, the clinical appearance of LP is correlated histologically to the presence of hyperkeratosis, and the microscopical rather than the clinical appearance of the lesion defines its malignant potential. As for oral leukoplakia, the clinical entity "leukoplakia" encompasses histologically heterogeneous lesions, ranging from simple hyperkeratosis to invasive carcinoma, as discussed later in detail [74,75]. It is the presence and grade of dysplasia that dictate its biological potential, i.e., its probability to develop into overt invasive cancer.

2. Etiology

Auerbach, who had previously demonstrated a similar etiological relationship with bronchial dysplasia, also firmly established the relationship between laryngeal dysplasia and cigarette smoke. His seminal study, involving serial sections from the entire larynx, established that both the number of cell rows in the basal layer of the epithelium and the percentage of atypical nuclei increased with the number of cigarettes smoked per day [76].

3. Clinical Findings

a. Incidence and Sites of Occurrence

These lesions have a striking predilection for males, with a M:F ratio ranging from 5:1 [18] to 7:1 [77] and 8:4 [78]. Most lesions occur in the fifth to seventh decade, with a mean of 52 [77] to 59 years [78]. A temporal gap exists between the occurrence of keratosis and CIS and invasive carcinoma, like for oral dysplasia. Miller and Fisher [79] and Bouquot *et al.* [80] observed that the peak age of incidence of CIS antedates that of SCC by 7 years: 55 vs 62. The existence of this gap indirectly supports the premalignant nature of dysplasia, analogous to oral leukoplakia and to other anatomical examples of premalignancy [67]. The great majority of LP occur in the true vocal cord: 33% in both, 35% in either left or right, and 11% in both vocal cords and interarytenoid areas [77].

V. CORRELATION BETWEEN CLINICAL AND HISTOLOGICAL CLASSIFICATIONS

The histological diagnosis of lesions presenting clinically as LP ranges from hyperplasia to invasive carcinoma. In the largest series of LP studied to date of 3256 cases, Waldron and Schaefer [66] found histological evidence of neoplastic changes (dysplasia or overt cancer) in 20% of the cases: 12.2% mild to moderate dysplasia, 4.5% severe dysplasia, and 3.1% invasive carcinoma. The remaining 80% of the lesions had varying combinations of hyperorthokeratosis and acanthosis, without dysplasia. Similar percentages of dysplasia were found by Banoczy and Csiba [69] in a series of 500 cases, although their overall incidence of invasive carcinoma in LP was found to be 9.6%.

Several authors have shown that lesions with heterogeneous appearance, with an alternation of white and red areas, are at increased risk of harboring dysplasia or invasive cancer compared to uniformly white lesions. Thus, Banoczy and Csiba [69] found the incidence of severe dysplasia/CIS to increase progressively, from 0.8% in LP simplex to 6.6% in

TABLE 4.5 Distribution of Dysplasia and Invasive Squamous Cell Carcinoma (SCC) in Oral Leukoplakia and Erythroplakia

Type of leukoplakia	Simple hyperplasia	Mild/moderate dysplasia	Severe dysplasia/CIS	SCC	Ref.
Leukoplakia	67–80%	12–19%	4–5%	3–10%	66,69
Leukoplakia simplex ($n=22$)		21 (17.5%)	1 (0.8%)		69
Leukoplakia verrucosa ($n=42$)		34 (28%)	8 (6.6%)		69
Leukoplakia erosiva ($n=56$)		42 (35%)	14 (12%)		69
Erythroplakia ($n=65$)		9%	40%	51%	81

LP verrucosa and 12% in LP erosiva. The incidence of mild–moderate dysplasia progressed, in the same lesions, from 17.5 to 28 to 35%, respectively [69]. However, it is lesions with the appearance of erythroplakia that harbor the highest percentage of high-grade dysplasia and overt cancer. In a series of 65 cases, Schaefer and Waldron [81] showed that 9% of erythroplakia show low-grade dysplasia, 40% high-grade dysplasia, and 51% invasive squamous cell carcinoma [81]. In the series of Mashberg ($n=158$), 89% of early, asymptomatic invasive squamous carcinoma cases and over 93% of CIS had an erythroplakia component [47]. SCC, compared to CIS, had a higher percentage of elevation (58% vs 35%), a granular or rough surface (59% vs. 35%), and was indurated more often (10% vs 0%). Data are summarized in Table 4.5.

VI. MALIGNANT PROGRESSION

A. Oral Cavity

The overall incidence of progression to invasive squamous cell carcinoma of premalignant lesions of the oral cavity ranges from 2.7 to 17.5% (see Table 4.6) in the studies with the longest follow-up and with the largest numbers of patients. As stressed previously, the progression rate is related to the degree of dysplasia, which in turn varies in relation to the site. Thus, in a series of 3256 cases, Waldron and Schafer [66] found the incidence of dysplasia and carcinoma *in situ* to be disproportionately high in the floor of mouth (42.9%; vs an 8% overall incidence of LP), tongue (24%, vs a 6.8% overall incidence of LP) and lips (24% vs an overall 10.3% incidence of LP). Thus, while all leukoplakias should be subject to histological examination, both the presence of redness (erythroplakic component) and the occurrence in "high-risk sites" should be of clinical concern. The incidence of progression to invasive carcinoma is also related to the clinical type of LP, consistent with their different association with high-grade dysplasia. None of the cases of LP simplex ($n=371$) was found to progress to cancer, whereas 5.5% of LP verrucosa ($n=183$) and 25.9% of LP erosiva ($n=116$) progressed during a mean observation time of 9.8 years [50] (see Table 4.7).

A higher risk for cancer in association with female sex has been reported by several authors: the overall incidence of malignant transformation in the series of Banoczy [50] is 8.8% in females vs 5.1% in males. In contrast, the M:F distribution of CIS and SCC was 3.2:1 and 1.9:1 in the whole series. An even higher difference in incidence was found for tongue cancer, 86.6% of which occurred in females. This site also showed a higher prevalence in the incidence of dysplasia among females [50]. A similar bias in the malignant transformation of females vs males (58% vs 42%) was observed by Silverman *et al.* [82]. Reasons for these gender-related differences are not known.

The etiological role of tobacco use in the development of SCC of the oral cavity has been well established. However, several authors have shown that LP occurring in nonsmokers has an excess risk of developing into SCC compared to LP arising in smokers [82–84].

It is worth stressing that most oral cancers arise *de novo*, without an associated precancerous lesion. While the association among leukoplakia, dysplasia, and the prospective

TABLE 4.6 Overall Progression Rate of Oral Leukoplakia to Invasive Cancer

N	Mean follow-up (years)	% Undergoing malignant transformation	Ref.
782	12	2.7	83
252	7.2	17.5	50
670	9.8	6	82

TABLE 4.7 Incidence of Dysplasia in Clinical Subtypes of Leukoplakia and their Frequency of Progression to SCC[a]

Type	N	Incidence of dysplasia	Progression to SCC
Leukoplakia simplex	371	22 (5.9%)	0
Leukoplakia verrucosa	183	42 (11.3%)	10 (5.5%)
Erythroplakia	116	56 (48%)	30 (25.9%)

[a]Mean follow-up 9.8 years; Ref. 82.

development of cancer is firmly established, examination of overt cancer only discloses coexisting leukoplakia or carcinoma *in situ* in a small subset of cases. A large retrospective study of oral and oropharyngeal cancers occurring over a span of 54 years in a small community in Minnesota addressed this issue retrospectively. Analysis of this set of 201 cases showed that only 7% of invasive cancers had adjacent CIS, whereas an additional 2% had severe epithelial dysplasia [85]. However, a limitation of this study is that it relied entirely on pathology reports and thus may have underestimated the percentage of dysplasia, secondary to sampling error and/or underreporting by the pathologist. In the same report, the authors show that cancers associated with leukoplakia are smaller, less invasive, and their histological grade lower than those without it [84]. Other authors had observed that the progression time from CIS into overt SCC cancer is extremely variable [86]. These findings highlight that, although likely to include a preneoplastic phase, the natural history of SCC may show significant patient-to-patient variations.

A distinctive type of hyperplasia, labeled proliferative verrucous leukoplakia, has been described [87,88]. This lesion was identified retrospectively and is characterized by a verrucous clinical appearance, an expanding and often multifocal growth pattern, and a high (up to 70%) rate of progression into invasive carcinoma [64]. The initial manifestation is usually that of simple hyperkeratosis. However, the lesion tends to recur and progress to dysplasia or invasive carcinoma, often in a multifocal distribution, retaining an exophytic-verrucous appearance and thus the diagnosis can reliably be made only retrospectively [87]. There is a M:F ratio of 4:1; the mean age of occurrence is 62 years, and the most frequent sites are the gingiva and tongue. The histological appearance is that of simple hyperkeratosis, dysplasia, or verrucous carcinoma [64,87]. Notably, 69% of the patients have no history of tobacco exposure [64]. In contrast, 78% show evidence of HPV 16 infection, highlighting the possible transforming role of this virus in this setting [89].

B. Larynx

The reported frequency of transformation of leukoplakia varies from 3.5 to 21%; the largest studies reported frequencies from 4.4 to 16% (see Table 4.8). Like for oral dysplasia, the rate of progression to overt cancer increases with the degree of dysplasia. Thus in the series of Blackwell, where 62 leukoplakias were studied for a mean follow-up of 74 months, the rate of progression was 0/6 in the absence of dysplasia, 12% for mild dysplasia (3/26), 33% (5/15) for moderate dysplasia, and 33% for severe dysplasia/CIS (5/15) [15]. Kambic *et al.* [18] found an overall incidence of SCC of 17/88 (19%): 12/17 in high-grade dysplasia, 3/17 in low-grade dysplasia, and 2/17 in simple hyperplasia. The rate of recurrence is also related to the severity of dysplasia: 53% (9/17) for CIS and 3/17 (18%) for moderate–severe dysplasia [75]. The progression rate of CIS to invasive carcinoma was found to be 63% in a group of 27 patients managed conservatively, after a mean follow-up time of 9 months [90]. In the series of Gillis *et al.* [91] progression to CIS or invasive carcinoma was observed in 3/7 patients with keratosis and in 5/12 cases of atypia with or without keratosis; progression to invasive SCC was observed in 3/8 cases of CIS. Norris and Peale [73] used the same terminology and found that the incidence of progression was related to the presence of atypia: only 1/30 keratosis without atypia progressed to SCC after 32 months. Of 86 cases of keratosis with atypia, 11 progressed, after an average of 22 months: 5 to SCC, 4 to CIS, and 2 to CIS with equivocal evidence of invasion. Hellquist *et al.* [78] found an overall incidence of progression to SCC of 8.7% ($n=161$). SCC developed in 2/98 (2%) of patients with slight dysplasia, 3/24 (12%) with moderate dysplasia, and 9/39 (23%) with severe dysplasia/CIS. Additionally, 5/98 cases with hyperplasia or mild dysplasia progressed to moderate or severe dysplasia; 3/24 moderate dysplasia progressed to severe dysplasia. The mean follow-up time was not indicated, but more than 86% of patients had more than 2 years follow-up and 57% more than 5 years. Crissman *et al.* [14] stressed that 36% of 25 patients with CIS had microinvasive carcinoma (Figs. 4.26 and 4.27) and another three developed invasive carcinoma in 6 to 8 years (data summarized in Table 4.9). In a large series of patients followed from 1.5 to 12 years and classified according to the Ljubjana classification scheme, 0.7% of simple ($n=380$), 1% of abnormal ($n=414$), and 9.5% of atypical hyperplasia ($n=105$) progressed to invasive carcinoma [18].

Other investigators have stressed the importance of specific histological parameters in predicting progression into invasive SCC. The occurrence of single cell intraepithelial keratinization (Figs. 4.17, 4.18, and 4.22) [14] pleomorphism, mitotic activity, and mucosa-associated

TABLE 4.8 Frequency of Progression of Laryngeal Leukoplakia/Dysplasia to Invasive Squamous Carcinoma

	Total number of patients	Percent progressing	Ref.
Sllamniku *et al.* (1989)	921	6.7	94
Crissman *et al.* (1988)	25	12	14
Bouquot *et al.* (1991)	108	16	80
Lundgren *et al.* (1987)	232	13	95
Plch *et al.* (1988)	227	4.4	96
McGavran *et al.* (1971)	84	3.5	77
Miller *et al.* (1971)	203	15.7	79
Blackwell *et al.* (1995)	62	21	15
Norris *et al.* (1963)	116	10	97
Hellquist *et al.* (1982)	161	8.7	78

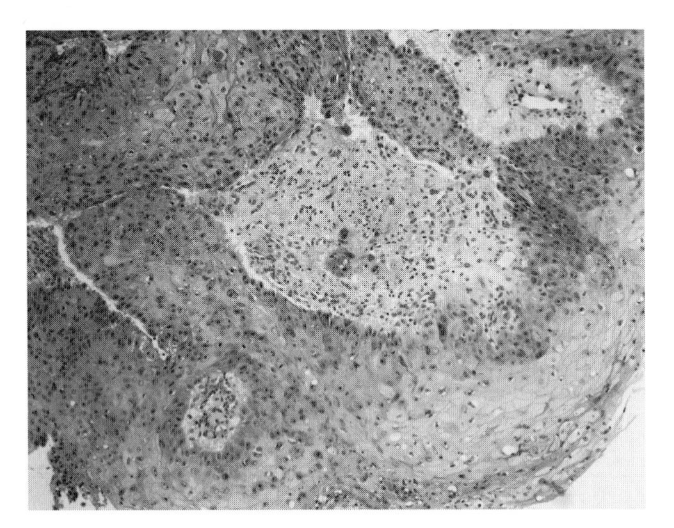

FIGURE 4.26 Microinvasive carcinoma at low power.

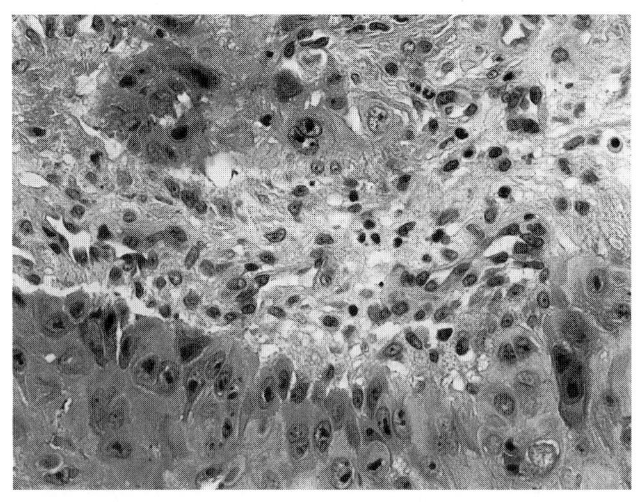

FIGURE 4.27 Microinvasive carcinoma at high power. Dysplastic cells, from a layer of basal cells showing marked pleomorphism, encroach upon the underlying submucosa. The basement membrane in this area has become blurred. Further down, a detached nest of pleomorphic cells, with individual cell keratinization, is present. These findings are compatible with early microscopic invasion.

inflammation [15] has been found to confer an increased likelihood for progression to SCC. In the series of Crissman *et al.* [14], dyskeratosis was further associated with an increased likelihood of recurrence.

Follow-up studies of dysplasia highlight that these lesions may recur not only in the same site, but also in anatomically separate foci, as either CIS or invasive carcinoma. Thus in the series of Gillis *et al.* [91], 13 of 57 patients treated by radiotherapy or surgery recurred or developed *de novo* as CIS or invasive SCC. In 2/42 patients, a second primary occurred. These data highlight the presence of multiple foci of transformed cells within the upper aerodigestive tract, stressing that the entire field is prone to develop cancer, as first described by Slaughter *et al.* [92].

References

1. Coltrera, M., Zarbo, R. J., Sakr, W., and Gown, A. M. (1992). Markers for dysplasia of the upper aerodigestive tract. *Am. J. Pathol.* **141**, 817–825.
2. Yashima, K., Maitra, A., Rogers, B. B., Timmons, C. F., Rathi, A., Pinar, H., Wright, W. E., Shay, J. W., and Gazdar, A. (1998). Expression of the RNA component of telomerase during human development and differentiation. *Cell Growth Differ.* 805–813.
3. Zidar, N., Cor, and Kambic, V. (1996). Expression of Ki-67 antigen and proliferative cell nuclear antigen in benign and malignant epithelial lesions of the larynx. *J. Laryngol. Otol.* **110**, 440–445.
4. World Health Organization Collaborating Centre for Oral Precancerous Lesions. (1978). Definition of leukoplakia and related lesions: An aid to studies on oral precancer. *Oral Surg.* **46**(4), 518–539.
5. Browne, R. M., and Potts, A. J. C. (1986). Dysplasia in salivary gland ducts in sublingual leukoplakia and erythroplakia. *Oral Surg. Oral Med. Oral Pathol.* **62**, 44–48.

TABLE 4.9 Incidence of Progression to SCC in Relation to Histology in Laryngeal Dysplasia

	Number progressing to SCC in each histological group			
No. dysplasia	**Mild**	**Moderate**	**Severe/CIS**	**Ref.**
0/6 (0%)	3/26 (12%)	5/15 (30%)	5/15 (30%)	15
	2/98 (2%)	3/24 (12.5%)	9/24 (38%)	78
			3/25 (12%)[a]	14
			17/27 (63%)	90
2/17 (12%)	3/17 (18%)	High-grade dysplasia 12/17 (71%)		18
1/30 (3%)	Atypia 5/86 (6%)[b]			97

[a]3 more revealed microinvasion on review.
[b]4 more progressed to CIS; 2 had CIS with equivocal invasion.

6. Sternberg, S. (1992). "Histology for Pathologists," pp. 451–455. Raven Press, New York.
7. Nasiell, M. (1996). Metaplasia and atypical metaplasia in the bronchial mucosa: A histopathological and cytopathological study. *Acta Cytol.* **10**, 421–427.
8. Stiblar-Martincic, D. (1997). Histology of laryngeal mucosa. *Acta Otolaryngol. Suppl. (Stockh).* **527**, 137–141.
9. Fechner, R. E., and Mills, S. E. (1996). Larynx and pharynx. *In* "Histology for Pathologistsl." (S. Sternberg, ed.), pp. 443–455. Lippincott-Raven, New York.
10. WHO International Histological Classification of Tumors. (1991). "Histologic Typing of Tumours of the Upper Respiratory Tract and Ear." (K. Shanmugaratnam, *et al.*, eds.), 2nd Ed. Springer, New York.
11. WHO International Histological Classification of Tumors. (1997). "Histological Typing of Cancer and Precancer of the Oral Mucosa." (J. J. Pindborg, P. A. Reichart, C. J. Smith, I. vander Waal, *et al.*, eds.), 2nd Ed. Springer, New York.
12. Kurman, R. J., Norris, H. J., and Wilkinson, E. (1992). "Atlas of Tumor Pathology. Tumors of the Cervix, Vagina, and Vulva," pp. 44–55. Armed Forces Institue of Pathology, Washington.
13. Hellquist, H., Cardesa, A., Gale, N., Kambic, V., and Michaels, L. (1999). Criteria for grading in the Ljubjana classification of epithelial hyerplastic laryngeal lesions: A study by members of the Working Group on epithelial Hyperplastic Laryngeal Lesions of the European Society of Pathologists . *Histopathology* **34**, 226–233.
14. Crissman, J. D., Zarbo, R. J., Drozdowicz, S., Jacobs, J., Ahmad, K., and Weaver, A. (1998). Carcinoma in situ and microinvasive squamous carcinoma of the laryngeal glottis. *Arch. Otolaryngol. Head Neck Surg.* **114**, 299–307.
15. Blackwell, K. E., Yao-Shi, F., and Calcaterra, T. C. (1995). Laryngeal dysplasia: A clinicopathologic study. *Cancer* **75**(2), 457–463.
16. Crissman, J. D., and Zarbo, R. J. (1989). Dysplasia, in situ carcinoma, and progression to invasive squamous cell carcinoma of the upper aerodigestive tract. *Am. J. Surg. Pathol.* **13**, 5–16.
17. Crissman, J. D., Visscher, D. W., and Sakr, W. (1993). Premalignant lesions of the upper aerodigestive tract: Pathologic classification. *J. Cell Biochem. Suppl.* **17F**, 49–56.
18. Kambic, V., Gale, N., and Ferluga, D. (1992). Laryngeal hyperplastic lesions, follow-up study and application of lectins and anticytokeratins for their evaluation. *Pathol. Res. Pract.* **188**, 1067–1077.
19. Kambic, V. (1997). Epithelial hyperplastic lesions: A challenging topic in laryngology. *Acta Otolaryngol.* **527**, 7–11.
20. Crissman, J. D. (1982). Laryngeal keratosis preceding laryngeal carcinoma: A report of four cases. *Arch Otolaryngol.* **108**, 445–448.
21. Crissman, J. D., and Zarbo, R. J. (1991). Quantitation of DNA ploidy in squamous intrepithelial neoplasia of the laryngeal glottis. *Arch. Otolaryngol. Head Neck Surg.* **117**, 182–188.
22. Kambic, V., and Gale, N. (1995). "Epithelial Hyperplastic Lesions of the Larynx." Elsevier, Amsterdam.
23. Michaels, L. (1997). The Kambic-Gale method of assessment of epithelial hyperplastic lesions of the larynx in comparison with the dysplasia grade method. *Acta Otolaryngol. Suppl.* **527**, 17–20.
24. Abbey, L.M., Kaugars, G. E., Gunsolley, J. C., Burns, J. C., Page, D. G., Svirsky, J. A., Eisenberg, E., Kruthchkoff, D. J., and Cushing, M. (1995). Intraexaminer and interexaminer reliability in the diagnosis of oral epithelial dysplasia. *Oral Surg. Oral Med. Oral Radiol. Endo.* **80**, 188–191.
25. Vogelsten, B., Fearon, E. R., Hamilton, S. R., Kern, S. E., Preisinger, A. C., Leppert, M., Nakamura, Y., White, R., Smits, A. M., and Bos, J. L. (1988). Genetic alterations during colorectal tumor development. *N. Engl. J. Med.* **319**(9), 525–532.
26. Lindberg, K. and Rheinwald, J. G. (1989). Suprabasal 40Kda keratin (K19) expression as an immunohistological marker of premalignany in oral epithelium. *Am J Pathol.* **134**, 89–98
27. Cattoretti, G., Becker, M. H., Key, G., Dichrow, M., Schluter, C., Galle, J., and Gerdes, J. (1992). Monoclonal antibody against recombinant parts of the Ki-67 antigen (MIB-1 and MIB-3) detect proliferating cells in microwave-processed formalin fixed paraffin sections. *J. Pathol.* **168**, 357–363.
28. Gerdes, J., Lemke, H., Baisch, Wacker, H., Schwab, U., and Stein, H. (1984). Cell cycle analysis of a cell proliferation-associated human nuclear antigen defined by the monoclonal antibody Ki-67. *J. Immunol.* **133**, 1710–1715.
29. Gallo, O., Franchi, A., Chiarelli, I., Porfirio, B., Grande, A., Simonetti, L., Bocciolini, C., and Fini-Storchi. (1997). Potential biomarkers in predicting progression of epitheliala hyperplastic lesions of the larynx. *Acta Otolaryngol. Suppl. (Stockh).* **527**, 30–38.
30. Silvestri, F., Bussani, R., Pavletic, N., Mannone, T., and Bosatra, A. (1997). From epithelial dysplasia to squamous carcinoma of the head and neck region: Evolutive and prognostic histopathological markers. *Acta Otolaryngol. Suppl. (Stockh).* **527**, 49–51.
31. Gale, N., Zidar, N., Kambic, V., Poljak, M., and Cor, A. (1997). Epidermal growth factor receptor, c-erbB-2 and p53 overexpressions in epithelial hypeplastic lesions of the larynx. *Acta Otolaryngol. Suppl. (Stockh).* **527**, 105–110.
32. Nadal, A., Campo, E., Pinto, J., Mallofre', C., Palacin, A., Arias, C., Traserra, J., and Cardesa, A. (1995), p53 expression in normal, dysplastic, and neoplastic epithelium: Absence of correlation with prognostic factors. *J. Pathol.* **175**, 181–188.
33. Cardesa, A., Nadal, A., Jares, P., Mallofre', C., Campo, E., and Traserra, J. (1997). Hyperplastic lesions of the larynx: Experience of the Barcelon group. *Acta Otloaryngol. Suppl. (Stockh).* **527**, 43–47.
34. Hellquist, H.B. (1997). Apoptosis in epithelial hyperplastic laryngeal lesions. *Acta Otolaryngol. Suppl. (Stockh).* **527**, 25–29.
35. Sheu, L., Chen, A., Meng, C., Chieh, K., Going, F., and Lee, W. (1997). Analysis of bcl-2 expression in normal, inflamed, dysplastic nasopharyngeal epithelia, and nasopharyngeal carcinoma. *Hum. Pathol.* **28**, 556–562.
36. Kok, K., Naylor, S. L., and Buys, C.H. (1997). Deletions of the short arm of chromosome 3 in solid tumors and the search for tumor suppressor genes. *Adv. Cancer Res.* **71**, 27–92.
37. Tseng, J. E., Kemp, B. L., Khuri, F. R., Kurie, J. M., Lee, J. S., Zhou, X., Hong, W. K., and Mao, L. (1999). Loss of FHIT is frequent in stage I non-small cell lung cancer and in the lungs of chronic smokers. *Cancer Res.* **59**, 4789–4803.
38. Sozzi, G., Sard, L., De Gregorio, L., Marchetti, A., Mussso, K., Buttitta, F., Tornielli, S., Pellegrini, S., *et al.* (1997). Association between cigarette smoking and FHIT gene alterations in lung cancer. *Cancer Res.* **57**, 2121–2133.
39. Nelson, H. H., Wiencke, J. K., Gunn, L., Wain, J. C., Christiani, D. C., and Kelsey, K. T. (1998). Chromosome 3p14 alterations in lung cancer: evidence that FHIT exon deletion is a target of tobacco carcinogens and asbestos. *Cancer Res.* **58**, 1804–1807.
40. Siprashvilli, Z., Sozzi, G., Barnes, L. D., McCue, P., Robinoson, A. K., Eryomin, V., Sard, L., Tagliabue, E., Greco, A., Fusetti, L., Schwartz, G., Pierotti, M. A., Croce, C. M., and Huebner, K. (1997). Replacement of FHIT in cancer cells suppresses tumorigenicity. *Proc. Natl. Acad. Sci. USA* **9**, 13771–13776.
41. Rosin, M.P., Cheng, X., Poh, C., Lam, W., Huang, Y., Lovas, J., Berean, K., Epstein, J. B., Priddy, R., Nhu, D. L., and Zhang, L. (2000). Use of allelic loss to predict malignant risk for low grade oral epithelial dysplasia. *Clin. Cancer Res.* 357–362.
42. Califano, J., van der Riet, P., Westra, W., Nawroz, H., Clayman, G., Piantadosi, S., Corio, R., Lee, D., Greenberg, D., Koch, W., and Sidransky, D. (1996). Genetic progression model for head and neck cancer: implications for field cancerization. *Cancer Res.* **56**, 2488–2492.

43. Sakr, W. A., Zarbo, R. J., Jacobs, J., and Crissman, J. D. (1987). Distribution of basement membrane in squamous cell carcinoma of the head and neck. *Hum. Pathol.* **18**, 1043–1050.
44. Pathmanathan, R., Prasad, U., Sadler, R., FLynn, K., and Raab-Traub, N. (1995). Clonal proliferations of cells infected with Epstein-Barr virus in preinvasive lesions related to nasopharyngeal carcinoma. *N. Engl. J. Med.* **333**(11), 724–726.
45. Becker, J., Leser, U., Marschall, M., Langford, A., Jilg, W., Gelderblom, H., Reichart, P., and Wolf, H. (1991). Expression of proteins coded by Epstein Barr virus transactivator genes depends on the differentiation of epithelial cells in oral hairy leukoplakia. *Proc. Natl. Acad. Sci. USA* **88**, 8322–8366.
46. Schwimmer, E. (1877). Die idiopatisches scleimhaut plaques der mundhohle (leukoplakia buccalis). *Arch. Dermatol. Syphilol.* **9**, 511–570.
47. Mashberg, A., Morrissey, J. B., and Garfinkel, L. (1973). A study of the appearance of early asymptomatic oral squamous cell carcinoma. *Cancer* **32**(6), 1436–1445.
48. Mashberg, A. (1978). Erythroplasia: The earliest sign of asymptomatic oral cancer. *J. Am. Dent. Assoc.* **96**, 615–620.
49. Shafer, W.G. (1975). Oral Carcinoma in situ. *Oral Surg.* **39**(2), 227–238.
50. Banoczy, J. (1977). Follow-up studies in oral leukoplakia. *J. Maxillofac. Surg.* **5**(1), 69–75.
51. Fenoglio Preiser Cecilia, M. (ed.) (1999). Gastrointestinal Pathology: An Atlas and Text, 2nd Ed. Lippincott Raven, New York.
52. Eisenberg, E., and Krutchkoff, D. J. (1992). Lichenoid lesions of oral mucosa. *Oral Surg. Oral Med. Oral Pathol.* **73**, 699–704.
53. Eisenberg, E. (1992). Lichen planus and oral cancer: Is there a connection between the two? *J. Am. Dent. Assoc.* **123**(5), 104–108.
54. Zhang, L., Cheng, X., Yong-hua, L., Poh, C., Zeng, T., Priddy, R., Lovas, J., Freedman, P., Daley, T., and Rosin, M.P. (2000). High frequency of allelic loss in dysplastic lichenoid lesions. *Lab. Invest.* **80**(2), 233–237.
55. Krutchkoff, D. J., and Eisenberg, E. (1985). Lichenoid dysplasia: A distinct histopathologic entity. *Oral Surg. Oral Med. Oral Pathol.* **30**, 308–315.
56. Juan Rosai, (ed.) (1996). "Ackermann's Surgical Pathology," 8th Ed.
57. Greenspan, D., Greenspan, J. S., Conant, M., Petersen, V., Silverman, S., Jr., and de Souza, Y. (1984). Oral "hairy" leukoplakia in male homosexuals: Evidence of association with both papillomavirus and a herpes-group virus. *Lancet* **2**(8407), 831–834.
58. Adler-Storthz, K., Ficarra, G., Woods, K. V., Gaglioti, D., Di Pietro, M., and Shillitoe, E. J. (1992). Prevalence of Epstein-Barr and HPV in the oral mucosa of HIV-infected patients. *J. Oral Pathol.* **21**, 164–170.
59. Goedert, J. J. (2000). The epidemiology of acquired immunodeficiency syndrome malignancies. *Semin. Oncol.* **27**(4), 390–401.
60. Franceschi, S., Dal Maso, L., Arniani, S., Crosignani, P., Vercelli, M., Simonato, L., Falcini, F., Zanetti, R., Barchielli, A., Serraino, D., and Rezza, G. (1998). Risk of cancer other than Kaposi's sarcoma and non-Hodgkin's lymphoma in persons with AIDS in Italy: Cancer and AIDS registry linkage study. *Br. J. Cancer* **78**(7), 966–970.
61. Beral, V., and Newton, R. (1998). Overview of the epidemiology of immunodeficiency-associated cancers. *J. Natl. Cancer Inst. Monogr.* **23**, 1–6.
62. Singh, B., Sabin, S., Rofim, O., Shaha, A., Har-El, G., and Lucente, F.E. (1999). Alterations in head and neck cancer occurring in HIV-infected patients: Results of a pilot, longitudinal, prospective study. *Acta Oncol.* **38**(8), 1047–1050.
63. Barry, B., and Gehanno, P. (1999). Squamous cell carcinoma of the ENT organs in the course of the HIV infection. *Ann. Otolaryngol. Chir. Cervicofac.* **116**(3), 149–153.
64. Silverman, S., Jr., and Gorsky, M. (1997). Proliferative verrucous leukoplakia: A follow-up study of 54 cases. *Oral Surg. Oral Med. Oral Pathol. Oral Radiol. Endod.* **84**, 154–157.
65. King, G. N., Healy, C. M., Glover, M. T., Kwan, J. T., Williams, D. M., Leigh, I. M., Worthington, H. V., and Thornhill, M. H. (1995). Increased prevalence of dysplastic and malignant lesions of the lips in renal transplant recipients. *N. Engl. J. Med.* **332**, 1052–1057.
66. Waldron, C. A., and Shafer, W. G. (1975). Leukoplakia revisited: A clinicopathologic study 3256 oral leukoplakias. *Cancer* **36**, 1386–1392.
67. Patten Stanley (1978). "Diagnostic Cytopathology of the Uterine Cervix," 2nd Ed. Karger, Basel.
68. Saccomanno, G., Archer, V. E., Auerbach, O., Saunders, R. P., and Brennan, L. M. (1974). Development of carcinoma of the lung as reflected in exfoliated cells. *Cancer* **33**, 256–270.
69. Banoczy, J., and Csiba, A. (1976). Occurrence of epithelial dysplasia in oral leukoplakia: Analysis and follow-up study of 120 cases. *Oral Surg. Oral Med. Oral Pathol.* **42**(6), 766–774.
70. Pindborg, J. J., Renstrup, G., and Poulsen, H. E. (1963). Silverman Studies in oral leukoplakias. V. Clinical and histological sign of malignancy. *Acta Odontol Scand.* **211**, 407–414.
71. Banoczy, J. (1983). Oral leukoplakia and other white lesions of the oral mucosa related to dermatological disorders. J. Cutan. Pathol. 10, 238–256.
72. Pierce, N. H. (1920). Leukoplakia laryngis. *Ann. Otol. Rhinol. Laryngol.* **29**, 33–56.
73. Norris, C. M., and Peale, A. R. (1963). Keratosis of the larynx. *J. Laryngol. Otol.* **77**, 635–647.
74. Goodman, M. L. (1984). Keratosis (leukoplakia) of the larynx. *Orthop. Clin. North Am.* **17**(1), 179–183.
75. Elman, A. J., Goodman, M., Wang, C. C., Pilch, B., and Busse, J. (1979). In situ carcinoma of the vocal cords. *Cancer* **43**, 2422–2428.
76. Auerbach, O., Hammond, E. C., and Garfinkel, L. (1970). Histologic changes in the larynx in relation to smoking habits. *Cancer* **23**, 92–104.
77. McGavran, M. H., Bauer, W. C., and Ogura, J. H. (1960). Isolated laryngeal keratosis: Its relation to carcinoma of the larynx based on a clinicopathologic study of 87 consecutive cases with long-term follow-up. *Laryngoscope* **70**, 932–951.
78. Hellquist, H., Lundgren, J., and Oloffson, J. (1982). Hyperplasia, keratosis, dysplasia and carcinoma in situ of the vocal cord: A follow-up study. *Clin. Otolaryngol.* **7**, 11–27.
79. Miller, A. H., and Fisher, H. R. (1971). Clues to the life history of carcinoma in situ of the larynx. *Laryngoscope* **81**, 1475–1480.
80. Bouquot, J. E., Kurland, L. T., and Weiland, L. H. (1991). Laryngeal keratosis and carcinoma in the Rochester, MN, population, 1935–1984. *Cancer Detect. Prev.* **15**, 83–91.
81. Shafer, W. G., and Waldron, C. A. (1975). Erythroplakia of the oral cavity. *Cancer* **36**, 1021–1028.
82. Silverman, S., Gorsky, M., and Lozada, F. (1984). Oral leukoplakia and malignant transformation: A follow-up study of 257 patients. *Cancer* **53**, 563–568.
83. Einhorn, J., and Wersal, J. (1967). Incidence of oral carcinoma in patients with leukoplakia of the oral mucosa. *Cancer* **20**(12), 2189–2193.
84. Roed-Petersen, B. (1971). Cancer development in oral leukoplakia: Follow up of 331 patients. *J. Dent. Res.* **50**, 711.
85. Bouquot, J. E., Weiland, L. H., and Kurland, L. T. (1988). Leukoplakia and carcinoma in situ synchronously associated with invasive oral/oropharyngeal carcinoma in Rochester, Minnesota, 1935–1984. *Oral Surg. Oral Med. Oral Pathol.* **65**, 199–207.
86. Ackermann, L. V., and McGavran, M. H. (1958). Proliferating benign and malignant lesions of the oral cavity. *J. Oral Surg.* **16**, 400–413.
87. Hansen, L. S., Olson, J. A., and Silverman, S. (1985). Proliferative verucous leukoplakia. *Oral Surg. Oral Med. Oral Pathol.* **60**, 285–298.
88. Zakrzewska, J. M., Lopes, V., Speight, P., and Hopper, C. (1996). Proliferative verrucous leukoplakia. *Oral Med. Oral Pathol. Oral Radio. Endod.* **82**, 392–401.
89. Palefsky, J. M., Silverman, S., and Abdel-Salaam, M. (1995). Association between proliferative verrucous leukoplakia and infection with human papillomvirus type 16. *J. Oral. Pathol. Med.* **24**, 193–197.
90. Hintz, B. L., Kagan, A. R., Nussbaum, H., Rao, A. R., Chan, P. Y., and Miles, J. (1981). A "watchful waiting" policy for in situ carcinoma of the vocal cords. *Arch. Otolaryngol.* **107**, 746–751.

91. Gillis, T. M., Incze, J., Strong, M. S., Vaughan, C. W., and Simpson, G.T. (1983). Natural history and management of keratosis, atypia, carcinoma in situ, and microinvasive cancer of the larynx. *Am. J. Surg.* **146**, 512–516.
92. Slaughter, D. P., Southwick, H. W., and Smejkal, W. (1953). Field cancerization in oral stratified squamous epithelium: Clinical implications of multicentric origin. *Cancer* **6**, 963–968.
93. Christensen, M. E., Therkildsen, M. H., Hansen, B. L., Albeck, H., Hansen, G. N., and Bretlau, P. (1992). Epidermal growth factor receptor expression on oral mucosa dysplastic epithelia and squamous cell carcinomas. *Eur. Arch. Otorhinolaryngol.* **249**(5), 243–247.
94. Sllamniku, B., Bauer, W., Painter, C., and Sessions, D. (1989). The transformation of laryngeal keratosis into invasive carcinoma. *Am. J. Otolaryngol.* **10**(1), 42–54.
95. Lungren, J., and Olofsson, J. (1987). Malignant tumours in patients with non-invasive squamous cell lesions of the vocal cords. *Clin. Otolaryngol.* **12**(1), 39–43.
96. Plch, J., Par, I., Navratilova, I., Blahova, M., and Zavadil, M. (1998). Long term follow-up study of laryngeal precancer. *Auris Nasus Larynx* **25**(4), 407–412.
97. Norris, C. M., and Peale, A. R. (1963). Keratosis of the larynx. *J. Laryng. Otol.* (August), 635–647.

P A R T II

BASIC MOLECULAR MECHANISMS

CHAPTER

5

Animal Models in Head and Neck Cancer

D. T. WONG, R. TODD, G. SHKLAR, and A. RUSTGI
Laboratory of Molecular Pathology
Harvard University
School of Dental Medicine
Boston, Massachusetts 02115

I. Introduction 57
II. History of Experimental Head and Neck Cancer Research 57
A. Animal Models for Head and Neck Cancer Research: The Value of the Hamster Buccal Pouch Experimental Model for the Study of Oral Cancer 58
III. Genetic Models for Head and Neck Cancer Research 59
A. Targeting Genes to Oral Cavity Epithelia with Promoters 59
B. Other Emerging Genetic Models of Head and Neck Cancer 60
C. Targeted Ablation of Genes in the Oral Cavity 61
D. Future Directions 61
IV. Integrative Human-Based Discoveries and Animal Model Testing for Head and Neck Cancer Research 61
References 61

I. INTRODUCTION

While the majority of head and neck cancers are of squamous epithelial origin, many other epithelial (such as the salivary gland cancers) and mesenchymal (osteosarcoma, fibrosarcoma, rhabodmyosarcoma) malignancies are rarer, but well documented. Therefore, there is no single animal model to study all forms of head and neck cancers. Even for the study of head and neck squamous cell carcinogenesis, no single animal model has emerged as a comprehensive resource for genetic, environmental, molecular, and biochemical analysis. A survey of the literature revealed that ~2067 publications since 1996 reported the use of various animal models for oral cancer research. These animal models included baboon (10), cat (82), dog (224), horse (18), monkey (78), owl (1), rabbit (110), snake (6), hamster (523), rat (438), and mouse (577). The use of rodents constituted ~74% of these studies. This chapter, primarily using oral cancer as an example, explores the use of animal models for head and neck cancer research from the following four perspectives. First, the history of experimental head and neck cancer research is outlined. Second, current commonly used animal models for head and neck cancer research are discussed. Third, emerging genetic models for head and neck cancer research are given. Fourth, an integrative human-based discovery and animal testing/validation approach to head and neck cancer research promises an accelerated means of identifying novel genetic pathways abrogated during head and neck carcinogenesis.

II. HISTORY OF EXPERIMENTAL HEAD AND NECK CANCER RESEARCH

The major form of head and neck cancer is squamous cell carcinoma of the oral cavity. Therefore, attempts to study the experimental pathology of head and neck cancer were focused on the development of experimental models for oral cancer. Because the primary risk factors for oral cancer development are tobacco and alcohol, chemical carcinogens were used to induce oral cancers in animal models. Other investigations attempted to develop experimental models for malignant tumors of salivary glands, larynx, and other head and neck sites.

Following the experimental production of skin cancer using cold tar [1] in 1918 by Yamagiwa and Ichikawa and the discovery of carcinogenic polycyclic hydrocarbons by Cook and co-workers [2], Salley [3] developed the first experimental model for oral cancer using the hamster buccal pouch and multiple paintings with the chemical carcinogen 7,12-dimethylbenz[*a*]anthracene (DMBA). Salley's work was extended by Morris [4], who studied the effects of age, sex, and concentration of chemical carcinogen in relation to the production of hamster buccal pouch tumors. Renstrup and associates [5] and Silberman and Shklar [6] were able to show that chronic irritation could augment chemical carcinogenesis in this oral cancer model. These studies confirmed Berenblum's [7] concept of cocarcinogenic action by local irritation and led to many investigations confirming the hamster buccal pouch cancer model as a suitable experimental model for oral cancer.

Chemically induced tumors of other oral sites in experimental animals have also been studied. However, the reliability of these models has not been comparable to that of the hamster buccal pouch model. The carcinomas have not been found to develop in a consistent pattern over time, as in the hamster pouch model. Levy, in 1958, first attempted to induce carcinoma of the tongue in mice by the implantation of methylcholanthrene. However, results were inconsistent, although some early evidence of carcinogenesis was found after 16 weeks.

Fujino and associates [9] were able to induce carcinomas of the tongue in 47% of mice after 25 weeks using applications of 4-nitro-quinoline-*N*-oxide (4NQO). Dachi [8] induced tongue carcinomas in 25% of animals using DMBA in dimethylsulfoxide. Fujita and co-workers [9] were the first to produce malignancies of tongue consistently using DMBA in acetone together with physical trauma. These studies were corroborated by Marefat and Shklar [10,11], who found that DMBA in acetone could induce lingual carcinomas in hamsters without the physical trauma, but that the malignancies took an extra 2–4 weeks to develop. This technique without the trauma has been used in many experimental procedures that study the biology of experimental tongue cancer and its prevention [12].

Experimental carcinogenesis at oral cavity sites other than the tongue and buccal pouch has also been induced and studied. Goldhaber [13] was able to induce malignancies in the labial vestibule of desalivated mice and concluded that carcinogenesis was facilitated by the collection of debris and resultant ulceration. Mesrobian and Shklar [14] were able to induce gingival carcinomas in hamsters using applications of DMBA powder covered with cyanoacrylate to prevent the carcinogen from being washed away by saliva.

Experimental models for other head and neck cancer sites have also been developed. Salivary gland tumors were induced by Steiner [15] in 1942 and later by Shafer [16] and Standish [17]. Cataldo and associates [18] found that pellets of DMBA powder implanted into the submandibular glands of rats resulted in the development of epithelial-lined cysts, with squamous cell carcinomas arising in the cyst walls. Cataldo and Shklar [19] found that the same procedure in hamsters resulted in the development of fibrosarcomas rather than epithelial cysts and carcinomas. Turbiner and Shklar [20] found that the rat salivary gland model could vary with different strains of animals, and this genetic difference in susceptibility to salivary gland carcinogenesis could help explain different experimental results in different experiments using the rat model. The rat salivary gland model was used to study various systemic influences causing enhancement of carcinogenesis [21] or retardation of carcinogenesis [22]. Experimental cancer of the larynx was studied by Bernfeld and colleagues [23] in experiments dealing with the inhalation of tobacco smoke. The experiments required susceptible inbred Syrian hamsters, as this type of malignancy is difficult to induce with tobacco smoke alone.

A. Animal Models for Head and Neck Cancer Research: The Value of the Hamster Buccal Pouch Experimental Model for the Study of Oral Cancer

The hamster buccal pouch experimental cancer model, based on the topical application of DMBA, is the most widely used experimental model for the study of oral cancer [24,25]. Squamous cell carcinomas develop slowly from an initial precancerous lesion similar to human dysplastic leukoplakia [26]. The malignancies gradually become invasive and have the potential of metastasis to regional (cervical) lymph nodes [27]. The tumors resemble their human counterparts both grossly and microscopically [28–30], as well as having similar metabolic markers [31,32] and oncogene expression [33–37]. Initiation and promotion of the tumors can be demonstrated nicely in this model [38,39]

Most of the research carried out on oral carcinogenesis in the hamster buccal pouch model has served to elucidate many aspects of cancer biology and to illustrate some of the major risk factors related to human oral cancer. Studies on hamster pouch tumors were among the first to illustrate the role of the immune system in the control of tumor development. Immunosuppressive agents, such as methotrexate, cortisone, and specific antilymphocyte serum, were found to enhance experimental oral carcinogenesis [40–42], whereas immunoenhancing agents, such as BCG and levamisole, were found to inhibit experimental oral carcinogenesis [43,44]. It was also found that peritoneal-derived macrophages, immune effectors, demonstrated a decrease in Fc and C_3 receptors in tumor-bearing animals, indicating a diminished tumor-killing capacity through antibody-dependent cellular cytotoxicity [45]. A decreased density of Langerhans cells (analogous to macrophages in the initiation of the immune response) and loss of their complex dendritic networks was also found in buccal pouches of carcinogen-treated hamsters [46]. Mast cells were found to gradually infiltrate and degranulate during experimental carcinogenesis, releasing their cytokines [47]. Eosinophil leukocytes were also found to infiltrate the tumor sites and release transforming growth factor α (TGF-α) [48].

It was also found that the major known risk factors in oral cancer, such as tobacco, alcohol, and chronic irritation [49], could be studied by experimental pathology and offer a clearer picture of mechanism than that obtained by epidemiology. Chronic mechanical irritation was found to enhance carcinogenesis by Renstrup and associates [5], whereas the chronic irritation induced by topically applied croton oil was found to, similarly, augment carcinogenesis in the hamster buccal pouch [6]. Alcohol was found to enhance experimental oral carcinogenesis and induced liver damage in the Syrian hamster [50]. Herpesvirus infection was also found to enhance experimental oral carcinogenesis [51].

The control of oral cancer and precancerous leukoplakia has been demonstrated experimentally in the hamster buccal pouch model using a variety of antioxidant micronutrients, such as retinoids, carotenoids, vitamin E, and glutathione. Combinations of micronutrients were shown to be more effective than single micronutrients and to exert a synergistic effect [52]. The nutrients were not only able to inhibit experimental carcinogenesis, but were able to totally prevent tumor development and were able to regress established squamous cell carcinomas [53–56]. The mechanisms of cancer control by micronutrients are gradually becoming clarified based on studies using the hamster buccal pouch experimental cancer model [57]. They involve common pathways of activity at the molecular level, including enhancement of the p53 suppressor gene [58], decreased expression of various oncogenes and growth factors, stimulation of the immune response through immune cytokines [59,60], and depression of tumor angiogenesis [61,62]. Investigations on tumor prevention and regression with micronutrients, utilizing the hamster model, have served as a basis for the development of human trials [58].

Investigations using the hamster oral cancer model have served to corroborate many significant concepts in the biology of cancer and have clarified many concepts dealing with the clinical management of oral cancer. For example, the outdated attitude that biopsy of oral cancer could be dangerous by causing cancer spread could be tested in the experimental model. Neither incision nor manipulation of established cancers in the hamster pouch led to increased spread or metastasis [63].

III. GENETIC MODELS FOR HEAD AND NECK CANCER RESEARCH

Animal models of head and neck cancer, especially oral cancer, have been used extensively to study chemical carcinogens as elaborated upon in the previous section. However, genetic approaches have lagged considerably until very recently through the employment of gene overexpression in transgenic mice and targeted gene ablation in embryonic stem cells of mice. The advantages of genetic models in mice are several, including the ability to recapitulate human carcinogenesis in large numbers of animals, dissect the molecular basis for the different stages of carcinogenesis, availability of mouse-specific molecular reagents, and application of novel chemopreventive and therapeutic approaches.

A. Targeting Genes to Oral Cavity Epithelia with Promoters

Because the oral cavity is lined by a stratified squamous epithelium, there are certain immediate considerations. This is the incorporation of promoters that have been used predominantly for skin models. These include viral promoters, such as human papillomavirus (HPV) E6 and E7, and cellular promoters, i.e., genes that are associated with basal and suprabasal cells. In the latter context, promoters for keratins 5 and 14 will lead to the targeting of genes to basal cells. Seven hundred base pairs of the 3′ downstream sequence was used to drive the expression of an intronless human K14 gene in transgenic mice, and the construct was expressed in a fashion analogous to the endogenous K14 gene: in the basal layer of stratified squamous epithelia [64]. Suprabasal promoters include those for keratins 4 and 13 (preferred partners in oral cavity, whereas in skin it is keratins 1 and 10) as well as for involucrin [65]. Six thousand base pairs of the 5′ upstream K5 sequence directed proper basal cell-specific expression in all stratified epithelia [66]. However, only 90 bp of the K5 promoter directs expression to stratified epithelia, with expression predominantly in the epidermis, hair follicles, and tongue.

Other promoters studied include adenosine deaminase (ADA), which is highly expressed in four tissues of the mouse: the maternal decidua, the fetal placenta, the keratinizing epithelium of the upper alimentary tract (tongue, esophagus, and forestomach), and the absorptive epithelium of the proximal small intestine [67,68]. However, ADA is produced at relatively low levels in all other tissues.

A viral promoter may prove to be of broad appeal in recapitulating oral carcinogenesis. The Epstein–Barr virus (EBV) contains nearly 170 kb of the genomic sequence. Most genes are involved in viral replication; however, some play a role in the ability of the virus to immortalize B lymphocytes. Nearly 800 bp of the ED-L2 promoter, located 3′ to the LMP-1 gene, regulates heterologous reporter genes in cell lines derived from the esophagus and oral cavity, but not of other cell lineages, including B lymphocytes (Fig. 5.1) [69]. In turn, the promoter is regulated transcriptionally in a complex interplay of tissue-specific and relatively tissue-restricted transcriptional factors [70–72]. Given these findings, the EBV ED-L2 promoter has been linked to the human cyclin D1 cDNA and utilized this transgene to generate founder lines. This transgene is transcribed specifically in the tongue, esophagus, and forestomach, all sharing a stratified squamous epithelium. The transgene protein product localizes to the basal and suprabasal compartments of these squamous epithelial tissues, and mice from different lines develop dysplasia, a precursor to carcinoma: mild dysplasia by 6–8 months of age and moderate–severe dysplasia by 16–18 months of age in contrast to age-matched wild-type littermates. Furthermore, the dysplastic phenotype is associated with increased cell proliferation based on PCNA overexpression and abnormalities in cyclin-dependent kinase 4 (cdk4), epidermal growth factor receptor (EGFR), and p53 [73]. In aggregate, these studies suggest that alterations in certain oncogenes and tumor suppressor genes occur early during oral carcinogenesis.

More recently, we have bred the ED-L2 cyclin D1 mice with p53-deficient mice. Importantly, the cyclin D1/p53 heterozygous mice develop severe dysplasia by 3–6 months in the oral–esophageal epithelium and histologic evidence of invasive cancer by 12 months (unpublished observations by A. Rustgi and colleagues). The cyclin D1/p53 null mice also develop severe dysplasia by 3–6 months in the oral–esophageal epithelium with suggestions of microinvasion; however, it proves to be difficult to observe these mice beyond about 6 months due to the fact that they succumb to systemic sarcomas or lymphomas, the expected phenotype of the p53 null alone genotype.

B. Other Emerging Genetic Models of Head and Neck Cancer

In addition to the ED-L2 transgenic mouse model, a number of mouse-based genetic models for head and neck cancer are currently being developed. These efforts are primarily ongoing in the molecular carcinogenesis units of the Oral and Pharyngeal Cancer Branch (OPCB) at the National Institute of Dental and Craniofacial Research under the leadership of J. Silvio Gutkind.

Adopting an approach that has been developed in Harold Varmus' laboratory while at NIH [74–76], which first targets the avian leucosis virus receptor, *tv*-a, to the target tissue of interest and then uses highly effective avian retroviruses to carry the transgene and target its expression *in vivo* in a tissue-specific manner. Investigators at the OPCB have created a transgenic mouse line carrying the *tv*-a receptor gene to the basal layer of stratified epithelium using the cytokeratin 5 (K5) promoter with the hope of being able to eventually induce tumor development on sequential or coexpression of known oncogenes and dominant-negative tumor suppressors. The advantages of this mouse genetic model include (1) the targeting tissue-specific transgene expression without the need to generate a new mouse line; (2) the ability

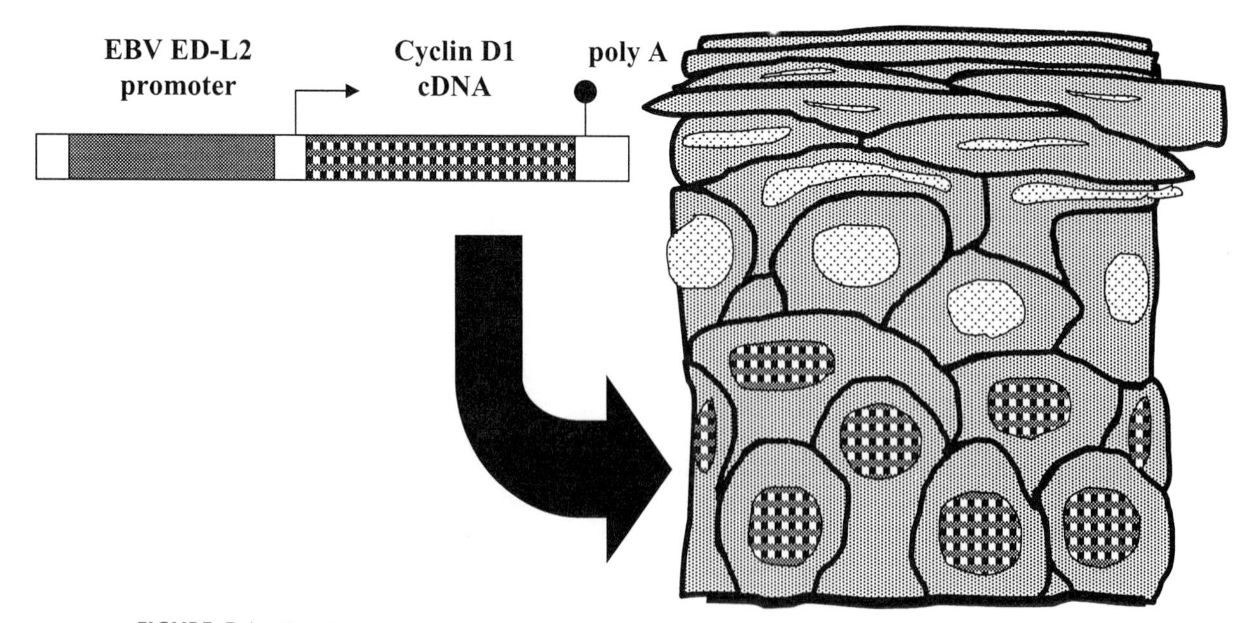

FIGURE 5.1 The EBV ED-L2 promoter targets cyclin D1 to the basal and suprabasal layers of oral–esophageal squamous epithelia in transgenic mice.

to sequentially target multiple transgene to the same host; (3) the ability to monitor the fate of individual, genetically modified cells; (4) the ability to combine oncogenic and tumor suppressor alterations in a specific tissue/cell type to provide insights regarding the effects of different mutations in carcinogenesis and recapitulating the multistep nature of tumorigenesis; and (5) the ability to test the role of newly discovered candidate molecules (e.g., by gene expression profiling analysis) in head and neck cancer carcinogenesis.

The OPCB has also generated a K5 transgenic mouse capable of inducing the expression of transgenes to the basal layer of stratified epithelium. This mouse line carries the tetracycline-inducible promoter system (*tet*-on receptor) with the eventual hope of being able to induce the expression of candidate oncogenes under the control of the tetracycline responsive promoter.

A third mouse model currently developed by this group utilizes the sprr3 promoter. This is a member of the small proline-rich family of proteins that is expressed preferentially in the oral epithelium. The sprr3 promoter is being explored to target transgenes (oncogenes and dominant tumor suppressors) to oral epithelium.

C. Targeted Ablation of Genes in the Oral Cavity

Given the relative tissue-specific expression of keratin 4, targeted disruption of this gene through homologous recombination in mouse embryonic stem cells has a striking phenotype of epithelial hyperplasia in the tongue and esophagus, suggesting an impairment in differentiation [77]. This is evident in K4 null mice as early as 2 months, but K4 heterozygous mice and wild-type mice are normal. No cancer develops, perhaps due to compensation of K4 loss by K13. Interestingly, mice lacking involucrin from embryonic stem cells develop normal tissue structures.

D. Future Directions

The ED-L2 promoter will be a powerful tool to target genes to the oral cavity, either singly or in combination in transgenic mice. Furthermore, this promoter may be useful in strategies for conditional knockout of oncogenes and tumor suppressor genes. Apart from these considerations, oral cavity-specific genes may emerge from ongoing efforts in microarray approaches with subsequent application in genetically engineered animal models.

Mouse genetic models will have a particularly important role in head and neck cancer research. These models will permit the site-specific interactions of oncogenes and tumor suppressors in the head and neck region. Another utility of these models will be to validate and test the role of newly discovered molecules in head and neck cancer carcinogenesis. This is the basis of the integrative human-based discovery and animal model testing for head and neck cancer research.

IV. INTEGRATIVE HUMAN-BASED DISCOVERIES AND ANIMAL MODEL TESTING FOR HEAD AND NECK CANCER RESEARCH

In addition to being able to importantly address the detailed molecular contribution of genes and molecules in carcinogenesis, the availability of genetic mouse models for head and neck cancer will dovetail very well with the current exploration of discoveries using genome-wide approaches to identify critical molecular targets in this cancer site. Technologies to monitor thousands of gene expression or through genome-wide allele typing procedures, coupled with advances in DNA sequencing, allow investigators to identify consistent genetic alterations in head and neck cancer. For epithelial cancers, this is further aided by recently developed laser capture microdissection technology, which permits the precise procurement of normal, premalignant, and cancer cells from the same histological section [78,79]. This ability to allow histologically homogeneous epithelial populations from the same patient to be compared genetically and biochemically is a significant technological advancement that will have important impact on cancer biology. Each of the identified molecular alterations, even whose identity and functional aspects are known, still needs to be tested and validated for their biological contribution in head and neck carcinogenesis. This is precisely how the genetic model for head and neck cancer will be of value: to validate and test the functional role of each altered target by overexpressing or expressing a dominant-negative interfering form of the molecule in head and neck cancer sites.

References

1. Yamagiwa, K., and Ichikawa, K. (1918). Experimental study of the pathogenesis of carcinoma. *J. Cancer Res.* **3**, 1–12.
2. Cook, J. W., *et al.*, (1932). The production of cancer by pure hydrocarbons. *Proc. R. Soc.* **B3**, 455–465.
3. Salley, J. J. (1954). Experimental carcinogenesis in the cheek pouch of the Syrian hamster. *J. Dent. Res.* **33**, 253–262.
4. Morris, A. L. (1961). Factors influencing experimental carcinogenesis in hamster cheek pouch. *J. Dent. Res.* **40**, 3–15.
5. Renstrup, G., Smulow, J., and Glickman, I. (1962). Effect of chronic mechanical irritation on chemically induced carcinogenesis in the hamster cheek pouch. *J. Am. Dent. Assoc.* **64**, 770–777.
6. Silberman, S., and Shklar, G. (1963). The effect of carcinogen (DMBA) applied to the hamster's bucal pouch in combination with croton oil. *Oral Surg.* **16**, 1344–1365.
7. Berenblum, I. (1941). The cocarcinogenic action of crotin resin. *Cancer Res.* **1**, 44–48.
8. Dachi, S. F. (1967). Experimental production of carcinomas of the hamster tongue. *J. Dent. Res.* **46**(6), 1480.

9. Fujita, K., *et al.* (1973). Experimental production of lingual carcinomas in hamsters: Tumor characteristics and site of formation. *J. Dent. Res.* **52**(6), 1176–1185.
10. Marefat, P., and Shklar, G. (1977). Experimental production of lingual leukoplakia and carcinoma. *Oral Surg. Oral Med. Oral Pathol.* **44**(4), 578–586.
11. Marefat, P., and Shklar, G. (1979). Lingual leukoplakia and carcinoma. An experimental model. *Prog. Exp. Tumor Res.* **24**, 259–268.
12. Shklar, G., *et al.* (1980). Retinoid inhibition of experimental lingual carcinogenesis: Ultrastructural observations. *J. Natl. Cancer Inst.* **65**(6), 1307–1316.
13. Goldhaber, P. (1957). The role of saliva and other environmental factors in oral carcinogenesis. *J. Am. Dent. Assoc.* **54**, 517–519.
14. Mesrobian, A. Z., and Shklar, G. (1969). Gingival carcinogenesis in the hamster, using tissue adhesives for circinogen fixation. *J. Periodontol.* **40**(10), 603–606.
15. Steiner, P. E. (1942). Comparative pathology of induced tumors of the salivary gland. *Arch. Pathol.* **34**, 613–624.
16. Shafer, W. G. (1962). Experimental salivary gland carcinogenesis. *J. Dent. Res.* **41**, 117–124.
17. Standish, M. (1957). Early histologic changes in induced tumors of the submaxillary glands of the rat. *Am. J. Pathol.* **33**, 671–689.
18. Cataldo, E., Shklar, G., and Chauncey, H. (1964). Experimental submaxillary gland tumors in rats: Histology and histochemistry. *Arch. Pathol.* **77**, 301–316.
19. Cataldo, E., and Shklar, G. (1964). Chemical carcinogenesis in the hamster submaxillary gland. *J. Dent. Res.* **43**, 568–579.
20. Turbiner, S., and Shklar, G. (1969). Variations in experimental carcinogenesis of submandibular gland in three strains of rats. *Arch. Oral Biol.* **14**(9), 1065–1071.
21. Turbiner, S., Shklar, G., and Cataldo, E. (1970). The effect of cold stress on chemical carcinogenesis of rat salivary glands. *Oral Surg. Oral Med. Oral Pathol.* **29**(1), 130–137.
22. Pisanty, S., Eisenberg, E., and Shklar, G. (1978). The effect of levamisole on experimental carcinogenesis of rat submandibular gland. *Arch. Oral Biol.* **23**(2), 131–135.
23. Bernfeld, P., Homburger, F., and Russfield, A. B. (1979). Cigarette smoke-induced cancer of the larynx in hamsters (CINCH): A method to assay the carcinogenicity of cigarette smoke. *Prog. Exp. Tumor Res.* **24**, 315–319.
24. Shklar, G. (1972). Experimental oral pathology in the Syrian hamster. *Prog. Exp. Tumor Res.* **16**, 518–538.
25. Gimenez-Conti, I. B., and Slaga, T. J. (1993). The hamster cheek pouch carcinogenesis model. *J. Cell. Biochem. Suppl.*, 83–90.
26. Santis, H., Shklar, G., and Chauncey, H. H. (1964). The histochemistry of experimentally induced leukoplakia and carcinoma of the hamster buccal pouch. *Oral Surg.* **17**, 207–218.
27. Shklar, G. (1984). Experimental pathology of oral cancer. *In* "Oral Cancer" (G. Shklar, ed.), pp. 41–54. Saunders, Philadelphia.
28. Salley, J. J. (1957). Histologic changes in the hamster cheek pouch during early hydrocarbon carcinogenesis. *J. Dent. Res.* **36**, 48–55.
29. Shklar, G., Eisenberg, E., and Flynn, E. (1979). Immunoenhancing agents and experimental leukoplakia and carcinoma of the hamster buccal pouch. *Prog. Exp. Tumor Res.* **24**, 269–282.
30. Malament, D. S., and Shklar, G. (1981). Inhibition of DMBA carcinogenesis of hamster buccal pouch by phenanthrene and dimethylnaphthalene. *Carcinogenesis* **2**, 723–729.
31. Shklar, G. (1965). Metabolic characteristics of experimental hamster pouch carcinomas. *Oral Surg.* **20**, 336–339.
32. Solt, D. B. (1981). Localization of gamma-glutamyl transpeptidase in hamster buccal pouch epithelium treated with 7,12-dimethylbenz[*a*]anthracene. *J. Natl. Cancer Inst.* **67**(1), 193–200.
33. Wong, D. T. W., and Biswas, D. K. (1987). Expression of c-erb B proto-oncogene during dimethylbenzanthracene-induced tumorigenesis in hamster cheek pouch. *Oncogene* **2**, 67–72.
34. Wong, D. T. W. (1988). Histone gene (H3) expression in chemically transformed oral keratinocytes. *Exp. Mol. Pathol.* **49**, 206–214.
35. Wong, D. T. W., *et al.* (1989). Detection of Ki-ras messenger RNA in normal and chemically transformed hamster oral keratinocytes. *Cancer Res.* **49**, 4562–4567.
36. Wong, D. T. W., *et al.* (1988). Transforming growth factor-α in chemically transformed hamster oral keratinocytes. *Cancer Res.* **48**, 3130–3134.
37. Husain, Z., *et al.* (1989). Sequential expression and cooperative interaction of c-Ha-ras and c-erbB genes in in vivo chemical carcinogenesis. *Proc. Natl. Acad. Sci. USA* **86**(4), 1264–1268.
38. Odukoya, O., and Shklar, G. (1982). Two-phase carcinogenesis in hamster buccal pouch. *Oral Surg.* **54**, 547–552.
39. Odukoya, O., and Shklar, G. (1984). Initiation and promotion in experimental oral carcinogenesis. *Oral Surg. Oral Med. Oral Pathol.* **58**(3), 315–320.
40. Shklar, G., Cataldo, E., and Fitzgerald, A. L. (1966). The effect of methotrexate on chemical carcinogenesis of hamster buccal pouch. *Cancer Res.* **26**, 2218–2224.
41. Shklar, G. (1966). Cortisone and hamster buccal pouch carcinogenesis. *Cancer Res.* **26**(12), 2461–2463.
42. Woods, D. A. (1969). Influence of antilymphocyte serum on DMBA induction of oral carcinoma. *Nature* **224**, 276.
43. Giunta, J. L., Reif, A. E., and Shklar, G. (1974). Bacillus Calmette-Guerin and anti-lymphocyte serum in carcinogenesis: Effects on hamster buccal pouch. *Arch. Path.* **98**, 237–240.
44. Eisenberg, E., and Shklar, G. (1977). Levamisole and hamster pouch carcinogenesis. *Oral Surg.* **43**, 562–571.
45. Antoniades, D. Z., Schwartz, J., and Shklar, G. (1984). The effect of chemically induced oral carcinomas on peritoneal macrophages. *J. Clin. Lab. Immunol.* **14**(1), 17–22.
46. Schwartz, J. L., *et al.* (1981). Distribution of Langerhans cells in normal and carcinogen-treated mucosa of buccal pouches of hamsters. *J. Dermatol. Surg. Oncol.* **7**, 1005–1010.
47. Flynn, E. A., Schwartz, J. L., and Shklar, G. (1991). Sequential mast cell infiltration and degranulation during experimental carcinogenesis. *J. Cancer Res. Clin. Oncol.* **117**, 1–8.
48. Elovic, A., *et al.* (1990). Production of transforming growth factor alpha by hamster eosinophils. *Am. J. Pathol.* **137**(6), 1425–1434.
49. Graham, S., *et al.* (1977). Dentition, diet, tobacco, and alcohol in the epidemiology of oral cancer. *J. Natl. Cancer Inst.* **59**(6), 1611–1618.
50. Freedman, A., and Shklar, G. (1978). Alcohol and hamster buccal pouch carcinogenesis. *Oral Surg.* **46**, 794–805.
51. Park, N. H., Sapp, J. P., and Herbosa, E. G. (1986). Oral cancer induced in hamsters with herpes simplex infection and simulated snuff dipping. *Oral Surg. Oral Medi. Oral Pathol.* **62**(2), 164–168.
52. Shklar, G., *et al.* (1993). The effectiveness of a mixture of beta-carotene, alpha-tocopherol, glutathione, and ascorbic acid for cancer prevention. *Nutr. Cancer* **20**(2), 145–151.
53. Shklar, G., *et al.* (1987). Regression by vitamin E of experimental oral cancer. *J. Natl. Cancer Inst.* **78**(5), 987–992.
54. Trickler, D., and Shklar, G. (1987). Prevention by vitamin E of experimental oral carcinogenesis. *J. Natl. Cancer Inst.* **78**(1), 165–169.
55. Suda, D., Schwartz, J., and Shklar, G. (1986). Inhibition of experimental oral carcinogenesis by topical beta carotene. *Carcinogenesis* **7**(5), 711–715.
56. Schwartz, J., and Shklar, G. (1988). Regression of experimental oral carcinomas by local injection of beta- carotene and canthaxanthin. *Nutr. Cancer* **11**(1), 35–40.
57. Shklar, G., and Oh, S. K. (2000). Experimental basis for cancer prevention by vitamin E. *Cancer Invest.* **18**(3), 214–222.
58. Shklar, G., and Schwartz, J. (1993). Oral cancer inhibition by micronutrients. The experimental basis for clinical trials. *Eur. J. Cancer B Oral Oncol.* **29B**(1), 9–16.

59. Schwartz, J. L., Sloane, D., and Shklar, D. (1989). Prevention and inhibition of oral cancer in the hamster buccal pouch model associated with carotenoid immune enhancement. *Tumour Biol.* **10**(6), 297–309.
60. Shklar, G., and Schwartz, J. (1988). Tumor necrosis factor in experimental cancer regression with alphatocopherol, beta-carotene, canthaxanthin and algae extract. *Eur. J. Cancer Clin. Oncol.* **24**(5), 839–850.
61. Shklar, G., and Schwartz, J. L. (1996). Vitamin E inhibits experimental carcinogenesis and tumour angiogenesis. *Eur. J. Cancer B Oral Oncol.* **32B**(2), 114–119.
62. Schwartz, J. L., and Shklar, G. (1996). Glutathione inhibits experimental oral carcinogenesis, p53 expression, and angiogenesis. *Nutr. Cancer* **26**(2), 229–236.
63. Shklar, G. (1968). The effect of manipulation and incision on experimental carcinoma of hamster buccal pouch. *Cancer Res.* **28**, 2180–2182.
64. Vassar, R., *et al.* (1989). Tissue-specific and differentiation-specific expression of a human K14 keratin gene in transgenic mice. *Proc. Natl. Acad. Sci. USA* **86**(5), 1563–1567.
65. Djian, P., Easley, K., and Green, H. (2000). Targeted ablation of the murine involucrin gene. *J. Cell Biol.* **151**(2), 381–388.
66. Byrne, C., and Fuchs, E. (1993). Probing keratinocyte and differentiation specificity of the human K5 promoter in vitro and in transgenic mice. *Mol. Cell. Biol.* **13**(6), 3176–3190.
67. Winston, J. H., *et al.* (1992). 5′ flanking sequences of the murine adenosine deaminase gene direct expression of a reporter gene to specific prenatal and postnatal tissues in transgenic mice. *J. Biol. Chem.* **267**(19), 13472–13479.
68. Xu, P. A., *et al.* (1999). Regulation of forestomach-specific expression of the murine adenosine deaminase gene. *J. Biol. Chem.* **274**(15), 10316–10323.
69. Nakagawa, H., *et al.* (1997). The targeting of the cyclin D1 oncogene by an Epstein-Barr virus promoter in transgenic mice causes dysplasia in the tongue, esophagus and forestomach. *Oncogene* **14**(10), 1185–1190.
70. Nakagawa, H., Inomoto, T., and Rustgi, A. K. (1997). A CACCC box like cis-regulatory element of an Epstein-Barr virus promoter interacts with a nuclear transcriptional factor in tissue-specific squamous epithelia. *J. Biol. Chem.* **272**, 16688–16699.
71. Jenkins, T. D., Nakagawa, H., and Rustgi, A. K. (1997). The keratinocyte specific Epstein-Barr virus ED-L2 promoter is regulated by phorbol 12-myristate 13-acetate through two cis-regulatory elements containing E-box and Krüpple-like factor motifs. *J. Biol. Chem.* **272**, 24433–24442.
72. Jenkins, T. D., *et al.* (1998). Transactivation of the human keratin 4 and Epstein-Barr virus ED-L2 promoters by gut-enriched Kruppel-like factor. *J. Biol. Chem.* **273**(17), 10747–10754.
73. Mueller, A., *et al.* (1997). A transgenic mouse model with cyclin D1 overexpression results in cell cycle, epidermal growth factor receptor, and p53 abnormalities. *Cancer Res.* **57**(24), 5542–5549.
74. Holland, E. C., and Varmus, H. E. (1998). Basic fibroblast growth factor induces cell migration and proliferation after glia-specific gene transfer in mice. *Proc. Natl. Acad. Sci. USA* **95**(3), 1218–1223.
75. Holland, E. C., *et al.* (1998). Modeling mutations in the G1 arrest pathway in human gliomas: Overexpression of CDK4 but not loss of INK4a-ARF induces hyperploidy in cultured mouse astrocytes. *Genes Dev.* **12**(23), 3644–3649.
76. Holland, E. C., *et al.* (1998). A constitutively active epidermal growth factor receptor cooperates with disruption of G1 cell-cycle arrest pathways to induce glioma-like lesions in mice. *Genes Dev.* **12**(23), 3675–3685.
77. Ness, S. L., *et al.* (1998). Mouse keratin 4 is necessary for internal epithelial integrity. *J. Biol. Chem.* **273**(37), 23904–23911.
78. Simone, N., *et al.* (1998). Laser capture microdissection: Opening the microscopic frontier to molecular analysis. *Trends Genet.* **14**, 253–294.
79. Simone, N. L., *et al.* (2000). Laser capture microdissection: Beyond functional genomics to proteomics. *Mol. Diagn.* **5**(4), 301–307.

CHAPTER

6

Adhesion Receptors in Oral Cancer Invasion

BARRY L. ZIOBER

Department of Otorhinolaryngology—Head and Neck Surgery
University of Pennsylvania
Philadelphia, Pennsylvania 19104

RANDALL H. KRAMER

Oral Cancer Research Center—Department of Stomatology,
School of Dentistry
University of California, San Francisco
San Francisco, California 94143

I. Introduction 65
II. Cell–Substrate Adhesion 66
  A. Collagen Receptors 66
  B. Fibronectin/Tenascin Receptors 67
  C. Laminin Receptors 68
III. Extracellular Matrix–Integrins Interactions and Signal Transduction 70
IV. Cadherins and Intracellular Adhesion 71
  A. Cadherins and Regulation of Invasion 72
  B. Integrin–Cadherin Cross Talk 72
  C. Regulation of Cadherin Expression 74
V. Conclusions 74
References 74

Squamous cell carcinoma of the oral cavity is characterized by local and distant metastasis, and it is this aspect of the disease that is the most challenging for developing clinical approaches to therapy. Invasion is a necessary component of metastasis, tumor cells infiltrate into adjacent tissues, degrading basement membranes and extracellular matrix, and disrupting tissue architecture. An important element of invasion is adhesion receptors, predominantly integrins, that modulate not only cell motility but also survival and proliferation. Additionally, a second class of receptors, the cadherins, form intercellular adhesions between carcinoma cells and are also relevant to the invasive process. Cell–cell adhesions are responsible for forming stratifying cell layers, influence the differentiated state of the tumor cells, and tend to restrain invasion. The process of squamous cell carcinoma invasion and dissemination requires active cell migration through the extracellular matrix with the simultaneous remodeling of intercellular adhesions. Observations suggest that it is the interplay and crosstalk between these two receptor families, integrins and cadherins, that modulate invasion. During tumor progression, the development of variant cells with alterations in the expression of adhesion receptors and their associated signaling pathways could result in the derivation of cells with a highly invasive and metastatic phenotype.

I. INTRODUCTION

Despite advancements in surgery, radiation, and chemotherapy, only about half of individuals diagnosed with oral cancer will survive 5 years [1,2]. The spread to regional lymph nodes in the neck, which can occur soon after the development of oral squamous cell carcinoma (SCC), is made possible by the highly invasive behavior of SCC and the rich lymphatic drainage from the oral cavity. While early detection is the most critical step at present to reduce the morbidity and mortality of oral cancer, it is essential that we gain a better understanding about the basic mechanisms of invasion that lead to metastasis. If alterations in specific genes can be identified in precancerous and malignant lesions, then enhanced diagnostic and treatment protocols might be developed to prevent the development and/or spread of these neoplasms [3,4]. A more refined approach may be forthcoming from chromosomal deletion studies that have reported differences in chromosomes between normal and malignant oral mucosal cells [5]. More effort is needed to identify pathways regulating the invasive tumor phenotype, including how specific adhesion receptors are involved. Integrins are clearly important in the invasive process, whereas intercellular cadherin receptors restrain invasion and promote a more differentiated phenotype. Additionally, these same receptors are known to transduce signals that not only mediate adhesion and motility, but also regulate cell survival and programmed cell death.

II. CELL–SUBSTRATE ADHESION

The interaction of normal epithelial cells with the extracellular matrix (ECM) is important for cell survival and proliferation. When denied ECM adhesion, normal cells and certain tumor cells can be driven into the G1 phase of the cell cycle, where they fail to proliferate [6]. SCC cells are known to be highly anchorage dependent and generally fail to grow in semisolid or suspension cultures, indicating the importance of interactions with ECM [7,8]. In addition, SCC cells, because of their unrestrained growth, their ability to invade surrounding tissues, and their propensity to metastasize, routinely generate secondary tumors in organs distant from the primary tumor. Thus, the constituents of the ECM and their receptors, the integrins, contribute both to the regulation of normal epithelial growth and differentiation and to the mechanisms of SCC tumor formation and metastasis.

As ligands for integrins, ECM molecules are essential in proper tissue development, adult tissue maintenance, wound healing, and oncogenesis [9]. The proliferation of stratifying normal epithelium is confined principally to the layer of cells that are attached to the basal lamina or basement membrane. This supportive matrix is composed of several components derived from keratinocytes and stromal cells or from the mutually cooperative interactions between these two cell populations. Changes in the composition of the ECM molecules, by altered expression, can contribute greatly to oncogenetic development, hyperplastic growth, and tumor development [10–12]. The major ECM molecules implicated in SCC development include collagens, fibronectin, tenascin, and laminin. Thus, changes in the ECM components or their receptors, the integrins, are essential for transformation of a premalignant squamous epithelium into a malignant lesion.

As a large family of transmembrane receptors, integrins are responsible for mediating the adhesive interactions of the cell with ECM macromolecules [13]. Integrins provide a linkage between the cell cytoskeleton and the extracellular environment. Each integrin is a heterodimer composed of a noncovalently associated α and β subunit. Transport of the integrin complex to the cell surface and the formation of the extracellular ligand-binding site located near the amino-terminal region of the integrin subunits require the pairing of α and β chains.

Common structural features that highlight the overall similarity of α subunits are revealed when sequences of the known subunits are aligned. The α subunits are glycoproteins containing a large, globular extracellular domain, a transmembrane region, and a short carboxyl-terminal cytoplasmic tail preceded by the conserved GFFKR sequence. Within the extracellular domains are seven homologous amino acid sequence repeats (numbered I–VII). Repeat domains include regions that are considered potential binding sites for divalent cations. When occupied with appropriate metals, these cation-binding sites stabilize the tertiary structure of the molecule, which is required for the interaction of the receptor with ligand as well as for the association of α and β subunits [13, 14].

In addition to providing adhesive and migratory functions, the integrin–ECM interactions, can regulate several cellular properties, including cell growth, apoptosis, angiogenesis, protease production, differentiation, and gene expression, all properties involved in malignant conversion and invasion [15–20]. Thus, changes in integrin profiles, which alter the sum of the signaling cascades within the cell, can have profound effects on cellular behavior. The major integrin receptors found in SCC cells include $\alpha2\beta1$, $\alpha3\beta1$, $\alpha5\beta1$, $\alpha6\beta1/\alpha6\beta4$, and αv complexes. Several studies have examined the expression of these integrin heterodimers in SCC. The most recent and pertinent studies are reviewed briefly here.

A. Collagen Receptors

The major collagenous components of the skin are collagen type I and type III, which make up primarily the dermis. Collagen type IV is the major collagen isoform comprising the basement membrane framework of the dermo-epidermal junction. Minor collagens making up the dermo-epidermal interface also include types VII and XVII [21]. Collagens, as with other ECM molecules, can promote cell adhesion, migration, differentiation, and growth [22].

Several studies have reported changes in collagen expression during SCC formation. For example, increased induction and deposition of collagen types I and III have been reported in well-differentiated SCC [23]. In contrast, loss of collagen VII was detected in highly differentiated tumors, whereas lack of collagen IV was associated with decreasing tumor cell differentiation [24]. Reduction of the type IV collagen, in particular, suggests that loss of the continuity of the subepithelial basement membrane is associated with the progressive nature of SCC [25]. This is an interesting observation in light of recent work demonstrating that unique domains of type IV collagen possess antiangiogenic and thus antitumor properties. Thus, the potential loss of type IV collagen may increase blood vessel formation, resulting in expanded tumor growth [22]. Whether this occurs in SCC is unknown.

Loss of collagen VII has also been suggested to serve as a potential marker for early invasive tumor growth [24]. In a related study, analysis of basement membrane antigens in SCC also revealed a reduction in collagens type IV and VII as compared to normal epithelium [26]. Interestingly, loss of collagen VII occurred before loss of collagen IV [27]. This was most apparent at the leading edge of the invasive tumor mass, emphasizing that there is an ordered decline of collagen isoforms. Such an ordered decline may provide a potential marker of tumor invasion [26]. The factor(s) responsible for these reductions is presently unknown. One study has noted that type VII collagen mRNA is synthesized at high levels in SCC tumors, but its protein distribution is impaired. This may explain the loss of this isoform in SCC [28]. However, metalloproteinases (MMP), particularly MMP-2, MMP-9, and MMP-13, are frequently expressed specifically in SCC and may account for a substantial reduction in these collagen isoforms [26,29–31].

The major collagen receptor is α2β1. However, α2β1 is not a dedicated collagen receptor, as it can also serve as a receptor for tenascin-C, laminin, and collagen [32,33]. A comparison of several metastatic SCC tumor cell lines with nonmetastatic cell lines revealed that α2β1 is strongly expressed in metastatic lines [34]. This was in agreement with immunohistochemical analysis of invasive and metastatic tumor biopsies, which also possessed a marked elevation of α2β1 [34,35]. In contrast, loss of this same receptor, as compared to normal epithelium, has the strongest correlation with SCC tumor progression. Furthermore, reduction of α2β1 has been suggested to serve as a prognostic evaluator for SCC [20,36]. As indicated earlier, several of the collagen ligands of α2β1 are downregulated in SCC. This may help explain the loss of this receptor in SCC. However, increased expression of α2β1 may be the result of a compensatory mechanism to overcome the loss of these collagen isoforms. Further studies are required to determine the role and function of this receptor in SCC development.

B. Fibronectin/Tenascin Receptors

Fibronectin is a large, alternatively spliced ECM molecule whose expression has been shown to directly influence the establishment and maintenance of the transformed phenotype [37]. Immunohistochemical analysis of 112 primary and 29 metastatic SCCs revealed that the expression of fibronectin is increased markedly in the tumor stroma, in particular at the invasive tumor site [38]. Oncofetal fibronectins, as demonstrated in a more recent study, were detected throughout the stromal compartment in association with SCC. The primary source of this fibronectin was determined to be stromal fibro/myofibroblasts [39]. Thus, in this study it was concluded that in SCC, conversion to a fetal ECM is occurring during tumor progression.

In addition to tumor specimens, several SCC cell lines have been shown to produce large amounts of fibronectin [40]. In contrasting reports, fibronectin was found to be absent in the later stages of SCC development, and this absence was associated with a decreasing degree of differentiation [27]. Furthermore, reverse transcriptase–polymerase chain reaction (RT–PCR) analysis has revealed complex patterns of fibronectin splicing in normal epithelium and SCC. The significance of such splicing patterns is unknown [41]. However, it is of interest to note that the high heparin-binding domain and the V region of fibronectin have been shown, *in vitro*, to play important roles in SCC invasion, motility, and spreading [42].

Like fibronectin, tenascin-C is a mosaic-like protein made up of several subunits. It is a large glycoprotein reported to possess antiadhesive modulating properties. In normal oral mucosa, tenascin-C expression is not detected [33]. However, during wound regeneration, epithelial cells do produce and secrete tenascin-C, which plays an important role in regulating cell migration and proliferation [43]. Analysis of tenascin-C in SCC has demonstrated that it is also upregulated, both in tumor specimens and in SCC cell lines [33]. This upregulation, as in wound-regenerating epithelial cells, is important for SCC tumor cell migration and suggests a role for tenascin-C in tumor invasion [33]. A more recent study indicates that this upregulation of tenascin-C tumor in ECM is dependent on factors provided by both SCC tumor cells and peritumor fibroblasts [44]. In support of this study, recent work, using *in situ* hybridization and immunohistochemistry, has revealed that carcinoma cells themselves can directly produce ECM components such as tenascin-C [45]. Thus, together these results suggest that the expression of tenascin-C in SCC appears to play a role in tumor cell migration and/or invasion, requiring both stromal cells and keratinocytes for its synthesis. The inappropriate secretion of tenascin-C by SCC may be due to improper regulation of the normal wound repair processes in the tumor cell.

The classic fibronectin receptor is α5β1. In normal and metaplastic epithelium, α5β1 expression is usually not detectable [46]. However, in the later stages of the murine skin carcinogenesis model, α5β1 is upregulated [47]. In neoplastic cells, α5β1 is expressed and contributes to SCC cell adhesion and migration on fibronectin [48]. Metastatic cell lines tended to show an increase in expression of this receptor that was paralleled in invasive and metastatic tumor specimens [34]. Thus, overall, α5β1 appears to be upregulated during SCC progression and likely contributes to the invasive process.

Several secondary fibronectin receptors are formed by the heterodimerization of αv with several β subunits, including β3, β5, and β6. Several other fibronectin receptors have not been detected in epithelial cells, including α4β1, α8β1, αvβ3, αvβ6, and αvβ8. αvβ1 is a low-affinity receptor whose expression can only be detected *in vitro* in cultured keratinocytes [49]. Some members of the αv family of integrins can also mediate adhesion and migration on vitronectin and tenascin [48]. However, information regarding vitronectin in SCC is limited and therefore is not discussed here.

αvβ5 can be detected in normal epithelium, with strong expression in the basal layers [50]. The ligands of αvβ5 include fibronectin and tenascin. The β3 and β6 partners of αv are not expressed in normal tissue. In SCC the staining of αvβ5 has been reported as variable within and between SCC tumor specimens. However, the least staining of αvβ5 was detected in poorly differentiated tumors [50]. To identify a functional property of αvβ5 in transformation, the αv subunit was transfected into a neoplastic keratinocyte cell line. This resulted in the pairing of αv with the β5 subunit. In these αv transfectants, anchorage-independent growth was suppressed and an increased capacity for terminal differentiation was observed [51]. These results suggest that loss of integrin receptors may be responsible for the abnormal properties of keratinoctyes in oral SCC [51].

In contrast to αvβ5, the β6 subunit is expressed in all carcinomas. However, like αvβ5, αvβ6 binds fibronectin and tenascin. That αvβ6 is not detectable in normal epithelium indicates that this subunit is neoexpressed and suggests that it is probably involved in oral SCC tumor progression [50]. Analysis of 11 SCC cell lines demonstrated that αvβ6 was upregulated in 8 (73%) of these lines [48]. Recent work has found that latent transforming growth factor (TGF)-β is a ligand for αvβ6 and that αvβ6-expressing cells can induce activation of TGF-β1 [52]. Because TGF-β is known to be a growth inhibitor of normal squamous epithelium and αvβ6 is upregulated in SCC, one would predict that SCC should be sensitive to the actions of TGF-β. However, several SCC cell lines are refractory to the effects of TGF-β, suggesting that T6F-β in these tumor cells is not available or is processed incorrectly, the TGF-β receptor(s) is absent or defective, or there are deficiencies in one of the many TGF-β downstream signaling factors [53]. In support of these observations, aberrant TGF-β I and II receptor expression has been shown to contribute to the pathogenesis of SCC [53,54].

C. Laminin Receptors

Laminins are large, heterodimeric extracellular glycoproteins composed of an α, a β, and a γ subunit. To date, 12 different laminin heterodimers have been identified [55]. We have identified several laminin isoforms in SCC cell lines, including laminins 2, 4, 10, and 11 [56]. However, the primary laminins expressed in stratified epithelium are laminin-5 and laminin-10/11.

Laminin-10/11 is composed of α5, β1/β2, and γ2. Because the laminin-10/11 subunits are expressed in SCC, one could hypothesize that this laminin isoform plays a significant role in SCC tumor progression [55]. In addition, keratinocytes of epidermal outgrowths that migrate into the wound bed express laminin-10/11 [57]. However, little is known about the involvement of laminin-10/11 in SCC except that SCC cells upregulate this laminin isoform and express the integrin receptors for laminin-10/11 [58].

One of the major matrix components of epithelial basement membranes is laminin-5. Laminin-5 is composed of three subunits: α3, β3, and γ2 [59]. An increasing body of evidence suggests that laminin-5 plays a significant role in SCC tumor development and invasion. Early studies indicated that laminin-5 is synthesized and secreted in several normal cells and tumor types, including mucosal keratinocytes and head and neck squamous cell carcinomas [60,61]. In these carcinomas, laminin-5 was overexpressed primarily at the invasive fronts [61]. It was determined, by immunohistochemistry and *in situ* hybridization, that laminin-5 in normal mucosa and lichen planus, was present as a thin, continuous line located in the basement membrane region. In epithelial dysplasia, this staining was discontinuous and more diffuse [62]. However, there was a striking, intense cytoplasmic staining of the carcinoma cells along the invasive border and the individual infiltrating carcinoma cells in invasive carcinomas and lymph node metastasis [62]. These observations indicate that in SCC tumor progression, synthesis and secretion of laminin-5 are altered, the most intense expression occurring at the invasive front and in the lymph node metastasis. Ono and colleagues [63] found a significant correlation between laminin-5-immunopositive SCC cells and increased infiltrative growth and poorer differentiation. In addition, these same researchers, by analyzing patient survival data, found that increased laminin-5 expression was significantly associated with poorer patient outcome. Overall, increased laminin-5 expression may provide a significant marker for the invasive phenotype and appears to be indicative of a poorer outcome for patients with SCC.

How laminin-5 induces migration, and thus the invasive phenotype, has been the focus of several interesting reports. It appears that proteolytic cleavage of the individual subunits comprising laminin-5 is important in regulating the migratory phenotype. First, cleavage of the γ2 subunit by bone morphogenetic protein-1 (BMP-1) facilitates the formation of basement membrane assembly, but not hemidesmosome assembly, suggesting that it may provide a more motility-inducing ECM environment [64]. BMP-1, also known as type I procollagen C-proteinase, is expressed in several SCC cell lines, implying a possible role for this proteinase in SCC tumor invasion [65]. Second, two separate reports have also demonstrated that the laminin-5 γ2 chain is cleaved [66,67]. However, instead of BMP-1, this cleavage occurs by MMP-2 and MT1-MMP, both of which are upregulated and expressed in SCC [26,30,31,68]. The resulting cleavage of laminin-5 by these MMPs in some cell types produces a more migration-inducing laminin substrate. In a third report, cleavage of the globular domain of the $\alpha 3$ subunit of laminin-5 converts this extracellular protein from a cell motility-inducing factor to a protein complex that can trigger formation of the stationary structures called hemidesmosomes, thus preventing cell movement [69]. Researchers have reported that laminin-5 can trigger the disruption of cell–cell junctions and induce scattering of oral SCC cells [70]. Together, these results implicate a substantial role for laminin-5 in SCC tumor progression and invasion. The involvement of laminin-5 in invasion appears to be dependent on how the molecule is cleaved proteolytically. More work is required to identify the regulatory factors responsible for laminin-5 cleavage and how these cleavage products regulate the static and dynamic mechanisms of cell movement. In addition, the proteolytic cascade for this processing needs to be clarified. Finally, it is interesting to note that laminin-5 was one of 13 genes found to be overexpressed in SCC using the new technique of subtraction hybridization coupled with microarray analysis, indicating the importance of this ECM molecule in SCC [71].

The major laminin receptors expressed in SCC include $\alpha 3\beta 1$, $\alpha 6\beta 1$, and $\alpha 6\beta 4$. $\alpha 3\beta 1$ is considered primarily a laminin-5 receptor [72,73]. It has also been implicated as a receptor for other ECM components, but the physiological significance of these weak interactions is unknown [74, and references therein]. Studies of $\alpha 3\beta 1$-null mice have revealed that this receptor is required for the establishment and/or maintenance of basement membrane integrity [74]. In contrast to another laminin-5 receptor, $\alpha 6\beta 4$, $\alpha 3\beta 1$ is considered to play an active role in cell spreading and motility on laminin-5 [60,75]. Adhesion of $\alpha 3\beta 1$ to laminin-5 stimulates a signaling pathway involving mitogen-activated protein kinase (MAPK), which can regulate cell growth, thus presumably contributing to the development of SCC [76].

$\alpha 3\beta 1$ has been shown to form a stable and specific association with the transmembrane-4 family member CD151. This association provides a linkage with phosphatidylinositol 4-kinase, which apparently regulates cell motility [77]. In addition, $\alpha 3\beta 1$ can pair with other tetraspanin proteins, which in turn stimulates the production of MMP-2 [78]. More recent results have shown that an $\alpha 3\beta 1$-dependent pathway is required for the sustained production of MMP-9 in keratinocytes [79]. Thus, it appears that $\alpha 3\beta 1$ can regulate the expression of proteases. These same $\alpha 3\beta 1$-linked expression pathways have not been fully explored in SCC, but they may be important in SCC development and tumor cell dissemination, especially as many SCC tumors produce both MMP-2 and MMP-9.

Like $\alpha 2\beta 1$, $\alpha 3\beta 1$ has been shown to be upregulated in several SCC tumor cell lines and biopsies [34]. In particular, $\alpha 3\beta 1$ (and $\alpha 2\beta 1$) is increased in the suprabasilar area during the development of SCC [80]. In contrast, a reduced expression of $\alpha 3\beta 1$ has been reported to correlate with poor histological differentiation in SCC [20]. Based on these studies, it is likely that $\alpha 3\beta 1$ plays a significant role in SCC tumor development and tumor invasion; however, more thorough studies are required.

$\alpha 6$ can pair with $\beta 1$ and $\beta 4$ in SCC cells, giving rise to two laminin receptors, $\alpha 6\beta 1$ and $\alpha 6\beta 4$. Because $\alpha 6$ combines preferentially with $\beta 4$, this is the dominant complex found. Nevertheless, the biological consequences of each form could be important in regulating the behavior of SCC cell lines. Like $\alpha 3\beta 1$, $\alpha 6\beta 4$ is abundant in keratinocytes and functions as a receptor for laminin-5. However, $\alpha 6\beta 4$ can efficiently stimulate migration on laminin-1 substrates as well [81]. Several studies have demonstrated that keratinocyte adhesion to laminin-5 is important in regulating the dermal–epidermal junction in skin. In contrast to $\alpha 3\beta 1$, $\alpha 6\beta 4$ was originally believed to be involved in the static structures called hemidesmosomes and to contribute little to cell migration. Extracellular laminin-5 anchoring molecules are linked directly with the keratin filament network within the cell by hemidesmosomes [74]. Mutations within either subunit of the $\alpha 6\beta 4$ heterodimer highlight the importance of this receptor in dermal–epidermal junction integrity [82,83]. Finally, during wound healing, the dermal–epidermal adhesive structure provided by $\alpha 6\beta 4$ is undesirable. Thus, for epithelial cells to migrate, they must first disassemble their hemidesmosomes [69]. SCC tumor invasion may represent a normal wound repair process, involving laminin-5 and its two primary receptors, $\alpha 6\beta 4$ and $\alpha 3\beta 1$, that is no longer properly controlled.

$\alpha 6$ has been reported to show a stronger expression in nonmetastatic cells than in metastatic SCC cell lines [34]. However, this subunit is upregulated in tumor biopsies from invasive and metastatic cases [34]. Analysis of $\alpha 6\beta 4$ levels in poorly differentiated SCCs revealed pronounced expression of this receptor along with laminin-5 at the invasive front [35]. Furthermore, in tumors of patients with SCC, $\alpha 6\beta 4$ levels have also been reported to be increased, localizing along the invasive border [84]. Earlier studies indicated that, $\alpha 6\beta 4$ expression in normal squamous epithelium is limited to contact sites of the basement membrane. In aggressive SCC, recurrent tumors, and metastatic tumor cell lines, this normal basal polarity of $\alpha 6\beta 4$ is lost and there is more intense suprabasal staining of $\alpha 6\beta 4$ [85]. This suggests that cytoplasmic or membrane components, possibly CD151 and bullous pemphigoid antigens 1 and 2, that function in $\alpha 6\beta 4$ polarity and hemidesmosome formation may be defective or lacking during SCC tumor progression [86,87].

An interesting question arises as to the regulation of $\alpha 6\beta 4$ in normal epithelial architecture versus tumor invasion: if $\alpha 6\beta 4$ is involved primarily in inhibiting cell migration by the formation of hemidesmosomes, why does increased expression of this receptor not inhibit SCC tumor invasion? Studies directed at understanding the signaling pathways elicited by $\alpha 6\beta 4$ (see later), including the signaling pathways that disrupt the hemidesmosome structure, the factors responsible for proper cell polarity, or the proteolytic regulation of the motility-inducing laminin-5 isoforms, are required to answer this question. Finally, it should be noted that the cytoplasmic interactions of α6β4 may also play important roles in whether this receptor functions in cell adhesion or motility. For example, BPAG1, which localizes to the inner surface of hemidesmosomal structures containing α6β4, positively influenced cell migration when removed by homologous recombination [88].

Decreased expression of α6β4, in direct contrast to the aforementioned reports, has also been described in SCC. Normally, α6β4 expression shows a basally polarized distribution. Lack of a restricted basal polarity of α6β4, by absence of β4 expression, has been suggested to be an early marker of oral malignancy [89]. Examination of normal and epithelial SCC tissue sections for α6 expression demonstrated that the staining intensity of this subunit was reduced significantly in SCC compared to normal epithelium [36]. Similarly, α6β4 was found to be downregulated in oral SCC, and this reduction in α6β4 correlated with poor histological differentiation [20]. Thus, loss of α6β4 and the absence of hemidesmosome formation may provide a more motile and invasive SCC tumor. This notion is supported by the fact that blocking antibodies to β4 can result in the stimulation of cell migration and an increase in MMP-2 activity [90]. Finally, transfection of the β4 subunit into a neoplastic keratinocyte cell line failed to restore differentiation capacity or proliferation properties, suggesting that α6β4 is not required for these properties in SCC [51].

III. EXTRACELLULAR MATRIX–INTEGRINS INTERACTIONS AND SIGNAL TRANSDUCTION

It is now well established that integrins, in addition to functioning in cell adhesion and migration, can function as signal transduction molecules that stimulate many intracellular pathways regulating cell differentiation, growth, apoptosis, motility, and protease production, to name a few [37]. The binding of integrins with their ECM ligands and the formation of integrin–cytoskeleton complexes in focal adhesions are regulated by complex signal transduction mechanisms involving a large number of proteins associated directly or indirectly with the integrin subunits themselves. One of the key signaling molecules involved in integrin-mediated cell signaling is focal adhesion kinase (FAK) [91]. Integrin ligation promotes the phosphorylation of FAK, which creates numerous binding sites for several proteins, such as p130Cas, Src family kinases, Grb-2, phosphatidylinositol 3′ kinase, and others. These in turn link integrin adhesion to several cellular pathways, e.g., the Ras/MAPK and C-Jun kinase pathways, which are ultimately responsible for cell survival, migration, and growth [92,93].

Multiple invasive and metastatic tumor cancers show overexpression of FAK [92,94]. In preinvasive lesions of oral cancer (carcinoma *in situ*) and in a large group of invasive SCCs, enhanced FAK immunostaining has also been found [92, and references therein]. It is believed that upregulation of FAK leads to increased cell migration and contributes to tumor cell survival under anchorage-independent conditions [92,93]. Such a population of tumor cells would have a strong tendency to invade and metastasize. The mechanism for the increased expression of FAK in SCC tumors is unknown. However, it has been discovered that in all cases of invasive SCC, the *fak* gene copy number was increased [94]. This same result was confirmed in frozen sections of SCC as well. The genomic location of the *fak* gene has been confirmed as chromosome 8 q [94]. Interestingly, this region of chromosome 8 is frequently overrepresented in SCCs, especially in metastatic carcinomas [95].

Ligation of ECM by integrins initiates several rearrangements in the cytoskeleton that are required for cell motility and thus invasion. The Rho family GTPases Cdc42, Rac, and Rho are major controllers of the actin cytoskeleton, regulating cell adhesion, migration, and invasion [96]. Binding of the integrin with ECM components results in the assembly of focal adhesion complexes, containing cytoskeletal proteins as well as several signal transduction molecules. It is still not clear what signaling molecule bridges integrin ligation and these GTPases. However, it has been suggested to be a kinase [97].

Classically, the integrin activation of Cdc42 leads to filopodia, that of Rac results in the formation of lamellipodia, and that of Rho causes stress fiber formation [96]. However, recent work has also established other functions for these GTPases. For example, activation of Rac is sufficient to cause the disruption of cadherin-dependent cell–cell adhesion in keratinocytes, which suggests that Rac is important in regulating cell–cell and cell–ECM adhesion during SCC development [98]. Cdc42 has also been implicated in regulating cell–cell adhesion and playing an important role in establishing the initial polarization in epithelial cells [96]. Rho is involved in the formation of stress fibers and focal adhesion complexes, both of which are necessary for the forward movement of cells. Rho has also been shown to moderate cell–cell adhesion. Integrin-mediated cell adhesion to the ECM can either stimulate or inhibit Rho, depending on several factors, including cell type, engagement of specific integrins, and time engaged [97]. For example, cross-linking α6β4 stimulates Rho activity, whereas β1 inhibits this activity. It is believed that the c-Src family of kinases is required for the inhibition of Rho [97]. Finally, for all of these GTPases, a fine balance among contractility, cell–cell adhesion, and GTPase concentration is necessary to determine if the cell is static or motile. Thus, given the importance of these GTPases in modulating the cellular events described, it is likely that changes in the activation and/or inactivation of these molecules can lead to an invasive and metastatic phenotype [96]. However, few studies have analyzed these GTPases in SCC.

The α6β4 integrin must first be released from stable hemidesmosome adhesion complexes before an SCC tumor cell can become motile and invasive. Thus, α6β4 and signaling events associated with this receptor have been the focus of several investigations. Initially, it was discovered that α6β4, when ligated to laminin, caused phosphorylation of Shc, activation of Ras, and stimulation of MAP kinases Erk and Jnk. These were the first findings coupling α6β4 to signaling pathways regulating cell proliferation [99,100]. In a related study, epidermal growth factor (EGF) was found to stimulate cell migration, and this migration induced α6β4 mobilization from hemidesmosomes [101]. This mobilization of α6β4 from hemidesmosomal components occurred by the activation of protein kinase C. Thus, ligation of α6β4 can stimulate cellular proliferation, whereas signals originating from other pathways that disrupt stable cell adhesion can lead to increased cell movement and contribute to the invasive properties of SCC.

By ligating with laminin, α6β4 can activate several pathways regulating carcinoma invasion. By itself, α6β4 can promote cell migration [102]. On laminin, α6β4 facilitated the translocation of the small GTPase RhoA to membrane ruffles. In general, these results established that an α6β4-mediated pathway can activate RhoA and function in promoting cell migration [102]. Work initiated by Shaw and co-workers [103] began to define a mechanism by which α6β4 potentiates carcinoma invasion. Tumor cell invasion stimulated by α6β4 is dependent on phosphoinositide-3 OH kinase (PI3K) activity and results in α6β4-specific formation of dynamic sites that are required for cell motility. Besides PI3K, it was determined that this invasive process also required another GTPase, Rac [103]. In addition, ligation of the α6β4 heterodimer protected the cells from apoptosis through a pathway that also involved PI3K/Akt [104]. It was determined subsequently that α6β4, by activating the Akt/PI3K pathway, can promote cell survival in p53-deficient carcinoma cells [105]. Substantial evidence has attributed loss of and/or mutations in p53 in the pathogenesis of SCC [106]. Finally, it has been indicated that the β4 subunit of α6β4 can stimulate specific signals responsible for cell motility, expression of MMP-2, and tumor invasion [90]. Collectively, these data indicate that α6β4 interacts with several pathways modulating cell growth, motility, survival, and proteinase production, suggesting an important role for this receptor in SCC development and invasion. Thus, α6β4 and its associated pathways may provide potential targets for the development of SCC therapies.

In addition to functioning as adhesion- or migration-regulating molecules, integrins can also activate several pathways involved in tumor progression. These pathways are dependent on particular integrin subunits and integrin cytoplasmic domains. For example, adhesion of α6β1 and α6β4 can result in tyrosine phosphorylation of distinct cellular proteins, suggesting activation of unique signaling pathways by these receptors [100,107]. In addition, the cytoplasmic domain of the α2 subunit can communicate with specific signaling molecules and pathways, including R-Ras and PI3K [108]. Thus, one particular integrin complex probably does not dictate the invasive tumor cell properties in SCC. The invasive phenotype is the result of the integrin composition present on the tumor cell surface and the summation of all the signaling events elicited by these integrins.

Several factors have been implicated in the signaling processes of integrins. For example, R-Ras, a member of the Ras family, is located in the cytoplasm and is required for the regulation of integrin activation and thus ligand binding [109]. R-Ras, when inhibited or not stimulated, may contribute to tumor cell invasion. This may be most important in SCCs displaying *ras* mutations. In addition, several recently identified integrin-associated proteins, e.g., p130Cas, PTEN, HEF1, Src, protein-tyrosine phosphatase-α, and others, have been reported to participate in many integrin-mediated events [110–113]. More work is required to determine their contribution to SCC and tumor invasion. However, by thorough examination of the molecular mechanisms of integrin signaling, novel treatment strategies for cancer cell growth, invasion, and metastasis are expected to be developed.

IV. CADHERINS AND INTERCELLULAR ADHESION

Oral SCC is formed from stratifed squamous epithelium and as such utilizes the E-cadherin system to maintain cell–cell adhesion. Cadherins are concentrated into junctional plaque structures that can be visualized by electron microscopy as zonula adherens and desmosomes [114,115]. The cadherin family has been organized into classical cadherins and desmosomes, but both groups of receptors share highly conserved transmembrane and extracellular regions and interact with a class of cytoplasmic link proteins termed catenins. Cadherin function requires interaction with the actin cytoskeleton. Consequently, cadherin function is regulated from the cytoplasmic side and depends on the interaction with catenins, which bind to the cytoplasmic domain of cadherins, thereby forming links to the cytoskeleton [reviewed in 116,117]. For adherens junctions, these link proteins are β-catenins and plakoglobin (γ-catenin). β-Catenin and plakoglobin, but not α-catenin, bind directly to the cytoplasmic N-terminal domain of E-cadherin; α-catenin can associate with the N-terminal regions of β-catenin and plakoglobin, which is essential in order for cadherins to interact with the cytoskeleton. The Armadillo repeat region of β-catenins and plakoglobin mediates their association with E-cadherin and their complexing with the adenomatous polyposis coli (APC) tumor suppressor protein. The association of

APC with β-catenin appears to regulate its stability. The Armadillo repeats in plakoglobin are also crucial for its binding to desmosomal cadherins (desmogleins and desmocollins). Finally, the amino-terminal region of α-catenin links cadherin–catenin complexes with the actin filaments via α-actinin (Fig. 6.1, see also color insert).

Catenins are now known to be key regulatory molecules that mediate the transduction of extracellular contacts between cadherins during epithelial reorganization and also provide for the linkage of cadherins to intracellular signaling pathways [reviewed in 118–120]. Importantly, catenins transmit signals that regulate gene expression. Thus, β-catenin binds both cadherin and other catenins, but when β-catenin is in excess, it binds the LEF family of transcription factors, and this complex of catenin and transcription factor is transported to the nucleus where it effects gene expression. In contrast, expression of the *wnt*-1 protooncogene increases the level and function of catenins, thereby stabilizing cadherin-mediated cell–cell adhesion. Wnt signaling involves GSK3-β, and through the phosphorylation of β-catenin, its own turnover is regulated. Also, axin binds β-catenin, APC, and GSK3-β directly, and this tetrameric complex regulates stabilization of β-catenin [121,122]. Finally, $p120^{ctn}$ or p120 contains Armadillo repeats, associates with E-cadherin–catenin complexes, is tyrosine phosphorylated in cells transformed with Src or in response to growth factors, and may regulate displacement of β-catenin from E-cadherin complexes [123,124]. In certain tumors, β-catenin appears to act as an oncogene. For example, in colon cancer, APC is mutated, resulting in the elevation of β-catenin, whereas normally APC and GSK3-β facilitate its degradation. However, in colon carcinoma cells and in melanoma cells [125,126], β-catenin is usually highly expressed and forms complexes with Tcf-4 or Lef-1 [127,128], thereby activating gene expression that regulates cell growth or suppresses programmed cell death.

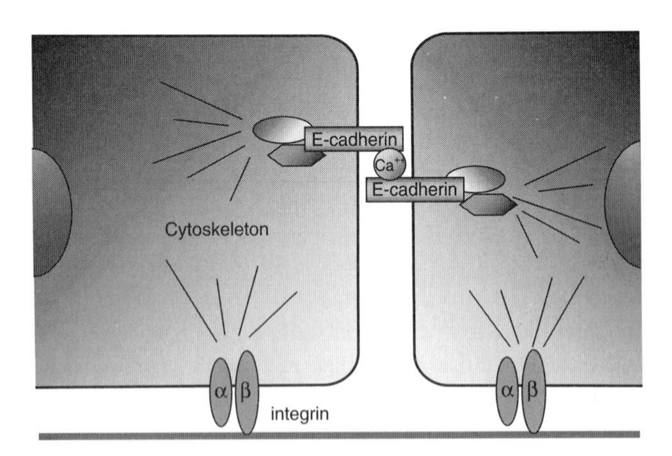

FIGURE 6.1 Integrin and cadherin adhesion receptors. (See also color insert.)

A. Cadherins and Regulation of Invasion

Cadherins are required for cells to remain tightly associated in normal and malignant epithelia, and in their absence the many other cell-junction proteins expressed in epithelial cells are generally not capable of supporting intercellular adhesion [115]. Cell dissociation and scattering, with loss of cell–cell adhesion and junctional communication, are required for the invasiveness and metastasis of malignant tumor cells. It is well established that the downregulation of cell–cell adhesion in tumor cells favors their dissemination [114]. E-cadherin is an important suppressor of epithelial tumor cell motility, invasion, and metastasis [129]. The loss of E-cadherin or its dysfunction leads to increased motility and invasiveness of carcinoma cells, and transfection of cadherin cDNA into deficient carcinoma cells can reverse the invasive phenotype and reduce tumorigenicity [130]. However, a number of tumors can disperse and invade, despite an abundant expression of cadherin molecules at their cell surface. In these cases, several defects in cadherin function have been identified that could account for the suppression of cadherin activity, including loss of α-catenin [131,132] and elevated tyrosine phosphorylation of β-catenin, $p120^{CAS}$, and cadherin [133–136]. In well- and moderately differentiated human SCC, studies have found that a modest but variable expression of E-cadherin is preserved as lesions advance through premalignant to invasive and metastatic stages [137–140]. However, in head and neck SCC, different levels of E-cadherin expression have been reported. Some of this variation in staining pattern may be due to the sensitivities of the different antibodies or methods used. In some studies with human head and neck SCC, a loss or decrease of E-cadherin expression was found during tumor progression, and metastatic lesions tended to have reduced levels of this adhesion receptor [141]. This suggests that in some cases, the tumor cells, possibly under stimulation by cytokines, could temporarily uncouple cadherins, thereby permitting distant metastasis followed by reexpression of the cadherin. Alternatively, the E-cadherin junctional complex may be present but somehow defective (due to dysfunctional catenins), allowing facile dissociation of the invasive cells following stimulation by effectors like integrin–ECM interaction and/or activation by cytokines (Fig 6.2).

B. Integrin–Cadherin Cross Talk

The integrin family of receptors, as detailed earlier, is clearly important in the process of invasion. Relatively little is known, however, about the interaction and cross talk between the two different adhesion receptor families, integrins and cadherins. Several studies now indicate an important dialogue between the two receptor types. For example, a number of studies have shown that cadherins can modulate or replace integrin function. In terminal differentiation of skin keratinocytes, cadherins play a role in the downregulation of integrin expression [142]. Cadherin-mediated adhesion can substitute for integrin-mediated, anchorage-dependent growth [8]. In another study, it was reported that both integrins and cadherins regulate contact-mediated inhibition of cell migration [143]. These reports suggest the possible existence of signaling pathways between integrin receptors and cadherin-mediated cell–cell contacts. Conversely, integrins can alter cadherin function. Thus, in migrating neural crest cells, β1 and β3 integrins elicit intracellular signals that regulate the surface distribution and activity of N-cadherin [144]. von Schlippe *et al.* [145] reported that the treatment of melanoma cell monolayers with blocking monoclonal antibody to αv integrin induced E-cadherin-mediated spheroid formation. In a different study, it was found that the reexpression of β1 integrin in integrin-null epithelial cell lines induced the disruption of polarity, intercellular adhesion, and cell scattering by the downregulation of cadherin and catenin function, which appeared to involve the activation of Rac1 and RhoA [146]. This phenomenon was not solely the result of stimulated cell motility; interaction of the β1 integrin with ligand was required. In addition, this study showed that β1 integrin-null cells formed abundant adhesions, whereas cells that had reexpressed the integrin exhibited a more scattered phenotype. In another study, Dufour *et al.* [147] found that the type of cadherin (cadherin-7 or N-cadherin) expressed defined whether fibronectin would induce cell motility and dispersion. Genda *et al.* [148] found that integrin β1 and β5 activity in HCC cells could inhibit cadherin-induced cell–cell adhesion and possibly involved c-Src. Kawano *et al.* [70] showed that oral SCC cells are able to form E-cadherin-dependent multicellular aggregates when plated onto non-adherent substrates. When these spheroids were confronted with ECM substrates, cell–cell adhesion was reversed. This indicates that the breakdown of cadherin-mediated cell–cell adhesion and active cell scattering by specific ECM ligands are triggered by intracellular signals elicited by cell-surface integrins. Taken together, these results suggest that several different integrin receptors can, after ligation with ligand, trigger the disruption of junctional adhesions and induce motility and invasion.

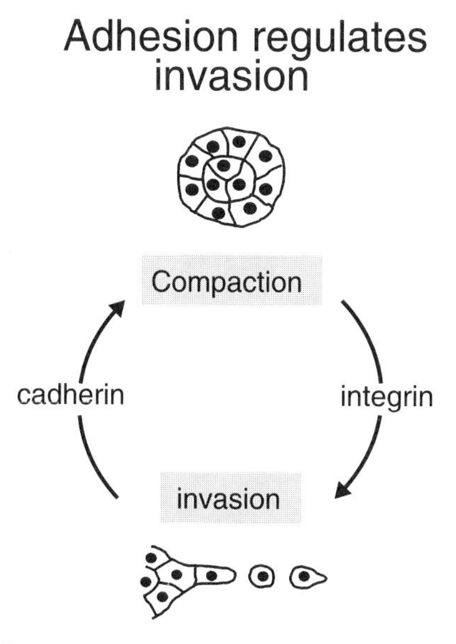

FIGURE 6.2 Adhesion receptors regulate tumor invasion. Local invasion of ECM via integrins initiates dissociation of invading cells, whereas formation of extensive cell–cell adhesions minimizes invasion and promotes a differentiated phenotype.

The mechanism by which each type of receptor regulates the other is not completely understood. Certainly the importance of the cytoskeleton in this process is appreciated in that both integrins and cadherins associate with and dynamically regulate the actin cytoskeleton. In the case of integrins, interfacing with actin is involved, with the formation of focal adhesion plaques that mediate attachment to the ECM. Similarly, cadherins form associations with actin filaments via the interaction with link proteins (catenins) during the formation of junctional adhesions such as zonula adherens [149]. Thus, one possible mechanism by which integrins and cadherins may cross talk is through their common interaction with the actin cytoskeleton [150]. Another potential mechanism may involve specific signaling pathways. For example, the adapter protein Shc, which is involved in growth factor activation by Ras, has been shown to associate with E-cadherin [151]. Additionally, cell–cell contact regulating proliferation has been linked to alterations in cyclin-dependent kinases [152,153]. Arregui *et al.* [154] showed that the nonreceptor tyrosine kinase Fer may be involved in mediating integrin–cadherin coordinate regulation. Using antennapedia fusion peptides, the authors found that sequences of the juxtamembrane domain of N-cadherin induced dysregulation of both cadherin and integrin function that involved dissociation of Fer from cadherin and transfer to integrin heterodimers.

The capacity of integrin and cadherin receptors to reversibly modulate cell phenotype is a phenomenon similar to epithelial–mesenchymal transition (EMT), a term that originally described discrete events occurring during developmental processes (e.g., gastrulation) [155]. However, various types of carcinomas frequently undergo a similar process in which differentiated epithelioid cells switch during tumor progression to a dedifferentiated fibroblastic phenotype [156,157]. EMT by carcinoma cells is an irreversible process and represents a discrete stage of tumor progression. Previous work has shown that in certain carcinomas, epithelial differentiation correlates with the level of cadherin expression [156]. Alternatively, the conversion can follow exposure of cells to certain growth factors and is not associated with an alteration in cadherin levels [158,159]. Kawano *et al.* [70] suggested a different variant of EMT whereby fibroblastoid cells can be converted to an epithelial phenotype by substrate deprivation and forced aggregation. After engagement of integrins with the ECM, the cells are restored to an epithelial monolayer as an intermediate and temporary stage, but eventually revert to their original fibroblastoid phenotype. This reversible modulation in cell behavior was associated with the regulation of E-cadherin levels.

C. Regulation of Cadherin Expression

The level of cadherin expression, and therefore important aspects of the cellular phenotype, is determined by a complicated series of regulatory events [reviewed in 160]. Mutational inactivation of E-cadherin has been identified in a number of human tumors, including gastric and breast carcinomas, and leads to loss of normal cell–cell adhesion. Also, in human tumors, transcriptional inactivation of cadherin expression can frequently occur. It is well known that E-cadherin expression can be regulated by methylation of the promoter and neighboring CpG regions, thereby causing loss of E-cadherin mRNA [161]. Cadherin levels can also be regulated by changes in the stability of the expressed protein. For example, cadherins become stabilized as they accumulate at junctional plaques, which is presumably related to their immobilization and withdrawal from the short-lived cytoplasmic pool of receptors [70]. In other systems, mature E-cadherin has been found to turn over rapidly, and its half-life has been estimated to be in the range of about 5–6 h [162]. Other studies based on morphometric analyses have suggested that cadherins may be stabilized after their incorporation into sites of cell–cell contact [161,163,164]. There are a number of ways by which cadherins could reinforce their stable association in these junctions. For example, as cadherin lattices are formed, stability could be enhanced as the cadherin–catenin complex associates with the actin cytoskeleton; additionally, the dimerization and lateral clustering of cadherins that follow junctional plaque formation may further stabilize receptor localization [165]. Several oral SCC tumor cell lines have been shown not to express E-cadherin [8,166], but the mechanism responsible for this has not been characterized. There appears to be a form of cadherin switching. For example, loss of E-cadherin can induce the upregulation of other cadherins, such as N-cadherin [167] or cadherin-11 [168]. In the case of oral SCC, N-cadherin is expressed in a subset of oral SCC cells and appears to induce altered morphology and invasion [166,169]. In different carcinomas, recent work suggests that N-cadherin regulates scatter response through the FGF-4 receptor [169,170].

V. CONCLUSIONS

Invasive squamous cell carcinoma is the most common malignant tumor of the upper aerodigestive tract. Despite advances in surgical and radiotherapeutic approaches for therapy, the overall prognosis of this cancer remains poor. These tumors are highly invasive, giving rise to both local and distant metastases. It is therefore important to understand the molecular events involved in the invasive process and the steps required for metastasis. Thus, unraveling in greater detail the complex mechanisms involved in cell–ECM and cell–cell adhesion, the common connections between the two adhesion systems, and the signaling pathways induced and regulated by each are new areas with the greatest potential for developing new treatment strategies. However, the multiplicity of adhesion receptors and the potentially diverse signaling pathways suggest that this approach will present major challenges. Powerful new techniques, such as DNA microarrays and tyrosine kinase profiling, should lead to the development of novel prognostic indicators and therapeutic techniques and should eventually advance tumor detection, staging, and treatment [171–174].

References

1. Greenlee, R. T., Murray, T., Bolden, S., and Wingo, P. (2000). Cancer statistics 2000. *CA Cancer J. Clin.* **50**, 7–33.
2. Fu, K. F., Silverman, S., and Kramer, A. M. (1998). Spread of tumor, staging, and survival. *In* "Oral Cancer" (S. Silverman, Jr., ed.), 4th Ed., pp. 67–74. Decker, Ontario.
3. Schoelch, M. L., Le, Q. T., Silverman, S., McMillan, A., Dekker, N. P., Fu, K. K., *et al.* (1999). Apoptosis-associated proteins and the development of oral squamous cell carcinoma. *Oral. Oncol.* **35**, 77–85.
4. Schoelch, M. L., Regezi, J. A., Dekker, N. P., Ng, O. L., McMillan, A., Ziober B.L., *et al.* (1999). Cell cycle proteins and the development of oral squamous cell carcinoma. *Eur. J. Cancer* **35**, 333–342.
5. Partridge, M., Emilion, G. G., Pateromichelakis, S., Phillips, E., and Langdon, J. D. (1999). The location of candidate tumour suppressor gene loci at chromosomes 3p, 8p and 9p for oral squamous cell carcinoma. *Int. J. Cancer* **83**, 318–326.
6. Frisch, S. M., and Ruoslahti, E. (1997). Integrins and anoikis. *Curr. Opin. Cell Biol.* **9**, 701–706.
7. Rheinwald, J. G., and Beckett, M. A. (1980). Defective terminal differentiation in culture as consistent and selectable character of malignant human keratinocytes. *Cell* **22**, 629–632.
8. Kantak, S., and Kramer, R. H. (1998). E-Cadherin regulates apoptosis and anchorage-independent growth in squamous cell carcinoma cells. *J. Biol. Chem.* **273**(27), 16953–16961.
9. Albelda, S. M., and Buck, C. A. (1990). Integrins and other cell adhesion molecules. *FASEB J.* **4**, 2868–2880.
10. Ginsberg, M. H., Loftus, J. C., D'Souza, S., and Plow, E. F. (1990). Ligand binding to integrins: Common and ligand specific recognition mechanisms. *Cell Differ. Dev.* **32**, 203–214.
11. Cheresh, D. A. (1992). Structural and biologic properties of integrin-mediated cell adhesion. *Clin. Lab. Med.* **12**, 217–236.
12. Bosman, F. T. (1993). Integrins: Cell adhesives and modulators of cell function. *Histochem. J.* **25**, 469–477.
13. Hynes, R. O. (1992). Integrins: Versatility, modulation, and signaling in cell adhesion. *Cell* **69**, 11–25.
14. Schwartz, M. A., and Ingber, D. E. (1994). Integrating with integrins. *Mol. Biol. Cell* **5**, 389–393.
15. Mizejewski, G. J. (1999). Role of integrins in cancer: Survey of expression patterns. *Proc. Soc. Exp. Biol. Med.* **222**,124–138.

16. Kanemoto, T., Reich, R., Royce, L., Greatorex, D., Adler, S. H., Shiraishi, N., *et al.* (1990). Identification of an amino acid sequence from the laminin A chain that stimulates metastasis and collagenase IV production. *Proc. Natl. Acad. Sci. USA* **87**, 2279–2283.
17. Mackay, A. R., Gomez, D. E., Nason, A. M., and Thorgeirsson, U. P. (1994). Studies on the effects of laminin, E-8 fragment of laminin and synthetic laminin peptides PA22-2 and YIGSR on matrix metalloproteinases and tissue inhibitor of metalloproteinase expression. *Lab. Invest.* **70**, 800–806.
18. Stack, M. S., Gray, R. D., and Pizzo, S. V. (1993). Modulation of murine B16F10 melanoma plasminogen activator production by a synthetic peptide derived from the laminin A chain. *Cancer Res.* **53**, 1998–2004.
19. Yamamura, K., Kibbey, M. C., and Kleinman, H. K. (1993). Melanoma cells selected for adhesion to laminin peptides have different malignant properties. *Cancer Res.* **53**, 423–428.
20. Bagutti, C., Speight, P. M., and Watt, F. M. (1998). Comparison of integrin, cadherin, and catenin expression in squamous cell carcinomas of the oral cavity. *J. Pathol.* **186**, 8–16.
21. Garrone, R., Lethias, C., and Guellec, D. L. (1997). Distribution of minor collagens during skin development. *Microsc. Res. Tech.* **38**, 407–412.
22. Maeshima, Y., Colorado, P. C., Torre, A., Holthaus, K. A., Grunkemeyer, J. A., Ericksen, M. B., *et al.* (2000). Distinct antitumor properties of a type IV collagen domain derived from basement membrane. *J. Biol. Chem.* **275**, 21340–21348.
23. Stenbeack, F., Maelinen, M. J., Jussila, T., Kauppila, S., Risteli, J., Talve, L., *et al.* (1999). The extracellular matrix in skin development: A morphological study. *J. Cutan. Pathol.* **26**, 327–338.
24. Hagedorn, H., Schreiner, M., Wiest, I., Teubel, J., Schleicher, E. D., and Nerlich, A. G. (1998). Defective basement membrane in laryngeal carcinomas with heterogeneous loss of distinct components. *Hum. Pathol.* **29**, 447–454.
25. Tosios, K. I., Kapranos, N., and Papanicolaou, S. I. (1998). Loss of basement membrane components laminin and type IV collagen parallels the procession of oral epithelial neoplasia. *Histopathology* **33**, 261–268.
26. Dumas, V., Kanitakis, J., Charvat, S., Euvard, S., Faure, M., and Claudy, A. (1999). Expression of basement membrane antigens and matrix metalloproteinases 2 and 9 in cutaneous basal and squamous cell carcinoma. *Anticancer Res.* **19**, 2929–2938.
27. Nerlich, A. G., Lebeau, A., Hagedorn, H. G., Sauer, U., and Schleicher, E. D. (1998). Morphological aspects of altered basement membrane metabolism in invasive carcinomas of the breast and the larynx. *Anticancer Res.* **18**, 3515–3520.
28. Kainulainen, T., Grenman, R., Oikarinen, A., Greenspan, D. S., and Salo, T. (1997). Distribution and synthesis of type VII collagen in oral squamous cell carcinoma. *J. Oral Pathol. Med.* **26**, 414–418.
29. Johansson, N., Vaalamo, M., Grénman, S., Hietanen, S., Klemi, P., Saarialho-Kere, U., and Kähäri, V. M. (1999). Collagenase-3 (MMP-13) is expressed by tumor cells in invasive vulvar squamous cell carcinomas. *Am. J. Pathol.* **154**, 469–480.
30. Kahari, V. M., and Saarialho-Kere, U. (1997). Matrix metalloproteinases in skin. *Exp. Dermatol.* **6**, 199–213.
31. Ziober, B. L., Turner, M. A., Palefsky, J. M., Banda, M. J., and Kramer, R. H. (2001). Type I collagen degradation by invasive oral squamous cell carcinoma. *Oral. Oncol.* **36**, 365–372.
32. Chan, B. M. C., and Hemler, M. E. (1993). Multiple functional forms of the integrin VLA-2 can be derived from a single α2 cDNA clone: Interconversion of forms induced by an anti-β1 antibody. *J. Cell Biol.* **120**, 537–543.
33. Ramos, D. M., Chen, B. L., Boylen, K. B., Stern, M., Kramer, R. H., Sheppard, D., *et al.* (1997). Stromal fibroblasts influence oral squamous cell carcinoma cells' interactions with tenascin-C and matrix synthesis. *Int. J. Cancer* **72**, 1–8.
34. Shinohara, K., Nakamura, S., Sasaki, M., Kurahara, S., Ikebe, T., Harada, T., *et al.* (1999). *Am. J. Clin. Pathol.* **111**, 75–88.
35. Thorp, A. K., Reibel, J., Schiedt, M., Sternersen, T. C., Therkildsen, M. H., Carter, W. G., *et al.* (1998). Can alterations in integrin and laminin-5 expression be used as markers of malignancy? *APMIS* **106**, 1170–1180.
36. Maragou, P., Bazopoulou-Kyrkanidou, E., Panotopoulou, E., Kakarantza-Angelopoulou, E., Sklavounou-Andrikopoulou, A., and Kotaridis, S. (1999). Alteration of integrin expression in oral squamous cell carcinomas. *Oral Dis.* **5**, 20–26.
37. Giancotti, F. G., and Ruoslahti, E. (1999). Integrin signaling. *Science* **285**, 1028–1032.
38. Harada, T., Shinohara, M., Nakamura, S., and Oka, M. (1994). An immunohistochemical study of the extracellular matrix in oral squamous cell carcinoma and its association with invasive and metastatic potential. *Virch. Arch.* **424**, 257–266.
39. Kosmel, H., Berndt, A., Strassburger, S., Borsi, L., Rousselle, P., Mandel, U., *et al.* (1999). Distribution of laminin and fibronectin isoforms in oral mucosa and oral squamous cell carcinoma. *Br. J. Cancer* **81**, 1071–1079.
40. Varani, J., Schuger, L., Fligiel, S. E., Inman, D. R., and Chakrabarty, S. (1991). Production of fibronectin by human tumor cells and interaction with exogenous fibronectin: Comparison of cell lines obtained from colon adenocarcinomas and squamous carcinomas of the upper aerodigestive tract. *Int. J. Cancer* **47**(3), 421–425.
41. Mighell, A. J., Thompson, J., Hume, W. J., Markham, A. F., and Robinson, P. A. (1997). RT-PCR investigation of fibronectin mRNA isoforms in malignant, normal and reactive oral mucosa. *Oral Oncol.* **33**, 155–162.
42. Kapila, Y. L., Niu, J., and Johnson, P. W. (1997). The high affinity heparin-binding domain and the V region of fibronectin mediate invasion of human oral squamous cell carcinoma cells in vitro. *J. Biol. Chem.* **272**,18932–18938.
43. Yoshimura, E., Majima, A., Sakakura, Y., Sakakura, T., and Yoshida, T. (1999). Expression of tenascin-C and the integrin alpha 9 subunit in regeneration of rat nasal mucosa after chemical injury: Involvement in migration and proliferation of epithelial cells. *Histochem. Cell Biol.* **111**, 259–264.
44. Ramos, D. M., Chen, B., Regezi, J., Zardi, L., and Pytela, R. (1998). Tenascin-C matrix assembly in oral squamous cell carcinoma. *Int. J. Cancer* **75**, 680–687.
45. Hindermann,W., Berndt, A., Borsi, L., Lou, X., Hyckel, P., Katenkamp, D., *et al.* (1999). Synthesis and protein distribution of unspliced large tenascin-C isoform in oral squamous cell carcinoma. *J. Pathol.* **189**, 475–480.
46. Vitolo, D., Ciocci, L., Ferrauti, P., Cicerone, E., Gallo, A., DeVincentiis, M., *et al.* (2000). α5 integrin distribution and TGFβ1 gene expression in supraglottic carcinoma: Their role in neoplastic local invasion and metastasis. *Head Neck* **22**, 438–456.
47. Gomez, M., and Cano, A. (1995). Expression of beta 1 integrin receptors in transformed mouse epidermal keratinocytes: Upregulation of alpha 5 beta 1 in spindle carcinoma cells. *Mol. Carcinogen.* **12**, 153–165.
48. Koivisto, L., Greman, R., Heino, J., and Larjava, H. (2000). Integrins alpha 5 beta 1, alphavbeta1, and alphavbeta6 collaborate in squamous cell spreading and migration on fibronectin. *Exp. Cell Res.* **255**, 10–17.
49. Koivisto, L., Larjava, K., Häkkinen, L., Uitto, V. J., Heino, J., and Larjava, H. (1999). Different integrins mediate cell spreading, haptotaxis and lateral migration of HaCaT keratinocytes on fibronectin. *Cell. Adhes. Commun.* **7**, 245–257
50. Jones, J., Watt, F., and Speight, P. M. (1997). Changes in the expression of alpha v integrins in oral squamous cell carcinomas. *J. Oral Pathol. Med.* **26**, 63–68.
51. Jones, J., Sugiyama, M., Giancotti, F., Speight, P. M., and Watt, F. M. (1996). Transfection of beta 4 integrin subunit into a neoplastic

keratinocyte line fails to restore terminal differentiation capacity or influence proliferation. *Cell Adhes. Commun.* **4**, 307–316.

52. Munger, J. S., Huang, X., Kawakatsu, H., Griffiths, M. J., Dalton, S. L., Wu, J., *et al.* (1999). The integrin alpha v beta 6 binds and activates latent TGF beta 1: A mechanism for regulating pulmonary inflammation and fibrosis. *Cell* **96**(3), 319–328.
53. Lange, D., Persson, U., Wollina, U., ten Dijke, P., Castelli, E., Heldin, C. H., *et al.* (1999). Expression of TGF-beta related Smad proteins in human epithelial skin tumors. *Int. J. Oncol.* **14**, 1049–1056.
54. Muro-Cacho, C. A., Anderson, M., Codero, J., and Muamoz-Antonia, T. (1999). Expression of transforming growth factor beta type II receptors in head and neck squamous cell carcinoma. *Clin. Cancer Res.* **5**, 1243–1248.
55. Colognato, H., and Yurchenco, P. D. (2000). Form and function: The laminin family heterotrimers. *Dev. Dynam.* **218**, 213–234.
56. Tanaka, H., and Kramer, R. H. (*in preparation*).
57. Nguyen, B. P., Gil, S. G., and Carter, W. G. (2000). Deposition of laminin 5 by keratinocytes regulates integrin adhesion and signaling. *J. Biol. Chem.* **275**, 31896–31907.
58. Kikkawa, Y., Sanzen, N., Fujiwara, H., Sonnenberg, A., and Sekiguchi, K. (2000). Integrin binding specificity of laminin 10/11 is recognized by α3β1, α6β1, and α6β4 integrins. *J. Cell Sci.* **113**, 869–876.
59. Rousselle, P., Lunstrum, G. P., Keene, D. R., and Burgeson, R. E. (1991). Kalinin: An epithelial-specific basement membrane adhesion molecule that is a component of anchoring filaments. *J. Cell. Biol.* **114**, 567–576.
60. Zhang, K., Kim, J. P., Woodley, D. T., Waleh, N. S., Chen, Y. Q., and Kramer, R. H. (1996). Restricted expression and function of laminin 1-binding integrins in normal and malignant oral mucosal keratinocytes. *Cell Adhes. Commun.* **4**, 159–174.
61. Pyke, C., Salo, S., Ralfkiar, E., Ramer, J., Dane, K., and Tryggvason, K. (1995). Laminin-5 is a marker of invading cancer cells in some carcinomas and is co-expressed with the receptor for urokinase plasminogen activator in budding cancer cells in colon adenocarcinomas. *Cancer Res.* **55**, 4132–4139.
62. Salo, T., Autio-Harmainen, H., Oikarinen, A., Salo, S., Tryggvason, K., and Salo, T. (1997). Altered distribution and synthesis of laminin-5 (kalinin) in oral lichen planus, epithelial dysplasias, and squamous cell carcinomas. *Br. J. Dermatol.* **136**, 331–336.
63. Ono, Y., Nakanishi, Y., Ino, Y., Niki, T., Yamada, T., Yoshimura, K., *et al.* (1999). Clinocopathologic significance of laminin-5 gamma2 chain expression in squamous cell carcinoma of the tongue: Immunohistochemical analysis of 67 lesions. *Cancer* **85**, 2315–2321.
64. Amano, S., Scott, I. C., Takahara, K., Koch, M., Gerecke, D. R., Keene, D. R., *et al.* (2000). Bone morphogenetic protein 1 (BMP-1) is an extracellular processing enzyme of the laminin 5 γ2 chain. *J. Biol. Chem.* **275**, 22728–22735.
65. Hatakeyama, S., Gao, Y. H., Ohara-Nemoto, Y., Kataoka, H., and Satoh, M. (1997). Expression of bone morphogenetic proteins of human neoplastic epithelial cells. *Biochem. Mol. Biol. Int.* **42**, 497–505.
66. Giannelli, G., Falk-Marziller, J., Schiraldi, O., Stetler-Stevenson, W. G., and Quaranta,V. (1998). Induction of cell migration by matrix metalloprotease-2 cleavage of laminin-5. *Science* **277**, 225–228.
67. Koshikawa, N., Ciannelli, G., Criulli, V., Miyazaki, K., and Quaranta, V. (2000). Role of cell surface metalloprotease MT1-MMP in epithelial cell migration over laminin-5. *J. Cell Biol.* **148**, 615–624.
68. Yoshizaki, T., Salo, H., Maruyama, Y., Murono, S., Furukawa, M., Parks, C. S., *et al.* (1997). Increased expression of membrane type 1-matrix metalloproteinase in head and neck carcinoma. *Cancer* **79**, 139–144.
69. Goldfinger, L. E., Hopkinson, S. B., deHart, G. W., Collawn, S., Couchman, J. R., and Jones, J. C. (1999). The α3 laminin subunit, α6β4 and α3β1 integrin coordinately regulate wound healing in cultured epithelial cells and in the skin. *J. Cell Sci.* **112**, 2615–2629.
70. Kawano, K., Kantak, S. S., Murai, M., Yao, C.C., and Kramer, R. H. (2001). Integrin alpha 3 beta 1 engagement disrupts intercellular adhesion. *Exp. Cell Res.* **262**, 180–196.
71. Villaret, D. B., Wang, T., Dillon, D., Xu, J., Sivam, D., Cheever, M. A., *et al.* (2000). Identification of genes overexpressed in head and neck squamous cell carcinoma using a combination of DNA subtraction and microarray analysis. *Laryngoscope* **110**, 374–381.
72. Baker, S. E., Hopkinson, S. B., Fitchman, M., Andreason, G. L., Frasier, F., Plopper, G., *et al.* (1996). Laminin-5 and hemidesmosomes; role of the alpha 3 chain subunit in hemidesmosome stability and assembly. *J. Cell Sci.* **109**, 2509–2520.
73. Mizushima, H., Takamura, H., Miyagai, Y., Kikkawa, K., Yamanaka, N., Yasumitsu, H., *et al.* (1997). Identification of integrin-dependent and independent cell adhesion domains in COOH-terminal globular region of laminin 5 α3 chain. *Cell Growth Differ.* **8**, 979–987.
74. DiPersio, C. M., Hodivala-Dilke, K. M., Jaenisch, R., Kreidberg, J. A., and Hynes, R. O. (1997). Alpha 3 beta 1 integrin is required for normal development of the epidermal basement membrane. *J. Cell Biol.* **137**(3), 729–742.
75. Carter, W. G., Kuar, P., Gil, S. G., Gahr, P. J., and Wayner, E. A. (1990). Distinct functions for integrins α3β1 in focal adhesions and α6β4/bullus antigen in a new stable anchoring contact (SAC) of keratinocytes: Relation to hemidesmosomes. *J. Cell Biol.* **111**, 3141–3154.
76. Gonzales, M., Haan, K., Baker, S. E., Fitchmun, M., Todorov, I., Weitaman, S., *et al.* (1999). A cell signal pathway involving laminin-5, alpha 3 beta 1 integrin, and mitogen-activated protein kinase can regulate epithelial cell proliferation. *Mol. Biol. Cell* **10**, 259–270.
77. Yauch, R. L., Berditchevski, F., Harler, M. B., Reichner, J., and Hemler, M. E. (1998). Highly stoichiometric, stable, and specific association of integrin alpha 3 beta 1 with CD151 provides a major link to phosphatidylinositol 4-kinase, may regulate cell migration. *Mol. Biol. Cell* **9**, 2751–2765.
78. Sugiura, T., and Berditchevski, F. (1999). Function of alpha 3 beta 1-tetraspanin protein complexes in tumor cell invasion: Evidence for the role of the complexes in production of matrix metalloproteinase 2 (MMP-2). *J. Cell Biol.* **146**, 1375–1389.
79. DiPersio, C. M., Shao, M., Di Costanzo, L. D., Kreidberg, J. A., and Hynes, R. O. (2000). Mouse keratinocytes immortalized with large T antigen acquire α3β1 integrin-dependent secretion of MMP-9/gelatinase B. *J. Cell Sci.* **113**, 2909–2921.
80. Van Waes, C. (1995). Cell adhesion and regulatory molecules involved in tumor formation, hemostasis, and wound healing. *Head Neck* **17**, 140–147.
81. Rabinovitz, I., and Mercurio, A. M. (1997). The integrin α6β4 functions in carcinoma cell migration on laminin-1 by mediating the formation and stabilization of actin-containing motility structures. *J. Cell Biol.* **139**, 1873–1884.
82. Georges-Labouesse, E., Messaddeq, N., Yehia, G., Cadalbert, L., Dierich, A., and LeMeur, M. (1996). Absence of integrin α6 leads to epidermolysis bullosa and neonatal death in mice. *Nature Genet.* **13**, 370–373.
83. Van der Neut, R., Krimpenfort, P., Calafat, J., Niessen, C. M., and Sonnenberg, A. (1996). Epithelial detachment due to absence of hemidesmosomes in integrin β4 null mice. *Nature Genet.* **13**, 366–369.
84. Van Waes, C., Surh, D. M., Cehn, Z., and Carey, T. E. (1997). Inhibition of integrin mediated cell adhesion of human head and neck squamous cell carcinoma to extracellular matrix laminin by monoclonal antibodies. *Int. J. Oncol.* **11**, 457–464.
85. Van Waes, C., Kozarsky, K. F., Warren, A. B., Kidd, L., Paugh, D., Liebert, M., *et al.* (1991). The antigen A9 associated with aggressive human squamous carcinoma is structurally and functionally similar to the newly defined integrin alpha 6 beta 4. *Cancer Res.* **51**, 2395–2402.
86. Lotus, S., Geuijen, C. A. W., Oomne, L. C. J. M., Calfat, J., Janssen, H., and Sonnenberg, A. (2000). The tetraspan moleculs CD151, a novel constituent of hemidesmosomes, associates with the integrin alpha 6

beta 4 and may regulate the spatial organization of hemidesmosomes. *J. Cell Biol.* **149**, 969–982.

87. Fung, L. F., Yeun, P. W., Wei, W., Kwan, H. S., and Tsao, S. W. (1998). Identification of genes differentially expressed in nasopharyngeal carcinomas by messenger RNA differential display. *Int. J. Oncol.* **13**, 85–89.
88. Guo, L., Degenstein, L., Dowling, J., Yu, Q. C., Wollmann, R., Perman, B., and Fuchs, E. (1995) Gene targeting of BPAG1: Abnormalities in mechanical strength and cell migration in stratified epithelia and neurologic degeneration. *Cell* **81**(2), 233–243.
89. Garzino-Demo, P., Carrozzo, M., Trusolino, L., Savia, P., Gandolfo, S., and Marchisio, P. C. (1998). Altered expression of alpha 6 integrin subunit in oral squamous cell carcinoma and oral potentially malignant lesions. *Oral Oncol.* **34**, 204–210.
90. Daemi, N., Thomasset, N., Lissitsky, J. C., Dumortier, J., Jacquier, M. F., Pourreyron, C., *et al.* (2000). Anti-beta 4 integrin antibodies enhance migratory and invasive abilities of human colon adenocarcinoma cells and their MMP-2 expression. *Int. J. Cancer* **85**, 850–858.
91. Ruest, P. J., Roy, S., Shi, E., Mernaugh, R. L., and Hanks, S. K. (2000). Phosphospecific antibodies reveal focal adhesion kinase activation loop phosphorylation in nascent and mature focal adhesions and requirement for the autophosphorylation site. *Cell Growth Differ.* **11**, 41–48.
92. Kornberg, L. J. (1998). Focal adhesion kinase and its potential involvement in tumor invasion and metastasis. *Head Neck* **20**, 745–752.
93. Almeida, E. A. C., Ilic, D., Han, Q., Haudk, C. R., Jin, F., Kawalatsu, H., *et al.* (2000). Matrix survival signaling: from fibronectin via focal adhesion kinase to c-Jun NH_2-terminal kinase. *J. Cell Biol.* **149**, 741–754.
94. Agochiya, M., Brunton, V. G., Owens, D. W., Parkinson, E. K., Paraskeva, C., Keith, W. N., *et al.* (1999). Increased dosage and amplification of the focal adhesion kinase gene in human cancer cells. *Oncogene* **18**, 5646–5653.
95. Welkoborsky, H. J., Bernauer, H. S., Riazimand, H. S., Jacob, R., Mann, W. J., and Hinni, M. L. (2000). Patterns of chromosomal aberrations in metastasizing and nonmetastasizing squamous cell carcinomas of the oropharynx and hypopharynx. *Ann. Otol. Rhinol. Laryngol.* **109**(4), 401–410.
96. Schmitz, A. A. P., Govek, E. E., Bottner, B., and van Aelst, L. (2000). Rho GTPases: Signaling, migration, and invasion. *Exp. Cell Res.* **261**, 1–12.
97. Sarita, S. K., and Burridge, K. (2000). Focal adhesions: A nexus for intracellular signaling and cytoskeletal dynamics. *Exp. Cell Res.* **261**, 25–36.
98. Braga, V. M. M., Betson, M., Li, X., and Lamarche-Vane, N. (2000). Activation of the small GTPase Rac is sufficient to disrupt cadherin-dependent cell-cell adhesion in normal human keratinocytes. *Mol. Biol. Cell* **11**, 3703-3721.
99. Mainiero, F., Murgia, C., Wary, K. K., Curatola, A. M., Pepe, A., Blumemberg, M., *et al.* (1997). The coupling of alpha 6 beta 4 integrin to Ras-Map kinase pathways mediated by Shc controls keratinocyte proliferation. *EMBO J.* **16**, 2365–2375.
100. Mainiero, F. A., Pepe, K. K., Wary, L., Spinardi, M., Mohammadi, J., Schlessinger, J., *et al.* (1995). Signal transduction by the $\alpha6\beta4$ integrin: Distinct $\beta4$ subunit sites mediate recruitment of Shc/Grb2 and associate with cytoskeleton of the hemidesmosomes. *EMBO J.* **14**, 4470–4481.
101. Rabinovitz, I., Toker, A., and Mercurio, A. M. (1999). Protein kinase C-dependent mobilization of the alpha 6 beta 4 integrin from hemidesmosomes and its association with actin rich cell protrusions drive the chemotactic migration of carcinoma cells. *J. Cell Biol.* **146**, 1147–1160.
102. O'Conner, K. L., Nguyen, B. K., and Mercurio, A. M. (2000). RhoA function in lamella formation and migration is regulated by the alpha 6 beta 4 integrin and cAMP metabolism. *J. Cell Biol.* **148**, 253–258.
103. Shaw, L. M., Rabinovitz, I., Wang, H. H., Toker, A., and Mercurio, A. M. (1997). Activation of phosphoinositide 3-OH kinase by alpha 6 beta 4 integrin promotes carcinoma invasion. *Cell* **91**, 949–960.
104. Tang, K., Nie, D., Cai, Y., and Honn, K. V. (1999). The beta 4 integrin subunit rescues A431 cells from apoptosis through a PI3K/Akt kinase signaling pathway. *Biochem. Biophys. Res. Commun.* **264**, 127–132.
105. Bachelder, R. E., Ribick, M. J., Marchetti, A., Falcioni, R., Soddu, S., Davis, K. R., *et al.* (1999). p53 inhibits alpha 6 beta 4 integrin survival signaling by promoting the caspase 3-dependent cleavage of AKT/PKB. *J. Cell Biol.* **147**, 1063–1072.
106. Pavelic, Z. P., and Gluckman, J. L. (1997). The role of p53 tumor supressor gene in human head and neck tumorigenesis. *Acta Otolaryngol. Suppl.* **527**, 21–24.
107. Jewell, K., Kapron-Bras, C., Jeevaratnam, P., and Dehar, S. (1995). Stimulating tyrosine phosphorylation of distinct proteins in response to antibody-mediated ligation and clustering of $\alpha3$ and $\alpha6$ integrins. *J. Cell Sci.* **108**, 1165–1174.
108. Keely, P. J., Rusyn, E. V., Cox, A. D., and Parise, L. V. (1999). R-Ras signals through specific integrin alpha cytoplasmic domains to promote migration and invasion of breast epithelial cells. *J. Cell Biol.* **145**, 1077–1088.
109. Wang, B., Zou, J. X., Ek-Rylander, B., and Rouslathi, E. (2000). R-Ras contains a proline-rich site that binds to SH3 domains and is required for integrin activation by R-Ras. *J. Biol. Chem.* **275**, 5222–5227.
110. Harder, K. W., Moller, N. P., Peacock, J. W., and Jirik, F. R. (1998). Protein-tyrosine phosphatase alpha regulates Src family kinases and alters cell-substratum adhesion. *J. Biol. Chem.* **273**, 31890–31900.
111. Kook, S., Shim, S. R., Choi, S. J., Ahnn, J., Kim, J. I., Eom, S. H., *et al.* (2000). Caspase-mediated cleavage of p130cas in etoposide-induced apoptotic rat-1 cells. *Mol. Biol. Cell.* **11**, 929–939.
112. Jin, F., Reynolds, A. B., Hines, M. D., Jensen, P. J., Johnson, K. R., and Wheelock, M. J. (1999). Src induces morphological changes in A431 cells that resemble epidermal differentiation through an SH3- and Ras-independent pathway. *J. Cell Sci.* **112**, 2913–2924.
113. Sakakibara, A., and Hattori, S. (2000). Chat, a Cas/HEF1-associated adaptor protein that integrates multiple signaling pathways. *J. Biol. Chem.* **275**, 6404–6410.
114. Takeichi, M. (1991). Cadherin cell adhesion receptors as a morphogenetic regulator. *Science* **251**, 1451–1455.
115. Gumbiner, B. M. (1996). Cell adhesion: The molecular basis of tissue architecture and morphogenesis. *Cell* **84**, 345–357.
116. Yap, A. S., Brieher, W.M, Pruschy, M., and Gumbiner, B. M. (1997). Lateral clustering of the adhesive ectodomain: A fundamental determinant of cadherin function. *Curr. Biol.* **7**(5), 308–315.
117. Barth, A. I. M., Näthke, I. S., and Nelson, W. J. (1997). Cadherins, catenins and APC protein: Interplay between cytoskeletal complexes and signaling pathways. *Curr. Opin. Cell Biol.* **9**, 683–690.
118. Behrens, J. (1999). Cadherins and catenins: Role in signal transduction and tumor progression. *Cancer Metast. Rev.* **18**(1), 15–30.
119. Vleminckx, K., and Kemler, R. (1999). Cadherins and tissue formation: Integrating adhesion and signaling. *Bioessays* **21**(3), 211–220.
120. Nollet, F., Berx, G., and van Roy, F. (1999) The role of the E-cadherin/catenin adhesion complex in the development and progression of cancer. *Mol. Cell. Biol. Res. Commun.* **2**, 77–85.
121. Kishida, S., Yamamoto, H., Ikeda, S., Kishida, M., Sakamoto, I., Koyama, S., *et al.* (1998). Axin, a negative regulator of the wnt signaling pathway, directly interacts with adenomatous polyposis coli and regulates the stabilization of beta-catenin. *J. Biol. Chem.* **273**(18), 10823–10826.
122. Hart, M. J., de los Santos, R., Albert, I. N., Rubinfeld, B., and Polakis, P. (1998). Downregulation of beta-catenin by human Axin and its association with the APC tumor suppressor, beta-catenin and GSK3 beta. *Curr. Biol.* **8**(10), 573–581.

123. Thoreson, M. A., Anastasiadis, P. Z., Daniel, J. M., Ireton, R. C., Wheelock, M. J., Johnson, K. R., *et al.* (2000). Selective uncoupling of p120(ctn) from E-cadherin disrupts strong adhesion. *J. Cell. Biol.* **148**(1), 189–202.
124. Anastasiadis, P. Z., and Reynolds, A. B. (2000). The p120 catenin family: Complex roles in adhesion, signaling and cancer. *J. Cell Sci.* **113**(Pt 8), 1319–1334.
125. Korinek, V., Barker, N., Morin, P. J., van Wichen, D., de Weger, R., Kinzler, K. W., *et al.* (1997). Constitutive transcriptional activation by a beta catenin-Tcf complex in APC-/- colon carcinoma. *Science* **275**, 1784–1787.
126. Morin, P. J., Sparks, A. B., Korinek, V., Barker, N., Clevers, H., Vogelstein, B., *et al.* (1997). Activation of β-catenin-tcf signaling in colon cancer mutations in β-catenin or APC. *Science* **275**, 1787–1789.
127. Roose, J., and Clevers, H. (1999). TCF transcription factors: Molecular switches in carcinogenesis. *Biochim. Biophys. Acta* **1424**(2–3), M23–M37.
128. Eastman, Q., and Grosschedl, R. (1999). Regulations of LEF-1/TCF transcription factors by Wnt and other signals. *Curr. Opin. Cell Biol.* **11**, 233–240.
129. Birchmeier, W., and Behrens, J. (1994). Cadherin expression in carcinomas, role in the formation of cell junctions and the prevention of invasiveness. *Biochem. Biophys. Acta* **1198**, 11–26.
130. Navarro, P., Gomez, M., Pizarro, A., Gamallo, C., Quintanilla, M., and Cano, A. (1991). A role for the E-cadherin cell-cell adhesion molecule during tumor progression of mouse epidermal carcinogenesis. *J. Cell. Biol.* **115**, 517–533.
131. Shimoyama, Y., Nagafuchi, A., Fujita, S., Gotoh, M., Takeichi, T., Tsukita, S., and Hirohashi, S. (1992). Cadherin dysfunction in a human cancer cell line, possible involvement of loss of α-catenin expression in reduced cell-cell adhesiveness. *Cancer Res.* **52**, 5770–5774.
132. Kadowaki, T., Shiozaki, H., Inoue, M., Tamura, S., Oka, H., Doki, Y., Iihara, K., Matsui, S., Iwazaki, T., Nagafuchi, A., Tsukita, S., Mori, T. (1994). E-cadherin and α-catenin expression in human esophageal cancer. *Cancer Res.* **54**, 291–296.
133. Matsuyoshi, N., Hamaguchi, M., Taniguchi, S., Nagafuchi, A., Tsukita, S., and Takeichi, M. (1992). Cadherin-mediated cell-cell adhesion is perturbed by v-src tyrosine phosphorylation in metastatic fibroblasts. *J. Cell Biol.* **118**, 703–714.
134. Hamaguchi, M., Matsuyoshi, N., Ohnishi, Y., Gotoh, B., Takeichi, M., and Nagai, Y. (1993) p60v-src causes tyrosine phosphorylation and inactivasion of the N-cadherin-catenin cell adhesion system. *EMBO J.* **12**, 307–314.
135. Behrens, J. (1999). Cadherins and catenins: Role in signal transduction and tumor progression. *Cancer Metast. Rev.* **18**(1), 15–30.
136. Reynolds, A. B., Daniel, J., AcCrea, P. D., Wheelock, M. J., Wu, J., and Zhang, Z. (1994). Identification of a new catenin, the tyrosine kinase substrate p120CAS associates with E-cadherin complexes. *Mol. Cell. Biol.* **14**, 8333–8342.
137. Mattijssen, V., Peters, H. M., Schalkwijk, L., Manni, J. J., van't Hof-Grootenboer, B., de Mulder, P. H., *et al.* (1993). E-cadherin expression in head and neck squamous-cell carcinoma is associated with clinical outcome. *Int. J. Cancer* **55**, 580–585.
138. Sakaki, T., Wato, M., Kaji, R., Mushimoto, K., Shirasu, R., and Tanaka, A. (1994). Correlation of E- and P-cadherin expression with differentiation grade and mode of invasion in gingival carcinoma. *Pathol. Int.* **44**, 280–286.
139. Andrews, N. A., Jones, A. S., Helliwell, T. R., and Kinsella, A. R. (1997). Expression of the E-cadherin-catenin cell adhesion complex in primary squamous cell carcinomas of the head and neck and their nodal metastases. *Br. J. Cancer* **75**, 1474–1480.
140. Bowie, G. L., Caslin, A. W., Roland, N. J., Field, J. K., Jones, A. S., and Kinsella, A. R. (1993). Expression of the cell-cell adhesion molecule E-cadherin in squamous cell carcinoma of the head and neck. *Clin. Otolaryngol.* **18**, 196–201.
141. Schipper, J. H., Frixen, U. H., Behrens, J., Unger, A., Jahnke, K., and Birchmeier, W. (1991). E-cadherin expression in squamous cell carcinomas of head and neck: Inverse correlation with tumor dedifferentiation and lymph node metastasis. *Cancer Res.* **51**, 6328–6337.
142. Hodivala, K. J., and Watt, F. M. (1994). Evidence that cadherins play a role in the downregulation of integrin expression that occurs during keratinocyte terminal differentiation. *J. Cell Biol.* **124**(4), 589–600.
143. Huttenlocher, A., Lakonishok, M., Kinder, M., Wu, S., Truong, T., Knudsen, K.A, *et al.* (1998). Integrin and cadherin synergy regulates contact inhibition of migration and motile activity. *J. Cell Biol.* **141**(2), 515–526.
144. Monier-Gavell, F., and Duband, J.-L. (1997). Cross talk between adhesion molecules, control of N-cadherin activity by intracellular signal elicited by beta 1 and beta 3 integrins in migrating neural crest cells. *J. Cell Biol.* **137**, 1666–1681.
145. von Schlippe, M., Marshall, J. F., Perry, P., Stone, M., Zhu, A. J., and Hart, I. R. (2000). Functional interaction between E-cadherin and αv-containing integrins in carcinoma cells. *J. Cell Sci.* **113**, 425–437.
146. Gimond, C., van der Flier, A., van Delft, S., Brakebusch, C., Kuikman, I., Collard, J. G., *et al.* (1999). Induction of cell scattering by expression of beta1 integrins in beta1-deficient epithelial cells requires activation of members of the rho family of GTPases and downregulation of cadherin and catenin function. *J. Cell Biol.* **147**(6), 1325–1340.
147. Dufour, S., Beauvais-Jouneau, A., Delouvee, A., and Thiery, J. P. (1999). Differential function of N-cadherin and cadherin-7 in the control of embryonic cell motility. *J. Cell Biol.* **146**(2), 501–516.
148. Genda, T., Sakamoto, M., Ichida, T., Asakura, H., and Hirohashi, S. (2000). Loss of cell-cell contact is induced by integrin-mediated cell-substratum adhesion in highly-motile and highly-metastatic hepatocellular carcinoma cells. *Lab. Invest.* **80**(3), 387–394.
149. Knudsen, K. A., Soler, A. P., Johnson, K. R., and Wheelock, M. J. (1995) Interaction of alpha-actinin with the cadherin/catenin cell-cell adhesion complex via alpha-catenin. *J. Cell Biol.* **130**, 67–77.
150. Yamada, K. M., and Geiger, B. (1997). Molecular interactions in cell adhesion complexes. *Curr. Opin. Cell Biol.* **9**, 76-85.
151. Xu, Y., Guo, D. F., Davidson, M., Inagami, T., and Carpenter, G. (1997). Interaction of the adaptor protein Shc and the adhesion molecule cadherin. *J. Biol. Chem.* **272**, 13463–13466.
152. St. Croix, B., Florenes, V. A., Rak, J. W., Flanagan, M., Bhattacharya, N., Slingerland, J. M., *et al.* (1996). Impact of the cyclin-dependent kinase inhibitor p27Kip1 on resistance of tumor cells to anticancer agents. *Nature Med.* **2**, 1204–1210.
153. St. Croix, B., and Kerbel, R. S. (1997). Cell adhesion and drug resistance in cancer. *Curr. Opin. Oncol.* **9**, 549–556.
154. Arregui, C., Pathre, P., Lilien, J., and Balsamo, J. (2000). The nonreceptor tyrosine kinase fer mediates cross-talk between N-cadherin and beta1-integrins. *J. Cell Biol.* **149**(6), 1263–1274.
155. Hay, E. D. (1995). An overview of epithelio-mesenchymal transformation. *Acta Anat.* **154**(1), 8–20.
156. Birchmeier, C., Birchmeier, W., and Brand-Saberi, B. (1996). Epithelial-mesenchymal transitions in cancer progression. *Acta Anat.* **156**(3), 217–226.
157. Perl, A. K., Wilgenbus, P., Dahl, U., Semb, H., and Christofori, G. (1998). A causal role for E-cadherin in the transition from adenoma to carcinoma. *Nature* **392**, 190–193.
158. Matsumoto, K., Matsumoto, K., Nakamura, T., and Kramer, R. H. (1994). Hepatocyte growth factor/scatter factor induces tyrosine phosphorylation of focal adhesion kinase (p125 FAK) and promotes migration and invasion by oral squamous cell carcinoma cells. *J. Biol. Chem.* **269**, 31807–31813.
159. Jouanneau, J., Tucker, G. C., Boyer, B., Valles, A. M., and Thiery, J. P. (1991). Epithelial cell plasticity in neoplasia. *Cancer Cells* **3**(12), 525–529.

160. Gumbiner, B. M. (2000). Regulation of cadherin adhesive activity. *J. Cell Biol.* **148**(3), 399–403.
161. Chang, H. W., Chow, V., Lam, K. Y., Wei, W. I., and Yuen, A. (2002). Loss of E-cadherin expression resulting from promoter hypermethylation in oral tongue carcinoma and its prognostic significance. *Cancer*, **94**, 386–392.
162. Troxell, M. L., Chen, Y.-T., Cobb, N., Nelson, W.J, and Marrs, J. A. (1999). Cadherin function in junctional complex rearrangement and posttranslational control of cadherin expression. *Am. J. Physiol.* **276**(45), C404–C418.
163. Adams, C. L., Nelson, W. J., and Smith, S. J. (1996). Quantitative analysis of cadherin-catenin-actin reorganization during development of cell-cell adhesion. *J. Cell Biol.* **135**, 1899–1911.
164. Angres, B., Barth A., and Nelson, W. J. (1996). Mechanism for transition from initial to stable cell-cell adhension, kinetic analysis of E-cadherin-mediated adhesion using a quantitative adhension assay. *J. Cell Biol.* **134**, 549–557.
165. Yap, A. S., Brieher, W. M., and Gumbiner, B. M. (1997). Molecular and functional analysis of cadherin-based adherens junctions. *Annu. Rev. Cell Dev. Biol.* **13**, 119–146.
166. Islam, S., Carey, T., Wolf, G., Wheelock, M., and Johnson, K. (1996). Expression of N-cadherin by human squamous carcinoma cells induces a scattered fibroblastic phenotype with disrupted cell-cell adhesion. *J. Cell Biol.* **135**, 1643–1654.
167. Pishvaian, M. J., Feltes, C. M., Thompson, P., Bussemakers, M. J., Schalken, J. A., and Byers, S. W. (1999). Cadherin-11 is expressed in invasive breast cancer cell lines. *Cancer Res.* **59**(4), 947–952.
168. Giroldi, L. A., Bringuier, P. P., Shimazui, T., Jansen, K., and Schalken, J. A. (1999). Changes in cadherin-catenin complexes in the progression of human bladder carcinoma. *Int. J. Cancer* **82**(1), 70–76.
169. Nieman, M. T., Prudoff, R. S., Johnson, K. R., and Wheelock, M. J. (1999). N-cadherin promotes motility in human breast cancer cells regardless of their E-cadherin expression. *J. Cell Biol.* **147**, 631–644.
170. Hazan, R. B., Phillips, G. R., Qiao, R. F., Norton, L., and Aaronson, S. A. (2000). Exogenous expression of N-cadherin in breast cancer cells induces cell migration, invasion, and metastasis. *J. Cell Biol.* **148**(4), 779–790.
171. Welford, S. M., Gregg, J., Chen, E., Garrison, D., Sorensen, P. H., Denny, C. T., *et al.* (1998). Detection of differentially expressed genes in primary tumor tissues using representational differences analysis coupled to microarray hybridization. *Nucleic Acids Res.* **26**, 3059–3065.
172. Zuber, J., Tchernitsa, O. I., Hinzmann, B., Schmitz, A. C., Grips, M., Hellriegel, M., *et al.* (2000). A genome-wide survey of RAS transformation targets. *Nature Genet.* **24**, 144–152.
173. Peale, F. V., Mason, K., Hunter, A. W., and Bothwell, M. (1998). Multiplex display polymerase chain reaction amplifies and resolves related sequences sharing a single moderately conserved domain. *Anal. Biochem.* **256**, 158–168.
174. Robinson, D, Hej, F., Pretlow, T., and Kung, H. J. (1996). A tyrosine kinase profile of prostate carcinoma. *Proc. Natl. Acad. Sci. USA* **93**, 5958–5962.

CHAPTER

7

Angiogenesis, Basic Mechanisms, and Role in Head and Neck Squamous Cell Carcinoma

ELENA TASSI and ANTON WELLSTEIN

Lombardi Cancer Center
Georgetown University
Washington, DC 20007

I. Epidemiology, Pathology, and Biology of Head and Neck Squamous Cell Carcinoma (HNSCC) 81
II. Basic Mechanisms of Angiogenesis 82
A. General Concepts 82
B. Mechanisms of Tumor Angiogenesis 83
C. Tumor Angiogenesis: The Angiogenic Switch 83
III. Angiogenesis in HNSCC 83
A. Clinical Evidences 83
B. Angiogenic Factors in HNSCC 85
IV. Antiangiogenic Therapies in HNSCC 91
V. Conclusions 92
References 92

In the past few years, tumor angiogenesis has been correlated to the progression and metastasis of different human tumors. Several studies have been performed in the attempt to define how angiogenesis contributes to the progression of head and neck squamous cell carcinoma (HNSCC). This review gives further insights into the understanding of the correlation between angiogenesis and HNSCC tumorigenesis. Analysis of the general mechanisms of tumor angiogenesis, clinical evidences, and the function of distinct proangiogenic factors involved in HNSCC tumor progression, in addition to the current antiangiogenic therapies, are discussed.

I. EPIDEMIOLOGY, PATHOLOGY, AND BIOLOGY OF HEAD AND NECK SQUAMOUS CELL CARCINOMA (HNSCC)

Head and neck cancers encompass a diverse group of uncommon tumors that are frequently aggressive in their biological behavior. The vast majority (95%) of head and neck tumors are squamous cell carcinomas (HNSCC), the most common type of cancer affecting the lining of the airways and upper digestive organs. HNSCC comprises a wide variety of epithelial malignant lesions affecting the nasal cavity and paranasal sinuses, nasopharynx, oral cavity, lips, alveolar ridge and retromolar trigone, floor of the mouth, tongue, hard and soft palat, tonsil, pharyngeal wall, larynx, hypopharynx, and salivary glands. Cancer in these locations originates from the cuboidal cells along the basement membrane of the mucosa and usually has profound impairing effects on breathing, speaking, and swallowing. The disease is characterized by local tumor aggressiveness, early recurrence, and high frequency of second primary tumors. Neoplasias of the head and neck region are relatively infrequent in comparison with cancer occurring in the breast, lung, prostate, and colon and represent approximately 5.6% of all tumors. However, nearly 46,000 new cases of head and neck cancer are diagnosed every year in the United States. It is more frequent in men, by a ratio of 5:1, but an increasing percentage is occurring in women. Moreover, the incidence increases with age, especially after the fifth or sixth decade. The occurrence of HNSCC correlates most closely with the use of tobacco. Alcohol consumption, by itself, has been proved to be a risk factor for the development of pharyngeal and laryngeal tumors, although is a less potent agent than tobacco. In addition, the abuse of both alcohol and tobacco appears to have a synergistic effect for HNSCC development and results in a multiplicative increase in risk. Other risk factors, such as ultraviolet light, radiation exposure, and viruses (Epstein–Barr virus, papilloma virus, herpes simplex virus), are associated with development of the HNSCC. Tumorigenesis can be viewed as results of a complex "yin-and-yang" balance of

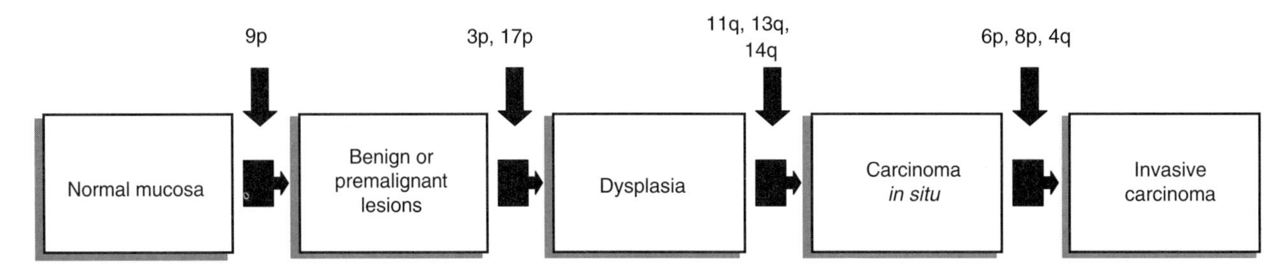

FIGURE 7.1 Genetic progression model for HNSCC. Genetic chromosomal aberrations during the progressive development of head and neck tumors have been determined by analysis of various premalignant and malignant HNSCC lesions. Adapted from Califano *et al.*, *Cancer Res.* **56**, 2488–2492 (1996).

tightly regulated oncogenes and tumor suppressor genes [1]. In this regard, cytogenetic and molecular studies have demonstrated that somatic mutations that activate oncogenes (e.g., *Ras*, *Myc*, *ErbB2*, *EGFR*, *bcl2*, *int-2*, *hst-1*, *ems-1*, *cyclinDI*), as well as point mutations, deletion, or hypermethylation that lead to tumor suppressor genes inactivation (e.g., *p16*, *TP53*, *PTEN*, *Rb*), are involved in and account for the development and progression of HNSCC [2–4].

Int-2 (*FGF3*) and *hst-1* (*FGF4*) oncogenes are fibroblast growth factor (FGF) family members and have been shown to be involved in the induction of cell proliferation and angiogenesis [5–8]. *Int-2* and *hst-1* have been found to be amplified and coamplified in 30–52% of HNSCC [9]. However, because they are not expressed in a significant percentage of HNSCC [10], their function in this disease still remains unclear. Analysis conducted on HNSCC specimens reported a consistent loss of hetero-zygosity (LOH) clustered around several chromosomal bands or regions, particularly 1p11-12, 1p22, 1q25, 3p11-q11, 5q10, 5q13, 8q10, 11q13, 10q23, and 17p31 [11,12]. Moreover, deletions at 3p13-p24, 5q12-q23, 8p22-p23, 9p21-p24, and 18q22-q23 [13] and LOH on chromosome arms 5q and 19p [14] frequently occur in HNSCC.

The LOH analysis of 87 head and neck tumors, including benign and premalignant lesions, led Califano *et al.* [15] to propose a preliminary progression model for tumorigenesis in HNSCC (Fig. 7.1). In accordance with this model, loss of chromosome 9p appears to be the earliest event in benign and preinvasive lesions, whereas 3p and 17p (p53 locus) deletions are observed in the dyplastic phase. Carcinoma *in situ* is the result of alterations of p16, p53 genes, and loss of 11q, 13q, and 14q and invasive lesions occur on 6p, 8p, and 4p chromosomal region loss. This review focuses on the tumor/stromal interaction evidenced as tumor angiogenesis.

II. BASIC MECHANISMS OF ANGIOGENESIS

A. General Concepts

Classically, angiogenesis is a process by which endothelial sprouts grew from preexisting postcapillary vessels [16]. A more complex view considers angiogenesis as a mechanism promoting the growth and the remodeling of a capillary network [17,18]. In the adult organism, most vasculature is quiescent and angiogenesis is limited generally to conditions of tissue repair and remodeling (menstruation and mammary gland involution), otherwise largely controlled by pathological conditions, such as wound healing, inflammation, and neoplastic growth [18–20]. Angiogenesis is initiated in response to specific stimuli, when the quiescent vasculature can become activated to grow new capillaries through a complex multistep process. The first step starts with vasodilatation, followed by the increase of vessel permeability, which leads to the formation of vascular fenestrations, vesiculovacuolar organelles, and redistribution of platelet endothelial cell adhesion molecule (PECAM)-1 and vascular endothelial (VE)-cadherin. Destabilization of the mature vessel allows endothelial cells to loosen interendothelial cell contacts and periendothelial cell support. In addition, proteolytic modification and degradation of the basement membrane surrounding a preexisting vessel driven by proteinases of the plasminogen activator, matrix metalloproteinases, collagenases, chymase, and heparanases, allow chemotactic endothelial cell migration into the surrounding stroma [21]. Dissolution of the extracellular matrix also results in the activation and release of a handful of angiogenic growth factors, such as basic fibro-blast growth factor (bFGF or FGF-2), acidic fibroblast growth factor (aFGF or FGF-1), and vascular endothelial growth factor (VEGF) that induce endothelial cell proliferation [18]. Consequently, endothelial cells converge and tightly assemble to give origin to a new lumen or they intercalate with endothelial cells in a preexisting vessel in a resulting capillary elongation. In the new lumen, endothelial cells undergo differentiation, thus acquiring specialized morphological characteristics that depend on the host tissue in which they localize [22]. The emerging vascular plexus undergoes final remodeling, whose molecular mechanisms are still undefined. In this last process, called "pruning," the newly formed capillary-like vessels are rearranged from an irregular into a structured network of branching vessels and capillary loops.

B. Mechanisms of Tumor Angiogenesis

In neoplastic tissues, endothelial cells present various morphological and physiological abnormalities and give origin to anomalous vasculature, characterized by tortuous, elongated, and dilated vessels, atypical enlargement of the vascular lumen, as well as excessive branching. The newly formed vessels are randomly fused either with arterioles or venules and create an atypical microcirculation. As a result, neoplastic regions are often hypoxic and acidic due to the chaotic and slow tumor blood flow [17,20,23–25]. In addition, tumor vessels are leaky, as they present endothelial fenestrae, vesicles and transcellular holes, widened interendothelial junctions, and a discontinuous or absent basement membrane [26–28]. Tumor vessels may also lack functional perivascular cells [29] and their walls can be formed by a mosaic of alternating endothelial and cancer cells (vasculogenic mimicry) [30]. Some of the original host vessels in the tumor disintegrate, are obstructed, or compressed. Of the remaining vessels, arteries seem to become permanently dilated and resistant to the invasive and destructive growth of tumor cells.

C. Tumor Angiogenesis: The Angiogenic Switch

Nearly a century ago Goldman first speculated that angiogenesis was required for tumors growth *in vivo* [31] and other following reports started confirming this hypothesis [32]. In 1968, two independent groups presented evidence that tumors could secrete a diffusible "angiogenic" factor, thus including an angiogenic response in a host tissue bed separated from the blood supply by filter membranes [33,34]. In 1971, Folkman stated several pioneering hypotheses regarding the requirement of new blood vessel formation for tumor growth and metastasis [35,36]. He proposed that most primary solid tumors, 1–2 mm^3 in size, probably originate as a vascular structures and go through a prolonged, and apparently dormant, growth due to a balance between mitogenesis and apoptosis. At these dimensions, tumor cells can still receive oxygen and nutrients by passive diffusion from blood vessels and survival is possible without a massive blood vessel supply. The duration of this prevascular phase may extend for years. These microscopic tumor masses can eventually activate an angiogenic process, the so-called "angiogenic switch," by inducing the sprouting of new capillaries from surrounding mature blood vessels, which will infiltrate the tumor mass, thereby potentially initiating the increase in the size of the tumor mass and its metastatic perfusion. The hypothesis of ectopic secretion from tumor cells of a proangiogenic factor [37,38] would account for the positive regulation of the angiogenic switch mechanism. It is now widely accepted that the "angiogenic switch" is "off" when the effect of proangiogenic molecules is counterweighted by that of antiangiogenic molecules, and it is "on" when the concentration of angiogenic inducers prevail on that of the angiogenic inhibitors.

Evidence that the angiogenic switch turns "on" during early stages preceding the appearance of solid tumor derives from studies conducted with three diverse transgenic mouse models [39–43] and with preneoplastic lesions associated to mammary duct [44–46] and uterine cervix [47,48] tumors. Thus activation of the angiogenic switch during early stages of tumor development suggests that the regulation of the angiogenic process is likely to be a rate-limiting step in the progression of numerous solid tumors [49].

In addition to this widely accepted angiogenic theory, a new angiogenic model of tumor growth has been suggested [50–52]. It is likely that many tumors, and metastases in particular, would develop without vascularization, especially when they grow within or metastasize toward vascularized tissues [53–55]. In these conditions, tumor cells appear to grow by coopting preexisting host vessels and initiate their development as well-vascularized small tumors [51].

III. ANGIOGENESIS IN HNSCC

A. Clinical Evidences

The requirement of solid tumors to stimulate an angiogenic response in the host has been shown to be a prerequisite for the development of tumor growth, invasion, and metastasis. Patients with HNSCC frequently are affected with tumor masses of several centimeters in size. Given the fact that only solid tumors as little as 1–2 mm^3 in size can grow in an avascular environment, several hypotheses have been raised regarding the capability of HNSCCs to initiate an angiogenic response.

The first experimental indication supporting a correlation between increased angiogenesis and tumor growth derives from the analysis of cutaneous metastasizing melanomas [56,57]. In addition, in 1991, Weidner *et al.* [58] reported that the density of capillary microvessels was elevated significantly in invasive breast carcinoma, thus being a closer predictor of metastasis than tumor size or grade. Several subsequent experimental studies and clinicopathological reports further confirmed existence of a tumor/blood vessel relationship in various types of solid tumors [46,59,60], such as nonsmall cell lung [61], prostate [62], testicular [63], colorectal, and gastric carcinomas [64], including squamous cell carcinoma of the head and neck.

The first evidence corroborating the induction of blood vessel formation in HNSCC arose from a study conducted by Petruzzelli *et al.* [65], in which specimens from HNSCC-affected patients were grafted onto the chick chorioallantoic membrane (CAM) of chick embryos. This resulted in a significant augmentation of the angiogenic response in the xenografts compared to normal control tissues (Fig. 7.2).

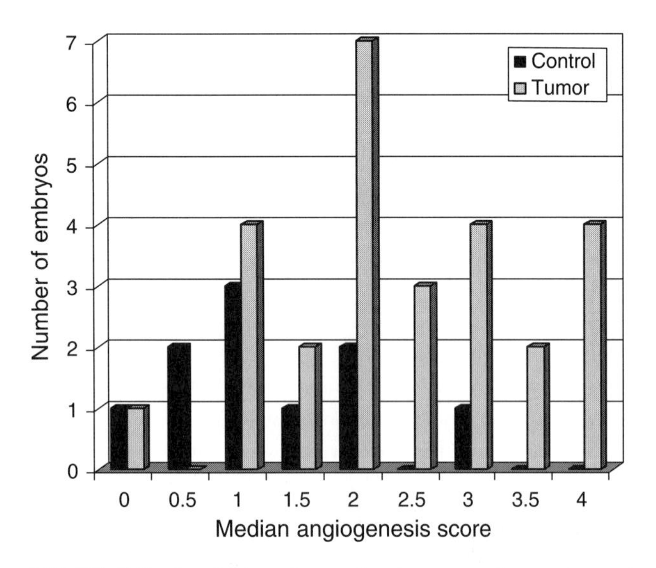

FIGURE 7.2 Distribution of control and tumor explant angiogenesis scores. Mann–Whitney rank sum analysis of median angiogenesis scores demonstrated significantly increased neovascularization induced by tumor explants ($p=0.01$). Adapted from Petruzzelli *et al.*, *Ann. Otol. Rhinol. Laryngol.* **102**, 215–221 (1993).

Moreover, the radial outgrowth ("spoke wheel") pattern of CAM vessels was identified in the tumor nodules and was lacking in the nontumor explants.

Is angiogenesis an effective marker in the progression of HNSCC? To date, several controversial observations emerged from various studies attempting to correlate the degree of neovascularization with prognosis, regional occurrence, and metastasis in the squamous cell carcinoma of the head and neck. During the past decades, several reports have been published evaluating the extent of microvessel growth in human neoplastic tissues derived from HNSCC. In general, microvessels are identified by immunohistochemical staining against specific markers for endothelial cells, such as von Willebrand factor, factor VIII, CD31, CD34, CD36, *Ulex europaeus* I agglutinin, or ABH blood group isoantigens [66–68]. Immunoreactive areas are then evalu-ated by determining the intensity of the staining using a 1 to 4+ grading range or by microvessel count per high-power field.

In 1988, Delides *et al.* [69] by analyzing 25 nasopharyngeal tumors histologically, observed the existence of a correlation between angiogenesis and response to irradiation and survival, thus proposing that tumors with a greater count of microvessels are likely to present a more effective therapeutic response [69]. Analogous findings have been reported in a study conducted by Zatterstrom *et al.* [70] in which 48 heterogeneous early and late stage HNSCC biopsies were examined by immunostaining with factor VIII monoclonal antiobodies. The significance of the tumor neovascularization in relation of clinical outcome in response to radiotherapy was confirmed (Table 7.1). Nonetheless, contradictory findings were reported in another study in which tumors from 42

TABLE 7.1 Response to Radiotherapy in Relation to Microvessel Density Score above versus below 4.75[a]

	Response		
Microvessel density score	Complete	Partial/no	Total
<4.75	6	7	13
>4.75	29	6	35

[a]Adapted from Zatterstrom *et al.*, *Head Neck* **17(4)**, 312–318 (1995).

patients affected by HNSCC in various areas of the aerodigestive tract were examined [71]. The authors were not able to find a statistically significant difference between rate of angiogenesis in the tumors with local recurrence and rate of metastasis. However, analysis of a small number of cases, as well as the heterogeneity of the origin sites of the malignancies, may account for the discrepancy of these results.

In 1993, Gasparini and colleagues [72] assessed that 70 patients with locally advanced HNSCC, presenting locoregional and distance metastases, displayed a significantly higher capillary density than those without metastatic diffusion. Increased angiogenesis was also correlated with relapse and metastasis in laryngeal squamous cell carcinoma [73]. Likewise, increased microvessel concentration was also found in premetastatic specimens from 66 patients diagnosed with squamous cell cancer of the oral cavity, thus suggesting that angiogenesis is likely to be an important independent predictor of nodal metastasis [74] (Fig. 7.3 and Table 7.2). However, opposing conclusions arose from a report by Leedy *et al.* [75], in which no correlation between tumor vascularity and nodal metastasis was determined in 57 patients affected by early (T1 and T2) tongue squamous cell carcinoma (Table 7.3). However, in an attempt to compare microvessel density in 33 late stage (T3 and T4) and metastatic versus early stage (T1 and T2) and nonmetastatic oral squamous cell carcinoma (OSCC) specimens, Alcalde and colleagues [76] observed a higher level of angiogenesis in the most advanced and invasive areas of the tumors. Likewise, another study confirmed the correlation between the count of microvessel to lymph node metastasis in 41 primary OSCC [77]. In contrast, two independent studies on T1 and T2–T4 OSCC, respectively, were performed by Gleich *et al.* [78,79]. In both analyses, the lack of a representative correlation between angiogenesis and either tumor aggressiveness or prognosis was detected, thereby concluding that oral tumor development may proceed independently from angiogenesis. However, it was also shown that the level of vascularity in oral carcinomas was significantly higher than that detected in normal oral tissue. Interestingly, a progressive increase in vascularity from normal to dysplastic to neoplastic tissues was found, thus suggesting that angiogenesis may be an early, rather than late, step in oral tumor progression [80] (Fig. 7.4).

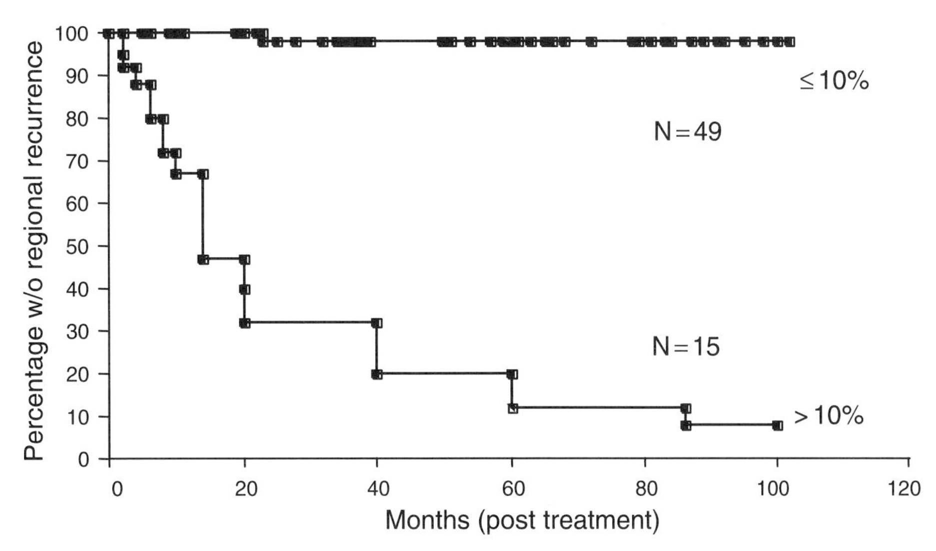

FIGURE 7.3 Regional recurrence and probability of patient being disease free within a given percentage staining group at time in months. At 30 months, the low-staining group will have a 98% probability of being disease free, decreasing to 30% in the high-staining group. Modified from Williams *et al.*, *Am. J. Surg.* **168**, 373–380 (1994).

In an attempt to correlate the number of microvessels with early recurrence or metastasis within 10 randomly selected late stage malignancies, Albo and colleagues [81] found an elevated angiogenic rate in the tumor masses, although a much greater microvessel count was detected consistently in the tissues immediately surrounding the neoplastic areas (Fig. 7.5). In addition, associated with the prominent vessel density, abundant inflammatory cells (macrophages and mast cells) were found in the peritu-moral tissues. Indeed, several studies promote the hypothesis that inflammatory cells may migrate to neoplastic areas, chemotactically attracted by tumor cells, thereby secreting proangiogenic factors [82]. Further conflicting data emerged from a report by Dray *et al.* [83] in which 108 patients with primary early stage HNSCC did not exhibit increased microvessel density in the neoplastic samples.

In light of these findings, angiogenesis indeed cannot conclusively be considered an independent prognostic marker in head and neck squamous cell carcinoma. However, it is worth taking into consideration that the intrinsic high vascular density of the head and neck region and the examination of a limited neoplastic area compared with the entire tumor, as well as flaws and limitations in the experimental techniques and patient recruitment, may account for the contradictory results of these studies.

B. Angiogenic Factors in HNSCC

The observation that tumors could be implanted either into an avascularized region, such as retina [84–86], or on a vascularized surface, such as the chick chorioallantoic membrane [82], and in both cases initiate the growth of new capillaries structures, implied the possibility that tumors could secrete diffusible factors responsible for the activation of an angiogenic process. As a consequence, angiogenic activators function by eliciting the sprouting of capillaries from dormant vasculature. Angiogenic factors are polypeptide growth factors whose pleiotropic effects have been shown to regulate tumor growth and development. *In vivo*, these factors may potentially act in an autocrine or paracrine manner in order to promote tumor formation. In addition, they are responsible for the regulation of secondary events in tumorigenesis, such as angiogenesis and immunomodulation.

TABLE 7.2 Correlation between Angiogenesis (above versus below 10%) and Regional Recurrence[a]

Angiogenesis (%)	No. of patients (%)	Regional recurrence (%)	*p* value
<10	49 (77)	1 (2)	<0.0001
>10	15 (23)	14 (93)	<0.0001

[a]Adapted from Williams *et al., Am. J. Surg.* **168**(5), 373–380 (1994).

TABLE 7.3 Comparison of Vascularity in Patients with Neck Metastasis with Those Without[a]

	Metastasis (+)	Metastasis (–)	*p* value
Vascularity grade	2.48	2.59	0.65
Vessels/x 200 field	59.8	61.5	0.8
Vessels/x 400 field	20.2	21.4	0.68

[a]Modified from Leedy *et al., Otol. Head Neck Surg.* **111**(4), 417–422 (1994).

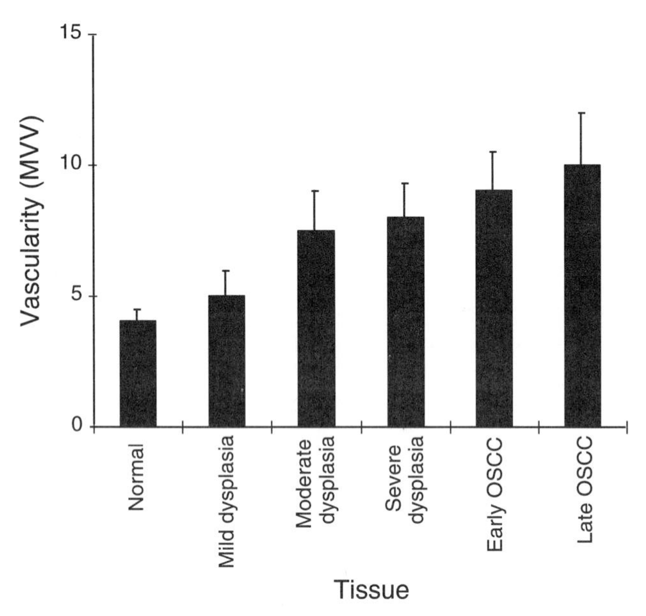

FIGURE 7.4 Changes in vascularity with disease progression. Vascularity was measured as microvascular volume (MVV). Adapted from Pazouki *et al*., *J. Pathol*. **183**, 39–43 (1997).

The application of experimental procedures, such as proliferation and chemotaxis assays of endothelial cells, allowed several groups of researchers to identify an increasing number of proangiogenic factors [49], which have been shown to modulate angiogenesis *in vivo* and *in vitro*. Among these, FGFs and VEGF are the most potent inducers of angiogenesis [87–92]. The angiogenic factors responsible for the progression of the squamous cell carcinoma of the head and neck have not been as well described as for other types of cancer [93]. Although an increasing number of reports have confirmed the correlation of angiogenesis with HNSCC development [65,72,74,76,80,81], how the angiogenic factors regulate the distinct steps of the angiogenic process in this type of cancer is not fully understood [93–95].

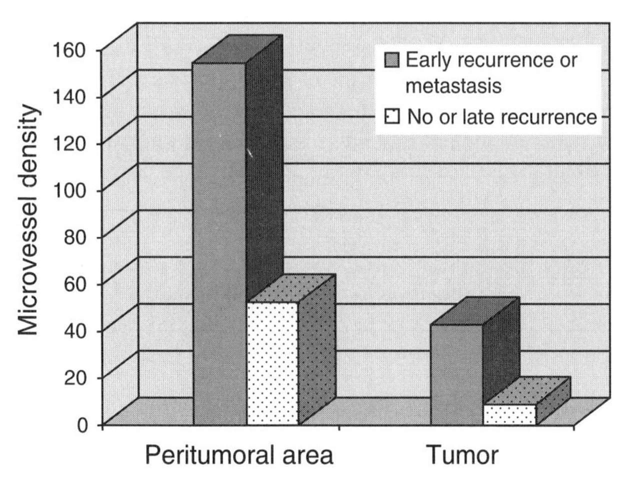

FIGURE 7.5 Microvessel count and clinical outcome in the tumor and in the tissue peripheral to the tumor. Adapted from Albo *et al*., *Ann. Plast. Surg*. **32**, 588–594 (1994).

Several proangiogenic factors and cytokines have been shown to be secreted by HNSCC cell lines, such as bFGF, insulin-like growth factor II (IGF-II), platelet-derived endothelial cell growth factor (PD-ECGF), transforming growth factor (TGF)-α [96,97], prostaglandin 2 (PGE_2), transforming growth factor (TGF)-β, and VEGF [95]. As a consequence, it has been demonstrated that supernatants from HNSCC cultured cell lines displayed stimulatory effects on endothelial cell growth into microvessel-like structures and on their motility in various *in vitro* models [94,95] (Fig. 7.6).

1. Basic Fibroblast Growth Factor

Basic fibroblast growth factor (bFGF or FGF-2) is one of 20 distinct members of the FGF family widely expressed in normal [98,99] and malignant [100–104] tissues, including squamous cell carcinoma of the head and neck [105]. bFGF is an 18-kDa protein shown to have biological effects in different cells and organ systems, including tumorigenesis and angiogenesis [106–108]. The biological response to bFGF is modulated by interaction with low-affinity cell surface and extracellular matrix heparan sulfate proteoglycans (HSPGs), which enable the growth factor to bind and activate its high-affinity tyrosine kinase receptors (FGFR1 and FGFR2), thereby forming a trimolecular active complex [109–112]. bFGF lacks the classic leader sequence, which targets intracellular proteins for secretion to the extracellular environment. Several reports indicate that bFGF release occurs via ER- and Golgi-independent passive processes, such as cell death, wounding, and chemical injury [87,113,114]. Nevertheless, other secretory mechanisms that account for bFGF release might exist. Digestion of HSPGs by heparinases or by glycosaminoglycan-degrading enzymes is an additional established mechanism for the solubilization of bFGF from the extracellular matrix [115–117]. It has been

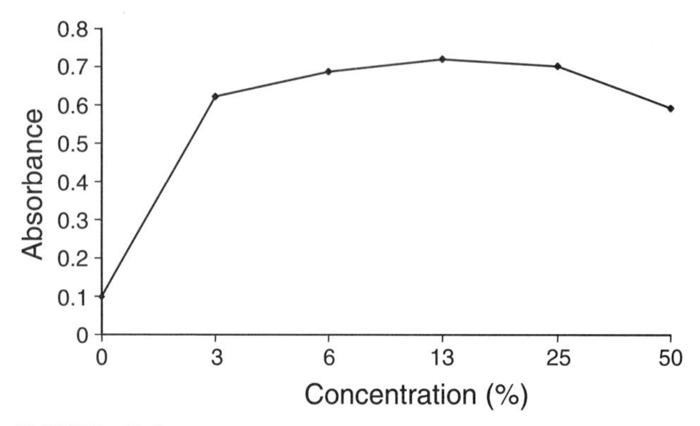

FIGURE 7.6 Head and neck squamous cell carcinoma-induced endothelial cell proliferation: dose response. TMB assay of endothelial cell proliferation in response to increasing concentrations of tumor supernatants. Adapted from Petruzzelli *et al*., *Head Neck* **19**, 57682 (1997).

proposed that the binding of bFGF to a secreted bFGF-binding protein, designated FGF-BP1 [118], might represent a novel mechanism for bFGF release from the extracellular matrix [119,120]. FGF-BPI has been shown to be upregulated in adult skin during early stages of carcinogenesis, as well as in some colon carcinoma and squamous cell carcinoma cell lines [120,121]. Moreover, the depletion of FGF-BP mRNA levels in ribozyme-targeted squamous cell carcinoma (SCC) and colon adenocarcinoma cell lines resulted in a significant reduction of tumor growth and angiogenic rate, thereby suggesting the role of FGF-BP as an angiogenic switch molecule in human tumors [121. Further studies will be needed to elucidate the role of FGF-BP1 in the progression of HNSCC. The release of bFGF *in vivo* may influence solid tumor growth and neovascularization by an autocrine [122–126] and paracrine [127] mode of action. Accordingly, neutralizing anti-bFGF antibodies affect tumor growth under defined experimental conditions [121].

bFGF has been shown to exert motogenic and mitogenic effects in fibroblasts and endothelial cells, and also to be a powerful inducer of angiogenesis *in vivo* [128,129]. The binding of bFGF to the cell surface receptor induces receptor tyrosine-kinase dimerization and autophosphorylation [130]. The phosphorylated FGFRs associate and subsequently activate SH2 domain-containing downstream signaling molecules, such as PLCγ [131,132] and Src [133,134]. Moreover, on ligand-dependent receptor autophosphorylation, adaptor proteins, such as Grb2 and Shc, link the FGFRs to the Ras/MAP kinase signaling cascade [135]. Grb2 and Shc form a complex with the GDP/GTP exchange factor Son of Sevenless (Sos) that results in the translocation of Ras to the plasma membrane and its further activation by the exchange of GDP for GTP by Sos. Thus, activated Ras leads to the consecutive activation of a cascade of protein kinases involving Raf, MEK, and p44/42^{MAPK}, also known as extracellular signal-regulated kinase 1 and 2 (ERK-1 and ERK-2) [135]. Furthermore, bFGF-dependent phosphorylation of FRS-2 (FGF receptor substrate 2), a membrane associated protein, has been shown to link FGFR to p44/42^{MAPK} [136].

The bFGF-dependent upregulation of p44/42^{MAPK} is likely to be the molecular mechanism by which the growth factor is likely to induce the mitogenesis of endothelial cells in the angiogenic process. However, whether the induction of endothelial cell migration occurs on the activation of the same or different signal transduction pathway(s) is still controversial [137].

To date, a handful of investigators have been attempting to correlate the overexpression of bFGF with the angiogenic phenotype of HNSCC. bFGF was found to be strongly expressed in very actively proliferating and highly oxygenated regions of murine SCCs [138,139], whereas it was not detected in tumor areas bordering on necrosis, which are poorly vascularized and hypoxic [140]. Shultze-Hector and Haghayegh [105] further confirmed these findings by demonstrating a correlation between bFGF production by tumor cells and vascular endothelial cell growth rate in murine HNSCC cell lines. Moreover, the staining of human HNSCC cell lines with polyclonal antibodies against bFGF revealed a highly heterogeneous distribution of the growth factor within the tumor that was associated with the inhomogeneous tumor cell prolif-eration throughout the viable tumor tissue [105] (Fig. 7.7). Dellacono *et al.* [141] showed in 45 HNSCC specimens that in the majority of samples, cells presented an intense positive staining for bFGF. In addition, in samples from patients affected by an early stage disease, bFGF expression was more elevated than in those with late stage HNSCC. These results also correlated with the expression of FGFR1 and FGFR2 in the same specimens, thus suggesting that tumor cells may upregulate FGFR expression in response to augmented bFGF levels in the surrounding tissues.

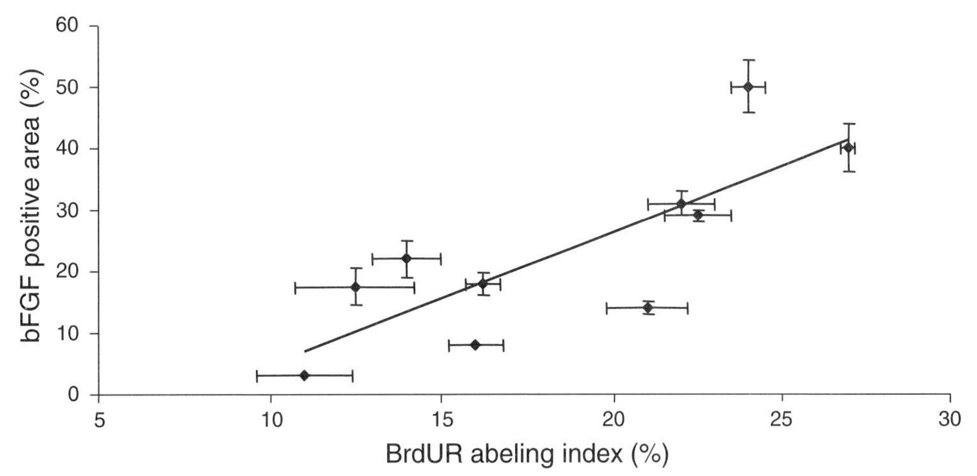

FIGURE 7.7 Relative extent of bFGF-positive tumor areas as a function of overall labeling index in 10 different human SCC. Adapted from Schultz-Hectpr amd Haghayegh, *Cancer Res.* **53**, 1444–1449 (1993).

Forootan and colleagues [144] analyzed the expression of bFGF in relation to the angiogenic degree in specimens from 51 patients affected by HNSCC of the tongue. Although bFGF was expressed heterogeneously in different regions of the tumor, in agreement with observations of others [105, 141–143], no significant correlation between raw or volume-weighted vessel counts and intensity of bFGF expression in primary carcinomas or in metastatic deposits was established. These latest data were in support of previous findings in which the high expression of bFGF in differentiated areas of 11 HNSCCs was not linked to the vascular count [145]. Contradictory results were reported in a simultaneous study conducted on 50 patients affected by oral cavity, oropharynx, larynx and hypopharynx primary HNSCC. A significantly higher expression of bFGF was found in the tumor samples compared to noral control tissues. This expresion was correlated with the mean number of microvessels, and increased vessel density in the tumors was associated with strong bFGF expression [146] (Fig. 7.8).

The ubiquitous presence of bFGF in squamous cell carcinomas of the head and neck underlines the importance of bFGF in the resulting growth of the carcinoma and its supporting stroma. However, even though the effect of bFGF on neoangiogenesis in HNSCC is still a conflicting topic, it is noteworthy that incongruities in the experimental results may arise not only from differences in the types of tumors being studied, but also in the specimens being selected, such as cultured cell lines or tumor tissues.

2. Vascular Endothelial Growth Factor

Vascular endothelial growth factor, also known as vascular permeability factor (VPF), is a multifunctional 46-kDa proangiogenic cytokine, with a pivotal role in the regulation of normal and pathological angiogenesis. VEGF was initially characterized for its specific and potent ability to promote endothelial cell proliferation in a paracrine mode of action [89,90,147]. In addition, VEGF can induce permeability and vascular leakage of plasma proteins. In particular, it has been demonstrated that extravasation of plasma fibrinogen alters the fibrin deposition in the extracellular matrix. As a consequence, the extracellular matrix is transformed into the mature stroma characteristic of tumors where increased proliferation of fibroblasts and endothelial cells occurs [28].

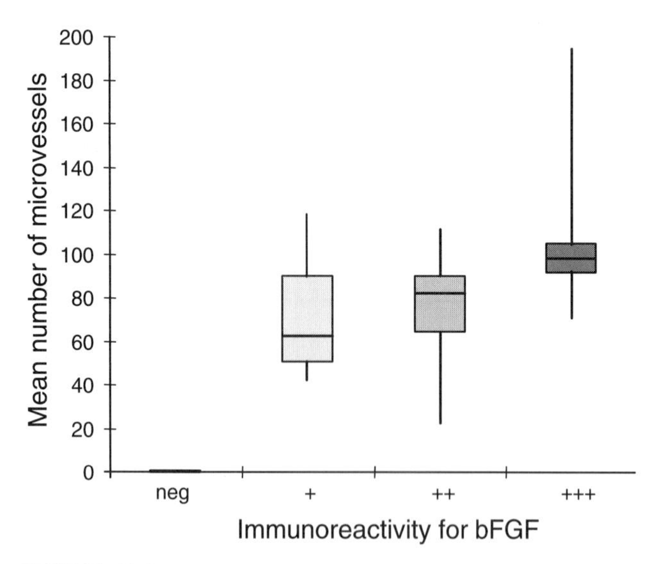

FIGURE 7.8 bFGF protein expression in HNSCC tumor tissue correlates with the mean number of microvessels per microscopic field. Adapted from Riedel *et al.*, *Head Neck* **22**, 183–189 (2000).

Disruption of a single VEGF allele or both VEGF alleles in mice results in embryonic lethality due to severe vasculature abnormalities or almost a complete absence of vasculature, respectively [148,149]. To date, five other members belonging to the VEGF family have been identified based on their homology to VEGF: placenta growth factor (PIGF), VEGF-B, VEGF-C, VEGF-D, and VEGF-E [150–152]. The different members of the VEGF family have an overlapping ability to bind to and activate their respective cell surface receptors. Three high-affinity vascular endothelial growth factor receptors (VEGFRs) have been cloned and characterized biochemically: VEGFR-1/Flt-1, VEGFR-2/Flk-1/KDR, and VEGFR-3/Flt-4 [153]. Studies have demonstrated that both VEGFR-2 and VEGFR-1 are essential for the normal development of embryonic vasculature. However, their respective roles in endothelial cell proliferation and differentiation appear to be distinct. The major growth and permeability functions of VEGF seem to be mediated exclusively by the engagement of VEGFR-2. In contrast, VEGFR-1 appears to exert an opposite negative role, either by sequestering VEGF and impairing its binding to VEGFR-2 or by directly suppressing VEGFR-2 signaling. In fact, VEGFR-2 knockout mice fail to develop a vasculature and have a limited number of endothelial cells, whereas VEGFR-1-deficient mice are characterized by the presence of disorganized tubules, resulting from the abnormal clustering of an excessive number of endothelial cells [91, 154, 155]. VEGFR-3 is mostly expressed in lymphatic vessels, and its contribution to the development of blood vessels is not yet fully defined [156].

VEGFRs are transmembrane receptor tyrosine kinases whose activation occurs on ligand binding, followed by receptor dimerization and activation by autophosphorylation of tyrosine residues in the cytoplasmic tail [157–159]. The activated receptor is likely to provide docking sites for SH2 domain-containing proteins and, upon the engagement of multiprotein aggregates, initiates a downstream signaling cascade, associated with endothelial cell proliferation and survival. Activated VEGFR-2, but not VEGFR-1, has been demonstrated to bind to Shc, Grb2, Nck, *ras* GTPase, P13-K, PLC-γ, and PKC, as well as to the protein phosphatases SHP-1 and SHP-2 [90,91,160,161]. Furthermore, phosphorylated VEGFR-2 leads to the activation of both p44/42^{MAPK} [162] and AKT/PKB [163] signal transduction pathways. The engagement of VEGFR-2 by its ligand results in diverse cellular biological responses, such as changes in cell morphology, actin reorganization, membrane ruffling, chemotaxis, and mitogenesis [161]. VEGF-dependent endothelial cell migration may occur focal adhesion kinase (FAK) and paxillin tyrosine phosphorylation and consequent activation [162]. Understanding of a VEGFR-1 downstream signal transduction pathway(s) is yet to be fully elucidated. However, its importance in the promotion of normal vascular development in the fetus [164], as well as the lethality of VEGFR-1-deficient mouse embryos [155], strongly suggests that this receptor may signal through novel mechanisms.

Numerous studies have demonstrated that VEGF mRNA is upregulated in the vast majority of human tumors, such as lung [165], thyroid [166], breast [44,167], gastrointestinal tract [168], kidney and bladder [169], ovary [170], and cervix uteri carcinomas [48], as well as angiosarcomas [171] and glioblastomas [172]. VEGF gene expression has been described to be upregulated by different mechanisms, among which oxygen tension seems to play a major role, both *in vitro* and *in vivo* [173–176]. Hypoxia-dependent VEGF gene transcription has been demonstrated to occur in both normal and tumor tissues [173] and to be mediated by the transcriptional activity of hypoxia-inducible factor-1 (HIF-1) [177]. Furthermore, upregulation of VEGF mRNA expression and increased VEGF secretion may be dependent on the effect of various mitogenic growth factors and cytokines, such as TGF-β [178], IL-1α, PGE$_2$ [179], IL-6 [180], and IGF-1 [181]. These factors, in addition to their direct mitogenic effect on malignant cells, may facilitate the tumor growth via increased VEGF production and secretion [91].

Several studies reported the elevated expression of VEGF mRNA in HNSCC. *In situ* hybridization analysis revealed a significantly greater expression of both VEGF and VEGFR-2 in 29 cases of oral cavity and larynx invasive squamous cell carcinoma compared to normal control tissues [182]. Similar results were obtained in an immunohistochemical analysis conducted on 63 specimens of various HNSCCs [183]. In addition, a study conducted on 156 patients revealed that high VEGF expression was detectable not only in the cancer cells, but also in the tumor-infiltrating inflammatory cells, such as plasma cells and macrophages, and in the cells of the tumor bordering tissues [184].

A correlation between VEGF expression with increased microvessel density, early recurrence, and worse prognosis has been observed in primary breast cancer [185, 186], lung [187], and gastric carcinoma [188]. The first direct evidence of a correlation between VEGF expression and increased microvessel density in HNSCC stems from a study conducted on a series of patients affected by different stages of head and neck lesions. In particular, the greatest number of microvessels was detected in premalignant lesions. Although increased angiogenesis was also evident in early and late stage HNSCC, compared to normal tissues, these findings suggest that VEGF is likely to regulate the early angiogenic steps of the tumor progression toward more aggressive phenotypes [189] (Fig. 7.9). Moreover, in 29 specimens from human nasopharyngeal carcinoma, a significant relationship between microvessel density and high VEGF expression was determined, thus suggesting the importance of VEGF-dependent angiogenesis in the occurrence of lymph node metastasis [190]. Likewise, the immunohistochemical analysis of 29 oral SCC specimens revealed a correlation between the intensity of VEGF expression with lymph node metastasis, but in contrast to the study by Wakisaka *et al.* [190], not with vessel density [191]. In addition a poor prognostic outcome is likely to be a direct consequence of high VEGF expression in HNSCC [192].

Wild-type p53 downregulates VEGF transcription, whereas p53 mutants have no effect on the promoter activity [193]. In this regard, the number of p53 mutations in VEGF-positive HNSCC has been found significantly higher than VEGF-negative tumors, thus suggesting the important role of p53 in the regulation of VEGF-dependent angiogenesis in HNSCC [194]. It has been demonstrated that tumor cells secrete VEGF to the bloodstream [195] and, as a consequence, an elevation in VEGF protein levels has been found in the serum of some cancer patients [187,196,197], as well as in patients affected by HNSCC [198,199] (Fig. 7.10). High serum levels of VEGF

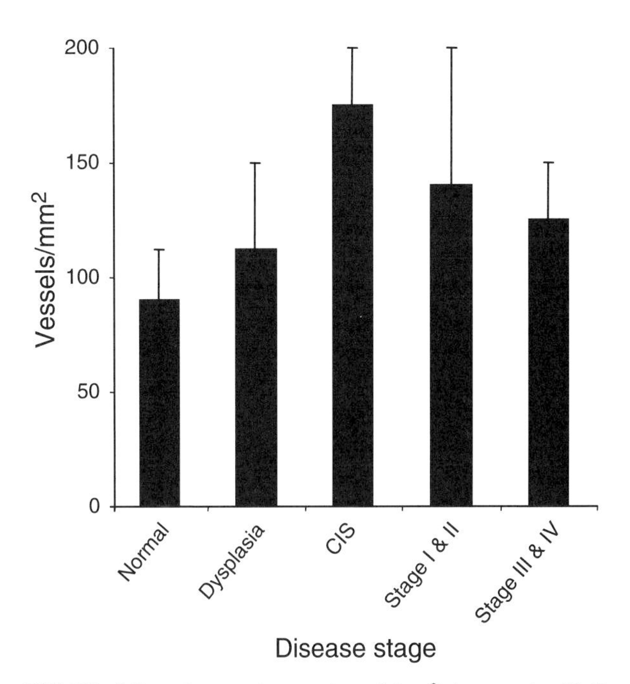

FIGURE 7.9 Microvessel count (vessels/mm^2) in normal epithelium, preinvasive [dysplasia and carcinoma *in situ* (CIS)], and invasive (stages I–IV) human HNSCC. Adapted from Sauter *et al.*, *Clin. Cancer Res.* **5**, 775–782 (1999).

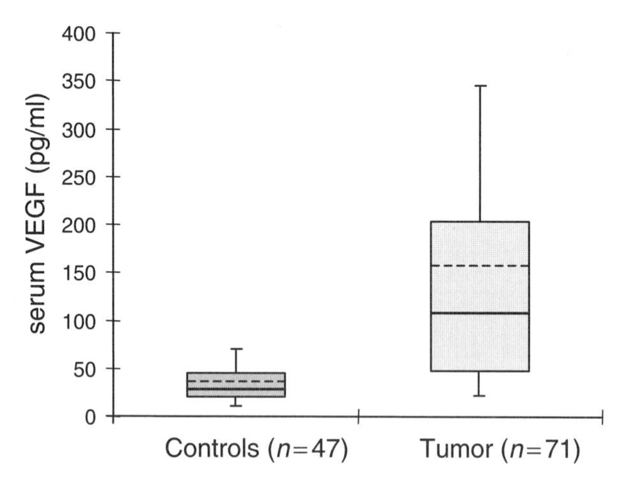

FIGURE 7.10 Comparison of mean and median serum VEGF concentrations in patients with HNSCC ($n=71$) and healthy controls ($n=47$). Modified from Riedel *et al.*, *Eur. Arch. Otorhinolaryngol.* **257**, 332–336 (2000).

can enhance not only endothelial cell proliferation, but also vascular permeability, thus contributing to tumor cell extravasation and development of metastatic disease. Moreover, it has been demonstrated that VEGF inhibits antigen-presenting dendritic cells [200]. Therefore, prolonged exposure of the immune system to high levels of circulating VEGF may impair the immune response in the host elicited by tumors.

It has been shown that HNSCC cells in culture secrete, in addition to VEGF, a high level of other angiogenic modulatory factors, such as TGFβ and PGE_2 in the conditioned medium. The HNSCC cell line-derived supernatant can stimulate both endothelial cell growth and motility. Neutralization of TGFβ resulted in an increased endothelial cell proliferation, thus suggesting that TGFβ may act as an antiproliferative factor counterbalancing the role of VEGF in the mitogenesis of endothelial cells. In contrast, PGE_2 was likely not to have any effect on endothelial cell mitogenesis. However, both HNSCC-derived TGFβ and PGE_2 were shown to induce endothelial cell motility, as the anti-TGFβ antibody and the PGE_2 inhibitor indomethacin blocked cell migration [94].

3. Platelet-Derived Endothelial Cell Growth Factor

A few studies investigated whether platelet-derived endothelial cell growth factor (PD-ECGF) is a positive regulator in the progression of neoangiogenesis in HNSCC. The platelet-derived endothelial cell growth factor (PD-ECGF) is a 45-kDa protein, identical to thymidine P_i deoxyribosyltransferase (TP^3), an enzyme involved in DNA synthesis [201,202]. PD-ECGF has been shown to promote the stimulation of endothelial cell chemotaxis *in vitro* and angiogenesis *in vivo* [203,204]. Correlation of PD-ECGF expression and angiogenesis has been reported in a wide variety of human tumors, such as ovary, breast, stomach, lung, and colon and rectum carcinomas [205–210]. To date, very few studies have investigated the effect of PD-ECGF on angiogenesis in HNSCC. Fujieda and co-workers [211] conducted an immunohistochemical analysis on 58 oral and oropharyngeal SCC specimens and demonstrated a high expression of PD-ECGF. Although increased microvessel density was observed in the tumors, no correlation between PD-ECGF expression score and vessel count was found. Similar findings were reported in the analysis of 95 HNSCC specimens, where PD-ECGF was found highly expressed in the tumors, but its expression was not associated with the angiogenic degree of the samples [212]. However, conflicting findings emerged from a study by Giatromanolaki *et al.* [213] in which microvessel density in locally advanced HNSCC was correlated with the over-expression of PD-ECGF. Moreover, the interrelation between angiogenesis and PD-ECGF was also associated with an increased accumulation of mutated p53 in the nuclei and the consequent inhibition of thrombospondin antiangiogenic activity.

4. Thrombospondin and p53

Mutation or loss of p53 tumor suppressor gene has been largely documented to be associated with the development and progression of HNSCC [223,224]. Furthermore, p53 has been shown to be involved in the control of angiogenesis by upregulating the transcription of thrombospondin (TSP), an inhibitor of angiogenesis [225,226]. Thus, a reduced TSP expression, dependent on *p53* alteration, can represent one of the mechanisms that may account for the increase of angiogenesis in oral squamous cell carcinoma. In addition, mutation of p53 inversely alters the expression of bax and bcl-2 apoptosis regulatory proteins, thereby triggering an antiapoptotic response and consequentially leading to an increase of tumor cell mitogenic rate [227].

5. Proinflammatory and Proangiogenic Cytokines

Patients affected by HNSCC often present an alteration in host immunity, inflammation, and angiogenesis [79,214,215]. It has been demonstrated that cytokines with proinflammatory, proangiogenic, and immunoregulatory activity are secreted by HNSCC and could contribute to the progression of the disease. Various cytokines, including IL-1α, IL-6, IL-8, and granulocyte macrophage-colony stimulating factor (GM-CSF), have been reported to be expressed in HNSCC cells [216–221]. In particular, the elevated coexpression of IL-8 and VEGF angiogenic factors in HNSCC samples has been proposed as an additional mechanism of angiogenesis in the progression of the disease [222]. However, whether the angiogenesis-mediated progression of HNSCC is dependent on the synergistic function of both IL-8 and VEGF or on the predominant action of either factor in a distinct stage of the disease has yet to be understood.

6. Nitric Oxide Synthase

High levels of nitric oxide synthase (NOS) have been found in solid tumors [228,229], including HNSCC [230], albeit NOS specific function in tumorigenesis has yet to be elucidated completely. It has been reported that NOS plays an important role in VEGF-dependent angiogenesis regulation [231,232]. In addition, NOS is likely to be responsible for the neovascularization in HNSCC. In fact, Gallo *et al.* [233] demonstrated that the progression of angiogenesis and consequently lymph node metastasis in 27 HNSCC are correlated with increase of NOS activity.

IV. ANTIANGIOGENIC THERAPIES IN HNSCC

Antiangiogenic inhibitors are used as part of therapeutic strategies addressed to arrest endothelial cell proliferation and differentiation in the tumor environment, thereby preventing cancer growth and metastasis. In tumor masses, one capillary vessel with a single endothelial cell is able to provide nutrients to approximately 100 cancer cells. For this reason, inhibition of endothelial cell growth might represent an amplification factor in halting the progression of tumor growth. In addition, because physiological angiogenesis is downregulated in adults, minimal side effects are obtained upon prolonged antiangiogenic treatments [234–237].

Neutralization of proangiogenic factor expression or activity, inhibition of the capillary basement membrane turnover, and inhibition of endothelial cell proliferation, migration, and differentiation are the diverse effects that the therapeutic application of specific antiangiogenic agents can target. Patients affected by HNSCC often develop multiple second primary tumors, which determine the failure of surgical, chemo-, and radiotherapeutic treatments. For this reason, various groups have investigated the efficacy of antiangiogenic factors in the prevention of HNSCC progression [238]. The most effective angiogenic inhibitors are summarized in Table 7.4.

TABLE 7.4 Classification of Angiogenic Inhibitors[a]

Agent	Reference
Agents neutralizing the expression or activity of factors stimulating angiogenesis	
Antibodies to bFGF	[260]
Antibodies to VEGF	[261]
Angiogenin antagonists	[262]
Suramin and analogs	[263–265]
Sulfonic derivatives of distamycin A	[266]
Recombinant platelet factor 4 (rPF4)	[267]
Angiostatin	[268]
Endostatin	[269]
Agents inhibiting turnover of capillary basement membrane or extracellular matrix	
Antibody to integrin avb3	[270]
Matrix metalloproteinase inhibitors	[271]
Minocycline	[272]
PAI-1	[273]
Aurintricaboxylic acid	[274]
Agents inhibiting endothelial cell growth, migration, and tube formation	
TNP-470 (AGM-1470)	[256]
Quinoline-3-carboxamide (linomide)	[275]
Tamoxifen	[276]
D-Penicillamine	[277]
Genistein	[278]
Thalidomide	[279]
Interferons	[280]
Interleukin-12	[281]
Retinoids	[247]
Vitamin D and analogs	[282,283]
Potential new antiangiogenic therapeutics	
Nitric oxide synthetase inhibitor	[284]
Thrombospondin antagonist	[285]
fps/fes inhibitor	[286]

[a]Adapted from Bhargava, P. Antiangiogenesis. *In* "Clinical Oncology" (M. D. Abeloff, J. O. Armitage, A. S. Lichter, and J. E. Niederhuber, eds.), 2nd Ed., pp. 243–250. Churchill Livingstone, New York (2000).

Retinoic acid (RA) has been investigated extensively for its ability to prevent HNSCC lesions. It has been demonstrated that RA modulates HNSCC tumor cell mitogenesis, differentiation, and apoptosis [239–242], although its maximal effectiveness is achieved when used at high dosage. In addition, several reports have documented RA antiangiogenic properties in different *in vivo* models [243–247]. However, the specific mechanism(s) by which RA reduces angiogenesis is largely unknown. One possibility is that RA directly exerts an inhibitor effect on tumor and endothelial cell proliferation or causes the production and release of endogenous antiangiogenic factors. Nevertheless, there is no indication of a direct downregulation of angiogenic growth factors, such as bFGF or VEGF, or their receptors by retinoids. Lingen *et al.* [219] reported that the treatment of oral squamous cell carcinoma cell lines with RA resulted in a switch from a potently angiogenic to an antiangiogenic phenotype. Two reports by our laboratory showed that long-term treatment of the ME-180 squamous cell carcinoma cell line with retinoic acid resulted in downregulation of the FGF-BP mRNA proangiogenic factor, both in cultured cells *in vitro* and *in vivo* in xenograft tumors [242,248]. Hence, negative regulation of the angiogenic modulator FGF-BP by retinoids may account for some of the RA-mediated inhibition of angiogenesis.

Similar to RA, interferon-α (IFN-α) retains an antiangiogenic effect, both *in vitro* and *in vivo* [249–251]. However, treatment of HNSCC patients with high doses of RA or IFN-α results in the development of acute toxicity [252]. Promising results emerged from phase I and II clinical trials in which the association of low doses of both RA and interferon-α was used to treat human advanced solid tumors, including HNSCC [253,254]. Lingen *et al.* [255] showed that the long-term use of both agents synergistically down-modulated angiogenesis and growth of HNSCC. In light of these findings, the synergistic and nontoxic combination of antiangiogenic molecules may represent a powerful therapeutic approach in HNSCC chemoprevention.

Several studies demonstrated the potent cytotoxic activity of TNP-470, a fumagallin derivative drug on endothelial cells. In addition, TPN-470-dependent inhibition of human tumor growth and metastasis has been demonstrated [256, 257]. However, contradictory results arose from independent studies attempting to investigate the antiangiogenic effects of TNP-470 on HNSCC progression. Gleich *et al.* [258] reported the failure of TNP-470 in reducing tumor growth in mice with implanted oral cancer. In contrast with this, Ueda *et al.* [259] described the effectiveness of TNP-470 in inhibiting oral SCC cell line growth. Moreover, the same authors showed that the tumor development in mice, as a result of oral SCC subcutaneous inoculation, was blocked dramatically by the antiangiogenic effect of TNP-470.

V. CONCLUSIONS

An increase in microvessel density has been demonstrated to be an important prognostic factor in the development of tumor growth, invasion, and metastasis. However, several controversial findings emerged from reports attempting to analyze the correlation between angiogenesis and progression of HNSCC. Taking into account the high expression of angiogenic factors in HNSCC specimens and in serum from HNSCC patients, as well as the success of antiangiogenic therapies, it is likely that angiogenesis plays a significant role in the development of head and neck squamous cell carcinoma.

References

1. Renan, M. J. (1993). How many mutations are required for tumorigenesis? Implications from human cancer data. *Mol. Carcinog.* **7**, 139.
2. Voravud, N., Charuruks, N., and Mutirangura, A. (1997). Squamous cell carcinoma of head and neck. *J. Med. Assoc. Thai* **80**, 207.
3. Shao, X., Tandon, R., Samara, G., Kanki, H., Yano, H., Close, L. G., Parsons, R., and Sato, T. (1998). Mutational analysis of the PTEN gene in head and neck squamous cell carcinoma. *Int. J. Cancer* **77**, 684.
4. Nagai, M. A. (1999). Genetic alterations in head and neck squamous cell carcinomas. *Braz. J. Med. Biol. Res.* **32**, 897.
5. Field, J. K. (1992). Oncogenes and tumour-suppressor genes in squamous cell carcinoma of the head and neck. *Eur. J. Cancer B Oral Oncol.* **28B**, 67.
6. Shackleford, G. M., MacArthur, C. A., Kwan, H. C., and Varmus, H. E. (1993). Mouse mammary tumor virus infection accelerates mammary carcinogenesis in Wnt-1 transgenic mice by insertional activation of int-2/Fgf-3 and hst/Fgf-4. *Proc. Natl. Acad. Sci. USA* **90**, 740.
7. Wellstein, A., Zugmaier, G., Califano, J. A. D., Kern, F., Paik, S., and Lippman, M. E. (1991). Tumor growth dependent on Kaposi's sarcoma-derived fibroblast growth factor inhibited by pentosan polysulfate. *J. Natl. Cancer Inst.* **83**, 716.
8. Costa, M., Danesi, R., Agen, C., Di Paolo, A., Basolo, F., Del Bianchi, S., and Del Tacca, M. (1994). MCF-10A cells infected with the int-2 oncogene induce angiogenesis in the chick chorioallantoic membrane and in the rat mesentery. *Cancer Res.* **54**, 9.
9. Muller, D., Millon, R., Velten, M., Bronner, G., Jung, G., Engelmann, A., Flesch, H., Eber, M., Methlin, G., and Abecassis, J. (1997). Amplification of 11q13 DNA markers in head and neck squamous cell carcinomas: Correlation with clinical outcome. *Eur. J. Cancer* **33**, 2203.
10. Williams, M. E., Gaffey, M. J., Weiss, L. M., Wilczynski, S. P., Schuuring, E., and Levine, P. A. (1993). Chromosome 11Q13 amplification in head and neck squamous cell carcinoma. *Arch. Otolaryngol. Head Neck Surg.* **119**, 1238.
11. Jin, Y., Mertens, F., Mandahl, N., Heim, S., Olegard, C., Wennerberg, J., Biorklund, A., and Mitelman, F. (1993). Chromosome abnormalities in eighty-three head and neck squamous cell carcinomas: Influence of culture conditions on karyotypic pattern. *Cancer Res.* **53**, 2140.
12. Cowan, J. M., Beckett, M. A., and Weichselbaum, R. R. (1993). Chromosome changes characterizing in vitro response to radiation in human squamous cell carcinoma lines. *Cancer Res.* **53**, 5542.
13. Van Dyke, D. L., Worsham, M. J., Benninger, M. S., Krause, C. J., Baker, S. R., Wolf, G. T., Drumheller, T., Tilley, B. C., and Carey, T. E. (1994). Recurrent cytogenetic abnormalities in squamous cell carcinomas of the head and neck region. *Genes Chromosomes Cancer* **9**, 192.
14. Ransom, D. T., Leonard, J. H., Kearsley, J. H., Turbett, G. R., Heel, K., Sosars, V., Hayward, N. K., and Bishop, J. F. (1996). Loss of heterozygosity studies in squamous cell carcinomas of the head and neck. *Head Neck* **18**, 248.
15. Califano, J., van der Riet, P., Westra, W., Nawroz, H., Clayman, G., Piantadosi, S., Corio, R., Lee, D., Greenberg, B., Koch, W., and Sidransky, D. (1996). Genetic progression model for head and neck cancer: Implications for field cancerization. *Cancer Res.* **56**, 2488.
16. Folkman, J., and Shing, Y. (1992). Angiogenesis. *J. Biol. Chem.* **267**, 10931.
17. Carmeliet, P. (2000). Mechanisms of angiogenesis and arteriogenesis. *Nature Med.* **6**, 389.
18. Hanahan, D., and Folkman, J. (1996). Patterns and emerging mechanisms of the angiogenic switch during tumorigenesis. *Cell* **86**, 353.
19. Salamonsen, L. A., and Woolley, D. E. (1996). Matrix metalloproteinases in normal menstruation. *Hum. Reprod.* **11**(Suppl. 2), 124.
20. Carmeliet, P., and Jain, R. K. (2000). Angiogenesis in cancer and other diseases. *Nature* **407**, 249.
21. Blood, C. H., and Zetter, B. R. (1990). Tumor interactions with the vasculature: Angiogenesis and tumor metastasis. *Biochim. Biophys. Acta* **1032**, 89.
22. Risau, W. (1998). Development and differentiation of endothelium. *Kidney Int. Suppl.* **67**, S3.
23. Helmlinger, G., Yuan, F., Dellian, M., and Jain, R. K. (1997). Interstitial pH and pO2 gradients in solid tumors in vivo: High-resolution measurements reveal a lack of correlation. *Nature Med.* **3**, 177.
24. Maniotis, A. J., Folberg, R., Hess, A., Seftor, E. A., Gardner, L. M., Pe'er, J., Trent, J. M., Meltzer, P. S., and Hendrix, M. J. (1999). Vascular channel formation by human melanoma cells in vivo and in vitro: Vasculogenic mimicry. *Am. J. Pathol.* **155**, 739.
25. Baish, J. W., and Jain, R. K. (2000). Fractals and cancer. *Cancer Res.* **60**, 3683.
26. Hobbs, S. K., Monsky, W. L., Yuan, F., Roberts, W. G., Griffith, L., Torchilin, V. P., and Jain, R. K. (1998). Regulation of transport pathways in tumor vessels: Role of tumor type and microenvironment. *Proc. Natl. Acad. Sci. USA* **95**, 4607.
27. Hashizume, H., Baluk, P., Morikawa, S., McLean, J. W., Thurston, G., Roberge, S., Jain, R. K., and McDonald, D. M. (2000). Openings

between defective endothelial cells explain tumor vessel leakiness. *Am. J. Pathol.* **156**, 1363.
28. Dvorak, H. F., Nagy, J. A., Feng, D., Brown, L. F., and Dvorak, A. M. (1999). Vascular permeability factor/vascular endothelial growth factor and the significance of microvascular hyperpermeability in angiogenesis. *Curr. Top. Microbiol. Immunol.* **237**, 97.
29. Benjamin, L. E., Golijanin, D., Itin, A., Pode, D., and Keshet, E. (1999). Selective ablation of immature blood vessels in established human tumors follows vascular endothelial growth factor withdrawal. *J. Clin. Invest.* **103**, 159.
30. Folberg, R., Hendrix, M. J., and Maniotis, A. J. (2000). Vasculogenic mimicry and tumor angiogenesis. *Am. J. Pathol.* **156**, 361.
31. Goldman, E. (1907). The growth of malignant disease in man and the lower animals with special reference to the vascular system. *Lancet* **2**, 1236.
32. Ide, A. G., Baker, N. H., and S. L. W. (1939). Vascularization of the Brown-Pearce rabbit epithelioma transplant as seen in the transparent ear chamber. *Am. J. Radiol.* **42**, 891.
33. Greenblatt, M., and Shubi, P. (1968). Tumor angiogenesis: Transfilter diffusion studies in the hamster by the transparent chamber technique. *J. Natl. Cancer Inst.* **41**, 111.
34. Ehrmann, R. L., and Knoth, M. (1968). Choriocarcinoma. Transfilter stimulation of vasoproliferation in the hamster cheek pouch. Studies by light and electron microscopy. *J. Natl. Cancer Inst.* **41**, 1329.
35. Folkman, J. (1971). Tumor angiogenesis: Therapeutic implications. *N. Engl. J. Med.* **285**, 1182.
36. Folkman, J. (1972). Anti-angiogenesis: New concept for therapy of solid tumors. *Ann. Surg.* **175**, 409.
37. Folkman, J., Merler, E., Abernathy, C., and Williams, G. (1971). Isolation of a tumor factor responsible or angiogenesis. *J. Exp. Med.* **133**, 275.
38. Folkman, J. (1974). Proceedings: Tumor angiogenesis factor. *Cancer Res.* **34**, 2109.
39. Hanahan, D. (1985). Heritable formation of pancreatic beta-cell tumours in transgenic mice expressing recombinant insulin/simian virus 40 oncogenes. *Nature* **315**, 115.
40. Kandel, J., Bossy-Wetzel, E., Radvanyi, F., Klagsbrun, M., Folkman, J., and Hanahan, D. (1991). Neovascularization is associated with a switch to the export of bFGF in the multistep development of fibrosarcoma. *Cell* **66**, 1095.
41. Arbeit, J. M., Munger, K., Howley, P. M., and Hanahan, D. (1994). Progressive squamous epithelial neoplasia in K14-human papillomavirus type 16 transgenic mice. *J. Virol.* **68**, 4358.
42. Arbeit, J. M. (1996). Transgenic models of epidermal neoplasia and multistage carcinogenesis. *Cancer Surv.* **26**, 7.
43. Arbeit, J. M., Olson, D. C., and Hanahan, D. (1996). Upregulation of fibroblast growth factors and their receptors during multi-stage epidermal carcinogenesis in K14-HPV16 transgenic mice. *Oncogene* **13**, 1847.
44. Brown, L. F., Berse. B., Jackman, R. W., Tognazzi, K., Guidi, A. J., Dvorak, H. F., Senger, D. R., Connolly, J. L., and Schnitt, S. J. (1995). Expression of vascular permeability factor (vascular endothelial growth factor) and its receptors in breast cancer. *Hum Pathol.* **26**, 86.
45. Guidi, A. J., Fischer, L., Harris, J. R., and Schnitt, S. J. (1994). Microvessel density and distribution in ductal carcinoma in situ of the breast. *J. Natl. Cancer Inst.* **86**, 614.
46. Weidner, N., Folkman, J., Pozza, F., Bevilacqua, P., Allred, E. N., Moore, D. H., Meli, S., and Gasparini, G. (1992). Tumor angiogenesis: A new significant and independent prognostic indicator in early-stage breast carcinoma. *J. Natl. Cancer Inst.* **84**, 1875.
47. Smith-McCune, K. K., and Weidner, N. (1994). Demonstration and characterization of the angiogenic properties of cervical dysplasia. *Cancer Res.* **54**, 800.
48. Guidi, A. J., Abu-Jawdeh, G., Berse, B., Jackman, R. W., Tognazzi, K., Dvorak, H. F., and Brown, L. F. (1995). Vascular permeability factor (vascular endothelial growth factor) expression and angiogenesis in cervical neoplasia. *J. Natl. Cancer Inst.* **87**, 1237.
49. Folkman, J., and D'Amore, P. A. (1996). Blood vessel formation: What is its molecular basis? *Cell* **87**, 1153.
50. Zagzag, D., Hooper, A., Friedlander, D. R., Chan, W., Holash, J., Wiegand, S. J., Yancopoulos, G. D., and Grumet, M. (1999). In situ expression of angiopoietins in astrocytomas identifies angiopoietin-2 as an early marker of tumor angiogenesis. *Exp. Neurol.* **159**, 391.
51. Holash, J., Maisonpierre, P. C., Compton, D., Boland, P., Alexander, C. R., Zagzag, D., Yancopoulos, G. D., and Wiegand, S. J. (1999). Vessel cooption, regression, and growth in tumors mediated by angiopoietins and VEGF. *Science* **284**, 1994.
52. Yancopoulos, G. D., Davis, S., Gale, N. W., Rudge, J. S., Wiegand, S. J., and Holash, J. (2000). Vascular-specific growth factors and blood vessel formation. *Nature* **407**, 242.
53. Wesseling, P., van der Laak, J. A., de Leeuw, H., Ruiter, D. J., and Burger, P. C. (1994). Quantitative immunohistological analysis of the microvasculature in untreated human glioblastoma multiforme: Computer-assisted image analysis of whole-tumor sections. *J. Neurosurg.* **81**, 902.
54. Holmgren, L., O'Reilly, M. S., and Folkman, J. (1995). Dormancy of micrometastases: Balanced proliferation and apoptosis in the presence of angiogenesis suppression. *Nature Med.* **1**, 149.
55. Pezzella, F., Pastorino, U., Tagliabue, E., Andreola, S., Sozzi, G., Gasparini, G., Menard, S., Gatter, K. C., Harris, A. L., Fox, S., Buyse, M., Pilotti, S., Pierotti, M., and Rilke, F. (1997). Non-small-cell lung carcinoma tumor growth without morphological evidence of neo-angiogenesis. *Am. J. Pathol*, **151**, 1417.
56. Srivastava, A., Laidler, P., Hughes, L. E., Woodcock, J., and Shedden, E. J. (1986). Neovascularization in human cutaneous melanoma: A quantitative morphological and Doppler ultrasound study. *Eur. J. Cancer Clin. Oncol.* **22**, 1205.
57. Srivastava, A., Laidler, P., Davies, R. P., Horgan, K., and Hughes, L. E. (1988). The prognostic significance of tumor vascularity in intermediate-thickness (0.76-4.0 mm thick) skin melanoma: A quantitative histologic study. *Am. J. Pathol.* **133**, 419.
58. Weidner, N., Semple, J. P., Welch, W. R., and Folkman, J. (1991). Tumor angiogenesis and metastasis: Correlation in invasive breast carcinoma. *N. Engl. J. Med.* **324**, 1.
59. Weidner, N., Carroll, P. R., Flax, J., Blumenfeld, W., and Folkman, J. (1993). Tumor angiogenesis correlates with metastasis in invasive prostate carcinoma. *Am. J. Pathol.* **143**, 401.
60. Folkman, J. (1995). Seminars in medicine of the Beth Israel Hospital, Boston: Clinical applications of research on angiogenesis. *N. Engl. J. Med.* **333**, 1757.
61. Macchiarini, P., Fontanini, G., Hardin, M. J., Squartini, F., and Angeletti, C. A. (1992). Relation of neovascularisation to metastasis of non-small-cell lung cancer. *Lancet* **340**, 145.
62. Wakui, S., Furusato, M., Itoh, T., Sasaki, H., Akiyama, A., Kinoshita, I., Asano, K., Tokuda, T., Aizawa, S., and Ushigome, S. (1992). Tumour angiogenesis in prostatic carcinoma with and without bone marrow metastasis: A morphometric study. *J. Pathol.* **168**, 257.
63. Ikinger, U., Terwey, B., Wurster, K., Manegold, C., and Moehring, K. (1985). Vascularization of malignant testicular tumors. *Urology* **26**, 41.
64. Moreira, L. F., Iwagaki, H., Hizuta, A., Sakagami, K., and Orita, K. (1992). Outcome in patients with early colorectal carcinoma. *Br. J. Surg.* **79**, 436.
65. Petruzzelli, G. J., Snyderman, C. H., Johnson, J. T., and Myers, E. N. (1993). Angiogenesis induced by head and neck squamous cell carcinoma xenografts in the chick embryo chorioallantoic membrane model. *Ann. Otol. Rhinol. Laryngol.* **102**, 215.
66. Mukai, K., Rosai, J., and Burgdorf, W. H. (1980). Localization of factor VIII-related antigen in vascular endothelial cells using an immunoperoxidase method. *Am. J. Surg. Pathol.* **4**, 273.

67. Holthofer, H., Virtanen, I., Kariniemi, A. L., Hormia, M., Linder, E., and Miettinen, A. (1982). Ulex europaeus I lectin as a marker for vascular endothelium in human tissues. *Lab. Invest.* **47**, 60.
68. Ramani, P., Birch, B. R., Harland, S. J., and Parkinson, M. C. (1991). Evaluation of endothelial markers in detecting blood and lymphatic channel invasion in pT1 transitional carcinoma of bladder. *Histopathology* **19**, 551.
69. Delides, G. S., Venizelos, J., and Revesz, L. (1988). Vascularization and curability of stage III and IV nasopharyngeal tumors. *J. Cancer. Res. Clin. Oncol.* **114**, 321.
70. Zatterstrom, U. K., Brun, E., Willen, R., Kjellen, E., and Wennerberg, J. (1995). Tumor angiogenesis and prognosis in squamous cell carcinoma of the head and neck. *Head Neck* **17**, 312.
71. Carrau, R. L., Barnes, E. L., Snyderman, C. H., Petruzzelli, G., Kachman, K., Rueger, R., D'Amico, F., and Johnson, J. T. (1995). Tumor angiogenesis as a predictor of tumor aggressiveness and metastatic potential in squamous cell carcinoma of the head and neck. *Invasion Metastasis* **15**, 197.
72. Gasparini, G., Weidner, N., Maluta, S., Pozza, F., Boracchi, P., Mezzetti, M., Testolin, A., and Bevilacqua, P. (1993). Intratumoral microvessel density and p53 protein: Correlation with metastasis in head-and-neck squamous-cell carcinoma. *Int. J. Cancer* **55**, 739.
73. Beatrice, F., Cammarota, R., Giordano, C., Corrado, S., Ragona, R., Sartoris, A., Bussolino, F., and Valente, G. (1998). Angiogenesis: Prognostic significance in laryngeal cancer. *Anticancer Res.* **18**, 4737.
74. Williams, J. K., Carlson, G. W., Cohen, C., Derose, P. B., Hunter, S., and Jurkiewicz, M. J. (1994). Tumor angiogenesis as a prognostic factor in oral cavity tumors. *Am. J. Surg.* **168**, 373.
75. Leedy, D. A., Trune, D. R., Kronz, J. D., Weidner, N., and Cohen, J. I. (1994). Tumor angiogenesis, the p53 antigen, and cervical metastasis in squamous carcinoma of the tongue. Otolaryngol. *Head Neck Surg.* **111**, 417.
76. Alcalde, R. E., Shintani, S., Yoshihama, Y., and Matsumura, T. (1995). Cell proliferation and tumor angiogenesis in oral squamous cell carcinoma. *Anticancer Res.* **15**, 1417.
77. Penfold, C. N., Partridge, M., Rojas, R., and Langdon, J. D. (1996). The role of angiogenesis in the spread of oral squamous cell carcinoma. *Br. J. Oral. Maxillofac. Surg.* **34**, 37.
78. Gleich, L. L., Biddinger, P. W., Pavelic, Z. P., and Gluckman, J. L. (1996). Tumor angiogenesis in T1 oral cavity squamous cell carcinoma: Role in predicting tumor aggressiveness. *Head Neck* **18**, 343.
79. Gleich, L. L., Biddinger, P. W., Duperier, F. D., and Gluckman, J. L. (1997). Tumor angiogenesis as a prognostic indicator in T2-T4 oral cavity squamous cell carcinoma: A clinical-pathologic correlation. *Head Neck* **19**, 276.
80. Pazouki, S., Chisholm, D. M., Adi, M. M., Carmichael, G., Farquharson, M., Ogden, G. R., Schor, S. L., and Schor, A. M. (1997). The association between tumour progression and vascularity in the oral mucosa. *J. Pathol.* **183**, 39.
81. Albo, D., Granick, M. S., Jhala, N., Atkinson, B., and Solomon, M. P. (1994). The relationship of angiogenesis to biological activity in human squamous cell carcinomas of the head and neck. *Ann. Plast. Surg.* **32**, 588.
82. Kessler, D. A., Langer, R. S., Pless, N. A., and Folkman, J. (1976). Mast cells and tumor angiogenesis. *Int. J. Cancer* **18**, 703.
83. Dray, T. G., Hardin, N. J., and Sofferman, R. A. (1995) Angiogenesis as a prognostic marker in early head and neck cancer. *Ann. Otol. Rhinol. Laryngol.* **104**, 724.
84. Stone, J., Itin, A., Alon, T., Pe'er, J., Gnessin, H., Chan-Ling, T., and Keshet, E. (1995). Development of retinal vasculature is mediated by hypoxia-induced vascular endothelial growth factor (VEGF) expression by neuroglia. *J. Neurosci.* **15**, 4738.
85. Alon, T., Hemo, I., Itin, A., Pe'er, J., Stone, J., and Keshet, E. (1995). Vascular endothelial growth factor acts as a survival factor for newly formed retinal vessels and has implications for retinopathy of prematurity. *Nature Med.* **1**, 1024.
86. Benjamin, L. E., Hemo, I., and Keshet, E. (1998). A plasticity window for blood vessel remodelling is defined by pericyte coverage of the preformed endothelial network and is regulated by PDGF- B and VEGF. *Development* **125**, 1591.
87. Bikfalvi, A., Klein, S., Pintucci, G., and Rifkin, D. B. (1997). Biological roles of fibroblast growth factor-2. *Endocr. Rev.* **18**, 26.
88. Powers, C. J., McLeskey, S. W., and Wellstein, A. (2000). Fibroblast growth factors, their receptors and signaling. *Endocr. Relat. Cancer* **7**, 165.
89. Leung, D. W., Cachianes, G., Kuang, W. J., Goeddel, D. V., and Ferrara, N. (1989). Vascular endothelial growth factor is a secreted angiogenic mitogen. *Science* **246**, 1306.
90. Ferrara, N., and Keyt, B. (1997). Vascular endothelial growth factor : Basic biology and clinical implications. *Exs.* **79**, 209.
91. Ferrara, N., and Davis-Smyth, T. (1997). The biology of vascular endothelial growth factor. *Endocr. Rev.* **18**, 4.
92. McMahon, G. (2000). VEGF receptor signaling in tumor angiogenesis. *Oncologist* **5**, 3.
93. Shemirani, B., and Crowe, D. L. (2000). Head and neck squamous cell carcinoma lines produce biologically active angiogenic factors. *Oral Oncol.* **36**, 61.
94. Benefield, J., Petruzzelli, G. J., Fowler, S., Taitz, A., Kalkanis, J., and Young, M. R. (1996). Regulation of the steps of angiogenesis by human head and neck squamous cell carcinomas. *Invasion Metastasis* **16**, 291.
95. Petruzzelli, G. J., Benefield, J., Taitz, A. D., Fowler, S., Kalkanis, J., Scobercea, S., West, D., and Young, M. R. (1997). Heparin-binding growth factor(s) derived from head and neck squamous cell carcinomas induce endothelial cell proliferations. *Head Neck* **19**, 576.
96. Thomas, G. R., Tubb, E. E., Sessions, R. B., and Cullen, K. J. (1993). Growth factor mRNA expression in human head and neck malignant tumors. *Arch. Otolaryngol. Head Neck Surg.* **119**, 1247.
97. Alcalde, R. E., Terakado, N., Otsuki, K., and Matsumura, T. (1997). Angiogenesis and expression of platelet-derived endothelial cell growth factor in oral squamous cell carcinoma. *Oncology* **54**, 324.
98. Cordon-Cardo, C., Vlodavsky, I., Haimovitz-Friedman, A., Hicklin, D., and Fuks, Z. (1990). Expression of basic fibroblast growth factor in normal human tissues. *Lab. Invest.* **63**, 832.
99. Baird, A., and Klagsbrun, M. (1991). The fibroblast growth factor family. *Cancer Cells* **3**, 239.
100. Morrison, R. S., Gross, J. L., Herblin, W. F., Reilly, T. M., LaSala, P. A., Alterman, R. L., Moskal, J. R., Kornblith, P. L., and Dexter, D. L. (1990). Basic fibroblast growth factor-like activity and receptors are expressed in a human glioma cell line. *Cancer Res.* **50**, 2524.
101. Gross, J. L., Morrison, R. S., Eidsvoog, K., Herblin, W. F., Kornblith, P. L., and Dexter, D. L. (1990). Basic fibroblast growth factor: A potential autocrine regulator of human glioma cell growth. *J. Neurosci. Res.* **27**, 689.
102. Luqmani, Y. A., Graham, M., and Coombes, R. C. (1992). Expression of basic fibroblast growth factor, FGFR1 and FGFR2 in normal and malignant human breast, and comparison with other normal tissues. *Br. J. Cancer* **66**, 273.
103. Nanus, D. M., Schmitz-Drager, B. J., Motzer, R. J., Lee, A. C., Vlamis, V., Cordon-Cardo, C., Albino, A. P., and Reuter, V. E. (1993). Expression of basic fibroblast growth factor in primary human renal tumors: Correlation with poor survival. *J. Natl. Cancer Inst.* **85**, 1597.
104. Yamanaka, Y., Friess, H., Buchler, M., Beger, H. G., Uchida, E., Onda, M., Kobrin, M. S., and Korc, M. (1993). Overexpression of acidic and basic fibroblast growth factors in human pancreatic cancer correlates with advanced tumor stage. *Cancer Res.* **53**, 5289.
105. Schultz-Hector, S., and Haghayegh, S. (1993). Beta-fibroblast growth factor expression in human and murine squamous cell carcinomas and

its relationship to regional endothelial cell proliferation. *Cancer Res.* **53**, 1444.

106. Shing, Y., Folkman, J., Sullivan, R., Butterfield, C., Murray, J., and Klagsbrun, M. (1984). Heparin affinity: Purification of a tumor-derived capillary endothelial cell growth factor. *Science* **223**, 1296.
107. Folkman, J., and Klagsbrun, M. (1987). Angiogenic factors. *Science* **235**, 442.
108. Basilico, C., and Moscatelli, D. (1992). The FGF family of growth factors and oncogenes. *Adv. Cancer Res.* **59**, 115.
109. Yayon, A., Klagsbrun, M., Esko, J. D., Leder, P., and Ornitz, D. M. (1991). Cell surface, heparin-like molecules are required for binding of basic fibroblast growth factor to its high affinity receptor. *Cell* **64**, 841.
110. Klagsbrun, M., and Baird, A. (1991). A dual receptor system is required for basic fibroblast growth factor activity. *Cell* **67**, 229.
111. Rapraeger, A. C., Krufka, A., and Olwin, B. B. (1991). Requirement of heparan sulfate for bFGF-mediated fibroblast growth and myoblast differentiation. *Science* **252**, 1705.
112. Rapraeger, A. C., Guimond, S., Krufka, A., and Olwin, B. B. (1994). Regulation of heparan sulfate in fibroblast growth factor signaling. *Methods Enzymol.* **245**, 219.
113. Mignatti, P., Morimoto, T., and Rifkin, D. B. (1992). Basic fibroblast growth factor, a protein devoid of secretory signal sequence, is released by cells via a pathway independent of the endoplasmic reticulum-Golgi complex. *J. Cell Physiol.* **151**, 81.
114. McNeil, P. L., Muthukrishnan, L., Warder, E., and D'Amore, P. A. (1989). Growth factors are released by mechanically wounded endothelial cells. *J. Cell Biol.* **109**, 811.
115. Bashkin, P., Doctrow, S., Klagsbrun, M., Svahn, C. M., Folkman, J., and Vlodavsky, I. (1989). Basif fibroblast growth factor binds to subendothelial extracellular matrix and is released by heparitinase and heparin-like molecules. *Biochemistry* **28**, 1737.
116. Saksela, O., and Rifkin, D. B. (1990). Release of basic fibroblast growth factor-neparan sulfate complexes from endothelial cells by plasminogen activator-mediated proteolytic activity. *J. Cell Biol.* **110**, 767.
117. Buczek-Thomas, J. A., and Nugent, M. A. (1999). Elastase-mediated release of heparan sulfate proteoglycans from pulmonary fibroblast cultures: A mechanism for basic fibroblast growth factor (bFGF) release and attenuation of bfgf binding following elastase-induced injury. *J. Biol. Chem.* **274**, 25167.
118. Wu, D. Q., Kan, M. K., Sato, G. H., Okamato, T., and Sato, J. D. (1991). Characterization and molecular cloning of a putative binding protein for heparin-binding growth factors. *J. Biol. Chem.* **266**, 16778.
119. Czubayko, F., Smith, R. V., Chung, H. C., and Wellstein, A. (1994). Tumor growth and angiogenesis induced by a secreted binding protein for fibroblast growth factors. *J. Biol. Chem.* **269**, 28243.
120. Kurtz, A., Wang, H. L., Darwiche, N., Harris, V., and Wellstein, A. (1997). Expression of a binding protein for FGF is associated with epithelial development and skin carcinogenesis. *Oncogene* **14**, 2671.
121. Czubayko, F., Liaudet-Coopman, E. D., Aigner, A., Tuveson, A. T., Berchem, G. J., and Wellstein, A. (1997). A secreted FGF-binding protein can serve as the angiogenic switch in human cancer. *Nature Med.* **3**, 1137.
122. Schulze-Osthoff, K., Risau, W., Vollmer, E., and Sorg, C. (1990). In situ detection of basic fibroblast growth factor by highly specific antibodies. *Am. J. Pathol.* **137**, 85.
123. Takahashi, J. A., Mori, H., Fukumoto, M., Igarashi, K., Jaye, M., Oda, Y., Kikuchi, H., and Hatanaka, M. (1990). Gene expression of fibroblast growth factors in human gliomas and meningiomas: Demonstration of cellular source of basic fibroblast growth factor mRNA and peptide in tumor tissues. *Proc. Natl. Acad. Sci. USA* **87**, 5710.
124. Zagzag, D., Miller, D. C., Sato, Y., Rifkin, D. B., and Burstein, D. E. (1990). Immunohistochemical localization of basic fibroblast growth factor in astrocytomas. *Cancer Res.* **50**, 7393.
125. Statuto, M., Ennas, M. G., Zamboni, G., Bonetti, F., Pea, M., Bernardello, F., Pozzi, A., Rusnati, M., Gualandris, A., and Presta, M. (1993). Basic fibroblast growth factor in human pheochromocytoma: A biochemical and immunohistochemical study. *Int. J. Cancer* **53**, 5.
126. Ohtani, H., Nakamura, S., Watanabe, Y., Mizoi, T., Saku, T., and Nagura, H. (1993). Immunocytochemical localization of basic fibroblast growth factor in carcinomas and inflammatory lesions of the human digestive tract. *Lab. Invest.* **68**, 520.
127. Laiho, M., and Keski-Oja, J. (1989). Growth factors in the regulation of pericellular proteolysis: A review. *Cancer Res.* **49**, 2533.
128. Moscatelli, D., Presta, M., and Rifkin, D. B. (1986). Purification of a factor from human placenta that stimulates capillary endothelial cell protease production, DNA synthesis, and migration. *Proc. Natl. Acad. Sci. USA* **83**, 2091.
129. Schweigerer, L., Neufeld, G., Friedman, J., Abraham, J. A., Fiddes, J. C., and Gospodarowicz, D. (1987). Capillary endothelial cells express basic fibroblast growth factor, a mitogen that promotes their own growth. *Nature* **325**, 257.
130. Klint, P., and Claesson-Welsh, L. (1999). Signal transduction by fibroblast growth factor receptors. *Front Biosci.* **4**, D165.
131. Burgess, W. H., Dionne, C. A., Kaplow, J., Mudd, R., Friesel, R., Zilberstein, A., Schlessinger, J., and Jaye, M. (1990). Characterization and cDNA cloning of phospholipase C-gamma, a major substrate for heparin-binding growth factor 1 (acidic fibroblast growth factor)-activated tyrosine kinase. *Mol. Cell. Biol.* **10**, 4770.
132. Mohammadi, M., Honegger, A. M., Rotin, D., Fischer, R., Bellot, F., Li, W., Dionne, C. A., Jaye, M., Rubinstein, M., and Schlessinger, J. (1991). A tyrosine-phosphorylated carboxy-terminal peptide of the fibroblast growth factor receptor (Flg) is a binding site for the SH2 domain of phospholipase C-gamma 1. *Mol. Cell. Biol.* **11**, 5068.
133. Zhan, X., Plourde, C., Hu, X., Friesel, R., and Maciag, T. (1994). Association of fibroblast growth factor receptor-1 with c-Src correlates with association between c-Src and cortactin. *J. Biol. Chem.* **269**, 20221.
134. Landgren, E., Blume-Jensen, P., Courtneidge, S. A., and Claesson-Welsh, L. (1995). Fibroblast growth factor receptor-1 regulation of Src family kinases. *Oncogene* **10**, 2027.
135. Lewis, T. S., Shapiro, P. S., and Ahn, N. G. (1998). Signal transduction through MAP kinase cascades. *Adv. Cancer Res.* **74**, 49.
136. Kouhara, H., Hadari, Y. R., Spivak-Kroizman, T., Schilling, J., Bar-Sagi, D., Lax, I., and Schlessinger, J. (1997). A lipid-anchored Grb2-binding protein that links FGF-receptor activation to the Ras/MAPK signaling pathway. *Cell* **89**, 693.
137. Boilly, B., Vercoutter-Edouart, A. S., Hondermarck, H., Nurcombe, V., and Le Bourhis, X. (2000). FGF signals for cell proliferation and migration through different pathways. *Cytokine Growth Factor Rev.* **11**, 295.
138. Hirst, D. G., and Denekamp, J. (1979). Tumour cell proliferation in relation to the vasculature. *Cell Tissue Kinet.* **12**, 31.
139. Sutherland, R. M. (1986). Importance of critical metabolites and cellular interactions in the biology of microregions of tumors. *Cancer* **58**, 1668.
140. Hirst, D. G., Hazlehurst, J. L., and Brown, J. M. (1985). Changes in misonidazole binding with hypoxic fraction in mouse tumors. *Int. J. Radiat. Oncol. Biol. Phys.* **11**, 1349.
141. Dellacono, F. R., Spiro, J., Eisma, R., and Kreutzer, D. (1997). Expression of basic fibroblast growth factor and its receptors by head and neck squamous carcinoma tumor and vascular endothelial cells. *Am. J. Surg.* **174**, 540.
142. Hughes, C. J., Reed, J. A., Cabal, R., Huvos, A. G., Albino, A. P., and Schantz, S. P. (1994). Increased expression of basic fibroblast growth factor in squamous carcinogenesis of the head and neck is less prevalent following smoking cessation. *Am. J. Surg.* **168**, 381.
143. Partridge, M., Kiguwa, S., Luqmani, Y., and Langdon, J. D. (1996) Expression of bFGF, KGF and FGF receptors on normal oral mucosa and SCC. *Eur. J. Cancer B Oral Oncol.* **32B**, 76.

144. Forootan, S. S., Ke, Y., Jones, A. S., and Helliwell, T. R. (2000). Basic fibroblast growth factor and angiogenesis in squamous carcinoma of the tongue. *Oral Oncol.* **36**, 437.
145. Janot, F., el-Naggar, A. K., Morrison, R. S., Liu, T. J., Taylor, D. L., and Clayman, G. L. (1995). Expression of basic fibroblast growth factor in squamous cell carcinoma of the head and neck is associated with degree of histologic differentiation. *Int. J. Cancer* **64**, 117.
146. Riedel, F., Gotte, K., Bergler, W., Rojas, W., and Hormann, K. (2000). Expression of basic fibroblast growth factor protein and its down-regulation by interferons in head and neck cancer. *Head Neck* **22**, 183.
147. Ferrara, N., and Henzel, W. J. (1989). Pituitary follicular cells secrete a novel heparin-binding growth factor specific for vascular endothelial cells. *Biochem. Biophys. Res. Commun.* **161**, 851.
148. Carmeliet, P., Ferreira, V., Breier, G., Pollefeyt, S., Kieckens, L., Gertsenstein, M., Fahrig, M., Vandenhoeck, A., Harpal, K., Eberhardt, C., Declercq, C., Pawling, J., Moons, L., Collen, D., Risau, W., and Nagy, A. (1996). Abnormal blood vessel development and lethality in embryos lacking a single VEGF allele. *Nature* **380**, 435.
149. Ferrara, N., Carver-Moore, K., Chen, H., Dowd, M., Lu, L., O'Shea, K. S., Powell-Braxton, L., Hillan, K. J., and Moore, M. W. (1996). Heterozygous embryonic lethality induced by targeted inactivation of the VEGF gene. *Nature* **380**, 439.
150. Eriksson, U., and Alitalo, K. (1999). Structure, expression and receptor-binding properties of novel vascular endothelial growth factors. *Curr. Top. Microbiol. Immunol.* **237**, 41.
151. Neufeld, G., Cohen, T., Gengrinovitch, S., and Poltorak, Z. (1999). Vascular endothelial growth factor (VEGF) and its receptors. *FASEB J.* **13**, 9.
152. Meyer, M., Clauss, M., Lepple-Wienhues, A., Waltenberger, J., Augustin, H. G., Ziche, M., Lanz, C., Buttner, M., Rziha, H. J., and Dehio, C. (1999). A novel vascular endothelial growth factor encoded by Orf virus, VEGF-E, mediates angiogenesis via signalling through VEGFR-2 (KDR) but not VEGFR-1 (Flt-1) receptor tyrosine kinases, *EMBO J.* **18**, 363.
153. Mustonen, T., and Alitalo, K. (1995). Endothelial receptor tyrosine kinases involved in angiogenesis. *J. Cell Biol.* **129**, 895.
154. Shalaby, F., Rossant, J., Yamaguchi, T. P., Gertsenstein, M., Wu, X. F., Breitman, M. L., and Schuh, A. C. (1995). Failure of blood-island formation and vasculogenesis in Flk-1-deficient mice. *Nature* **376**, 62.
155. Fong, G. H., Rossant, J., Gertsenstein, M., and Breitman, M. L. (1995). Role of the Flt-1 receptor tyrosine kinase in regulating the assembly of vascular endothelium. *Nature* **376**, 66.
156. Taipale, J., Makinen, T., Arighi, E., Kukk, E., Karkkainen, M., and Alitalo, K. (1999). Vascular endothelial growth factor receptor-3. *Curr. Top. Microbiol. Immunol.* **237**, 85.
157. Wiesmann, C., Fuh, G., Christinger, H. W., Eigenbrot, C., Wells, J. A., and de Vos, A. M. (1997). Crystal structure at 1.7 A resolution of VEGF in complex with domain 2 of the Flt-1 receptor. *Cell* **91**, 695.
158. Hubbard, S. R. (1999). Structural analysis of receptor tyrosine kinases. *Prog. Biophys. Mol. Biol.* **71**, 343.
159. Schlessinger, J. (2000). Cell signaling by receptor tyrosine kinases. *Cell* **103**, 211.
160. Guo, D., Jia, Q., Song, H. Y., Warren, R. S., and Donner, D. B. (1995). Vascular endothelial cell growth factor promotes tyrosine phosphorylation of mediators of signal transduction that contain SH2 domains: Association with endothelial cell proliferation. *J. Biol. Chem.* **270**, 6729.
161. Kroll, J., and Waltenberger, J. (1997). The vascular endothelial growth factor receptor KDR activates multiple signal transduction pathways in porcine aortic endothelial cells. *J. Biol. Chem.* **272**, 32521.
162. Abedi, H., and Zachary, I. (1997). Vascular endothelial growth factor stimulates tyrosine phosphorylation and recruitment to new focal adhesions of focal adhesion kinase and paxillin in endothelial cells. *J. Biol. Chem.* **272**, 15442.
163. Gerber, H. P., McMurtrey, A., Kowalski, J., Yan, M., Keyt, B. A., Dixit, V., and Ferrara, N. (1998). Vascular endothelial growth factor regulates endothelial cell survival through the phosphatidylinositol 3′-kinase/Akt signal transduction pathway: Requirement for Flk-1/KDR activation. *J. Biol. Chem.* **273**, 30336.
164. Hiratsuka, S., Minowa, O., Kuno, J., Noda, T., and Shibuya, M. (1998). Flt-1 lacking the tyrosine kinase domain is sufficient for normal development and angiogenesis in mice. *Proc. Natl. Acad. Sci. USA* **95**, 9349.
165. Mattern, J., Koomagi, R., and Volm, M. (1996). Association of vascular endothelial growth factor expression with intratumoral microvessel density and tumour cell proliferation in human epidermoid lung carcinoma. *Br. J. Cancer* **73**, 931.
166. Viglietto, G., Maglione, D., Rambaldi, M., Cerutti, J., Romano, A., Trapasso, F., Fedele, M., Ippolito, P., Chiappetta, G., Botti, G., *et al.* (1995). Upregulation of vascular endothelial growth factor (VEGF) and downregulation of placenta growth factor (PIGF) associated with malignancy in human thyroid tumors and cell lines. *Oncogene* **11**, 1569.
167. Yoshiji, H., Gomez, D. E., Shibuya, M., and Thorgeirsson, U. P., (1996). Expression of vascular endothelial growth factor, its receptor, and other angiogenic factors in human breast cancer. *Cancer Res.* **56**, 2013.
168. Brown, L. F., Berse, B., Jackman, R. W., Tognazzi, K., Manseau, E. J., Senger, D. R., and Dvorak, H. F. (1993). Expression of vascular permeability factor (vascular endothelial growth factor) and its receptors in adenocarcinomas of the gastrointestinal tract. *Cancer Res.* **53**, 4727.
169. Brown, L. F., Berse, B., Jackman, R. W., Tognazzi, K., Manseau, E. J., Dvorak, H. F., and Senger, D. R. (1993). Increased expression of vascular permeability factor (vascular endothelial growth factor) and its receptors in kidney and bladder carcinomas. *Am. J. Pathol.* **143**, 1255.
170. Olson, T. A., Mohanraj, D., Carson, L. F., and Ramakrishnan, S. (1994). Vascular permeability factor gene expression in normal and neoplastic human ovaries. *Cancer Res.* **54**, 276.
171. Hashimoto, M., Ohsawa, M., Ohnishi, A., Naka, N., Hirota, S., Kitamura, Y., and Aozasa, K. (1995). Expression of vascular endothelial growth factor and its receptor mRNA in angiosarcoma. *Lab. Invest.* **73**, 859.
172. Plate, K. H., Breier, G., Weich, H. A., and Risau, W. (1992). Vascular endothelial growth factor is a potential tumour angiogenesis factor in human gliomas in vivo. *Nature* **359**, 845.
173. Shweiki, D., Itin, A., Soffer, D., and Keshet, E. (1992). Vascular endothelial growth factor induced by Hypoxia may mediate hypoxia-initiated angiogenesis. *Nature* **359**, 843.
174. Minchenko, A., Bauer, T., Salceda, S., and Caro, J. (1994). Hypoxic stimulation of vascular endothelial growth factor expression in vitro and in vivo. *Lab. Invest.* **71**, 374.
175. Brogi, E., Wu, T., Namiki, A., and Isner, J. M. (1994). Indirect angiogenic cytokines upregulate VEGF and bFGF gene expression in vascular smooth muscle cells, whereas hypoxia upregulates VEGF expression only. *Circulation* **90**, 649.
176. Shima, D. T., Adamis, A. P., Ferrara, N., Yeo, K. T., Yeo, T. K., Allende, R., Folkman, J., and D'Amore, P. A. (1995). Hypoxic induction of endothelial cell growth factors in retinal cells: Identification and characterization of vascular endothelial growth factor (VEGF) as the mitogen, *Mol. Med.* **1**, 182.
177. Forsythe, J. A., Jiang, B. H., Iyer, N. V., Agani, F., Leung, S. W., Koos, R. D., and Semenza, G. L. (1996). Activation of vascular endothelial growth factor gene transcription by hypoxia-inducible factor 1. *Mol. Cell. Biol.* **16**, 4604.
178. Pertovaara, L., Kaipainen, A., Mustonen, T., Orpana, A., Ferrara, N., Saksela, O., and Alitalo, K. (1994). Vascular endothelial growth factor is induced in response to transforming growth factor-beta in fibroblastic and epithelial cells. *J. Biol. Chem.* **269**, 6271.

179. Ben-Av, P., Crofford, L. J., Wilder, R. L., and Hla, T. (1995). Induction of vascular endothelial growth factor expression in synovial fibroblasts by prostaglandin E and interleukin-1: A potential mechanism for inflammatory angiogenesis. *FEBS Lett.* **372**, 83.
180. Cohen, T., Nahari, D., Cerem, L. W., Neufeld, G., and Levi, B. Z. (1996). Interleukin 6 induces the expression of vascular endothelial growth factor. *J. Biol. Chem.* **271**, 736.
181. Warren, R. S., Yuan, H., Matli, M. R., Ferrara, N., and Donner, D. B. (1996). Induction of vascular endothelial growth factor by insulin-like growth factor 1 in colorectal carcinoma. *J. Biol. Chem.* **271**, 29483.
182. Denhart, B. C., Guidi, A. J., Tognazzi, K., Dvorak, H. F., and Brown, L. F. (1997). Vascular permeability factor/vascular endothelial growth factor and its receptors in oral and laryngeal squamous cell carcinoma and dysplasia. *Lab. Invest.* **77**, 659.
183. Eisma, R. J., Spiro, J. D., and Kreutzer, D. L. (1997). Vascular endothelial growth factor expression in head and neck squamous cell carcinoma. *Am. J. Surg.* **174**, 513.
184. Salven, P., Heikkila, P., Anttonen, A., Kajanti, M., and Joensuu, H. (1997). Vascular endothelial growth factor in squamous cell head and neck carcinoma: Expression and prognostic significance. *Mod. Pathol.* **10**, 1128.
185. Toi, M., Hoshina, S., Takayanagi, T., and Tominaga, T. (1994). Association of vascular endothelial growth factor expression with tumor angiogenesis and with early relapse in primary breast cancer. *Jpn. J. Cancer Res.* **85**, 1045.
186. Gasparini, G. (2000). Prognostic value of vascular endothelial growth factor in breast cancer. *Oncologist* **5**, 37.
187. Salven, P., Ruotsalainen, T., Mattson, K., and Joensuu, H. (1998). High pre-treatment serum level of vascular endothelial growth factor (VEGF) is associated with poor outcome in small-cell lung cancer. *Int. J. Cancer* **79**, 144.
188. Maeda, K., Chung, Y. S., Ogawa, Y., Takatsuka, S., Kang, S. M., Ogawa, M., Sawada, T., and Sowa, M. (1996). Prognostic value of vascular endothelial growth factor expression in gastric carcinoma. *Cancer* **77**, 858.
189. Sauter, E. R., Nesbit, M., Watson, J. C., Klein-Szanto, A., Litwin, S., and Herlyn, M. (1999). Vascular endothelial growth factor is a marker of tumor invasion and metastasis in squamous cell carcinomas of the head and neck. *Clin. Cancer Res.* **5**, 775.
190. Wakisaka, N., Wen, Q. H., Yoshizaki, T., Nishimura, T., Furukawa, M., Kawahara, E., and Nakanishi, I. (1999). Association of vascular endothelial growth factor expression with angiogenesis and lymph node metastasis in nasopharyngeal carcinoma. *Laryngoscope* **109**, 810.
191. Moriyama, M., Kumagai, S., Kawashiri, S., Kojima, K., Kakihara, K., and Yamamoto, E. (1997). Immunohistochemical study of tumor angiogenesis in oral squamous cell carcinoma. *Oral Oncol.* **33**, 369.
192. Mineta, H., Miura, K., Ogino, T., Takebayashi, S., Misawa, K., Ueda, Y., Suzuki, I., Dictor, M., Borg, A., and Wennerberg, J. (2000). Prognostic value of vascular endothelial growth factor (VEGF) in head and neck squamous cell carcinomas. *Br. J. Cancer* **83**, 775.
193. Mukhopadhyay, D., Tsiokas, L., and Sukhatme, V. P. (1995) Wild-type p53 and v-Src exert opposing influences on human vascular endothelial growth factor gene expression. *Cancer Res.* **55**, 6161.
194. Riedel, F., Gotte, K., Schwalb, J., Schafer, C., and Hormann, K. (2000). Vascular endothelial growth factor expression correlates with p53 mutation and angiogenesis in squamous cell carcinoma of the head and neck. *Acta Otolaryngol.* **120**, 105.
195. Dvorak, H. F., Brown, L. F., Detmar, M., and Dvorak, A. M. (1995). Vascular permeability factor/vascular endothelial growth factor, microvascular hyperpermeability, and angiogenesis. *Am. J. Pathol.* **146**, 1029.
196. Salven, P., Manpaa, H., Orpana, A., Alitalo, K., and Joensuu, H. (1997). Serum vascular endothelial growth factor is often elevated in disseminated cancer. *Clin. Cancer Res.* **3**, 647.
197. Linder, C., Linder, S., Munck-Wikland, E., and Strander, H. (1998). Independent expression of serum vascular endothelial growth factor (VEGF) and basic fibroblast growth factor (bfgf) in patients with carcinoma and sarcoma. *Anticancer Res.* **18**, 2063.
198. Homer, J. J., Anyanwu, K., Ell, S. R., Greenman, J., and Stafford, N. D. (1999). Serum vascular endothelial growth factor in patients with head and neck squamous cell carcinoma. *Clin. Otolaryngol.* **24**, 426.
199. Riedel, F., Gotte, K., Schwalb, J., Wirtz, H., Bergler, W., and Hormann, K. (2000). Serum levels of vascular endothelial growth factor in patients with head and neck cancer. *Eur. Arch. Otorhinolaryngol.* **257**, 332.
200. Gabrilovich, D. I., Chen, H. L., Girgis, K. R., Cunningham, H. T., Meny, G. M., Nadaf, S., Kavanaugh, D., and Carbone, D. P. (1996). Production of vascular endothelial growth factor by human tumors inhibits the functional maturation of dendritic cells. *Nature Med.* **2**, 1096.
201. Friedkin, M., and Roberts, D. (1954). The enzymatic synthesis of nucleotides. II. Thymidine and related pyrimidine nucleosides. *J. Biol. Chem.* **207**, 257.
202. Zimmerman, M., and Seidenberg, J. (1964). Deoxyribosyl transfer. I. Thymidine phosphorylase and nucleoside deoxyribosyltransferase in normal and malignant tissues. *J. Biol. Chem.* **239**, 2618.
203. Miyazono, K., Okabe, T., Urabe, A., Takaku, F., and Heldin, C. H. (1987). Purification and properties of an endothelial cell growth factor from human platelets. *J. Biol. Chem.* **262**, 4098.
204. Ishikawa, F., Miyazono, K., Hellman, U., Drexler, H., Wernstedt, C., Hagiwara, K., Usuki, K., Takaku, F., Risau, W., and Heldin, C. H. (1989). Identification of angiogenic activity and the cloning and expression of platelet-derived endothelial cell growth factor. *Nature* **338**, 557.
205. Reynolds, K., Farzaneh, F., Collins, W. P., Campbell, S., Bourne, T. H., Lawton, F., Moghaddam, A., Harris, A. L., and Bicknell, R. (1994). Association of ovarian malignancy with expression of platelet-derived endothelial cell growth factor, *J. Natl. Cancer Inst.* **86**, 1234.
206. Toi, M., Hoshina, S., Taniguchi, T., Yamamoto, Y., Ishitsuka, H., and Tominaga, T. (1995). Expression of platelet-derived endothelial cell growth factor/thymidine phosphorylase in human breast cancer. *Int. J. Cancer* **64**, 79.
207. Takebayashi, Y., Miyadera, K., Akiyama, S., Hokita, S., Yamada, K., Akiba, S., Yamada, Y., Sumizawa, T., and Aikou, T. (1996). Expression of thymidine phosphorylase in human gastric carcinoma. *Jpn. J. Cancer Res.* **87**, 288.
208. Koukourakis, M. I., Giatromanolaki, A., Kakolyris, S., O'Byrne, K. J., Apostolikas, N., Skarlatos, J., Gatter, K.C., and Harris, A. L. (1998). Different patterns of stromal and cancer cell thymidine phosphorylase reactivity in non-small-cell lung cancer: Impact on tumour neoangiogenesis and survival. *Br. J. Cancer* **77**, 1696.
209. Takebayashi, Y., Yamada, K., Maruyama, I., Fujii, R., Akiyama, S., and Aikou, T. (1995). The expression of thymidine phosphorylase and thrombomodulin in human colorectal carcinomas. *Cancer Lett.* **92**, 1.
210. Takebayashi, Y., Akiyama, S., Akiba, S., Yamada, K., Miyadera, K., Sumizawa, T., Yamada, Y., Murata, F., and Aikou, T. (1996). Clinicopathologic and prognostic significance of an angiogenic factor, thymidine phosphorylase, in human colorectal carcinoma, *J. Natl. Cancer Inst.* **88**, 1110.
211. Fujieda, S., Sunaga, H., Tsuzuki, H., Tanaka, N., and Saito, H. (1998). Expression of platelet-derived endothelial cell growth factor in oral and oropharyngeal carcinoma. *Clin. Cancer Res.* **4**, 1583.
212. Fukuiwa, T., Takebayashi, Y., Akiba, S., Matsuzaki, T., Hanamure, Y., Miyadera, K., Yamada, Y., and Akiyama, S. (1999). Expression of thymidine phosphorylase and vascular endothelial cell growth factor in human head and neck squamous cell carcinoma and their different characteristics. *Cancer* **85**, 960.

213. Giatromanolaki, A., Fountzilas, G., Koukourakis, M. I., Arapandoni, P., Theologi, V., Kakolyris, S., Georgoulias, V., Harris, A. L., and Gatter, K. C. (1998). Neo-angiogenesis in locally advanced squamous cell head and neck cancer correlates with thymidine phosphorylase expression and p53 nuclear oncoprotein accumulation. *Clin. Exp. Metastasis* **16**, 665.
214. Wolf, G. T., Hudson, J. L., Peterson, K. A., Miller, H. L., and McClatchey, K. D. (1986). Lymphocyte subpopulations infiltrating squamous carcinomas of the head and neck: Correlations with extent of tumor and prognosis. Otolaryngol. *Head Neck Surg.* **95**, 142.
215. Young, M. R., Wright, M. A., Lozano, Y., Prechel, M. M., Benefield, J., Leonetti, J. P., Collins, S. L., and Petruzzelli, G. J. (1997). Increased recurrence and metastasis in patients whose primary head and neck squamous cell carcinomas secreted granulocyte–macrophage colony-stimulating factor and contained $CD34^+$ natural suppressor cells. *Int. J. Cancer* **74**, 69.
216. Mann, E. A., Spiro, J. D., Chen, L. L., and Kreutzer, D. L. (1992). Cytokine expression by head and neck squamous cell carcinomas. *Am. J. Surg.* **164**, 567.
217. Yamamura, M., Modlin, R. L., Ohmen, J. D., and Moy, R. L. (1993). Local expression of antiinflammatory cytokines in cancer. *J. Clin. Invest.* **91**, 1005.
218. Cohen, R. F., Contrino, J., Spiro, J. D., Mann, E. A., Chen, L. L., and Kreutzer, D. L. (1995). Interleukin-8 expression by head and neck squamous cell carcinoma. Arch. Otolaryngol. *Head Neck Surg.* **121**, 202.
219. Lingen, M. W., Polverini, P. J., and Bouck, N. P. (1996). Retinoic acid induces cells cultured from oral squamous cell carcinomas to become anti-angiogenic. *Am. J. Pathol.* **149**, 247.
220. Chen, Z., Colon, I., Ortiz, N., Callister, M., Dong, G., Pegram, M. Y., Arosarena, O., Strome, S., Nicholson, J. C., and Van Waes, C. (1998). Effects of interleukin-l alpha, interleukin-1 receptor antagonist, and neutralizing antibody on proinflammatory cytokine expression by human squamous cell carcinoma lines. *Cancer Res.* **58**, 3668.
221. Woods, K. V., El-Naggar, A., Clayman, G. L., and Grimm, E. A. (1998). Variable expression of cytolines in human head and neck squamous cell carcinoma cell lines and consistent expression in surgical specimens. *Cancer Res.* **58**, 3132.
222. Eisma, R. J., Spiro, J. D., and Kreutzer, D. L. (1999). Role of angiogenic factors: Coexpression of interleukin-8 and vascular endothelial growth factor in patients with head and neck squamous carcinoma. *Laryngoscope* **109**, 687.
223. Boyle, J. O., Hakim, J., Koch, W., van der Riet, P., Hruban, R. H., Roa, R. A., Correo, R., Eby, Y. J., Ruppert, J. M., and Sidransky, D. (1993). The incidence of p53 mutations increases with progression of head and neck cancer. *Cancer Res.* **53**, 4477.
224. Hegde, P. U., Brenski, A. C., Caldarelli, D. D., Hutchinson, J., Panje, W. R., Wood, N. B., Leurgans, S., Preisler, H. D., Taylor, S. G. T., Caldarelli, L., and Coon, J. S. (1998). Tumor angiogenesis and p53 mutations: Prognosis in head and neck cancer, *Arch. Otolaryngol. Head Neck Surg.* **124**, 80.
225. Dameron, K. M., Volpert, O. V., Tainsky, M. A., and Bouck, N. (1994). Control of angiogenesis in fibroblasts by p53 regulation of thrombospondin-1. *Science* **265**, 1582.
226. Dameron, K. M., Volpert, O. V., Tainsky, M. A., and Bouck, N. (1994). The p53 tumor suppressor gene inhibits angiogenesis by stimulating the production of thrombospondin. *Cold Spring Harb. Symp. Quant. Biol.* **59**, 483.
227. Ravi, D., Ramadas, K., Mathew, B. S., Nalinakumari, K. R., Nair, M. K., and Pillai, M. R. (1998). Angiogenesis during tumor progression in the oral cavity is related to reduced apoptosis and high tumor cell proliferation. *Oral Oncol.* **34**, 543.
228. Thomsen, L. L., Lawton, F. G., Knowles, R. G., Beesley, J. E., Riveros-Moreno, V., and Moncada, S. (1994) Nitric oxide synthase activity in human gynecological cancer. *Cancer Res.* **54**, 1352.
229. Thomsen, L. L., Miles, D. W., Happerfield, L., Bobrow, L. G., Knowles, R. G., and Moncada, S. (1995). Nitric oxide synthase activity in human breast cancer. *Br. J. Cancer* **72**, 41.
230. Rosbe, K. W., Prazma, J., Petrusz, P., Mims, W., Ball, S. S., and Weissler, M. C. (1995). Immunohistochemical characterization of nitric oxide synthase activity in squamous cell carcinoma of the head and neck. Otolaryngol. *Head Neck Surg.* **113**, 541.
231. Morbidelli, L., Chang, C. H., Douglas, J. G., Granger, H. J., Ledda, F., and Ziche, M. (1996). Nitric oxide mediates mitogenic effect of VEGF on coronary venular endothelium. *Am. J. Physiol.* **270**, H411.
232. Ziche, M., Morbidelli, L., Choudhuri, R., Zhang, H. T., Donnini, S., Granger, H. J., and Bicknell, R. (1997). Nitric oxide synthase lies downstream from vascular endothelial growth factor-induced but not basic fibroblast growth factor-induced angiogenesis. *J. Clin. Invest.* **99**, 2625.
233. Gallo, O., Masini, E., Morbidelli, L., Franchi, A., Fini-Storchi, I., Vergari, W. A., and Ziche, M. (1998). Role of nitric oxide in angiogenesis and tumor progression in head and neck cancer. *J. Natl. Cancer Inst.* **90**, 587.
234. Brower, V. (1999). Tumor angiogenesis: New drugs on the block. Nature Biotechnol. **17**, 963.
235. Giavazzi, R., Albini, A., Bussolino, F., DeBraud, F., Presta, M., Ziche, M., and Costa, A. (2000). The biological basis for antiangiogenic therapy. *Eur. J. Cancer* **36**, 1913.
236. Kerbel, R. S. (2000). Tumor angiogenesis: Past, present and the near future. *Carcinogenesis* **21**, 505.
237. Bhargava, P. (2000). Antiangiogenesis. *In* "Clinical Oncology" (M. D. Abeloff, J. O. Armitage, A. S. Lichter, and J. E. Niederhuber, eds.), 2nd Ed., p. 243. Churchill Livingstone, New York.
238. Lingen, M. W. (1999) Angiogenesis in the development of head and neck cancer and its inhibition by chemopreventive agents. *Crit. Rev. Oral. Biol. Med.* **10**, 153.
239. Zou, C. P., Clifford, J. L., Xu, X. C., Sacks, P. G., Chambon, P., Hong, W. K., and Lotan, R. (1994). Modulation by retinoic acid (RA) of squamous cell differentiation, cellular RA-binding proteins, and nuclear RA receptors in human head and neck squamous cell carcinoma cell lines. *Cancer Res.* **54**, 5479.
240. Shalinsky, D. R., Bischoff, E. D., Gregory, M. L., Gottardis, M. M., Hayes, J. S., Lamph, W. W., Heyman, R. A., Shirley, M. A., Cooke, T. A., Davies, P. J., *et al* (1995). Retinoid-induced suppression of squamous cell differentation in human oral squamous cell carcinoma xenografts (line 1483) in athymic nude mice. *Cancer Res.* **55**, 3183.
241. Sacks, P. G., Harris, D., and Chou, T. C. (1995). Modulation of growth and proliferation in squamous cell carcinoma by retinoic acid: A rationale for combination therapy with chemotherapeutic agents. *Int. J. Cancer* **61**, 409.
242. Liaudet-Coopman, E. D., Berchem, G. J., and Wellstein, A. (1997). In vivo inhibition of angiogenesis and induction of apoptosis by retinoic acid in squamous cell carcinoma. Clin. *Cancer Res.* **3**, 179.
243. Oikawa, T., Hirotani, K., Nakamura, O., Shudo, K., Hiragun, A., and Iwaguchi, T. (1989). A Highly potent antiangiogenic activity of retinoids. *Cancer Lett.* **48**, 157.
244. Arensman, R. M., and Stolar, C. J. (1979). Vitamin A effect on tumor angiogenesis. *J. Pediatr. surg.* **14**, 809.
245. Majewski, S., Szmurlo, A., Marczak, M., Jablonska, S., and Bollag, W. (1993). Inhibition of tumor cell-induced angiogenesis by retinoids, 1,25-dihydroxyvitamin D3 and their combination. *Cancer Lett.* **75**, 35.
246. Majewski, S., Szmurlo, A., Marczak, M., Jablonska, S., and Bollag, W. (1994). Synergistic effect of retinoids and interferon alpha on tumor-induced angiogenesis: Anti-angiogenic effect on HPV-harboring tumor-cell lines. *Int. J. Cancer* **57**, 81.
247. Majewski, S., Marczak, M., Szmurlo, A., Jablonska, S., and Bollag, W. (1995). Retinoids interferon alpha, 1,25-dihydroxyvitamin D3 and their combination inhibit angiogenesis induced by non-HPV-harboring

tumor cell lines: RAR alpha mediates the antiangiogenic effect of retinoids. *Cancer Lett.* **89**, 117.

248. Liaudet-Coopman, E. D. E., and Wellstein, A. (1996). Regulation of gene expression of a binding protein for fibroblast growth factors by retinoic acid. *J. Biol. Chem.* **271**, 21303.
249. Maheshwari, R. K., Srikantan, V., Bhartiya, D., Kleinman, H. K., and Grant, D. S. (1991). Differential effects of interferon gamma and alpha on in vitro model of angiogenesis. *J. Cell. Physiol.* **146**, 164.
250. Ruszczak, Z., Detmar, M., Imcke, E., and Orfanos, C. E. (1990). Effects of rIFN alpha, beta, and gamma on the morphology, proliferation, and cell surface antigen expression of human dermal microvascular endothelial cells in vitro. *J. Invest. Dermatol.* **95**, 693.
251. Saiki, I., Sato, K., Yoo, Y. C., Murata, J., Yoneda, J., Kiso, M., Hasegawa, A., and Azuma, I. (1992). Inhibition of tumor-induced angiogenesis by the administration of recombinant interferon-gamma followed by a synthetic lipid-A subunit analogue (GLA-60). *Int. J. Cancer* **51**, 641.
252. Hong, W. K., Lippman, S. M., Itri, L. M., Karp, D. D., Lee, J. S., Byers, R. M., Schantz, S. P., Kramer, A. M., Lotan, R., Peters, L. J., *et al.* (1990). Prevention of second primary tumors with isotretinoin in squamous-cell carcinoma of the head and neck. *N. Engl. J. Med.* **323**, 795.
253. Lippman, S. M., Glisson, B. S., Kavanagh, J. J., Lotan, R., Hong, W. K., Paredes-Espinoza, M., Hittelman, W. N., Holdener, E. E., and Krakoff, I. H. (1993). Retinoic acid and interferon combination studies in human cancer. *Eur. J. Cancer.* **29**A, S9.
254. Moore, D. M., Kalvakolanu, D. V., Lippman, S. M., Kavanagh, J. J., Hong, W. K., Borden, E. C., Paredes-Espinoza, M., and Krakoff, I. H. (1994). Retinoic acid and interferon in human cancer: Mechanistic and clinical studies. *Semin. Hematol.* **31**, 31.
255. Lingen, M. W., Polverini, P. J., and Bouck, N. P. (1998). Retinoic acid and interferon alpha act synergistically as antiangiogenic and antitumor against human head and neck squamous cell carcinoma. *Cancer Res.* **58**, 5551.
256. Ingber, D., Fujita, T., Kishimoto, S., Sudo, K., Kanamaru, T., Brem, H., and Folkman, J. (1990). Synthetic analogues of fumagillin that inhibit angiogenesis and suppress tumour growth. *Nature* **348**, 555.
257. Yanase, T., Tamura, M., Fujita, K., Kodama, S., and Tanaka, K. (1993). Inhibitory effect of angiogenesis inhibitor TNP-470 on tumor growth and metastasis of human cell lines in vitro and in vivo. *Cancer Res.* **53**, 2566.
258. Gleich, L. L., Zimmerman, N., Wang, Y. O., and Gluckman, J. L. (1998). Angiogenic inhibition for the treatment of head and neck cancer. *Anticancer Res.* **18**, 2607.
259. Ueda, N., Kamata, N., Hayashi, E., Yokoyama, K., Hoteiya, T., and Nagayama, M. (1999). Effects of an anti-angiogenic agent, TNP-470, on the growth of oral squamous cell carcinomas. *Oral Oncol.* **35**, 554.
260. Hori, A., Sasada, R., Matsutani, E., Naito, K., Sakura, Y., Fujita, T., and Kozai, Y. (1991). Suppression of solid tumor growth by immunoneutralizing monoclonal antibody against human basic fibroblast growth factor. *Cancer Res.* **51**, 6180.
261. Kim, K. J., Li, B., Winer, J., Armanini, M., Gillett, N., Phillips, H. S., and Ferrara, N. (1993). Inhibition of vascular endothelial growth factor-induced angiogenesis suppresses tumour growth in vivo. *Nature* **362**, 841.
262. Olson, K. A., Fett, J. W., French, T. C., Key, M. E., and Vallee, B. L. (1995). Angiogenin antagonists prevent tumor growth in vivo. *Proc. Natl. Acad. Sci. USA* **92**, 442.
263. Braddock, P. S., Hu, D. E., Fan, T. P., Stratford, I. J., Harris, A. L., and Bicknell, R. (1994). A structure-activity analysis of antagonism of the growth factor and angiogenic activity of basic fibroblast growth factor by suramin and related polyanions. *Br. J. Cancer* **69**, 890.
264. Takano, S., Gately, S., Neville, M. E., Herblin, W. F., Gross, J. L., Engelhard, H., Perricone, M., Eidsvoog, K., and Brem, S. (1994). Suramin, an anticancer and angiosuppressive agent, inhibits endothelial cell binding of basic fibroblast growth factor, migration, proliferation, and induction of urokinase-type plasminogen activator. *Cancer Res.* **54**, 2654.
265. Myers, C., Cooper, M., Stein, C., LaRocca, R., Walther, M. M., Weiss, G., Choyke, P., Dawson, N., Steinberg, S., Uhrich, M. M., *et al.* (1992) Suramin: A novel growth factor antagonist with activity in hormone- refractory metastatic prostate cancer. *J. Clin. Oncol.* **10**, 881.
266. Tanaka, N. G., Sakamoto, N., Inoue, K., Korenaga, H., Kadoya, S., Ogawa, H., and Osada, Y. (1989). Antitumor effects of an antiangiogenic polysaccharide from an Arthrobacter species with or without a steroid. *Cancer Res.* **49**, 6727.
267. Maione, T. E., Gray, G. S., Petro, J., Hunt, A. J., Donner, A. L., Bauer, S. I., Carson, H. F., and Sharpe, R. J. (1990). Inhibition of angiogenesis by recombinant human platelet factor-4 and related peptides. *Science* **247**, 77.
268. O'Reilly, M. S., Holmgren, L., Shing, Y., Chen, C., Rosenthal, R. A., Moses, M., Lane, W. S., Cao, Y., Sage, E. H., and Folkman, J. (1994). Angiostatin: A novel angiogenesis inhibitor that mediates the suppression of metastases by a Lewis lung carcinoma. *Cell* **79**, 315.
269. O'Reilly, M. S., Boehm, T., Shing, Y., Fukai, N., Vasios, G., Lane, W. S., Flynn, E., Birkhead, J. R., Olsen, B. R., and Folkman, J. (1997). Endostatin: An endogenous inhibitor of angiogenesis and tumor growth. *Cell* **88**, 277.
270. Brooks, P. C., Montgomery, A. M., Rosenfeld, M., Reisfeld, R. A., Hu, T., Klier, G., and Cheresh, D. A. (1994). Integrin alpha v beta 3 antagonists promote tumor regression by inducing apoptosis of angiogenic blood vessels. *Cell* **79**, 1157.
271. Brown, P. D., and Giavazzi, R. (1995). Matrix metalloproteinase inhibition: A review of anti-tumour activity. *Ann. Oncol.* **6**, 967.
272. Tamargo, R. J., Bok, R. A., and Brem, H. (1991). Angiogenesis inhibition by minocycline. *Cancer Res.* **51**, 672.
273. Mueller, B. M., Yu, Y. B., and Laug, W. E. (1995). Overexpression of plasminogen activator inhibitor 2 in human melanoma cells inhibits spontaneous metastasis in scid/scid mice. *Proc. Natl. Acad. Sci. USA* **92**, 205.
274. Gagliardi, A. R., and Collins, D. C. (1994) Inhibition of angiogenesis by aurintricarboxylic acid. *Anticancer Res.* **14**, 475.
275. Vukanovic, J., Passaniti, A., Hirata, T., Traystman, R. J., Hartley-Asp, B., and Isaacs, J. T. (1993). Antiangiogenic effects of the quinoline-3-carboxamide linomide. *Cancer Res.* **53**, 1833.
276. Gagliardi, A., and Collins, D. C. (1993). Inhibition of angiogenesis by antiestrogens, *Cancer Res.* **53**, 533.
277. Matsubara T., Saura, R., Hirohata, K., and Ziff, M. (1989). Inhibition of human endothelial cell proliferation in vitro and neovascularization in vivo by D-penicillamine. *J. Clin. Invest.* **83**, 158.
278. Fotsis, T., Pepper, M., Adlercreutz, H., Fleischmann, G., Hase, T., Montesano, R., and Schweigerer, L. (1993). Genistein, a dietary-derived inhibitor of in vitro angiogenesis. *Proc. Natl. Acad. Sci. USA* **90**, 2690.
279. D'Amato, R. J., Loughnan, M. S., Flynn, E., and Folkman, J. (1994). Thalidomide is an inhibitor of angiogenesis. *Proc. Natl. Acad. Sci. USA* **91**, 4082.
280. Sidky, Y. A., and Borden, E. C. (1987). Inhibition of angiogenesis by interferons: Effects on tumor- and lymphocyte-induced vascular responses. *Cancer Res.* **47**, 5155.
281. Voest, E. E., Kenyon, B. M., O'Reilly, M. S., Truitt, G., D'Amato, R. J., and Folkman, J. (1995). Inhibition of angiogenesis in vivo by interleukin 12, *J. Natl. Cancer Inst.* **87**, 581.
282. Oikawa, T., Hirotani, K., Ogasawara, H., Katayama, T., Nakamura, O., Iwaguchi, T., and Hiragun, A. (1990). Inhibition of angiogenesis by vitamin D3 analogues. *Eur. J. Pharmacol.* **178**, 247.
283. Shokravi, M. T., Marcus, D. M., Alroy, J., Egan, K., Saornil, M. A., and Albert, D. M. (1995). Vitamin D inhibits angiogenesis in transgenic murine retinoblastoma. *Invest. Ophthalmol. Vis. Sci.* **36**, 83.

284. Ziche, M., Morbidelli, L., Masini, E., Amerini, S., Granger, H. J., Maggi, C. A., Geppetti, P., and Ledda, F. (1994). Nitric oxide mediates angiogenesis in vivo and endothelial cell growth and migration in vitro promoted by substance P. *J. Clin. Invest.* **94**, 2036.
285. Weinstat-Saslow, D. L., Zabrenetzky, V. S., VanHoutte, K., Frazier, W. A., Roberts, D. D., and Steeg, P. S. (1994). Transfection of thrombospondin 1 complementary DNA into a human breast carcinoma cell line reduces primary tumor growth, metastatic potential, and angiogenesis, *Cancer Res.* **54**, 6504.
286. Greer, P., Haigh, J., Mbamalu, G., Khoo, W., Bernstein, A., and Pawson, T. (1994). The Fps/Fes protein-tyrosine kinase promotes angiogenesis in transgenic mice. *Mol. Cell. Biol.* **14**, 6755.

C H A P T E R

8

Cell Cycle Regulatory Mechanisms in Head and Neck Squamous Cell Carcinoma

W. ANDREW YEUDALL and KATHARINE H. WRIGHTON

Molecular Carcinogenesis Group
Head and Neck Cancer Program
Guy's King's and St. Thomas' Schools of Medicine and Dentistry
King's College London
London SE1 9RT, United Kingdom

I. Introduction 101
II. Progression through the Mammalian Cell Cycle 101
A. Regulation of Cyclin-Dependent Kinase (CDK) Activity 101
B. Function of CDK Complexes 103
III. Cyclin-Dependent Kinase Inhibitors (CKIs) 104
A. Cip/Kip Family 104
B. Ink4 Family 106
C. p12^{DOC-1} 108
IV. Alterations in Cell Cycle Regulators 108
A. p53 108
B. G1/S Deregulation 108
C. Cell Cycle Regulation and Therapy 109
References 110

I. INTRODUCTION

Precise temporal and spatial regulation of cellular proliferation and programmed cell death is important for normal development, including cell and tissue growth and differentiation. Upsetting the fine balance of these processes, as a result of genetic and/or biochemical abnormalities, may have serious consequences for the organism in terms of disease states, including neoplasia. It is, therefore, of considerable importance to understand (1) the molecular mechanisms that regulate cell proliferation and cell death and (2) how the disruption of regulatory pathways plays a role in tumor development and progression. This knowledge allows us to predict disease course and to target specific biochemical mediators for therapeutic purposes.

II. PROGRESSION THROUGH THE MAMMALIAN CELL CYCLE

In order to complete a round of cell division, proliferating cells must grow and copy their DNA content. Thus, cells progress in a cyclical fashion through a series of well-recognized steps (Fig. 8.1). After dividing, cells enter a gap phase (G1), followed by a phase of DNA synthesis (S). Subsequent to this is a second gap (G2), after which the cells enter mitosis (M). Mitosis can be further subdivided into prophase, metaphase, anaphase, and telophase, at which point two daughter cells are formed. Cells that are not actively proliferating exit the cycle into quiescence (G0).

Progression through the cell cycle is driven by the assembly and disassembly of a series of protein kinase complexes (Table 8.1), which phosphorylate their substrates on specific serine and/or threonine residues. These enzyme complexes consist of a catalytic subunit, the cyclin-dependent kinase (CDK); a regulatory cyclin subunit; and proliferating cell nuclear antigen (PCNA), a component of the DNA polymerase δ enzyme complex. CDK complexes may also include, perhaps paradoxically, one of the cyclin kinase inhibitors (CKIs; see Table 8.2 and later), such as p21 (WAF1/Cip1/Sdi1) or p27 (Kip1).

A. Regulation of Cyclin-Dependent Kinase (CDK) Activity

CDK activity is regulated by a number of mechanisms, which probably reflects the importance of maintaining exquisite control of cell proliferation in response to a continuously

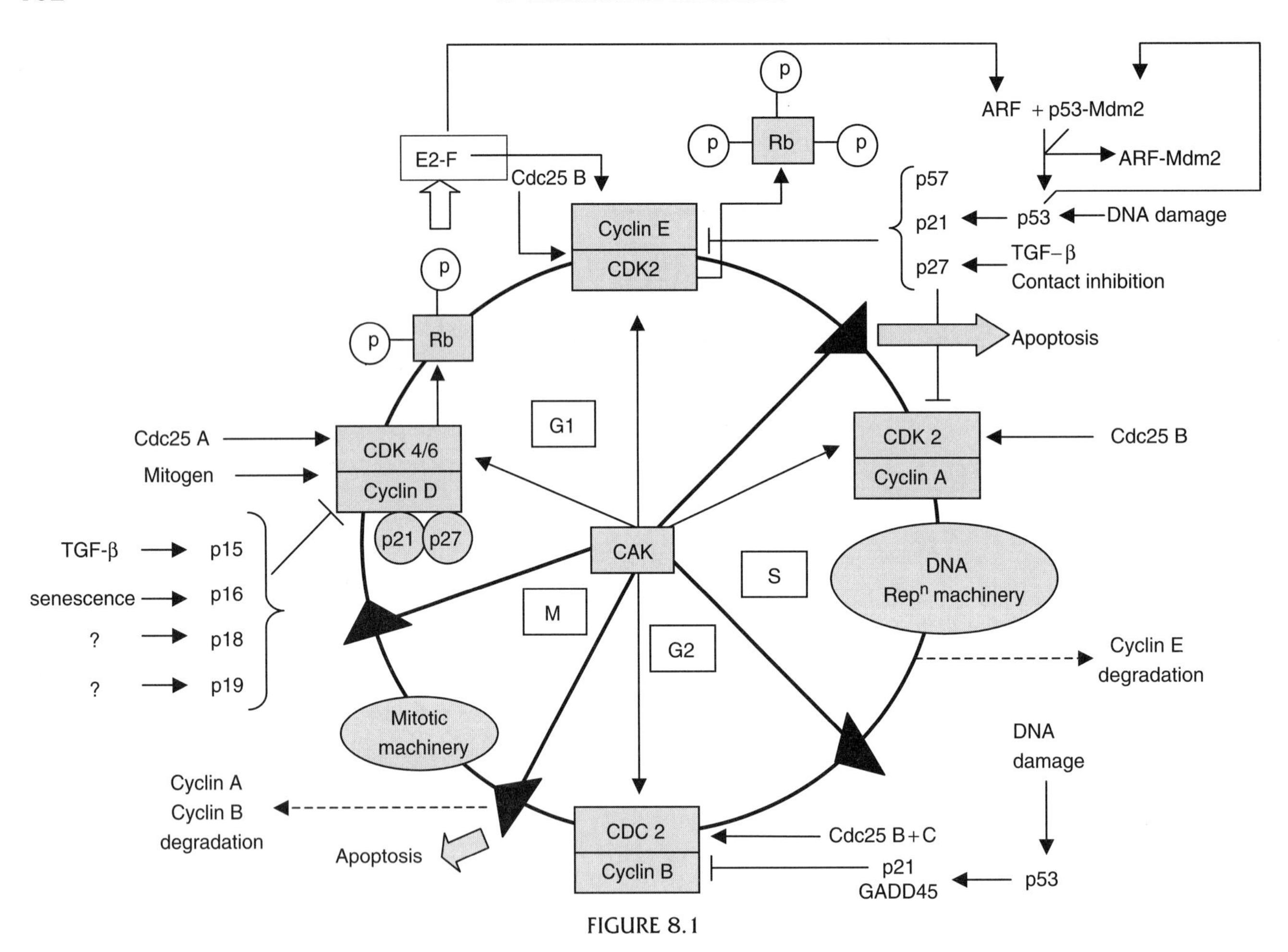

FIGURE 8.1

changing environment. First, to enable an active complex to form, the CDK must associate with its cognate cyclin, e.g., cyclin A with CDK2 or CDC2; cyclin B with CDC2; D-type cyclins with CDK4 and CDK6; and cyclin E with CDK2. This may be regulated by the availability of the cyclin molecule within the cell. Transcription and translation of cyclin genes are temporally regulated, such that cyclin proteins are synthesized only at specific times during the cell cycle. Cyclins are also subject to regulated proteolysis within the cell. This is mediated by ubiquitin-dependent proteosomes and is dependent on a specific amino acid motif—the destruction box—present within the cyclin molecule. Ubiquitin-mediated degradation of cyclins is also dependent, in some cases, on concomitant phosphorylation of the cyclin. For instance, destruction of cyclin E requires its phosphorylation by CDK2, whereas D-type cyclins are also phosphorylated prior to proteolysis. The availability of cyclins to complex with CDKs may additionally be regulated by subcellular compartmentalization. Thus, during S phase and G2, cyclin B is confined to the cytoplasm but translocates rapidly to the nucleus immediately prior to mitosis.

TABLE 8.1 Relationships of Cyclin-Dependent Kinase (CDK) and Cyclin Subunits

CDK	CYCLIN	Cycle stage and role of active CDK/cyclin complex	Cycle outcome
CDK4/6	Cyclin D 1, 2, 3	Acts mid-G1 to phosphorylate Rb	G1 and S progression
CDK2	Cyclin E	Acts late in G1 to phosphorylate proteins for G1/S transition	G1/S transition
CDK2	Cyclin A	Phosphorylates DNA replication machinery in S phase	DNA synthesis
CDC2	Cyclin B 1, cyclin A	Phosphorylates histones / lamins for G2/M transition	Mitosis
CDK7	Cyclin H	Forms complex also known as CAK. Acts throughout cell cycle to phosphorylate CDKs	CDK activation

TABLE 8.2 Relationships between CDKs and CDK Inhibitors (CKIs)

CDK–Cyclin	CKI	Initiator of CKI activity
CDK4/6–cyclin D	p16/Ink4A	Senescence
	p15/Ink4B	TGF-β
	p18/InkC	Development; ? others
	p19/InkD	Development; ? others
	p21	Mitogens; DNA damage
	p27	TGF-β; cell contact
CDK2–cyclin E/A	p21	DNA damage
	p27	TGF-β/senescence
	p57	?
CDC2–cyclin	p21	DNA damage (p53)

CDK activity is also regulated by the phosphorylation status of the kinase itself. One important site is a conserved threonine residue, e.g., T160 in CDK2, which is a target for CDK-activating kinase (CAK; CDK7-cyclin H-mat1) [1,2]. Phosphorylation of this residue is thought to improve binding between the kinase and its substrate, although it has no direct effect on the ATP-binding site or the catalytic site.

CDKs are also subject to inhibitory phosphorylations on conserved threonine and tyrosine residues (T14, Y15), which must be removed to enable full activation of the kinase. This is carried out by members of the cdc25 phosphatase family, of which three isoforms of cdc25 (A, B, and C) have been identified in humans. It is likely that the inhibitory phosphorylation of CDKs plays a major role in regulating cell cycle progression. cdc25A and cdc25B are expressed during G1 and at the G1/S-phase boundary. Thus, cdc25A activity is probably required for the onset of S phase, as it is phosphorylated and activated by cyclin E–CDK2 complexes, which are active at the G1/S transition. Further evidence exists to implicate cdc25A in activation of the G1 kinases CDK4 and CDK6, as transforming growth factor β (TGFβ) is thought to inhibit their activity by decreasing the levels of cdc25A that are present in the cell [3]. cdc25C is thought to be the isoform that is active during mitosis, targeting CDC2 complexes. It is now known that regulation of the level of cdc25A is a crucial component of the DNA damage response of human cells. When cells were exposed to ultraviolet light or ionizing radiation, cdc25A was destroyed rapidly by a ubiquitin-proteasome-dependent mechanism, resulting in the persistent inhibitory phosphorylation of CDK2, thereby preventing entry into S phase [4]. This process appears to occur independently of the activity of the p53 tumor suppressor protein, a major regulator of the G1 DNA damage checkpoint (reviewed in Giaccia and Kastan [5]). However, it was shown that the kinase Chk1, which is capable of phosphorylating all three cdc25 isoforms, is activated in response to ultraviolet irradiation and increases its association with cdc25A. Blocking Chk1 activity using chemical inhibitors was found to overcome UV-induced degradation of cdc25A [4], suggesting that Chk1 is, indeed, an upstream regulator of cdc25A in the G1 DNA damage response.

CDK activity is subject to further regulation by association with members of two classes of proteins: the Cip/Kip family and the Ink4 family of CKIs. The Cip/Kip family consists of p21 (also known as WAF1 [6]; Cip1 [7]; sdi1 [8]; Cap20 [9], p27 (Kip1 [10–12]), and p57 (Kip2 [13,14]). These molecules are capable of inhibiting the function of G1/S (CDK2–cyclin E) and S-phase (CDK2–cyclin A) complexes, thereby blocking progression through G1 into S phase. Initially, it was also thought that Cip/Kip CKIs inhibited cyclin D-dependent kinase activity. However, it is now believed that these molecules act as positive regulators of CDK4 and CDK6 complexes. In contrast, Ink4 CKIs show considerable specificity for CDK4 and CDK6, and hence negatively influence progression through G1. The function of CKIs and their deregulation in cancer are discussed in detail later.

B. Function of CDK Complexes

In response to mitogenic signaling, primarily via the Ras-Raf-MEK-ERK cascade, transcription of D-type cyclins is induced [15–20], and the assembly of CDK4/6-cyclin D complexes is facilitated in an ERK-dependent manner [21,22]. Further mitogenic signaling through phosphatidylinositol 3-OH kinase (PI-3K) to protein kinase B (PKB/AKT) maintains cyclin D in the nucleus by preventing its phosphorylation by glycogen synthase kinase-3β (GSK-3β) and thus its subsequent export and proteolysis [23].

The major substrate of G1 CDK–cyclin D complexes is the product of the retinoblastoma gene Rb, which becomes phosphorylated on serine and threonine residues during mid-G1. Rb is one of the family of "pocket proteins," which together with its related proteins p107 and p130 play a central role in regulating cell cycle progression. In its underphosphorylated state, Rb is able to bind and sequester members of the E2-F family of transcription factors (reviewed by Weinberg [24]). However, phosphorylation by CDKs results in a conformational change in Rb, which probably opens it up to subsequent phosphorylation in late G1 by CDK2–cyclin E. The net result of this is to facilitate the release of E2-F, which then coordinates the transcription of genes whose products are required for DNA synthesis and, thus, progression through S phase (reviewed by Nevins [25]), including cyclins E and A. Indeed, forced expression of E2-F in quiescent cells can drive them into S phase [26–28] and may overcome antiproliferative signals that normally result in G1 arrest [29]. This results in increased levels and availability of A and E cyclins, which are necessary for the activation of CDK2. CDK2 activation is further enhanced by the sequestration of Cip/Kip CKIs by cyclin D–CDK complexes, thereby relieving their inhibitory effects on CDK2. Indeed, Cip/Kip CKIs are thought to act as assembly factors for cyclin D–CDK complexes, as they bind to both kinase and cyclin subunits and enhance their association [30].

Further mechanisms that enable the cell to progress into and through the S phase include phosphorylation of p27/Kip1 by CDK2, and its subsequent targeting for degradation by the ubiquitin-proteasome machinery [31,32], and maintenance of Rb in a hyperphosphorylated state. These relieve, to a considerable extent, the dependence of the cell on mitogens, although this requirement is probably regained as a result of destruction of cyclin E and cyclin A during S phase and G2, respectively [33]. Increased proteasome-mediated degradation of p27 has been reported in aggressive colorectal carcinomas and may be another mechanism to abrogate growth repression in at least a subset of cancers [34].

Upon cessation of mitogenic stimulation, D cyclins are degraded rapidly (see earlier discussion), releasing Cip/Kip proteins, which are then available to bind and inhibit CDK2 complexes and mediate cell cycle arrest in G1 [10–12]. At least in the case of p27/Kip1, this occurs as a result of the CKI disrupting the ATP-binding site of the CDK subunit [35]. In contrast, complexes containing equimolar amounts of cyclin D, CDK4/6, and p21/p27 are catalytically active [30,36]. The importance of the CDK4–p27 interaction for cell cycle progression is underlined elsewhere [37], which demonstrated a delayed S-phase entry in fibroblasts from CDK4$^{-/-}$ mice, whereas this was partially restored in cells derived from CDK4$^{-/-}$ p27$^{-/-}$ animals.

III. CYCLIN-DEPENDENT KINASE INHIBITORS (CKIs)

A. Cip/Kip Family

As mentioned previously, Cip/Kip CKIs act as assembly factors for G1 CDK–cyclin D complexes, while they also inhibit CDK2–cyclin E/A. p21/Cip1 was isolated independently by a number of groups using different approaches. For instance, Cip1 was identified as a CDK2-interacting protein using a yeast interaction screen [7], as a transcriptional target and mediator of G1 arrest by the tumor suppressor protein p53 (WAF1—wild-type p53 activated fragment) [6], and as a molecule whose expression was upregulated in senescent cells [8]. The gene encoding p21 is located on chromosome 6p21.2 in humans [6].

The discovery that p21 expression is upregulated in response to wild-type p53 activity is of considerable importance. It is now clear that p21 is responsible to a large extent for the G1 arrest that occurs in response to DNA-damaging agents, including ultraviolet or γ-irradiation, and commonly used cancer chemotherapeutics such as adriamycin, 5-fluorouracil, bleomycin, and cisplatin [38]. Thus, as p53 is inactivated at high frequency in a broad spectrum of human tumors [39], the p21-dependent G1 cell cycle arrest in response to DNA damage is abrogated, and cells continue to proliferate and duplicate their DNA content without repairing the damage. Furthermore, overexpression of molecules that induce the transcription of S-phase genes, such as E2-F (mentioned earlier) or B-myb [40], may enable tumor cells to overcome p21-mediated growth arrest. p21 is also induced during p53-dependent apoptosis, but not during apoptosis that occurs by p53-independent mechanisms [41]. Other studies have shown that p21 can block programmed cell death in a number of cell systems, including differentiating muscle cells [42] and prostaglandin-treated colon cancer cells [43]. However, no effect on apoptosis was observed when p21 was expressed ectopically in a neural cell line [44], suggesting that p21-dependent regulation of cell death pathways may be cell type specific or may depend on other factors, such as endogenous p53 status or growth factor signaling. Indeed, p21 has been shown to inhibit the kinases c-jun NH_2-terminal kinase (JNK) and p38, both of which have been implicated in cell death in response to stress signals [45].

Although, mutation of p53 is common in human cancer, some tumors retain one or more wild-type alleles. As p21 is a critical mediator of p53-dependent growth arrest, it might, therefore, be predicted that tumors expressing a functional p53 would subvert this mechanism by expressing aberrant forms of p21. However, from a number of studies carried out to date, it would appear that intragenic mutation of p21 is relatively rare in human cancers [46–48], although polymorphisms have been identified [49,50], some of which are associated with an increased risk of neoplasia [51]. This may indicate a fundamental requirement for p21 in cellular growth control and development, although it should be noted that p21$^{-/-}$ mice are viable [52], albeit more susceptible to the effects of DNA-damaging agents.

As p21 plays a central role in regulating cell cycle progression, it is perhaps not surprising that factors other than wild-type p53 are capable of inducing p21 expression. The first evidence for p53-independent expression of p21 came from a series of experiments by Michieli *et al.* [53], who demonstrated that growth factors, including platelet-derived growth factor (PDGF) and fibroblast growth factor (FGF), could induce p21 in p53-null fibroblasts, although no increase in p21 mRNA was observed when these cells were exposed to ionizing radiation, indicating that p53-dependent induction of p21 was compromised. Indeed, the elevation of p21 expression in response to mitogenic agents is entirely consistent with its role as an assembly factor for cyclin D–CDK4/6 complexes. Further underlining this concept is a report [54] demonstrating that antisense oligonucleotides targeting p21 result in decreased cyclin D–CDK4 complexes, whereas cyclin E–CDK2 is unaffected, and that p21 is required for the proliferation of vascular smooth muscle cells in response to PDGF. A number of other effectors of p21 expression have been identified. These include transforming growth factor α (TGFα), TGFβ, and activin A [55–59], Smad proteins, which signal downstream of TGFβ/activin [60,61], IL-1 [62], tumor necrosis factor α (TNFα [63]), interferon γ (IFNγ [64]), epidermal growth factor (EGF [65,66]), and steroids such as progesterone [67]. Interestingly, it has been shown that members of the jun family of transcription factors (c-Jun, junB, junD, ATF-2) cooperate with the transcription factor Sp1 to transactivate transcription from the p21 promoter [68]. Thus, this likely provides an important mechanism for integrating signals from a number of biochemical pathways to regulate cell cycle progression and/or cell death.

Removal of cells from cycle into a quiescent state is an important aspect of terminal differentiation. Thus, inhibition of CDK activity is necessary for this to occur, and CKIs, including p21, are now known to play a role in this process. For example, p21 expression is induced in response to the differentiation programs of HL60 promyelocytic cells by agents such as butyrate, retinoids, and dimethyl sulfoxide [69,70]. Furthermore, A431 squamous carcinoma cells undergo a reversible differentiation in culture when treated with EGF [71]. It is now known that this occurs as a result of elevated p21 expression [65,66] and is mediated primarily by a STAT-1-dependent pathway [64]. Additional studies on differentiating muscle cells have shown that p21 levels are elevated during terminal differentiation in response to MyoD and that ectopic expression of p21 in this system enhances muscle-specific gene expression [72–74]. However, the precise role of p21 in the process of terminal differentiation remains unclear, as more recent evidence suggests (1) that while p21 is increased in postmitotic cells, the levels may in fact decrease further along the differentiation pathway and (2) that ectopic expression of p21 at this stage can inhibit expression of differentiation markers [75].

Aside from its role as a CDK inhibitor, p21 has also been shown to inhibit DNA replication *in vitro* [76]. In a cell-free system, p21 was found to interact directly with PCNA, in the absence of cyclin–CDK complexes. This interaction blocks PCNA function to act as a processivity factor for DNA polymerases δ and ε *in vitro* [77] and is also important in preventing S-phase progression [78], inhibiting endoreduplication [79], and enhancing DNA repair [80,81]. The domain of p21 that participates in PCNA binding is located in the carboxyl terminus of the protein [82]. Binding of p21 to PCNA is regulated by phosphorylation within the carboxyl terminus of p21 at S146 and/or T145 [83], which blocks the interaction of these molecules. However, the kinase(s) responsible for this posttranslational modification of p21 remains to be identified. It is interesting to note that while p21 inhibits PCNA-dependent DNA replication, it does not interfere with the function of PCNA in nucleotide excision repair [84,85]. Thus, in response to DNA damage, p53 induces expression of p21 to halt cell cycle progression and DNA replication, but does not hinder repair of the damaged genetic material.

The second member of the p21-related proteins to be identified was p27/Kip1, which was isolated by three independent laboratories [10–12,86]. The gene encoding p27 is situated on human chromosome 12p12.3-pter [87–89] and shares 44% identity with p21 in the amino-terminal region of the protein.

In a similar manner to p21, p27 acts as an assembly factor for cyclin D–CDK complexes and is present as a component of such complexes in proliferating cells. p27 levels are relatively high in quiescent cells and are titrated out by assembling cyclin D–CDK complexes during G1. This relieves the inhibitory activity of p27 on cyclin E–CDK2, enabling its activation in late G1, whereas cyclin D2–CDK4 is not inhibited by this CKI [36]. p27 also facilitates the nuclear accumulation of cyclin D–CDK complexes, as does p21.

A role for p27 in programmed cell death in some cell types has been demonstrated. When p27 is transiently overexpressed in transformed epithelial cells or fibroblasts, the cells undergo apoptosis [90,91]. In contrast, lack of p27 in mesangial cells results in mitogen-depleted apoptosis [92], whereas ectopic expression in myeloid leukemic cells has been shown to block cell death [93]. In a subsequent study of lymphoid apoptosis, downregulation of p27 was found to occur by two mechanisms: one dependent on caspase-8 activity in response to Fas activation and a second caspase-independent mechanism in etoposide or cycloheximide-treated cells [94]. Thus, the contribution of p27 to programmed cell death may depend not only on the type of cell, but also the initiating stimulus.

Deregulation of p27 activity is likely to play a role in malignant disease. Haploinsufficiency for p27 results in greatly increased susceptibility of affected animals to develop tumors following exposure to chemical carcinogens or ionizing radiation [95]. Consistent with this is the finding that several human tumor types contain reduced levels of p27, including breast and colon cancers [34,96,97], and that this is indicative of more aggressive disease. Furthermore, overexpression of Myc is a frequent alteration found in cancer. One major mechanisms whereby Myc may contribute to increased cellular proliferation is through induction of cyclin D1 and D2, leading to sequestration of p27 [98,99] and release of the block on cyclin A/E–CDK2 activities. In addition, it has been suggested that overlap exists between p27- and pRb-regulated pathways, as pRb haploinsufficiency on a p27-null background leads to more rapid tumor development and more aggressive lesions than either defect alone [100].

Diminished levels of p27 are likely to contribute to the malignant phenotype of other human tumors. It has been shown that the BCR-ABL oncoprotein, which is expressed in many leukemic cell lines as a result of chromosomal translocation, results in a decrease in p27 expression through a PI3-K/AKT-dependent mechanism [101]. In a separate study, increased activity of AKT was found to enhance prostate cancer progression by blocking p27 expression [102]. This is likely to be a direct effect of AKT to inhibit the activity of AFX/Forkhead transcription factors, which have been shown to transactivate p27 transcriptionally [103]. In addition, it is probable that neoplasms in which function of the PTEN tumor suppressor protein has been lost (see Tamura *et al.* [104] for review) will also contain low levels of p27, as PTEN negatively regulates AKT activity. Furthermore, oncogenic signaling from HER-2/neu results in decreased cellular levels of p27 by promoting its export into the cytoplasm and thus enhancing ubiquitin-mediated degradation [105]. Hence, downregulation of p27 is likely to be of importance in many human cancers, including breast (mentioned earlier), ovarian, lung, gastric, and some oral tumors [106,107].

A third member of the Cip/Kip family of CKIs is p57/Kip2, which was cloned by virtue of its homology with p21 [13] and as a cyclin D interacting protein using a yeast two-hybrid screen [14]. Murine p57 contains an amino-terminal CDK inhibitory domain, a proline-rich domain, an acidic domain, and a carboxy-terminal domain that shares homology with p27. In addition, the carboxy-terminal domain contains a putative site for phosphorylation by CDKs and is thought to play a role in nuclear targeting [13]. Whereas the human p57 retains amino- and carboxy-terminal regions, the internal domains are replaced with a number of proline-alanine repeats [14]. From *in situ* hybdridization studies of murine embryogenesis, p57 is highly expressed in terminally differentiated cells such as brain, heart, muscle, eye, and lungs, suggesting that it plays a central role in regulating cell cycle exit during development. This has been confirmed by the results of *in vivo* studies of p57-null mice [108,109]. The majority of these animals die as neonates, harboring a number of developmental defects, such as cleft palate and gastrointestinal abnormalities. The phenotype of p57 knockout mice also includes short limbs, likely as a result of a failure to exit the cell cycle during chondrocyte differentiation and, consequently, failure to undergo endochondral ossification [108,109].

The gene encoding human p57, *KIP2*, is located on chromosome 11p15.5 [14], which may be altered in a number of tumor types and in a familial syndrome—Beckwith-Wiedemann syndrome (BWS)—which is characterized by a number of developmental abnormalities related to increased growth and which carries an elevated risk of childhood tumors [110]. *KIP2* is paternally imprinted, which is consistent with the suggestion that BWS is carried maternally, thus suggesting that $p57^{KIP2}$ may play a central role in this syndrome [109]. However, subsequent studies have provided little evidence for p57 alterations in cancer, including Wilms' tumors and sarcomas [111–114] or squamous cell carcinomas [115], although decreased expression of $p57^{KIP2}$ has been observed in some bladder carcinomas [116].

Expression of $p57^{KIP2}$ is capable of suppressing cellular transformation by MYC and RAS. This function has been shown to be dependent on its cyclin–CDK binding and also on its ability to bind PCNA through the C-terminal domain, which results in blocks on DNA replication and S-phase entry [117]. Perhaps unsurprisingly, ectopic expression of $p57^{KIP2}$ inhibits the growth of tumor cell lines in culture and may induce a senescent phenotype [118]. However, the precise contribution of $p57^{KIP2}$ to the development of human cancer remains unclear.

B. Ink4 Family

The Ink4 family of CKIs consists of four members: p16/Ink4A, p15/Ink4B, p18/Ink4C, and p19/Ink4D. These molecules contain multiple ankyrin repeats and show specificity for G1 CDKs, namely CDK4 and CDK6. p16/Ink4A was first observed as a 16-kDa component of cyclin D–CDK4 immunoprecipitates [119]. Binding of Ink4 CKIs to the CDK results in inhibition of kinase activity, in contrast to the effects of Cip/Kip proteins. Indeed, INK4 proteins probably compete with Cip/Kip CKIs for CDK4 subunits and block binding of Cip/Kips and cyclins. This will have the effect of making more Cip/Kip proteins available to bind cyclin E–CDK2, thus blocking exit from G1 into S phase.

The gene encoding p16 is located on the short arm of chromosome 9 in humans, at 9p21. This site is a major tumor suppressor locus in a wide range of human cancers, as determined by a number of studies of loss of heterozygosity [120–126]. As a negative regulator of cell growth, p16 seemed to be a likely candidate tumor suppressor, and although initial opinion was sceptical [127], this has now been confirmed by a broad volume of work. Two initial studies demonstrated homozygous deletion of the p16 locus in cell lines from a wide range of tumor cell lines, including lung, breast, bone, epidermis, ovary, lymphoreticular malignancies, and melanoma [128,129]. Furthermore, in samples where one allele was retained, the remaining allele frequently contained missense or nonsense mutations. Other subsequent studies of primary tumor material were unable to detect mutation and deletion of p16 at the high frequency observed in cell lines [130–132], casting doubt on the previous findings. However, it has now been established that p16 expression may be silenced in tumors, primarily by methylation of the promoter [133]. Thus, inactivation of p16 by deletion, mutation, or transcriptional silencing is likely to play a major role in tumorigenesis by deregulating CDK activity during G1. Indeed, loss of p16 is functionally equivalent to inactivation of Rb as, in such cases, Rb remains in a hyperphosphorylated (nonsuppressive) state [24]. The relationship between p16 loss and Rb function is underlined by several studies. In an analysis of human lung cancers, an inverse correlation was observed between these two molecules, such that 89% of small cell tumors retained expression of p16, but either lost Rb or expressed an aberrant form, whereas the remainder lacking p16 retained a normal Rb protein [134]. Conversely, in nonsmall cell lung cancers where Rb is generally normal, p16 was found to be absent or otherwise inactivated, except in the minority of lesions where Rb was deleted or mutated. These data were further extended by subsequent studies [135,136]. p16 was found to compete with cyclin D for binding to G1 CDKs [136], and in cells where Rb was inactivated, no association between CDK4 and D cyclins was observed. Cell cycle-dependent expression of p16 was reported by Tam *et al.* [137], with highest levels being found during the S phase, suggesting a requirement to inhibit G1 CDKs and perhaps enhance the progression through the S phase by forcing Cip/Kip proteins onto cyclin E–CDK2 complexes. It has also been reported that the p16 promoter is a target for JunB and that overexpression of JunB results in inhibition of cyclin D-associated kinase activity and prolongation of G1 [138].

p16 may also play a role in the development of some inherited cancers. For example, the genetic locus for familial melanoma (MLM) maps to chromosome 9p21 [139], and it is likely that p16 is a candidate melanoma susceptibility gene, as germline alterations have been found in a high proportion of cases [140], although other workers have reported alterations at lower frequency [141]. It is notable, however, that pancreatic adenocarcinomas, which also show 9p losses, exhibit p16 alterations at high frequency [142], suggestive of a role in tumor development, and that this type of malignancy is very common in familial melanoma cohorts [140].

Further direct evidence for the role of p16 as a tumor suppressor has come from studies where p16 was reintroduced into p16-null cells. Expression of p16 was found to block Ras-induced S-phase entry in rat embryo fibroblasts [143], whereas the growth in culture of glioma cells lacking endogenous p16 was reduced markedly by the expression of p16 [144]. In addition, gene inactivation studies in mice have revealed that loss of p16 in the germline greatly enhances susceptibility of animals to spontaneous tumor development, whereas cells cultured from these mice can be fully transformed by an activated ras oncogene, unlike normal cells [145]. While this latter study also inactivated the adjacent p19ARF sequence (see later), it seems likely that, when taken together with all the other available evidence, p16 is indeed a *bona fide* tumor suppressor gene. Furthermore, loss of p16, together with expression of an activated ras oncogene in melanocytes, results in increased melanoma formation after a short latency [146].

Available evidence suggests that CDK4, cyclin D, p16, and Rb are components of the same growth regulatory pathway that functions to control progression through G1 into the S phase. Initially, it was thought that p16-mediated growth arrest was entirely dependent on the presence of a functional Rb protein, as $Rb^{-/-}$ cells are insensitive to p16-induced growth arrest [147–149]. However, experiments using cells cultured from mice in which the Rb-related protein p107 and p130 have been inactivated have shown that Rb alone is insufficient for p16-mediated growth arrest [150], suggesting that while these molecules may share some functions, nonredundant activities are required for G1 arrest by p16.

It is now well recognized that p16 plays a critical role in cellular senescence. In one study, transfer of chromosome 9 was shown to induce a senescent phenotype in neoplastic cell lines [151], whereas loss of heterozygosity at 9p21 was found in immortal but not senescent cell lines derived from malignant tumors [152]. Further work demonstrated that immortal keratinocytes suppressed expression of p16 by a number of mechanisms, including transcriptional silencing and gene deletion [153], whereas normal p16 expression levels were observed in the majority of senescent cultures. Other studies in fibroblasts showed that p16 accumulated during senescence, likely due to increased p16 mRNA stability rather than increased transcription [154]. Thus, p16 is an attractive candidate for a tumor suppressor on chromosome 9p.

It has been recognized that the INK4A locus, as well as encoding the p16 protein, is capable of utilizing an upstream exon 1b together with an alternative reading frame within exon 2, thereby directing synthesis of the protein p19/ARF (alternative reading frame) in mouse and p14/ARF in humans [155–158]. Like p16, ARF is also a tumor suppressor [156,159]. However, its mode of action is distinct from that of p16 or any of the other INK4 proteins. Whereas p16 acts on the pRb pathway by inhibiting CDK4/6, ARF acts on the p53 pathway by interacting with MDM2, thereby blocking degradation of p53 and resulting in p53 stabilization and activation [160–162]. Furthermore, it has been shown that binding of ARF to MDM2 results in sequestration of MDM2 in the nucleolus [163]. This prevents shuttling of MDM2 between the nucleus and cytoplasm (to where it must export p53 for degradation), thereby leaving p53 free within the nucleus to transactivate transcription of responsive genes [164]. Interestingly, ARF is required for p53 activation by proliferative signals such as Myc, but not for activation in response to ultraviolet or ionizing radiation [165] (reviewed in Sherr [166] and Prives [167]). In addition, ARF is under transcriptional control of E2-F, thus providing an additional mechanism whereby cells that have deregulated G1 progression (e.g., through loss of p16, pRb, or elevation of CDK activity) and an elevated pool of free E2-F are still subject to G1 arrest [168]. An additional mechanism has been described for growth control by ARF, as it appears that ARF can target certain E2-F species for degradation [169]. Notably, germline deletion of *CDKN2* exon 1b sequences in a melanoma–astrocytoma syndrome family has been reported [170], which results in loss of ARF function but not p16 or p15 and implicates ARF inactivation as a predisposing factor for this disease. Taken together, it is clear that deletion of all or part of the *INK4A/ARF* locus, therefore, has considerable impact on two of the major growth suppressive pathways inactivated in human cancers.

Other members of the Ink4 family of CDK inhibitors are $p15^{INK4b}$ [171,172], $p18^{INK4c}$ [172,173], and $p19^{INK4d}$ [173,174]. p15 is encoded by a gene also situated on chromosome 9p21, proximal to that which encodes p16, and therefore represents an additional candidate for a tumor suppressor at this locus. Several studies have now presented evidence that p15 function is lost in some human tumors and cell lines, including nonsmall cell lung cancer [175], leukemias and lymphomas [176], and glioblastomas [177]. Although a few samples harbored intragenic mutations within the p15-coding region [175,176], homozygous deletion was found to be more common. Thus, loss of 9p may represent an efficient method through which cells deregulate their growth control mechanisms by deleting both p16 and p15 genes in one event.

$p18^{INK4c}$ maps to human chromosome 1p32, a locus that shows structural abnormalities in a range of different malignancies [172]. p18 has been shown to associate strongly with cyclin D–CDK6 complexes *in vitro* and *in vivo* and to inhibit phosphorylation of pRb [172]. Furthermore, ectopic expression of p18 is able to suppress the growth of cells that express a functional pRb, a similar finding to that observed for p16 [143]. The fourth member of the Ink4 family is $p19^{INK4d}$ [173,174]. In humans, the gene encoding $p19^{INK4d}$ is located on chromosome 19. Thus far, no overt phenotypes have been reported as a result of gene inactivation studies in mice. Both p18 and p19 are expressed during development, in contrast to p16 and p15.

C. $p12^{DOC-1}$

A novel hamster cDNA encoding an 87 amino acid polypeptide has been isolated, which exhibits properties of tumor suppression when overexpressed in malignant hamster oral keratinocytes [178]. The gene encoding doc-1 (deleted in oral cancer) showed frequent loss of heterozygosity and reduced expression during tumor progression. Human *DOC-1* is located on chromosome 12q24, encodes a 115 amino acid product, and is expressed in nuclei and cytoplasm of normal basal keratinocytes and in nuclei of suprabasal cells [179]. Furthermore, human oral tumors did not stain for the $p12^{DOC-1}$ protein. $p12^{DOC-1}$ has been identified as a binding partner for CDK2, which negatively regulates CDK2 activity by targeting it for degradation and/or sequestering monomeric CDK2, thereby resulting in reduced cell cycle traverse [180]. Thus, $p12^{DOC-1}$ is an attractive candidate for a tumor suppressor protein. In addition, ectopic expression of doc-1 in malignant hamster keratinocytes induces apoptosis [181].

IV. ALTERATIONS IN CELL CYCLE REGULATORS

A. p53

As in many other cancers, loss or mutation of the *P53* gene occurs at high frequency and is thought to play a role in the malignant progression of oral epithelia. Indeed, a plethora of reports are present in the literature as testimony to this. Studies of oral squamous cell carcinoma cell lines and carcinoma tissues have shown point mutations and deletions, which result in amino acid substitution, in-frame deletion, or premature termination of translation [182–187]. Furthermore, mutation of p53 in head and neck squamous cell carcinoma (HNSCC) results in abrogation of its ability to transactivate transcription of responsive promoters [188] and loss of the G1 DNA damage checkpoint [189], whereas aberrant forms of p53 can reduce the activity of coexpressed wild-type p53 [188]. Thus, in some cases, p53 mutation can contribute to development of a tumorigenic phenotype and may be implicated in metastasis [190]. It has been suggested that positive immunostaining for p53 in histologically normal mucosa adjacent to squamous cell carcinomas may be indicative of potential areas of local recurrence following surgical treatment [191], which may have profound implications for aggressive follow-up. However, some premalignant lesions have been shown to lack p53 alterations [192]; thus in some cases, progression to malignancy may be independent of p53 status or may involve mutation at a later stage of tumor development.

B. G1/S Deregulation

As noted earlier, several proteins are known to participate in regulation of the G1 phase of the cell cycle. Thus, there are a number of ways through which G1 deregulation may occur in cancer. Elevated expression of the cyclin D1 protein has been found in several tumor types. One mechanism by which this occurs is through amplification of the locus on chromosome 11q13, which has been demonstrated in malignancies of breast [193,194], esophagus [195,196], lung [197], and head and neck [198,199]. In addition, increased transcription from the cyclin D1 promoter, in the absence of gene amplification, results in more cyclin D protein being available to bind G1 CDKs and enable their activation. Bartkova *et al.* [200] reported elevated expression of cyclin D1 in HNSCC tumor specimens using immunohistochemical techniques and showed that different patterns and levels of D-type cyclins were present in HNSCC tumor cells. Furthermore, they demonstrated an absolute requirement for cyclin D1 for cell cycle progression, suggesting that its overexpression may play a crucial role in carcinogenesis. In a series of HNSCC cell lines, the enzymatic activity of CDK4, CDK6, and CDK2 was found to be markedly elevated, compared to that of normal keratinocytes [201]. This is most likely due to elevated G1 CDK expression in a number of cases, loss of p16/Ink4A expression, and, in the case of CDK2, elevated cyclin E expression. In a separate study, elevated CDK6 activity was found in oral cancer cells, and this kinase was found to be associated with p18/Ink4C [202]. Taken together, these reports suggest that CDK6 may play an important role in the development and/or maintenance of cellular transformation during oral carcinogenesis. Overexpression of cyclins A and B has also been demonstrated in a series of floor-of-mouth carcinomas, and thus is likely to result in further deregulation of S phase and G2/M [203].

Additional confirmation of cyclin D1 overexpression in head and neck cancer has been provided by genome-wide expression studies using microarray technology [204]. In addition, these authors demonstrated upregulation of cyclin H, the binding partner for the CDK-activating kinase, which had not been reported previously in these lesions and which may represent a mechanism through which cancer cells can deregulate the activity of multiple cell cycle kinases. A role for cyclin D1 overexpression in early oral carcinogenesis is suggested by the results of studies that demonstrated cyclin D1 gene amplification [205] and protein overexpression [205,206] in premalignant oral lesions. Thus, although loss of pRb expression appears to occur only at low frequency in HNSCC, despite deletion or loss of chromosome 13q [207], pRb-regulated pathways are compromised in these tumors.

It is also likely that alterations in cell cycle regulatory proteins may be of some prognostic value. For instance, elevated expression of cyclin D1, detected immunohistochemically, has been associated with a decrease in the overall survival of HNSCC patients [208,209]. Furthermore, 80% of patients with tumors that were cyclin D1 negative were found to have a 5-year disease-free interval compared to only 47% of patients whose lesions showed positivity for cyclin D1 [208].

There is now a considerable body of evidence to support the loss of p16/Ink4A as a critical step in the development of HNSCC [210–212]. This may occur through chromosomal deletion (in which case the expression of p15Ink4B is also likely to be abrogated), by intragenic mutation, or by transcriptional silencing as a result of promoter methylation. Indeed, preferential inactivation of p16 over p15 may emphasize its importance as the major tumor suppressor on chromosome 9p21 that is important in the genesis of head and neck squamous malignancies [153,201]. Furthermore, in a study of 148 tongue carcinomas, loss of p16 expression was predictive of a poor outcome, whereas p16 loss together with cyclin D1 overexpression indicated a considerably worse survival at 5 years [209]. Loss of p16 expression may occur as an early step in tumorigenesis, as it has been shown to be absent in around 80% of premalignant lesions in one study [213], and frequent loss of exon 1α and, in some cases, exon 2 has been reported in dysplasias and carcinomas [214]. Loss of p27 in oral cancers has also been reported [215–217] and may be a predictor of poor prognosis [217].

C. Cell Cycle Regulation and Therapy

Genetic modification of tumor cells as a treatment modality is an attractive option for dealing with head and neck cancers, as the accessibility of lesions for direct delivery of therapeutic reagents, and the potential for preserving vital surrounding structures, is excellent. Loss of expression of cell cycle inhibitors in cancer makes these suitable candidates for gene replacement strategies, and several studies have utilized this approach in head and neck cancer. As mentioned previously, alteration of p53 in HNSCC is common. Thus, it is not surprising that gene therapy to replace wild-type p53 in tumor cells has shown some promise. Recombinant adenoviruses carrying a wild-type p53 gene are capable of inhibiting the growth of HNSCC cell lines in culture [218] and in preclinical animal models [219–221] and can induce apoptosis of transduced cells [222]. Furthermore, topical application may be a viable option, as multiple layers of epithelial cells are transduced, at least in a cell culture model [223]. These studies have now been translated into clinical trials, with p53 adenoviruses being used as adjuvants to surgical treatment of HNSCC [224].

Other cell cycle regulators are also candidates for use as therapeutic reagents. Overexpression of E2-F in HNSCC cells has been shown to induce apoptosis [225]. Furthermore, p21/WAF1 is a mediator of the growth suppressive effects of both wild-type p53 and TGF-β [56,57,226], both of which may be inactivated in HNSCC [227–229], and replacement of p21 has been found to result in up to 70% inhibition of tumor cell growth in a preclinical animal model [230]. However, cotransduction of tumor cells with p21 and p16/Ink4A did not provide improved growth suppression compared with the delivery of p16 alone [231]. This implies that inhibition at multiple points throughout the cell cycle is unnecessary and that inhibition during G1 is sufficient to cause growth arrest.

As an alternative strategy for inhibiting G1 progression, the use of antisense cyclin D1 constructs has been tested in HNSCC cells [232]. Interestingly, in addition to suppressing cell growth and tumorigenicity, cells became more sensitive to the effects of the chemotherapeutic cisplatin. This may open the door for the design of combination therapy regimes, which couple modulating cell cycle progression and induction of apoptosis with either radiotherapy or chemotherapy [189,233] or with small molecule or peptide inhibitors of cyclin–CDK complexes [234,235].

Acknowledgments

We are grateful to Dr. Andrew Sunters for helpful comments on the manuscript and to the Concern Foundation for support.

References

1. Makela, T. P., Tassan, J. P., Nigg, E. A., Frutiger, S., Hughes, G. J., and Weinberg, R. A. (1994). A cyclin associated with the CDK-activating kinase MO15. *Nature* **371**, 254–257.
2. Nigg, E. A. (1995). Cyclin-dependent kinase 7: At the cross-roads of transcription, DNA repair and cell cycle control? *Curr. Opin. Cell Biol.* **8**, 312–317.
3. Iavarone, A., and Massague, J. (1997). Repression of the CDK activator Cdc25A and cell-cycle arrest by cytokine TGF-beta in cells lacking the CDK inhibitor p15. *Nature* **387**, 417–422.
4. Mailand, N., Falck, J., Lukas, C., Syljuasen, R. G., Welcker, M., Bartek, J., and Lukas, J. (2000). Rapid destruction of human Cdc25A in response to DNA damage. *Science* **288**, 1425–1429.
5. Giaccia, A. J., and Kastan, M. B. (1998). The complexity of p53 modulation: Emerging patterns from divergent signals. *Genes Dev.* **12**, 2973–2983.
6. El-Deiry, W. S., Tokino, T., Velculescu, V. E., Levy, D. B., Parsons, R., Trent, J. M., Lin, D., Mercer, W. E., Kinzler, K. W., and Vogelstein, B. (1993). WAF1, a potential mediator of p53 tumor suppression. *Cell* **75**, 817–825.
7. Harper, J. W., Adami, G. R., Wei, N., Keyomarsi, K., and Elledge, S. J. (1993). The p21 Cdk-interacting protein Cip1 is a potent inhibitor of G1 cyclin-dependent kinases. *Cell* **75**, 805–816
8. Noda, A., Ning, Y., Venable, S. F., Pereira-Smith, O. M., and Smith, J. R. (1994). Cloning of senescent cell-derived inhibitors of DNA synthesis using an expression screen. *Exp. Cell Res.* **211**, 90–98.
9. Gu, Y., Turek, C. W., and Morgan, D. O. (1993). Inhibition of CDK2 activity *in vivo* by an associated 20K regulatory subunit. *Nature* **366**, 707–710.
10. Polyak, K., Lee, M. H., Erdjument-Bromage, H., Koff, A., Roberts, J. M., Tempst, P., and Massague, J. (1994). Cloning of p27Kip1, a cyclin-dependent kinase inhibitor and a potential mediator of extracellular antimitogenic signals. *Cell* **78**, 59–66.
11. Polyak, K., Kato, J. Y., Solomon, M. J., Sherr, C. J., Massague, J., Roberts, J. M., and Koff, A. (1994). p27Kip1, a cyclin-Cdk inhibitor, links transforming growth factor-beta and contact inhibition to cell cycle arrest. *Genes Dev.* **8**, 9–22.
12. Toyoshima, H., and Hunter, T. (1994). p27, a novel inhibitor of G1 cyclin-Cdk protein kinase activity, is related to p21. *Cell* **78**, 67–74.
13. Lee, M.-H., Reynisdottir, I., and Massague, J. (1995). Cloning of $p57^{KIP2}$, a cyclin-dependent kinase inhibitor with unique domain structure and tissue distribution. *Genes Dev.* **9**, 639–649.
14. Matsuoka, S., Edwards, M. C., Bai, C., Parker, S., Zhang, P., Baldini, A., Harper, J. W., and Elledge, S. J. (1995). $p57^{KIP2}$, a structurally distinct member of the $p21^{CIP1}$ Cdk inhibitor family, is a candidate tumor suppressor gene. *Genes Dev.* **9**, 650–662.
15. Filmus, J., Robles, A. I., Shi, W., Wong, M. J., Colombo, L. L., and Conti, C. J. (1994). Induction of cyclin D1 overexpression by activated ras. *Oncogene* **9**, 3627–33.
16. Lavoie, J. N., L'Allemain, G., Brunet, A., Muller, R., and Pouyssegur, J. (1996). Cyclin D1 expression is regulated positively by the p42/p44MAPK and negatively by the p38/HOGMAPK pathway. *J. Biol. Chem.* **271**, 20608–20616.
17. Winston, J. T., Coats, S. R., Wang, Y. Z., and Pledger, W. J. (1996). Regulation of the cell cycle machinery by oncogenic ras. *Oncogene* **12**, 127–134.
18. Kerkhoff, E., and Rapp, U. R. (1997). Induction of cell proliferation in quiescent NIH 3T3 cells by oncogenic c-Raf-1. *Mol. Cell. Biol.* **17**, 2576–2586.
19. Aktas, H., Cai, H., and Cooper, G. M. (1997). Ras links growth factor signaling to the cell cycle machinery via regulation of cyclin D1 and the Cdk inhibitor p27KIP1. *Mol. Cell. Biol.* **17**, 3850–3857.
20. Weber, J. D., Raben, D. M., Phillips, P. J., and Baldassare, J. J. (1997). Sustained activation of extracellular-signal-regulated kinase 1 (ERK1) is required for the continued expression of cyclin D1 in G1 phase. *Biochem. J.* **326**, 61–68.
21. Peeper, D. S., Upton, T. M., Ladha, M. H., Neuman, E., Zalvide, J., Bernards, R., DeCaprio, J. A., and Ewen, M. E. (1997). Ras signalling linked to the cell-cycle machinery by the retinoblastoma protein. *Nature* **386**, 177–181.
22. Cheng, M., Sexl, V., Sherr, C. J., and Roussel, M. F. (1998). Assembly of cyclin D-dependent kinase and titration of p27Kip1 regulated by mitogen-activated protein kinase kinase (MEK1). *Proc. Natl. Acad. Sci. USA*. **95**, 1091–1096.
23. Diehl, J. A., Cheng, M., Roussel, M. F., and Sherr, C. J. (1998). Glycogen synthase kinase-3 beta regulates cyclin D1 proteolysis and subcellular localization. *Genes Dev.* **12**, 3499–3511.
24. Weinberg, R. A. (1995). The retinoblastoma protein and cell cycle control. *Cell* **81**, 323–330.
25. Nevins, J. R. (1998). Toward an understanding of the functional complexity of the E2F and retinoblastoma families. *Cell Growth Differ.* **9**, 585–593.
26. Johnson, D. G., Schwarz, J. K., Cress, W. D., and Nevins, J. R. (1993). Expression of transcription factor E2F1 induces quiescent cells to enter S phase. *Nature* **365**, 349–352.
27. Shan, B., and Lee, W. H. (1994). Deregulated expression of E2F-1 induces S-phase entry and leads to apoptosis. *Mol. Cell. Biol.* **14**, 8166–8173.
28. Qin, X. Q., Livingston, D. M., Kaelin, W. G. Jr., and Adams, P. D. (1994). Deregulated transcription factor E2F-1 expression leads to S-phase entry and p53-mediated apoptosis. *Proc. Natl. Acad. Sci. USA* **91**, 10918–10922.
29. Zhang, H. S., Postigo, A. A., and Dean, D. C. (1999). Active transcriptional repression by the Rb-E2F complex mediates G1 arrest triggered by p16INK4a, TGFbeta, and contact inhibition. *Cell* **97**, 53–61.
30. LaBaer, J., Garrett, M. D., Stevenson, L. F., Slingerland, J. M., Sandhu, C., Chou, H. S., Fattaey, A., and Harlow, E. (1997). New functional activities for the p21 family of CDK inhibitors. *Genes Dev.* **11**, 847–862.
31. Pagano, M., Tam, S. W., Theodoras, A. M., Beer-Romero, P., Del Sal, G., Chau, V., Yew, P. R., Draetta, G. F., and Rolfe, M. (1995). Role of the ubiquitin-proteasome pathway in regulating abundance of the cyclin-dependent kinase inhibitor p27. *Science* **269**, 682–685.
32. Montagnoli, A., Fiore, F., Eytan, E., Carrano, A. C., Draetta, G. F., Hershko, A., and Pagano, M. (1999). Ubiquitination of p27 is regulated by Cdk-dependent phosphorylation and trimeric complex formation. *Genes Dev.* **13**, 1181–1189.
33. King, R. W., Deshaies, R. J., Peters, J.-M., and Kirschner, M. W. (1996). How proteolysis drives the cell cycle. *Science* **274**, 1652–1659.
34. Loda, M., Cukor, B., Tam, S. W., Lavin, P., Fiorentino, M., Draetta, G. F., Jessup, J. M., and Pagano, M. (1997). Increased proteasome-dependent degradation of the cyclin-dependent kinase inhibitor p27 in aggressive colorectal carcinomas. *Nat. Med.* **3**, 231–234.
35. Russo, A. A., Jeffrey, P. D., Patten, A. K., Massague, J., and Pavletich, N. P. (1996). Crystal structure of the p27Kip1 cyclin-dependent-kinase inhibitor bound to the cyclin A-Cdk2 complex. *Nature* **382**, 325–31.
36. Blain, S. W., Montalvo, E., and Massague, J. (1997). Differential interaction of the cyclin-dependent kinase (Cdk) inhibitor p27Kip1 with cyclin A-Cdk2 and cyclin D2-Cdk4. *J. Biol. Chem.* **272**, 25863–25872.
37. Tsutsui, T., Hesabi, B., Moons, D. S., Pandolfi, P. P., Hansel, K. S., Koff, A., and Kiyokawa, H. (1999). Targeted disruption of CDK4 delays cell cycle entry with enhanced p27(Kip1) activity. *Mol. Cell. Biol.* **19**, 7011–7019.
38. Deng, C., Zhang, P., Harper, J. W., Elledge, S. J., and Leder, P. (1995). Mice lacking $p21^{CIP1/WAF1}$ undergo normal development, but are defective in G1 checkpoint control. *Cell* **82**, 675–684.
39. Harris, C. C. (1993). p53: At the crossroads of molecular carcinogenesis and risk assessment. *Science* **262**, 1980.
40. Lin, D., Fiscella, M., O'Connor, P. M., Jackman, J., Chen, M., Luo, L. L., Sala, A., Travali, S., Appella, E., and Mercer, W. E. (1994).

Constitutive expression of B-myb can bypass p53-induced Waf1/Cip1-mediated G1 arrest. *Proc. Natl. Acad. Sci. USA* **91**, 10079–10083.

41. El-Deiry, W. S., Harper, J. W., O'Connor, P. M., Velculescu, V. E., Canman, C. E., Jackman, J., Pietenpol, J. A., Burrell, M., Hill, D. E., Wang, Y., Wiman, K. G., Mercer, W. E., Kastan, M. B., Kohn, K. W., Elledge, S. J., Kinzler, K. W., and Vogelstein, B. (1994). WAF1/CIP1 is induced in p53-mediated G1 arrest and apoptosis. *Cancer Res.* **54**, 1169–1174.
42. Wang, J., and Walsh, K. (1996). Resistance to apoptosis conferred by Cdk inhibitors during myocyte differentiation. *Science* **273**, 359–361.
43. Gorospe, M., Wang, X., Guyton, K. Z., and Holbrook, N. J. (1996). Protective role of p21(Waf1/Cip1) against prostaglandin A2-mediated apoptosis of human colorectal carcinoma cells. *Mol. Cell. Biol.* **16**, 6654–6660.
44. Erhardt, J. A., and Pittman, R. N. (1998). p21WAF1 induces permanent growth arrest and enhances differentiation, but does not alter apoptosis in PC12 cells. *Oncogene* **16**, 443–451.
45. Shim, J., Lee, H., Park, J., Kim, H., and Choi, E. J. (1996). A non-enzymatic p21 protein inhibitor of stress-activated protein kinases. *Nature* **381**, 804–806.
46. Shiohara, M., El-Deiry, W. S., Wada, M., Nakamaki, T., Takeuchi, S., Yang, R., Chen, D. L., Vogelstein, B., and Koeffler, H. P. (1994). Absence of WAF1 mutations in a variety of human malignancies. *Blood* **1[illegible]**, 3781–3784.
47. Gao, X., Chen, Y. Q., Wu, N., Grignon, D. J., Sakr, W., Porter, A. T., and Honn, K. V. (1995). Somatic mutations of the WAF1/CIP1 gene in primary prostate cancer. *Oncogene* **11**, 1395–1398.
48. Bhatia, K., Fan, S., Spangler, G., Weintraub, M., O'Connor, P. M., Judde, J. G., and Magrath, I. (1995). A mutant p21 cyclin-dependent kinase inhibitor isolated from a Burkitt's lymphoma. *Cancer Res.* **55**, 1431–1435.
49. Chedid, M., Michieli, P., Lengel, C., Huppi, K., and Givol, D. (1994). A single nucleotide substitution at codon 31 (Ser-Arg) defines a polymorphism in a highly conserved region of the p53-inducible gene WAF1/CIP1. *Oncogene* **9**, 3021–3024.
50. Li, Y.-J., Laurent-Puig, P., Salmon, R. J., Thomas, G., and Hamelin, R. (1995). Polymorphisms and probable lack of mutation in the WAF1-CIP1 gene in colorectal cancer. *Oncogene* **10**, 599–601.
51. Bahl, R., Arora, S., Nath, N., Mathur, M., Shukla, N. K., and Ralhan, R. (2000). Novel polymorphism in p21waf1/cip1 cyclin-dependent kinase inhibitor gene: Association with human esophageal cancer. *Oncogene* **19**, 323–328.
52. Brugarolas, J., Chandrasekaran, C., Gordon, J. I., Beach, D., Jacks, T., and Hannon, G. J. (1995). Radiation-induced cell cycle arrest compromised by p21 deficiency. *Nature* **377**, 552–557.
53. Michieli, P., Chedid, M., Lin, D., Pierce, J. H., Mercer, W. E., and Givol, D. (1994). Induction of WAF1/CIP1 by a p53-independent pathway. *Cancer Res.* **54**, 3391–3395.
54. Weiss, R. H., Joo, A., and Randour, C. (2000). p21(Waf1/Cip1) is an assembly factor required for platelet-derived growth factor-induced vascular smooth muscle cell proliferation. *J. Biol. Chem.* **275**, 10285–10290.
55. Shiohara, M., Akashi, M., Gombart, A. F., Yang, R., and Koeffler, H. P. (1996). Tumor necrosis factor alpha: Posttranscriptional stabilization of WAF1 mRNA in p53-deficient human leukemic cells. *J. Cell. Physiol.* **166**, 568–576.
56. Datto, M. B., Li, Y., Panus, J. F., Howe, D. J., Xiong, Y., and Wang, X.-F. (1995). Transforming growth factor beta induces the cyclin-dependent kinase inhibitor p21 through a p53-independent mechanism. *Proc. Natl. Acad. Sci. USA* **92**, 5545–5549.
57. Li, C.-Y., Suardet, L., and Little, J. B. (1995). Potential role of WAF1/Cip1/p21 as a mediator of TGF-β cytoinhibitory effect. *J. Biol. Chem.* **270**, 4971–4974.
58. Zauberman, A., Oren, M., and Zipori, D. (1997). Involvement of p21(WAF1/Cip1), CDK4 and Rb in activin A mediated signaling leading to hepatoma cell growth inhibition. *Oncogene* **15**, 1705–1711.
59. Li, J. M., Datto, M. B., Shen, X., Hu, P. P., Yu, Y., and Wang, X. F. (1998). Sp1, but not Sp3, functions to mediate promoter activation by TGF-beta through canonical Sp1 binding sites. *Nucleic Acids Res.* **26**, 2449–2456.
60. Hunt, K. K., Fleming, J. B., Abramian, A., Zhang, L., Evans, D. B., and Chiao, P. J. (1998). Overexpression of the tumor suppressor gene Smad4/DPC4 induces p21waf1 expression and growth inhibition in human carcinoma cells. *Cancer Res.* **58**, 5656–5661.
61. Moustakas, A., and Kardassis, D. (1998). Regulation of the human p21/WAF1/Cip1 promoter in hepatic cells by functional interactions between Sp1 and Smad family members. *Proc. Natl. Acad. Sci. USA* **95**, 6733–6738.
62. Osawa, Y., Hachiya, M., Koeffler, H. P., Suzuki, G., and Akashi, M. (1995). IL-1 induces expression of WAF1 mRNA in human fibroblasts: Mechanisms of accumulation. *Biochem. Biophys. Res. Commun.* **216**, 429–437.
63. Akashi, M., Hachiya, M., Osawa, Y., Spirin, K., Suzuki, G., and Koeffler, H. P. (1995). Irradiation induces WAF1 expression through a p53-independent pathway in KG-1 cells. *J. Biol. Chem.* **270**, 19181–19187.
64. Chin, Y. E., Kitagawa, M., Su, W. C., You, Z. H., Iwamoto, Y., and Fu, X. Y. (1996). Cell growth arrest and induction of cyclin-dependent kinase inhibitor p21 WAF1/CIP1 mediated by STAT1. *Science* **272**, 719–722.
65. Fan, Z., Lu, Y., Wu, X., DeBlasio, A., Koff, A., and Mendelsohn, J. (1995). Prolonged induction of p21Cip1/WAF1/CDK2/PCNA complex by epidermal growth factor receptor activation mediates ligand-induced A431 cell growth inhibition. *J. Cell. Biol.* **131**, 235–242.
66. Jakus, J., and Yeudall, W. A. (1996). Growth inhibitory concentrations of EGF induce p21 (WAF1/Cip1) and alter cell cycle control in squamous carcinoma cells. *Oncogene* **12**, 2369–2376.
67. Owen, G. I., Richer, J. K., Tung, L., Takimoto, G., and Horwitz, K. B. (1998). Progesterone regulates transcription of the p21(WAF1) cyclin-dependent kinase inhibitor gene through Sp1 and CBP/p300. *J. Biol. Chem.* **273**, 10696–10701.
68. Kardassis, D., Papakosta, P., Pardali, K., and Moustakas, A. (1999). c-Jun transactivates the promoter of the human p21(WAF1/Cip1) gene by acting as a superactivator of the ubiquitous transcription factor Sp1. *J. Biol. Chem.* **274**, 29572–29581.
69. Steinman, R. A., Hoffman, B., Iro, A., Guillouf, C., Liebermann, D. A., and El-Houseini, M. E. (1994). Induction of p21 (WAF1/CIP1) during differentiation. *Oncogene* **9**, 3389–3396.
70. Zhang, W., Grasso, L., McClain, C. D., Gambel, A. M., Cha, Y., Travali, S., Deisseroth, A. B., and Mercer, W. E. (1995). p53-independent induction of WAF1/CIP1 in human leukemia cells is correlated with growth arrest accompanying monocyte/macrophage differentiation. *Cancer Res.* **55**, 668–674.
71. MacLeod, C. L., Luk, A., Castagnola, J., Cronin, M., and Mendelsohn, J. (1986). EGF induces cell cycle arrest of A431 human epidermoid carcinoma cells. *J. Cell. Physiol.* **127**, 175–182.
72. Parker, S. B., Eichele, G., Zhang, P., Rawls, A., Sands, A. T., Bradley, A., Olson, E. N., Harper, J. W., and Elledge, S. J. (1995). p53-independent expression of $p21^{Cip1}$ in muscle and other terminally differentiating cells. *Science* **267**, 1024–1027.
73. Halevy, O., Novitch, B. G., Spicer, D. B., Skapek, S. X., Rhee, J., Hannon, G. J., Beach, D., and Lassar, A. B. (1995). Correlation of terminal cell cycle arrest of skeletal muscle with induction of p21 by MyoD. *Science* **267**, 1018–1021.
74. Skapek, S. X., Rhee, J., Spicer, D. B., and Lassar, A. B. (1995). Inhibition of myogenic differentiation in proliferating myoblasts by cyclin D1-dependent kinase. *Science* **267**, 1022–1024.
75. Di Cunto, F., Topley, G., Calautti, E., Hsaio, J., Ong, L., Seth, P. K., and Dotto, G. P. (1998). Inhibitory function of p21 Cip1/WAF1 in

differentiation of primary mouse keratinocytes independent of cell cycle control. *Science* **280**, 1069–1072.
76. Waga, S., and Stillman, B. (1998). Cyclin-dependent kinase inhibitor p21 modulates the DNA primer-template recognition complex. *Mol. Cell. Biol.* **18**, 4177–4187.
77. Waga, S., Hannon, G. J., Beach, D., and Stillman, B. (1994). The p21 inhibitor of cyclin-dependent kinases controls DNA replication by interaction with PCNA. *Nature* **369**, 574–578.
78. Rousseau, D., Cannella, D., Boulaire, J., Fitzgerald, P., Fotedar, A., and Fotedar, R. (1999). Growth inhibition by CDK-cyclin and PCNA binding domains of p21 occurs by distinct mechanisms and is regulated by ubiquitin-proteasome pathway. *Oncogene* **18**, 4313–4325.
79. Stewart, Z. A., Leach, S. D., and Pietenpol, J. A. (1999). p21(Waf1/Cip1) inhibition of cyclin E/Cdk2 activity prevents endoreduplication after mitotic spindle disruption. *Mol. Cell. Biol.* **19**, 205–215.
80. McDonald, E. R., III, Wu, G. S., Waldman, T., and El-Deiry, W. S. (1996). Repair Defect in p21 WAF1/CIP1 –/– human cancer cells. *Cancer Res.* **56**, 2250–2255.
81. Savio, M., Stivala, L. A., Scovassi, A. I., Bianchi, L., and Prosperi, E. (1996). p21waf1/cip1 protein associates with the detergent-insoluble form of PCNA concomitantly with disassembly of PCNA at nucleotide excision repair sites. *Oncogene* **13**, 1591–1598.
82. Chen, J., Jackson, P. K., Kirschner, M. W., and Dutta, A. (1995). Separate domains of p21 involved in the inhibition of Cdk kinase and PCNA. *Nature* **374**, 386–388.
83. Scott, M. T., Morrice, N., and Ball, K. L. (2000). Reversible phosphorylation at the C-terminal regulatory domain of p21(Waf1/Cip1) modulates proliferating cell nuclear antigen binding. *J. Biol. Chem.* **275**, 11529–11537.
84. Li, R., Waga, S., Hannon, G. J., Beach, D., and Stillman, B. (1994). Differential effects by the p21 CDK inhibitor on PCNA-dependent DNA replication and repair. *Nature* **371**, 534–537.
85. Shivji, M. K., Grey, S. J., Strausfeld, U. P., Wood, R. D., and Blow, J. J. (1994). Cip1 inhibits DNA replication but not PCNA-dependent nucleotide excision-repair. *Curr. Biol.* **4**, 1062–1068.
86. Slingerland, J. M., Hengst, L., Pan, C. H., Alexander, D., Stampfer, M. R., and Reed, S. I. (1994). A novel inhibitor of cyclin-Cdk activity detected in transforming growth factor beta-arrested epithelial cells. *Mol. Cell. Biol.* **14**, 3683–3694.
87. Bullrich, F., MacLachlan, T. K., Sang, N., Druck, T., Veronese, M. L., Allen, S. L., Chiorazzi, N., Koff, A., Heubner, K., Croce, C. M., and Giordano, A. (1995). Chromosomal mapping of members of the cdc2 family of protein kinases, cdk3, cdk6, PISSLRE, and PITALRE, and a cdk inhibitor, p27Kip1, to regions involved in human cancer. *Cancer Res.* **55**, 1199–1205.
88. Pietenpol, J. A., Bohlander, S. K., Sato, Y., Papadopoulos, N., Liu, B., Friedman, C., Trask, B. J., Roberts, J. M., Kinzler, K. W., Rowley, J. D., and Vogelstein, B. (1995). Assignment of the human p27Kip1 gene to 12p13 and its analysis in leukemias. *Cancer Res.* **55**, 1206–1210.
89. Ponce-Castaneda, M. V., Lee, M. H., Latres, E., Polyak, K., Lacombe, L., Montgomery, K., Mathew, S., Krauter, K., Sheinfeld, J., Massague, J., and Cordon-Cardo, C. (1995). p27Kip1: Chromosomal mapping to 12p12-12p13.1 and absence of mutations in human tumors. *Cancer Res.* **55**, 1211–1214.
90. Craig, C., Wersto, R., Kim, M., Ohri, E., Li, Z., Katayose, D., Lee, S. J., Trepel, J., Cowan, K., and Seth, P. (1997). A recombinant adenovirus expressing p27Kip1 induces cell cycle arrest and loss of cyclin-Cdk activity in human breast cancer cells. *Oncogene* **14**, 2283–2289.
91. Schreiber, M., Muller, W. J., Singh, G., and Graham, F. L. (1999). Comparison of the effectiveness of adenovirus vectors expressing cyclin kinase inhibitors p16INK4A, p18INK4C, p19INK4D, p21(WAF1/CIP1) and p27KIP1 in inducing cell cycle arrest, apoptosis and inhibition of tumorigenicity. *Oncogene* **18**, 1663–1676.
92. Hiromura, K., Pippin, J. W., Fero, M. L., Roberts, J. M., and Shankland, S. J. (1999). Modulation of apoptosis by the cyclin-dependent kinase inhibitor p27(Kip1). *J. Clin. Invest.* **103**, 597–604.
93. Eymin, B., Haugg, M., Droin, N., Sordet, O., Dimanche-Boitrel, M. T., and Solary, E. (1999). p27Kip1 induces drug resistance by preventing apoptosis upstream of cytochrome c release and procaspase-3 activation in leukemic cells. *Oncogene* **18**, 1411–1418.
94. Frost, V., and Sinclair, A. J. (2000). p27KIP1 is down-regulated by two different mechanisms in human lymphoid cells undergoing apoptosis. *Oncogene* **19**, 3115–3120.
95. Fero, M. L., Randel, E., Gurley, K. E., Roberts, J. M., and Kemp, C. J. (1998). The murine gene p27Kip1 is haplo-insufficient for tumor suppression. *Nature* **396**, 177–180.
96. Catzavelos, C., Tsao, M. S., DeBoer, G., Bhattacharya, N., Shepherd, F. A., and Slingerland, J. M. (1999). Reduced expression of the cell cycle inhibitor p27Kip1 in nonsmall cell lung carcinoma: A prognostic factor independent of Ras. *Cancer Res.* **59**, 684–688.
97. Porter, P. L., Malone, K. E., Heagerty, P. J., Alexander, G. M., Gatti, L. A., Firpo, E. J., Daling, J. R., and Roberts, J. M. (1997). Expression of cell-cycle regulators p27Kip1 and cyclin E, alone and in combination, correlate with survival in young breast cancer patients. *Nature Med.* **3**, 222–225.
98. Bouchard, C., Thieke, K., Maier, A., Saffrich, R., Hanley-Hyde, J., Ansorge, W., Reed, S., Sicinski, P., Bartek, J., and Eilers, M. (1999). Direct induction of cyclin D2 by Myc contributes to cell cycle progression and sequestration of p27. *EMBO J.* **18**, 5321–5333.
99. Perez-Roger, I., Kim, S. H., Griffiths, B., Sewing, A., and Land, H. (1999). Cyclins D1 and D2 mediate myc-induced proliferation via sequestration of p27(Kip1) and p21(Cip1). *EMBO J.* **18**, 5310–5320.
100. Park, M. S., Rosai, J., Nguyen, H. T., Capodieci, P., Cordon-Cardo, C., and Koff, A. (1999). p27 and Rb are on overlapping pathways suppressing tumorigenesis in mice. *Proc. Natl. Acad. Sci. USA* **96**, 6382–6387.
101. Gesbert, F., Sellers, W. R., Signoretti, S., Loda, M., and Griffin, J. D. (2000). BCR/ABL regulates expression of the cyclin-dependent kinase inhibitor p27Kip1 through the phosphatidylinositol 3-Kinase/AKT pathway. *J. Biol. Chem.* **275**, 39223–39230.
102. Graff, J. R., Konicek, B. W., McNulty, A. M., Wang, Z., Houck, K., Allen, S., Paul, J. D., Hbaiu, A., Goode, R. G., Sandusky, G. E., Vessella, R. L., and Neubauer, B. L. (2000). Increased AKT activity contributes to prostate cancer progression by dramatically accelerating prostate tumor growth and diminishing p27Kip1 expression. *J. Biol. Chem.* **275**, 24500–24505.
103. Dijkers, P. F., Medema, R. H., Pals, C., Banerji, L., Thomas, N. S., Lam, E. W., Burgering, B. M., Raaijmakers, J. A., Lammers, J. W., Koenderman, L., and Coffer, P. J. (2000). Forkhead transcription factor FKHR-L1 modulates cytokine-dependent transcriptional regulation of p27(KIP1). *Mol. Cell. Biol.* **20**, 9138–9148.
104. Tamura, M., Gu, J., Tran, H., and Yamada, K. M. (1999). PTEN gene and integrin signaling in cancer. *J. Natl. Cancer Inst.* **91**, 1820–1828.
105. Yang, H. Y., Zhou, B. P., Hung, M. C., and Lee, M. H. (2000). Oncogenic signals of HER-2/neu in regulating the stability of the cyclin-dependent kinase inhibitor p27. *J. Biol. Chem.* **275**, 24735–24739.
106. Slamon, D. J., Godolphin, W., Jones, L. A., Holt, J. A., Wong, S. G., Keith, D. E., Levin, W. J., Stuart, S. G., Udove, J., Ullrich, A, *et al.* (1989). Studies of the HER-2/neu proto-oncogene in human breast and ovarian cancer. *Science* **244**, 707–712.
107. Beckhardt, R. N., Kiyokawa, N., Xi, L., Liu, T. J., Hung, M. C., el-Naggar, A. K., Zhang, H. Z., and Clayman, G. L. (1995). HER-2/neu oncogene characterization in head and neck squamous cell carcinoma. *Arch. Otolaryngol. Head Neck Surg.* **121**, 1265–1270.
108. Yan, Y., Frisen, J., Lee, M.-H., Massague, J., and Barbacid, M. (1997). Ablation of the CDK inhibitor p57Kip2 results in increased apoptosis

and delayed differentiation during mouse development. *Genes Dev.* **11**, 973–983.

109. Zhang, P., Liegeois, N. J., Wong, C., Finegold, M., Hou, H., Thompson, J. C., Silverman, A., Harper, J. W., DePinho, R. A., and Elledge, S. J. (1997). Altered cell differentiation and proliferation in mice lacking p57KIP2 indicates a role in Beckwith–Weidemann syndrome. *Nature* **387**, 151–158.
110. Weidemann, H. R. (1983). Tumours and hemihypertrophy associated with Weidemann–Beckwith syndrome. *Eur. J. Pediatr.* **141**, 129.
111. Reid, L. H., Crider-Miller, S. J., West, A., Lee, M.-H., Massague, J., and Weissman, B. E. (1996). Genomic organization of the human $p57^{KIP2}$ gene and its analysis in the G401 Wilms' tumor assay. *Cancer Res.* **56**, 1214–1218.
112. Orlow, I., Iavarone, A., Crider-Miller, S. J., Bonilla, F., Latres, E., Lee, M.-H., Gerald, W. H., Massague, J., Weissman, B. E., and Cordon-Cardo, C. (1996). Cyclin-dependent kinase inhibitor $p57^{KIP2}$ in soft tissue sarcomas and Wilms' tumors. *Cancer Res.* **56**, 1219–1221.
113. Taniguchi, T., Okamoto, K., and Reeve, A. E. (1997). Human p57(KIP2) defines a new imprinted domain on chromosome 11p but is not a tumour suppressor gene in Wilms' tumour. *Oncogene* **14**, 1201–1206.
114. Anderson, J., Gordon, A., McManus, A., Shipley, J., and Pritchard-Jones, K. (1999). Disruption of imprinted genes at chromosome region 11p15.5 in paediatric rhabdomyosarcoma. *Neoplasia* **1**, 340–348.
115. Lai, S., Goepfert, H., Gillenwater, A. M., Luna, M. A., and El-Naggar, A. K. (2000). Loss of imprinting and genetic alterations of the cyclin-dependent kinase inhibitor p57KIP2 gene in head and neck squamous cell carcinoma. *Clin. Cancer Res.* **6**, 3172–3176.
116. Oya, M., and Schulz, W. A. (2000). Decreased expression of p57(KIP2) mRNA in human bladder cancer. *Br. J. Cancer* **83**, 626–631.
117. Watanabe, H., Pan, Z. Q., Schreiber-Agus, N., DePinho, R. A., Hurvitz, J., and Xiong, Y. (1998). Suppression of cell transformation by the cyclin-dependent kinase inhibitor p57KIP2 requires binding to proliferating cell nuclear antigen. *Proc. Natl. Acad. Sci. USA* **95**, 1392–1397.
118. Tsugu, A., Sakai, K., Dirks, P. B., Jung, S., Weksberg, R., Fei, Y. L., Mondal, S., Ivanchuk, S., Ackerley, C., Hamel, P. A., and Rutka, J. T. (2000). Expression of p57(KIP2) potently blocks the growth of human astrocytomas and induces cell senescence. *Am. J. Pathol.* **157**, 919–932.
119. Serrano, M., Hannon, G. J., and Beach, D. (1993). A new regulatory motif in cell-cycle control causing specific inhibition of cyclin D/CDK4. *Nature* **366**, 704–707.
120. Diaz, M. O., Rubin, C. M., Harden, A., Ziemin, S., Larson, R. A., Le Beau, M. M., Rowley, J. D. (1990). Deletions of interferon genes in acute lymphoblastic leukemia. *N. Eng. J. Med.* **332**, 77–82.
121. Fountain, J. W., Karayiorgou, M., Ernstoff, M. S., Kirkwood, J. M., Vlock, D. R., Titus-Ernstoff, L., Bouchard, B., Vijayasaradhi, S., Houghton, A. N., Lahti, J., *et al.* (1992). Homozygous deletions within human chromosome band 9p21 in melanoma. *Proc. Natl. Acad. Sci. USA* **89**, 10557–10561.
122. Olopade, O. I., Buchhagen, D. L., Malik, K., Sherman, J., Nobori, T., Bader, S., Nau, M. M., Gazdar, A. F., Minna, J. D., and Diaz, M. O. (1993). Homozygous loss of the interferon genes defines the critical region on 9p that is deleted in lung cancers. *Cancer Res.* **53**, 2410–2415.
123. Cairns, P., Shaw, M. E., and Knowles, M. A. (1993). Initiation of bladder cancer may involve deletion of a tumor suppressor gene on chromosome 9. *Oncogene* **8**, 1083–1085.
124. Nawroz, H., van der Riet, P., Hruban, R. H., Koch, W., Ruppert, J. M., and Sidransky, D. (1994). Allelotype of head and neck squamous cell carcinoma. *Cancer Res.* **54**, 1152–1155.
125. Ah-See, K. W., Cooke, T. G., Pickford, I. R., Soutar, D., and Balmain, A. (1994). An allelotype of squamous cell carcinoma of the head and neck using microsatellite markers. *Cancer Res.* **54**, 1617–1621.
126. Cairns, P., Tokino, K., Eby, Y., and Sidransky, D. (1994). Homozygous deletions of 9p21 in primary human bladder tumors detected by comparative multiplex polymerase chain reaction. *Cancer Res.* **54**, 1422–1424.
127. Bonetta, L. (1994). Open questions on p16. *Nature* **370**, 180.
128. Kamb, A., Gruis, N. A., Weaver-Feldhaus, J., Liu, Q., Harshman, K., Tavtigian, S. V., Stockert, E., Day, R. S., 3rd, Johnson, B. E., and Skolnick, M. H. (1994). A cell cycle regulator potentially involved in genesis of many tumor types. *Science* **264**, 436–440.
129. Nobori, T., Miura, K., Wu, D. J., Lois, A., Takabayashi, K., and Carson, D. A. (1994). Deletions of the cyclin-dependent kinase-4 inhibitor gene in multiple human cancers. *Nature* **368**, 753–756.
130. Cairns, P., Mao, L., Merlo, A., Lee, D. J., Schwab, D., Eby, Y., Tokino, K., van der Riet, P., Blaugrund, J. E., and Sidransky, D. (1994). Rates of p16 (MTS1) mutations in primary tumors with 9p loss. *Science* **265**, 415–416.
131. Spruck, C. H., Gonzalez-Zulueta, M., Shibata, A., Simoneau, A. R., Lin, M. F., Gonzales, F., Tsai, Y. C., and Jones, P. A. (1994). p16 gene in uncultured tumors. *Nature* **370**, 183–184.
132. Kamb, A., Liu, Q., Harshman, K., Tavtigian, S., Cordon-Cardo, C., and Skolnick, M. H. (1994). Rates of p16 (MTS1) mutations in primary tumors with 9p loss. *Science* **265**, 416–417.
133. Merlo, A., Herman, J. G., Mao, L., Lee, D. J., Gabrielson, E., Burger, P. C., Baylin, S. B., and Sidransky, D. (1995). 5′ CpG island methylation is associated with transcriptional silencing of the tumor suppressor p16/CDKN2/MTS1 in human cancers. *Nature Med.* **1**, 686–692.
134. Otterson, G. A., Kratzke, R. A., Coxon, A., Kim, Y. W., and Kaye, F. J. (1994). Absence of p16INK4 protein is restricted to the subset of lung cancer lines that retains wildtype RB. *Oncogene* **9**, 3375–3378.
135. Aagaard, L., Lukas, J., Bartkova, J., Kjerulff, A. A., Strauss, M., and Bartek, J. (1995). Aberrations of p16Ink4 and retinoblastoma tumour-suppressor genes occur in distinct sub-sets of human cancer cell lines. *Int. J. Cancer.* **61**, 115–120.
136. Parry, D., Bates, S., Mann, D. J., and Peters, G. (1995). Lack of cyclin D-Cdk complexes in Rb-negative cells correlates with high levels of p16INK4/MTS1 tumour suppressor gene product. *EMBO J.* **14**, 503–511.
137. Tam, S. W, Shay, J. W., and Pagano, M. (1994). Differential expression and cell cycle regulation of the cyclin-dependent kinase inhibitor $p16^{Ink4}$. *Cancer Res.* **54**, 5816–5820.
138. Passegue, E., and Wagner, E. F. (2000). JunB suppresses cell proliferation by transcriptional activation of $p16^{INK4a}$ expression. *EMBO J.* **19**, 2969–2979.
139. Cannon-Albright, LA, Goldgar, D. E., Meyer, L. J., Lewis, C. M., Anderson, D. E., Fountain, J. W., Hegi, M. E., Wiseman, R. W., Petty, E. M., Bale, A. E., *et al.* (1992). Assignment of a locus for familial melanoma, *MLM*, to chromosome 9p13-22. *Science* **258**, 1148–1152.
140. Hussussian, C. J., Struewing, J. P., Goldstein, A. M., Higgins, P. A. T., Ally, D. S., Sheahan, M. D., Clark, W. H. Jr., Tucker, M. A., and Dracopoli, N. C. (1994). Germline p16 mutations in familial melanoma. *Nature Genet.* **8**, 15–21.
141. Kamb, A., Shattuck-Eidens, D., Eeles, R., Liu, Q., Gruis, N. A., Ding, W., Hussey, C., Tran, T., Miki, Y., Weaver-Feldhaus, J., *et al.* (1994). Analysis of the p16 gene (CDKN2) as a candidate for the chromosome 9p melanoma susceptibility locus. *Nature Genet.* **8**, 21–26.
142. Caldas, C., Hahn, S. A., da Costa, L. T., Redston, M. S., Schutte, M., Seymour, A. B., Weinstein, C. L., Hruban, R. H., Yeo, C. J., and Kern, S. E. (1994). Frequent somatic mutations and homozygous deletions of the p16 (MTS1) gene in pancreatic adenocarcinoma. *Nature Genet.* **8**, 27–32.
143. Serrano, M., Gomez-Lahoz, E., DePinho, R. A., Beach, D., and Bar-Sagi, D. (1995). Inhibition of ras-induced proliferation and cellular transformation by $p16^{ink4}$. *Science* **267**, 249–252.

144. Arap, W., Nishikawa, R., Furnari, F. B., Cavenee, W. K., and Huang, H.-J. S. (1995). Replacement of the p16/CDKN2 gene suppresses human glioma cell growth. *Cancer Res.* **55**, 1351–1354.
145. Serrano, M., Lee, H., Chin, L., Cordon-Cardo, C., Beach, D., and DePinho, R. A. (1996). Role of the INK4a locus in tumor suppression and cell mortality. *Cell* **85**, 27–37.
146. Chin, L., Pomerantz, J., Polsky, D., Jacobson, M., Cohen, C., Cordon-Cardo, C., Horner II, J. W., and DePinho, R. A. (1997). Cooperative effects of *INK4a* and *ras* in melanoma susceptibility in vivo. *Genes Dev.* **11**, 2822–2834.
147. Koh, J., Enders, G. H., Dynlacht, B. D., and Harlow, E. (1995). Tumor-derived p16 alleles encoding proteins defective in cell cycle inhibition. *Nature* **375**, 506–510.
148. Lukas, J., Parry, D., Aagard, L., Mann, D. J., Bartkova, J., Strauss, M., Peters, G., and Bartek, J. (1995). Retinoblastoma protein dependent inhibition by the tumor suppressor p16. *Nature* **375**, 503–506.
149. Medema, R. H., Herrera, R. E., Lam, F., and Weinberg, R. A. (1995). Growth suppression by p16ink4 requires functional retinoblastoma protein. *Proc. Natl. Acad. Sci. USA* **92**, 6289–6293.
150. Bruce, J. L., Hurford, R. K., Classon, M., Koh, J., and Dyson, N. (2000). Requirements for cell cycle arrest by $p16^{INK4A}$. *Mol. Cell* **6**, 737–742.
151. Porterfield, B. W., Diaz, M. O., Rowley, J. O., and Olopade, O. I. (1992). Induction of senescence in two neoplastic cell lines by microcell transfer of human chromosome 9. *Proc. Am. Assoc. Cancer Res.* **33**, 73.
152. Loughran, O., Edington, K. G., Berry, I. J., Clark, L. J., and Parkinson, E. K. (1994). Loss of heterozygosity of chromosome 9p21 is associated with the immortal phenotype of neoplastic human head and neck keratinocytes. *Cancer Res.* **54**, 5045–5049.
153. Loughran, O., Malliri, A., Owens, D., Gallimore, P. H., Stanley, M. A., Ozanne, B., Frame, M. C., and Parkinson, E. K. (1996). Association of CDKN2A/p16INK4A with human head and neck keratinocyte replicative senescence: Relationship of dysfunction to immortality and neoplasia. *Oncogene* **13**, 561–568.
154. Hara, E., Smith, R., Parry, D., Tahara, H., Stone, S., and Peters, G. (1996). Regulation of p16CDKN2 expression and its implications for cell immortalization and senescence. *Mol. Cell. Biol.* **16**, 859–867.
155. Stone, S., Jiang, P., Dayananth, P., Tavtigian, S. V., Katcher, H., Parry, D., Peters, G., and Kamb, A. (1995). Complex structure and regulation of the *P16 (MTS1)* locus. *Cancer Res.* **55**, 2988–2994.
156. Quelle, D. E., Zindy, F., Ashmun, R. A., and Sherr, C. J. (1995). Alternative reading frames of the *INK4a* tumor suppressor gene encode two unrelated proteins capable of inducing cell cycle arrest. *Cell* **83**, 993–1000
157. Mao, L., Merlo, A., Bedi, G., Shapiro, G. I., Edwards, C. D., Rollins, B. J., and Sidransky, D. (1995). A novel $p16^{INK4a}$ transcript. *Cancer Res.* **55**, 2995–2997.
158. Duro, D., Bernard, O., Della Valle, V., Berger, R., and Larsen, C.-J. (1995). A new type of $p16^{INK4/MTS1}$ gene transcript expressed in B-cell malignancies. *Oncogene* **11**, 21–29.
159. Kamijo, T., Bodner, S., van de Kamp, E., Randle, D. H., and Sherr, C. J. (1999). Tumor spectrum in ARF-deficient mice. *Cancer Res.* **59**, 2217–2222.
160. Pomerantz, J., Schreiber-Agus, N., Liegeois, N. J., Silverman, A., Alland, L., Chin, L., Potes, J., Chen, K., Orlow, I., Lee, H.-W., Cordon-Cardo, C., and DePinho, R. (1998). The Ink4a tumor suppressor gene product, p19ARF, interacts with MDM2 and neutralizes MDM2's inhibition of p53. *Cell* **92**, 713–723.
161. Zhang, Y., Xiong, Y., and Yarbrough, W. G. (1998). ARF promotes MDM2 degradation and stabilizes p53: ARF-INK4a locus deletion impairs both the Rb and p53 tumor suppressor pathways. *Cell* **92**, 725–734.
162. Stott, F. J., Bates, S., James, M. C., McConnell, B. B., Starborg, M., Brookes, S., Palmero, I., Ryan, K., Hara, E., Vousden, K. H., and Peters, G. (1998). The alternative product from the human CDKN2A locus, p14(ARF), participates in a regulatory feedback loop with p53 and MDM2. *EMBO J.* **17**, 5001–5014.
163. Weber, J. D., Taylor, L. J., Roussel, M. F., Sherr, C. J., and Bar-Sagi, D. (1999). Nucleolar Arf sequesters Mdm2 and activates p53. *Nature Cell Biol.* **1**, 20–26.
164. Tao, W., and Levine, A. J. (1999). $P19^{ARF}$ stabilizes p53 by blocking nucleo-cytoplasmic shuttling of Mdm2. *Proc Natl. Acad. Sci. USA* **96**, 6937–6941.
165. Kamijo, T., Zindy, F., Roussel, M. F., Quelle, D. E., Downing, J. R., Ashmun, R. A., Grosveld, G., and Sherr, C. J. (1997). Tumor suppression at the mouse INK4a locus mediated by the alternative reading frame product p19ARF. *Cell* **91**, 649–659.
166. Sherr, C. J. (1998). Tumor surveillance via the ARF-p53 pathway. *Genes Dev.* **12**, 2984-2991.
167. Prives, C. (1998). Signaling to p53: Breaking the MDM2-p53 circuit. *Cell* **95**, 5–8.
168. Bates, S., Phillips, A. C., Clark, P. A., Stott, F., Peters, G., Ludwig, R. L., and Vousden, K. H. (1998). p14ARF links the tumour suppressors RB and p53. *Nature* **395**, 124–125.
169. Martelli, F., Hamilton, T., Silver, D. P., Sharpless, N. E., Bardessy, N., Rokas, M., DePinho, R. A., Livingston, D. M., and Grossman, S. R. (2001). p19ARF targets certain E2F species for degradation. *Proc. Natl. Acad. Sci. USA* **98**, 4455–4460.
170. Randerson-Moor, J. A., Harland, M., Williams, S., Cuthbert-Heavens, D., Sheridan, E., Aveyard, J., Sibley, K., Whitaker, L., Knowles, M., Bishop, J. N., and Bishop, D. T. (2001). A germline deletion of $p14^{ARF}$ but not *CDKN2A* in a melanoma–neural system tumour syndrome family. *Hum. Mol. Genet.* **10**, 55–62.
171. Hannon, G. J., and Beach, D. (1994). $p15^{INK4B}$ is a potential effector of TGF-β-induced cell cycle arrest. *Nature* **371**, 257–260.
172. Guan, K.-L., Jenkins, C. W., Li, Y, Nichols, M. A., Wu, X., O'Keefe, C. L., Matera, A. G., Xiong, Y. (1994). Growth suppression by p18, a $p16^{INK4}$/MTS1 and $p14^{INK4B}$/MTS2-related CDK6 inhibitor, correlates with wild-type pRb function. *Genes Dev.* **9**, 2939–2952.
173. Hirai, H., Roussel, M. F., Kato, J.-Y., Ashmun, R. A., and Sherr, C. J. (1995). Novel INK4 proteins, p19 and p18, are specific inhibitors of the cyclin D-dependent kinases CDK4 and CDK6. *Mol. Cell. Biol.* **15**, 2672–2681.
174. Chan, F. K. M., Zhang, J., Cheng, L., Shapiro, D. N., and Winoto, A. (1995). Identification of human and mouse p19, a novel CDK4 and CDK6 inhibitor with homology to $p16^{ink4}$. *Mol. Cell. Biol.* **15**, 2682–2688.
175. Okamoto, A., Hussain, S. P., Hagiwara, K., Spillare, E. A., Rusin, M. R., Demetrick, D. J., Serrano, M., Hannon, G. J., Shiseki, M., Zariwala, M., Xiong, Y., Beach, D. H., Yokota, J., and Harris, C. C. (1995). Mutations in the $p16^{INK4}$/MTS1/CDKN2, $p15^{INK4B}$/MTS2, and p18 genes in primary and metastatic lung cancer. *Cancer Res.* **55**, 1448–1451.
176. Otsuki, T., Clark, H. M., Wellmann, A., Jaffe, E. S., and Raffeld, M. (1995). Involvement of CDKN2 ($p16^{INK4A}$/MTS1) and $p15^{INK4B}$/MTS2 in human leukemias and lymphomas. *Cancer Res.* **55**, 1436–1440.
177. Jen, J., Harper, J. W., Bigner, S. H., Bigner, D. D., Papadopoulos, N., Markowitz, S., Willson, J. K., Kinzler, K. W., and Vogelstein, B. (1994). Deletion of p16 and p15 genes in brain tumors. *Cancer Res.* **54**, 6353–6358.
178. Todd, R., McBride, J., Tsuji, T., Donoff, R. B., Nagai, M., Chou, M. Y., Chiang, T., and Wong, D. T. (1995). Deleted in oral cancer-1 (doc-1), a novel oral tumor suppressor gene. *FASEB J.* **9**, 1362–1370.
179. Tsuji, T., Duh, F. M., Latif, F., Popescu, N. C., Zimonjic, D. B., McBride, J., Matsuo, K., Ohyama, H., Todd, R., Nagata, E., Terakado, N., Sasaki, A., Matsumura, T., Lerman, M. I., and Wong, D. T. (1998). Cloning, mapping, expression, function, and mutation analyses of the human ortholog of the hamster putative tumor suppressor gene Doc-1. *J. Biol Chem.* **273**, 6704–6709.

180. Shintani, S., Ohyama, H., Zhang, X., McBride, J., Matsuo, K., Tsuji, T., Hu, M. G., Hu, G., Kohno, Y., Lerman, M., Todd, R., and Wong, D. T. W. (2000). p12 DOC-1 Is a novel cyclin-dependent kinase 2-associated protein. *Mol. Cell. Biol.* **20**, 6300–6307.
181. Cwikla, S. J., Tsuji, T., McBride, J., Wong, D. T., and Todd, R. (2000). doc-1-mediated apoptosis in malignant hamster oral keratinocytes. *J. Oral Maxillofac. Surg.* **58**, 406–414.
182. Gusterson, B. A., Anbazhagan, R., Warren, W., Midgely, C., Lane, D. P., O'Hare, M., Stamps, A., Carter, R., and Jayatilake, H. (1991). Expression of p53 in premalignant and malignant squamous epithelium. *Oncogene* **6**, 1785–1789.
183. Somers, K. D., Merrick, M. A., Lopez, M. E., Incognito, L. S., Schechter, G. L. and Casey, G. Frequent *p53* mutations in head and neck cancer. *Cancer Res.* **52**, 5997–6000 (1992).
184. Sakai, E., and Tsuchida, N. (1992). Most human squamous cell carcinomas in the oral cavity contain mutated p53 tumor-suppressor genes. *Oncogene* **7**, 927–933.
185. Brachman, D. G., Graves, D., Vokes, E., Beckett, M., Haraf, D., Montag, A., Dunphy, E., Mick, R., Yandell, D., and Weichselbaum, R. R. (1992). Occurrence of p53 gene deletions and human papilloma virus infection in human head and neck cancer. *Cancer Res.* **52**, 4832–4836.
186. Burns, J. E., Baird, M. C., Clark, L. J., Burns, P. A., Edington, K., Chapman, C., Mitchell, R., Robertson, G., Soutar, D. and Parkinson, E. K. (1993). Gene mutations and increased levels of p53 protein in human squamous cell carcinomas and their cell lines. *Br. J. Cancer* **67**, 1274–1284.
187. Yeudall, W. A., Paterson, I. C., Patel, V., and Prime, S. S. (1995). Presence of human papillomavirus sequences in tumour-derived human oral keratinocytes expressing mutant p53. *Eur. J. Cancer* **31B**, 136–143.
188. Yeudall, W. A., Jakus, J., Ensley, J. F. and Robbins, K. C. (1997). Functional characterization of p53 molecules expressed in human squamous cell carcinomas of the head and neck. *Mol. Carcinogen.* **18**, 89–96.
189. Patel, V., Ensley, J. F., Gutkind, J. S., and Yeudall, W. A. (2000). p53-independent apoptosis induced by γ-irradiation and bleomycin in head and neck squamous carcinoma cells. *Int. J. Cancer* **88**, 737–743.
190. Cardinali, M., Kratochvil, F. J., Ensley, J. F., Robbins, K. C., and Yeudall, W. A. (1997). Functional characterization *in vivo* of mutant p53 molecules derived from squamous cell carcinomas of the head and neck. *Mol. Carcinog.* **18**, 78–88.
191. Cruz, I. B., Meijer, C. J. L. M., Snijders, P. J. F., Snow, G. B., Walboomers, J. M. M., and van der Waal, I. (2000). P53 immunoexpression in non-malignant oral mucosa adjacent to oral squamous cell carcinoma: potential consequences for clinical management. *J. Pathol.* **191**, 132–137.
192. Burns, J. E., Clark, L. J., Yeudall, W. A., Mitchell, R. J., Mackenzie, K., Chang, S. E., and Parkinson, E. K. (1994). The p53 status of cultured premalignant oral keratinocytes. *Br. J. Cancer* **70**, 571–575.
193. Fantl, V., Richards, M. A., Smith, R., Lammie, G. A., Johnstone, G., Allen, D., Gregory, W., Peters, G., Dickson, C., and Barnes, D. M. (1990). Gene amplification on chromosome band 11q13 and estrogen receptor status in breast cancer. *Eur. J. Cancer* **26**, 423–429.
194. Lammie, G. A., and Peters, G. (1991). Chromosome 11q13 abnormalities in human cancer. *Cancer Cells* **3**, 413–420.
195. Kitagawa, Y., Ueda, M., Ando, N., Shinozawa, Y., Shimizu, M., and Abe, O. (1991). Significance of *int-2/hst-1* coamplification as a prognostic factor in patients with esophageal squamous cell carcinoma. *Cancer Res.* **51**, 1504–1508.
196. Wagata, T., Ishizaki, K., Imamura, M., Shimada, Y., Ikenaga, M., and Tobe, T. (1991). Deletion of 17p and amplification of the *int-2* gene in esophageal carcinomas. *Cancer Res.* **51**, 2113–2117.
197. Berenson, J. R., Koga, H., Yang, J., Pearl, J., Holmes, E. C., Figlin, R., and Group, L. C. S. (1990). Frequent amplification of the *bcl-1* locus in poorly differentiated squamous cell carcinoma of the lung. *Oncogene* **5**, 1343–1348.
198. Berenson, J. R., Yang, J., and Mickel, R. (1989). Frequent amplification of the *bcl-1* locus in head and neck squamous cell carcinomas. *Oncogene* **4**, 1111–1116.
199. Somers, K. D., Cartwright, S. L., and Schechter, G. L. (1990). Amplification of the int-2 gene in human head and neck squamous cell carcinomas. *Oncogene* **5**, 915–920.
200. Bartkova, J., Lukas, J., Muller, H., Strauss, M., Gusterson, B., and Bartek, J. (1995). Abnormal patterns of D-type cyclin expression and G1 regulation in human head and neck cancer. *Cancer Res.* **55**, 949–956.
201. Patel, V., Jakus, J., Harris, C., Ensley, J. F., Robbins, K. C., and Yeudall, W. A. (1997). Altered expression and activity of G1 cyclin dependent kinases characterize squamous cell carcinomas of the head and neck. *Int. J. Cancer* **73**, 551–555.
202. Timmermann, S., Hinds, P. W., and Munger, K. (1997). Elevated activity of cyclin-dependent kinase 6 in human squamous cell carcinoma lines. *Cell Growth Differ.* **8**, 361–370.
203. Kushner, J., Bradley, G., Young, B., and Jordan, R. C. (1999). Aberrant expression of cyclin A and cyclin B1 proteins in oral carcinoma. *J. Oral. Pathol. Med.* **28**, 77–81.
204. Leethanakul, C., Patel, V., Gillespie, J., Pallente, M., Ensley, J. F., Koontongkaew, S., Liotta, L. A., Emmert-Buck, M., and Gutkind, J. S. (2000). Distinct pattern of expression of differentiation and growth-related genes in squamous cell carcinomas of the head and neck revealed by the use of laser capture microdissection and cDNA arrays. *Oncogene* **19**, 3220–3224.
205. Rousseau, A., Lim, M. S., Lin, Z., and Jordan, R. C. K. (2001). Frequent cyclin D1 gene amplification and protein overexpression in oral epithelial dysplasias. *Oral Oncol.* **37**, 268–275.
206. Castle, J. T., Cardinali, M., Kratochvil, F. J., Abbondanzo, S. L., Kessler, H. P., Auclair, P. L., and Yeudall, W. A. (1999). p53 and cyclin D1 staining patterns of malignant and premalignant oral lesions in age-dependent populations. *Oral Surg. Oral Med. Oral Pathol.* **88**, 326–332.
207. Yoo, G. H., Xu, H. J., Brennan, J. A., Westra, W., Hruban, R. H., Koch, W., Benedict, W. F., and Sidransky, D. (1994). Infrequent inactivation of the retinoblastoma gene despite frequent loss of chromosome 13q in head and neck squamous cell carcinoma. *Cancer Res.* **54**, 4603–4606.
208. Michalides, R., van Veelen, N., Hart, A., Loftus, B., Wientjens, E., and Balm, A. (1995). Overexpression of cyclin D1 correlates with recurrence in a group of forty-seven operable squamous cell carcinomas of the head and neck. *Cancer Res.* **55**, 975–978.
209. Bova, R. J., Quinn, D. I., Nankervis, J. S., Cole, I. E., Sheridan, B. F., Jensen, M. J., Morgan. G. J., Hughes, C. J., and Sutherland, R. L. (1999). Cyclin D1 and p16INK4A expression predict reduced survival in carcinoma of the anterior tongue. *Clin. Cancer Res.* **5**, 2810–2819.
210. Yeudall, W. A., Crawford, R. Y., Ensley, J. F., and Robbins, K. C. (1994). *MTS1/CDK4I* is altered in cell lines derived from primary and metastatic oral squamous cell carcinoma. *Carcinogenesis* **15**, 2683–2686.
211. Reed, A. L., Califano, J., Cairns, P., Westra, W. H., Jones, R. M., Koch, W., Ahrendt, S., Eby, Y., Sewell, D., Nawroz, H., Bartek, J., and Sidransky, D. (1996). High frequency of *p16* (CDKN2/MTS-1/INK4A) inactivation in head and neck squamous cell carcinoma. *Cancer Res.* **56**, 3630–3633.
212. Wu, C. L., Roz, L., McKown, S., Sloan, P., Read, A. P., Holland, S., Porter, S., Scully, C., Paterson, I., Tavassoli, M., and Thakker, N. (1999). DNA studies underestimate the major role of CDKN2A inactivation in oral and oropharyngeal squamous cell carcinomas. *Genes Chrom. Cancer* **25**, 16–25.
213. Sartor, M., Steingrimsdottir, H., Elamin, F., Gaken, J., Warnakulasuriya, S., Partridge, M., Thakker, N., Johnson, N. W., and

Tavassoli, M. (1999). Role of p16/MTS1, cyclin D1 and RB in primary oral cancer and oral cancer cell lines. *Br. J. Cancer* **80**, 79–86.

214. Shahnavaz, S. A., Bradley, G., Regezi, J. A., Thakker, N., Gao, L., Hogg, D., and Jordan, R. K. C. (2001). Patterns of *CDKN2* gene loss in sequential oral epithelial dysplasias and carcinomas. *Cancer Res.* **61**, 2371–2375.
215. Jordan, R. C., Bradley, G., and Slingerland, J. (1998). Reduced levels of the cell-cycle inhibitor p27Kip1 in epithelial dysplasia and carcinoma of the oral cavity. *Am. J. Pathol.* **152**, 585–590.
216. Ito, R., Yasui, W., Ogawa, Y., Toyosawa, S., Tahara, E., and Ijuhin, N. (1999). Reduced expression of cyclin-dependent kinase inhibitor p27(Kip1) in oral malignant tumors. *Pathobiology* **67**, 169–173.
217. Mineta, H., Miura, K., Suzuki, I., Takebayashi, S., Amano, H., Araki, K., Harada, H., Ichimura, K., Wennerberg, J. P., and Dictor, M. R. (1999). Low p27 expression correlates with poor prognosis for patients with oral tongue squamous cell carcinoma. *Cancer* **85**, 1011–1017.
218. Liu, T. J., Zhang, W. W., Taylor, D. L., Roth, J. A., Goepfert, H., and Clayman, G. L. (1994). Growth suppression of human head and neck cancer cells by the introduction of a wild-type p53 gene via a recombinant adenovirus. *Cancer Res.* **54**, 3662–3667.
219. Clayman, G. L., el-Naggar, A. K., Roth, J. A., Zhang, W. W., Goepfert, H., Taylor, D. L., and Liu, T. J. (1995). In vivo molecular therapy with p53 adenovirus for microscopic residual head and neck squamous carcinoma. *Cancer Res.* **55**, 1–6.
220. Clayman, G. L., Liu, T. J., Overholt, S. M., Mobley, S. R., Wang, M., Janot, F., and Goepfert, H. (1996). Gene therapy for head and neck cancer: Comparing the tumor suppressor gene p53 and a cell cycle regulator WAF1/CIP1 (p21). *Arch. Otolaryngol. Head Neck Surg.* **122**, 489–493.
221. Overholt, S. M., Liu, T. J., Taylor, D. L., Wang, M., El-Naggar, A. K., Shillitoe, E., Adler-Storthz, K., John, L. S., Zhang, W. W., Roth, J. A., and Clayman, G. L. (1997). Head and neck squamous cell growth suppression using adenovirus-p53-FLAG: A potential marker for gene therapy trials. *Clin. Cancer Res.* **3**, 185–191.
222. Liu, T. J., el-Naggar, A. K., McDonnell, T. J., Steck, K. D., Wang, M., Taylor, D. L., and Clayman, G. L. (1995). Apoptosis induction mediated by wild-type p53 adenoviral gene transfer in squamous cell carcinoma of the head and neck. *Cancer Res.* **55**, 3117–3122.
223. Eicher, S. A., Clayman, G. L., Liu, T. J., Shillitoe, E. J., Storthz, K. A., Roth, J. A., and Lotan, R. (1996). Evaluation of topical gene therapy for head and neck squamous cell carcinoma in an organotypic model. *Clin. Cancer Res.* **2**, 1659–1664.
224. Clayman, G. L., Frank, D. K., Bruso, P. A., and Goepfert, H. (1999). Adenovirus-mediated wild-type p53 gene transfer as a surgical adjuvant in advanced head and neck cancers. *Clin. Cancer Res.* **5**, 1715–1722.
225. Liu, T. J., Wang, M., Breau, R. L., Henderson, Y., El-Naggar, A. K., Steck, K. D., Sicard, M. W., and Clayman, G. L. (1999). Apoptosis induction by E2F-1 via adenoviral-mediated gene transfer results in growth suppression of head and neck squamous cell carcinoma cell lines. *Cancer Gene Ther.* **6**, 163–171.
226. Malliri, A., Yeudall, W. A., Nikolic, M., Crouch, D. H., Parkinson, E. K., and Ozanne, B. (1996). Sensitivity to TGF-β1 induced growth arrest is common in human squamous cell carcinoma cell lines: C-MYC down regulation and p21^{WAF1} induction are important early events. *Cell Growth Diff.* **7**, 1291–1304.
227. Prime, S. S., Matthews, J. B., Patel, V., Game, S. M., Donnelly, M., Stone, A., Paterson, I. C., Sandy, J. R., and Yeudall, W. A. (1994). TGF-β receptor regulation mediates the response to exogenous ligand but is independent of the degree of cellular differentiation in human oral keratinocytes. *Int. J. Cancer* **56**, 406–412.
228. Garrigue-Antar, L., Munoz-Antonia, T., Antonia, S. J., Gesmonde, J., Vellucci, V. F., and Reiss, M. (1995). Missense mutations of the transforming growth factor β type II receptor in human head and neck squamous carcinoma cells. *Cancer Res.* **55**, 3982–3987.
229. Lesaca, E. E., Ensley, J. F., and Yeudall, W. A. (1998). Cellular factors may enable squamous carcinoma cells to overcome TGFβ-mediated repression of CDK2 activity. *Oral Oncol.* **34**, 52–57.
230. Cardinali, M., Jakus, J., Robbins, K. C., Ensley, J. F., Shah, S., and Yeudall, W. A. (1998). p21^{WAF1} retards the growth of human squamous carcinoma cells *in vivo*. *Oral Oncol* **34**, 211–218.
231. Mobley, S. R., Liu, T. J., Hudson, J. M., and Clayman, G. L. (1998). In vitro growth suppression by adenoviral transduction of p21 and p16 in squamous cell carcinoma of the head and neck: A research model for combination gene therapy. *Arch. Otolaryngol. Head Neck Surg.* **124**, 88–92.
232. Nakashima, T., and Clayman, G. L. (2000). Antisense inhibition of cyclin D1 in human head and neck squamous cell carcinoma. *Arch. Otolaryngol. Head Neck Surg.* **126**, 957–961.
233. Patel, V., Senderowicz, A. M., Pinto, D., Igishi, T., Raffeld, M., Quintanilla-Martinez, L., Ensley, J. F., Sausville, E. A., and Gutkind, J. S. (1998). Flavopiridol, a novel cyclin-dependent kinase inhibitor, suppresses the growth of head and neck squamous cell carcinomas by inducing apoptosis. *J. Clin. Invest.* **102**, 1674–1681.
234. Fahraeus, R., Paramio, J. M., Ball, K. L., Lain, S., and Lane, D. P. (1996). Inhibition of pRb phosphorylation and cell-cycle progression by a 20-residue peptide derived from p16$^{CDKN2/INK4A}$. *Curr. Biol.* **6**, 84–91.
235. Soni, R., O'Reilly, T., Furet, P., Muller, L., Stephan, C., Zumstein-Mecker, S., Fretz, H., Fabbro, D., and Chaudhuri, B. (2001). Selective *in vivo* and *in vitro* effects of a small molecule inhibitor of cyclin-dependent kinase 4. *J. Natl. Cancer Inst.* **93**, 436–446.

CHAPTER

9

Oncogenes and Tumor Suppressor Genes in Oral or Head and Neck Squamous Cell Carcinoma

CRISPIAN SCULLY
Eastman Dental Institute for Oral Health Care Sciences
University College London
University of London
London WC1X 8LD, United Kingdom

J. K. FIELD
Molecular Genetics and Oncology Group
Clinical Dental Sciences
University of Liverpool and
Roy Castle International Centre for Lung Cancer Research
Liverpool L69 3BX, United Kingdom

HIDEKI TANZAWA
Department of Oral Surgery
Chiba University
Chiba 260-8670, Japan

I. Introduction 117
II. Cancer 118
III. Cell Regulation, Oncogenes, and Tumor Suppressor Genes 118
A. Cell Signaling 118
B. Cell Cycle 119
IV. Tumor Suppressor Genes 120
V. Detection of Individuals at Risk 125
VI. Molecular Diagnosis 125
VII. Staging 125
VIII. Prognostication 125
A. Fractional Allelic Loss 126
B. Identification of Second Primary and Recurrent Tumors 126
C. The Future 126
References 126

I. INTRODUCTION

Carcinogenesis is the result of a series of genetic mutations resulting in unregulated growth of a clone of cells and the development of a malignant lesion that is largely monoclonal. Cancer is cell proliferation caused by cell dysregulation as a consequence of disruption to cell signaling, the cell growth cycle, or mechanisms that normally repair cell damage or eliminate dysfunctional cells. Chromosomal (genetic) damage affects these regulatory processes via transcribed proteins.

Oral squamous cell carcinoma (oral cancer) and many squamous cell carcinomas of the head and neck (SCCHN) arise as a consequence of multiple molecular events induced by the effects of carcinogens from habits such as tobacco, influenced by environmental factors, possibly viruses in some instances, against a background of heritable susceptibility. Consequent genetic damage affects many chromosomes and genes, and it is the accumulation of these genetic changes that appears to lead to carcinoma in some instances, sometimes via a clinically evident premalignant, or potentially malignant, lesion. Enhanced function of oncogenes, impaired function of tumor suppressor genes (TSGs) or their products, or increased telomerase activity is involved.

There is about to be an explosion in the understanding of this area with the advent of a number of new technologies. DNA technology, especially allelic imbalance [loss of heterozygosity (LOH)] studies, has identified changes particularly in certain chromosomes shown in Table 9.1. Overexpression of oncogenes, especially those on chromosome 11 (PRAD-1 in particular) and 17 (H-*ras*) and TSGs, particularly 3p—especially 3p14.2 (FHIT), 3p24, and 3p21.3, where the TSGs involved, are as yet unidentified; 9p21 where p16 (INK4A/MTS-1) is the main target TSG; and 17p13 where p53 is the major target TSG, have been incriminated, as discussed elsewhere [1–5]. These findings already offer the possibility of the development of a number of advances in risk assessment [3–5], diagnosis, prognostication, and management [6,7] mainly via dideoxynucleotide sequencing and molecular probing [8,9] and in association with the use of other biomarkers [10–12]. More recent advances in technology, such as the use of laser capture microdissection and cDNA arrays, place this area at a watershed, when many new findings are about to appear [13].

TABLE 9.1 Chromosomes Affected by Loss of Heterozygosity in SCCHN[a]

Chromosome affected by LOH	Approximate % SCCHN affected
3p13-14	60
4p	43
5q12-q22	30
8p	65
9p22-p24	43
10p	39
13q12-q24	30
18q	60
21	52

[a]From Jones *et al.* [256].

TABLE 9.2 Candidate Biomarkers for Oral Carcinoma

Gene/location	Alteration	Clinical significance
p53	Mutation/ overexpression	Nodal metastasis, recurrence
p16	Mutation/LOH	Poor prognosis
p21	Overexpression	Aggressive phenotype
p27	Decreased expression	Tumor progression
PRAD-1	Amplification	Poor prognosis
hst-1	Amplification	Advanced tumor, metastasis
bcl-1	Amplification	Advanced tumor, metastasis
H-ras	Overexpression	Favorable prognosis
EGFR	Overexpression	Tumor size; TNM staging
bcl-2	Overexpression	Favorable prognosis
Bax	Decreased	Poor prognosis expression
3p	LOH	Poor prognosis, nodal metastasis
3p13.21.24-26	AI	Poor prognosis
5q31.2	LOH	Histological grade
8q24	LOH	Tumor size, histological grade
13q14.3	LOH	Nodal metastasis
16q24	LOH	Nodal metastasis
18q/DCC region	LOH/AI	Poor prognosis
18q21	Mutation	Histological grade
21q	AI	TNM staging

LOH, loss of heterozygosity; AI, allelig imbalance.

II. CANCER

Cancer arises from damage by a number of mutagens to the DNA of genes located at various points on the short (p) or long (q) arms of a number of chromosomes. Mutagens include chemicals (such as carcinogens), physical agents (such as ionizing radiation), and biological agents (such as microorganisms), and some mutations arise "spontaneously."

The consequence of chromosomal (genetic) damage is cell dysregulation with disruption in cell signaling, cell cycle and growth control, and/or cell damage repair mechanisms via proteins transcribed from the responsible genes. It is the accumulation of such genetic changes that appears to lead, in some instances, to such cell dysregulation that growth becomes autonomous and invasive mechanisms develop, leading to carcinoma.

Cancer is associated with aneuploidy, reflecting complex karyotypes arising from multiple genetic events, and with micronuclei-chromosome or chromatid fragments formed in proliferating cells from chromosome nondysjunction resulting from DNA damage [14]. Molecular genetic alterations include point mutations, amplifications, rearrangements, and deletions (Table 9.2).

III. CELL REGULATION, ONCOGENES, AND TUMOR SUPPRESSOR GENES

A. Cell Signaling

Cell signaling is essential for messages to be passed from the plasma membrane to the nucleus. Examples of signal mechanisms include the products of genes such as Harvey *ras* (H-*ras*) and epidermal growth factor receptor (EGFR).

1. H-*ras*

H-*ras*, one of the first protooncogenes that caught the attention of molecular biologists interested in cell signaling, cell growth control, and cancer, is only one of many genes involved in cell signaling. Located on chromosome 17p, it is responsible for a 21-kDa protein p21, an enzyme with guanosine triphosphatase activity (a GTPase) that acts as a signal transduction protein, transmitting mitogenic signals from the cell surface to the cell interior. p21 is inherently a GTPase, normally of short-lived activity. H-*ras* mutations can result in continued activity of p21, disturbing cell signaling and thus growth control. H-*ras* may thus act as an oncogene or possibly a TSG.

H-*ras* mutations have been found in SCCHN and oral cancers, mainly in patients in the developing world, particularly from India [15–18] and in advanced tumors [19], mainly at codons 12 or 61, but also 13 [20]. The high incidence of H-*ras* mutations in oral cancer in India is independent of p53 mutations (see later) and is mutually exclusive [21]. Although *ras* overexpression has been seen in a large percentage of tumors [22], no association was found between LOH at H-*ras* and overexpression of the gene [23].

Mutations in codon 12 of H-*ras* in SCCHN and oral carcinomas are, however, rare in the Western world, where they are seen in only about 5% [23–28].

There is interest in the possible interactions between human papillomaviruses (HPV), which have potentially oncogenic early genes such as E6 and E7, and H-*ras*. In squamous carcinomas, H-*ras* is mutated in about 22%, but only half of these samples contain HPV [29].

Other *ras* genes are infrequently mutated in oral carcinoma [18,19,30].

2. EGFR

The epidermal growth factor receptor gene (erbB1), which maps to 7p13-q22, encodes a cell-signaling receptor, which is a transmembrane tyrosine-specific phosphokinase that binds ligands, including epidermal growth factor (EGF), transforming growth factor α (TGFα), amphiregulin, heparin-binding EGF-like growth factor, betacellulin, cripto, and epiregulin, and can activate intracellular signaling via protein tyrosine kinase [31]. Activation of EGFR results in phosphorylation of the tyrosine residues of several proteins.

EGFR is overexpressed in epithelium adjacent to [32], and in, SCCHN [33] and is an early change. The reported proportion of oral carcinomas that overexpress EGFR varies widely, and some authors have even reported a decreased expression [34,35]. EGFR expression correlates with matrix-metalloproteinase 3 expression, which contributes to invasion and metastasis [36].

3. c-*myc*

The c-*myc* oncogene, on chromosome 8q24, may be amplified and overexpressed in oral carcinomas [20,37–41] and in about 7% of SCCHN [42].

4. PRAD-1

Chromosome 11q13 genes include *cyclin D1* (PRAD-1 or bcl-1), *hst-1* (FGF4), *int2* (FGF3), and *EMS1* (cortactin). PRAD-1 codes a 295-kDa protein that has sequence homology to the cyclin D1 gene, which regulates the cell cycle by binding and activating p34 CDK-4 and CDK-6 kinases. It can be detected in SCCHN [43,44] and can cooperate with H-*ras* in carcinogenesis [45]. Cyclin D1 is amplified in oral carcinomas [28,46–53] and there may be gene amplifications [49,54].

4. hst-1 and Int-2

As to hst-1, results are equivocal, with some workers reporting amplification [55], and others denying it [56]. The same applies to Int-2, where some have found it amplified in SCCHN [57], and others have not [56], and there appears to be little transcription of these genes.

5. HER-2/neu

Studies of the chromosome 17 oncogene HER-2/neu, also termed c-erbB-2, which codes a protein P185/HER-2 with tyrosine kinase activity, have produced conflicting results, although studies suggest that it may be overexpressed in oral cancer [33,58].

6. Bcl-2

The Bcl-2 protooncogene, on chromosome 18q21, is involved in the regulation of apoptosis, is regulated by p53, and shows an inverse relationship in expression [59]. High p53 protein leads to high Bax and low Bcl-2. Bcl-2 inhibits the apoptosis that results from treatment with radiation/ chemotherapy. Some have reported a reduced expression of Bcl-2 in oral carcinomas [60,61], an increased Bax expression [62], and a decreased ratio of bcl-2/Bax with increased apoptosis [61]. Others have found that nearly one-quarter overexpress Bcl-2 [50,63], but the expression of these apoptosis-related genes may influence prognostication. Bcl-X is expressed early in malignant transformation, with variable bcl-2 expression [64].

B. Cell Cycle

The cell cycle consists of a stationary G0 phase, which, upon receipt of appropriate cell signals, is followed by entry into G1, S, and G2 before mitosis (M). The cycle is regulated by genes in a highly sophisticated fashion via a number of proteins [65]. The transition from one phase to the next is checked at a "checkpoint" before the cycle can progress.

Checkpoint genes not only control transition between phases of the cycle, but also coordinate progression with cell signals. When DNA is damaged, there may be arrest of the cycle, which can facilitate DNA repair. G1 arrest prevents replication of a damaged DNA template, and G2 arrest allows the segregation of damaged chromosomes. Checkpoint genes can modulate regulation of the integrity of the genome if the DNA damage is such as not to immediately cause cell death.

1. Cyclins and Kinases

Key players in cell cycle regulation include a group of heterodimeric protein kinases comprising a cyclin (the regulatory element) together with a cyclin-dependent kinase (CDK) as the catalyst. These protein kinases dephosphorylate the DNA-binding retinoblastoma protein (pRb) (pp105 or pp110), thereby releasing critical transcription factors, such as those of the E2F family. For example, progression from the stationary G0 phase of the cell cycle through the G phase is mediated by two CDK complexes (CDK4–cyclin D, and CDK2–cyclin E) [66].

Progression depends on the timely formation and disassociation of protein kinase complexes, each of which consists of a regulatory cyclin, an enzyme (a cyclin-dependent kinase), and proliferating cell nuclear antigen (PCNA). The complexes are regulated by phosphorylation/dephosphorylation. Cyclins, and cyclin-dependent kinases (CDKs), are thus the most important positive regulators of cell cycle progression. Negative regulators include products of some TSGs, such as inhibitors of CDK (CDK inhibitors), of which the p16 gene is the best known [66].

Cyclins can phosphorylate and thereby inactivate Rb, whereas cyclin-dependent kinase inhibitors (CKIs) can reverse this inactivation. Cyclin D1 (bcl-1 or PRAD-1), mapped to 11q13, regulates the cell cycle by involvement in control of the transit from G1 to S phase, becoming associated with CDK4 or CDK6 and then resulting in threonine or serine kinase activity leading to transition from G1 to S phases.

IV. TUMOR SUPPRESSOR GENES

The main factors regulating cell cycle progression therefore include cyclins, cyclin kinases, and inhibitory Rb, p21, p27, and p53 proteins. Overexpression of oncogenes, especially those on chromosome 11 (PRAD-1, Int-2, hst-1, and Bcl-1 in particular) and 17 (H-*ras*) has been implicated in the carcinogenesis of SCCHN [67], as has increased telomerase activity [68–72] and TSGs, which have been identified in LOH studies in oral and SCCHN, particularly in chromosomes 3, 9, 11, and 17. Regions identified most commonly have included 3p, especially 3p24, 3p21.3, and 3p14, where the TSGs are as yet unidentified; 9p21, where p16 (INK4A/MTS-1) is the main target; and 17p13, where p53 is the prime target. Many other TSGs may also be involved. DNA normally has two copies (alleles) of every sequence. If these alleles are different, the sequence is termed "polymorphic" in the population and "heterozygous" in the individual. A change from the heterozygous state to a hemizygous state is termed loss of heterozygosity. LOH indicates consistent allelic loss at marker loci in a specific genomic region and generally indicates genetic instability and the inactivation of specific tumor suppressor genes within the region of allele loss. LOH appears to represent the second genetic inactivation step in the complete loss of a TSG locus. Analyses of chromosomal allelic losses (LOH), suggestive of microsatellite instability (MI) [73], have thus allowed for the identification of chromosomal regions harboring tumor suppressor genes, and such analyses have a major advantage over cytogenetic techniques, as the resolution is much better, especially if closely spaced microsatellite markers are used.

Studies into genetic aberrations have also been facilitated by other new DNA technology [6], which can detect mutations in oncogenes or tumour suppressor genes, or microsatellites. DNA survives well and can be amplified by the polymerase chain reaction (PCR), and thus very small amounts of DNA are required. RNA can also be examined but is less stable so that it is best converted to cDNA using reverse transcriptase (RT-PCR). DNA alterations can also be detected by restriction landmark genomic scanning (RLGS) [74]. Allelotyping techniques, including the use of polymeric microsatellite markers and restriction fragment length polymorphism (RFLP), enable determination of LOH, indicative of allelic imbalance (AI), to be examined. AI has been considered an early event in oral carcinogenesis [75–77], especially on chromosome arms 3p, 9p, 11q, 13q, and 17p (50–70%), as outlined earlier, but it is possible that genetic alterations at other sites may play a role. LOH on chromosomes 3p, 9p, 11q, 13q, and 17p on oral cancer suggests a possible pathway of progression in oral carcinogenesis involving a generalized increased rate of errors during DNA replication, and defective repair of DNA, as shown by MIN. The occurrence of multiple areas of allelic loss on several chromosomes, together with the sequential loss of several tumor suppressor genes during experimental carcinogenesis [78–81], is entirely consistent with the hypothesis that oral carcinogenesis involves multiple molecular steps (presumably over a fairly prolonged period) [1,6,82]. The steps involved appear similar independent of the age of the patient [83], but may differ between users and nonusers of tobacco [84].

It is clear that no one locus is responsible for the carcinogenic progression, however, few studies have tackled this question on a genome wide level, preferring to concentrate on a specific gene or chromosomal region of interest. The problem with the latter approach is conceptualizing the results in the wider arena. One of the methods used to overcome this drawback is the use of a range of microsatellite markers covering the majority of the chromosomal arms at multiple loci and using the concept of fractional allele loss (FAL). This approach has been used successfully in cancers such as colon and bladder cancer, as well as in squamous cell carcinoma of the head and neck [85]. The latter study was reviewed in detail by Scully and Field [2], and the most important correlation found was between FAL status and survival thus emphasizing the relationship between the most important clinical parameter in cancer, together with an overreaching genetic measure of damage. Thereby, FAL becomes a very important genetic technique by which to address other salient questions surrounding the molecular progression of cancer, i.e., is there only one major genetic pathway in carcinogenesis based on allelic imbalance data? Nunn *et al.* [77] addressed this question. The availability of an extremely large data set on allelic imbalance in squamous cell carcinomas of the head and neck allowed this group to question whether allelic imbalance at specific loci on 3p, 9p, and 17p is solely responsible for head and neck carcinogenesis. It is of note that many of the SCCHN specimens in the Nunn *et al.* [77] were of oral origin and no significant difference was found between the tumor sites and FAL status. The SCCHN specimens were subdivided into high Fal (HFAL), median FAL (MFAL), and low FAL (LFAL) groups on the basis of global SCCHN genome data available to this research group. SCCHN tumors in this series were then further analyzed with a large number of microsatellites on 3p, 9p, and 17p in order to determine if any particular group of tumors had a predominance of LOH at all or any combination of these regions. Results of this analysis demonstrated a very clear grouping of allelic imbalance on chromosomes 3p, 9p, and 17p depending on the FAL score. A significantly higher level of allelic imbalance was found in the HFAL group of SCCHN specimens than in the LFAL group ($P < 0.005$). In particular, it was found that a subgroup of patients with LFAL did not show allelic imbalance on 3p, 9p, or 17p but demonstrated allelic imbalance on other chromosome arms, which are most likely involved in the initiation and progression of SCCHN. These findings provide evidence for a second genetic pathway in the initiation of SCCHN (i.e., loci on; 2p, 7p, 8q, 11q, 13q, 17q, 18p, 18q, and 19q), and LOH on 3p, 9p, and 17p cannot be considered to represent the only genetic pathway in SCCHN. These results emphasize the limitations of certain studies where only a small set of microsatellite markers are used in the analysis and then attempting to generalize the results to a role in molecular progression. The quantity of work cannot be underestimated in this type of analysis, and thus many groups have to concentrate on very limited regions of the genome. However, the advent of chip technologies and automated sequencing techniques will open up possibilities for many more in-depth analyses of allelic imbalance, which may elucidate the molecular progression of carcinogenesis [86]. Changes in 9p21 and in 3p appear early, followed by p53 changes (Fig. 9.1), but it is the accumulation of, rather than the sequence of, genetic events that appears to determine progression to malignancy [82].

Fluorescence *in situ* hybridization (FISH) and comparative genome hybridization (CGH) have also proved significant advances. FISH uses a probe from specific chromosomal regions to hybridize and identify specific areas of amplification or deletion, allowing direct visualization of chromosomal abnormalities in interphase cells. CGH involves mixing tumor and normal DNA and hybridizing to normal metaphase chromosomal spreads to detect mutations or deletions [87,88]. Deletions or amplifications are seen as changes in the ratio of the intensities of the two labels along the target chromosome. The *cancer genome anatomy project* will undoubtedly uncover many more new genes implicated [89].

1. Chromosome 3

Aberrations have been identified in chromosome 3 in oral cancer [40,90–95], and the short arm of chromosome 3 (3p) is often deleted in SCCHN [96–97]. LOH has been identified at 3p in 52% of oral carcinomas and mapped to three distinct regions of loss, 3p13-p21.1, 3p21.3-p23, and 3p25 [98], which appear to overlap with those described by others [90], and mainly at D3S1293 (3p24-p25), D3S1079, and D3S659 (3p13) [91] and at D3F15S2 (3p21) [82,92]. The region close to 3p14 is a fragile site on 3p. LOH has been found in 74% of SCCHN at 3p14-cen, 3p21.3, and 3p24-ter (*retinoic acid receptor β:RARβ*) [90]. At least two or even three TSGs may be involved on 3p, at least in SCCHN [90,91].

The *fragile histidine triad* (FHIT) gene, localized to 3p14.2, is altered in SCCHN with decreased or aberrant protein [99,100], but no mutations or deletions [101–103]. The FHIT protein has dinucleoside triphosphate hydrolase activity.

3p21 is the region with the highest rate of allelic deletion [101], and TSGs at 3p21.2-p21.3, and 3p25 in particular may be implicated [102]. 3p21 contains several TSGs, including D8 (CHCM; *ubiquitin-activating enzyme*), ACY 1 (*aminoacylase*), APEH (D3S48E), and PTP gamma (*protein tyrosine phosphatase*). The MMR gene hMLH1 involved in DNA repair, a gene located in 3p21.3, which is the human

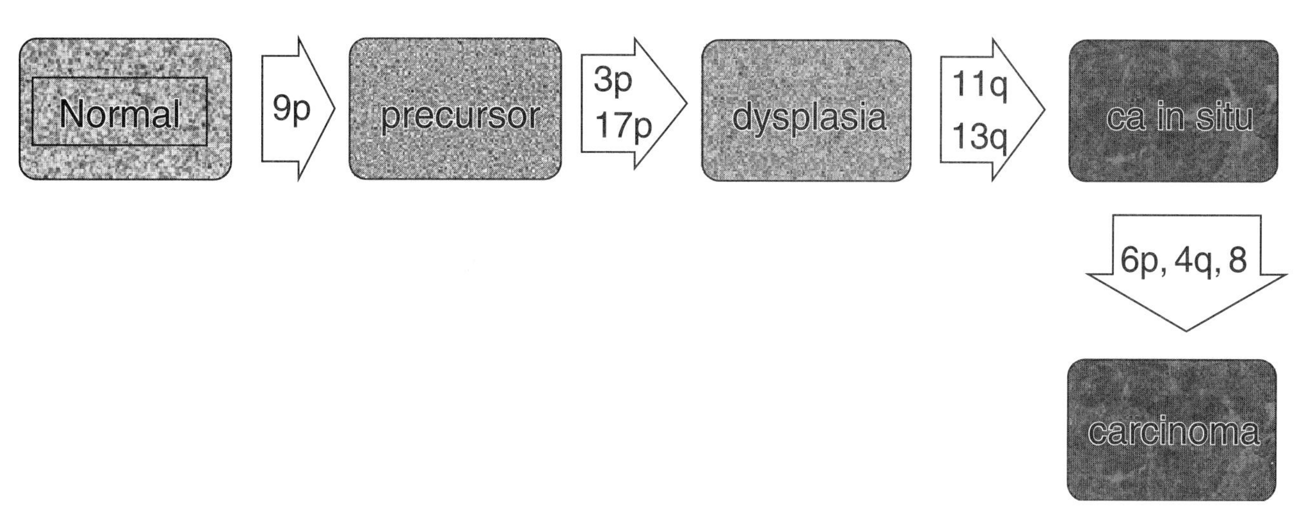

FIGURE 9.1 Carcinogenesis.

homologue of the *ribosomal protein L14* (RPL14) gene may be the subject of transcriptional loss [104], and the *arginine-rich protein* gene at 3p21.1 is also mutated in SCCHN [105].

Whether the *TGFβ type II receptor* gene on 3p22 is involved remains unclear, although mutations have been detected [106] and there is decreased expression on SCCHN cells [107]. The *von-Hippel Lindau* (VHL) TSG at 3p25, however, appears uninvolved [103,108,109].

2. Chromosome 4

Chromosome 4 anomalies include allelic instability [110], and MIN has been found in 31%, at the long arm of chromosome 4 (4q) at 4q25 [110,111] and 14q31-32.1 [82].

3. Chromosome 5

Chromosome 5 anomalies have been recorded in 25 to 73% of oral cancers [40,93,112–114] at D5S178 [114]. LOH involves 5q, and at 5q21-22, there is the APC (*adenomatous polyposis coli*) locus [97,115–116]. 5p aberrations have also been found [117].

4. Chromosome 6

LOH has been found in 21–38% oral cancers on 6p and 23–25% on 6q [40,82,94]. Chromosome 6 bears MHC genes on 6p21 and several other genes, including *tumor necrosis factor* genes TNF-α (cachectin) and TNF-β (lymphotoxin), and heat shock protein 70. However, allelism at TNF does not appear to influence oral cancer [118].

P21 (WAF1/CIP1/Sdil) (which is distinct from the ras p21), a general inhibitor of CDKs, whose gene is at 6p.21.2, is regulated by p53 in response to DNA damage and interacts with proliferating cell nuclear antigen (PCNA) to block its DNA synthesis function [119]. A major effect of p53 is to block the cell cycle by stimulating p21, which may thus lead to G1 arrest, and is a major factor in p53-mediated cell growth arrest, apoptosis, and senescence.

P21 polymorphisms and/or abnormal protein has been detected in oral and SCCHN [120–124] but is not necessarily dependent on p53 status [123–126].

5. Chromosome 7

Deletions of chromosome 7 and LOH involving 7q or 7p have been seen in SCCHN. Genotypic changes in chromosome 7 have also been described in oral cancer [33,127–129] and tissue adjacent to oral cancer [130].

A putative TSG localized to 7q31.1 has also been reported, as well as the PP5/TFPI-2 gene, which has proteinase-inhibitory activity for thrombin, plasmin, and trypsin, secreted by various tumors [131].

6. Chromosome 8

Chromosome 8p deletions [132] and LOH have been reported in 40% of SCCHN [82] and oral cancers [94,133–135], at 8p11-8p21 [133], 8p22 [133], and especially 8p23 [133,135,136], suggesting the presence of two or more TSGs [103], but these are as yet unidentified.

7. Chromosome 9

There is LOH at 9p21-p22 in up to 72% of tumors [82,137–139]. This is the most commonly reported chromosomal defect in SCCHN [140] in at least 5 markers [94], with a higher rate of LOH than for other chromosomes. The specific genes involved are unclear, but the CDKN2 gene (MTS-1), which encodes p16, a cyclin-dependent kinase inhibitor, has assumed considerable importance.

Several regions appear to be involved in oral carcinoma [103], especially the area concentrated between D9S161 (9p21) and D9S156 (9p23-9p22) [40,137]. Chromosome 9p21-22 appears to contain several TSGs, including *CDKN2* (p16), another TSG close to the *methylthioadenosine phosphorylase* (MTAP) gene, one at 9p22, the *interferon α cluster* gene (IFNA) [141], and possibly p15, located only 28 kb from p16 [142]. The genes for cyclin-dependent kinase-4 (CDK-4) inhibitors p15 and p16, and also p16β, which inhibit progression through the G1 phase of the cell cycle, map to chromosome 9p21.

The phosphorylation of pRb is regulated positively by a G1 cyclin termed cyclin D1/CDK4 and is regulated negatively by CKIs such as p16 (CDKN2/MTS-1/INK4A), in part via cyclin D cyclin-dependent kinase (CDK) 4 complexes, and pRb is also regulated by p15 and p18. Thus a major effect of p16 is to halt progression of the cell growth cycle at G1. The CDKN2 locus on chromosome 9p21 actually encodes two proteins: p16 and p16β [p19(ARF)]. The protein p16β, related to p16, also mediates G1 arrest, but in this instance by destabilizing MDM2, a protein that binds to p53 protein (see later), causing its degradation and thereby loss of cell cycle control leading to unregulated cell growth.

Interestingly, p16 changes correlate with tobacco and betel use [143] and appear early in neoplasia [144]. However, sequence analysis of the p16 gene in SCCHN with LOH on 9p has indicated that point mutations or insertions are relatively uncommon in tumors [47,145–149], although they are common in cell lines from SCCHN [150]. G-T transversions in particular, with some G-A transitions involving p16 (see later), have been found more in oral cancer cell cultures (40%) than in tumors (16%) [150].

However, the absence of mutations in tumors with 9p21 LOH does *not* necessarily indicate that p16 is not the target gene. Small homozygous deletions represent an important mechanism of inactivation of 9p21 in many tumor types, including SCCHN, and can be seen in tumor, dysplastic tissue, and histologically normal mucosa from patients with SCCHN [96,142,151]. Fine mapping of these deletions implicates a minimal region that includes p16 but excludes p15 (p15/INK4B) [145], although p15 may be involved in a minority [142].

Even where there are no deletions, p16 is frequently inactivated through an alternative mechanism [52,53,141, 151–153], which is by methylation of the 5′ CpG-rich region of p16, resulting in a complete block of gene transcription—*transcriptional silencing* [151,154]. Methylation inactivation has been found in up to 30% of oral and SCCHN, suggesting that inactivation of the promoter region could be important in the genesis and promotion of SCCHN [142,148,149,151,153,155,156].

There is altered expression of CDKs in tumor cells [157]. Cyclins can be disturbed by p16 abberrations and also by cyclin D1 (PRAD) amplification [47,48,52] independently of p16 inactivation [52]. The p16/cyclin D/CDK/pRb pathway appears to be as crucial to carcinogenesis as is p53 [158] (see later), and both paths appear to interplay [159], although they act separately as well [123].

LOH has not been found in 9q in oral cancer [94,139], but has, however, been found in 35% of SCCHN at 9q31-q34 and 9q22.1-q32 [93] and in 70% LOH at 9p22-q23.3, suggesting that this may be additional TSGs on chromosome 9 [94]. Indeed, a number of putative TSGs are located on 9q, including those related to Gorlin's syndrome (basal cell naevus syndrome: 9q31) and the Ferguson–Smith syndrome (multiple self-healing squamous epitheliomata: 9q22-q33).

8. Chromosome 10

Deletions of the long arm of chromosome 10 have been described in many tumor types, and frequent LOH has been identified in SCCHN on 10q, suggesting the presence of a TSG [160]. The *PTEN/MMAC1* gene has been identified as a possible candidate TSG located at 10q23 [161,162] with deletions demonstrable in about 10% SCCHN [163] but few mutations [160,164].

9. Chromosome 11

LOH has been reported in over 60% SCCHN at 11q [82] and in 50% of oral cancers [82,93,94], and rearrangements have been described [165]. LOH appears to represent amplification of cyclin D1 (PRAD-1) [166], the main protooncogene involved in oral carcinogenesis (see earlier discussion). Although the main changes reported have been in cyclin D1 (PRAD), at least two putative TSGs on 11q23 and q25 may also be associated with the development of oral SCC [167]. Deletions at 11q13 are common in SCCHN [49,97,168], especially at 11q13-qter [168]. Studies have also shown the frequent loss of imprinting of growth-promoting (IGF2) and control (H19) genes on chromosome 11p15 [169].

10. Chromosome 12

A minority of SCCHN appear to have LOH on 12p and 12q [40,94]. P27 (Kip1), which maps to chromosome 12p12-12p13.1, plays a role in G1 arrest induced by TGFβ in normal cells, regulating proliferation by binding and inhibiting the G1 cyclin–CDK complexes. P27 expression is correlated with Bax expression [170]. The p27 protein expression is reduced in oral carcinoma [123,170–174] due to increased degradation by ubiquitin-proteasome [172].

It is noteworthy that a putative TSG, termed Doc-1 (*deleted in oral cancer*), has also been characterized and mapped on 12q24 [175], but at present, however, no intragenic mutation of the gene has been found.

11. Chromosome 13

Some 50% of patients with SCCHN have LOH involving 13q [82,94,139,176,177], and multiple regions show deletions [178]. Many of the aberrations have been seen at or close to the 13q32-ter or 13q14.2-q14.3 sites [177,179], 13q34, 13q14.3, and 13q12.1 [178], which are close to the *retinoblastoma gene* (Rb). Rb is a key component of the G1 checkpoint and can be affected by dephosphorylation, which renders it active, leading to G1 arrest, and by phosphorylation, which inactivates it. Rb dephosphorylation results in the release of transcription factors such as E2F-1, essential to the expression of S-phase cell regulatory genes.

Loss of the retinoblastoma gene was said to be uncommon in SCCHN [176,177] and oral carcinomas [28,97,179] suggesting that Rb played a minor role [144], but there are now several reports of altered retinoblastoma protein pRb in oral cancer [143,144,180]. However, there may be other TSGs at 13q14, such as the *leukaemia associated gene* 1 (*leu* 1) at 13q14.3-q21 [178], and aberrations upstream of Rb appear common [53,179].

12. Chromosome 16

A minority of SCCHN have demonstrated LOH on 16p or 16q [40,93,94], and translocations have been found at 16q22 in a few [181]. The *E-cadherin* gene, which is at 16q21.1, is not mutated, but, if silenced transcriptionally by hypermethylation, appears to result in the acquisition of an invasive phenotype [182].

13. Chromosome 17

The p53 gene, so called because it produces a 53-kDa nuclear phosphoprotein, is located on chromosome 17p and exerts important effects on cell growth. p53 is, with p73 and p63, part of a family of TSGs [183]. It acts as a transcription factor and, by activating transcription of a number of target genes, controls cell cycle progression, acting as a

G1 checkpoint control, and regulates DNA repair, apoptosis, and differentiation. p53 also controls cell cycle progression via the regulation of other proteins, including p21 [Waf1/Cip1] and MDM2, and also GADD45, Bax, and Bcl-2. p53 blocks the cell cycle by switching on the GADD45 gene (growth arrest and DNA inducible) in cells exposed to stress. The GADD45 protein product binds PCNA and inhibits entry into the cell cycle S phase.

Mutations or deletions of the p53 gene, or inactivated protein, can alter p53 activity and lead to disturbed cell cycle control, as the half-life of the mutant p53 protein is up to 6–8 h compared to that of the wild type, which is only up to 20 min. p53 can also be inactivated by interaction with other proteins such as MDM2, methylation, or sequestration [184,185]. p53 is a TSG with one of the highest frequency of mutations in cancers of any gene thus far studied. p53 has 11 exons, and exons 5 to 8 are the most highly conserved (codons 126–306) and contain the majority of the mutations within the p53 gene.

Investigation into p53 expression in SCCHN has demonstrated that approximately 60% of these tumors have *immunohistochemically* detectable p53, thus suggesting gene mutations [186,187]. Antibody positivity has thus frequently been taken to indicate p53 mutation; conversely, negativity has sometimes been regarded as excluding mutations. However, the frequency of detection of p53 mutations based on immunohistochemical studies is not necessarily accurately representative of the state of p53. Antigen retrieval increases the sensitivity of p53 immunoreactivity significantly, but there is no association between p53 protein overexpression and gene mutations [188–190]. Absence of reactivity with p53 antibodies does not exclude alterations in the gene; frameshift as well as nonsense mutations may be found in antibody-negative tumors, as the antibodies fail to detect truncated p53 proteins [191,192]. Conversely, there may be overexpression of p53 protein without apparent gene mutations [193–195], overexpression in completely benign oral lesions [196], and discordance between p53 mutation and LOH [77,197]. Thus in a sizable number of cases, p53 protein expression does not reflect the mutational status of the gene [190].

With these caveats in mind, it is interesting that p53 mutations have been demonstrated in up to two-thirds (from 12 to 100%) of oral and SCCHN [191,198–210], mainly in basal epithelia and at the advancing front of the tumor [191,211]. LOH is reported in 50–55% in some studies [82,93,94,139,212,213]. However, others have not found LOH, but simply deletions or rearrangements in a minority of tumors [92]. A detailed study of SCCHN has shown LOH on 17p in 50%, often involving the TP53 locus (42%), but more frequently involving 17p13 [82], the p53 locus, or the CHRNB1 locus (acetylcholine receptor subunit B1) at 17p11.1-p12 [213]. p53 mutations precede histologic changes [214], suggesting the p53 changes are early events in carcinogenesis and may be found independently of p16 mutations [123].

In carcinomas, p53 mutations are often the same in the primary tumor and in nodal metastases [207,215–217] and in immediately adjacent epithelium [207,217,218], as well as in recurrences [219]. The molecular events, however, differ in tumor distant epithelium [220,221] and in second primary tumors [215,216,222], suggesting that genetic changes occur independently at different sites in the upper aerodigestive tract. More recently, some have even found discordance between p53 in the primary tumour and metastasis from SCCHN [223].

In SCCHN in Western societies, p53 overexpression correlates with mutagen-induced chromosome fragility [224] and has been demonstrated by certain authors to be related to the use of tobacco [201,225–227] and alcohol consumption [226] in both carcinomas and in premalignant lesions of the head and neck [228,229]. Studies indicate that p53 mutations are found mainly in smokers/former smokers [86] and in younger rather than older patients in some [230] but not all studies [180] and are rarely found in nonsmoking/nondrinking patients younger than 40 years [231].

p53 mutations, although common in SCCHN from Western countries, are uncommon in those from the developing world [120,193,194,206, 232–236]. Some two-thirds of p53 mutations in head and neck cancers are at guanine (G) nucleotides [217,237], implicating carcinogens from tobacco smoke, as guanine residues are the preferential targets for chemical carcinogen-induced damage [238]. In a large series of SCCHN, most p53 mutations were G:C–A:T (31%), G:C–T:A (18%), or deletions/inversions (19%) [217,239]. Mutations in p53 in oral dysplastic and malignant lesions are mostly G–A transitions in Japan and Switzerland, G–T transversions in the United Kingdom/United States [191,216,217,239], and transitions in India [206] but a range of alterations, including transversions, transitions, deletions, and polymorphisms, can be seen [240].

Most p53 mutations have been in exons 2–11, mainly at exons 4–9, in codons 238–248 (exon 7), and 278–281 (exon 8) [187]. Hot spots for mutations are predominantly in codons 220, 245–248, 278–281 [186,191,216,239,241,242], 149 and 274 [206], 157, 175, 186, 248, 273, and 282 [243]. In oral carcinoma patients, most of whom were women with no obvious risk factors for cancer, a specific exon 8 14-bp deletion in codons 287–292, which results in a truncated p53 protein lacking the C terminal, has been implicated [244].

p53 expression may be altered both by p53 gene mutation as a consequence of carcinogens, radiation, or other mutagens or by other mechanisms, such as degradation of the p53 protein by viral or other proteins [245]. Significant in this respect may be viral proteins such as human papillomavirus (HPV) type 16 E6 protein and cellular proteins such as MDM-2 (murine double minute-2) [246–249], p21, GADD45, cyclins, CDKs, and PCNA [205].

Germline abnormalities in p53 appear to underlie some cancer-prone families [250], including cases of Li–Fraumeni syndrome [251], and with less penetrance in some families with SCCHN [252]. 17q24 is the locus for the tylosis oesophageal cancer gene (*TOC*) with patients also being predisposed to oral leukoplakia (Clarke–Howel–Evans–McConnell syndrome) [253]. Also on chromosome 17q, at 17q21.3, is the putative metastasis supressor gene NM23, which is expressed in oral carcinoma [254].

14. Chromosome 18

LOH has been described in SCCHN mostly at 18q [255–257], with the highest single incidence of LOH being at 18q21.1-21.3 [256], but some have reported LOH at 18p in oral cancer [258]. Putative TSGs include DCC, DPC4/Smad4, and MADR2/Smad2. LOH at the DCC (*deleted in colon cancer*) gene at 18q21.1 is only found in 12–24% of SCCHN, suggesting that the D18S35 and DCC loci are different and that the D18S35 marker is mapping a second TSG [97,255]. Furthermore, neither DCC nor DPC4 (*deleted in pancreatic carcinoma*) appears incriminated in SCCHN [97,259] or at least plays only a minor role [257]. Another TSG at 18q21 may be incriminated [257], such as *MADR2* or a distinct other gene.

15. Chromosome 21

LOH and MI are found on chromosome 21q in oral cancer, and three putative TSGs may be present [260].

16. Chromosome 22

LOH has been detected in 73% of oral carcinomas, mainly at 22q13, but not at the NF2 gene [261].

V. DETECTION OF INDIVIDUALS AT RISK

Screening of smokers and drinkers for p53 expression or mutations alone cannot yet be considered specific or sensitive enough for the detection of individuals at risk for cancer. p53 mutations are found mainly in smokers/former smokers [86], but are not found in nonsmoking/nondrinking patients younger than 40 [231]. However, no correlation has been found between exposure to these carcinogens and fractional allelic loss (FAL) [2,85]. p53 mutations are also uncommon in betel use in several studies (discussed earlier), although in some studies, betel users have had p53 mutations [235,236], mostly in users of tobacco [221]. p53 overexpression has been demonstrated to be related to the use of tobacco [22,201,226,227,262] and alcohol consumption [226] in carcinomas of the head and neck [228,229], but others have found either no relationship with tobacco use [216,263,264] or tobacco or alcohol [265] or, indeed, an inverse relationship [204,207].

Inherited p53 mutations in some families (Li–Fraumeni syndrome) underlie some cases of SCCHN [266]. Inherited p16 mutations in some families underlie some cases of head and neck cancer [158,267,268].

VI. MOLECULAR DIAGNOSIS

A number of molecular methods are available to aid the diagnosis of cancer [269]. The potential for detecting markers of carcinogenesis in exfoliated cells is attractive [270]. p53 mutations can be detected in oral epithelial [271] and salivary cells [272] from patients with SCCHN, and there are detectable changes in oral exfoliated cells from high-risk individuals in chromosomes 3p, 9p, and 17p [273]. p53 antibodies can be found in serum [274,278] and saliva [278], but only in a minority of patients, and therefore cannot be considered diagnostically useful, although it is possible antibody detection could be helpful in follow-up.

VII. STAGING

Molecular assays can augment clinical and conventional histopathological staging, as they can detect cancer cells when they comprise less than 0.01% of the total cell population [9,279]. The detection of p53 mutations at tumor resection margins heralds future recurrence and has been demonstrated in 38% of light microscopically tumor-free margins [279].

VIII. PROGNOSTICATION

Cervical lymph node metastases have been demonstrated by p53 changes in 21% of histopathologically tumor-free nodes [279], showing that about 80% of patients have a more advanced stage of cancer than hitherto suspected [279]. Molecular probing also examines a larger volume of the lymph node tissue, reducing sampling errors present in conventional histopathological examinations.

Chromosomal imbalances with overrepresentation affecting 2q12, 3q21-29, 6p21.1, 11q13, 14q23, 14q24, 14q31, 14q32, 15q24, 16q22, and deletions of 8p21-22 and 18q11.2 correlate with a shorter disease-free interval and survival, with a special significance identified in 3q21-29, 11q13, and 8p21-22 [280]. LOH in 3p SCCHN may be strongly associated with prognosis [91,226] and correlate with nodal metastases [226]. The majority of stage II–III oral carcinomas [98] and TNM stage IV SCCHN [226] show LOH at 3p, in contrast to TNM stage I tumors [98], and there is a correlation with poor survival [281]. Allelic imbalances

(AI) at 3p24-26, 3p21, and 3p13 appear to predict poor prognosis [281].

p16 deletions appear crucial for malignant progression [283], and chromosome 9p21 deletions appear to influence tumor recurrence [96], possibly through p16 mutations [284]. AI at 9p21 appears to predict poor prognosis [282].

p53 expression is more frequent in poorly differentiated SCCHN compared with differentiated tumors [285], and may correlate with metastasis [275,286,287] and poor prognosis [288–291], although not all would agree [191,195,218,228,263,264,274,292–299,300–302]. p53 gene mutations rather than protein overexpression may predict tumor recurrence [303,304].

p53 levels in serum may be more powerful predictors of relapse and death than tissue immunodetection [275].

A. Fractional Allelic Loss

Single genetic alterations do not appear to affect survival, as discussed earlier and elsewhere [305]. However, when multiple aberrations are detected, survival is affected. Accumulated genetic damage, as provided by allelotype analysis using FAL, provides a useful molecular indicator of the behavior of head and neck carcinomas. FAL for a particular tumor is defined as the number of chromosomal arms on which allelic loss is observed, divided by the number of chromosomal arms for which allelic markers are informative [306]. AI at key loci has been used to identify a FAL that predicts recurrence and reduced survival and thus appears to predict poor prognosis in oral and head and neck cancers [85,282]. FAL above 0.22 correlates with local metastasis and decreased survival [85].

The first comprehensive allelotype of SCCHN, which analyzed FAL values for 52 carcinomas with LOH information on nine or more (9–39) chromosome arms, found a positive correlation between FAL and tumor grade, and nodal involvement at pathological examination [85]. The FAL was also related to clinical outcome; a FAL > median value correlated with poor survival even when previously untreated tumors were analyzed separately. Analysis of 40 advanced tumors also demonstrated a correlation between FAL and prognosis.

A more recent slightly smaller study of FAL in oral cancer showed that AI at 3p22-26, 3p14.3-12.1, and 9p21 was a better prognosticator than the TNM system [282], and a case–control study confirmed that a microsatellite assay can identify persons at risk of developing oral SSC in a field of cancerization [307].

B. Identification of Second Primary and Recurrent Tumors

Molecular analyses can differentiate recurrences from second primary tumors [146,303,304]. It is clear that multifocal tumors in the head and neck may harbor identical genetic changes, suggesting that a single progenitor clone can populate different contiguous areas of the upper aerodigestive tract via re-epithelialization or intramucosal migration [308].

The site of origin of unknown primary tumors in patients presenting with cervical node metastasis can be identified by molecular analyses of p53 [309].

Gene mutations in p53 have also enabled lung metastases to be distinguished from second primary tumors in the aerodigestive tract [310].

C. The Future

There has been a significant publication regarding prognostic factors in colorectal cancer based on a consensus statement from the College of American Pathologists in 1999.

Those of us working in oral and SCCHN cancer have much to learn from the paper as it indicates the way the field has to move in order to take molecular biomarkers into clinical practice. It is of note that certain molecular biomarkers are now considered to be very important in determining colonic cancer prognosis. These include LOH at the 18q (DCC gene allelic loss) and a high degree of microsatellite instability in determining important factors in category IIB colorectal cancer. The fact that molecular biomarkers have now been included in prognostication indicates a major change in thinking. Are we ready to take up this challenge in oral and SCCHN cancer?

References

1. Todd, R., Donoff, R. B., and Wong, D. T. (1997). The molecular biology of oral carcinogenesis; toward a tumor progression model. *J. Oral Maxillofac. Surg.* **55**, 613–623.
2. Scully, C., and Field, J. K. (1997). Genetic aberrations in squamous cell carcinoma of the head and neck (SCCHN), with reference to oral carcinoma. *Int. J. Oncol.* **10**, 5–21.
3. Scully, C., Field, J. K., and Tanzawa, H. (2000). Genetic aberrations in oral or head and neck squamous cell carcinoma (SCCHN). 1. Carcinogen metabolism, DNA repair and cell cycle control. *Oral Oncol.* **36**, 256–263.
4. Scully, C., Field, J. K., and Tanzawa, H. (2000). Genetic aberrations in oral or head and neck squamous cell carcinoma (SCCHN). 2. Chromosomal aberrations. *Oral Oncol.* **36**, 311–327.
5. Scully, C., Field, J. K., and Tanzawa, H. (2000). Genetic aberrations in oral or head and neck squamous cell carcinoma (SCCHN). 3. Clinicopathological applications. *Oral Oncol.* **36**, 404–413.
6. Sidransky, D. (1995). Molecular genetics of head and neck cancer. *Curr. Opin. Oncol.* **7**, 229–233.
7. Bradford, C. R. (1999) Predictive factors in head and neck cancer. *Hematol. Oncol. Clin. North Am.* **13**, 777–785.
8. Mao, L., Lee, D. J., Tockman, M. S., Erozan, Y. S., Askin, F., and Sidransky, D. (1994). Microsatellite alterations as clonal markers in the detection of human cancers. *Proc. Natl. Acad. Sci. USA* **91**, 9871–9875.
9. Brennan, J. A., and Sidransky, D. (1996). Molecular staging of head and neck squamous carcinoma. *Cancer Metastasis Rev.* **15**, 3–10.

10. Scully, C., and Burkhardt, A. (1993). Tissue markers of potentially malignant oral epithelial lesions. *J. Oral Pathol. Med.* **22**, 246–256.
11. Papadimitrakopoulou, V. A., Shin, D. M., and Hong, W. K. (1996). Molecular and cellular biomarkers for field cancerization and multistep process in head and neck tumorigenesis. *Cancer Metastasis Rev.* **15**, 53–76.
12. Oh, Y., and Mao, L. (1997). Biomarkers in head and neck carcinoma. *Curr. Opin. Oncol.* **9**, 247–256.
13. Leethanakul, C., Patel, V., Gillespie, J., Pallente, M., Ensley, J. F., Koontongkaew, S., Liotta, L. A., Emmert-Buck, M., and Gutkind, J. S. (2000). Distinct pattern of expression on differentiation and growth-related genes in squamous cell carcinomas of the head and neck revealed by the use of laser capture microdissection and cDNA arrays. *Oncogene* **19**, 3220–3224.
14. Stich, H. F. (1987). Micronucleated exfoliated cells as indicators for genotoxic damage and as markers in chemoprevention trials. *J. Nutr. Growth Cancer* **4**, 9–18.
15. Saranath, D., Panchal, R. G., Nair, R., Mehta, A. R., Sanghavi, V., Sumegi, J., Klein, G., and Deo, M. G. (1989). Oncogene amplification in squamous cell carcinoma of the oral cavity. *Jpn. J. Cancer Res.* **80**, 430–437.
16. Saranath, D., Bhoite, L. T., Mehta, A. R., Sanghavi, V., and Deo, M. G. (1991). Loss of allelic heterozygosity at the harvey *ras* locus in human oral carcinomas. *J. Cancer Res. Clin. Oncol.* **117**, 484–488.
17. Saranath, D., Chang, S. E., Bhoite, L. T., *et al.* (1991). High frequency mutation in codons 12 and 61 of H-ras oncogene in chewing tobacco-related human oral carcinoma in India. *Br. J. Cancer* **63**, 573–578.
18. Das, N., Majumder, J., and DasGupta, U. B. (2000). ras gene mutations in oral cancer in eastern India. *Oral Oncol.* **36**, 76–80.
19. McDonald, J. S., Jones, H., Pavelic, Z. P., Pavelic, L. J., Stambrook, P. J., and Gluckman, J. L. (1994). Immunohistochemical detection of the H-ras, K-ras, and N-ras oncogenes in squamous cell carcinoma of the head and neck. *J. Oral Pathol. Med.* **23**, 342–346.
20. Tadokoro, K., Ueda, M., Ohshima, T., Fujita, K., Rikimaru, K., Takashashi, N., Enomoto, S., and Tschuchida, N. (1989). Activation of oncogenes in human oral cancer cells: A novel codon 13 mutation of c-H-ras-1 and concurrent amplifications of c-erbB-1 and c-myc. *Oncogene* **4**, 499–505.
21. Munirajan, A. K., Mohanprasad, B. K., Shanmugam, G., and Tsuchida, N. (1998). Detection of a rare point mutation at codon 59 and relatively high incidence of H-ras mutation in Indian oral cancer. *Int. J. Oncol.* **13**, 971–974.
22. Field, J. K., Spandidos, D. A., and Stell, P. M. (1992). Overexpression of the p53 gene in head and neck cancer, linked with heavy smoking and drinking. *Lancet* **339**, 502–503.
23. Sheng, Z. M., Barrois, M., Klijanienko, J., Micheau, C., Richard, J. M., Riou, G. (1990). Analysis of the c-Ha-ras-1 gene for deletion, mutation, amplification and expression in lymph node metastases of human head and neck carcinomas. *Br. J. Cancer* **62**, 398–404.
24. Chang, S. E., Bhatia, P., Johnson, N. W., Morgan, P. R., McCormick, F., and Young, B. (1991). Ras mutations in United Kingdom examples of oral malignancies are infrequent. *Int. J. Cancer* **48**, 409–412.
25. Warnakulasuriya, K. A., Chang, S. E., and Johnson, N. W. (1992). Point mutations in Ha-ras oncogene are detectable in formalin-fixed tissues of oral squamous cell carcinomas, but are infrequent in British cases. *J. Oral Pathol. Med.* **21**, 225–229.
26. Clark, L. J., Edington, K., Swan, I. R., McLay, K. A., Newlands, W. J., Wills, L. C., Young, H. A., Johnston, P. W., Mitchell, R., Robertson, G., *et al.* (1993). The absence of Harvey *ras* mutations during development and progression of squamous cell carcinomas of the head and neck. *Br. J. Cancer* **68**, 617–620.
27. Yeudall, W. A., Torrance, L. K., Elsegood, K. A., Speight, P., Scully, C., and Prime, S. S. (1993). *Ras* gene point mutation is a rare event in premalignant tissues and malignant cells and tissues from oral mucosal lesions. *Eur. J. Cancer B Oral Oncol.* **29B**, 63–67.
28. Xu, J., Gimenez-Conti, I. B., Cunningham, J. E., Collet, A. M., Luna, M. A., Lanfranchi, H. E., Spitz, M. R., and Conti, C. J. (1998). Alterations of p53, cyclin D1, Rb, and H-ras in human oral carcinomas related to tobacco use. *Cancer* **83**, 204–212.
29. Anderson, J. A., Irish, J. C., McLachlin, C. M., and Ngan, B. Y. (1994). H-ras oncogene mutation and human papillomavirus infection in oral carcinomas. *Arch. Otolaryngol. Head Neck Surg.* **120**, 755–760.
30. Kuo, M. Y., Jeng, J. H., Chiang, C. P., and Hahn, L. J. (1994). Mutations of Ki-ras oncogene codon 12 in betel quid chewing-related human oral squamous cell carcinoma in Taiwan. *J. Oral Pathol. Med.* **23**, 70–74.
31. Todd, R., and Wong, D. T. (1999). Epidermal growth factor receptor (EGFR) biology and human oral cancer. *Histol. Histopathol.* **14**, 491–500.
32. Grandis, J. R., and Tweardy, D. J. (1993). Elevated levels of transforming growth factor alpha and epidermal growth factor receptor messenger RNA are early markers of carcinogenesis in head and neck cancer. *Cancer Res.* **53**, 3579–3584.
33. Ibrahim, S. O., Lillehaug, J. R., Johannessen, A. C., Liavaag, P. G., Nilsen, R., and Vasstrand, E. N. (1999). Expression of biomarkers (p53, transforming growth factor alpha, epidermal growth factor receptor, c-erbB-2/neu and the proliferative cell nuclear antigen) in oropharyngeal squamous cell carcinomas. *Oral Oncol.* **35**, 302–313.
34. Kunikata, M., Yamada, K., Yamada, T., Mori, M., Tatemoto, Y., and Osaki, T. (1992). Epidermal growth-factor receptor in squamous-cell carcinoma and other epidermal lesions of squamous origin: An immunohistochemical study. *Acta Histochem. Cytochem.* **25**, 387–394.
35. Sakai, H., Kawano, K., Okamura, K., and Hashimoto, N. (1990). Immunohistochemical localization of c-*myc* oncogene product and EGF receptor in oral squamous cell carcinoma. *J. Oral Pathol. Med.* **19**, 1–4.
36. Kusukawa, J., Harada, H., Shima, I., Sasaguri, Y., Kameyama, T., and Morimatsu, M. (1996). The significance of epidermal growth factor receptor and matrix metalloproteinase-3 in squamous cell carcinoma of the oral cavity. *Eur. J. Cancer B Oral Oncol.* **32B**, 217–221.
37. Riviere, A., Wilckens, C., and Loning, T. (1990). Expression of c-erbB2 and c-myc in squamous epithelia and squamous cell carcinomas of the head and neck and the lower female genital tract. *J. Oral Pathol. Med.* **19**, 408–413.
38. Leonard, J. H., Kearsley, J. H., Chenevix-Trench, G., and Hayward, N. K. (1991). Analysis of gene amplification in head and neck squamous cell carcinoma. *Int. J. Cancer* **48**, 511–515.
39. Eversole, L. R., and Sapp, J. P. (1993). c-myc oncoprotein expression in oral precancerous and early cancerous lesions. *Eur. J. Cancer B Oral Oncol.* **29B**, 131–135.
40. Field, J. K. (1995). The role of oncogenes and tumour-suppressor genes in the aetiology of oral, head and neck squamous cell carcinoma. *J. R. Soc. Med.* **88**, 35P–39P.
41. Rodrigo, J. P., Lazo, P. S., Ramos, S., Alverez, I., and Suarez, C. (1996). MYC amplification in squamous cell carcinomas of the head and neck. *Arch. Otolaryngol. Head Neck Surg.* **122**, 504–507.
42. Muller, D., Millon, R., Lidereau, R., Engelmann, A., Bronner, G., Flesch, H., Eber, M., Methlin, G., and Abecassis, J. (1994). Frequent amplification of 11q13 DNA markers is associated with lymph node involvement in human head and neck squamous cell carcinomas. *Eur. J. Cancer B Oral Oncol.* **30B**, 113–120.
43. Bartkova, J., Lukas, J., Muller, H., Strauss, M., Gusterson, B., and Bartek, J. (1995). Abnormal patterns of D-type cyclin expression and G1 regulation in human head and neck cancer. *Cancer Res.* **55**, 949–956.
44. Michalides, R., van Veelen, N., Hart, A., Loftus, B., Wientjens, E., and Balm, A. (1995). Overexpression of cyclin D1 correlates with recurrence in a group of forty-seven operable squamous cell carcinomas of the head and neck. *Cancer Res.* **55**, 975–978.
45. Lovec, H., Sewing, A., Lucibello, F. C., Muller, R., and Moroy, T. (1994). Oncogenic activity of cyclin D1 revealed through cooperation

with Ha-*ras*: Link between cell cycle control and malignant transformation. *Oncogene* **9**, 323–326.
46. Kotelnikov, V. M., Coon, J. S. IV, Mundle, S., Kelanic, S., LaFollette, S., Taylor, S. N., Hutchinson, J., Panje, W., Caldarelli, D. D., and Preisler, H. D. (1997). Cyclin D1 expression in squamous cell carcinomas of the head and neck and in oral mucosa in relation to proliferation and apoptosis. *Clin. Cancer Res.* **3**, 95–101.
47. Olshan, A. F., Weissler, M. C., Pei, H., Conway, K., Anderson, S., Fried, D. B., *et al.* (1997). Alterations of the p16 gene in head and neck cancer: Frequency and association with p53, PRAD-1 and HPV. *Oncogene* **14**, 811–818.
48. Kyomoto, R., Kumazawa, H., Toda, Y., Sakaida, N., Okamura, A., Iwanaga, M., Shintaku, M., Yamashita, T., Hiai, H., and Fukumoto, M. (1997). Cyclin-D1-gene amplification is a more potent prognostic factor than its protein overexpression in human Head-and-neck squamous-cell carcinoma. *Int. J. Cancer* **74**, 576–581.
49. Jin, Y., Hoglund, M., Jin, C., Martins, C., Wennerberg, J., Akervall, J., Mandahl, N., Mitelman, F., and Mertens, F. (1998). FISH characterization of head and neck carcinomas reveals that amplification of band 11q13 is associated with deletion of distal 11q. *Genes Chromosomes Cancer* **22**, 312–320.
50. Staibano, S., Mignogna, M. D., LoMuzio, L., DiAlberti, L., DiNatale, E., Lucariello, A., Mezza, E., Bucci, E., and DeRosa, G. (1998). Overexpression of cyclin-D1, bcl-2, and bax proteins, proliferating cell nuclear antigen (PCNA) and DNA-ploidy in squamous cell carcinoma of the oral cavity. *Hum. Pathol.* **29**, 1189–1194.
51. Kuo, M. Y., Lin, C. Y., Hahn, L. J., Cheng, S. J., and Chiang, C. P. (1999). Expression of cyclin D1 is correlated with poor prognosis in patients with areca quid chewing-related oral squamous cell carcinomas in Taiwan. *J. Oral Pathol. Med.* **28**, 165–169.
52. Okami, K., Reed, A. L., Cairns, P., Koch, W. M., Westra, W. H., Wehage, S., Jen, J., and Sidransky, D. (1999). Cyclin D1 amplification is independent of p16 inactivation in head and neck squamous cell carcinoma. *Oncogene* **18**, 3541–3545.
53. Sartor, M., Steingrimsdottir, H., Elamin, F., Gaken, J., Warnakulasuriya, S., Partridge, M., Thakker, N., Johnson, N. W., and Tavassoli, M. (1999). Role of p16/MTS1, cyclin D1 and RB in primary oral cancer and oral cancer cell lines. *Br. J. Cancer* **80**, 79–86.
54. Fortin, A., Guerry, M., Guerry, R., Talbot, M., Parise, O., Schwaab, G., Bosq, J., Bourhis, J., Salvatori, P., Janot, F., and Busson, P. (1997). Chromosome 11q13 gene amplifications in oral and oropharyngeal carcinomas; no correlation with subclinical lymph node invasion and disease recurrence. *Clin. Cancer Res.* **3**, 1609–1614.
55. Volling, P., Jungehulsing, M., Jucker, M., Stutzer, H., Diehl, V., and Tesch, H. (1993). Coamplification of the hst and bcl-1 oncogenes in advanced squamous cell carcinomas of the head and neck. *Eur. J. Cancer* **29A**, 383–389.
56. Lucas, J. M., Mountain, R. E., Gramza, A. W., Schuller, D. E., Wilkie, N. M., and Lang, J. C. (1994). Expression of cyclin D1 in squamous cell carcinomas of the head and neck. *Int. J. Oncol.* **5**, 469–472.
57. Zhou, D. J., Casey, G., and Cline, M. J. (1988). Amplification of human int-2 in breast cancers and squamous carcinomas. *Oncogene* **2**, 279–282.
58. Ibrahim, S. O., Vasstrand, E. N., Liavaag, P. G., Johannessen, A. C., and Lillehaug, J. R. (1997). Expression of c-erbB proto-oncogene family members in squamous cell carcinoma of the head and neck. *Anti Cancer Res.* **17**, 4539–4546.
59. Kanekawa, A., Tsuji, T., Oga, A., Sasaki, K., and Shinozaki, F. (1999). Chromosome 17 abnormalities in squamous cell carcinoma of the oral cavity, and its relationship with p53 and Bcl-2 expression. *Anticancer Res.* **19**, 81–86.
60. Birchall, M. A., Schock, E., Harmon, B. V., and Gobe, G. (1997). Apoptosis, mitosis, PCNA and bcl-2 in normal, leukoplakic and malignant epithelia of the human oral cavity: Prospective, in vivo study. *Oral Oncol.* **33**, 419–425.
61. Loro, L. L., Vintermyr, O. K., Liavaag, P. G., Jonsson, R., and Johannessen, A. C. (1999). Oral squamous cell carcinoma is associated with decreased bcl-2/bax expression ratio and increased apoptosis. *Hum. Pathol.* **30**, 1097–1105.
62. Ito, T., Fujieda, S., Tsuzuki, H., Sunaga, H., Fan, G., Sugimoto, C., Fukuda, M., and Saito, H. (1999). Decreased expression of Bax is correlated with poor prognosis in oral and oropharyngeal carcinoma. *Cancer Lett.* **140**, 81–91.
63. Kannan, K., Latha, P. N., and Shanmugam, G. (1998). Expression of bcl-2 oncoprotein in Indian oral squamous cell carcinomas. *Oral Oncol.* **34**, 373–376.
64. Schoelch, M. L., Le, Q. T., Silverman, S., Jr., McMillan, A., Dekker, N. P., Fu, K. K., Ziober, B. L., and Regezi, J. A. (1999). Apoptosis-associated proteins and the development of oral squamous cell carcinoma. *Oral Oncol.* **35**, 77–85.
65. Schoelch, M. L., Regezi, J. A., Dekker, N. P., Ng, I. O., McMillan, A., Ziober, B. L., *et al.* (1999). Cell cycle proteins and the development of oral squamous cell carcinoma. *Oral Oncol.* **35**, 333–342.
66. Roberts, J. M. (1999). Evolving ideas about cyclins. *Cell* **98**, 129–132.
67. Scully, C. (1993). Oncogenes, tumor suppressors and viruses in oral squamous cell carcinoma. *J. Oral Pathol. Med.* **22**, 337–347.
68. Mao, L., El-Naggar, A. K., Fan, Y. H., Lee, J. S., Lippman, S. M., Kayser, S., Lotan, R., and Hong, W. K. (1996).Telomerase activity in head and neck squamous cell carcinoma and adjacent tissues. *Cancer Res.* **56**, 5600–5604.
69. Mutirangura, A., Supiyaphun, P., Trirekapan, S., Sriuranpong, V., Sakuntabhai, A., Yenrudi, S., and Voravud, N. (1996). Telomerase activity in oral leukoplakia and head and neck squamous cell carcinoma. *Cancer Res.* **56**, 3530–3533.
70. Shay, J.W., and Bacchetti, S. (1997). A survey of telomerase activity in human cancer. *Eur. J. Cancer* **33**, 787–791.
71. Chang, L. Y., Lin, S. C., Chang, C. S., Wong, Y. K., Hu, Y. C., and Chang, K. W. (1999). Telomerase activity and in situ telomerase RNA expression in oral carcinogenesis. *J. Oral Pathol. Med.* **28**, 389–396.
72. Liao, J., Mitsuyasu, T., Yamane, K., and Ohishi, M. (2000). Telomerase activity if oral and maxillofacial tumors. *Oral Oncol.* **36**, 347–352.
73. Ishwad, C. S., Ferrell, R. E., Rossie, K. M., *et al.* (1995). Microsatellite instability in oral cancer. *Int. J. Cancer* **64**, 332–335.
74. Yamamoto, K., Konishi, N., Inui, T., Nakamura, M., Hiasa, Y., Kirita, T., and Sugimura, M. (1999). DNA alterations in human oral squamous cell carcinomas detected by restriction landmark genomic scanning. *J. Oral Pathol. Med.* **28**, 102–106.
75. Emilion, G., Langdon, J. D., Speight, P., and Partridge, M. (1996). Frequent gene deletions in potentially malignant oral lesions. *Br. J. Cancer* **73**, 809–813.
76. Roz, L., Wu, C. L., Porter, S., Scully, C., Speight, P., Read, A., Sloan, P., and Thakker, N. (1996). Allelic imbalance on chromosome 3p in oral dysplastic lesions: an early event in oral carcinogenesis. *Cancer Res.* **56**, 1228–1231.
77. Nunn, J., Scholes, A. G., Liloglou, T., Nagini, S., Jones, A. S., Vaughan, E. D., Gosney, J. R., Rogers, S., Fear, S., Field, J. K. (1999). Fractional allele loss indicates distinct genetic populations in the development of squamous cell carcinoma of the head and neck (SCCHN). *Carcinogenesis* **20**, 2219–2228.
78. Moroco, J. R., Solt, D. B., and Polverini, P. J. (1990). Sequential loss of suppressor genes for three specific functions during in vivo carcinogenesis. *Lab. Invest.* **63**, 298–306.
79. Suzui, M., Yoshimi, N., Tanaka, T., and Mori, H. (1995). Infrequent Ha-ras mutations and absence of Ki-ras, N-ras and p53 mutations in 4-nitroquinoline 1-oxide-induced rat oral lesions. *Mol. Carcinog* **14**, 294–298.
80. Shintani, S., Matsumura, T., Alcalde, R.E., and Yoshihama, Y. (1996). Sequential expression of myc-, ras-, oncogene products and EGF receptor during DMBA-induced tongue carcinogenesis. *Int. J. Oncol.* **8**, 821–826.

81. Chang, K. W., Lin, S.-C., Koos, S., Pather, K., and Solt, D. (1996). p53 and Ha-ras mutations in chemically induced hamster buccal pouch carcinomas. *Carcinogenesis* **17**, 595–600.
82. Califano, J., van der Riet, P., Westra, W., Nawroz, H., Clayman, G., Piantadosi, S., Corio, R., Lee, B., Greenberg, B., Koch, W., and Sidransky, D. (1996). Genetic progression model for head and neck cancer; implications for field cancerization. *Cancer Res.* **56**, 2488–2492.
83. Jin, Y. T., Myers, J., Tsai, S. T., Goepfert, H., Batsakis, J. G., and El-Naggar, A. K. (1999). Genetic alterations in oral squamous cell carcinoma of young adults. *Oral Oncol.* **35**, 251–256.
84. Koch, W. M., and McQuone, S. (1997).Clinical and molecular aspects of squamous cell carcinoma of the head and neck in the nonsmoker and nondrinker. *Curr. Opin. Oncol.* **9**, 257–261.
85. Field, J. K., Kiaris, H., Risk, J. M., Tsiriyotis, C., Adamson, R., Zoumpourlis, V., *et al.* (1995). Allelotype of squamous cell carcinoma of the head and neck: Fractional allele loss correlates with survival. *Br. J. Cancer* **72**, 1180–1188.
86. Liloglou, T., Scholes, A. G., Spandidos, D. A., Vaughan, E. D., Jones, A. S., and Field, J. K. (1997). p53 mutations in squamous cell carcinoma of the head and neck predominate in a subgroup of former and present smokers with a low frequency of genetic instability. *Cancer Res.* **57**, 4070–4074.
87. Hermsen, M. A., Joenje, H., Arwert, F., Braakhuis, B. J. M., Baak, J. P. A., Westerveld, A., and Slater, R. (1997). Assessment of chromosomal gains and losses in oral squamous cell carcinoma by comparative genomic hybridisation. *Oral Oncol.* **33**, 414–418.
88. Wolff, E., Girod, S., Liehr, T., Vorderwulbecke, U., Ries, J., Steininger, H., and Gebhart, E. (1998). Oral squamous cell carcinomas are characterised by a rather uniform pattern of genomic imbalances detected by comparative genomic hybridisation. *Oral Oncol.* **34**, 186–190.
89. Shillitoe, E. J., May, M., Patel, V., Lethanakul, C., Ensley, J. F., Strausberg, R. L., and Gutkind, J. S. (2000). Genome-wide analysis of oral cancer. Early results from the Cancer Genome Anatomy Project. *Oral Oncol.* **36**, 8–16.
90. Maestro, R., Gasparotto, D., Vukosavljevic, T., Barzan, L., Sulfaro, S., and Boiocchi, M. (1993). Three discrete regions of deletion at 3p in head and neck cancers. *Cancer Res.* **53**, 5775–5779.
91. Field, J. K., Tsiriyotis, C., Zoumpourlis, V., Howard, P., Spandidos, D. A., and Jones, A. S. (1994). Allele loss on chromosome 3 in squamous-cell carcinoma of the head and neck correlates with poor clinical prognostic indicators. *Int. J. Oncol.* **4**, 543–549.
92. Partridge, M., Kiguwa, S., and Langdon, J.D. (1994). Frequent deletion of chromosome 3p in oral squamous cell carcinoma. (1994). *Eur. J. Cancer B Oral Oncol.* **30B**, 248–251.
93. Ah-See, K. W., Cooke, T. G., Pickford, I. R., Soutar, D., and Balmain, A. (1994). Allelotype of squamous carcinoma of the head and neck using microsatellite markers. *Cancer Res.* **54**, 1617–1621.
94. Nawroz, H., van der Riet, P., Hruban, R. H., Koch, W., Ruppert, J. M., and Sidransky, D. (1994). Allelotype of head and neck squamous cell carcinoma. *Cancer Res.* **54**, 1152–1155.
95. Ishwad, C. S., Ferrell, R. E., Rossie, K. M., *et al.* (1996). Loss of heterozygosity of the short arm of chromosomes 3 and 9 in oral cancer. *Int. J. Cancer* **69**, 1–4.
96. Lydiatt, W. M., Davidson, B. J., Schantz, S. P., Caruana, S., and Chaganti, R. S. (1998). 9p21 deletion correlates with recurrence in head and neck cancer. *Head Neck* **20**, 113–118.
97. Venugopalan, M., Wood, T. F., Wilczynski, S. P., Sen, S., Peters, J., Ma, G. C., Evans, G. A., and Srivatsan, E. S. (1998). Loss of heterozyosity in squamous ceil carcinomas of the head and neck defines a tumour suppressor gene region on 11q13. *Cancer Genet. Cytogenet.* **104**, 124–132.
98. Wu, C. L., Sloan, P., Read, A. P., Harris, R., and Thakker, N. (1994). Deletion mapping on the short arm chromosome 3 in squamous cell carcinoma of the oral cavity. *Cancer Res.* **54**, 6484–6488.
99. Virgilio, L., Schuster, M., Gollin, S. M., Veronese, M. L., Ohta, M., *et al.* (1996). FHIT gene alterations in head and neck squamous cell carcinomas. *Proc. Natl. Acad. Sci. USA* **93**, 9770–9775.
100. Kisielewski, A. E., Xiao, G. H., Liu, S. C., Klein-Szanto, A. J., Novara, M., Sina, J., Bleicher, K., Yeung, R. S., and Goodrow, T. L. (1998). Analysis of the FHIT gene and its product in squamous cell carcinomas of the head and neck. *Oncogene* **17**, 83–91.
101. Gonzalez, M. V., Pello, M. F., Ablanedo, P., Suarez, C., Alvarez, V., and Coto, E. (1998). Chromosome 3p loss of heterozygosity and mutation analysis of the FHIT and β-cat genes in squamous cell carcinoma of the head and neck. *J. Clin. Pathol.* **51**, 520–524.
102. Uzawa, N., Yoshida, M. A., Hosoe, S., Oshimura, M., Amagasa, T., and Ikeuchi, T. (1998). Functional evidence for involvement of multiple putative tumor suppressor genes on the short arm of chromosome 3 in human oral squamous cell carcinogenesis. *Cancer Genet. Cytogenet.* **107**, 125–131.
103. Partridge, M., Emilion, G., Pateromichelakis, S., Phillips, E., and Langdon, J. (1999). Location of candidate tumour suppressor gene loci at chromosomes 3p, 8p, and 9p for oral squamous cell carcinomas. *Int. J. Cancer* **83**, 318–325.
104. Shriver, S. P., Shriver, M. D., Tirpak, D. L., Bloch, L. M., Hunt, J. D., Ferrell, R. E., and Siegfried, J. M. (1998). Trinucleotide repeat length variation in the human ribosomal protein L14 gene (RPL14): Localisation to 3p21.3 and loss of heterozygosity in lung and oral cancers. *Mutat. Res.* **406**, 9–23.
105. Shridar, R., Shridar, V., Rivard, S., Siegfried, J. M., Pietraszkiewicz, H., Ensley, J., Pauley, R., Grignon, D., Sakr, W., Miller, O. J., and Smith, D. I. (1996). Mutations in the arginine-rich protein gene in lung, breast, and prostate cancers, and in squamous cell carcinomas of the head and neck. *Cancer Res.* **56**, 5576–5578.
106. Garrigue-Antar, L., Munoz-Antonia, T., Antonia, S. J., Gesmonde, J., Velluci, V. F., and Reiss, M. (1995). Missense mutations of the transforming growth factor β type II receptor in human head and neck squamous carcinoma cells. *Cancer Res.* **55**, 3982–3987.
107. Eisma, R. J., Spiro, J. D., von Biberstein, S. E., Lindquist, R., and Kreutzer, D. L. (1996). Decreased expression of TGF receptors on head and neck SqCC tumor cells. *Am. J. Surg.* **172**, 641–645.
108. Waber, P. G., Lee, N. K., and Nisen, P. D. (1996). Frequent allelic loss at chromosome arm 3p is distinct from genetic alterations of the Von-Hippel Lindau tumour suppressor gene in head and neck cancer. *Oncogene* **12**, 365–369.
109. Uzawa, K., Suzuki, H., Yokoe, H., Tanzawa, H., and Sato, K. (1995). Mutational state of p16/CDKN2 and VHL genes in squamous cell carcinoma of the oral cavity. *Int. J. Oncol.* **7**, 895–899.
110. Pershouse, M. A., El-Naggar, A. K., Hurr, K., Lin, H., Yung, W. K., and Steck, P. A. (1997). Deletion mapping of chromosome 4 in head and neck squamous cell carcinoma. *Oncogene* **14**, 369–373.
111. Wang, X. L., Uzawa, K., Imai, F. L., and Tanzawa, H. (1999). Localisation of a novel tumor suppressor gene associated with human oral cancer on chromosome 4q25. *Oncogene* **18**, 823–825.
112. Largey, J. S., Meltzer, S. J., Sauk, J. J., Hebert, C. A., and Archibald, D. W. (1994). Loss of heterozygosity involving the APC gene in oral squamous cell carcinomas. *Oral Surg. Oral Med. Oral Pathol.* **77**, 260–263.
113. Uzawa, K., Yoshida, H., Susuki, H., Tanzawa, H., Shimazaki, J., Seino, S., and Sato, K. (1994). Abnormalities of the adenomatous polyposis coli gene in human oral squamous cell carcinolma. *Int. J. Cancer* **58**, 814–817.
114. Wang, X. L., Uzawa, K., Nakanishi, H., Tanzawa, H., and Sato, T. (1997). Allelic imbalance on the long arm of chromosome 5 in oral squamous cell carcinoma. *Int. J. Oncol.* **10**, 535–538.
115. Huang, J. S., Chiang, C. P., Kok, S. H., Kuo, Y. S., and Kuo, M. Y. (1997). Loss of heterozygosity of APC and MCC genes in oral squamous cell carcinomas in Taiwan. *J. Oral Pathol. Med.* **26**, 322–326.

116. Mao, E. J., Schwartz, S. M., Daling, J. R., and Beckmann, A. M. (1998). Loss of heterozygosity at 5q21-22 (adenomatous polyposis coli gene region) in oral squamous cell carcinoma is common and correlated with advanced disease. *J. Oral Pathol. Med.* **27**, 297–302.
117. Martins, C., Jin, Y., Jin, C., Wennerberg, J., Hoglund, M., and Mertens, F. (1999). Fluorescent *in situ* hybridisation (FISH) characterization of pericentromeric breakpoints on chromosome 5 in head and neck squamous cell carcinomas. *Eur. J. Cancer* **35**, 498–501.
118. Matthias, C., Jahnke, V., Fryer, A., Strange, R., Ollier, W., and Hajeer, A. (1998). Influence of tumour necrosis factor microsatellite polymorphisms on susceptibility to head and neck cancer. *Arch. Otolaryngol.* **118**, 284–288.
119. Harada, K., and Ogden, G. R. (2000). An overview of the cell cycle arrest protein $p21^{WAF1}$. *Oral Oncol.* **36**, 3–7.
120. Heinzel, P. A., Balaram, P., and Berbard, H. U. (1996). Mutations and polymorphisms in the p53, p21 and p16 genes in oral carcinomas of Indian betel quid chewers. *Int. J. Cancer* **68**, 420–423.
121. Erber, R., and Klein, W. (1997). Aberrant p21 protein accumulation in Head-and-neck cancer. *Int. J. Cancer* **74**, 383–389.
122. van Oijen, M. G., Tilanus, M. G., Medema, R. H., and Slootweg, P. J. (1998). Expression of p21 (Waf1/Cip1) in head and neck cancer in relation to proliferation, differentiation, p53 status and cyclin D1 expression. *J. Oral Pathol. Med.* **27**, 367–375.
123. Warnakulasuriya, K. A., Tavassoli, M., and Johnson, N. W. (1998). Relationship of p53 overexpression to other cell cycle regulatory proteins in oral squamous cell carcinoma. *J. Oral Pathol. Med.* **27**, 376–381.
124. Kudo, Y., Takata, T., Ogawa, I., Sato, S., and Nikai, H. (1999). Expression of p53 and p21CIP1/WAF1 proteins in oral epithelial dysplasia and squamous cell carcinomas. *Oncol. Rep.* **6**, 539–545.
125. Agarwal, S., Mathur, M., Shukla, N. K., and Ralhan, R. (1998). Expression of cyclin-dependent kinase inhibitor p21 waf1/cip1 in premalignant and malignant oral lesions; relationship with p53 status. *Oral Oncol.* **34**, 353–360.
126. Yook, J. I., and Kim, J. (1998). Expression of p21WAF1/CIP1 is unrelated to p53 tumour suppressor gene status in oral squamous cell carcinomas. *Oral Oncol.* **34**, 198–203.
127. Voravud, N., Shin, D. M., Ro, J. Y., Lee, J. S., Hong, W. K., and Hittelman, W. N. (1993). Increased polysomies of chromosomes 7 and 17 during head and neck multistage tumorigenesis. *Cancer Res.* **53**, 2874–2883.
128. Rubin Grandis, J., Chakraborty, A., Melhem, M. F., Zeng, Q., and Tweardy, D. J. (1997). Inhibition of epidermal growth factor receptor gene expression and function decreases proliferation of head and neck squamous carcinoma but not normal mucosal epithelial cells. *Oncogene* **15**, 409–416.
129. Maiorano, E., Favia, G., Maisonneuve, P., and Viale, G. (1998). Prognostic implications of epidermal growth factor receptor immunoreactivity in squamous cell carcinoma of the oral mucosa. *J. Pathol.* **185**, 167–174.
130. Shin, D. M., Ro, J. Y., Hong, W. K., and Hittelman, W. N. (1994). Dysregulation of epidermal growth factor receptor expression in premalignant lesions during head and neck tumorigenesis. *Cancer Res.* **54**, 3153–3159.
131. Wang, X. L., Uzawa, K., Miyakawa, A., Shiiba, M., Watanabe, T., Sato, T., Miya, T., Yokoe, H., and Tanzawa, H. (1998).Localisation of a tumour-suppressor gene associated with human oral cancer on 7q31.1. *Int. J. Cancer* **75**, 671–674.
132. Patel, V., Yeudall, W. A., Gardner, S., Mutlu, S., Scully, C., and Prime, S. S. (1993). Consistent chromosomal anomalies in keratinocyte cell lines derived from untreated malignant lesions of the oral cavity. *Genes Chromosomes Cancer* **7**, 109–115.
133. Wu, C. L., Roz, L., Sloan, P., Read, A. P., Holland, S., Porter, S. R., Scully, C., Speight, P. M., and Thakker, N. (1997). Deletion mapping defines three discrete areas of allelic imbalance on chromosome arm 8p in oral and oropharyngeal squamous cell carcinomas. *Genes Chromosomes Cancer* **20**, 347–353.
134. El-Naggar, A. K., Coombes, M. M., Batsakis, J. G., Hong, W. K., Goepfert, H., and Kagan, J. (1998). Localisation of chromosome 8p regions involved in early tumorigenesis of oral and laryngeal squamous carcinoma. *Oncogene* **16**, 2983–2987.
135. Ishwad, C. S., Shuster, M., Bockmuhl, U., Thakker, N., Shah, P., Toomes, C., Dixon, M., Ferrell, R. E., Gollin, S. M. (1999). Frequent allelic loss and homozygous deletion in chromosome band 8p23 in oral cancer. *Int. J. Cancer* **80**, 25–31.
136. Ono, K., Miyakawa, A., Fukuda, M., Shiiba, M., Uzawa, K., Watanabe, T., Miya, T., Yokoe, H., Imai, Y., and Tanzawa, H. (1999). Allelic loss on the short arm of chromosome 8 in oral squamous cell carcinoma. *Oncol. Rep.* **6**, 785–789.
137. van der Riet, P., Nawroz, H, Hruban, R. H., Corio, R., Tokino, K., Koch, W., and Sidransky, D. (1994). Frequent loss of chromosome 9p21-22 early in head and neck cancer progression. *Cancer Res.* **54**, 1156–1158.
138. El-Naggar, A. K., Hurr, K., Batsakis, J. G., Luna, M. A., Goepfert, H., and Huff, V. (1995). Sequential loss of heterozygosity at microsatellite motifs in preinvasive and invasive head and neck squamous carcinoma. *Cancer Res.* **55**, 2656–2659.
139. Lydiatt, W. M., Davidson, B. J., Shah, J., Schantz, S. P., and Chaganti, R. S. (1994). The relationship of loss of heterozygosity to tobacco exposure and early recurrence in head and neck squamous-cell carcinoma. *Am. J. Surg.* **168**, 437–440.
140. Miracca, E. C., Kowalski, L. P., and Nagai, M. A. (1999). High prevalence of p16 genetic alterations in head and neck tumours. *Br. J. Cancer* **81**, 677–683.
141. Nakanishi, H., Wang, X. L., Imai, F. L., Kato, J., Shiiba, M., Miya, T., Imai, Y., and Tanzawa, H. (1999). Localisation of a novel tumour suppressor gene loci on chromosome 9p21-22 in oral cancer. *Anticancer Res.* **19**, 29–34.
142. Zhao, Y., Zhang, S., Fu, B., and Xiao, C. (1999). Abnormalities of tumor suppressor genes p16 and p15 in primary maxillofacial squamous cell carcinomas. *Cancer Genet. Cytogenet.* **112**, 26–33.
143. Pande, P., Mathur, M., Shukla, N. K., and Ralhan, R. (1998). pRb and p16 protein alterations in human oral tumorigenesis. *Oral Oncol.* **34**, 396–403.
144. El-Naggar, A. K., and Clayman, G. (1999). Expression of p16, Rb, and cyclin D1 gene products in oral and laryngeal squamous carcinomas; Biological and clinical implications. *Hum. Pathol.* **30**, 1013–1018.
145. Cairns, P., Mao, L., Merlo, A., Lee, D. J., Schwab, D., Eby, Y., *et al.* (1994). Rates of p16 (MTS1) mutations in primary tumours with 9p loss. *Science* **265**, 415–417.
146. Zhang, L., Epstein, J., Band, P., Berean, K., Hay, J., Cheng, X., and Rosin, M. P. (1999). Local tumor recurrence or emergence of a new primary lesion? A molecular analysis. *J. Oral Pathol. Med.* **28**, 381–384.
147. Nakanishi, H., Uzawa, K., Yokoe, H., Miya, T., Wang, X. L., Watanabe, T., *et al.* (1997). Rare mutations of the growth suppressor genes involved in negative regulation of the cell cycle: p15, p16 and p18 genes in oral squamous cell carcinoma. *Int. J. Oncol.* **11**, 1129–1133.
148. Gonzalez, M. V., Pello, M. F., Lopez-Larrea, C., Suarez, C., Menendez, M. J., and Coto, E. (1997). Deletion and methylation of the tumor suppressor gene p16/CDKN2 in primary head and neck squamous cell carcinoma. *J. Clin. Pathol.* **50**, 509–512.
149. Riese, U., Dahse, R., Fiedler, W., Theuer, C., Koscielny, S., Ernst, G., Beleites, E., Claussen, U., and von Eggeling, F. (1999). Tumor suppressor gene p16(CDKN2A) mutation status and promoter inactivation in head and neck cancer. *Int. J. Mol. Med.* **4**, 61–65.
150. Zhang, S. Y., Klein-Szanto, A. J., Sauter, E. R., Shafarenko, M., Mitsunga, S., Nobori, T., Carson, D. A., Ridge, J. A., and

Goodrow, T. L. (1994). Higher frequency of alterations in the p16/CDKN2 gene in squamous cell carcinoma cell lines than in primary tumors of the head and neck. *Cancer Res.* **54**, 5050–5053.

151. Reed, A. L., Califano, J., Cairns, P., Westra, W. H., Jones, R. M., Koch, W., Ahrendt, S., Eby, Y., Sewell, D., Nawroz, H., Bartek, J., and Sidransky, D. (1996). LHigh frequency of p16 (CDKN2/MTS-1/INK4A) inactivation in head and neck squamous carcinoma. *Cancer Res.* **56**, 3630–3633.
152. Lang, J. C., Tobin, E. J., Knobloch, T. J., *et al.* (1998). Frequent mutation of p16 in squamous cell carcinoma of the head and neck. *Laryngoscope* **108**, 923–928.
153. Wu, C. L., Roz, L., McKown, S., Sloan, P., Read, A. P., Holland, S., Porter, S., Scully, C., Paterson, I., Tavassoli, M., and Thakker, N. (1999). DNA studies underestimate the major role of CDKN2A inactivation in oral and oropharyngeal squamous cell carcinomas. *Genes Chromosomes Cancer* **25**, 16–25.
154. Merlo, A., Herman, J. G., Mao, L., Lee, D. J., Gabrielson, E., Burger, P. C., Baylin, S. B., and Sidransky, D. (1995). 5′ CpG island methylation is associated with transcriptional silencing of the tumour suppressor p16/CDKN2/MTS1 in human cancers. *Nature Med.* **1**, 686–692.
155. El-Naggar, A. K., Lai, S., Clayman, G., Lee, J. K., Luna, M. A., Goepfert, H., *et al.* (1997). Methylation, a major mechanism of p16/CDKN2 gene inactivation in head and neck squamous carcinoma. *Am. J. Pathol.* **151**, 1767–1774.
156. Tao, J., Wu, B., Fang, W., *et al.* (1997). Mutation and methylation of CDKN2 gene in human head and neck carcinomas. *J. Clin. Pathol.* **26**, 152–154.
157. Patel, V., Jakus, J., Harris, C. M., Ensley, J. F., Robbins, K. C., and Yeudall, W. A. (1997). Altered expression and activity of G1/S cyclins and cyclin-dependent kinases characterise squamous cell carcinomas of the head and neck. *Int. J. Cancer* **73**, 551–555.
158. Liggett, W. H., and Sidransky, D. (1998). Role of the p16 tumour suppressor gene in cancer. *J. Clin. Oncol.* **16**, 1197–1206.
159. Sandig, V., Brand, K., Herwig, S., Lukas, J., Bartek, J., and Strauss, M. (1997). Adenovirally transferred p16 INK4/CDKNZ and p53 genes cooperate to induce apoptotic tumour cell death. *Nature Med* **3**, 313–319.
160. Gasparotto, D., Vukosavljevic, T., Piccinin, S., Barzan, L., Sulfaro, S., Armellin, M., Boiocchi, M., and Maestro, R. (1999). Loss of heterozygosity at 10q in tumors of the upper respiratory tract is associated with poor prognosis. *Int. J. Cancer* **84**, 432–436.
161. Henderson, Y. C., Wang, E., and Clayman, G. L. (1998). Genotypic analysis of tumor suppressor genes PTEN/MMAC1 and p53 in head and neck squamous cell carcinomas. *Laryngoscope* **108**, 1553–1556.
162. Steck, P. A., Pershouse, M. A., Jasser, S. A., Yung, W. K., Lin, H., Ligon, A. H., Langford, L. A., Baumgard, M. L., Hattier, T., Davis, T., Frye, C., Hu, R., Swedlund, B., Teng, D. H., and Tavtigian, S. V. (1997). Identification of a candidate tumour suppressor gene, MMAC1, at chromosome 10q23.3 that is mutated in multiple advanced cancers. *Nature Genet.* **15**, 356–362.
163. Okami, K., Wu, L., Riggins, G., Cairns, P., Goggins, M., Evron, E., Halachmi, N., Ahrendt, S. A., Reed, A. L., Hilgers, W., Kern, S. E., Koch, W. M., Sidransky, D., and Jen, J. (1998). Analysis of PTEN/MMAC1 alterations in aerodigestive tract tumors. *Cancer Res.* **58**, 509–511.
164. Chen, Q., Luo, G., Li, B., and Samaranayake, L. P. (1999). Expression of p16 and CDK4 in oral premalignant lesions and oral squamous cell carcinomas: A semi-quantitative immunohistochemical study. *J. Oral Pathol. Med.* **28**, 158–164.
165. Zaslav, A. L., Stamberg, J., Steinberg, B. M., Lin, Y. J., and Abramson, A. (1991). Cytogenetic analysis of head and neck carcinomas. *Cancer Genet. Cytogenet.* **56**, 181–187.
166. Lese, C. M., Rossie, K. M., Appel, B. N., Reddy, J. K., Johnson, J. T., Myers, E. N., and Gollin, S. M. (1995). Visualisation of INT2 and HST1 amplification in oral squamous cell carcinomas. *Genes Chromosomes Cancer* **12**, 288–295.
167. Uzawa, K., Suzuki, H., Komiya, A., Nakanishi, H., Ogawara, K., Tanzawa, H., and Sato, K. (1996). Evidence for two distinct tumor-suppressor gene loci on the long arm of chromosome 11 in human oral cancer. *Int. J. Cancer* **67**, 510–514.
168. Mertens, F., Johansson, B., Hoglund, M., and Mitelman, F. (1997). Chromosomal imbalance maps of malignant solid tumour; a cytogenetic survey of 3,185 neoplasms. *Cancer Res.* **57**, 2765–2780.
169. El-Naggar, A. K., Lai, S., Tucker, S. A., Clayman, G. L., Goepfert, H., Hong, W. K., and Huff, V. (1999). Frequent loss of imprinting at the IGF2 and H19 genes in head and neck squamous carcinoma. *Oncogene* **18**, 7063–7069.
170. Fujieda, S., Inuzuka, M., Tanaka, N., Sunaga, H., Fan, G. K., Ito, T., Sugimoto, C., Tsuzuki, H., and Saito, H. (1999). Expression of p27 is associated with Bax expression and spontaneous apoptosis in oral and oropharyngeal carcinoma. *Int. J. Cancer* **84**, 315–320.
171. Jordan, R. C. K., Bradley, G., and Slingerland, J. (1998). Reduced levels of the cell-cycle inhibitor p27 Kip 1 in epithelial dysplasia and carcinoma of the oral cavity. *Am. J. Pathol.* **152**, 585–590.
172. Kudo, Y., Takata, T., Yasui, W., Ogawa, I., Miyauchi, M., Takekoshi, T., Tahara, E., and Nikai, H. (1998). Reduced expression of cyclin-dependent kinase inhibitor p27 Kip 1 is an indicator of malignant behavior in oral squamous cell carcinoma. *Cancer* **83**, 2447–2455.
173. Mineta, H., Miura, K., Suzuki, I., Takebayashi, S., Amano, H., Araki, K., Harada, H., Ichimura, K., Wennerberg, J. P., and Dictor, M. R. (1999). Low p27 expression correlates with poor prognosis for patients with oral tongue squamous cell carcinoma. *Cancer* **85**, 1011–1017.
174. Venkatesan, T. K., Kuropkat, C., Caldarelli, D. D., Panje, W. R., Hutchinson, J. C., Jr., Chen, S., and Coon, J. S. (1999). Prognostic significance of p27 expression on carcinoma of the oral cavity and oropharynx. *Laryngoscope* **109**, 1329–1333.
175. Tsuji, T., Duh, F. M., Latif, F., Popescu, N. C., Zimonjic, D. B., McBride, J., Matsuo, K., Ohyama, H., Todd, R., Nagata, E., Terakado, N., Sasaki, A., Matsumura, T., Lerman, M. I., and Wong, D. T. (1998). Cloning, mapping, expression, function, and mutation analyses of the human ortholog of the hamster putative tumor suppressor gene Doc-1. *J. Biol. Chem.* **273**, 6704–6709.
176. Yoo, G. H., Xu, H. J., Brennan, J. A., Westra, W., Hruban, R. H., Koch, W., Benedict, W. F., and Sidransky, D. (1994). Infrequent inactivation of the retinoblastoma gene despite frequent loss of chromosome 13q in head and neck squamous cell carcinoma. *Cancer Res.* **54**, 4603–4606.
177. Maestro, R., Piccinin, S., Doglioni, C., *et al.* (1996). Chromosome 13q deletion mapping in head and neck squamous cell carcinomas: Identification of two distinct regions of preferential loss. *Cancer Res.* **56**, 1146–1150.
178. Gupta, V. K., Schmidt, A. P., Pashia, M. E., Sunwoo, J. B., and Scholnick, S. B. (1999). Multiple regions of deletion on chromosome arm 13q in Head-and-neck squamous-cell carcinoma. *Int. J. Cancer* **84**, 453–457.
179. Ogawara, K., Miyakawa, A., Shiba, M., Uzawa, K., Watanabe, T., Wang, X. L., Sato, T., Kubosawa, H., Kondo, Y., and Tanzawa, H. (1998). Allelic loss of chromosome 13q14.3 in human oral cancer: Correlation with lymph node metastasis. *Int. J. Cancer* **79**, 312–317.
180. Regezi, J. A., Dekker, N. P., McMillan, A., Ramirez-Amador, V., Meneses-Garcia, A., Ruiz-Godoy Rivera, L. M., Chrysomali, E., and Ng, I. O. L. (1999). p53, p21, Rb and MDM2 proteins in tongue carcinoma from patients <35 versus >75 years. *Oral Oncol.* **35**, 379–383.
181. Owens, W., Field, J. K., Howard, P. J., and Stell, P. M. (1992). Multiple cytogenetic aberrations in squamous cell carcinomas of the head and neck. *Eur. J. Cancer B Oral Oncol.* **28B**, 17–21.
182. Saito, Y., Takazawa, H., Uzawa, K., Tanzawa, H., and Sato, K. (1998). Reduced expression of E-cadherin in oral squamous cell carcinoma: Relationship with DNA methylation of 5′ CpG island. *Int. J. Oncol.* **12**, 293–298.

183. Chen, X. (1999). The p53 family: Same response, different signals? *Mol. Med. Today* **5**, 387–392.
184. Chang, F., Syrjanen, S., and Syrjanen, K. (1995). Implications of the p53 tumour suppressor gene in clinical oncology. *J. Clin. Oncol.* **13**, 1009–1022.
185. Ralhan, R., Sandhya, A., Meera, M., Bohdan, W., and Nootan, S. K. (2000). Induction of MDM2-P2 transcripts correlates with stabilized wild-type p53 in betel- and tobacco-related human oral cancer. *Am. J. Pathol.* **157**, 587–596.
186. Field, J. K., Pavelic, Z. P., Spandidos, D. A., Stambrook, P. J., Jones, A. S., and Gluckman, J. L. (1993). The role of the p53 tumour suppressor gene in squamous-cell carcinoma of the head and neck. *Arch. Otolaryngol. Head Neck Surg.* **119**, 1118–1122.
187. Raybaud-Diogene, H., Tetu, B., Morency, R., Fortin, A., and Monteil, R. A. (1996). p53 overexpression in head and neck squamous cell carcinoma: review of the literature. *Eur. J. Cancer B Oral Oncol.* **32B**, 143–149.
188. Calzolari, A., Chiarelli, I., Bianchi, S., Messerini, L., Gallo, O., Porfirio, B., and Mattiuz, P.L. (1997) Immunohistochemical vs molecular biology methods; complementary techniques for effective screening of p53 alterations in head and neck cancer. *Am. J. Clin. Pathol.* **107**, 7–11.
189. van Heerden, W. F. P., van Rensburg, E. J., Hemmer, J., Raubenheimer, E. J, and Engelbrecht, S. (1998). Correlation between p53 gene mutation, p53 protein labeling and PCNA expression in oral squamous cell carcinomas. *Anticancer Res.* **18**, 237–240.
190. Taylor, D., Koch, W. M., Zahurak, M., Shah, K., Sidransky, D., and Westra, W.H. (1999). Immunohistochemical detection of p53 protein accumulation in head and neck cancer: Correlation with p53 gene alterations. *Hum. Pathol.* **30**, 1221–1225.
191. Burns, J. E., Baird, M. C., Clark, L. J., Burns, P. A., Edington, K., Chapman, C., Mitchell, R., Robertson, G., Soutar, D., and Parkinson, E. K. (1993). Gene mutations and increased levels of p53 protein in human squamous cell carcinomas and their cell lines. *Br. J. Cancer* **67**, 1274–1284.
192. Chen, Y. T., Xu, L., Massey, L., Zlotolow, I. M., Huvos, A. G., Garin-Chesa, P., and Old, L. J. (1994). Frameshift and nonsense p53 mutations in squamous-cell carcinoma of head and neck: Non-reactivity with three anti-p53 monoclonal antibodies. *Int. J. Oncol.* **4**, 609–614.
193. Ranasinghe, A., MacGeoch, C., Dyer, S., Spurr, N., and Johnson, N. W. (1993). Some oral carcinoma from Sri Lanka betel/tobacco chewers overexpress p53 oncoprotein but lack mutations in exons 5-9. *Anticancer Res.* **13**, 2065–2068.
194. Ranasinghe, A. W., Warnakulasuriya, K. A. A. S., and Johnson, N. W. (1993). Low prevalence of expression of p53 oncoprotein in oral carcinomas from Sri Lanka associated with betel and tobacco chewing. *Eur. J. Cancer B Oral Oncol.* **29B**, 147–150.
195. Xu, L., Chen, Y. T., Huvos, A. G., Zlotolow, I. M., Rettig, W. J., Old, L. J., and Garin-Chesa, P. (1994). Overexpression of p53 protein in squamous cell carcinomas of head and neck without apparent gene mutations. *Diagn. Mol. Pathol.* **3**, 83–92.
196. Dowell, S. P., and Ogden, G. R. (1996). The use of antigen retrieval for immunohistochemical detection of p53 over-expression in malignant and benign oral mucosa: A cautionary note. *J. Oral Pathol. Med.* **25**, 60–64.
197. Partridge, M., Kiguwa, S., Emilion, G., Pateromichelakis, S., A'Hern, R., and Langdon, J. D. (1999). New insights into p53 protein stabilisation in oral squamous cell carcinoma. *Oral Oncol.* **35**, 45–55.
198. Field, J. K., Spandidos, D. A., Malliri, A., Gosney, J. R., Yiagnisis, M., and Stell, P. M. (1991). Elevated p53 expression correlates with a history of heavy smoking in squamous cell carcinoma of the head and neck. *Br. J. Cancer* **64**, 573–577.
199. Brachman, D. G., Graves, D., Vokes, E., Beckett, M., Haraf, D., Montag, A., Dunphy, E., Mick, R., Yandell, D., and Weichselbaum, R. R. (1992). Occurrence of p53 gene deletions and human papilloma virus infection in human head and neck cancer. *Cancer Res.* **52**, 4832–4836.
200. Langdon, J. D., and Partridge, M. (1992). Expression of the tumour suppressor gene p53 in oral cancer. *Br. J. Oral Maxillofac. Surg.* **30**, 214–220.
201. Ogden, G. R., Kiddie, R. A., Lunny, D. P., and Lane, D. P. (1992). Assessment of p53 protein expression in normal, benign and malignant oral mucosa. *J. Pathol.* **166**, 389–394.
202. Chang, Y., Lin, Y., Tsai, C., Shu, C., Tsai, M., Choo, K., and Liu, S. (1992). Detection of mutations in p53 gene in human head and neck carcinomas by single strand conformation polymorphism analysis. *Cancer Lett.* **67**, 167–174.
203. Warnakulasuriya, K. A., and Johnson, N. W. (1992). Expression of p53 mutant nuclear phosphoprotein in oral carcinoma and potentially malignant oral lesions. *J. Oral Pathol. Med.* **21**, 404–408.
204. Matthews, J. B., Scully, C., Jovanovic, A., van der Waal, I., Yeudall, W. A., and Prime, S. S. (1993). Relationship of tobacco/alcohol use to p53 expression in patients with lingual squamous cell carcinomas. *Eur. J. Cancer B Oral Oncol.* **29B**, 285–289.
205. Matsumura, T., Yoshihama, Y., Kimura, T., Shintani, S., and Alcalde, R. E. (1996). p53 and MDM2 expression in oral squamous cell carcinoma. *Oncology* **53**, 308–312.
206. Munirajan, A. K., Tutsumi-Ishii, Y., Hohanprasad, B. K., Hirano, Y., Munakata, N., Shanmugan, G., and Tsuchida, N. (1996). p53 gene mutations in oral carcinomas from India. *Int. J. Cancer* **66**, 297–300.
207. Yan, J. J., Tzeng, C. C., and Jin, Y. T. (1996). Overexpression of p53 protein in squamous cell carcinoma of buccal mucosa and tongue in Taiwan: An immunohistochemical and clinicopathological study. *J. Oral Pathol. Med.* **25**, 55–59.
208. Steingrimsdottir, H., Penhallow, J., Farzaneh, F., Johnson, N., and Tavassoli, M. (1997). Detection of p53 mutations in oral cancer samples using a sensitive PCR-based method. *Biochem. Soc. Trans.* **25**, 315–318.
209. Nagai, M. A., Miracca, E. C., Yamamoto, L., Moura, R. P., Simpson, A. J., Kowalski, L. P., and Brentani, R. R. (1998). TP53 genetic alterations in head-and-neck carcinomas from Brazil. *Int. J. Cancer* **76**, 13–18.
210. Kannan, K., Munirajan, A. K., Krishnamurthy, J., Bhuvarahamurthy, V., Mohanprasad, B. K., Panishankar, K. H., Tsuchida, N., and Shanmugam, G. (1999). Low incidence of p53 mutations in betel quid and tobacco chewing-associated oral squamous cell carcinoma from India. *Int. J. Oncol.* **15**, 1133–1136.
211. Gusterson, B. A., Anbahaghan, R., Warren, W., *et al.* (1991). Expression of p53 in premalignant and malignant squamous epithelium. *Oncogene* **6**, 1785–1789.
212. Caamano, J., Zhang, S. Y., Rosvold, E. A., Bauer, B., and Klein-Szanto, A. J. (1993). p53 alterations in human squamous cell carcinomas and carcinoma cell lines. *Am. J. Pathol.* **142**, 1131–1139.
213. Adamson, R., Jones, A. S., and Field, J. K. (1994). Loss of heterozygosity studies on chromosome 17 in head and neck cancer using microsatellite markers. *Oncogene* **9**, 2077–2082.
214. Regezi, J. A., Zarbo, R. J., Regev, E., Pisanty, S., Silverman, S., and Gazit, D. (1995). p53 protein expression in sequential biopsies of oral dysplasias and in situ carcinomas. *J. Oral Pathol. Med.* **24**, 18–22.
215. Chung, K. Y., Mukhopadyay, T., Kim, J., Casson, A., Ro, J. Y., Goepfert, H., Hong, W.K., and Roth, J. A. (1993). Discordant p53 gene mutations in primary head and neck cancers and corresponding second primary cancers of the upper aerodigestive tract. *Cancer Res.* **53**, 1676–1683.
216. Zariwala, M., Schmid, S., Pfaltz, M., Ohgaki, H., Kleihues, P., and Schafer, R. (1994). p53 gene mutations in oropharyngeal carcinomas: A comparison of solitary and multiple primary tumours and lymph-node metastases. *Int. J. Cancer* **56**, 807–811.
217. Fogel, S., Ahomadegbe, J. C., Barrois, M., Le Bihan, M. L., Douc-Rasy, S., Duvillard, P., *et al.* (1996). Incidence elevee des mutations p53 dans les tumeurs primitives et metastases des voies aerodigestives

superieures et frequente en proteine dans les epitheliums normaux. *Bull. Cancer* **83**, 227–233.

218. Shintani, S., Yoshihama, Y., Emilio, A. R., and Matsumura, T. (1995). Overexpression of p53 is an early event in the tumorigenesis of oral squamous cell carcinomas. *Anticancer Res.* **15**, 305–308.
219. Gasparotto, D., Maestro, R, Barzan, L., *et al.* (1995). Recurrences and second primary tumors in the head and neck region: Differentiation by p53 mutation analysis. *Ann. Oncol.* **6**, 933–939.
220. Nees, M., Homann, N., Discher, H., Andl, T., Enders, C., Herold-Mende, C., Schulmann, A., and Bosch, F. X. (1993). Expression of mutated p53 occurs in tumor-distant epithelia of head and neck cancer patients: A possible molecular basis for the development of multiple tumors. *Cancer Res.* **53**, 4189–4196.
221. Kaur, J., Srivastava, A., and Ralhan, R. (1994). Overexpression of p53 protein in betel- and tobacco-related human oral dysplasia and squamous cell carcinoma in India. *Int. J. Cancer* **58**, 340–345.
222. Koch, W. M., Boyle, J. O., Mao, L., Hakim, J., Hruban, R. H., and Sidransky, D. (1994). p53 gene mutations as markers of tumour spread in synchronous oral cancers. *Arch. Otolaryngol. Head Neck Surg.* **120**, 943–947.
223. Kropveld, A., van Mansfeld, A. D., Nabben, N., Hordijk, G. J., and Slootweg, P. J. (1996). Discordance of p53 status in matched primary tumours and metastases in head and neck squamous cell carcinoma patients. *Eur. J. Cancer B Oral Oncol.* **32B**, 388–393.
224. Gallo, O., Bianchi, S., Giovannucci-Uzzielli, M. L., Santoro, R., Lenzi S., Salimbeni, C., Abbruzzese, M., and Alajmo, E. (1995). p53 oncoprotein overexpression correlates with mutagen-induced chromosome fragility in head and neck cancer patients with multiple malignancies. *Br. J. Cancer* **71**, 1008–1012.
225. Field, J. K. (1992). Oncogenes and tumour-suppressor genes in squamous cell carcinoma of the head and neck. *Eur. J. Cancer B Oral Oncol.* **28B**, 67–76.
226. Field, J. K., Zoumpourlis, V., Spandidos, D. A., and Jones, A. S. (1994). p53 expression and mutations in squamous cell carcinoma of the head and neck: Expression correlates with the patients use of tobacco and alcohol. *Cancer Detect. Prev.* **18**, 197–208.
227. Lazarus, P., Stern, J., Zwiebel, N., Fair, A., Richie, J. P., Jr., and Schantz, S. (1996). Relationship between p53 mutation incidence in oral cavity squamous cell carcinomas and patient tobacco use. *Carcinogenesis* **17**, 733–739.
228. Brennan, J. A., Boyle, J. O., Koch, W. M., Goodman, S. N., Hruban, R. H., Eby, Y. J., Couch, M. J., Forastiere, A. A., and Sidransky, D. (1995). Association between cigarette smoking and mutation of the p53 gene in squamous cell carcinoma of the head and neck. *N. Engl. J. Med.* **332**, 712–717.
229. Lazarus, P., Garewal, H. S., Sciubba, J., Zwiebel, N., Calcagnotto, A., Fair, A., Schaefer, S., and Richer, J. P., Jr. (1995). A low incidence of p53 mutations in pre-malignant lesions of the oral cavity from non-tobacco users. *Int. J. Cancer* **60**, 458–463.
230. Koch, W. M., Patel, H., Brennan, J., Boyle, J. O., and Sidransky, D. (1995). Squamous cell carcinoma of the head and neck in the elderly. *Arch. Otolaryngol. Head Neck Surg.* **121**, 262–265.
231. Sorensen, D. M., Lewark, T. M., Haney, J. L., Meyers, A. D., Krause, G., and Franklin, W. A. (1997). Absence of p53 mutations in squamous carcinoas of the tongue in nonsmoking and nondrinking patients younger than 40 years. *Arch. Otolaryngol. Head Neck Surg.* **123**, 503–506.
232. Wong, Y. K., Liu, T. Y., Chang, K. W., Lin, S. C., Chao, T. W., Li, P. L., and Chang, C.S. (1998). p53 alterations in betel quid and tobacco associated oral suqamous cell carcinomas from Taiwan. *J. Oral Pathol. Med.* **27**, 243–248.
233. Thomas, S., Brennan, J., Martel, G., Frazer, I., Montesano, R., Sidransky, D., and Hollstein, M. (1994). Mutations in the conserved regions of p53 are infrequent in betel-associated oral cancers from Papua New Guinea. *Cancer Res.* **54**, 3588–3593.
234. Baral, R. N., Patnaik, S., and Das, B. R. (1998). Co-overexpression of p53 and c-myc proteins linked with advanced stages of betel and tobacco-related oral squamous cell carcinomas from eastern India. *Eur. J. Oral Sci.* **106**, 907–913.
235. Chiba, I., Muthumala, M., Yamazaki, Y., Uz Zaman, A., Iizuka, T., Amamiya, A., Shibata, T., Kashiwazaki, H., Sugiura, C., and Fukuda, H. (1998). Characteristics of mutations in the p53 gene of oral squamous-cell carcinomas associated with betel-quid chewing in Sri Lanka. *Int. J. Cancer* **77**, 839–842.
236. Kaur, J., Srivastava, A., and Ralhan, R. (1998). Prognostic significance of p53 protein over-expression in betel and tobacco related oral oncogenesis. *Int. J. Cancer* **79**, 370–375.
237. Boyle, J. O., Hakim, J., Koch, W., van der Riet, P., Hruban, R. H., Roa, R. A., Correo, R., Eby, Y. J., Ruppert, M., and Sidransky, D. (1993). The incidence of p53 mutations increases with progression of head and neck cancer. *Cancer Res.* **53**, 4477–4480.
238. Kriek, E., Engelse, L. D., Scherer, E., and Westra, J. G. (1984). Formation of DNA modifications by chemical carcinogens. Identification, localization and quantification. *Biochim. Biophys. Acta* **738**, 181–201.
239. Greenblatt, M. S., Bennett, W. P., Hollstein, M., and Harris, C. C. (1994). Mutations in the p53 tumour suppressor gene: Clues to cancer etiology and molecular pathogenesis. *Cancer Res.* **54**, 4855–4878.
240. Olshan, A. F., Weissler, M. C., Pei, H., and Conway, K. (1997). p53 mutations in head and neck cancer: New data and evaluation of mutational spectra. *Cancer Epidemiol. Biomark. Prev.* **6**, 499–504.
241. Sakai, E., and Tsuchida, N. (1992). Most human squamous cell carcinomas in the oral cavity contain mutated p53 tumour-suppressor genes. *Oncogene* **7**, 927–933.
242. Somers, K. D., Merrick, M. A., Lopez, M. E., Incognito, L. S., Schechter, G. L., and Casey, G. (1992). Frequent p53 mutations in head and neck cancer. *Cancer Res.* **52**, 5997–6000.
243. Ibrahim, S. O., Vasstrand, E. N., Johannessen, A. C., Idris, A. M., Magnusson, B., Nilsen, R., and Lillehaug, J. R. (1999). Mutations of the p53 gene in oral squamous cell carcinomas from Sudanese dippers of nitrosamine-rich Toombak and non-snuff-dippers from the Sudan and Scandinavia. *Int. J. Cancer* **81**, 527–534.
244. Nylander, K., Schildt, E. B., Eriksson, M., Magnusson, A., Mehle, C., and Roos, G. (1996). A non-random deletion in the p53 gene in oral squamous cell carcinoma. *Br. J. Cancer* **73**, 1381–1386.
245. Lee, G., Hong, H.J., You, Y.-O., Bang, E.-H., Kook, J.-K., and Min, B.-M. (1998). Effect of p53 gene transfer on the cell proliferation and cell cycle progression in a human oral cancer cell line with p53 mutation. *Int. J. Oral Biol.* **23**, 189–199.
246. Mietz, J. A., Unger, T., Huibregtse, J. M., and Howley, P. M. (1992). The transcriptional transactivation function of wild-type p53 is inhibited by SV40 large T-antigen and by HPV-16 E6 oncoprotein. *EMBO J.* **11**, 5013–5020.
247. Hubbert, N. L., Sedman, S. A., and Schiller, J. T. (1992). Human papillomavirus type 16 E6 increases the degradation rate of p53 in human keratinocytes. *J. Virol.* **66**, 6237–6241.
248. Momand, J., Zambetti, G. P., Olson, D. C., George, D., and Levine, A. J. (1992). The mdm-2 oncogene product forms a complex with the p53 protein and inhibits p53-mediated transactivation. *Cell* **69**, 1237–1245.
249. Gujuluva, C., Sin, K. H., and Park, N. H. (1996). Role of HPV in tumorigenesis of oral keratinocytes: Implication of p53, $p21^{WAF1/CIP1}$, gadd45, cyclins, cyclin-dependent kinases and PCNA in oral cancer. *Int. J. Oncol.* **8**, 21–28.
250. Wang, Q., Lasset, C., Sobol, H., and Ozturk, M. (1996). Evidence of a hereditary p53 syndrome in cancer-prone families. *Int. J. Cancer* **65**, 554–557.
251. Birch, J. M., *et al.* (1994). Prevalence and diversity of constitutional mutations in the p53 gene among 21 Li-Fraumeni families. *Cancer Res.* **54**, 1298–1304.

252. Gallo, O., Sardi, I., Pepe, G., Franchi, A., Attanasio, M., Giusti, B. Bocciolini, C., and Abbate, R. (1999). Multiple primary tumors of the upper aerodigestive tract; is there a role for constitutional mutations in the p53 gene? *Int. J. Cancer* **82**, 180–186.
253. Stevens, H. P., Kelsell, D. P., Bryant, S. P., Bishop, D. T., Spurr, N. K., Weissenbach, J., Marger, D., Marger, R. S., and Leigh, I. M. (1996). Linkage of an American pedigree with palmoplantar keratoderma and malignancy (palmoplantar ectodermal dysplasia type 111) (17q24): Literature survey and proposed updated classification of the keratodermas. *Arch. Dermatol.* **132**, 640–651.
254. Lo Muzio, L., Mignogna, M. D., Pannone, G., Staibano, S., Procaccini, M., Serpico, R., DeRosa, G., and Scully, C. (1999). The NM23 gene and its expression in oral squamous cell carcinoma. *Oncol. Rep.* **6**, 747–751.
255. Rowley, H., Jones, A. S., and Field, J. K. (1995). Chromosome 18: A possible site for a tumour suppressor gene deletion in squamous cell carcinoma of the head and neck. *Clin. Otolaryngol.* **20**, 266–271.
256. Jones, J. W., *et al.* (1997). Frequent loss of heterozygosity on chromosome arm 18q in squamous cell carcinomas. *Arch. Otolaryngol. Head Neck Surg.* **123**, 610–614.
257. Watanabe, T., Wang, X. L., Miyakawa, A., Shiiba, M., Imai, Y., Sato, T., and Tanzawa, H. (1997). Mutational state of tumour suppressor genes (DCC, DPC4) and alteration on chromosome 18q21 in human oral cancer. *Int. J. Oncol.* **11**, 1287–1290.
258. Cowan, J. M., Beckett, M. A., Ahmed-Swan, S., and Weichselbaum, R. R. (1992). Cytogenetic evidence of the multistep origin of head and neck squamous cell carcinomas. *J. Natl. Cancer Inst.* **84**, 793–797.
259. Kim, S. K., Fan, Y., Papadimitrakopoulou, V., Clayman, G., Hittleman, W. N., Hong, W. K., Lotan, R., and Mao, L. (1996). DPC4, a candidate tumour suppressor gene, is altered infrequently in head and neck squamous cell carcinoma. *Cancer Res.* **56**, 2519–2521.
260. Yamamoto, N., Uzawa, K., Miya, T., Watanabe, T., Yokoe, H., Shibahara, T., Noma, H., and Tanzawa, H. (1999). Frequent allellic loss/imbalance on the long arm of chromosome 21 in oral cancer: Evidence for three discrete tumour suppressor gene loci. *Oncol. Rep.* **6**, 1223–1227.
261. Miyakawa, A., Wang, X. L., Nakanishi, H., Imai, F. L., Shiiba, M., Miya, T., Imai, Y., and Tanzawa, H. (1998). Allelic loss on chromsome 22 in oral cancer: Possibility of the existence of a tumour suppressor gene on 22q13. *Int. J. Oncol.* **13**, 705–709.
262. Yin, X. Y., Smith, M. L., Whiteside, T. L., Johnson, J. T., Herberman, R. B., and Locker, J. (1993). Abnormalities in the p53 gene in tumors and cell lines of human squamous cell carcinomas of the head and neck. *Int. J. Cancer* **54**, 322–327.
263. Hogmo, A., Munck-Wikland, E., Kuylenstierna, R., Lindholm, J., and Auer, G. (1994). Nuclear- DNA content and p53 immunostaining in oral squamous-cell carcinoma: An analysis of a consecutive 10 year material. *Int. J. Oncol.* **5**, 915–920.
264. Nakanishi, Y., Noguchi, M., Matsuno, Y., Saikawa, M., Mukai, K., Shimosato, Y., and Hirohashi, S. (1995). p53 expression in multicentric squamous cell carcinoma and surrounding squamous epithelium of the upper aerodigestive tract. *Immuno Cancer* **75**, 1657–1662.
265. Franceschi, S., Gloghini, A., Maestro, R., *et al.* (1995). Analysis of the p53 gene in relation to tobacco and alcohol in cancers of the upper aero-digestive tract. *Int. J. Cancer* **60**, 872–876.
266. Trizna, Z., and Schantz, S. P. (1992). Hereditary and environmental factors associated with risk and progression of head and neck cancer. *Otolaryngol. Clin. North Am.* **25**, 1089–1103.
267. Sun, S., Pollock, P. M., Liu, L., Karimi, S., Jothy, S., Milner, B. J., *et al.* (1997). CDKN2A mutation in a non-FAMMM kindred with cancers at multiple sites results in a functionally abnormal protein. *Int. J. Cancer* **73**, 531–536.
268. Yarbrough, W. G., Apelikova, O., Pei, H., Olshan, A. F., and Liu, E. T. (1996). Familial tumour syndrome associated with a germline nonfunctional p16INK4a allele. *J. Natl. Cancer Inst.* **88**, 1489–1490.
269. Cairns, P., and Sidransky, D. (1999). Molecular methods for the diagnosis of cancer. *Biochim. Biophys. Acta* **1423**, C11–C18.
270. Welkoborsky, H. J., Hinni, M., Dienes, H. P., and Mann, W. J. (1995). Potential early markers of carcinogenesis in the mucosa of the head and neck using exfoliative cytology. *Ann. Otol. Rhinol. Laryngol.* **104**, 503–510.
271. Huang, M. F., Chang, Y. C., Liao, P. S., Huang, T. H., Tsay, C. H., and Chou, M. Y. (1999). Loss of heterozygosity of p53 gene of oral cancer detected by exfoliative cytology. *Oral Oncol.* **35**, 296–301.
272. Boyle, J. O., Mao, L., Brennan, J. A., Koch, W. M., Eisele, D. W., Saunders, J. R., and Sidransky, D. (1994). Gene mutations in saliva as molecular markers for head and neck squamous cell carcinomas. *Am. J. Surg.* **168**, 429–432.
273. Rosin, M. P., Epstein, J. B., Berean, K., Durham, S., Hay, J., Cheng, X., Zeng, T., Huang, Y., and Zhang, L. (1997). The use of exfoliative cell samples to map clonal genetic alterations in the oral epithelium of high-risk patients. *Cancer Res.* **57**, 5258–5260.
274. Lavieille, J. P., Lubin, R., Soussi, T., Reyt, E., Brambilla, C., and Riva, C. (1996). Analysis of p53 antibody response in patients with squamous cell carcinomas of the head and neck. *Anticancer Res.* **16**, 2385–2388.
275. Bourhis, J., Bosq, J., Wilson, G. D., Bressac, B., Talbot, M., Leridant, A. M., Dendale, R., Janin, N., Armand, J. P., Luboinski, B., Malaise, E. P., Wibault, P., and Eschwege, F. (1994). Correlation between p53 gene expression and tumor-cell proliferation in oropharyngeal cancer. *Int. J. Cancer* **57**, 458–462.
276. Friedrich, R. E., Bartel-Friedrich, S., Plambeck, K., Bahlo, M., and Klapdor, R. (1997). p53 auto-antibodies in the sera of patients with oral squamous cell carcinoma. *Anticancer Res.* **17**, 3183–3184.
277. Maass, J. D., Gottschlich, S., Goeroegh, T., Lippert, B. M., and Werner, J. A. (1997). Head and neck cancer and p53-immunogenicity. *Anticancer Res.* **17**, 2873–2874.
278. Tavassoli, M., Brunel, N., Maher, R., Johnson, N. W., and Soussi, T. (1998). p53 antibodies in the saliva of patients with squamous cell carcinoma of the oral cavity. *Int. J. Cancer* **78**, 390–391.
279. Brennan, J. A., Mao, L., Hruban, R. H., Boyle, J.O. *et al.* (1995). Molecular assessment of histopathological staging in squamous cell carcinoma of the head and neck. *N. Engl. J. Med.* **332**, 429–435.
280. Bockmuhl, U., Schluns, K., Kuchler, I., Petersen, S., and Petersen, I. (2000). Genetic imbalances with impact on survival in head and neck cancer patients. *Am. J. Pathol.* **157**, 369–375.
281. Partridge, M., Emilion, G., and Langdon, J.D. (1996). LOH at 3p correlates with a poor survival in oral squamous cell carcinoma. *Br. J. Cancer* **73**, 366–371.
282. Partridge, M., Emilion, G., Pateromichelakis, S., A'Hern, R., Lee, G., Phillips, E., and Langdon, J. (1999). The prognostic significance of allelic imbalance at key chromosomal loci in oral cancer. *Br. J. Cancer* **79**, 1821–1827.
283. Yeudall, W. A., Crawford, R. Y., Ensley, J. D., and Robbins, K. C. (1994). MTS1/CDK4I is altered in cell lines derived from primary and metastatic oral squamous cell carcinoma. *Carcinogenesis* **15**, 2683–2686.
284. Danahey, D. G., Tobin, E. J., Schuller, D. E., Bier-Laning, C. M., Weghorst, C. M., and Lang, J. C. (1999). p16 mutation frequency and clinical correlation in head and neck cancer. *Acta Otolaryngol* **119**, 285–288.
285. Nagai, M. A., Miracca, E. C., Yamamoto, L., Kowalski, L. P., and Brentani, R. R. (1995). TP53 mutations in upper aerodigestive squamous cell carciomas from a group of Brazilian patients. *Am. J. Surg.* **170**, 492–494.
286. Gasparini, G., Weidnere, N., Maluta, S., *et al.* (1993). Intratumoral microvessel density and p53 protein: Correlation with metastasis in head and neck squamous cells carcinoma. *Int. J. Cancer* **55**, 739–744.
287. Osaki, T., Kimura, T., Tatemoto, Y., Dapeng, L., Yoneda, K., and Yamamoto, T. (2000). Diffuse mode of tumor cell invasion and

expression of mutant p53 protein but not of p21 protein are correlated with treatment failure in oral carcinomas and their metatstatic foci. *Oncology* **59**, 36–43.

288. Field, J. K., Malliri, A., Butt, S. A., Gosney, J. R., Phillips, D., Spandidos, D. A., *et al.* (1993). p53 over-expression in end-stage squamous-cell carcinoma of the head and neck prior to chemotherapy treatment-expression correlates with a very poor clinical outcome. *Int. J. Oncol.* **3**, 431–435.
289. Gluckman, J. L., Stambrook, P. J., and Pavelic, Z. P. (1994). Prognostic significance of p53 protein accumulation in early stage T1 oral cavity cancer. *Eur. J. Cancer B Oral. Oncol.* **30B**, 281.
290. Warnakulasuriya, K. A., and Johnson, N. W. (1994). Association of overexpression of p53 oncoprotein with the state of cell proliferation in oral carcinoma. *J. Oral Pathol. Med.* **23**, 246–250.
291. Shin, D. M., Lee, J. S., Lippman, S. M., Lee, J. J., Tu, Z. N., Choi, G., *et al.* (1996). p53 expression: Predicting recurrence and second primary tumors in head and neck squamous cell carcinoma. *J. Natl. Cancer Inst.* **88**, 519–529.
292. Pavelic, Z. P., Gluckman, J. L., Gapany, M., Reising, J., Craven, J. M., Kelley, D. J., Pavelic, L., Gapany, S., Biddinger, P., and Strambrook, P. J. (1992). Improved immunohistochemical detection of p53 protein in paraffin-embedded tissues reveals elevated levels in most head and neck and lung carcinomas: Correlation of clinicopathological parameters. *Anticancer Res.* **12**, 1389–1394.
293. Nishioka, H., Hiasa, Y., Hayashi, I., Kitahori, Y., Konishi, N., and Sugimura, M. (1993). Immunohistochemical detection of p53 oncoprotein in human oral squamous cell carcinomas and leukoplakias: Comparison with proliferating cell nuclear antigen staining and correlation with clinicopathological findings. *Oncology* **50**, 426–429.
294. Sauter, E. R., Ridge, J. A., Gordon, J., and Elsenberg, B. L. (1992). p53 over-expression correlates with increased survival in patients with squamous carcinoma of the tongue base. *Am. J. Surg.* **164**, 651–653.
295. Gapany, M., Pavelic, Z. P., Gapany, S. R., *et al.* (1993). Relationship between immunohistochemically detectable p53 protein and prognostic factors in head and neck tumors. *Cancer Detect. Prevent.* **17**, 379–386.
296. Leedy, D. A., Trune, D. R., Kronz, J. D., Weidner, N., and Cohen, J. I. (1994). Tumor angiogenesis, the p53 antigen and cervical metastasis in squamous cell carcinoma of the tongue. *Otolaryngol. Head Neck Surg.* **111**, 417–422.
297. Mukhopadhyay, D., Chatterjee, R., and Chakraborty, R. N. (1994). Association of p53 expression with cytokinetics and HPV capsid antigen prevalence in oral carcinomas. *Cancer Lett.* **87**, 99–105.
298. Schipper, J. H., and Kelker, W. (1994). Die expression der tumorsuppressor-gene p53 und Rb bei patienten mit plattenepihtelkarzinomen im Kopf/Halsbereich. *HNO* **42**, 270–274.
299. Ahomadegbe, J. C., Barrois, M., Fogel, S. *et al.* (1995). High incidence of p53 alterations (mutation, deletion, overexpression) in head and neck primary tumors and metastases; absence of correlation with clinical outcome. Frequent protein overexpression in normal epithelium and in early non-invasive lesions. *Oncogene* **10**, 1217–1227.
300. Nylander, K., Stenling, R., Gustafsson, H., Zackrisson, B., and Roos, G. (1995). p53 expression and cell proliferation in squamous cell carcinomas of the head and neck. *Cancer* **75**, 87–93.
301. Stoll, C., Baretton, G., and Lohrs, U. (1998). The influence of p53 and associated factors on the outcome of patients with oral squamous cell carcinoma. *Virch. Arch.* **433**, 427–433.
302. Dahse, R., Fiedler, W., von Eggeling, F., Schimmel, B., Koscielny, S., Beleites, E., Claussen, U., and Ernst, G. (1999). p53 genotyping: An effective concept for molecular testing of head and neck cancer? *Int. J. Mol. Med.* **4**, 279–283.
303. Ma, L., Ronai, A., Riede, U. N., and Kohler, G. (1998). Clinical implications of screening p53 gene mutations in head and neck squamous cell carcinomas. *J. Cancer Res. Clin. Oncol.* **124**, 389–396.
304. Mineta, H., Borg, A., Dictor, M., Wahlberg, P., Akervall, J., and Wennerberg, J. (1998). p53 mutation, but not p53 overexpression, correlates with survival in head and neck squamous cell carcinoma. *Br. J. Cancer* **78**, 1084–1090.
305. Gleich, L. L., Li, Y. Q., Wang, X., Stambrook, P. J., and Gluckman, J. L. (1999). Variable genetic alterations and survival in head and neck cancer. *Arch. Otolaryngol. Head Neck Surg.* **125**, 949–952.
306. Vogelstein, B., Fearon, E. R., Kern, S. E., Hamilton, S. R., Preisingerm, A. C., Nakamura, Y., and White, R. (1989). Allelotype of colorectal carcinomas. *Science* **244**, 207–211.
307. Partridge, M., Pateromichelakis, S., Phillips, E., Emilion, G. G., A'Hern, R. P., and Langdon, J. D. (2000). A case-control study confirms that microsatelite assay can identify patients at risk of developing oral squamous cell carcinoma within a field of cancerization. *Cancer Res.* **60**, 3893–3898.
308. Scholes, A. J., Woolgar, J. A., Boyle, M. A., Brown, J. S., Vaughan, E. D., Hart, C. A., Jones, A. S., and Field, J. K. (1998). Synchronous oral carcinomas: Independent or common clonal origin? *Cancer Res.* **58**, 2003–2006.
309. Califano, J., Westra, W. H., Koch, W., Meininger, G., Reed, A., Yip, L., Boyle, J. O., Lonardo, F., Sidransky, D. (1999). Unknown primary head and neck squamous cell carcinoma; molecular identification of the site of origin. *J. Natl. Cancer Inst.* **91**, 599–604.
310. Leong, P. P., Rezai, B., Koch, W. M., Reed, A., Eisele, D., Lee, D. J., Sidransky, D., Jen, J., and Westra, W. H. (1998). Distinguishing second primary tumors from lung metastases in patients with head and neck squamous cell carcinoma. *J. Natl. Cancer Inst.* **90**, 972–977.

CHAPTER

10

Proteolysis in Carcinogenesis

THOMAS H. BUGGE

Oral and Pharyngeal Cancer Branch
National Institute of Dental and Craniofacial Research
National Institute of Health
Bethesda, Maryland 20892

I. Tumor-Associated Protease Systems 137
A. The Plasminogen Activation System 137
B. Matrix Metalloproteinases, ADAMs, and ADAMTS 138
C. Lysosomal Proteases: Cathepsins 139
D. The Coagulation Cascade 139
E. "Novel" Tumor-Associated Serine Proteases 139
II. Molecular Targets for Proteases in Tumor Progression 140
A. Zymogen Activation 140
B. Degradation of Tissue Barriers 141
C. Generation of a Provisional Matrix for Tumor Cell Survival, Adhesion, and Proliferation 142
D. Proteases as Regulators of Tumor Cell Growth, Survival, and Motility 142
III. Extracellular Proteolysis in Head and Neck Cancer Progression 143
References 144

Proteases have attracted substantial attention as potential targets for cancer therapy, including the treatment of head and neck cancer. This chapter gives an overview of the variety of secreted and pericellular proteases that have been associated with the dissemination of human tumors and reviews data that link these proteases to tumor progression. It describes the work carried out to unravel the complex functions of proteases in malignancy, to identify their relevant molecular targets, and describes the broadening concepts regarding the roles of pericellular proteolysis in tumor biology.

I. TUMOR-ASSOCIATED PROTEASE SYSTEMS

A. The Plasminogen Activation System

The association of the plasminogen activation (PA) system with tumor progression is particularly well documented and is the subject of long-standing investigation [1–3]. The PA system is a complex system of serine proteases, protease inhibitors, and protease receptors that governs the conversion of the abundant protease zymogen, plasminogen (Plg), to the active serine protease, plasmin. Plg is produced predominantly by the liver and is present in high concentration in plasma and most extravascular fluids [1]. Plasmin is formed by the proteolytic cleavage of Plg by either of two Plg activators, the urokinase Plg activator (uPA) and the tissue Plg activator (tPA), of which uPA, until today, has attracted the most attention in the context of tumor progression. uPA is a 52-kDa serine protease that is secreted as an inactive single chain proenzyme (pro-uPA) and is efficiently converted to active two-chain uPA by plasmin, when bound to its cellular receptor, the uPA receptor (uPAR) [1,4]. Two-chain uPA, in turn, is a potent activator of Plg. Both pro-uPA and Plg are catalytically inactive proenzymes, and the mechanism of initiation of uPA-mediated Plg activation is not fully understood (Section II,A). Plg activation by uPA is regulated by two physiological inhibitors, Plg activator inhibitor-1 and -2 (PAI-1 and PAI-2), each forming a 1:1 complex with uPA [5–7]. Plasmin generated by the cell surface PA system is relatively protected from its primary physiological inhibitor α_2-antiplasmin [4,8,9], leading to the suggestion that the plasmin-mediated degradation of extracellular matrix (ECM) is strictly associated with the cell surface.

Extensive direct and indirect evidence has accumulated over more than two decades for the involvement of the PA system in proteolysis associated with tumor invasion and metastasis [1,2,3,10]. Histological analysis of both human tumors and experimental mouse tumors have consistently demonstrated expression of the components of the PA system specifically at sites of active invasion [11–14]. uPAR and uPA are overexpressed with remarkable consistency in malignant human tumors, including cancers of the head and neck [15], colon [12], breast [14, 16], bladder [17], liver [18], lung [19], pancreas [20], and ovaries [21], as well as both monocytic and myelogenous leukemias [22,23]. *In situ* hybridization and immunohistochemical studies of various human tumors have also demonstrated that cancer cells typically express uPAR, whereas pro-uPA may be expressed by either cancer cells or adjacent stromal cells [12,24,25]. Furthermore, Plg activators, Plg activator receptor, and inhibitors are all excellent prognostic markers in human cancer [2,26]. Thus, extensive epidemiological studies have revealed that high expression of Plg activator, uPAR, or PAI-1 in human breast [27–29], colorectal [30], lung [31,32], prostate [33], and gastric cancer [34,35] is associated with a poor prognosis.

A direct causal relationship between Plg activator expression and tumor progression has also been demonstrated in a number of studies using the chick chorioallantoic membrane metastasis model and the mouse spontaneous and experimental metastasis models. These studies have employed a variety of strategies to inhibit Plg activator function, including antisense RNA techniques [36,37], anticatalytic PA antibodies [38–42], natural and synthetic PA inhibitors or uPAR antagonists [43–47], and, lately, genetically engineered mice with targeted deficiencies in components of the PA system [48–50]. Collectively, these studies have firmly established the PA system as an important contributor to the dissemination of malignant tumors.

B. Matrix Metalloproteinases, ADAMs, and ADAMTS

The matrix metalloproteinase (MMP) family is a large group of secreted proteases that require zinc for catalytic activity [51]. This family is expanding rapidly but currently has approximately 26 members [52,53]. MMPs can be organized by their domain structure into five distinct classes. All MMPs contain an N-terminal signal peptide, a prodomain conferring latency to the enzyme, and a catalytic domain. The smallest members of the MMP family, the matrilysins (MMP-7 and MMP-26), contain only these domains. Collagenase-1 (MMP-1), collagenase-3 (MMP-13), neutrophil collagenase (MMP-8), stromelysin-1 (MMP-3), stromelysin-2 (MMP-10), and macrophage metalloelastase (MMP-12) contain an additional proline-rich hinge region and a hemopexin domain that is involved in substrate binding. Stromelysin-3 (MMP-11) contains, in addition to these domains, a short furin activation sequence immediately following the prodomain. The two gelatinases, gelatinase A (MMP-2) and gelatinase B (MMP-9), also contain an additional fibronectin-like domain required for the binding of the two enzymes to gelatin. Four members of the MMP family, membrane type (MT)1-MMP (MMP-14), MT2-MMP (MMP-15), MT3-MMP (MMP-16), and MT5-MMP (MMP-24), are type I transmembrane proteins attached to the cell surface via a C-terminal membrane-spanning domain. Two MMPs, MT4-MMP (MMP-17) and MT6-MMP (MMP-25), are associated with the cell surface by a glycosylphosphatidylinositol anchor, and one member (MMP-23) is a type II transmembrane protein attached to the cell via an N-terminal signal anchor [52,53]. All membrane-bound MMPs contain the furin activation sequence originally identified in stromelysin-3, suggesting that they are secreted in a catalytically active form. The principal inhibitors of MMPs are the widely expressed tissue inhibitors of metalloproteinases (TIMPs) that currently comprise four members (TIMP-1 through 4). These inhibitors form a noncovalent 1:1 complex with activated MMPs. Each TIMP is generally capable of inhibiting most MMPs, although some specificity of inhibition has been observed [52,54].

Closely related to MMPs are the more recently identified families of ADAMs and ADAMTS proteases. ADAMs proteases are all cell surface type I transmembrane multidomain proteins that have N-terminal pro- and metalloprotease domains similar to MMPs, followed by disintegrin, cysteine-rich, EGF-like, transmembrane, and cytoplasmic domains [55]. ADAMTS are secreted proteins containing pro- and metalloprotease domains followed by thrombospondin type I motifs [56].

The MMP family was first linked to tissue remodeling during tumor progression by their consistent upregulation in tumor tissue [review in 52,57]. The association of MMPs with cancer was further strengthened by the demonstration that the level of MMPs and TIMPs in tumors correlates with clinical outcome [58–63]. More recently, the causal relation between MMP expression and tumor progression was demonstrated directly by the use of highly specific MMP inhibitors and by the use of transgenic mice with specific MMP deficiencies or overexpression of MMPs [64–67]. Few studies have so far directly addressed the functions of ADAM and ADAMTS proteases in cancer progression.

However, the limited insight gained so far into the biological functions of these protease families strongly suggests that members of both protease families will also be found to be important modulators of tumor progression. ADAM proteases are localized on the cell surface, and at least one member (ADAM-17) has "sheddase" activity and is required for the proteolytic liberation of growth factors from membrane-bound precursors and the inactivation of growth factor receptors by the shedding of their ectodomains from the cell surface (Section II,D,1) [68,69]. ADAMTS are deposited in the ECM and appear to be involved in the maturation, modification, and degradation of components of the ECM during development and homeostasis [70].

C. Lysosomal Proteases: Cathepsins

Cathepsins are a group of lysosomal cysteine (cathepsins B, H, L, S, C, S, K, O, F, V, X, and W) or aspartic proteases (cathepsin D) that are involved predominantly in phagolysosomal protein degradation. Lysosomal cysteine proteinases are papain-like enzymes with a molecular mass of 20–35 kDa [71]. Cathepsins are synthesized as proenzymes that undergo proteolytic activation in the acidic environment of the late endosome or lysosome. Most of the cathepsins are expressed ubiquitously (cathepsins B, D, L, S, C, O, F, and X), but a few demonstrate strict tissue specificity, such as cathepsin K, which is expressed exclusively by osteoclasts, cathepsin V, which is expressed in testes and thymus, and cathepsin W, which is expressed in a subset of T lymphocytes [71,72]. The principal inhibitors of cysteine cathepsins are cystatins, including stefins, cystatins, and kininogens [71].

The altered expression of lysosomal proteases during malignant progression was first observed for cathepsin B in human breast cancer [73]. Since then, changes in the localization, level of expression, or activity of cathepsin B, H, and L and their inhibitors, cystatin C, stefin A, and stefin B, have been documented in a variety of other human tumors, including head and neck, lung, ovary, brain, colorectal, gastric, bladder, and melanoma. Furthermore, cathepsin and cathepsin inhibitors are prognostic factors, with high expression levels in tumors and biological fluids being associated with poor survival [71,72,74,75].

The association between tumor progression and altered expression of the lysosomal cathepsins and their inhibitors was established early. Nevertheless, the possibility of a direct involvement of cathepsins in tumor-mediated ECM degradation remained controversial due to the presumed exclusively intracellular location of these proteases and their requirement of acidic pH for function. The realization that phagolysosomal ECM degradation may represent a rate-limiting step in cell migration [76–78], that cathepsins can redistribute from the lysosome to the cell surface during malignant progression [74,79–81], and that cathepsin K is a highly efficient fibrillar collagenase during osteoclast-mediated bone resorption [82,83] has strengthened the notion of cathepsins as critical contributors to tumor progression. Experiments using transgenic mice with an altered expression of cathepsins to unravel their specific functions in tumor progression are now in progress [84].

D. The Coagulation Cascade

Vascular patency is governed by the coagulation cascade, a sophisticated system of circulating serine proteases and cell-associated receptors. The central step in the coagulation pathway is generation of the serine protease thrombin from the inactive protease zymogen prothrombin by proteolytic cleavage by factor Xa. Under physiological conditions, thrombin formation is triggered by vascular injury, facilitating the contact of factor VII with a cellular receptor, tissue factor (TF), and the sequential proteolytic activation of coagulation factors VII to VIIa, X to Xa, and V to Va [85]. Thrombin has pleiotropic functions mediated through the cleavage and activation of other serine proteinases, transglutaminase, and thrombin activated, G-protein coupled receptors. Most tumor cells, as well as nonneoplastic stromal cells, possess strong procoagulant activities that promote the local formation of thrombin in the absence of vascular injury [86,87], and many tumor cells and tumor stromal cells are equipped with thrombin-activated receptors. The causal relation between the activation of the coagulation cascade and tumor progression is now well established. For example, pretreatment of tumor cells or systemic treatment of experimental animals with a variety of highly specific inhibitors of TF, factor Xa, or thrombin dramatically reduces pulmonary metastasis of a wide range of tumor types [88–93]. The specific mechanisms by which the coagulation cascade promotes tumor dissemination is currently being elucidated (Section II,C).

E. "Novel" Tumor-Associated Serine Proteases

In addition to the well-characterized protease families described in Sections II,A–II,D, a rapidly expanding group of recently identified secreted or cell surface-associated serine proteases, of mostly unknown physiological function, are gathering increasing attention in the context of tumor progression. As was the case with the discovery of several MMPs, some of these proteases were originally identified as a direct consequence of their overexpression by human tumors or tumor cell lines [94–96], and their specific association with tumor progression and their value as prognostic factors are currently under evaluation. This heteroge neous group of novel tumor-associated proteases currently includes the secreted proteases neuropsin (also known as ovasin, bsp1, TAGD-14 and KLK-8) [97–100] and neurosin (also known as zyme, protease M, BSP, BSSP, and myelencephalon-specific protease) [101–106], the stratum corneum chymotryptic enzyme [107], the type II transmembrane proteases hepsin [108], TMPRSS2 [109], TMPRSS3 [96], MT-SP1 (also known as ST-14, matriptase, epithin, TAGD-15, and prostamin) [94,110–113], and the type I transmembrane proteases testisin (also known as eosinophil serine protease) [114,115] and fibroblast activation protein (FAP, also known as seprase) [116,117]. Neuropsin and neurosin, named after their preferential expression in the central nervous system, are overexpressed dramatically in ovarian cancer [94,95,99,118]. Indeed, neurosin has been reported to be a reliable new serum biomarker of ovarian carcinoma [118]. Neurosin is also expressed in cultured breast cancer lines [103] and neuropsin in squamous cell carcinoma (K. List and T. Bugge, unpublished data). Ovarian cancer cells express high levels of testisin and stratum corneum chymotryptic enzyme, a protease whose normal expression is restricted to the cornifying layer of the skin [119,120]. Hepsin has been shown to be overexpressed with remarkable consistency in human prostate cancer, and high levels of hepsin correlated with poor survival [121–124]. Hepsin is also overexpressed in renal carcinoma cells and in ovarian cancer, a tumor speculated to disseminate in the abdomen through activation of the coagulation cascade and deposition of an adhesive fibrin matrix on the peritoneal wall [125]. In this respect, it is interesting to note that hepsin is an activator of factor VII and can initiate the extrinsic pathway of blood coagulation on the cell surface, leading to the deposition of peritumoral fibrin [126]. TMPRSS3 is overexpressed in pancreatic cancer [96], and FAP is highly expressed by tumor cells or in the stromal fibroblasts of a wide variety of human tumors, including tumors of the bone, breast, colorectal, lung, melanoma, and soft tissue sarcomas [127–129].

Overexpression of the type II transmembrane serine proteases, hepsin, TMPRSS3, and MT-SP1 by human tumors is particularly noteworthy. The prototypic member of this class of proteases, enteropeptidase, is the physiological activator of trypsin in the digestive tract, and in this capacity is the initiator of a proteolytic cascade reaction that leads to the activation of procarboxypeptidase, chymotrypsinogen, and proelastase [130,131]. The novel type II transmembrane serine proteases could be speculated to have similar functions in tumor progression. Thus, MT-SP1, which is expressed in a number of epithelial tumors [132], (T. Bugge and K. List, unpublished data), has been demonstrated to be a highly efficient activator of pro-uPA and, thus, to be a candidate initiator of the PA cascade (Section II,A) [133,134]. Future studies will undoubtedly rapidly expand our knowledge of the expression and function of these novel serine proteases in tumor progression.

II. MOLECULAR TARGETS FOR PROTEASES IN TUMOR PROGRESSION

A. Zymogen Activation

The majority of extracellular and membrane-associated proteases are synthesized as catalytically inactive proenzymes (zymogens) that are activated by the endoproteolytic cleavage of one ore more peptide bonds. *In vivo*, these protease zymogens are organized into complex hierarchies that undergo sequential proteolytic activation. These cascade reactions eventually lead to the activation of the "effector" proteases that modify the extracellular environment to promote tumor progression (Sections II,B–II,D). Thus, the specific molecular targets for many, if not most, proteases relevant for tumor dissemination are likely to be other protease zymogens. Elucidation of the complex pathways of activation of functionally important proteolytic cascades, such as the PA system, MMPs, and the coagulation cascade during tumor progression, is an area of long-standing investigation that is seminal to the understanding of proteases in tumor biology, and the design of therapeutic strategies aimed at protease inhibition.

Nevertheless, many aspects of protease activation are still poorly understood. The principal causes are the considerable technical difficulties associated with studying protease activation *in vivo*, the uncertainty as to the interpretation of data obtained with purified components *in vitro*, and the apparent existence of multiple context and cell type-specific pathways of zymogen activation. However, several lines of evidence have converged to suggest that the initiation of proteolytic cascades in both neoplastic and nonneoplastic processes is predominantly a cell surface-associated event and involves the binding of one or more protease zymogens to specific high-affinity cell surface receptors [10,85,135]. On the cell surface, protease zymogens are in a spatially favorable orientation for activation and are often protected against specific soluble inhibitors. In the PA system, one cascade-initiating pathway is believed to involve the concomitant binding of pro-uPA to its high-affinity cell surface receptor, uPAR, and of Plg to as yet uncharacterized cell surface receptors. The concomitant binding of pro-uPA and Plg to the cell surface strongly potentiates uPA-mediated Plg activation and plasmin-mediated activation of pro-uPA, leading to a powerful feedback loop that results in productive plasmin formation [4,10]. How this feedback loop is initiated is unclear, but may include the activation of pro-uPA by other proteases besides plasmin or the activation of Plg by a very low intrinsic activity of pro-uPA. The transmembrane serine protease MT-SP1, true tissue kallikrein, hepatocyte growth factor activator, and cathepsin B are all candidate initiators of the pro-uPA cascade [133,134,136–138]. Activation of the MMP family is also increasingly believed to be cell surface dependent. MMP-2, a collagenase that is associated intimately with tumor angiogenesis, is activated on the cell surface by the membrane-anchored MMP-14 in a process that requires juxtaposition of the two proteases by TIMP-2 [139]. MMP-14 has also been proposed to be involved in the activation of several other members of the MMP family on the cell surface [140]. Potential interactions that lead to the localization of MMPs to the cell surface include MMP-1 binding to integrin $\alpha_2\beta_1$, MMP-13 binding to urokinase Plg activator receptor-associated protein/Endo 180, and MMP-9 binding to CD44 [141–142a]. Generation of the metastasis-promoting protease thrombin is also tightly linked to the surface of tumor (or tumor stromal) cells and is initiated by the binding of coagulation factor VII to its high-affinity cell surface receptor, TF, leading to the sequential proteolytic activation of coagulation factors VII to VIIa, X to Xa, and V to Va, and prothrombin to thrombin (Section I,D). Tumor cells may also utilize alternative cell surface-associated pathways for thrombin generation, including the direct cleavage and activation of factor X by cancer procoagulant, a membrane-associated, vitamin K-dependent cysteine protease [86,143,144], or activation of factor VII by hepsin, a novel transmembrane serine protease (Section I,E) [126].

MMPs, cathepsins, Plg activators, and related serine proteases can mutually activate each others zymogens *in vitro*. The extent to which such "cross talk" is taking place *in vivo* is still not entirely understood. Many studies have focused on the role of plasmin in the activation of MMPs, and numerous experiments have demonstrated the participation of plasmin in the activation of proMMP-1, -2, -3, -9, -10, -12, and –13 *in vitro* [145]. A less clear picture has emerged from cell-based assays. Plasmin has been reported to be both a poor activator and a very efficient activator of MMP-9 [146–148]. The genetic deficiency of uPA abrogates plasmin-mediated MMP-13 activation by cultured macrophages [149]. MMP-13 can also be activated by plasmin in fibroblast cultures, but MMP-13 activated this way is degraded rapidly [150]. In contrast, keratinocytes appear to be completely dependent of plasmin for the activation of MMP-13 and subsequent degradation of fibrillar collagen, strongly implicating the PA system in MMP activation (S. Netzel-Arnett, D. Mitola, H. Birkedal Hansen, and T. Bugge, unpublished data). Other serine proteases may also participate in MMP activation. Genetic studies have shown that MMP-2 is activated by the mast cell-specific serine protease, chymase, during human papillomavirus-16 E6/E7-induced squamous carcinoma, providing direct evidence for a pivotal role of serine proteinases in MMP activation during tumor progression [151].

B. Degradation of Tissue Barriers

In order for a solid tumor to metastasize successfully to distant organs, it must be able to degrade, or trigger, the degradation of two basement membranes during the steps of extravasation and intravasation. The tumor must also be able to transverse a complex meshwork of interstitial matrix that is rich in cross-linked glycoproteins. In addition to having the capacity to negotiate the typical array of basement membrane and interstitial matrix glycoproteins (e.g., laminin, fibronectin, entactin, and collagen), tumors must also be able to degrade a matrix highly abundant in cross-linked fibrin, which is generated by the strong procoagulant activity inherent to tumor cells and by the vascular damage caused by tissue invasion (Section II,C). As has been recognized for more than half a century, it is largely inconceivable that these tasks can be accomplished successfully in the absence of appropriate matrix degrading proteases [1, 152, and references therein]. Intense efforts have been made to identify the critical matrix-degrading proteases in this process due to the obvious clinical importance of confining tumors to their site of origin. As described in Section I, solid tumors overexpress a wide variety of secreted or cell-associated proteases, most of which are endowed with the capacity to degrade at least some components of the basement membrane and interstitial matrix *in vitro*. The development of efficient, specific, and bioavailable synthetic inhibitors of matrix-degrading proteases, as well as several transgenic mouse lines deficient in matrix-degrading proteases, has provided novel powerful tools to directly assess the contribution of individual matrix-degrading proteases to cancer invasion and metastasis *in vivo*. Preliminary studies with transgenic mice and synthetic protease inhibitors have largely confirmed the involvement of matrix-degrading proteases in tumor dissemination [48–50,64,67]. Surprisingly, however, demonstration of an absolute requirement of a specific tumor-associated protease, or even a protease family, for basement membrane and interstitial matrix degradation during tumor invasion *in vivo* has proven to be remarkably difficult. Negating protease activity to transplanted or genetically induced tumors by gene targeting or through pharmacological inhibition retards, but does not prevent, invasion and metastasis of the tumor to distant sites and, by inference, does not completely abolish basement membrane and interstitial matrix degradation [49,50,64,67]. For example, polyomavirus middle T-induced mammary adenocarcinoma expresses high amounts of Plg activator at the invasive edge, but retains some of its metastatic capacity even in mice with a complete deficiency in Plg. Likewise, a transplanted Lewis lung carcinoma expresses high levels of several MMPs and Plg activator, but is very metastatic both in Plg-deficient mice and in mice treated with a broad- spectrum MMP inhibitor [49,64]. Furthermore, MMP-9-deficient mice have a reduced incidence of papillomavirus-induced squamous carcinoma, but the few tumors ensuing in MMP-9-deficient mice are, if anything, more, rather than less, invasive than MMP-9 sufficient tumors [67]. There may be several possible explanations for these somewhat surprising findings, all of which are not mutually exclusive. (a) Tumor cells, or tumor-associated stromal cells, may be endowed with several parallel proteolytic pathways, each being independently capable of efficiently dissolving basement membrane and interstitial matrix. Inhibition of just one, or a few, of these pathways will be insufficient to prevent tumor-associated matrix dissolution and metastatic dissemination [153]. (b) Cancer invasion and metastasis involve multiple, discrete steps, such as intravasation, arrest, extravasation, growth, and development of a supporting neovasculature. The dissolution of basement membrane and interstitial matrix degradation, although essential to the metastatic process, may simply not be a particularly rate-limiting step in the complex cascade of events that precedes final metastatic dissemination. Thus, even a profound reduction in the rate of dissolution of a basement membrane achieved by gene knockouts or by pharmacological protease inhibition may not manifest itself in a dramatic reduction in distant metastasis. (c) Given the complex, multistep nature of metastatic process, it is also conceivable that nonspecific, or even specific, protease inhibition may impair some of the steps of the metastatic pathway, while accelerating others, with a modest reduction in distant metastasis representing the net result of systemic protease inhibition, masking even a profound effect of protease ablation on matrix degradation.

C. Generation of a Provisional Matrix for Tumor Cell Survival, Adhesion, and Proliferation

The interaction of tumor cells with ECM is essential for adhesion, migration, invasion, and survival. Most, if not all, solid tumors directly induce the formation of a provisional extracellular fibrin-rich matrix at the site of active invasion. This is accomplished through activation of the coagulation cascade, which ultimately leads to the conversion of fibrinogen to cross-linked fibrin by the action of thrombin and factor XIII (Section I,D). The TF/factor VII pathway appears to be principally responsible for the activation of coagulation, but other tumor cell surface proteases, such as hepsin and cancer procoagulant, may also trigger coagulation by the direct cleavage of coagulation factors VII and X, respectively (Section II,A). Fibrin and its various proteolytic derivatives have been shown to display an extraordinary range of biological activities, strongly implying a pivotal role of provisional fibrin matrices in tumor biology. Fibrin matrices promote the migration of both tumor cells and tumor stromal cells, such as endothelial cells, macrophages, and fibroblasts [87,154–156]. Tumor and stromal cells interact directly with fibrinogen and fibrin via $\alpha v\beta 3$ and $\alpha m\beta 2$ integrin receptors, as well as nonintegrin receptors of the ICAM family [154,157–159]. Furthermore, highly purified fibrin gels can induce neovascularization directly when implanted into experimental animals [160,161], and fibrin degradation products, released during fibrin turnover, have been shown to display powerful chemotactic, immune-modulatory, as well as angiogenic, properties [161a–164].

The ability of tumor cells to induce the rapid formation of a provisional fibrin matrix also appears to play a critical role during hematogenous metastasis. The bloodstream is a particularly inhospitable environment for tumor cells, where they are subject to destruction by shear stress, anoikis, and immune-mediated killing. However, tumor cells form fibrin-rich platelet–tumor cell aggregates almost immediately after their initial adhesion to the microvascular endothelium of their target organs through the local activation of the coagulation cascade [165–167]. Multiple studies have converged to demonstrate that tumor cell survival and subsequent metastasis formation are critically dependent on this process and that specific inhibitors of TF, factor Xa, or thrombin are among the most powerful antimetastatic agents known [88–93].

D. Proteases as Regulators of Tumor Cell Growth, Survival, and Motility

Traditionally, tumor-associated proteolysis has been studied almost exclusively in the context of degradation of extracellular matrix barriers due to the obvious necessity of such events taking place during invasion and metastasis. During the last decades, however, it has been gradually realized that pericellular proteolysis has multiple other indispensable roles in virtually all aspects of the normal life of a cell, regulating such key processes as growth, differentiation, adhesion, migration, and programmed cell death by the modification of the extracellular environment [168]. It is, therefore, not surprising that genetic studies of neoplastic progression have demonstrated conclusively that extracellular proteases have profound roles even in early stage tumorigenesis. For example, transgenic overexpression of MMPs in the mammary gland can directly drive malignant progression, whereas loss of MMP expression reduces the formation of premalignant lesions in mice prone to multiple intestinal neoplasia [65,66]. These revelations have led to a gradual shift in the paradigm of tumor-associated proteases as mere obliterators of matrix in the destructive path of invading cancer cells. Today, a more holistic view is favored in which pericellular proteases can act as tumor progression factors in all stages of malignant progression by creating a local environment that is conducive to the growth, survival, and progression of a tumor through the modulation of growth factor pathways, cell–cell adhesions, and cell–matrix adhesions. This paradigm shift has also led to intensified studies of the function of matrix-associated proteases in the early stages of malignant progression prior to the establishment of the invasive phenotype. These studies are clearly in their infancy, and the exact protease targets remain to be firmly established. However, a range of possible candidate substrates is now emerging from biochemical, cell biological, and genetic studies of both physiological and neoplastic tissue remodeling [169].

1. Growth Factor Availability

Many growth factors need proteolytic processing for activation and biological activity. Latency is maintained by membrane anchorage, association with the extracellular matrix, or the tight binding to specific latency-maintaining proteins. Members of the transforming growth factor-β, fibroblast growth factor, and epidermal growth factor ligand families all require proteolytic release for biological function [170]. It is more than likely that tumor-associated proteases play an important role in regulating the activity of these growth factors during tumor progression, as illuminated by several studies [52,169,170]. For example, degradation of proteoglycans by a variety of proteases that are overexpressed in tumors releases fibroblast growth factor and transforming growth factor-β from the extracellular matrix [171]. The insulin-like growth factor-binding protein, which confers latency to the insulin-like growth factor, is a direct target of MMPs during transgene-induced hepatocellular carcinoma. Thus, overexpression of TIMP-1 impairs the release of active insulin-like growth factor by preventing the degradation of insulin-like growth factor-binding protein-3 by MMPs and suppresses hyperplasia and tumorgenesis induced by SV40 T antigen expression by hepatocytes [172]. Likewise, in a transgenic model of pancreatic cancer, systemic MMP inhibitor treatment or genetic ablation of MMP-9 suppresses the proteolytic release of the vascular endothelial growth factor, thereby strongly impairing tumor vascularization [173].

2. Matrix Modification

Proteases can modify the extracellular matrix to facilitate cell migration. Indeed, proteolytic cleavage of collagen is a prerequisite for keratinocyte migration on this substrate. Although the mechanism is not entirely elucidated, it appears that limited proteolysis exposes hidden binding sites for integrins and integrin ligand engagement induces signal transduction that promotes cell migration [174,175]. Likewise, limited cleavage of laminin 5 by secreted or membrane-bound MMPs stimulates keratinocyte migration, possibly through the exposure of cryptic promigratory sites [176,177]. The exposure of novel cell–matrix interaction sites through limited proteolysis of the ECM may also promote tumor cell survival directly. For example, collagen cleavage by MMP-2 has been reported to expose cryptic avβ3 integrin-binding sites, thereby suppressing the apoptosis of melanoma cells through integrin signaling [178].

3. Cell–Cell and Cell–Matrix Adhesion Receptors

Dissolution of cell–cell and cell–matrix adhesions is essential for tumor cell invasion and appears to be at least partially achieved through the proteolytic degradation of adhesion receptors. Several classes of cell adhesion molecules are susceptible to cleavage by proteases *in vitro* and in cell culture systems, and studies of nontransformed cells have revealed that the proteolysis of adhesion receptors can have dramatic effects on cell behavior. Overexpression of MMP-3 in normal mammary epithelial cells results in the cleavage of E-cadherin, triggering a progressive phenotypic conversion that includes cytoskeletal reorganization, growth factor and protease expression, and the acquisition of an invasive phenotype [179]. Similar findings have been reported for immortalized fibroblasts [180]. Downregulation of MMP activity by TIMPs or synthetic MMP inhibitors stabilizes cadherin-mediated cell–cell contacts and promotes contact inhibition, whereas diminution of TIMP activity reduces cell–cell contacts and impairs contact inhibition [180].

III. EXTRACELLULAR PROTEOLYSIS IN HEAD AND NECK CANCER PROGRESSION

The contribution of extracellular proteolysis to the progression of squamous cell carcinomas of the head and neck has been studied less extensively than carcinomas of the lung, breast, and colon. However, the expression, prognostic significance, and causal involvement of proteases in the progression of carcinomas of the head and neck appear to be similar to other types of human cancer. Plg activators, and members of the MMP and cathepsin families, have all been documented to be consistently overexpressed in the stromal or tumor compartment of squamous cell carcinomas, as analyzed by *in situ* hybridization, reverse transcription polymerase chain reaction (RT-PCR), immunohistochemistry, quantitative enzyme-linked immunosorbant assay (ELISA), or laser capture microdissection of surgically resected tumors. This includes uPA, tPA, MMP-2, -3, -7, -9, -10, -11, -13, and -14 and cathepsins B, D, H, and L [15,181–196] (A. Curino, V. Patel, S. Gutkind, and T. Bugge, unpublished data). A direct correlation between the high expression of uPA, MMP-2, and cathepsins B and L and poor prognosis has been established for head and neck cancer, suggesting a causal involvement of these proteases in tumor progression [191,197,198]. Furthermore, the direct association between Plg activator and MMP expression and squamous cell carcinoma progression has been established directly in carcinogen or oncogene-induced tumor models using mice with targeted deletions in protease genes [48,67]. Surprisingly, in both cases, the incidence of squamous cell carcinoma was reduced dramatically by deletion of the protease, suggesting a role of the MMP and PA systems already in early stage carcinogenesis.

References

1. Danø, K., Andreasen, P. A., Grøndahl-Hansen, J., Kristensen, P., Nielsen, L. S., and Skriver, L. (1985). Plasminogen activators, tissue degradation, and cancer. *Adv. Cancer Res.* **44**, 139–266.
2. Andreasen, P. A., Kjøller, L., Christensen, L., and Duffy, M. J. (1997). The urokinase-type plasminogen activator system in cancer metastasis: A review. *Int. J. Cancer* **72**, 1–22.
3. Andreassen, P. A., Egelund, R., and Petersen, H. H. (2000). The plasminogen activation system in tumor growth, invasion, and metastasis. *Cell. Mol. Life Sci.* **57**, 25–40.
4. Ellis, V., Behrendt, N., and Danø, K. (1991). Plasminogen activation by receptor-bound urokinase: A kinetic study with both cell-associated and isolated receptor. *J. Biol. Chem.* **266**, 12752–12758.
5. Cubellis, M. V., Andreasen, P., Ragno, P., Mayer, M., Danø, K., and Blasi, F. (1989). Accessibility of receptor-bound urokinase to type-1 plasminogen activator inhibitor. *Proc. Natl. Acad. Sci. USA* **86**, 4828–4832.
6. Ellis, V., Wun, T. C., Behrendt, N., Ronne, E., and Danø, K. (1990). Inhibition of receptor-bound urokinase by plasminogen-activator inhibitors. *J. Biol. Chem.* **265**, 9904–9908.
7. Baker, M. S., Bleakley, P., Woodrow, G. C., and Doe, W. F. (1990). Inhibition of cancer cell urokinase plasminogen activator by its specific inhibitor PAI-2 and subsequent effects on extracellular matrix degradation. *Cancer Res.* **50**, 4676–4684.
8. Plow, E. F., Freaney, D. E., Plescia, J., and Miles, L. A. (1986). The plasminogen system and cell surfaces: Evidence for plasminogen and urokinase receptors on the same cell type. *J. Cell Biol.* **103**, 2411–2420.
9. Hall, S. W., Humphries, J. E., and Gonias, S. L. (1991). Inhibition of cell surface receptor-bound plasmin by alpha 2-antiplasmin and alpha 2-macroglobulin. *J. Biol. Chem.* **266**, 12329–12336.
10. Danø, K., Rømer, J., Nielsen, B. S., Bjorn, S., Pyke, C., Rygaard, J., and Lund, L. R. (1999). Cancer invasion and tissue remodeling: Cooperation of protease systems and cell types. *APMIS* **107**, 120–127.
11. Skriver, L., Larsson, L. I., Keilberg, V., Nielsen, L. S., and Andreassen, P. B. (1984). Immunocytochemical localization of urokinase-type plasminogen activator in Lewis lung carcinoma. *J. Cell Biol.* **99**, 752–757.
12. Pyke, C., Kristensen, P., Ralfkiaer, E., Grøndahl-Hansen, J., Eriksen, J., Blasi, F., and Danø, K. (1991). Urokinase type plasminogen activator is expressed in stromal cells and its receptor in cancer cells at invasive foci in human colon cancer. *Am. J. Pathol.* **138**, 1059–1067.
13. Pyke, C., Kristensen, P., Ralfkiaer, E., Eriksen, J., and Danø, K. (1991). The plasminogen activation system in human colon cancer: Messenger RNA for the inhibitor PAI-1 is located in endothelial cells in the tumor stroma. *Cancer Res.* **51**, 4067–4071.
14. Pyke, C., Graem, N., Ralfkier, W., Rønne, E., Høyer-Hansen, G., Brunner, N., and Danø, K. (1993). The receptor for urokinase is present in tumor-associated macrophages in ductal breast carcinoma. *Cancer Res.* **53**, 1911–1915.
15. Schmidt, M., Schler, G., Gruensfelder, P., Muller, J., and Hoppe, F. (2000). Urokinase receptor up-regulation in head and neck squamous cell carcinoma. *Head Neck* **22**, 498–504.
16. Dublin, E., Hanby, A., Patel, N. K., Liebman, R., and Barnes, D. (2000). Immunohistochemical expression of uPA, uPAR, and PAI-1 in breast carcinoma. Fibroblastic expression has strong associations with tumor pathology. *Am. J. Pathol.* **157**, 1219–1227.
17. Hudson, M. A., and McReynolds, L. M. (1997). Urokinase and the urokinase receptor: Association with *in vitro* invasiveness of human bladder cancer cell lines. *J. Natl. Cancer Inst.* **89**, 709–717.
18. De Petro, G., Tavian, D., Copeta, A., Portolani, N., Giulini, S. M., and Barlati, S. (1998). Expression of urokinase-type plasminogen activator (u-PA), u-PA receptor, and tissue-type PA messenger RNAs in human hepatocellular carcinoma. *Cancer Res.* **58**, 2234–2239.
19. Morita, S., Sato, A., Hayakawa, H., Ihara, H., Urano, T., Takada, Y., and Takada, A. (1998). Cancer cells overexpress mRNA of urokinase-type plasminogen activator, its receptor and inhibitors in human non-small-cell lung cancer tissue: Analysis by Northern blotting and *in situ* hybridization. *Int. J. Cancer* **78**, 286–292.
20. Taniguchi, T., Kakkar, A. K., Tuddenham, E. G., Williamson, R. C., and Lemoine, N. R. (1998). Enhanced expression of urokinase receptor induced through the tissue factor-factor VIIa pathway in human pancreatic cancer. *Cancer Res.* **58**, 4461–4467.
21. Sier, C. F., Stephens, R., Bizik, J., Mariani, A., Bassan, M., Pedersen, N., Frigerio, L., Ferrari, A., Danø, K., Brunner, N., and Blasi, F. (1998). The level of urokinase-type plasminogen activator receptor is increased in serum of ovarian cancer patients. *Cancer Res.* **58**, 1843–1849.
22. Plesner, T., Ralfkiaer, E., Wittrup, M., Johnsen, H., Pyke, C., Pedersen, T. L., Hansen, N. E., and Danø, K. (1994). Expression of the receptor for urokinase-type plasminogen activator in normal and neoplastic blood cells and hematopoietic tissue. *Am. J. Clin. Pathol.* **102**, 835–841.
23. Lanza, F., Castoldi, G. L., Castagnari, B., Todd, R. F. III, Moretti, S., Spisani, S., Latorraca, A., Focarile, E., Roberti, M. G., and Traniello, S. (1998). Expression and functional role of urokinase-type plasminogen activator receptor in normal and acute leukaemic cells. *Br. J. Haematol.* **103**, 110–123.
24. Pyke, C., Ralfkiaer, E., Ronne, E., Hoyer-Hansen, G., Kirkeby, L., and Danø, K. (1994). Immunohistochemical detection of the receptor for urokinase plasminogen activator in human colon cancer. *Histopathology* **24**, 131–138.
25. Nielsen, B. S., Sehested, M., Timshel, S., Pyke, C., and Danø, K. (1996). Messenger RNA for urokinase plasminogen activator is expressed in myofibroblasts adjacent to cancer cells in human breast cancer. *Lab. Invest.* **74**, 168–177.
26. Danø, K., Behrendt, N., Brunner, N, Ellis, V., Ploug, M., and Pyke, C. (1994). The urokinase receptor: Protein structure and role in plasminogen activation and cancer invasion. *Fibrinolysis* **9**, 189–203.
27. Duffy, M. J., Reilly, D., O'Sullivan, C., O'Higgins, N., Fennely, J. J., and Andreassen, P. A. (1990). Urokinase-plasminogen activator, a new and independent prognostic marker in breast cancer. *Cancer Res.* **50**, 6827–6829.
28. Foekens, J. A., Schmitt, M., van Putten, W. L. J., Peters, H. A., Bontenbal, M., Janicke, F., and Klijn, J. G. M. (1992). Prognostic value of urokinase-type plasminogen activator in 671 primary breast cancer patients. *Cancer Res.* **52**, 6101–6105.
29. Grøndahl-Hansen, J., Christensen, I. J., Rosenquist, C., Brunner, N., Mouridsen, H. T., Danø, K., and Blichert-Toft, M. (1993). High levels of urokinase-type plasminogen activator and its inhibitor PAI-1 in cytosolic extracts of breast carcinomas are associated with poor prognosis. *Cancer Res.* **53**, 2513–2521.
30. Ganesh, S., Siene, C. E. M., Heerding, M. M., Griffioen, G., Lamers, C. B. H. W., and Verspaget, H. W. (1994). Diagnostic relevance of plasminogen activators and their inhibitors in colorectal cancer. *Cancer Res.* **54**, 4065–4071.
31. Oka, T., Ishida, T., Nishino, T., and Sugimachi, K. (1991). Immunohistochemical evidence of urokinase-type plasminogen activator in primary and metastatic tumors of pulmonary adenocarcinoma. *Cancer Res.* **51**, 3522–3525.
32. Pedersen, H., Grøndahl-Hansen, J., Francis, D., Østerlind, K., Hansen, H. H., Danø, K., and, Brunner, N. (1994). Prognostic impact of urokinase, urokinase receptor and type 1 plasminogen activator inhibitor in squamous and large cell lung cancer tissue. *Cancer Res.* **54**, 4671–4675.
33. Heinert, G., Kircheimer, J. C., Pfluger, H., and Binder, B. R. (1988). Urokinase-type plasminogen activator as a marker for the formation of distant metastasis in prostate carcinomas. *J. Urol.* **140**, 1466–1469.
34. Heiss, M. M., Allgayer, H., Gruetzner, K. U., Funke, I., Babic, R., Jauch, K. W., and Schildberg, F. W. (1995). Individual development and uPA-receptor expression of disseminated tumor cells in bone

marrow: A reference to early systemic disease in solid tumors. *Nature Med.* **1**, 1035–1039.
35. Nekarda, H., Siewert, J. R., Schmitt, M., and Ulm, K. (1994). Tumor-associated proteolytic factors uPA and PAI-1 and survival in totally resected gastric cancer. *Lancet* **343**, 117.
36. Kook, Y. H., Adamski, J., Zelent, A., and Ossowski, L. (1994). The effect of antisense inhibition of urokinase receptor in human squamous cell carcinoma on malignancy. *EMBO J.* **13**, 3983–3991.
37. Yu, H., and Schultz, R. M. (1990). Relationship between secreted urokinase plasminogen activator activity and metastatic potential in murine B16 cells transfected with human urokinase sense and antisense genes. *Cancer Res.* **50**, 7623–7633.
38. Hearing, V. J., Law, L. W., Corti, A., Apella, E., and Blasi, F. (1988). Modulation of metastatic potential by cell surface urokinase of murine melanoma cells. *Cancer Res.* **48**, 1270–1278.
39. Kobayashi, H., Gotoh, J., Shinohara, H., Moniwa, N., and Terao, T. (1994). Inhibition of the metastasis of Lewis lung carcinoma by antibody against urokinase-type plasminogen activator in the experimental and spontaneous metastasis model. *Thromb. Haemostasis* **71**, 474–480.
40. Ossowski, L., and Reich, E. (1983). Antibodies to plasminogen activator inhibit human tumor metastasis. *Cell* **35**, 611–619.
41. Ossowski, L. (1988). Plasminogen activator dependent pathways in dissemination of human tumors in the chick embryo. *Cell* **52**, 321–328.
42. Ossowski, L., Russo-Payne, H., and Wilson, E. L. (1991). Inhibition of urokinase-type plasminogen activator by antibodies: The effect on dissemination of a human tumor in the nude mouse. *Cancer Res.* **51**, 274–281.
43. Crowley, C. W., Cohen, R. L., Lucas, B. K., Liu, G., Shuman, M. A., and Levinson, A. D. (1993). Prevention of metastasis by inhibition of the urokinase receptor. *Proc. Natl. Acad. Sci. USA* **90**, 5021–5025.
44. Kobayashi, H., Gotoh, J., Fujie, M., Shinohara, H., Moniwa, N., and Terao, T. (1994). Inhibition of metastasis of Lewis lung carcinoma by a synthetic peptide within growth factor-like domain of urokinase in the experimental and spontaneous metastasis model. *Int. J. Cancer* **57**, 727–733.
45. Min, Y. M., Doyle, L. V., Vitt, C. R., Zandonella, C. L., Stratton-Thomas, J. R., Shuman, M. A., and Rosenberg, S. (1996). Urokinase receptor antagonists inhibit angiogenesis and primary tumor growth in syngeneic mice. *Cancer Res.* **56**, 2428–2433.
46. Mueller, B. M., Yu, Y. B., and Laug, W. E. (1995). Overexpression of plasminogen activator inhibitor 2 in human melanoma cells inhibits spontaneous metastasis in *scid/scid* mice. *Proc. Natl. Acad. Sci. USA* **92**, 205–209.
47. Rabbani, S. A., Harakidas, P., Davidson, D. J., Henkin, J., and Mazar, A. P. (1995). Prevention of prostate-cancer metastasis *in vivo* by a novel synthetic inhibitor of urokinase-type plasminogen activator (uPA). *Int. J. Cancer* **63**, 840–845.
48. Shapiro, R. L., Duquette, J. G., Roses, D. F., Nunes, I., Harris, M. N., Kamino, H., Wilson, E. L., and Rifkin, D. B. (1996). Induction of primary cutaneous melanocytic neoplasms in urokinase-type plasminogen activator (uPA)-deficient and wild-type mice: Cellular blue nevi invade but do not progress to malignant melanoma in uPA-deficient animals. *Cancer Res.* **56**, 3597–3604.
49. Bugge, T. H., Kombrinck, K. W., Xiao, Q., Holmback, K., Daugherty, C. C., Witte, D. P., and Degen, J. L. (1997). Growth and dissemination of Lewis lung carcinoma in plasminogen-deficient mice. *Blood* **90**, 4522–4431.
50. Bugge, T. H., Lund, L. R., Kombrinck, K. K., Nielsen, B. S., Holmback, K., Drew, A. F., Flick, M, J., Witte, D. P., Danø, K., and Degen, J. L. (1998). Reduced metastasis of Polyoma virus middle T antigen-induced mammary cancer in plasminogen-deficient mice. *Oncogene* **16**, 3097–3104.
51. Birkedal-Hansen, H., Moore, W. G., Bodden, M. K., Windsor, L. J., Birkedal-Hansen, B., DeCarlo, A., and Engler, J. A. (1993). Matrix metalloproteinases: A review. *Crit. Rev. Oral Biol Med.* **4**, 197–250.
52. McCawley, L. J., and Matrisian, L. M. (2000). Matrix metalloproteinases: Multifunctional contributors to tumor progression. *Mol. Med. Today* **6**, 149–156.
53. Pei, D., Kang, T., and Qi, H. (2000). Cysteine array matrix metalloproteinase (CA-MMP)/MMP-23 is a type II transmembrane matrix metalloproteinase regulated by a single cleavage for both secretion and activation. *J. Biol. Chem.* **275**, 33988–33997.
54. Brew, K., Dinakarpandian, D., and Nagase, H. (2000). Tissue inhibitors of metalloproteinases: Evolution, structure and function. *Biochim. Biophys. Acta* **1477**, 267–283.
55. Primakoff, P., and Myles, D. G. (2000). The ADAM gene family: Surface proteins with adhesion and protease activity. *Trends Genet.* **16**, 83–87.
56. Tang, B. L. (2001). ADAMTS: A novel family of extracellular matrix proteases. *Int. J. Biochem. Cell Biol.* **33**, 33–44.
57. Werb, Z., Vu, T. H., Rinkenberger, J. L., and Coussens, L. M. (1999). Matrix-degrading proteases and angiogenesis during development and tumor formation. *APMIS* **107**, 11–18.
58. Murray, G. I., Duncan, M. E., O'Neil, P., Melvin, W. T., and Fothergill, J. E. (1996). Matrix metalloproteinase-1 is associated with poor prognosis in colorectal cancer. *Nature Med.* **2**, 461–462.
59. Grignon, D. J., Sakr, W., Toth, M., Ravery, V., Angulo, J., Shamsa, F., Pontes, J. E., Crissman, J. C., and Fridman, R. (1996). High levels of tissue inhibitor of metalloproteinase-2 (TIMP-2) expression are associated with poor outcome in invasive bladder cancer. *Cancer Res.* **56**, 1654–1659.
60. Sier, C. F., Kubben, F. J., Ganesh, S., Heerding, M. M., Griffioen, G., Hanemaaijer, R., van Krieken, J. H., Lamers, C. B., and Verspaget, H. W. (1996). Tissue levels of matrix metalloproteinases MMP-2 and MMP-9 are related to the overall survival of patients with gastric carcinoma. *Br. J. Cancer* **74**, 413–417.
61. Gohji, K., Fujimoto, N., Fujii, A., Komiyama, T., Okawa, J., and Nakajima, M. (1996). Prognostic significance of circulating matrix metalloproteinase-2 to tissue inhibitor of metalloproteinases-2 ratio in recurrence of urothelial cancer after complete resection. *Cancer Res.* **56**, 3196–3198.
62. Talvensaari-Mattila, A., Paakko, P., Hoyhtya, M., Blanco-Sequeiros, G., and Turpeenniemi-Hujanen, T. (1998). Matrix metalloproteinase-2 immunoreactive protein: A marker of aggressiveness in breast carcinoma. *Cancer* **83**, 1153–1162.
63. Kanayama, H., Yokota, K., Kurokawa, Y., Murakami, Y., Nishitani, M., and Kagawa, S. (1998). Prognostic values of matrix metalloproteinase-2 and tissue inhibitor of metalloproteinase-2 expression in bladder cancer. *Cancer* **82**, 1359–1366.
64. Anderson, I. C., Shipp, M. A., Docherty, A. J. P., and Teicher, B. A. (1996). Combination therapy including a gelatinase inhibitor and cytotoxic agent reduces local invasion and metastasis of murine Lewis lung carcinoma. *Cancer Res.* **56**, 715–718.
65. Wilson, C. L., Heppner, K. J., Labosky, P. A., Hogan, B. L., and Matrisian, L. M. (1997). Intestinal tumorigenesis is suppressed in mice lacking the metalloproteinase matrilysin. *Proc. Natl. Acad. Sci. USA* **94**, 1402–407.
66. Sternlicht, M. D., Lochter, A., Sympson, C. J., Huey, B., Rougier, J. P., Gray, J. W., Pinkel, D., Bissell, M. J., and Werb, Z. (1999). The stromal proteinase MMP3/stromelysin-1 promotes mammary carcinogenesis. *Cell* **98**, 137–146.
67. Coussens, L. M., Tinkle, C. L., Hanahan, D., and Werb. Z. (2000) MMP-9 supplied by bone marrow-derived cells contributes to skin carcinogenesis. *Cell* **103**, 481–490.
68. Peschon, J. J., Slack, J. L., Reddy, P., Stocking, K. L., Sunnarborg, S. W., Lee, D. C., Russell, W. E., Castner. B. J., Johnson, R. S., Fitzner, J. N., Boyce, R. W., Nelson, N., Kozlosky, C. J., Wolfson, M. F., Rauch, C. T., Cerretti, D. P., Paxton, R. J., March, C. J., and Black, R. A. (1998). An essential role for ectodomain shedding in mammalian development. *Science* **282**, 1281–1284.
69. Blobel, C. P. (2000). Remarkable roles of proteolysis on and beyond the cell surface. *Curr. Opin. Cell Biol.* **12**, 606–612.

70. Kaushal, G. P., and Shah, S. V. (2000). The new kids on the block: ADAMTSs, potentially multifunctional metalloproteinases of the ADAM family. *J. Clin. Invest.* **105**, 1335–1337.
71. Turk, B., Turk, D., and Turk, V. (2000). Lysosomal cysteine proteases: More than scavengers. *Biochim. Biophys. Acta* **1477**, 98–111.
72. Koblinski, J. E., Ahram, M., and Sloane, B. F. (2000). Unraveling the role of proteases in cancer. *Clin. Chim. Acta* **291**, 113–135.
73. Poole, A. R., Tiltman, K. J., Recklies, A. D., and Stoker, T. A. (1978). Differences in secretion of the proteinase cathepsin B at the edges of human breast carcinomas and fibroadenomas. *Nature* **273**, 545–547.
74. Frosch, B. A., Berquin, I., Emmert-Buck, M. R., Moin, K., and Sloane. B. F. (1999). Molecular regulation, membrane association and secretion of tumor cathepsin B. *APMIS* **107**, 28–37.
75. Kos, J., Werle, B., Lah, T., and Brunner, N. (2000). Cysteine proteinases and their inhibitors in extracellular fluids: Markers for diagnosis and prognosis in cancer. *Int. J. Biol. Markers* **15**, 84–89.
76. Everts, V., van der Zee, E., Creemers, L., and Beertsen, W. (1996). Phagocytosis and intracellular digestion of collagen, its role in turnover and remodelling. *Histochem. J.* **28**, 229–245.
77. Arora, P. D., Manolson, M. F., Downey, G. P., Sodek, J., and McCulloch, C. A. (2000). A novel model system for characterization of phagosomal maturation, acidification, and intracellular collagen degradation in fibroblasts. *J. Biol. Chem.* **275**, 35432–35441.
78. Segal, G., Lee, W., Arora, P. D., McKee, M., Downey, G., and McCulloch, C. A. (2001). Involvement of actin filaments and integrins in the binding step in collagen phagocytosis by human fibroblasts. *J. Cell Sci.* **114**, 119–129.
79. Sloane, B. F., Rozhin, J., Johnson, K., Taylor, H., Crissman, J. D., and Honn, K. V. (1986). Cathepsin B: Association with plasma membrane in metastatic tumors. *Proc. Natl. Acad. Sci. USA* **83**, 2483–2487.
80. Rozhin, J., Wade, R. L., Honn, K. V., and Sloane, B. F. (1989). Membrane-associated cathepsin L: A role in metastasis of melanomas. *Biochem. Biophys. Res. Commun.* **164**, 556–561.
81. Sameni, M., Moin, K., and Sloane, B. F. (2000). Imaging proteolysis by living human breast cancer cells. *Neoplasia* **2**, 496–504.
82. Gelb, B. D., Shi, G. P., Chapman, H. A., and Desnick, R. J. (1996). Pycnodysostosis, a lysosomal disease caused by cathepsin K defi ciency. *Science* **273**, 1236–1238.
83. Saftig, P., Hunziker, E., Wehmeyer, O., Jones, S., Boyde, A., Rommerskirch, W., Moritz, J. D., Schu, P., and von Figura, K. (1998). Impaired osteoclastic bone resorption leads to osteopetrosis in cathepsin-K-deficient mice. *Proc. Natl. Acad. Sci. USA.* **95**, 13453–13458.
84. Giranda, V. L., and Matrisian, L. M. (1999). The protease consortium: an alliance to advance the understanding of proteolytic enzymes as therapeutic targets for cancer. *Mol. Carcinog.* **26**, 139–142.
85. Davie, E. W., Fujikawa, K., and Diesel, W. (1991). The coagulation cascade: Initiation, maintenance, and regulation. *Biochemistry* **30**, 10363–10370.
86. Donati, M. B. (1995). Cancer and thrombosis: From Phlegmasia Alba Dolens to transgenic mice. *Thrombosis Haemostasis* **74**, 278–281.
87. Dvorak, H. F., Nagy, J. A., Berse, B., Brown, L. F., Yeo, K.-T, Yeo, T.-K., Dvorak, A. M., van de Water, L., Siousat, T. M., and Senger, D. R. (1992). Vascular permeability factor, fibrin, and the pathogenesis of tumor stroma formation. *Ann. N. Y. Acad. Sci.* **667**, 101–111.
88. Colucci, M., Delaini, F., de Bellis Vitti, G., Locati, D., Poggi, A., Semeraro, N., and Donati, M. B. (1983). Warfarin inhibits both procoagulant activity and metastatic capacity of Lewis lung carcinoma cells: Role of vitamin K deficiency. *Biochem. Pharmacol.* **32**, 1689–1691.
89. Esumi, N., Fan, D., and Fidler, I. J. (1991). Inhibition of murine melanoma experimental metastasis by recombinant desulfatohirudin, a highly specific thrombin inhibitor. *Cancer Res.* **51**, 4549–4556.
90. Fischer, E. G., Ruf, W. R., and Mueller, B. M. (1995). Tissue factor-initiated thrombin generation activates the signaling thrombin receptor on malignant melanoma cells. *Cancer Res.* **55**, 1629–1632.
91. Mueller, B. M., Reisfeld, R. A., Edginton, T. S., and Ruf, W. (1992). Expression of tissue factor by melanoma cells promotes efficient metastasis. *Proc. Natl. Acad. Sci. USA* **89**, 11832–11836.
92. Tuszynski, G. P., Gasic, T. B., and Gasic, G. J. (1987). Isolation and characterization of antistasin. *J. Biol. Chem.* **262**, 9718–9723.
93. Palumbo, J. S., Kombrinck, K. W., Drew, A. F., Grimes, T. S., Kiser, J. H., Degen, J. L., and Bugge, T. H. (2000). Fibrinogen is an important determinant of the metastatic potential of circulating tumor cells. *Blood* **96**, 3302–3309.
94. Tanimoto, H., Underwood, L. J., Wang, Y., Shigemasa, K., Parmley, T. H., and O'Brien, T. J. (2001). Ovarian tumor cells express a transmembrane serine protease: A potential candidate for early diagnosis and therapeutic intervention. *Tumour Biol.* **22**, 104–114.
95. Underwood, L. J., Shigemasa, K., Tanimoto, H., Beard, J. B., Schneider, E. N., Wang, Y., Parmley, T. H., and O'Brien, T. J. (2000). Ovarian tumor cells express a novel multi-domain cell surface serine protease. *Biochim Biophys Acta* **1502**, 337–350.
96. Wallrapp, C., Hahnel, S., Muller-Pillasch, F., Burghardt, B., Iwamura, T., Ruthenburger, M., Lerch, M. M., Adler, G., and Gress, T. M. (2000). A novel transmembrane serine protease (TMPRSS3) overexpressed in pancreatic cancer. *Cancer Res.* **60**, 2602–2606.
97. Chen, Z. L., Yoshida, S., Kato, K., Momota, Y., Suzuki, J., Tanaka, T., Ito, J., Nishino, H., Aimoto, S., and Kiyama, H. (1995). Expression and activity-dependent changes of a novel limbic-serine protease gene in the hippocampus. *J. Neurosci.* **15**, 5088–5097.
98. Davies, B. J., Pickard, B. S., Steel, M., Morris, R. G., and Lathe, R. (1998). Serine proteases in rodent hippocampus. *J. Biol Chem.* **273**, 23004–23011.
99. Underwood, L. J., Tanimoto, H., Wang, Y., Shigemasa, K., Parmley, T. H., and O'Brien, T. J. (1999). Cloning of tumor-associated differentially expressed gene-14, a novel serine protease overexpressed by ovarian carcinoma. *Cancer Res.* **59**, 4435–4439.
100. Magklara, A., Scorilas, A., Katsaros, D., Massobrio, M., Yousef, G. M., Fracchioli, S., Danese, S., and Diamandis, E. P. (2001). The human KLK8 (neuropsin/ovasin) gene: Identification of two novel splice variants and its prognostic value in ovarian cancer. *Clin. Cancer Res.* **7**, 806–811.
101. Yamashiro, K., Tsuruoka, N., Kodama, S., Tsujimoto, M., Yamamura, Y., Tanaka, T., Nakazato, H., and Yamaguchi, N. (1997). Molecular cloning of a novel trypsin-like serine protease (neurosin) preferentially expressed in brain. *Biochim. Biophys. Acta* **350**, 11–14.
102. Little, S. P., Dixon, E. P., Norris, F., Buckley, W., Becker, G. W., Johnson, M., Dobbins, J. R., Wyrick, T., Miller, J. R., MacKellar, W., Hepburn, D., Corvalan, J., McClure, D., Liu, X., Stephenson, D., Clemens, J., and Johnstone, E. M. (1997). Zyme, a novel and potentially amyloidogenic enzyme cDNA isolated from Alzheimer's disease brain. *J. Biol. Chem.* **272**, 25135–25142.
103. Anisowicz, A., Sotiropoulou, G., Stenman, G., Mok, S. C., and Sager, R. (1996). A novel protease homolog differentially expressed in breast and ovarian cancer. *Mol. Med.* **5**, 624–636.
104. Matsui, H., Kimura, A., Yamashiki, N., Moriyama, A., Kaya, M., Yoshida, I., Takagi, N., and Takahashi, T. (2000). Molecular and biochemical characterization of a serine proteinase predominantly expressed in the medulla oblongata and cerebellar white matter of mouse brain. *J. Biol. Chem.* **275**, 11050–11057.
105. Meier, N., Dear, T. N., and Boehm, T. A. (1999). Novel serine protease overexpressed in the hair follicles of nude mice. *Biochem. Biophys. Res. Commun.* **258**, 374–378.
106. Scarisbrick, I. A., Towner, M. D., and Isackson, P. J. (1997). Nervous system-specific expression of a novel serine protease: Regulation in the adult rat spinal cord by excitotoxic injury. *J. Neurosci.* **17**, 8156–8168.
107. Hansson, L., Stromqvist, M., Backman, A., Wallbrandt, P., Carlstein, A., and Egelrud, T. (1994). Cloning, expression, and characterization of stratum corneum chymotryptic enzyme: A skin-specific human serine proteinase. *J. Biol. Chem.* **269**, 19420–19426.

108. Leytus, S. P., Loeb, K. R., Hagen, F. S., Kurachi, K., and Davie, E. W. (1988). A novel trypsin-like serine protease (hepsin) with a putative transmembrane domain expressed by human liver and hepatoma cells. *Biochemistry* **27**, 1067–1074.
109. Paoloni-Giacobino, A., Chen, H, Peitsch, M. C., Rossier, C., and Antonarakis, S. E. (1997). Cloning of the TMPRSS2 gene, which encodes a novel serine protease with transmembrane, LDLRA, and SRCR domains and maps to 21q22.3. *Genomics* **44**, 309–320.
110. Zhang, Y., Cai, X., Schlegelberger, B., and Zheng, S. (1998). Assignment of human putative tumor suppressor genes ST13 (alias SNC6) and ST14 (alias SNC19) to human chromosome bands 22q13 and 11q24—>q25 by *in situ* hybridization. *Cytogenet. Cell Genet.* **83**, 56–57.
111. Takeuchi, T., Shuman, M. A. and Craik, C. S. (1999). Reverse biochemistry: Use of macromolecular protease inhibitors to dissect complex biological processes and identify a membrane-type serine protease in epithelial cancer and normal tissue. *Proc. Natl. Acad. Sci. USA* **96**, 11054–11061.
112. Lin, C. Y., Anders, J., Johnson, M., Sang, Q. A., and Dickson, R. B. (1999). Molecular cloning of cDNA for matriptase, a matrix-degrading serine protease with trypsin-like activity. *J. Biol. Chem.* **274**, 18231–18236.
113. Kim, M. G., Chen, C., Lyu, M. S., Cho, E. G., Park, D., Kozak, C., and Schwartz, R. H. (1999). Cloning and chromosomal mapping of a gene isolated from thymic stromal cells encoding a new mouse type II membrane serine protease, epithin, containing four LDL receptor modules and two CUB domains. *Immunogenetics* **49**, 420–428.
114. Hooper, J. D., Nicol, D. L., Dickinson, J. L., Eyre, H. J., Scarman, A. L., Normyle, J. F., Stuttgen, M. A., Douglas, M. L., Loveland, K. A., Sutherland, G. R., and Antalis, T. M. (1999). Testisin, a new human serine proteinase expressed by premeiotic testicular germ cells and lost in testicular germ cell tumors. *Cancer Res.* **59**, 3199–3205.
115. Inoue, M., Kanbe, N., Kurosawa, M., and Kido, H. (1998). Cloning and tissue distribution of a novel serine protease esp-1 from human eosinophils. *Biochem. Biophys. Res. Commun.* **18**, 307–312.
116. Niedermeyer, J., Scanlan, M. J., Garin-Chesa, P., Daiber, C., Fiebig, H. H., Old, L. J., Rettig, W. J., and Schnapp, A. (1997). Mouse fibroblast activation protein: Molecular cloning, alternative splicing and expression in the reactive stroma of epithelial cancers. *Int. J. Cancer* **71**, 383–389.
117. Goldstein, L. A., Ghersi. G., Pineiro-Sanchez, M. L., Salamone, M., Yeh, Y., Flessate, D., and Chen, W. T. (1997). Molecular cloning of seprase: A serine integral membrane protease from human melanoma. *Biochim. Biophys. Acta* **1361**, 11–19.
118. Diamandis, E. P., Yousef, G. M., Soosaipillai. A. R., and Bunting, P. (2000). Human kallikrein 6 (zyme/protease M/neurosin): A new serum biomarker of ovarian carcinoma. *Clin. Biochem.* **33**, 579–583.
119. Shigemasa, K., Underwood, L. J., Beard, J., Tanimoto, H., Ohama, K., Parmley, T. H., and O'Brien, T. J. (2000). Overexpression of testisin, a serine protease expressed by testicular germ cells, in epithelial ovarian tumor cells. *J. Soc. Gynecol. Investig.* **7**, 358–362.
120. Tanimoto, H., Underwood, L. J., Shigemasa, K., Yan Yan, M. S., Clarke, J., Parmley, T. H., and O'Brien, T. J. (1999). The stratum corneum chymotryptic enzyme that mediates shedding and desquamation of skin cells is highly overexpressed in ovarian tumor cells. *Cancer* **86**, 2074–2082.
121. Dhanasekaran, S. M., Barrette, T. R., Ghosh, D., Shah, R., Varambally, S., Kurachi, K., Pienta, K. J., Rubin, M. A., and Chinnaiyan, A. M. (2001) Delineation of prognostic biomarkers in prostate cancer. *Nature* **412**, 822–826.
122. Welsh, J. B., Sapinoso, L. M., Su, A. I., Kern, S. G., Wang-Rodriguez, J., Moskaluk, C. A., Frierson, H. F., Jr, and Hampton, G. M. (2001). Analysis of gene expression identifies candidate markers and pharmacological targets in prostate cancer. *Cancer Res.* **15**, 5974–5978.
123. Magee, J. A., Araki, T., Patil, S., Ehrig, T., True, L., Humphrey, P. A., Catalona, W. J., Watson, M. A., and Milbrandt, J. (2001). Expression profiling reveals hepsin overexpression in prostate cancer. *Cancer Res.* **61**, 5692–5696.
124. Luo, J., Duggan, D. J., Chen, Y., Sauvageot, J., Ewing, C. M., Bittner, M. L., Trent, J. M., and Isaacs, W. B. (2001). Human prostate cancer and benign prostatic hyperplasia: molecular dissection by gene expression profiling. *Cancer Res.* **61**, 4683–4688.
125. Nagy, J. A., Meyers, M. S., Masse, E. M., Herzberg, K. T., and Dvorak, H. F. (1995). Pathogenesis of ascites tumor growth: Fibrinogen influx and fibrin accumulation in tissues lining the peritoneal cavity. *Cancer Res.* **55**, 369–375.
126. Kazama, Y., Hamamoto, T., Foster, D. C., and Kisiel, W. (1995). Hepsin, a putative membrane-associated serine protease, activates human factor VII and initiates a pathway of blood coagulation on the cell surface leading to thrombin formation. *J. Biol. Chem.* **270**, 66–72.
127. Garin-Chesa, P., Old, L. J., and Rettig, W. J. (1990). Cell surface glycoprotein of reactive stromal fibroblasts as a potential antibody target in human epithelial cancers. *Proc. Natl. Acad. Sci. USA* **87**, 7235–7239.
128. Rettig, W. J., Su, S. L., Fortunato, S. R., Scanlan, M. J., Raj, B. K., Garin-Chesa, P., Healey, J. H., and Old, L. J. (1994). Fibroblast activation protein: Purification, epitope mapping and induction by growth factors. *Int. J. Cancer* **58**, 385–392.
129. Mersmann, M., Schmidt, A., Rippmann, J. F., Wuest, T., Brocks, B., Rettig, W. J., Garin-Chesa, P., Pfizenmaier, K., and Moosmayer, D. (2001). Human antibody derivatives against the fibroblast activation protein for tumor stroma targeting of carcinomas. *Int. J. Cancer* **92**, 240–248.
130. Mann, N. S., and Mann, S. K. (1994). Enterokinase. *Proc. Soc. Exp. Biol. Med.* **206**, 114–118.
131. Hooper, J. D., Clements, J. A., Quigley, J. P., and Antalis, T. M. (2001). Type II transmembrane serine proteases: Insights into an emerging class of cell surface proteolytic enzymes. *J. Biol. Chem.* **276**, 857–860.
132. Oberst, M., Anders, J., Xie, B., Singh, B., Ossandon, M., Johnson, M., Dickson, R. B., and Lin, C. Y. (2001). Matriptase and HAI-1 are expressed by normal and malignant epithelial cells *in vitro* and *in vivo*. *Am. J. Pathol.* **158**, 1301–1311.
133. Lee, S. L., Dickson, R. B., and Lin, C. Y. (2000). Activation of hepatocyte growth factor and urokinase/plasminogen activator by matriptase, an epithelial membrane serine protease. *J. Biol. Chem.* **275**, 36720–36725.
134. Takeuchi, T., Harris, J. L., Huang, W., Yan, K. W., Coughlin, S. R., and Craik, C. S. (2000). Cellular localization of membrane-type serine protease 1 and identification of protease-activated receptor-2 and single-chain urokinase-type plasminogen activator as substrates. *J. Biol. Chem.* **275**, 26333–26342.
135. Murphy, G., Stanton, H., Cowell, S, Butler, G., Knauper, V., Atkinson, S., and Gavrilovic, J. (1999). Mechanisms for promatrix metalloproteinase activation. *APMIS* **107**, 38–44.
136. List, K., Jensen, O. N., Bugge, T. H., Lund, L. R., Ploug, M., Danø, K., and Behrendt, N. (2000). Plasminogen-independent initiation of the pro-urokinase activation cascade *in vivo*: Activation of pro-urokinase by glandular kallikrein (mGK-6) in plasminogen-deficient mice. *Biochemistry* **39**, 508–515.
137. Romisch, J., Vermohlen, S., Feussner, A., and Stohr, H. (1999). The FVII activating protease cleaves single-chain plasminogen activators. *Haemostasis* **29**, 292–299.
138. Kobayashi, H., Schmitt, M., Goretzki, L., Chucholowski, N., Calvete, J., Kramer, M., Gunzler, W. A., Janicke, F., and Graeff, H. (1991). Cathepsin B efficiently activates the soluble and the tumor cell receptor-bound form of the proenzyme urokinase-type plasminogen activator (Pro-uPA). *J. Biol. Chem.* **266**, 5147–5152.

139. Strongin, A. Y., Collier, I., Bannikov, G., Marmer, B. L., Grant, G. A., and Goldberg, G. I. (1995). Mechanism of cell surface activation of 72-kDa type IV collagenase: Isolation of the activated form of the membrane metalloprotease. *J. Biol. Chem.* **270**, 5331–5338.
140. Cowell, S., Knauper, V., Stewart, M. L., DOrtho, M. P., Stanton, H., Hembry, R. M., Lopez-Otin, C., Reynolds, J. J., and Murphy, G. (1998). Induction of matrix metalloproteinase activation cascades based on membrane-type 1 matrix metalloproteinase: Associated activation of gelatinase A, gelatinase B and collagenase 3. *Biochem. J.* **331**, 453–458.
141. Dumin, J. A., Dickeson, S. K., Stricker, T. P., Bhattacharyya-Pakrasi, M., Roby, J. D., Santoro, S. A., and Parks, W. C. (2001). Pro-collagenase-1 (matrix metalloproteinase-1) binds the alpha(2)beta(1) integrin upon release from keratinocytes migrating on type I collagen. *J. Biol. Chem.* **276**, 29368–29374.
142. Barmina, O. Y., Walling, H. W., Fiacco, G. J., Freije, J. M., Lopez-Otin, C., Jeffrey, J. J., and Partridge, N. C. (1999). Collagenase-3 binds to a specific receptor and requires the low density lipoprotein receptor-related protein for internalization. *J. Biol. Chem.* **274**, 30087–30093.
142a. Yu, Q., and Stamenkovic, I. (1999). Localization of matrix metalloproteinase 9 to the cell surface provides a mechanism for CD44-mediated tumor invasion. *Genes Dev.* **13**, 35–48.
143. Ruf, R., and Mueller, B. M. (1996). Tissue factor in cancer angiogenesis and metastasis. *Curr. Opin. Hematol.* **3**, 379–384.
144. Zacharski, L. R., Wojtukiewicz, M. Z., Costantini, V., Ornstein, D. L., and Memoli, V. A. (1992). Pathways of coagulation/fibrinolysis activation in malignancy. *Semin. Thromb. Hemost.* **18**, 104–116.
145. Woessner, J. F., and Nagase, H. (1996). "Matrix Metalloproteinases and TIMPSs." Oxford Univ. Press, Oxford.
146. Lijnen, H. R., Silence, J., Van Hoef, B., and Collen, D. (1998). Stromelysin-1 (MMP-3)-independent gelatinase expression and activation in mice. *Blood* **91**, 2045–2053.
147. Ramos-DeSimone, N., Hahn-Dantona, E., Sipley, J., Nagase, H., French, D. L., and Quigley, J. P. (1999). Activation of matrix metalloproteinase-9 (MMP-9) via a converging plasmin/stromelysin-1 cascade enhances tumor cell invasion. *J. Biol. Chem.* **274**, 13066–13076.
148. Mazzieri, R., Masiero, L., Zanetta, L., Monea, S., Onisto, M., Garbisa, S., and Mignatti, P. (1997). Control of type IV collagenase activity by components of the urokinase-plasmin system: A regulatory mechanism with cell-bound reactants. *EMBO J.* **16**, 2319–2332.
149. Carmeliet, P., Moons, L., Lijnen, R., Baes, M., Lemaitre, V., Tipping, P., Drew, A., Eeckhout, Y., Shapiro, S., Lupu, F., and Collen, D. (1997). Urokinase-generated plasmin activates matrix metalloproteinases during aneurysm formation. *Nature Genet.* **17**, 439–444.
150. Knauper, V., Will, H., Lopez-Otin, C., Smith, B., Atkinson, S. J., Stanton, H, Hembry, R. M., and Murphy, G. (1996). Cellular mechanisms for human procollagenase-3 (MMP-13) activation: Evidence that MT1-MMP (MMP-14) and gelatinase a (MMP-2) are able to generate active enzyme. *J. Biol. Chem.* **271**, 17124–17131.
151. Coussens, L. M., Raymond, W. W., Bergers, G., Laig- Webster. M., Behrendtsen, O., Werb, Z., Caughey, G. H., and Hanahan, D. (1999) Inflammatory mast cells up-regulate angiogenesis during squamous epithelial carcinogenesis. *Genes Dev.* **13**, 1382–1397.
152. Liotta, L. A., Thorgeirsson, U. P., and Garbisa, S. (1982). Role of collagenases in tumor cell invasion. *Cancer Metastasis Rev.* **1**, 277–288.
153. Lund, L. R., Rømer, J., Bugge, T. H., Nielsen, B. S., Frandsen, T. L., Degen, J. L., Stephens, R. W., and Danø, K. (1999). Functional overlap between two classes of matrix-degrading proteases in wound healing. *EMBO J.* **18**, 4645–4656.
154. Altieri, D. C., Mannucci, P. M., and Capitanio, A. M. (1986). Binding of fibrinogen to human monocytes. *J. Clin. Invest.* **78**, 968–976.
155. Dejana, E., Languino, L. R., Polentarutti, N., Balconi, G., Ryckewaert , J. J., Larrieu, M. J., Donati, M. B., Mantovani, A., and Marguerie, G. (1985). Interaction between fibrinogen and cultured endothelial cells: Induction of migration and specific binding. *J. Clin. Invest.* **75**, 11–18.
156. Sherman, L. A., and Lee, J. (1977). Specific binding of soluble fibrin to macrophages. *J. Exp. Med.* **145**, 76–85.
157. Katagiri, Y., Hiroyama, T., Akamatsu, N., Suzuki, H., Yamazaki, H., and Tanoue, K. (1995). Involvement of alpha v beta 3 integrin in mediating fibrin gel retraction. *J. Biol. Chem.* **270**, 1785–1790.
158. Languino, L. R., Duperray, A., Joganic, K. J., Fornaro, M., Thornton, G. B., and Altieri, D. C. (1995). Regulation of leukocyte-endothelium interaction and leukocyte transendothelial migration by intercellular adhesion molecule 1-fibrinogen recognition. *Proc. Natl. Acad. Sci. USA* **92**, 1505–1509.
159. Simon, D. I., Ezratty, A. M., Francis, S. A., Rennke, H., and Loscalzo, J. (1993). Fibrin(ogen) is internalized and degraded by activated human monocytoid cells via Mac-1 (CD11b/CD18): A nonplasmin fibrinolytic pathway. *Blood* **82**, 2414–2422.
160. Dvorak, H. F., Dvorak, A. M., Manseau, E. J., Wiberg, L., and Churchill, W. H. (1979). Fibrin gel investment associated with line 1 and line 10 solid tumor growth, angiogenesis, and fibroplasia in guinea pigs: Role of cellular immunity, myofibroblasts, microvascular damage, and infarction in line 1 tumor regression. *J. Natl. Cancer Inst.* **62**, 1459–1472.
161. Dvorak, H. F., Harvey, V. S., Estrella, P., Brown, L. F., McDonagh, J., and Dvorak, A. M. (1987). Fibrin containing gels induce angiogenesis: Implications for tumor stroma generation and wound healing. *Lab. Invest.* **57**, 673–686.
161a. Plow, E. F., and Edginton, T. S. (1986). Lymphocyte suppressive peptides are derived predominantly from the Aα chain. *J. Immunol.* **137**, 1910–1915.
162. Robson, S. C., Saunders, R., Purves, L. R., de Jager, C., Corrigal, A., and Kirch, R. E. (1993). Fibrin and fibrinogen degradation products with an intact D-domain C terminal gamma chain inhibit an early step in accessory cell-dependent lymphocyte mitogenesis. *Blood* **81**, 3006–3014.
163. Skogen, W. F., Senoir, R. M., Griffin, G. L., and Wilner, G. D. (1988). Fibrinogen-derived peptide Bb1-42 is a multidomain neutrophil chemoattractant. *Blood* **71**, 1475–1479.
164. Thompson, W. D., Smoth, E. B., Stirk, C. M., and Wang, J. (1993). Fibrin degradation products in growth stimulatory extracts of pathological lesions. *Blood Coagul. Fibrinolysis* **4**, 113–115.
165. Cavenaugh, P. G., Sloane, B. F., and Honn, K. V. (1988). Role of the coagulation system in tumor-cell-induced platelet aggregation and metastasis. *Haemostasis* **18**, 37–46.
166. Chew, E. C., and Wallace, A. C. (1976). Demonstration of fibrin in early stages of metastasis. *Cancer Res.* **36**, 1904–1909.
167. Crissman, J. D., Hatfield, J. S., Menter, D. G., Sloane, B. S., and Honn, K. V. (1988). Morphological studies of the interaction of intravascular tumor cells with endothelial cells and subendothelial matrix. *Cancer Res.* **48**, 4065–4072.
168. Werb, Z. (1997). ECM and cell surface proteolysis: Regulating cellular ecology. *Cell* **91**, 439–442.
169. McCawley, L. J., and Matrisian, L. M. (2001). Matrix metalloproteinases: They're not just for matrix anymore! *Curr. Opin. Cell Biol.* **13**, 534–540.
170. Rifkin, D. B., Mazzieri, R., Munger, J. S., Noguera, I., and Sung, J. (1999). Proteolytic control of growth factor availability. *APMIS* **107**, 80–85.
171. Whitelock, J. M., Murdoch, A D., Iozzo, R V., and Underwood, P. A. (1996). The degradation of human endothelial cell-derived perlecan and release of bound basic fibroblast growth factor by stromelysin, collagenase, plasmin, and heparanases. *J. Biol. Chem.* **271**, 10079–10086.
172. Martin, D. C., Fowlkes, J. L., Babic, B., and Khokha, R. (1999). Insulin-like growth factor II signaling in neoplastic proliferation is

blocked by transgenic expression of the metalloproteinase inhibitor TIMP-1. *J. Cell Biol.* **146**, 881–892.
173. Bergers, G., Brekken, R., McMahon, G., Vu, T. H., Itoh, T., Tamaki, K., Tanzawa, K., Thorpe, P., Itohara, S., Werb, Z., and Hanahan, D. (2000). Matrix metalloproteinase-9 triggers the angiogenic switch during carcinogenesis. *Nature Cell Biol.* **2**, 737–744.
174. Pilcher, B. K., Dumin, J. A., Sudbeck, B. D., Krane, S. M., Welgus, H. G., and Parks, W. C. (1997). The activity of collagenase-1 is required for keratinocyte migration on a type I collagen matrix. *J. Cell Biol.* **137**, 1445–1457.
175. Messent, A. J., Tuckwell, D. S., Knauper, V., Humphries, M. J., Murphy, G., and Gavrilovic, J. (1998). Effects of collagenase-cleavage of type I collagen on alpha2beta1 integrin-mediated cell adhesion. *J. Cell Sci.* **111**, 1127–1135.
176. Giannelli, G., Falk-Marzillier, J., Schiraldi, O., Stetler-Stevenson, W. G., and Quaranta, V. (1997). Induction of cell migration by matrix metalloprotease-2 cleavage of laminin-5. *Science* **277**, 225–228
177. Koshikawa, N., Giannelli, G., Cirulli, V., Miyazaki, K., and Quaranta, V. (2000). Role of cell surface metalloprotease MT1-MMP in epithelial cell migration over laminin-5. *J. Cell Biol.* **148**, 615–624.
178. Montgomery, A. M., Reisfeld, R. A., and Cheresh, D. A. (1994). Integrin alpha v beta 3 rescues melanoma cells from apoptosis in three-dimensional dermal collagen. *Proc. Natl. Acad. Sci. USA.* **91**, 8856–8860.
179. Lochter, A., Galosy, S., Muschler, J., Freedman, N., Werb, Z., and Bissell, M. J. (1997). Matrix metalloproteinase stromelysin-1 triggers a cascade of molecular alterations that leads to stable epithelial-to-mesenchymal conversion and a premalignant phenotype in mammary epithelial cells. *J. Cell Biol.* **139**, 1861–1872.
180. Ho, A. T., Voura, E. B., Soloway, P. D., Watson, K. L., and Khokha, R. (2001). MMP inhibitors augment fibroblast adhesion through stabilization of focal adhesion contacts and upregulation of cadherin function. *J. Biol. Chem.* **276**, 40215–40224.
181. Rømer, J., Pyke, C., Lund, L. R., Ralfkiaer, E., and Danø, K. (2001). Cancer cell expression of urokinase-type plasminogen activator receptor mRNA in squamous cell carcinomas of the skin. *J. Invest. Dermatol.* **116**, 353–358.
182. Schmidt, M., and Hoppe, F. (1999). Increased levels of urokinase receptor in plasma of head and neck squamous cell carcinoma patients. *Acta Otolaryngol.* **119**, 949–953.
183. Yasuda, T., Sakata, Y., Kitamura, K., Morita, M., and Ishida, T. (1997). Localization of plasminogen activators and their inhibitor in squamous cell carcinomas of the head and neck. *Head Neck* **19**, 611–616.
184. Itaya, T. (1996). Relationship between head and neck squamous cell carcinomas and fibrinolytic factors: Immunohistological study. *Acta Otolaryngol. Suppl.* **525**, 113–119.
185. Bjorlin, G., Ljungner, H., Wennerberg, J., and Astedt, B. (1987). Plasminogen activators in human xenografted oro-pharyngeal squamous cell carcinomas. *Acta Otolaryngol.* **104**, 568–572.
186. Miyajima, Y., Nakano, R., and Morimatsu, M. (1995). Analysis of expression of matrix metalloproteinases-2 and -9 in hypopharyngeal squamous cell carcinoma by *in situ* hybridization. *Ann. Otol. Rhinol. Laryngol.* **104**, 678–684.
187. Okada, A., Bellocq, J. P., Rouyer, N., Chenard, M. P., Rio, M.C., Chambon, P., and Basset, P. (1995). Membrane-type matrix metalloproteinase (MT-MMP) gene is expressed in stromal cells of human colon, breast, and head and neck carcinomas. *Proc. Natl. Acad. Sci. USA* **92**, 2730–2734.
188. Birkedal-Hansen, B., Pavelic, Z. P., Gluckman, J. L., Stambrook, P., Li, Y. Q., and Stetler-Stevenson, W. G. (2000). MMP and TIMP gene expression in head and neck squamous cell carcinomas and adjacent tissues. *Oral Dis.* **6**, 376–382.
189. Kusukawa, J. (1993). Expression of matrix metalloproteinase-2 related to lymph node metastasis of oral squamous cell carcinoma: A clinicopathologic study. *Am. J. Clin. Pathol.* **99**, 18–23.
190. Muller, D., Wolf, C., Abecassis, J., Millon, R., Engelmann, A., Bronner, G., Rouyer, N., Rio, M. C., Eber, M., and Methlin, G. (1993). Increased stromelysin 3 gene expression is associated with increased local invasiveness in head and neck squamous cell carcinomas. *Cancer Res.* **53**, 165–169.
191. Budihna, M., Strojan, P., Smid, L., Skrk, J., Vrhovec, I., Zupevc, A., Rudolf, Z., Zargi, M., Krasovec, M., Svetic, B., Kopitar-Jerala, and N., Kos, J. (1996). Prognostic value of cathepsins B, H, L, D and their endogenous inhibitors stefins A and B in head and neck carcinoma. *Biol. Chem. Hoppe Seyler* **377**, 385–390.
192. Johansson, N. (1997). Expression of collagenase-3 (matrix metalloproteinase-13) in squamous cell carcinomas of the head and neck. *Am. J. Pathol.* **151**, 499–508.
193. Imanishi, Y. (2000). Clinical significance of expression of membrane type 1 matrix metalloproteinase and matrix metalloproteinase-2 in human head and neck squamous cell carcinoma. *Hum. Pathol.* **31**, 895–904.
194. Kawada, A., Hara, K., Kominami, E., Kobayashi, T., Hiruma, M., and Ishibashi, A. (1996). Cathepsin B and D expression in squamous cell carcinoma. *Br. J. Dermatol.* **135**, 905–910.
195. Zeillinger, R., Eder, S., Schneeberger, C., Ullrich, R., Speiser, P., Swoboda, H. (1996). Cathepsin D and PAI-1 expression in human head and neck cancer. *Anticancer Res.* **16**, 449–453.
196. Kos, J. (1995). Lysosomal proteases cathepsins D, B, H, L and their inhibitors stefins A and B in head and neck cancer. *Biol. Chem. Hoppe Seyler* **376**, 401–405.
197. Yoshizaki, T., Maruyama, Y., Sato, H., and Furukawa, M. (2001). Expression of tissue inhibitor of matrix metalloproteinase-2 correlates with activation of matrix metalloproteinase-2 and predicts poor prognosis in tongue squamous cell carcinoma. *Int. J. Cancer* **95**, 44–50.
198. Strojan, P., Strojan, P., Budihna, M., Smid, L., Vrhovec, I., and Skrk, J. (2000). Urokinase-type plasminogen activator, plasminogen activator inhibitor type 1 and cathepsin D: Analysis of their prognostic significance in squamous cell carcinoma of the head and neck. *Anticancer Res.* **20**, 3975–3981.

C H A P T E R

11

Papillomaviruses in Head and Neck Squamous Cell Carcinoma

S. KIM* and E. J. SHILLITOE†

*Department of Otolaryngology and Communication Sciences and
†Microbiology and Immunology, SUNY Upstate Medical University, Syracuse, New York 13210

I. Introduction 151
II. Human Papillomaviruses (HPVs) 152
A. Virus Life Cycle 152
B. Virion and Genomic Structure 152
III. Interaction between Cellular Host Factors and E6 of High-Risk HPVs 153
IV. Interaction between Cellular Host Factors and E7 of High-Risk HPVs 154
V. Methods of Detection of HPVs 155
VI. HPV and Normal Mucosa of the Upper Aerodigestive Tract 155
VII. HPV and Benign Lesions of the Head and Neck 156
A. Benign Oral Lesions 156
B. Laryngeal Papillomatosis 156
VIII. HPVs and Premalignant Lesions of the Head and Neck 157
A. Premalignant Lesions of the Oral Cavity 157
B. Premalignant Lesions of the Larynx 158
C. Schneiderian Papilloma of the Nasal Cavity 158
IX. HPV and SCCA of Head and Neck 159
A. Detection of HPV in HNSCC 159
B. Does HPV Infection Increase the Risk of HNSCC? 160
C. Do HPV-Positive Tumors Constitute a Distinct and Separate Clinical Entity? 161
X. Conclusions 161
References 162

Viral etiologies have been implicated in the pathogenesis of many types of cancer. Although the human papillomavirus (HPV) appears to be involved in anogenital cancer, its involvement in head and neck squamous cell carcinoma (HNSCC) has been controversial. Certain HPV viral proteins show oncogenic potential *in vitro* via well-established interactions with cellular factors, but the *in vivo* implications have yet to be demonstrated. The prevalence of HPV in HNSCC also appears to be similar to the prevalence of HPV in normal mucosa of the upper aerodigestive tract. However, HPV is found in tonsillar carcinomas at a rate significantly higher than in normal mucosa, and high-risk HPV-16 appears to be the predominant type. Epidemiological studies also suggest that the risk of developing oral cancer is increased in the setting of HPV infection and certain sexual practices and history. Although much data supporting the association of HPV with HNSCC remain circumstantial, a growing body of evidence lends support for a causative relationship between HPV and HNSCC.

I. INTRODUCTION

Viral etiologies have been implicated in the pathogenesis of many different types of human cancer. The human papillomaviruses (HPVs) have been implicated in the carcinogenesis of human uterine cervical and anogenital cancer. In the case of uterine cervical cancer, HPV infection appears to be a necessary but insufficient etiological agent, as HPV DNA is found in up to 90% of the cervical biopsies positive for squamous cell carcinoma [1]. It is also found with high frequency in carcinomas of the anus, vulva, and penis [2]. In the region of the upper aerodigestive tract, HPVs are the causative agent of respiratory papillomatosis. However, their association with squamous cell carcinoma of the head and neck region (HNSCC) is less certain, as HPV DNA is found in only 15–40% of oral carcinomas. Although HPVs have demonstrated the ability to induce oncogenic transformation of cells *in vitro*, the causal relationship between

HPV and HNSCC remains controversial. This chapter examines epidemiological data suggesting a causal relationship between HPV and HNSCC, along with a review of the transforming interaction between host cellular proteins and HPV viral proteins.

II. HUMAN PAPILLOMAVIRUSES (HPVs)

A. Virus Life Cycle

Papillomaviruses are small (55 nm), nonenveloped, double-stranded DNA viruses belonging to the papovavirus family. A subset of human papillomaviruses such as type 1 (HPV-1), HPV-4, HPV-60, and HPV-65 cause strictly cutaneous lesions of keratinizing squamous epithelium. Others are associated with benign lesions of mucous membrane (e.g., condylomata accuminata and respiratory papillomatosis) and show a low probability of malignant transformation. These low-risk viruses include type-6 (HPV-6), HPV-11, and HPV-13 [3]. Other HPV types are found predominantly in malignant tumors of the uterine cervix, anogenital area, and the oral cavity. A specific subset of HPVs, including types 16, 18, 31, 33, and 39, is detected in up to 99% of anogenital preneoplastic lesions and squamous cell carcinomas [4]. These are termed "high-risk" HPVs. Their genomes are integrated into the host DNA and are transcriptionally active in both tumors and tumor-derived cell lines [4,5]. Based on this evidence, the International Agency for Research on Cancer (IARC, 1997) has classified HPV-16 and -18 as carcinogenic in humans (group 1), HPV-31 and -33 as probably carcinogenic in humans (group 2A), and some of the remaining HPV types as possibly carcinogenic in humans (group 2B) [6].

Most papillomaviruses have a strong tropism for squamous epithelial cells and induce papillomas in many higher vertebrate species. HPV infection of epithelial cells occurs as the result of self-inoculation from the genital area, sexual contact with an infected partner, or by inoculation from the maternal genital tract during birth [7,8]. The initial infection of the epidermal layer occurs in the basal and parabasal keratinocytes where viral DNA can be demonstrated by *in situ* hybridization [9]. The initial infection may either lead to the production of papillomas as a result of active viral replication or result in a latent infection in which the tissue remains histologically normal. It is also possible that the viruses can be cleared by the immune system of the host prior to the onset of clinically apparent HPV infection.

Latent HPV infection of clinically normal appearing mucosa of the upper aerodigestive tract has been well established. The detection rate of HPV in samples of normal oral mucosa ranges from 0 to 60% [10]. Activation of the dormant HPV in latent infections occurs via two mechanisms. Injury to the epithelial layer in the vicinity of infected tissue can result in HPV activation, presumably due to the resulting inflammation, which in turn stimulates cell proliferation [11–14]. Immunosuppression of the host can also lead to activation of latent infection [15]. Patients undergoing immunosuppression following renal transplant have been known to develop cutaneous papillomas [16].

In the case of active infection, viral replication does not occur until the HPV-containing cells reach terminal differentiation in the stratum spinosum and stratum granulosum layer [6]. The exact viral–host interaction that defines this restriction is not yet fully understood. The necessity for terminal differentiation of host cells for viral replication limits the amount of viral production, as these host cells are ultimately destined for growth arrest. To overcome this situation, the HPV viral genome induces transformation of the host cells and maintains continued replication of terminally differentiated keratinocytes. This alteration in the differentiation pattern of keratinocytes results in thickening of the squamous layer typical of papillomas. Dysplasia is rarely seen in papillomas. Instead, a common finding is cells, referred to as *koilocytes*, in the upper layer of papillomas, which display an area of clearing around the nucleus. Parakeratosis and the presence of multinucleated cells often accompany this change (Fig. 11.1, see also color insert). In the final stage of viral production, the infectious viral particles are most likely shed, packaged within the desquamated squames rather than as free particles [17].

B. Virion and Genomic Structure

HPVs consist of circular, double-strand DNA encased in a small, nonenveloped, icosahedral capsid. The capsid

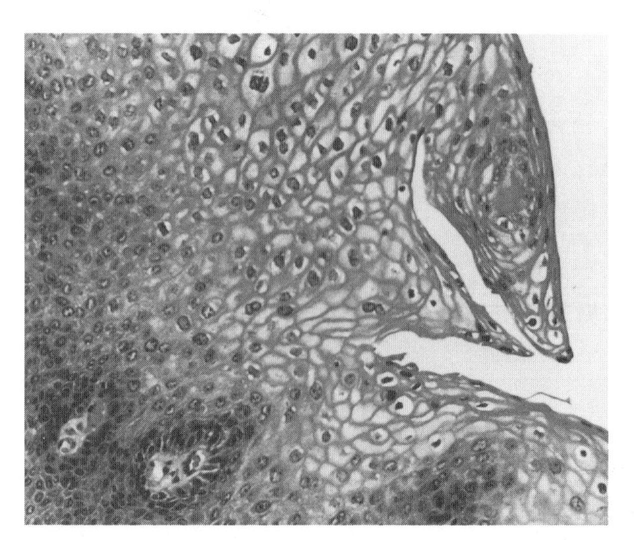

FIGURE 11.1 Histopathological appearance of laryngeal papillomatosis demonstrating the thickening of the epithelium around a fibrovascular core, oval nuclei in the superficial layer, and cytoplasmic vacuolization (Koilocytosis). Courtesy of Dr. D. M. Humphrey. (See also color insert.)

consists of at least two structural proteins encoded by the L1 and L2 reading open reading frames of the viral genome. Production of these structural proteins, as well as packaging of the viral genome, occurs in terminally differentiated keratinocytes of the superficial epithelium. Replication of the viral genome occurs in the nucleus of squamous epithelial cells.

Analysis of the human papillomavirus genome indicates that all open reading frames are located on one strand of viral DNA. The coding strand contains approximately 10 open reading frames, which have been classified as either early (E) or late (L) genes. Early genes are expressed during the early phase of infections and encode proteins that regulate gene expression and replication. This group of genes also confers the oncogenic transforming property on HPVs. The two late genes, L1 and L2, encode for structural proteins and are expressed in productively infected cells during active virion production.

The El protein has helicase activity that is important in viral replication [9]. The E2 protein appears to be a DNA-binding protein that is involved in the modulation of viral transcription [6]. The E2 protein binds to specific binding sites within the viral genome [18]. However, it has also been shown to transactivate heterologous promoters that do not contain E2-binding sites [19,20]. The E5 protein has been found to interact with membrane-bound growth factors [21] and displays *in vitro* transforming activities [22]. The function of the E4 protein is not yet elucidated, but its mRNA is the most abundant viral transcript in papillomas [23]. The transforming property of high-risk HPVs, however, appears to localize mostly to E6 and E7 proteins. Analysis of primary anogenital and cervical SCCA has revealed that only the E6 and E7 proteins are expressed with consistency in these tumors [24,25]. Although E7 shows the predominant transforming activity in rodent cells [26,27,28], interaction between E6 and E7 is necessary for the efficient transformation of human keratinocytes [29,30]. It should be emphasized that the E6 and E7 proteins of low-risk HPV such as HPV-6 or HPV-11 show very poor transforming ability in *in vitro* assays [31,32]. This difference is most likely due to intrinsic differences in protein function rather than in levels of expression [33].

The high-risk HPV E6 protein exerts an oncogenic effect by inhibiting the infected cells from entering DNA damage-induced apoptosis. The HPV E6 protein has been shown to complex with the P53 tumor suppressor protein with high affinity [34]. This interaction results in the ubiquination of P53 and its subsequent degradation [35,36]. The level of P53 protein is reduced in HPV-16 E6-immortalized cells [37] and these cells fail to undergo G1 cell cycle arrest following DNA damage [38].

The HPV-16 E7 protein has been shown to interact with the retinoblastoma tumor suppressor gene product, PRb [39]. This tumor suppressor protein inhibits cell growth and replication in part by binding to the E2F family of transcription factors and thereby inhibiting their activity. Binding of HPV-16 E7 to PRb results in the release of active E2F transcription factors from PRb [40,41] which in turn induces the expression of genes important in cell division control. This chapter elaborates further the interactions between cellular host factor and E6 and E7 proteins of high-risk HPV.

III. INTERACTION BETWEEN CELLULAR HOST FACTORS AND E6 OF HIGH-RISK HPVs

The E6 protein of high-risk HPVs has been shown to interact with a number of cellular proteins. Central to the transforming property of the E6 protein is its interaction with the cellular P53 protein. This interaction is mediated by E6-associated protein (E6-AP) ligase [36]. Binding of E6 to P53 results in ubiquination of the P53 protein and its subsequent degradation [36]. The level of P53 is decreased in E6-immortalized cells, and these cells fail to undergo cell cycle arrest following DNA damage [37]. This decrease in the intracellular level of P53 has also been shown to increase mutagenesis in human cells *in vitro* as well as result in chromosomal instability [38,42–44]. Consistent with these data is the observation that 10–30% of patients who underwent irradiation of laryngeal papillomas eventually developed squamous cell carcinomas [45,46]. However, Kyono *et al.* [47] reported that the decrease in the intracellular P53 level is not a requirement for immortalization by E6 [47]. Furthermore, SCCs of the oral cavity positive for the presence of HPV often show a normal or increased level of P53 [48,49]. E6 proteins of low-risk HPV-6 and HPV-11 show a significantly lower affinity for the P53 protein and are defective for the ability to promote its degradation [50].

Oda *et al.* [51] showed that the intracellular level of a Src-family tyrosine kinase, Blk, is regulated by E6-AP. Binding of E6-AP to Blk leads to its ubiquitination and subsequent degradation. The consequence of E6 production on the E6-APBlk interaction is not yet defined, but it is possible that the sequestration of intracellular E6-AP to p53 in the presence of E6 may result in a longer half-life for Blk. This stabilization of Blk may result in the stimulation of mitotic activity.

Telomerase of human cervical keratinocytes has been shown to be activated by the HPV-16 E6 protein [52]. Telomeres, located on the ends of chromosomal DNA, shorten with each mitosis. This gradual truncation of chromosomal DNA with age is thought to contribute to senescence. Activation of telomerase results in preservation of the telomeres and is a characteristic of most cell lines and tumors. Expression of HPV-E6 in keratinocytes does lead to an extended life span but is not sufficient for immortalization.

Another interesting observation is the loss of expression of P16 in human uroepithelial cells that were transformed by HPV-16 E6 [53]. P16 blocks cell cycle progression and has been found to be markedly elevated in senescent human uroepithelial cells [53]. P16 inhibits the phosphorylation of PRb by cyclin D/cyclin-dependent kinase complexes. Phosphorylation of PRb results in the release of the E2F-1 transcription factor and subsequent cell cycle progression. In contrast, human uroepithelial cells transformed by HPV-16 E7 show a high level of P16 [53]. The elevated level of P16 in E7-transformed cells is most likely due to the direct inhibition of Rb by E7, which bypasses the action of P16. These interactions suggest a convergence point between the separate transforming mechanisms of the high-risk HPV E6 and E7 proteins.

Ronco *et al.* [54] showed that HPV-16 E6 binds to interferon regulatory factor-3 (IRF-3). IRF-3 is produced in the presence of double-stranded RNA or viral infection and induces the transcription of interferon-β (IF-β). More importantly, the binding of HPV-16 E6 to IRF-3 significantly impaired this transactivation of IF-β promoters. This suppression of IF-β expression may allow the human papillomavirus a means to circumvent the antiviral activity of the infected cell.

The E6 protein of high-risk HPVs has been found to interact with other cellular proteins as well. E6TP1, a novel putative GTPase-activating protein [80], focal adhesion protein paxillin [56,57], calcium-binding protein ERC 55 [58], and the human homologue of the *Drosophila* discs large tumor suppressor protein [59] have all been shown to interact with E6. However, the *in vivo* implication of these interactions has yet to be elucidated.

The interaction between P53 and high-risk HPV E6 protein appears to be crucial to the transforming property of E6 proteins. The enhanced degradation of P53 results in tumorigenesis and antiapoptosis conditions for the infected cell. However, the expression of E6 protein can also lead to the immortalization of the host cells by modification of a number of different endogenous pathways, such as the loss of P16 expression and the activation of telomerases. It appears that the transforming effect of the E6 protein does not localize to one specific interaction but is due to a number of pleiotropic interactions.

IV. INTERACTION BETWEEN CELLULAR HOST FACTORS AND E7 OF HIGH-RISK HPVs

A key aspect of the transforming property of high-risk HPV E7 protein is its interaction with the PRb [40]. Binding of E7 to PRb results in release of the E2F family of growth factors. E2F directly regulates the transcription of a diverse set of genes involved in DNA replication and cell growth, including B-*myb*, cdc2, c-*myc*, and cyclin A [60,61]. Furthermore, a fusion protein consisting of the N-terminal half of the HPV-16 E7 protein and the full-length HPV E6 protein was shown to promote the *in vitro* degradation of PRb by ubiquination [62]. Therefore, degradation of PRb with a subsequent release of E2F transcription factors may contribute to the transforming activity of E7 of HPV-16 and -18. The E7 protein has been shown to bind to other transcription factors, such as the TATA box-binding protein [63] and c-jun [64], but these interactions need to be elucidated further.

As is the case for E6, the E7 proteins of high-risk HPVs and low-risk HPVs are quite similar in amino acid composition and structural organization, yet differ in their transforming potentials. The E7 proteins of HPV-16 and -18 bind to PRb with higher affinity than the E7 proteins of HPV-6 and -11. The *in vitro* transforming activity of HPV E7 is thus associated with its efficiency of binding to PRb [65].

The E7 protein of high-risk HPVs also interacts with a number of other cellular host factors. Of particular interest is the binding of high-risk HPV E6 with Mi2B, a component of a histone deacetylase complex involved in chromatin remodeling and histone deacetylation activity [66]. Gene expression can be regulated *via* the differential acetylation of histones within nucleosome [67]. Acetylation of lysine residues within the histones results in weakening of the DNA–histone interaction and leads to a more open chromatin structure. This allows easier access of the transcription machinery to the DNA strands. This finding suggests that the E7 protein exerts its transforming effect partly by influencing the histone deacetylation pathway. Relevant to this finding is the observation that the E7 protein also binds to histone H1 kinase at the G2/M phase of the cell cycle [68]. Although the *in vivo* significance of this finding is as yet unclear, mutant E7 lacking in histone H1 kinase-binding activity was transformation defective.

In a study by Funk *et al.* [69], HPV-16 E7 was shown to affect directly the regulation of the cell cycle by binding to, and inhibiting the activity of, P21. P21 is induced by cellular DNA damage and is a critical determinant of G1 arrest in response to DNA damage [70]. It inhibits cyclin-dependent kinase (CDK) activity and proliferating cell nuclear antigen (PCNA)-dependent DNA replication by binding to CDK/cyclin complexes and to PNCA [71,72]. Funk and colleagues also showed that low-risk HPV (HPV-6) E7 had significantly lower affinity for P21 than HPV-16 E7 [69]. By abrogating the inhibitory effect of P21, HPV-16 E7 can release infected cells from cell cycle arrest. The inability to respond to cell cycle arrest signals can also promote genetic instability during the progression of HPV-induced neoplasm. Similar to the high-risk HPV E6 protein, expression of the E7 protein enhances the mutagenic potential of the infected cells, as well as enhancing the integration of foreign DNA into the host cell DNA [42,44].

Other less-defined interactions between E7 and host cellular factors are the S4 subunit of the 26 S proteosome [73], M2 pyruvate kinase [74], and hTid-1, a homologue of the *Drosophila* tumor suppressor protein Tid56 [75].

V. METHODS OF DETECTION OF HPVs

Methods for detecting HPV have been reviewed [76]. Conventional light microscopy (LM) is able to detect the active HPV infection by visualizing the presence of koilocytosis along with parakeratosis and the presence of mutinucleated cells. However, these histological findings are not absolutely specific of HPV infection and can result in false positives. Although the presence of HPV particles can be visualized directly by electron microscopy (EM), both LM and EM are unable to detect HPV in latent infections and are unable to differentiate HPV types.

Immunohistochemistry utilizes labeled antibodies against the capsid protein of HPV to detect the presence of HPV viral particles. The assay is performed directly on tissue sections, which may be fresh or paraffin embedded. Antibodies against the capsid protein usually react with all subtypes and do not allow detection of latent infections. HPV types can be differentiated by using type-specific antibodies raised against HPV oncoproteins. However, this also does not allow detection of other HPV types.

In situ hybridization utilizes radioactively or biotin-labeled complementary DNA probes against viral DNA or RNA. This method can be used directly on tissue sections and allows for correlation between the presence of HPV and histological findings. By targeting specific viral RNA transcripts, the level of expression of a particular viral gene may also be examined. The relatively moderate sensitivity of *in situ* hybridization requires 50 to 100 copies of viral DNA per cell in multiple cells [77]. For this reason, *in situ* hybridization cannot detect latent infections.

Southern blot hybridization offers higher sensitivity and can detect 1 to 10 copies of viral DNA per cell [78]. Type-specific HPV detection is achieved under stringent conditions, but HPV types different from the probe can be detected by decreasing the stringency of the reaction. This method is best used on fresh tissue and is less effective in retrospective studies on archived material.

The polymerase chain reaction (PCR) offers the highest sensitivity and can detect as low as one copy of HPV DNA molecule per cell. This sensitivity allows the detection of latent as well as active infections. Because of such high sensitivity, the presence of minute quantities of HPV DNA, which may not be meaningful clinically or biologically, will still result in positive detection by PCR. The use of HPV type-specific primers also allows the identification of specific HPV types. However, some primers (i.e., primers targeting the L1 region of HPV genome) can also amplify regions of the human genome [76]. Confirming the identity of the PCR product after amplification (with direct sequencing, restriction endonuclease digestion, or Southern blot) will prevent this situation. With *in situ* PCR, the amplification reaction is performed directly on tissue sections, either fresh or paraffin embedded. This technique allows the use of PCR on archived specimens for retrospective study. The correlation between HPV signals and histological findings can be examined as well. The disadvantage of *in situ* PCR is the possibility of false positives due to its high sensitivity.

VI. HPV AND NORMAL MUCOSA OF THE UPPER AERODIGESTIVE TRACT

The rate of HPV detection in normal mucosa of the upper aerodigestive tract reported in literature ranges widely from 0 to 60% when all the subsites of head and neck are considered (Table 11.1). Miller and White [10] retrospectively reviewed 15 reports of HPV detection in normal oral tissue. The mean prevalence of HPV in normal oral mucosa was found to be 13.5% (standard deviation ±19.4%, range 0 to 60%). When the review was limited to reports that utilized PCR as the method of detection, the detection rate increased to 25.4% [10]. In one of the larger studies involving normal oral mucosal specimens from 97 volunteers, Ostwald *et al.* [79] showed HPV DNA in only 1% (1/97) of the subjects using PCR with a consensus primer for HPV type 6/11, -16, and -18. Bouda *et al.* [80]

TABLE 11.1 Detection of HPV in Normal Tissues of the Upper Aerodigestive Tract

Sites	HPV positive	Methods[a]	Reference
Oral cavity	1/97 (1%)	PCR	Ostwald *et al.* [79]
Oral cavity	0/16 (0%)	PCR	Bouda *et al.* [80]
Oral cavity	36/60 (60%)	PCR	Lawton *et al.* [81]
Oral cavity	21/48 (44%)	PCR	Jalal *et al.* [82]
Nasal cavity	0/21 (0%)	PCR, ISH	Buchwald *et al.* [84]
Nasal cavity	1/61 (1.6%)	PCR	Eike *et al.* [86]
Larynx	4/12 (30%)	PCR	Nunez *et al.* [87]

[a]PCR, polymerase chain reaction; ISH, *in situ* hybridization.

examined normal oral mucosa of 16 volunteers using nested PCR with primers designed for detection of HPV-6, -11, -16, -18, -31, and -33. None of the 16 volunteers examined were shown to harbor HPV DNA. In contrast, Lawton and co-workers [81] were able to detect HPV in clinically normal oral mucosa of 60% of 60 subjects using PCR. HPV-16 was the prevalent type. Jalal *et al.* [82] also employed PCR to detect HPV-16 in 21 out of 48 healthy young volunteers. Furthermore, the prevalence of HPV in clinically normal oral mucosa of patients with genital papilloma appears to be similar to that found in normal subjects. In a study by Kellkowski *et al.* [83], oral mucosa of 309 patients with genital HPV infections were analyzed by dot blot hybridization using labeled DNA probes for HPV-2, -6, -7, -11, -13, and -16. HPV infection was found in 3.8% of 309 women. Of the patients with a positive finding, only 2 had clinical oral lesions indicative of HPV infection.

HPV is rarely demonstrated in normal nasal mucosa [84,85]. Buchwald *et al.* [84] examined mucosal biopsies taken from 21 otherwise healthy individuals who underwent surgery for a nasal fracture or deviated nasal septum. Neither polymerase chain reaction nor *in situ* hybridization was able to detect the presence of HPV in any of the specimens. In a similar study, Eike *et al.* [86] detected HPV in only 1 of 61 nasal smear biopsies by PCR in otherwise healthy individuals.

Nunez *et al.* [87] examined the rate of HPV detection in normal laryngeal mucosa. Twelve clinically normal larynges were obtained during autopsy and were examined with PCR using primers for HPV-11, -16, and -18. The presence of HPV was demonstrated in 4 of 12 (40%) specimens and all were HPV-11.

The detection rate of HPV appears to vary from 2 to 40% depending on the anatomic location. There also appears to be a wide variance in the detection rate within individual anatomic locations. Such variance in rate of detection is most likely due to the difference in methods of detection, the difference in the study population, and the small number of specimens examined. Regardless, detection of HPV in normal oral mucosa serves to emphasize the need for caution when examining the epidemiological association between HPV and lesions of head and neck.

VII. HPV AND BENIGN LESIONS OF THE HEAD AND NECK

A. Benign Oral Lesions

Benign lesions of the oral cavity include squamous papilloma, verruca vulgaris, condyloma acuminatum, and focal epithelial hyperplasia. These lesions have a very low probability of transformation into malignant entities.

The incidence of oral squamous papilloma is less than 0.1% and occurs usually on the soft palate or the uvula. These lesions occur mostly in the third to fifth decades of life. A histology of papillomas shows a projection of squamous epithelium surrounding fibrovascular cores. These lesions are clearly associated with HPV and mainly demonstrate HPV-6 [88–90] and HPV-11 [88,89,91]. HPV-16 has also been reported in oral papillomas [92].

Focal epithelial hyperplasia (FEH), also called Heck's disease, is a benign familial disorder characterized by multiple papillomatous lesions of the oral cavity. The lesions of focal epithelial hyperplasia show a hyperplastic epithelium with keratosis and broadening of the rete ridges (so called "Bronze Age axe" appearance). HPV-13 and -32 have been isolated from papillomatous lesions of FEH [93–95]. These HPV subtypes are rarely found in other oral lesions and appear to be specific etiological factors for FEH in genetically predisposed individuals [92].

Oral verrucae vulgaris are the mucosal counterpart of warts found on keratinizing squamous epithelium. These lesions are found most frequently on lip, labial mucosa, and the tongue. Histology shows hyperkeratosis with elongated rete ridges that are bent inward at the margins of the lesion. HPV-2 and -4 are found in verruca vulgaris and are most likely the causative agent of these lesions [96,97].

Condyloma accuminata are usually seen in the anogenital region and have a more cauliflower-like appearance than papillomas. In the oral cavity, these lesions are found on buccal and labial mucosa, tongue, palate, and gingiva. As in the condyloma accuminata of the anogenital region, HPV-6 and -11 are the most frequently isolated genotypes.

B. Laryngeal Papillomatosis

Laryngeal papillomatosis is the most common laryngeal tumor found in children and can involve the trachea, bronchi, and the lungs. Treatment is primarily surgical but recurrence is common. This may be due to the presence of latent HPV infections in the adjacent, normal mucosa [98,99].

In contrast to other benign or premalignant lesions of the head and neck, a definite causal relationship can be established for laryngeal papillomatosis and human papillomavirus type-6 and -11. Lack *et al.* [100] and Costa *et al.* [101] were initially able to demonstrate the presence of HPV in laryngeal papillomas with immunohistochemistry. The detection rate ranged from 48 to 50% in these studies. However, later studies utilizing *in situ* hybridization or more sensitive PCR methods have detected HPV-6 and -11 in 90 to 100% of both adult and juvenile laryngeal papillomatosis (Table 11.2) [98,102–108]. Studies by Gomez *et al.* [108] and Pou *et al.* [109] were able to demonstrate the presence of HPV-6/11 in 100% of specimens using PCR. With such a high detection rate, there is little doubt that HPV-6 and -11 are the causative agents of laryngeal papillomatosis.

TABLE 11.2 Detection of HPV in Laryngeal Papillomatosis

No. of specimens	HPV positive	Methods[a]	Reference
26	26/26 (100%)	SBH	Abramson *et al.* [102]
10	10/10 (100%)	ISH	Terry *et al.* [103]
14	13/14 (93%)	SBH	Corbitt *et al.* [104]
13	13/13 (100%)	PCR	Levi *et al.* [105]
8	8/8 (100%)	ISH	Tsutsumi *et al.* [106]
23	23/24 (96%)	ISH	Rimell *et al.* [107]
11	10/11 (91%)	PCR	Rihkanen *et al.* [98]
20	20/20 (100%)	PCR	Gomez *et al.* [108]

[a]PCR, polymerase chain reaction; ISH, *in situ* hybridization; SBH, Southern blot hybridization.

High-risk HPV-16 is found infrequently in laryngeal papillomatosis and may increase the risk of malignant degeneration. Pou *et al.* [109] examined the records and biopsy specimens of 29 patients with laryngeal papillomatosis. Detection of HPV on 24 of 29 paraffin-embedded specimen was performed using PCR. Twenty-one of 24 patients were infected with HPV-6 and the other 3 demonstrated the presence of HPV-11 or -16. Furthermore, 2 of 3 patients who eventually developed squamous cell carcinoma of the larynx were infected with HVP-11 or -16. This study also noted a more aggressive clinical course when the laryngeal lesions were coinfected with other viruses, such as Epstein–Barr virus, cytomegalovirus, or herpes simplex virus.

VIII. HPVs AND PREMALIGNANT LESIONS OF THE HEAD AND NECK

A. Premalignant Lesions of the Oral Cavity

Premalignant lesions of the oral cavity include leukoplakia, erythroplakia, and lichen planus. Leukoplakia is a clinical term that describes a white patch in the mucous membrane of the mouth, the tongue, or the lip. It is usually considered the result of chronic inflammation resulting in the proliferation of epithelial and connective tissue. Histological examination shows a variety of morphologies, including hyperkeratosis, parakeratosis, dyskeratosis, and carcinoma *in situ*. The most common histological finding is benign epithelial hyperplasia, which is found in 80% of cases [110]. Epithelial dysplasia is demonstrated in 20% of the cases, and the incidence of carcinoma *in situ* is noted to be approximately 2%. Erythroplakia describes slightly raised, friable, red granular lesions on the mucosal membrane of the oral cavity usually found on the anterior tonsillar pillar or the retromolar trigone [111]. These lesions have a higher malignant potential than leukoplakia. Lichen planus is also a disorder of the skin and mucous membrane with a malignant potential. In one study, Vas Kovskaia and Abramova [112] followed 725 patients with lichen planus of the oral cavity and labial mucosa for up to 32 years and noted malignant degeneration in 4% of the patients.

Several studies have demonstrated the presence of HPV in these premalignant lesions [80,92,113–115]. Kashima *et al.* [113] utilized Southern transfer hybridization and reverse blot hybridization to detect HPV-11 in 1 out of 30 patients with leukoplakia of the oral cavity. In addition, samples of oral mucosa from 22 patients with lichen planus were analyzed using immunoperoxidase staining with antibody to the HPV capsid antigen. The presence of HPV was detected in 4 of 22 specimen.

Investigations using more sensitive techniques of HPV detection have demonstrated a higher detection rate. Nielsen *et al.* [114] examined 49 patients with oral premalignant lesions and 20 control patients for the presence of HPV with immunohistochemical staining, *in situ* hybridization, and PCR analyzed by Southern blot hybridization with an HPV-16-specific probe. The overall HPV detection rate was 40.8% in premalignant lesions and 0% in control subjects. Furthermore, HPV was found in 62.5% of verrucous leucoplakias, 50% of the erythroplakias, 45.5% of the homogeneous leukoplakias, 33.3% of erythroleukoplakias, and 12.5% of nodular leukoplakias.

Bouda *et al.* [80] used a highly sensitive nested PCR technique to detect HPV in 30 of 34 leukoplakia specimens. None of the normal specimens was found to be infected. Further analysis using genotype-specific PCR showed at least one high-risk type (HPV-16, -18, or -33) in 98% of the infected specimens. HPV-16 was the prevalent type, being found in 71% of the infected specimens. However, no correlation was found between the HPV type and the degree of dysplasia. Positivity for HPV was also independent of the tobacco history of the analyzed group of patients. Sand *et al.* [115] analyzed a group of 22 patients with lichen planus and 7 patients with leukoplakia of the oral cavity. Six of 22 (27.3%) patients with lichen planus were HPV positive by PCR. Five of the 6 HPV-positive patients demonstrated the presence of HPV-18. Two (26.9%) of 7 specimens of leukoplakia were positive for both HPV-11 and -16. None of the 12 controls was positive for HPV. This study was also unable to find a correlation between HPV infection and the use of alcohol and tobacco.

Although a subset of the normal population does carry latent HPV infections, each of the studies presented earlier included a group of control specimens that did not demonstrate the presence of HPV. The presence of HPV in premalignant lesions of the oral cavity but not in normal oral mucosa is strongly suggestive of a relationship between HPV and dysplastic transformation of oral mucosa.

B. Premalignant Lesions of the Larynx

As in the oral cavity, leukoplakia of the larynx is considered a precursor of malignant lesions. Studies concerning the detection of HPV in premalignant lesions of the larynx are both limited and conflicting. In a study involving 115 biopsy samples of laryngeal keratosis, the presence of HPV could not be detected in any of the specimens by *in situ* hybridization [116]. Fouret *et al.* [117] screened 57 cases of paraffin-embedded laryngeal premalignant lesions with immunohistochemistry and PCR. In this study, HPV was detected in 6 of 57 specimen and all were of type 16. Azzimonti *et al.* [118] also examined 50 paraffin-embedded tissues containing laryngeal precancerous lesions with PCR, as well as immunohistochemistry with antibody against the L1 protein. The presence of HPV DNA was demonstrated in 28 of 50 specimen (56%), including 9 of 12 cases with mild dysplasia (64%), 3 of 6 cases with moderate dysplasia (50%), and 7 of 11 cases with severe dysplasia (64%). The remaining 11 cases of HPV-positive specimens were from 21 cases with keratosis and without any dysplasia. The presence of HPV was independent of the grade of dysplasia in the premalignant lesion.

C. Schneiderian Papilloma of the Nasal Cavity

Papillomas arising from the pseudostriated columnar epithelium of the nasal cavity are referred to as Schneiderian papillomas and can be subtyped histologi cally as inverting, fungiform, or cylindric cell papilloma. Malignant change is found in approximately 10% of inverting papillomas.

As with laryngeal papillomatosis, HPV-6 and -11 are found most frequently in Schneiderian papillomas. The rate of HPV detection in Schneiderian papilloma ranges from 3 to 90% [84,119–124]. This wide variance may be due to differences in population examined, the method utilized, and small numbers of samples. In one of the larger studies, the incidence and subtypes of HPV in Schneiderian papillomas were analyzed by Weiner *et al.* [123] with the use of PCR. Of the 88 tumor specimen analyzed, 69 were the inverting subtype, 17 the fungiform subtype, and 2 the cylindric subtype. The presence of HPV was demonstrated in 5 of 69 (6.8%) inverting papillomas, 17 of 17 (100%) fungiform papillomas, and 0 of 2 cylindric cell papillomas. Three of the HPV-positive inverting papillomas showed HPV-11 and two showed HPV-16. HPV-6b and -11 accounted for all cases of fungiform papillomas. Buchwald *et al.* [85] and Sarkar *et al.* [119] (Table 11.3) also demonstrated the preferential association between HPV and fungiform subtypes. In all three studies, the prevalent HPV types were HPV-6 and -11.

Hwang *et al.* [124] analyzed paraffin-embedded samples of inverting papillomas using PCR with type-specific primers for HPV-6, -11, -16, -18, and -33. Of the 36 cases of inverting papillomas, 2 cases of HPV-11 and 1 of HPV-6 were detected. HPV-16 was also found in 2 of 5 inverting papilloma specimens with coexisting squamous cell carcinoma. It was further noted that inverted papillomas recurred in 2 (66%) of 3 cases positive for HPV and 2 (6%) of 33 cases negative for HPV. This association between the presence of HPV and the higher risk of recurrence after surgical excision was also noted by Beck *et al.* [121]. In this study, the tumor samples and records of 32 patients who underwent resection of inverting papillomas were examined. HPV was found in 20 (63%) of 32 specimens by PCR. Thirteen of 15 patients with HPV-positive tumors eventually developed recurrence, whereas recurrence was noted in none of the 10 patients with HPV-negative tumors. These finding suggest that the presence of HPV in inverted papilloma may predispose the lesion to a higher rate of recurrence.

TABLE 11.3 Detection of HPV in Schneiderian Papilloma

Inverting papilloma	Fungiform papillom	Cylindrical cell papilloma	Reference
5/69 (8%)	17/17 (100%)	0/2 (0%)	Sarkar *et al.* [119][a]
HPV 6/11	HPV 6/11	—	
3/52 (6%)	11/16 (69%)	0/5 (0%)	Buchwald *et al.* [85][a]
HPV 6/11	HPV 6/11	—	
0/24 (0%)	1/9 (11%)	0/2 (0%)	Weiner *et al.* [123][b]
—	HPV 6/11	—	

[a]PCR and *in situ* hybridization performed on paraffin-embedded specimens.
[b]PCR performed on paraffin-embedded specimens.

IX. HPV AND SCCA OF HEAD AND NECK

A. Detection of HPV in HNSCC

Several investigators have examined the detection rate of HPV in HNSCC. When SCCA of the oral cavity, oropharynx, hypopharynx, larynx, and the esophagus were combined, the prevalence appears to range from 7.9 to 25% (Table 11.4) [125–129]. HPV-16 was detected most frequently (64 to 100% of the HPV isolates). In one of the larger studies, Gillison *et al.* [125] analyzed 253 specimens of HNSCC using PCR, Southern blot hybridization, and *in situ* hybridization. HPV was detected in 62 of 253 (25%) cases, with HPV-16 accounting for 90% of the HPV-positive tumors.

The prevalence of HPV in SCCA of nasal cavity, nasopharynx, and paranasal sinuses varies from 0 to 66% with a mean prevalence of 17% [130–134] (Table 11.5). Again, HPV-16 was the predominant HPV type in these neoplasms. Therefore, a specific high-risk HPV type appears to be the prevalent type found in HNSCC. The prevalence of HPV in squamous cell carcinoma of head and neck, however, does not appear to be significantly higher than the prevalence of HPV in normal mucosa of the upper aerodigestive tract.

Nevertheless, when the prevalence of HPV is compared among squamous cell carcinoma of different anatomic sites within the head and neck, there is a significantly preferential association of HPV for tonsillar SCC. Brandsma and Abramson [129] were the first to compare the prevalence of HPV between various anatomic sites within the head and neck. In this study, 101 cases of HNSCC and controls consisting of 116 cases of biopsy specimens from benign disease or structural abnormalities were analyzed for the presence of HPV using Southern blot hybridization for HPV-11, -16, and -18. When all the anatomic sites were considered, the prevalence of HPV was 8.6 and 1% among cancerous and control specimens, respectively. However, the prevalence rate of HPV in tonsillar SCC specimens was significantly higher at 29%. All HPV-positive lesions showed HPV-16.

Paz *et al.* [128], who analyzed 167 cases of HNSCC for the presence of HPV using Southern blotting and PCR, also demonstrated this predilection of HPV for the tonsillar region. The overall prevalence rate of HPV detection in all anatomic sites of head and neck combined was 15%, whereas SCC from tonsillar fossa demonstrated a detection rate of 60%. HPV-16 was the most prevalent type, accounting for 76% of HPV positivity. In the study by Gillison *et al.* [125], HPV was found in 32 of 34 (94%) tonsillar carcinoma, where the prevalence for all anatomic sites of head and neck combined was 25%. HPV-16 accounted for 90% of HPV-positive tumors in this series.

Several studies have examined the detection rate of HPV specifically in tonsillar carcinoma [138–140]. In a study by Snijders *et al.* [135], the presence of HPV DNA was assessed in 10 cases of tonsillar carcinoma, as well as 7 control cases of tonsillectomy due to recurrent tonsillitis. Type-specific PCR was utilized with primers specific for types 6, 11, 16, 18, 31, and 33. HPV was detected in 10 of 10 (100%) tonsillar carcinomas and in none of the control tonsillectomy specimens. Four carcinomas contained HPV-16, 4 contained HPV-33, and one contained an HPV-16/33 double infection. This study also utilized RT-PCR and RNA *in situ* hybridization to detect viral RNA transcripts in tumor cells, indicating that the detection of HPV represents an active infection. Wilczynski *et al.* [136] utilized type-specific PCR to detect HPV in 22 cases of tonsillar carcinoma. HPV was detected in 14 of 22 (64%) tonsillar carcinoma specimens, and 12 of the 14 HPVs were of type 16.

These findings suggest very strongly that there is a predilection of HPV for tonsillar carcinoma (Table 11.6). The high-risk HPV-16 accounts for the majority of HPV detected in HNSCC, including tonsillar carcinoma, where it accounts for approximately 80% of HPV positivity. The basis for this preferential association between high-risk HPV and tonsillar carcinoma is not yet understood. There may be intrinsic differences in the intracellular or extracellular environment of the tonsillar region when compared with neighboring sites of head and neck. It is also possible that the tonsillar region is more prone to mucosal injury, facilitating viral access to the basal and parabasal layers of the mucosa and promoting the activation of latent HPV infections.

TABLE 11.4 Detection of HPV in Oral Cavity, Oropharynx, Hypopharynx, Larynx, and Esophagus

No. of tumors	HPV-positive tumors	HPV 16	Reference
253	62/253 (25%)	56/62 (90%)	Gillison *et al.* [125][a,b]
248	64/248 (25.8%)	41/64 (64%)	Schwartz *et al.* [127][a,c]
167	25/167 (15%)	19/25 (76%)	Paz *et al.* [128][a,d]
101	8/101 (7.9%)	8/8 (100%)	Brandsma and Abramson [129][a,e]

[a]HPV-16 is the most frequently detected type.
[b]PCR, *in situ* hybridization, and Southern blot performed on fresh-frozen specimens.
[c]PCR performed on fresh-frozen specimens.
[d]PCR and Southern blot performed on fresh-frozen specimens.
[e]*In situ* hybridization and Southern blot performed on fresh-frozen specimens.

TABLE 11.5 Detection of HPV in Nasal Cavity, Nasopharynx, and Paranasal Sinuses

No. of tumors	HPV-positive tumors	HPV types					Reference
		6	11	16	18	33	
6	4/6 (66%)	0	1	3	NE[a]	NE	Syrjanen *et al.* [131][b]
49	8/49 (14%)	NE	NE	7	1	NE	Furuta *et al.* [130][c]
8	1/8 (12%)	1	0	0	0	0	Judd *et al.* [134][d]
16	0/16 (0%)	NE	NE	0	0	NE	Dickens *et al.* [133][e]
15	4/15 (26%)	0	1	3	NE	NE	Hording *et al.* [132][f]

[a]Not examined.
[b]Light microscopy, *in situ* hybridization, and Southern blot hybridization performed on paraffin-embedded specimens.
[c]PCR performed on paraffin-embedded specimens.
[d]PCR and *in situ* hybridization performed on paraffin-embedded specimens.
[e]PCR and Southern blot hybridization performed on paraffin-embedded specimens.
[f]PCR performed on paraffin-embedded specimens.

B. Does HPV Infection Increase the Risk of HNSCC?

The association between HPV infection and the risk of developing HNSCC was examined by Smith *et al.* [126]. The authors analyzed 93 cases of an oral SCC and gender frequency-matched control group of 205 patients with no history of oral cancer. The presence of HPV among the cases and controls was determined by PCR analysis of exfoliated oral epithelium. The overall prevalence of HPV among the control group was 15%, with HPV-16 accounting for 42.9% of the HPV-positive tumors. As expected, the current use of alcohol and tobacco increased the risk of oral SCC [odds ratio (OR) = 2.57; 95% confidence interval (CI): 1.22–5.42 and OR = 2.63; 95% CI = 1.22–5.71, respectively]. However, when adjusted for known risk factors of oral SCC, multivariate analysis showed that the HPV status was also associated with a statistically significant increase in the risk of oral cancer independent of age, gender, or alcohol and tobacco use (OR = 3.70; 95% CI: 1.47–9.32; $P < 0.05$). The risk of oral SCC associated with HPV status was somewhat higher than that noted for tobacco use and alcohol consumption.

The association between sexual history and the risk of oral SCC was examined by Schwartz *et al.* [127]. It is a well-established fact that certain sexual histories or practices increase the risk of cervical cancer due to the sexual transmission of HPV. This study examined 284 cases of patients with oral SCC and an age–gender-matched control group of 477 subjects with no history of oral SCC. A detailed sexual history was obtained from both groups, including age at first intercourse, lifetime number of opposite sex partners, history of having performed oral sex, and prior diagnosis of genital warts. Oral SCC specimens were analyzed for the presence of HPV using type-specific PCR, and the control group was

TABLE 11.6 Detection of HPV in Tonsillar Carcinomas

No. of tumors	HPV + tumors	HPV type								Reference
		6	11	16	18	31	33	59	Unknown	
7	2/7 (29%)	NE[a]	0	2	0	NE	NE	NE	0	Brandsma and Abramson [129][b]
28	6/28 (21%)	0	0	6	NE	NE	NE	NE	0	Niedobitek *et al.* [137][c]
10	10/10 (100%)	0	0	6	4	0	0	NE	0	Snijder *et al.* [135][d]
15	9/15 (60%)	0	0	8	0	0	0	NE	1	Paz *et al.* [128][e]
22	14/22 (64%)	0	NE	12	0	0	1	1	0	Wilczynski *et al.* [136][f]
32	32/34 (94%)	—	—	—	—	—	—	—	—	Gillison *et al.* [125][g,h]

[a]Not examined.
[b]*In situ* hybridization and Southern blot performed on fresh-frozen specimens.
[c]*In situ* hybridizaiton, Southern blot, and immunohistochemical staining on paraffin-embedded specimens.
[d]PCR, Southern blot hybridization, RT-PCR, and RNA *in situ* hybridization on fresh-frozen specimens.
[e]PCR and Southern blot performed on fresh-frozen specimens.
[f]PCR and *in situ* hybridization performed on fresh-frozen and paraffin-embedded specimens.
[g]HPV types detected in tonsillar carcinoma not reported, but HPV-16 accounted for 90% of HPV-positive tumors for all subsites of head and neck.
[h]PCR, *in situ* hybridization, and Southern blot performed on fresh-frozen specimens.

assayed using type-specific PCR on exfoliated oral epithelium. Seropositivity to the HPV-16 capsid protein was also measured as an indicator of HPV exposure. When adjusted for age, tobacco use, and alcohol consumption, the risk of oral SCC in males was increased with decreasing age at first sexual intercourse, increasing number of lifetime opposite sex partners, and history of genital warts (Table 11.7). These patterns were not seen among women. The association between HPV status and increased risk of oral cancer was strongest for tumors containing HPV-16. For example, the OR for case subjects with ≥15 sex partners and HPV-16-positive tumors was 2.5 (95% CI=1.1–5.6), whereas the OR was 0.9 (95% CI=0.2–4.3) for subjects with ≥15 sex partners and tumors containing HPV-6 or -11. Therefore, sexual risk factors similar to those for women with cervical cancer appear to be involved in men and oral cancer as well.

Although a further study is warranted, these findings suggest that the presence of HPV in the upper aerodigestive tract is not an incidental finding but a significant risk factor in the pathogenesis of oral cancer.

C. Do HPV-Positive Tumors Constitute a Distinct and Separate Clinical Entity?

If HPV is involved in the pathogenesis of HNSCC, it is possible that the carcinogenic mechanism involved may represent a pathway distinct from that utilized by other known risk factors of HNSCC. Gillison *et al.* [125] retrospectively analyzed 253 cases of newly diagnosed or recurrent HNSCC by type-specific PCR, *in situ* hybridization, and Southern blot hybridization. The patients were followed for a median of 30 months. HPV-positive tumors were found predominantly in the oropharynx (OR=9.7, 95% CI=4.2–22). As noted earlier, the tonsillar area was involved most frequently within the oropharynx (OR=9.1, 95% CI=4.6–18). More interestingly, for oropharyngeal tumors only, HPV-positive tumors were noted to have distinct clinical behavior and morphological appearance when compared to HPV-negative tumors. At the time of presentation, HPV-positive tumors were noted to be more poorly differentiated (OR=3.5, 95% CI=1.4–8.8) and more likely to have a poorly differentiated, nonkeratinizing, basaloid morphology than HPV-negative tumors (OR=19.8, 95% CI=5.3–7.4). The poorly differentiated appearance of HPV-positive tumors was also noted by Wilczynzki *et al.* [136] among HPV-positive tonsillar carcinoma. Of note, this nonkeratinizing, basaloid morphology is also found commonly among anogenital carcinomas [138,139].

TABLE 11.7 Increased Risk of Oral Cancer with Certain Sexual History[a]

	Odds ratio (95% CI)
Lifetime number of opposite sex partners	
1.0[b]	1.0
2–4	1.3 (0.6–2.0)
5–14	1.5 (0.7–3.1)
≥15	2.3 (1.1–5.0)
Age at first regular sexual intercourse	
≥25[b]	1.0
20–24	1.7 (0.8–3.5)
18–19	1.6 (0.7–3.7)
≤18	3.4 (1.5–7.5)
Prior history of genital warts	
No[b]	1.0
Yes	2.2 (1.0–4.9)

[a]From Schwartz *et al.* [127].

[b]Reference group for odds ratio calculation.

A more intriguing finding in this study was the improved prognosis among patients with HPV-positive, oropharyngeal tumors compared with HPV-negative, oropharyngeal tumors. After adjusting for factors associated with poor prognosis (lymph node status, age, and alcohol consumption), patients with HPV-positive, oropharyngeal tumors had an approximately 40% reduction in the risk of death from all causes when compared with HPV-negative counterparts [hazard ratio (HR)=0.60, 95% CI=0.35–1.0]. When analyzed for risk of death from cancer, the authors noted a 60% reduction in the risk of death in HPV-positive, oropharyngeal tumors when compared with HPV-negative, oropharyngeal tumors (HR=0.40, 95% CI=0.19–0.84).

The association of HPV-positive SCCA with a particular anatomic location, characteristic morphology, and decreased risk of death suggests that HPV-positive, oropharyngeal cancers comprise a distinct clinical and pathological entity with improved prognosis. This observation not only strengthens the causal association between HPV and HNSCC, but also may lead to improved treatment formulation for HPV-positive oropharyngeal tumors.

X. CONCLUSIONS

Although much circumstantial evidence currently exits, the literature still lacks unequivocal evidence linking HPV with HNSCC. When all anatomic sites are considered, the prevalence of HPV in HNSCC does not appears to be higher than the prevalence of HPV in clinically normal mucosa. Detection of HPV also often requires a highly sensitive PCR method, which is associated with a high false-positive rate. Although HPV viral proteins have been shown to interact *in vitro* with host cellular factors in an oncogenic manner, an *in vivo* model has yet to be developed.

However, it is difficult to ignore the highly preferential association of HPV with tonsillar carcinoma, as well as the finding of high-risk HPV type 16 in these tumors. These findings are observed consistently in the literature.

The tonsillar region contains epithelial elements, as well as lymphoid tissue, and may form an intra- or extracellular environment distinct from neighboring anatomic sites. It is also possible that this area is more predisposed to epithelial injury, which than facilitates the subsequently viral infection.

Epidemiological findings also lend further support to a causal relationship between HPV and HNSCC, particularly for oropharyngeal tumors: (i) the risk of oral cancer has been shown to be increased in the presence of HPV infection, (ii) certain sexual risk factors known to be associated with cervical cancer for women have been shown to the associated with oral cancer in men, and (iii) HPV-positive oropharyngeal tumors appear to represent a distinct clinical and histopathological identity with improved prognosis. Further studies involving 1700 head and neck tumor specimens are currently ongoing under the direction of World Health Organization's International Agency for Research on Cancer [140]. A 5-year prospective study involving 1000 patients is also in the planning stage at Johns Hopkins Oncology Center. Data from these studies should further elucidate the role of HPV in HNSCC [140].

An etiological association between HPV and oral cancer implies that preventive interventions can be designed. A vaccine against HPV might be designed that might decrease the incidence of oral cancer. A trial of an HPV vaccine is currently being planned [140]. Furthermore, pharmacological agents that block the actions of E6 and E7 protein may be designed for the treatment of HPV-positive oral cancers. If the improved prognosis of HPV-positive oropharyngeal tumors is confirmed by further studies, then testing for HPV may also become a routine procedure in staging of these tumors.

References

1. Eversole, L. R., Laipis, P. J., Merrell, P., and Choi, E. (1987). Demonstration of human papillomavirus DNA in oral condyloma accuminatum. *J. Oral Pathol.* **16**, 266–272.
2. zur Hausen, H., and de Villiers, E. M. (1994). Human papillomaviruses. *Annu. Rev. Microbiol.* **48**, 427–447.
3. zur Hausen, H. (1991). Human papillomaviruses in the pathogenesis of anogenital cancer. *Virology* **184**, 9–13.
4. Shirasawa, H., Tomita, Y., Sekiya, S., Takamizawa, H., and Simizu, B. (1987). Integration and transcription of human papillomavirus type 16 and 18 sequences in cell lines derived from cervical carcinomas. *J. Gen. Virol.* **68**, 583–591.
5. Shirasawa, H., Tomita, Y., Kubota, K., Kasai, T., Sekiya, S., Takamizawa, H., and Simizu, B. (1988). Transcriptional differences of the human papillomavirus type 16 genome between precancerous lesions and invasive carcinomas. *J. Virol.* **62**, 1022–1027.
6. Sugarman, P. B., and Shillitoe, E. J. (1997). The high risk human papillomaviruses and oral cancer: Evidence for and against a causal relationship. *Oral Dis.* **3**, 130–147.
7. Yeudall, W. A. (1992). Human papillomaviruses and oral neoplasia. *Eur. J. Cancer Oral Oncol.* **28B**, 61–66.
8. Puranen, M., Yliskoski, M., Saarikoski, S., Syrjanen, K., and Syrjanen, S. (1996). Vertical transmission of human papillomavirus from infected mothers to their newborn babies and persistence of virus in childhood. *Am. J. Obstet. Gynecol.* **174**, 694–699.
9. Schneider, A., Oltersdorf, T., Schneider, V., and Gissmann, L. (1987). Distribution pattern of human papilloma virus 16 genome in cervical neoplasia by molecular in situ hybridization of tissue sections. *Int. J. Cancer* **39**, 717–721.
10. Miller, C. S., and White, D. K. (1996). Human papillomavirus expression in oral mucosa, premalignant conditions, and squamous cell carcinoma: A retrospective review of the literature. *Oral Med. Oral Surg. Oral Pathol. Oral Radiol. Endod.* **82**, 57–68.
11. Amtmann, E., Volm, M., and Wayss, K. (1984). Tumour induction in the rodent *Mastomys natalensis* by activation of endogenous papilloma virus genomes. *Nature* **308**, 291–292.
12. Siegsmund, M., Wayss, K., and Amtmann, E. (1991). Activation of latent papillomavirus genomes by chronic mechanical irritation. *J. Gen. Virol.* **72**, 2787–2789.
13. Campo, M. S., Jarrett, W. F., O'Neil, W., and Barron, R. J. (1994). Latent papillomavirus infection in cattle. *Res. Vet. Sci.* **56**, 151–157.
14. Amella, C. A., Lofgren, L. A., Ronn, A. M., Nouri, M., Shikowitz, M. J., and Steinberg, B. M. (1994). Latent infection induced with cottontail rabbit papillomavirus. A model for human papillomavirus latency. *Am. J. Pathol.* **144**, 1167–1171.
15. Benton, C., Shahidullah, H., and Hunter, J. A. A. (1992). Human papillomavirus in the immunosuppressed. *Papillomavirus Rep.* **3**, 23.
16. Shamanin, V., zur Hausen, H., Lavergne, D., Proby, C. M., Leigh, I. M., Neumann, C., Hamm, H., Goos, M., Haustein, U. F., Jung, E. G., Plewig, G., Wolff, H., and de Villiers, E. M. (1996). Human papillomavirus infections in nonmelanoma skin cancers from renal transplant recipients and nonimmunosuppressed patients. *J. Natl. Cancer. Inst.* **88**, 802–811.
17. Steinberg, B. M. (1999). Viral etiologies of head and neck cancer *In* "Head and Neck Cancer: A Multidisciplinary Approach" (L. B. Harrison, R. B. Sessions, and W. K. Hong, eds.), p.36. Lippincott-Raven, Philadelphia.
18. Ustav, M., and Stenlund, A. (1991). Transient replication of BPV-1 requires two viral polypeptides encoded the E1 and E2 open reading frames. *EMBO J.* **10**, 449–457.
19. Haugen, T. H., Cripe, T. P., Ginder, G. D., Karin, M., and Turek, L. P. (1987). Trans-activation of an upstream early gene promoter of bovine papilloma virus-1 by a product of the viral E2 gene. *EMBO J.* **6**, 145–152.
20. Haugen, T. H., Turek, L. P., Mercurio, F. M., Cripe, T. P., Olson, B. J., Anderson, R. D., Seidl, D., Karin, M., and Schiller, J. (1988). Sequence-specific and general transcriptional activation by the bovine papillomavirus-1 E2 trans-activator requires an N-terminal amphipathic helix-containing E2 domain. *EMBO J.* **7**, 4245–4253.
21. Leechanachai, P., Banks, L., Moreau, F., and Matlashewski, G. (1992). The E5 gene from human papillomavirus type 16 is an oncogene which enhances growth factor-mediated signal transduction to the nucleus. *Oncogene* **7**, 19–25.
22. Schiller, J. T., Vass, W. C., Vousden, K. H., and Lowy, D. R. (1986). E5 open reading frame of bovine papillomavirus type 1 encodes a transforming gene. *J. Virol.* **57**, 1–6.
23. Stoler, M. H., Whitbeck, A., Wolinsky, S. M., Broker, T. R., Chow, L. T., Howett, M. K., and Kreider, J. W. (1990). Infectious cycle of human papillomavirus type 11 in human foreskin xenografts in nude mice. *J. Virol.* **64**, 3310–3318.
24. Schneider-Gadicke, A., and Schwarz, E. (1986). Different human cervical carcinoma cell lines show similar transcription patterns of human papillomavirus type 18 early gene. *EMBO J.* **5**, 2285–2292.
25. Smotkin, D., and Wettstein, F. O. (1986). Transcription of human papillomavirus type 16 early genes in a cervical cancer and a cancer-derived cell line and identification of the E7 protein. *Proc. Natl. Acad. Sci. USA* **83**, 4680–4684.
26. Kanda, T., Furuno, A., and Yoshiike, K. (1988). Human papillomavirus type 16 open reading frame E7 encodes a transforming gene for rat 3Y1 cells. *J. Virol.* **62**, 610–613.

27. Phelps, W. C., Yee, C. L., Munger, K., and Howley, P. M. (1988). The human papillomavirus type 16 E7 gene encodes transactivation and transforming functions similar to those of adenovirus E1A. *Cell* **53**, 539–547.
28. Vousden, K. H., Doniger, J., DiPaolo, J. A., and Lowy, D. R. (1988). The E7 open reading frame of human papillomavirus type 16 encodes a transforming gene. *Oncogene Res.* **3**, 167–175.
29. Hawley-Nelson, P., Vousden, K. H., Hubbert, N. L., Lowy, D. R., and Schiller, J. T. (1989). HPV 16 E6 and E7 proteins cooperate to immortalize human foreskin keratinocytes. *EMBO J.* **8**, 3905–3910.
30. Munger, K., Phelps, W. C., Bubb, V., Howley, P. M., and Schlegel, R. (1989). The E6 and E7 genes of the human papillomavirus type 16 together are necessary and sufficient for transformation of primary human keratinocytes. *J. Virol.* **63**, 4417–4421.
31. Storey, A., Pim, D., Murray, A., Osborn, K., Banks, L., and Crawford, L. (1988). Comparison of the in vitro transforming activities of human papillomavirus types. *EMBO J.* **7**, 1815–1820.
32. Woodworth, C. D., Doniger, J., and DiPaolo, J. A. (1989). Immortalization of human foreskin keratinocytes by various human papillomavirus DNAs corresponds to their association with cervical carcinoma. *J. Virol.* **63**, 159–164.
33. Barbosa, M. S., Vass, W. C., Lowy, D. R., and Schiller, J. T. (1991). In vitro biological activities of the E6 and E7 genes vary among human papillomaviruses of different oncogenic potential. *J. Virol.* **65**, 292–298.
34. Werness, B. A., Levine, A. J., and Howley, P. M. (1990). Association of human papillomavirus type 16 and 18 E6 proteins with p53. *Science* **248**, 76–79.
35. Scheffner, M., Werness, B. A., Huibregtse, J. M., Levine, A. J., and Howley, P. M. (1990). The E6 oncoprotein encoded by human papillomavirus types 16 and 18 promotes the degradation of p53. *Cell* **63**, 1129–1136.
36. Scheffner, M., Huibregtse, J. M., Vierstra, R. D., and Howley, P. M. (1993). The HPV-16 E6 and E6-AP complex functions as a ubiquitin-protein ligase in the ubiquitination of p53. *Cell* **75**, 495–505.
37. Band, V., De Caprio, J. A., Delmolino, L., Kulesa,V., and Sager, R. (1991). Loss of p53 protein in human papillomavirus type 16 E6-immortalized human mammary epithelial cells. *J. Virol.* **65**, 6671–6676.
38. Kessis, T. D., Slebos, R. J., Nelson, W. G., Kastan, M. B., Plunkett, B. S., Han, S. M., Lorincz, A. T., Hedrick, L., and Cho, K. R. (1993). Human papillomavirus 16 E6 expression disrupts the p53-mediated cellular response to DNA damage. *Proc. Natl. Acad. Sci. USA* **90**, 3988–3992.
39. Dyson, N., Howley, P. M., Munger, K., and Harlow, E. (1989). The human papilloma virus-16 E7 oncoprotein is able to bind to the retinoblastoma gene product. *Science* **243**, 934–937.
40. Morris, J. D., Crook, T., Bandara, L. R., Davies, R., LaThangue, N. B., and Vousden, K. H. (1993). Human papillomavirus type 16 E7 regulates E2F and contributes to mitogenic signaling. *Oncogenes* **8**, 893–898.
41. Pagano, M., Durst, M., Joswig, S., Draetta, G., and Jansen-Durr, P. (1992). Binding of the human E2F transcription factor to the retinoblastoma protein but not to cyclin A is abolished in HPV-16-immortalized cells. *Oncogene* **7**, 1681–1686.
42. White, A. E., Livianos, E. M., and Tlsty, T. D. (1994). Differential disruption of genomic integrity and cell cycle regulation in normal human fibroblasts by the HPV oncoproteins. *Genes Dev.* **8**, 666–677.
43. Havre, P. A., Yuan, J., Hedrick, L., Cho, K. R., and Glazer, P. M. (1995). p53 inactivation by HPV 16 E6 results in increased mutagenesis in human cells. *Cancer Res.* **55**, 4420–4244.
44. Kessis, T. D., Connolly, D. C., Hedrick, L., and Cho, K. R. (1996). Expression of HPV 16 E6 or E7 increases integration of foreign DNA. *Oncogene* **13**, 427–431.
45. Galloway, T. C., Soper, G. R., and Elsen, G. (1960). Carcinomas of the larynx after irradiation of papilloma. *Arch. Otolaryngol. Head Neck Surg.* **72**, 289–294.
46. Majoros, M., Devine, K. D., and Parkhill, E. M. (1963). Malignant transformation of benign laryngeal papillomas in children after radiation therapy. *Surg. Clin. North Am.* **43**, 1049–1061.
47. Kyono, T., Foster, S. A., Koop, I., McDougall, J. K., Galloway, D. A., and Klingelhutz, A. J. (1998). Both Rb/p16INK4a inactivation and telomerase activity are required to immortalize human epithelial cells. *Nature* **396**, 84–88.
48. Majoros, M., Devine, K. D., and Parkhill, E. M. (1963). Malignant transformation of benign laryngeal papillomas in children after radiation therapy. *Surg. Clin. North Am.* **43**, 1049–1061.
49. Shindoh, M., Chiba, I., Yasuda, M., Saito, T., Funaoka, K., Kohgo, T., Amemiya, A., Sawada, Y., and Fujinaga, K. (1995). Detection of human papillomavirus DNA sequences in oral squamous cell carcinomas and their relation to p53 and proliferating cell antigen expression. *Cancer* **76**, 1513–1521.
50. Crook, T., Tidy, J. A., and Vousden, K. H. (1991). Degradation of p53 can be targeted by HPV E6 sequences distinct from those required for p53 binding and trans–activation. *Cell* **67**, 547–556.
51. Oda, H., Kumar, S., and Howley, P. M. (1999). Regulation of the Src family tyrosine kinase Blk through E6AP-mediated ubiquitination. *Proc. Natl. Acad. Sci. USA* **96**, 9557–9562.
52. Klingelhutz, A. J., Foster, S. A., and McDougall, J. K. (1996). Telomerase activation by the E6 gene product of papillomavirus type 16. *Nature* **380**, 79–82.
53. Reznikoff, C. A., Yeager, T. R., Belair, C. D., Savelieva, E., Puthenveettil, J. A., and Stadler, W. M. (1996). Elevated p16 at senescence and loss of p16 at immortalization in human papillomavirus 16 E6, but not E7, transformed human uroepithelial cells. *Cancer Res.* **56**, 2886–2890.
54. Ronco, L. V., Karpova, A. Y., Vidal, M., and Howley, P. M. (1998). Human papillomavirus 16 E6 oncoprotein binds to interferon regulatory factor-3 and inhibits its transcriptional activity. *Genes Dev.* **12**, 2061–2072.
55. Gao, Q., Srinivasan, S., Boyer, S. N., Wazer, D. E., and Band, V. (1999). The E6 oncoproteins of high-risk papillomaviruses bind to a novel putative GAP protein, E6TP1, and target it for degradation. *Mol. Cell. Biol.* **19**, 733–744.
56. Tong, X., and Howley, P. M. (1997). The bovine papillomavirus E6 oncoprotein interact with paxillin and disrupts the actin cytoskeleton. *Proc. Natl. Acad. Sci. USA* **94**, 4412–4417.
57. Vande Pol, S. B., Brown, M. C., and Turner, C. E. (1998). Association of bovine papillomavirus type 1 E6 oncoprotein with the focal adhesion protein paxillin through a conserved protein interaction motif. *Oncogene* **16**, 43–52.
58. Chen, J. J., Reid, C. E., Band, V., and Androphy, E. J. (1995). Interaction of papillomavirus E6 oncoprotein with a putative calcium-binding protein. *Science* **269**, 529–531.
59. Kiyono, T., Hiraiwa, A., Fujita, M., Hayashi, Y., Akiyama, T., and Ishibashi, M. (1997). Binding of high-risk human papillomavirus E6 oncoproteins to the human homologue of the Drosophila discs large tumor suppressor protein. *Proc. Natl. Acad. Sci. USA* **94**, 11612–11616.
60. Lam, E. W., Morris, J. D., Davies, R., Crook, T., Watson, R. J., and Vousden, K. H. (1994). HPV 16 E7 oncoprotein deregulates B-myb expression: Correlation with targeting of p107/E2F complexes. *EMBO J.* **13**, 871–878.
61. Sugerman, P. B., Joseph, B. K., and Savage, N. W. (1995). The role of oncogenes, tumour suppressor genes and growth factors in oral squamous cell carcinoma: A case of apoptosis versus proliferation. *Oral Dis.* **1**, 172–188.
62. Scheffner, M., Munger, K., Huibregtse, J. M., and Howley, P. M. (1992). Targeted degradation of the retinoblastoma protein by human papillomavirus E7-E6 fusion proteins. *EMBO J.* **11**, 2425–2431.
63. Massimi, P., Pim, D., Storey, A., and Banks, L. (1996). HPV-16 E7 and adenovirus E1a complex formation with TATA box binding protein is enhanced by casein kinase II phosphorylation. *Oncogene* **12**, 2325–2330.

64. Nead, M. A., Baglia, L. A., Antinore, M. J., Ludlow, J. W., and McCance, D. J. (1998). Rb binds c-Jun and activates transcription. *EMBO J.* **17**, 2342–2352.
65. Heck, D. V., Yee, C. L., Howley, P. M., and Munger, K. (1992). Efficiency of binding the retinoblastoma protein correlates with the transforming capacity of the E7 oncoproteins of the human papillomaviruses. *Proc. Natl. Acad. Sci. USA* **89**, 4442–4446.
66. Brehm, A., Nielsen, S. J., Miska, E. A., McCance, D. J., Reid, J. L., Bannister, A. J., and Kouzarides, T. (1999). The E7 oncoprotein associates with Mi2 and histone deacetylase activity to promote cell growth. *EMBO J.* **18**, 2449–2458.
67. Struhl, K. (1998). Histone acetylation and transcriptional regulatory mechanisms. *Genes Dev.* **12**, 599–606.
68. Davies, R., Hicks, R., Crook, T., Morris, J., and Vousden, K. (1993). Human papillomavirus type 16 E7 associates with a histone H1 kinase and with p107 through sequences necessary for transformation. *J. Virol.* **67**, 2521–2528.
69. Funk, J. O., Waga, S., Harry, J. B., Espling, E., Stillman, B., and Galloway, D. A. (1997). Inhibition of CDK activity and PCNA-dependent DNA replication by p21 is blocked by interaction with the HPV-16 E7 oncoprotein. *Genes Dev.* **11**, 2090–2100.
70. Dulic, V., Kaufmann, W. K., Wilson, S. J., Tlsty, T. D., Lees, E., Harper, J. W., Elledge, S. J., and Reed, S. I. (1994). p53-dependent inhibition of cyclin-dependent kinase activities in human fibroblasts during radiation-induced G1 arrest. *Cell* **76**, 1013–1023.
71. Gu, Y., Turck, C. W., and Morgan, D. O. (1993). Inhibition of CDK2 activity in vivo by an associated 20K regulatory subunit. *Nature* **366**, 707–710.
72. Warbrick, E., Lane, D. P., Glover, D. M., and Cox, L. S. (1995). A small peptide inhibitor of DNA replication defines the site of interaction between the cyclin-dependent kinase inhibitor $p21^{WAF1}$ and proliferating cell nuclear antigen. *Curr. Biol.* **5**, 275–282.
73. Berezutskaya, E., and Bagchi, S. (1997). The human papillomavirus E7 oncoprotein functionally interacts with S4 subunit of the 26 S proteasome. *J. Biol. Chem.* **272**, 30135–30140.
74. Zwerschke, W., Mazurek, S., Massimi, P., Banks, L., Eigenbrodt, E., and Jansen-Durr, P. (1999). Modulation of type M^2 pyruvate kinase activity by the human papillomavirus type 16 E7 oncoprotein. *Proc. Natl. Acad. Sci. USA* **96**, 1291–1296.
75. Schilling, B., De-Medina, T., Syken, J., Vidal, M., and Munger, K. (1998). A novel human DnaJ protein, hTid-1, a homolog of the Drosophila tumor suppressor protein Tid56, can interact with the human papillomavirus type 16 E7 oncoprotein. *Virology* **247**, 74–85.
76. Scully, C., Prime, S., and Maitland, N. (1985). Papillomaviruses: Their possible role in oral disease. *Oral Surg. Oral Med. Oral Pathol.* **60**, 166–174.
77. Steinberg, B. M. (1999). Viral etiologies of head and neck cancer *In* "Head and Neck Cancer: A Multidisciplinary Approach" (L. B. Harrison, R. B. Sessions, and W. K. Hong, eds.), p. 39. Lippincott-Raven, Philadelphia.
78. de Villiers, E. M. (1989). Papillomaviruses in cancers and papillomas of the aerodigestive tract. *Biomed. Pharmacother.* **43**, 31–36.
79. Ostwald, C., Muller, P., Barten, M., Rutsatz, K., Sonnenburg, M., Milde-Langosch, K., and Loning, T. (1994). Human papillomavirus DNA in oral squamous cell carcinomas and normal mucosa. *J. Oral Pathol. Med.* **23**, 220–225.
80. Bouda, M., Gorgoulis, V. G., Kastrinakis, N. G., Giannoudis, A., Tsoli, E., Danassi-Afentaki, D., Foukas, P., Kyroudi, A., Laskaris, G., Herrington, C. S., and Kittas, C. (2000). "High risk" HPV types are frequently detected in potentially malignant and malignant oral lesions, but not in normal oral mucosa. *Mod. Pathol.* **13**, 644–653.
81. Lawton, G., Thomas, S., Schonrock, J., Monsour, F., and Frazer, I. (1992). Human papillomaviruses in normal oral mucosa: A comparison of methods for sample collection. *J. Oral Pathol. Med.* **21**, 265–269.
82. Jalal, H., Sanders, C. M., Prime, S. S., Scully, C., and Maitland, N. J. (1992). Detection of human papilloma virus type 16 DNA in oral squames from normal young adult. *J. Oral Pathol. Med.* **21**, 465–470.
83. Kellokoski, J., Syrjanen, S., Yliskoski, M., and Syrjanen, K. (1992). Dot blot hybridization in detection of human papillomavirus (HPV) infection in the oral cavity of women with genital HPV infections. *Oral Microbiol. Immunol.* **7**, 19–23.
84. Buchwald, C., Franzmann, M. B., Jacobsen, G. K., and Lindeberg, H. (1994). Human papillomavirus and normal nasal mucosa: Detection of human papillomavirus DNA in normal nasal mucosa biopsies by polymerase chain reaction and in situ hybridization. *Laryngoscope* **104**, 755–757.
85. Buchwald, C., Franzmann, M. B., Jacobsen, G. K., and Lindeberg, H. (1995). Human papillomavirus (HPV) in sinonasal papillomas: A study of 78 cases using in situ hybridization and polymerase chain reaction. *Laryngoscope* **105**, 66–71.
86. Eike, A., Buchwald, C., Rolighed, J., and Lindeberg, H. (1995). Human papillomavirus (HPV) is rarely present in normal oral and nasal mucosa. *Clin. Otolaryngol.* **20**, 171–173.
87. Nunez, D. A., Astley, S. M., Lewis, F. A., and Wells, M. (1994). Human papilloma virus: A study of their prevalence in normal larynx. *J. Laryngol. Otol.* **108**, 319–320.
88. de Villiers, E. M., Neumann, C., Le, J. Y., Weidauer, H., and zur Hausen, H. (1986). Infection of the oral mucosa with defined types of human papillomaviruses. *Med. Microbiol. Immunol.* **174**, 287–294.
89. Syrjanen, S. M., Syrjanen, K.J., and Lamberg, M. A. (1986). Detection of human papillomavirus DNA in oral mucosal lesions using in situ DNA-hybridization applied on paraffin sections. *Oral Surg. Oral Med. Oral Pathol.* **62**, 660–667.
90. Naghashfar, Z., Sawada, E., Kutcher, M.J, Swancar, J., Gupta, J., Daniel, R., Kashima, H., Woodruff, J. D., and Shah, K. (1985). Identification of genital tract papillomaviruses HPV-6 and HPV-16 in warts of the oral cavity. *J. Med. Virol.* **17**, 313–324.
91. Loning, T., Reichart, P., Staquet, M. J., Becker, J., and Thivolet, J. (1984). Occurrence of papillomavirus structural antigens in oral papillomas and leukoplakias. *J. Oral Pathol.* **13**, 155–165.
92. Al-Bakkal, G., Ficarra, G., McNeill, K., Eversole, L. R., Sterrantino, G., and Birek, C. (1999). Human papilloma virus type 16 E6 gene expression in oral exophytic epithelial lesions as detected by *in situ* rtPCR. *Oral Surg. Oral Med. Oral Pathol. Oral Radiol. Endod.* **87**, 197–208.
93. Beaudenon, S., Praetorius, F., Kremsdorf, D., Lutzner, M., Worsaae, N., Pehau-Arnaudet, G., and Orth, G. (1987). A new type of human papillomavirus associated with oral focal epithelial hyperplasia. *J. Invest. Dermatol.* **88**, 130–135.
94. Pfister, H., Hettich, I., Runne, U., Gissman, L., and Chilf, G. N. (1983). Characterization of human papillomavirus type 13 from focal epithelial hyperplasia Heck lesions. *J. Virol.* **47**, 363–366.
95. Hernandez-Jauregui, P., Eriksson, A., Tamayo Perez, R., Pettersson, U., and Moreno-Lopez, J. (1987). Human papillomavirus type 13 DNA in focal epithelial hyperplasia among Mexicans. *Arch. Virol.* **93**, 131–137.
96. Alder-Storthz, K., Newland, J. R., Tessin, B. A., Yeudall, W. A., and Shilllitoe, E. J. (1986). Identification of human papillomavirus types in oral verruca vulgaris. *J. Oral Pathol.* **15**, 230–233.
97. Alder-Storthz, K., Newland, J. R., Tessin, B. A., Yeudall, W. A., and Shillitoe, E. J. (1986). Human papillomavirus type 2 DNA in oral verrucous carcinoma. *J. Oral Pathol.* **15**, 472–475.
98. Rihkanen, H., Aaltonen, L. M., and Syrjanen, S. M. (1993). Human papillomavirus in laryngeal papillomas and in adjacent normal epithelium. *Clin. Otolaryngol.* **18**, 470–474.
99. Erisen, L., Fagan, J. J., and Myers, E. N. (1996). Late recurrences of laryngeal papillomatosis. *Arch. Otolaryngol. Head Neck Surg.* **122**, 942–944.

100. Lack, E. E., Jenson, A. B., Smith, H. G., Healy, G. B., Pass, F., and Vawter, G. F. (1980). Immunoperoxidase localization of human papillomavirus in laryngeal papillomas. *Intervirology* **14**, 148–154.
101. Costa, J., Howley, P. M., Bowling, M. C., Howard, R., and Bauer, W. C. (1981). Presence of human papilloma viral antigens in juvenile laryngeal papilloma. *Am. J. Clin. Pathol.* **75**, 194–197.
102. Abramson, A. L., Steinberg, B. M., and Winkler, B. (1987). Laryngeal papillomatosis: Clinical, histopathologic and molecular studies. *Laryngoscope* **97**, 678–685.
103. Terry, R. M., Lewis, F. A., Griffiths, S., Wells, M., and Bird, C. C. (1987). Demonstration of human papillomavirus type 6 and 11 in juvenile laryngeal papillomatosis by in-situ DNA hybridization. *J. Pathol.* **153**, 245–248.
104. Corbitt, G., Zarod, A. P., Arrand, J. R., Longson, M., and Farrington, W. T. (1988). Human papillomavirus (HPV) genotypes associated with laryngeal papilloma. *J. Clin. Pathol.* **41**, 284–288.
105. Levi, J. E., Delcelo, R., Alberti, V. N., Torloni, H., and Villa, L. L. (1989). Human papillomavirus DNA in respiratory papillomatosis detected by in situ hybridization and the polymerase chain reaction. *Am. J. Pathol.* **135**, 1179–1184.
106. Tsutsumi, K., Nakajima, T., Gotoh, M., Shimosato, Y., Tsunokawa, Y., Terada, M., Ebihara, S., and Ono, I. (1989). In situ hybridization and immunohistochemical study of human papillomavirus infection in adult laryngeal papillomas. *Laryngoscope* **99**, 80–85.
107. Rimell, F., Maisel, R., and Dayton, V. (1992). In situ hybridization and laryngeal papillomas. *Ann. Otol. Rhinol. Laryngol.* **101**, 119–126.
108. Gomez, M. A., Drut, R., Lojo, M. M., and Drut, R. M. (1995). Detection of human papillomavirus in juvenile laryngeal papillomatosis using polymerase chain reaction. *Medicina* **55**, 213–217.
109. Pou, A. M., Rimell, F. L., Jordan, J. A., Shoemaker, D. L., Johnson, J. T., Barua, P., Post, J. C., and Ehrlich, G. D. (1995). Adult respiratory papillomatosis: Human papillomavirus type and viral coinfections as predictors of prognosis. *Ann. Otol. Rhinol. Laryngol.* **104**, 758–762.
110. Rodriguez-Perez, I., and Banoczy, J. (1982). Oral leukoplakia: A histopathological study. *Acta Morphol. Hung.* **30**, 289–298.
111. Sharma, P. K., Schuller, D. E., and Baker, S. R. (1998). Malignant neoplasms of the oral cavity. *In* "Otolaryngology—Head and Neck Surgery" (C. W. Cummings, ed.), 3rd Ed., p. 1425. Mosby, St. Louis.
112. Vas Kovskaia, G. P., and Abramova, E. I. (1981). Cancer development from lichen planus on the oral and labial mucosa. *Stomatologiia (Mosk)* **60**, 46–48.
113. Kashima, H. K., Kutcher, M., Kessis, T., Levin, L. S., de Villiers, E. M., and Shah, K. (1990). Human papillomavirus in squamous cell carcinoma, leukoplakia, lichen planus, and clinically normal epithelium of the oral cavity. *Ann. Otol. Rhinol. Laryngol.* **99**, 55–61.
114. Nielsen, H., Norrild, B., Vedtofte, P., Praetorius, F., Reibel, J., and Holmstrup, P. (1996). Human papillomavirus in oral premalignant lesions. *Eur. J. Cancer Oral Oncol.* **32B**, 264–270.
115. Sand, L., Jalouli, J., Larsson, P. A., and Hirsch, J. M. (2000). Human papilloma viruses in oral lesion. *Anticancer Res.* **20**, 1183–1188.
116. Gallo, O., Bianchi, S., Giannini, A., Boccuzzi, S., Calzolari, A., and Fini-Storchi, O. (1994). Lack of detection of human papillomavirus (HPV) in transformed laryngeal keratosis by in situ hybridization (ISH) technique. *Acta Oto-Laryngologica* **114**, 213–217.
117. Fouret, P., Dabit, D., Sibony, M., Alili, D., Commo, F., Saint-Guily, J. L., and Callard, P (1995). Expression of p53 protein related to the presence of human papillomavirus infection in precancerous lesions of the larynx. *Am. J. Pathol.* **146**, 599–604.
118. Azzimonti, B., Hertel, L., Aluffi, P., Pia, F., Monga, G., Zocchi, M., Landolfo, S., and Gariglio, M. (1999). Demonstration of multiple HPV types in laryngeal premalignant lesions using polymerase chain reaction and immunohistochemistry. *J. Med. Virol.* **59**, 110–116.
119. Sarkar, F. H., Visscher, D. W., Kintanar, E. B., Zarbo, R. J., and Crissman, J. D. (1992). Sinonasal Schneiderian papillomas: Human papillomavirus typing by polymerase chain reaction. *Mod. Pathol.* **5**, 329–332.
120. Fu, Y. S., Hoover, L., Franklin, M., Cheng, L., and Stoler, M. H. (1992). Human papillomavirus identified by nucleic acid hybridization in concomitant nasal and genital papillomas. *Laryngoscope* **102**, 1014–1019.
121. Beck, J. C., McClatchey, K. D., Lesperance, M. M., Esclamado, R. M., Carey, T. E., and Bradford, C. R. (1995). Presence of human papillomavirus predicts recurrence of inverted papilloma. *Otolaryngol. Head Neck Surg.* **113**, 49–55.
122. Beck, J. C., McClatchey, K. D., Lesperance, M. M., Esclamado, R. M., Carey, T. E., and Bradford, C. R. (1995). Human papillomavirus types important in progression of inverted papilloma. *Otolaryngol. Head Neck Surg.* **113**, 558–563.
123. Weiner, J. S., Sherris, D., Kasperbauer, J., Lewis, J., Li, H., and Persing, D. (1999). Relationship of human papillomavirus to schneiderian papillomas. *Laryngoscope* **109**, 21–26.
124. Hwang, C. S., Yang, H. S., and Hong, M. K. (1998). Detection of human papillomavirus (HPV) in sinonasal inverted papillomas using polymerase chain reaction (PCR). *Am. J. Rhinol.* **12**, 363–366.
125. Gillison, M. L., Koch, W. M., Capone, R. B., Spafford, M., Westra, W. H., Wu, L., Zahurak, M. L., Daniel, R. W., Viglione, M., Symer, D. E., Shah, K. V., and Sidransky, D. (2000). Evidence for a causal association between human papillomavirus and a subset of head and neck cancer. *J. Natl. Cancer Inst.* **92**, 709–720.
126. Smith, E. M., Hoffman, H. T., Summersgill, K. S., Kirchner, H. L., Turek, L. P., and Haugen, T. H. (1998). Human papillomavirus and risk of oral cancer. *Laryngoscope* **108**, 1098–1103.
127. Schwartz, S. M., Daling, J. R., Doody, D. R., Wipf, G. C., Carter, J. J., Madeleine, M. M., Mao, E. J., Fitzgibbons, E. D., Huang, S., Beckmann, A. M., McDougall, J. K., and Galloway, D. A. (1998). Oral cancer risk in relation to sexual history and evidence of human papillomavirus infection. *J. Natl. Cancer Inst.* **90**, 1626–1636.
128. Paz, I. B., Cook, N., Odom-Maryon, T., Xie, Y., and Wilczynski, S. P. (1997). Human papillomavirus (HPV) in head and neck cancer. An association of HPV16 with squamous cell carcinoma of Waldeyer's tonsillar ring. *Cancer* **79**, 595–604.
129. Brandsma, J. L., and Abramson, A. L. (1989). Association of papillomavirus with cancers of the head and neck. *Arch. Otolaryngol. Head Neck Surg.* **115**, 621–625.
130. Furuta, Y., Takasu, T., Asai, T., Shinohara, T., Sawa, H., Nagashima, K., and Inuyama, Y. (1992). Detection of human papillomavirus DNA in carcinomas of the nasal cavities and paranasal sinuses by polymerase chain reaction. *Cancer* **69**, 353–357.
131. Syrjanen, S., Happonen, R. P., Virolainen, E., Siivonen, L., and Syrjanen, K. (1987). Detection of human papillomavirus (HPV) structural antigens and DNA types in inverted papillomas and squamous cell carcinomas of the nasal cavities and paranasal sinuses. *Acta Oto-Laryngologica* **104**, 334–341.
132. Hording, U., Nielsen, H. W., Daugaard, S., and Albeck, H. (1994). Human papillomavirus types 11 and 16 detected in nasopharyngeal carcinomas by the polymerase chain reaction. *Laryngoscope* **104**, 99–102.
133. Dickens, P., Srivastava, G., and Liu, Y. T. (1992). Human papillomavirus 16/18 and nasopharyngeal carcinoma. *J. Clin. Pathol.* **45**, 81–82.
134. Judd, R., Zaki, S. R., Coffield, L. M., and Evatt, B. L. (1991). Human papillomavirus type 6 detected by the polymerase chain reaction

in invasive sinonasal papillary squamous cell carcinoma. *Arch. Pathol. Lab. Med.* **115**, 1150–1153.

135. Snijders, P. J., Cromme, F. V., van den Brule, A. J., Schrijnemakers, H. F., Snow, G. B., Meijer, C. J., and Walboomers, J. M. (1992). Prevalence and expression of human papillomavirus in tonsillar carcinoma, indicating a possible viral etiology. *Int. J. Cancer* **51**, 845–850.
136. Wilczynski, S. P., Lin, B. T., Xie, Y., and Paz, I. B. (1998). Detection of human papillomavirus DNA and oncoprotein over expression are associated with distinct morphological patterns of tonsillar squamous cell carcinoma. *Am. J. Pathol.* **152**, 145–156.
137. Niedobitek, G., Pitteroff, S., Herbst, H., Shepherd, P., Finn, T., Anagnostopoulos, I., and Stein, H. (1990). Detection of human papillomavirus type 16 DNA in carcinomas of the palatine tonsil. *J. Clin. Pathol.* **43**, 918–921.
138. Frisch, M., Fenger, C., van den Brule, A. J., Sorensen, P., Meijer, C. J., Walboomers, J. M., Adami, H. O., Melbye, M., and Glimelius, B. (1999). Variants of squamous cell carcinoma of the anal canal and perianal skin and their relation to human papillomaviruses. *Cancer Res.* **59**, 753–757.
139. Gregoire, L., Cubilla, A. L., Reuter, V. E., Haas, G. P., and Lancaster, W. D. (1995). Preferential association of human papillomavirus with high-grade histologic variants of penile-invasive squamous cell carcinoma. *J. Natl. Cancer Inst.* **87**, 1705–1709.
140. McNeil, C. (2000). HPV in oropharyngeal cancers: New data inspires hope for vaccines. *J. Natl. Cancer Inst.* **92**, 680–681.

C H A P T E R

12

Clinical Correlations of DNA Content Parameters, DNA Ploidy, and S-Phase Fraction in Head and Neck Cancer

Z. MACIOROWSKI
Institute Curie
Paris 75005, France

JOHN F. ENSLEY
Karmanos Cancer Institute
Wayne State University
Detroit, Michigan 48201

I. Introduction 167
II. Technical Considerations 168
  A. Tissue Preparation 168
  B. Tissue Source 168
  C. Standards 168
  D. Methods of Analysis (Flow vs Image vs Laser Scanning Cytometry) 168
III. DNA Content Parameters for Squamous Cell Carcinomas of the Head and Neck 169
  A. General Considerations 169
  B. Presenting Clinical Parameters 169
  C. Outcome Following Surgery 169
  D. Response to Radiotherapy 174
  E. Response to Chemotherapy 176
IV. DNA Content Parameters and Nonsquamous Cell Carcinomas of the Head and Neck 176
  A. Thyroid Carcinomas 176
  B. Salivary Gland Tumors 176
V. Multiparameter Studies 177
VI. Conclusions 178
  References 179

Patients with squamous cell carcinoma of the head and neck (SCCHN) vary tremendously in their natural history and response to treatment. In the past, most of these patients presented with disease that was incurable or likely to fail conventional surgery and/or radiotherapy. Advances in treatment, including concurrent chemoradiation and sequential combination chemotherapy–concurrent chemoradiation, have resulted in improved survival. In addition, conventional therapy results in unacceptable functional and cosmetic defects in the many of the patients who are cured. Advances in organ preservation strategies now provide alternatives to standard therapy, but not all patients are candidates for these approaches. Patients who have similar tumor burdens and presenting clinical and morphological features may differ widely in their clinical course and response to treatment. They are currently stratified into groups using clinical tumor node metastasis (TNM) staging and a spectrum of tumor/stromal histological characteristics, which are insufficient for predicting natural history and treatment outcome for individual patients.

I. INTRODUCTION

Complements to the clinical and pathological classification of squamous cell carcinoma of the head and neck (SCCHN) tumors are needed to better stratify these patients for treatment regimens and for the identification of new treatments for patients with incurable disease and those who fail local therapy. Effective cytotoxic treatment regimens are now curing patients with advanced stages, as well as preventing recurrences and reducing the undesirable effects of conventional therapy in patients with early stage cancers [1–3]. Overall (OR) and complete response (CR) rates of 80–90 and 35–54%, respectively, have been reported in advanced tumors [4], even in N3 stage cancers, where up to 30% of patients were reported to have achieved a CR with cytotoxic therapy [2,5–8]. Thus, knowledge of which tumors or tumor components will respond to the different treatment modalities is essential for the design of more effective cytotoxic regimens for resistant tumors, reduction of cytotoxic regimen intensity for advanced tumors that are responsive, and proper selection of early stage tumors unlikely to respond to conventional therapy.

Large hyperchromatic nuclei have long been a hallmark of malignancy [9]. This histopathologic feature reflects massive rearrangements of the chromatin structure and changes in the chromatin content and occurs in a large proportion of human solid tumors. Quantification of the amount of DNA per cell, using flow cytometry or image analysis, allows an objective assessment of the degree of aberration of the chromatin content, as well as an estimation of the number of cells actively dividing, i.e., in the DNA synthesis phase (S-phase) of the cell cycle [10]. The amount of DNA per cell is usually expressed as the DNA index (DI), i.e., ratio of the DNA content of the G0G1 phase tumor cell to that of a normal G0G1 phase diploid cell, which is considered to have a DI of 1. Cellular DNA content parameters have become important as prognostic indicators for hematological malignancies and solid tumors. Aneuploid DNA content (DNA index of >1.1) has been predictive of poor survival in surgically treated, early stage tumors of the breast [11], bladder [12,13], prostate [14], kidney [15], colon [16], and lung [17]. Investigations in childhood lymphoblastic leukemia [18], advanced lymphoma [19], medulloblastomas [20], neuroblastomas [21], breast cancer [22], advanced colon cancer [23], prostate cancer [24], sarcomas [25], and head and neck cancer [26] have indicated that aneuploid tumors or aneuploid subpopulations are more sensitive than diploid populations to cytotoxic therapy. This chapter focuses on DNA content parameters in SCCHN, their correlation with clinical and histological parameters, and with response to the different treatment modalities currently in use.

II. TECHNICAL CONSIDERATIONS

Twenty years of literature and a large body of work now exist concerning DNA content parameters in head and neck cancer. Most of the published studies have been retrospective, using paraffin-embedded material and either flow cytometry or image analysis to measure DNA content. DNA ploidy has emerged as a robust prognostic indicator, despite the differences in techniques used, tumor site, selection of patient population, and treatment variations.

A. Tissue Preparation

Tissue disaggregation method and the type of material available greatly affect the quality of the final preparation, proportions, and representativity of normal and tumor subpopulations, and thus ploidy and cell cycle determinations. Tissue disaggregation methods need to be tailored to tumor type and material, ensuring no loss of cell populations due to an inadequate release of tumor cells with tight intercellular connections or important stromal components, nor destruction of more fragile populations by strong enzymatic treatments. The disaggregation technique used should be assessed in terms of total cell yield, representativity of aneuploid populations, cellular viability and integrity of cell surface and cytoplasmic antigens, quality of DNA histogram and correlation with prognosis, response to therapy, or classic clinicopathologic parameters [27–29].

B. Tissue Source

Tissue source may be fresh or frozen, biopsy or surgical resection material, fine needle samples, or paraffin-embedded material. Fresh biopsy or surgical material must be treated soon after reception, but the whole cells with intact cytoplasm thus available allow numerous subsequent fixation and staining techniques for simultaneous analysis of cellular antigens or processes in combination with the DNA content parameters. Frozen and paraffin-embedded material, however, yield mainly nuclei, and in the case of paraffin-embedded material, many sliced nuclei, which precludes analysis of cytoplasmic or cell surface antigens of interest. Paraffin-embedded material generally produces poorer quality histograms than fresh or frozen, with debris levels that sometimes make S-phase analysis unreliable.

A series of rules and recommendations published in 1995 by the consensus comittee on DNA flow cytometry [30] addressed the problems encountered in the identification of diploid and aneuploid populations, acceptable peak coeffecient of variation, number of acquired cells necessary for adequate analysis, and acceptable debris and aggregate levels. The currently available analysis programs have powerful algorithms to automatically identify all the peaks in a histogram, subtract debris and aggregates, and calculate the best S-phase fit, but the results must be read carefully and critically by an experienced human eye.

C. Standards

Ideally, normal tissue of the same type as the tumor from the same patient should be used as a diploid control. In reality, this is often difficult to obtain, and human lymphocytes fixed and stained in the same way as the tumor are often used. Paraffin-embedded tumors present an added difficulty in that variations in fixation time before embedding routinely seen in pathology laboratories affect peak position. Establishment of the position of a diploid peak is often not possible, and the lowest peak in the histogram is generally taken to be the diploid peak.

D. Methods of Analysis (Flow vs Image vs Laser Scanning Cytometry)

Ploidy determination is possible using image analysis, laser scanning cytometry, or flow cytometry. Flow cytometry allows rapid analysis of a large number of cells, thus generating enough data for statistically valid calculations of cell cycle fractions, and the possibility of simultaneous analysis of other cellular antigens of interest. The image analysis of ploidy content is much slower, but has the advantage that cellular morphology, and sometimes also tissue architecture, is maintained, often allowing the disinction of normal vs tumor cells.

III. DNA CONTENT PARAMETERS FOR SQUAMOUS CELL CARCINOMAS OF THE HEAD AND NECK

A. General Considerations

A wide range of ploidy values have been reported in tumors from patients with SCCHN, as summarized in Tables 12.1–12.3. The average percentage of aneuploid tumors for these studies is about 60%, with a range of 25 to 81%. No consistent differences can be seen that relate these variations to source of tissue or technique used, i.e., paraffin vs fresh or frozen material. Most of the studies used paraffin-embedded samples and showed an average value for all stages and sites combined of 57% aneuploidy ranging from 15 to 80%, whereas fresh or frozen tissue showed average values of 57 and 51% and ranged from 42 to 81% and from 30 to 71%. Average values for early stage tumors (Table 12.1) were 47%, ranging from 30 to 64%. Recurrent tumors were reported to have lower rates of aneuploidy than untreated tumors (54% vs 67%) [31]. Metastatic lymph nodes were found to have lower rates of aneuploidy than the corresponding primary tumors (59% vs 73%) [32]. Where S-phase fractions were calculated, aneuploidy was found to correlate with a high S-phase [31,33,34]. Gandour-Edwards *et al.* [35] compared flow and image analysis on the same specimens and found no difference between the two methods. Berlinger *et al.* [36] compared fresh vs fixed specimens and found similar results with both methods. Studies comparing ploidy in biopsy vs surgical resection material have shown that in 97% of cases, biopsies yield the same results as the corresponding surgical specimens [26,37]. A striking factor in reviewing these studies is the variation in the quality of the DNA histogram, which has a major impact on the ability to discriminate the presence of aneuploid peaks. Peak coefficient of variation, which gives a good indication of histogram quality, and cell yields are not often reported, and often sample histograms are not shown, making evaluation of final results and comparison between studies difficult. Technical aspects of tissue preparation and staining will affect the DNA histogram, representativity of aneuploid and diploid populations produced, and thus the proportion of aneuploid tumors, as discussed earlier. Patient selection may also have an impact on aneuploidy, as some studies have reported a correlation between aneuploidy and stage [38,39].

B. Presenting Clinical Parameters

Aneuploidy has been found to correlate with prognostically poor presenting clinical parameters: high clinical stage, increased tumor size, high pathological grade, node positivity, and extension to bone. As can be seen in Tables 12.1–12.5, correlations varied from study to study. In a study of 38 laryngeal *in situ* tumors using image analysis, Munck-Wikland [40] found that 79% had high-degree DNA aberrations of greater than 2.5c DNA content (2c = normal diploid DNA content); 33% of these lesions progressed to cancer, whereas none of the lesions with low-degree DNA aberrations progressed to cancer. In a large number of studies, aneuploidy was found to correlate with the presence and number of histologically positive nodes. This was true for T1 and T2 tumors [41,42], as well as late stage cancers [39,43–45]. Stage IV patients with N3 tumors, however, were reported to be diploid more often (57%) than stage IV patients with N0-2 nodes (24%) [31]. Aneuploidy was correlated with an increased tumor size in many studies [45,46]. Aneuploidy was also associated with histologically high-grade undifferentiated tumors [47–49]. Detailed examination of histologic parameters has shown that aneuploid tumors exhibit a high nuclear grade, small cord or single cell pattern of invasion, decreased stromal or desmoplastic response, and high mitotic index [50]. Trends were also seen for aneuploid tumors to invade small capillary structures. Diploid tumors showed well-defined tumor stromal interfaces, an increased desmoplastic response, a low nuclear grade, a mitotic index, and a low inflammatory response.

C. Outcome Following Surgery

Nearly all studies comparing DNA content parameters and outcome after surgery as initial therapy in patients with early stage SCCHN report a survival advantage for patients with DNA diploid tumors. In a study of 49 stage I oral cancers treated by local surgical excision, Hogmo and colleagues [51] reported 56% of the aneuploid tumors recurred vs 33% of the diploid tumors. A study by Mishra and Mohanty [52] of 78 T1 and T2 buccal mucosal cancers with curative resections showed a local-regional recurrence rate of 72% in aneuploid tumors vs 6% in diploid tumors. Ploidy was a highly significant predictor of disease-free survival (DFS): a 4-year DFS of 28% for patients with aneuploid tumors vs 94% for patients with diploid tumors. Munck-Wikland *et al.* [53] reported increased local-regional recurrence in aneuploid tumors in 47 T1N0 tongue cancers following surgery. In a mixed stage and site study by Kokal and co-workers [54], 50 of 76 patients were treated by surgery alone, and survival advantages were seen for diploid tumors at all stages; 79% recurrence rates in aneuploid tumors vs 0% in diploid tumors for stage I and II, and 69% vs 8% for stages II and IV. In a large study by

TABLE 12.1 Stage I and II SCCHN DNA Content Studies

Material/ analysis	#	Site	Aneuploid (%)	Treatment	Aneuploid: Clinical correlation	Response	Recurrence	Survival	Reference
Paraffin image	49	Oral cavity, Tongue	64	Surgery	NA	NA	Increased in aneuploid	NA	Hogmo *et al.* [51]
Paraffin flow	78	Buccal mucosa	39	Surgery	High stage	NA	Increased locoregional in aneuploid	Decreased in aneuploid	Mishra and Mohanty [52]
Paraffin flow	72	Larynx	46	Radiotherapy	NA	NA	Increased local recurrence in diploid	No difference	Toffoli *et al.* [56]
Fresh flow	36	Tongue	42	Surgery	Node + Also in T1-2 only	NA	NA	No difference	Saito *et al.* [41]
Paraffin image	28	Glottis	NA	Radiotherapy	NA	NA	Increased in aneuploid	NA	Munck-Wikland *et al.* [58]
Paraffin Image	47	Tongue	NA	Surgery	NA	NA	Increased in locoregional aneuploid	NA	Munck-Wikland *et al.* [53]
Paraffin flow	60	Tongue	30	Not specified	NA	NA	No difference	DFS, No difference	Farrar *et al.* [103]
Paraffin flow	15	Tongue	67	Surgery	NA	NA	NA	5-year DFS decreased in aneuploid	Sickle-santello *et al.* [104]
Paraffin image	50	Oral cavity	44	Surgery and/or radiotherapy	Node+	Aneuploid inc. response to radiotherapy	Increased in aneuploid	5 year decreased in aneuploid	Tytor *et al.* [42]

TABLE 12.2 Stage III and IV SCCHN DNA Content Studies

Material/analysis	#	Site	Aneuploid (%)	Treatment	Aneuploid: Clinical correlation	Response	Recurrence	Local control	Survival	Reference
Paraffin flow	22	Maxillary sinus	25	Surgery Radiotherapy Chemotherapy	NA	NA	NA	NA	Increased in aneuploid	Halverson *et al.* [70]
Paraffin flow	94	Oral cavity Oropharynx	48	Surgery/ Radiotherapy	No correlation	NA	NA	NA	Increased in aneuploid (trend)	Syms *et al.* [66]
Frozen image	94	Larynx	40	Surgery/ Radiotherapy	Node+	NA	Increased distant in aneuploid	NA	DFS decreased in aneuploid	Wolf *et al.* [44]
Paraffin flow	143	Multisite	53	Radiotherapy	NA	NA	Increased in aneuploid	Decreased in aneuploid	No difference	Fu *et al.* [59]
Paraffin image	50	Larynx	NA	Induction chemotherapy	NA	CR aneuploid only	No difference	NA	No difference	Gregg *et al.* [71]
Fresh image	88	Larynx	55	Surgery Radiotherapy	No correlation	NA	Increased in aneuploid	NA	DFS decreased in aneuploid	Truelson *et al.* [63]
Paraffin flow	100	Larynx	50	Surgery Radiotherapy	No correlation	NA	NA	NA	No difference	Cooke *et al.* [105]
Fresh flow	237	Multisite	70	Surgery Chemotherapy	N3 decreased aneuploidy	NA	Decreased aneuploidy in recurrents	NA	NA	Ensley *et al.* [26]
Paraffin flow	76	Multisite	66	Surgery 23+ radiotherapy	Node+ high stage	NA	Increased in aneuploid	NA	OS, DFS decreased in aneuploid	Kokal *et al.* [54]
Fresh flow	155	Multisite	69	NA	High grade invasive high mitosis	NA	NA	NA	NA	Sakr *et al.* [50]
Paraffin flow	51	Nasopharynx	16	Radiotherapy	NA	NA	Increased LR and distant in aneuploid	NA	OS DFS increased in aneuploids and high proliferation	Yip *et al.* [67]
Fresh Flow	–	Multisite	66	Surgery Radiotherapy Chemotherapy	NA	Increased in aneuploid	NA	NA	Increased in aneuploid	Ensley *et al.* [26]
Paraffin flow	105	Multisite	54	Cisplatinum	NA	No difference	NA	NA	Aneuploid increased survival treated vs untreated	Cooke *et al.* [73]
Paraffin flow	31	Nasopharynx	25	Radiotherapy or Radiotherapy/ Surgery or Radiotherapy/ Chemotherapy	NA	NA	NA	NA	Decreased in aneuploid	Cheng *et al.* [69]

TABLE 12.3 Stage I–IV SCCHN DNA Content Studies

Material/analysis	#	Site	Aneuploid (%)	Treatment	Aneuploid: Clinical correlation	Response	Recurrence	Survival	Reference
Paraffin image and flow	28	Oral cavity, Pharynx	54	Surgery, radiotherapy for persistant disease	No correlation	NA	Increased in aneuploid	Decreased in aneuploid	Raybaud 2000 [60]
Paraffin flow	93	Oral cavity, Tongue	55	Surgery radiotherapy	High stage High grade Node+	NA	Increased in aneuploid	OS + RFS decreased in aneuploid	Rubio-Bueno 1998 [39]
Paraffin image	52	Oral cavity	60	Surgery radiotherapy	Size, Node+	NA	Increased locoregional in aneuploid	5 year OS Decreased in aneuploid	Schimming *et al.* [45]
Fresh flow	429	Oral cavity	81	Surgery	High Grade Size Node+	NA	Increased locoregional in aneuploid	OS Decreased in aneuploid	Hemmer *et al.* [46]
Paraffin image and flow	26	Multisite	73	Surgery	Node+	NA	NA	Decreased in aneuploid	Gandour Edwards *et al.* [35]
Fresh flow	49	Multisite	47	Surgery + radiotherapy	Node+	NA	Increased locoregional and distant in aneuploid	No difference	Lampe [43]
Fresh flow	40	Oral cavity	58	Surgery	Node + High grade	NA	NA	NA	Chen *et al.* [48]
Paraffin flow	38	Tongue	37	Radiotherapy/surgery or surgery/radiotherapy	NA	NA	NA	DFS OS Decreased in aneuploid	Gomez *et al.* [62]
Paraffin flow	172	Oral cavity	67	Surgery/Radiotherapy	High grade	NA	Increased in aneuploid	Node negative DFS OS Increased in aneuploid	Kearsley *et al.* [47]
Paraffin flow	133	Larynx	57	Surgery radiotherapy chemotherapy	High grade	NA	NA	OS Decreased in node negative aneuploid S phase>15%	Rua *et al.* [49]

Paraffin flow	55	Nasopharynx	60	Radiotherapy or chemotherapy	NA	NA	NA	OS No difference	Costello *et al.* [68]
Paraffin image	45	Multi site	76	Radiotherapy Surgery Chemotherapy	Node + High stage	NA	NA	DFS Decreased in aneuploid	Holm [38]
Frozen flow	72	Multi site	51	Radiotherapy	High grade	Cr decreased in aneuploids	NA	OS Node + Decreased in aneuploid	Zatterstrom *et al.* [61]
Paraffin flow	296	Multi site	64	Surgery 44/296 radiotherapy	Node+	NA	Local and disatant mets increased in aneuploid	Surgery: decreased in aneuploid Radiotherapy: increased in aneuploid for all stages	Guo *et al.* [55]
Frozen flow	200	Multi site	71	Surgery	Node + High stage	NA	NA	NA	Wennenberg *et al.* [106
Fresh flow	497	Multi site	66 primary 50 mets	Not specified	NA	NA	NA	NA	Burgio *et al.* [32]
Paraffin image	140	Oral cavity	51	Surgery/radiotherapy/ chemotherapy Stage III IV: radiotherapy/ chemotherapy	Node+ Size	Aneuploid response to radiotherapy	No difference	I, II: decreased in aneuploid III, : IV increased in aneuploid due to therapy	Tytor *et al.* [57]
Fresh flow	110	Oral cavity	73	Surgery	High grade Node+	NA	NA	NA	Hemmer *et al.* [107]
Fresh flow	131	Maxillary sinus	67	Surgery, radiotherapy	NA	NA	NA	OS decreased in aneuploid Diploid low S phase: Increased survival	Muller *et al.* [33]
Paraffin flow	55	Nasopharynx	60	Radiotherapy or chemotherapy	NA	NA	NA	No difference	Costello *et al.* [68]

TABLE 12.4 Thyroid Malignancies DNA Content Studies

#	Type	Aneuploid (%)	Aneuploid: Clinical correlation	Survival	Reference
125	All	Papillary 24 Follicular 56 Medullary 57	Age Infiltrating CA	Decreased in aneuploid	Joensuu *et al.* [78]
121	All	7	NA	Decreased in aneuploid	Tsuchiya *et al.* [79]
25	Medullary	50	NA	Decreased in aneuploid	Schroder 1988 [86]
119	Medullary	23	NA	Decreased in aneuploid	Hay *et al.* [86]
211	Medullary	28	NA	Decreased in aneuploid	Ekman *et al.* [84]
247	Medullary	28	NA	Decreased in aneuploid	Bergohlm *et al.* [87]
20 adenoma 20 carcinoma	Follicular	29	NA	Decreased in aneuploid	Backdahl 1986 [108]
29 adenoma 11 carcinoma	Follicular	14 AD 28 CA	NA	No difference	Hruban *et al.* [82]
60 adenoma 64 carcinoma	Follicular	Adenoma 25 Carcinoma 55	Cannot distinguish malignant vs benign	No difference: Ploidy Decreased in high S phase	Grant *et al.* [83]
150	Papillary	12	Poorly differentiated	Decreased in aneuploid	Hrafinkelsson *et al.* [80]
40	Papillary	20	Invasive	NA	Stern 1997 [81]
11 adenoma 6 carcinoma	Hurthle cell	Adenoma 55 Carcinoma 67	Cannot distinguish Malignant vs benign	No difference	Bronner *et al.* [76]
36	Hurthle cell	19	Cannot distinguish malignant vs benign	Decreased in aneuploid	Mcleod *et al.* [77]
16	Parathyroid	4% adenomas 31% carcinomas	NA	Increased recurrence in aneuploid	Obara *et al.* [89]
39	Parathyroid	67	NA	Decreased time to recurrence in aneuploid	Sandelin *et al.* [90]

Hemmer *et al.* [46], of 429 surgically treated oral cancers, ploidy was predictive of local-regional recurrence and recurrence-free survival and was the only predictor of overall and disease-free survival for the node-negative group. In a mixed stage and site study by Guo *et al.* [55] of patients treated with surgery alone, survival advantages were seen for diploid tumors overall and for each stage: 32% of patients with aneuploid tumors vs 49% of patients with diploid tumors survived 5 years. A subgroup of 44 patients in this study received adjuvant radiotherapy, and their survival rates were very different: 62% of patients with aneuploid tumors survived 5 years compared to 28% of patients with diploid tumors.

In our series of 200 patients with advanced stage III and IV SCCHN treated initially with surgery, DNA content discriminated a 2:1 survival difference for patients with DNA diploid tumor specimens [26,28]. We analyzed The North American Head and Neck Cancer Intergroup Study 0034 for stage III and IV patients with margin negative resections for DNA content and showed essentially the same 2:1 survival advantage for patients with DNA diploid tumors (unpublished data). We also demonstrated a highly significant relationship between DNA ploidy status and individual features of tumor histopathology and the tumor–stromal relationship [50]. In particular, an association with the presence of DNA aneuploid clones and loss of tumor–stromal border, local immune and desmoplastic response, and increased mitotic index was demonstrated. This may explain the relative inability of the surgeon to obtain and the pathologist to detect negative margin surgical resections in patients with DNA aneuploid tumors.

D. Response to Radiotherapy

The correlation of DNA content parameters and the response to conventionally administered radiotherapy have yielded contradictory results. A number of studies have found that aneuploid tumors respond better to radiotherapy, whereas others have found that diploid tumors show a better response and increased survival. In a study of 72 T1 and 2 larnygeal tumors treated with radiotherapy alone, Toffoli and co-workers [56] found increased local regional persistence and recurrence in diploid tumors. Tytor and colleagues [42] found that aneuploid tumors in 24 T1 and 2 oral cancers treated by radiotherapy showed a nonsignificant trend toward an increased response to radiotherapy: 33% of aneuploid tumors recurred or persisted vs 67% of diploid tumors; however, a nonsignificant 5-year survival advantage was seen for diploid tumors. In another study of stage III and IV oral cancers treated with preoperative radiotherapy, the same group [57] found that 85% of diploid tumors vs 31% of aneuploid tumors had residual tumor at surgery.

TABLE 12.5 Salivary Gland\Malignancies DNA Content Studies

#	Type	Aneuploid (%)	Aneuploid: Clinical correlation	Recurrence	Survival	Reference
24	Adenoid cystic	33	Advanced stage Solid architecture	Increased in aneuploid	Decreased in aneuploid	Tytor *et al.* [91]
51	Adenoid cystic	24	Advanced stage High grade	Increased in aneuploid	NA	Franzen *et al.* [92]
26	Adenoid cystic	38	Solid architecture Advanced stage Node+	NA	NA	Luna *et al.* [93]
52	Adenoid cystic	87 nondiploid	NA	NA	DFS Decreased in nondiploid	Hamper *et al.* [94]
48	Mucoepidermoid	67	NA	NA	Decreased in aneuploid and high S phase	Hicks *et al.* [95]
15	Acinic cell	53	NA	NA	Decreased in aneuploid	El-Naggar *et al.* [98]
45	Acinic cell	42	NA	No difference	No difference	Timon *et al.* [97]
30	All	58	NA	NA	No difference	Felix *et al.* [99]
16	All	50	Advanced stage High S phase	Increased in aneuploid	NA	Tytor *et al.* [91]
55	All	22	Size, grade, node+	NA	Decreased in aneuploid	Carillo *et al.* [100]
37	All	24	NA	Increased in aneuploid	Decreased in aneuploid	Hamper *et al.* [101]

No survival advantage was seen for diploid over aneuploid tumors. However, Munck-Wikland *et al.* [58] looked at 28 T1 N0 glottic tumors treated with radiotherapy and found increased recurrence in aneuploid tumors. In 143 stage III and IV tumors from multiple sites, Fu and colleagues [59] found an increased local regional and distant recurrence in aneuploid tumors after radiotherapy, but no difference between diploid and aneuploid tumors for survival. In a study of 56 oral cavity and pharyngeal tumors, Raybaud and co-workers [60] found aneuploid tumors to be radioresistant: 50% of aneuploid tumors recurred vs 11% of diploid tumors.

In patients treated mainly with radiotherapy or radiotherapy and surgery, Zatterstrom *et al.* [61] found complete responses in 86% of diploid tumors vs 59% of aneuploid tumors, with median survival times of >81 and 16 months, respectively. Muller and colleagues [33] found that in maxillofacial tumors treated by surgery followed by radiotherapy, aneuploid tumors had a mean survival of 9.5 months vs 12 months in diploid tumors. Other studies where treatment was a mixture of surgery and radiotherapy showed increased recurrence and decreased survival for aneuploid tumors or found no difference in survival between diploid and aneuploid tumors [39,47,62,63]. As discussed earlier, however, Guo *et al.* [55] found a 5-year survival advantage for aneuploid vs diploid tumors treated by surgery and radiotherapy. In Holm's [38] series, similar CR rates were reported for diploid and aneuploid tumors following preoperative radiotherapy: 73% vs 65%, respectively, although a larger (50% vs 38%) but not significant proportion of patients with diploid tumors survived compared to aneuploid tumors. An analysis of these data revealed that 80% of the diploid tumors received 4000 cGy preoperative radiation and surgery, whereas many of the aneuploid tumors received less radiation. Goldsmith's [64] report of 48 patients with advanced laryngeal SCC treated with post-operative radiation and surgery indicated that patients with aneuploid tumors responded better than diploid tumors and were associated with a statistically significant survival advantage at 18 months. Franzen *et al.* [65] reported that 73% of oral SCCHN with pretreatment aneuploid specimens achieved histological CR following radiotherapy compared with only 11% of patients with pretreatment diploid specimens. In a study of 94 late stage oral and oropharyngeal tumors treated with surgery and postoperative radiotherapy, Syms and colleagues [66] found a trend for survival advantage for aneuploid tumors.

Studies on radiation-treated nasophayngeal cancers have also shown varied results. Yip *et al.* [67] studied 51 late stage NPC in a Chinese population, with an aneuploidy rate of 16% treated with radiotherapy, and found that aneuploid tumors had a higher rate of local regional and distant recurrences and a decreased 12-year survival compared to diploid tumors. A high S phase was also prognostic for decreased survival. In an Australian population of 55 NPC, Costello *et al.* [68] found an aneuploidy rate of 60% and no 5-year survival advantage for diploid tumors over aneuploid tumors. Histology of the tumors was also very different: the Chinese population was either PDSCC or UCNT, whereas only 35% of the Australian population was UNCT or PDSCC, 65% were keratinizing SCC. A third study by

Cheng *et al.* [69] of 41 stage III and IV NPC patients treated with radiotherapy, divided equally among WHO histological classifications 1, 2, and 3, showed 25% aneuploidy and decreased 2 and 5-year survival for aneuploid tumors. Preliminary analysis of the North American Head and Neck Cancer Intergroup Int 0099 nasopharyngeal study also showed that the DNA ploidy status of tumor specimens was associated with a substantial difference in patient survival: DNA diploid 61% versus 38 % for specimens with DNA aneuploid content (unpublished data).

E. Response to Chemotherapy

Studies correlating DNA content and the response to chemotherapy indicate either that survival is similar for diploid and aneuploid tumors or, more often, that aneuploid tumors show an increased response to chemotherapy and a survival advantage over diploid tumors. In a study by Halvorson *et al.* [70] of 22 T3 and T4 tumors of the maxillary sinus, of which 18 were treated with preoperative radiation and chemotherapy, a survival advantage was seen for aneuploid tumors: all 4 patients with aneuploid tumors were alive after a mean follow-up of 6 years vs 10 of the 18 patients with diploid tumors. In 50 stage III and IV laryngeal tumors treated with cisplatinum–5-fluorouracil (5-FU) induction chemotherapy followed by radiotherapy in the Veterans Administration Laryngeal Preservation Study, Gregg and colleagues [71] found that a high DNA index was associated with a clinical complete response. No patient with a low DNA index achieved a complete response. Survival analysis showed no difference between diploid and aneuploid tumors. Similarly, Tenneval *et al.* [72] found that a better response to cisplatinum–5-FU treatment was associated with aneuploidy, 12% of diploid tumors vs 39% of aneuploid tumors achieved a complete response. Cooke and co-workers [73] found that cisplatinum treatment of advanced recurrent SCCHN prolonged the mean survival time for patients with aneuploid tumors from 55 days untreated to 224 days for treated patients, whereas for patients with diploid tumors, the increase was only from 74 to 118 days. In a study by Rua *et al.* [49] of 133 laryngeal cancers, 73 of which were treated with radiotherapy and chemotherapy, no survival advantage was seen for diploid over aneuploid tumors. In a study of 200 patients treated with cisplatinum containing combination chemotherapy, Ensley *et al.* [26] found that 66% of the tumors were aneuploid pretherapy but only 9% were aneuploid posttherapy. Eighty percent of the patients with aneuploid tumors achieved complete clinical remission, and 43% of these were also histologically negative. Only 2% of the patients with diploid tumors achieved complete clinical remission, and none of these were histologically negative. These studies indicate that the aneuploid components of tumors are susceptible to cytotoxic treatment and that this choice of treatment can counteract the survival advantage seen for diploid tumors over aneuploid tumors treated by surgery alone.

IV. DNA CONTENT PARAMETERS AND NONSQUAMOUS CELL CARCINOMAS OF THE HEAD AND NECK

A. Thyroid Carcinomas

DNA content parameters in thyroid tumors have also shown correlations with aggressiveness, relapse, and survival. Percentage aneuploidy also varies between studies and tumor types, with a range of 7 to 93%. Papillary carcinomas show less aneuploidy (0 to 25%) than follicular (27 to 93%) or medullary (28 to 57%) carcinomas. Although the aneuploidy rates are higher in carcinomas than in adenomas, attempts to use aneuploidy to distinguish benign from malignant tumors in follicular [74,75] and Hurthle cell tumors [76,77] have been unsuccessful. Aneuploidy has, however, shown prognostic correlations with recurrence and survival. In a study of papillary, follicular, and medullary tumors, Joensuu *et al.* [78] found that aneuploidy was associated with increased age of the patient, infiltrating carcinomas, and poor survival. In a study of 121 tumors, Tsuchiya and colleagues [79] found a 5-year survival rate of 97% of patients with diploid tumors vs 37% of patients with aneuploid tumors. In papillary carcinomas, Hrafinkelsson *et al.* [80] found aneuploidy and high S phase to be unfavorable prognostic factors. In a study of 40 papillary carcinomas, Stern *et al.* [81] found that 40% of invasive carcinomas were aneuploid vs 0% of noninvasive carcinomas. In follicular carcinomas, correlations were more variable, with some studies showing aneuploidy associated with decreased survival, whereas others were inconclusive [82,83]. Studies of medullary tumors showed aneuploidy to be prognostic of poor survival [84,85]. In a study of 119 medullary cancers, Hay and co-workers [86] found 10-year mortality rates of 10% for patients with diploid tumors vs 49% of patients with aneuploid tumors. A study of 211 medullary cancers by Bergholm *et al.* [87] showed 10-year survival rates of 67% for diploid tumors vs 30% for aneuploid tumors. Aneuploidy has also been found to predict recurrence and distant metastases in Hurthle cell carcinomas [77,88]. Studies of parathyroid tumors all showed a correlation of aneuploidy with recurrence. Obara and co-workers [89] found that 80% of aneuploid vs 31% of diploid tumors recurred, and Sandelin *et al.* [90] found that aneuploidy correlated with time to recurrence.

B. Salivary Gland Tumors

In salivary gland tumors, aneuploidy correlated with advanced stage and grade, as well as with increased recurrence and decreased survival. Aneuploidy rates were lower than SCHNN, generally from 20 to 40%. In adenoid cystic carcinomas, Tytor *et al.* [91] found that aneuploidy correlated with advanced clinical stage and solid architecture. Seventy-five percent of aneuploid tumors recurred vs 19% of diploid tumors, and cumulative survival was worse for aneuploid tumors. In studies of adenoid cystic tumors, Franzen and co-workers [92] showed treatment failure to be increased in aneuploid tumors, 100% of anuploid tumors recurred vs 14% of diploid tumors. Luna *et al.* [93] found aneuploidy to be associated with solid architecture, lymph node metatases, and clinical stage, and Hamper *et al.* [94] found shorter survival times for aneuploid adenoid cystic tumors. Studies of mucoepidermoid tumors showed that aneuploidy correlated with poor prognosis and decreased survival [95,96]. Reports on acinic cell tumors are mixed. Two studies by Hamper *et al.* [94,96] and one by Timon *et al.* [97] showed no correlation between aneuploidy and prognosis, whereas a study by El-Naggar and colleagues [98] showed 100% of patients with diploid tumors vs 50% of patients with aneuploid tumors alive after 10 years of follow-up. In studies with mixed typed of salivary gland tumors, results were similar: aneuploidy correlated with advanced stage, recurrence, and mortality [91,99–101].

V. MULTIPARAMETER STUDIES

Simultaneous analysis of several cellular or molecular genetic markers in parallel with DNA content parameters is also possible using flow or image cytometry and multiplies the information possible on a single sample. Slide-based image analysis or laser scanning cytometric assays allow the restaining of the slides and reanalysis of the same cells. Examples of double staining are shown in Figs. 12.1 and 12.2 (see also color insert). Simultaneous analysis of abnormal keratin expression and DNA content in advanced SCCHN tumors (Fig. 12.1) allows the discrimination of tumor

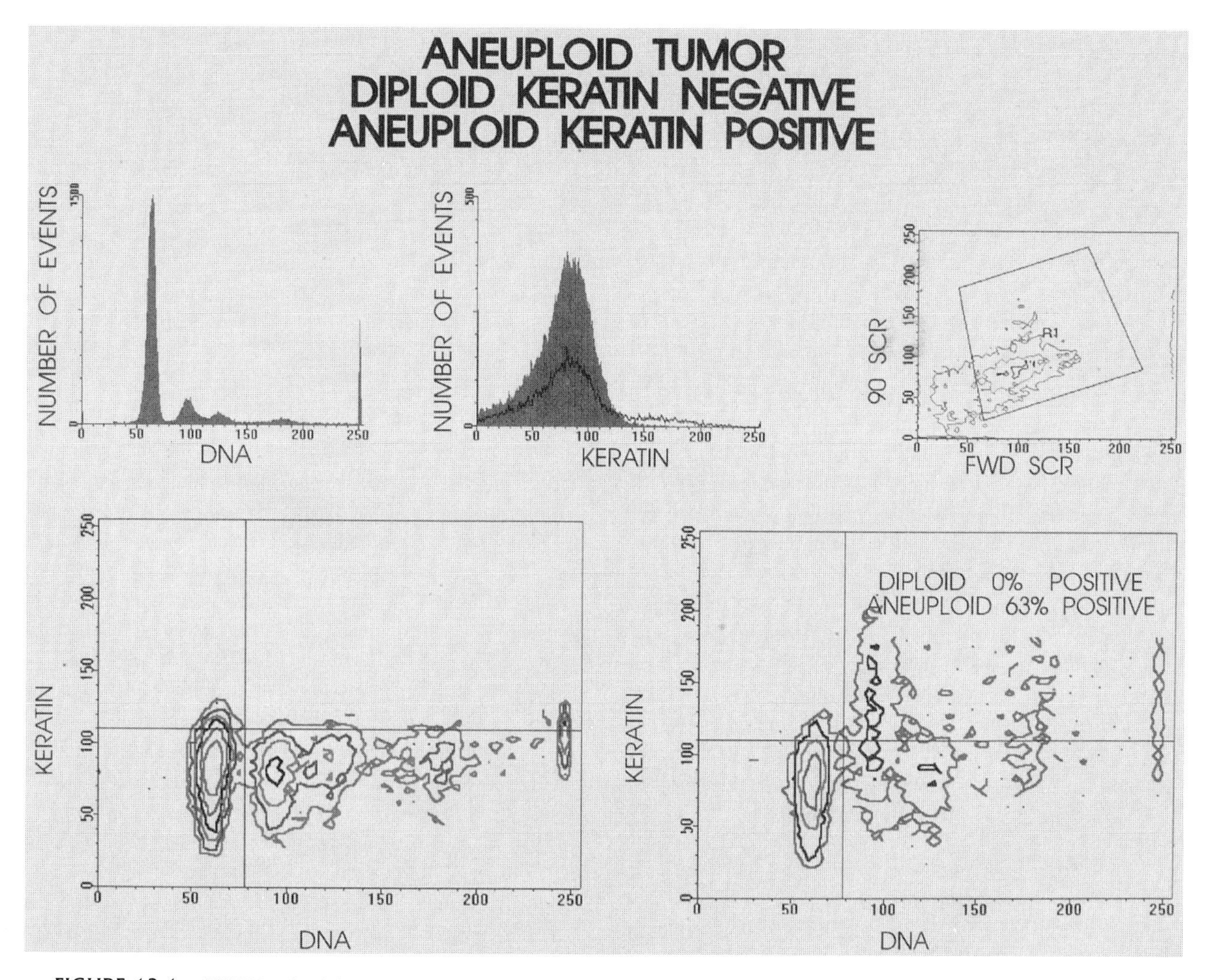

FIGURE 12.1 SCCHN stained simultaneously for DNA content (propidium iodide) and keratin (FITC-CAM 5.2). (See also color insert.)

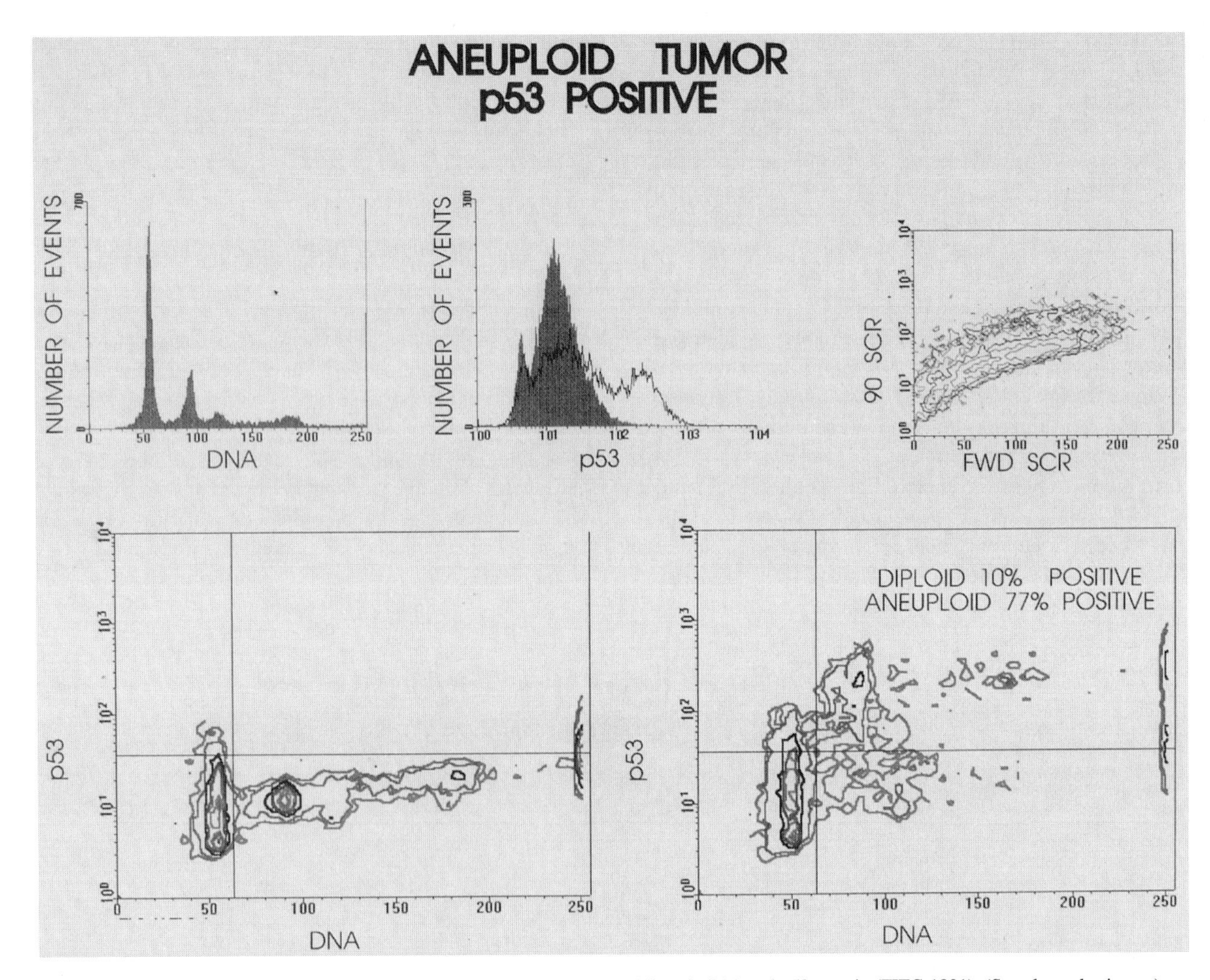

FIGURE 12.2 SCCHN stained simultaneously for DNA content (propidium iodide) and p53 protein (FITC-1801). (See also color insert.)

subpopulations in mixed diploid and aneuploid tumors. In a study of 130 SCCHN patients [102], 92% of pure diploid tumors showed malignant keratin expression, but in 47% of specimens containing diploid and aneuploid populations, the diploid component was negative for malignant keratin expression. These analyses demonstrate the existence and allow for the identification of pure aneuploid SCCHN tumors, which are most likely to show a complete response to chemotherapy. Abnormal p53 expression is another marker of interest in SCCHN, and multiparametric analysis, which we developed and validated in over 50 SCCHN tumor specimens [102], allows the correlation of p53 expression with ploidy or cell cycle information, as seen in Fig. 12.2.

VI. CONCLUSIONS

DNA content parameters in squamous and nonsquamous carcinomas of the head and neck have demonstrated utility in determining the natural history and the prognosis of the tumor in terms of response to therapy and survival. Diploid tumors are associated with distinct tumor–stromal interfaces, allowing the achievement of negative margins following surgery. Relapse-free and overall survival after surgery is superior for diploid tumors compared to aneuploid tumors. Aneuploid tumors, however, demonstrate noncohesive growth characteristics and poor tumor–stromal barriers and achievement of clean margins is difficult. The increased sensitivity of aneuploid tumors to cytotoxic therapy reduces the survival advantage generally seen for diploid tumors. It is therefore possible to envision therapeutic regimens tailored to individual tumors using these parameters, thus increasing survival and reducing the associated functional and cosmetic defects in patients unresponsive to conventional treatment. The response to treatment may be monitored by assessing levels of apoptosis induction along with the disappearance of aneuploid populations. Multiparametric analysis of new cellular and genetic markers of interest, e.g., cell cycle-related proteins such as cyclins, angiogenesis or metastatic markers such as tissue proteases, or fluorescence *in situ* hybridization (FISH) staining to identify early changes in diploid tumors, may help further identify subgroups sensitive to new therapies.

References

1. Ensley, J., Pietraszkiewicz, H., Maciorowski, Z., Hassan, M., Kish, J., Tapazaglou, E., Jacobs, J., Weaver, A., Atkinson, D., Binns, P., and Al-Sarraf, M. (1988). Comparative techniques and clinical correlations of cell cultures from squamous cell carcinomas of the head and neck. *In* "Head and Neck Oncology Research: Proceedings of the Second International Research Conference on Head and Neck Cancer" (G. T. Wolf and T. Carey, eds.), pp. 35–41. Kugler, Amsterdam.
2. Ensley, J., Kish, J., Tapazaglou, E., and Al-Sarraf, M. (1987). Strategies for improvement in complete response rates with adjuvant combined chemotherapy in advanced head and neck cancer. *In* "Adjuvant Therapy of Cancer" (V. S. Salmon, ed.), pp. 101–111. Decker, Philadelphia.
3. Ensley, J., Kish, J., and Al-Sarraf, M. (1986). The development of optimum induction chemotherapy regimens for advanced head and neck cancer: The Wayne State University Experience. *In* "Neo-Adjuvant Chemotherapy: First International Congress" (C. Jacquillat, M. Weil, and D. Khayat, eds.), pp. 353–361. John Libbey, London.
4. Al-Sarraf, M., Kish, J. A., and Ensley, J. F. (1991). The Wayne State University experience with adjuvant chemotherapy of head and neck cancer. *Hematol. Oncol. Clin. North Am.* **5**, 687–700.
5. Kish, J., Ensley, J., and Al-Sarraf, M. (1987). Clinical results of chemotherapy in recurrent head and neck carcinoma. *In* "Scientific and Clinical Perspective of Head and Neck Cancer Management Strategies for Cure" (J. Jacobs, M. Al-Sarraf, J. Crissman, and F. Valeriote, eds.), pp. 199–211. Elsevier, New York.
6. Ensley, J., Kish, J., Tapazaglou, E., Jacobs, J., Weaver, A., Atkinson, D., Ahmed, K., Heilbrun, L., and Al-Sarraf, M. (1988). The justification and strategies for the continued intensification of induction regimens in advanced untreated head and neck cancer. *In* "Head and Neck Oncology Research: Proceedings of the Second International Research Conference on Head and Neck Cancer" (G. Wolf and T. Carey, eds.), pp. 313–323. Kugler, Amsterdam.
7. Al-Sarraf, M., Pajak, T., Marcial, V., Mowry, P., Cooper, J., Stetz, J., Ensley, J., and Velez-Garcia, E. (1987). Concurrent radiotherapy and chemotherapy with cisplatin in inoperable squamous cell carcinoma of the head and neck: An RTOG Study. *Cancer* **59**, 259–265.
8. Ensley, J., Kish, J., Tapazaglou, E., Jacobs, J., Weaver, A., Atkinson, D., Ahmed, K., Mathog, B., and Al-Sarraf, M. (1988). An intensive 5-course alternating combination chemotherapy induction regimen employed in patients with advanced unresectable head and neck cancer. *J. Clin. Oncol.* **6**, 1147–1154.
9. Broders, A. (1926). Carcinoma: Grading and practical application. *Arch. Pathol.* **2**, 376–381.
10. Barlogie, B., Raber, M. N., Schumann, J., Johnson, T. S., Drewinko, B., Swartzendruber, D. E., Gohde, W., Andreeff, M., and Freireich, E. J. (1983). Flow cytometry in clinical cancer research. *Cancer Res.* **43**, 3982–3997.
11. McDivitt, R. W., Stone, K. R., Craig, R. B., Palmer, J. O., Meyer, J. S., and Bauer, W. C. (1986). A proposed classification of breast cancer based on kinetic information derived from a comparison of risk factors in 168 primary operable breast cancers. *Cancer* **57**, 269–276.
12. Frankfurt, O. S., and Huben, R. P. (1984). Clinical applications of DNA flow cytometry for bladder tumors. *Urology* **23**, 29–34.
13. Jacobsen, A. B., Pettersen, E. O., Amellem, O., Berner, A., Ous, S., and Fossa, S. D. (1992). The prognostic significance of deoxyribonucleic acid flow cytometry in muscle invasive bladder carcinoma treated with preoperative irradiation and cystectomy. *J. Urol.* **147**, 34–37.
14. Zetterberg, A., and Esposti, P. L. (1980). Prognostic significance of nuclear DNA levels in prostatic carcinoma. *Scand. J. Urol. Nephrol. Suppl.* **55**, 53–58.
15. Otto, U., Baisch, H., Huland, H., and Kloppel, G. (1984). Tumor cell deoxyribonucleic acid content and prognosis in human renal cell carcinoma. *J. Urol.* **132**, 237–239.
16. Wolley, R. C., Schreiber, K., Koss, L. G., Karas, M., and Sherman, A. (1982). DNA distribution in human colon carcinomas and its relationship to clinical behavior. *J. Natl. Cancer Inst.* **69**, 5–22.
17. Volm, M., Hahn, E., Mattern, J., and Vogt-Moykopf, I. (1988). Independent flow cytometric prognostic factors for the survival of patients with non-small cell lung cancer: A five year followup. *Proc. AACR* **29**, 26.
18. Look, A. T., Roberson, P. K., Williams, D. L., Rivera, G., Bowman, W. P., Pui, C. H., Ochs, J., Abromowitch, M., Kalwinsky, D., Dahl, G. V., *et al.* (1985). Prognostic importance of blast cell DNA content in childhood acute lymphoblastic leukemia. *Blood* **65**, 1079–1086.
19. Woodbridge, T., Grierson, H., Pierson, J., Pauza, M., Collins, M., Armitage, J., Weisenburger, D., and Purtilo, D. (1987). DNA aneuploidy and low proliferative activity predict a favorable clinical outcome in diffuse large-cell carcinoma. *Proc. AACR* **28**, 131.
20. Tomita, T., Yasue, M., Engelhard, H. H., McLone, D. G., Gonzalez-Crussi, F., and Bauer, K. D. (1988). Flow cytometric DNA analysis of medulloblastoma: Prognostic implication of aneuploidy. *Cancer* **61**, 744–749.
21. Oppedal, B. R., Storm-Mathisen, I., Lie, S. O., and Brandtzaeg, P. (1988). Prognostic factors in neuroblastoma: Clinical, histopathologic, and immunohistochemical features and DNA ploidy in relation to prognosis. *Cancer* **62**, 772–780.
22. Briffod, M., Spyratos, F., Tubiana-Hulin, M., Pallud, C., Mayras, C., Filleul, A., and Rouesse, J. (1989). Sequential cytopunctures during preoperative chemotherapy for primary breast carcinoma: Cytomorphologic changes, initial tumor ploidy, and tumor regression. *Cancer* **63**, 631–637.
23. Meyer, J. S., and Prioleau, P. G. (1981). S-phase fractions of colorectal carcinomas related to pathologic and clinical features. *Cancer* **48**, 1221–1228.
24. Frankfurt, O. S., Chin, J. L., Englander, L. S., Greco, W. R., Pontes, J. E., and Rustum, Y. M. (1985). Relationship between DNA ploidy, glandular differentiation, and tumor spread in human prostate cancer. *Cancer Res.* **45**, 1418–1423.
25. Look, A. T., Douglass, E. C., and Meyer, W. H. (1988). Clinical importance of near-diploid tumor stem lines in patients with osteosarcoma of an extremity. *N. Engl. J. Med.* **318**, 1567–1572.
26. Ensley, J., Pietraszkiewicz, H., Pear, A., Sakr, W., and Maciorowski, Z. (1994). Clinical Application of DNA content parameters in head and neck cancers. *Cancer Mol. Biol.* **1**, 9–18.
27. Hitchcock, C., and Ensley, J. (1993). Technical considerations for dissociation of fresh and archival tumors. *In* "Clinical Flow Cytometry Principles and Application" (K. Bauer, R. Duque, and V. Shankey, eds.), pp. 93–109. Williams and Wilkins, Baltimore.
28. Ensley, J. F., and Maciorowski, Z. (1994). Clinical applications of DNA content parameters in patients with squamous cell carcinomas of the head and neck. *Semin. Oncol.* **21**, 330–339.
29. Ensley, J. F., Maciorowski, Z., Hassan, M., Pietraszkiewicz, H., Sakr, W., and Heilbrun, L. K. (1993). Variations in DNA aneuploid cell content during tumor dissociation in human colon and head and neck cancers analyzed by flow cytometry. *Cytometry* **14**, 550–558.
30. Shankey, T. V., Rabinovitch, P. S., Bagwell, B., Bauer, K. D., Duque, R. E., Hedley, D. W., Mayall, B. H., Wheeless, L., and Cox, C. (1993). Guidelines for implementation of clinical DNA cytometry: International Society for Analytical Cytology. *Cytometry* **14**, 472–477.

31. Ensley, J. F., Maciorowski, Z., Hassan, M., Pietraszkiewicz, H., Heilbrun, L., Kish, J. A., Tapazoglou, E., Jacobs, J. R., and Al-Sarraf, M. (1989). Cellular DNA content parameters in untreated and recurrent squamous cell cancers of the head and neck. *Cytometry* **10**, 334–338.
32. Burgio, D. L., Jacobs, J. R., Maciorowski, Z., Alonso, M. M., Pietraszkiewicz, H., and Ensley, J. F. (1992). DNA ploidy of primary and metastatic squamous cell head and neck cancers. *Arch. Otolaryngol. Head Neck Surg.* **118**, 185–187.
33. Muller, R. P., Addicks, H. W., and Meier, E. M. (1985). Pretherapeutic flow cytometric DNA investigations in radiotherapy patients with maxillo-facial carcinomas. *Int. J. Radiat. Oncol. Biol. Phys.* **11**, 1613–1619.
34. Johnson, T. S., Williamson, K. D., Cramer, M. M., and Peters, L. J. (1985). Flow cytometric analysis of head and neck carcinoma DNA index and S-fraction from paraffin-embedded sections: Comparison with malignancy grading. *Cytometry* **6**, 461–470.
35. Gandour-Edwards, R. F., Donald, P. J., Yu, T. L., Howard, R. R., and Teplitz, R. L. (1994). DNA content of head and neck squamous carcinoma by flow and image cytometry. *Arch. Otolaryngol. Head Neck Surg.* **120**, 294–297.
36. Berlinger, N. T., Malone, B. N., and Kay, N. E. (1987). A comparison of flow cytometric DNA analyses of fresh and fixed squamous cell carcinomas. *Arch. Otolaryngol. Head Neck Surg.* **113**, 1301–1306.
37. Hemmer, J., Kraft, K., and Polackova, J. (1998). Representativity of incisional biopsies for the assessment of flow cytometric DNA content in head and neck squamous cell carcinoma. *Pathol. Res. Pract.* **194**, 105–109.
38. Holm, L. E. (1982). Cellular DNA amounts of squamous cell carcinomas of the head and neck region in relation to prognosis. *Laryngoscope* **92**, 1064–1069.
39. Rubio Bueno, P., Naval Gias, L., Garcia Delgado, R., Domingo Cebollada, J., and Diaz Gonzalez, F. J. (1998). Tumor DNA content as a prognostic indicator in squamous cell carcinoma of the oral cavity and tongue base. *Head Neck* **20**, 232–239.
40. Munck-Wikland, E., Kuylenstierna, R., Lindholm, J., and Auer, G. (1991). Image cytometry DNA analysis of dysplastic squamous epithelial lesions in the larynx. *Anticancer Res.* **11**, 597–600.
41. Saito, T., Sato, J., Satoh, A., Notani, K., Fukuda, H., Mizuno, S., Shindoh, M., and Amemiya, A. (1994). Flow cytometric analysis of nuclear DNA content in tongue squamous cell carcinoma: Relation to cervical lymph node metastasis. *Int. J. Oral. Maxillofac. Surg.* **23**, 28–31.
42. Tytor, M., Franzen, G., Olofsson, J., Brunk, U., and Nordenskjold, B. (1987). DNA content, malignancy grading and prognosis in T1 and T2 oral cavity carcinomas. *Br. J. Cancer* **56**, 647–652.
43. Lampe, H. B. (1993). DNA analysis of head and neck squamous cell carcinoma by flow cytometry. *Laryngoscope* **103**, 637–644.
44. Wolf, G.T., Fisher, S. G., Truelson, J. M., and Beals, T. F. (1994). DNA content and regional metastases in patients with advanced laryngeal squamous carcinoma: Department of Veterans Affairs Laryngeal Study Group. *Laryngoscope* **104**, 479–483.
45. Schimming, R., Hlawitschka, M., Haroske, G., and Eckelt, U. (1998). Prognostic relevance of DNA image cytometry in oral cavity carcinomas. *Anal. Quant. Cytol. Histol.* **20**, 43–51.
46. Hemmer, J., Nagel, E., and Kraft, K. (1999). DNA aneuploidy by flow cytometry is an independent prognostic factor in squamous cell carcinoma of the oral cavity. *Anticancer Res.* **19**, 1419–1422.
47. Kearsley, J. H., Bryson, G., Battistutta, D., and Collins, R. J. (1991). Prognostic importance of cellular DNA content in head-and-neck squamous-cell cancers: A comparison of retrospective and prospective series. *Int. J. Cancer* **47**, 31–37.
48. Chen, R. B., Suzuki, K., Nomura, T., and Nakajima, T. (1993). Flow cytometric analysis of squamous cell carcinomas of the oral cavity in relation to lymph node metastasis. *J. Oral Maxillofac. Surg.* **51**, 397–401.
49. Rua, S., Comino, A., Fruttero, A., Cera, G., Semeria, C., Lanzillotta, L., and Boffetta, P. (1991). Relationship between histologic features, DNA flow cytometry, and clinical behavior of squamous cell carcinomas of the larynx. *Cancer* **67**, 141–149.
50. Sakr, W., Hussan, M., Zarbo, R. J., Ensley, J., and Crissman, J. D. (1989). DNA quantitation and histologic characteristics of squamous cell carcinoma of the upper aerodigestive tract. *Arch. Pathol. Lab. Med.* **113**, 1009–1014.
51. Hogmo, A., Kuylenstierna, R., Lindholm, J., and Munck-Wikland, E. (1999). Predictive value of malignancy grading systems, DNA content, p53, and angiogenesis for stage I tongue carcinomas. *J. Clin. Pathol.* **52**, 35–40.
52. Mishra, R. C., and Mohanty, S. (1996). Study of tumour ploidy in early stages of buccal mucosa cancer. *Eur. J. Surg. Oncol.* **22**, 58–60.
53. Munck-Wikland, E., Kuylenstierna, R., Lind, M., Lindholm, J., Nathanson, A., and Auer, G. (1992). The prognostic value of cytometric DNA analysis in early stage tongue cancer. *Eur. J. Cancer B Oral Oncol.* **28B**, 135–138.
54. Kokal, W. A., Gardine, R. L., Sheibani, K., Zak, I. W., Beatty, J. D., Riihimaki, D. U., Wagman, L. D., and Terz, J. J. (1988). Tumor DNA content as a prognostic indicator in squamous cell carcinoma of the head and neck region. *Am. J. Surg.* **156**, 276–280.
55. Guo, Y. C., DeSanto, L., and Osetinsky, G. V. (1989). Prognostic implications of nuclear DNA content in head and neck cancer. *Otolaryngol. Head Neck Surg.* **100**, 95–98.
56. Toffoli, G., Franchin, G., Barzan, L., Cernigoi, C., Carbone, A., Sulfaro, S., Franceschi, S., and Boiocchi, M. (1995). Brief report: Prognostic importance of cellular DNA content in T1-2 N0 laryngeal squamous cell carcinomas treated with radiotherapy. *Laryngoscope* **105**, 649–652.
57. Tytor, M., Franzen, G., and Olofsson, J. (1989). DNA ploidy in oral cavity carcinomas, with special reference to prognosis. *Head Neck* **11**, 257–263.
58. Munck-Wikland, E., Fernberg, J. O., Kuylenstierna, R., Lindholm, J., and Auer, G. (1993). Proliferating cell nuclear antigen (PCNA) expression and nuclear DNA content in predicting recurrence after radiotherapy of early glottic cancer. *Eur. J. Cancer B Oral Oncol.* **29B**, 75–79.
59. Fu, K. K., Hammond, E., Pajak, T. F., Clery, M., Doggett, R. L., Byhardt, R. W., McDonald, S., and Cooper, J. S. (1994). Flow cytometric quantification of the proliferation-associated nuclear antigen p105 and DNA content in advanced head & neck cancers: Results of RTOG 91-08. *Int. J. Radiat. Oncol. Biol. Phys.* **29**, 661–671.
60. Raybaud, H., Fortin, A., Bairati, I., Morency, R., Monteil, R. A., and Tetu, B. (2000). Nuclear DNA content, an adjunct to p53 and Ki-67 as a marker of resistance to radiation therapy in oral cavity and pharyngeal squamous cell carcinoma. *Int. J. Oral Maxillofac. Surg.* **29**, 36–41.
61. Zatterstrom, U. K., Wennerberg, J., Ewers, S. B., Willen, R., and Attewell, R. (1991). Prognostic factors in head and neck cancer: Histologic grading, DNA ploidy, and nodal status. *Head Neck* **13**, 477–487.
62. Gomez, R., el-Naggar, A. K., Byers, R. M., Garnsey, L., Luna, M. A., and Batsakis, J. G. (1992). Squamous carcinoma of oral tongue: Prognostic significance of flow-cytometric DNA content. *Mod. Pathol.* **5**, 141–145.
63. Truelson, J. M., Fisher, S. G., Beals, T. E., McClatchey, K. D., and Wolf, G. T. (1992). DNA content and histologic growth pattern correlate with prognosis in patients with advanced squamous cell carcinoma of the larynx: The Department of Veterans Affairs Cooperative Laryngeal Cancer Study Group. *Cancer* **70**, 56–62.
64. Goldsmith, M. M., Cresson, D. S., Postma, D. S., Askin, F. B., and Pillsbury, H. C. (1986). Significance of ploidy in laryngeal cancer. *Am. J. Surg.* **152**, 396–402.
65. Franzen, G., Olofsson, J., Tytor, M., Klintenberg, C., and Risberg, B. (1987). Preoperative irradiation in oral cavity carcinoma: A study with special reference to DNA pattern, histological response and prognosis. *Acta Oncol.* **26**, 349–355.

66. Syms, C. A., III, Eibling, D. E., McCoy, J. P., Jr., Barnes, L., Emanuel, B., Fowler, C., Wagner, R., and Johnson, J. T. (1995). Flow cytometric analysis of primary and metastatic squamous cell carcinoma of the oral cavity and oropharynx. *Laryngoscope* **105**, 149–155.
67. Yip, T. T., Lau, W. H., Chan, J. K., Ngan, R. K., Poon, Y. F., Lung, C. W., Lo, T. Y., and Ho, J. H. (1998). Prognostic significance of DNA flow cytometric analysis in patients with nasopharyngeal carcinoma. *Cancer* **83**, 2284–2292.
68. Costello, F., Mason, B. R., Collins, R. J., and Kearsley, J. H. (1990). A clinical and flow cytometric analysis of patients with nasopharyngeal cancer. *Cancer* **66**, 1789–1795.
69. Cheng, D. S., Campbell, B. H., Clowry, L. J., Hopwood, L. E., Murray, K. J., Toohill, R. J., and Hoffmann, R. G. (1990). DNA content in nasopharyngeal carcinoma. *Am. J. Otolaryngol.* **11**, 393–397.
70. Halvorson, D. J., Day, S., Christian, D. R., Jr., and Porubsky, E. S. (1999). Flow cytometry and squamous cell carcinoma of the maxillary sinus: A possible prognostic indicator for multimodality intervention. *Oncology* **56**, 248–252.
71. Gregg, C. M., Beals, T. E., McClatchy, K. M., Fisher, S. G., and Wolf, G. T. (1993). DNA content and tumor response to induction chemotherapy in patients with advanced laryngeal squamous cell carcinoma. *Otolaryngol. Head Neck Surg.* **108**, 731–737.
72. Tennvall, J., Albertsson, M., Biorklund, A., Wennerberg, J., Anderson, H., Andersson, T., Elner, A., and Mercke, C. (1991). Induction chemotherapy (cisplatin + 5-fluorouracil) and radiotherapy in advanced squamous cell carcinoma of the head and neck. *Acta Oncol.* **30**, 27–32.
73. Cooke, L. D., Cooke, T. G., Bootz, F., Forster, G., Helliwell, T. R., Spiller, D., and Stell, P. M. (1990). Ploidy as a prognostic indicator in end stage squamous cell carcinoma of the head and neck region treated with cisplatinum. *Br. J. Cancer* **61**, 759–762.
74. Lukacs, G. L., Balazs, G., Zs-Nagy, I., and Miko, T. (1994). Clinical meaning of DNA content in the long term behaviour of follicular thyroid tumours: A 12-year follow up. *Eur. J. Surg.* **160**, 417–423.
75. Zedenius, J., Auer, G., Backdahl, M., Falkmer, U., Grimelius, L., Lundell, G., and Wallin, G. (1992). Follicular tumors of the thyroid gland: Diagnosis, clinical aspects and nuclear DNA analysis. *World J. Surg.* **16**, 589–594.
76. Bronner, M. P., Clevenger, C. V., Edmonds, P. R., Lowell, D. M., McFarland, M. M., and LiVolsi, V. A. (1988). Flow cytometric analysis of DNA content in Hurthle cell adenomas and carcinomas of the thyroid. *Am. J. Clin. Pathol.* **89**, 764–769.
77. McLeod, M. K., Thompson, N. W., Hudson, J. L., Gaglio, J. A., Lloyd, R. V., Harness, J. K., Nishiyama, R., and Cheung, P. S. (1988). Flow cytometric measurements of nuclear DNA and ploidy analysis in Hurthle cell neoplasms of the thyroid. *Arch. Surg.* **123**, 849–854.
78. Joensuu, H., Klemi, P., Eerola, E., and Tuominen, J. (1986). Influence of cellular DNA content on survival in differentiated thyroid cancer. *Cancer* **58**, 2462–2467.
79. Tsuchiya, A., Sekikawa, K., Ando, Y., Suzuki, S., Kimijima, I., and Abe, R. (1990). Flow cytometric DNA analysis of thyroid carcinoma. *Jpn. J. Surg.* **20**, 510–514.
80. Hrafinkelsson, J., Stal, O., Enestrom, S., Jonasson, J. G., Bjornsson, J., Olafsdottir, K., and Nordenskjold, B. (1988). Cellular DNA pattern, S-phase frequency and survival in papillary thyroid cancer. *Acta Oncol.* **27**, 329–333.
81. Stern, Y., Lisnyansky, I., Shpitzer, T., Nativ, O., Medalia, O., Feinmesser, R., and Aronson, M. (1997). Comparison of nuclear DNA content in locally invasive and noninvasive papillary carcinoma of the thyroid gland. *Otolaryngol. Head Neck Surg.* **117**, 501–503.
82. Hruban, R. H., Huvos, A. G., Traganos, F., Reuter, V., Lieberman, P. H., and Melamed, M. R. (1990). Follicular neoplasms of the thyroid in men older than 50 years of age: A DNA flow cytometric study. *Am. J. Clin. Pathol.* **94**, 527–532.
83. Grant, C. S., Hay, I. D., Ryan, J. J., Bergstralh, E. J., Rainwater, L. M., and Goellner, J. R. (1990). Diagnostic and prognostic utility of flow cytometric DNA measurements in follicular thyroid tumors. *World J. Surg.* **14**, 283–289; discussion 289–290.
84. Ekman, E. T., Bergholm, U., Backdahl, M., Adami, H. O., Bergstrom, R., Grimelius, L., and Auer, G. (1990). Nuclear DNA content and survival in medullary thyroid carcinoma: Swedish Medullary Thyroid Cancer Study Group. Cancer 65, 511–517.
85. Schroder, S., and Dralle, H. (1989). Prognostic factors in medullary thyroid carcinomas. Horm. Metab. Res. Suppl. 21, 26–28.
86. Hay, I. D., Ryan, J. J., Grant, C. S., Bergstralh, E. J., van Heerden, J. A., and Goellner, J. R. (1990). Prognostic significance of nondiploid DNA determined by flow cytometry in sporadic and familial medullary thyroid carcinoma. Surgery 108, 972–979; discussion 979–980.
87. Bergholm, U., Bergstrom, R., and Ekbom, A. (1997). Long-term follow-up of patients with medullary carcinoma of the thyroid. Cancer 79, 132–138.
88. Galera-Davidson, H., Bibbo, M., Bartels, P. H., Dytch, H. E., Puls, J. H., and Wied, G. L. (1986). Correlation between automated DNA ploidy measurements of Hurthle-cell tumors and their histopathologic and clinical features. Anal. Quant. Cytol. Histol. 8, 158–167.
89. Obara, T., Fujimoto, Y., Kanaji, Y., Okamoto, T., Hirayama, A., Ito, Y., and Kodama, T. (1990). Flow cytometric DNA analysis of parathyroid tumors: Implication of aneuploidy for pathologic and biologic classification. Cancer 66, 1555–1562.
90. Sandelin, K., Auer, G., Bondeson, L., Grimelius, L., and Farnebo, L. O. (1992). Prognostic factors in parathyroid cancer: A review of 95 cases. World J. Surg. 16, 724–731.
91. Tytor, M., Gemryd, P., Wingren, S., Grenko, R. T., Lundgren, J., Lundquist, P. G., and Nordenskjold, B. (1993). Heterogeneity of salivary gland tumors studied by flow cytometry. Head Neck 15, 514–521.
92. Franzen, G., Klausen, O. G., Grenko, R. T., Carstensen, J., and Nordenskjold, B. (1991). Adenoid cystic carcinoma: DNA as a prognostic indicator. Laryngoscope 101, 669–673.
93. Luna, M. A., el-Naggar, A., Batsakis, J. G., Weber, R. S., Garnsey, L. A., and Goepfert, H. (1990). Flow cytometric DNA content of adenoid cystic carcinoma of submandibular gland: Correlation of histologic features and prognosis. Arch. Otolaryngol. Head Neck Surg. 116, 1291–1296.
94. Hamper, K., Lazar, F., Dietel, M., Caselitz, J., Berger, J., Arps, H., Falkmer, U., Auer, G., and Seifert, G. (1990). Prognostic factors for adenoid cystic carcinoma of the head and neck: A retrospective evaluation of 96 cases. J. Oral Pathol. Med. 19, 101–107.
95. Hicks, M. J., el-Naggar, A. K., Byers, R. M., Flaitz, C. M., Luna, M. A., and Batsakis, J. G. (1994). Prognostic factors in mucoepidermoid carcinomas of major salivary glands: A clinicopathologic and flow cytometric study. Eur. J. Cancer B Oral Oncol. 30B, 329–334.
96. Hamper, K., Mausch, H. E., Caselitz, J., Arps, H., Berger, J., Askensten, U., Auer, G., and Seifert, G. (1990). Acinic cell carcinoma of the salivary glands: The prognostic relevance of DNA cytophotometry in a retrospective study of long duration (1965–1987). Oral Surg. Oral Med. Oral Pathol. 69, 68–75.
97. Timon, C. I., Dardick, I., Panzarella, T., Patterson, B., Thomas, M. J., Ellis, G. L., and Gullane, P. J. (1994). Acinic cell carcinoma of salivary glands: Prognostic relevance of DNA flow cytometry and nucleolar organizer regions. Arch. Otolaryngol. Head Neck Surg. 120, 727–733.
98. el-Naggar, A. K., Batsakis, J. G., Luna, M. A., McLemore, D., and Byers, R. M. (1990). DNA flow cytometry of acinic cell carcinomas of major salivary glands. J. Laryngol. Otol. 104, 410.
99. Felix, A., El-Naggar, A. K., Press, M. F., Ordonez, N. G., Fonseca, I., Tucker, S. L., Luna, M. A., and Batsakis, J. G. (1996). Prognostic significance of biomarkers (c-erbB-2, p53, proliferating cell nuclear antigen, and DNA content) in salivary duct carcinoma. Hum. Pathol. 27, 561–566.

100. Carrillo, R., Batsakis, J. G., Weber, R., Luna, M. A., and el-Naggar, A. K. (1993). Salivary neoplasms of the palate: A flow cytometric and clinicopathological analysis. J. Laryngol. Otol. 107, 858–861.
101. Hamper, K., Brugmann, M., Caselitz, J., Arps, H., Berger, J., Askensten, U., Auer, G., and Seifert, G. (1989). Prognosis of salivary adenocarcinomas: A retrospective study of 52 cases with special regard to cytochemically assessed nuclear DNA content. *Virch. Arch. A Pathol. Anat. Histopathol.* **416**, 57–64.
102. Ensley, J. F. (1996). The clinical application of DNA content and kinetic parameters in the treatment of patients with squamous cell carcinomas of the head and neck. *Cancer Metastasis Rev.* **15**, 133–141.
103. Farrar, W. B., Sickle-Santanello, B. J., Keyhani-Rofagha, S., DeCenzo, J. F., and O'Toole, R. V. (1989). Follow-up on flow cytometric DNA analysis of squamous cell carcinoma of the tongue. *Am. J. Surg.* **157**, 377–380.
104. Sickle-Santanello, B. J., Farrar, W. B., Dobson, J. L., O'Toole, R. V., and Keyhani-Rofagha, S. (1986). Flow cytometric analysis of DNA content as a prognostic indicator in squamous cell carcinoma of the tongue. *Am. J. Surg.* **152**, 393–395.
105. Cooke, L. D., Cooke, T. G., Forster, G., Helliwell, T. R., and Stell, P. M. (1991). Cellular DNA content and prognosis in surgically treated squamous carcinoma of the larynx. *Br. J. Cancer* **63**, 1018–1020.
106. Wennerberg, J., Baldetorp, B., and Wahlberg, P. (1998). Distribution of non-diploid flow-cytometric DNA indices and their relation to the nodal metastasis in squamous cell carcinomas of the head and neck. *Invasion Metastasis* **18**, 184–191.
107. Hemmer, J., Schon, E., Kreidler, J., and Haase, S. (1990). Prognostic implications of DNA ploidy in squamous cell carcinomas of the tongue assessed by flow cytometry. *J. Cancer Res. Clin. Oncol.* **116**, 83–86.
108. Backdahl, M., Auer, G., Forsslund, G., Granberg, P. O., Hamberger, B., Lundell, G., Lowhagen, T., and Zetterberg, A. (1986). Prognostic value of nuclear DNA content in follicular thyroid tumours. *Acta Chir. Scand.* **152**, 1–7.

PART III

PREVENTION AND DETECTION

CHAPTER

13

Smoking Behavior and Smoking Cessation among Head and Neck Cancer Patients

ROBERT A. SCHNOLL
Fox Chase Cancer Center
Cheltenham, Pennsylvania 19012

CARYN LERMAN
University of Pennsylvania Health Sciences
Philadelphia, Pennyslvania 19104

I. Introduction 185
II. Adverse Effects of Smoking for Head and Neck Cancer Patients 186
III. Smoking Rates among Head and Neck Cancer Patients 187
IV. Why Head and Neck Cancer Patients May Have Trouble Quitting Smoking 188
- A. Demographic and Medical Correlates 188
- B. Psychological Factors among Head and Neck Cancer Patients 189
- C. Psychological Correlates of Smoking in the General Population 190
- D. Genes and Smoking Behavior 191

V. State of the Science in Smoking Cessation Interventions 192
- A. Types of Smoking Cessation Approaches 192
- B. Smoking Cessation Intervention Studies with Cancer Patients 193
- C. Recommendations for Smoking Cessation Interventions for Cancer Patients 194

VI. Directions for Future Research 195
References 196

Primary prevention in the form of smoking cessation among asymptomatic populations is the most optimal approach to decreasing head and neck cancer morbidity and mortality. However, in light of data showing that continued tobacco use among head and neck cancer patients can influence treatment outcome, smoking cessation interventions for head and neck cancer patients are becoming an important priority. Despite the provision of cessation messages often conveyed to patients by their healthcare providers, more than one-third of head and neck cancer patients continue to smoke during and following treatment. Fortunately, the past decade has witnessed a substantial growth in the understanding of the factors that influence smoking behavior, as well as the availability of effective smoking cessation interventions. This chapter (1) reviews the rationale for integrating smoking cessation treatments into the standard medical care for head and neck cancer patients, (2) identifies the potential determinants of smoking behavior in this population, including psychological, demographic, medical, and genetic factors, (3) discusses the range of empirically based smoking cessation interventions and provides recommendations for smoking cessation treatment in this population, and (4) provides suggestions for future research.

I. INTRODUCTION

That tobacco use is a primary cause of cancers of the head and neck is well documented [1,2]. Indeed, it is widely recognized that there is a dose–response relationship between the duration of tobacco use and the risk of oral, oropharangeal, laryngeal, and hypopharangeal cancer [3,4]. However, research conducted over the past decade has also demonstrated that survival duration, risk of recurrence and vulnerability to a second primary neoplasm, and treatment efficacy are affected by continued smoking after head and neck cancer diagnosis and treatment [5–7]. Such data, when combined with the high incidence of second primary tumors among initially cured head and neck cancer patients [8], underscore the need for integrating smoking cessation treatment into the standard medical care for head and neck cancer patients.

Today, those interested in developing a smoking cessation treatment program for head and neck cancer patients are fortunate to have a sizable body of literature to rely on for understanding what factors—be they medical, demographic, psychological, or even genetic—to consider when devising such an intervention. In addition, an impressive number of studies concerning the effectiveness and comparative efficacy of a broad range of smoking cessation intervention approaches have been conducted, culminating in the recently published Agency for Healthcare Research and Quality Smoking Cessation Clinical Practice Guideline (AHRQ; 9). To this point, only a small amount of research has been conducted on tobacco use by head and neck cancer patients [5,10]. While barriers to cessation identified among the general population of smokers may apply to cancer patients, certain factors that influence cessation may be either particularly important (i.e., depression) or unique (i.e., disease-related factors) for this population of smokers. Consideration of these important or unique issues concerning tobacco use among cancer patients when designing smoking cessation interventions that are tailored to the needs of this population of smokers may improve the efficacy of these treatments. In turn, the establishment of smoking cessation services for head and neck cancer patients can be expected to yield important health benefits for head and neck cancer patients, including improved survival rates and enhanced quality of life [7].

This chapter provides a (1) rationale for the provision of smoking cessation treatment for head and neck cancer patients, (2) brief overview of the literature on the determinants of smoking behavior, including psychological, demographic, medical, and genetic factors, and (3) description of the range of evidence-based smoking cessation treatment options that are available for use with head and neck cancer patients, as well as recommendations for clinical treatments for this population of smokers. In addition, we highlight several directions for future research in the area of smoking behavior among head and neck cancer patients. Overall, this chapter was tailored to answer the question: Based on what is known about what drives tobacco use, what is the best approach to treating nicotine addiction among head and neck cancer patients?

II. ADVERSE EFFECTS OF SMOKING FOR HEAD AND NECK CANCER PATIENTS

The growing interest in smoking cessation among head and neck cancer patients is borne out of data suggesting that survival duration is influenced by whether the patient ceases to smoke once diagnosed or continues to smoke during or following treatment [5–7]. Browman *et al.* [11] examined the relationship between continued smoking and survival among 115 patients with stage III or IV squamous cell carcinoma of the head and neck. Thirty-nine percent of patients who smoked lived at least 2 years after diagnosis, compared with 66% of nonsmokers. Overall, the median survival time for smokers was 16 months versus 30 months for nonsmokers. Likewise, De Boer *et al.* [12] examined longitudinal predictors of survival in a sample of 133 head and neck cancer patients. After controlling for the type of medical treatment received, tumor stage, and age, a history of smoking between diagnosis and treatment was a significant predictor of disease mortality. In particular, patients who continued to smoke following their diagnosis were almost twice as likely to have been deceased at the 6-year follow-up compared to patients who abstained. Finally, Stevens *et al.* [13] reported that of 269 patients recovering from treatment for head and neck cancer, the mean survival time was significantly lower for patients who continued to smoke than for patients who quit smoking following their diagnosis. It is important to note, however, that another study with 208 patients with head and neck cancer failed to replicate this relationship between continued smoking and reduced survival [14].

The link between survival duration and smoking behavior is likely mediated by the influence of smoking on the risk of recurrence and/or a second primary tumor [15,16]. In the study conducted by Stevens and colleagues [13], patients who continued to smoke were twice as likely to experience a disease recurrence compared with patients who abstained following treatment. In addition, Day *et al.* [17] examined the effects of continued smoking on risk for a second primary tumor among a large group of patients with primary oral or pharyngeal cancer. Current smokers were almost four times as likely to develop a second primary tumor compared to former smokers, and this risk increased with the duration and intensity of smoking. Likewise, Hiyama *et al.* [18], in a study with laryngeal cancer patients, showed that risk of a second primary tumor increased among smokers and was proportional to the level of exposure (i.e., heavy smokers were at greater risk than light smokers). Data also suggest that smoking may have an adverse effect on the response to medical treatment. In the study by Browman *et al.* [11], 45% of smokers experienced a complete tumor response following radiotherapy (i.e., no evidence of disease 13 weeks after completion of treatment) vs 75% of patients who discontinued smoking. In addition, Fountzilas *et al.* [19], in a study that evaluated correlates of tumor response to chemotherapy, reported that patients who were nonsmokers were significantly more likely to exhibit a complete tumor response than smokers.

Continued smoking has also been associated with adverse health consequences relevant to cancer patients with advanced disease and a poor life expectancy, namely a greater frequency of physical symptoms associated with treatment, which can reduce the quality of life substantially [5,20]. Continued smoking by head and neck cancer patients has been associated with increased bronchial secretion, impaired pulmonary function, loss of taste, and dry mouth [21]. Dry mouth, in particular, has been shown to (1) greatly increase a patients' risk for esophageal injury by decreasing the removal of acid from the esophagus [22] and (2) increase the potential for mucousitis—a condition that can interfere with nutritional intake and increase the risk for infection [23]. Rugg *et al.* [24] assessed the duration of mucousitis in patients receiving radiation therapy for advanced head and neck cancer. Smokers averaged 21 weeks of mucousitis, whereas nonsmokers averaged 13 weeks. Continued smoking in this population has been linked to an increased risk of deep venous thrombosis, pulmonary embolism, and impaired wound healing following surgery [21,25,26]. In addition, head and neck cancer patients who continue to smoke following medical therapy are at heightened risk for soft tissue and bone necrosis [27] and have a greater difficulty regaining the quality of their voice following surgery [28].

Thus, continuing to smoke following diagnosis and/or treatment is a prognostic factor in the recovery from head and neck cancer. Continued smoking can also derail recovery from surgery (i.e., prolong wound healing, delay voice recovery, bone and tissue death) and exacerbate side effects of radiotherapy and chemotherapy (i.e., mucousitis). Further, continued smoking can increase the patient's risk for noncancer-related illnesses, such as chronic obstructive pulmonary disease, coronary heart disease, and stroke [29,30]. In addition, continued smoking is related to high levels of alcohol consumption, which can independently increase the risk for adverse health consequences, including decreased survival, following head and neck cancer diagnosis and treatment [17,19,31]. Therefore, quitting smoking following a diagnosis of head and neck cancer may be an important way to increase survival rates from this disease, while at the same time improving the quality of life of patients and survivors.

III. SMOKING RATES AMONG HEAD AND NECK CANCER PATIENTS

Despite the documented health advantage that smoking cessation can afford cancer patients, many continue to smoke following diagnosis and treatment. Two studies have evaluated the rate of smoking in the general population of cancer patients. In one study with close to 6000 patients, Spitz *et al.* [32] reported that 30% of male patients and 29% of female patients were self-classified smokers. In a second study with about 700 patients surveyed from six cancer centers, 18% of patients were current smokers [33]. The smoking rate among head and neck cancer patients may be particularly high compared to the rate of tobacco use among the general population of cancer patients and compared to the rate of tobacco use in the general (noncancer) population; two studies indicate that 55–69% of head and neck cancer patients are current smokers [31,34]. Additionally, although relapse may be delayed among head and neck cancer patients, about 25% of patients who quit following their diagnosis return to smoking within 1 year [35]. Finally, many head and neck cancer patients have a history of heavy alcohol use in addition to their tobacco use [7], and Christensen *et al.* [36] found that, on average, their sample of head and neck cancer patients consumed close to 14 drinks per week. Further, three studies have indicated an existing substance abuse disorder in 18–45% of head and neck cancer patients [31,37,38], with alcohol use increasing over time [35].

In prior studies with head and neck cancer patients, there is wide variability in the prevalence rates of continued smoking, with a range of 25–90% [6,15,32,34–36,39,40,41]. Much of this variability can be attributed to differences across studies, such as inconsistent follow-up intervals, divergent methods for assessing quitting, reductions in baseline smoking rates in the population over time, and medical and demographic differences among the samples [34]. Early studies [40] and those with longer follow-up intervals [15] reported the highest smoking rates. More recent studies, which defined smoking as taking a single puff in the past 30 days and followed patients up to 2 years after diagnosis, have reported rates of continued smoking to be 25–40% [35,39,41]. In an ongoing prospective study with head and neck cancer patients at Fox Chase Cancer Center, preliminary findings indicate that about 30% of patients continue to smoke [42], which converges well with a recent review of this literature, which suggested that the rate of continued smoking in this population is about 33% [7].

Nevertheless, prospective studies in this area are the exception and only two of these studies utilized samples exceeding 100 patients. Retrospective or cross-sectional studies can be biased by the fallibility of patient recall and/or the tendency for patients who smoke to be underrepresented due to early mortality. Cross-sectional studies can also yield erroneous prevalence rates if the research overlooks when the measurement occurs; it is common for patients to quit at diagnosis or in response to surgical treatment. However, these patients may relapse quickly following the completion of treatment or surgical recovery. Thus, the prevalence rate can vary as a function of when the measurement of smoking behavior takes place. An additional concern is variability in the denominator for calculating smoking rates. It is often unclear whether the rate of smoking reported is based on considering all head and neck cancer patients or just those with a history of smoking. If the number of patients using tobacco is compared to all head and neck patients, the rate will be much lower compared to if the rate is compared to only those head and neck patients who smoked prior to diagnosis. Likewise, variability in the definition of "smoked prior to diagnosis" can affect prevalence estimates; if this variable is defined as "smoked in the preceding 12 months," the rate may be larger than if this variable is defined as "a history of smoking."

For the purposes of assessing estimates of continued smoking among cancer patients, only patients with a history of smoking during the 6 months prior to diagnosis should be included (and serve as the denominator of the estimate). The goal of this assessment is to determine what proportion of patients quit in response to their diagnosis. Patients who have been abstinent for more than 6 months have likely quit for reasons other than the diagnosis and have a low likelihood for relapse [43] and, therefore, should not be included in the analyses. In addition, biochemical verification of smoking status may be especially important given the possibility of deception with a large degree of imposed pressure on the patient to quit.

As such, larger prospective studies with biochemical verification may be needed before a definitive sense of the rate of continued smoking in this population is ascertained. Future studies should follow a cohort of newly diagnosed patients who smoked at any time during the 6 months preceding their diagnosis. In addition, novel measurement strategies of smoking may be needed, such as the time line follow-back method. This approach to assessing continuous abstinence is reliable for retrospective assessments of alcohol use [44], psychoactive drug use [45], and smoking [46]. In short, this method consists of interviewing the participant to record smoking practices on each day preceding the interview date back to the date of diagnosis. With more rigorous research and measurement strategies, which consider factors that can confound findings, a more accurate estimation of continued smoking rates among cancer patients could be obtained.

IV. WHY HEAD AND NECK CANCER PATIENTS MAY HAVE TROUBLE QUITTING SMOKING

There is now sizable literature concerning determinants of smoking behavior for the general population of smokers [47]. In contrast, few studies have assessed correlates of smoking among cancer patients, and prospective studies are lacking. Understanding what factors influence smoking behavior among this population is essential for developing and implementing smoking cessation interventions for head and neck cancer patients. The next few sections review the available literature concerning demographic, medical, psychological, and genetic correlates of smoking. Because few studies have examined psychological correlates and no study has evaluated genetic factors associated with smoking among head and neck cancer patients, we rely mostly on literature formulated with the general population of smokers to review the potential role that these two factors play in determining the smoking behavior of head and neck cancer patients.

A. Demographic and Medical Correlates

Few consistent findings have emerged with regard to demographic variables as predictors of smoking behavior among head and neck cancer patients. While older cancer patients were more likely to remain abstinent in one study [32], they were less likely in another [48]. In yet a third study, there was no significant difference in mean age between patients who continued to smoke and those who abstained [35]. Likewise, while higher education has been linked with greater abstinence by some [32,34], this relationship has not been replicated by others [40]. Further, four studies found no differences in cessation rates between men and women [32,34–36].

Smoking history variables (e.g., amount smoked, duration smoked, level of nicotine dependence) have also been evaluated as correlates of continued smoking among head and neck cancer patients. Amount smoked, often assessed as number of cigarettes smoked prior to diagnosis, has not been associated with the likelihood of continued smoking [32,34–36,48]. Likewise, duration of smoking does not appear to be associated with continued smoking either [34,36], with one study suggesting—counter to what one would expect—that patients who started smoking more recently were more likely to be a continuous smoker [35]. In contrast, nicotine addiction appears to be associated with the smoking behavior of head and neck cancer patients, and many head and neck cancer patients resemble heavily addicted smokers from the general population. Gritz and colleagues [35,48] reported that, according to a single item indication of nicotine dependence (i.e., "smoking within 30 min of waking"), 70–75% of head and neck cancer patients were classified as strongly addicted to nicotine. This compares to 90% of heavy smokers about to participate in a smoking cessation intervention [49] and greatly exceeds the rate of 40% found among light smokers in the general population [50]. In our ongoing prospective study of smoking behavior among head and neck cancer patients, we have found that 68% of patients indicate that they smoke within 30 min of waking up [51]. We have also found that patients with higher levels of nicotine dependence are more likely to continue to smoke after diagnosis [51], a finding reported by others as well [35,48].

Alcohol use has also been examined as a correlate of smoking behavior among head and neck cancer patients, as many head and neck cancer patients abuse (or at least heavily use) alcohol, and the use of alcohol and tobacco has been consistently linked in studies in the general population. The comorbidity between tobacco and alcohol use is extensive, with 80% of alcoholics being smokers and 30% of smokers being alcoholics [52,53]. However, studies with head and neck cancer patients have not yielded consistent findings in this area. Vander Ark *et al.* [41] found that head and neck cancer patients who were heavy drinkers were at a greater risk of continuing to smoke than patients who were light or moderate drinkers; however, this relationship was not found in other studies [32,35,36].

Disease-related variables may influence smoking behavior and cessation among head and neck cancer patients. A history of more extensive medical treatment (i.e., surgery and adjunctive therapies versus surgery only, or surgery versus only radiation) has been associated with a greater rate of continued abstinence [34,35,41,48]. Likewise, the cancer site can influence cessation rates. Two studies found that laryngeal head and neck cancer patients were more likely to abstain from tobacco use than patients with oral cavity cancer [32,34,41]. Further, in our ongoing research [42] we found that time since diagnosis is associated with the ability to quit. We have found that smokers report a longer duration of time between diagnosis and assessment than abstainers. One interpretation of these data may be that patients in late stages of recovery (i.e., completed treatment) minimize the seriousness of their diagnosis and are thus less committed to quitting. Alternatively, smoking rates may be lower during treatment due to the greater likelihood of physical disability, which makes it impossible to smoke. Regardless, this finding underscores that risk of relapse increases once treatment is complete. In contrast, findings concerning the association between disease stage and smoking behavior have not been consistent; a higher disease stage was associated with a greater likelihood for abstinence by Ostroff *et al.* [34], but not by others [35,36,41].

B. Psychological Factors among Head and Neck Cancer Patients

Although a sizable literature concerning psychological correlates of smoking among the general population of smokers has accumulated, few studies have assessed psychological correlates of smoking behavior among head and neck cancer patients. In general, studies have focused on quit motivation, beliefs about quitting, and emotional distress as correlates of smoking and cessation behavior in this population.

1. Motivation to Quit

An early study by Gritz *et al.* [48] found that motivation to quit, assessed using a stage-based measure derived from the transtheoretical model [43] at a baseline assessment, predicted smoking status among head and neck cancer patients assessed at a 12-month follow-up. In particular, 11/24 (46%) of patients who indicated at baseline that they were *not* prepared to quit within the next 12 months (i.e., precontemplators) were able to maintain abstinence over the 12-month study compared to 26/44 (59%) patients who indicated at baseline that they were prepared to quit within the next 12 months (i.e., contemplators). More recently, Gritz and colleagues [35] found that 10/21 (48%) head and neck cancer patients classified as pre-contemplators continued to smoke after their diagnosis and treatment compared to 11/37 (30%) patients classified as contemplators. Such data parallel findings from the general population of smokers [43]. Further, our longitudinal study with head and neck and lung cancer patients found that higher baseline levels of quit motivation predicted abstinence assessed 3 months later [42].

2. Beliefs about Quitting

A study by Christensen *et al.* [36] examined whether perceived control (i.e., believing that future health outcomes are affected by one's behavior) and self-blame (i.e., attributing the cause of the diagnosis to one's behavior) influenced smoking behavior among head and neck cancer patients. A significant interaction effect between these two correlates of smoking behavior was detected; i.e., a greater proportion of individuals with high levels of self-blame *and* low levels of perceived control continued to smoke versus patients with high levels of self-blame *but* high levels of perceived control (64% vs 25%, respectively). Thus, self-blame is adaptive, in the context of smoking among head and neck cancer patients, only insofar as patients also believe that they can do something about regaining their health.

In addition, Gritz *et al.* [35] found that patients who were able to maintain continual abstinence at follow-up assessments reported higher levels of quitting self-efficacy (i.e., confidence in their ability to quit) versus patients who continued to smoke. In our ongoing study of correlates of smoking behavior among cancer patients we have also found that higher levels of self-efficacy predict a greater likelihood of quitting [42]. However, we have also found that other beliefs are associated with smoking and cessation behavior among cancer patients. Our cross-sectional analyses indicate that lower levels of quitting pros (i.e., advantages of quitting) and risk perceptions (i.e., risk of recurrence) and higher quitting cons (i.e., disadvantages of quitting) and fatalistic beliefs (i.e., believing that there is no use in quitting) are associated with a greater likelihood of continued smoking [42]. Our prospective analyses, moreover, indicate that higher baseline levels of quitting self-efficacy, pros of quitting, and risk perceptions, as well as lower levels of cons of quitting, predict quitting assessed at a 3-month follow-up [42].

3. Emotional Distress

Available data also point to a link between higher levels of emotional distress and a lower likelihood of maintaining abstinence following diagnosis and treatment. Gritz *et al.* [35] reported that emotional distress (e.g., symptoms of depression and anxiety), measured by the Profile of Moods

States [54], was not related to the probability of a smoking relapse among head and neck cancer patients. However, our research has indicated that patients who continue to smoke exhibit higher levels of emotional distress concerning quitting and their cancer [42], a finding that converges well with data accumulated in the general population of smokers.

Taken together, the available (but limited) literature on smoking among head and neck cancer patients suggests that patients with low levels of quit motivation and self-efficacy and patients with high self-blame–low perceived control are at greatest risk of continuing to smoke following their diagnosis and treatment. The patient's decisional balance (awareness of the pros of quitting and minimizing the cons of quitting), perceived vulnerability to the adverse health effects of continued smoking (e.g., risk of recurrence), and emotional distress might also play a role in determining ability to quit smoking.

C. Psychological Correlates of Smoking in the General Population

The aforementioned findings from studies with head and neck cancer patients converge well with data accrued from studies with the general population of smokers. These studies are reviewed in the following section.

1. Motivation to Quit

In research conducted with the general population of smokers, prospective studies show that higher levels of motivation to quit smoking are a consistent predictor of smoking cessation [55], and treatment studies show that initial motivation to quit smoking predicts success in a smoking cessation intervention [56]. In a study with close to 5000 smokers, a stage measure of readiness to quit was used to demonstrate that a greater likelihood of attempting to quit smoking was predicted by higher levels of quit motivation [57].

2. Beliefs about Quitting

One of the most consistent predictors of the ability to quit smoking is quitting self-efficacy [58,59]. A higher level of quitting self-efficacy among adult smokers predicts greater interest and enrollment in smoking cessation treatments [60], as well as a greater likelihood that individuals will maintain abstinence following cessation treatment [61–66]. Quitting self-efficacy is considered one of the most central components of a successful smoking cessation intervention [58,67,68]. Likewise, higher levels of perceived control are associated with greater levels of quit motivation and a greater number of quit attempts [69–73].

In addition, there is a sizable literature concerning the impact of risk perceptions on smoking and cessation behavior [74,75]. A lack of awareness of the adverse effects of smoking may prevent smokers from quitting. About 60% of smokers indicate that they are at no greater risk of developing cancer than those who have never smoked [76–78], a phenomenon referred to as an "optimistic bias" [79]. In turn, as awareness of the adverse health effects of smoking increases, interest in quitting and the likelihood of actual cessation grow [74]. Perceived susceptibility to cancer has been shown to be a significant predictor of the probability that smokers will enroll in smoking cessation programs [80], and increases in risk perceptions have been observed among those who are able to quit smoking following a cessation program [81].

An additional belief about quitting that has been examined as a correlate of smoking cessation in the general population is decisional balance—the cognitive process of weighing the relative pros and cons of a given behavioral option [82]. Numerous studies have shown that higher levels of the pros of quitting (e.g., "quitting will improve my health") and lower levels of the cons of quitting (e.g., "I like the image of a smoker") are associated with a greater likelihood of smoking abstinence [83–88]. In other words, those who fail to recognize that quitting smoking is beneficial (e.g., improved health) or focus on the negative outcomes of quitting (e.g., depression) are less likely to maintain abstinence after treatment.

3. Emotional Distress

In studies with the general population of smokers, the relationship between depression and smoking [89] and between depression and cessation behavior [90] is well established. Among smokers trying to quit, a lower rate of cessation has been associated with signs of depressive symptoms [91]. Anda *et al.* [92] reported that smokers who exhibited clinical depression at a baseline assessment were 40% less likely to have quit at a 9-year follow-up compared to nondepressed smokers. Finally, several studies have shown that smokers with a history of major depression are less likely to be successful in a smoking cessation intervention than smokers without such a history [93,94]. Kinnunen *et al.* [91] found that nondepressed smokers showed a higher rate of absitinence (37%) than depressed smokers (29.5%) after treatment with a nicotine replacement therapy (gum). Likewise, in a prospective study, Rausch *et al.* [95] found that baseline assessments of depressive symptoms differentiated between those who maintained abstinence after a behavioral cessation program and those who relapsed.

The role of depression in determining smoking behavior among head and neck cancer patients may be particularly important to consider given the high rate of depression in this population [7]. For many patients, the shock, sadness, and worry associated with the initial diagnosis leads to clinical depression and suicidal ideation, especially when treatment involves disfiguring surgery. Disfigurement and loss of speech, taste, and smell can threaten self-image and self-identity, thereby increasing the risk for depression and suicidal ideation [96]. Indeed, head and neck cancer patients are at high risk for suicide [97], and early cross-sectional studies indicated that close to 40% exhibit clinical depression as defined by DSM-III criteria [98]. A more recent cross-sectional study reported that about 17% of head and neck cancer patients manifested adjustment disorder or major depression assessed using the Structured Clinical Interview for the DSM-IIIR (SCID; 99). Four recent prospective studies, using a self-report measure to assess depression (Center for Epidemiologic Studies–Depression; CES-D), found that 26–28% of patients exhibit depressive symptoms immediately after diagnosis [100–103]. Rates of depression in these studies diminished slightly to 22–28% 6 months later. In a review of this literature, Moadel *et al.* [7] indicated that about one-third of head and neck cancer patients manifest long-term psychiatric problems, including depression.

One plausible theoretical model explaining the link between depression and smoking is the self-medication hypothesis [104–106]. This theory contends that smokers use nicotine to increase their sense of physical arousal and mental alertness and to diminish feelings of sadness, worry, or tension. Several studies have documented the cognitive and affective consequences of nicotine, including a euphoria feeling comparable to the effects of other drugs of abuse such as cocaine, antidepressant qualities, and enhanced memory, attention, concentration, and learning capacity [89,90]. Additional support for this model is derived from studies that have delineated the cascade of neurobiological effects from nicotine administration within the central and peripheral nervous system (e.g., release of dopamine), which underlie enhanced cognitive functioning and improved affect [89,107,108].

D. Genes and Smoking Behavior

There is sizable variability in people's ability to quit smoking. Some people can quit and maintain abstinence easily, whereas others battle their addiction forever. Much of this variability is attributable to the psychological factors reviewed earlier. However, several genetic polymorphisms—individually or in combination—may account for some of this variability as well [109,110]. One line of research has focused on genes in the dopamine pathway, which is considered a key neurotransmitter in the reward mechanism of addictive behaviors [111–112]. The hypothesis is that certain individuals may have a harder time quitting smoking because they possess a genetic profile that yields lower rates of endogenous dopamine activity [108].

Initial research in this area focused on the role of polymorphisms on dopamine *receptor* genes, including *DRD1* [113], *DRD2* [114], and *DRD4* [115] receptors. Individuals who carry the more rare *DRD2-A1* allele [116] or the closely linked *DRD2-B1* allele [117] have been shown to have a higher likelihood of smoking. The *DRD2-A1* allele has been shown to be associated with a reduced density of dopamine D2 receptors [113]. However, in a family-based study consisting of 138 nuclear families with at least one offspring who is a habitual smoker, Bierut *et al.* [118] reported no significant difference in the frequency between DRD2 alleles transmitted and not transmitted to smokers [119].

Genetic variation in the dopamine transporter gene (*SLC6A3*) has been examined as a correlate of smoking behavior. These studies assessed the relationship between the presence vs the absence of the nine-repeat allele of this gene (9/9 or 9/* vs */*, where the asterisk refers to alleles other than nine) and smoking behavior. The presence of the nine-allele may decrease levels of dopamine transporter, thereby increasing levels of dopamine [111]. Thus, it has been hypothesized that individuals with the nine-repeat genotypes may be less likely to smoke and may have an easier time quitting versus those manifesting *SLC6A3**/** genotypes. Using a sample of 289 smokers and 233 nonsmokers, Lerman *et al.* [120] found evidence for an interaction between *SLC6A3* and *DRD2*. Smokers with the nine-repeat genotype of *SLC6A3* and the *DRD2-A2* genotype were significantly less likely to be smokers. In addition, smokers with the *SLC6A3-9* genotype reported a longer average quit duration (retrospectively) compared with smokers without this genotype (472 days vs 230 days). In a second study, Sabol *et al.* [121] examined the link between *SLC6A3* genotypes and smoking behavior in over 1000 smokers, quitters, and nonsmokers. Smokers were less likely to have the *SLC6A3-9* genotype than former smokers (42% vs 52%), although smokers and nonsmokers were no different, suggesting that *SLC6A3-9* may influence quitting behavior rather than smoking initiation per se. Jorm *et al.* [122], however, failed to replicate an association between the *SLC6A3* genotypes and smoking in an Australian sample.

Genes important in serotonin regulation are also relevant to smoking because of the role of serotonin in depression and anxiety, two traits linked with smoking behavior [123,124]. Although there is no evidence for a main effect on smoking status of a polymorphism in the serotonin transporter gene [125], preliminary evidence shows that this gene modifies the effect of anxiety-related traits on smoking behavior [126,127]. The tryptophan hydroxylase (*TPH*) gene codes for a rate-limiting enzyme in the biosynthesis of serotonin. Preliminary data suggest that individuals who are homozygous for the more rare *A* allele of *TPH* are more likely to initiate smoking and to start smoking at an earlier age [128,129].

Investigations of other candidate genes have been less fruitful. There was an initial finding of a modest association of smoking with a mutation in a gene governing nicotine metabolism (*CYP2A6*; 130). However, other groups of researchers were not able to confirm this association [131,132]. Although genes regulating nicotine receptor function would be prime candidates for smoking risk, data on *functional* genetic variation in humans are not yet available.

Although research on the genetics of complex behaviors, such as smoking, is receiving increasing attention, there are several limitations to be overcome. One factor contributing to the lack of replication has been the reliance on case–control (association) data. In case–control studies, there is the possibility of population stratification due to the fact that cases (smokers) and controls (nonsmokers) may have been drawn from populations with different ethnicities. Because the frequencies of these polymorphisms may vary across ethnic groups and because ethnic/cultural factors may influence smoking practices, ethnic admixture can bias case–control studies. Family-based analyses, such as the transmission disequilibrium test (TDT), can minimize bias due to ethnic admixture [133]. A second limitation of previous research in the area of genetics and smoking has been the reliance on crude assessments of smoking status (e.g., smoker vs nonsmoker). Because complex behaviors like smoking are likely caused by multiple genes acting in conjunction with environmental factors, it is critical that the phenotypes for smoking be well defined. Future studies to understand the genetic basis of smoking behavior in cancer patients must use multiple measures of nicotine dependence, as well as longitudinal phenotypes, such as the rate of progression to nicotine dependence and smoking persistence. Further, large sample sizes will be needed to provide confirmatory data regarding genetic effects and to measure important medical, psychological, and socioenvironmental covariates.

V. STATE OF THE SCIENCE IN SMOKING CESSATION INTERVENTIONS

Psychosocial and genetic factors may each play an important role in determining smoking cessation among head and neck cancer patients, as well as with the general population of smokers. In addition, emotional distress may be especially important to consider when designing a smoking cessation intervention for head and neck cancer patients. These findings have important implications for the selection of appropriate smoking cessation interventions for head and neck cancer patients who continue to smoke. Fortunately, there is a range of evidence-based smoking cessation intervention methods that have been tested in the general population.

A. Types of Smoking Cessation Approaches

Relevant smoking cessation interventions can be grouped into four main categories: (1) self-help, (2) clinic based, (3) physician based or (4) pharmacologic [5,134,135]. Certain treatment approaches, however, show comparatively greater effectiveness at promoting cessation.

1. Self-Help

Upward of 80% of smokers report using a self-help method [136], including "cold-turkey" or manuals (e.g., *Clear Horizons*; 137), so these methods have relatively high participation rates (i.e., 5% of all smokers) [138]. However, these methods yield low abstinence rates, somewhere around 10% [135,139–141]. The AHRQ review of self-help methods showed that they are no more effective at generating quitting than no intervention, as they resulted in a cessation rate of about 14.4% versus the 14.3% rate produced by no self-help [142]. In addition, given the greater propensity toward symptoms of depression among head and neck cancer patients, self-help approaches are unlikely to be very effective for promoting cessation in this population.

2. Clinic Based

This type of intervention, which typically includes individual or group behavioral counseling that offers social support, education, and/or problem-solving/skills training, yields substantially higher cessation rates [135]. Information about the harms of smoking, the benefits of quitting, self-efficacy training, and skills training to deal with depression and high-risk situations (e.g., being with smokers, bars and parties) are common elements of this approach. A review of 58 studies by the AHRQ [142] found that individual behavioral counseling increased cessation rates to about 17% versus a no contact control group. Further, a dose–response relationship between the intensity and the duration of contacts of counseling and treatment efficacy has been found [142]: clinic-based interventions that include over 1.5 h of total contact time or include more than four sessions increase cessation rates to about 25%. Thus, clinic-based interventions are seen as an integral dimension of a smoking cessation program for any population of smokers [142] and are likely to be effective for head and neck cancer patients.

3. Physician Based

With upward of 70% of all smokers visiting their primary-care physician at least once each year, physician-delivered smoking cessation interventions represent a cost-effective method for disseminating smoking cessation messages and services to a large number of smokers [143]. Public health advocacy agencies have tried to encourage physicians to implement smoking cessation interventions by providing clinical treatment guidelines and instructions [144], and oncologists can play an important role in improving quit rates among their patients [142]. Reviews of physician-based treatments indicate that such interventions produce modest quit rates of about 10% [145,146]. Further, studies show that only about 50% of smokers have their smoking status and interest in cessation assessed by physicians in a primary-care setting [147]. Therefore, although physician-based interventions can reach a substantial number of smokers, additional research is needed to (1) train physicians in providing effective smoking cessation treatments and (2) promote adherence among physicians to the assessment and treatment of smoking behavior. Regular access to head and neck cancer patients by physicians certainly suggests that they can play a role in treating nicotine addiction in this population of smokers.

4. Pharmacologic Interventions

Many new pharmacologic treatments for nicotine addiction have been developed over the past decade, including nicotine replacement therapies (NRTs), such as the transdermal patch, and antidepressant medications, such as buproprion (Zyban). The "patch" is considered one of the more effective NRTs available, as it is associated with fewer compliance problems, produces fewer side effects, and requires less clinician involvement for patient training [148]. Meta-analytic studies show that the patch can double cessation rates when compared with placebo and can significantly increase cessation rates when added to behavioral counseling [144]. As such, the AHRQ [142] recommends that all smokers should be offered some form of NRT unless there is a clear medical contraindication. NRTs are not, however, recommended as stand-alone treatments, but rather are prescribed in conjunction with clinic-based therapies [134].

A great deal of attention has been paid to the use of bupropion for treating nicotine addiction, and this antidepressant has been examined for smoking cessation in four studies. Two studies conducted by Ferry and colleagues [149,150] with smokers recruited from a veteran's administration hospital found that bupropion outperformed placebo, with both conditions receiving behavioral counseling. A third study assessed the dose–response effect of bupropion [151]. In terms of continuous abstinence, 24% of participants receiving the 300-mg dosage were abstinent at a 1-year follow-up compared to 18% for the 150-mg dosage, 14% for the 100-mg dosage, and 10.5% for placebo. Only the comparison between the 300-mg and placebo conditions was significant. In the fourth study, Jorenby *et al.* [152] examined the comparative and additive effectiveness of bupropion and the transdermal nicotine patch. Heavy smokers were randomized to receive a placebo, bupropion, the patch, or both bupropion and the patch. The treatment period lasted for 9 weeks and all participants received weekly clinic-based counseling sessions (e.g., awareness of smoking triggers and weight management). Assessment of point-prevalence abstinence at a 6-month follow-up showed the following cessation rates: 19% (placebo), 21% (patch), 35% (bupropion), and 39% (patch plus bupropion). Only bupropion alone and in combination with the patch outperformed placebo. Further, bupropion alone and in combination with the patch outperformed the patch alone, but there was no difference between bupropion alone and bupropion in combination with the patch. These effects were maintained at 12 months.

Therefore, clinic-based interventions, NRT, and bupropion appear to be viable options for promoting abstinence from tobacco use among head and neck cancer patients. The use of bupropion and clinic-based interventions targeted toward mood management may be especially relevant for head and neck cancer patients in light of their levels of nicotine addiction and depressive symptoms. Finally, because the oncologist or members of the health-care team routinely have access to all cancer patients (and therefore all patients who smoke), their involvement in an effort to provide smoking cessation treatment could be beneficial.

B. Smoking Cessation Intervention Studies with Cancer Patients

Smoking cessation interventions provided in a medical context can be uniquely effective at promoting cessation, as hospitalizations or physician visits represent a "window of opportunity" where patients are particularly motivated to quit smoking and especially receptive to smoking cessation interventions [153,154]. Indeed, a cancer diagnosis can increase the salience of the adverse health effects of smoking, in turn increasing the patient's readiness and intention to quit [6]. Nevertheless, only four randomized or quasi experimental studies of smoking cessation treatments in the oncologic context have been conducted.

The first study randomized 186 squamous cell head and neck cancer patients to either "usual care" physician quit advice (e.g., risks of continued smoking and benefits of cessation) or to an "enhanced" physician quit advice intervention that included expression of confidence in the patient's ability to quit, a written quit date contract, tailored self-help booklets, and booster advice to remain abstinent [48]. The results showed extremely high continuous quit rates for both conditions at a 12-month follow-up (usual care=77%; intervention=64%). Cinciripini and colleagues [5] suggest that the strong physician quit advice provided in both study conditions resulted in contamination across the groups and the absence of a treatment effect. These extremely high quit rates may also be attributable to the inclusion of ex-smokers who had not smoked for up to 1 year prior to study recruitment as participants.

In a second study, Stanislaw and Wewers [155] randomized cancer patients with various tumor sites to either a usual care or an intervention that consisted of three in-person sessions and five follow-up phone contacts. The in-person sessions comprised identifying the patients' smoking habits, reviewing the health benefits of cessation, providing self-help materials, aiding the patient in recognizing triggers and selecting alternative behaviors for smoking, and giving the patient an audio tape of relaxation exercises. The follow-up calls encouraged abstinence and discussed relapse. Control patients received routine care (quit advice). At 5 weeks, the quit rate (confirmed biochemically) for the experimental group was 75% (9/12) versus 43% (6/14) for the control group, although the small sample led to a nonsignificant comparison.

In the third study, Griebel *et al.* [156] randomized surgical cancer patients with various types of malignancies to either usual care or to a one-time smoking cessation consultation with an oncology nurse. The nurse discussed the risks of continued smoking and the benefits of cessation, the importance of setting a quit date, how to manage withdrawal symptoms using relaxation techniques, and how to use a self-help guide. Participants in the experimental condition received five weekly telephone contacts to either congratulate and reinforce abstinence or to offer assistance for relapse. After 6 weeks, 21% (3/14) of experimental patients exhibited cotinine-validated abstinence versus 14% (2/14) of the patients in the usual care condition. Again, the small sample size may have led to a nonsignificant comparison of quit rates.

Finally, the fourth study [157] examined the effectiveness of a nurse-managed smoking cessation treatment derived from the Agency for Health Care Policy and Research, now known as the AHRQ, using a quasiexperimental design. Fourteen patients (with different types of cancer) received the intervention, which consisted of quit advice, guidance in the establishment of a quit date, the teaching of behavioral and cognitive strategies to manage the stress of abstinence, the provision of support and encouragement, the use of self-help guides, the recommendation to use nicotine replacement therapy and bupropion, and follow-up telephone contacts. The comparison group consisted of 11 patients who received "usual care" prior to the initiation of the intervention (i.e., quit advice and the availability of self-help literature). At 6 months, 71% (10/14) of intervention patients were abstinent (confirmed biochemically) versus 55% (6/11) of control patients. As with the previous studies described, a small sample size limited the power of this study to show statistical significance between the groups.

C. Recommendations for Smoking Cessation Interventions for Cancer Patients

The necessary components of an effective smoking cessation intervention for head and neck cancer patients can be identified by blending physician-based, pharmacologic, and clinic-based interventions and by designing the clinic-based dimension of the intervention based on empirical data concerning determinants of cessation behavior among head and neck cancer patients. Based on these principles, we offer the following recommendations.

1. Integrate Smoking Assessment and Smoking Cessation Treatment into the Standard Medical Care Provided to Cancer Patients

This is both an infrastructure recommendation and a clinical recommendation. First, the ideal treatment program is one that is seen as part of the overall coordinated effort of care for the patient. The program should be integrated into the treatment facility. Thus, intervention specialists should be identified and resources allocated to establishing an in-house facility. Second, all members of the health-care team should be involved in smoking cessation treatment. Oncologists and nurses should be trained and encouraged to identify and refer patients for smoking cessation. A formal system may need to be established in order to ensure the identification of all patients who smoke. Patients can complete a form on their initial visit, which includes questions pertaining to their smoking history and behavior. A trained nurse or health educator can review these forms in order to identify—and then recruit—eligible patients into the smoking cessation program. In addition, chart reminders can be used to cue nurses and oncologists to refer eligible patients and brochures can be placed in clinic areas. This sort of approach can capitalize on the unique advantage of the physician-based interventions: the recruitment of a large number of smokers into a cessation program [142,143].

2. Ensure Adequate Assessment and Referral of Patients

It is critical that smokers be identified and referred to treatment. One challenge to recruiting patients is deciding on criteria for program entry. Because patients often quit at the time of diagnosis, but relapse shortly there after, programs may need to use looser criteria for program entry. So, rather than requiring that patients be current daily smokers, it may be better to allow patients who have smoked in the past 3 or 6 months into the program. In this way, one can recruit patients who exhibit a high probability for relapse once treatment is complete.

3. Include Pharmacologic Treatments into a Comprehensive Cessation Program

Because head and neck cancer patients report a high level of nicotine addiction, some form of NRT appears especially warranted. Considering the high rate of depression manifested by head and neck cancer patients, other pharmacologic interventions, such as the use of antidepressants, may be uniquely effective with this population as well. Thus, NRT or bupropion should be seen as a necessary component of a smoking cessation program for cancer patients [142]. Importantly, there are several psychiatric and medical contraindications for use of bupropion, so a psychiatric and medical assessment is critical prior to use of this medication.

4. Address Alcohol Use

The unusually high rate of alcohol use and abuse among head and neck cancer patients calls for the need to assess for the presence of this condition. Consideration for the integration of clinical strategies to address this potential barrier to cessation (and an independent factor predictive of a negative prognosis) is needed.

5. Establish a Standardized Clinic-Based Behavioral Counseling Component to the Intervention Program

This component should assist the patient to (1) develop motivation to quit and quitting self-efficacy, (2) articulate the adverse effects of continued smoking (e.g., reduced treatment efficacy, increased risk for recurrence), (3) overcome self-blame or at least develop some sense of personal control over one's prognosis, (4) overcome the cons to quitting (e.g., the idea of giving up a way to manage weight or manage emotions), and (5) manage emotional distress that often accompanies attempts to quit as well as the diagnosis itself. The AHRQ recommends four to seven individual or group sessions to attain these objectives and provides an outline of strategies to address each of these treatment elements [10,142].

6. Include Relapse Prevention Techniques

Head and neck cancer patients appear highly likely to relapse once treatment is complete. Smoking cessation interventions should include long-term follow-up and booster sessions and provide patients with specific skills to avoid relapse, such as identifying and learning how to deal with triggers [142]. It may even be more appropriate for cessation programs to target program enrollment toward patients who have completed treatment.

Finally, in the future it may possible to tailor interventions to the patients' genetic and/or psychological profile. Although we consider this to be premature given the available literature, the coming decade may yield data concerning the benefits of tailoring smoking cessation interventions in this manner. Individual differences are likely to moderate responsiveness to smoking cessation interventions [158]. For instance, ongoing studies are examining the efficacy of targeting more intensive interventions, including pharmacologic treatments such as bupropion, to individuals with "high-risk" genotypes or to those who manifest chronic or high levels of depression (with less intensive interventions provided to those with "protective" genes or manifesting low depressive symptoms). This area of research may indeed demonstrate the important benefits in terms of increased abstinence from tobacco use of "treatment matching."

VI. DIRECTIONS FOR FUTURE RESEARCH

The work that has already been accomplished has laid the groundwork for many new exciting avenues of research. To conclude, we identify areas in need of additional research.

Longitudinal studies are needed to provide additional data on the rates and patterns of smoking among head and neck cancer patients and to identify predictors of continuous smoking. Much of this research is cross-sectional and utilizes small samples. In addition, several determinants of smoking and cessation behavior have yet to be evaluated sufficiently among head and neck cancer patients, including decisional balance, emotional distress, risk perceptions, and genetic factors. We will not know the full scope of the public health problem, nor will we completely understand what differentiates between patients who continue to smoke and those who are able to abstain until large prospective studies are completed in this area. Importantly, gene–gene or gene–environment interactions can be tested only with sufficiently large samples.

Additional studies are needed to evaluate alternative smoking cessation treatments for this population. Only four studies, manifesting methodological shortcomings such as small samples and less than ideal internal validity, have been published in this area. Further research is needed to establish a literature to rely upon for the designing and integration of smoking cessation treatment programs within cancer treatment facilities. Unanswered questions in this area include what form should tobacco treatment programs for cancer patients take? Should interventions focus more on one dimension than another (i.e., addressing depression vs the pros of quitting)? Should smoking and alcohol use be addressed simultaneously? Finally, what are the mechanisms through which treatments are successful?

Finally, related to the need for additional treatment studies, future research is needed to evaluate the effectiveness of tailored treatments with this population, as well as interventions that could include the patient's relatives who smoke. Because head and neck cancer patients manifest unique characteristics that may influence smoking, such as high rates of depression, this population is ideal for assessing the role of individual differences in moderating responsiveness to specific types of smoking cessation treatments, such as bupropion. Further, the patient's diagnosis may also represent a "teachable moment" for the patient's relatives who smoke. Because these relatives may be momentarily more receptive to smoking cessation interventions in light of the diagnosis of their relative [159], studies are needed to examine the effectiveness of smoking cessation interventions for the patient's relatives who smoke as well. Such an approach to disseminating a smoking cessation program may represent an especially effective way to reach a large number of smokers who may not be otherwise targeted by a smoking cessation treatment program [159].

More attention has been allocated to the smoking behavior of cancer patients. With the growing survival rate for most cancers, the need to promote smoking abstinence among those already diagnosed with cancer has become much more relevant and important. Indeed, increasing the rate of cessation among patients with head and neck cancer will likely improve survival rates and enhance the quality of life for these patients. However, a substantial amount of additional research in this area is needed before we are ready to integrate cessation programs into the landscape of today's comprehensive cancer centers. With the growth in the literature concerning smoking behavior among cancer patients seen over the past decade, there is good reason to believe that the next decade will see major advances toward addressing the issues outlined in this chapter and a resolution to the ultimate question of how best to promote permanent abstinence from tobacco use among cancer patients.

Acknowledgments

Support to Dr. Schnoll for the preparation of this manuscript was provided by National Institutes of Health Grants R25 CA57708, RO3 CA88610, and P30 CA06927, and support to Dr. Lerman for the preparation of this manuscript was provided by National Institutes of Health Grants RO1 CA63562, RO1 CA63563, and P50 CA/DA84718.

We acknowledge Ms. Pamela Bradley, who provided technical assistance in the preparation of this chapter.

References

1. Clayman, G. L., Lippman, S. M., Laramore, G. E., and Hong, W. K. (2000). Neoplasms of the head and neck. *In* "Cancer Medicine" (J. F. Holland *et al.*, eds.), 5th Ed., pp. 1174–1226. Decker, Hamilton, Ontario, Canada.
2. USDHHS (2000). "Reducing Tobacco Use: A Report of the Surgeon General-Executive Summary." USDHHS, Centers for Disease Control and Prevention, National Center for Chronic Disease Prevention and Health Promotion, Office on Smoking and Health, Atlanta, CA.
3. Maier, H., and Weidauer, H. (1995). Alcohol drinking and tobacco smoking are the chief risk factors for ENT tumors: Increased incidence of mouth cavity, pharyngeal, and laryngeal carcinomas. *Fortshr. Med.* **113**, 157–160.
4. Zheng, W., McLaughlin, J. K., Chow, W. H., Chien, H. T., and Blot, W. J. (1993). Risk factors for cancers of the nasal cavity and paranasal sinuses among white men in the United States. *Am. J. Epidemiol.* **138**, 965–972.
5. Cinciripini, P. M., Gritz, E. R., Tsoh, J. Y., and Skaar, K. L. (1998). Smoking cessation and cancer prevention. *In* "Psycho-Oncology" (J. C. Holland, ed.). Oxford Univ. Press, New York.
6. Gritz, E. R. (1991). Smoking and smoking cessation in cancer patients. *Br. J. Addict.* **86**, 549–554.
7. Moadel, A. B., Ostroff, J. S., and Schantz, S. P. (1998). Head and neck Cancer. *In* "Psycho-Oncology" (J. C. Holland, ed.). Oxford Univ. Press, New York.
8. Lippman, S. M., and Hong, W. K. (1992). Retinoid chemoprevention of upper aerodigestive tract carcinogensis. *In* "Important Advances in Oncology" (V. T. Devita, S. Hellman, and S. A. Rosenberg, eds.), pp. 93–109. Lippincott, Philadelphia.
9. Fiore, M. C., Bailey, W. C., Cohen, S. J., *et al.* (2000). "Treating Tobacco Use and Dependence." USDHHS Public Health Service, Rockville, MD.
10. Hecht, J. P., Emmons, K. M., Brown, R. A., Everett, K. D., Farrell, N. C., Hitchcock, P., and Sales, S. D. (1994). Smoking interventions for patients with cancer: Guidelines for nursing practice. *Oncol. Nurs. For.* **21**, 1657–1666.
11. Browman, G. P., Wong, G., Hodson, I., Sathya, J., Russell, R., McAlpine, L., Skingley, P., and Levine, M. N. (1993). Influence of cigarette smoking on the efficacy of radiation therapy in head and neck cancer. *N. Engl. J. Med.* **328**, 59–63.
12. De Boer, M. F., Van den Borne, B., Pruyn, J. F. A., Ryckman, R. M., Volovics, L., Knegt, P. P., Meeuwis, C. A., Mesters, I., and Verwoerd, C. D. (1998). Psychosocial and physical correlates of survival and recurrence in patients with head and neck carcinoma: Results of a 6-year longitudinal study. *Cancer* **83**, 2567–2579.
13. Stevens, M. H., Gardner, J. W., Parkin, J. L., and Johnson, L. P. (1983). Head and neck cancer survival and lifestyle change. *Arch. Otolaryngol.* **109**, 746–749.
14. de Graeff, A., de Leeuw, J. R., Ros, W. J., Hordijk, G. J., Blijham, G. H., and Winnubst, J. A. (2001). Sociodemographic factors and quality of life as prognostic indicators in head and neck cancer. *Eur. J. Cancer* **37**, 332–339.
15. Moore, C. (1971). Cigarette smoking and cancer of the mouth, pharynx and larynx. *JAMA* **218**, 553–558.
16. Silverman, S., Gorsky, M., and Greenspan, D. (1983). Tobacco usage in patients with head and neck carcinomas: A follow-up study on habit changes and second primary oral/orophayngeal cancers. *J. Am. Dent. Assoc.* **106**, 33–35.
17. Day, G. L., Blot, W. J., Shore, R. E., McLaughlin, J. K., Austin, D. F., Greenberg, R. S., Liff, J. M., Preston-Martin, S., Sarker, S., Schoenberg, J. B., and Fraumerni, J. F., Jr. (1994). Second cancers following oral and pharyngeal cancers: role of tobacco and alcohol. *JNCI* **86**, 131–137.
18. Hiyama, T., Sato, T., Yoshino, K., Tsukuma, H., Hanai, A., and Fujimoto, I. (1992). Second primary cancer following laryngeal cancer with special reference to smoking habits. *Japan J. Cancer Res.* **83**, 334–339.
19. Fountzilas, G., Kosmidis, P., Beer, M., Sridhar, K. S., Banis, K., Vristios, A., and Daniilidis, J. (1992). *Ann. Oncol.* **3**, 553–558.
20. Des Rochers C., Dische, S., and Saunders, M. I. (1992). The problem of cigarette smoking in radiotherapy for cancer in the head and neck. *Clin. Oncol.* **4**, 214–216.
21. Benowitz, N. L. (1998). Pharmacologic aspects of cigarette smoking and nicotine addiction. *N. Engl. J. Med.* **319**, 1318–1330.
22. Korsten, M. A., Rosman, A. S., Fishbein, S., Shlein, R. D., Goldberg, H. E., and Biener, A. (1991). Chronic xerostomia increases esophageal acid exposure and is associated with esophageal injury. *Am. J. Med.* **90**, 701–706.
23. McCarthy, G. M., Awde, J. D., Ghandi, H., Vincent, M., and Kocha, W. I. (1998). Risk factors associated with mucositis in cancer patients receiving 5-fluorouracil. *Oral Oncol.* **34**, 484–490.
24. Rugg, T., Saunders, M. L., and Dische, S. (1990). Smoking and mucosal reactions to radiotherapy. *Br. J. Radiat.* **63**, 554–556.
25. Dresler, C. M., Roper, C., Patterson, G. A., and Cooper, J. D. (1993). Effect of physician advice on smoking cessation in patients undergoing thoracotomy. *ABS Chest* **104**, 18.
26. USDHHS (1988). The health consequences of smoking. *In* "Nicotine Addiction: A Report of the Surgeon General." USDHHS, Public Health

Service, Center for Disease Control, Centers for Health Promotion and Education, Office on Smoking and Health, Rockville, MD.

27. Whittet, H. B., Lund, J. V., Brockbank, M., and Feyerbend, C. (1991). Serum cotinine as an objective marker for smoking habit in head and neck malignancy. *J. Laryngol. Otol.* **105**, 1039–1039.
28. Karim, A. B., Snow, G. B., Siek, H. T., and Njo, K. H. (1983). The quality of voice in patients irradiated for laryngeal carcinoma. *Cancer* **51**, 47–49.
29. Taylor, S. E. (1995). "Health Psychology," 3rd Ed. Mcgraw-Hill, New York.
30. USDHHS (1990). "The Health Benefits of Smoking Cessation: A Report of the Surgeon General." U.S. Government Printing Office, Washington D.C.
31. Deleyiannis, F. W., Thomas, D. B., Vaughan, T. L., and Davis, S. (1996). Alcoholism: Independent predictor of survival in patients with head and neck cancer. *J. Natl. Cancer Inst.* **88**, 542–549.
32. Spitz, M. R., Fueger, J. J., Chamberlain, R. M., *et al.* (1990). Cigarette smoking patterns in patients after treatment of upper aerodigestive tract cancers. *J. Cancer Educ.* **5**, 109–113.
33. Orleans, C. T., Rotberg, H., Quade, D., and Lees, P. (1990). A hospital quit-smoking consult service: Pilot evaluation and intervention model. *Prev. Med.* **19**, 198–212.
34. Ostroff, J. S., Spiro, R. H., and Kraus, D. H. (1995) Prevalence and predictors of continued tobacco use after treatment of patients with head and neck cancer. *Cancer* **75**, 569–576.
35. Gritz, E. R., Schacherer, C., Koehly, L., Nielsen, I. R., and Abemayor, E. (1999). Smoking withdrawal and relapse in head and neck cancer patients. *Head Neck* **21**, 420–427.
36. Christensen, A. J., Moran, P. J., Ehlers, S. L., Raichle, K., Karnell, L., and Funk, G. (1999). Smoking and drinking behavior in patients with head and neck cancer: Effects of behavioral self-blame and perceived control. *J. Behav. Med.* **22**, 407–418.
37. Bronheim, H., Strain, J. J., and Biller, H. F. (1991). Psychiatric aspects of head and neck surgery, 1. New surgical techniques and psychiatric consequences. *Gen Hosp. Psychiatry* **13**, 165–176.
38. Huber, F. T., Bartels, H., and Siewert, J. R. (1990). Treatment of postoperative alcohol withdrawal syndrome after esophageal resection. *Langenbecks Arch. Chir. Suppl. II Verh. Dtsch. Ges. Chir.* 1141–1143.
39. Hasuo, S., Koyama, Y., Kinoshita, N., Tanaka, H., Ajiki, W., Yoshino, K., Furukawa, H., and Oshima, A. (1998). Smoking behavior and cognition for smoking cessation after diagnosis of head and neck cancer or stomach cancer. *Nippon Koshu Eisei Zasshi* **45**, 732–739.
40. Silverman, S., Jr., and Griffith, M. (1972). Smoking characteristics of patients with oral carcinoma and the risk for second oral primary carcinoma. *J. Am. Dent. Assoc.* **85**, 637–40.
41. Vander Ark, W., DiNardo, L. J., and Oliver, D. S. (1997). Factors affecting smoking cessation in patients with head and neck cancer. *Laryngoscope* **107**, 888–892.
42. Schnoll, R. A., Malstrom, M., James, C., Rothman, R. L., Miller, S. M., Ridge, J. A., Movsas, B., Langer, C., Unger, M., and Goldberg, M. (2001). Longitudinal predictors of continued smoking among cancer patients. *Proc. Am. Soc. of Prev. Oncol.*
43. Prochaska, J. O., and Velicer, W. F. (1997). The transtheoretical model of health behavior change. *Am. J. Health Prom.* **12**, 38–48.
44. Sobell, L. C., Brown, J., Leo, G. I., and Sobell, M. B. (1996). The reliability of the Alcohol Timeline Followback when administered by telephone and by computer. *Drug Alcohol Depend.* **42**, 49–54.
45. Fals-Stewart, W., O'Farrell, T. J., Freitas, T. T., McFarlin, S. K., and Rutigliano, P. (2000). The timeline followback reports of psychoactive substance use by drug-abusing patients: Psychometric properties. *J. Consult. Clin. Psychol.* **68**, 134–144.
46. Gariti, P. W., Alterman, A. I., Ehrman, R. N., and Pettinati, H. M. (1998). Reliability and validity of the aggregate method of determining number of cigarettes smoked per day. *Am. J. Addict.* **7**, 283–287.
47. Engstrom, P. E., Clapper, M., Schnoll, R. A., and Orleans, C. T. (2000). Prevention of tobacco-related cancers. *In* "Cancer Medicine" (J. F. Holland *et al.*, eds.), 5th Ed., pp. 127–140. Decker, Hamilton, Ontario, Canada.
48. Gritz, E. R., Carr, C. R., Rapkin, D., Abermayor, E., Chang, L. C., Wong, W., Belin, T. R., Calcaterra, T., Robbins, T., Chonkich, G., Beumer, J., and Ward, P. H. (1993). Predictors of long-term smoking cessation in head and neck cancer patients. *Cancer Epidem. Biomark.* **2**, 261–270.
49. Payne, T. J., Smith, P. O., McCracken, L. M., McSherry, W. C., and Antony, M. M. (1994). Assessing nicotine dependence: A comparison of the Fagerstrom Tolerance Questionnaire (FTQ) with the Fagerstrom Test for Nicotine Dependence (FTND) in a clinical sample. *Add Behav.* **19**, 307–317.
50. Etter, J. F., Vu Duc, T., and Perneger, T. V. (1999). Validity of the Fagerstrom test for nicotine dependence and of the heaviness of smoking index among relatively light smokers. *Addicson.* **94**, 269–281.
51. Schnoll, R. A., Miller, S., Malstrom, M., James, C. L., Unger, M., Langer, C., Ridge, J. A., Goldberg, M., and Movsas, B. (2000). Predictors and rates of smoking among cancer patients. *Proc. Am. Soc. Prev. Oncol.* Bethesda, MD.
52. Miller, N. S., and Gold, M. S. (1998). Comorbid cigarette and alcohol addiction: Epidemiology and treatment. *J. Addict. Dis.* **17**, 55–66.
53. Sher, J. J. Gotham, H. J., Erickson, D. J., and Wood, P. K. (1996). A prospective, high risk study of the relationship between tobacco dependence and alcohol use disorders. *Alcohol Clin. Exp. Res.* **20**, 485–492.
54. McNair, P. M., Lorr, M., and Droppleman, L. (1992). "Profile of Mood States Manual." Educational and Industrial Testing Services, San Diego, CA.
55. Osler, M., and Prescott, E. (1998). Psychosocial, behavioral, and health determinants of successful smoking cessation: A longitudinal study of Danish adults. *Tob. Cont.* **7**, 262–267.
56. Sciamanna, C. N., Hoch, J. S., Duke, G. C., Fogle, M. N., and Ford, D. E. (2000). Comparison of five measures of motivation to quit smoking among a sample of hospitalized smokers. *J. Gen. Intern. Med.* **15**, 16–23.
57. Fava, J. L., Ruggiero, L., and Grimley, D. M. (1998). The development and structural confirmation of the Rhode Island Stress and Coping Inventory. *J. Behav. Med.* **21**, 601–611.
58. Berarducci, A., and Lengacher, C. A. (1998). Self-efficacy: An essential component of advanced-practice nursing. *Nurs. Conn.* **11**, 55–67.
59. Shiffman, S., Balabanis, M. H., Paty, J. A., Engberg, J., Gwaltney, C. J., Liu, K. S., Gnys, M., Hickcox, M., and Paton, S. M. (2000). Dynamic effects of self-efficacy on smoking lapse and relapse. *Health Psychol.* **19**, 315–323.
60. Pohl, J. M., Martinelli, A., and Antonakos, C. (1998). Predictors of participation in a smoking cessation intervention group among low-income women. *Addict Behav.* **23**, 699–704.
61. Gulliver, S. B., Hughes, J. R., Solomon, L. J., and Dey, A. N. (1995). An investigation of self-efficacy, partner support and daily stresses as predictors of relapse to smoking in self-quitters. *Addiction* **90**, 767–772.
62. Matheny, K. B., and Weatherman, K. E. (1998). Predictors of smoking cessation and maintenance. *J. Clin. Psychol.* **54**, 223–225.
63. Rosal, M. C., Ockene, J. K., Hurley, T. G., Alan, K., and Hebert, J. R. (1998). Effectiveness of nicotine-containing gum in the physician-delivered smoking intervention study. *Prev. Med.* **27**, 262–267.
64. Scholte, R. H., and Breteler, M. H. (1997). Withdrawal symptoms and previous attempts to quit smoking: Associations with self-efficacy. *Subst. Use Misuse* **32**, 133–148.
65. Stuart, K., Borland, R., and McMurray, N. (1994). Self-efficacy, health locus of control, and smoking cessation. *Addict Behav.* **19**, 1–12.

66. Woodby, L. L., Windsor, R. A., Snyder, S. W., Kohler, C. L., and Diclemente, C. C. (1999). Predictors of smoking cessation during pregnancy. *Addiction* **94**, 283–292.
67. Halpern, M. T., Schmier, J. K., Ward, K. D., and Klesges, R. C. (2000). Smoking cessation in hospitalized patients. *Respir. Care* **45**, 330–336.
68. Meland, E., Maeland, J. G., and Laerum, E. (1999). The importance of self-efficacy in cardiovascular risk factor change. *Scand. J. Public Health* **27**, 11–17.
69. Bursey, M., and Craig, D. (2000). Attitudes, subjective norm, perceived behavioral control, and intentions related to adult smoking cessation after coronary artery bypass graft surgery. *Public Health Nurs.* **17**, 460–467.
70. Hanson, M. J. (1997). The theory of planned behavior applied to cigarette smoking in African-American, Puerto Rican, and non-Hispanic white teenage females. *Nurs. Res.* **46**, 155–162.
71. Hill, A. J., Boudreau, F., Amyot, E., Dery, D., and Godin, G. (1997). Predicting the stages of smoking acquisition according to the theory of planned behavior. *J. Adolesc. Health* **21**, 107–115.
72. Nguyen, M. N., Beland, F., and Otis, J. (1998). Is the intention to quit smoking influenced by other heart-healthy lifestyle habits in 30- to 60-year-old men? *Addict. Behav.* **23,** 23–30.
73. Norman, P., Conner, M., and Bell, R. (1999). The theory of planned behavior and smoking cessation. *Health Psychol.* **18**, 89–94.
74. Weinstein, N. D. (1998). Accuracy of smokers' risk perceptions. *Ann. Behav. Med.* **20**, 135–140.
75. USDHHS (1989). "Reducing the Health Consequences of Smoking." U.S. Department of Health and Human Services, Public Health Service, Rockville, MD
76. Ayanian, J. Z., and Cleary, P. D. (1999). Perceived risks of heart disease and cancer among cigarette smokers. *JAMA* **281**, 1019–1021.
77. McCoy, S. B., Gibbons, F. X., Reix, T. J., Gerrard, M., Luus, C. A., and Sufka, A. V. (1992). Perceptions of smoking risk as a function of smoking status. *J. Behav. Med.* **15**, 469–488.
78. Strecher, V. J., Kreuter, M., and Korbin, S. C. (1995). Do cigarette smokers have unrealistic perceptions of their heart attack, cancer, and stroke risks? *J. Behav. Med.* **18**, 45–54.
79. Weinstein, N. D. (1987). Unrealistic optimism about susceptibility to health problems: Conclusions from a community-wide sample. *J. Behav. Med.* **10**, 481–500.
80. Kviz, F. J., Crittenden, K. S., Belzer, L. J., and Warnecke, R. B. (1991). Psychosocial factors and enrollment in a televised smoking cessation program. *Health Educ. Quart.* **18**, 445–461.
81. Halpern, M. T., and Warner, K. E. (1994). Differences in former smokers' beliefs and health status following smoking cessation. *Am. J. Prev. Med.* **10**, 31–37.
82. Prochaska, J. O. (1994). Strong and weak principles for progressing from precontemplation to action on the basis of twelve problem behaviors. *Health Psychol.* **13**, 47–51.
83. Ahijevych, K., and Parsley, L. A. (1999). Smoke constituent exposure and stage of change in black and white women cigarette smokers. *Addict Behav.* **24**, 115–120.
84. Dijkstra, A., de Vries, H., and Bakker, M. (1996). Pros and cons of quitting, self-efficacy, and the stages of change in smoking cessation. *J. Consult. Clin. Psychol.* **64**, 758–763.
85. Etter, J. F., and Perneger, T. V. (1999). Associations between the stages of change and the pros and cons of smoking in a longitudinal study of Swiss smokers. *Addict. Behav.* **24**, 419–424.
86. Fava, J. L., Velicer, W. F., and Prochaska, J. O. (1995). Applying the transtheoretical model to a representative sample of smokers. *Addict. Behav.* **20**, 189–203.
87. Herzog, T. A., Abrams, D. B., Emmons, K. M., Linnan, L. A., and Shadel, W. G. (1999). Do processes of change predict smoking stage movements? A prospective analysis of the transtheoretical model. *Health Psychol.* **18**, 369–375.
88. King, T. K., Marcus, B. H., Pinto, B. M., Emmons, K. M., and Abrams, D. B. (1996). Cognitive-behavioral mediators of changing multiple behaviors: Smoking and a sedentary lifestyle. *Prev. Med.* **25**, 684–691.
89. Quattrocki, E., Baird, A., and Yurgelun-Todd, D. (2000). Biological aspects of the link between smoking and depression. *Harv. Rev. Psychiatry* **8**, 99–110.
90. Hall, S. M., Munoz, R. F., Reus, V. I., and Sees, K. L. (1993). Nicotine, negative affect, and depression. *J. Consult. Clin. Psychol.* **61**, 761–767.
91. Kinnunen, T., Doherty, K., Militello, F. S., and Garvey, A. J. (1996). Depression and smoking cessation: Characteristics of depressed smokers and effects of nicotine replacement. *J. Consult. Clin. Psychol.* **64**, 791–798.
92. Anda, R. F., Williamson, D. F., Escobedo, L. G., Mast, E. E., and Giovino, G. A. (1990). Depression and the dynamics of smoking: A national perspective. *JAMA* **264**, 1541–1545.
93. Covey, J. C. (1990). Depressive mood, the single-parent home, and adolescent smoking. *Am. J. Pub. Health* **80**, 1330–1333.
94. Glassman, A. H. (1993). Cigarette smoking: Implications for psychiatric illness. *Am. J. Psychiatry* **150**, 546–553.
95. Rausch, J. L., Nichinson, B., Lamke, C., and Matloff, J. (1990). Influence of negative affect on smoking cessation treatment outcome: A pilot study. *Br. J. Addict.* **85**, 929–933.
96. Kugaya, A., Akechi, T., Okamura, H., Mikami, I., and Uchitomi, Y. (1999). Correlates of depressed mood in ambulatory head and neck cancer patients. *Psycho-Oncol.* **8**, 494–499.
97. Breitbart, W. (1995). Identifying patients at risk for, and treatment of major complications of cancer. *Supp. Care. Cancer* **3**, 45–60.
98. Morton, R. P., Davies, A. D. M., Baker, J., Baker, G., and Stell, P. M. (1984). Quality of life in treated head and neck cancer patients: A preliminary report. *Clin. Otolaryngol.* **9**, 181–85.
99. Kugaya, A., Akechi, T., Okulama, T., Nakano, T., Mikami, I., Okamura, H., and Uchitomi, Y. (2000). Prevalence, predictive factors, and screening for psychologic distress in patients with newly diagnosed head and neck cancer. *Cancer* **88**, 2817–2823.
100. de Graeff, A., de Leeuw, R. J., Ros, W. J. G., Hordijk, G. J., Batterman, J. J., Blijham, G. H., and Winnubst, J. A. M. (1999). A prspective study on the quality of life of laryngeal cancer patients treated with radiotherapy. *Head Neck* **21**, 291–196.
101. de Graeff, A., de Leeuw, R. J., Ros, W. J. G., Hordijk, G.-J., Blijham, G. H., and Winnubst, J. A. M. (2000). Pretreatment factors predicting quality of life after treatment for head and neck cancer. *Head Neck* **22**, 398–407.
102. de Leeuw, R. J., de Graeff, A., Ros, W. J. G., Hordijk, G. J., Blijham, G. H., and Winnubst, J. A. M. (2000). Negative and positive influences of social support on depression in patients with head and neck cancer: A prospective study. *Psycho-Oncol.* **9**, 20–28.
103. de Leeuw, R. J., de Graeff, A., Ros, W. J. G., Blijham, G. H., Hordijk, G. J., and Winnubst, J. A. M. (2000). Prediction of depressive symptomatoloty after treatment of head and neck cancer: The influence of pre-treatment physical and depressive symptoms, coping, and social support. *Head Neck* **22**, 799–807.
104. Carmody, T. P. (1989). Affect regulation, nicotine addiction, and smoking cessation. *J. Psychoact. Drugs* **21**, 331–342.
105. Hughes, J. R. (1988). Clonidine, depression, and smoking cessation. *JAMA* **259**, 2901–2902.
106. Lerman, C., Audrain, J., Orleans, C. T., Boyd, R., Gold, K., Main, D., and Caporaso, N. (1996). Investigation of mechanisms linking depressed mood to nicotine dependence. *Addict. Behav.* **21**, 9–19.
107. Cinciripini, P. M., Hecht, S. S., Henningfield, J. E., Manley, M. W., and Kramer, B. (1997). Tobacco addiction: Implication for treatment and cancer prevention. *JNCI* **89**, 1852–1867.
108. Henningfield, J. E., Schuh, L. M., and Jarvik, M. E. (1995). Pathophysiology of tobacco dependence. *In* "Psychopharmacology: The Fourth Generation of Progress" (F. E. Bloom and D. J. Kupfer, eds.), pp. 1715–1730. Raven Press, New York.

109. Pomerleau, O. F., and Kardia, S. L. (1999). Introduction to the featured section: Genetic research on smoking. *Health Psychol.* **18**, 3–6.
110. Noble, E. P. (1998). The *DRD2* gene, smoking, and lung cancer. *JNCI* **90**, 343–347.
111. Heinz, A., Goldman, D., Jones, D. W., Palmour, R., Hommer, D., Gorey, J. G., Lee, K. S., Linnoila, M., and Weinberger, D. R. (2000). Genotype influences in vivo dopamine transporter availability in human striatum. *Neuropsychopharmacology* **22**, 133–139.
112. Bergen, A. W., and Caporaso, N. (1999). Cigarette smoking. *JNCI* **91**, 1365–1375.
113. Comings, D. E., Gade, R., Wu, S., Chiu, C., Dietz, G., Muhleman, D., Saucier, G., Ferry, L., Rosenthal, R. J., Lesier, H. R., Rugle, L. J., and MacMurray, D (1997). Studies of the potential role of dopamine D_1 receptor gene in addictive behaviors. *Mol. Psychiatry* **2**, 44–56.
114. Comings, D. E., Ferry, L., Bradshaw-Robinson, S., Burchette, R., Chiu, C., and Muhleman, D. (1996). The dopamine D_2 receptor (DRD2) gene: A genetic risk factor in smoking. *Pharmacogenetics* **6**, 73–79.
115. Sheilds, P. G., Lerman, C., Audrain, J., Bowman, E. D., Main, D., Boyd, N. R., and Caporaso, N. E. (1998). Dopamine D_4 receptors and the risk of cigarette smoking in African Americans and Caucasians. *Cancer Epidemiol. Biomark. Prev.* **7**, 453–458.
116. Noble, E. P., St. Jeor, S. T., Syndulko, K., St. Jeor, S. C., Fitch, R. J., Brunner, R. L., and Sparkes, R. S. (1994). D_2 dopamine receptor gene and cigarette smoking: A reward gene? *Med. Hypoth.* **42**, 257–260.
117. Spitz, M. R., Shi, H., Yang, F., Suchanek Hudmon, K., Jiang, H., Chamberlain, R. M., Amos, C. I., Wan, Y., Cinciripini, P., Hong, W. K., and Wu, X. (1998). Case-control study of D_2 receptor gene and smoking status in lung cancer patients. *JNCI* **90**, 358–363.
118. Bierut, L. J., Rice, J. P., Edenberg, H. J., Goate, A., Foroud, T., Cloninger, C. R., Begleiter, H., Conneally, P. M., Crowe, R. R., Hesselbrock, V., Li, T. K., Nurnberger, J. I., Jr., Porjesz, B., Schuckit, M. A., and Reich, T. (2000). Family-based study of the association of the dopamine D2 receptor gene (DRD2) with habitual smoking. *Am. J. Med. Genet.* **14**, 299–302.
119. Singleton, A. B., Thomson, J. H., Morris, C. M., Court, J. A., Lloyd, S., and Cholerton, S. (1998). Lack of association between the dopamine D2 receptor gene allele DRD2*A1 and cigarette smoking in a United Kingdom population. *Pharmacogenetics* **8**, 125–128.
120. Lerman, C., Caporaso, N. E., Audrain, J., Main, D., Bowman, E. D., Lockshin, B., Boyd, N. R., and Sheilds, P. (1999). Evidence suggesting the role of specific genetic factors in cigarette smoking. *Health Psychol.* **18**, 14–20.
121. Sabol, S., Nelson, M. L., Fisher, C., Gunzerath, L., Brody, C. L., Hu, S., Sirota, L. A., Marcus, S., Greenberg, B. D., Lucas, F. R., Benjamin, J., Murphy, D. L., and Hamer, D. H. (1999). A genetic association for cigarette smoking behavior. *Health Psychol.* **18**, 7–13.
122. Jorm, A. F., Henderson, A. S., Jacomb, P., Christensen, H., Korten, A. E., Rodgers, B., Tan, X., and Easteal, S. (2000). Association of smoking and personality with a polymorphism of the dopamine transporter gene: Results from a community survey. *Am. J. Med. Gen.* **12**, 331–334.
123. Gilbert, D. G. (1995). Personality, psychopathology, tobacco use, and individual differences in effects of nicotine. *In* "Smoking: Individual Differences, Psychopathology, and Emotion" (K. P. Baker and H. Seltzer, eds.), pp. 149–176. Taylor and Francis, Washington, DC.
124. Audrain, J., Lerman, C., Gomez-Caminero, A., Boyd, N. R., and Orleans, C. T. (1998). The role of trait anxiety in nicotine dependence. *J. Appl. Biobehav. Res.* **3**, 29–42.
125. Lerman, C., Shields, P. G., Audrain, J., Main, D., Cobb, B., Boyd, N. R., and Caporaso, N. (1998). The role of the serotonin transporter gene in cigarette smoking. *Cancer Epidemiol. Biomark. Prev.* **7**, 253–255.
126. Hu, S., Brody, C. L., Fisher, C., Gunzerath, L., Nelson, M. L., Sabol, S., Sirota, L., Marcus, S. E., Greenberg, B. D., Murphy, D. L., and Hamer, D. H. (2000). Interaction between the serotonin transporter gene and neuroticism in cigarette smoking behavior. *Mol. Psychiatry* **5**, 181–88.
127. Lerman, C., Caporaso, N. E., Audrain, J., Main, D., Boyd, N. R., and Sheilds, P. (2000). Interacting effects of the serotonin transporter gene and neuroticism in smoking practices and nicotine dependence. *Mol. Psychiatry* **5**, 189–192.
128. Lerman, C., Caporaso, N. E., Bush, A., Zheng, Y-L., Audrain, J., Main, D., and Shields, P. G. (2002). Tryptophan hyroxylase gene variant and smoking behavior. *Am. J. Med. Gen.* **105**, 513–520.
129. Sullivan, P. F., Jiang, X., Neale, M. C., Kendler, K. S., and Straub, R. E. (2002). Association of the tryptophan hyroxylase gene with smoking initiation but not progression to nicotine dependence. *Am. J. Med. Gen.* **105**, 479–484.
130. Pianezza, M. L., Sellers, E. M., and Tyndale, R. F. (1998). Nicotine metabolism defect reduces smoking. *Nature* **393**, 750.
131. Oscarson, M., Gullsten, H., Rautio, A., Bernal, M. L., Sinues, B., Dahl, M. L., Stengard, J. H., Pelkonen, O., Raunio, H., and Ingelman-Sundberg, M. (1998). Genotyping of human cytochrome P450 2A6 (CYP2A6), a nicotine C-oxidase. *FEBS Lett.* **438**, 201–205.
132. London, S. J., Idle, J. R., Daly, A. K., and Coetzee, G. A. (1999). Genetic variation of CYP2A6, smoking, and risk of cancer. *Lancet* **353**, 898–899.
133. Spielman, R. S., McGinnis, R. E., and Ewens, W. J. (1993). Transmission test for linkage disequilibrium: The insulin gene region and insulin-dependent diabetes mellitus (IDDM). *Am. J. Hum. Gen.* **52**, 506–516.
134. Cinciripini, P. M., and McClure, J. B. (1998). Smoking cessation: Recent developments in behavioral and pharmacologic interventions. *Oncology (Huntingt)* **12**, 249–256.
135. Wetter, D. W., Fiore, M. C., Gritz, E. R., Lando, H. A., Stitzer, M. L., Hasselblad, V., and Baker, T. B. (1998). The Agency for Health Care Policy and Research findings and implications for psychologists. *Am. Psychol.* **53**, 657–669.
136. Fiore, M. C., Novotny, T. E., Pierce, J. P., Giovino, G. A., Hatziandreu, E. J., Newcomb, P. A., Surawicz, T. S., and Davis, R. M. (1990). Methods used to quit smoking in the United States. Do cessation programs help? *JAMA* **263**, 2760–2765.
137. Orleans, C. T., Rotberg, H., Quade, D. and Lees, P. (1990). Hospital quite smoking consult service: Pilot evaluation and intervention model. *Prev. Med.* **19**, 198–212.
138. Abrams, D. B., Orleans, C. T., Niaura, R. S., Goldstein, M. G., Prochaska, J. O., and Velicer, W. (1996). Integrating individual and public health perspectives for treatment of tobacco dependence under managed health care: A combined stepped-care and matching model. *Ann. Behav. Med.* **18**, 290–304.
139. Curry, S. J. (1993). Self-help interventions for smoking cessation. *J. Consul. Clin. Psychol.* **61**, 790–803.
140. Glynn, T., Boyd, G. M., and Gruman, J. C. (1990). Essential elements of self-help/minimal intervention strategies for smoking cessation. *Health Educ. Quart.* **17**, 329–345.
141. Lando, H. A., Pirie, P. L., McGovern, P. G., Pechacek, T. F., Swim, J., and Loken, B. (1991). A comparison of self-help approaches to smoking cessation. *Addict. Behav.* **16**, 183–193.
142. Fiore, M. C., Bailey, W. C., Cohen, S. J., *et al.* (2000). "Treating Tobacco Use and Dependence." USDHHS, Public Health Service, Rockville, MD.
143. Kottke, T. E. (1998). Observing the delivery of smoking-cessation interventions. *Am. J. Prev. Med.* **14**, 71–72.
144. Fiore, M. C., Bailey, W. C., Cohen, S. J., *et al.* (1996). "Smoking Cessation: Information for Specialists. Clinical Practice Guideline." USDHHS, Public Health Service, Agency for Health Care Policy and Research and Centers for Disease Control and Prevention, Rockville, MD.
145. Glynn, T. (1988). Physicians and a smoke-free society. *Arch. Intern. Med.* **148**, 1013–1016.

146. Kottke, T., Battista, R. N., DeFriese, G. H., and Brekke, M. L. (1988). Attributes of successful smoking cessation interventions in medical practice: A meta-analysis of 39 controlled trials. *JAMA* **259**, 2883–2889.
147. Goldstein, M. G., Niaura, R., Willey-Lessne, C., DePue, J., Eaton, C., Rakowski, W., and Dube, C. (1997). Physicians counseling smokers. A population-based survey of patients' perceptions of health care provider-delivered smoking cessation interventions. *Arch. Intern. Med.* **23**, 1313–1319.
148. Anderson, C. B., and Wetter, D. W. (1997). Behavioral and pharmacologic approaches to smoking cessation. *Cancer Met. Rev.* **16**, 393–404.
149. Ferry, L. H., Robbins, A. S., Scariati, P. D., *et al.* (1992). Enhancement of smoking cessation using the antidepressant bupropion hydrochloride. *Circulation* **86**, I-671. [Abstract]
150. Ferry, L. H., and Burchette, R. J. (1994). Efficacy of bupropion for smoking cessation in non-depressed smokers. *J. Addict. Dis.* **13**, 249.
151. Hurt, R. D., Sachs, D. P., Glover, E. D., Offord, K. P., Johnston, J. A., Dale, L. C., Khayrallah, M. A., Schroeder, D. R., Glover, P. N., Sullivan, C. R., Croghan, I. T., and Sullivan, P. M. (1997). A comparison of sustained-release bupropion and placebo for smoking cessation. *N. Engl. J. Med.* **23**, 1195–1202.
152. Jorenby, D. E., Leischow, S. J., Nides, M. A., Rennard, S. I., Johnston, J. A., Hughes, A. R., Smith, S. S., Muramoto, M. L., Daughton, D. M., Doan, K., Fiore, M. C., and Baker, T. B. (1999). A controlled trial of sustained-release bupropion, a nicotine patch, or both for smoking cessation. *N. Engl. J. Med.* **340**, 685–691.
153. Emmons, K. M., and Goldstein, M. G. (1992). Smokers who are hospitalized: A window of opportunity for cessation interventions. *Prev. Med.* **21**, 262–269.
154. Orleans, C. T., Kristeller, J. L., and Gritz, E. R. (1993). Helping hospitalized smokers quit: New directions for treatment and research. *J. Consult. Clin. Psychol.* **61**, 778–789.
155. Stanislaw, A. E., and Wewers, M. E. (1994). A smoking cessation intervention with hospitalized surgical cancer patients: A pilot study. *Cancer Nurs.* **17**, 81–86.
156. Griebel, B., Wewers, M. E., and Baker, C. A. (1998). The effectiveness of a nurse-managed minimal smoking-cessation intervention among hospitalized patients with cancer. *Oncol. Nurs. Forum* **25**, 897–902.
157. Browning, K. K., Ahijevych, K. L., Ross, P., Jr, and Wewers, M. E. (2000). Implementing the Agency for Health Care Policy and Research's Smoking Cessation Guideline in a lung cancer surgery clinic. *Oncol. Nurs. Forum* **27**, 1248–1254.
158. Glassman, A. H. (1998). Psychiatry and cigarettes. *Arch. Gen. Psychiatry* **55**, 692–693.
159. Schilling, A., Conaway, M. R., Wingate, P. J., Atkins, J. N., Berkowitz, I. M., Clamon, G. H., DiFino, S. M., and Vinciguerra, V. (1997). Recruiting cancer patients to participate in motivating their relatives to quit smoking: A cancer control study of the Cancer and Leukemia Group B (CALGB 9072). *Cancer* **79**, 152–60.

CHAPTER

14

Nutrients, Phytochemicals, and Squamous Cell Carcinoma of the Head and Neck

OMER KUCUK and ANANDA PRASAD

Karmanos Cancer Institute
Wayne State University
Detroit, Michigan 48201

I. Introduction 201
II. Tobacco and Nutrients 202
III. Alcohol and Nutrients 203
IV. Cancer Treatment and Nutrients 203
V. Nutritional Consequences of Radiation and Chemotherapy 203
VI. Effects of Selected Micronutrients on Radiation and Chemotherapy Toxicity 204
A. Vitamin E 204
B. Zinc 204
C. Gastrointestinal Toxicity 205
D. Nephrotoxicity 205
VII. Nutrients and Antitumor Effects of Radiation/Chemotherapy 205
VIII. Nutrients and Pharmaceuticals in the Prevention of Radiation and Chemotherapy Toxicity 206
A. Selenium 206
B. Zinc 207
C. Phytochemicals 207
IX. Conclusions 208
References 208

Although squamous cell cancer of the head and neck is associated with tobacco use, micronutrients and phytochemicals in the diet play an important role in the etiology and progression of this deadly disease. Furthermore, these compounds may have significant interactions with cancer therapies, such as radiation and chemotherapy, which may be antagonistic, additive, or synergistic. Some micronutrients and/or phytochemicals may prevent the toxicity of chemotherapy and/or radiation therapy while improving its efficacy. Clinical studies should be conducted to investigate the potential role of micronutrients and phytochemicals in the prevention and/or treatment of squamous cell cancer of the head and neck.

I. INTRODUCTION

Squamous cell carcinoma of the head and neck (SCCHN) is strongly associated with tobacco use. However, chromosome instability and defective DNA repair may underlie susceptibility to environmental carcinogenesis [1]. Hsu [2] suggested that chromosome fragility in the general population exists to varying degrees and indicates genetic instability and that individuals most genetically sensitive to carcinogens are more likely to develop cancer. The number of bleomycin-induced chromosomal breaks in cultured peripheral blood lymphocytes may be a measure of an individual's "mutagen sensitivity," i.e., susceptibility to environmental cancers [1,3]. Hsu *et al.* [3] found wide interindividual variability in chromatid breakage rates, i.e., genotoxicity induced by bleomycin. Approximately 12% of normal persons were regarded as bleomycin sensitive, whereas nearly 50% of patients with cancer in the upper aerodigestive tract were found to be sensitive [1]. In a prospective study of mutagen sensitivity in patients with upper aerodigestive tract cancer, Schantz *et al.* [4] found that mutagen sensitivity correlated with the risk of developing second malignant tumors in patients cured of head and neck cancer. A case–control study showed that persons with untreated upper aerodigestive tract cancer express greater sensitivity than controls when their cells are exposed to bleomycin *in vitro* [5]. These findings have been confirmed by Cloos *et al.* [6], who found increased mutagen sensitivity in head and neck squamous cell carcinoma patients, particularly in those with multiple primary tumors.

Mutagen sensitivity can be modulated *in vitro* [7] and *in vivo* [8] by various nutrients. The antioxidant micronutrients α-tocopherol and ascorbic acid have been shown to protect against carcinogen-induced chromosomal breakage [7–10]. Because vegetable and fruit consumption has been found to be protective against the development of upper aerodigestive tract cancers [11,12], certain antioxidant micronutrients in the diet may provide protection against DNA-damaging carcinogens. Kucuk *et al.* [13] investigated the intraindividual variation in mutagen sensitivity and its possible correlation with plasma nutrient levels in a group of 25 healthy individuals in Hawaii. Mutagen sensitivity, as assessed by bleomycin-induced chromosomal breaks in cultured peripheral blood lymphocytes and plasma nutrient levels, was measured monthly for 11 months. Significant inverse correlations were found between mutagen sensitivity scores and the plasma levels of α-carotene ($r=-0.64$), total carotenoids ($r=-0.41$), and ascorbic acid ($r=-0.40$). There were also significant inverse associations between monthly mean plasma levels of α-carotene ($r=-0.58$), β-carotene ($r=-0.76$), and total carotenoids ($r=-0.72$) and monthly mean chromosomal breaks. In contrast, there was a significant positive correlation between monthly mean plasma triglyceride level ($r=0.60$) and monthly mean mutagen sensitivity. These results suggest that mutagen sensitivity could potentially be reduced by dietary modifications or by supplementing certain micronutrients.

Epidemiological data consistently show an inverse relationship between cancer risk and dietary intake of vegetables and fruits or their antioxidant micronutrients [12]. Most cancer patients have low micronutrient levels at presentation. Cancer is a disease of aging and micronutrient deficiencies common among older individuals. Monget *et al.* [14] found that serum concentrations of most micronutrients had an inverse association with age and most nursing home residents had low serum levels of vitamin C, zinc, and selenium. Micronutrient deficiency may also be present in pediatric cancer patients. Donma *et al.* [15] found reduced hair zinc levels in children with malignancies compared to healthy children and children with cancers in remission.

Negri *et al.* [16] investigated the relationship between selected micronutrients and oral and pharyngeal cancer risk using data from a case–control study conducted in Italy and Switzerland. Cases were 754 incident, histologically confirmed oral cancers (344 of the oral cavity and 410 of the pharynx) and controls were 1775 subjects with no history of cancer admitted to hospitals in the same catchment areas. Dietary habits were investigated using a validated food-frequency questionnaire. Odds ratios (ORs) were computed after allowance for age, sex, center, education, occupation, body mass index, smoking and drinking habits, and nonalcohol energy intake. ORs were 0.95 for retinol, 0.61 for carotene, 0.91 for lycopene, 0.83 for vitamin D, 0.74 for vitamin E, 0.63 for vitamin C, 0.82 for thiamine, 0.87 for riboflavin, 0.59 for vitamin B_6, 0.61 for folic acid, 0.62 for niacin, 0.91 for calcium, 0.88 for phosphorus, 0.65 for potassium, 0.82 for iron, 0.67 for nonalcohol iron, and 0.89 for zinc. When the combined intake of vitamins C and E and carotene was considered, the protective effect of each nutrient was more marked or restricted to subjects with low intake of the other two. The association with vitamin C and carotene was independent of smoking and drinking habits, whereas that with vitamin E was less evident in those heavily exposed to alcohol or tobacco. In general, the more a micronutrient was correlated to total vegetable and fruit intake, the stronger its protective effect against oral cancer.

Alhasan *et al.* [17] showed that a soy isoflavone, genistein, inhibited cell proliferation, caused cell cycle arrest at the S/G2–M phase, and induced apoptosis in the squamous cell carcinoma cell line HN4. These effects appeared to be dose and time dependent, and specific for tumor cells, because genistein did not affect normal keratinocytes. Alhasan *et al.* [18] also observed that these changes were accompanied by the downregulation of Cdk1 and CyclinB1 and the upregulation of p21WAF1, which may be responsible for the induction of cell cycle arrest and apoptosis. Evidence for the induction of apoptosis was supported by the appearance of a DNA ladder and the cleavage of poly-ADP-ribose polymerase (PARP), the hallmark of apoptosis. This was also accompanied by the upregulation of Bax, with modest downregulation of Bcl-2 protein expression, which changes the balance between pro- and antiapoptosis molecules in favor of proapoptosis. Furthermore, they also observed downregulation and degradation of Cdc25C, which is a marker of cell proliferation and plays an important role in cyclin B–Cdk1 complex activation. The downregulation followed by the degradation of Cdc25C is an indicator of G2/M arrest and antiproliferation effects of genistein. These results suggest that genistein may have a role in the prevention and/or treatment of SCCHN.

II. TOBACCO AND NUTRIENTS

Tobacco use is a major risk factor for SCCHN. Tobacco use has consistently been associated with increased oxidative stress and decreased serum antioxidant micronutrient levels. Pamuk *et al.* [19] reported on the relationship between current cigarette smoking and serum concentrations of vitamins C, E, and A and five carotenoids in 91 low-income, African-American women. Among smokers, serum concentrations of α-carotene, β-carotene, cryptoxanthin, and lycopene averaged 71–79% of the concentrations among nonsmokers. Mean serum concentrations of vitamins C and E and lutein/zeaxanthin were only slightly lower among smokers relative to nonsmokers. Among current smokers, mean serum concentrations of all five carotenoids had an inverse correlation with the amount smoked. Ross *et al.* [20] determined plasma concentrations of carotenoids, ascorbic acid, α-tocopherol and γ-tocopherol in plasmas from 50 smokers and 50 age-matched never-smoker Scottish men. Significantly less α-carotene, β-carotene, cyrptoxanthin, and ascorbic acid were found in smokers compared to persons who never smoked. Pakrashi and Chatterjee [21] measured the prostatic excretion of zinc in ejaculates of 29 tobacco smokers, 25 tobacco chewers, and 30 nonusers of tobacco. They found reduced levels of zinc in the ejaculates of tobacco smokers compared to tobacco chewers and tobacco nonusers. It has been postulated that smoking results in the depletion of antioxidant micronutrients by generating oxidative stress. However, a low dietary intake of antioxidant micronutrients by smokers may also be a factor in the observed inverse association. For example, Faruque *et al.* [22] observed a lower dietary intake of vitamin C, carotenoids, and zinc and a lower plasma level of vitamin C in 44 male students who smoked compared to 44 male nonsmoker students.

Similar to SCCHN, lung cancer is also caused by smoking. Because smoking generates oxidative stress and leads to decreased serum levels of β-carotene, supplementing the diet with β-carotene in smokers to prevent lung cancer is reasonable. However, a large clinical study conducted in Finland found just the opposite [23]. This unexpected result might have been due to the paradoxical prooxidant effect of β-carotene in lungs where the oxygen tension is high. Studies have shown that β-carotene, at high concentrations, has a prooxidant effect when the oxygen pressure is also high [24]. Therefore, in the lungs, where the oxygen tension is high, administering large doses of β-carotene may lead to oxidative DNA damage and a higher incidence of cancer. Furthermore, β-carotene and tobacco smoke interact to increase AP-1 production in ferret lungs [25], which may also explain the increased risk of lung cancer with β-carotene supplementation in current smokers but not in nonsmokers.

III. ALCOHOL AND NUTRIENTS

Alcohol consumption has also been associated with increased oxidative stress and decreased micronutrient levels. Alcohol and tobacco in combination may result in even more severe micronutrient deficiencies compared to either one used alone. Tsubono *et al.* [26] examined the associations of smoking and alcohol with plasma levels of β-carotene, α-carotene, lutein, lycopene, and zeaxanthin in 634 healthy men aged 40–49 years. After controlling for age, serum cholesterol, serum triglycerides, body mass index, green vegetables, yellow vegetables, and fruits, there was a significant inverse association between smoking and alcohol consumption and plasma levels of β-carotene and α-carotene; only smoking reduced the level of lutein, and neither smoking nor alcohol significantly reduced the level of lycopene or zeaxanthin. In a population-based sample of 400 individuals, Brady *et al.* [27] found an association between smoking and alcohol consumption and lower serum levels of α-carotene, β-carotene, β-cryptoxanthin, and lutein/zeaxanthin. In addition, lower serum lycopene was associated with older age. Lecomte *et al.* [28] measured plasma carotenoid levels in 118 healthy men consuming low or moderate alcohol and 95 alcoholics. β-Carotene, α-carotene, zeaxanthin/lutein, lycopene, and β-cryptoxanthin levels were significantly lower in alcoholics and 21 days after withdrawal plasma levels of all carotenoids increased. However, Leo *et al.* [29] did not find a significant difference in the levels of carotenoids, retinol, and α-tocopherol from oropharyngeal mucosa samples of 11 chronic alcoholics with oropharyngeal cancer and 11 control subjects.

IV. CANCER TREATMENT AND NUTRIENTS

Nutritional status is known to profoundly impact treatment morbidity, efficacy, and overall prognosis in cancer patients [30–37]. Various prognostic nutritional indices have been developed to predict treatment complications and overall survival [30,31,34,37]. Radiation and chemotherapy are better tolerated and may be more effective in nutritionally sound individuals [30,31]. For example, head and neck cancer patients with poor nutritional status are at increased risk for postoperative wound breakdown and infections, fistula formation, and flap loss [30,34]. These patients frequently present with significant weight loss and chronic protein-calorie malnutrition, which may be exacerbated by an acute weight loss due to decreased intake secondary to tumor-induced dysphagia [30–34]. Approximately 30–40% of patients with advanced stage head and neck cancer have severe malnutrition and an additional 20–30% have moderate malnutrition at the time of presentation [30–34]. Olmedilla *et al.* [38] found that the plasma levels of carotenoids, retinol, and vitamin E were significantly lower in patients who had laryngectomy for laryngeal cancer compared to control subjects. After commercial enteral formula feeding, carotenoid levels decreased further and retinol and tocopherol levels increased; however, the levels remained lower than the controls for all micronutrients [38]. Postoperative alterations of the upper aerodigestive tract may further compromise intake, increase metabolic demands, and compound the nutritional stress [32,35]. Because there are no known zinc stores in the human body, zinc deficiency sets in quickly with malnutrition in these patients [39].

V. NUTRITIONAL CONSEQUENCES OF RADIATION AND CHEMOTHERAPY

Both radiation therapy and chemotherapy have been associated with increased oxidative stress, which may further deplete tissue levels of antioxidant micronutrients, particularly in smokers and in the presence of inadequate dietary intake. Faber *et al.* [40] measured lipid peroxidation, plasma glutathione and glutathione peroxidase activity, and plasma micronutrient levels in cancer patients before and after doxorubicin-containing chemotherapy. The concentration of lipid peroxidation products, measured as thiobarbituric acid-reactant materials, in the plasma of cancer patients was higher than in controls and the level was increased further after chemotherapy. These results indicated that cancer patients had increased oxidative stress at presentation, which was further aggravated by doxorubicin treatment. Cancer patients had lower levels of glutathione, glutathione peroxidase, selenium, and zinc levels, but these were not modified further by chemotherapy. Torii *et al.* [41] reported that doxorubicin treatment caused cardiomyopathy and increased lipid peroxidation and lower α-tocopherol levels in the myocardium of spontaneously hypertensive rats.

Radiation therapy of malignancies in the head and neck area results in a marked reduction in the saliva flow rate and alterations in saliva composition within the first week of therapy and impairs saliva flow throughout the duration of therapy. Decreased secretion of saliva may lead to symptoms such as oral pain and burning sensations, loss of taste and appetite, and an increased incidence of oral disease. These symptoms may affect eating and increase the risk of inadequate nutritional intake. Backstrom *et al.* [42] investigated the average nutritional intake of 24 patients treated for malignancies in the head and neck region who had dry mouth symptoms that had persisted for at least 4 months after the completion of radiation therapy. The average caloric intake was 1925 calories in the irradiated patients with dry mouth symptoms compared to 2219 calories in the age- and sex-matched controls. The average intakes of vitamin A, β-carotene, vitamin E, vitamin B_6, folic acid, iron, and zinc were significantly lower in irradiated patients than in controls.

VI. EFFECTS OF SELECTED MICRONUTRIENTS ON RADIATION AND CHEMOTHERAPY TOXICITY

A. Vitamin E

Many toxicities associated with chemotherapy and/or radiation therapy may be preventable by vitamin E administration. Radioprotection by vitamin E may be due in part to its antioxidant effects [43] and in part to its immunostimulatory and immunoprotective effects [44–47]. Vitamin E was found to be effective in preventing chemotherapy-induced oral mucositis [48,49] and protected against doxorubicin cardiotoxicity without compromising the effectiveness of the drug [50]. The efficacy of vitamin E in the selective protection of murine erythroid progenitor cells from chemotherapy toxicity [51], as well as prevention of severe toxicity caused by tumor necrosis factor [52], has been reported. A selective antitumor effect of α-tocopherol against murine leukemia cells while protecting the murine bone marrow against doxorubicin toxicity has also been reported [53]. Srinivasan and Weiss [54] showed that pretreatment with α-tocopherol protected mice against radiation lethality, and the effect of another radioprotective agent, WR-3689, was enhanced when given in combination with vitamin E. Nattakom *et al.* [55] reported the case of a 44-year-old woman who developed severe venoocclusive disease after bone marrow transplantation and was treated with vitamin E and glutamine, which resulted in complete resolution of the clinical and biochemical signs of severe hepatic dysfunction asssociated with this disease.

In addition to its cardioprotective effect, α-tocopherol pretreatment prevented the development of doxorubicin-induced focal glomerulosclerosis and renal failure in an animal model [56]. Topical application of vitamin E was also very effective in promoting the healing of skin wounds caused by doxorubicin-induced skin necrosis [57]. In animals given an oral or topical vitamin E preparation prior to treatment with doxorubicin, dermal incision wounds healed much better compared to control animals, suggesting that vitamin E may play an important role in postoperative wound healing, especially in doxorubicin-impaired wounds [58].

Oral administration of vitamin E concurrently with intravenous chemotherapy with cyclophosphamide, methotrexate, and 5-fluorouracil (CMF) in rats protected the intestinal membranes against chemotherapy-induced toxicity [59]. CMF-induced decreases in intestinal basolateral membrane levels of ATPases, alkaline phosphatase, 5′-nucleotidase, and sulfhydryl groups and an increase in malondialdehyde levels were restored to normal by coadministration of vitamin E. Vascular endothelial damage induced by intravenous cisplatin administration was prevented by vitamin E treatment in rats [60]. In the cisplatin plus vitamin E group, there was restoration of cisplatin-induced morphological changes in the endothelium, as well as restoration of superoxide dismutase and Na/K-ATPase levels.

B. Zinc

In a review article, Sorenson [61] listed copper, iron, manganese, and zinc complexes as protective agents against radiation-induced immunosuppression, cell damage, and death in lethally irradiated animals. Srivastava *et al.* [62] observed decreased cisplatin-induced nephrotoxicity and gastrointestinal toxicity in animals given zinc before chemotherapy.

Radiation therapy to the head and neck region frequently results in xerostomia and lack of taste. Abnormalities of taste have been related to a deficiency of zinc in humans by several investigators [63,64]. Decreased taste acuity (hypogeusia) has been observed in zinc-deficient subjects, such as patients with liver disease, malabsorption syndrome, chronic uremia, and after burns and administration of penicillamine. Chronically debilitated patients, such as cancer patients, also develop hypogeusia. In a double blind study, Mahajan *et al.* [64] showed that zinc was effective in improving taste acuity in subjects with chronic uremia.

Another neurosensory disorder, abnormal dark adaptation, has been also related to a deficiency of zinc in humans [65]. Zinc supplementation to zinc-deficient sickle cell anemia patients is known to correct this abnormality [65]. Decreased dark adaptation has been identified as the dose-limiting toxicity for fenretinide (4-hydroxyphenylretinamide), a cancer chemopreventive retinoid currently under clinical investigation. Clinical trials should be conducted with a combination of zinc and fenretinide to determine if the combination has reduced toxicity and enhanced chemopreventive activity.

C. Gastrointestinal Toxicity

Antioxidant micronutrients have been found to prevent gastrointestinal toxicities of radiation and chemotherapy. Mills [66] reported that β-carotene decreases the oral mucositis that is induced by chemotherapy and radiation therapy. Klimberg *et al.* [67] observed a protective effect of glutamine on the small bowel mucosa of rats receiving abdominal radiation. Carroll *et al.* [68] found that pretreatment with a variety of antioxidant compounds and micronutrients, including ribose-cystein, amifostine, glutamine, vitamin E, and magnesium chloride/ATP, afforded protection against radiation-induced small bowel and large bowel injury in rats. Current studies are investigating the efficacy of glutamine in preventing oral mucositis induced by radiation and chemotherapy in patients with SCCHN.

D. Nephrotoxicity

Cisplatin is a drug that is active against head and neck cancer; however, severe nephrotoxicity and neurotoxicity are dose-limiting side effects of this drug and may result in significant morbidity. Administration of cisplatin with pre- and posttreatment hydration and mannitol-induced diuresis lowers the concentration of cisplatin in the kidneys and reduces its nephrotoxicity. An alternative approach to protect against the side effects of cisplatin is provided by chemoprotectors. Selenium has been reported to reduce cisplatin-induced nephrotoxicity [69,70], in addition to its well-known chemopreventive properties [71]. Sodium selenite has been shown to protect rodents against cisplatin nephrotoxicity without reducing the antitumor activity of the drug [70]. Reactions between cisplatin and nucleophilic metabolites of selenite may be responsible for the protective effect of selenite against cisplatin nephrotoxicity [72]. Vermeulen *et al.* [73] concluded that sodium selenite protected rodents against cisplatin-induced nephrotoxicity without influencing the systemic availability of cisplatin.

Sadzuka *et al.* [74,75] demonstrated that cisplatin-induced nephrotoxicity was closely associated with an increase in lipid peroxidation and a decrease in the activity of enzymes that protect against lipid peroxidation. Pretreatment with α-tocopherol and glutathione significantly decreased the amount of lipid peroxides produced in the kidney by the administration of cisplatin [76]. Sugihara *et al.* [77,78] found that α-tocopherol prevented lipid peroxidation and nephrotoxicity induced by cisplatin in rodents. Bogin *et al.* [79] reported that pretreatment with a combination of cysteine and α-tocopherol is protective against the nephrotoxicity and biochemical changes induced by the administration of cisplatin in rats.

VII. NUTRIENTS AND ANTITUMOR EFFECTS OF RADIATION/CHEMOTHERAPY

The mechanism of action of radiation therapy and some chemotherapeutic agents involves the generation of toxic oxygen-free radicals. Supplementing patients with antioxidant micronutrients during therapy may potentially interfere with the antitumor effects of the treatment. However, many of the antioxidants have been found to prevent treatment toxicity without reducing the efficacy of radiation or chemotherapy. Furthermore, certain micronutrients have antitumor effects, inhibit cancer cell proliferation, and induce cancer cell differentiation.

Vitamin E inhibits growth and causes morphological changes in several tumor cell lines in tissue culture [80,81]. Animal studies and clinical trials have also shown chemopreventive [82,83] and antineoplastic activities [84,85] of vitamin E. A number of experimental studies further suggest that vitamin E can enhance the growth inhibitory effect of various cancer treatment modalities such as radiation, chemotherapy, and hyperthermia [80]. At some doses, vitamin E enhanced tumor killing by irradiation [86]. Prasad *et al.* [87] observed growth inhibitory effects of vitamin C alone, vitamin E alone, and a combination of vitamin C, vitamin E, β-carotene, and 13-*cis*-retinoic acid when SK-30 melanoma cells were exposed to these agents *in vitro*. Furthermore, ascorbic acid, alone or in combination with β-carotene, vitamin E, and 13-*cis*-retinoic acid, enhanced the growth-inhibitory effect of cisplatin, dacarbazine, tamoxifen, and interferon-α 2b.

Certain micronutrients have cancer chemopreventive properties and thus may play a role in the prevention of radiation- and chemotherapy-induced cancers. Krishnaswamy *et al.* [88] observed a 57% complete remission rate of oral preneoplastic lesions in 150 subjects given a multivitamin capsule containing vitamin A, riboflavin, zinc, and selenium twice weekly for 1 year. Satoh *et al.* [89] reported that an induction of increased pulmonary metallothionein by zinc or bismuth could prevent the development of lung cancer in mice receiving repeated injections of cisplatin and melphalan. Zinc administration potentiated the radioprotective effect of diltiazem in mice given a lethal dose of radiation [90]. A combination of zinc aspartate with amifostine, an antioxidant compound, conferred protection against lethal effects of radiation, as well as against the development of radiation-induced lymphomas in mice [91].

Oral administration of vitamins A and E in conjunction with FEMTX (fluorouracil, epirubicin, methotraxate) chemotherapy in patients with unresectable or metastatic gastric cancer does not appear to reduce the antitumor activity of the chemotherapeutic agents [92]. Glutathione administration protected rodents against the renal and lethal toxicity of cisplatin, but did not interfere with the antitumor activity of the drug [93]. In small clinical studies, protective effects of reduced glutathione against the renal toxicity of cisplatin [94,95] without interfering with its antitumor activity [95] were reported. Di Re *et al.* [96] confirmed these results in a larger series of 40 patients with ovarian cancer treated with high-dose cisplatin and cyclophosphamide given with pre- and posttreatment glutathione protection showing a significant protection against renal toxicity with no effect on antitumor activity.

Rouse *et al.* [97] hypothesized that intravenous glutamine protects liver cells from oxidant injury by increasing the intracellular glutathione content and that supplemental oral glutamine would increase the therapeutic index of methotrexate by improving host tolerance through changes in glutathione metabolism. Provision of glutamine-rich diet to rats with implanted fibrosarcomas treated with methotrexate decreased tumor glutathione and increased the antitumor effect of methotrexate, while maintaining or increasing host glutathione stores [97]. Significantly decreased glutathione levels in tumor cells correlated with their susceptibility to methotrexate and tumor shrinkage in animals that received a combination of glutamine and methotrexate. Thus oral glutamine supplementation enhanced the sensitivity of the tumor cells while protecting normal cells from the effects of methotrexate chemotherapy.

However, zinc administration has been shown to interfere with the antitumor activity of cisplatin by increasing its detoxification through increased metallothionein synthesis [98]. In C3H mice inoculated with bladder tumor (MBT-2), cisplatin and zinc sulfate were administered and a reduction in both renal toxicity and antitumor activity of cisplatin was observed [98].

VIII. NUTRIENTS AND PHARMACEUTICALS IN THE PREVENTION OF RADIATION AND CHEMOTHERAPY TOXICITY

Floersheim [90] reported protection of mice against a lethal dose of radiation by diltiazem, and synergistic effects occurred by combining diltiazem with zinc. In another study, Floersheim *et al.* [91] found that small doses of zinc and amifostine provided additive protection against radiation lethality in mice.

The phosphorothioate amifostine has been approved for clinical use in the prevention of cisplatin toxicity. Currently, its use in the prevention of radiation toxicity and toxicities of other chemotherapeutic agents is under investigation. The usefulness of phosphorothioates is limited by their toxicity, and it has been proposed that combining other agents with phosphorothioates may improve their efficacy and/or lower their toxicity [99]. Because zinc aspartate and a combination of zinc aspartate with amifostine have shown better protective effects for normal tissue compared to tumor tissue [100–102], further clinical studies should test the low-dose amifostine and zinc combination to determine if radiation protection can be achieved without the side effects of high-dose amifostine.

Micronutrient supplementation may prevent adverse effects of cancer chemotherapy and radiation therapy without interfering with their antitumor effects. This protection may result in lower treatment-associated morbidity and mortality and better quality of life for cancer patients. There may even be potentiation of the antineoplastic effect of radiation and chemotherapy by micronutrients, which may lead to increased tumor responsiveness to treatments. Furthermore, micronutrients lack intrinsic toxicity, which is a major problem with some of the pharmaceutical compounds approved for clinical use as toxicity preventive agents. However, currently there are insufficient clinical data to support the use of micronutrients as adjunct to radiation or chemotherapy. Future clinical intervention trials should investigate the potential benefits and risks of micronutrient supplementation during cancer therapy.

A. Selenium

Kiremidjian *et al.* [103] conducted a randomized, double-blind placebo-controlled study to determine whether the oral intake of 200 mcg/day of sodium selenite will abrogate depressed immune function of patients receiving therapy for squamous cell carcinoma of the head and neck. Subjects were given one selenium/placebo tablet/day for 8 weeks, beginning on the day of their first treatment for the disease (e.g., surgery, radiation, or surgery and radiation), and their immune functions were monitored. Supplementation with selenium during therapy resulted in a significantly enhanced cell-mediated immune responsiveness, as reflected in the ability of the patient's lymphocytes to respond to stimulation with mitogen, to generate cytotoxic lymphocytes, and to destroy tumor cells. The enhanced responsiveness was evident during therapy and following the conclusion of therapy. In contrast, patients in the placebo arm of the study showed a decline in immune responsiveness during therapy, which was followed, in some patients, by an enhancement, but the responses of the group remained significantly lower than baseline values. Data also show that at baseline, patients entered in the study had significantly lower plasma Se levels than healthy individuals, and patients in stage I–II disease had significantly higher plasma selenium levels than patients in stage III–IV disease.

B. Zinc

Zinc deficiency causes a profound reduction in the activity of a thymic hormone, thymulin. Prasad *et al.* [104] found decreased production of interleukin(IL)-2 and interferon-γ by TH1 cells, reduced natural killer (NK) cell activity, and decreased recruitment of T-cell precursors in zinc-deficient subjects. Mocchegiani *et al.* [105] observed a significant increase or stabilization in the body weight of AIDS patients who received zinc supplement in addition to AZT, associated with an increase in CD4 cells and plasma thymulin and a decrease in the frequency of opportunistic infections. Abdulla *et al.* [106] observed that plasma zinc was decreased and that the copper:zinc ratio in the plasma was significantly higher in 13 patients with SCCHN in comparison to healthy controls. Patients who showed a marked decrease in plasma zinc levels died within 12 months. The authors suggested that the plasma zinc and copper/zinc ratio may be of value as a potential screening and predicting test in patients with head and neck cancer. However, Garofalo *et al.* [107] observed (1) no significant difference in serum zinc and (2) no diagnostic or prognostic value in these parameters in patients with head and neck cancer.

Zinc deficiency is known to cause weight loss, abnormal cellular immune functions, hypogeusia, and difficulty in wound healing, all of which commonly occur in malnourished head and neck cancer patients. Wound healing is, in many respects, analogous to growth, and inasmuch as zinc deficiency affects growth adversely, it is not surprising that zinc deficiency also causes impaired wound healing [108–110] and that zinc supplementation promotes wound healing in zinc-deficient subjects [108–110]. Zinc-deficient rats show a significant reduction in total collagen, a reduction in the total dry weight of the sponge connective tissue and the non-collagenous protein content, a decrease in RNA/DNA, and a depletion of polyribosomes in sponge connective tissue, suggesting that the proliferation of fibroblasts is impaired as a result of zinc deficiency [111].

Prasad *et al.* [112] described the zinc levels in plasma, lymphocytes, and granulocytes in zinc-deficient and zinc-sufficient subjects with head and neck cancer and healthy volunteers. By cellular zinc criteria, a mild deficiency of zinc was observed in 25% of the normal healthy volunteers and 48% of the head and neck cancer subjects. Productions of IL-2 and tumor necrosis factor (TNF)- α were decreased significantly in zinc-deficient subjects in both groups (cancer and healthy volunteers), whereas the productions of IL-4, IL-5, and IL-6 were not affected by zinc status. The mean IL-4 production in cancer patients was higher than in noncancer subjects, but statistically the difference was not significant. In zinc-deficient subjects of both groups, the production of IL-1β was increased significantly in comparison to zinc-sufficient subjects. NK cell activity was decreased in zinc-deficient subjects in comparison to zinc-sufficient subjects in both groups. The ratios of $CD4^+/CD8^+$ and $CD4^+CD45RA^+/CD4^+CD45R0^+$ cells were decreased in zinc-deficient subjects. Fifty-seven percent (27/47) of patients were classified as nutritionally deficient (NUTR-), whereas 43% (20/47) were nutritionally sufficient (NUTR+) based on their prognostic nutritional index (PNI) indices. Zinc status was inversely associated with tumor size ($p=0.002$), disease stage ($p=0.04$), and unplanned hospital days ($p=0.04$). Zinc and nutrition interaction was significant for postoperative febrile days ($p=0.03$) and for disease-free interval ($p=0.01$). Fifty percent of the morbidities (pulmonary and nonpulmonary) were due to infectious episodes. Thus, they observed that the zinc status of head and neck cancer patients affects significantly cell-mediated immune functions and clinical morbidities. Results showed that the functions of TH1 cells were compromised, as evidenced by the decreased production of IL-2 and IFN-γ in zinc-deficient head and neck cancer patients, whereas the TH2 cytokines were unaffected. NK cell lytic activity was also decreased in zinc-deficient patients. Thus, there is an imbalance between the functions of TH1 and TH2 cells, which may have been responsible for cell-mediated immune function disorders in zinc-deficient cancer patients. Further research must be carried out in order to document the effect of zinc supplementation in zinc-deficient patients with squamous cell carcinoma of head and neck.

Zinc deficiency and cell-mediated immune dysfunction are present in a large percentage of head and neck cancer patients at initial presentation [112]. Zinc deficiency is associated with an increased tumor size, overall stage of the cancer, and unplanned hospitalizations [112]. The disease-free interval is longest for the group with zinc-sufficient and nutrition-sufficient status. If these results are confirmed in larger studies, zinc supplementation may be recommended for head and neck cancer patients at presentation to reduce treatment- and disease-related morbidity, to improve immune function, to delay disease recurrence, and to prevent second primary tumors.

C. Phytochemicals

Because major factors associated with cancer are prooxidant in nature, it is hypothesized that the administration of antioxidant supplements would prevent cancer. An increased consumption of fruits and vegetables, which contain numerous antioxidant micronutrients, has consistently been associated with a lower risk of cancer in epidemiological studies. However, vegetables and fruits contain numerous cancer-preventive compounds with different mechanisms of action, and they are all taken together in small quantities as a part of a complex diet. It is therefore inappropriate to extrapolate from epidemiological studies and conclude that just because a micronutrient has an inverse association in epidemiological studies it would result in cancer risk reduction when taken as a dietary supplement. Clinical studies are needed to investigate the mechanisms of action, as well as efficacy and toxicity of each micronutrient, alone and in combination with other micronutrients. Multiple chemopreventive agents, when taken together, may have synergistic or antagonistic interactions or no interactions with each other. Examples of promising nutritional chemopreventive compounds in clinical trials include vitamin E, selenium, lycopene, folic acid, and soy isoflavones. The importance of conducting clinical chemoprevention trials has become very clear recently, when several clinical trials showed that the agents hypothesized to prevent cancer did exactly the opposite [23,113]. A large chemoprevention study conducted to determine whether β-carotene and/or α-tocopherol would prevent lung cancer showed that β-carotene supplementation increased the risk of lung cancer [23]. These unexpected results highlight the importance of conducting well-designed, prospective, randomized clinical trials before making recommendations to the public regarding the use of supplements.

Richardson *et al.* [114] assessed the prevalence of supplement use among cancer patients attending outpatient clinics at the University of Texas M.D. Anderson Cancer Center, Houston, Texas. Of the 453 participants, 62.6% used vitamins and herbs. The authors concluded that "given the number of patients combining vitamins and herbs with conventional treatments, the oncology community must improve patient–provider communication, offer reliable information to patients, and initiate research to determine possible drug–herb–vitamin interactions." Nam *et al.* [115] determined the prevalence and patterns of the use of complementary and alternative medicine (CAM) therapies among patients with and those at high risk for prostate cancer. Of the patients presenting to urology clinics and the support group, 27.4 and 38.9% with and 25.8 and 80% at high risk for prostate cancer, respectively, used some form of complementary therapy. Because some CAM therapies used by prostate cancer patients may include herbs, such as PC-SPES [116], and phytochemicals, such as lycopene [117] and soy isoflavones [118] with potent biological effects, it is very important to obtain accurate information from patients regarding their CAM use.

IX. CONCLUSIONS

Micronutrients and phytochemicals have significant roles in the initiation and progression of SCCHN. These compounds may have a role in the prevention of *de novo* cancer in high-risk populations, such as tobacco smokers, as well as in the prevention of disease progression or relapse. In addition, simultaneous use of these compounds with radiation and chemotherapy may improve the efficacy of treatment while decreasing its toxicity. However, there are no data from randomized clinical trials showing their efficacy and safety in this population. Clinical studies investigating the use of these agents in the prevention and/or treatment of SCCHN should be high priority.

References

1. Hsu, T. C., Spitz, M. R., and Schantz, S. P. (1991). Mutagen sensitivity: A biological marker of cancer susceptibility. *Cancer Epidemiol. Biomark. Prev.* **1**, 83–89.
2. Hsu, T. C. (1983). Genetic instability in the human population: a working hypothesis. *Hereditas* **98**, 1–9.
3. Hsu, T. C., Johnston, D. A., Cherry, L. M., Ramkisson, D., Schantz, S. P., Jessup, J. M., Winn, R. J., Shirley, L., and Furlong, C. (1989). Sensitivity to genotoxic effects of bleomycin in humans: Possible relationship to environmental carcinogenesis. *Int. J. Cancer* **43**, 403–409.
4. Schantz, S. P., Spitz, M. R., and Hsu, T. C. (1990). Mutagen sensitivity in head and neck cancer patients: A biologic marker for risk of multiple primary malignancies. *J. Natl. Cancer Inst.* **82**, 1773–1775.
5. Spitz, M. R., Fueger, J. J., Halabi, S., Schantz, S. P., Sample, D., and Hsu, T. C. (1993). Mutagen sensitivity in upper aerodigestive tract cancer: A case-control analysis. *Cancer Epidemiol. Biomark. Prev.* **2**, 329–333.
6. Cloos, J., Braakhuis, B. J. M., Steen, I., Copper, M. P., DeVries, N., Nauta, J. J. P., and Snow, G. P. (1994). Increased mutagen sensitivity in head-and-neck squamous-cell carcinoma patients, particularly those with multiple primary tumors. *Int. J. Cancer* **56**, 816–819.
7. Trizna, Z., Schantz, S. P., and Hsu, T. C. (1991). Effects of N-acetyl-L-cysteine and ascorbic acid on mutagen-induced chromosomal sensitivity in patients with head and neck cancers. *Am. J. Surg.* **162**, 294–298.
8. Pohl, H., and Reidy, J. (1989). Vitamin C intake influences the bleomycin-induced chromosome damage assay: Implications for detection of cancer susceptibility and chromosome breakage syndromes. *Mutat. Res.* **224**, 247–252.
9. Trizna, Z., Hsu, T. C., and Schantz, S. P. (1992). Protective effects of vitamin E against bleomycin-induced genotoxicity in head and neck cancer patients in vitro. *Anticancer Res.* **2**, 325–328.
10. Shamberger, R. L., Baughman, F. F., Kalchert, S. L., *et al.* (1973). Carcinogen-induced chromosomal breakage decreased by antioxidants. *Proc. Natl. Acad. Sci. USA* **70**, 1461–1463.
11. LaVecchia, C., Negri, E., D'Avanzo, B., Franchesci, S., Decarli, A., and Boyle, P. (1990). Dietary indicators of laryngeal cancer risk. *Cancer Res.* **50**, 4497–4500.
12. Block, G., Patterson, B., and Subar, A. (1992). Fruits, vegetables, and cancer prevention: A review of epidemiological evidence. *Nutr. Cancer* **18**, 1–29.
13. Kucuk, O., Pung, A., Franke, A., Custer, L., Wilkens, L., LeMarchand, L., Higuchi, C., Cooney, R., and Hsu, T. C. (1995). Variability of mutagen sensitivity in healthy adults: Correlations with plasma nutrient levels. *Cancer Epidemiol. Biomark. Prev.* **4**, 217–221.

14. Monget, A. L., Galan, P., Preziosi, P., Keller, H., Bourgeois, C., Arnaud, J., Favier, A., and Hercberg, S. (1996). Micronutrient status in elderly people. *Int. J. Vit. Nutr. Res.* **66**, 71–76.
15. Donma, M. M., Donma, O., and Tas, M. A. (1993). Hair zinc and copper concentrations and zinc: Copper ratios in pediatric malignancies and healthy children from southeastern Turkey. *Biol. Trace Element Res.* **36**, 51–63.
16. Negri, E., Franceschi, S., Bosetti, C., Levi, F., Conti, E., Parpinel, M., and La Vecchia, C. (2000). Selected micronutrients and oral and pharyngeal cancer. *Int. J. Cancer* **86**, 122–127.
17. Alhasan, S. A., Pietrasczkiwicz, H., Alonso, M. D., Ensley, J., and Sarkar, F. H. (1999). Genistein-induced cell cycle arrest and apoptosis in a head and neck squamous cell carcinoma cell line. *Nutr. Cancer* **34**, 12–19.
18. Alhasan, S. A., Ensley, J. F., and Sarkar, F. H. (2000). Genistein induced molecular changes in a squamous cell carcinoma of the head and neck cell line. *Int. J. Oncol.* **16**, 333–338.
19. Pamuk, E. R., Byers, T., Coates, R. J., Vann, J. W., Sowell, A. L., Gunter, E. W., and Glass, D. (1994). Effects of smoking on serum nutrient concentrations in African-American women. *Am. J. Clin. Nutr.* **59**, 891–895.
20. Ross, M. A., Crosley, L. K., Brown, K. M., Duthie, S. J., Collins, A. C., Arthur, J. R., and Duthie, G. G. (1995). Plasma concentrations of carotenoids and antioxidant vitamins in Scottish males: Influences of smoking. *Eur. J. Clin. Nutr.* **49**, 861–865.
21. Pakrashi, A., and Chatterjee, S. (1995). Effect of tobacco consumption on the function of male accessory sex glands. *Int. J. Androl.* **18**, 232–236.
22. Faruque, M. O., Khan, M. R., Rahman, M. M., and Ahmed, F. (1995). Relationship between smoking and antioxidant micronutrient status. *Br. J. Nutr.* **73**, 625–632.
23. Albanes, D., Heinonen, O. P., Taylor, P. R., Virtamo, J., Edwards, B. K., Rautalahti, M., Hartman, A. M., Palmgren, J., Freedman, L. S., Haapakoski, J., Barrett, M. J., Pietinen, P., Malila, N., Tala, E., Liippo, K., Salomaa, E. R., Tangrea, J. A., Teppo, L., Askin, F. B., Taskinen, E., Erozan, Y., Greenwald, P., and Huttunen, J. K. (1996). Alpha-Tocopherol and beta-carotene supplements and lung cancer incidence in the alpha-tocopherol, beta-carotene cancer prevention study: Effects of baseline characteristics and study compliance. *J. Natl. Cancer Inst.* **88**, 1560–1570.
24. Palozza, P., Luberto, C., Calviello, G., Ricci, P., and Bartoli, G. M. (1997). Antioxidant and prooxidant role of beta–carotene in murine normal and tumor thymocytes: Effects of oxygen partial pressure. *Free Radic. Biol. Med.* **22**, 1065–1073.
25. Wang, X. D., Liu, C., Bronson, R. T., Smith, D. E., Krinsky, N. I., and Russell, M. (1999). Retinoid signaling and activator protein-1 expression in ferrets given beta-carotene supplements and exposed to tobacco smoke. *J. Natl. Cancer Inst.* **91**, 60–66.
26. Tsubono, Y., Tsugane, S., and Gey, K. F. (1996). Differential effects of cigarette smoking and alcohol consumption on the plasma levels of carotenoids in middle aged Japanese men. *Jap. J. Cancer Res.* **87**, 563–569.
27. Brady, W. E., Mares-Perlman, J. A., Bowen, P., and Stacewicz-Sapuntzakis, M. (1996). Human serum carotenoid concentrations are related to physiologic and lifestyle factors. *J. Nutr.* **126**, 129–137.
28. Lecomte, E., Grolier, P., Herbeth, B., Pirollet, P., Musse, N., Paille, F., Braesco, V., Siest, G., and Artur, Y. (1994). The relationship of alcohol consumption to serum carotenoid and retinol levels: Effects of withdrawal. *Int. J. Vit. Nutr. Res.* **64**, 170–175.
29. Leo, M. A., Seitz, H. K., Maier, H., and Lieber, C. S. (1995). Carotenoid, retinoid and vitamin E status of the oropharyngeal mucosa in alcoholics. *Alcohol Alcoholism* **30**, 163–170.
30. Goodwin, W. J., and Torres, J. (1984). The value of prognostic nutritional index in the management of patients with carcinoma of the head and neck. *Otolaryngol. Head Neck Surg.* **6**, 932–937.
31. Brooks, G. B. (1985). Nutritional status: A prognostic indicator in head and neck cancer. *Otolaryngol. Head Neck Surg.* **93**, 69–74.
32. Williams, E. F., and Meguid, M. M. (1989). Nutritional concepts and considerations in head and neck surgery. *Head Neck Surg.* **11**, 393–399.
33. Bassett, M. R., and Dobie, R. A. (1983). Patterns of nutritional deficiency in head and neck cancer. *Otolaryngol. Head Neck Surg.* **91**, 119–125.
34. Hooley, R., Levine, H., Flores, T. C., Wheeler, T., and Steiger, E. (1983). Predicting postoperative complications using nutritional assessment. *Arch. Otolaryngol.* **109**, 83–85.
35. Sobol, S. M., Conoyer, J. M., and Sessions, D. G. (1979). Enteral and parenteral nurition in patients with head and neck cancer. *Ann. Otolaryngol.* **88**, 495–501.
36. Mullen, J. L., Gertner, M. H., Buzby, G. P., Goodhart, G. L., and Rosato, E. F. (1979). Implications of malnutrition in the surgical patient. *Arch. Surg.* **114**, 121–125.
37. Buzby, G. P., Mullen, J. L., Matthews, D. C., Hobbes, C. I., and Rosato, E. F. (1980). Prognostic nutritional index in gastrointestinal surgery. *Am. J. Surg.* **139**, 160–167.
38. Olmedilla, B., Granado, F., Blanco, I., and Rojas-Hidalgo, E. (1996). Evaluation of retinol, alpha- tocopherol, and carotenoids in serum of men with cancer of the larynx before and after commercial enteral formula feeding. *J. Parenteral Enteral Nutr.* **20**, 145–149.
39. Prasad, A. S. (1993). "Biochemistry of Zinc." Plenum Press, New York.
40. Faber, M., Coudray, C., Hida, H., Mousseau, M., and Favier, A. (1995). Lipid peroxidation products, and vitamin and trace element status in patients with cancer before and after chemotherapy, including Adriamycin: A preliminary study. *Biol. Trace Element Res.* **47**, 117–123.
41. Torii, M., Ito, H., and Suzuki, T. (1992). Lipid peroxidation and myocardial vulnerability in hypertrophied SHR myocardium. *Exp. Mol. Pathol.* **57**, 29–38.
42. Backstrom, I., Funegard, U., Andersson, I., Franzen, L., and Johansson, I. (1995). Dietary intake in head and neck irradiated patients with permanent dry mouth symptoms. *Oral Oncol. Eur. J. Cancer* **31B**, 253–257.
43. Machlin, L. J., and Bendich, A. (1987). Free radical tissue damage; protective role of antioxidant micronutrients. *FASEB J.* **1**, 441–445.
44. Roy, R. M., and Petrella, M. (1987). Humoral immune response of mice injected with tocopherol after exposure to X-irradiation. *Immunopharmacol. Immunotoxicol.* **9**, 47–50.
45. Roy, R. M., Petrella, M., and Shateri, H. (1988). Effect of administering tocopherol after irradiation on survival and proliferation of murine lymphocytes. *Pharmacol. Ther.* **39**, 393–395.
46. Srinivasan, V., Jacobs, A. J., Simpson, S. A., and Weiss, J. F. (1983). Radioprotection by vitamin E: effects on hepatic enzymes, delayed type hypersensitivity, and postirradiation survival of mice. *In* "Modulation and Mediation of Cancer by Vitamins" (F. L. Meyskens and K. N. Prasad, eds.), pp.119–131. Karger, Basel.
47. Tengerdy, R. P., Mathias, M. M., and Nockels, C. F. (1981). Vitamin E, immunity and disease resistance. *Adv. Exp. Med. Biol.* **135**, 27–42.
48. Lopez, I., Goudou, C., Ribrag, V., Sauvage, C., Hazebroucq, G., and Dreyfus, F. (1994). Treatment of mucositis with vitamin E during administration of neutropenic antineoplastic agents. *Ann. Med. Interne (Paris)* **145**, 405–408.
49. Wadleigh, R., Redman, R., Cohen, M., Krasnow, S., Anderson, A., and Graham, M. (1990). Vitamin E in the treatment of chemotherapy-induced mucositis. *Proc. Am. Soc. Clin. Oncol.* **9**, A1237.
50. Myers, C., McGuire, W., and Young, R. (1976). Adriamycin amelioration of toxicity by alpha-tocopherol. *Cancer. Treat. Rep.* **60**, 961–962.
51. Gogu, S. R., Green, G., Gupta, V., and Agarwal, K. C. (1990). Selective protection of murine erythroid progenitor cells with vitamin E from drug induced toxicity. *Proc. Am. Assoc. Cancer Res.* **31**, 404.
52. Satomi, N., Sakurai, A., Haranaka, R., and Haranaka, K. (1988). Preventive effects of several chemicals against lethality of recombinat human tumor necrosis factor. *J. Biol. Response Mod.* **7**, 54–64.

53. Fariss, M. W., Fortuna, M. B., Everett, C. K., Smith, J. D., Trent, D. F., and Djuric, Z. (1994). The selective antiproliferative effects of alpha-tocopheryl hemisuccinate and cholesteryl hemisuccinate on murine leukemia cells result from the action of the intact compounds. *Cancer Res.* **54**, 3346–3351.
54. Srinivasan, V., and Weiss, J. F. (1992). Radioprotection by vitamin E: Injectable vitamin E administered alone or with WR-3689 enhances survival of irradiated mice. *Int. J. Radiat. Oncol. Biol. Phys.* **23**, 841–845.
55. Nattakom, T. V., Charlton, A., and Wilmore, D. W. (1995). Use of vitamin E and glutamine in the successful treatment of severe veno-occlusive disease following bone marrow transplantation. *Nutr. Clin. Practice* **10**, 16–18.
56. Washio, M., Nanishi, F., Okuda, S., Onoyama, K., and Fujishima, M. (1994). Alpha-tocopherol improves focal glomerulosclerosis in rats with Adriamycin-induced progressive renal failure. *Nephron* **68**, 347–352.
57. Lucero, M. J., Vigo, J., Rabasco, A. M., Sanchez, J. A., and Martin, F. (1993). Protection by alpha-tocopherol against skin necrosis induced by doxorubicin hydrochloride. *Pharmazie* **48**, 772–775.
58. Bauer, G., O'Connell, S., and Devereux, D. (1994). Reversal of doxorubicin-impaired wound healing using Triad compound. *Am. Surgeon* **60**, 175–179.
59. Subramaniam, S., Subramaniam, S., and Devi, C. S. (1995). Vitamin E protects intestinal basolateral membrane from CMF-induced damages in rat. *Indian J. Physiol. Pharmacol.* **39**, 263–266.
60. Ito, H., Okafuji, T., and Suzuki, T. (1995). Vitamin E prevents endothelial injury associated with cisplatin injection into the superior mesenteric artery in rats. *Heart Vessels* **10**, 178–184.
61. Sorenson, J. R. (1992). Essential metalloelement metabolism and radiation protection and recovery. *Radiat. Res.* **132**, 19–29.
62. Srivastava, R. C., Farookh, A., Ahmad, N., Misra, M., Hasan, S. K., and Husain, M. M. (1995). Reduction of cis-platinum induced nephrotoxicity by zinc histidine complex: The possible implication of nitric oxide. *Biochem. Mol. Biol. Int.* **36**, 855–862.
63. Henkin, R. J., Aamodt, R. L., Aggarwal, R. P., and Foster, D. M. (1982). The role of zinc in taste and smell. *In* "Clinical, Biochemical and Nutritional Aspects of Trace Elements" (A. S. Prasad, ed.), pp. 161–188. A. R. Liss, New York.
64. Mahajan, S. K., Prasad, A. S., Lambujon, J., Abbasi, A. A., Briggs, W. A., and McDonald, F. D. (1980). Improvement of uremic hypogeusia by zinc: A double-blind study. *Am. J. Clin. Nutr.* **33**, 1517–1521.
65. Warth, J. A., Prasad, A. S., Zwas, F., and Frank, R. N. (1981). Abnormal dark adaptation in sickle cell anemia. *J. Lab. Clin. Med.* **98**, 189–194.
66. Mills, E. E. (1988). The modifying effect of beta-carotene on radiation and chemotherapy induced oral mucositis. *Br. J. Cancer* **57**, 416–417.
67. Klimberg, V. S., Souba, W. W., Dolson, D. J., Salloum, R. M., Hautamaki, R. D., Plumley, D. A., Mendenhall, W. M., Bova, F. J., Khan, S. R., Hackett, R. L., Bland, K. I., and Copeland, E. M., III (1990). Prophylactic glutamine protects the intestinal mucosa from radiation injury. *Cancer* **66**, 62–68.
68. Carroll, M. P., Zera, R. T., Roberts, J. C., Schlafmann, S. E., Feeney, D. A., Johnston, D. R., West, M. A., and Bubrick, M. P. (1995). Efficacy of radioprotective agents in preventing small and large bowel radiation injury. *Dis. Colon Rectum* **38**, 716–722.
69. Meyer, K. B., and Madias, N. E. (1994). Cisplatin nephrotoxicity. *Min Electrol Metab.* **20**, 201–213.
70. Baldew, G. S., Van den Hamer, C. J. A., Los, G., Vermeulen, N. P. E., De Goeij, J. J. M., and McVie, J. G. (1989). Selenium-induced protection against cis-diamminedichloroplatinum(II) nephrotoxicity in mice and rats. *Cancer Res.* **49**, 3020–3023.
71. El-Bayoumy, K. (1991). The role of selenium in cancer prevention. *In* "Practice of Oncology" (V. T. DeVita, S. Hellman, and S. S. Rosenberg, eds.), 4th Ed., pp. 1–15. Lippincott, Philadelphia, PA.
72. Baldew, G. S., Mol, J. G. J., De Kanter, F. J. J., Van Baar, B., De Goeij, J. J. M., and Vermeulen, N. P. E. (1991). The mechanism of interaction between cisplatin and selenite. *Biochem. Pharmacol.* **41**, 1429–1437.
73. Vermeulen, N. P. E., Baldew, G. S., Los, G., McVie, J. G., and De Goeij, J. J. M. (1993). Reduction of cisplatin nephrotoxicity by sodium selenite: Lack of interaction at the pharmacokinetic level of both compounds. *Drug Metab. Disp.* **21**, 30–36.
74. Sadzuka, Y., Shoji, T., and Takino, Y. (1991). Change of lipid peroxide levels in rat tissues after cisplatin administration. *Toxicol. Lett.* **57**, 159–166.
75. Sadzuka, Y., Shoji, T., and Takino, Y. (1992). Effect of cisplatin on the activities of enzymes which protect against lipid peroxidation. *Biochem. Pharmacol.* **43**, 1872–1875.
76. Sadzuka, Y., Shoji, T., and Takino, Y. (1992). Mechanism of the increase in lipid peroxide induced by cisplatin in the kidneys of rats. *Toxicol. Lett.* **62**, 293–300.
77. Sugihara, K., Nakano, S., Koda, M., Tanaka, K., Fukuishi, N., and Gemba, M. (1987). Stimulatory effect of cisplatin on production of lipid peroxidation in renal tissues. *Jpn. J. Pharmacol.* **43**, 247–252.
78. Sugihara, K., Nakano, S., and Gemba, M. (1987). Effect of cisplatin on in vitro production of lipid peroxides in rat kidney cortex. *Jpn. J. Pharmacol.* **44**, 71–76.
79. Bogin, E., Marom, M., and Levi, Y. (1994). Changes in serum, liver and kidneys of cisplatin-treated rats: Effects of antioxidants. *Eur. J. Clin. Chem. Clin. Biochem.* **32**, 843–851.
80. Prasad, K. N., and Rama, B. N. (1984). Modification of the effect of pharmacological agents, ionizing radiation and hyperthermia on tumor cells by vitamin E. *In* "Vitamins, Nutrition and Cancer" (K. N. Prasad, ed.), pp. 76–104. Karger, Basel.
81. El Attar, T. M., and Lin, H. S. (1993). Inhibition of human oral squamous carcinoma cell (SCC-25) proliferation by prostaglandin E2 and vitamin E succinate. *J. Oral Pathol. Med.* **22**, 425–427.
82. Boone, C. W., Kelloff, G. J., and Malone, W. E. (1990). Identification of candidate cancer chemopreventive agents and their evaluation in animal models and human clinical trials: A review. *Cancer Res.* **50**, 2–9.
83. Knekt, P., Aromaa, A., Maatela, J., Aaran, R. K., Nikkari, T., Hakama, M., Hakulinen, T., Peto, R., and Teppo, L. (1991). Vitamin E and cancer prevention. Am. J. Clin. Nutr. 53, 283s–286s.
84. Shklar, G., Schwartz, J., Trickler, D. P., and Ninklan, K. (1987). Regression by vitamin E of experimental oral cancer. J. Natl. Cancer Inst. 78, 987–989.
85. Helson, L. A. (1984). A phase I study of vitamin E and neuroblastoma. *In* "Vitamins, Nutrition and Cancer" (K. N. Prasad, ed.), pp. 274–281. Karger, Basel.
86. Kagerud, A., and Peterson, H. I. (1981). Tocopherol in irradiation of experimental neoplasms. *Acta Radiol. Oncol.* **20**, 97–100.
87. Prasad, K. N., Hernandez, C., Edwards-Prasad, J., Nelson, J., Borus, T., and Robinson, W. A. (1994). Modification of the effect of tamoxifen, cisplatin, DTIC, and interferon-alpha 2b on human melanoma cells in culture by a mixture of vitamins. *Nutr. Cancer* **22**, 233–245.
88. Krisnaswamy, K., Prasad, M. P., Krishna, T. P., Annapurna, V. V., and Reddy, G. A. (1995). A case study of nutrient intervention of oral precancerous lesions in India. *Eur. J. Cancer B Oral Oncol.* **31B**, 41–48.
89. Satoh, M., Kondo, Y., Mita, M., Nakagawa, I., Naganuma, A., and Imura, N. (1993). Prevention of carcinogenicity of anticancer drugs by metallothionein induction. *Cancer Res.* **53**, 4767–4768.
90. Floersheim, G. L. (1993). Radioprotective effects of calcium antagonists used alone or with other types of radioprotectors. *Radiat. Res.* **133**, 80–87.
91. Floersheim, G. L., Christ, A., Koenig, R., Racine, C., and Gudat, F. (1992). Radiation-induced lymphoid tumors and radiation lethality are inhibited by combined treatment with small doses of zinc aspartate and WR2721. *Int. J. Cancer* **52**, 604–608.
92. Pyrhonen, S., Kuitunen, T., Nyandoto, P., and Kouri, M. (1995). Randomised comparison of fluorouracil, epidoxorubicin and methotrexate (FEMTX) plus supportive care with supportive care alone in patients with non-resectable gastric cancer. *Br. J. Cancer* **71**, 587–591.

93. Zunino, F., Pratesi, G., Micheloni, A., Cavalletti, E., Sala, F., and Tofanetti, O. (1989). Protective effect of reduced glutathione against cisplatin-induced renal and systemic toxicity and its influence on the therapeutic activity of the antitumor drug. *Chem. Biol. Interactions* **70**, 89–101.
94. Bohm, S., Oriana, S., Spatti, G. B., Tognella, S., Tedeschi, M., Zunino, F., and Di Re, F. (1988). A clinical study of reduced glutathione as a protective agent against cisplatin-induced toxicity. *In* "Platinum and Other Metal Compounds in Cancer Chemotherapy" (M. Nicolini, ed.), pp. 456–459. Martinus Nijhoff Publishers, Boston.
95. Oriana, S., Bohm, S., Spatti, G., Zunino, F., and Di Re, F. (1987). A preliminary clinical experience with reduced glutathione as a protector against cisplatin toxicity. *Tumori* **73**, 337–340.
96. Di Re, F., Bohm, S., Oriana, S., Spatti, G. B., and Zunino, F. (1990). Efficacy and safety of high-dose cisplatin and cyclophosphamide with glutathione protection in the treatment of bulky advanced epithelial ovarian cancer. *Cancer Chemother. Pharmacol.* **25**, 355–360.
97. Rouse, K., Nwoked, E., Woodliff, J. E., Epstein, J., and Klimberg, V. S. (1995). Glutamine enhances selectivity of chemotherapy through changes in glutathione metabolism. *Ann. Surg.* **221**, 420–426.
98. Satoh, M., Kloth, D. M., Kadhim, S. A., Chin, J. L., Naganuma, A., Imura, N., and Cherian, M. G. (1993). Modulation of both cisplatin nephrotoxicity and drug resistance in murine bladder tumor by controlling metallothionein synthesis. *Cancer Res.* **53**, 1829–1832.
99. Weiss, J. F., Kumar, K. S., Walden, T. L., Neta, R., Landauer, M. R., and Clark, E. P. (1990). Advances in radioprotection through the use of combined agent regimens. *Int. J. Radiat. Biol.* **57**, 709–722.
100. Floersheim, G. L., and Floersheim, P. (1986). Protection against ionizing radiation and synergism with thiols by zinc aspartate. *Br. J. Radiol.* **59**, 597–602.
101. Floersheim, G. L., and Bieri, A. (1990). Further studies on selective radioprotection by organic zinc salts and synergism of zinc aspartate with WR-2721. *Br. J. Radiol.* **63**, 468–475.
102. Floersheim, G. L., Chiodetti, N., and Bieri, A. (1988). Differential radioprotection of bone marrow and tumor cells by zinc aspartate. *Br. J. Radiol.* **61**, 501–508.
103. Kiremidjian-Schumacher, L., Roy, M., Glickman, R., Schneider, K., Rothstein, S., Cooper, J., Hochster, H., Kim, M., and Newman, R. (2000). Selenium and immunocompetence in patients with head and neck cancer. *Biol. Trace Elem. Res.* **73**, 97–111.
104. Prasad, A. S., Beck, F. W. J., Grabowski, S. M., Kaplan. J., and Mathog, R. H. (1997). Zinc deficiency: Changes in cytokine production and T-cell subpopulations in patients with head and neck cancer and in non-cancer subjects. *Proc. Asso. Am. Physicians* **109**, 68–77.
105. Mocchegiani, E., Veccia, S., Ancarani, F., Scalise, G., and Fabris, N. (1995). Benefit of oral zinc supplementation as an adjunct to zidovudine (AZT) therapy against opportunistic infections in AIDS. *Int. J. Immunopharmacol.* **17**, 719–727.
106. Abdulla, M., Biorklund, A., Mathur, A., and Wallenius, K. (1979). Zinc and copper levels in whole blood and plasma from patients with squamous cell carcinomas of head and neck. *J. Surg. Oncol.* **12**, 107–113.
107. Garofalo, J. A., Erlandson, E., Strong, E. W., Lesser, M., Gerold, F., Spiro, R., Schwartz, M., and Good, R. A. (1980). Serum zinc, serum copper, and the Cu/Zn ratio in patients with epidermoid cancers of the head and neck. *J. Surg. Oncol.* **15**, 381–386.
108. Wacker, W. E. C. (1976). Role of zinc in wound healing: A critical review. *In* "Trace Elements in Human Health and Disease" (A. S. Prasad, ed.), Vol. I, pp. 107–113. Academic Press, New York.
109. Strain, W. H., Pories, W. J., and Hinshaw, J. R. (1960). Zinc studies in skin repair. *Surg. Forum* **11**, 291–292.
110. Pories, W. J., and Strain, W. H. (1966). Zinc and wound healing. *In* "Zinc Metabolism" (A. S. Prasad, ed.), pp. 378–394. Charles C. Thomas, Springfield, IL.
111. Fernandez–Madrid, F., Prasad, A. S., and Oberleas, D. (1973). Effect of zinc deficiency on nucleic acid, collagen and non-collagenous protein of the connective tissue. *J. Lab. Clin. Med.* **82**, 951–961.
112. Prasad, A. S., Beck, F. W. J., Doerr, T. D., Shamsa, F. H., Penny, H. S., Marks, S. C., Kaplan, J., Kucuk, O., and Mathog, R. H. (1998). Nutritional status of head and neck cancer patients; zinc deficiency and cell mediated immune functions. *J. Am. Coll. Nutr.* **5**, 409–418.
113. Omenn, G. S., Goodman, G. E., Thornquist, M. D., Balmes, J., Cullen, M. R., Glass, A., Keogh, J. P., Meyskens, F. L. Jr, Valanis, B., Williams, J. H. Jr, Barnhart, S., Cherniack, M. G., Brodkin, C. A., and Hammar, S. (1996). Risk factors for lung cancer and for intervention effects in CARET, the Beta-Carotene and Retinol Efficacy Trial. *J. Natl. Cancer Inst.* **88**, 1550–1559.
114. Richardson, M. A., Sanders, T., Palmer, J. L., Greisinger, A., and Singletary, S. E. (2000). Complementary/alternative medicine use in a comprehensive cancer center and the implications for oncology. *J. Clin. Oncol.* **18**, 2505–2514.
115. Nam, R. K., Fleshner, N., Rakovitch, E., Klotz, L., Trachtenberg, J., Choo, R., Morton, G., and Danjoux, C. (1999). Prevalence and patterns of the use of complementary therapies among prostate cancer patients: An epidemiological analysis. *J. Urol.* **161**, 1521–1524.
116. DiPaola, R. S., Zhang, H., Lambert, G. H., Meeker, R., Licitra, E., Rafi, M. M., Zhu, B. T., Spaulding, H., Goodin, S., Toledano, M. B., Hait, W. N., and Gallo, M. A. (1998). Clinical and biologic activity of an estrogenic herbal combination (PC-SPES) in prostate cancer. *N. Engl. J. Med.* **339**, 785–791.
117. Kucuk, O., Sarkar, F., Sakr, W., Djuric, Z., Khachik, F., Pollak, M., Bertram, J., Grignon, D., Banerjee, M., Crissman, J., Pontes, E., and Wood, D. P., Jr. (2001). Lycopene supplementation in men with localized prostate cancer: Modulation of biomarkers and clinical endpoints. *Cancer Epidemiol. Biomark. Prev.* **10**, 861–868.
118. Davis, J. N., Muqim, N., Bhuiyan, M., Kucuk, O., Pienta, K. J., and Sarkar, F. H. (2000). Inhibition of prostate specific antigen expression by genistein in prostate cancer cells. *Int. J. Oncol.* **16**, 1091–1097.

CHAPTER

15

Molecular Epidemiology of Head and Neck Cancer

QINGYI WEI,* PETER SHIELDS,† ERICH M. STURGIS,* and MARGARET R. SPITZ*

*Department of Epidemiology, The University of Texas M. D. Anderson Cancer Center, Houston, Texas 77030 and

†Cancer Genetics and Epidemology Program, Lombardi Cancer Center, Georgetown University Medical Center, Washington, DC 20007

I. Introduction 213
II. Epidemiology of Head and Neck Cancer 213
  A. Smokeless Tobacco Products 214
  B. Molecular and Genetic Changes in Squamous Cell Carcinoma of the Head and Neck 214
III. Molecular Epidemiology of Head and Neck Cancer 215
IV. Xenobiotic Metabolism of Carcinogens in Head and Neck Cancer 215
V. DNA Repair Phenotype and Risk of Head and Neck Cancer 216
  A. Host–Cell Reactivation Assay for NER 217
  B. Mutagen Sensitivity Assay 217
  C. ^{32}P-Postlabeling Assay of DNA Adducts 218
  D. Assays for DNA Repair Gene Transcript (mRNA) Levels 218
VI. Polymorphisms in DNA Repair Genes 219
VII. Polymorphisms in Cell Cycle Genes 220
References 221

I. INTRODUCTION

Oropharyngeal cancers, also known as head and neck cancers, are almost always squamous cell cancers and include cancers arising in the oral cavity, tongue, pharynx, and larynx. About 40,000 persons will be diagnosed with squamous cell carcinoma of the head and neck (SCCHN) each year, 12,000 of which are fatal [1]. There is a male:female ratio of about 2:1. Only a fraction of individuals exposed to tobacco smoke and alcohol develop SCCHN, suggesting that there are differences in individual susceptibility to carcinogenesis and the impact of gene–environment interactions. Tobacco carcinogens undergo a series of metabolic activation and detoxification steps that determine the internal dose of exposure and ultimately impact the level of DNA damage incurred. Both endogenous and exogenous exposures to carcinogens or genotoxic agents cause cell cycle delays [2] that allow cells to repair such DNA damage. Therefore, cellular DNA repair capacity (DRC) is central to maintaining genomic integrity and normal cellular functions [3]. Molecular epidemiology studies of tobacco-induced carcinogenesis have been reviewed comprehensively [4–6]. Studies have shown that polymorphisms of genes controlling drug metabolisms [7–10] and DNA repair [11–13] may contribute to a variation of tobacco-induced carcinogenesis in the general population. This chapter focuses on molecular epidemiological studies on the roles of carcinogen metabolism and DNA repair in susceptibility to SCCHN.

II. EPIDEMIOLOGY OF HEAD AND NECK CANCER

The major risk factors for oropharyngeal cancers are tobacco (cigarettes and smokeless tobacco) and alcohol use. There is a dose-response effect for both smoking and alcohol use; together the two agents act synergistically [14–26]. The attributable risk for alcohol and/or tobacco use is about 75–80% for men and 52–61% for women [14,15]. There is some evidence for a weak familial association in smokers [27]. Some studies suggest that tobacco consumption is more likely than alcohol consumption to give rise to precursor lesions [28,29] and to cancer [20,30], although the evidence for this comes from small studies. Talamini and co-workers [26] studied 60 nonsmoking drinkers and 32 nondrinking smokers and compared them to controls. Depending on the amount of drinks per week, the odds ratio (OR) reached 5.3 [95% confidence interval (CI) = 1.1, 24.8] in nonsmokers and 7.2 (95% CI = 1.1–46) in smokers. Three published studies that report data by gender all indicate that there is an increased risk for women compared to men, especially at the highest levels of smoking [14,15,22].

Cigarette type and SCCHN risk have not been studied extensively. Black tobacco is more harmful than blond tobacco, as are hand-rolled cigarettes compared to manufactured cigarettes [31]. While there is no clear difference between filtered and nonfiltered cigarettes [14,15] one study indicated a lower risk for "low tar" cigarettes compared to "high tar" [18]. Where studies are available, there are no differences in risk for similar levels of smoking in Caucasian Americans compared to African Americans [14,32], although one study suggested that African Americans were at a lower risk, but there was no breakdown by smoking and drinking categories.

Smoking cessation changes the risk of oropharyngeal cancers. In one study, cancer of the larynx was found to be markedly less likely among ex-smokers than among current cigarette smokers [33]. In a relatively large case–control study in Brazil [34], 784 cases of mouth, pharynx, and larynx cancers were compared to 1578 noncancer controls: compared to never smokers, former smokers of >20 years had an OR = 2.0 (95% CI 1.0–3.8) for all types combined; lower risks for mouth (OR = 1.6) and pharyngeal cancer (OR = 1.5) and a higher risk for laryngeal cancer (OR = 3.6). The benefits of quitting were greatest for cigarettes and lesser for cigars and pipes.

A. Smokeless Tobacco Products

Smokeless tobacco products include chewing tobacco, dry snuff (used in the nasal cavity), wet snuff (a moist wad of tobacco, usually placed between the lips and gums), and nass (a mixture of tobacco, lime, ash, and cotton oil). The use of smokeless tobacco varies greatly around the world, e.g., there are particularly high rates in Scandinavia (where a popular form of snuff is known as snus), India, southeastern Asia, Sudan, and parts of the United States. Smokeless tobacco products from these different regions are produced differently and have different levels of carcinogens [35,36]. In the United States, there has been a sharp increase in the use of these products since the 1980s, particularly among athletes and high school and college students [37]. Data from the 1986 to 1987 National Survey of Oral Health in U.S. school children examined relationships between smokeless tobacco and alcohol and the presence of oral soft tissue lesions [38]. In the study sample of over 17,000 children between the ages of 12 and 17, 1.5% had smokeless tobacco lesions of the mouth. Factors associated with these lesions included male gender, white race, current snuff use, and current chewing tobacco use, with snuff having the highest risk (OR = 18.4).

A "snuff pouch keratosis" in the mouth is common among users of snuff, with a prevalence of 1.6 per 1000 adults in a population-based study of 23,616 white American adults over the age of 35 [39]. Early lesions can be commonly found in adolescent snuff users [38]. There are other reports linking smokeless tobacco products with precancerous lesions of the oral cavity. In India, chewing tobacco is associated with erythroplakia [40], and oral dysplasia [29]. Stopping the use of smokeless tobacco results in the disappearance of oral leukoplakia [41].

The type of tobacco in smokeless tobacco products carries different risks for SCCHN [42–47], although not all studies are supportive of this conclusion. In the United States, Winn and co-workers reported a 4.2-fold increased risk (95% CI = 2.6–6.7) in Southern white women who exclusively use snuff [46]. In contrast, an analysis of the relationship between smokeless tobacco and cancer of the oral cavity in the National Mortality Followback Study did not detect an increased risk [48]. Toombak in the Sudan carries a very high risk [49], whereas the use of snus in Sweden generally is not associated with SCCHN [24,49,50]. Snus is not fermented and so has a much lower level of *N*-nitrosamines [51] and has a lower genotoxic potential [52], which might explain the lack of increased risk. A large number of studies in India, including cohort, case–control, and intervention studies, support the association between oral cancer and smokeless tobacco, and these studies are consistent, strong, coherent, and temporally plausible [43,44,49].

B. Molecular and Genetic Changes in Squamous Cell Carcinoma of the Head and Neck

Preneoplastic lesions, which include keratosis, dysplasia, carcinoma *in situ*, and microinvasive cancer, are considered a sequential continuum [53]. Keratosis is the commonest oral lesion, occurring as white (leukoplakia) or red (erythroplakia) patches, and is present in 1 to 10% of adults [54]. Some molecular evidence exists that premalignant lesions are the direct precursors of invasive lesions [55]. Cessation of smoking does not remove the potential for progression of the disease and all patients must be followed indefinitely [53]. Excellent reviews of the molecular changes present in oropharyngeal cancer have been published [54,56]. Many of the molecular changes present in smoking-related upper aerodigestive tract tumors, including lung and oropharyngeal cancers, are similar and commence during multistage pathogenesis [54,56]. The changes include frequent losses at chromosome arms 3p, 9p, 17p, 5q, and 8p, aneuploidy, *p53* gene mutations, and expression abnormalities of the TGF-α-signaling pathway, activation of, telomerase, downregulation of RAR-β, and inactivation of the *p16* gene [57–60]. Deregulation of the cell cycle is related to the degree of tobacco exposure [61,62].

III. MOLECULAR EPIDEMIOLOGY OF HEAD AND NECK CANCER

The study of genetic susceptibilities can improve the accuracy of estimates for association from carcinogen exposure [63]. Tobacco toxicants affect people to a variable degree. There is large interindividual variation in cellular responses, e.g., in metabolism and detoxification of toxicants, and DNA repair. As other cellular responses to DNA damage are identified (e.g., cell cycle delays, heat shock), interindividual variation in risk is likely to be discovered for these as well. Interindividual effects in cellular responses could be due to genetically determined differences in enzyme expression, kinetics, or stability. Also, induction of enzymes from previous exposures or comorbidity may also contribute to cancer risk, and induction has a genetic component.

Disease risk from genetic variation can range from small to large, depending on the genetic penetrance. Highly penetrant cancer susceptibility genes cause familial cancers, but account for less than 1% of all cancers [64]. Low penetrant genes cause common sporadic cancers and have large public health consequences [65] because they have high prevalences.

Genetic susceptibility can be assessed either phenotypically (measuring the resultant enzymatic function) or genotypically (determining the genetic code). Phenotypic assays may include determining enzymatic activity by administering probe drugs to individuals and measuring blood levels or urinary metabolites, assessing carcinogen metabolic capacity in cultured lymphocytes, or establishing the ratios of endogenously produced substances, such as estrogen metabolite ratios. Some extensively studied phenotypes in relation to smoking risk include aryl hydrocarbon hydroxylase activity [66]. In general, it is preferable to use a genetic-based assay to assess cancer risk because DNA is easier to obtain and the assays are technically simpler. However, phenotypes represent a multigenic trait and may not be characterized adequately by only one genetic assay. Therefore, there is a role for both genetic- and phenotype-based assays in research studies.

How and why people smoke cigarettes affect their level of exposure and consequently their cancer risk. The greatest contributor to smoking addiction is the availability of tobacco and our cultural acceptance of tobacco smoking. Genetics plays a lesser role. The tobacco smoking epidemic has only occurred since the mid-1920s, and it is unlikely that our genetics have evolved in that amount of time. Nonetheless, twin studies indicate a substantial genetic role for both smoking initiation and smoking persistence [67–69]. People smoke in ways that will maintain a desired blood nicotine level. Nicotine in turn stimulates reward mechanisms in the brain. Presynaptic nicotinic acetylcholine receptors stimulate the secretion of dopamine into neuronal synapses. There are also effects on other pathways, such as those that involve serotonin.

For dopamine, synaptic dopamine stimulates dopamine receptors, where five subtypes have been identified, which are considered to be D1- or D2-like. Synaptic dopamine levels are governed by the presynaptic dopamine transporter protein. In humans, different types of data support the link between nicotine and dopamine. The genes that code for dopamine receptors (e.g., *DRD2*, *DRD4*), dopamine transporter reuptake (*SL6A3*), and dopamine synthesis (e.g., dopamine hydroxylase, tyrosine hydroxylase, tryptophan hydroxylase, catechol-*O*-methyltransferase, monoamine oxidase) are polymorphic. Some of the polymorphisms result in altered protein function. Persons with higher levels of synaptic dopamine or "more stimulation" of dopamine receptors may have less rewarding effects from nicotine and so would be less likely to become smokers and would quit more easily. For example, in a study of 500 smokers and nonsmokers, several candidate genes have been implicated [70], whereas our other studies of candidate genes have yielded null results [71]. Other investigators have also reported supporting evidence [72–74].

IV. XENOBIOTIC METABOLISM OF CARCINOGENS IN HEAD AND NECK CANCER

Oropharyngeal epithelium can metabolically activate tobacco smoke carcinogens, which cause DNA damage [75,76]. The highest levels of *CYP1A1* expression have been reported in these tissues compared to others [76]. *NAT1*, but not *NAT2*, activity is present, and there is some evidence that *CYP2C* plays an important role in these tissues. Aromatic DNA and 4-ABP adducts have been detected in laryngeal tissues, which were higher in smokers compared to nonsmokers [77,78].

N-Nitroso compounds in smokeless tobacco have been demonstrated to cause cancers of the mouth and lip, nasal cavity, esophagus, stomach, and lungs in laboratory animals. Urinary metabolites of tobacco specific nitrosomes (TSNs) have been measured in persons using smokeless tobacco products, and higher levels were associated with oral leukoplakia, indicating higher use of the products [50]. Hemoglobin adducts to these carcinogens are measurable in the blood of smokeless tobacco users [79] and thus may be useful biomarkers for measuring exposure levels among users. Hemoglobin adduct levels have been found to be higher in snuff users compared to nonusers [79]. Other exposures that occur with the use of smokeless tobacco products include compounds that cause oxidative DNA damage [80].

Several studies have indicated that there is increased risk for oropharyngeal cancers, which have a heritable trait demonstrated by genetic polymorphisms, although which markers play the greatest role is not yet known [81–85] and there is some evidence for a greater effect in persons with lower levels of smoking [86]. In one study, heritable traits in carcinogen metabolism increased the frequency of *p53* mutations [87].

The *p53* gene is commonly mutated in cancers associated with smokeless tobacco products, and while some differences in spectra have been reported for different regions in the world, no hot spots or patterns have been shown consistently compared to oral cavity cancers related to smoking. Mutational spectra of *p53* in oropharyngeal cancers are similar to those in the lung [88], although some disagree [89]. Mutations occur more commonly in smokers than in nonsmokers [57,88,90,91]. In a study by Brennan and co-workers [57] the frequency of *p53* mutations for tobacco and alcohol users was higher than for either of these exposures alone. In one study, heritable traits in carcinogen metabolism increased the frequency of *p53* mutations [87].

V. DNA REPAIR PHENOTYPE AND RISK OF HEAD AND NECK CANCER

In the process of evolution, species of all living organisms have developed sophisticated DNA repair pathways and mechanisms (Table 15.1) to battle genomic insults from environmental hazards in order to survive and maintain genomic integrity. The DNA repair capacity appears to meet the challenge from the natural environment; e.g., human skin repair capacity can just meet the repair demand from sunlight exposure at midday [92]. Overloaded DNA damage leads to either cell death or mutant cancerous cells that have escaped from repair systems. It has been reported that more than 150 human genes are involved in various repair pathways and that the number is likely to increase when the human genome project refines its published draft of the human genome [93]. These repair genes are grossly categorized into four important and well-characterized repair pathways: base excision repair (BER), nucleotide excision repair (NER), mismatch repair (MMR), and homologous recombinational repair (RCR).

Assays that measure cellular DNA repair are now being applied in population studies to investigate the association between DNA repair and susceptibility to cancer. Generally, cellular responses to DNA damage fall into three major categories: direct reversal of damage, e.g., enzymatic photoreactivation; excision of damage by BER or NER; and postreplication repair, namely MMR and RCR [3]. While the presence of only one unrepaired DNA lesion can block the transcription of an essential gene [94,95], there is a wide range of repair ability in the general population [96,97], with xeroderma pigmentation (XP) patients representing the

TABLE 15.1 Some Important Genes in Major Human DNA Repair Pathways[a]

Type	Genes involved	Damage involved
Base excision repair (BER)	DNA ligase (LIG3) DNA glycosylase (*MBD4, MPG, MYH, NTH1, OGG1, SMUG1, TDG, UNG*), *APE1, APE2, XRCC1* *ADPRT, ADPRTL2, ADPRTL3*	Single base damage repair
Nucleotide excision repair (NER)	*XPA, XPC XPF/ERCC4, XPE, XPG/ERCC5, ERCC1* *CSB/ERCC6, CSA/CKN1, XAB2* *TFIIH (XPB/ERCC3 XPD/ERCC2, GTF2H1, GTF2H2\1, GTF2H3, GTF2H4, CDK7, CCNH, MNAT), DDB1, DDB2, MMS19, CENN2* *RAD23A, RAD23B,* *LIG1 RPA1, RPA2, RPA3*	Bulky nucleotide damage, including ultraviolet photoproducts and chemical carcinogen-induced adducts
Mismatch repair (MMR)	*MSH2, MHS3, MSH6, MSH4, MSH5, MLH1, MLH3, PMS1, PMS2, PMS2L3, PMS2L4*	Base mismatch
Recombinational repair (RCR)	*RAD50, RAD51, RAD51B, RAD51C, RAD51D, RAD54L, RAD54B RAD52, DMC1, MRE11A, NBS1, ERCC1, XPF/ERCC4, XRCC2, XRCC3, XRCC4, XRCC5, XRCC6* *XRCC7, XRCC8, BRCA1, BRCA2*	Double strand breaks V(D)J recombination

[a]See review by Wood *et al.* [93].

extreme end of the repair spectrum [98]. Because there is a shortage of target tissues for laboratory experiments, peripheral blood lymphocytes have been used extensively as surrogate tissue [96,99].

A. Host–Cell Reactivation Assay for NER

While there are many assays that measure the efficiency of multiple steps of excision repair individually, the ability to test the whole pathway is needed for population studies, in which time, cost, and repeatability of measurements are major concerns. Therefore, the host cell reactivation (HCR) assay, which measures the expression level of a damaged reporter gene as a marker of repair proficiency in the host cell, is the assay of choice [100,101]. The HCR assay uses undamaged cells, is relatively fast, and is an objective way of measuring repair [100]. In the assay, a damaged nonreplicating recombinant plasmid (pCMV*cat*) harboring a chloramphenicol acetyltransferase reporter gene is introduced by transfection into primary lymphocytes. Reactivated chloramphenicol acetyltransferase enzyme activity is measured as a function of nucleotide excision repair of the damaged bacterial gene [100]. Both lymphocytes [96] and skin fibroblasts [102] from patients who have basal cell carcinoma but not XP have lower excision-repair rates of a UV-damaged reporter gene than individuals without cancer. Their finding suggests that the repair capacity of lymphocytes can be considered a reflection of an individual's overall repair capacity.

To investigate whether differences in DRC for repairing tobacco carcinogen-induced DNA damage are associated with differential susceptibility to tobacco-related cancer [97,103] the HCR assay was used with benzo[a]pyrene diol epoxid (BPDE)-damaged plasmids in both an initial pilot study (51 patients and 56 frequency-matched controls) and a subsequent large hospital-based case–control study of lung cancer (316 each lung cancer patients and controls). A statistically significantly lower DRC was observed in cases compared with controls and was associated with a greater than a twofold increased risk of lung cancer [97]. Compared to the highest DRC quartile in controls, suboptimal DRC was associated with adjusted risks for lung cancer in a dose–response fashion ($P_{trend} < 0.001$). Cases who were younger at diagnosis (<60 years), female, lighter smokers, or who reported a family history of cancer exhibited the lowest DRC and the highest lung cancer risk among their subgroups, suggesting that these subgroups may be especially susceptible to lung cancer [96]. A low DRC found in women in this study is consistent with epidemiological findings that women are at a higher risk of tobacco-induced cancer as compared to men [104–106]. Using the same assay, Cheng and colleagues [107] investigated the role of DRC in head and neck cancer, and again DRC was significantly lower in cases (8.6%) than in controls (12.4%) with a similar dose–response trend; i.e., those in the middle and lowest tertiles of DRC had a more than a two- and fourfold increased risk, respectively. These results suggest that suboptimal DRC may contribute to the susceptibility to tobacco carcinogenesis.

B. Mutagen Sensitivity Assay

This is another functional assay that measures quantitatively chromatid breaks in response to *in vitro* exposure to carcinogens in short-term cultures of peripheral blood lymphocytes. Several case–control [108–110] and cohort studies [111,112] have suggested that induced and spontaneous lymphocytic chromosome aberrations can be used as markers of susceptibility. The implications of chromosomal aberrations and genomic instability in carcinogenesis of the head and neck have been reviewed comprehensively elsewhere [113,114].

In the general population, the frequency of spontaneous chromosome aberrations is low [115], and classic cytogenetic assays assessing these types of aberrations may not be applicable to epidemiological studies requiring a large number of samples. Therefore, Hsu and co-workers [116,117] developed an assay for mutagen sensitivity to measure genetic susceptibility by estimating the frequency of *in vitro* bleomycin-induced breaks in short-term lymphocyte cultures. Bleomycin is considered radiomimetic (i.e., it causes the generation of free oxygen radicals), which is relevant to tobacco-induced carcinogenesis because numerous compounds in tobacco condensate may generate free oxygen radicals that can induce single- and double-stranded breaks. Mutagen sensitivity has consistently been shown in case–control studies to be a significant independent predictor of the risk for lung and upper aerodigestive tract cancers [109,110,116,118,119]. Lighter smokers and former smokers appear to be at higher risk than heavier smokers, as are younger patients. In upper aerodigestive tract cancer, bleomycin sensitivity was highest in subjects under age 30 [120]. These results suggest that this assay may serve as a biomarker for susceptibility to tobacco-related cancer.

The bleomycin assay has since been modified by using BPDE as the test mutagen [121], and BPDE sensitivity has also been associated with a significantly elevated risk for head and neck cancer [122] and lung cancer [123]. In a pilot case–control study of 60 SCCHN patients and 112 healthy controls, the high frequency of BPDE-induced chromatid breaks was associated with more than a twofold increased risk of head and neck cancer, and there was a dose–response relationship between the frequency of BPDE-induced chromatid breaks and risk, suggesting that the BPDE-induced breaks/cell (b/c) value is a significant risk factor for head and neck cancer [122]. In lung cancer, risks were higher for lighter smokers and younger patients. There was also a dose–response relationship between the quartiles of numbers of BPDE-induced breaks and lung cancer risk, and risk was the highest in subjects who were sensitive to both BPDE and bleomycin [123]. Findings that cancer is more likely to develop at younger ages in people who have the sensitive phenotype support the hypothesis that BPDE-induced chromatid breaks are a marker for genetic susceptibility to tobacco-induced carcinogenesis.

It has been suggested that the mutagen sensitivity assay indirectly measures the effectiveness of one or more DNA repair mechanisms [124]. A correlation between the cellular DNA repair capacity measured by the host cell reactivation assay and the frequency of mutagen-induced *in vitro* chromatid breaks has been reported [11a,121]. Mutagen sensitivity may also involve an inherent chromatin alteration that permits more efficient translation of DNA damage into chromosome damage after exposure to a mutagen [126]. Although the mechanism underlying the association between induced chromosomal aberrations and susceptibility to cancer remains to be unraveled, tobacco smoke causes both oxidative damage and bulky adducts. Defects in both BER and NER mechanisms may therefore increase the risk of smoking-related cancer dramatically.

C. ^{32}P-Postlabeling Assay of DNA Adducts

A relatively large variation is observed in the level of persistent DNA adducts *in vivo* that are believed to be related to smoking [127,128]. Although this variation could be partly due to experimental methodology, it may reflect true biological variation that is a valid phenotypic marker for the joint effect of host metabolic activities and DNA repair in response to carcinogen exposure [129]. Using the ^{32}P-postlabeling assay developed by Reddy *et al.* [130], Phillips *et al.* [131] noted a linear relationship between the levels of aromatic DNA adducts in human lung and the number of cigarettes smoked per day. While some studies have failed to find a correlation between lymphocyte adduct levels and smoking habits [132,133], one study reported a significant difference between levels in smokers and nonsmokers [99].

A large *in vivo* variation of adduct levels may be driven by variation in the activities of enzymes involved in carcinogen bioactivation [134,135], such as CYP1A1 [136], which can be induced by smoking in the target tissues [137]. To tackle this problem, Li *et al.* [129] developed a new assay of *in vitro* induction of carcinogen–DNA adducts by an ultimate carcinogen. In this assay, stimulated lymphocytes were treated with BPDE [the ultimate carcinogen of B(a)P, which does not need bioactivation]. Therefore, the variation in levels of BPDE-induced DNA adducts should reflect only genetic variation in phase II enzymes and DRC. However, phase II enzymes have little effect, if any, on the *in vitro* formation of adducts in this assay because of the relatively large dose (4 μM) of BPDE used and the rapid binding of BPDE to DNA, which peaks within 15 min [138]. The ultimate carcinogen generates *in vitro* adduct levels that are a hundredfold higher than *in vivo* adduct levels; furthermore, the variation in such induced adduct levels is within a hundredfold rather than a thousandfold as often seen *in vivo*.

In a pilot study of 91 patients with squamous cell carcinomas of the head and neck and 115 controls, Li *et al.* [139] found that levels of BPDE–DNA adducts were significantly higher in cases than in controls (mean ± SD, 76.8 ± 77.4 and 47.1 ± 48.0/10^7 nucleotides, respectively; $P < 0.001$). Using the median level of control values (35/10^7) as the cutoff point, about 66% of cases were distributed above this level. High levels of BPDE-induced DNA adducts were associated with more than a twofold increased risk. There was a statistically significant dose–response relationship between the quartile levels of BPDE-induced DNA adducts and the risk of head and neck cancer (trend test, $P = 0.003$), suggesting that this biomarker may have the potential to complement with other biomarkers in identifying individuals at an increased risk of developing tobacco-related cancers. Indeed, similar findings were also observed in lung cancer studies [129,140].

D. Assays for DNA Repair Gene Transcript (mRNA) Levels

While the DRC phenotype can be affected by genetic polymorphisms of the genes that participate in the repair pathway, epigenetic factors could also influence the repair outcome. For instance, the expression levels of repair genes may be affected epigenetically. To investigate the variation in expression levels, a multiplex reverse transcriptase polymerase chain reaction (RT-PCR) assay has been used to measure the levels of several DNA repair gene transcripts relative to those of a ubiquitous housekeeping gene [141]. In this technique, transcripts from several repair genes and the β-actin gene are amplified simultaneously, and the transcript levels are quantified in relation to the β-actin level by a computerized densitometry analysis of gel electrophoresis of the multiplex RT-PCR products. This assay is also flexible in grouping into one experiment the genes involved in the same repair pathway such as MMR [142] or NER [143].

Using this multiplex RT-PCR assay, Wei *et al.* [144] simultaneously evaluated the relative expression levels of five MMR genes (*hMSH2*, *hMLH1*, *hPMS1*, *hPMS2*, and h*GTBP/hMSH6*) in the peripheral lymphocytes of 78 patients with head and neck cancer and 86 controls. The relative MMR gene expression was not correlated with disease stage or tumor site in the cases or with smoking and alcohol use in the controls, but increased with age in both cases and controls. The mean expression levels of *hMLH1, hPMS1*, and *hGTBP/hMSH6* were significantly lower in cases than in controls. A low expression of *hMLH1* was associated with more than a fourfold increased risk, and a low expression of *hGTBP/hMSH6* was associated with more than a twofold increased risk. Cheng *et al.* [145] used this assay to measure the relative expression levels of five NER genes [*ERCC1, XPB/ERCC3, XPG/ERCC5, CSB/ERCC6*, and *XPC* (ERCC, excision repair cross-complementing; CSB, Cockayne's syndrome complementary group B)] in phytohemagglutinin-stimulated peripheral lymphocytes obtained from 75 lung cancer patients and 95 controls. They found a 12.2 and 12.5% decrease in the baseline expression levels of *XPG/ERCC5* and *CSB/ERCC6*, respectively, in cases compared with controls ($P < 0.01$). When the median expression level in the controls was used as the cutoff point, lung cancer patients were significantly more likely than controls to have reduced expression levels of XPG/ERCC5 and CSB/ERCC6, which were associated with more than a twofold increased risk of lung cancer. There was also a dose–response relationship between reduced expression levels and increased lung cancer risk (trend test: $P < 0.01$). These results suggest that individuals with low expression levels of DNA repair genes may be at a higher risk of tobacco-related cancer.

VI. POLYMORPHISMS IN DNA REPAIR GENES

Genetic polymorphisms of DNA repair genes may also contribute to variation in DNA repair capacity. Clearly, functional (phenotypic) studies of DNA repair in individuals with various genotypes of DNA repair genes are needed. However, it will be difficult to detect subtle differences in DNA repair capacity due to a single polymorphism of a single gene in a very complex pathway. The entire coding regions of several DNA repair genes on chromosome 19, i.e., three NER genes (*ERCC1, XPD/ERCC2*, and *XPF/ERCC4*), one RCR gene (*XRCC3*), and one BER gene (*XRCC1*), have been resequenced in 12 normal individuals [146]. Of these, 7 variants of *ERCC1*, 17 of *XPD/ERCC2*, 6 of *XPF/ERCC4*, 4 of *XRCC3*, and 12 of *XRCC1* were identified. Among these variants, 4 variants of *XPD/ERCC2*, 3 variants of *XRCC1*, 1 variant of *XRCC3*, and 1 variant of *XPF/ERCC4* result in an amino acid sequence change. Later, another 6 variants of *XPF/ERCC4* were identified in 38 individuals [147]; 2 variants of *XPA* (chromosome 9) and 2 *XPB/ERCC3* (chromosome 2) were identified in 35 individuals; and 2 variants of *XPC* (chromosome 3) [148] and 3 variants of *XPG/ERCC5* (chromosome 13) [149] were identified. Although the significance of these variants is largely unknown, the implication is that variants that cause amino acid substitutions may have an impact on the function of the proteins and therefore on the efficiency of DNA repair. Those variants that do not cause an amino acid change may also have an impact on DNA repair function because they may lie in introns that regulate splicing, cause mRNA instability, or may be linked to genetic changes in other genes. Therefore, the impact of these polymorphisms on disease outcomes is important to ultimately understanding their functional relevance.

The XPD protein is an evolutionarily conserved helicase, a subunit of transcription factor IIH (TFIIH) that is essential for transcription and NER [150]. Mutations in *XPD* prevent its protein from interacting with *p44*, another subunit of TFIIH [151], and cause decreased helicase activity, resulting in a defect in NER. Mutations at different sites result in distinct clinical phenotypes [152]. *XPD* is also thought to be involved in repairing genetic damage induced by tobacco carcinogens [125].

Several *XPD* polymorphisms have been identified in the coding regions with a relative high frequency [146,153]. These common (allele frequencies >0.20) polymorphisms include C22541A (156Arg) of exon 6 and C35326T (711Asp) without amino acid changes and G23592A (Asp312Asn) of exon 10 and A35931C (Lys751Gln) of exon 23 with amino acid changes. The Lys751Gln polymorphism is located about 50 bases upstream from the poly(A) signal and therefore may alter XPD protein function. In a study of 31 women, those with the 751Gln/Gln genotype were found to have higher levels of chromatid aberrations induced by X-ray [154]. However, this finding was not confirmed in another study that measured the frequency of smoking-induced sister chromatid exchanges and polyphenol DNA adducts ($n = 61$) [155].

In a case–control study of 189 head and neck patients and 496 cancer-free controls, Sturgis *et al.* [125] found that the frequency of the 22541 AA homozygous genotype was lower in cases (15.9%) than in controls (20.4%), but this difference was not statistically different. The frequency of the 751Gln/Gln homozygous genotype was higher in cases (16.4%) than in controls (11.5%) and was associated with a borderline increased risk (OR = 1.55). The risk was higher in older subjects (OR = 2.22), current smokers (OR = 1.83), and current drinkers (OR = 2.59). Furthermore, the 751Gln variant was found to be associated with more than a twofold risk of melanoma [156], and the 751 Lys→Gln substitution may affect protein function [153]. The Asp312Asn variant was found to be associated with nearly a twofold lung cancer risk in two independent studies [157,158]. The *XPD* C22541A and C35326T polymorphisms are silent, resulting in no amino acid substitutions [146], and were not found to be associated with an increased risk of cancer [11,159]. However, it is possible that such a sequence variation could affect RNA stability or otherwise disturb protein synthesis [153].

Several polymorphisms of *XRCC1* have also been identified [146]. These include polymorphisms resulting in a nonconservative amino acid substitution at C26304T of codon 194 (Arg194Trp) in exon 6; G27466A of codon 280 (Arg280His) in exon 9, and G28152A of codon 399 (Arg399Gln) in exon 10. Although the functional relevance of these variants is unknown, codon 399 is within the *XRCC1* BRCT (breast cancer susceptibility protein-1) domain (codons 314–402) [160], which is highly homologous to

BRCA1 (a gene also involved in DNA repair), containing a binding site for poly(ADP-ribose) polymerase (PARP) [161]. Because the role of *XRCC1* in BER involves bringing together DNA polymerase β (β-pol), DNA ligase III (LIG III), and PARP at the site of DNA damage [162–164], the codon 399 variant could have an impact on repair activity. The codon 194 polymorphism resides in the linker regions of the *XRCC1* N-terminal domain separating the helix 3 and β-pol involved in binding a single nucleotide gap DNA substrate [165]. Lunn *et al.* [166] reported that the codon 399 variant was associated with higher levels of both aflatoxin B1–DNA adducts and glycophorin A variants in a normal population, suggesting that this variant may be an adverse genotype. However, few studies have studied the associations between polymorphisms of the DNA repair gene *XRCC1* and the risk of cancer.

In a case–control study, Sturgis *et al.* [125] reported that 88.7% of 203 head and neck cancer cases and 85.6% of 424 controls lacked the codon 194 Trp variant, with a significant risk for oral cavity and pharynx cancers of 2.46. Thirty-two cases (15.8%) and 46 controls (10.8%) were homozygous for the codon 399 Gln variant (adjusted OR = 1.59) for all cases. Furthermore, when the two genotypes were combined, the adjusted OR was 1.51 for either risk genotype and 2.02 for both risk genotypes. In addition, the codon 399 Arg/Gln or Gln/Gln genotypes were associated with a risk of breast cancer in African Americans and gastric cancer in a Chinese population [166a,166b]; the codon 280 Arg/His and His/His genotypes were associated with a risk of lung cancer in a Chinese population [167]; and both the 194Trp and the 399Gln variant alleles were associated with an increased risk of colon cancer in an Egyptian population [168]. However, some of these polymorphisms were not found to be associated with an increased risk of cancer in other studies [158,159,169]. Despite some conflicting reports, the variants of *XRCC1* and their impact on cancer risk have generated much interest in the scientific community [170].

The *hOGG1* gene is localized on chromosome 3p25 and encodes two forms of protein that result from an alternative splicing of a single messenger RNA [171]. The α-hOGG1 protein has a nuclear localization, whereas the β-hOGG1 is targeted to the mitochondrion. The α-hOGG1 protein is a DNA glycosylase/AP lyase that excises 8-OH-G and Fapy-G from γ-irradiated DNA. There are both somatic and polymorphic mutations of the *hOGG1* gene in lung and kidney tumors [172,173]. The mutant forms α-hOGG1-Gln46 and α-hOGG1-His154 are defective in their catalytic capacities, especially for 8-OH-Gua [174]. A polymorphism at codon 326 (Ser326Cys) leads to hOGG1-Ser326 and hOGG1-Cys326 proteins [172]. Activity in the repair of 8-hydroxyguanine appears to be greater with the Ser326 protein than with the Cys326 protein. Because tobacco carcinogens produce 8-hydroxyguanine residues, the capacity to repair these lesions can be involved in cancer susceptibility.

Sugimura *et al.* [175,175a] found that those with the hOOG1 Cys/Cys of codon 326 were at an increased risk of squamous cell and nonadenocarcinoma lung cancer compared to those with the Ser/Cys or those with the Ser/Ser genotypes combined (ORs of 3.0 and 2.2, respectively). The distributions of this polymorphism varied for different populations (Chinese, Japanese, Micronesians, Melanesians, Hungarians, and Australian Caucasians), with a much lower prevalence of the Cys allele in the latter three populations. Another population-based study of 128 lung cancer cases and 268 controls identified a polymorphic allele 3 in hMMH/OGG1 exon 1 and found it to be significantly related to risk of adenocarcinoma of the lung (OR = 3.2) among Japanese [176]. This polymorphism was also identified in European patients with head and neck or kidney cancer not associated with increased risk [177]. Polymorphisms involving intron 4 and exon 7 were present in 30% of 33 patients. Loss of heterozygosity (LOH) at the *hOGG1* gene locus was found to be a very common occurrence in lung tumorigenesis [178]. However, the *hOGG1* polymorphisms studied were not found to be major contributors to individual lung cancer susceptibility in Caucasians, although it is reported to be associated with esophageal cancer in a Chinese population [179]. Correlations between DNA repair genotype and phenotype are needed, and confirmatory large, well-designed case–control or cohort studies will be required to confirm the impact of these genetic variants and cancer risk.

VII. POLYMORPHISMS IN CELL CYCLE GENES

In response to DNA damage, the cell cycle becomes arrested under the control of many cell cycle-related genes, one of which is cyclin D1 (*CCND1*), which plays a critical role in cell cycle control. Studies show that cyclin D1 overexpression and loss of p16INK4A expression predict early relapse and reduced survival in squamous cell carcinoma of the anterior tongue [180]. The G→A polymorphism (G870A) of exon 4 of *CCND1* creates an alternate splicing site of its mRNA, encoding a protein with an altered carboxy-terminal domain. It has been suggested that DNA damage in cells with the A allele bypasses the G1/S checkpoint of the cell cycle more easily than damage in cells without the A allele. In a hospital-based case–control study of 233 newly diagnosed SCCHN patients and 248 noncancer controls, Zheng *et al.* [181] found that compared with the wild-type *CCND1* GG, the *CCND1* AA genotype was associated with an almost twofold increased risk for SCCHN. Among the cases, the mean age of onset was 59.0, 56.8, and 55.5 years for the GG, GA, and AA genotypes, respectively. In the stratification analysis, the *CCND1* AA variant genotype was associated with a more than a threefold increased risk in individuals who were ≤45 years old, females, nonsmokers, and nonalcohol users. These results suggest that the *CCND1* polymorphism is associated with early onset of SCCHN and contributes to susceptibility to SCCHN. However, polymorphisms of p16 have no effect on the risk of SCCHN [182] because p16 abnormalities are likely mediated by other mechanisms such as methylation status [183] in the etiology of SCCHN.

In summary, DNA repair is critical in maintaining genomic integrity and genetic stability. Ample evidence shows that etiologic or susceptibility factors for SCCHN include suboptimal DNA repair, resulting in higher levels of tobacco-induced DNA adducts, higher frequencies of chromosomal aberrations, and likely high frequencies of mutations in oncogenes or tumor suppresser genes. The susceptible phenotype of suboptimal DNA repair will be evaluated further by genotyping the genes involved in the specific repair pathway. The functional relevance of genetic polymorphisms offers biological plausibility in using genotypes to predict DRC, and are amenable to high-throughput analysis, thereby adding in molecular epidemiological studies. Some genetic variants in cell cycle genes may also be relevant in genetic susceptibility to SCCHN. Ultimately, the goal of such work is to reliably estimate an individual's risk of developing SCCHN based on their genetic profile. Such a risk assessment would allow for improved primary prevention, such as individualized tobacco and alcohol cessation consulting/treatment, early detection through a more aggressive screening practice, chemoprevention through diet modification, and pharmaceutical and secondary prevention for detecting and preventing secondary primary malignancies.

Acknowledgments

The authors thank Mrs. Joanne Sider for assistance in preparing the manuscript. This investigation was supported in part by National Institute of Health Grants CA 55769 (to M.R.S.) and ES11740 (to Q.W.) and by National Institute of Environmental Health Sciences Grant ES07784.

References

1. Greenlee, R. T., Murray, T., Bolden, S., and Wingo, P. A. (2000). Cancer statistics, 2000. *CA Cancer J. Clin.* **50**, 7–33.
2. Hartwell, L. H., and Weinert, T. A. (1989). Checkpoints: Controls that assure the order of cell cycle events. *Science* **246**, 629–633.
3. Sancar, A. (1995). DNA repair in humans. *Annu. Rev. Genet.* **29**, 69–105.
4. Shields, P. G., and Harris, C. C. (2000). Cancer risk and low-penetrance susceptibility genes in gene–environment interactions. *J. Clin. Oncol.* **18**, 2309–2315.
5. Perera, F. P. (2000). Molecular epidemiology: On the path to prevention? *J. Natl. Cancer Inst.* **92**, 602–612.
6. Christiani, D. C. (2000). Smoking and the molecular epidemiology of lung cancer. *Clin. Chest Med.* **21**, 87–93.
7. Rebbeck, T. R. (1997). Molecular epidemiology of the human glutathione S-transferase genotypes GSTM1 and GSTT1 in cancer susceptibility. *Cancer Epidemiol. Biomark. Prev.* **6**, 733–743.
8. Autrup, H. (2000). Genetic polymorphisms in human xenobiotic metabolizing enzymes as susceptibility factors in toxic response. *Mutat. Res.* **464**, 65–76.
9. Bartsch, H., Nair, U., Risch, A., Rojas, M., Wikman, H., and Alexandrov, K. (2000). Genetic polymorphism of CYP genes, alone or in combination, as a risk modifier of tobacco-related cancers. *Cancer Epidemiol. Biomark. Prev.* **9**, 3–28.
10. Nair, U., and Bartsch, H. (2001). Metabolic polymorphisms as susceptibility markers for lung and oral cavity cancer. *IARC Sci. Publ.* **154**, 271–290.
11. Sturgis, E. M., Castillo, E. J., Li, L., *et al.* (1999). Polymorphisms of DNA repair gene XRCC1 in squamous cell carcinoma of the head and neck. *Carcinogenesis* **20**, 2125–2129.

11a. Sturgis, E. M., Clayman, G. L., Guan, Y., Guo, Z., and Wei, Q. (1999). DNA repair in lymphoblastoid cell lines from patients with head and neck cancer. *Arch. Otolaryngol. Head Neck Surg.* **125**, 185–190.

12. Sturgis, E. M., Zheng, R., Li, L., Castillo, E. J., Eicher, S. A., Chen, M., Strom, S. S., Spitz, M. R., and Wei, Q. (2000). XPD/ERCC2 polymorphisms and risk of head and neck cancer: A case–control analysis. *Carcinogenesis* **21**, 2219–2223.
13. Berwick, M., and Vineis, P. (2000). Markers of DNA repair and susceptibility to cancer in humans: An epidemiologic review. *J. Natl. Cancer Inst.* **92**, 874–897.
14. Blot, W. J., *et al.* (1988). Smoking and drinking in relation to oral and pharyngeal cancer. *Cancer Res.* **48**, 3282–3287.
15. Hayes, R. B., *et al.* (1999). Tobacco and alcohol use and oral cancer in Puerto Rico. *Cancer Causes Control* **10**, 27–33.
16. Iribarren, C., Tekawa, I. S., Sidney, S., and Friedman, G. D. (1999). Effect of cigar smoking on the risk of cardiovascular disease, chronic obstructive pulmonary disease, and cancer in men. *N. Engl. J. Med.* **340**, 1773–1780.
17. Keller, A. Z., and Terris, M. (1965). The association of alcohol and tobacco with cancer of the mouth and pharynx. *Am. J. Public Health Nation. Health* **55**, 1578–1585.
18. La Vecchia, C., *et al.* (1990). Type of cigarettes and cancers of the upper digestive and respiratory tract. *Cancer Causes Control* **1**, 69–74.
19. Lewin, F., *et al.* (1998). Smoking tobacco, oral snuff, and alcohol in the etiology of squamous cell carcinoma of the head and neck: A population-based case-referent study in Sweden. *Cancer* **82**, 1367–1375.
20. Macfarlane, G. J., *et al.* (1995). Alcohol, tobacco, diet and the risk of oral cancer: A pooled analysis of three case-control studies. *Eur. J. Cancer B Oral Oncol.* **31B**, 181–187.
21. Mashberg, A., Boffetta, P., Winkelman, R., and Garfinkel, L. (1993). Tobacco smoking, alcohol drinking, and cancer of the oral cavity and oropharynx among U.S. veterans. *Cancer* **72**, 1369–1375.
22. Muscat, J. E., Richie, J. P., Jr., Thompson, S., and Wynder, E. L. (1996). Gender differences in smoking and risk for oral cancer. *Cancer Res.* **56**, 5192–5197.
23. Sanderson, R. J., de Boer, M. F., Damhuis, R. A., Meeuwis, C. A., and Knegt, P. P. (1997). The influence of alcohol and smoking on the incidence of oral and oropharyngeal cancer in women. *Clin. Otolaryngol.* **22**, 444–448.
24. Schildt, E. B., Eriksson, M., Hardell, L., and Magnuson, A. (1998). Oral snuff, smoking habits and alcohol consumption in relation to oral cancer in a Swedish case-control study. *Int. J. Cancer* **77**, 341–346.
25. Takezaki, T., *et al.* (1996). Tobacco, alcohol and dietary factors associated with the risk of oral cancer among Japanese. *Jpn. J. Cancer Res.* **87**, 555–562.
26. Talamini, R., *et al.* (1998). Cancer of the oral cavity and pharynx in nonsmokers who drink alcohol and in nondrinkers who smoke tobacco. *J. Natl. Cancer Inst.* **90**, 1901–1903.

27. Goldstein, A. M., *et al.* (1994). Familial risk in oral and pharyngeal cancer. *Eur. J. Cancer B Oral Oncol.* **30B**, 319–322.
28. Jaber, M. A., Porter, S. R., Gilthorpe, M. S., Bedi, R., and Scully, C. (1999). Risk factors for oral epithelial dysplasia: The role of smoking and alcohol. *Oral Oncol.* **35**, 151–156.
29. Kulasegaram, R., Downer, M. C., Jullien, J. A., Zakrzewska, J. M., and Speight, P. M. (1995). Case-control study of oral dysplasia and risk habits among patients of a dental hospital. *Eur. J. Cancer B Oral Oncol.* **31B**, 227–231.
30. Elwood, J. M., Pearson, J. C., Skippen, D. H., and Jackson, S.M. (1984). Alcohol, smoking, social and occupational factors in the aetiology of cancer of the oral cavity, pharynx and larynx. *Int. J. Cancer* **34**, 603–612.
31. De Stefani, E., Boffetta, P., Oreggia, F., Mendilaharsu, M., and Deneo-Pellegrini, H. (1998). Smoking patterns and cancer of the oral cavity and pharynx: A case-control study in Uruguay. *Oral Oncol.* **34**, 340–346.
32. Day, G. L., *et al.* (1993). Racial differences in risk of oral and pharyngeal cancer: Alcohol, tobacco, and other determinants. *J. Natl. Cancer Inst.* **85**, 465–473.
33. Public Health Service (1964). Smoking and health. Report of the Advisory Committee to the Surgeon General of the Public Health Service. PHS publication No. 1103.
34. Schlecht, N. F., *et al.* (1999). Interaction between tobacco and alcohol consumption and the risk of cancers of the upper aero-digestive tract in Brazil. *Am. J. Epidemiol.* **150**, 1129–1137.
35. Gupta, P. C., Murti, P. R., and Bhonsle, R. B. (1996). Epidemiology of cancer by tobacco products and the significance of TSNA. *Crit. Rev. Toxicol.* **26**, 183–198.
36. Hoffmann, D., and Djordjevic, M. V. (1997). Chemical composition and carcinogenicity of smokeless tobacco. *Adv. Dent. Res.* **11**, 322–329.
37. Christen, A. G. (1980). Tobacco chewing and snuff dipping. *N. Engl. J. Med.* **302**, 818.
38. Tomar, S. L., Winn, D. M., Swango, P. A., Giovino, G. A., and Kleinman, D. V. (1997). Oral mucosal smokeless tobacco lesions among adolescents in the United States. *J. Dent. Res.* **76**, 1277–1286.
39. Kimura, S., Umeno, M., Skoda, R. C., Meyer, U. A., and Gonzalez, F. J. (1989). The human debrisoquine 4-hydroxylase (CYP2D) locus: Sequence and identification of the polymorphic CYP2D6 gene, a related gene, and a pseudogene. *Am. J. Hum. Genet.* **45**, 889–904.
40. Carter, A. M., Catto, A. J., and Grant, P. J. (1999). Association of the alpha-fibrinogen Thr312Ala polymorphism with poststroke mortality in subjects with atrial fibrillation. *Circulation* **99**, 2423–2426.
41. Campisi, R., Czernin, J., Schoder, H., Sayre, J. W., and Schelbert, H. R. (1999). L-Arginine normalizes coronary vasomotion in long-term smokers. *Circulation* **99**, 491–497.
42. Brinton, L. A., *et al.* (1984). A case-control study of cancers of the nasal cavity and paranasal sinuses. *Am. J. Epidemiol.* **119**, 896–906.
43. Carpenter, C. L., Jarvik, M. E., Morgenstern, H., McCarthy, W. J., and London, S. J. (1999). Mentholated cigarette smoking and lung-cancer risk. *Ann. Epidemiol.* **9**, 114–120.
44. Jacob, P., III, Yu, L., Shulgin, A. T., and Benowitz, N. L. (1999). Minor tobacco alkaloids as biomarkers for tobacco use: Comparison of users of cigarettes, smokeless tobacco, cigars, and pipes. *Am. J. Public Health* **89**, 731–736.
45. Rao, D. N., Ganesh, B., Rao, R. S., and Desai, P. B. (1994). Risk assessment of tobacco, alcohol and diet in oral cancer: A case-control study. *Int. J. Cancer* **58**, 469–473.
46. Winn, D. M., *et al.* (1981). Snuff dipping and oral cancer among women in the southern United States. *N. Engl. J. Med.* **304**, 745–749.
47. Wynder, E. L., Bross, I. J., and Feldman R. M. (1957). A study of the etiological factors in cancer of the mouth. *Cancer* **10**, 1300–1323.
48. Sterling, T. D., Rosenbaum, W. L., and Weinkam, J. J. (1992). Analysis of the relationship between smokeless tobacco and cancer based on data from the National Mortality Followback Survey. *J. clin. Epidemiol.* **45**, 223–231.
49. Idris, A. M., *et al.* (1998). The Swedish snus and the Sudanese toombak: Are they different? *Oral Oncol.* **34**, 558–566.
50. Kresty, L. A., *et al.* (1996). Metabolites of a tobacco-specific nitrosamine, 4-(methylnitrosamino)-1-(3-pyridyl)-1-butanone (NNK), in the urine of smokeless tobacco users: Relationship between urinary biomarkers and oral leukoplakia. *Cancer Epidemiol. Biomark. Prev.* **5**, 521–525.
51. Hautanen, A., *et al.* (1999). Joint effects of an aldosterone synthase (CYP11B2) gene polymorphism and classic risk factors on risk of myocardial infarction. *Circulation* **100**, 2213–2218.
52. Waldum, H. L., *et al.* (1996). Long-term effects of inhaled nicotine. *Life Sci.* **58**, 1339–1346.
53. Gillis, T. M., Incze, J., Strong, M. S., Vaughan, C. W., and Simpson, G. T. (1983). Natural history and management of keratosis, atypia, carcinoma-in situ, and microinvasive cancer of the larynx. *Am. J. Surg.* **146**, 512–516.
54. Mao, L., and El-Naggar, A. (1999). Molecular changes in the multistage pathogenesis of head and neck cancer. *In* "Molecular Pathology of Early Cancer" (S. Srivastava, D. E., Henson, and A. F., Gazdar, eds.). IOS Press, Amsterdam.
55. Califano, J., *et al.* (2000). Genetic progression and clonal relationship of recurrent premalignant head and neck lesions. *Clin. Cancer Res.* **6**, 347–352.
56. Sidransky, D. (1997). Molecular biology of head and neck tumors. *In* "Cancer: Principles and Practice of Oncology" (V. T. DeVita, S. Hellman, Jr., and S. A. Rosenberg, eds.), 5th Ed. Lippincott, Philadelphia.
57. Brennan, J. A., *et al.* (1995). Association between cigarette smoking and mutation of the p53 gene in squamous-cell carcinoma of the head and neck. *N. Engl. J. Med.* **332**, 712–717.
58. Field, J. K., *et al.* (1995). Allelotype of squamous cell carcinoma of the head and neck: Fractional allele loss correlates with survival [published erratum appears in *Br. J. Cancer* **74**, 1153 (1996)]. *Br. J. Cancer* **72**, 1180–1188.
59. Izzo, J. G., *et al.* (1998). Dysregulated cyclin D1 expression early in head and neck tumorigenesis: In vivo evidence for an association with subsequent gene amplification. *Oncogene* **17**, 2313–2322.
62. Gallo, O., Bianchi, S., and Porfirio, B. (1995). Bcl2 overexpression and smoking history in head and neck cancer. *J. Natl. Cancer Inst.* **87**, 1024–1025.
60. Picard, E., *et al.* (1999). Expression of retinoid receptor genes and proteins in non-small-cell lung cancer. *J. Natl. Cancer Inst.* **91**, 1059–1066.
61. Davidson, B. J., Lydiatt, W. M., Abate, M. P., Schantz, S. P., and Chaganti, R. S. (1996). Cyclin D1 abnormalities and tobacco exposure in head and neck squamous cell carcinoma. *Head Neck* **18**, 512–521.
63. Khoury, M. J., and Wagener, D. K. (1995). Epidemiological evaluation of the use of genetics to improve the predictive value of disease risk factors [published erratum appears in *Am. J. Hum. Genet.* **58**(1), 253 (1996)]. *Am. J. Hum. Genet.* **56**, 835–844.
64. Fearon, E. R. (1997). Human cancer syndromes: Clues to the origin and nature of cancer. *Science* **278**, 1043–1050.
65. Shields, P. G., and Harris, C. C. (2000). Cancer risk and low penetrance susceptibility genes in gene-environment interactions. *J. Clin. Oncol.* **18**, 2309–2315.
66. Kellermann, G., Shaw, C. R., and Luyten-Kellerman, M. (1973). Aryl hydrocarbon hydroxylase inducibility and bronchogenic carcinoma. *N. Engl. J. Med.* **289**, 934–937.
67. Carmelli, D., Swan, G. E., Robinette, D., and Fabsitz, R. (1992). Genetic influence on smoking: A study of male twins. *N. Engl. J. Med.* **327**, 829–833.
68. Heath, A. C., *et al.* (1993). Genetic contribution to risk of smoking initiation: Comparisons across birth cohorts and across cultures, *J. Subst. Abuse* **5**, 221–246.

69. Heath, A. C., and Martin, N. G. (1993). Genetic models for the natural history of smoking: Evidence for a genetic influence on smoking persistence. *Addict. Behav.* **18**, 19–34.
70. Shields, P. G., *et al.* (1998). Dopamine D4 receptors and the risk of cigarette smoking in African Americans and Caucasians. *Cancer Epidemiol. Biomark. Prev.* **7**, 453–458.
71. Lerman, C., *et al.* (1997). Lack of association of tyrosine hydroxylase genetic polymorphism with cigarette smoking. *Pharmacogenetics* **7**, 521–524.
72. Comings, D. E., *et al.* (1996). The dopamine D2 receptor (DRD2) gene: A genetic risk factor in smoking. *Pharmacogenetics* **6**, 73–79.
73. Noble, E. P., *et al.* (1994). D2 dopamine receptor gene and cigarette smoking: A reward gene? *Med. Hypotheses* **42**, 257–260.
74. Spitz, M. R., *et al.* (1998). A case-control study of the dopamine D2 receptor gene and smoking status in lung cancer patients. *J. Natl. Cancer Inst.* **90**, 358–363.
75. Badawi, A. F., Stern, S. J., Lang, N. P., and Kadlubar, F. F. (1996). Cytochrome P-450 and acetyltransferase expression as biomarkers of carcinogen-DNA adduct levels and human cancer susceptibility. *Prog. Clin. Biol. Res.* **395**, 109–140.
76. Kabat, G. C., Morabia, A., and Wynder, E. L. (1991). Comparison of smoking habits of blacks and whites in a case-control study. *Am. J. Public Health* **81**, 1483–1486.
77. Flamini, G., *et al.* (1998). 4-Aminobiphenyl-DNA adducts in laryngeal tissue and smoking habits: An immunohistochemical study. *Carcinogenesis* **19**, 353–357.
78. Szyfter, K., *et al.* (1994). Aromatic DNA adducts in larynx biopsies and leukocytes. *Carcinogenesis* **15**, 2195–2199.
79. Carmella, S. G., *et al.* (1990). Mass spectrometric analysis of tobacco-specific nitrosamine hemoglobin adducts in snuff dippers, smokers, and nonsmokers. *Cancer Res.* **50**, 5438–5445.
80. Nair, J., Ohshima, H., Nair, U. J., and Bartsch, H. (1996). Endogenous formation of nitrosamines and oxidative DNA-damaging agents in tobacco users. *Crit. Rev. Toxicol.* **26**, 149–161.
81. Cullen, P., Schulte, H., and Assmann, G. (1997). The Munster Heart Study (PROCAM): Total mortality in middle-aged men is increased at low total and LDL cholesterol concentrations in smokers but not in nonsmokers. *Circulation* **96**, 2128–2136.
82. Helbock, H. J., *et al.* (1998). DNA oxidation matters: The HPLC-electrochemical detection assay of 8-oxo-deoxyguanosine and 8-oxo-guanine. *Proc. Natl. Acad. Sci. USA* **95**, 288–293.
83. LeVois, M. F. (1997). Environmental tobacco smoke and coronary heart disease. *Circulation* **96**, 2086–2089.
84. Rebbeck, T. R. (1997). Molecular epidemiology of the human glutathione S-transferase genotypes GSTM1 and GSTT1 in cancer susceptibility. *Cancer Epidemiol. Biomark. Prev.* **6**, 733–743.
85. Sumida, H., *et al.* (1998). Does passive smoking impair endothelium-dependent coronary artery dilation in women? *J. Am. Coll. Cardiol.* **31**, 811–815.
86. Jourenkova, N., *et al.* (1998). Larynx cancer risk in relation to glutathione S-transferase M1 and T1 genotypes and tobacco smoking. *Cancer Epidemiol. Biomark. Prev.* **7**, 19–23.
87. Lazarus, P., *et al.* (1998). p53, but not p16 mutations in oral squamous cell carcinomas are associated with specific CYP1A1 and GSTM1 polymorphic genotypes and patient tobacco use. *Carcinogenesis* **19**, 509–514.
88. Liloglou, T., *et al.* (1997). p53 mutations in squamous cell carcinoma of the head and neck predominate in a subgroup of former and present smokers with a low frequency of genetic instability. *Cancer Res.* **57**, 4070–4074.
89. Olshan, A. F., Weissler, M. C., Pei, H., and Conway, K. (1997). p53 mutations in head and neck cancer: New data and evaluation of mutational spectra. *Cancer Epidemiol. Biomark. Prev.* **6**, 499–504.
90. Field, J. K., Zoumpourlis, V., Spandidos, D. A., and Jones, A. S. (1994). p53 expression and mutations in squamous cell carcinoma of the head and neck: Expression correlates with the patients' use of tobacco and alcohol. *Cancer Detect. Prev.* **18**, 197–208.
91. He, J., *et al.* (1999). Passive smoking and the risk of coronary heart disease: A meta-analysis of epidemiologic studies. *N. Engl. J. Med.* **340**, 920–926.
92. Setlow, R. B. (1982). DNA repair, aging, and cancer. *Natl. Cancer Inst. Monogr.* **60**, 249–255.
93. Wood, R. D., Mitchell, M., Sgouros, J., and Lindahl, T. (2001). Human DNA repair genes. *Science* **291**, 1284–1289.
94. Protic-Sabljic, M., and Kraemer, K. H. (1985). One pyrimidine dimer inactivates expression of a transfected gene in xeroderma pigmentosum cells. *Proc. Natl. Acad. Sci. USA* **82**, 6622–6626.
95. Koch, K. S., Fletcher, R. G., Grond, M. P, Inyang, A. I., Lu, X. P., Brenner, D. A., *et al.* (1993). Inactivation of plasmid reporter gene expression by one benzo(a)pyrene diol-epoxide DNA adduct in adult rat hepatocytes. *Cancer Res.* **53**, 2279–2286.
96. Wei, Q., Matanoski, G. M., Farmer, E. R., Hedayati, M. A., Grossman, L. (1993). DNA repair and aging in basal cell carcinoma: A molecular epidemiology study. *Proc. Natl. Acad. Sci. USA* **90**(4), 1614–1618.
97. Wei, Q., Cheng, L., Amos, C. I., *et al.* (2000). Repair of tobacco carcinogen-induced DNA adducts and lung cancer risk: A molecular epidemiological study. *J. Natl. Cancer Inst.* **92**, 1764–1772.
98. Hoeijmakers, J. H. J. (1994). Human nucleotide excision repair syndromes: Molecular clues to unexpected intricacies. *Eur. J. Cancer* **30**, 1912–1921.
99. Gupta, R. C., Earley, K., and Sharma, S. (1988). Use of human peripheral blood lymphocytes to measure DNA binding capacity of chemical carcinogens. *Proc. Natl. Acad. Sci. USA* **85**, 3513.
100. Athas, W. F., Hedayati, M., Matanoski, G. M., *et al.* (1991). Development and field-test validation of an assay for DNA repair in circulating lymphocytes. *Cancer Res.* **51**, 5786–5793.
101. Slebos, R. J., and Taylor, J. A. (2001). A novel host cell reactivation assay to assess homologous recombination capacity in humans. *Biochem. Biophys. Res. Commun.* **281**, 212–219.
102. Alcalay, J., Freeman, S. E., Goldberg, L. H., and Wolf, J. E. (1990). Excision repair of pyrimidine dimers induced by simulated solar radiation in the skin of patients with basal cell carcinoma. *J. Invest. Dermatol.* **95**, 506–509.
103. Wei, Q., Spitz, M. R., Gu, J., *et al.* (1996). DNA repair capacity correlates with mutagen sensitivity in lymphoblastoid cell lines. *Cancer Epidemiol. Biomark. Prev.* **5**, 199–204.
104. Brownson, R. C., Chang, J. C., and Davis, J. R. (1992). Gender and histologic type variations in smoking-related risk of lung cancer. *Epidemiology* **3**, 61–64.
105. Risch, H. A., Howe, G. R., Jain, M., Burch, J. D., Holowaty, E. J., and Miller, A. B. (1993). Are female smokers at higher risk for lung cancer than male smokers? A case-control analysis by histologic type. *Am. J. Epidemiol.* **138**, 281–293.
106. Zang, E. A., and Wynder, E. L. (1996). Differences in lung cancer risk between men and women: Examination of the evidence. *J. Natl. Cancer Inst.* **88**, 183–192.
107. Cheng, L., Eicher, S. A., Guo, Z., Hong, W. K., Spitz, M. R., and Wei, Q. (1998). Reduced DNA repair capacity in head and neck cancer patients. *Cancer Epidemiol. Biomark. Prev.* **7**, 465–468.
108. Spitz, M. R., Hoque, A., Trizna, Z., *et al.* (1994). Mutagen sensitivity as a risk factor for second malignant tumors following upper aerodigestive tract malignancies. *J. Natl. Cancer Inst.* **86**, 1681–1684.
109. Strom, S. S., Wu, X., Sigurdson, A. J., *et al.* (1995). Lung cancer, smoking patterns, and mutagen sensitivity in Mexican-Americans. *Monogr. Natl. Cancer Inst.* **18**, 29–33.
110. Wu, X., Delclos, G. L., Annegers, F. J., *et al.* (1995). A case-control study of wood-dust exposure, mutagen sensitivity, and lung-cancer risk. *Cancer Epidemiol. Biomark. Prev.* **4**, 583–588.
111. Hagmar, L., Brogger, A., Hansteen, I., *et al.* (1994) Cancer risk in humans predicted by increased levels of chromosome aberrations in

lymphocytes: Nordic study group on the health risk of chromosome damage. *Cancer Res.* **54**, 2919–2922.

112. Bonassi, S., Abbondandolo, A., Camurri, L., *et al.* (1995). Are chromosome aberrations in circulating lymphocytes predictive of future cancer onset in humans? Preliminary results of an Italian cohort study. *Cancer Genet. Cytogenet.* **79**, 133–135.
113. Field, J. K. (1996). Genomic instability in squamous cell carcinoma of the head and neck. *Anticancer Res.* **16**, 2421–2431.
114. Scully, C., Field, J. K., and Tanzawa, H. (2000). Genetic aberrations in oral or head and neck squamous cell carcinoma 2: Chromosomal aberrations. *Oral Oncol.* **36**, 311–327.
115. Dutrillaux, B. (1995). Pathways of chromosome alteration in human epithelial cancers. *Adv. Cancer Res.* **67**, 59–82.
116. Hsu, T. C., Johnston, D. A., Cherry, L. M., *et al.* (1989). Sensitivity to genotoxic effects of bleomycin in humans: Possible relationship to environmental carcinogenesis. *Int. J. Cancer* **43**, 403–409.
117. Hsu, T. C., Spitz, M. R., and Schantz, S. P. (1991). Mutagen sensitivity: A biologic marker of cancer susceptibility. *Cancer Epidemiol. Biomark. Prev.* **1**, 83–89.
118. Spitz, M. R., Fueger, J. J., Beddingfield, N. A., *et al.* (1989). Chromosome sensitivity to bleomycin-induced mutagenesis, an independent risk factor for upper aerodigestive tract cancers. *Cancer Res.* **49**, 4626–4628.
119. Spitz, M. R., Hsu, T. C., Wu, X. F., Fueger, J. J., Amos, C. I., and Roth, J. A. (1995). Mutagen sensitivity as a biologic marker of lung cancer risk in African Americans. *Cancer Epidemiol. Biomark. Prev.* **4**, 99–103.
120. Schantz, S. P., Hsu, T. C., Ainslie, N., and Moser, R. P. (1989). Young adults with head and neck cancer express increased susceptibility to mutagen-induced chromosome damage. *JAMA* **262**, 3313–3315.
121. Wei, Q., Cheng, L., Hong, W. K., *et al.* (1996). Reduced DNA repair capacity in lung cancer patients. *Cancer Res.* **56**, 4103–4107.
122. Wang, L. E., Sturgis, E. M., Eicher, S. A., Spitz, M. R., Hong, W. K., and Wei, Q. (1998). Mutagen sensitivity to benzo(a)pyrene diol epoxide and the risk of squamous cell carcinoma of the head and neck. *Clin. Cancer Res.* **4**, 1773–1778.
123. Wu, X., Gu, J., Amos, C. I., Jiang, H., Hong, W. K., and Spitz, M. R. (1998). A parallel study of in vitro sensitivity to benzo[a]pyrene diol epoxide and bleomycin in lung cancer cases and controls. *Cancer* 1118–1127.
124. Hsu, T. C. (1983). Genetic instability in the human population: A working hypothesis. *Hereditas* **98**, 1–9.
125. Sturgis, E. M., Spitz, M. R., and Wei, Q. (1998). DNA repair and genomic instability in tobacco induced malignancies of the lung and upper aerodigestive tract. *Environ. Carcinogen. Ecol.* **C16**, 1–30.
126. Pandita, T. K., and Hittelman, W. N. (1995). Evidence of a chromatin basis for increased mutagen sensitivity associated with multiple primary malignancies of the head and neck. *Int. J. Cancer* **61**, 738–743.
127. Everson, R. B., Randerath, E., Santella, R. M., Cefalo, R. C., Avitts, T. A., and Randerath, K. (1986). Detection of smoking-related covalent DNA adducts in human placenta. *Science* **231**, 54–57.
128. Geneste, O., Camus, A. M., Castegaro, M., *et al.* (1991). Comparison of pulmonary DNA adduct levels, measured by 32P-postlabelling and aryl hydrocarbon hydroxylase activity in lung parenchyma of smokers and ex-smokers. *Carcinogenesis* **12**, 1301–1305.
129. Li, D., Wang, M., Cheng, L., Spitz, M. R., Hittelman, W. N., and Wei, Q. (1996). In vitro induction of benzo(a)pyrene diol epoxide-DNA adducts in peripheral lymphocytes as a susceptibility marker for human lung cancer. *Cancer Res.* **56**, 3638–3641.
130. Reddy, M. V., and Randerath, K. (1986). Nuclease P1-mediated enhancement of sensitivity of 32P-postlabeling test for structurally diverse DNA adducts. *Carcinogenesis* **7**, 1543–1557.
131. Phillips, D. H., Hewer, A., and Grover, P. L. (1986). Aromatic DNA adducts in human bone marrow and peripheral blood leukocytes. *Carcinogenesis* **7**, 2071–2075.
132. Savela, K., and Hemminki, K. (1991). DNA adducts in lymphocytes and granulocytes of smokers and nonsmokers detected by the 32P-postlabelling assay. *Carcinogenesis* **12**, 503–508.
133. van Schooten, F. J., Hillebrand, M. J., van Leeuwen, F. E., *et al.* (1992). Polycyclic aromatic hydrocarbon-DNA adducts in white blood cells from lung cancer patients: No correlation with adduct levels in lung. *Carcinogenesis* **13**, 987–993.
134. Gelboin, H. V. (1980). Benzo[*a*]pyrene metabolism, activation, and carcinogenesis: Role and regulation of mixed function oxidases and related enzymes. *Physiol. Rev.* **60**, 1107–1166.
135. Gonzalez, M. V., Alvarez, V., Pello, M. F., Menendez, M. J., Suarez, C., and Coto, E. (1998). Genetic polymorphism of N-acetyltransferase-2, glutathione S-transferase-M1, and cytochromes P450IIE1 and P450IID6 in the susceptibility to head and neck cancer. *J. Clin. Pathol.* **51**, 294–298.
136. Eaton, D. L., Gallagher, E. P., Bammler, T. K., and Kunze, K. L. (1995). Role of cytochrome P4501A2 in chemical carcinogenesis: Implications for human variability in expression and enzyme activity. *Pharmacogenetics* **5**, 259–274.
137. Bartsch, H., Rojas, M., Alexandrov, K., *et al.* (1995). Metabolic polymorphism affecting DNA binding and excretion of carcinogens in humans. *Pharmacogenetics* **5**, S84–S90.
138. Krolewski, B., Little, J. B., and Reynolds, R. J. (1988). Effect of duration of exposure to benzo(a)pyrene diol-epoxide on neoplastic transformation, mutagenesis, cytotoxicity, and total covalent binding to DNA of rodent cells. *Teratog. Carcinog. Mutagen* **8**, 127–136.
139. Li, D., Firozi, P. F., Wang, *et al.* (2001). In vitro BPDE-induced DNA adducts in peripheral lymphocytes as a risk factor for squamous cell carcinoma of the head and neck. *Int. J. Cancer* **93**, 436–440.
140. Li, D., Firozi, P. F., Wang, L. E., Bosken, C. H., Spitz, M. R., Hong, W. K., and Wei, Q. (2001). Sensitivity to DNA damage induced by benzo(a)pyrene diol epoxide and risk of lung cancer: A case-control analysis. *Cancer Res.* **61**(4), 1445–1450.
141. Wei, Q., Xu, X., Cheng, L., Legerski, R. J., and Ali-Osman, F. (1995). Simultaneous amplification of four DNA repair genes and beta-actin in human lymphocytes by multiplex reverse transcriptase-PCR. *Cancer Res.* **55**, 5025–5029.
142. Wei, Q., Guan, Y., Cheng, L., Radinsky, R., Bar-Eli, M., Tsan, R., Li, L., and Legerski, R. J. (1997). Expression of five selected human mismatch repair genes simultaneously detected in normal and cancer cell lines by a nonradioactive multiplex reverse transcription-polymerase chain reaction. *Pathobiology* **65**, 293–300.
143. Cheng, L., Guan, Y., Li, L., Legerski, R. J., Einspahr, J., Bangert, J., Alberts, D. S., and Wei, Q. (1999). Expression in normal human tissues of five nucleotide excision repair genes measured simultaneously by multiplex reverse transcription-polymerase chain reaction. *Cancer Epidemiol. Biomark. Prev.* **8**, 801–807.
144. Wei, Q., Eicher, S. A., Guan, Y., *et al.* (1998). Reduced expression of hMLH1 and hGTBP/hMSH6: A risk factor for head and neck cancer. *Cancer Epidemiol. Biomark. Prev.* **7**, 309–314.
145. Cheng, L., Spitz, M. R., Hong, W. K., and Wei, Q. (2000). Reduced expression levels of nucleotide excision repair genes in lung cancer: A case-control analysis. *Carcinogenesis* **21**, 1527–1530.
146. Shen, M. R., Jones, I. M., and Mohrenweiser, H. (1998). Nonconservative amino acid substitution variants exist at polymorphic frequency in DNA repair genes in healthy humans. *Cancer Res.* **58**, 604–608.
147. Fan, F., Liu, C., Tavare, S., and Arnheim, N. (1999). Polymorphisms in the human DNA repair gene XPF. *Mutat. Res.* **406**, 115–120.
148. Khan, S. G., Metter, E. J., Tarone, R. E., *et al.* (2000). A new xeroderma pigmentosum group C poly(AT) insertion/deletion polymorphism. *Carcinogenesis* **21**, 1821–1825.
149. Emmert, S., Schneider, T. D., Khan, S. G., and Kraemer, K. H. (2001). The human XPG gene: Gene architecture, alternative splicing

and single nucleotide polymorphisms. *Nucleic Acids Res.* **29**, 1443–1452.

150. Coin, F., Marinoni, J. C., Rodolfo, C., Fribourg, S., Pedrini, A. M., and Egly, J. M. (1998). Mutations in the XPD helicase gene result in XP and TTD phenotypes, preventing interaction between XPD and the p44 subunit of TFIIH. *Nature Genet.* **20**, 184–188.
151. Reardon, J. T., Ge, H., Gibbs, E., Sancar, A., Hurwitz, J., and Pan, Z. Q. (1996). Isolation and characterization of two human transcription factor IIH (TFIIH)-related complexes: ERCC2/CAK and TFIIH. *Proc. Natl. Acad. Sci. USA* **93**, 6482–6487.
152. Taylor, E. M., Broughton, B. C., Botta, E., *et al.* (1997). Xeroderma pigmentosum and trichothiodystrophy are associated with different mutations in the XPD (ERCC2) repair/transcription gene. *Proc. Natl. Acad. Sci. USA* **94**, 8658–8663.
153. Broughton, B. C., Steingrimsdottir, H., and Lehmann, A. R. (1996). Five polymorphisms in the coding sequence of the xeroderma pigmentosum group D gene. *Mutat. Res.* **362**, 209–211.
154. Lunn, R. M., Helzlsouer, K. J., Parshad, R., *et al.* (2000). XPD polymorphisms: Effects on DNA repair proficiency. *Carcinogenesis* **21**, 551–555.
155. Duell, E. J., Wiencke, J. K., Cheng, T. J., *et al.* (2000). Polymorphisms in the DNA repair genes XRCC1 and ERCC2 and biomarkers of DNA damage in human blood mononuclear cells. *Carcinogenesis* **21**, 965–971.
156. Tomescu, D., Kavanagh, G., Ha, T., Campbell, H., and Melton, D. W. (2001). Nucleotide excision repair gene XPD polymorphisms and genetic predisposition to melanoma. *Carcinogenesis* **22**, 403–408.
157. Spitz, M. R., Wu, X., Wang, Y., Wang, L. E., Shete, S., Amos, C. I., Guo, Z., Lei, L., Mohrenweiser, H., and Wei, Q. (2001). Modulation of nucleotide excision repair capacity by XPD polymorphisms in lung cancer patients. *Cancer Res.* **61**, 1354–1357.
158. Butkiewicz, D., Rusin, M., Enewold, L., Shields, P. G., Chorazy, M., and Harris, C. C. (2001). Genetic polymorphisms in DNA repair genes and risk of lung cancer. *Carcinogenesis* **22**, 593–597.
159. Matullo, G., Guarrera, S., Carturan, S., *et al.* (2001). DNA repair gene polymorphisms, bulky DNA adducts in white blood cells and bladder cancer in a case-control study. *Int. J. Cancer.* **92**, 562–567.
160. Zhang, X., Morera, S., Bates, P. A., *et al.* (1998). Structure of an XRCC1 BRCT domain: A new protein-protein interaction module. *EMBO J.* **17**, 6404–6411.
161. Masson, M., Niedergang, C., Schreiber, V., Muller, S., Demarcia, J. M., and Demurcia G. (1998). XRCC1 is specifically associated with PARP polymerase and negatively regulates its activity following DNA damage. *Mol. Cell. Biol.* **18**, 3563–3571.
162. Caldecott, K. W., McKeown, C. K., Tucker, J. D., Ljungquist, S., and Thompson, L. H. (1994). An interaction between the mammalian DNA repair protein XRCC1 and DNA ligase III. *Mol. Cell. Biol.* **14**, 68–76.
163. Kubota, Y., Nash, R. A., Klungland, A., Schar, P., Barnes, D. E., and Lindahl, T. (1996). Reconstitution of DNA base excision-repair with purified human proteins: Interaction between DNA polymerase beta and the XRCC1 protein. *EMBO J.* **15**, 6662–6670.
164. Cappelli, E., Taylor, R., Cevasco, M., Abbondandolo, A., Caldecott, K., and Forsina, G. (1997). Involvement of XRCC1 and DNA ligase gene products in DNA base excision repair. *J. Biol. Chem.* **272**, 23970–23975.
165. Marintchev, A., Mullen, M. A., Maciejewski, M. W., Pan, B., Gryk, M. R., and Mullen, G. P. (1999). Solution structure of the single-strand break repair protein XRCC1 N-terminal domain. *Nature. Struct. Biol.* **6**, 884–893.
166. Lunn, R. M., Langlois, R. G., Hsieh, L. L., Thompson, C. L., and Bell, D. A. (1999). XRCC1 polymorphisms: Effects on aflatoxin B1-DNA adducts and glycophorin A variant frequency. *Cancer Res.* **59**, 2557–2561.

166a. Shen, H., Sturgis, E. M., Khan, S. G., *et al.* (2001). An intronic poly(AT) polymorphism of the DNA repair gene XPC and risk of squamous cell carcinoma of the head and neck: A case-control study. *Cancer Res.* **61**, 3321–3325.

166b. Shen, H., Xu, Y., Qian, Y., *et al.* (2000). Polymorphisms of the DNA repair gene XRCC1 and risk of gastric cancer in a Chinese population. *Int. J. Cancer* **88**, 601–606.

167. Ratnasinghe, D., Yao, S. X., Tangrea, J. A., *et al.* (2001). Polymorphisms of the DNA repair gene XRCC1 and lung cancer risk. *Cancer Epidemiol. Biomark. Prev.* **10**, 119–123.
168. Abdel-Rahman, S. Z., Soliman, A. S., Bondy, M. L., *et al.* (2000). Inheritance of the 194Trp and the 399Gln variant alleles of the DNA repair gene XRCC1 are associated with increased risk of early-onset colorectal carcinoma in Egypt. *Cancer Lett.* **9**, 79–86.
169. Stern, M. C., Umbach, D. M., van Gils, C. H., Lunn, R. M., and Taylor J. A. (2001). DNA repair gene XRCC1 polymorphisms, smoking, and bladder cancer risk. *Cancer Epidemiol. Biomark. Prev.* **10**, 125–131.
170. Kaiser, J. (2001). Diversity in mending DNA damage. In toxicologists hit the west coast. *Science* **292**, 837–838.
171. Boiteux, S., and Radicella, J. P. (2000). The human OGG1 gene: Structure, functions, and its implication in the process of carcinogenesis. *Arch. Biochem. Biophys.* **377**, 1–8.
172. Kohno, T., Shinmura, K., Tosaka, M., *et al.* (1998). Genetic polymorphisms and alternative splicing of the hOGG1 gene, that is involved in the repair of 8-hydroxyguanine in damaged DNA. *Oncogene* **16**, 3219–3225.
173. Chevillard, S., Radicella, J. P., Levalois, C., *et al.* (1998). Mutations in OGG1, a gene involved in the repair of oxidative DNA damage, are found in human lung and kidney tumours. *Oncogene* **16**, 3083–3086.
174. Audebert, M., Radicella, J. P., and Dizdaroglu, M. (2000). Effect of single mutations in the OGG1 gene found in human tumors on the substrate specificity of the OGG1 protein. *Nucleic Acids Res.* **28**(14), 2672–2678.
175. Sugimura, H., Kohno, T., Wakai, K., *et al.* (1999). hOGG1 Ser326Cys polymorphism and lung cancer susceptibility. *Cancer Epidemiol. Biomark. Prev.* **8**(8), 669–674.

175a. Sugimura, H., Wakai, K., Genka, K., *et al.* (1998). Association of Ile462Val (Exon 7) polymorphism of cytochrome P450 IA1 with lung cancer in the Asian population: Further evidence from a case-control study in Okinawa. *Cancer Epidemiol. Biomark. Prev.* **7**, 413–417.

176. Ishida, T., Takashima, R., Fukayama, M., *et al.* (1999). New DNA polymorphisms of human MMH/OGG1 gene: Prevalence of one polymorphism among lung-adenocarcinoma patients in Japanese. *Int. J. Cancer* **80**, 18–21.
177. Blons, H., Radicella, J. P., Laccoureye, O., *et al.* (1999). Frequent allelic loss at chromosome 3p distinct from genetic alterations of the 8-oxoguanine DNA glycosylase 1 gene in head and neck cancer. *Mol. Carcinog.* **26**, 254–260.
178. Wikman, H., Risch, A., Klimek, F., *et al.* (2000). hOGG1 polymorphism and loss of heterozygosity (LOH): Significance for lung cancer susceptibility in a Caucasian population. *Int. J. Cancer* **88**, 932–937.
179. Xing, D. Y., Tan, W., Song, N., and Lin, D. X. (2001). Ser326Cys polymorphism in hOGG1 gene and risk of esophageal cancer in a Chinese population. *Int. J. Cancer* **95**, 140–143.
180. Bova, R. J., Quinn, D. I., Nankervis, J. S., *et al.* (1999). Cyclin D1 and p16INK4A expression predict reduced survival in carcinoma of the anterior tongue. *Clin. Cancer Res.* **5**, 2810–2819.
181. Zheng, Y., Shen, H., Sturgis, E. M., Wang, L. E., Eicher, S. A., Strom, S. S., Frazier, M. L., Spitz, M. R., and Wei, Q. (2001). Cyclin D1 polymorphism and risk for squamous cell carcinoma of the head and neck: A case-control study. *Carcinogenesis*, **22**, 1195–1199.

182. Zheng, Y., Shen, H., Sturgis, E. M., Wang, L. E., Shete, S., Spitz, M. R., and Wei, Q. (2002). Haplotypes of two variants in p16 (CDKN2/MTS1/INK4a) Exon 3 and risk of squamous cell carcinoma of the head and neck: A case-control study. *Cancer Epidemiol. Biomarkers Prev.* **11**, 640–645.

183. Rosas, S. L., Koch, W., da Costa Carvalho, M. G., Wu, L., Califano, J., Westra, W., Jen, J., and Sidransky, D. (2001). Promoter hypermethhylation patterns of p16, O6-methylguanine-DNA-methyltransferase, and death-associated protein kinase in tumors and saliva of head and neck cancer patients. *Cancer Res.* **61**, 939–942.

CHAPTER

16

Head and Neck Field Carcinogenesis

WALTER N. HITTELMAN

Department of Experimental Therapeutics
The University of Texas M. D. Anderson Cancer Center
Houston, Texas 77030

I. Introduction 227
II. Clinical Evidence for Head and Neck Field Cancerization 229
  A. Role of Tobacco, Alcohol Exposure, and Viral Exposure 229
  B. Second Primary Tumor Development 229
  C. Premalignant Lesions and Risk of Tumor Development 230
III. Histopathologic Evidence of Field Cancerization 230
  A. Animal Models of Head and Neck Carcinogenesis 230
  B. Histologic Changes in the Carcinogen-Exposed Epithelium 231
  C. Histologic Changes in the Field of Head and Neck Tumors 231
IV. Genetic Evidence for Field Cancerization 231
  A. DNA Damage Resulting from Carcinogen Exposure 231
  B. Chromosome Changes in the Carcinogen-Exposed Field 232
  C. Evidence for Chromosome Instability in the Head and Neck Epithelium 233
  D. *In Situ* Hybridization Evidence for Clonal Outgrowths during Field Cancerization 234
  E. Molecular Genetic Evidence for Clonal Outgrowths during Field Cancerization 235
V. Phenotypic Changes Associated with Field Carcinogenesis 236
VI. Clinical Implications of Field Cancerization 236
  A. Head and Neck Tumor Risk Estimation 237
  B. Assessment of Response to Chemopreventive Intervention 237
  C. Implications for Management of Recurrent Disease or Second Primaries 238
  References 238

Head and neck tumorigenesis has been proposed to represent a field cancerization process. The whole epithelium is exposed to carcinogenic insult (e.g., tobacco products) and cofactors (e.g., alcohol), resulting in chronic tissue damage and wound repair. DNA damage created by carcinogens is translated into more permanent genetic changes during proliferation, resulting in the presence of genomic damage throughout the exposed field. With chronic injury, genomic alterations accumulate, leading to multifocal clonal outgrowths that grow out at the expense of normal epithelium and continue to evolve toward malignancy. The field cancerization process in the head and neck epithelium is manifest by clinically detectable lesions (e.g., leukoplakia) prior to cancer development, progressive histologic changes, and alterations of genetic and phenotypic markers in the epithelial cells. With continued genetic evolution throughout the epithelial field, multiple, genetically distinct primary tumors may develop in synchrony or sequentially over time. This chapter reviews evidence for the field cancerization process during multistep head and neck tumorigenesis and discusses the implications of such a process in the clinical management of individuals at high risk for first and second head and neck primary tumors.

I. INTRODUCTION

Head and neck cancer remains a significant public health problem throughout the world. It is estimated that in the year 2002 in the United States there will be approximately 28,900 new cases of cancer of the oral cavity and pharynx and approximately 8900 new cases of laryngeal cancer [1]. This accounts for approximately 5% of the malignancies. However, head and neck cancer can account for up to 40% of malignancies in the Far East and India [2]. Despite improvements in surgery, radiotherapy, and chemotherapy, the 5-year survival rates have improved only marginally over the past 20–30 years. New approaches need to be developed that can reduce both the incidence and the mortality of head and neck cancer [3]. The new strategies can take the form of preventive measures that interfere with the head and neck tumorigenesis process or therapeutic measures that interfere with growth, survival, and metastatic processes associated with tumor progression [4,5]. The development of targeted approaches for both prevention and therapy will require improved understanding of the process of head and neck tumor development, as well as further characterization of the specific molecular events that underlie the tumor phenotype.

Tumors of the head and neck have been proposed to reflect a "field cancerization" process whereby the tissue region is thought to be exposed to carcinogenic insult (e.g., tobacco products and alcohol) and is at increased risk for multistep tumor development [6–10]. As shown in Fig. 16. 1, the working model for head and neck cancer development is that exposure to carcinogens leads to the accumulation of genetic changes throughout the directly exposed tissue, as well as in tissues exposed systemically to the metabolic products of the carcinogens. Genomic damage in the form of DNA adducts or their processed repair intermediates (e.g., DNA single and double strand breaks, abasic DNA residues) associated with the carcinogenic exposure leads to cellular and tissue damage and various types of tissue response. One component of the tissue reaction might be an inflammatory response, leading to the influx of cells such as granulocytes and macrophages that can produce free radicals as part of their normal function. This would lead to further damage to the genome of the epithelial cells. Another component of this tissue response may be a healing reaction, leading to increased levels of cellular proliferation. With cellular proliferation, damage to the DNA of the cells, if not repaired prior to DNA synthesis, is translated during DNA synthesis into more permanent types of genomic change such as DNA point mutation, deletion, insertion, and recombination, which can then be transmitted to the offspring of the damaged cells. When changes occur in regions of the genome that influence the survival of the cell following injury or influence the regulatory circuits of the cell, these genetically altered cells may preferentially expand within the exposed epithelium at the expense of their epithelial neighbors. With continued carcinogenic exposure and resultant chronic tissue healing, clonal outgrowths of these initiated cells may continue to accumulate new genetic hits and lead to the development of subclonal outgrowths with more evolved genotypes (Fig. 16. 2). Eventually, one or more of these clones may collect sufficient numbers of genetic hits to influence all the pathways necessary for tumor development, invasion, and metastasis.

The concept of field cancerization in the setting of head and neck cancer is important because it has critical implications for

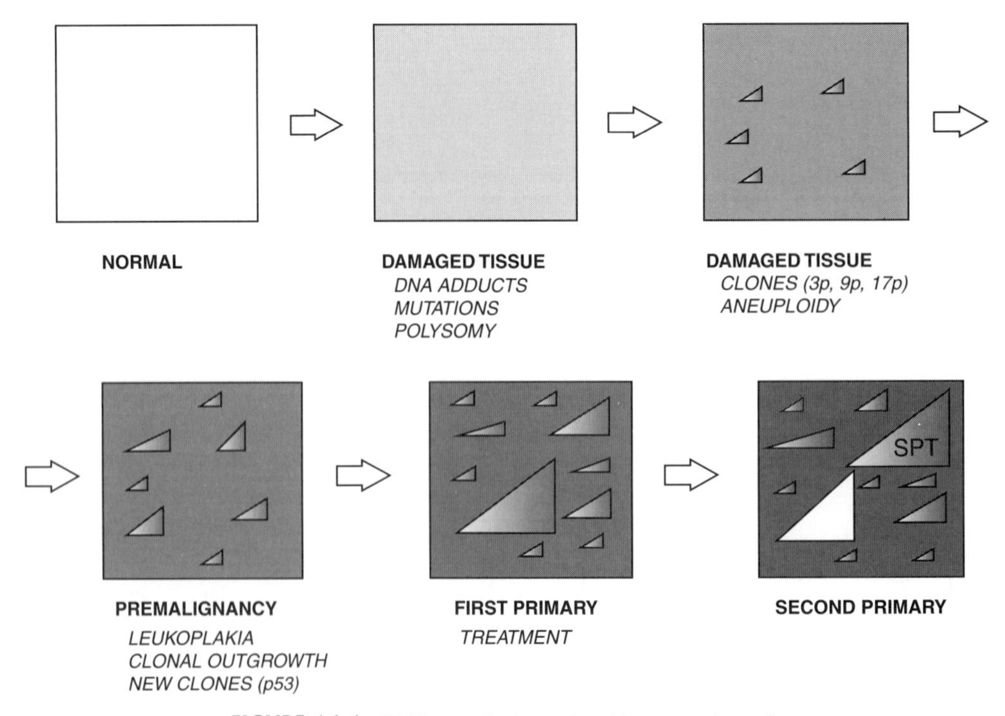

FIGURE 16-1 Field cancerization and multistep tumorigenesis.

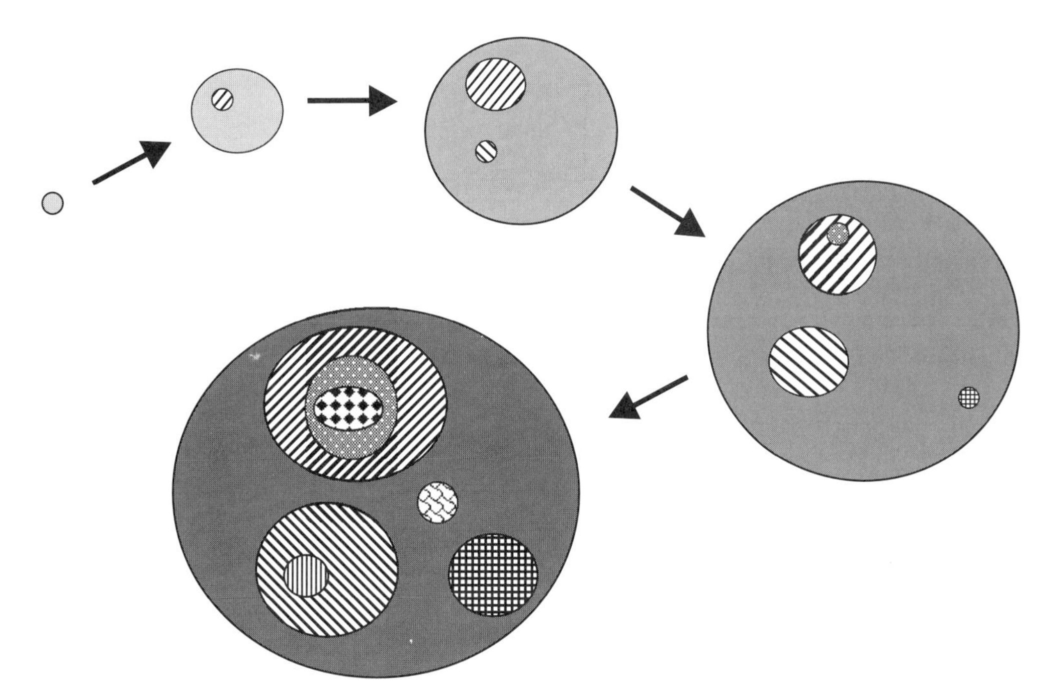

FIGURE 16-2 Multiple focal clonal evolution during head and neck tumorigenesis.

assessing the risk of first and second primary tumors and thus can have a profound impact on the clinical management of individuals with premalignant lesions, of individuals who have been definitively treated for a first primary tumor and remain at high risk for developing a second primary tumor, and of individuals presenting with either a recurrence or a second primary tumor. The goal of this chapter is to review the clinical, pathologic, and genetic evidence for the concept of head and neck field cancerization and then discuss its implications for the clinical management of individuals at increased risk for developing first or second head and neck cancers.

II. CLINICAL EVIDENCE FOR HEAD AND NECK FIELD CANCERIZATION

A. Role of Tobacco, Alcohol Exposure, and Viral Exposure

The clinical manifestations of a field cancerization and multistep head and neck tumorigenesis process can be seen in multiple ways. First, head and neck cancer has been shown to be highly associated with tobacco and alcohol exposure [11–14]. In fact, it is estimated that 50–70% of deaths resulting from oral and laryngeal cancer can be attributed to tobacco smoking [15]. The risk of developing head and neck cancer increases as a function of both the intensity (e.g., packs per day) and the duration (e.g., pack years) of tobacco exposure [16,17] and decreases gradually following cessation of tobacco exposure [18].

While the majority of head and neck cancers is attributable to tobacco and alcohol exposure, nearly one-fourth to one-third of head and neck cancers occur in nonsmokers. The etiologic agents or intrinsic factors responsible for head and neck cancer in nonsmokers are less understood. However, they may also influence the head and neck region in a field cancerization manner. For example, mucosa of the upper aerodigestive tract can be infected by human papillomaviruses (HPV), albeit in a potentially more focal fashion than that affected by tobacco exposure [19]. While the reported frequency is highly variable depending on the detection technique, more than 50% of head and neck cancers, especially in the tonsil and base of tongue, have been reported to harbor (HPV) [20,21]. It has been shown that (1) high-risk HPV16 is present in 90% of HPV-positive head and neck cancer cases [22] and (2) that the gene products of HPV16 are reported to be especially potent with regard to interfering with cell cycle regulation, altering cellular response to injury, promoting genomic instability, and facilitating immortalization [23]. These findings suggest that HPV16 can facilitate the acquisition of genetic events associated with head and neck tumorigenesis or can act in combination with exposure of the head and neck mucosa to carcinogens [24].

B. Second Primary Tumor Development

A second clinical feature that suggests head and neck field cancerization is the frequent multifocal nature of the disease [9,25]. While it is not infrequent that multiple sites of disease are concurrently detected, the more frequent observation is the high frequency of second primary development in individuals whose first head and neck primary has been definitively treated. For clinical purposes, second primary tumors have been defined using the criteria developed by Warren and Gates whereby tumors were called second primary tumors if they exhibited a different histology, if they showed a similar histology but occurred more than 3 years after therapy for the first primary, or if they were separated from the initial primary tumor by more than 2 cm [26]. More recently, the validity of this definition has been examined through molecular classification of the first and second tumors, which will be discussed in more detail later. Nevertheless, using clinical criteria as the basis for defining a second primary, individuals who suffer a first primary tumor in the head and neck region have at least a 40% chance of developing a second aerodigestive tract tumor in their lifetime at a rate of approximately 3.5–4% per year [27–31].

A recent analysis of second primary development in individuals participating in a head and neck retinoid chemoprevention trial provided additional insights regarding head and neck field carcinogenesis [31]. First, the majority (70%) of the second primary tumors involved tobacco-related sites. In fact, the highest frequency of second primary tumors in tobacco-associated sites occurred in current smokers (4.2% per year) followed by that in former smokers (3.2% per year) followed by that in nonsmokers (1.9% per year). Second, the site and frequency of the second primary tumor development were related to the site of the first primary, perhaps providing insight to the individual nature of the epithelial fields most affected by carcinogen exposure. In this case, the rates of second primary tumor development in tobacco-related sites were highest for pharyngeal primaries (5.1% per year) when compared to oral cavity (4.3% per year) and larynx (2.7% per year). For example, second primary tumors arising following an oral cavity first primary occurred frequently in the oral cavity, whereas those arising following a laryngeal first primary occurred frequently in the lung.

It is notable that nonsmokers also show a defined rate of second primary tumor development, albeit lower than that found associated with current or former tobacco exposure. This might suggest that the nature of the field in nonsmokers is more defined than that in current or former smokers. Along these lines, when the rates of disease-specific survival were compared between individuals with HPV-positive and HPV-negative head and neck squamous cell tumors using a Kaplan–Meier plot, the two curves appeared to separate at around 18 months with survival being higher for those having HPV-positive tumors [21]. It was therefore hypothesized that the risk of second primaries might be lower in individuals with HPV-positive tumors due to more localized HPV infections and perhaps a more limited epithelial field susceptible for the development of a second primary tumor.

C. Premalignant Lesions and Risk of Tumor Development

A third clinical feature consistent with the notion of head and neck field cancerization is the finding of multifocal foci of intraepithelial neoplasia in the carcinogen-exposed tissue. The presence of such lesions in the exposed epithelium is associated with an increased risk for developing head and neck cancer. For example, oral leukoplakia and oral erythroplakia are common lesions observed in individuals exposed to tobacco or other products containing carcinogenic compounds (e.g., betel quid) [32]. Overall, while the risk of developing oral cancer over 10 years in individuals with oral leukoplakia is approximately 10%, this frequency approaches 40% when leukoplakia lesions are more histologically advanced [33,34]. Of importance, only approximately half of oral cancers develop at the site of leukoplakia. This would suggest that the increased cancer risk associated with intraepithelial neoplasia is for the whole field that is carcinogen exposed.

A similar observation is seen with intraepithelial neoplasia of the larynx. While tobacco exposure is the most common factor leading to tissue changes in the larynx, gastroesophageal reflux is also known to contribute in association with exposure of the laryngeal epithelium to low pH and bile acids [35–39]. As with the oral cavity, the presence of advanced histologic lesions in the larynx is a significant risk factor for developing laryngeal cancer, with 40–50% of the cases developing tumors within 3–4 years [40–43].

III. HISTOPATHOLOGIC EVIDENCE OF FIELD CANCERIZATION

A. Animal Models of Head and Neck Carcinogenesis

Exposure of animal model epithelial tissues to carcinogens demonstrates the presence of histopathologic changes that evolve during multistage carcinogenesis. Frequently, these epithelial changes can be observed throughout the entire exposed epithelium, even though tumors may evolve in only a few foci. One animal model system that demonstrates this nicely for the head and neck is the Syrian hamster buccal pouch carcinogenesis model [44]. In this model, the hamster buccal pouch is painted with 7,12-dimethylbenz[a]anthracene (DMBA) three times a week and this is associated with a progressive histologic change from normal to hyperplasia to dysplasia with or without papilloma to squamous cell carcinoma. Interestingly, histologic changes can also be viewed in the contralateral buccal pouch tissue, possibly due to the transfer of carcinogen from one side to the other or global effects of an inflammatory stimulus associated with tissue injury.

A second animal model that has provided insight into the nature of head and neck field cancerization and multistep tumorigenesis involves painting the rodent oral epithelium with 4-nitroquinoline-1-oxide [45,46]. Again, the treated epithelium shows progressive histologic changes prior to cancer development throughout the exposed epithelium. One important finding from these studies is that field changes can be seen in the painted epithelium as well as in the underlying connective tissue, including the generation and infiltration of inflammatory cells and changes in vascularization within the stroma. These findings might support the hypothesis that epithelial tumorigenesis involves more than simply an accumulation of genetic changes in the epithelium due to direct carcinogenic exposure. That is, it likely involves an interactive process between the damaged epithelium and the underlying stroma that may be responding to the damaged epithelium by eliciting a wound healing process [47,48]. In some experimental models, the growth of epithelial cells on top of stromal components derived from more advanced lesions is associated with abnormal epithelial cell growth and differentiation [49]. These models and cells derived from these tissues have provided the opportunity to explore the early genotypic and phenotypic changes that occur in the field of cancerization prior to the development of cancer at one or more foci within the epithelial field [50].

B. Histologic Changes in the Carcinogen-Exposed Epithelium

Careful histopathologic analysis of the aerodigestive tract of current and former smokers demonstrates the presence of histologic changes throughout the exposed tissue. For example, Auerbach and colleagues [51] took lobectomy and pneumonectomy specimens of smokers and individuals with lung cancer and carried out histopathological evaluations of the epithelial linings throughout the lung. Over 90% of the tissue sections derived from light smokers and nearly all the tissue sections derived from heavy smokers and from individuals with lung cancer showed some form of epithelial change, including loss of cilia, basal cell hyperplasia, and/or carcinoma *in situ*.

Similar histologic changes are seen throughout the upper aerodigestive tract of individuals with exposure to tobacco products and alcohol [35]. In many cases, the clinically evident premalignant lesions, such as leukoplakia and erythroplakia, show the greatest degree of histologic change. These changes appear to be highly related to carcinogenic exposure. For example, the risk of vocal cord metaplasia and dysplasia is much higher in smokers compared to nonsmokers, and the size of the affected area increases with aging, tobacco, and alcohol exposure [42–54]. However, the finding that the risk of vocal cord dysplasia is higher in blue collar workers than in white collar workers, after stratification of smoking habits, suggests that other etiologic agents may influence histological changes in the epithelial field [55].

C. Histologic Changes in the Field of Head and Neck Tumors

Careful analyses of head and neck squamous cell carcinomas frequently show premalignant lesions that appear to undergo a continuous histologic progression from histologically normal epithelium to hyperplasia to dysplasia to carcinoma *in situ* to invasive carcinoma. This finding provides both a visualization of the multistep tumorigenesis process and a setting in which to more fully explore the timing of specific genotypic and phenotypic changes associated with multistep tumorigenesis. At the same time, head and neck tumor specimens also exhibit premalignant lesions that are not contiguous with the primary tumor. As discussed later, molecular genetic analyses of these various lesions indicate that they may be genetically distinct from the primary tumor.

IV. GENETIC EVIDENCE FOR FIELD CANCERIZATION

A. DNA Damage Resulting from Carcinogen Exposure

If whole tissue fields receive carcinogen exposure and are at increased risk for developing tumors, one would expect to see evidence of DNA damage throughout the field. Cigarette smoke includes a complex mixture of more than 4000 chemical compounds that are delivered to tissue in both gas and tar phases. Of these, more than 40 have been identified as carcinogens. In some cases, the chemicals are procarcinogens [e.g., benzo(a)pyrene and nitrosamines] that need to be activated to induce DNA damage. In other cases, the compounds are directly reactive unless detoxified by the host. One form of DNA damage that is caused by tobacco exposure is the DNA adduct, whereby a chemical moiety becomes covalently attached to the DNA. Evidence for the presence of DNA adducts in the tobacco-exposed head and neck tissue field has been measured a number of ways. For example, a number of investigators have utilized the ^{32}P-postlabeling technique to detect DNA adducts in normal-appearing mucosa of the larynx and oral cavity [55–58]. The levels of these adducts are found to be correlated with tobacco exposure, as well as the expression of metabolic activating enzymes such as P450 2C, 3A4, and/or 1A1. Of interest, similar levels of adducts can be detected in oral biopsies as buccal mucosa scrapings from smokers, and these levels are significantly higher than that found in similar samples of nonsmokers [59].

While the ^{32}P-postlabeling technique is very sensitive in that it can detect one adducted nucleotide in a background of up to 10^{10} normal nucleotides [60], the procedure still requires significant amounts of tissue and its use is limited when only small biopsies are available. More recently, some investigators have turned to more sensitive assays systems that can be accomplished on a cell-by-cell basis. For example, antibodies have been developed that can recognize specific types of DNA adducts, including polycyclic aromatic hydrocarbon–DNA (PAH–DNA) adducts, 4-aminobiphenyl–DNA adducts, and malonaldehyde–DNA adducts [61–64]. While all of these studies show higher adduct levels in smokers when compared to nonsmokers, considerable variation in adduct levels was found between individuals, suggesting either differences in their levels of exposure or interindividual differences in their capability to activate or detoxify carcinogens or to repair DNA adducts caused by carcinogens. Of importance to the notion of field carcinogenesis, DNA adducts could be detected in various oral cavity sites within the same individual (e.g., floor of mouth and buccal mucosa), and the levels of adducts within the individual were related to the degree of ongoing smoking exposure (e.g., amount of tar and number of cigarettes smoked per day) [61]. However, the levels of adducts detected were not related to the degree of smoking history (e.g., pack-years). Thus, the tissue half-life of DNA adducts may be relatively short. Also, when cells proliferate, the DNA adducts get diluted because they do not reproduce themselves during DNA replication. As a result, the measurement of DNA adducts might reflect the level of ongoing DNA damage more than the levels of accumulated levels of DNA damage.

B. Chromosome Changes in the Carcinogen-Exposed Field

Carcinogenic exposure to tissue can induce cell death in damaged cells, cell turnover, and subsequent wound healing. Associated with the wound healing process is an increase in the proliferative component of the damaged tissue. When cells attempt to replicate DNA containing DNA adducts, the result is the induction of DNA mutations and chromosome damage in the form of broken or rearranged chromosomes. When cells containing chromosome damage attempt to divide, chromosome regions no longer attached to kinetochores tend to lag behind during anaphase and may not be incorporated into the main nucleus during telophase. The result of this is the formation of micronuclei. Indeed, exfolliated cells from the oral cavity of carcinogen-exposed individuals demonstrate increased frequencies of micronuclei [65]. Interestingly, the frequency of micronuclei is increased greatly (upward toward eightfold) in individuals exposed to both tobacco products and alcohol when compared to each agent alone [66]. The reason for this is not well understood. However, it has been suggested that ethanol exposure can have an inhibitory effect on the repair of DNA damage induced by carcinogens [67,68]. In addition, exposure of cells *in vitro* to cigarette smoke condensate can also cause abnormalities in the centrosomal location in mitotic cells, perhaps leading to subsequent chromosome nondisjunction [69].

Again, because micronuclei are formed during mitosis, they do not accumulate as cell populations undergo multiple cell cycles. For example, a dramatic increase in micronuclei frequency can be detected in exfolliated cells of the oral mucosa during radiation therapy; however, the micronuclei frequency decreases significantly within a couple weeks of therapy completion.

In an attempt to explore more permanent genetic changes in the head and neck epithelial field exposed to carcinogens, investigators have attempted to examine mitotic cells for evidence of changes in chromosome number as well as in chromosome structure. Unfortunately, the fraction of actively dividing cells present in the epithelium is small and thus mitotic figures of exfolliated cells or cells derived from biopsies are few in number [70]. However, if these cells are placed in culture and analyzed after a short period of *in vitro* growth, chromosome changes can be visualized in epithelial cells derived from squamous cell carcinomas of the head and neck and skin [71–74]. An interesting finding from these studies was the diversity of clonal populations that could be identified within the same cell culture, suggesting the presence of several distinct populations within the sample. It is still not clear whether these separate clones were derived from tumor cells or from premalignant outgrowths in the field of the tumors or from other cell types contained within the specimen [75]. In any event, these findings suggest that chromosome changes are ongoing throughout the exposed epithelial field.

In order to overcome the problem of the paucity of mitotic cells in fresh tissue specimens, our own laboratory turned to a technique called premature chromosome condensation (PCC) that permits the visualization of chromosomes while cells are still in interphase [76]. Using this technique on tissues derived at different stages of tumorigenesis in the DMBA-treated hamster cheek pouch model, we demonstrated two important findings. First, distinct chromosome changes were observed in the premalignant phase and the number of chromosome changes was found to accumulate with continued carcinogenic exposure [77]. Second, chromosome damage was observed throughout the field of carcinogen exposure, not only in the specific site that ultimately developed tumors. The PCC technique was also applied to a limited number of cases where cells were dissociated from fresh leukoplakia biopsies. Of interest, there was a wide variation in chromosome copy numbers per cell suggestive of a profound ongoing chromosome instability process in the premalignant field [78].

C. Evidence for Chromosome Instability in the Head and Neck Epithelium

While the PCC technique is useful in situations where fresh target cells are plentiful, it is not easily adaptable to tissue biopsies of carcinogen-exposed head and neck epithelium. The technique of *in situ* hybridization (ISH) involves the use of labeled DNA probes that recognize chromosome-specific repetitive target sequences, chromosome single gene copy sequences, sequences along the whole chromosome length, or sequences in chromosome segments [79–81]. The attractiveness of this technology is that it permits genetic analysis of nondividing cells such as those found in exfoliated populations, touch preparations, dissociated tissues, or on tissue sections of archival material [82]. The visual output of this procedure can be obtained using fluorescence or immunocytochemical reactions.

To examine the nature of chromosome instability in the field of head and neck tumors, archival surgical specimens were sought that exhibited a contiguous histologic progression from normal adjacent epithelium through hyperplasia to dysplasia to invasive cancer. *In situ* hybridization using probes for highly repeated sequences located in the centromere regions of specific chromosomes was then used on tissue sections of these specimens to detect chromosome polysomy, or the presence of cells with three or more chromosome copies [83]. Of interest to the notion of field cancerization, chromosome polysomy could be detected in 35% of the cases of apparently normal epithelium adjacent to the tumor. Moreover, the frequency of chromosome polysomy was found to increase with evidence of histologic progression through the transition from normal epithelium to hyperplasia (65% of cases) to dysplasia (95% of cases) to invasive cancer. In this setting, each chromosome probe used (i.e., probes for chromosomes 7, 9, 11, and 17) showed the same trend toward increasing chromosome instability with histologic progression. A similar finding was reported in a series of 16 squamous cell carcinomas of the head and neck using probes for chromosomes 1, 3, 7, 8, 9, and 17 [84]. Of interest, synchronous low-grade and high-grade dysplastic lesions showed discordant chromosome signatures, suggesting a multifocal pattern of chromosome instability. In addition, exfolliated cells from normal-appearing mucosa in the field of head and neck tumors also showed evidence of chromosome instability [85]. Thus, global genomic instability may occur very early in the carcinogen-exposed epithelial field, likely a downstream consequence of carcinogen-induced DNA damage and subsequent wound healing processes.

In the case of normal epithelium and premalignant lesions adjacent to tumors, one can think of these regions as being epithelium at 100% cancer risk. Thus, one might expect to see the highest degrees of chromosome instability. It was therefore of interest to determine whether chromosome polysomy could be detected in the epithelium of individuals thought to be at increased risk for developing head and neck cancer. Pilot *in situ* hybridization studies on oral leukoplakia lesions using probes for chromosomes 7 and 17 showed a wide degree of variation in chromosome polysomy levels between different subjects. While dysplastic lesions showed higher levels of chromosome polysomy than hyperplastic lesions, some hyperplastic lesions showed high levels of chromosome polysomy [86].

These *in situ* hybridization studies have been expanded in association with chemoprevention trials using the reversal of leukoplakia and histologic progression as end points. While similar observations have been made, several findings relating to field cancerization have been made. For example, many biopsy samples show heterogeneous degrees of histologic progression within the same sample. While, in general, dysplastic lesions show higher levels of chromosome polysomy than hyperplastic lesions, hyperplastic lesions in the same field as dysplastic lesions show higher levels of chromosome polysomy than hyperplastic lesions in a biopsy without evidence of more histologically advanced sites [87]. Thus, the measurement of chromosome instability seems to provide unique global information about the tissue field that may not be detected by histology.

This observation is important because the sampling of tissue fields for malignancy-associated changes must be considered somewhat random. Despite the presence of a clinically evident leukoplakia lesion, one cannot predict the future location of a cancer. Thus, it is desirable to have a technology that can provide relevant information about events such as genomic instabiity that drive the tumorigenesis process throughout the field.

The idea that measures of chromosome instability provide additional information to histology is further supported when advanced premalignant lesions of the oral cavity were compared to those in the larynx. While severely dysplastic lesions showed higher levels of chromosome polysomy than lesions with moderate or mild dysplasia in both tissue sites, lesions in the oral cavity showed significantly higher levels of chromosome polysomy than those lesions in the larynx [88]. It is therefore possible that different tissue sites respond differently in terms of histological change for the same degree of ongoing genetic insult to the epithelial field.

Another setting where chromosome instability measurements add information to histology is in the setting where individuals have decreased their exposure to carcinogenic insult (e.g., following smoking cessation). Biopsies from individuals who still show clinical manifestations of leukoplakia despite smoking cessation show higher chromosome polysomy levels in their oral epithelium than those from current smokers with leukoplakia [89]. This would suggest that ongoing tissue damage and wound healing due to carcinogenic exposure may provide a reactive tissue component that contributes to apparent histologic progression.

Much of the discussion so far has focused on extrinsic carcinogen exposure as the driving force for the generation of chromosome instability in the epithelium at risk. However, some individuals develop head and neck cancer as well as head and neck cancer intraepithelial neoplasia without known risk factors such as tobacco and alcohol exposure. In some cases, this may be associated with HPV infection. However, there still appears to be groups of individuals that develop head and neck lesions with no known risk factors. The underlying forces driving tumorigenesis are not well understood. One possibility is that factors intrinsic to the host are responsible, including intrinsic defects in DNA repair, cell cycle regulation, or tissue homeostasis. The importance of these intrinsic factors in head and neck tumorigenesis is discussed extensively in another chapter of this volume. Interestingly, however, *in situ* hybridization analyses of leukoplakia samples from individuals without a smoking and drinking history show relatively high levels of chromosome polysomy [88]. Thus, the notion of field cancerization can still be applied to epithelial tissues without known etiologic factors.

One attractive feature of the *in situ* hybridization technique is that it permits a genetic analysis while still retaining tissue architecture. This allows one to examine cofactors that might influence the degree of ongoing genetic instability in the tissue because one can examine adjacent tissue sections for possible interacting events in the same tissue region. For example, leukoplakia samples can exhibit manifestations of HPV infection (e.g., koilocytosis) in the tissue sections. By carrying out a spatial examination of chromosome polysomy within the multilayered epithelium, it was found that lesions showing HPV manifestations exhibited retention of chromosomally altered cells into the parabasal and superficial layers of the epithelium [87]. Thus it is possible that HPV infection may enhance the survival and retention of damaged cells in the epithelium and enhance the accumulation of genetic changes in the exposed field. In a similar type of analysis, it was shown that epithelial regions showing altered p53 expression showed higher levels of chromosome polysomy than epithelial regions showing no abnormal p53 protein levels [89]. This association was clearly visible within tissue sections showing a transition from normal to abnormal levels of p53 protein levels. This might suggest that alterations in p53 function enhance genomic instability induced by carcinogenic exposure, possibly by inhibiting the turnover of damaged cells in the epithelium.

D. *In Situ* Hybridization Evidence for Clonal Outgrowths during Field Cancerization

When cell populations growing *in vitro* on culture dishes are treated with agents that induce cellular death, the dying cells frequently fall off the culture dish and the surviving cells repopulate the surface of the dish in a clonal manner. Chronic cell injury and subsequent cell repopulation might then lead to the accumulation of additional genetic changes in the population that permit the development and outgrowth of cells that have preferential survival capabilities. One might imagine that a similar phenomenon could occur in the head and neck epithelium with chronic carcinogenic exposure that takes place over years, resulting in waves of tissue injury followed by wound healing.

The *in situ* hybridization studies described previously indicated that head and neck tumors and premalignant lesions demonstrated significant levels of chromosome polysomy. However, it was not known whether these cells with abnormal numbers of chromosomes were distributed randomly throughout the carcinogen-exposed epithelial field or were clustered as a result of selective outgrowth of cells (i.e., clones) with abnormal numbers of chromosomes. To address this issue, our laboratory used an image analysis system associated with the light microscope to record the coordinates and chromosome copy number of each cell analyzed in the epithelial regions of biopsy-derived tissue sections. This permitted the creation of a spatial genetic map representation of the epithelium. Because of nuclear truncation associated with cutting tissue sections, the number of chromosome copies per cell was underrepresented. Therefore, it was difficult to easily visualize outgrowths of monosomic (one chromosome copy per cell), trisomic (three copies per cell), and tetrasomic (four copies per cell) clones. However, using special gating techniques (similar to that used in flow cytometry analysis) and nearest neighbor analysis, localized chromosome indices within the epithelium could be determined. When this localized chromosome index was three-halves that found in normal lymphocytes, the epithelial cell region was considered as part of a trisomic clone of cells. When this localized chromosome index was one-half that of lymphocytes, it was considered part of a monosomic clone.

When this genetic mapping approach was applied to tissue sections exhibiting an apparent contiguous transition from normal epithelium through premalignant lesions to invasive head and neck cancer, the increase in polysomic cells during the transition was found to be partly associated with an increased number of cells involved in clonal outgrowths. Several interesting patterns were discovered with this spatial type of clonal analysis. First, when different chromosome probes were utilized on adjacent tissue sections, distinct regions of clonal outgrowths could be visualized. Similar types of results were also seen in other epithelial regions chronically exposed to carcinogens and/or environmental factors [90]. This would suggest that the chronically exposed epithelium consists of a mosaic of clones and subclones of varying size. Second, this approach allowed the detection of monosomic clones, as well as trisomic and tetrasomic clones. In the setting of analysis of thin tissue sections (e.g., 4-μm sections), there is an underrepresentation of chromosome copy numbers, as many of the nuclei are cut during the sectioning, leaving only partial nuclei in the tissue section. As a result, there is a copy number frequency shift toward lower values such that the normal average chromosome copy number decreases from 2 copies to about 1.2–1.4 copies on the average. The nearest neighborhood analysis can take this into account by looking for regions of cells that exhibit half this normal average number of chromosome copies. Thus, regions of monosomic populations can be identified within the epithelium and increase the sensitivity for detecting clonal outgrowths in the chronically damaged epithelium.

A third beneficial result became apparent while carrying out these spatial analyses of chromosome copy numbers. When adjacent sections to the chromosome analyses were examined spatially for other types of genotypic/phenotypic alterations that might interact functionally with the cellular response to carcinogenic damage, it was found that some types of changes facilitated clonal outgrowth in the damaged epithelium. For example, abnormal increases in cyclin D1 expression were found to be associated with an increase in generalized genetic instability. However, when overexpression of cyclin D1 occurred at the same time as loss of heterozygosity at chromosome 9p21, tetrasomic clones appeared in the epithelium [91]. Cyclin D1 overexpression also appeared to upregulate the expression of fragile sites in the chromosome 11q13 region (detected by *in situ* hybridization using probes that flank the fragile site) [92] and facilitate the development and outgrowth of clones containing chromosome 11q13 gene amplification (including cyclin D1) [93,94]. Thus, as genetic and phenotypic events accumulate in the tissue over time, different types of changes may interact functionally to increase the degree of genetic instability (per unit degree of carcinogenic exposure) or to facilitate the outgrowth of clones either by permitting damaged cells to survive damage or to overcome senescence.

This *in situ* hybridization mapping technique has also been applied to intraepithelial neoplastic lesions of subjects at increased risk for head and neck cancer due to clinical manifestations of leukoplakia/erythroplakia of the oral cavity or larynx. Evidence for multifocal clonal outgrowths is also found in these subjects' risk epithelia and can be found very early in the histologic progression pathway [87]. The frequency of cells involved in clonal outgrowths in these precursor lesions is found to increase with histologic transition from normal-appearing epithelium to hyperplasia to mild to moderate to severe dysplasia. However, there is a high degree of variability of clonal frequencies found within individuals' biopsies that have similar histologic appearances. Again, factors such as overexpression of cyclin D1 and loss of p16 expression are associated with genetic instability and clonal outgrowths in these precursor lesions [95].

E. Molecular Genetic Evidence for Clonal Outgrowths during Field Cancerization

Throughout the years, successive technological innovations have permitted descriptions of specific molecular and phenotypic events associated with oral cancer development. For example, cytogenetic studies, starting with metaphase analysis and moving to G banding, fluorescence *in situ* hybridization, spectral karyotyping, and comparative genomic hybridization (CGH) have now identified a large number of specific chromosome changes in tumor specimens [96–98]. Molecular analyses (e.g., microsatellite analyses, mutation analyses) improved the delineation of chromosome regions frequently altered or deleted and likely to harbor tumor suppressor genes [99–102]. More recent studies have indicated that alteration of the remaining function or expression of the wild type allele may occur through genetic (e.g., mutation) or epigenetic (e.g., promoter methylation) pathways [103,104]. One important take-home lesson is that an important driving force of the head and neck tumorigenesis process must be the induction of random chromosome instability, which may impact specific chromosome regions important in the development of the malignant phenotype [105,106].

Examinations of premalignant epithelial regions in the field of head and neck cancers have demonstrated the presence of some of these same genetic and epigenetic changes [107,108]. Some genetic events (e.g., 9p21 and 3p LOH) are found frequently in hyperplastic and mild dysplasia, whereas other events (e.g., LOH at 4q, 8p, 11q, and 17p LOH) are observed more frequently in more advanced lesions [109–111]. More recently, some of these same molecular events have been detected in precursor lesions (e.g., oral lichen planus, leukoplakia, and erythroplakia) in individuals without detectable head and neck cancer [95,112–116]. Thus, it is clear from this molecular genetic information and from the *in situ* hybridization studies described earlier that clonal outgrowths can occur early during the head and neck tumorigenesis process.

While more detailed descriptions of the molecular changes associated with head and neck tumorigenesis and the implications of their timing during cancer development are provided elsewhere in this volume, these molecular events have proven useful in characterizing the nature of head and neck field cancerization. Early cytogenetic studies on short-term cultures of head and neck tumors demonstrated the presence of multiple clonal populations within the same specimen [71–74]. Thus there was some question whether head and neck tumors were multiclonal or whether some genetically distinct premalignant cells in the specimen were able to proliferate *in vitro* and contribute unique cytogenetic information [117].

With the advent of molecular technologies that could be applied to small samples (e.g., microdissection of cells from the epithelium and polymerase chain reaction amplification of informative genetic regions), the question of unifocal versus multifocal head and neck tumors could be better addressed. Early studies compared first and second primary tumors or synchronous tumors one molecular marker at a time (e.g., p53 mutation) and found evidence for gene discordance between tumors [118,119]. When multiple primary head and neck cancers were examined with multiple molecular markers to define a more global genotype of the lesions, there was still considerable evidence for the discordance of tumors, suggesting multifocality of genetically distinct tumors [120–122].

However, other investigators have found that some of the apparently molecularly distinct tumors still contain some common molecular events [123–125]. Moreover, close molecular examination of the apparently normal margins of head and neck tumors had provided evidence for the presence of small number of cells that carry some of the same genetic changes found in the primary tumor [126,127]. Thus, the alternative hypothesis suggests that a single progenitor clone may get established early during the disease process and migrate through the epithelium and then discordantly evolve along distinct molecular pathways [128]. In this latter hypothesis, the head and neck cancerization process still involves the development of multifocal patterns of clones and subclonal evolution throughout the epithelial field [90]. However, if initiated epithelial cell migration occurs early during development, it is possible that a second tumor could arise in a distant site in the epithelial field and still be considered part of the same original clone. More recently, hybrid hypotheses have been proposed whereby some second primaries are thought to arise from a totally distinct molecular pathway, some are late recurrences or metastases of the same primary tumor, and some are derived from the same genetically altered precursor clone but have subsequently evolved along a different molecular pathway toward cancer. This latter category has been termed "SFT" or second field cancer [129,130].

V. PHENOTYPIC CHANGES ASSOCIATED WITH FIELD CARCINOGENESIS

Chronic exposure of the head and neck epithelium to carcinogens leads to a cyclical pattern of tissue damage and rehealing processes, as well as an accumulation of genetic changes that drive multistep tumorigenesis throughout the exposed field. As a result, the tissue undergoes considerable phenotypic change, some of which might be important to the tumorigenesis process and some of which may simply reflect a response to tissue injury. Phenotypic changes that are important in the tumorigenesis process might be proposed to be those that are important for developing the tumor phenotype, i.e., altering pathways that influence inappropriate cell growth, immortalization, differentiation, cell loss, ability to take over the epithelium, migrate, invade, and metastasize, and develop new vasculature [131].

One approach to identifying phenotypic changes that might be important for head and neck tumor development is to examine phenotypic changes that occur in the field of head and neck tumors. For example, immunohistochemical examination of tissue sections of tumors and their adjacent normal and premalignant lesions has demonstrated an altered phenotypic expression of molecules that influence many of these tumor-associated processes. For example, altered patterns of proliferation can be detected early during the tumorigenesis process, even in normal-appearing epithelium adjacent to the tumor [132,133]. Moreover, the levels of abnormal proliferation are found to increase as the tissue passes through the histologic progression to cancer. The driving forces for abnormal epithelial proliferation in the epithelial field are not known; however, proliferation changes are associated with an increased expression of gene products that are known to influence the initiation of proliferation, such as growth factor receptors and their ligands [134–137], products that drive cells through the cell cycle, such as cyclin D1 and DNA replication factors [93,138,139], and abnormal expression of products such as p53 that might be important in controlling growth and permitting cell death [140–142]. Another important phenotypic event that must occur during tumorigenesis is immortalization. In some cases, this may be facilitated by HPV infection in the epithelial field. Other important events in immortalization may be the activation of telomerase and the inactivation of p16, both of which apparently can occur very early during the multistep tumorigenesis process [143–145].

Many of these same types of changes can be found in lesions in the cancerization field of individuals at risk for developing head and neck cancer. While some of these phenotypic events might be important for particular functions, such as continued proliferation, their abnormal expression might also play a role in driving the tumorigenesis process itself. For example, overexpression of cyclin D1 can lead to genetic instability and facilitate gene amplification and clonal outgrowth [91–93,95]. Similarly, dysfunctional p53 expression can also facilitate the accumulation of genetic damage, as well as permit the survival of damaged cells [89]. Thus, despite a common exposure to carcinogen, the rate of multifocal progression at different regions of the cancerization field may differ according to the molecular and phenotypic events that have already accumulated at that site.

VI. CLINICAL IMPLICATIONS OF FIELD CANCERIZATION

The notion of head and neck field cancerization has important clinical implications in several settings, including first and second cancer risk estimation, assessment of response to chemopreventive intervention, and clinical management of secondary or recurrent head and neck primaries.

A. Head and Neck Tumor Risk Estimation

One of the problems facing the clinical management of individuals exhibiting so-called premalignant lesions is that it is difficult to identify which lesions are simply reactive and are not destined to progress to cancer and which lesions are likely to progress to cancer. While lesions demonstrating higher histologic progression are associated with an increased cancer risk, the prediction of cancer risk by histologic status is limited in accuracy. Moreover, it is known that nearly half of head and neck cancers develop at sites away from the premalignant lesions. These factors create two problems. First, the future site of the tumor is unknown and therefore one does not know where to biopsy. Second, a careful analysis of the target lesion for specific changes that represent late events during tumorigenesis might not provide a proper risk assessment if the future tumor site is not biopsied.

The notion that head and neck tumors arise in a field of tumorigenesis may provide an alternative approach for risk assessment. Studies in *in vitro* and *in vivo* model systems, as well as in the clinical setting, have shown a dose response between carcinogen exposure and tumor incidence. While it is true that specific changes have to occur in a particular epithelial site in order to develop the tumor phenotype, the studies described in this chapter suggest that molecular and phenotypic changes are occurring throughout the carcinogenic field. Thus, it might be predicted that those tissues carrying the greatest burden of genetic and phenotypic change might be at the highest risk for developing a cancer somewhere in the exposed epithelial field.

A number of studies have provided some support for this notion. For example, individuals whose oral and laryngeal premalignant lesions show the highest degree of generalized genetic instability, as assessed by chromosome *in situ* hybridization, have been shown to exhibit significantly higher rates of progression to head and neck cancer [86,88,146]. Similarly, individuals whose tissue fields exhibit altered p53 and cyclin D1 expression (in part due to the presence of a cyclin D1 polymorphism), both known to affect the accumulation of genetic instability, have also been reported to have an increased rate of head and cancer development [95,147–149]. Similarly, as described in more depth elsewhere in this volume, studies examining clonal outgrowths in the head and neck epithelium indicate significant potential for the assessment of head and neck cancer risk. For example, DNA content measurements of cells in oral leukoplakia have shown that individuals whose lesions show unstable aneuploid DNA develop oral squamous cell carcinomas at a much higher rate than lesions showing stable tetraploidy or stable diploidy [150]. Similarly, those individuals who show increased genetic instability and clonal outgrowth (in the form of increased frequencies of LOH) in biopsied lesions show an increased head and neck tumor risk [95,151–153].

If markers of epithelial field carcinogenesis can be used to stratify individuals with premalignant lesions according to cancer risk, this would be useful for identifying individuals who would best benefit by different chemopreventive strategies. For example, individuals with a very high risk for developing head and neck tumors might benefit from more aggressive chemoprevention approaches, such as combination treatments or even gene therapy strategies (e.g., targeting p53). The narrowing of these studies to individuals at highest risk would increase the efficacy of these prevention strategies, especially if cancer incidence is the primary end point. These subjects might also benefit from more frequent surveillance in the hope of early cancer detection. However, individuals whose epithelial fields show measures of decreased risk might benefit from strategies that slow the acquisition of new genetic changes (e.g., tobacco and alcohol cessation and antioxidants).

If measured changes in the head and neck epithelium can be used for assessing the risk of developing a first primary tumor, one might suspect that similar measurements might prove useful for predicting the likelihood of developing a second head and neck primary tumor following definitive treatment of the first primary. Along these lines, overexpression of p53 in the first primary tumor as well as overexpression in normal-appearing epithelia in the field of the first primary tumor has been associated, but not uniformly, with an increased incidence of second primaries [154–157]. Similarly, overexpression of glutathione *S*-transferase in the normal epithelium in the field of the first primary predicts second primary tumor development [158].

B. Assessment of Response to Chemopreventive Intervention

The genetic and phenotypic changes seen in the cancerization field can be used as intermediate markers of response during chemoprevention studies [159,160]. For example, premalignant lesions of the head and neck frequently show loss of expression of the retinoic acid receptor β [161]. Because retinoids have been shown to upregulate this receptor *in vitro*, one potential biomarker of chemopreventive efficacy is the reexpression of retinoic acid receptor β [162]. Retinoid trials in oral leukoplakia have suggested that individuals whose lesion show abnormal p53 expression exhibit a decreased response [163]. This has led to the generation of new treatment strategies using combinations of agents, including 13-*cis*-retinoic acid, interferon α, and α-tocopherol [164]. Measurements of the epithelial field during these trials have provided unique information. First, those individuals whose epithelial fields show the highest genetic instability and p53 expression show decreased response rates to treatment [165]. Second, downregulation of cyclin D1 is a necessary but not sufficient condition for response [149]. Third, while the treatment can result in the disappearance of histologically abnormal epithelium, clinically evident disease, and improved organ function (e.g., voice quality in the larynx), the original clonal outgrowths remain in the epithelial field [166,167]. This would suggest that the reversal of phenotypic abnormalities might precede the disappearance of genotypic changes during response. The clinical implication of this finding is that subjects might benefit from prolonged chemopreventive intervention in order to reverse the clonal outgrowth and evolution that had been established through years of chronic exposure of the epithelial field to carcinogens and their cofactors.

C. Implications for Management of Recurrent Disease or Second Primaries

The clinical approach to the management of recurrence and or metastases is generally different from that of a second primary tumor. In the case of recurrent disease or metastases, the biology of the recurring disease might be considered similar to that of the primary disease. If recurrence indicates that the tumor was somewhat refractory to the first treatment approach, then subsequent treatments might include additional approaches or modalities than that used for the primary tumor. Moreover, the next treatment would tend to involve a more systemic approach to disease control, especially if the second event is a metastasis. However, if the second primary has evolved through a distinct molecular pathway compared to the first primary, then one might consider treatments similar to that appropriate for a first primary tumor (e.g., surgical excision alone with or without radiotherapy). In addition, if a second primary tumor has occurred, one would seriously consider adjuvant chemopreventive measures to prevent additional, genetically distinct tumors from subsequently arising in the cancerization field.

As discussed earlier, it is not always possible to definitively distinguish recurrent disease from second primary tumors based solely on clinical criteria. For this reason, there is increased interest for molecularly characterizing similarities and differences between the first primary cancer and the second malignant event. If the second event shows many of the same molecular genetic changes as the first primary, then one would expect that it is either a recurrence or a metastasis. However, if the second event shares few molecular changes in common with the first primary tumor, then the second event likely represents the field cancerization process whereby multiple clones have evolved independently to the invasive state. If the second event contains both similar and different molecular changes from the first primary, it is not clear whether it will behave as a second primary or a recurrence or metastasis. It will be important to carry out retrospective studies combining molecular characterization of both the first primary and the second event with careful analysis of clinical outcome, especially in the setting where chemopreventive intervention or adjuvant treatment was undertaken in an attempt to prevent second primaries as well as prevent recurrences [31,168].

Acknowledgments

Supported by NIH DE 13157, Public Health Service Grant CA-68089 and NCI CA-52051, and EDRN 86390. W.N.H. is a Sophie Caroline Steves Professor in Cancer Research.

References

1. American Cancer Society, Cancer Facts & Figures 2002.
2. Saranath, D., Bhoite, L. T., and Deo, M. G. (1993). Molecular lesions in human oral cancer: The Indian scene. *Oral Oncol. Eur. J. Cancer* **29B**, 107–112.
3. DeVesa, S. S., Blot, W. J., Stone, B. J., Miller, B. A., Tarone, R. E., and Fraumeni, J. F. (1995). Recent cancer trends in the United States. *J. Natl. Cancer Inst.* **87**, 175–182.
4. Meyskens, F. L., Jr. (1990). Coming of age: The chemoprevention of cancer. *N. Engl. J. Med.* **323**, 825–827.
5. Vokes, E. E., Weischelbaum, R. R., Lippman, S. M., and Hong, W. K. (1993). Head and neck cancer. *N. Engl. J. Med.* **328**, 184–194.
6. Choi, S. Y., and Kahlo, H. (1991). Effect of cigarette smoking and alcohol consumption in the aetiology of cancer of the oral cavity, pharynx, and larynx. *Int. J. Epidemiol.* **20**, 878–885.
7. Mashberg, A., Boffetta, P., Winkelman, R., and Garfinkel, L. (1993). Tobacco smoking, alcohol drinking, and cancer of the oral cavity and oropharynx among U.S. veterans. *Cancer* **72**, 1369–1375.
8. Day, G. L., Blot, W. J., Shore, R. E., McLaughlin, J. K., Austin, D. F., Greenberg, R. S., Liff, J. M., Preston-Martin, S., Sarkar, S., Schoenberg, J. B., and Fraumeni, J. F., Jr. (1994). Second cancers following oral and pharyngeal cancers: Role of tobacco and alcohol. *J. Natl. Cancer Inst.* **86**, 131–137.
9. Slaughter, D. L., Southwick, H. W., and Smejkal, W. (1953). "Field cancerization" in oral stratified squamous epithelium: Clinical implications of multicentric origin. *Cancer* **6**, 963–968.
10. Farber, E. (1984). The multistep nature of cancer development. *Cancer Res.* **44**, 4217–4223.
11. Jacobs, C. D. (1990). Etiologic considerations for head and neck squamous cancers. *In* "Carcinomas of the Head and Neck: Evaluation and Management" (C. Jacobs, ed.), pp. 265–282. Kluwer Academic, Boston.
12. Hoffman, D., and Hecht, S. S. (1985). Nicotine-derived N-nitrosamines and tobacco-related cancer: Current status and future directions. *Cancer Res.* **45**, 935–944.
13. Doll, R., and Peto, R. (1981). The causes of cancer: Quantitative estimates of avoidable risks of cancer in the United States today. *J. Natl. Cancer Inst.* **66**, 1191–1308.
14. McCoy, G. D., and Wynder, E. L. (1979). Etiological and preventive implications in alcohol carcinogenesis. *Cancer Res.* **39**, 2844–2850.
15. Wynder, E. L., and Gori, G. B. (1977). Contribution of the environment to cancer: An epidemiologic exercise. *J. Natl. Cancer Inst.* **58**, 825–832.
16. Choi, S. Y., and Kahyo, H. (1991). Effect of cigarette smoking and alcohol consumption in the aetiology of cancer of the oral cavity, pharynx and larynx. *Int. J. Epidemiol.* **20**, 878–885.

17. Franceschi, S., Levi, F., La Vecchia, C., Conti, E., Dal Maso, L., Barzan, L., and Talamini, R. (1999). Comparison of the effect of smoking and alcohol drinking between oral and pharyngeal cancer. *Int. J. Cancer* **83**, 1–4.
18. Blot, W. J., McLaughlin, J. K., Winn, D. M., Austin, D. F., Greenberg, R. S., Preston-Martin, S., Bernstein, L., Schoenberg, J. B., Stemhagen, A., and Fraumeni, J. F., Jr. (1988). Smoking and drinking in relation to oral and pharyngeal cancer. *Cancer Res.* **48**, 3282–3287.
19. Franceschi, S., Munoz, N., Bosch, X. F., Snijders, P. J., and Walboomers, J. M. (1996). Human papillomavirus and cancers of the upper aerodigestive tract: A review of epidemiological and experimental evidence. *Cancer Epidemiol. Biomark. Prev.* **5**, 567–575.
20. Gillison, M. L., and Shah, K. V. (2001). Human papillomavirus-associated head and neck squamous cell carcinoma: Mounting evidence for an etiologic role for human papillomavirus in a subset of head and neck cancers. *Curr. Opin. Oncol.* **13**, 183–188.
21. Gillison, M. L., Koch, W. M., Capone, R. B., Spafford, M., Westra, W. H., Wu, L., Zahurak, M. L., Daniel, R. W., Viglione, M., Symer, D. E., Shah, K. V., and Sidransky, D. (2000). Evidence for a causal association between human papillomavirus and a subset of head and neck cancers. *J. Natl. Cancer Inst.* **92**, 709–720.
22. Mork, J., Lie, A. K., Glattre, E., Hallmans, G., Jellum, E., Koskela, P., Moller, P., Pukkala, E., Schiller, J. T., Youngman, L., Lehtinen, M., and Dillner, J. (2001). Human papillomavirus infection as a risk factor for squamous cell carcinoma of the head and neck. *N. Engl. J. Med.* **344**, 1125–1131.
23. Munger, K. (2002). The role of human papillomaviruses in human cancers. *Front. Biosci.* **7**, D641–D649.
24. Li, S.-L., Kim, M. S., Cherrick, H. M., Doniger, J., and Park, N.-H. (1992). Sequential combined tumorigenic effect of HPV-16 and chemical carcinogens. *Carcinogenesis* **13**, 1981–1987.
25. Gluckman, J. O., Crissman, J. D., and Donegan, J. O. (1980). Multicentric squamous-cell carcinoma of the upper aerodigestive tract. *Head Neck Surg.* **3**, 90–96.
26. Warren, S. and Gates, O. (1932). Multiple primary malignant tumors: A survey of the literature and statistical study. *Am. J. Cancer* **51**, 1358–1414.
27. Cahan, W. G., Castro, E. B., Posen, P. P., and Strong, E. W. (1976). Separate primary carcinomas of the esophagus and head and neck regions in the same patient. *Cancer* **37**, 85–89.
28. Vikram, B. (1984). Changing pattern of failure on advanced head and neck cancer. *Arch. Otolaryngol.* **110**, 564–565.
29. Lippman, S. M., and Hong, W. K. (1989). Second malignant tumors in head and neck squamous cell carcinoma: The overshadowing threat for patients with early-stage disease. *Int. J. Radiat. Oncol. Biol. Phys.* **17**, 691–694.
30. Day, G. L., Blot, W. J., Shore, R. E., McLaughlin, J. K., Sustin, D. F., Greenberg, R. S., Liff, J. M., Preston-Martin, S., Sarkar, S., Schoenberg, J. B., and Fraumeni, J. F., Jr. (1994). Second cancers following oral and pharyngeal cancers: Role of tobacco and alcohol. *J. Natl. Cancer Inst.* **86**, 131–137.
31. Khuri, F. R., Kim, E. S., Lee, J. J., Winn, R. J., Benner, S. E., Lippman, S. M., Fu, K. K., Cooper, J. S., Vokes, E. E., Chamberlain, R. M., Williams, B., Pajak, T. F.,Goepfert, H., and Hong, W. K. (2001). The impact of smoking status, disease stage, and index tumor site on second primary tumor incidence and tumor recurrence in the head and neck retinoid chemoprevention trial. *Cancer Epidemiol. Biomark. Prev.* **10**, 823–829.
32. Axell, T., Pindborg, J. J., Smith, C. J., and van der Waal, I. (1996). Oral white lesions with special reference to precancerous and tobacco related lesions: Conclusions of an international symposium held in Uppsala, Sweden, May 18–21 1994. International Collaborative Group on Oral White Lesions. *J. Oral Pathol. Med.* **25**, 49–54.
33. Lumerman, H., Freedman, P., and Kerpel, S. (1995). Oral epithelial dysplasia and the development of invasive squamous carcinoma. *Oral Surg. Oral. Med. Oral Pathol.* **79**, 321–329.
34. Silverman, S., Jr., Gorosky, M., and Lozada, F. (1984). Oral leukoplakia and malignant transformation: A follow-up study of 257 patients. *Cancer* **53**, 563–568.
35. Auerbach, O., Hammond, E. C., and Garfinkel, L. (1970). Histologic changes in the larynx in relation to smoking habits. *Cancer* **25**, 92–104.
36. Blackwell, K. E., Calcaterra, T. C., and Fu, Y. S. (1995). Laryngeal dysplasia: Epidemiology and treatment outcome. *Ann. Otol. Rhinol. Laryngol.* **104**, 596–602.
37. Morrison, M. D. (1988). Is chronic gastroesophageal reflux a causative factor in glottic carcinoma? *Otolaryngol. Head Neck Surg.* **99**, 370–373.
38. Ward, P. H., and Hanson, D. G. (1988). Reflux as an etiological factor of carcinoma of the laryngopharynx. *Laryngoscope* **98**, 1195–1199.
39. Koufman, J. A. (1991). The otolaryngologic manifestations of gastroesophageal reflux disease (GERD: A clinical investigation of 225 patients using ambulatory 24-hour pH monitoring and an experimental investigation of the role of acid and pepsin in the development of laryngeal injury. *Laryngoscope* **101**(Suppl. 53), 1–78.
40. Hojslet, P. E., Nielsen, V. M., and Palvio, D. (1989). Premalignant lesions of the larynx. A follow-up study. *Acta Oto-Laryngol.* **107**, 150–155.
41. Zeitels, S. M. (1995). Premalignant epithelium and microinvasive cancer of the vocal fold: The evolution of phonomicrosurgical management. *Laryngoscope* **105**, 1–51.
42. Hellquist, H., Lundgren, J., and Olofsson, J. (1982). Hyperplasia, keratosis, dysplasia and carcinoma in situ of the vocal cords: A follow-up study. *Clin. Otolaryngol.* **7**, 11–27.
43. Puttney, F. J., and O'Keefe, J. J., (1953). The clinical significance of keratosis of the larynx as a premalignant lesion. *Arch. Otolaryngol.* **62**, 348–357.
44. Shklar, G. (1984). Experimental pathology of oral cancer. *In* "Oral Cancer" (G. Shklar, ed.), pp. 41–54. Saunders, Philadelphia.
45. Steidler, N. E., and Reade, P. C. (1984). Experimental induction of oral squamous cell carcinomas in mice with 4-nitroquinoline-1-oxide. *Oral Surg. Oral Med. Oral Path.* **57**, 524–531.
46. Prime, S. S., Malamos, D., Rosser, T. J., and Scully, C. (1986). Oral epithelial atypia and acantholytic dyskeratosis in rats painted with 4-nitroquinoline N-oxide. *J. Oral. Path.* **15**, 280–283.
47. Bissell, M. J., and Radisky, D. (2001). Putting tumours in context. *Nature Rev. Cancer* **1**, 46–54.
48. Tlsty, T. D. (2001). Stromal cells can contribute oncogenic signals. *Semin. Cancer Biol.* **11**, 97–104.
49. Radisky, D., Hagios, C., and Bissell, M. J. (2001). Tumors are unique organs defined by abnormal signaling and context. *Semin. Cancer Biol.* **11**, 87–95.
50. Sacks, P. G. (1996). Cell, tissue and organ culture as *in vitro* models to study the biology of squamous cell carcinomas of the head and neck. *Cancer Metastas. Rev.* **15**, 27–51.
51. Auerbach, O., Stout, A. P., Hammond, E. C., and Garfinkel, L. (1961). Changes in bronchial epithelium in relation to cigarette smoking and in relation to lung cancer. *N. Engl. J. Med.* **265**, 253–267.
52. Muller, K. M., and Krohn, B. R. (1980). Smoking habits and their relationship to precancerous lesions of the larynx. *J. Cancer Res. Clin. Oncol.* **96**, 211–217.
53. Grasl, M. C., Neuwirth-Riedl, K., Vutuc, C., Horak, F., Vorbeck, F., and Banyai, M. (1990). Risk of vocal chord dysplasia in relation to smoking, alcohol intake, and occupation. *Eur. J. Epidemiol.* **6**, 45–48.
54. Stell, P. M., and Watt, J. (1984). Squamous metaplasia of the subglottic space and its relation to smoking. *Ann. Otol. Rhinol. Laryngol.* **93**, 124–126.
55. Randerath, K., Reddy, M. V., and Gupta, R. C. (1981). ^{32}P-labeling test for DNA damage. *Proc. Natl. Acad. Sci. USA* **78**, 6126–6129.

56. Hirabayashi, H., Koshii, K., Uno, K., Ohgaki, H., Nakasone, Y., Fujisawa, T., Shono, N., Hinohara, T., and Hirabayashi, K. (1990). Laryngeal epithelial changes on effects of smoking and drinking. *Auris Nasus Larynx* **17**, 105–114.
57. Jones, N. J., McGregor, A. D., and Waters, R. (1993). Detection of DNA adducts in human oral tissue: Correlation of adduct levels with tobacco smoking and differential enhancement of adducts using the butanol extraction and nuclease P1 versions of ^{32}P-postlabeling. *Cancer Res.* **53**, 1522–1528.
58. Nath, R. G., Ocando, J. E., Guttenplan, J. B., and Chung, F. L. (1998). 1,N2-propanodeoxyguanosine adducts: Potential new biomarkers of smoking-induced DNA damage in human oral tissue. *Cancer Res.* **58**, 581–584.
59. Stone, J. G., Jones, N. J., MeGregor, A. D., and Waters, R. (1995). Development of a human biomonitoring assay using buccal mucosa: Comparison of smoking-related DNA adducts in mucosa versus biopsies. *Cancer Res.* **55**, 1267–1270.
60. Gupta, R. C. (1985). Enhanced sensitivity of ^{32}P-postlabeling analysis of aromatic carcinogen:DNA adducts. *Cancer Res.* **45**, 5656–5662.
61. Besarati Nia, A., Van Straaten, H. W., Godschalk, R. W., Van Zandwijk, N., Balm, A. J., Kleinjans, J. C., and Van Schooten, F. J. (2000). Immunoperoxidase detection of polycyclic aromatic hydrocarbon-DNA adducts in mouth floor and buccal mucosa cells of smokers and nonsmokers. *Environ. Mol. Mutagen* **36**, 123–133.
62. Zhang, Y. J., Hsu, T. M., and Santella, R. M. (1995). Immunoperoxidase detection of polycyclic aromatic hydrocarbon-DNA adducts in oral mucosa of smokers and nonsmokers. *Cancer Epidemiol. Biomark. Prev.* **4**, 133–138.
63. Romano, G., Mancini, R., Fedele, P., Curigliano, G., Flamini, G., Giovagnoli, M. R., Malara, N., Boninsegna, A., Vecchione, A., Santella, R. M., and Cittadini, A. (1997). Immunohistochemical analysis of 4-aminobiphenyl-DNA adducts in oral mucosal cells of smokers and nonsmokers. *Anticancer Res.* **17**, 2827–2830.
64. Zhang, Y., Chen, S. Y., Hsu, T., and Santella, R. M. (2002). Immunohistochemical detection of malondialdehyde-DNA adducts in human oral mucosa cells. *Carcinogenesis* **23**, 207–211.
65. Stich, H. F., and Rosin, M. P. (1984). Micronuci in exfolliated human cells as a tool for studies in cancer risk and intervention. *Cancer Lett.* **22**, 241–253.
66. Stich, H. F., and Rosin, M. P. (1983). Quantitating the synergistic effect of smoking and alcohol consumption with the micronucleus test on human buccal mucosa cells. *Int. J. Cancer* **31**, 305–308.
67. Hsu, T. C., Furlong, C., and Spitz, M. R. (1991). Ethyl alcohol as a cocarcinogen with special reference to the aerodigestive tract: A cytogenetic study. *Anticancer Res.* **11**, 1097–1101.
68. Hsu, T. C., and Furlong, C. (1991). The role of ethanol in oncogenesis of the upper aerodigestive tract; inhibition of DNA repair. *Anticancer Res.* **11**, 1995–1998.
69. Hsu, T. C., Cherry, L. M., Bucana, C., Shirley, L. R., and Gairola, C. G. (1991). Mitosis-arresting effects of cigarette smoke condensate on human lymphoid cell lines. *Mutat. Res.* **259**, 67–78.
70. Teyssier, J. R. (1989). The chromosomal analysis of solid tumors: A triple challenge. *Cancer Genet. Cytogenet.* **37**, 103–125.
71. Jin, Y. S., Heim, S., Mandahl, N., Biorklund, A., Wennerberg, J., and Mitelman, F. (1988). Multiple apparently unrelated clonal chromosome abnormalities in a squamous cell carcinoma of the tongue. *Cancer Genet. Cytogenet.* **32**, 93–100.
72. Heim, S., Mandahl, N., and Mitelman, F. (1988). Genetic convergence and divergence in tumor progression. *Cancer Res.* **48**, 5911–5916.
73. Jin, Y. S., Heim, S., Mandahl, N., Biorklund, A., Wennerberg, J. and Mitelman, F. (1990). Unrelated clonal chromosomal aberrations in carcinomas of the oral cavity. *Genes Chromosomes Cancer* **1**, 209–215.
74. Jin, Y. S., Heim, S., Mandahl, N., Biorklund, A., Wennerberg, J., and Mitelman F. (1990). Multiple clonal chromosome aberrations in squamous cell carcinomas of the larynx. *Cancer Genet. Cytogenet.* **44**, 209–216.
75. Mertens, F., Jin, Y., Heim, S., Mandahl, N., Jonsson, N., Mertens, O., Persson, B., Salemark, L., Wennerberg, J., and Mitelman, F. (1992). Clonal structural chromosome aberrations in nonneoplastic cells of the skin and upper aerodigestive tract. *Genes Chromosomes Cancer* **4**, 235–240.
76. Hittelman, W. N. (1982). Premature chromosome condensation in the diagnosis of malignancies. *In* "Premature Chromosome Condensation" (R. Johnson, P. N. Rao, and K. Sperling, eds.), pp. 309–358. Academic Press, New York.
77. Hittelman, W. N., Lee, J. S., Cheong, N., Shin, D. M., and Hong, W. K. (1991). The chromosome view of "field cancerization" and multi-step carcinogenesis: Implications for chemopreventive approaches. *In* "Chemoimmuno Prevention of Cancer" (V. Pastorino and W. K. Hong, eds.), pp. 41–47. Georg Thieme Verlag, Stuttgard.
78. Sacks, P. G., Hong, W. K., and Hittelman, W. N. (1991). *In vitro* studies of the premalignant process: Initial culture of oral premalignant lesions. *Cancer Bull.* **43**, 485–489.
79. Pinkel, D., Straume, T., and Gray, J. W. (1986). Cytogenetic analysis using quantitative, high-sensitivity, fluorescence hybridization. *Proc. Natl. Acad. Sci. USA* **83**, 2934–2938.
80. Lichter, P., Cremer, T., Borden, J., Manuelidis, L., and Ward, D. C. (1988). Delineation of individual human chromosomes in metaphase and interphase cells by in situ suppression hybridization using recombinant DNA libraries. *Hum. Genet.* **80**, 224–234.
81. Cremer, T., Lichter, P., Borden, J., Ward, D. C., and Manuelidis, L. (1988). Detection of chromosome aberrations in metaphase and interphase tumor cells by in situ hybridization using chromosome-specific library probes. *Hum. Genet.* **80**, 235–246.
82. Hopman, A. H., Voorter, C. E., and Ramaekers, F. C. (1994). Detection of genomic changes in cancer by in situ hybridization. *Mol. Biol. Rep.* **19**, 31–44.
83. Voravud, N., Shin, D. M., Ro, J. Y., Lee, J. S., Hong, W. K., and Hittelman, W. N. (1993). Increased polysomies of chromosomes 7 and 17 during head and neck multistage tumorigenesis. *Cancer Res.* **53**, 2874–2883.
84. Ai, H., Barrera, J. E., Meyers, A. D., Shroyer, K. R., and Varella-Garcia, M. (2001). Chromosomal aneuploidy precedes morphological changes and supports multifocality in head and neck lesions. *Laryngoscope* **111**, 1853–1858.
85. Barrera, J. E., Ai, H., Pan, Z., Meyers, A. D., and Varella-Garcia, M. (1998). Malignancy detection by molecular cytogenetics in clinically normal mucosa adjacent to head and neck tumors. *Arch. Otolaryngol. Head Neck Surg.* **124**, 847–851.
86. Lee, J. S., Kim, S. Y., Hong, W. K., Lippman, S. M., Ro, J. Y., Gay, M. L., and Hittelman, W. N. (1993). Detection of chromosomal polysomy in oral leukoplakia, a premalignant lesion. *J. Natl. Cancer Inst.* **85**, 1951–1954.
87. Kim, J., Shin, D. M., El-Naggar, A., Lee, J. S., Corrales, C., Lippman, S. M., Hong, W. K., and Hittelman, W. N. (2001). Chromosome polysomy and histological characteristics in oral premalignant lesions. *Cancer Epidemiol. Biomark. Prev.* **10**, 319–325.
88. Hittelman, W. N., Papadimitrakopoulou, V., El-Naggar, A. K., Clayman, G., Myers, J., Lee, J. S., Corrales, C., Lee, J. J., Hong, W. K., and Shin, D. M. (2000). High genetic instability is associated with poor therapeutic outcome following biochempreventive treatment of advanced premalignant lesions (APL) of the larynx and oral cavity. *Proc. AACR* **41**, 223.
89. Shin, D. M., Charuruks, N., Lippman, S. M., Lee, J. J., Ro, J. Y., Hong, W. K., and Hittelman, W. N. (2001). p53 protein accumulation and genomic instability in head and neck multistep tumorigenesis. *Cancer Epidemiol. Biomark. Prev.* **10**, 603–609.
90. Hittelman, W. N. (1999). Clones and subclones in the lung cancer field. *J. Natl. Cancer Inst.* **91**, 1796–1799.
91. Izzo, J., Papadimitrakopoulou, V., El-Naggar, A., Lee, J. J., Hong, W. K., and Hittelman, W. N. (1999). Detection of clonal genetic

events preceding tetraploidization and cyclin D1 (CCND1) gene amplification during head and neck (HNSCC) tumorigenesis. *Proc. AACR* **40**, 945.
92. Koster, M. I., Izzo, J., El-Naggar, A., Hong, W. K., and Hittelman, W. N. (1999). Expression of fragile sites (FRA) on chromosome 11q13 is an early event during head and neck (HNSCC) tumorigenesis, preceding gene amplification. *Proc. AACR* **40**, 4547.
93. Izzo, J. G., Papadimitrakopoulou, V., Li, X. Q., Ibarguen, H., Lee, J. S., Ro, J. Y., Hong, W. K., and Hittelman, W. N. (1998). Ectopic overexpression of cyclin D1 in head and neck tumorigenesis: In vitro evidence for an early event, enhancing gene amplification. *Oncogene* **17**, 2313–2322.
94. Roh, H. J., Shin, D. M., Lee, J. S., Ro, J. Y., Tainsky, M. A., Hong, W. K., and Hittelman, W. N. (2000). Visualization of the timing of gene amplification during multistep head and neck tumorigenesis. *Cancer Res.* **60**, 6496–6502.
95. Papadimitrakopoulou, V. A., Izzo, J., Mao, L., Keck, J., Hamilton, D., Shin, D. M., El-Naggar, A., den Hollander, P., Liu, D., Hittelman, W. N., and Hong, W. K. (2001). Cyclin D1 and p16 alterations in advanced premalignant lesions of the upper aerodigestive tract: Role in response to chemoprevention and cancer development. *Clin. Cancer Res.* **7**, 3127–3124.
96. Van Dyke, D. L., Worsham, M. J., Benninger, M. S., Krause, C. J., Baker, S. R., Wolf, G. T., Drumheller, T., Tilley, B. C., and Carey, T. E. (1994). Recurrent cytogenetic abnormalities in squamous cell carcinomas of the head and neck region. *Genes Chromosomes Cancer* **9**, 192–206.
97. Gollin, S. M. (2001). Chromosomal alterations in squamous cell carcinomas of the head and neck: Window to the biology of disease. *Head Neck* **23**, 238–253.
98. Bockmuhl, U., Wolf, G., Schmidt, S., Schwendel, A., Jahnke, V., Dietel, M., and Petersen, I. (1998). Genomic alterations associated with malignancy in head and neck cancer. *Head Neck* **20**, 145–151.
99. Nawroz, H., van der Riet, P., Hruban, R., Koch, W., Ruppert, J., and Sidransky, D. (1994). Allelotype of head and neck squamous cell carcinoma. *Cancer Res.* **54**, 1152–1155.
100. Ah-See, K. W., Cooke, T. G., Pickford, I. R., Soutar, D., and Balmain, A. (1994). An allelotype of squamous carcinoma of the head and neck using microsatellite markers. *Cancer Res.* **54**, 1617–1621.
101. El Naggar, A. K., Coombes, M. M., Batsakis, J. G., Hong, W. K., Goepfert, H., and Kagan, J. (1998). Localization of chromosome 8p regions involved in early tumorigenesis of oral and laryngeal squamous carcinoma. *Oncogene* **16**, 2983–2987.
102. Cairns, P., Polascik, T., Eby, Y., Tokino, K., Califano, J., Merlo, A., Mao, L., Herath, J., Jenkins, R., Westra, W., *et al.* (1995). Frequency of homozygous deletion at p16/CDKN2 in primary human tumours. *Nature Genet.* **11**, 210–212.
103. Baylin, S. B., Herman, J. G., Graff, J. R., Vertino, P. M., and Issa, J. P. (1998). Alterations in DNA methylation: A fundamental aspect of neoplasia. *Adv. Cancer Res.* **72**, 141–196.
104. Sanchez-Cespedes, M., Esteller, M., Wu, L., Nawroz-Danish, H., Yoo, G. H., Koch, W. M., Jen, J., Herman, J. G., and Sidransky, D. (2000). Gene promoter hypermethylation in tumors and serum of head and neck cancer patients. *Cancer Res.* **60**, 892–895.
105. Lengauer, C., Kinzler, K. W., and Vogelstein, B. (1998). Genetic instabilities in human cancers. *Nature* **396**, 643–649.
106. Saunders, W. S., Shuster, M., Huang, X., Gharaibeh, B., Enyenihi, A. H., Petersen, I., and Gollin, S. M. (2000). Chromosomal instability and cytoskeletal defects in oral cancer cells. *Proc. Natl. Acad. Sci. USA* **97**, 303–308.
107. Sidransky, D. (1995). Molecular genetics of head and neck cancer. *Curr. Opin. Oncol.* **7**, 229–233.
108. Mao, L. (2000). Can molecular assessment improve classification of head and neck premalignancy? *Clin. Cancer Res.* **6**, 321–322.
109. El Naggar, A. K., Lai, S., Luna, M. A., Zhou, X. D., Weber, R. S., Goepfert, H., and Batsakis, J. G. (1995). Sequential p53 mutation analysis of pre-invasive and invasive head and neck squamous carcinoma. *Int. J. Cancer* **64**, 196–201.
110. El Naggar, A. K., Hurr, K., Batsakis, J. G., Luna, M. A., Goepfert, H., and Huff, V. (1995). Sequential loss of heterozygosity at microsatellite motifs in preinvasive and invasive head and neck squamous carcinoma. *Cancer Res.* **55**, 2656–2659.
111. Boyle, J. O., Hakim, J., Koch, W., van der Riet, P., Hruban, R. H., Roa, R. A., Correo, R., Eby, Y. J., Ruppert, J. M., and Sidransky, D. (1993). The incidence of p53 mutations increases with progression of head and neck cancer. *Cancer Res.* **53**, 4477–4480.
112. Zhang, L., Michelsen, C., Cheng, X., Zeng, T., Priddy, R., and Rosin, M. P. (1997). Molecular analysis of oral lichen planus: A premalignant lesion? *Am. J. Pathol.* **8**, 323–327.
113. Califano, J., van der Riet, P., Westra, W., Nawroz, H., Clayman, G., Piatadosi, S., Corio, R., Lee, D., Greenberg, B., Koch, W., and Sidransky, D. (1996). Genetic progression model for head and neck cancer: Implications for field cancerization. *Cancer Res.* **56**, 2488–2492.
114. Roz, L., Wu, C. L., Porter, S., Scully, C., Speight, P., Read, A., Sloan, P., and Thakker, N. (1996). Allelic imbalance on chromosome 3p in oral dysplastic lesions: An early event in oral carcinogenesis. *Cancer Res.* **56**, 1228–1231.
115. Mao, L., Lee, J. S., Fan, Y. H., Ro, J. Y., Batsakis, J. G., Lippman, S., Hittelman, W., and Hong, W. K. (1996). Frequent microsatellite alterations at chromosomes 9p21 and 3p14 in oral premalignant lesions and their value in cancer risk assessment. *Nature Med.* **2**, 682–685.
116. Emilion, G., Langdon, J. D., Speight, P., and Partridge, M. (1996). Frequent gene deletions in potentially malignant oral lesions. *Br. J. Cancer* **73**, 809–813.
117. Akin, N. B., and Baker, M. C. (1991). Squamous cell carcinomas of the head, neck, and skin. Monoclonal or polyclonal origin? *Cancer Genet. Cytogenet.* **54**, 135–136.
118. Chung, K. Y., Mukhopadhyay, T., Kim, J., Casson, A., Ro, J. Y., Goepfert, H., Hong, W. K., and Roth, J. A. (1993). Discordant p53 gene mutations in primary head and neck cancers and corresponding second primary cancers of the upper aerodigestive tract. *Cancer Res.* **53**, 1676–1683.
119. Kanjilal, S., Strom, S. S., Clayman, G. L., Weber, R. S., el-Naggar, A. K., Kapur, V., Cummings, K. K., Hill, L. A., Spitz, M. R., Kripke, M. L., and Ananthaswamy, H. N. (1995). p53 mutations in nonmelanoma skin cancer of the head and neck: Molecular evidence for field cancerization. *Cancer Res.* **55**, 3604–3609.
120. Scholes, A. G., Woolgar, J. A., Boyle, M. A., Brown, J. S., Vaughan, E. D., Hart, C. A., Jones, A. S., and Field, J. K. (1998). Synchronous oral carcinomas: Independent or common clonal origin? *Cancer Res.* **58**, 2003–2006.
121. Nunn, J., Scholes, A. G., Liloglou, T., Nagini, S., Jones, A. S., Vaughan, E. D., Gosney, J. R., Rogers, S., Fear, S., and Field, J. K. (1999). Fractional allele loss indicates distinct genetic populations in the development of squamous cell carcinoma of the head and neck (SCCHN). *Carcinogenesis* **20**, 2219–2228.
122. Jang, S. J., Chiba, I., Hirai, A., Hong, W. K., and Mao, L. (2001). Multiple oral squamous epithelial lesions: Are they genetically related? *Oncogene* **20**, 2235–2242.
123. Bedi, G. C., Westra, W. H., Gabrielson, E., Koch, W., and Sidransky, D. (1996). Multiple head and neck tumors: Evidence for a common clonal origin. *Cancer Res.* **56**, 2484–2487.
124. Califano, J., Leong, P. L., Koch, W. M., Eisenberger, C. F., Sidransky, D., and Westra, W. H. (1999). Second esophageal tumors in patients with head and neck squamous cell carcinoma: An assessment of clonal relationships. *Clin. Cancer Res.* **5**, 1862–1867.
125. Califano, J., Westra, W. H., Meininger, G., Corio, R., Koch, W. M., and Sidransky, D. (2000). Genetic progression and clonal relationship

of recurrent premalignant head and neck lesions. *Clin. Cancer Res.* **6**, 347–352.

126. Brennan, J. A., Mao, L., Hruban, R. H., Boyle, J. O., Eby, Y. J., Koch, W. M., Goodman, S. N., and Sidransky, D. (1995). Molecular assessment of histopathological staging in squamous-cell carcinoma of the head and neck. *N. Engl. J. Med.* **332**, 429–435.
127. Partridge, M., Li, S. R., Pateromichelakis, S., Francis, R., Phillips, E., Huang, X. H., Tesfa-Selase, F., and Langdon, J. D. (2000). Detection of minimal residual cancer to investigate why oral tumors recur despite seemingly adequate treatment. *Clin. Cancer Res.* **6**, 2718–2725.
128. Califano, J., van der Riet, P., Westra, W., Nawroz, H., Clayman, G., Piantadosi, S., Corio, R., Lee, D., Greenberg, B., Koch, W., and Sidransky, D. (1996). Genetic progression model for head and neck cancer: Implications for field cancerization. *Cancer Res.* **56**, 2488–2492.
129. Hittelman, W. N. (1999). Molecular cytogenetic evidence for multistep tumorigenesis: Implications for risk assessment and early detection. *In* "Molecular Pathology of Early Cancer" (S. Srivastava, D. E. Henson, and A. Gazdar, eds.), pp. 385–404. IOS Press, Van Diemanstratt, Netherlands.
130. Braakhuis, B. J., Tabor, M. P., Rene Leemans, C., van Der Waal, I., Snow, G. B., and Brakenhoff, R. H. (2002). Second primary tumors and field cancerization in oral and oropharyngeal cancer: Molecular techniques provide insights and definitions. *Head Neck* **24**, 198–206.
131. Hanahan, D., and Weinberg, R. A. (2000). The hallmarks of cancer. *Cell* **100**, 57–70.
132. Shin, D. M., Voravud, N., Ro, J. Y., Lee, J. S., Hong, W. K., and Hittelman, W. N. (1993). Sequential upregulation of proliferating cell nuclear antigen in head and neck tumorigenesis: A potential biomarker. *J. Natl. Cancer Inst.* **85**, 971–978.
133. Kotelnikov, V. M., Coon, J. S., Taylor, S., Hutchinson, J., Panje, W, Caldareill, D D., LaFollette, S., and Preisler H. D. (1996). Proliferation of epithelia of noninvolved mucosa in patients with head and neck cancer. *Head Neck* **18**, 522–528.
134. Shin, D. M., Ro, J. Y., Hong, W. K., and Hittelman, W. N. (1994). Dysregulation of epidermal growth factor receptor expression in premalignant lesions during head and neck tumorigenesis. *Cancer Res.* **54**, 3153–3159.
135. Grandis, J. R., Tweardy, D. J., and Melhem, M. F. (1998). Asynchronous modulation of transforming growth factor alpha and epidermal growth factor receptor protein expression in progression of premalignant lesions to head and neck squamous cell carcinoma. *Clin. Cancer Res.* **4**, 13–20.
136. Lango, M. N., Dyer, K. F., Lui, V. W., Gooding, W. E., Gubish, C., Siegfried, J. M., and Grandis J. R. (2002). Gastrin-releasing peptide receptor-mediated autocrine growth in squamous cell carcinoma of the head and neck. *J. Natl. Cancer Inst.* **94**, 375–383.
137. Nakayama, H., Ikebe, T., Beppu, M., and Shirasuna, K. (2001). High expression levels of nuclear factor kappaB, IkappaB kinase alpha and Akt kinase in squamous cell carcinoma of oral cavity. *Cancer* **92**, 3037–3044.
138. Bartkova, J., Lukas, J., Muller, H., Strauss, M., Gusterson, B., and Bartek, J. (1995). Abnormal patterns of D-type cyclin expression and G1 regulation in human head and neck cancer. *Cancer Res.* **55**, 949–956.
139. Kodani, I., Shomori, K., Osaki, M., Kuratate, I., Ryoke, K., and Ito, H. (2001). Expression of minichromosome maintenance 2 (MCM2), Ki-67, and cell-cycle-related molecules, and apoptosis in the normal-dysplasia-carcinoma sequence of the oral mucosa. *Pathobiology* **69**, 150–158.
140. Nees, M., Homann, N., Discher, H., Andl, T., Enders, C., Herold-Mende, C., Schuhmann, A., and Bosch, F. X. (1993). Expression of mutated p53 occurs in tumor-distant epithelia of head and neck cancer patients: A possible molecular basis for the development of multiple tumors. *Cancer Res.* **53**, 4189–4196.
141. Zhang, L., Rosin, M., Priddy, R., and Xiao, Y. (1993). p53 expression during multistage human oral carcinogenesis. *Int. J. Oncol.* **3**, 735–739.
142. Shin, D. M., Kim, J., Ro, J. Y., Hittelman, J., Roth, J. A., Hong, W. K., and Hittelman, W. N. (1994). Activation of p53 gene expression in premalignant lesions during head and neck tumorigenesis. *Cancer Res.* **54**, 321–326.
143. Mao, L., El-Naggar, A. K., Fan, Y. H., Lee, J. S., Lippman, S. M., Kayser, S., Lotan, R., and Hong, W. K. (1996). Telomerase activity in head and neck squamous cell carcinoma and adjacent tissues *Cancer Res.* **56**, 5600–5604.
144. Kim, H. R., Christensen, R., Park, N. H., Sapp, P., Kang, M. K., and Park, N. H. (2001). Elevated expression of hTERT is associated with dysplastic cell transformation during human oral carcinogenesis in situ. *Clin. Cancer Res.* **7**, 3079–3086.
145. Papadimitrakopoulou, V., Izzo, J., Lippman, S. M., Lee, J. S., Fan, Y. H., Clayman, G., Ro, J. Y., Hittelman, W. N., Lotan, R., Hong, W. K., and Mao, L. (1997). Frequent inactivation of p16INK4a in oral premalignant lesions. *Oncogene* **14**, 1799–1803.
146. Lee, J. J., Hong, W. K., Hittelman, W. N., Mao, L., Lotan, R., Shin, D. M., Benner, S. E., Xu, X.-C., Lee, J. S., Papadimitrakoupoulou, V. M., Geyer, C., Perez, C., Martin, J. W., El-Naggar, A. K., and Lippman, S. M. (2000). Predicting cancer development in oral leukoplakia: Ten years of translational research. *Clin. Cancer Res.* **6**, 1702–1710.
147. Uhlman, D. L., Adams, G., Knapp, D., Aeppli, D. M., and Niehans, G. (1996). Immunohistochemical staining for markers of future neoplastic progression in the larynx. *Cancer Res.* **56**, 2199–2205.
148. Cruz, I. B., Snijders, P. J., Meijer, C. J., Braakhuis, B. J., Snow, G. B., Walboomers, J. M., and van der Waal, I. (1998). p53 expression above the basal cell layer in oral mucosa is an early event of malignant transformation and has predictive value for developing oral squamous cell carcinoma. *J. Pathol.* **184**, 360–368.
149. Izzo, J. G., Papadimitrakopoulou, V. A., Den Hollander, P. L. C., Liu, D., El-Naggar, A., Hong, W. K., and Hittelman, W. N. (2002). Cyclin D1 A870G polymorphism enhances genomic instability during head and neck tumorigenesis. *Proc. AACR* **43**, 293.
150. Sudbo, J., Kildal, W., Risberg, B., Koppang, H. S., Danielsen, H. E., and Reith, A. (2001). DNA content as a prognostic marker in patients with oral leukoplakia. *N. Engl. J. Med.* **344**, 1270–1278.
151. Sudbo, J, Kildal, W., Johannessen, A. C., Koppang, H. S., Sudbo, A., Danielsen, H. E., Risberg, B., and Reith, A. (2002). Gross genomic aberrations in precancers: Clinical implications of a long-term follow-up study in oral erythroplakias. *J. Clin. Oncol.* **20**, 456–462.
152. Rosin, M. P., Cheng, X., Poh, C., Lam, W. L., Huang, Y., Lovas, J., Berean, K., Epstein, J. B., Priddy, R., Le, N. D., and Zhang, L. (2000). Use of allelic loss to predict malignant risk for low-grade oral epithelial dysplasia. *Clin. Cancer Res.* **6**, 357–62.
153. Partridge, M., Pateromichelakis, S., Phillips, E., Emilion, G. G., A'Hern, R. P., and Langdon, J. D. (2000). A case-control study confirms that microsatellite assay can identify patients at risk of developing oral squamous cell carcinoma within a field of cancerization. *Cancer Res.* **60**, 3893–3898.
154. Shin, D. M., Lee, J. S., Lippman, S. M., Lee, J. J., Tu, N., Choi, G., Heyne, K., Shin, H. J. C., Ro, J. Y., Goepfert, H., Hong, W. K., and Hittelman, W. N. (1996). p53 expression predicts early recurrence and second primary tumors in head and neck squamous cell carcinoma. *J. Natl. Cancer Inst.* **88**, 519–529.
155. Homann, N., Nees, M., Conradt, C., Dietz, A., Weidauer, H., Maier, H., and Bosch, F. X. (2001). Overexpression of p53 in tumor-distant epithelia of head and neck cancer patients is associated with an increased incidence of second primary carcinoma. *Clin. Cancer Res.* **7**, 290–296.
156. Cruz, I. B, Meijer, C. J., Snijders, P. J, Snow, G. B., Walboomers, J. M., and van Der Waal, I. (2000). p53 immunoexpression in non-malignant

oral mucosa adjacent to oral squamous cell carcinoma: Potential consequences for clinical management. *J. Pathol.* **191**, 132–137.

157. Bongers, V., Snow, G. B., van der Waal, I., and Braakhuis, B. J. (1995). Value of p53 expression in oral cancer and adjacent normal mucosa in relation to the occurrence of multiple primary carcinomas. *Eur. J. Cancer B Oral. Oncol.* **31**, 392–395.
158. Bongers, V., Snow, G. B., de Vries, N., Cattan, A. R., Hall, A. G., van der Waal, I., and Braakhuis B. J. (1995). Second primary head and neck squamous cell carcinoma predicted by the glutathione S-transferase expression in healthy tissue in the direct vicinity of the first tumor. *Lab. Invest.* **73**, 503–510.
159. Lippman, S. M., Lee, J. S., Lotan, R., Hittelman, W., Wargovich, M. J. and Hong, W. K. (1990). Biomarkers as intermediate endpoints in chemoprevention trials. *J. Natl. Cancer Inst.* **82**, 555–560.
160. Papadimitrakopoulou, V. A., Shin, D. M., and Hong, W. K. (1996). Molecular and cellular biomarkers for field cancerization and multistep process in head and neck tumorigenesis. *Cancer Metastas. Rev.* **15**, 53–76.
161. Xu, X. C., Ro, J. Y., Lee, J. S., Shin, D. M., Hong, W. K., and Lotan, R. (1994). Differential expression of nuclear retinoid receptors in normal, premalignant, and malignant head and neck tissues. *Cancer Res.* **54**, 3580–3587.
162. Lotan, R., Xu, X. C., Lippman, S. M., Ro, J. Y., Lee, J. S., Lee, J. J., and Hong, W. K. (1995). Suppression of retinoic acid receptor-beta in premalignant oral lesions and its up-regulation by isotretinoin. *N. Engl. J. Med.* **332**, 1405–1410.
163. Shin, D. M, Xu, X. C., Lippman, S. M., Lee, J. J., Lee, J. S., Batsakis, J. G., Ro, J. Y., Martin, J. W., Hittelman, W. N., Lotan, R., and Hong, W. K. (1997). Accumulation of p53 protein and retinoic acid receptor beta in retinoid chemoprevention. *Clin. Cancer Res.* **3**, 875–880.
164. Papadimitrakopoulou, V. A., Clayman, G. L., Shin, D. M., Myers, J. N., Gillenwater, A. M., Goepfert, H., El-Naggar, A. K., Lewin, J. S., Lippman, S. M., and Hong W. K. (1999). Biochemoprevention for dysplastic lesions of the upper aerodigestive tract. *Arch. Otolaryngol. Head Neck Surg.* **125**, 1083–1089.
165. Shin, D. M., Mao, L., Papadimitrakopoulou, V. M., Clayman, G., El-Naggar, A., Shin, H. J., Lee, J. J., Lee, J. S., Gillenwater, A., Myers, J., Lippman, S. M., Hittelman, W. N., and Hong, W. K. (2000). Biochemopreventive therapy for patients with premalignant lesions of the head and neck and p53 gene expression. *J. Natl. Cancer Inst.* **92**, 69–73.
166. Mao L., El-Naggar A. K., Papadimitrakopoulou, V., Shin, D. M., Shin, H. C., Fan, Y., Zhou, X., Clayman, G., Lee, J. J., Lee, J. S., Hittelman, W. N., Lippman, S. M., and Hong, W. K. (1998). Phenotype and genotype of advanced premalignant head and neck lesions after chemopreventive therapy. *J. Natl. Cancer Inst.* **90**, 1545–1551.
167. Hittelman, W. N., Shin, D. M., Mao, L., Papadimitrakopoulou, V. A., Lee, J., El-Naggar, A. K., Lippman, S. M., Clayman, G., Myers, J. N., Shin, H. J., Lee, J. S., and Hong, W. K. (2000). Pretreatment biomarkers predict response of advanced head and neck premalignant lesions to biochemoprevention. *Proc. ASCO* **19**, 414a.
168. Shin, D. M., Khuri, F. R., Murphy, B., Garden, A. S., Clayman, G., Francisco, M., Liu, D., Glisson, B. S., Ginsberg, L., Papadimitrakopoulou, V., Myers, J., Morrison, W., Gillenwater, A., Ang, K. K., Lippman, S. M., Goepfert, H., and Hong, W. K. (2001). Combined interferon-alfa, 13-cis-retinoic acid, and alpha-tocopherol in locally advanced head and neck squamous carcinoma: Novel bioadjuvant phase II trial. *J. Clin. Oncol.* **19**, 3010–3017.

CHAPTER

17

Molecular Markers of Oral Premalignant Lesion Risk

MIRIAM P. ROSIN,*,† LEWEI ZHANG,‡ and CATHERINE POH†,‡

*British Columbia Cancer Agency/Cancer Research Centre, Vancouver, British Columbia, Canada V5Z 4E6 and
†School of Kinesiology, Simon Fraser University, Burnaby, British Columbia, Canada V5A 1S6 and
‡ Faculty of Dentistry, University of British Columbia, Vancouver, British Columbia, Canada V6T 1Z3

I. Introduction 245
II. Oral Premalignant Lesions (OPLs): Traditional and Evolving Definitions 246
III. Lack of a Reliable System for Predicting Progression Risk of OPLs 247
IV. Lack of a Consensus on Treatment for OPLs 247
V. Use of Microsatellite Analysis to Identify Oral Mucosal Regions at Risk for Progression 248
A. Loss of Heterozygosity Assessment in Oral Squamous Cell Carcinoma (SCCs) and OPLs 248
B. Prediction of Risk of Progression for OPLs 248
C. Prediction of Recurrence of SCC in Oral Cancer Patients 250
D. Use of Microsatellite Analysis to Identify Type 1 and 2 OPLs 251
VI. Use of Molecular Markers to Manage OPLs 251
A. Staging of OPLs 251
B. Use of Molecular Markers to Make Treatment Decisions 252
C. Use of Molecular Markers to Assess Treatment Efficacy 252
VII. Exfoliated Cell Sampling: A Noninvasive Method for Identification, Risk Prediction, and Management of OPLs 253
VIII. The Need for Longitudinal Studies 255
References 256

The application of molecular technology to samples will revolutionize the way in which we detect and manage oral mucosa regions with increased cancer risk. This chapter discusses briefly the limitations of current approaches for identifying oral premalignant lesions (OPLs). We suggest that the old definition of OPLs be extended to include not only *morphologically* but also *genetically* altered tissue in which cancer is more likely to occur. As an example of how molecular markers can have an impact on clinical activities, this chapter describes data obtained on OPLs using microsatellite analysis for loss of heterozygosity (LOH). This approach can be used to (1) identify those lesions that appear clinically and histologically benign, yet contain high-risk genetic alterations; (2) provide an early indication of recurrence of a cancer at a former oral cancer site; (3) construct risk models and staging systems that can be used to triage patients into different treatment groups, thus sparing low-risk cases and focusing intervention on high-risk lesions; and (4) judge efficacy of treatment of OPLs. In addition, the potential use of exfoliated cells as a source for DNA for microsatellite analysis is discussed, with an indication of how such a noninvasive approach can be used to extend our ability to identify and manage regions of high risk. In summary, in the future we are likely to see a gradual extension of the classical clinical and histological approaches to following OPLs to an inclusion of an evaluation of molecular markers. The role of scientists in translational research will be to determine how best to integrate these different disciplines.

I. INTRODUCTION

Oral squamous cell carcinoma (SCC) is believed to progress through sequential stages of premalignancies to invasive cancer. Once invasion takes place, prognosis is poor and mortality and morbidity rates are dismal. Five-year survival for oral SCC has remained at less than 50% for the last three decades [1]. Even when treatment is effective, it is often at a cost of serious cosmetic and/or functional compromise.

These statistics support the need for earlier diagnosis of the disease and the development of better approaches to identifying mucosa at a premalignant stage. Such efforts are timely, given the possibility of applying current technological advances to the identification of genetic change in clinical specimens. We know that cancer development is driven by the accumulation of alterations to critical control genes and that these in turn are associated with the clinical and histological changes that occur in a tissue during carcinogenesis. The question to be asked is how to begin to integrate such genetic information into clinical practice, to use it to improve our ability to detect regions in the oral mucosa with increased cancer risk. This chapter presents a brief summary of the traditional approaches used to identify oral premalignant lesions (OPLs), the limitations of these approaches, and suggestions on how molecular techniques could facilitate the identification and management of such lesions.

II. ORAL PREMALIGNANT LESIONS (OPLs): TRADITIONAL AND EVOLVING DEFINITIONS

Historically, the definition of an oral premalignant lesion dates back to 1978, when it was defined by the World Health Organization (WHO) as a morphologically altered tissue in which cancer is more likely to occur than in its apparently normal counterpart [2]. At the clinical level, OPLs present most frequently as leukoplakia or occasionally as erythroplakia [3]. The WHO [2] defines leukoplakia in the oral cavity as a white patch or plaque of oral mucosa that cannot be characterized clinically or pathologically as any other diagnosable disease and is not removed by rubbing. Erythroplakia are fiery red patches that again cannot be characterized clinically or pathologically as any other definable lesion [3]. Both of these definitions rely heavily on the ability of a clinician to exclude other oral lesions [4]. This can be a challenge, as a number of very common oral conditions can present clinically as white or red patches. For example, oral frictional keratosis and lichen planus are often indistinguishable clinically from leukoplakia. Also, mucositis of various causes can resemble erythroplakia.

The greatest problem lies in the ability to exclude reactive lesions. The subjectivity of such exclusions is probably largely responsible for the wide variation in rates of malignant progression for oral leukoplakia in different studies (0.15 to 50%) [5–7]. Studies with a high proportion of reactive lesions (e.g., frictional keratosis) would have a lower rate of cancer transformation. This difficulty in lesion differentiation is also probably responsible for the fact that the most common sites for leukoplakia are different from those in tumors in patients in developed countries. Although most leukoplakia are on the buccal and alveolar mucosae [8–11], the majority of dysplastic leukoplakia (up to 93%) and 80–90% of tumors preferentially develop on the floor of mouth and ventrolateral tongue [5,9,10,12–15]. This apparent contradiction can be partially explained by the fact that both buccal and alveolar mucosae are prone to trauma and to the occurrence of frictional keratosis.

One additional point is worrisome with respect to our reliance on the presence of leukoplakia or erythroplakia as the first indication of an area at risk for malignancy. Increasing evidence suggests that histological dysplasia might not always manifest clinical change. The white mucosa patches seen clinically are a result of the thickening of the oral mucosal epithelial lining, a condition associated with epithelial hyperplasia (hyperkeratosis and/or epithelial acanthosis). Dysplastic lesions present frequently as white patches because epithelial hyperplasia commonly accompanies dysplasia. However, this may not always be so, as dysplastic changes (loss of differentiation, increased mitosis, and increased nuclear-cytoplasmic ratio) have been reported in "normal" appearing tissue that lies outside of clinical premalignant and malignant lesions [16,17]. With progression of the premalignant lesions, dysplastic epithelium can become atrophic. Such lesions will appear as red patches because submucosal vascularity reflects better through the thinned epithelium. Again, the presence of alterations in epithelial thickness (i.e., hyperplasia or atrophy) is detected clinically as a lesion.

Evidence also suggests that tissue can be normal both clinically and histologically, yet contain genetic changes found frequently in premalignant and malignant lesions. These morphologically innocuous but genetically abnormal cells may have an increased cancer risk [18–26] (see Section VI). Such clones of abnormal cells may be responsible for cases of "de nova" development of SCC. In the Western world, a significant proportion of oral SCCs appear to develop without preceding leukoplakia [27]. For instance, a study from the Netherlands showed that only about 50% of oral SCCs were associated with or preceded by leukoplakia [28]. Another indication of such a possibility lies in the sudden appearance of second primary cancers at distant sites in a patient who already has had one oral SCC. Such occurrences have been reported without preceding leukoplakia at the new cancer site and are attributed to a field effect on cancerization.

Taken together, the aforementioned data suggest that we may need to revise the old definition of OPLs to include not only *morphologically*, but also *genetically* altered tissue in which cancer is more likely to occur. This would allow us to identify and study a broader range of alterations that could put oral mucosa at risk. Figure 17.1 shows a scenario that includes four types of such alterations. Type 1 has only genetic changes and can only be detected molecularly. Type 2 shows both genetic alterations and dysplasia and is detectable by both molecular and histological techniques. Types 3 and 4 consist of lesions that have

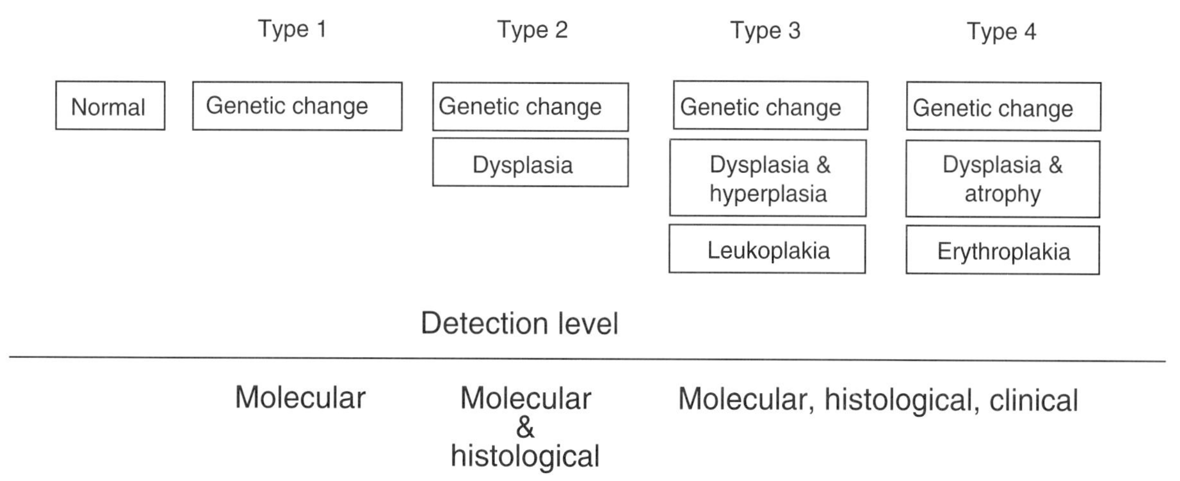

FIGURE 17.1 Detection limits for different approaches to assessing alterations in the oral cavity.

genetic, histological, and clinical alterations and hence can be detected with each of these approaches. The value in adding molecular markers to the evaluation of types 3 and 4 lies not in the identification of the lesion, but rather in the prediction of risk. This is a major problem in the current management of OPLs and is discussed later.

III. LACK OF A RELIABLE SYSTEM FOR PREDICTING PROGRESSION RISK OF OPLs

OPLs consist of a heterogeneous group of lesions that vary in potential for malignant transformation with only a fraction eventually progressing into cancer [5,6,29,30]. This heterogeneity partly results from inclusion of various reactive lesions as true leukoplakia and partly from the fact that premalignant lesions at different stages of progression (different risks) are all lumped together.

Histological criteria (the presence and degree of dysplasia) represent the current gold standard for judging the malignant risk of OPLs [6,31–34]. Histology has a relatively good predictive value for high-grade premalignant/preinvasive lesions (severe dysplasia and *CIS*). Many believe that these lesions, if untreated, have a high chance of progression into invasive lesions [33], with some believing that progression for such lesions is inevitable [35]. However, determination of prognosis for lesions with histological changes that are less than severe is more problematic. The majority of OPLs without dysplasia or with low-grade dysplasia (mild and moderate dysplasia) will not progress into cancer [6,33,36–38] and histology alone does not clearly differentiate between those that will and will not progress. Unfortunately, as a group, these lesions represent the bulk of leukoplakia and account for the majority of cases that later progress to cancer [6,9].

An increased malignant risk of OPLs is also associated with a number of other clinical factors. These include a nonhomogeneous clinical type, location at certain high-risk sites (floor of mouth, lateroventral tongue, and soft palate), a large size or long duration, presence of symptoms, history of cancer, absence of apparent etiologies, a female gender, and *Candida* infection [14,39–41].

There have been efforts to establish a staging system for oral leukoplakia that takes into consideration these different risk factors [29,30,42]. Such studies all use histology, clinical appearance, and size of the lesions as risk determinants for staging. Site of the lesion has also been proposed, but only in a single study [29]. Nonetheless, no studies have been done to confirm the risk prediction value of these systems, and currently no reliable means exist to predict the malignant risk of a given lesion.

IV. LACK OF A CONSENSUS ON TREATMENT FOR OPLs

The management of OPLs is not surprisingly a topic of heated debate given the problems in identifying and in predicting the risk for such lesions [5,31,38,43–46]. There is general agreement that high-grade OPLs (severe dysplasia or *CIS*) should be removed, in most cases, with a wide margin. However, there is no consensus for leukoplakia without dysplasia or with low-grade dysplasia. In fact, many suggest that such lesions could be monitored safely [36,38,44,47]. Even among those who choose to treat, there is no agreement on the method of treatment. Treatments vary widely from the use of anti-inflammatory agents, vitamins, or chemotherapy with bleomycin to surgical removal by scalpel, laser, cryoprobe, or electrosurgery. When lesions are excised, there is also a wide variation in the width of normal looking mucosa included as a margin. In some cases, there is no margin, in others a narrow margin (1–2 mm), and still others a slightly wider margin (≥3 mm). A high rate of recurrence (10–35%) suggests that the removal of such lesions is frequently inadequate [5,48].

V. USE OF MICROSATELLITE ANALYSIS TO IDENTIFY ORAL MUCOSAL REGIONS AT RISK FOR PROGRESSION

A. Loss of Heterozygosity Assessment in oral squamous cell carcinoma (SCCs) and OPLs

Early indications are that the incorporation of molecular techniques into clinical practice will eventually revolutionize the way we deal with patients. Such tools will both facilitate the identification of mucosa at risk for progression and provide information that will assist clinicians in designing and managing the treatment of such lesions. The remainder of this chapter focuses specifically on one assay, microsatellite analysis, which is receiving much attention as an early tool for such studies.

Microsatellite analysis is one of the more popular approaches available for detecting genetic alterations in premalignant lesions. This polymerase chain reaction-based protocol measures alterations in chromosomal regions that contain short tandem nucleotide repeats (microsatellites) [49]. The high degree of polymorphism in these regions allows one to detect both maternal and paternal alleles for most individuals and to determine whether lesions show an alteration in the quantity of one of the alleles, called a loss of heterozygosity (LOH) or allelic imbalance (AI). LOH is usually assumed to represent the deletion (or inactivation) of a known or putative tumor suppressor gene located near the microsatellite marker. For example, LOH at 9p21 corresponds to a loss of *p16* (*CDKN2/MTS1*), a cyclin-dependent kinase inhibitor involved in cell cycle regulation [7]; loss at 17p13, especially within the *TP53* locus, may be associated with inactivation of the *p53* gene; and finally LOH at 3p14 may involve the *FHIT* gene, although the latter association is still controversial.[1]

[1]Although LOH is usually equated with deletion of a tumor suppressor gene, occasionally this alteration might signal an amplification event, as chromosome gains in a tumor or premalignant lesion can also manifest as a relative change in gene dosage (or imbalance) when compared with normal DNA. For example, the LOH at 11q13 at the *bcl-1/int-2* locus is thought to represent the amplification of a region containing the protooncogene *cyclin D1* [50].

One of the main advantages of microsatellite analysis is that it requires only small quantities of DNA, and is thus suitable for use with the typically small OPLs, yet provides information of significant biological and clinical value. LOH is a common event in head and neck and oral SCCs, with numerous studies reporting its occurrence at specific chromosomal regions: 3p, 4q, 5q, 7q, 8p, 9p, 10q, 11q, 13q, 14q, 17p, 18q, and 22q [51–66]. There have been fewer studies on LOH in OPLs, probably because of the difficulty in working with these small lesions and the fact that such lesions are not as readily accessible as SCCs in the larger institutes in which such research often is conducted. Nevertheless, results from these studies clearly show that LOH is also a frequent event in OPLs, occurring in many of the same regions as in SCCs [52,67–72].

Among the earliest studies of LOH in head and neck OPLs was that of Califano *et al.* [51]. That study examined LOH patterns (10 loci examined) in a wide spectrum of lesions, including oral hyperplasia, dysplasia, CIS, and SCC, to determine whether there was an association of such alterations with the histological progression of the disease. The study showed that LOH occurred early in disease development. Loss of at least one locus occurred in nearly all samples of dysplasia and CIS but was present in less than one-third of hyperplasias. Although the authors suggested that it was the accumulation and not necessarily the order of genetic events that determined progression, they also indicated that some losses were more likely to occur early in the development of the disease and others at later stages. For example, LOH at 9p was the earliest event, associated with transition from normal to benign hyperplasia; LOH at 3p and 17p were more often associated with development of dysplasia; whereas CIS and SCC were characterized by additional deletions on 4q, 6p, 8p, 11q, 13q, and 14q.

B. Prediction of Risk of Progression for OPLs

The aforementioned studies suggest that LOH is a common event in OPLs and can occur early in carcinogenesis. Can this assay be used to improve our ability to predict risk of malignant transformation for OPLs? This possibility is supported by several studies.

In 1996, Mao *et al.* [70] examined 84 oral leukoplakia samples from 37 patients enrolled in a chemoprevention study. The samples were analyzed for LOH at 9p21 and 3p14. LOH at either or both loci was observed in 19 of these patients and this loss was strongly associated with cancer progression. Seven (37%) of the 19 positive cases later developed SCC. In contrast, only 1 of 18 cases (6%) without LOH progressed to cancer.

Partridge and co-workers also looked for an association between LOH pattern and progression. In an early study in that laboratory, they reported an association between multiple LOH in OPLs and increased progression risk [73]. This result was again observed in a more recent study from that laboratory [74] involving 78 OPLs, diagnosed histologically with hyperplasia or dysplasia, all from patients with no prior history of oral cancer. In half of the patients the lesions progressed, usually at or adjacent to the original site. Progressing lesions were characterized by the presence of multiple regions of LOH, with loss of 9p or 3p present in 94% of lesions.

We have completed a retrospective study that restricted the focus to OPLs without—or with minimal—dysplasia [72]. These are the lesions that are the most difficult for clinicians to manage. One hundred and sixteen biopsies of OPLs were examined to see if a correlation existed between LOH at 7 chromosome arms (3p, 4q, 8p, 9p, 11q, 13q, and 17p) and progression risk. None of the patients had a history of cancer prior to the studied hyperplasia or mild/moderate dysplasia. Twenty-nine of the 116 OPLs progressed to cancer. The progressing lesions showed not only a significantly higher number of LOHs, but also characteristic LOH patterns (Fig. 17.2) [72]. LOH at 3p and/or 9p was present in 97% of progressing lesions, suggesting that loss at these arms may be a progression prerequisite. However, because many nonprogressing lesions also showed losses at 3p and/or 9p, such loss alone is probably insufficient for malignant transformation. Indeed, cases with LOH at these arms but no others showed only a 3.8-fold increase in relative risk for cancer development. In contrast, individuals with additional losses at 4q, 8p, 11q, 13q, or 17p showed a 33-fold increase in relative cancer risk. In nonprogression cases, additional losses were uncommon. These results suggest that LOH patterns may differentiate three progression risk groups: low, with retention of 3p and 9p; intermediate, with losses at 3p and/or 9p; and high, with losses at 3p and/or 9p plus losses at 4q, 8p, 11q, 13q, or 17p [72,75].

We have also explored one other situation in which the microsatellite assay might facilitate clinical management of OPLs, which is in predicting risk for lesions at "low-risk" oral sites. As stated previously in Section I, among Western populations such lesions include those on the buccal and alveolar mucosa, which have a much lower risk of progression than OPL on the floor of the mouth and the ventrolateral tongue, high-risk (HR) sites [5,9,12–15,76–78]. We hypothesized that if the location of leukoplakia is a critical prognostic factor regardless of the degree of dysplasia, leukoplakias with a similar degree of dysplasia from HR and LR sites should be genetically different. To test this hypothesis, we analyzed 127 oral dysplasias from HR and LR intraoral sites for LOH at loci at 3p14, 9p21, and 17p11.2–13.1 (Table 17.1). Results showed that dysplasias from HR sites contained significantly higher LOH frequencies than those from LR sites (% with any loss, $P=0.0004$; % with multiple losses, $P=0.0001$; loss on 3p and/or 9p, $P=0.0005$, and % loss on each of the arms, $P<0.05$) [79].

These data support a more aggressive intervention for lesions at HR sites, regardless of the level of dysplasia observed in the lesion. For example, a hyperkeratotic lesion with mild epithelial dysplasia in the floor of mouth is considered by many to have enough risk to warrant complete

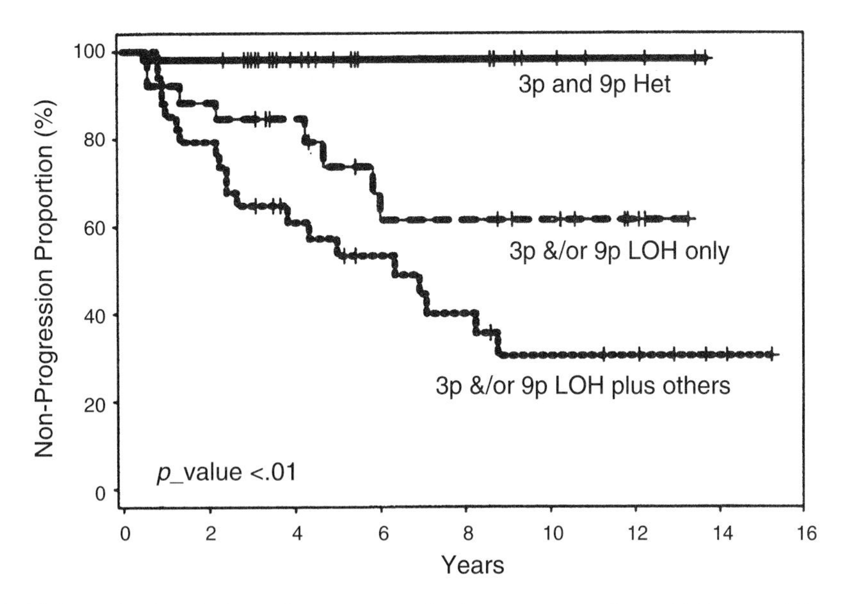

FIGURE 17.2 Probability of primary OPL not progressing into cancer, according to LOH pattern. All samples were assayed for LOH on seven arms (3p, 9p, 4q, 8p, 11q, 13q, and 17p). The lowest risk (RR = 1) was associated with retention of both 3p and 9p ($n=56$). LOH at 3p and/or 9p in the absence of LOH at any other arm had an intermediate level of risk (RR = 3.75; $n=26$). The highest risk was associated with LOH at 3p and/or 9p in the presence of LOH at any other arm (RR = 33.4; $n=34$). Adapted from Rosin *et al.* [72].

TABLE 17.1 LOH Frequencies in OPLs at High- and Low-Risk Sites[a]

No.	High risk[b]	Low risk[c]	P value
No. of cases	71	56	—
No. with any loss[d]	56 (79)[e]	27 (48)	0.0004
With multiple loss[f]	38 (54)	11 (20)	0.0001
LOH on 3p and /or 9p	53 (75)	24 (43)	0.0005

[a]From Zhang *et al.* [79].
[b]High-risk site (floor of mouth, ventral lateral tongue, soft palate–anterior–pillar–retromolar complex).
[c]Low-risk (rest of oral cavity).
[d]LOH present at 3p, 9p, or 17p.
[e]Loss/informative cases (% loss).
[f]LOH present at more than one of the three arms.

removal if the lesion persists after the elimination of possible causative factors [80]. Our molecular data support this intervention. In contrast, a mild dysplasia from the buccal mucosa is regarded to have significantly less malignant risk and is often left for further observation. Our data suggest that it is important to assess such lesions for their genetic profile. Although the majority may have little risk, a fraction (with high-risk genetic profiles) should be managed more aggressively.

Interestingly, when the degree of dysplasia was considered, LOH frequencies showed significant increases for HR compared to LR sites with mild and moderate dysplasias, but not with severe dysplasias/CIS [79]. Mucosa from LR sites, such as buccal mucosa and gingiva, is much more likely to be traumatized (e.g., biting) than mucosa from HR sites, such as the floor of mouth and ventral tongue. Injury due to trauma or local inflammation can induce atypical changes in epithelia that resemble low-grade dysplasias. Such changes are not believed to signal malignant risk. Thus the genetic damage associated with progression would be present more often at HR sites, where trauma is less likely to induce histological alterations. With an increasing degree of dysplasia, it is less likely that dysplastic changes are confused with reactive epithelial atypical changes. Microsatellite analysis might provide us with a valuable tool for differentiating reactive lesions at LR sites from "true" premalignancies.

Another example of the use of microsatellite analysis to differentiate between reactive lesions and high-risk OPL is with verrucous leukoplakia [81]. This lesion appears morphologically benign with little or no dysplasia, yet has a high cancer risk (reported malignant transformation rate of 70% [82]). The lack of dysplasia in verrucous leukoplakia has made it difficult to differentiate such lesions from reactive hyperplasia, often delaying diagnosis of the condition. We have shown that despite a benign appearance, such lesions possess multiple LOH alterations that are similar to those observed in severe dysplasia and CIS. The finding of such high-risk LOH patterns in verrucous leukoplakia may account for the high probability of progression. This finding also has important clinical implications, as it suggests that microsatellite analysis might facilitate differentiation between reactive hyperplasia and verrucous leukoplakia, significantly improving prognosis for patients with the latter lesions.

C. Prediction of Recurrence of SCC in Oral Cancer Patients

Molecular markers may also play a role in the early identification of high-risk OPLs in cancer patients after treatment. Such identification could significantly improve the prognosis of patients with oral SCC, as the high mortality associated with this disease is at least partially due to the frequent development of recurrent or secondary primary tumors [83–85].

At present, many lesions still escape identification and go on to become cancer despite vigorous monitoring of patients. Clinical and histological changes are not reliable indicators of malignant risk for OPLs in former cancer patients. The radiation and extensive surgery received by such patients induce reactive white and red lesions at previous cancer site that are not differentiated easily from true premalignant changes. Histologically, treatment effects also generate reactive changes resembling low-grade dysplasia. The situation is further complicated by the reluctance of clinicians to repeatedly biopsy such fragile sites, hence delaying the diagnosis of high-risk changes. Molecular markers that can identify high-risk lesions could revolutionize follow-up.

We did a retrospective study on biopsies from 68 patients with prior oral SCC who subsequently developed a leukoplakia at the former cancer site. The objective of the study was to determine whether microsatellite analysis of a leukoplakia at a former cancer site could predict the probability of development into a recurrent cancer. We were unable to differentiate cases that developed into a recurrent cancer on the basis of the more traditional predictors of recurrence (i.e., stage and grade of the primary cancer and presence or degree of dysplasia in the leukoplakia). However, LOH patterns were predictive of risk. The relative risk of recurrence was 23.5-fold higher when leukoplakia had a loss at 3p or 9p and was 26.8-fold higher when that loss occurred in combination with LOH at 4q, 8p, 11q, 13q, or 17p (unpublished data).

If this pattern of loss is confirmed in a prospective study, it would suggest the possibility of active intervention whenever a leukoplakia at a former cancer site contained a loss at either 3p or 9p. This would facilitate management of these lesions greatly.

D. Use of Microsatellite Analysis to Identify Type 1 and 2 OPLs

Section I suggested that genetic alterations occur outside of regions with identifiable clinical and/or histological change and that such alterations could signal a significant risk for such mucosa. There are now several studies using microsatellite analysis that support this possibility.

In one study, microsatellite analysis was performed on biopsies of normal or minimally altered bronchial epithelium of 54 chronic smokers (40 current smokers and 14 former smokers) [86]. LOH was present in the majority of these samples, occurring at 3p, 9p, and 17p in 75, 57, and 18%, respectively, of informative lesions studied. In contrast, LOH was not observed in samples from 5 nonsmoker controls, with the exception of a single case with LOH at 3p only.

A second study showed similar results [25]. Multiple biopsies were taken under fluorescence bronchoscopy of normal or minimally altered bronchial epithelium of 54 chronic smokers. Twenty-six of these biopsies were also histologically normal. Twelve (46%) of the 26 samples were shown to contain LOH. LOH was also present in dysplastic lesions taken from these same patients. None of the samples from 21 nonsmoker controls contained LOH. Thus, respiratory tracts of long-term heavy smokers appear to contain multiple abnormal premalignant genetic clones with or without accompanying morphological changes.

In another study, LOH was analyzed in nasopharyngeal biopsies taken from patients from southern China, where there is a high incidence of nasopharyngeal carcinoma [22]. LOH at 3p was observed in 74% of 23 biopsies from clinically and histologically normal-looking epithelia and in 75% of 8 dysplastic biopsies. In contrast, a low frequency of loss was observed for biopsies from subjects coming from northern or central China, where there is a low incidence of nasopharyngeal carcinoma.

Califano *et al.* [87] presented further evidence supporting the presence of high-risk genetic profiles in clinically and histologically normal tissue. In this study, microsatellite analysis was performed on metastatic tumors of 18 patients with unknown primary head and neck SCC. In such cases the tumor presents as a cervical lymph node metastasis without identification of the primary tumor. The LOH patterns of these metastatic tumors were compared with those in histologically benign surveillance biopsy specimens from the oropharyngeal mucosa, taken from sites mostly likely to harbor an occult primary tumor. In 10 (55%) of the 18 patients, alterations in biopsies of these normal-looking mucosae were the same as those in the metastatic tumors. Futhermore, 3 of the 10 patients subsequently developed SCC in or near the sites of morphologically normal-looking but genetically abnormal mucosae [87].

Finally, using molecular markers, many studies have demonstrated genetic changes in apparently normal-looking tissue adjacent to premalignant or malignant lesions that are similar to the lesions themselves [51,88–96] (unpublished data from this laboratory). This again supports the spread of abnormal cells beyond a clinical lesion. Indeed the molecular change does not necessarily have to be adjacent to a tumor, but in some cases can be quite distant from the tumor site [18–21,23] (unpublished data from this laboratory).

VI. USE OF MOLECULAR MARKERS TO MANAGE OPLs

A. Staging of OPLs

Because oral cancer develops along complex molecular pathways, it is unlikely that a single marker or a class of markers will be able to predict the outcome of every oral leukoplakia [97] nor will these markers replace pathological indicators in the near future. The value of histology as an indicator of cancer risk is timetested not only in the oral cavity or other head and neck regions, but also in other sites, including the uterine cervix, lung, breast, skin, and esophagus [6,32,98–102]. More likely, the future will see a gradual extension of the classical histological and clinical approaches to include an evaluation of molecular markers.

In support of this probability is a report by Lee *et al.* [103] that summarized 10 years of translational research on OPLs at MD Anderson. This report concluded that histology (presence of pronounced dysplasia) and molecular markers (LOH, *p53* mutation, and chromosome polysomy) were independent risk predictors. An increase in abnormality in either histology or genetic markers was associated with an increase in cancer risk. More recently, another study reported an association of ploidy in dysplastic lesions with progression risk for OPL [97,104]. The lowest risk was associated with diploid dysplastic leukoplakias (3% progressed into oral SCC), tetraploid lesions had an intermediate risk (60% showed progression), and aneuploid dysplasias showed the greatest risk (84% progressed). With time, more such studies will accrue. A future goal for individuals involved in translational research will be to determine which combinations of histological and molecular markers best predict risk.

How can such information best be handled? One possibility is to develop staging systems that incorporate different risk factors into risk models. One such model is presented in Table 17.2. For the sake of simplicity, this hypothetical model only includes pathological and molecular markers (clinical features, such as appearance, site, and size of the lesion, are not included, but should be part of an eventual screening system). In this model, the majority of OPLs

TABLE 17.2 Staging of Oral Premalignant Lesions with Both Pathologic and Genetic Indices[a]

Stage (cancer risk)	Pathology (P)[b] and genetic (G)[c] patterns
1	P_1+G_1
2	P_1+G_2
	P_2+G_1
	P_2+G_2
3	P_3+G_1
	P_3+G_2
	P_1+G_3
	P_2+G_3
	P_3+G_3

[a]Adapted from Zhang and Rosin [104a].

[b]Pathological indices: P_1, no dysplasia or mild dysplasia; P_2, moderate dysplasia and P_3, severe dysplasia/*CIS*.

[c]Genetic patterns, categorized by increasing risk: G_1, low risk; G_2, intermediate risk; and G_3 high risk.

would be placed into stage 1, with the lowest probability of progressing to cancer. Such lesions would contain both low-risk pathology (P_1) and genetic patterns (G_1). Emergence of intermediate-risk patterns in either histology (P_2) or genetic profile (G_2) would place a lesion into stage 2. Stage 3 would contain lesions with a high-risk genetic or histology pattern. It should be noted that the greatest impact of such a staging system would be on lesions in stage 3. Such lesions would now include cases with a relatively benign phenotype (without or with minimal dysplasia) but a high-risk genotype (e.g., P_1+G_3).

B. Use of Molecular Markers to Make Treatment Decisions

The value of the aforementioned approach would be to better target intervention. For example, stage 1 lesions (the bulk of OPLs) might be left untreated, perhaps with periodic monitoring for clinical or molecular change, thus sparing large numbers of patients from invasive procedures. Stage 2 OPLs might be treated conservatively or be given a chemopreventive regime. In contrast, lesions in stage 3 would require a more aggressive treatment involving either excision with wide margins or, for more diffuse lesions, some form of drug therapy. Significant toxicity might be justified for such cases.

A retrospective study in our laboratory supports this possibility. Sixty-six OPLs with low-grade dysplasia were investigated for their response to treatment by surgery and/or laser. The clinical history and LOH patterns of the lesions were known. Twenty-four of these lesions were not removed clinically (only received an incisional diagnostic biopsy). Ten of these 24 lesions showed a low-risk LOH pattern (i.e., stage 1 of the aforementioned scenario). Only 1 (10%) progressed into cancer. In contrast, the remaining 14 lesions had intermediate or high-risk LOH patterns (stages 2 and 3). Eight of these lesions (58%) developed SCC [105].

That study also showed that even when excised, the treatment of lesions with high-risk LOH patterns was frequently inadequate. Eleven such dysplastic lesions were studied. All of the lesions were judged to be clinically excised but went on to become cancer. Histological evidence for removal was not available, as most lesions were treated with laser. Fifty biopsies from the 11 progressing dysplasias (sampled over time) were analyzed by microsatellite analysis of loci on 7 chromosome arms (3p, 4q, 8p, 9p, 11q, 13q, and 17p). Nine of the 11 had LOH patterns that suggested the presence of the same clones (allelic specific mutations) in successive biopsies, indicating incomplete removal of these lesions, despite a clinical impression of removal.

Would a wide excision margin have improved outcome for these lesions? When a tumor is removed surgically, a wide normal-looking mucosa margin (1.5–2 cm) is taken to ensure removal of the lesion. Such treatment has a high cure rate for early stage tumors. However, in contrast to the aggressive treatment of oral SCC, surgical removal of OPLs is generally done with a small margin at best (1–3 mm). Increasing margin width could have a significant impact on high-risk OPLs.

However, it is possible that at least a portion of such lesions have extensive spread of premalignant clones outside of the clinical lesion. In such cases, removal may indeed be futile, and systemic treatment such as chemotherapy or chemoprevention might be more appropriate. Identification of such hidden widespread clones with focal clinical lesions could be facilitated by using molecular techniques.

C. Use of Molecular Markers to Assess Treatment Efficacy

Not only could molecular markers help us establish appropriate treatment plans for OPLs (whom to treat and how to treat), but, as indicated earlier, they have potential value as indicators of the efficacy of treatment. Such a use for molecular markers should be extended.

For example, a study by Brennan *et al.* [106] showed that the presence of genetically abnormal clones (cells with *p53* mutation) in margins of surgical resections of oral SCCs predicted tumor recurrence, even when such margins appeared to be histologically clean. The same principle could be applied to OPLs. Microsatellite analysis could be used to judge treatment efficacy [105].

In support of this possibility is a report of a chemoprevention trial of patients with OPLs treated with *N*-(4-hydroxyphenyl)retinamide (4-HPR). Efficacy of treatment was judged not only with clinical and histological end points, but also with microsatellite analysis. The study showed that "scars" of genetically altered clones with LOH often could be detected in lesions after treatment even when they appeared to have regressed both clinically and histologically. The subsequent reappearance of such lesions once therapy was halted appeared to be associated with the presence of these genetic clones [71,107,108].

VII. EXFOLIATED CELL SAMPLING: A NONINVASIVE METHOD FOR IDENTIFICATION, RISK PREDICTION, AND MANAGEMENT OF OPLs

One of the main barriers to the use of molecular techniques to identify and manage OPLs is the requirement for biopsies to provide the specimens for analysis. A possible solution to this problem is to perform molecular analysis on exfoliated cells collected noninvasively from a tissue. There are currently two types of studies aimed at exploring such approaches. In one, cells are collected "globally" from a targeted organ in the form of sputum (for the lung), urine (for the bladder), stools (for the bowels), saliva, and mouth rinse (for the oral cavity). The other approach uses cells collected from specific sites of interest within a tissue, such as with scrapes of leukoplakia within an oral cavity.

A number of studies have investigated the value of studying exfoliated cells obtained globally from a targeted organ. Such studies have shown that molecular changes in cells isolated from sputum and urine are similar to those observed in lung and bladder cancers, suggesting that the identification of such changes in exfoliated cells might serve as an early indication of the presence of a tumor in these organs [70,109]. Similar studies have been done on patients with head and neck SCC and have found that *p53* mutations and microsatellite alterations (LOH or microsatellite instability) in cells obtained from saliva match alterations present in tumor biopsies [110,111]. More recently, these studies have been extended to an assessment of saliva DNA from head and neck cancer patients for epigenetic alterations that may play a role in cancer development. Aberrant DNA methylation patterns were detected in exfoliated cell samples that were identical to those present in the tumors [112]. These results are extremely promising.

Our laboratory has investigated whether exfoliated cells taken from a specific site in the oral cavity could reveal molecular changes present at that site. In an early study, we compared LOH patterns in biopsies of oral SCCs to those obtained from exfoliated cell samples collected by scrapping the same SCC prior to biopsy [113]. Results showed that LOH patterns in exfoliated cells are highly representative of those in biopsies.

We are currently confirming the aforementioned results and also extending them to include OPLs. To date, we have analyzed and compared LOH frequencies at 3p14 and 9p21 (four primers for each location) in scrapes and concurrent biopsies of the same lesion from 68 clinical lesions (23 SCCs, 18 OPLs in patients with history of oral cancer, and 27 primary OPLs) (unpublished data). As shown in Table 17.3, 51 (75%) of the 68 lesions examined demonstrated LOH in their biopsies. Forty-four (86%) of the 51 positive cases showed identical LOH patterns in the scrapes to those seen in biopsies. Of significance, LOH was not detected in any exfoliated cell samples from lesions in which biopsies were negative for LOH, suggesting that the approach is very specific (no false-positive scrapes).

These data support the use of this approach for screening and follow-up of OPLs. However, it should be noted that the procedure is still in its infancy and that modifications are required to improve the sensitivity of the approach. Such modifications could incorporate other molecular end points into the analysis; however, given the limited amount of DNA in these samples, such a change would have to be accompanied by a modification of the technique toward a higher throughput analysis so that multiple regions of alteration could be examined at the same time. Such technical alterations are now being explored.

Even with the current scrape assay we are already able to suggest some ways in which the approach could be used to improve the clinical management of OPLs. One potential use is as an early screen of lesions where there is insufficient clinical information to warrant a decision to biopsy. For example, clinicians are often reluctant to biopsy a homogeneous leukoplakia located at a low-risk site because only a small proportion of these lesions have a malignant risk. However, loss at 3p14 and 9p21 in a scrape of such a lesion would support the biopsy of this lesion, irrespective of clinical appearance of the lesion or its history, given the increased risk for progression or recurrence associated with this pattern of alteration [70,72]. Thus, scrapes could be used to triage patients with clinically innocuous lesions, preventing

TABLE 17.3 Loss of Heterozygosity at 3p14 and 9p21 in Exfoliated Cells and Concurrent Biopsies

Sample	Total cases	Biopsies positive for LOH	Scrapes positive for LOH	LOH in both biopsies and scrapes
SCC	23	23/23 (100%)[a]	20/23 (87%)	20/23 (87%)
OPL after SCC	18	11/18 (61%)	10/18 (56%)	10/11 (91%)
OPL, no SCC history	27	17/27 (63%)	14/27 (52%)	14/17 (82%)
All cases	68	51/68 (75%)	44/68 (65%)	44/51 (86%)

[a]Loss at 3p14 and/or 9p21/informative cases (% loss).

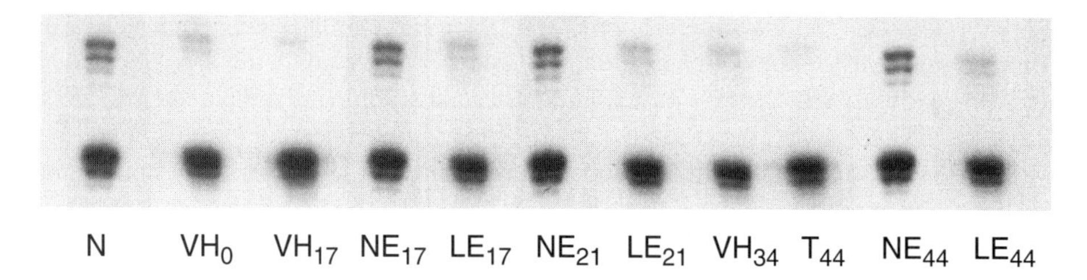

FIGURE 17.3 LOH analysis of primary OPL during progression into SCC. This 55-year-old male is a former smoker (from age 16 to 40) who developed a verrucous leukoplakia on the left buccal mucosa extending to the lower gum of the molar area. The index (VH_0) and repeat biopsies VH_{17} and VH_{34} taken 17 and 34 months thereafter (subscripts indicate time from index biopsy) all showed a hyperplasia with verrucous configuration. A verrucous carcinoma was diagnosed 44 months after the initial biopsy. DNA was isolated from stroma (N), verrucous hyperplasia (VH), and tumor (T) cells microdissected from lesion biopsies and from exfoliative cells from clinically normal oral mucosa (NE) and the lesion (LE). A loss of the upper allele at *D3S1300* was observed in all biopsies and lesion scrapes but was absent in the normal stroma and scrapes of the normal mucosa.

unnecessary biopsy in the majority of such cases, yet providing an opportunity to spot potentially high-risk lesions.

Another potential role for exfoliated cell analysis is to identify temporal patterns of genetic alterations during follow-up, to serve as early predictors of risk of progression for primary OPLs or of tumor recurrence or formation of second primaries in oral cancer patients. Figure 17.3 shows LOH analysis of a primary OPL with a high-risk LOH pattern (LOH on 3p and 4q, see Section IV,B). This lesion progressed into cancer 44 months after the initial assessment, despite laser ablation at 34 months. LOH was present in biopsies and scrapes on multiple occasions before the development of the carcinoma.

In a second example, Fig. 17.4 shows LOH patterns of biopsies and scrapes of a patient who developed a recurrent cancer. At 6 months after treatment, no apparent clinical changes were observed at the cancer site. However, microsatellite analysis of a scrape from the site showed the same allelic loss at 9p that had been present in the initial *CIS*. The same loss was seen in scrapes 9 and 14 months posttreatment, but there was still no apparent clinical lesion. Morphological changes were not seen until 25 months posttreatment. At this point, the lesion was biopsied and showed mild dysplasia, again with the same LOH pattern observed in previous scrapes. The lesion went on to become a recurrent tumor 53 months after treatment of the primary SCC, with all of the remaining scrapes and biopsies showing the same allelic loss. As exemplified by this case, molecular analysis of scrapes could enable us to identify high-risk lesions at sites of previous cancer months before the appearance of morphological changes and could guide treatment of such lesions, thus preventing tumor recurrence.

Finally, scrapes may play a role in assessing normal-looking mucosa adjacent to clinically visible high-risk OPLs to determine how far such lesions have spread. For those lesions shown to be discrete, surgical removal of the clinical lesion could be performed with margins that extend into regions that show no molecular change. However, genetic

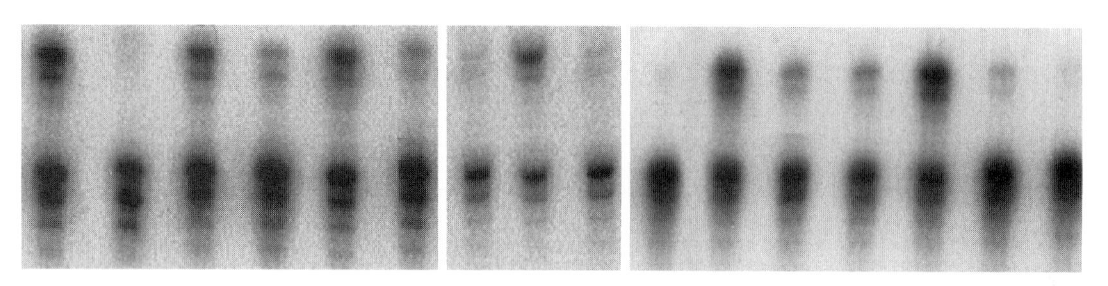

FIGURE 17.4 LOH patterns of biopsies and scrapes obtained from a former cancer site of a 62-year-old male smoker. This patient had a CIS (T_0) on the right lateral border of the tongue that was removed by excision. The original CIS had a loss of the upper allele on both *IFNA* (not shown) and *D9S171*. The same loss was observed in scrapes 6, 9, 14, 25, and 38 months and in biopsies 25, 38, 46, and 53 months after the initial CIS. The biopsies were diagnosed histologically as mild dysplasia (D_{25}), severe dysplasia (D_{39}), mild dysplasia (D_{46}), and finally SCC (T_{53}). N, normal genomic DNA from connective tissue. LE, scrape from lesion site. NE, scrapes of normal epithelium contralateral to the lesion site. Subscripts indicate time (months) from initial biopsy of CIS (T_0).

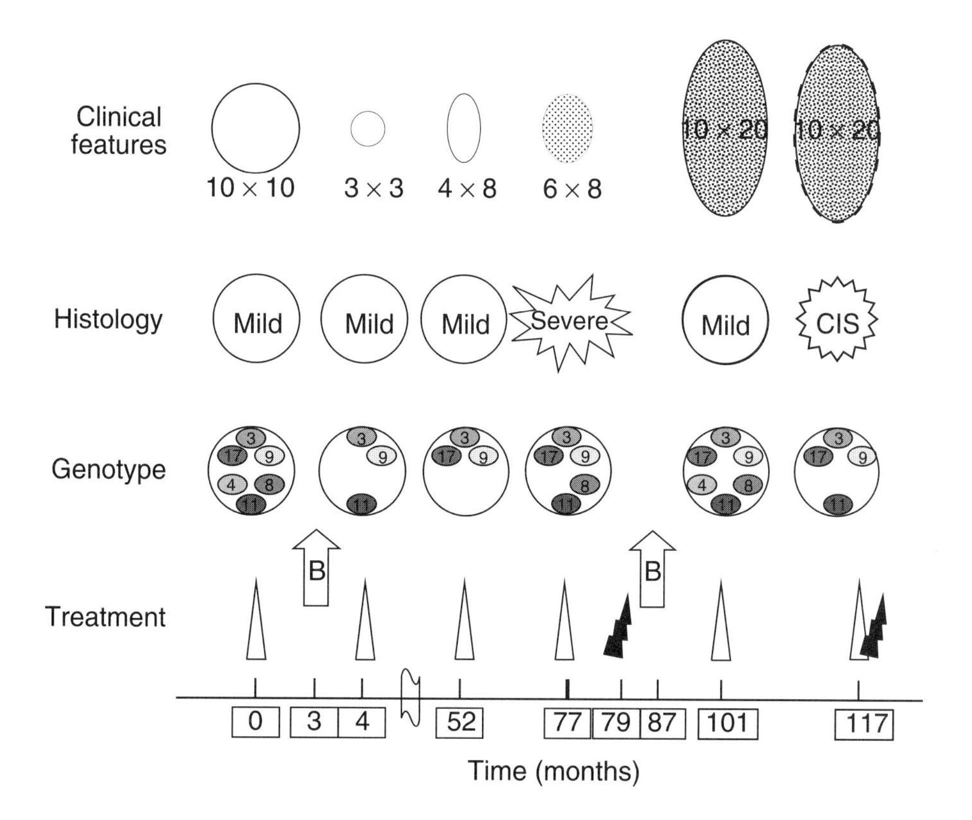

FIGURE 17.5 Temporal changes of clinical, histological, and molecular indices in an OPL that developed into CIS. The index biopsy of this lesion had a low clinicopathological risk (<2 cm in size, homogeneous in appearance, and mild dysplasia; for clinicopathological staging criteria, see Axell *et al.* [30]), but a high-risk genotype. Bleomycin treatment decreased the size of the lesion, but it persisted. At month 52, the lesion still maintained a low clinicopathological but high molecular risk. However, only a portion of the allelic losses seen in the index biopsy were present, probably due to the selective pressure of treatment on subclones within the initial lesion. At month 77, almost 6.5 years after molecular indication of high risk, the lesion finally showed clinicopathological progression to a high-risk lesion (nonhomogeneous and sever dysplaisa). The molecular risk remained high with emergence of a clone(s) that showed more of the allelic losses that were present in the initial lesion. The lesion was subsequently excised but recurred and was treated by topical bleomycin at month 87. At month 101, the lesion presented clinicopathologically as an intermediate risk lesion (2 cm in size, nonhomogeneous) with the same high-risk molecular clone as that seen in the index biopsy. The lesion progressed into a CIS at month 117. Clinical features: Numbers represent lesion size (diameter in mm); hollow circle represents homogeneous appearance; dot-filled circles represent nonhomogeneous appearance. Genotype: Numbers represent chromosomal arms with LOH. Treatment: Gray triangle represents incisional biopsy; hollow arrow with letter B represents topical bleomycin treatment; black jagged triangle represents excision.(see also color insert.)

analysis of scrapes could identify widespread lesions that are not amenable to surgical removal. Identification of these lesions could prevent unnecessary repeat surgeries and provide a rationale for other treatment, such as aggressive topical chemotherapy of the whole oral cavity, despite cost and potential side effects.

VIII. THE NEED FOR LONGITUDINAL STUDIES

Although data provided here are promising, proof of the actual prognostic value of molecular markers requires long-term longitudinal studies. In October 1999, we began accruing patients into a National Institute of Dental and Craniofacial Research-funded prospective study that will examine temporal patterns of allelic loss in biopsies and scrapes collected serially over time in 300 patients. The study has two arms. The first includes 150 patients that have been treated for oral cancer and who are at elevated risk for local recurrence and for development of second primaries. The second includes 150 patients with oral dysplasia, but no history of oral cancer. The study will follow all clinical lesions identified at entry into the study and those that develop during the course of the study. Among those currently enrolled, 53% of the cancer patients and 42% of dysplasia patients have >2 lesions at entry (24 and 14% have 3 or more lesions). Follow-up time will be up to 7 years in duration. To date we have recruited 110 patients with a history of oral SCC and 60 patients with primary dysplastic leukoplakia.

This will be one of the first studies to follow molecular changes in such lesions over an extended period. In addition, the study will collect information on clinical, histological, and molecular alterations for all lesions every 6 months. This will allow us to look for associations of the more traditional indicators of risk (clinical, histological) and molecular patterns that might better predict outcome. An example of the types of information that can be obtained with such analysis is shown in Fig. 17.5 (see also color insert). At the same time, information on alterations to smoking and alcohol habits or interim treatment will be collected to allow us to control for confounding effects produced by these alterations on markers and outcome.

Data generated by this study will lead to a better understanding of the natural history of the development of aerodigestive tract cancer by providing insight into the evolution of genetic clones in such lesions over time. It will allow us to determine what impact temporal shifts in genetic clones have on the phenotypic behavior of cells in a lesion and on its ultimate outcome. Finally, and more specifically, the study will test the three risk profiles identified in our earlier retrospective studies to determine their validity as prognostic indicators.

Acknowledgments

This work was supported by Grant 1 R01 DE13124-01 from the National Institute of Dental and Craniofacial Research and by a grant from the National Science Engineering Research Council (NSERC) of Canada.

References

1. Greenlee, R. T., Hill-Harmon, M. B., Murray, T., and Thun, M. (2001). Cancer statistics, 2001. *CA Cancer J. Clin.* **51**, 15–36.
2. World Health Organization (WHO) Collaborating Center for Oral Precancerous Lesions (1978). Definition of leukoplakia and related lesions: An aid to studies on oral precancer. *Oral Surg. Oral Med. Oral Pathol.* **46**, 518–539.
3. Pindborg, J. J., Reichart, P. A., Smith, C. J., and van der Wall, I. (1997). "Histological Typing of Cancer and Precancer of the Oral Mucosa," 2nd Ed. Springer, New York.
4. Bouquot, J. E., and Whitaker, S. B. (1994). Oral leukoplakia: Rationale for diagnosis and prognosis of its clinical subtypes or phases. *Quintessence Int.* **25**, 133–140.
5. Lumerman, H., Freedman, P., and Kerpel, S. (1995). Oral epithelial dysplasia and the development of invasive squamous cell carcinoma. *Oral Surg. Oral Med. Oral Pathol. Oral Radiol. Endod.* **79**, 321–329.
6. Silverman, S., Gorsky, M., and Lozada, F. (1984). Oral leukoplakia and malignant transformation: A follow-up study of 257 patients. *Cancer* **53**, 563–568.
7. Papadimitrakopoulou, V. A., and Hong, W. K. (1997). Retinoids in head and neck chemoprevention. *Proc. Soc. Exp. Boil. Med.* **216**, 238–290.
8. Renstrup, G. (1958). Leukoplakia of the oral cavity. *Acta Odonto. Scand.* **16**, 99.
9. Waldron, C. A., and Shafer, W. G. (1975). Leukoplakia revisited: A clinicopathologic study of 3256 oral leukoplakias. *Cancer.* **36**, 1386–1392.
10. Bouquot, J. E., and Gorlin, R. J. (1986). Leukoplakia, lichen planus and other oral keratosis in 23,616 white Americans over the age of 35 years. *Oral Surg. Oral Med. Oral Pathol.* **61**, 373–381.
11. Banoczy, J., and Rigo, O. (1991). Prevalence study of oral precancerous lesions within a complex screening system in Hungary. *Community Dent. Oral Epidemiol.* **19**, 265–267.
12. Mashberg, A., and Meyers, H. (1976). Anatomical site and size of 222 early asymptomatic oral squamous cell carcinomas: A continuing prospective study of oral cancer. *Cancer* **37**, 2149–2157.
13. Mashberg, A., Merletti, F., Boffetta, P., Gandolfo, S., Ozzello, F., Fracchia, F., and Terracini, B. (1989). Appearance, site of occurrence, and physical and clinical characteristics of oral carcinoma in Torino, Italy. *Cancer* **63**, 2522–2527.
14. Kramer, I. R., El-Labban, N., and Lee, K. W. (1978). The clinical features and risk of malignant transformation in sublingual keratosis. *Br. Dent. J.* **144**, 171–180.
15. Schell, H., and Schonberger, A. (1987). Site-specific incidence of benign and precancerous leukoplakias and cancers of the oral cavity. *Z. Hautkr.* **62**, 798–804.
16. El-Husseiny, G., Kandil, A., Jamshed, A., Khafaga, Y., Saleem, M., Allam, A., Al-Rajhi, N., Al-Amro, A., Rostom, A. Y., Abuzeid, M., Otieschan, A., and Flores, A. D. (2000). Squamous cell carcinoma of the oral tongue: An analysis of prognostic factors. *Br. J. Oral Maxillofac. Surg.* **38**, 193–199.
17. Incze, J., Vaughan, C. W., Jr., Lui, P., Strong, M. S., and Kulapaditharom, B. (1982). Premalignant changes in normal appearing epithelium in patients with squamous cell carcinoma of the upper aerodigestive tract. *Am. J. Surg.* **144**, 401–405.
18. Muto, S., Horie, S., Takahashi, S., Tomita, K., and Kitamura, T. (2000) Genetic and epigenetic alterations in normal bladder epithelium in patients with metachronous bladder cancer. *Cancer Res.* **60**, 4021–4025.
19. Grandis, J. R., Falkner, D. M., Melhem, M. F., Gooding, W. E., Drenning, S. D., and Morel, P. A. (2000). Human leukocyte antigen class I allelic and haplotype loss in squamous cell carcinoma of the head and neck: Clinical and immunogenetic consequences. *Clin. Cancer Res.* **6**, 2794–2802.
20. Man, Y. G., Martinez, A., Avis, I. M., Hong, S. H., Cuttita, F., Venzon, D. J., and Mulshine, J. L. (2000). Phenotypically different cells with heterogeneous nuclear ribonucleoprotein A2/B1 overexpression show similar genetic alterations. *Am. J. Resp. Cell Mol. Biol.* **23**, 636–645.
21. Wistuba, I. I., Berry, J., Behrens, C., Maitra, A., Shivapurkar, N., Milchgrub, S., Mackay, B., Minna, J. D., and Gazdar, A. F. (2000). Molecular changes in the bronchial epithelium of patients with small cell lung cancer. *Clin. Cancer Res.* **6**, 2604–2610.
22. Chan, A. S., To, K. F., Lo, K. W., Mak, K. F., Pak, W., Chiu, B., Tse, G. M., Ding, M., Li, X., Lee, J. C., and Huang, D. P. (2000). High frequency of chromosome 3p deletion in histologically normal nasopharyngeal epithelia from southern Chinese. *Cancer Res.* **60**, 5365–5370.
23. Lakhani, S. R., Chaggar, R., Davies, S., Jones, C., Collins, N., Odel, C., Stratton, M. R., and O'Hare, M. J. (1999). Genetic alterations in "normal" luminal and myoepithelial cells of the breast. *J. Pathol.* **189**, 496–503.
24. Mao, L., Lee, J. S., Kurie, J. M., Fan, Y. H., Lippman, S. M., Lee, J. J., Ro, J. Y., Broxson, A., Yu, R., Morice, R. C., Kemp, B. L., Khuri, F. R., Walsh, G. L., Hittelman, W. N., and Hong, W. K. (1997). Clonal genetic alterations in the lungs of current and former smokers. *J. Natl. Cancer Inst.* **89**, 857–862.
25. Wistuba, I. I., Lam, S., Behrens, C., Virmani, A. K., Fong, K. M., LeRiche, J., Samet, J. M., Srivastava, S., Minna, J. D., and Gazdar, A. F. (1997). Molecular damage in the bronchial epithelium of current and former smokers. *J. Natl. Cancer Inst.* **89**, 1366–1373.
26. Franklin, W. A., Gazdar, A. F., Haney, J., Wistuba, I. I., La Rosa, F. G., Kennedy, T., Ritchey, D. M., and Miller, Y. E. (1997). Widely dispersed p53 mutation in respiratory epithelium: A novel mechanism for field carcinogenesis. *J. Clin. Invest.* **100**, 2133–2137.

27. Sciubba, J. J. (1995). Oral leukoplakia. *Crit. Rev. Oral Biol. Med.* **6**, 147–160.
28. Schepman, K., der Meij, E., Smeele, L., and der Waal, I. (1999). Concomitant leukoplakia in patients with oral squamous cell carcinoma. *Oral Dis.* **5**, 206–209.
29. Schepman, K. P., and van der Waal, I. (1995). A proposal for a classification and staging system for oral leukoplakia: A preliminary study. *Eur. J. Cancer B Oral Oncol.* **31B**, 396–398.
30. Axell, T., Pindborg, J. J., Smith, C. J., and van der Waal, I. (1996). Oral white lesions with special reference to precancerous and tobacco-related lesions: Conclusions of an international symposium held in Uppsala, Sweden, May 18–21 1994. International Collaborative Group on Oral White Lesions. *J. Oral Pathol. Med.* **25**, 49–54.
31. Scully, C. (1995). Oral precancer: Preventive and medical approaches to management. *Oral Oncol. Eur. J. Cancer* **31**, 16–26.
32. Bosatra, A., Bussani, R., and Silvestri, F. (1997). From epithelial dysplasia to squamous carcinoma in the head and neck region: An epidemiological assessment. *Acta Otolaryngol.* **527**, 47–48.
33. Wright, J. M. (1998). A review and update of oral precancerous lesions. *Tex. Dent. J.* **115**, 15–19.
34. Scully, C., and Cawson, R. (1996). Potentially malignant oral lesions. *J. Epidem. Biostat.* **2**, 3–12.
35. Reid, B. C., Winn, D. M., Morse, D. E., and Pendrys, D. G. (2000). Head and neck in situ carcinoma: Incidence, trends, and survival. *Oral Oncol.* **36**, 414–420.
36. Schepman, K. P., van der Meij, E. H., Smeele, L. E. and van der Waall, I. (1998). Malignant transformation of oral leukoplakia: A follow-up study of a hospital-based population of 166 patients with oral leukoplakia from the Netherlands. *Oral Oncol.* **34**, 270–275.
37. Bouquot, J. (1999). Oral cancers with leukoplakia. *Oral Dis.* **5**, 183–184.
38. Mincer, H. H., Coleman, S. A., and Hopkins, K. P. (1972). Observations on the clinical characteristics of oral lesions showing histologic epithelial dysplasia. *Oral Surg. Oral Med. Oral Pathol.* **33**, 389–399.
39. van der Waal, I., Schepman, K. P., van der Meij, E. H., and Smeele, L. E. (1997). Oral leukoplakia: A clinicopathological review. *Oral Oncal.* **33**, 291–301.
40. Cloos, J., Steen, I., Timmerman, A. J., van der Schans, G. P., Snow, G. B., and Braakhuis, B. J. (1996). DNA-damage processing in blood lymphocytes of head-and-neck-squamous-cell-carcinoma patients is dependent on tumor site. *Int. J. Cancer* **68**, 26–29.
41. Axell, T., Holmstrup, P., Kramer, I. R. H., Pindorg, J. J., and Shear, M. (1984). International seminar on oral leukoplakia and associated lesions related to tobacco habits. *Commun. Dent. Oral Epidemiol.* **12**, 145–154.
42. van der Waal, I., Schepman, K. P., and van der Meij, E. H. (2000). A modified classification and staging system for oral leukoplakia. *Oral Oncol.* **36**, 264–266.
43. Tradati, N., Grigolat, R., Calabrese, L., Costa, L., Giugliano, G., Morelli, F., Scully, C., Boyle, P., and Chiesa, F. (1997). Oral leukoplakia: To treat or not? *Oral Oncol.* **33**, 317–321.
44. McCartan, B. (1998). Malignant transformation of leukoplakia. *Oral Surg. Oral Med. Oral Pathol. Oral Radiol. Endod.* **85**.
45. Allen, C. M. (1998). Malignant transformation of leukoplakia. *Oral Surg. Oral Med. Oral Pathol. Oral Radiol. Endod.* **85**, 348–349.
46. Banoczy, J., and Csiba, A. (1976). Occurrence of epithelial dysplasia in oral leukoplakia: Analysis and follow-up study of 12 cases. *Oral Surg. Oral. Med. Oral Pathol.* **42**, 766–774.
47. Einhorn, J., and Wersall, J. (1967). Incidence of oral carcinoma in patients with leukoplakia of the oral mucosa. *Cancer* **20**, 2189–2193.
48. Chiesa, F., Boracchi, P., Tradati, N., Rossi, N., Costa, L., Giardini, R., Marazza, M., and Zurrida, S. (1993). Risk of preneoplastic and neoplastic events in operated oral leukoplakias. *Eur. J. Cancer B Oral Oncol.* **29B**, 23–28.
49. Weber, J. L., and May P. E. (1989). Abundant class of human DNA polymorphisms which can be typed using the polymerase chain reaction. *Am. J. Hum. Genet.* **44**, 388–396.
50. Lese, C. M., Rossie, KI. M., Appel, B. N., Reddy, J. K., Johnson, J. T., Myers, E. N., and Gollin, S. M. (1995). Visualization of Int2 and Hst1 amplification in oral squamous cell carcinomas. *Genes Chromosomes Cancer* **12**, 288–295.
51. Califano, J., van der Riet, P., Westra, W., Nawroz, H., Clayman, G., Piantadosi, S., Corio, R., Lee, D., Greenberg, B., Koch, W., and Sidransky, D. (1996). Genetic progression model for head and neck cancer: Implications for field cancerization. *Cancer Res.* **56**, 2488–2492.
52. el-Naggar, A. K., Hurr, K., Batsakis, J. G., Luna, M. A., Goepfert, H., and Huff, V. (1995). Sequential loss of heterozygosity at microsatellite motifs in preinvasive and invasive head and neck squamous carcinoma. *Cancer Res.* **55**, 2656–2659.
53. Pershouse, M. A., El-Naggar, A. K., Hurr, K., Lin, H., Yung, W. K., and Steck, P. A. (1997). Deletion mapping of chromosome 4 in head and neck squamous cell carcinoma. *Oncogene* **14**, 369–373.
54. Shah, S. I., Yip, L., Greenberg, B., Califano, J. A., Chow, J., Eisenberger, C. F., Lee, D. J., Sewell, D. A., Reed, A. L., Lango, M., Jen, J., Koch, W. M., and Sidransky, D. (2000). Two distinct regions of loss on chromosome arm 4q in primary head and neck squamous cell carcinoma. *Arch. Otolaryngol. Head Neck Surg.* **126**, 1073–1076.
55. Uzawa, K., Yoshida, H., Suzuki, H., Tanzawa, H., Shimazaki, J., Seino, S., and Sato, K. (1994). Abnormalities of the adenomatous polyposis coli gene in human oral squamous-cell carcinoma. *Int. J. Cancer* **58**, 814–817.
56. Resto, V. A., Caballero, O. L., Buta, M. R., Westra, W. H., Wu, L., Westendorf, J. M., Jen, J., Hieter, P., and Sidransky, D. (2000). A putative oncogenic role for MPP11 in head and neck squamous cell cancer. *Cancer Res.* **60**, 5529–5535.
57. Field, J. K., Kiaris, H., Risk, J. M., Tsiriyotis, C., Adamson, R., Zoumpourlis, V., Rowley, H., Taylor, K., Whittaker, J., and Howard, P. (1995). Alleotype of squamous cell carcinoma of the head and neck: Fractional allele loss correlates with survival. *Br. J. Cancer* **72**, 1180–1118.
58. Partridge, M., Kiguwa, S., and Langdon, J. D. (1994). Frequent deletion of chromosome 3p in oral squamous cell carcinoma. *Eur. J. Cancer B Oral Oncol.* **30B**, 248–251.
59. Partridge, M., Emilion G., and Langdon J. D. (1996). LOH at 3p correlates with a poor survival in oral squamous cell carcinoma. *Br. J. Cancer* **73**, 366–371.
60. Partridge, M., Emilion, G., Pateromichelakis, S., Phillips, E., and Langdon, J. (1999). Location of candidate tumour suppressor gene loci at chromosomes 3p, 8p and 9p for oral squamous cell carcinomas. *Int. J. Cancer* **83**, 318–325.
61. Okami, K., Wu, L., Riggins, G., Cairns, P., Goggins, M., Evron, E., Halachmi, N., Ahrendt, S. A., Reed, A. L., Hilgers, W., Kern, S. E., Koch, W. M., Sidransky, D., and Jen, J. (1998). Analysis of PTEN/NMAC1 alterations in aerodigestive tract tumors. *Cancer Res.* **58**, 509–511.
62. Lee, D. J., Koch, W. M., Yoo, G., Lango, M., Reed, A., Califano, J., Brennan, J. A., Westra, W. H., Zahurak, M., and Sidransky, D. (1997). Impact of chromosome 14q loss on survival in primary head and neck squamous cell carcinoma. *Clin. Cancer Res.* **3**, 501–505.
63. Papadimitrakopoulou, V. A., Oh, Y., El-Naggar, A., Izzo, J., Clayman, G., and Mao, L. (1998). Presence of multiple incontiguous deleted regions at the long arm of chromosome 18 in head and neck cancer. *Clin. Cancer Res.* **4**, 539–544.
64. Mao, L.(1998). Tumor suppressor genes: Does FHIT fit? *J. Natl. Cancer Inst.* **90**, 412–414.
65. Ishwad, C.S., Shuster, M., Bockmuhl, U., Thakker, N., Shah, P., Toomes, C., Dixon, M., Ferrell, R. E., and Gollin, S. M. (1999). Frequent allelic loss and homozygous deletion in chromosome band 8p23 in oral cancer. *Int. J. Cancer* **80**, 25–31.

66. Miyakawa, A., Wang, X. L., Nakanishi, H., Imai, F. L., Shiiba, M., Miya, T., Imai, Y., and Tanzawa, H. (1998) Allelic loss on chromosome 22 in oral cancer: Possibility of the existence of a tumor suppressor gene on 22q13. *Int. J. Oncol.* **13**, 705–709.
67. Emilion, G., Langdon, J. D., Speight, P., and Partridge, M. (1996). Frequent gene deletions in potentially malignant oral lesions. *Br. J. Cancer* **73**, 809–813.
68. Roz, L., Wu, C. L., Porter, S., Scully, C., Speight, P., Read, A., Sloan, P., and Thakker, N. (1996). Allelic imbalance on chromosome 3p in oral dysplastic lesions: An early event in oral carcinogenesis. *Cancer Res.* **56**, 1228–1231.
69. van der Riet, P., Nawroz, H., Hruban, R. H., Corio, R., Tokino, K., Koch, W., and Sidransky, D. (1994). Frequent loss of chromosome 9p21-22 early in head and neck cancer progression. *Cancer Res.* **54**, 1156–1158.
70. Mao, L., Lee, J. S., Fan, Y. H. Ro, J.Y., Batsakis, J. G., Lippman, S., Hittelman, W., and Hong, W. K. (1996). Frequent microsatellite alterations at chromosomes 9p21 and 3p14 in oral permalignant lesions and their value in cancer risk assessment. *Nature Med.* **2**, 682–685.
71. Mao, L., el-Naggar, A. K., Papadimitrakopoulou, V., Shin, D. M., Shin, H. C., Fan, Y., Zhou, X., Clayman, G., Lee, J. J., Lee, J. S., Hittelman, W. N., Lippman, S. M., and Hong.W. K. (1998). Phenotype and genotype of advanced premalignant head and neck lesions after chemopreventive therapy. *J. Natl. Cancer Inst.* **90**, 1545–1551.
72. Rosin, M. P., Cheng, X., Poh, C., Lam, W. L., Huang, Y., Lovas, J., Berean, K., Epstein, J. B., Priddy, R., Le, N. D., and Zhang, L. (2000). Use of allelic loss to predict malignant risk for low-grade oral epithelial dysplasia. *Clin. Cancer Res.* **6**, 357–362.
73. Partridge, M., Emilion, G., Pateromichelakis, S., A'Hern, R., Phillips, E., and Langdon, J. (1998). Allelic imbalance at chromosomal loci implicated in the pathogenesis of oral precancer, cumulative loss and its relationship with progression to cancer. *Oral Oncol.* **34**, 77–83.
74. Partridge, M., Pateromichelakis, S., Phillips, E., Emilion, G. G., A'Hern, R. P., and Langdon, J. D. (2000). A case-control study confirms that microsatellite assay can identify patients at risk of developing oral squamous cell carcinoma within a field of cancerization. *Cancer Res.* **60**, 3893–3898.
75. Mao, L. (2000). Can molecular assessment improve classification of head and neck premalignancy? *Clin. Cancer Res.* **6**, 321–322.
76. Boffetta, P., Mashberg, A., Winkelmann, R., and Garfinkel, L. (1992). Carcinogenic effect of tobacco smoking and alcohol drinking on anatomic sites of the oral cavity and oropharynx. *Int. J. Cancer* **52**, 530–533.
77. Kramer, I. R. (1980). Oral leukoplakia, *J. R. Soc. Med.* **73**, 765–767.
78. Moore, C., and Catlin, D. (1967). Anatomic origins and locations of oral cancer. *Am. J. Surg.* **114**, 510–513.
79. Zhang, L., Cheung, K.-J., Cheng, X., Poh, C. F., Priddy, R. W., Epstein, J., and Rosin, M. P. (2001). Increased genetic damage in oral leukoplakia from high-risk sites: Potential impact on staging and clinical management. *Cancer* **91**, 2148–2155.
80. Mashberg, A., and Samit, A. (1995). Early diagnosis of asymptomic oral and oropharyngeal squamous cancers. *CA Cancer J. Clin.* **45**, 328–351.
81. Poh, C. F., Zhang, L., Lam, W. L., Zhang, X., An, D., Chau, C., Priddy, R., Epstein, J., and Rosin, M. P. (2001). A high frequency of allelic loss in oral verrucous lesions may explain malignant risk. *Lab. Invest.* **81**, 629–634.
82. Silverman, S. J., and Gorsky, M. (1997). Proliferative verrucous leukoplakia: A follow-up study of 54 cases. *Oral Surg. Oral Med. Oral Pathol.* **84**, 154–157.
83. Clayman, G. L., Lippman, S. M., Laramore, G. E., and Hong, W. K. (1996). Head and neck cancer. In "Cancer Medicine" (J. F. Holland, E. Frei, R. C. J. Bast, D. W. Kufe, D. L. Morton, and R. Weichselbaum, eds.) pp. 1645–1709. William & Wilkins, Philadelphia.
84. Vikram, B. (1994). Changing patterns of failure in advanced head and neck cancer. *Arch. Otolaryn. Head Neck Surg.* **110**, 564–565.
85. Parker, S. L., Tong, T., Bolden, S., and Wingo, P. A. (1996). Cancer statistics, 1996. *CA Cancer J. Clin.* **46**, 5–27.
86. Mao, L., Lee, J. S., Kurie, J. M., Fan, Y. H., Lippman, S. M., Lee, J. J., Ro, J. Y., Broxson, A. Yu, R., Morice, R. C., Kemp, B. L., Khuri, F. R., Walsh, G. L., Hittelman, W. N., and Hong, W. K. (1997). Clonal genetic alterations in the lungs of current and former smokers. *J. Natl. Cancer Inst.* **89**, 857–862.
87. Califano, J., Westra, W. H., Koch, W., Meininger, G., Reed, A., Yip, L., Boyle, J. O., Lonardo, F., and Sidransky, D. (1999). Unknown primary head and neck squamous cell carcinoma: Molecular identification of the site of origin. *J. Natl. Cancer Inst.* **91**, 599–604.
88. Lydiatt, W. M., Anderson, P. E., Bazzana, T., Casale, M., Hughes, C. J., Huvos, A. G., Lydiatt, D. D., and Schantz, S. P. (1998). Molecular support for field cancerization in the head and neck. *Cancer* **82**, 1376–1380.
89. Hittelman, W. N., Voravud, N., Shin, D. M., Lee, J. S., Ro, J. Y., and Hong. W. K. (1993). Early genetic changes during upper aerodigestive tract tumorigenesis. *J. Cell Biochem. Suppl.* 233–236.
90. Kishimoto, Y., Sugio, K., Hung, J. Y., Virmani, A. K., McIntire, D. D., Minna, J. D., and Gazdar, A. F. (1995). Allele-specific loss in chromosome 9p loci in preneoplastic lesions accompanying non-small-cell lung cancers. *J. Natl. Cancer Inst.* **87**, 1224–1229.
91. Smith, A. L., Hung, J., Walker, L., Rogers, T. E., Vuitch, F., Lee, E., and Gazdar, A. F. (1996). Extensive areas of aneuploidy are present in the respiratory epithelium of lung cancer patients *Br. J. Cancer* **73**, 203–209.
92. Deng, G., Lu, Y., Zlotnikov, G., Thor, A. D., and Smith, H. S. (1996). Loss of heterozygosity in normal tissue adjacent to breast carcinomas. *Science* **274**, 2057–2059.
93. Hittelman, W. N., Kim, H. J., Lee, J. S., Shin, D. M., Lippman, S. M., Kim, J., Ro, J. Y., and Hong, W. K. (1996). Detection of chromosome instability of tissue fields at risk: In situ hybridization. *J. Cell Biochem. Suppl.* **25**, 57–62.
94. Dolan, K., Garde, J., Walker, S. J., Sutton, R., Gosney, J., and Field, J. K. (2000). Histological and molecular mapping of adenocarcinoma of the oesophagus and gastroesophageal junction: Loss of heterozygosity occurs in histologically normal epithelium in the oesophagus and stomach. *Oncol. Rep.* **7**, 521–528.
95. Heinmoller, E., Dietmaier, W., Zirngibl, H., Heinmoller, P., Scaringe, W., Jauch, K. W., Hofstadter, F., and Ruschoff, J. (2000). Molecular analysis of microdissected tumors and preneoplastic intraductal lesions in pancreatic carcinoma. *Am. J. Pathol.* **157**, 83–92.
96. Carlson, J. A., Healy, K., Tran, T. A., Malfetano, J., Wilson, V. L., Rohwedder, A., and Ross, J. S. (2000). Chromosome 17 aneusomy detected by fluorescence in situ hybridization in vulvar squamous cell carcinomas and synchronous vulvar skin. *Am. J. Pathol.* **157**, 973–983.
97. Lippman, S. M., and Hong, W. K. (2001). Molecular markers of the risk of oral cancer. *N. Engl. J. Med.* **344**, 1323–1326.
98. Boone, C. W., Kelloff, G. J., and Steele, V. E. (1992). Natural history of intraepithelial neoplasia in humans with implications for cancer chemoprevention strategy. *Cancer Res.* **52**, 1651–1659.
99. Shekhar, M. P., Nangia-Makker, P., Wolman, S. R., Tait, L., Heppner, G. H., and Visscher, D. W. (1998). Direct action of estrogen on sequence of progression of human preneoplastic breast disease. *Am J. Pathol.* **152**, 1129–1132.
100. Pinto, A. P., and Crum, C. P. (2000). Natural history of cervical neoplasia: Defining progression and its consequence. *Clin. Obst. Gynecol.* **43**, 352–362.
101. Braithwaite, K. L., and Rabbits, P. H. (1999). Multi-step evolution of lung cancer. *Sem. Cancer Biol.* **9**, 255–265.
102. Geboes, K. (2000). Barrette's esophagus: The metaplasia-dysplasia-carcinoma sequence: morphological aspects. *Acta Gastroenter. Belgica* **63**, 13–17.

103. Lee, J. J., Hong, W. K., Hittelman, W. N., Mao, L., Lotan, R., Shin, D. M., Benner, S. E., Xu, X. C., Lee, J.S., Papadimitrakopoulou, V. M., Geyer, C., Perez, C., Martin, J. W., El-Naggar, A. K., and Lippman, S. M. (2000). Predicting cancer development in oral leukoplakia: Ten years of translational research. *Clin. Cancer Res.* **6**, 1702–1710.

104. Sudbo, J., Kildal, W., Risberg, B., Koppang, H. S., Danielsen, H. E., and Reith, A. (2001). DNA content as a prognostic marker in patients with oral leukoplakia. *N. Engl. J. Med.* **344**, 1270–1278.

104a. Zhang, L., and Rosin, M. P. (2001). Loss of heterozygosity: A potential tool in management of oral premalignant lesion? *J. Oral Pathol. Med.* **30**, 513–520.

105. Zhang, L., Poh, C. F., Lam, W. L., Epstein, J. B., Cheng, X., Zhang, X., Priddy, R., Lovas, J., Le, N. D., and Rosin, M. P. (2001). Impact of localized treatment in reducing risk of progression of low-grade oral dysplasia: Molecular evidence of incomplete resection. *Oral Oncol.* **37**, 505–512.

106. Brennan, J. A., Mao, L., Hruban, R. H., Boyle, J. O., Eby, Y. J., Koch, W. M., Goodman, S. N., and Sidransky, D. (1995). Molecular assessment of histopathological staging in squamous cell carcinoma of the head and neck. *Engl. J. Med.* **332**, 429–435.

107. Westra, W. H., and Sidransky, D. (1998). Phenotypic and genotypic disparity in premalignant lesions: Of calm water and crocodiles. *JNCI* **90**, 1500–1501.

108. Hong, W. K., Endicott, J., Itri, L. M., Doos, W., Batsakis, J. G., and Bell, R. (1986). 13-cis-retinoic acid in the treatment of oral leukoplakia. *N. Engl. J. Med.* **315**, 1501–1505.

109. Mao, L., Hruban, R. H., Boyle, J. O., Tockman, M., and Sidransky, D. (1994). Detection of oncogene mutations in sputum precedes diagnosis of lung cancer. *Cancer Res.* **54**, 1634–1637.

110. Spafford, M. F., Koch, W. M., Reed, A. L., Califano, J. A., Xu, L. H., Eisenberger, C. F., Yip, L., Leong, P. L., Wu, L., Liu, S. X., Jeronimo, C., Westra, W. H., and Sidransky, D. (2000). Detection of head and neck squamous cell carcinoma among exfoliated oral mucosal cells by microsatellite analysis. *Clin. Cancer Res.* **6**, 4171–4175.

111. Boyle, J. O., Mao, L., Brennan, J. A., Koch, W. M., Eisele, D. W., Saunders, J. R., and Sidransky, D. (1994). Mutations in saliva as molecular markers for head and neck squamous cell carcinomas. *Am. J. Surg.* **168**, 429–432.

112. Rosas, S. L., Koch, W., da Costa Carvalho, M. G., Wu, L., Califano, J., Westra, W., Jen, J., and Sidransky, D. (2001). Promoter hypermethylation patterns of p16, O6-methylguanine-DNA-methyltransferase, and death-associated protein kinase in tumors and saliva of head and neck cancer patients. *Cancer Res.* **61**, 939–942.

113. Rosin, M. P., Epstein, J. B., Berean, K., Durham, S., Hay, J., Cheng, X., Zeng, T., Huang, Y., and Zhang L. (1997). The use of exfoliative cell samples to map clonal genetic alterations in the oral epithelium of high-risk patients. *Cancer Res.* **57**, 5258–5260.

C H A P T E R

18

Chemoprevention of Oral Premalignant Lesions

SUSAN T. MAYNE,* BRENDA CARTMEL,* and DOUGLAS E. MORSE†

*Department of Epidemiology and Public Health, Yale University School of Medicine, New Haven, Connecticut 06520 and

†New York University College of Dentistry, New York, New York 10010

I. Oral Premalignant Lesions: Definition and Significance 261
II. Chemoprevention Trials: General Design and Outcome Assessment 262
III. Review of Trials 263
A. Retinoids 263
B. β-Carotene 264
C. Vitamin E and Selenium 265
D. Other Agents 265
IV. Validity of Oral Precancerous Lesions in Predicting Efficacy of Agents for Oral Cancer Prevention 266
V. Summary and Conclusions 267
References 268

Oral premalignant lesions have served as a widely used model system in chemoprevention research. Chemoprevention of these lesions has direct clinical relevance and can also be used to screen agents that may have efficacy in the prevention of oral, pharyngeal, or other cancers. This chapter introduces the reader to various types of oral premalignant lesions, discusses design and interpretation issues of relevance to chemoprevention trials in this clinical setting, and reviews the results of chemoprevention trials of oral premalignancy. Several agents/classes of agents have been evaluated for chemopreventive efficacy in oral precancerous lesions and are discussed herein, including various retinoids, β-carotene, the algae *Spirulina fusiformis,* vitamin E, selenium, tea, protease inhibitors (Bowman–Birk inhibitor concentrate), and bleomycin. Some of these agents have established efficacy in terms of regressing oral precancerous lesions, demonstrating the "proof of principle" behind chemoprevention of oral premalignancy. However, these same agents have limitations to their widespread use for this purpose. The chapter concludes with a discussion of the validity of oral precancerous lesions in predicting the efficacy of agents for oral cancer prevention and discusses challenges that lie ahead with regard to identifying suitable agents with clear evidence of efficacy but without significant toxicity.

I. ORAL PREMALIGNANT LESIONS: DEFINITION AND SIGNIFICANCE

Many oral and pharyngeal carcinomas are preceded by precancerous lesions that present as red, white, or red and white patches or plaques that will not rub off and cannot be characterized clinically or pathologically as any other disease [1]. White oral lesions meeting these criteria are termed oral leukoplakia; those that are red are referred to as erythroplakia; and those that are both red and white are termed oral erythroleukoplakia. Although each of these lesions has been classified as precancerous, it is well established that lesions with a red component are most likely to undergo malignant transformation or to exhibit either invasive or *in situ* carcinoma [2,3]. Small hyperplastic leukoplakia lesions have a 30–40% spontaneous regression rate and less than a 5% risk of malignant transformation, whereas erythroleukoplakia and dysplastic leukoplakia lesions are associated with a low rate of spontaneous regression and a 30–40% long-term risk of oral cancer [4]. A mainstay in the prevention of oral and pharyngeal cancer is, therefore, the early identification and removal of such lesions.

It is estimated that approximately 75% of all oral and pharyngeal cancer cases in the United States are attributable to smoking and drinking [5]. Smoking, and to some extent, drinking have also been implicated as risk factors for oral precancer [6–9]. Consequently, the reduction or elimination of smoking and heavy drinking represent another critical strategy for preventing oral and pharyngeal cancer.

Although the removal of precancerous lesions and the elimination of known risk factors are viable and important strategies for preventing oral and pharyngeal cancer, they do suffer from important limitations. First, precancerous lesions that have been removed can recur, thereby requiring subsequent biopsies and close follow-up. Further, because lesions can be large or multifocal, their complete removal may sometimes be deemed inappropriate, particularly when incisional biopsy reveals a histopathologic picture suggesting a low risk of transformation. An additional limitation of these prevention strategies is that persons who have been exposed previously to risk factors for oral/pharyngeal cancers are likely to have cells that have undergone initiation and promotion and are at risk of proceeding further down the path of oral carcinogenesis.

One additional strategy for the prevention of oral and pharyngeal cancer that has the potential to address these limitations, at least in part, is the use of chemopreventive agents. Cancer chemopreventive agents are specific nutrients or other chemicals that are used to suppress or reverse carcinogenesis and to prevent the development of invasive cancer [10]. These agents, while potentially useful for the general population, may have particular utility for persons who are presumed to be at an elevated risk of oral and pharyngeal cancer, due either to the presence of a precancerous lesion or a significant history of risk factor exposure.

Oral precancerous lesions have been relatively well studied in chemoprevention research. Oral precancerous lesions are used as intermediate end point biological markers in order to help screen potential chemopreventive agents that might have utility in the prevention of oral and pharyngeal cancers. Also, because treatment strategies for large or diffuse, multifocal oral precancerous lesions are limited, successful chemopreventive agents for these lesions would have direct clinical relevance.

II. CHEMOPREVENTION TRIALS: GENERAL DESIGN AND OUTCOME ASSESSMENT

Prior to reviewing the results of chemoprevention trials of various agents in oral precancerous lesions, it is necessary to understand how these trials are conducted, how efficacy is assessed, and the resulting implications for interpreting the results of these types of trials.

Previous trials investigating the use of chemopreventive agents in treating potentially precancerous oral lesions have focused primarily on individuals with oral leukoplakia. Oral leukoplakia serves as a useful model because it is often readily visible and accessible and can therefore be measured, photographed, and biopsied. Because the term "oral leukoplakia" represents a clinical diagnosis, however, the underlying histopathological picture can vary widely from lesion to lesion, with a concomitantly wide range in malignant potential. Consequently, chemoprevention trials of oral premalignancy often present information on or, in some instances, restrict their eligibility criteria to specific histopathologic diagnoses.

Clearly, patients who have had full excision of suspicious lesions are generally not good candidates for recruitment for efficacy trials, although such patients could be enrolled into trials aimed at maintenance of remissions (e.g., see Costa *et al.* [11]). In contrast, patients with multifocal lesions or lesions not fully excised would be suitable candidates for recruitment.

Prior to placing patients on an intervention, baseline evaluation of the lesion must be performed. The baseline evaluation is predicated on the response criteria selected for the trial, which can vary across trials. For example, efficacy evaluation in most trials includes a visual examination of the lesion. Responses are then quantified by bidimensional measurements of the lesions pre- and postintervention and/or by the use of intraoral photography of the lesions. The histological response is also sometimes used as a criterion in order to evaluate efficacy; this approach requires oral biopsies pre- and postintervention. While this approach may seem more objective than visual estimation of change in the size of a lesion, it must be recognized that this approach has limitations as well. The baseline biopsy, for example, is generally done on the most suspicious looking lesion/part of a lesion. It is quite possible that the initial biopsy might remove areas of the most severe dysplasia, with the result that lesions appear to improve histologically at the time of the second biopsy. Histological or visual responses could be susceptible to observer bias in estimating response in a single-arm trial. Also, spontaneous regression of oral leukoplakia lesions is known to occur in the absence of intervention. These limitations all point to the need for placebo controls and blinding of study subjects and of observers (double blinding) in order to evaluate the effect of a given intervention more objectively. For these reasons, the greatest emphasis is given to randomized, placebo-controlled trials of oral leukoplakia/oral dysplasia (see Section III).

While visual and histological evaluations of lesions have been the primary response criteria used in chemoprevention trials of oral premalignancy, other intermediate end points, such as micronuclei in exfoliated oral mucosal cells and biomarkers of cell proliferation, have also been evaluated (e.g., see Li *et al.* [12]). A promising research area of relevance here is the molecular assessment of risk of oral cancer in patients with leukoplakia. That is, molecular evaluation of biopsied tissue has been shown to be of value in predicting which lesions will progress to oral carcinomas [13,14]. Molecular markers may have utility in monitoring risk for patients with oral premalignancies and in evaluating efficacy in chemoprevention trials. The ability of a chemopreventive agent to reverse a molecular alteration that is strongly associated with the development of invasive malignancy might better predict efficacy in the prevention of oral cancer than visual changes in lesion size. Thus, it is likely that molecular end points will be commonly incorporated as part of the outcome assessment in future chemoprevention trials of oral leukoplakia.

III. REVIEW OF TRIALS

A. Retinoids

Retinoids, including vitamin A, were the first class of chemopreventive agents to be widely studied in the reversal of oral premalignant lesions. Many single-arm trials of vitamin A and certain retinoids were completed in the mid-1970s. These early single-arm trials will not be reviewed here, but are important as they set the foundation for newer, placebo-controlled trials that are considered here. These single-arm trials identified several aspects of retinoid chemoprevention that remain relevant today. That is, they established the concepts that retinoids had efficacy in the reversal of oral precancerous lesions, that retinoid-related toxicity caused some patients to discontinue therapy, and that relapses commonly occurred within 2–3 months after the cessation of therapy [15].

One of the first randomized, placebo-controlled studies to emerge following the early trials was a 3-month study of 13-*cis*-retinoic acid (2 mg/kg/day), which was published in 1986 [16]. The complete plus partial response rate in the 44 evaluable patients was 67% in the retinoid arm and 10% in the placebo arm ($p = 0.0002$). The histopathologic improvement rate (e.g., reversal of dysplasia) was also higher in the retinoid arm (54% vs 10%, $p = 0.01$). There were two major problems, however, with this high-dose, short-term approach. First, high-dose 13-*cis*-retinoic acid toxicity was substantial and not acceptable in this clinical setting. Second, over half of the responders recurred or developed new lesions within 3 months of stopping the intervention.

Based on the results of this placebo-controlled trial, a randomized maintenance trial was designed [17]. In this trial, patients initially received a 3-month induction course of high-dose 13-*cis*-retinoic acid (1.5 mg/kg/day), followed by a 9-month maintenance course with low-dose 13-*cis*-retinoic acid (0.5 mg/kg/day) or β-carotene (30 mg/day) in patients with responding or stable lesions during the high-dose induction phase. The induction-phase response rate was 55% (95% CI = 42–67%). During the maintenance phase, 2 (8%) of the patients in the retinoid group progressed versus 16 (55%) in the β-carotene group ($p < 0.001$). Toxic effects of low-dose 13-*cis*-retinoic acid maintenance therapy were mild, although greater than for β-carotene, with no patients discontinuing therapy because of toxicity.

Another trial of 13-*cis*-retinoic acid used topical application rather than oral dosing of the retinoid [18]. In this double-blind study, 10 patients with oral leukoplakia were randomly assigned to daily treatments (three topical applications) with 0.1% 13-*cis*-retinoic acid gel or placebo gel for 4 months. Patients who originally received placebo crossed over to the intervention for an additional 4 months. Following completion of topical retinoid treatment, all patients showed a significant improvement in size and clinical appearance (change in color from white to pink) of the oral lesions. In contrast, no patients showed improvements following completion of placebo treatment. Of 9 evaluable patients who completed the retinoid intervention, 1 had a complete response and 8 had partial responses (50% or greater reduction in the size of the lesion). No side effects were observed with this topical intervention.

Fenretinide [*N*-4-hydroxyphenyl retinamide (4HPR)], another promising retinoid, is also being evaluated in oral premalignancy [11]. Patients with oral premalignant lesions were treated with laser resection to remove lesions and were then randomized to 1 year of systemic 4HPR maintenance therapy (200 mg/day) or no intervention with a planned 4-year follow-up after completing the intervention. Recruitment was initiated in 1988 with a target sample size of 190, and recruitment was closed in 1994 with 174 patients enrolled (170 evaluable). A 3-day drug holiday at the end of each month was prescribed to avoid the adverse effects (night blindness) of lowering serum retinol by 4HPR treatment. There were 11 recurrences in each group, but only 3 new occurrences in the 4HPR group versus 12 in the control group, for an overall failure of 14/84 (17%) in the retinoid group versus 23/86 (27%) in the control group. Two other randomized studies involving retinol [19,20] and another involving the retinoid *N*-4-(hydroxycarbophenyl) retinamide (4HCR [21]) have been reported (see Table 18.1); all observed significant retinoid chemopreventive activity.

Thus, there is a consistent body of literature documenting that a variety of different retinoids have efficacy in the regression of oral premalignant lesions. However, as noted in the early trials, this literature clearly indicates that lesions return rapidly upon cessation of intervention, requiring the use of long-term preventive therapy. Consequently, a current goal with regard to both retinoids and other chemopreventive agents is to identify efficacious agents with minimal or no toxicity, allowing for the possibility of long-term preventive therapy.

Of concern with regard to the possible long-term use of 13-*cis*-retinoic acid for preventive therapy are results from a second primary lung cancer prevention trial of this agent [22]. This trial investigated the use of 13-*cis*-retinoic acid (30 mg/day) for the prevention of second primary tumors in patients with stage I nonsmall cell lung cancer. This retinoid

TABLE 18.1 Randomized Trials of Chemopreventive Agents in Oral Leukoplakia

Population	Intervention[b]	*N*	End point	Result	Statistical significance	Reference
United States	13cRA (2 mg/kg/day)	24	Regression	67%	$p=0.0002$	Hong *et al.* [16]
	Placebo	20		10%		
United States	13cRA (1.5 mg/kg/day) initially; stable or improved	66	Regression	55%		Lippman *et al.* [17]
	13cRA (0.5 mg/kg/day)	24	Progression after 13cRA induction	8%	$p<0.001$	
	β-carotene (30 mg/day)	29		55%		
Italy	13cRA gel (0.1%) vs placebo gel	10	Regression	100%	*p* not reported	Piattelli *et al.* [18]
	Placebo group crossed over at end of initial intervention period			0% placebo		
Italy[a]	4HPR (200 mg/day)	84	Recurrence + new lesion	17%	*p* not reported	Costa *et al.* [11]
	No treatment control	86		27%		
India	Vitamin A (300,000 IU/week)	50	Complete regression	52%	$p<0.0001$	Sankaranarayanan *et al.* [19]
	BC (360 mg/week)	55		33%		
	Placebo	55		10%		
India	Vitamin A (200,000 IU/week)	21	Complete regression	57%	$p<0.05$	Stich *et al.* [20]
	Placebo	33		3%		
China	4HCR (40 mg/day)	31	Complete regression	87%	$p<0.01$	Han *et al.* [21]
	Placebo	30		17%		
India (betel nut chewers)	BC (180 mg/week)	27	Complete regression	14.8%	p=NS	Stich *et al.* [26]
	BC and retinol (100,000 IU/week)	51		27.5%	$p<0.05$	
	Placebo	33		3%		
United States	BC (60 mg/day) initially	54	Regression	52%		Garewal *et al.* [32]
	Responders: BC (60 mg/day)	11	Relapse in responders	18% (failure)	p=NS	
	Placebo	12		17%		
Uzbekistan	BC (40 mg/day) + retinol (100,000 IU/week) + vitamin E (80 mg/week)	384	Leukoplakia prevalence	OR = 0.62 (0.39–0.98)	$p<0.05$	Zaridze *et al.* [33]
	Placebo	291				
India	*Spirulina fusiformis* (1 g/day)	60	Complete regression	45%	$p<0.0001$	Mathew *et al.* [37]
	Placebo (nonconcurrent)	55		7%		
China	Green tea (3 g mixed tea oral capsule + topical)	29	Partial regression (30% or more decrease in size)	37.9%	$p<0.05$	Li *et al.* [12]
	Placebo capsule and vehicle topical control	30		10.0%		
Canada	Bleomycin 1% in DMSO	10	% reduction in lesion size	81%	$p=0.001$	Epstein *et al.* [48]
	DMSO	12		21%		

[a]Ongoing; interim report.
[b]BC, β-carotene; DMSO, dimethyl sulfoxide; NS, not significant; OR, odds ratio; 4HPR, *N*-4-hydroxyphenyl retinamide; 4HCR, *N*-4-hydroxycarbophenyl retinamide; 13cRA, 13-*cis*-retinoic acid.

did not improve the overall rates of second primary tumors, recurrences or mortality. However, of potential concern was the observation that 13-*cis*-retinoic acid increased the risk of recurrence in smokers (hazard ratio for treatment by current versus never-smoking status = 3.11, 95% CI 1.00–9.71). Thus, current smokers should probably be advised not to use 13-*cis*-retinoic acid as a long-term chemopreventive agent, with the exception of smokers participating in other trials [23] with appropriate monitoring for long-term safety and toxicity.

B. β-Carotene

β-Carotene is structurally related to retinoids, being a precursor of vitamin A. Unlike the retinoids, β-carotene does not exhibit retinoid-like side effects such as mucocutaneous toxicity, making it attractive for evaluation as a long-term preventive therapy. Studies in the 1980s in populations at high risk of oral cancer (tobacco chewers, betel quid chewers) demonstrated that supplemental β-carotene and retinol reduced the frequency of oral micronuclei significantly [24–26]. These trials, suggesting beneficial effects of β-carotene on markers of DNA damage, set the stage for subsequent trials of this agent in oral premalignant lesions. At least nine trials have investigated the effects of supplemental β-carotene, alone or in combination with other agents, on the regression of oral leukoplakia. Five nonrandomized studies [27–31] and the nonrandomized induction phase of a randomized trial [32] reported response rates ranging from 44 to 97%. The response rates from these uncontrolled studies, however, must be interpreted cautiously for the reasons alluded to earlier. Four placebo-controlled trials of β-carotene and oral leukoplakia are available. Stich *et al.* [26] reported that the combination of β-carotene (180 mg/week) plus retinol (100,000 IU/week) produced complete remissions in 27.5%, β-carotene alone in 14.8%, and placebo in 3.0% of subjects (partial remissions were not reported) with a 6-month intervention. Sankaranarayanan *et al.* [19] got better response rates with a longer duration of intervention (12 months), consisting of 33% complete regression with β-carotene (360 mg/week) and 52% with retinyl palmitate (300,000 IU/week) versus 10% in the placebo arm. In a trial in Uzbekistan [33], 6 months of treatment with the combination of retinol (100,000 IU/week), β-carotene (40 mg/day), and vitamin E (80 mg/week) led to a significant reduction in the prevalence odds ratio of oral leukoplakia (OR=0.62; 95% CI=0.39–0.98). The risk of progression or no change versus regression was also reduced by 40% by this intervention, although it was not statistically significant (OR=0.60, 95% CI=0.23–1.63). A trial with even a longer duration of intervention is available: Garewal *et al.* [32] gave β-carotene (60 mg/day) to 54 patients with oral leukoplakia for 6 months and then randomized responders to continue supplementing with β-carotene (same dose) or placebo for another 12 months (total duration of supplementation up to 18 months). Of those who initially responded to β-carotene, 2 of 11 in the β-carotene arm and 2 of 12 in the placebo arm subsequently relapsed.

These trials indicate that supplemental forms of β-carotene have the ability to regress oral precancerous lesions. Of concern with regard to the possible long-term use of β-carotene for preventive therapy are results from two primary lung cancer prevention trials involving β-carotene [34,35]. Lung cancer incidence and mortality were increased significantly rather than reduced in smokers who received supplemental β-carotene for several years. These results, coupled with those seen for 13-*cis*-retinoic acid, suggest that chemoprevention in current smokers may be particularly challenging.

Evidence shows that the lung cancer-promoting effect of β-carotene in smokers is dose dependent [36] and, to date, is only observed with purified supplements of all-*trans* β-carotene. Given this, results of another trial of a carotenoid-containing plant extract are of interest. This trial [37] evaluated the chemopreventive efficacy of the blue-green algae *Spirulina fusiformis*, which is known to be a concentrated natural source of carotenoids and other micronutrients. The trial was done in tobacco chewers in Kerala, India, with 60 subjects given algae (lyophilized powder, 1 g/day for 12 months). These subjects were compared to 55 subjects from the placebo arm of another randomized trial (nonconcurrent control). The complete regression rate was 45% (20 of 44 evaluable) in subjects given the algae as compared to 7% (3 of 43 evaluable, $p<0.0001$) of subjects in a nonconcurrent placebo arm. The ability of the algae to regress lesions varied with the type of lesion, ranging from 57% for subjects with homogeneous leukoplakia to 25% for persons with erythroleukoplakia to 0% for persons with ulcerated and nodular lesions. Similar to what is observed in other trials, recurrent lesions commonly developed following the cessation of supplementation. As the algae *Spirulina* contains many natural compounds, including carotenoids, vitamin E, and other micronutrients, it is not clear which of these agents (or combinations of agents) are responsible for the observed efficacy. Nonetheless, there is considerable interest in these types of interventions given the unexpected adverse effects observed for high-dose single agents as described earlier.

C. Vitamin E and Selenium

Vitamin E (400 IU twice daily) was evaluated by itself in a single-arm, 24-week trial in oral leukoplakia in which both clinical (46%) and histological (21%) responses were observed [38]. Similarly, selenium (300 μg selenium/day for three 4-week cycles) has also been evaluated in a single-arm trial done in Italy [39]. A 38.8% response rate was observed following supplementation. While these trials suggest efficacy, both were uncontrolled and thus need to be interpreted cautiously.

D. Other Agents

In addition to the nutrients/nutrient derivatives described earlier, there are several other chemopreventive agents for oral precancerous lesions under evaluation. They are described briefly here.

1. Tea/Tea Components

It is well known that tea and tea components have the ability to inhibit cancer formation in various animal carcinogenesis models [40]. The chemopreventive activity of tea and its components in humans, however, is less clear. Tea contains polyphenolic compounds known as catechins. The most abundant catechin in tea is (−)-epigallocatechin-3-gallate (EGCG), which purportedly has numerous biochemical effects, including, but not limited to, antioxidant activity, inhibition of AP-1 activity, and inhibition of angiogenesis [40]. Given the documented antitumor activity in animal models and the ability to monitor EGCG levels in blood as a biomarker of adherence, it is not surprising that there is interest in tea as a possible chemopreventive agent for oral leukoplakia.

To date, a feasibility study of green tea in 28 patients with precancerous lesions has been completed in India [41], and results of one randomized trial of tea in oral leukoplakia patients from China are now available [12]. In the trial, 59 patients with oral leukoplakia were randomized to an intervention of mixed tea (encapsulated product developed by the authors; 3 g oral administration plus topical treatment) or a control group (placebo capsule and glycerin topical treatment) for 6 months. At the end of intervention, the lesion size decreased in 37.9% of 29 treated patients and increased in 3.4% as compared to decreases in 10.0% of 30 control patients and increases in 6.7% ($p < 0.01$). Persons in the treatment group also had a lower frequency of micronucleated exfoliated oral mucosal cells and less evidence of oral mucosal cell proliferation, as assessed by proliferating cell nuclear antigen (PCNA). These results are promising, but confirmation in other populations and with other tea preparations is needed.

2. Protease Inhibitors

Certain protease inhibitors are capable of preventing carcinogenesis in a wide variety of *in vivo* and *in vitro* model systems [42]. One particular protease inhibitor, derived from soybeans, is known as the Bowman–Birk inhibitor (BBI) and has been studied extensively [42]. A concentrate of BBI, known as Bowman–Birk inhibitor concentrate (BBIC), was approved by the FDA in 1992 as an investigational new drug. This approval is required prior to studying the chemopreventive efficacy of BBIC (or any new drug) in human populations. Human trials of BBIC ensued, including a recently completed phase IIa trial involving oral leukoplakia.

In this trial, 33 patients with oral leukoplakia were given BBIC for 1 month and 32 completed the intervention [43]. BBIC was dispensed as a powder that was reconstituted with a commercial saliva substitute, and patients were instructed to hold the BBIC suspension in the mouth for 1 min and then swallow. Clinical response was based on measurements of lesion areas and analysis of blinded clinical judgments of intraoral photographs. Two patients had complete responses and 8 had partial responses, resulting in an overall clinical response rate of 31%. The mean pretreatment total lesion area decreased from 615 to 438 mm^2 after treatment ($p<0.004$). As discussed earlier, because leukoplakia lesions can regress spontaneously in the absence of intervention, results from any single-arm trial need to be interpreted cautiously, but these results support further evaluation of BBIC in randomized trials. Of note, two patients were known to have subsequently developed oral cancers within 16 months of completing the BBIC trial, although only 2 of the initial 32 subjects had evidence of dysplasia (not those who developed malignancies). The authors concluded that it was extremely unlikely that BBIC acted as a procarcinogen, but this observation suggests that future trials of BBIC should ideally have multiyear intervention and follow-up.

3. Bleomycin

The previous sections emphasized the use of nutrients, nutrient derivatives, or diet-derived substances, usually consumed orally, as chemopreventive agents. A different strategy for chemoprevention involves the topical application of bleomycin, a drug that is better known for its use in cancer chemotherapy. Bleomycins are glycopeptides originally extracted from a strain of *Streptomyces verticillus* [44] and are cytotoxic antibiotics. The primary action of bleomycin is to produce single- and double-strand DNA breaks [44].

Single-arm trials in the 1980s first investigated the chemopreventive activity of bleomycin, administered topically as a solution with dimethyl sulfoxide (DMSO), in patients with oral precancerous lesions [45–47]. Clinical responses observed in these pilot studies served as the basis for a subsequent randomized trial of 1% bleomycin in DMSO versus DMSO alone [48]. Treatments were applied topically for 5 min once daily for 14 consecutive days, with posttreatment biopsies at 4 weeks following treatment. Twenty-two patients were randomized in this trial. Treated patients had a significant decrease in the size of the lesions ($p=0.001$) and had a nonsignificant histological reduction in dysplasia ($p=0.094$). Longer term follow-up of this same intervention in 19 treated patients, all of whom had dysplastic leukoplakia, has also been done (single arm trial). Partial responses or better were achieved in 94% of patients [49]. After a mean follow-up of 3.4 years, 31.6% of patients had no clinically visible lesions, 47.4% had clinically benign lesions, but 11% developed malignant transformation [49]. The authors concluded that topical bleomycin may have value in inhibiting tumor progression in this setting, but close follow-up for progression remains critical.

IV. VALIDITY OF ORAL PRECANCEROUS LESIONS IN PREDICTING EFFICACY OF AGENTS FOR ORAL CANCER PREVENTION

Because oral leukoplakia lesions can undergo malignant transformation, agents that either reduce lesion size or lead to histological improvements may have direct clinical benefits. This implies, however, that chemopreventive efficacy in oral leukoplakia predicts efficacy in the prevention of oral cancers, a premise that may not be true. It could be argued, for example, that lesions that respond might not be the types of lesions that are most likely to undergo malignant transformation. Thus, an examination of the validity of oral precancerous lesion response in predicting efficacy in oral cancer prevention is needed.

Direct evidence for the validity of oral precancerous lesions in predicting the efficacy of agents for oral cancer prevention would ideally come from large, randomized trials of oral cancer prevention that collect data on intermediate end point responses, including oral leukoplakias. To date, however, primary prevention trials of oral cancer have not been done due to the large sample sizes that would be required (tens of thousands of randomized individuals). Trials aimed at the prevention of second malignant tumors of the mouth and throat are feasible, and a few have been completed or are underway. However, most of these trials have not collected data on both precancerous lesions and malignancies within the same trial, primarily due to cost constraints. Nonetheless, data are now available from a few randomized trials of various retinoids and the carotenoid β-carotene that bear on the issue of the validity of oral precancerous lesions with regard to predicting efficacy in oral cancer prevention.

The first randomized trial aimed at the prevention of second primary cancers of the mouth and throat to be completed was a trial of high-dose 13-*cis*-retinoic acid versus placebo in 103 head and neck cancer patients [50]. Following definitive local therapy of stage I–IV disease, patients were assigned randomly to 12 months of 13-*cis*-retinoic acid (50–100 $mg/m^2/day$) or placebo. At a median followup of 32 months, there were no significant differences in primary disease recurrence (local, regional, or distant) or survival. The rate of second primary tumors, however, was significantly lower in the retinoid arm than in the placebo group, with second primary tumors developing in 2 (4%) of the retinoid-treated patients compared with 12 (24%) of the placebo-treated patients ($p=0.005$). This trial was reanalyzed with a longer median follow-up of 55 months [51]. With additional follow-up, each group had one more second primary tumor in the aerodigestive tract, resulting in a cumulative total of 3 of the retinoid group and 13 of the placebo group ($p=0.008$). As discussed earlier (see Section III,A), 13-*cis*-retinoic acid has shown consistent chemopreventive efficacy in the setting of oral precancerous lesions; therefore, the observed reduction in second primary tumors provides some assurance that responses in oral precancerous lesions predict responses for second primary tumors.

Supplemental β-carotene has also shown consistent chemopreventive efficacy in the setting of oral precancerous lesions (see Section III,B). To date, only one chemoprevention trial of supplemental β-carotene for the prevention of second primary tumors of the mouth and throat is available [52]. After definitive local therapy, 264 patients with stage I/II head and neck cancer were randomized to either supplemental β-carotene (50 mg/day) or identical placebo. The median follow-up was 51.1 months from the date of randomization. Persons randomized to supplemental β-carotene had a 31% decrease in second head and neck cancer risk (RR=0.69); however, the result was not statistically significant (95% CI 0.39–1.25). Thus, responses of premalignant lesions to 13-*cis*-retinoic acid and β-carotene are consistent with lower rates of second head and neck tumors developing in patients who received these agents.

A third randomized prevention trial studied the retinoid etretinate in 316 patients with stage I to III squamous cell carcinoma of the oral cavity or oropharynx [53]. In this trial, there was no difference in the rate of second primary tumor development between intervention and control arms. Trials of etretinate in oral precancerous lesions are not available. The lack of efficacy in this trial could be attributed to the fact that etretinate is not transcriptionally active [4]. This trial highlights the notion that the efficacy observed for one or more structurally related compounds (vitamin A, certain retinoids) should not be generalized to other related compounds in the absence of supporting clinical data.

V. SUMMARY AND CONCLUSIONS

Oral precancerous lesions have been relatively well studied in chemoprevention research. Oral precancerous lesions are used as intermediate end point biological markers in order to help screen potential chemopreventive agents that might have utility in the prevention of oral and pharyngeal cancers. Also, because treatment strategies for large or diffuse, multifocal oral precancerous lesions are limited, successful chemopreventive agents for these lesions would have direct clinical relevance. Clinical trials aimed at evaluating the chemopreventive efficacy of candidate compounds need to carefully consider how to classify responses, given the many issues in response assessment as detailed in this chapter. Single-arm trials have value for identifying agents with potential efficacy, but further evaluation in randomized trials is critical. To date, randomized trials of several agents are available, with efficacy clearly indicated for certain retinoids and supplemental β-carotene, demonstrating the "proof of principle" behind the chemoprevention of oral premalignancy. The most widely studied retinoid (13-*cis*-retinoic acid) and carotenoid (β-carotene), however, have been shown to interact with tobacco smoke to increase the risk of lung cancer (recurrences for 13-*cis*-retinoic acid; primary lung cancers for β-carotene). Given this, the long-term preventive therapy of these agents for oral precancerous lesions, at least in smokers, is unlikely.

Limited data from randomized trials also suggest efficacy for tea/tea components, and topically applied bleomycin, and single arm trials of vitamin E, selenium, and BBIC have also suggested further evaluation of these agents. The ultimate use of any agent for long-term chemoprevention in this clinical setting will require evidence of efficacy in oral precancerous lesions and in oral cancers. Also, a lack of evidence of acute toxicity and of adverse effects on other cancers/mortality should also be required prior to widespread use of any agent for chemoprevention.

References

1. WHO Collaborating Centre for Precancerous Lesions (1978). Definition of leukoplakia and other related lesions: An aid to studies on oral precancer. *Oral Surg.* **46**, 518–539.
2. Silverman, S., Jr., Gorsky, M., and Lozada, F. (1984). Oral leukoplakia and malignant transformation: A follow-up study of 257 patients. *Cancer (Phila.)* **53**, 563–568.
3. Mashberg, A., Morrissey, J. B., and Garfinkel, L. (1973). A Study of the appearance of early asymptomatic oral squamous cell carcinoma. *Cancer* **32**, 1436–1445.
4. Mayne, S. T., and Lippman, S. M. (2001). Cancer prevention: Diet and chemopreventive agents. Retinoids, carotenoids, and micronutrients. *In* "Principles and Practice of Oncology" (V. T. DeVita, Jr., S. Hellman, and S. A. Rosenberg, eds.), 6th Ed., pp. 575–590. Lippincott Williams & Wilkins, Philadelphia.
5. Blot, W. J., McLaughlin, J. K., Winn, D. M., Austin, D. F., Greenberg, R. S., Preston-Martin, S., Bernstein, L., Schoenberg, J. B., Stemhagen, A., and Fraumeni, J. F., Jr. (1988). Smoking and drinking in relation to oral and pharyngeal cancer. *Cancer Res.* **48**, 3282–3287.
6. Baric, J. M., Alman, J. E., Feldman, R. S., and Chauncey, H. H. (1976). Influence of cigarette, pipe, and cigar smoking, removable partial dentures, and age on oral leukoplakia. *Oral Surg.* **42**, 766–774.
7. Morse, D. E., Katz, R. V., Pendrys, D. G., Holford, T. R., Kruchkoff, D. J., Eisenberg, E., Kosis, D., and Mayne, S. T. (1996). Smoking and drinking in relation to oral epithelial dysplasia. *Cancer Epidemiol. Biomark. Prev.* **5**, 769–777.
8. Hashibe, M., Sankaranarayanan, R., Thomas, G., Kuruvilla, B., Mathew, B., Somanathan, T., Parkin, D. M., and Zhang, Z. F. (2000). Alcohol drinking, body mass index and the risk of oral leukoplakia in an Indian population. *Int. J. Cancer* **88**, 129–134.
9. Hashibe, M., Mathew, B., Kuruvilla, B., Thomas, G., Sankaranarayanan, R., Parkin, D. M., and Zhang, Z. F. (2000). Chewing tobacco, alcohol, and the risk of erythroplakia. *Cancer Epidemiol. Biomark. Prev.* **9**, 637–638.
10. Lippman, S. M., Benner, S. E., and Hong, W. K. (1994). Cancer chemoprevention. *J. Clin. Oncol.* **12**, 851–873.
11. Costa, A., De Palo, G., Decensi, A., Formelli, F., Chiesa, F., Nava, M., Camerini, T., Marubini, E., and Veronesi, U. (1995). Retinoids in cancer chemoprevention: Clinical trials with the synthetic analogue fenretinide. *Ann. N. Y. Acad. Sci.* **768**, 148–162.
12. Li, N., Sun, Z., Han, C., and Chen, J. (1999). The chemopreventive effects of tea on human oral precancerous mucosa lesions. *Exp. Biol. Med.* **220**, 218–224.
13. Lippman, S. M., and Hong, W.K. (2001). Molecular markers of the risk of oral cancer. *N. Engl. J. Med.* **344**, 1323–1326.
14. Sudbo, J., Kildal, W., Risberg, B., Koppang, H. S., Danielsen, H. E., and Reith, A. (2001). DNA content as a prognostic marker in patients with oral leukoplakia. *N. Engl. J. Med.* **344**, 1270–1278.
15. Lippman, S. M., Kessler, J. F., and Meyskens, F. L., Jr. (1987). Retionids as preventive and therapeutic anticancer agents (part II). *Cancer Treatment Rep.* **71**, 493–515.
16. Hong, W., Endicott, J., Itri, L. M., Doos, W., Batsakis, J. G., Bell, R., Fofonoff, S., Byers, R., Atkinson, E. N., Vaughan, C., Toth, B. B., Kramer, A., Dimery, I. W., Skipper, P., and Strong, S. (1986). 13-cis retinoic acid in the treatment of oral leukoplakia. *N. Engl. J. Med.* **315**, 1501–1505.
17. Lippman, S. M., Batsakis, J. G., Toth, B. B., Weber, R. S., Lee, J. J., Martin, J. W., Hays, G. L., Goepfert, H., and Hong, W. K. (1993). Comparison of low-dose isotretinoin with beta carotene to prevent oral carcinogenesis. *N. Engl. J. Med.* **328**, 15–20.
18. Piattelli, A., Fioroni, M., Santinelli, A., and Rubini, C. (1999). bcl-2 expression and apoptotic bodies in 13-*cis*-retinoic acid (isotretinoin)-topically treated oral leukoplakia: A pilot study. *Oral Oncol.* **35**, 314–320.
19. Sankaranarayanan, R., Mathew, B., Varghese, C., Sudhakaran, P. R., Menon, V., Jayadeep, A., Nair, M. K., Mathews, C., Mahalingam, T. R., Balaram, P., and Nair, P. P. (1997). Chemoprevention of oral leukoplakia with vitamin A and beta carotene: An assessment. *Oral Oncol.* **33**, 231–236.
20. Stich, H. F. Hornby, A. P., Mathew, B., Sankaranarayanan, R., and Nair, M. K. (1988). Response of oral leukoplakias to the administration of vitamin A. *Cancer Lett.* **40**, 93–101.
21. Han, J., Jiao, L., Lu, Y., Sun, Z., Gu, Q. M., and Scanlon, K. J. (1990). Evaluation of *N*-4-(hydroxycarbophenyl) retinamide as a cancer prevention agent and as a cancer chemotherapeutic agent. *In Vivo* **4**, 153–160.
22. Lippman, S. M., Lee, J. J., Karp, E. E., Vokes, E. E., Benner, S. E., Goodman, G. E., Khuri, F. R., Marks, R., Winn, R. J., Fry, W., Graziano, S. L., Ganadara, D. R., Okawara, G., Woodhouse, C. L., Williams, B., Perez, C., Kim, H. W., Lotan, R., Roth, J. A., and Hong, W. K. (2001). Randomized phase III intergroup trial of isotretinoin to prevent second primary tumors in stage I non-small-cell lung cancer. *J. Natl. Cancer Inst.* **93**, 605–618.
23. Khuri, F. R., Kim, E. S., Lee, J. J., Winn, R. J., Benner, S. E., Lippman, S. M., Fu, K. K., Cooper, J. S., Vokes, E. E., Chamberlain, R. M., Williams, B., Pajak, T. F., Goepfert, H., and Hong, W. K. (2001). The impact of smoking status, disease stage, and index tumor site on second primary tumor incidence and tumor recurrence in the head and neck retinoid chemoprevention trial. *Cancer Epidemiol. Biomark. Prev.* **10**, 823–829.
24. Stich, H. F., Rosin, M. P., and Vallejera, M. O. (1984). Reduction with vitamin A and beta-carotene administration of proportion of micronucleated buccal mucosal cells in Asian betel nut and tobacco chewers. *Lancet* **1**, 1204–1206.
25. Stich, H. F., Hornby, A. P., and Dunn, B. P. (1985). A pilot beta-carotene intervention trial with Inuits using smokeless tobacco. *Int. J. Cancer* **36**, 321–327.
26. Stich, H. F., Rosin, M. P., Hornby, P., Mathew, B., Sankaranarayanan, R., and Nair, M. K. (1988). Remission of oral leukoplakias and micronuclei in tobacco/betel quid chewers treated with beta-carotene and with beta-carotene plus vitamin A. *Int. J. Cancer* **42**, 195–199.
27. Garewal, H. S. Meyskens, F. L., Killen, D., Reeves, D., Kiersch, T. A., Elletson, H., Strosberg, A., King, D., and Steinbronn, K. (1990). Response of oral leukoplakia to beta-carotene. *J. Clin. Oncol.* **8**, 1715–1720.
28. Toma, S., Benso, S., Albanese, E., Palumbo, R., Cantoni, E., Nicolo, G., and Mangiante, P. (1992). Treatment of oral leukoplakia with beta-carotene. *Oncology* **49**, 77–81.
29. Malaker, K., Anderson, B. A., Beecroft, W. A., and Hodson, D. I. (1991). Management of oral mucosal dysplasia with beta-carotene and retinoic acid: A pilot cross-over study. *Cancer Detect. Prev.* **15**, 335–340.
30. Kaugars, G. E., Silverman, S., Jr., Lovas, J. G. L., Brandt, R. B., Riley, W. T., Dao, Q., Singh, V. N., and Gallo, J. (1994). A clinical trial of antioxidant supplements in the treatment of oral leukoplakia. *Oral Surg. Oral Med. Oral Pathol.* **78**, 462–468.
31. Barth, T. J., Zoller, J., Kubler, A., Born, I. A., and Osswald, H. (1997). Redifferentiation of oral dysplastic mucosa by the application of the antioxidants beta-carotene, alpha-tocopherol and vitamin C. *Int. J. Vit. Nutr. Res.* **67**, 368–376.
32. Garewal, H. S., Katz, R. V., Meyskens, F., Pitcock, J., Morse, D., Friedman, S., Peng, Y., Pendrys, D. G., Mayne, S., Alberts, D., Kiersch, T., and Graver, E. (1999). Beta-carotene produces sustained remissions in patients with oral leukoplakia: Results of a multicenter prospective trial. *Arch. Otolaryngol. Head Neck Surg.* **125**, 1305–1310.

33. Zaridze, D., Evstifeeva, T., and Boyle, P. (1993). Chemoprevention of oral leukoplakia and chronic esophagitis in an area of high incidence of oral and esophageal cancer. *Ann. Epidemiol.* **3**, 225–234.
34. The Alpha-Tocopherol, Beta Carotene Cancer Prevention Study Group (1994). The effect of vitamin E and beta carotene on the incidence of lung cancer and other cancers in male smokers. *N. Engl. J. Med.* **330**, 1029–1035.
35. Omenn, G. S., Goodman, G. E., Thornquist, M. D., Balmes, J., Cullen, M. R., Glass, A., Keogh, J. P., Meyskens, F. L., Jr., Valanis, B., Williams, J. H., Barnhart, S., and Hammar, S. (1996). Effects of a combination of beta carotene and vitamin A on lung cancer and cardiovascular disease. *N. Engl. J. Med.* **334**, 1150–1155.
36. Mayne, S. T. (1998). Beta-carotene, carotenoids, and cancer prevention. *Principles Pract. Oncol. Updates* **12**, 1–15.
37. Mathew, B., Sankaranarayanan, R., Nair, P. P., Varghese, C., Somanathan, T., Amma, B. P., Amma, N. S., and Nair, M. K. (1995). Evaluation of chemoprevention of oral cancer with *Spirulina fusiformis*. *Nutr. Cancer* **24**, 197–202.
38. Benner, S. E., Winn, R. J., Lippman, S. M., Poland, J., Hansen, K. S., Luna, M. A., and Hong, W. K. (1993). Regression of oral leukoplakia with alpha-tocopherol: A community clinical oncology program chemoprevention study. *J. Natl. Cancer Inst.* **85**, 44–47.
39. Toma, S., Micheletti, A., Giacchero, A., Coialbu, T., Collecchi, P., Esposito, M., Rotondi, M., Albanese, E., and Cantoni, E. (1991). Selenium therapy in patients with precancerous and malignant oral cavity lesions: preliminary results. *Cancer Detect. Prev.* **15**, 491–494.
40. Yang, C. S., Chung, J. Y., Yang, G.-Y., Chhabra, S. K., and Lee, M.-J. (2000). Tea and tea polyphenols in cancer prevention. *J. Nutr.* **130**, 472S–478S.
41. The Indian-US Head and Neck Cancer Cooperative Group (1997). Green tea and leukoplakia. *Am. J. Surg.* **174**, 552–555.
42. Kennedy, A. R. (1998). Chemopreventive agents: Protease inhibitors. *Pharmacol. Ther.* **78**, 167–209.
43. Armstrong, W. B., Kennedy, A. R., Wan, X. S., Taylor, T. H., Nguyen, Q. A., Jensen, J., Thompson, W., Lagerberg, W., and Meyskens, F. L., Jr. (2000). Clinical modulation of oral leukoplakia and protease activity by Bowman-Birk inhibitor concentrate in a phase IIa chemoprevention trial. *Clin. Cancer Res.* **6**, 4684–4691.
44. Cheson, B. D. (2001). Pharmacology of cancer chemotherapy: Miscellaneous chemotherapeutic agents. *In* "Principles and Practice of Oncology" (V. T. DeVita, Jr., S. Hellman, and S. A. Rosenberg, eds.) 6th Ed., pp. 452–459. Lippincott Williams & Wilkins, Philadelphia.
45. Hammersley, N., Ferguson, M. M., and Rennie, J. S. (1985). Topical bleomycin in the treatment of oral leukoplakia: A pilot study. *Br. J. Oral Maxillofac. Surg.* **23**, 251–258.
46. Malmstrom, M., Hietanen, J., Sane, J., and Sysmalainen, M. (1988). Topical treatment of oral leukoplakia with bleomycin. *Br. J. Oral Maxillofac. Surg.* **26**, 491–498.
47. Wong, F., Epstein, J., and Millner, A. (1989). Treatment of oral leukoplakia with topical bleomycin: A pilot study. *Cancer* **64**, 361–365.
48. Epstein, J. B., Wong, F. L. W., Millner, A., and Le, N. D. (1994). Topical bleomycin treatment of oral leukoplakia: A randomized double-blind clinical trial. *Head Neck* **16**, 539–544.
49. Epstein, J. B., Gorsky, M., Wong, F. L. W., and Millner, A. (1998). Topical bleomycin for the treatment of dysplastic oral leukoplakia. *Cancer* **83**, 629–634.
50. Hong, W. K. Lippman, S. M., Itri, L. M., Karp, D. D., Lee, J. S., Byers, R. M., Schantz, S. P., Kramer, A. M., Lotan, R., Peters, L. J., Dimery, I. W., Brown, B. W., and Goepfert, H. (1990). Prevention of second primary tumors with 13cRA in squamous-cell carcinoma of the head and neck. *N. Engl. J. Med.* **323**, 795–801.
51. Benner, S. E., Winn, R. J., Lippman, S. M., Poland, J., Hansen, K. S., Luna, M. A., and Hong, W. K. (1993). Regression of oral leukoplakia with alpha-tocopherol: A community clinical oncology program chemoprevention study. *J. Natl. Cancer Inst.* **85**, 44–47.
52. Mayne, S. T., Cartmel, B., Baum, M., Shor-Posner, G., Fallon, B. G., Briskin, K., Bean, J., Zheng, T., Cooper, D., Friedman, C., and Goodwin, W. J., Jr. (2001). Randomized trial of supplemental beta-carotene to prevent second head and neck cancer. *Cancer Res.* **61**, 1457–1463.
53. Bolla, M., Lefur, R., Ton Van, J., Domenge, C., Badet, J. M., Koskas, Y., and Laplanche, A. (1994). Prevention of second primary tumours with etretinate in squamous cell carcinoma of the oral cavity and oropharynx. Results of a multicentric double-blind randomized study. *Eur. J. Cancer* **30A**, 767–772.

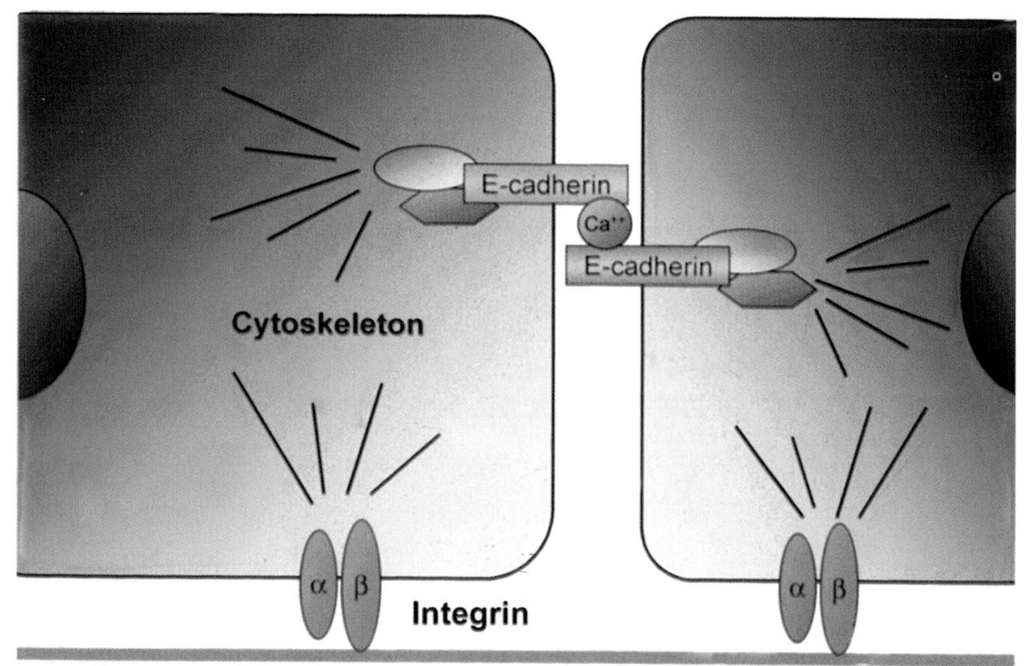

FIGURE 6.1 Integrin and cadherin adhesion receptors.

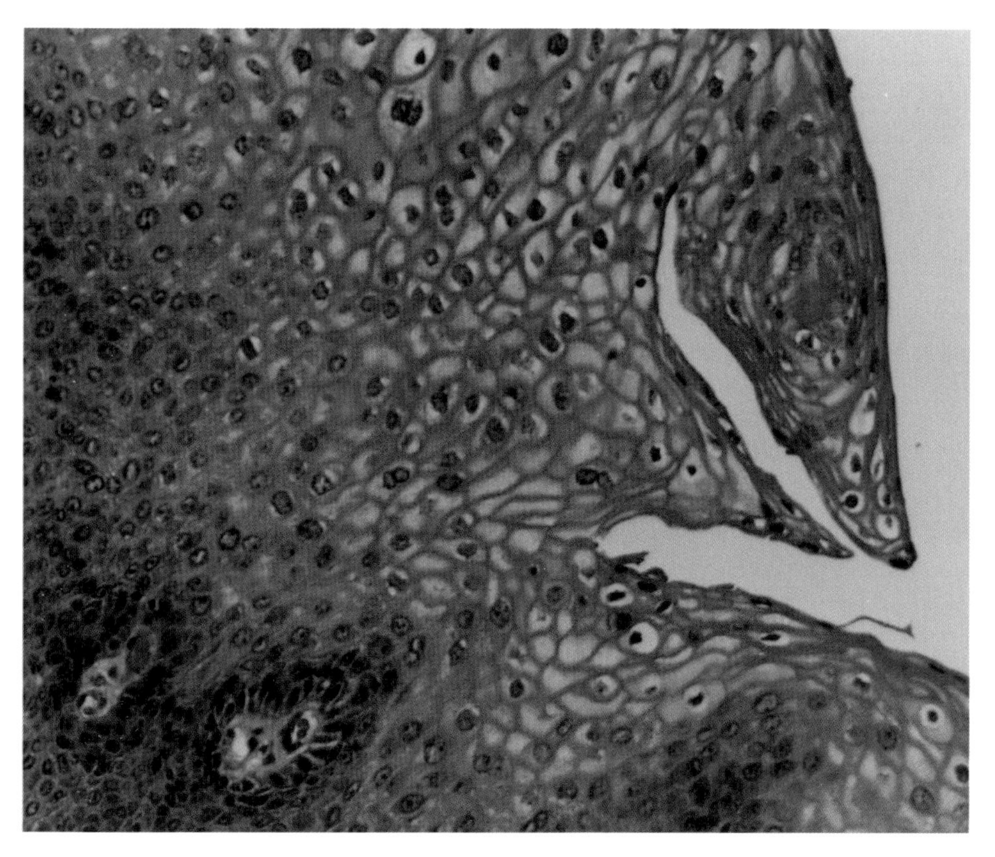

FIGURE 11.1 Histopathological appearance of laryngeal papillomatosis demonstrating the thickening of the epithelium around a fibrovascular core, oval nuclei in the superficial layer, and cytoplasmic vacuolization (Koilocytosis). Courtesy of Dr. D. M. Humphrey.

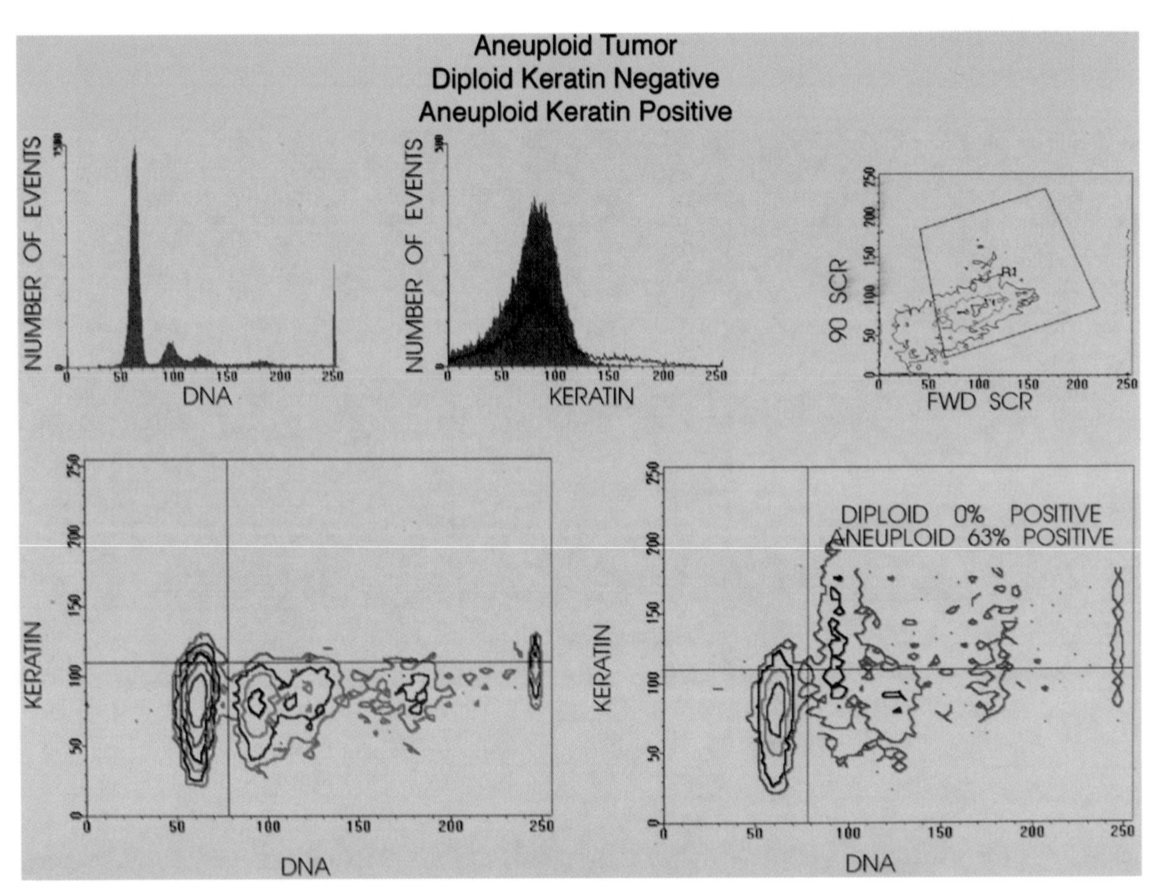

FIGURE 12.1 SCCHN stained simultaneously for DNA content (propidium iodide) and keratin (FITC-CAM 5.2).

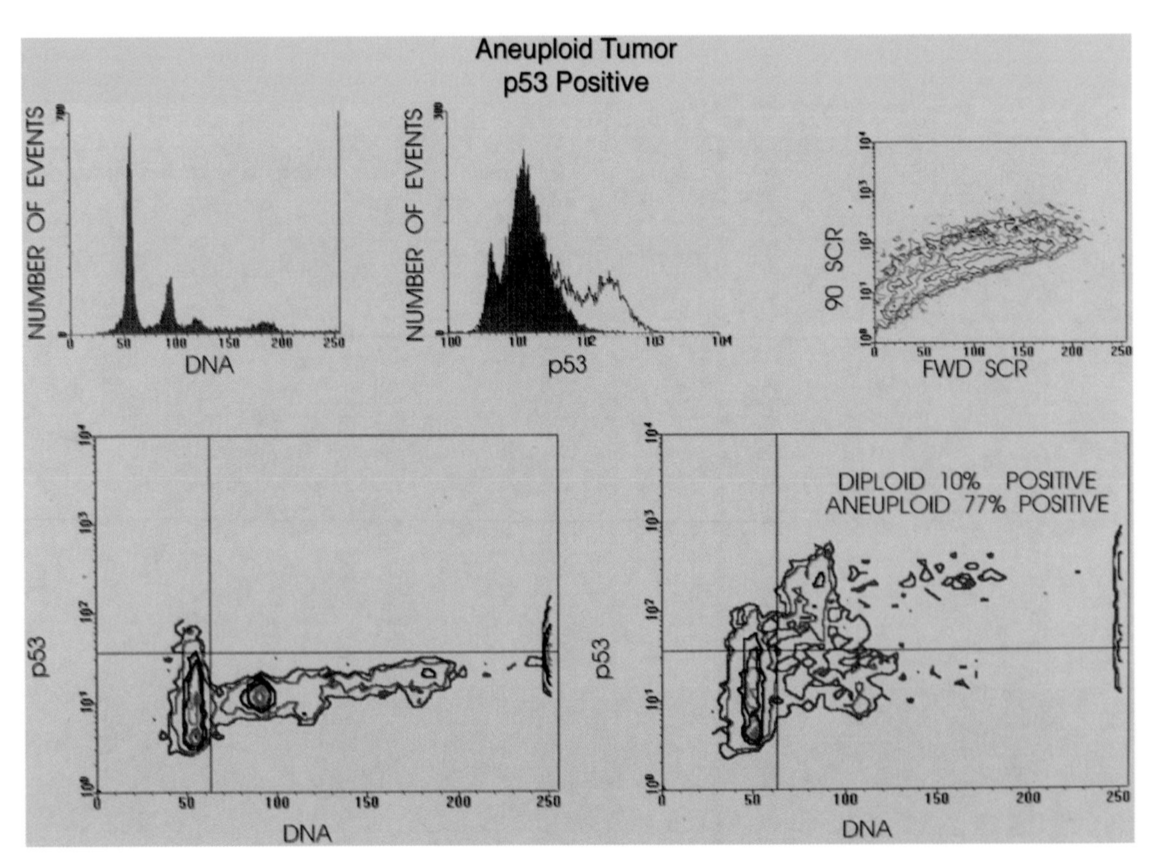

FIGURE 12.2 SCCHN stained simultaneously for DNA content (propidium iodide) and p53 protein (FITC-1801).

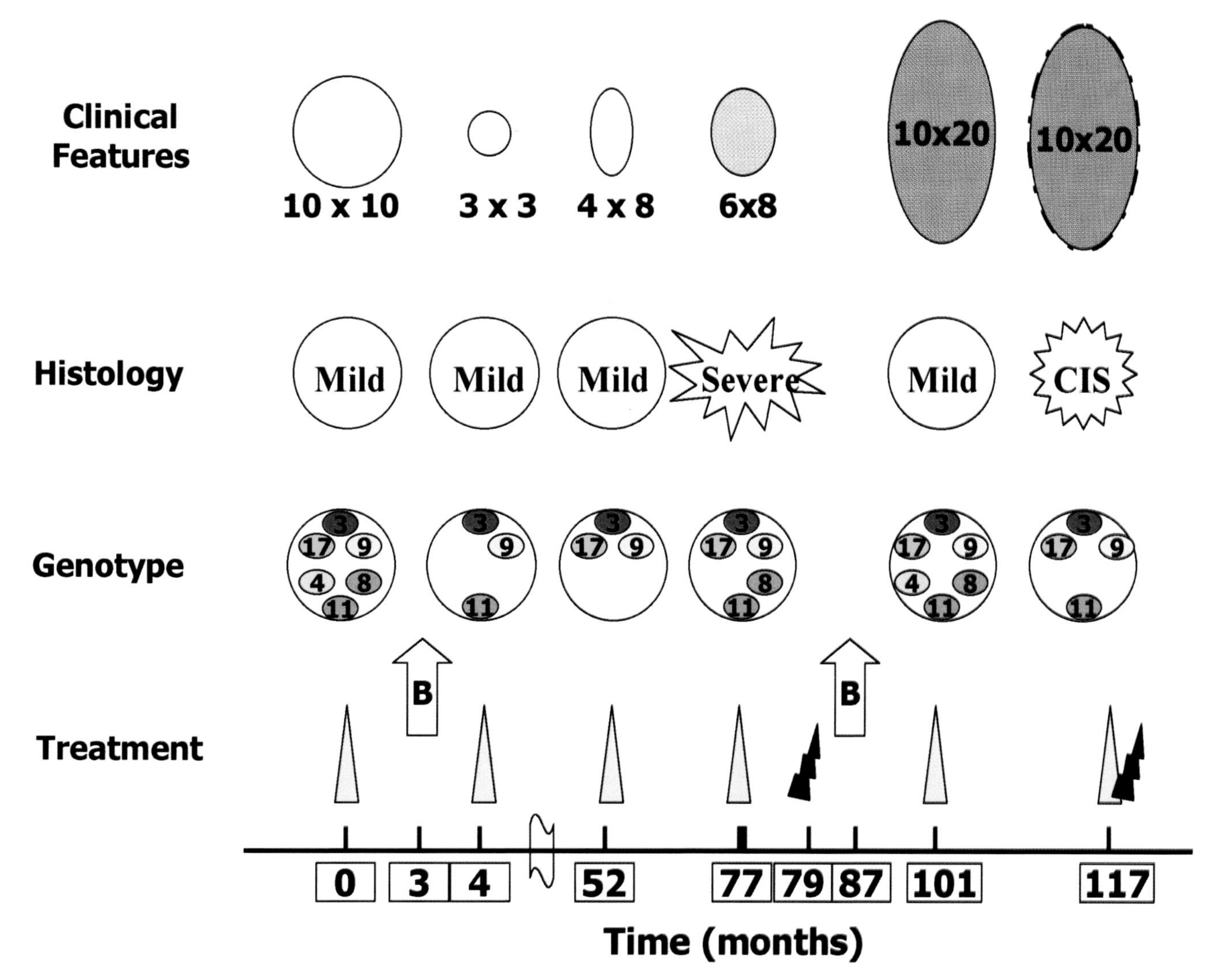

FIGURE 17.5 Temporal changes of clinical, histological, and molecular indices in an OPL that developed into CIS. The index biopsy of this lesion had a low clinicopathological risk (< 2 cm in size, homogeneous in appearance, and mild dysplasia, for clinicopathological staging criteria, see Axell *et al.* [30]), but a high-risk genotype. Bleomycin treatment decreased the size of the lesion, but it persisted. At month 52, the lesion still maintained a low clinicopathological but high molecular risk. However, only a portion of the allelic losses seen in the index biopsy were present, probably due to the selective pressure of treatment on subclones within the initial lesion. At month 77, almost 6.5 years after molecular indication of high risk, the lesion finally showed clinicopathological progression to a high-risk lesion (nonhomogeneous and severe dysplasia). The molecular risk remained high with emergence of a clone(s) that showed more of the allelic losses that were present in the initial lesion. The lesion was subsequently excised but recurred and was treated by topical bleomycin at month 87. At month 101, the lesion presented clinicopathologically as an intermediate-risk lesion (2 cm in size, nonhomogeneous) with the same high-risk molecular clone as that seen in the index biopsy. The lesion progressed into a CIS at month 117. Clinical features: Numbers represent lesion size (diameter in mm); hollow circle represents homogeneous appearance; dot-filled circles represent nonhomogeneous appearance. Genotype: Numbers represent chromosomal arms with LOH. Treatment: Gray triangle represents incisional biopsy; hollow arrow with letter B represents topical bleomycin treatment; black jagged triangle represents excision.

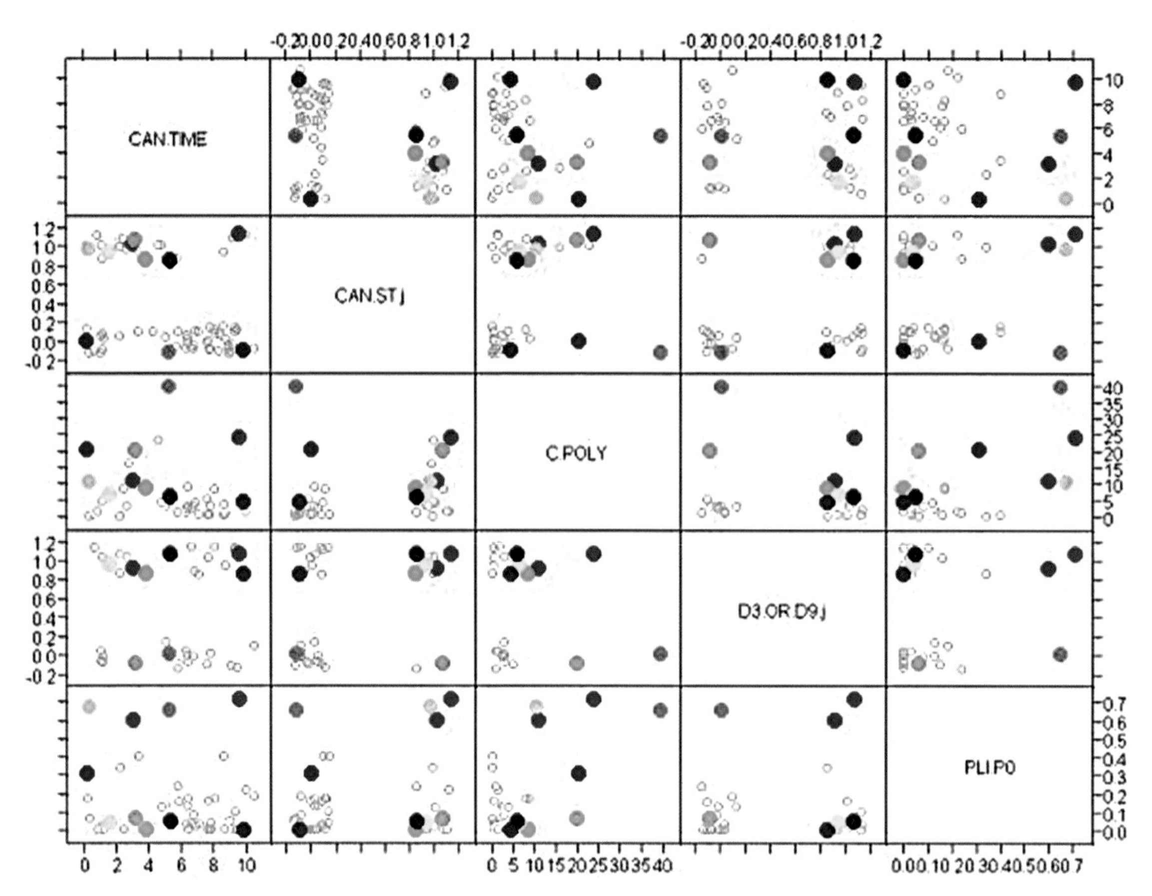

FIGURE 20.6 Scatter plot matrix of time (CAN.TIME), cancer status (CAN.ST), chromosomal polysomy (C.POLY), LOH at 3p or 9p, and p53.

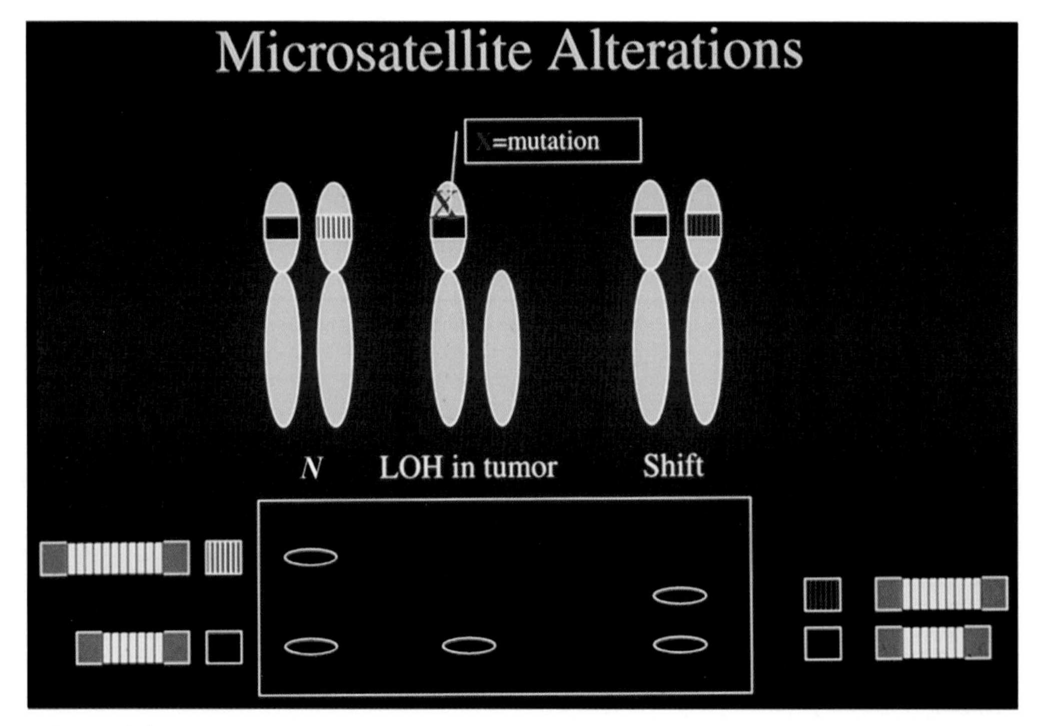

FIGURE 21.1 Schema for altered band pattern on PAGE for tumor-associated loss of heterozygosity (LOH) and microsatellite shift. *N* = normal (control) DNA displaying polymorphism of the hypothetical microsatellite. With LOH, a deletion of one chromosomal arm results in absence of the band for the microsatellite allele located on the deleted segment. A shift pattern occurs when the tumor cells display a new form of the microsatellite containing a novel number of repeats.

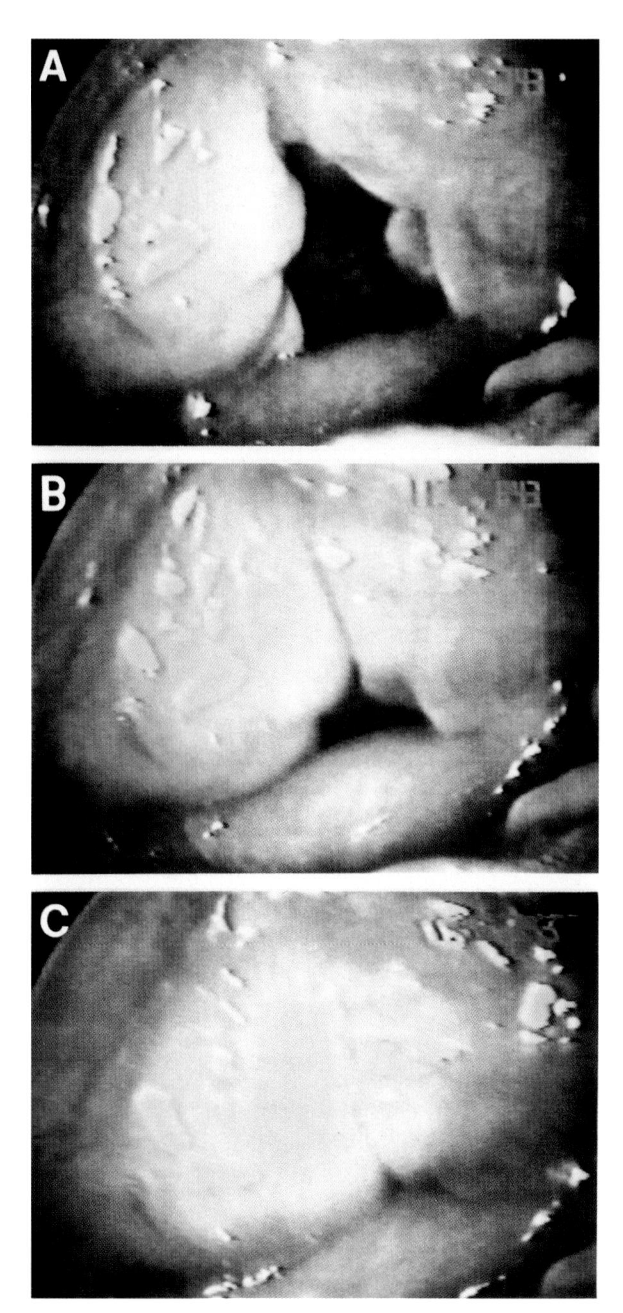

FIGURE 26.4 Postoperative appeareance of the larynx after supracricoid partial laryngectomy with cricohyoidoepiglottopexy in (A) respiration (arytenoids open), (B) phonation beginning (arytenoids partially closed), and (C) phonation (arytenoids closed). Published with permission from G.S. Weinstein and O. Laccourreye (1994). Supracricoid laryngectomy with cricohyoidoepiglottopexy. *Otolaryngol. Head Neck Surg.* **111**, 684–685.

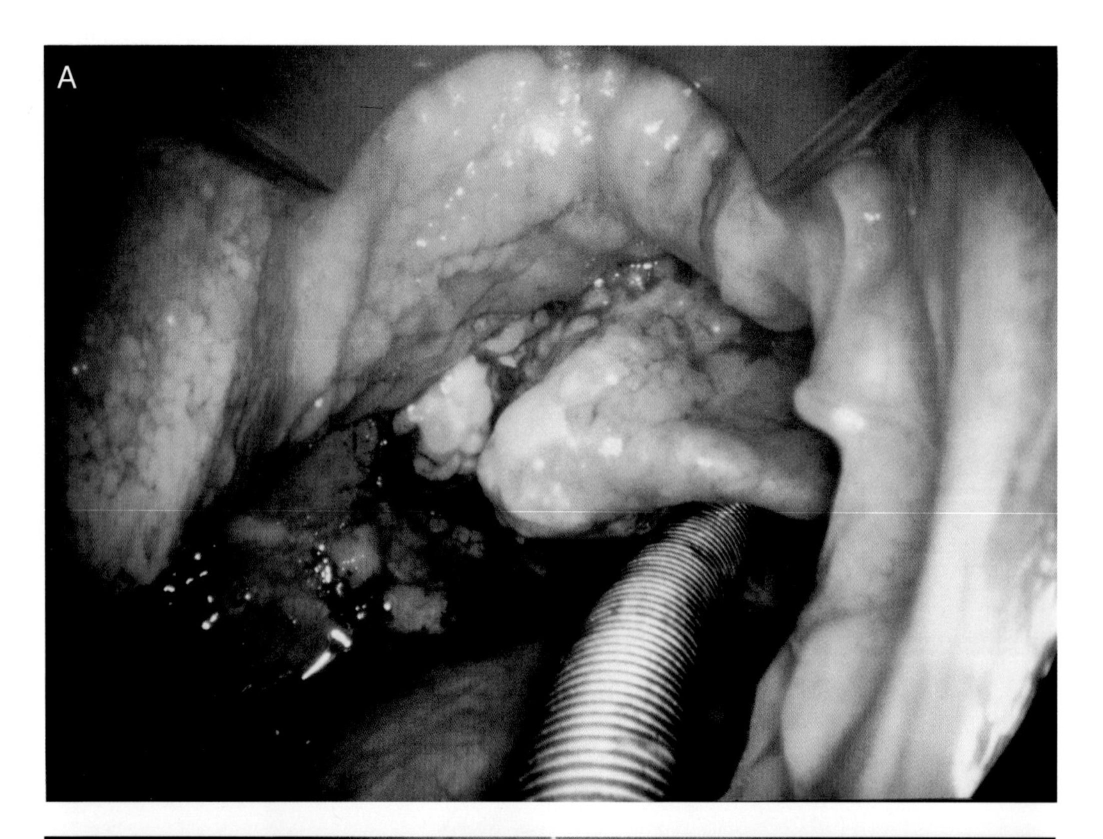

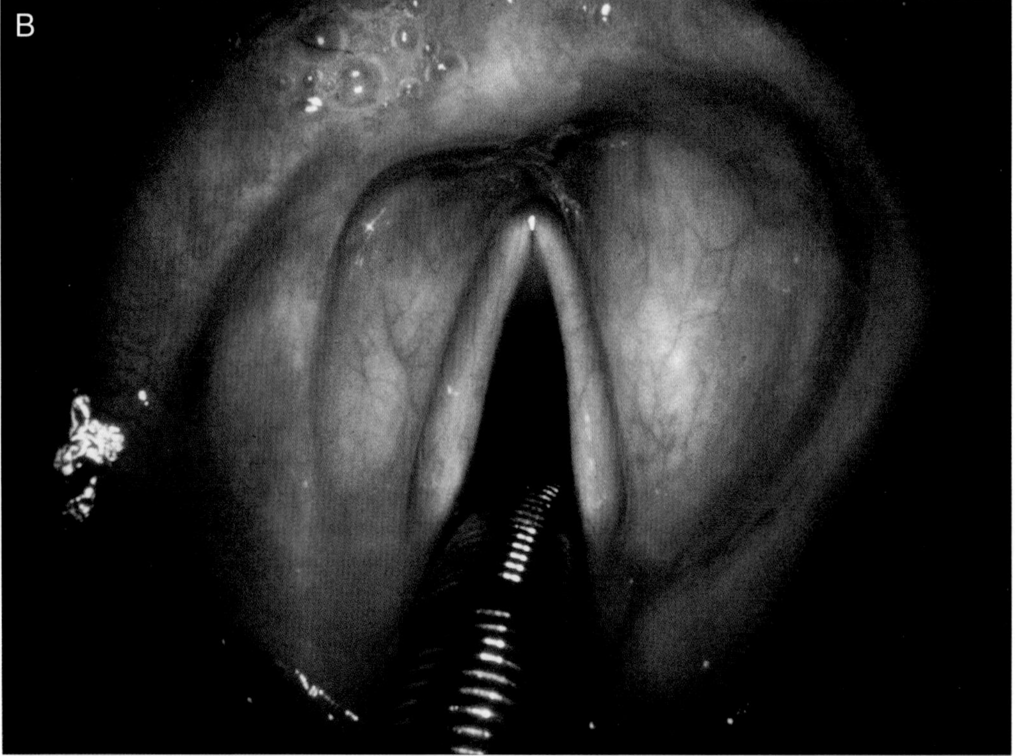

FIGURE 26.5 (A) A carcinoma of the epiglottis and the left aryepiglottic fold in an 89-year-old patient before laser resection (initially palliatively indicated). No tracheostomy was performed. (B) The larynx 1 year later without evidence of disease. Published with permission from "Organ Preservation Surgery for Laryngeal Carcinoma," G.S. Weinstein, O. Laccourreye, D. Brasnu, and H. Laccourreye, Singular Publishing, San Diego, 1999.

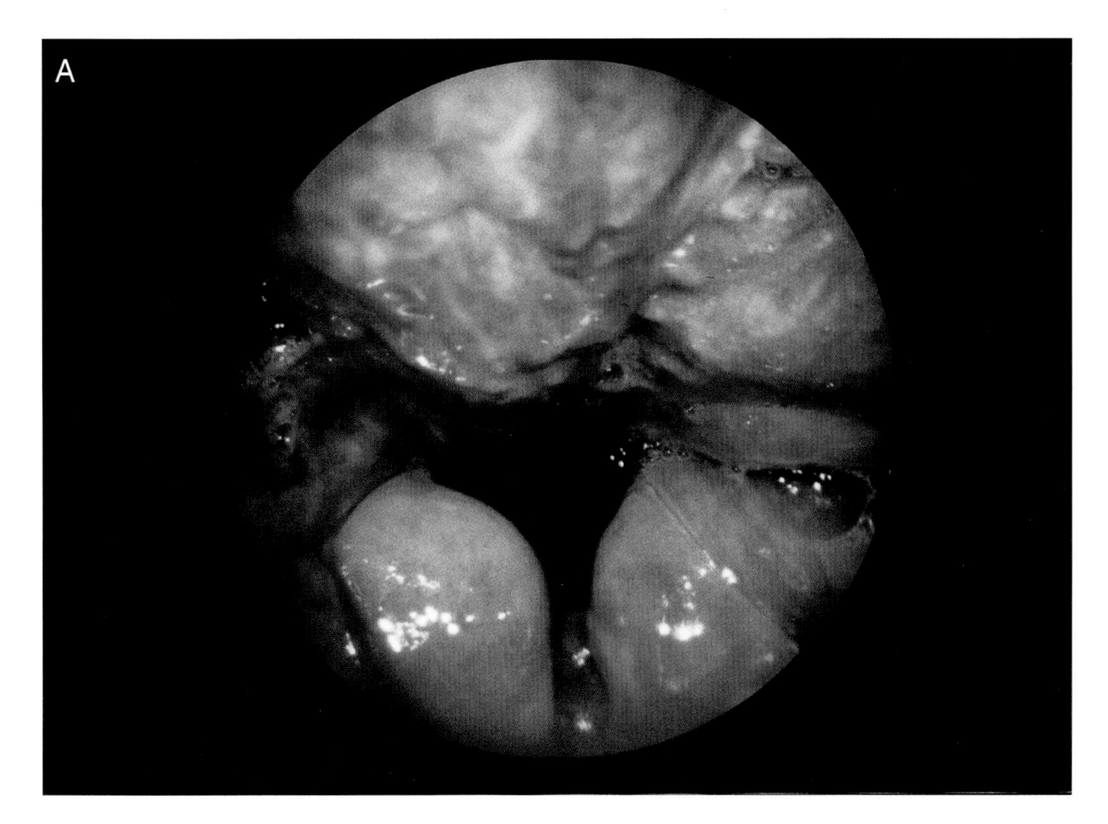

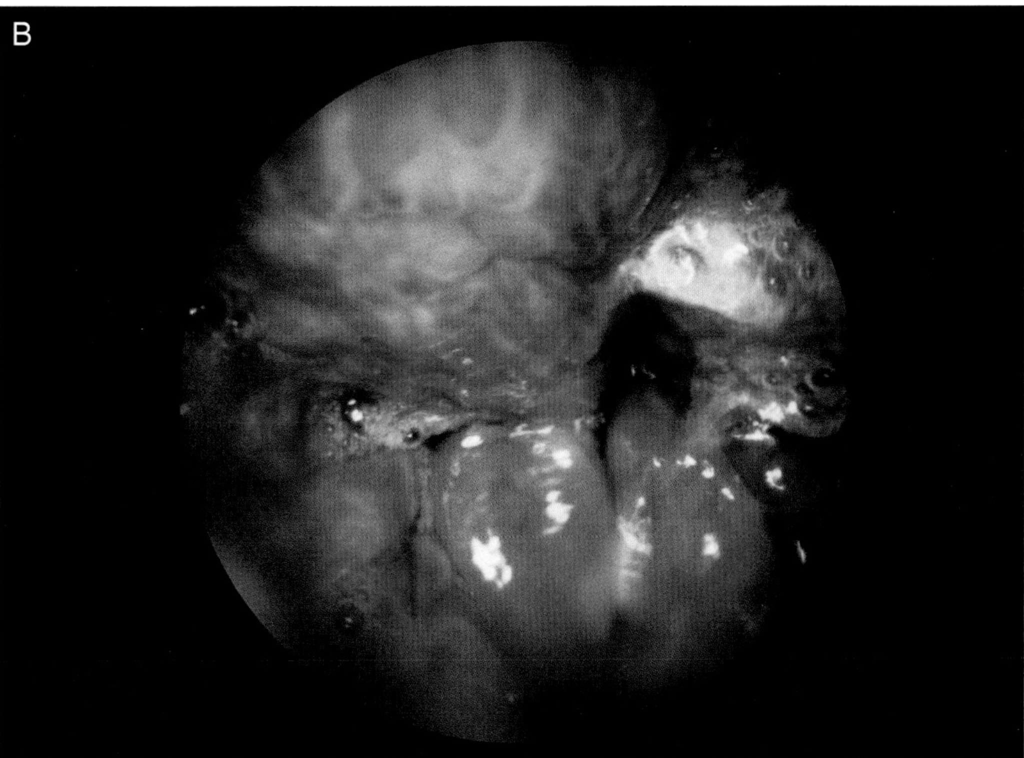

FIGURE 26.6 Postoperative appearance of the neolarynx after supracricoid partial laryngectomy with cricohyoidopexy sparing both arytenoids and (A) arytenoids open and (B) arytenoids closed. Published with permission from "Organ Preservation Surgery for Laryngeal Carcinoma," G.S. Weinstein, O. Laccourreye, D. Brasnu, and H. Laccourreye, Singular Publishing, San Diego, 1999.

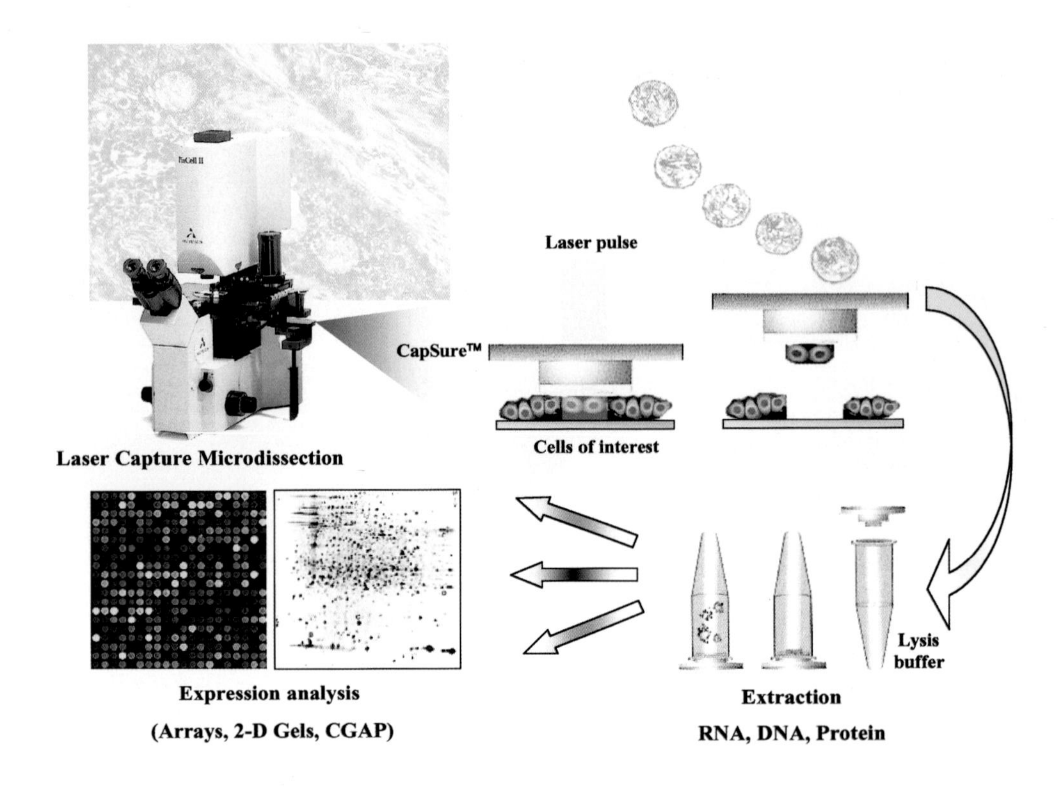

FIGURE 34.1 Application of laser capture microdissection (LCM). The procedure of LCM provides a quick and reliable method for the procurement of a pure population of cells from their native tissues. The process involves using the handle of the transport arm to manipulate the platform holding the vacuum-secured glass slide until an area of interest is determined with the microscope. A pulse of laser beam is then used to capture areas of tissues of interest onto ethylene vinyl acetate (EVA)-coated caps (CapSure). These caps are then transferred after LCM to Eppendorf tubes containing appropriate lysis buffer for extraction of macromolecules (RNA, DNA, and proteins). Using this method, frozen tissue sections can be microdissected immediately after staining, and thus a pure population of cells of interest (>95% purity) can be concentrated rapidly with maximum preservation of macromolecules. Adapted from Leethanakul *et al.* [67].

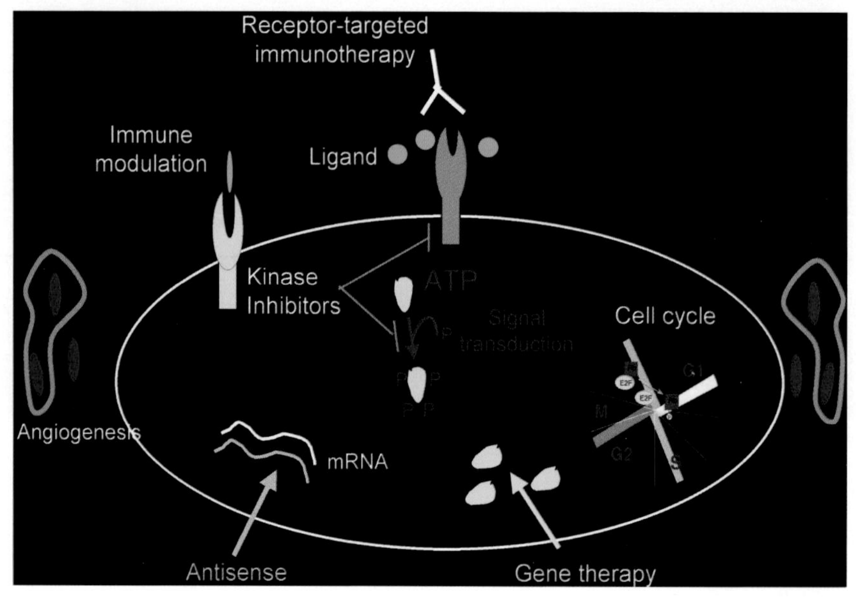

FIGURE 36.1 Gene therapy and immune modulation.

C H A P T E R

19

Aerodigestive Tract Chemoprevention Trials and Prevention of Second Primary Tumors

EDWARD S. KIM and FADLO R. KHURI

Department of Thoracic/Head and Neck Medical Oncology
The University of Texas M. D. Anderson Cancer Center
Houston, Texas 77030

I. Introduction 271
II. Epidemiology 272
III. Natural History 272
IV. Risk Factors 273
A. Alcohol and Tobacco 273
B. Viral Infection 274
C. Genetics 275
V. Biology of Head and Neck Cancer 275
A. Field Cancerization 275
B. Multistep Carcinogenesis 275
VI. Chemoprevention 276
VII. Chemoprevention Trials 276
A. Oral Premalignancy 276
B. Second Primary Tumors 279
VIII. Biochemoprevention 281
IX. Summary: Future Directions 281
References 281

I. INTRODUCTION

Head and neck cancers comprise a majority of the upper aerodigestive tract (UADT) epithelial malignancies. These tumors are typically of squamous cell histology and originate along the epithelial lining of the UADT. Damage to the epithelium results from chronic carcinogen exposure via inhalation or diet, especially from alcohol and tobacco use. These factors are common to head and neck cancer development worldwide. Prevention strategies have included primary prevention of tobacco and alcohol use. These campaigns, however, have not succeeded in improving the overall mortality, especially in developing countries.

The complexity of head and neck cancer treatment relates to the substantial progress that has been made in the areas of surgery, concomitant chemoradiotherapy, and intensity modulated radiotherapy over the past few decades. The 5-year survival rate has improved only slightly from the 1960s. The high mortality rate of this cancer makes it a worldwide public health menace. This disease devastates functional abilities, including problems with speech, swallowing, and cosmetic appearance, all of which lead to the subsequent loss of self-esteem; hence, making it a particularly troubling disease. In fact, even those patients who are fortunate enough to be cured of this illness often succumb to a second smoking-related cancer at some point in their lifetime. The problem of second, metachronous primary cancers in this patient group continues, with some studies indicating that these cancers are the major determinants of overall prognosis in patients definitively treated for early stage disease.

It has been shown that second primary tumors (SPTs) help determine survival among patients with advanced head and neck cancer who did not experience primary tumor recurrence within 2 years after definitive treatment [1]. Subsequently, larger studies [2,3] assessed the impact of second malignancies in patients who had prior head and neck cancer. A review of head and neck cancer patients treated on various Radiation Therapy Oncology Group (RTOG) protocols showed that despite improvements in local control, overall survival did not improve due to the development of second malignancies.

Novel approaches to diagnosis and treatment may emerge from the increased understanding of the biology of head and neck cancer. Chemoprevention methods, using the model of UADT malignancy, are still experimental but may one day prove useful for population-wide recommendations. Prevention strategies other than primary prevention with tobacco and alcohol cessation include prevention of recurrence and second primary tumors.

Large trials have been undertaken to study the major cancer sites, including oral cavity, hypopharynx, and larynx. These trials analyze rates of tumor incidence, recurrence, or SPT incidence and require large amounts of time and money. Development of validated intermediate biological markers may obviate the need for large, time-consuming trials. This chapter focuses on the development of new treatment strategies and intermediate markers for head and neck cancer. The major cancer sites discussed include the oral cavity, pharynx, and larynx. The establishment of effective preventive measures and early diagnosis of these lesions will certainly be necessary before eradication of this disease becomes reality.

II. EPIDEMIOLOGY

UADT cancers represent 3% of all U.S. cancers. An estimated 40,000 new cases of head and neck cancer were diagnosed in 2001 in the United States alone, resulting in more than 12,000 deaths [4]. Despite significant improvements in diagnosis, local management, and chemotherapy of head and neck cancer, there has been no significant increase in long-term survival since the early 1970s.

The 5-year survival rate is 40% for head and neck cancer patients in the United States and other developed countries. This is comparable to the 5-year survival rate in the 1960s, despite advances in detection, surgery, and chemotherapy. Typically, patients with head and neck squamous cell carcinoma (HNSCC) are over age 50 and incidence increases with age. Ninety percent of head and neck cancers occur in individuals with known chronic carcinogen exposure, predominantly tobacco and related substances such as betel leaf. This relationship of tobacco exposure and disease development has been clearly demonstrated as a dose–response relationship with greater risk proportional to duration and intensity of exposure [5,6]. Alcohol is a closely linked carcinogenic agent and seems to potentiate tobacco's effects on the aerodigestive epithelium [7].

However, other factors must be considered as etiologic agents in head and neck cancer other than alcohol and tobacco, as tobacco users far outnumber cases of this disease. Genetic susceptibility is becoming better understood and may provide an explanation. Mutagen sensitivity, a marker of underlying DNA repair deficiency, for example, is associated with an increased risk of head and neck cancer [8]. Ataxia telangiectasia, Fanconi's anemia, and xeroderma pigmentosum are syndromes characterized by mutagen sensitivity and have all been associated with oral cavity cancers [9].

Worldwide, UADT cancers account for roughly 10% of all cancers. The male-to-female ratio is approximately 4:1. Oral and pharyngeal cancers have decreased in incidence in white men and white women under the age of 65 over the past 2 decades. Geographical variation plays a role in head and neck cancer rate and incidence. For example, rates of laryngeal cancer in males 70–74 years old are two to six times higher in Bombay than in Scandinavia. Asians appear to have a higher oral cavity cancer incidence because of increased use of betel nut mixtures and smokeless tobacco [10]. In the United States, variations in the incidence of oral cancer during 1950–1969 were observed [11]. Males in urban areas had a higher rate of UADT cancer attributed to alcohol and tobacco consumption; in contrast, women in southern regions had increased oral cancer due to snuff use [12].

III. NATURAL HISTORY

Malignancies of the UADT are predominantly of epithelial mucosal origin. The region provides a unique opportunity to monitor cancer development. Many lesions are clinically apparent by direct visual inspection, indirect laryngoscopy, or palpation. Premalignant lesions include leukoplakia, defined as a white mucosal patch, or erythroplakia, a red mucosal patch, neither of which can be scraped off. These lesions can represent hyperplasia, dysplasia, or carcinoma *in situ*. These lesions are commonly observed in individuals with known carcinogen exposure and are associated with a high rate of progression to invasive cancer, usually over several years. These lesions have served as a model for the development of treatment for intraepithelial neoplasia. Squamous cell carcinomas of the head and neck have a high propensity for metastasis, especially to regional lymph nodes. Advanced disease ultimately leads to distant spread to other organs, commonly the lungs.

Diagnosing head and neck cancers at earlier stages is the key to treatment success. Lesions within the oral cavity may be diagnosed early because of their overt visual appearance. Other lesions, such as laryngeal tumors, frequently produce symptoms such as hoarseness during the early stages. However, there are regions in the UADT that are not readily visualized or do not produce early symptoms such as tumors arising from the pyriform sinus and nasopharnyx. These tumors are often not diagnosed early and are usually locally advanced or metastatic upon presentation.

For early stage squamous cell lesions of the head and neck, current therapy, including surgery and/or radiation, is effective. Locally advanced disease is cured much less frequently. Metastatic disease involving distant organs is not curable with current therapy. Although early stage UADT cancers can be "cured" by conventional therapy, these patients' course is complicated by the development of second primary tumors. These SPTs occur most commonly within the UADT, lungs, and esophagus.

Prevention and early detection of head and neck squamous cell carcinomas are very important. The natural history of these UADT tumors progressing from premalignant lesions, to invasive cancers, to occurrence of second primary tumors has led investigators to explore methods to prevent these cancers. Individuals at risk include those with known carcinogen exposure, especially from tobacco and alcohol consumption. The UADT provides an easily accessible region for following detectable lesions, especially leukoplakia or erythroplakia.

Studies of the molecular biology of HNSCC development may detect early genetic alterations that are not clinically apparent. This could lead to possible agents effective in blocking the process of carcinogenesis via chemopreventive measures or dietary modifications, before invasive cancer develops. These genetic alterations or biomarkers could provide a means for mass screening programs identifying individuals with premalignant lesions, early stage lesions, or otherwise at high risk of developing head and neck cancer. This could lead to earlier, more effective treatments to prevent cancer or cure early stage disease. Screening strategies that would detect head and neck cancer and decrease mortality are still being developed and guidelines are not yet established.

IV. RISK FACTORS

A. Alcohol and Tobacco

Cigarette smoking is the predominant cause of cancer of the lung, oral cavity, larynx, and pharynx [13,14]. As early as 1928, it has been reported that there is a higher incidence of smoking among patients with cancer than among controls [15] and in 1939 that a relationship between smoking and lung cancer was observed [16]. There are clearly strong dose–response carcinogenic relationships with tobacco [17,18] as studies document 5- to 25-fold increased risks of developing cancer for the heaviest smokers as compared with nonsmokers [18] and a higher risk of UADT cancer with unfiltered cigarette use. However, it has been demonstrated that overall risk decreases after successively longer periods of time of smoking cessation [19–21]. In heavy smokers, many years must pass before the risk decreases, but still falls short of those people who never smoked [19]. The risk of bronchogenic carcinoma appears less for cigar and pipe smokers, although these forms of tobacco are clearly associated with UADT cancers. The pooling of saliva, which contains carcinogens, in gravity-dependent areas such as along the lateral and ventral surfaces of the tongue and in the floor of the mouth may account for the increased frequency of oral squamous cell carcinomas in these locations [22].

There are an estimated 1 billion smokers worldwide, one-third of which reside in China. The incidence of smokers in the United States and other developed countries is decreasing [23]. From 1970 to 1985, cigarette consumption decreased in most developed countries. However, overall worldwide consumption increased 7%. This was due to the dramatic increase in cigarette consumption in third-world countries such as those in Latin America, Africa, and Asia [24]. The major carcinogenic activity of cigarette smoke resides in the tar fraction. More than 3000 chemicals have been identified in tobacco smoke. Polynuclear aromatic hydrocarbons such as benzopyrenes account for most tumorigenic activity [25]. Upon activation by specific enzymes in human tissue, the compounds become mutagenic. Binding of these activated carcinogens to macromolecules such as DNA may induce point mutations such as those involving the K-ras oncogene [26]. Nitrosamines are also present and have a propensity to form tumors in the upper airways of animals. Although cigarette smoking is the most prevalent form of tobacco consumption and has been associated with the greatest number of head and neck cancer cases, other patterns of tobacco use are also associated with significant risk.

Smokeless tobacco continues to be a growing international problem, as rates in the United States remain largely unchanged. Chewing tobacco is associated with the development of oral cavity cancer and is strongly associated with the development of oral leukoplakia, a premalignant lesion of the oral cavity. High school students have been observed to develop these lesions. The increased use by younger people most likely accounts for the recent excess of oral cancer mortality in this age group [27–32]. The use of smokeless tobacco also carries a risk of cheek and gum cancer and, to a lesser extent, pharyngeal cancer in long-term users [33]. The major carcinogens in smokeless tobacco are polonium 210 and tobacco-specific nitrosamines [34]. As the production (up 42% over the past two decades) and sales of smokeless tobacco continue to increase, the incidence pattern of squamous cell carcinoma of the oral cavity will change. In other areas of the world, such as India and Indonesia, tobacco is used in different forms, such as betel nut quids. Squamous cell cancer of the oral cavity accounts for 50% of all cancers in Bombay where this form of tobacco use is more prevalent. There is also a strong association between development of premalignant lesions of the oral cavity such as leukoplakia and use of this form of tobacco-containing quids [32,35–37].

UADT cancer rates in the United States have changed over time as tobacco-use habits have shifted. Among American women and nonwhite men, head and neck squamous cell cancer incidence and mortality rates have increased since the early 1950s, most strikingly for oral cancer (sevenfold increase in women). Regional differences have become apparent, as women in the southeastern United

States had a 30% greater risk of oral and pharyngeal cancer than women from northern urban areas [31,33]. Analysis of the impact of specific patterns of tobacco consumption and the development of head and neck cancer has been complicated by the fact that many individuals use tobacco in several forms. This is further confounded by a history of concomitant alcohol exposure.

Alcohol has been implicated as a cofactor in head and neck cancer development. It seems to have a synergistic effect with tobacco consumption acting in a multiplicative fashion [20,38]. Independently, however, alcohol appears to have only a modest effect on the increased risk of oropharyngeal and laryngeal cancers [39,40]. This risk is lower than that associated with tobacco alone. Investigations into whether type of alcoholic beverage consumed determines risk remain unclear. Ethanol itself seems to be the important factor [41,42]. The importance of alcohol consumption and the development of premalignant lesions such as oral leukoplakia are still unclear [43–46]. Cancer risks appear highest for heavy drinkers, independent of tobacco use. Also, drinkers who developed head and neck squamous cell carcinoma were more likely to develop oropharyngeal rather than laryngeal cancer, suggesting that despite field cancerization, individual sites may vary in their susceptibility and exposure to alcohol as a carcinogen [47]. The role of alcohol as an agent in carcinogenesis of the UADT, independent and in association with tobacco, still requires further investigation.

Despite aggressive campaigns for smoking cessation, especially for young people, tobacco-related malignancy remains a worldwide epidemic. Cigarette smoking, although socially acceptable in the past, today has found itself a target for increasing regulation. Concern about the health risks from exposure to passive smoke has contributed to these changes. As up to 75% of all UADT cancer cases are attributed to tobacco and/or alcohol, reduction in exposure could have a significant impact on cancer medicine, but will require both changes in personal lifestyle and governmental legislation [20].

B. Viral Infection

Viruses have been associated with UADT cancers. However, their role in carcinogenesis is not clear [48]. Papillomavirus infections and cancer development were first described by Shope in 1933 [49]. He observed that infection with the cottontail rabbit papillomavirus, a deoxyribonucleic acid (DNA) virus, led to invasive epithelial neoplasms. Infections usually result in benign, self-limiting warts or epithelial tumors [50–52]. Human papillomavirus (HPV) infections are associated with about 10% of the worldwide cancer burden, most of which are anogenital cancers [53,54]. The frequency of HPV in benign and precancerous lesions of the UADT has ranged from 18.5 to 35.9%. Its overall prevalence in head and neck tumors using polymerase chain reaction (PCR) was 34.5%, with the majority of cases containing the "high-risk" types 16 (40%) and 18 (11.9%) [54]. The causal role for HPV in tumorigenesis remains unclear. Cell culture studies have shown that high-risk HPVs have the ability to immortalize and transform epithelial tissue, whereas HPV-6 and 11, associated with benign lesions, do not [55–62]. Expression of the open reading frames E6 and E7 of HPV-16 or 18 is sufficient for immortalization [63–65]. Interaction between these HPV proteins and cellular proteins is believed to have a role in transformation. HPV-16 E7 forms a complex with the Rb tumor suppressor protein *in vivo* and *in vitro* [66,67]. The binding of Rb, which prevents abnormal cell proliferation, suppresses this activity. Other cell cycle regulators have been observed to form complexes with HPV-16 E7, including p107, p130, and cyclin-dependent kinase 2–cyclin A [68–70]. These interactions add to abnormal cell cycle progression and transformation.

HPV-16 and -18 E6 protein has been described to complex with p53, another tumor suppressor protein, leading to p53 degradation [71,72]. p53 induces cell cycle inhibition when DNA damage increases, thus allowing for repair. However, when p53 is suppressed when complexed to the E6 protein, DNA replication continues before repair occurs. This leads to accumulation of mutations, which could contribute to malignant conversion. Mechanisms other than the presence of HPV-16 and -18 proteins are involved in tumorigenesis, including carcinogen exposure [73–76].

Epstein–Barr virus (EBV), a member of the Herpesvirus family and the cause of infectious mononucleosis, was first reported in 1885 and infects more than 95% of the adult human population [77]. EBV has been clearly linked to nasopharyngeal carcinoma. Undifferentiated nasopharyngeal cancer, which is most prevalent in high-risk populations such as the Cantonese people of southern China, has been clearly linked to EBV and not to smoking [78]. In contrast, nasopharyngeal cancer in low-risk regions, such as the United States, has been linked predominantly to smoking [79]. EBV DNA is found in tissue from all nasopharyngeal cancer types, including premalignant. However, its expression is much lower in squamous cell carcinoma than in undifferentiated tumors. Studies show that immunoglobulin A (IgA)-early antigen and viral capsid antigen (VCA) of EBV are the most specific markers and IgG-VCA the most sensitive markers for nasopharyngeal cancer. Cell-free EBV DNA has been detected in the plasma and serum of patients with nasopharyngeal cancer in which significantly elevated levels were found in patients who eventually relapsed. Low levels of serum EBV DNA were observed in patients who remained in remission [80]. Antibodies against Epstein–Barr virus nuclear antigen-1(EBNA-1), a protein expressed consistently in EBV-infected cells and in EBV-related malignant tissues, may prove to be efficacious in developing serological markers for screening people at high risk for nasopharyngeal cancer [81]. EBV has been reported to occur in mesopharyngeal and hypopharyngeal carcinomas [82] and in oral hairy leukoplakia [83], and also to be associated with higher levels of transforming growth factor-β-1 in nasopharyngeal carcinoma [84]. Development of biomarkers and implementation of screening programs for EBV-associated cancers continue to progress, but further investigations are needed.

Herpes simplex virus-1 (HSV-1) has also been implicated as a possible etiologic agent in head and neck cancer. Studies have shown that patients with oral squamous cell carcinoma (SCC) have increased levels of HSV-1 IgA and IgM antibodies that are of prognostic significance [85]. However, other laboratory work has demonstrated the presence of HSV-1 gene products in isolated cases of oral SCC [86]; thus, HSV-1 may have a role in head and neck cancer tumorigenesis as research continues.

C. Genetics

Genetic risk factors for UADT cancers were first suggested when these cancers were found in atypical patients, such as nonusers of tobacco and alcohol, patients with early age of onset, or those with unusual tumor sites [87]. The rate of accumulation of genetic damage reflects the interaction between cells and carcinogens and the capacity of the cells to repair DNA damage [88]. Genetic-based research in tobacco-related carcinogenesis has focused on human leukocyte antigen (HLA) phenotypes, polymorphic variability of metabolic enzymes, and chromosome fragility [89–92]. Studies have demonstrated that in untreated UADT cancer, chromosome fragility, measured by bleomycin mutagen sensitivity, is a strong independent risk factor for UADT cancer [93]. It also has synergism with tobacco in causing UADT cancer. Chromosome fragility also correlated with the development of second primary tumors following UADT cancer [94].

V. BIOLOGY OF HEAD AND NECK CANCER

A. Field Cancerization

In field cancerization, diffuse epithelial injury results from carcinogen exposure in the upper aerodigestive tract; genetic changes and the presence of premalignant and malignant lesions in one region of the field are associated with an increased risk of cancer developing throughout the entire field. In 1953, Slaughter *et al.* [95] first described field cancerization associated with squamous cell carcinoma of the head and neck. Their studies included histological examination of 783 resected head and neck cancer specimens. Epithelium from sites beyond the original invasive cancer was found to be abnormal in every case, with abnormalities including epithelial hyperplasia and hyperkeratinization, dyskaryosis, and carcinoma *in situ*. In addition, 88 patients (11.2%) were found to have multiple, distinct invasive cancers within the surgical resection specimen. The authors suggested that oral cavity squamous cell carcinoma originated from epithelium that had been "preconditioned by an as-yet unknown carcinogenic agent" and resulted in an irreversible change that made development of cancer inevitable. Thus, a "preconditioned" or "condemned" epithelium could conceivably become activated or break down into cancer at multiple points producing separate tumors. The interaction of host susceptibility and carcinogen exposure (i.e., tobacco and/or alcohol) leads to the variation in cancer susceptibility and presentation. Development of SPTs confers a greater risk of mortality on early stage treated head and neck cancer patients. The concepts of field cancerization and a condemned epithelium explain the biological basis for UADT cancers and why they recur or persist despite therapeutic intervention.

The diffuse epithelial damage from carcinogen exposure certainly could be prevented by carcinogen avoidance. Because the region of exposure in the UADT is vast, screening and early detection become more difficult once exposure has occurred. Screening procedures may identify prevalent cases at an earlier stage in their natural history and would offer an opportunity for curative therapy. However, because the entire field has been carcinogen exposed, effective local therapy will not eliminate the risk of a second cancer developing. Areas of epithelium, which appear grossly normal, may already have phenotypic and genotypic changes, which predict an increased risk of developing UADT cancer.

B. Multistep Carcinogenesis

The multistep carcinogenesis theory describes a stepwise accumulation of alterations, both genotypic and phenotypic. Arresting one or several of the steps may impede the development of cancer. This has been particularly well described in the UADT by studies that focus on oral premalignant lesions (leukoplakia and erythroplakia) and the associated increased risk of progression to cancer. The complex fundamental biology of head and neck squamous cell cancer remains poorly understood despite intensive study. Epithelial carcinogenesis, based on animal studies, is divided into three phases: initiation, promotion, and progression. DNA damage occurs during initiation and the mutation becomes fixed after several cellular divisions. Carcinogen exposure strongly influences this step as these chemical events occur very rapidly. During promotion, phenotypic clonal expansion of the molecularly damaged cell occurs, leading to hyperplasia. This process is slow and occurs over a long period of time. Progression is the most complex step, as genetic and phenotypic changes occur with rapid cellular expansion [96]. Primary prevention efforts such as avoidance of carcinogens affect the initiation phase. Chemoprevention targets the promotion and progression phases as premalignant lesions evolve during these periods [97]. An important goal of clinical and basic research is to establish markers of these carcinogenic steps, thus helping to identify individuals at high risk for UADT cancers. These "intermediate markers" could not only be used as screening devices, but also as indicators for early evaluation of chemopreventive agent efficacy [98].

Longitudinal studies have shown that patients with a prior treated cancer have the single highest risk for development of second primary tumors [99,100], including patients with laryngeal cancers who were found to have a lifetime risk for SPTs of 25–40% in the SEER database.

The theory of clonal origin of metachronous or synchronous primary cancers remains controversial. Studies have focused on whether these SPTs are associated synchronously or metachronously with the index primary tumor. Theories generated from the field cancerization model were whether the areas of abnormality involved separate, independent clones with a unique set of genetic alterations or if they were genetically related and derived from a single cellular clone. Observations have led to a belief that a common, clonal origin of the histopathologically distinct areas in premalignant lesions exists. Additional genetic losses were associated with a more malignant phenotype [101]. For example, discordant p53 mutations found in multiple tumors in the same field were a result of genetic events such as 3p and 9p21 loss preceding p53 mutation [102]. Thus, as the clonal population expands, genetic heterogeneity occurs. This may explain the histopathologic abnormalities found in primary and second primary tumors. Genetic alterations present in preneoplastic and neoplastic lesions from HNSCC biopsies were used to evaluate a temporal order of events [103]. By using microsatellite analysis of minimal regions of loss on the 10 most frequently lost chromosomal arms in HNSCC, 3 patients had a clear genetic progression with new lesions of LOH correlating with histologic progression. Further studies evaluated paired tumors from patients with HNSCC and a solitary lung squamous cell carcinoma for LOH at chromosomal arms 3p and 9p [104]. Of 16 patients, 10 patients were found to have concordant patterns of loss at all chromosomal loci, whereas 3 patients had discordant patterns. These studies suggest a common clone and that squamous SPTs of the lung were most likely late primary tumor metastases as opposed to metachronous independent SPTs.

VI. CHEMOPREVENTION

Chemoprevention is defined as the use of specific natural or synthetic chemical agents to reverse, suppress, or prevent carcinogenic progression to invasive cancer [105,106]. Cancer chemoprevention is a rapidly developing field that approaches carcinogenesis from a different perspective. Previously, early detection techniques were employed to reduce morbidity and mortality with respect to cancer treatment. In lung cancer, this included early chest X-ray and sputum cytology analysis in individuals at high risk. Despite these early detection techniques, overall mortality did not improve. Chemoprevention bridges basic biologic research with clinical chemical intervention and attempts to halt the process of carcinogenesis. Chemoprevention trials are based on the hypothesis that interruption of the biological processes involved in carcinogenesis will inhibit this process and, in turn, reduce cancer incidence. This hypothesis provides a framework for the design and evaluation of chemoprevention trials, including the rationale for the selection of agents that are likely to inhibit biological processes and the development of intermediate markers associated with carcinogenesis. Trials in UADT have been encouraging, and this approach is being studied actively as a strategy to prevent head and neck cancer. Clinical studies have focused on reversing premalignant lesions such as oral leukoplakia and development of invasive cancers in high-risk individuals and in those with a previous history of head and neck cancer (Table 19.1). These trials are guided by the concepts of field cancerization and multistep carcino genesis. Because large randomized trials are difficult and expensive, the pursuit of intermediate biomarkers is paramount. These trials may serve as a model for chemoprevention strategies and increase the understanding of head and neck cancer biology with hope that findings may eventually be translated into the treatment of other body systems.

VII. CHEMOPREVENTION TRIALS

Cancer chemoprevention is still investigational, although its role in oncologic practice continues to expand. Prevention in cancer has become more prominent as frustration over the failures of current therapeutic modalities has grown. A variety of chemopreventive agents have been studied in over 40 randomized trials since 1990. Some clinical activity has been demonstrated, proving the potential utility of this method of cancer prevention. Still, large randomized trials are needed before chemoprevention agents can be fully integrated into standard oncologic practice. Major trials are listed in Tables 19.1 and 19.2.

A. Oral Premalignancy

Leukoplakia and erythroplakia are the predominant premalignant lesions in oral cancer [107–109]. Leukoplakia, a white patch that cannot be scraped off, is clearly associated with oral cancer development [110]. Erythroplakia, a red, velvety lesion in the oral mucosa, is more often associated

TABLE 19.1 Selected Randomized Head and Neck Chemoprevention Trials

Study	Agent	Sample size	Result
Hong 1986	Isotretinoin (1–2 mg/kg/day)	44	Positive
Stich 1988	Vitamin A (200,000 U/week)	65	Positive
Han 1990	Retinamide (40 mg/day)	61	Positive
Lippman 1993	Isotretinioin (0.5 mg/kg/day)	70	Positive
Costa 1994	Fenretinide (200 mg/day)	153	Positive
Hong 1990/1994	Isotretinoin (50–100 mg/m^2/day)	103	Positive
Bolla 1994	Etretinate (25–50 mg/day)	316	Positive
Garewal 1999	β-Carotene (60 mg/day)	50	Positive
Van Zandwijk 2000[a]	Retinyl palmitate (300,000 IU/day) *N*-Acetylcysteine (600 mg/day)	2592	Negative
RTOG[b] 91–15	Low-dose 13-cRA	1384	Ongoing

[a]Study included head and neck (60%) and lung (40%) cancer.
[b]Radiation Therapy Oncology Group.

with *in situ* or invasive carcinoma [111]. Both of these lesions harbor different degrees of histology, including hyperplasia and dysplasia, and are found in the United States predominantly in tobacco users. Standard therapy for these lesions includes surgical removal or laser excision. These lesions have shown propensity for relapse or development in new lesions, despite surgical intervention, most commonly in those patients with tobacco, cigarette and alcohol histories [112]. Spontaneous regression occurs in 10–20% of lesions. In the largest U.S. series, Silverman *et al.* [113] followed 257 leukoplakia patients for 7.2 years. Malignant transformation occurred in 45 of these patients (17.5%) who developed squamous cell carcinoma of the oral cavity. Cancer development was fourfold higher in erythroplakia as compared with leukoplakia, and marginal improvement of these lesions occurred with smoking cessation. These facts support the notion of field cancerization, that oral leukoplakia is a marker of this damage, and that systemic treatment is required for treatment. Because of the accessibility of these lesions and the capability to monitor them safely, they serve as an excellent model for chemoprevention and biomarker studies.

Because of its effects on epithelial differentiation, vitamin A has been studied as both a topical and a systemic treatment for oral leukoplakia. These trials, which were first initiated in the 1950s, demonstrated regression of oral leukoplakia and also documented the toxic effects associated with vitamin A [114–117]. As studies continued, the balance between toxicity and efficacy of other natural and synthetic vitamin A agents was critical for the development of chemoprevention trials in oral cancer. Stich *et al.* [118] reported a 57%

TABLE 19.2 Selected Studies of Natural Agent Chemoprevention in Oral Premalignancy

Investigator	Agent(s)	Sample size	Clinical responses (%)
Wulf, 1957	Vitamin A	20	90
Silverman *et al.*, 1963	Vitamin A	16	44
Silverman *et al.*, 1965	Vitamin A	6	83
Ryssel *et al.*, 1971	β-All-*trans*-retinoic acid	10	70
Koch, 1978, 1981	β-All-*trans*-retinoic acid	27	59
	13-*cis*-Retinoic acid	24	87
	Etretinate	24	92
	Etretinate (oral, topical)	24	83
	Etretinate	21	71
Stich *et al.*, 1988	β-Carotene	27	15
	β-Carotene and vitamin A	51	28
	Placebo	33	3
Garewal *et al.*, 1990	β-Carotene	24	71
Toma *et al.*, 1990	β-Carotene	24	27
	Selenium	25	33
Malaker *et al.*, 1991	β-Carotene	18	44
Benner *et al.*, 1993	β-Tocopherol	43	46
Kaugars *et al.*, 1994	β-Carotene, vitamins E,C	79	57

complete clinical response rate using vitamin A to reverse leukoplakia. Interest in other vitamin A agents and their role in cancer treatment was explored. Stich *et al.* [119] pioneered a series of trials testing β-carotene in high-risk groups, namely betel nut and snuff users. Micronucleated buccal mucosa cells were studied, introducing the concept of early intermediate end points as markers of activity. These investigators studied 130 patients who received placebo, β-carotene alone, or β-carotene plus retinol. They found that the combination of β-carotene plus retinol was twice as active in inducing remission in leukoplakia [120]. Single-arm trials of β-carotene followed, which reported decreasing response rates with increasing doses, but these studies were not randomized and their responses were not documented histologically [121–123]. However, these trials have been extremely important in developing potentially useful chemopreventive regimens against carcinogenesis. Other agents have been studied in the model of human oral leukoplakia, including α-tocopherol (vitamin E) and selenium. A nonrandomized trial in oral leukoplakia using selenium [124] produced a 33% response rate. A trial of α-tocopherol [125] produced a 46% response rate. Continued studies will help devise combination regimens that may be more efficacious than single agents [120]. As the result of successful treatment of leukoplakia lesions with vitamin A treatment, studies of other retinoids, including *trans*-retinoic acid, etretinate, 13-*cis*-retinoic acid (13-cRA), and *N*-4-(hydroxycarbophenyl) retinamide, have been performed. Clinical responses have been reported with these agents [117,126–132].

In 1986, a landmark trial by Hong *et al.* [133] reported several important aspects of retinoid treatment in oral premalignancy. This pioneering randomized, double-blind, placebo-controlled trial followed 44 patients treated with high-dose 13-cRA (1 to 2 mg per kilogram per day) for 3 months and followed for 6 months. Clinical responses, including both complete or partial response, were observed in 67% of those treated with the high-dose retinoid and in 10% of those given placebo ($p=0.0002$). Histologic responses, including reversal of dysplasia, occurred in 54% of retinoid-treated patients and in 10% of patients given placebo ($p=0.01$). Several important facts arose from this short high-dose retinoid regimen. First, toxicity was dose related and was reversible with drug cessation. Eighty-eight percent of patients receiving 2-mg/kg experienced moderate to severe cheilitis, dry skin, and peeling and 76% had conjunctivitis. Forty-seven percent of patients required dose reduction to 1 mg/kg/day and many still experienced mild skin toxicities. This group also had significantly lower response rates as compared with the 2 mg/kg/day dose group. Second, remission was short-lived after therapy was stopped, as more than 50% of participants relapsed within 3 months after drug cessation. These findings suggested that long-term administration of chemopreventive drugs was needed to confer protection from cancer development.

The study that followed addressed these problems of relapse and toxicity. This trial examined the effect of high-dose 13-cRA (1.5 mg/kg/day) induction for 3 months followed by maintenance therapy with low-dose 13-cRA (0.5 mg/kg/day) or β-carotene (30 mg/day) [134]. Low-dose 13-cRA was used to keep patients in remission after induction, but to avoid side effects with higher dose 13-cRA. Epidemiologic evidence suggested efficacy with β-carotene. Induction therapy produced a response rate of greater than 50% and reversal of dysplasia in 43%, which was consistent with the previous trial. However, in the maintenance phase, low-dose 13-cRA was much more effective than β-carotene in maintaining remissions. Relapse occurred in 8% of patients treated with low-dose 13-cRA, whereas 55% of β-carotene-treated patients experienced relapse ($p \leq 0.01$). 13-cRA treatment was also associated with a reduction in micronuclei frequency and with histological responses. The low-dose schedule significantly reduced the toxic effects of the drug, although they were still greater than those in the β-carotene group. This study confirmed that high-dose 13-cRA induction followed by low-dose maintenance therapy is effective in preventing recurrence with minimal toxicity.

These results led to the formation of a currently ongoing randomized trial in oral premalignancy. This study compares low-dose 13-cRA (0.5 mg/kg/day) for 1 year and then 0.25 mg/kg/day for the next 2 years in combination with β-carotene (50 mg per day) and retinyl palmitate (25,000 IU per day) for a total of 3 years. This study will attempt to identify an appropriate maintenance dose-maximizing efficacy while minimizing toxicity. Intermediate markers will also be assessed. However, the β-carotene dose was subsequently removed after results from the α-tocopherol, β-carotene (ATBC) study [135] and the β-carotene and retinol efficacy trial (CARET) [136] reported adverse outcomes with β-carotene supplementation in smokers as well as its lack of efficacy from the Physicians' Health Study [123].

The α-tocopherol, β-carotene (ATBC) cancer prevention study was a randomized, double-blind, placebo-controlled, primary prevention trial in which 29,133 Finnish male smokers received either α-tocopherol (50 mg a day) alone, β-carotene (20 mg a day) alone, both α-tocopherol and β-carotene, or a placebo [135]. Male participants were between 50 and 69 years of age and all smoked five or more cigarettes a day. Patients received follow-up observations for 5 to 8 years. Lung cancer incidence, the primary end point, did not change with the addition of α-tocopherol alone, nor did the overall mortality rate. However, both groups who received β-carotene supplementation (alone or with α-tocopherol) had an 18% increase in the incidence of lung cancer. There appeared to be a stronger adverse effect from β-carotene in the men who smoked more than 20 cigarettes a day. This trial raised the serious issue that pharmacologic doses of β-carotene could potentially be harmful in active smokers.

The β-carotene and retinol efficacy trial (CARET) confirmed the results of the Finnish trial. This randomized, double-blind, placebo-controlled trial tested the combination of 30 mg of β-carotene and 25,000 IU of retinyl palmitate against a placebo in 18,314 men and women aged 50 to 69 years who were at high risk for lung cancer [136]. A majority of participants (14,254) had a smoking history of at least 20 pack-years and were either current smokers or recent ex-smokers. Extensive occupational exposure to asbestos was noted in 4060 men. This trial was stopped after 21 months because there was evidence of no benefit or of possible harm. Lung cancer incidence, the primary end point, increased 28% in the active intervention group. The overall mortality rate also increased 17% in this group. Given these results, in addition to those of the ATBC trial, high-dose β-carotene is not recommended for patients at high risk who continue to smoke.

The Physicians Health Study, a randomized, double-blind, placebo-controlled trial, studied 22,071 healthy male physicians. Half of the participants (11,036) received 50 mg of β-carotene on alternate days and the other half (11,035) received placebos. The use of supplemental β-carotene showed virtually no adverse or beneficial effects on cancer incidence or overall mortality rate during a 12-year follow-up period [137].

Analysis of subgroups of the previously mentioned studies, especially the ATBC and CARET studies, has provided few explanations for the increase in lung cancer incidence. It seems that β-carotene has a harmful effect only in heavy smokers at high risk or people who have been exposed to asbestos. Current recommendations are for these people to avoid β-carotene in large doses [138].

Chemoprevention and treatment trials of oral prema lignancy will continue as head and neck cancer continues to be a challenge. Use of natural and synthetic agents may indeed reduce the risk of cancer in high-risk individuals. The optimal dosage and maintenance schedules still need further clarification. Through this approach we hope to identify accurate biomarkers and to establish effective treatment regimens for head and neck carcinogenesis.

B. Second Primary Tumors

Head and neck cancer patients who have been treated successfully remain at a significantly increased risk for developing additional neoplasms within the UADT and lungs [5,139–144]. The concept of multistep field cancerization explains the development of multiple independent tumor sites within the aerodigestive tract. In fact, despite occurring in all treatment stages of head and neck cancer, SPTs have the greatest impact on patients treated for early stage disease (stage I or II), which is usually curative [145,146]. The lifetime risk of developing a second primary tumor in head and neck squamous cell cancer is 20% and the annual rate is 4–6%. One study in oral cancer reported rates of 3.6% per year [147]. SPTs are the major cause of death after curative surgery in head and neck cancer and are the leading cause of death in early stage disease, more so than recurrence [1,3,145–148]. Retinoids have been proven active in oral premalignancy. These facts have provided the basis for chemoprevention trials in head and neck cancer evaluating SPTs and the impact of smoking.

Variations in reported occurrence rates depend on the population studied and the methods used for diagnosing an SPT. Currently, the Warren–Gates criteria [149], published in the 1930s are used to diagnose an SPT. The definition of SPT is as follows: (1) it is a new cancer of a different histological type; (2) it is a cancer, regardless of site, that occurs after more than 3 years; (3) in the head and neck, the lesion is separated from the initial primary tumor by more than 2 cm of clinically normal epithelium; and (4) in the lung, if the cancer is of squamous cell histological type and develops within 3 years, it presents as a solitary mass, the patient must be free of local or regional disease, and there must be changes consistent with dysplasia or carcinoma *in situ* within the bronchial epithelium. Using these criteria, the risk of local recurrence seems to decline over time, whereas the risk of SPT is constant for the first 8 years following initial head and neck cancer [1].

Because of the morbidity associated with the development of SPTs in head and neck cancer, the first phase III adjuvant chemoprevention trial was performed by Hong *et al.* [150] in 1990. This randomized, placebo-controlled, double-blind study followed 103 patients with stage I through IV (M0) head and neck cancer who were randomized to receive high-dose 13-cRA (100 mg/m^2/day) or placebo for 1 year after definitive local therapy. The dosage of 13-cRA was reduced to 50 mg/m^2/day after 13 of the first 44 patients experienced intolerable side effects. The primary end points were primary recurrence and SPT development. In the two treatment arms, there was no difference in local recurrence development or distant metastases. However, patients treated with 13-cRA had a dramatically lower incidence of SPTs. Of the 103 patients followed for a median of 42 months, SPTs developed in 6% (3 of 49) of those in the 13-cRA arm, whereas 28% (14 of 51) developed SPTs in the placebo arm. Consistent with field carcinogenesis, most of the SPTs developed in the UADT, esophagus, and lung in the placebo group (14 of 17) and were histologically squamous cell type. Additionally, none of the patients receiving 13-c-RA developed an SPT during the year of active treatment. Despite only 47% of the 13-cRA treatment arm patients completing the therapy as prescribed, the reduction in SPT development was still significant.

Based on the important findings of the previous study, a randomized chemoprevention trial in head and neck squamous cell cancer designed to prevent SPT development was instituted through the University of Texas M. D. Anderson Cancer Center and its affiliated Community Clinical Oncology Program (CCOP) and the Radiation Therapy Oncology Group (RTOG) [151,152]. This randomized, double-blind trial was launched in 1991 and studied the effect of low-dose 13-cRA to prevent second primary tumors in patients definitively treated for stage I or II squamous cell carcinoma for up to 3 years before participation (T1N0M0 or T2N0M0). Patients received 30 mg per day of 13-cRA or placebo for 3 years and were followed for an additional 4 years. This study recently completed accrual with 1190 randomized and 1384 registered patients. Second primary tumor incidence according to prior tumor stage as well as related to smoking status (current, former, never) has been reported. The annual primary tumor recurrence rate was 2.8% and the SPT occurrence rate was 5.1% annually. Stage II HNSCC had a higher rate of second primary tumor development than stage I. Additionally, active smokers had a significantly higher recurrence rate than former and never smokers (4.3% vs 3.3% vs 1.9%). This prospective study demonstrated for the first time the impact of active smoking status and second primary tumor development. The SPT rate was significantly higher in smokers vs nonactive smokers ($p=0.018$) and was marginally significant between former and never smokers ($p=0.11$). The site of SPT also differed by the primary index tumor. Patients with primary laryngeal cancers were most likely to develop SPT in either the lung or the larynx, whereas patients with oral cavity primaries were most likely to develop second primaries in either the oral cavity or the lung. Finally, patients with an index primary tumor of the pharynx developed SPTs in the lung, oral cavity, pharynx, or esophagus. Compared to previous trials, SPTs occurred more frequently than expected at the index primary sites of the oral upper aerodigestive tract. The lower dose of 13-cRA was also well tolerated with few grade 3 toxicities. This trial is scheduled to be unblinded in 2002 [153–156].

Additional completed studies in head and neck chemoprevention are listed in Tables 19.1 and 19.2. Other major phase III studies include EUROSCAN and the US-Intergroup NCI 91-0001 trial. EUROSCAN, a randomized adjuvant chemoprevention study of the European Organization for Research and Treatment of Cancer (EORTC) Head/Neck and Lung Cancer Groups, studied the effects of vitamin A (retinyl palmitate) and *N*-acetylcysteine in patients with early stage head and neck and lung cancer [155,157]. In the trials, 2592 patients with cancers of the larynx (Tis–T3, N0–N1), oral cavity (Tis–T2, N0–N1), and NSCLC (T1–T2, N0–N1) received retinyl palmitate (300,000 IU a day in year 1; 150,000 IU a day in year 2), *N*-acetylcysteine (600 mg a day for 2 years), both drugs, or placebo. There were no end point differences among the three active treatment arms and the placebo group in terms of lung cancer incidence, occurrence of second primary cancer, and survival. There was a statistically significant difference in time to development of SPTs within the carcinogen-exposed field ($p=0.045$) in favor of the retinoid-treated group. The majority (93%) of the patient population were considered regular smokers with at least half with greater than 43 pack-years of tobacco exposure. Problems with the study included differences in medication adherence across the three treatment groups and the testing of *N*-acetylcysteine, a drug with little established efficacy in chemoprevention in 1300 patients at risk.

US-Intergroup NCI 91-0001, a randomized, double-blind study using low-dose 13-cRA after complete resection of stage I NSCLC, completed accrual in April 1997 with 1486 participants [158]. The study objectives were to evaluate the efficacy of 13-cRA in reducing the incidence of second primary tumors in patients after complete resection of stage I NSCLC, to look at the qualitative and quantitative toxicity of daily low-dose 13-cRA (30 mg/day), and to compare the overall survival rates of patients receiving 13-cRA and those receiving a placebo. Randomization of 1304 patients was completed in June 1997. Patients were required to have complete resection of primary stage I NSCLC (postoperative T1 or T2,N0) 6 weeks to 3 years prior to registering. One thousand one hundred sixty-six patients with pathologic stage I NSCLC (6 weeks to 3 years from definitive resection and no prior radiotherapy or chemotherapy) were evaluated. Patients took the study drug for 3 years and were stratified at randomization by tumor stage, histology, and smoking status. After a median follow-up of 3.5 years, there were no statistically significant differences between the placebo and isotretinoin arms with respect to the time to SPTs, recurrences, or mortality. Multivariate analyses showed that the rate of SPTs was not affected by any stratification factor. The recurrence rate was affected by tumor stage [HR for T2 vs T1 = 1.77 (95% CI = 1.35 to 2.31)] and a treatment by smoking interaction [HR for treatment by current vs never-smoking status = 3.11 (95% CI = 1.00 to 9.71)]. Mortality was affected by tumor stage, histology, and a treatment by smoking interaction. The authors concluded isotretinoin did not improve overall survival rates of SPTs, recurrences, or mortality in stage I NSCLC and possibly, in subset analyses, that isotretinoin was harmful in current smokers and beneficial in never smokers.

Researchers at Yale University in a randomized, double-blind, placebo-controlled trial studied the efficacy of β-carotene (50 mg per day) in reducing local recurrence and SPTs in head and neck cancer [159]. Two hundred sixty-four patients, some recruited from the state tumor registry, with curatively treated early stage squamous cell carcinoma of the oral cavity, pharynx, or larynx were randomized. Patients were assigned to receive 50 mg of β-carotene per day or placebo and were followed for 90 months for the development of SPTs and local recurrences. After a median follow-up of 51 months, there was no difference between the two groups in the time to failure (SPTs and local recurrences). In site-specific analyses, supplemental β-carotene had no significant effect on second head and neck cancer or lung cancer. Total mortality was not significantly affected by the drug intervention. Based on the point estimates, the authors concluded that a nonsignificant but possible decrease in the risk of second head and neck cancer, as well as a possible increase in lung cancer risk, was suggested.

VIII. BIOCHEMOPREVENTION

Biochemoprevention is currently being studied as another method to prevent treatment failure. The combination of retinoids and interferons have single agent and when combined, synergistic effects in modulating proliferation, differentiation, and apoptosis. A phase II study tested the combination of interferon-α (IFN-α), 13-cRA, and α-tocopherol (AT) as adjuvant treatment in patients with definitively treated locally advanced HNSCC [160]. Patients were treated with IFN-α (3×10^6 IU/m^2, sc injection, three times a week), 13-cRA (50 mg/m^2/day orally, daily), and AT (1200 IU/day, orally, daily) for 12 months. Of 45 patients, 11 were stage III and 34 were stage IV. Thirty-eight (86%) of 44 patients completed the 12-month treatment. At a median of 24 months follow-up, local/regional recurrence was 9%, local/regional recurrence and distant metastases was 5%, and second primary tumors 2% (nonaerodigestive tract). Median 1- and 2-year survival rates were 98 and 91% with disease-free survival rates of 91 and 84%, respectively. This biologic combination was generally well tolerated and seems promising as adjuvant treatment for locally advanced HNSCC. A phase III study is ongoing. If positive, this approach will set a new standard of care in definitively treated advanced HNSCC patients.

As technology and medicine continue to develop, more precise techniques will appear. Currently, definitions regarding the classification of SPTs are the Warren–Gates criteria, published in the 1930s [142]. Intermediate biomarkers have become an important and exciting subject in the field of chemoprevention and head and neck cancer. Synchronous and metachronous tumors are differentiated via time of occurrence and distance from the index primary as well as histology. A more precise molecular classification method is needed and is currently being developed to better identify and differentiate the molecular profiles of SPT and recurrence. As this classification method becomes elucidated, it may open new avenues into the treatment and prevention of cancer.

IX. SUMMARY: FUTURE DIRECTIONS

Squamous cell carcinomas of the head and neck are a major public health problem in the United States. Exposure to tobacco and the tobacco-related illness that follows contributes to the horrible morbidity and mortality. Smoking cessation campaigns for current smokers and prevention of youth smokers may have a profound impact on the incidence of head and neck cancer. Behavior modification targeting both tobacco and alcohol use is needed. Patients with early stage cancers can still be cured as compared to later stage disease patients; however, screening strategies, which are cost-effective and efficacious, have yet to be developed.

Field carcinogenesis, or the concept of diffusely "condemned" carcinogen-exposed epithelium, has led to the development of intervention strategies. Chemoprevention efforts have impacted the treatment of premalignant lesions such as leukoplakia and the prevention of invasive disease in high-risk patients, as well as second primary tumors. Agents used in chemoprevention continue to be tested as the understanding of the biology of tumorigenesis continues to be elucidated. As ongoing chemoprevention studies continue, effective strategies in the prevention of head and neck cancer may be established. Dietary interventions continue to be studied and may influence further trials in reducing head and neck cancer incidence.

The treatment of head and neck cancer, like other cancers, is characterized by feelings of both frustration and hope. Chemopreventive agents have demonstrated efficacy thus far and hope to define their future role in treating and, more importantly, preventing head and neck cancers in high-risk individuals. Second primary tumors have emerged as an increasingly important problem, despite curative local therapy, underscoring the principle of field cancerization. Chemopreventive agents have impacted this arena as well and as further studies are performed, their role in preventing SPTs will be further defined. Small molecule compounds that target specific receptors or mutations may play a significant role as their side effect profiles are tolerable. Development of a risk model is important to help guide and tailor therapy for patients with various risk profiles, thus allowing stratification based on risk factors. A multidisciplinary approach from clinicians and basic researchers is needed to study the biology of head and neck cancer before chemoprevention can be incorporated into a societal standard of care.

References

1. Vikram, B. (1984). Changing patterns of failure in advanced head and neck cancer. *Arch. Otolaryngol.* **110**, 564–565.
2. Cooper, J. S., Pajak, T. K., Rubin, P., *et al.* (1989). Second malignancies in patients who have head and neck cancer: Incidence, effect on survival and implications based on the RTOG experience. *Int. J. Radiat. Oncol. Biol. Phys.* **17**, 449–456.
3. Licciardello, J. T., Spitz, M. R., and Hong, W. K. (1989). Multiple primary cancers in patients with cancer of the head and neck: Second cancer of the head and neck, esophagus and lung. *Int. J. Radiat. Oncol. Biol. Phys.* **17**, 467–476.

4. Greenlee, R. T., Hill-Harmon, M. B., Murray, T., and Thun, M. (2001). Cancer statistics, 2001. *CA Cancer J. Clin.* **1**, 15–36.
5. Wynder, E. L., and Stellman, S. D. (1979). Impact of long-term filter cigarette usage on lung and larynx cancer risk: A case-control study. *J. Natl. Cancer Inst.* **62**, 471.
6. Spitz, M. R., Fueger, J. J., Goepfert, H., Hong, W. K., and Newell, G. R. (1988). Squamous cell carcinoma of the upper aerodigestive tract: A case comparison analysis. *Cancer (Phila.)* **61**, 203.
7. Mashberg, A., Garfinkel, L., and Harris, S. (1981). Alcohol as a primary risk factor in oral squamous carcinoma. *CA Cancer J. Clin.* **31**, 146.
8. Schantz, S. P., and Hsu, T. C. (1989). Head and neck cancer patients express increased clastogen-induced chromosome fragility. *Head Neck* **11**, 337.
9. German, J. (ed). (1983). "Chromosome Mutation and Neoplasia." Alan R. Liss, New York.
10. Rothman, K. J., Cann, C. I., Flanders, D., and Fried, M. P. (1980). Epidemiology of laryngeal cancer. *Epidemiol. Rev.* **2**, 195–209.
11. Blot, W. J., and Fraumeni, J. F., Jr. (1977). Geographic patterns of oral cancer in the United States: Etiologic implications. *J. Chronic. Dis.* **30**, 745–757.
12. Winn, D. M., Blot, W. J., Shy, C. M., *et al.* (1981). Snuff dipping and oral cancer among women in the southern United States. *N. Engl. J. Med.* **304**, 745–749.
13. Wynder, E. L., and Stellman, S. D. (1977). Comparative epidemiology of tobacco-related cancers. *Cancer Res.* **37**, 4608.
14. Decker, J., and Goldstein, J. C. (1982). Risk factors in head and neck cancer. *N. Engl. J. Med.* **306**, 1151–1155.
15. Lombard, H. L., and Doering, C.R. (1928). Cancer studies in Massachusetts: Habits, characteristics, and environment of individuals with and without cancer. *N. Engl. J. Med.* **198**, 481.
16. Ochsner, A., and DeBakey, M. E. (1939). Primary pulmonary malig nancy: Treatment by total pneumonectomy: Analysis of 79 collected cases and presentation of 7 personal cases. *Surg. Gynecol. Obstet.* **68**, 435.
17. Williams, R. R., and Horn, J. W. (1977). Association of cancer sites with tobacco and alcohol consumption and socioeconomic status of patients: Interview study from the Third National Cancer Survey. *J. Natl. Cancer Inst.* **58**, 525.
18. Rothman, K. J., Cann, C. I., Flanders, D., *et al.* (1980). Epidemiology of laryngeal cancer. *Epidemiol. Rev.* **2**, 195.
19. Cann, C. L., Fried, M. P., and Rothman, K. J. (1985). Epidemiology of squamous cell cancer of the head and neck. *Otolaryngol. Clin. North Am.* **18**, 367–388.
20. Blot, W. J., McLaughlin, J. K., Winn, D. M., *et al.* (1988). Smoking and drinking in relation to oral and pharyngeal cancer. *Cancer Res.* **48**, 3282.
21. Wynder, E. L., Covey, L. S., Mabuchi, K., *et al.* (1976). Environmental factors in cancer of the larynx: A second look. *Cancer* **38**, 1591.
22. Moore, C., and Catlin, D. (1967). Anatomic origins and locations of oral cancer. *Am. J. Surg.* **114**, 510.
23. Hecht, S. S. (1999). Tobacco smoke carcinogens and lung cancer. *J. Natl. Cancer Inst.* **91**, 1194–1210.
24. Masironi, R., and Rothwell, K. (1988). Trends in cigarette smoking in the world. *World Health Stat. Q* **41**, 228.
25. Florin, I., Ruthberg, L., Curvall, M., *et al.* (1980). Screening of tobacco smoke constituents for mutagenicity using the Ames' test. *Toxicology* **18**, 219.
26. Slebos, R. J., Dalesio, O., Mooi, W. J., *et al.* (1990). Mutational activation of the K-ras oncogene is associated with smoking in adenocarcinoma of the lung. *N. Engl. J. Med.* **323**, 561–565.
27. Chen, K., Katz, R. V., and Krutchkoff, D. J. (1990). Intraoral squamous cell carcinoma: Epidemiologic patterns in Connecticut from 1935 to 1985. *Cancer* **66**, 1288–1296.
28. Cullen, J. W., Blot, W., Henningfield, J., *et al.* (1986). Health consequences of using smokeless tobacco: Summary of the Advisory Committee's report to the Surgeon General. *Public Health Rep.* **101**, 355–373.
29. Bouquot, J. E., Kurtland, L. T., and Wellend, L. H. (1988). Carcinoma in situ of the upper aerodigestive tract: Incidence, time trends, and follow-up in Rochester, Minnesota, 1935–1984. *Cancer* **61**, 1691–1698.
30. Depue, R. R. (1986). Rising mortality from cancer of the tongue in young white males. *N. Engl. J. Med.* **315**, 647.
31. Devesa, S. S., Blot, W. J., Stone, B. J., Miller, B. A., Tarone, R. E., and Fraumeni, J. F. (1995). Recent cancer trends in the United States. *J. Natl. Cancer Inst.* **87**, 175–182.
32. Squier, C. A. (1984). Smokeless tobacco and oral cancer: A cause for concern? *CA Cancer J. Clin.* **34**, 242.
33. Winn, D. M., Blot, W. J., Shy, C. M., *et al.* (1981). Snuff dipping and oral cancer among women in the southern United States. *N. Engl. J. Med.* **304**, 745.
34. Hoffman, D., Harley, N. H., Fisenne, I., *et al.* (1986). Carcinogenic agents in snuff. *J. Natl. Cancer Inst.* **76**, 435.
35. Muir, C. S., Kirk, R. (1960). Betel, tobacco, and cancer of the mouth. *Br. J. Cancer* **14**, 597–608.
36. Malaowalla, A. M., Silverman, S., Jr., Mani, N. J., *et al.* (1976). Oral cancer in 57,518 industrial workers of Gujarat, India: A prevalence and follow-up study. *Cancer* **37**, 1882–1886.
37. Mehta, F. S., Gupta, P. C., and Pindborg, J. J. (1981). Chewing and smoking habits in relation to precancer and oral cancer. *J. Cancer Res. Clin. Oncol.* **99**, 35–39.
38. Mashberg, A., Garfinkel, L., and Harris, S. (1981). Alcohol as a primary risk factor in oral squamous carcinoma. *CA Cancer J. Clin.* **31**, 146.
39. Tuyns, A. J., Esteve, J., Raymond, L., *et al.* (1988). Cancer of the larynx/hypopharynx, tobacco and alcohol: IARC international case-control study in Turin and Varese (Italy), Zaragosa and Navarra (Spain), Geneva (Switzerland), and Calvados (France). *Int. J. Cancer* **41**, 483–491.
40. Spitz, M. R., Fueger, J. J., Geopfert, H., Hong, W. K., and Newell, G. R. (1988). Squamous cell carcinoma of the upper aerodigestive tract: A case comparison analysis. *Cancer* **61**, 203.
41. Kabat, G. C., and Wynder, E. L. (1989). Type of alcoholic beverage and oral cancer. *Int. J. Cancer* **43**, 190–194.
42. Falk, R. T., Pickle, L. W., and Brown, L. M. (1989). Effect of smoking and alcohol consumption on laryngeal cancer risk in coastal Texas. *Cancer Res.* **49**, 4024–4029.
43. Graham, S., Dayal, H., Rohrer, T., *et al.* (1977). Dentition, diet, tobacco, and alcohol in the epidemiology of oral cancer. *J. Natl. Cancer Inst.* **59**, 1611.
44. Rothman, K., and Keller, A. (1972). The effect of joint exposure to alcohol and tobacco on risk of cancer of the mouth and pharynx. *J. Chronic. Dis.* **25**, 711.
45. Mashberg, A., Garfinkel, L., and Harris, S. (1981). Alcohol as a primary risk factor in oral squamous cell carcinoma. *CA Cancer J. Clin.* **31**, 146.
46. Wynder, E. L., and Stellman, S. D. (1977). Comparative epidemiology of tobacco-related cancers. *Cancer Res.* **37**, 4608.
47. Brugere, J., Guenel, P., LeClerc, A., and Rodriguez, J. (1986). Differential effects of tobacco and alcohol in cancer of the pharynx and mouth. *Cancer* **57**, 391–395.
48. Snijders, P. J., van den Brule, A. J., Meijer, C. J., and Walboomers, J. M. (1994). Papillomaviruses and cancer of the upper digestive and respiratory tracts. *Curr. Top. Microbiol. Immunol.* **186**, 177–198.
49. Shope, R. E., and Hurst, E. W. (1933). Infectious papillomatosis of rabbits. *J. Exp. Med.* **58**, 607–623.
50. Howley, P. M., and Schlegel, R. (1988). The human papillomaviruses: An overview. *Am. J. Med.* **85**(2A), 155–158.
51. zur Hausen, H. (1987). Papillomaviruses in human cancer. *Appl. Pathol.* **5**, 19–24.

52. Demoubren, W. A. G. E. (1932). Infectious oral papillomatosis of dogs. *Am. J. Pathol.* 843–852.
53. Bosch, F. X., Manos, M. M., Munoz, N., *et al.* (1995). Prevalence of human papillomavirus in cervical cancer: A worldwide perspective. International biological study on cervical cancer (IBSCC) Study Group. *J. Natl. Cancer Inst.* **87**(11), 796–802.
54. McKaig, R. G., Baric, R. S., and Olshan, A. F. (1998). Human papillomavirus and head and neck cancer: Epidemiology and molecular biology. *Head Neck* 250–265.
55. Woodworth, C. D., Bowden, P. E., Doniger, J., *et al.* (1988). Characterization of normal human exocervical epithelial cells immortalized in vitro by papillomavirus types 16 and 18 DNA. *Cancer Res.* **48**, 4620.
56. Pirisi, L., Yasumoto, S., Feller, M., Doniger, J., and DiPaolo, J. A. (1987). Transformation of human fibroblasts and keratinocytes with human papillomavirus type 16 DNA. *J. Virol.* **61**, 1061.
57. Durst, M., Dzarlieva-Petrusevska, R. T., Boukamp, P., Fusenig, N. E., Glissmann, L. (1987). Molecular and cytogenetic analysis of immortalized human primary keratinocytes obtained after transfection with human papillomavirus type 16 DNA. *Oncogene* **1**, 251.
58. Park, N. H., Min, B. M., Li, S. L., Huang, M. Z., Cherick, H. M., and Doniger, J. (1991). Immortalization of normal human keratinocytes with type 16 human papillomavirus. *Carcinogenesis* **12**, 1627.
59. Band, V., Zajchowski, D., Kulesa, V., and Sager, R. (1990). Human papillomavirus DNAs immortalize normal human mammary epithelial cells and reduce their growth factor requirements. *J. Virol.* **87**, 463.
60. Kaur, P., and McDougall, J. K. (1988). Characterization of primary human keratinocytes transformed by human papillomavirus type 18. *J. Virol.* **62**, 1917.
61. Sclegel, R., Phelps, W. C., Zhang, Y.-L., and Barbosa, M. (1988). Quantitative keratinocyte assay detects two biological activities of human papillomavirus DNA and identifies viral types associated with cervical carcinoma. *EMBO J.* **7**, 3181.
62. Pecoraro, G., Morgan, D., and Defendi, V. (1989). Differential effects of human papillomavirus type 6, 16, and 18 DNAs on immortalization and transformation of human cervical epithelial cells. *Proc. Natl. Acad. Sci. USA* **86**, 563.
63. Barbosa, M. S., and Schlegel, R. (1989). The E6 and E7 genes of HPV-18 are sufficient for inducing two-stage in vitro transformation of human keratinocytes. *Oncogene* **4**, 1529.
64. Munger, K., Phelps, W. C., Bubb, V., Howley, P. M., and Schlegel, R. (1989). The E6 and E7 genes of human papillomavirus type 16 together are necessary and sufficient for transformation of primary human keratinocytes. *J. Virol.* **63**, 4417.
65. Hudson, J. B., Bedell, M. A., McCance, D. J., and Laimins, L. A. (1990). Immortalization and altered differentiation of human keratinocytes in vitro by the E6 and E7 open reading frames of human papillomavirus type 18. *J. Virol.* **64**, 519.
66. Dyson, N., Howley, P. M., Munger, K., and Harlow, E. (1989). The human papilloma virus-16 E7 oncoprotein is able to bind to the retinoblastoma gene product. *Science* **243**, 934.
67. Munger, K., Werness, B. A., Dyson, N., Phelps, W. C., Harlow, E., and Howley, P. M. (1989). Complex formation of human papillomavirus E7 proteins with the retinoblastoma tumor suppressor gene product. *EMBO J.* **8**, 4099.
68. Arroyo, M., Bagchi, S., and Raychaudhuri, P. (1993). Association of the human papillomavirus type 16 E7 protein with the S-phase-specific E2F-cyclin A complex. *Mol. Cell. Biol.* **13**, 6537.
69. Dyson, N., Guida, P., Munger, K., and Harlow, E. (1992). Homologous sequences in adenovirus E1A and the human papillomavirus E7 proteins mediate interaction with the same set of cellular proteins. *J. Virol.* **66**, 6893.
70. Tommasino, M., Adamczewski, J. P., Carlotti, F. *et al.* (1993). HPV 16 E7 protein associates with the protein kinase $p33^{CDK2}$ and cyclin A. *Oncogene* **8**, 195.
71. Werness, B. A., Levine, A. J., and Howley, P. M. (1990). Association of human papillomavirus types 16 and 18 E6 proteins with p53. *Science* **248**, 76.
72. Huibregtse, J. M., Scheffner, M., and Howley, P. M. (1993). Localization of the E6-AP regions that direct human papillomavirus E6 binding, association with p53, and ubiquitination of associated proteins. *Mol. Cell. Biol.* **13**, 4918.
73. Kim, M. S., Shin, K.-H., Baek, J.-H., Cherrick, H. M., Park, N.-H. (1993). HPV-16, tobacco-specific N-nitrosamine, and N-methyl-N′-nitro-N-nitrosoguanidine in oral carcinogenesis. *Cancer Res.* **53**, 4811.
74. Peoraro, G., Lee, M., Morgan, D., and Defendi, V. (1990). Evolution of in vitro transformation and tumorigenesis of HPV 16 and HPV 18 immortalized primary cervical epithelial cells. *Am. J. Pathol.* **138**, 1.
75. Hurlin, P. J., Kaur, P., Smith, P. P., Perez-Reyes, N., Blanton, R. A., and McDougall, J. K. (1991). Progression of human papillomavirus type 19-immortalized human keratinocytes to a malignant phenotype. *Proc. Natl. Acad. Sci. USA* **88**, 570.
76. Garrett, L. R., Perez-Reyes, N., Smith, P. P., and McDougall, J. K. (1993). Interaction of HPV-18 and nitrosomethylurea in the induction of squamous cell carcinoma. *Carcinogenesis* **14**, 329.
77. Hemminki, K., and Dong, C. (1999). Detection of Epstein-Barr virus in invasive breast cancers. *J. Natl. Cancer Inst.* **91**(24), 2126.
78. Henderson, B. E., Louie, E., Soo H. J., *et al.* (1977). Risk factors associated with nasopharyngeal carcinoma. *N. Engl. J. Med.* **295**, 1101–1106.
79. Nam, J., McLaughlin, J. K., and Blot W. J. (1992). Cigarette smoking, alcohol, and nasopharyngeal carcinoma: A case-control study among U.S. whites. *J. Natl. Cancer Inst.* **84**, 619–622.
80. Lo, Y. M., Chan, L. Y., Chan, A. T. *et al.* (1999). Quantitative and temporal correlation between circulating cell-free Epstein-Barr virus DNA and tumor recurrence in nasopharyngeal carcinoma. *Cancer Res.* **59**(21), 5452–5455.
81. Chen, M. R., Yang, J. F., Hsu, T. Y. *et al.* (1996). Use of bacterially expressed GST/EBNA-1 fusion proteins for detection of antibodies in sera from patients with nasopharyngeal carcinoma and healthy donors. *Chung Hua Min Kuo Wei Sheng Wu Chi Mien I Hsueh Tsa Chih* **29**(2), 65–79.
82. Shimakage, M., Sasagawa, T., Yoshino, K. *et al.* (1999). Expression of Epstein-Barr virus in mesopharyngeal and hypopharyngeal carcinomas. *Hum. Pathol.* **30**(9), 1071–1076.
83. Wurapa, A. K., Luque, A. E., and Menegus, M. A. (1999). Oral hairy leukoplakia: A manifestation of primary infection with Epstein-Barr virus? *Scand. J. Infect. Dis.* **31**(5), 505–506.
84. Xu, J., Menezes, J., Prasad, U., and Ahmad, A. (1999). Elevated serum levels of transforming growth factor β1 in Epstein-Barr virus-associated nasopharyngeal carcinoma patients. *Int. J. Cancer* **84**, 396–399.
85. Shillitoe, E. J., Greenspan, D., Greenspan, J. S., and Silverman, S., Jr. (1986). Five-year survival of patients with oral cancer and its associations with antibody to herpes simplex virus. *Cancer* **58**, 2256.
86. Shillitoe, E. J., Hwang, C. B., Silverman, S., Jr., and Greenspan, J. S. (1986). Examination of oral cancer tissue for the presence of the proteins ICP4, ICP5, ICP6, ICP8, and gB of herpes simplex virus type 1. *J. Natl. Cancer Inst.* **76**, 371.
87. Lund, V. J., and Howard, D. J. (1990). Head and neck cancer in the young: A prognostic conundrum? *J. Laryngol. Otol.* **104**, 544–548.
88. Harris, C. C. (1989). Interindividual variation among humans in carcinogen metabolism, DNA adduct formation and DNA repair. *Carcinogenesis* **10**, 1563–1566.
89. Houck, J. R., Romano, P. J., Bartholomew, M., *et al.* (1992). Do histocompatibility antigens influence the risk of head and neck carcinoma? *Cancer* **69**, 2327–2332.
90. Schantz, S. P., Hsu, T. C., Ainslie, N., and Moser, R. P. (1989). Young adults with head and neck cancer express increased susceptibility to mutagen-induced chromosome damage. *JAMA* **262**, 3313–3315.

91. Shields, P. G., and Harris, C. C. (1991). Molecular epidemiology and the genetics of environmental cancer. *JAMA* **266**, 681–687.
92. Heckbert, S. R., Weiss, N. S., Hornung, S. K., *et al.* (1992). Glutathione S-transferase and epozide hydroxylase activity in human leukocytes in relation to the risk of lung cancer and other smoking-related cancers. *J. Natl. Cancer Inst.* **84**, 414–422.
93. Spitz, M. R., Fueger, J. J., Beddingfield, N. A., *et al.* (1989). Chromosome sensitivity to bleomycin-induced mutagenesis: An independent risk factor for upper aerodigestive tract cancers. *Cancer Res.* **49**, 4626–4628.
94. Schantz, S. P., Spitz, M. R., and Hsu, T. C. (1990). Mutagen sensiti vity in patients with head and neck cancers: A biologic marker for risk of multiple primary malignancies. *J. Natl. Cancer Inst.* **82**, 1773–1775.
95. Slaughter, D. P., Southwick, H. W., and Smejkal, W. (1953). Field cancerization in oral stratified squamous epithelium. *Cancer* **6**(5), 963–968.
96. Meyskens, F. L., Jr. (1997). Micronutrients. *In* "Cancer: Principles and Practice of Oncology" (V. T. DeVita, S. Hellman, and S. A. Rosenberg eds.), pp. 573–584. Lippincott-Raven, Philadelphia.
97. Clayman, G. L., Lippman, S. M., Laramore, G. E., and Hong, W. K. (1997). Head and neck cancer. *In* "Cancer Medicine" (J. F. Holland, R. C. Bast, D. L. Morton, E. Frei, D. W. Kufe, and R. R. Weichselbaum, eds.), pp. 1645–1710. Williams and Wilkins, Baltimore.
98. Lippman, S. M., Lee, J. S., Lotan, R., Hittelman, W., Wargovich, M. J., and Hong, W. K. (1990). Biomarkers as intermediate endpoints in chemoprevention trials. *J. Natl. Cancer Inst.* **82**, 555.
99. Gao, X., Fisher, S. G., Mohideen, N., *et al.* (2000). Second primary cancers in patients with laryngeal cancer: A population-based study. *Proc. Am. Soc. Clin. Oncol.* **19**, 414a. [Abstract]
100. Tucker, M. A., Murray, N., Shaw, E. G., *et al.* (1997). Second primary cancers related to smoking and treatment of small-cell lung cancer. *J. Natl. Cancer Inst.* **89**, 1782–1788.
101. Sporn, M. B., and Roberts, A. B. (1985). What is a retinoid? *Ciba Found. Symp.* **113**, 1–5.
102. Califano, J., van der Riet, P., Clayman, G., *et al.* (1996). A genetic progression model for head and neck cancer; implications for field cancerization. *Cancer Res.* **56**, 2488–2492.
103. Bedi, G. C., Westra, W. H., Gabrielson, E., *et al.* (1996). Multiple head and neck tumors: Evidence for a common clonal origin. *Cancer Res.* **56**, 2484–2487.
104. Leong, P. P., Rezai, B., Koch, W. M., *et al.* (1998). Distinguishing second primary tumors from lung metastases in patients with head and neck squamous cell carcinoma. *J. Natl. Cancer Inst.* **90**, 972–977.
105. Chung, K. Y., Mukhopadhyay, T., Kim, J., *et al.* (1993). Discordant p53 mutations in primary head and neck cancer and corresponding second primary cancers of the upper aerodigestive tract. *Cancer Res.* **53**, 1676–1683.
106. Benner, S. E., Lippman, S. M., Hong, and W. K. (1992). Prevention of second head and neck cancers. *Semin. Radiat. Oncol.* **2**, 206–212.
107. Shklar, G. (1986). Oral leukoplakia. *N. Engl. J. Med.* **315**, 1544–1545.
108. Mashberg, A., and Samit, A. M. (1989). Early detection, diagnosis, and management of oral and oropharyngeal cancer. *CA Cancer J. Clin.* **39**, 67–88.
109. Waldron, C. A., and Shafer, W. G. (1975). Leukoplakia revisited: A clinicopathologic study 3256 oral leukoplakias. *Cancer* **36**, 1386–1392.
110. WHO Collaborating Centre for Oral Precancerous Lesions (1978). Definition of leukoplakia and related lesions: An aid to studies on oral precancer. *Oral Surg. Oral Med. Oral Pathol.* **46**, 518–539.
111. Silverman, S., and Shillitoe, E. J. (1990). Etiology and predisposing factors. *In* "Oral Cancer" (S. Silverman ed.), 3rd Ed., pp. 7–39. American Cancer Society, Atlanta, GA.
112. Chiesa, F., Tradati, N., Marazza, M., *et al.* (1992). Prevention of local relapses and new localizations of oral leukoplakias with synthetic retinoids renretinide (4-HPR): Preliminary results. *Oral Oncol. Eur. J. Cancer* **28B**, 97–102.
113. Silverman, S., Gorsky, M., and Lozada, F. (1984). Oral leukoplakia and malignant transformation: A follow-up study of 257 patients. *Cancer* **53**, 563–568.
114. Wulf, K. (1957). Zur vitamin A behandlung der leukoplkien. *Arch. Klin. Exp. Derm.* **206**, 495–498.
115. Silverman, S., Renstrup, G., and Pindborg, J. J. (1963). Studies in oral leukoplakias. III. Effects of vitamin A comparing clinical, histopathologic, cytologic, and hematologic responses. *Acta Odont. Scand.* **21**, 271–292.
116. Silverman, S., Eisenberg, E., and Restrup, G. (1965). A study of the effects of high-doses of vitamin A on oral leukoplakia (hyperkeratosis), including toxicity, liver function, and skeletal metabolism. *J. Oral Ther. Pharmacol.* **2**, 9–23.
117. Koch, H. F. (1981). Effect of retinoids on precancerous lesions of oral mucosa. *In* "Retinoids, Advances in Basic Research and Therapy" (C. E. Orfanos, O. Braun-Falco, E. M. and Farber, eds.), pp. 307–312. Springer-Verlag, Berlin.
118. Stich, H. F., Hornby, A. P., Mathew, B., *et al.* (1988). Response of oral leukoplakias to the administration of vitamin A. *Cancer Lett.* **40**, 93–101.
119. Stich, H. F., Rosin, M. P., and Vallejera, M. O. (1984). Reduction with vitamin A and beta-carotene administration of the proportion of micronucleated buccal cells in Asian betel nut and tobacco chewers. *Lancet* **1**, 1204–1206.
120. Stich, H. F., Rosin, M. P., Hornby, A. P., *et al.* (1988). Remission of oral leukoplakias and micronuclei in tobacco/betel quid chewers treated with beta-carotene and with beta-carotene plus vitamin A. *Int. J. Cancer* **42**, 195–199.
121. Garewal, H. S., Meyskens, F. L., Killen, D., *et al.* (1990). Response of oral leukoplakia to beta-carotene. *J. Clin. Oncol.* **8**, 1715–1720.
122. Malaker, K., Anderson, B. J., Beecroft, W. A., and Hodson, D. I. (1992). Management of oral mucosal dysplasia with beta-carotene retinoic acid: A pilot cross-over study. *Cancer Detect. Prev.* **15**, 335–340.
123. Toma, S., Benso, S., Albanese, E., *et al.* (1992). Treatment of oral leukoplakia with beta carotene. *Oncology* **42**, 77–81.
124. Toma, S., Coialbu, T., Collecchi, P., *et al.* (1990). Aspetti biologici e prospettive applicative della chemioprevenzione nel cancro delle vie aerodigestive superiori. *Acta. Otorhinol. Ital.* **10**, 41–54.
125. Benner, S. E., Winn, R. J., Lippman, S. M., *et al.* (1993). Regression of oral leukoplakia with alpha-tocopherol: A Community Clinical Oncology Program (CCOP) chemoprevention study. *J. Natl. Cancer Inst.* **85**, 44–47.
126. Ryssel, H. J., Brunner, K. W., and Bollag, W. (1971). Die perorale Anwedung von Vitamin-A-Saure bie Leukoplakien, Hyperkeratosen und Plattenepithelkarzinomen: Ergebnisse und Vertaglichkeit. *Schweiz Med. Wochenschr.* **101**, 1027–1030.
127. Stuttgen, G. (1975). Oral vitamin A acid therapy. *Acta Derm. Venereol. Suppl.* **74**, 174–179.
128. Raque, C. J., Biondo, R. V., Keeran, M. G., *et al.* (1975). Snuff dippers keratosis (snuff-induced leukoplakia). *South Med. J.* **68**, 565–568.
129. Koch, H. F. (1978). Biochemical treatment of precancerous oral lesions: The effectiveness of various analogues of retinoic acid. *J. Oral Maxillofac. Surg.* **6**, 59–63.
130. Cordero, A. A., Allevato, M. A. J., Barclay, C. A., *et al.* (1981). Treatment of lichen planus and leukoplakia with the oral retinoid RO 10-9359. *In* "Retinoids, Advances in Basic Research and Therapy" (C. E. Orfanos, O. Braun-Falco, and E. M. Farber eds.), pp. 273–278. Springer-Verlag, Berlin.
131. Shah, J. P., Strong, E. W., DeCosse, J. J., *et al.* (1983). Effect of retinoids on oral leukoplakia. *Am. J. Surg.* **146**, 466–470.
132. Han, J., Jiao, L., Lu, Y., *et al.* (1990). Evaluation of N-4-(hydroxycarbophenyl)retinamide as a cancer agent. *In Vivo* **4**, 153–160.
133. Hong, W. K., Endicott, J., Itri, L. M., *et al.* (1986). 13-cis-retinoic acid in the treatment of oral leukoplakia. *N. Engl. J. Med.* **315**, 1501–1505.

134. Lippman, S. M., Batsakis, J. G., Toth, B. B., *et al.* (1993). Comparison of low-dose isotretinoin with beta-carotene to prevent oral carcinogenesis. *N. Engl. J. Med.* **328**, 15–20.
135. The Alpha-Tocopherol, Beta Carotene Cancer Prevention Study Group (1994). The effect of vitamin E and beta-carotene on the incidence of lung cancer and other cancers in male smokers. *N. Engl. J. Med.* **330**, 1029–1035.
136. Omenn, G. S., Goodman, G. E., Thornquist, M. D., *et al.* (1996). Effects of a combination of beta-carotene and vitamin A on lung cancer and cardiovascular disease. *N. Engl. J. Med.* **334**, 1150–1155.
137. Hennekans, C. H., Buring, J. E., Manson, J. E., *et al.* (1996). Lack of long-term supplementation with beta-carotene on the incidence of malignant neoplasms and cardiovascular disease. *N. Engl. J. Med.* **334**, 1145–1149.
138. Goodman, M., Morgan, R. W., Ray, R., *et al.* (1999). Cancer in asbestos-exposed occupational cohorts: A meta-analysis. *Cancer Causes Control* **10**(5), 453–465.
139. Boice, J. D., and Fraumeni, J. F. (1985). Second cancer following cancer of the respiratory system in Connecticut, 1935–1982. *Natl. Cancer Inst. Monogr.* **68**, 83–98.
140. Gluckman, J. L., and Crissman, J. D. (1983). Survival rates in 548 patients with multiple neoplasms of the upper aerodigestive tract. *Laryngoscope* **93**, 71–74.
141. De Vries, N., and Snow, G. B. (1986). Multiple primary tumours in laryngeal cancer. *J. Laryngol. Otol.* **100**, 915–918.
142. Yellin, A., Hill, L. R., and Benfield, J. R. (1986). Bronchogenic carcinoma associated with upper aerodigestive cancers. *J. Thorac. Cardiovasc. Surg.* **91**, 674–683.
143. Vokes, E. E., Weichselbaum, R. R., Lippman, S. M., and Hong, W. K. (1993). Head and neck cancer. *N. Engl. J. Med.* **328**, 184–193.
144. Lippman, S. M., and Hong, W. K. (1993). Not yet standard: Retinoids versus second primary tumors. *J. Clin. Oncol.* **11**, 1204–1207.
145. Lippman, S. M., and Hong, W. K. (1989). Second malignant tumors in head and neck squamous cell carcinoma: The overshadowing threat for patients with early stage disease. *Int. J. Radiat. Oncol. Biol. Phys.* **17**, 691–694.
146. Larson, J. T., Adams, G. L., and Fattah, H. A. (1990). Survival statistics for multiple primaries in head and neck cancer. *Otolaryngol. Head Neck Surg.* **103**, 14–24.
147. Tepperman, B. S., and Fitzpatrick, P. J. (1981). Second respiratory and upper digestive tract cancers after oral cancer. *Lancet* **9**, 547–549.
148. McDonald, S., Haie, C., Rubin, P., Nelson, D., and Divers, L. D. (1989). Second malignant tumors in patients with laryngeal carcinoma: Diagnosis, treatment and prevention. *Int. J. Radiat. Oncol. Biol. Phys.* **17**, 457–465.
149. Warren, S., and Gates, O. (1932). Multiple primary malignant tumors: A survey of the literature and statistical study. *Am. J. Cancer* **16**(4), 1358–1403.
150. Hong, W. K., Lippman, S. M., Itri, L. M., *et al.* (1990). Prevention of second primary tumors with isotretinoin in squamous-cell carcinoma of the head and neck. *N. Engl. J. Med.* **323**, 795–801.
151. Benner, S. E., Lippman, S. M., and Hong, W. K. (1992). Current status of chemoprevention of head and neck cancer. *Oncology* **6**, 61–66.
152. Benner, S. E., Pajak, T. F., Stetz, J., *et al.* (1994). Toxicity of isotretinoin in a chemoprevention trial to prevent second primary tumors following head and neck cancer. *J. Natl. Cancer Inst.* **86**, 1799–1801.
153. Khuri, F. R., Lee, J. J., Winn, R. J., *et al.* (1999). Interim analysis of randomized chemoprevention trial of HNSCC. *In* "Programs and Abstracts of the American Society of Clinical Oncology," p. 1503. [Abstract]
154. Kim, E. S., Khuri, F. R., Lee, J. J., Winn, R., Cooper, J. M., Fu, K., Pajak, T. F., Lippman, S. M., Vokes, E. E., Broxson, A., Williams, B., and Hong, W. K. (2000). Second primary tumor incidence related to primary index tumor and smoking status in a randomized chemoprevention study of head and neck squamous cell cancer. *In* Proceedings of the American Society of Clinical Oncology [Abstract]
155. de Vries, N., van Zandwijk, N., and Pastorino, U. (1999). Chemoprevention of head and neck and lung (pre)cancer. *Recent Results Cancer Res.* **151**, 13–25.
156. Khuri, F. R., Kim, E. S., Lee, J. J., *et al.* (2001). The impact of smoking status, disease stage, and index tumor site on second primary tumor incidence and tumor recurrence in the head and neck retinoid chemoprevention trial. *Cancer Epidemiol. Biomark. Prev.* **10**(8), 823–829.
157. de Vries, N., van Zandwijk, N., and Pastorino, U. (1994). Chemoprevention of second primary tumours in head and neck cancer in Europe: EUROSCAN. *Eur. J. Cancer B Oral Oncol.* **30B**(6), 367–368.
158. Lippman, S. M., Lee, J. J., Karp, D. D., *et al.* (2001). Randomized phase III intergroup trial of isotretinoin to prevent second primary tumors in stage I non-small cell lung cancer. *J. Natl. Cancer Inst.* **93**(8), 605–618.
159. Mayne, S. T., Cartmel, B., Baum, M., *et al.* (2001). Randomized trial of supplemental beta-carotene to prevent second head and neck cancer. *Cancer Res.* **61**(4), 1457–1463.
160. Shin, D. M., Khuri, F. R., Murphy, B., *et al.* (2001). Combined interferon-alfa, 13 cis-retinoic acid, and alpha-tocopherol in locally advanced head and neck squamous cell carcinoma: Novel bioadjuvant phase II trial. *J. Clin. Oncol.* **19**(12), 3010–3017.

CHAPTER

20

Statistical Methods for Biomarker Analysis for Head and Neck Carcinogenesis and Prevention

J. JACK LEE

Department of Biostatistics
The University of Texas M. D. Anderson Cancer Center
Houston, Texas 77030

I. Introduction 287
II. Type of Biomarker Measures 288
III. Standard Statistical Methods for Analyzing Biomarkers 289
IV. Distribution of Continuous Biomarkers 290
V. Choosing Cut Points for Continuous Biomarkers 291
VI. Analysis of Surrogate End Point Biomarkers 292
VII. Predicting Cancer Development Using Clinical, Histological, Epidemiologic, and Multiple Biomarker Information 295
A. Single-Covariate Cox Model Analysis on Medical, Demographic, and Epidemiologic Variables 295
B. Combined Predictive Effect of Histology and Prior Cancer History on Cancer Development 296
C. Biomarkers as Predictors for Cancer: One Covariate Case 297
D. Biomarkers as Predictors for Cancer: Multiple Covariates Case 298
E. Combined Biomarker Score and Patient Characteristics as Predictors for Cancer: Multiple Covariates Case 298
F. Sensitivity Analysis on Patients with Missing Biomarker Information 300
G. Recursive Partitioning for the Classification of Cancer Risk: Multiple Covariates Case 300
VIII. Summary and Design Considerations for Biomarker-Integrated Translational Studies 301
References 302

I. INTRODUCTION

Although the etiology of cancer is not fully understood, head and neck cancer provides an excellent model system for biomarker-integrated carcinogenesis and cancer prevention studies. The advantage of using the head and neck model system for biomarker research includes (1) well-known risk factors, such as the use of tobacco, alcohol, or, particularly in southeast Asia, betel nut chewing; (2) easily identifiable premalignant lesions, such as oral leukoplakia, oral erythroplakia, and hoarseness-related laryngeal lesions; (3) a well-established histological progression model from normal to hyperplasia, mild dysplasia, moderate dysplasia, severe dysplasia, to carcinoma *in situ*, and cancer; and (4) readily accessible anatomy for biopsy to obtain tissue samples. Similar to most other malignancies, head and neck cancer takes many years to develop. This long duration gives a window of opportunity to monitor, intervene, delay, or reverse the carcinogenesis. Several active agents for treating head and neck premalignant lesions, such as 13-*cis*-retinoic acid (13cRA), *N*-(4-hydroxyphenyl) retinamide (4-HPR), interferon α, and their combinations, have been reported in the literature [1–3]. With successful chemoprevention strategies and accessible tissues for biopsy to evaluate the treatment effect, the head and neck cancer model system provides an ideal setting for biomarker-integrated, translation chemoprevention studies [4–6].

The advance of biological science has fueled the discovery and characterization of many molecular, cellular, and genetic biomarkers in the past decades. In addition to the standard medical, demographic, epidemiologic, and histological information, increased knowledge of biomarkers, such as p53 protein and mutation, DNA content, chromosome polysomy, loss of heterozygosity (LOH), proliferating cell nuclear antigen (PCNA) or Ki-67, tumor growth factor (TGF)-α, retinoic acid nuclear acid receptor-β (RAR-β), DNA repair capacity, micronuclei, p16 methylation, and mutagen sensitivity assay, has enriched the field of cancer research tremendously [7–20]. The three main roles of biomarkers are (1) as intermediate or surrogate end points, (2) for risk assessment, and (3) as covariates or prognostic factors. This chapter describes the type of biomarkers and statistical methods applicable to different types of biomarkers. Examples are given to illustrate statistical methods.

To facilitate the discussion, various statistical methods will be illustrated in the following sections by using data obtained from a retinoid chemoprevention trial in patients with oral premalignant lesions (OPLs). While the main result of this OPL trial was reported in the literature [1], we include a brief summary, as follows. From 1988 to 1991, 70 advanced OPL patients were enrolled in a chemoprevention trial with a 3-month high-dose 13cRA induction (1.5 mg/kg/day) followed by a 9-month maintenance treatment with either low-dose 13cRA (0.5 mg/kg/day) or β-carotene (30 mg/day). Biopsies were taken at baseline and at 3-month and 12-month visits. Due to patient refusal, loss to follow-up, and limited amount of tissues, biomarkers have not been analyzed on all participants, but only on the available tissue samples. Because we use data from an updated database, the result reported herein may be slightly different from what was reported previously in the literature.

II. TYPE OF BIOMARKER MEASURES

Knowing the type of biomarker is essential for choosing the proper statistical method for biomarker analysis. Biomarkers can be measured qualitatively, quantitatively, or semiquantitatively. Qualitative measures can produce either dichotomous or categorical outcomes (Table 20.1). One of the simplest and most commonly used measures is to summarize the biomarker by a dichotomous or binary outcome, such as (+) or (–) p16 methylation, or the presence or absence of a p53 mutation. Qualitative biomarker measures can also produce categorical variables, such as the result of LOH analysis that can be described as noninformative, retention, loss, or shift. The possible outcomes in categorical variables do not necessarily have a natural order to their relationships.

Immunohistochemistry (IHC) staining is a widely applied method that stains cells with a specific monoclonal antibody of interest. Typical results of IHC can be summarized in semiquantitative measures, such as the staining intensity or percent positivity in the cells counted. Staining intensity is summarized either by reading the intensity in the majority of cells or in the highest stained region as 0, 1+, 2+, or 3+, which makes it an ordinal categorical variable. The percentage of cells stained positive can be considered as a continuous variable if the denominator is large and the value is away from 0 or 100%. True quantitative measures, such as the serum or tissue 4-HPR levels in a chemoprevention trial, produce continuous variables. Table 20.1 shows that continuous variables can be regrouped into ordinal categorical or dichotomous variables. Polytomous categorical variables, either ordinal or not, can be reduced further to dichotomous variables by combining the groups. Qualitative variables are often classified as categorical variables because no particular order may exist. However, quantitative variables typically are considered as continuous variables.

Table 20.2 gives an example of IHC staining for a certain marker, such as the p53 protein, in five subjects by the percentage of cells stained. Subject A has 100% of the cells stained negative. Subject B has 80% of the cells stained negative and 20% of the cells stained 1+, etc. There are at least four different methods for summarizing data.

1. By the positivity or negativity of the staining. In Table 20.2, there is only one subject with negative staining of all cells (subject A). The other four subjects have some cells with positive staining.
2. By the level of staining intensity. If the highest staining intensity is considered, the maximum staining intensity is 0 in subject A, 1+ in subject B, 2+ in subject C, and 3+ in subjects D and E, respectively.
3. By percentage of cells stained positive (labeling index, LI). The percentage positivity in each of the five

TABLE 20.1 Different Types of Biomarker Measures and the Resulting Outcome Variables

Type of biomarker measures	Type of outcome variables			
	Dichotomous	Polytomous-categorical	Polytomous-ordinal categorical	Continuous
Qualitative	Presence/absence of a certain trait or mutation	Different cell type: round, spindle, irregular, etc. LOH: noninformative, retention, loss, or shift		
Semiquantitative	Positive/negative staining		Staining intensity: 0, 1+, 2+, 3+	Percent positivity: % of cells stained (+)
Quantitative	Grouped marker: high/low		Grouped marker: high/medium/low	Serum or tissue 4-HPR levels

TABLE 20.2 Example of Immunohistochemistry Staining by Percentage of Cells Stained in Five Subjects

Intensity	Subject A	B	C	D	E
0+	100	80	80	20	0
1+	0	20	0	30	10
2+	0	0	20	40	10
3+	0	0	0	10	80

TABLE 20.3 Summary of Immunohistochemistry Staining in Five Subjects

Intensity	Positivity (%) 0–5%	6–25%	26–50%	51–75%	76–100%
0+					A
1+		B			
2+		C	D		
3+					E

subjects is 0, 20, 20, 80, and 100%, corresponding to a LI of 0, 0.2, 0.2, 0.8, and 1.0, respectively.

4. By the weighted mean index (WMI). The weighted mean index computes the weighted mean of the percentage of cell stained positive, weighted by intensity. The WMI is 0, 0.2, 0.4, 1.4, and 2.7, respectively.

When the intensity and percentage positivity are considered together, e.g., by examining the intensity and percent positivity in the majority of positively stained cells, Table 20.3 shows that subjects can be classified into groups described as "no expression" (no shading—subject A), "low expression" (with light gray shading—subjects B and C), or "high expression" (with dark gray shading—subjects D and E). The same result can be achieved by grouping subjects based on their WMI with the cut points of 0 and 1. Subjects with WMI = 0 belong to the no expression group, a value of $0 < \text{WMI} \leq 1$ ranks subjects in the low expression group, and WMI > 1 places subjects in the high expression group.

The choice of the cut point may vary from marker to marker, but a consistent grouping method needs to be sought and the results need to be validated by other studies. The choice of a cut-off point will be discussed further in Section V.

III. STANDARD STATISTICAL METHODS FOR ANALYZING BIOMARKERS

Table 20.4 lists the standard statistical methods, which can be applied to biomarker analysis. The association between two categorical variables, e.g., histology (hyperplasia or dysplasia) and p53 protein expression (low or high) in patients with oral leukoplakia, can be summarized in a

TABLE 20.4 Standard Statistical Methods for Biomaker Analysis

Statistical method	Type of data: Categorical	Continuous non Gaussian distributed	Continuous Gaussian distributed	Survival
Describe one group	Proportion	Median, percentiles	Mean (SD)	Kaplan–Meier curve
Compared one group to a preset value	χ2-square test, Fisher's exact test	Sign test or signed rank test	One-sample t test	Test for median survival or survival rate at a fixed time
Compared two paired groups	McNemar's test	Sign test or signed rank test	Paired t test	Methods for censored paired data
Compared paired data with several measurements	Cochran–Mantel–Haenszel test	Friedman's test	Repeated measures ANOVA	Methods for censored paired data
Compared two unpaired groups	χ^2 test, Fisher's exact test	Wilcoxon rank sum test	Two-sample t test	Log-rank test
Compared several unmatched groups	χ^2 test, Fisher's exact test	Kruskal–Wallis test	One-way ANOVA	Log-rank test
Model multiple prognostic variables	Logistic regression	Nonparametric regression	Multiple regression	Cox regression
Association among variables	χ^2 test, Fisher's exact test	Spearman correlation	Pearson correlation	Correlation in parametric survival models

2×2 contingency table. A χ^2 test can be applied if the sample size is large such that the test statistics can be approximated accurately by the χ^2 distribution. If the sample size is small, especially when the expected count of any cell is less than five under the null hypothesis of no association, Fisher's exact test should be applied.

When a biomarker is measured over time, e.g., when the p53 status is measured at baseline and after treatment, it results in paired data. Because p53 status is measured twice within the same patient, correlation between the measurements exists and appropriate statistical methods for analyzing paired data should be applied accordingly. When the correlation is properly accounted for, paired design can be statistically more powerful than the unpaired design because each patient serves as his/her own control.

IV. DISTRIBUTION OF CONTINUOUS BIOMARKERS

The first step of analyzing a continuous biomarker is to characterize its distribution by plotting data and computing summary statistics. Several functions and procedures are readily available in standard statistical software packages, such as SAS, S-PLUS, and SPSS. In addition, BLiP plot—a versatile, one-dimensional distribution plot written in S-PLUS—was developed to provide standard plots as well as customized plots [21]. We use the baseline chromosome polysomy in the OPL study for illustration. Chromosome polysomy (CP) is defined as the percentage of cells having three or more copies of chromosome 9 in the lesion by the technique of chromosome *in situ* hybridization (CISH) [10].

Forty patients had baseline CP measured in the OPL study. The summary statistics of CP are listed as follows.

N	40
Minimum	0.2
First quartile	1.375
Median	3.25
Third quartile	8.35
Maximum	39.5
Mean	6.642
SD	8.393

Figure 20.1 shows distribution plots using the BLiP function. Figure 20.1A gives the histogram of CP by an interval

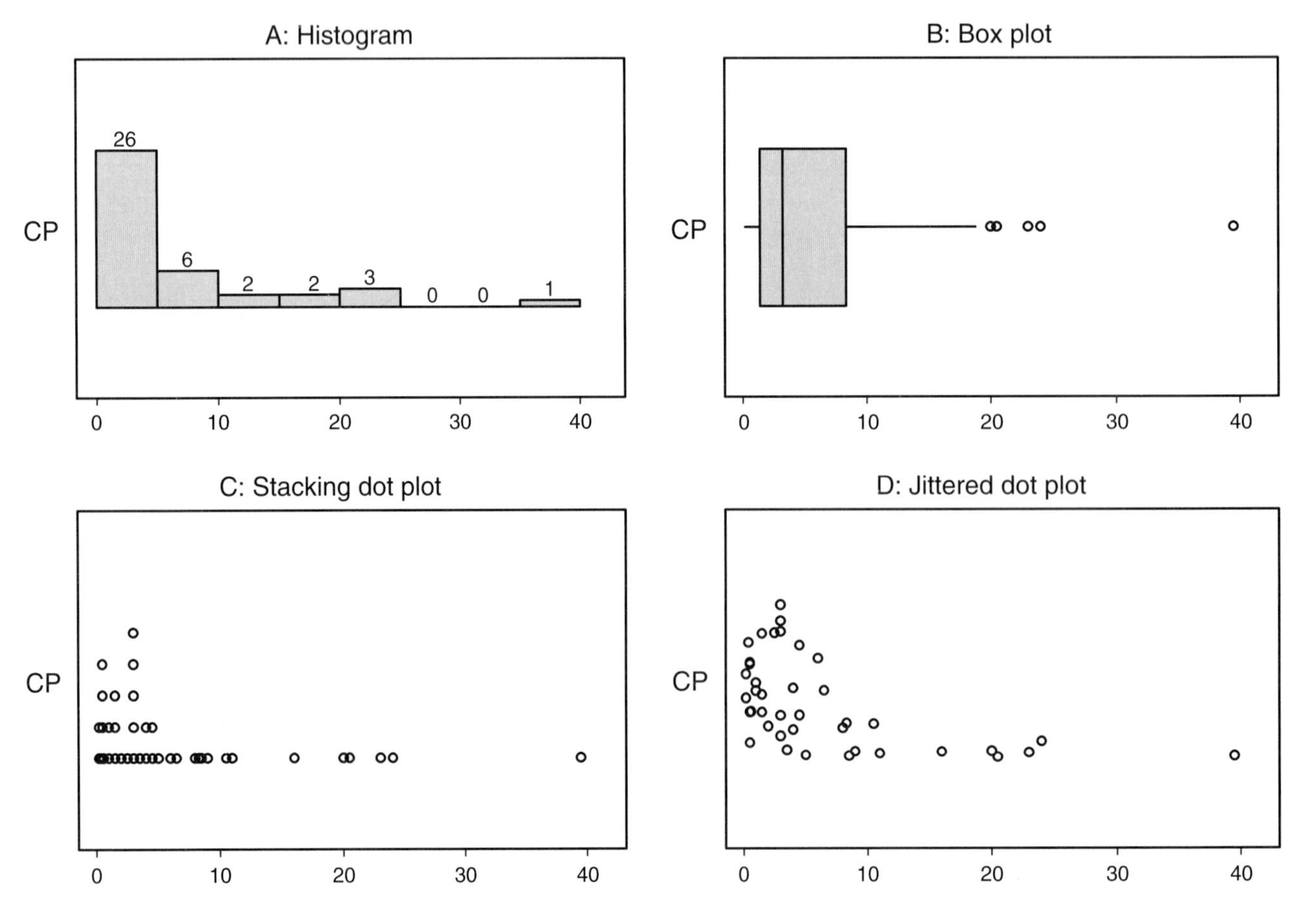

FIGURE 20.1 Distribution plots for chromosome polysomy (CP) in 40 patients with oral leukoplakia.

of 5. Most patients have low CP values with 26 between 0 and 5, and 6 between 5 and 10, etc. Only 1 patient has CP between 35 and 40. Figure 20.1B shows the box plot of the data. The middle shaded box indicates the range of the middle 50% of the data. The lower end, middle bar, and upper end of the box correspond to the 25th, 50th (median), and 75th percentiles. The lines or whiskers extending from either end of the box cover the vast majority of the data (about 99% if data are Gaussian distributed), with outliers indicated by separate dots. With a moderate or small sample size, it is beneficial to see the exact value of each data point. Figures 20.1C and 20.1D show two dot plots by stacking the points or jittering the points on the vertical axis according to the range of the density plot. The stacking dot plot is particularly useful when ties in the data are observed. The jittered plot separates the ties by jittering points. Jittered plots show the shape of the distribution and also identify the individual data point. While histograms and box plots present the information on grouped data, and dot plots reveal information of individual data, all four plots show clearly that the distribution of CP is concentrated heavily on the left (small values) and skewed to the right. Combining both the sumary statistics and the distribution plot can give a comprehensive assessment of data distribution and can be used for data screening and outlier identification. This procedure should be applied routinely as the first step of data analysis before model fitting and hypothesis testing.

V. CHOOSING CUT POINTS FOR CONTINUOUS BIOMARKERS

Many standard statistical methods, such as *t* test, ANOVA, and linear regression analysis, require that data are Gaussian distributed. Although the mean of a variable converges to Gaussian distribution as the sample size increases by the central limit theorem, Gaussian approximation may not be satisfactory with limited sample size. When the distribution is highly skewed, such as the CP example given earlier, outliers with extreme values become highly influential and can bias the statistical inference. One approach is to use nonparametric methods. Another common practice is to group the continuous variable into several categories, e.g., normal/abnormal, low/high, or low/medium/high groups, and analyze data based on the grouped variable. One main advantage of using the grouped variable is to gain robustness in inference making. Working with grouped data can also simplify the analysis and make the result easily understandable. For example, after grouping, all data can be shown in the contingency table format and analyzed accordingly. The ratio of odds of developing disease between abnormal versus normal marker groups, i.e., the odds ratio, can be easier understood and more meaningful than the regression coefficient obtained from analysis based on a non-Gaussian continuous variable. In addition, with grouped data, Kaplan–Meier survival curves can be plotted for different biomarker groups to provide a visual aid on the impact of the biomarker. One disadvantage of analysis based on grouped data is the loss of efficiency. When the Gaussian assumption is met, analysis based on continuous data typically will be more powerful than analysis based on grouped data. The choice of grouping data or not involves a trade-off in robustness versus efficiency. It also depends on the purpose of the study and the ultimate inference the investigator wants to make.

If grouping data is desirable, the next obvious step is to determine the number of groups and cut points. The choice may be arbitrary, but continuous biomarkers are typically grouped into two groups (negative/positive, normal/abnormal, low/high), three groups (low/medium/high), or four groups (four quartiles). Before any grouping takes place, one should first examine the distribution of the biomarker and its relationship with the primary variable(s) of interest by various distribution plots, scatter plots, trellis plots, residual plots, or martingale residual plots for survival data [22–25]. If the relationship between the outcome variable and the biomarker is monotone, the choice of number of groups depends on the inference objective. If the relationship is not monotone, careful evaluation is required to determine the underlying mechanism before deciding on whether or how to categorize the continuous variables. The common practice of categorizing data is as follows.

1. Use a commonly accepted standard cutoff when it is available. For example, a total blood cholesterol level of less than 200 mg/dl or a PSA blood level of less than 4 ng/ml is generally considered as the threshold for normal. Using the standard cutoff produces a consistent grouping scheme and makes it easy to compare results across studies.
2. Use percentiles in data. For example, case–control studies often use median, tertiles, or quartiles in the control group as cutoff values. If the analysis method is defined prospectively, it can remove the potential bias and subjectivity that occurred in the post-hoc data analysis. The percentile method, however, is still data dependent. Every study may have different cutoff values, which makes comparison between studies difficult.
3. Choose the optimal cut points. One big temptation in data analysis is to select the method yielding the most significant test result, i.e., the smallest *P* value. For example, Fig. 20.2 shows the result of using chromosome polysomy to predict cancer development in 40 leukoplakia patients. If CP is dichotomized into high and low groups, the bottom panel shows the number of patients in the high CP group (on the *y* axis) by the corresponding CP cutoff value (on the *x* axis). The top panel shows the *P* value of the log-rank test between the resulting high and low CP groups for each cut point. As can be seen, choosing a cutoff CP of 3 to 8 and 9 to 10 produces significant results at 5% level. The optimal cutoff

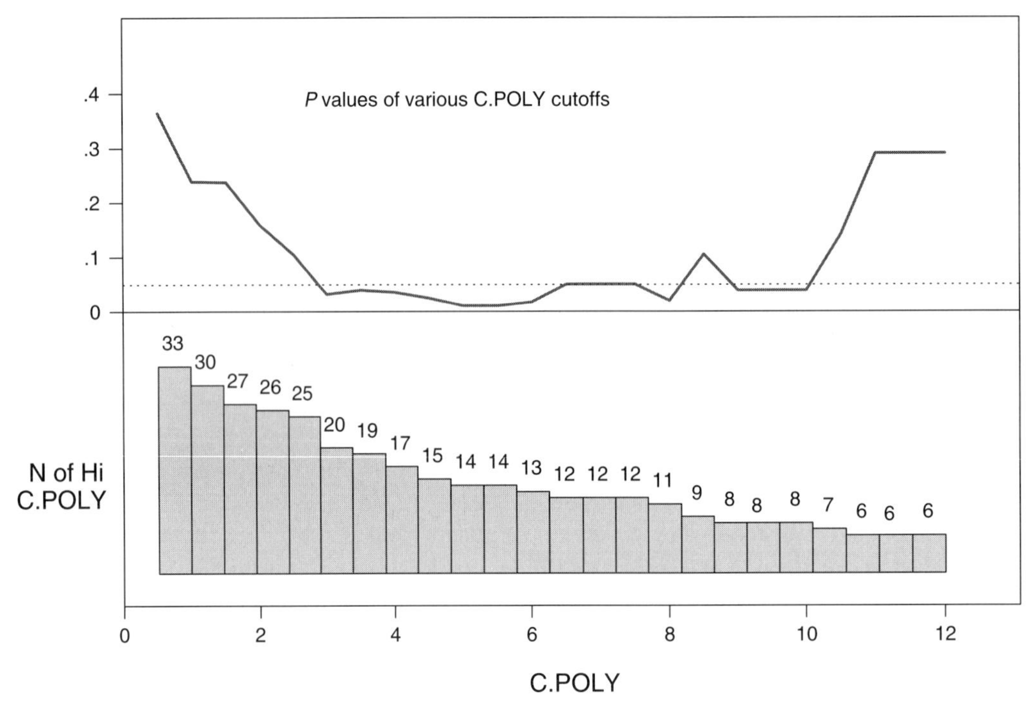

FIGURE 20.2 Choosing the cutoff value for dichotomize chromosome polysomy (CP) into low and high groups and the corresponding test result for comparing the time to cancer by CP groups using the log-rank test.

lies between 5 and 6, which produces a *P* value of less than 0.01. It has been reported in the literature that the optimal cutoff point approach is associated with a considerably inflated type I error rate [26]. Inflation of the type I error is due to repeated significance testing. Table 20.5 shows a well-known result that the type I error rate increases as the number of tests increases. Corrections were proposed to adjust the *P* value based on the optimal cut points [26]. By definition, the optimal cut point approach is data dependent. When the sample size is small, the result can be highly variable and vulnerable to random variations in data.

Whenever possible, a prespecified, standard, consistent method for defining cutoff points should be used. A haphazard, liberal choice for the cut point can severely damage the validity of statistical inference. It is also a main source of confusing and, sometimes, conflicting results reported in the literature. Sensitivity analysis by varying the cut points or by cross validation should be applied to examine the internal consistency in data. Analysis based on the post-hoc choice of cut points will need to be validated by prospectively defined cut points in future studies.

VI. ANALYSIS OF SURROGATE END POINT BIOMARKERS

This section, illustrates the use of standard statistical methods in identifying surrogate end point biomarkers (SEB) in cancer prevention studies. The previously reported OPL trial [1] will be taken as an example for assessing the use of RAR-β as a surrogate end point biomarker.

Four key aspects need to be evaluated for a biomarker to be considered as a good SEB candidate.

1. *Differential expression between normal and premalignant lesions or cancer.* A biomarker can be considered a surrogate for cancer development if it is directly involved or indirectly linked to the specific process of carcinogenesis.

TABLE 20.5 Repeated Significance Testings

Number of tests	Overall type I error
1	0.05
2	0.08
3	0.11
4	0.13
5	0.14
10	0.19
20	0.25
50	0.32
100	0.37
1000	0.53
∞	1.00

TABLE 20.6 Differential Expression of RAR-β between Normal and Oral Premalignant Lesions (OPLs)

	Normal	OPL	Total
RAR-β (−)	0 (0%)	28 (52%)	28
RAR-β (+)	7 (100%)	26 (48%)	33
Total	7 (100%)	54 (100%)	61

One indication is to show the association between biomarker and disease progression. Table 20.6 shows that RAR-β was expressed in seven out of seven normal subjects. However, only 48% of the 54 patients with OPLs expressed the RAR-β. The loss of RAR-β expression (in 52% of the patients) is significant in OPLs compared to normals (Fisher's exact test, $P=0.013$). We chose the Fisher's exact test because the expected number of RAR-β (−) and normal under the null hypothesis of no association is less than 5. Specifically, when there is no association between RAR-β expression and histology, the expected number of RAR-β (−) and normal histology is as follows:

$$\begin{aligned}&\text{Probability(RAR-}\beta\text{ negative and normal histology)}\times 61\\&=\text{probability(RAR-}\beta\text{ negative)}\\&\quad\times\text{probability(normal histology)}\times 61\\&=28/61\times 7/61\times 61=3.2.\end{aligned}$$

2. *Biomarker can be modulated by preventive treatment of choice*. After establishing the differential expression of a biomarker between normal and OPLs, the next step is to examine whether the biomarker can be modulated by the treatment of choice. If so, the biomarker can be used for evaluating treatment efficacy. Table 20.7 shows the result of 40 patients with OPLs treated with 13cRA. Data show an impressive upregulation of RAR-β expression from 42% (17/40) before the treatment to 88% (35/40) after the treatment. Specifically, 4 patients showed consistent RAR-β (−) and 16 patients had consistent RAR-β (+) expression before and after treatment; 19 patients had negative RAR-β expression at baseline but positive RAR-β expression after treatment and only 1 patient had a loss of RAR-β expression after treatment. McNemar's test shows that RAR-β is upregulated significantly after 13cRA treatment ($P<0.0001$). It is important to recognize that data of RAR-β expression before and after treatment is paired data and should be analyzed accordingly using McNemar's test. A common mistake is to take the same data but list them as shown in Table 20.8, and then use χ^2 test to test the change in RAR-β expression. One immediate indication that something is wrong is that the total sample size is inflated incorrectly (doubled).

TABLE 20.7 Upregulation of RAR-β by 13cRA

	After treatment		
Before treatment	RAR-β (−)	RAR-β (+)	Total
RAR-β (−)	4	19	23 (58%)
RAR-β (+)	1	16	17 (42%)
Total	5 (12%)	35 (88%)	40

TABLE 20.8 Upregulation of RAR-β by 13cRA (Incorrect Presentation of Data)

	Time		
RAR-β status	Before treatment	After treatment	Total
(−)	23	5	28
(+)	17	35	52
Total	40	40	80

3. *Biomarker modulation correlates with short-term response*. After observing the biomarker modulation by treatment, the next question to ask is, "What does it mean?" Does biomarker modulation mean anything in preventing cancer? Now we see that RAR-β expression can be upregulated by 13cRA treatment, but how does it relate to clinical findings? Table 20.9 shows the relationship between the change in RAR-β expression and a short-term (3-month) clinical response. Twenty-two out of 28 patients (79%) with RAR-β upregulation demonstrated a clinical response, whereas only 4 out of 12 (33%) patients without RAR-β upregulation had a clinical response. The association between RAR-β upregulation and clinical response is statistically significant (Fisher's exact test, $P=0.011$, χ^2 test, $P=0.006$).

4. *Biomarker change over time correlates with cancer development*. In addition to the correlation of biomarker modulation and short-term response, it is important to confirm that the worsening or progression of a biomarker measure is indicative of the ultimate development of cancer. With a median follow-up of 7 years, 22 of 70 (31.4%) patients in the OPL trial developed cancers in the upper aerodigestive tract. Figure 20.3A shows the Kaplan–Meier cancer-free survival plot with its 95% confidence interval (CI). The time to cancer development ranges from 2.6 months to more than 10 years after registration in the trial [27,28]. Figure 20.3B showed the distribution of time to

TABLE 20.9 RAR-β Upregulation Is Associated with Clinical Response

	Treatment result		
RAR-β Expression	No response	Response	Total
Decrease/no change	8 (67%)	4 (33%)	12 (100%)
Increase	6 (21%)	22 (79%)	28 (100%)
Total	14	26	40

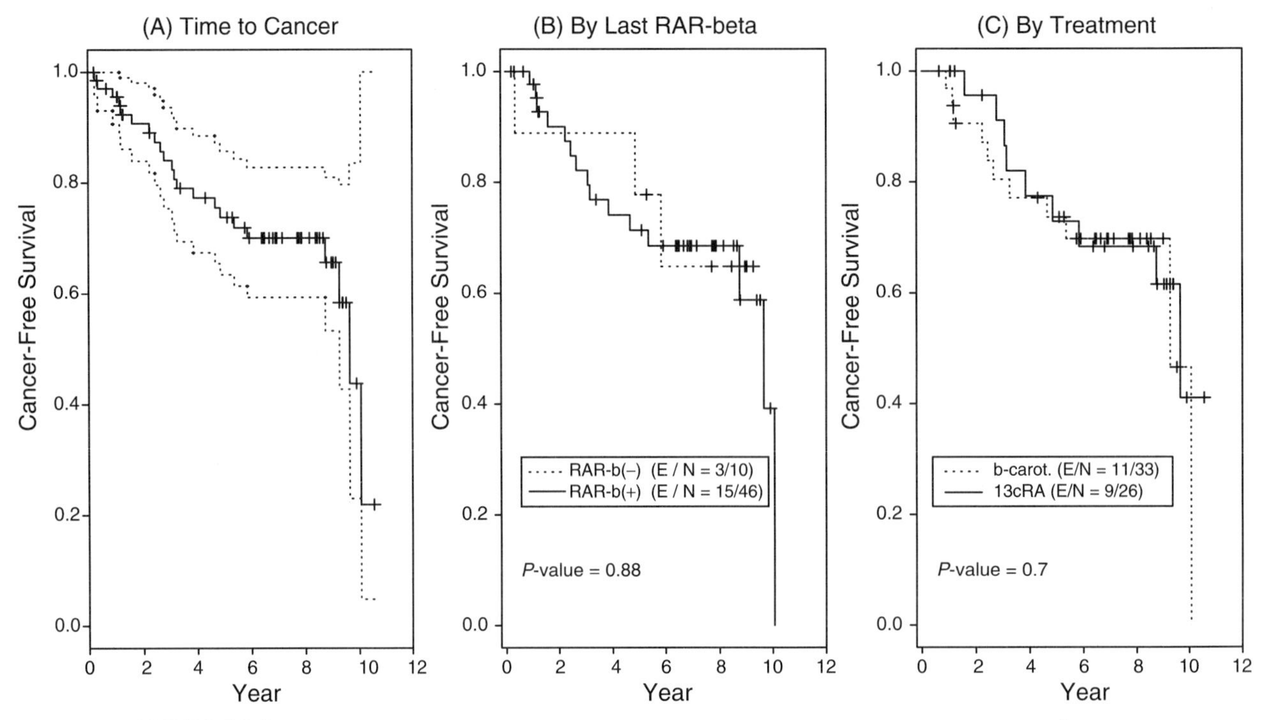

FIGURE 20.3 Kaplan–Meier plots for cancer-free survival with 95% confidence interval, by the last RAR-β status, and by treatment.

cancer development by the last measure of RAR-β expression. We chose the last measure of RAR-β expression because the RAR-β expression changed after treatment. Three out of 10 patients with a negative last RAR-β expression developed cancer, and 15 out of 46 with a positive last RAR-β expression developed cancer. There was no difference in time to cancer development between the two groups (log-rank test, $P = 0.88$).

There are two explanations for the lack of association of the last RAR-β expression with the prediction of cancer. First, in most cases, the last RAR-β expression was measured 12 months after trial registration, but the cancer developed considerably longer after that. We do not have the RAR-β expression status right before the development of cancer, nor do we have the RAR-β expression in the tumor tissues. Without this information, the relationship of loss of RAR-β expression and cancer development could not be assessed more definitely. Second, although the loss of RAR-β expression in the tumor has been well established in head and neck cancer, approximately 40% of patients developed cancer via other mechanisms that were independent of the loss of retinoid acid receptor expression. Further discussion of incorporating biomarkers for cancer risk modeling is given in the next section.

Long-term follow-up is required to study the relationship between biomarker and cancer development. For example, with a mean follow-up of 103 months, it has been reported that DNA ploidy is a prognostic marker for cancer development in oral leukoplakia patients [11]. Repeated biopsies are required to capture the change of biomarker over time. Survival analysis methods should be applied for time-to-event data. As shown earlier, the survival curves are typically computed using Kaplan–Meier product limit estimators. The difference in survival curves can be tested by the log-rank test or Gehan–Wilcoxon test. The Cox proportional hazards model can be applied to model the survival time by incorporating multiple covariates. With repeated measured data, the biomarker change over time can be analyzed by the Cox model with time-dependent covariates [22,23].

More statistical methods for defining, evaluating, and analyzing the surrogate end point biomarkers can be found in the following papers. Prentice [29] first gave a rigorous statistical definition of SEB that a surrogate for a true end point should yield a valid test of the null hypothesis of no association between the treatment and the true response. It implies that the SEB should essentially be in the disease progression pathway and that the treatment should affect the outcome by modulating the SEB. More general models were proposed to estimate the proportion of the treatment effect on the outcome explained by the SEB [30,31] and the association among treatment, SEB, and the clinical end point [32]. A latent disease model assumes that the treatment effect on both the SEB and the clinical end point is mediated through a latent variable [33]. De Gruttola *et al.* [34] gave a summary of considerations in evaluating the SEB in clinical trials.

VII. PREDICTING CANCER DEVELOPMENT USING CLINICAL, HISTOLOGICAL, EPIDEMIOLOGIC, AND MULTIPLE BIOMARKER INFORMATION

Using the previously presented chemoprevention trial with 70 OPL patients, our primary goal is to develop a model for predicting cancer development. As indicated earlier, 22 ofthe 70 patients (31.4%) developed cancers of the upper digestive tract with a median follow-up of 7 years. Over 20 medical/demographic and epidemiologic variables were collected in the study. There were also more than 10 biomarkers measured on available tissue samples at baseline and, for some markers, at additional follow-up visits. Not every patient had a biopsy. Missing data on biomarkers also resulted from exhausting tissue samples due to the limited amount of tissues procured. We will illustrate a successful strategy of comprehensive cancer risk assessment amid all these complications.

A. Single-Covariate Cox Model Analysis on Medical, Demographic, and Epidemiologic Variables

The Cox proportional hazards model is a very general semiparametric method for analyzing time-to-event data. Table 20.10 summarizes the prognostic value of medical, demographic, and epidemiologic variables when each variable is considered by itself for predicting cancer development in the Cox model. We found that the two strongest predictors for cancer are prior cancer history and histology. New cancers developed in 7 of 11 patients (63.6%) with a prior history of cancer compared to only 15 of 59 patients (25.4%) with no prior cancer history. The risk ratio was 1.89 ($P=0.009$) with a 95% confidence interval of (1.17, 3.05). Similarly, the cancer risk of OPLs with moderate or severe dysplasia was 2.3 times (95% CI: 1.39 to 3.81) higher than the cancer risk of OPLs with hyperplasia or mild dysplasia ($P=0.001$).

TABLE 20.10 Analysis of Cancer Risk by Patient Characteristics, Medical, Demographic, and Epidemiologic Variables Using the Cox Proportional Hazards Model

Variable	Category	Cancer	*N*	Risk ratio (95% confidence interval)	*P* value
Sex	Female	15	37	1.33 (0.84, 2.09)	0.22
	Male	7	33		
Age	> 60	16	35	1.61 (1.01, 2.58)	0.05
	≤ 60	6	35		
Smoking	Current	11	39	0.98 (0.64, 1.50)	0.93
	Former, never	11	31		
Alcohol	Current	16	49	1.27 (0.79, 2.04)	0.32
	Former, never	6	21		
Chewing tobacco	Yes	1	10	0.57 (0.21, 1.56)	0.27
	No	21	60		
Site	Tongue/floor of mouth	11	27	1.14 (0.74, 1.75)	0.56
	Others	11	43		
Cancer history	Yes	7	11	1.89 (1.17, 3.05)	0.009
	No	15	59		
Histology	Moderate/severe dysplasia	6	9	2.30 (1.39, 3.81)	0.001
	Hyperplasia/ mild dysplasia	16	61		
Erythroplakia	Yes	11	30	1.27 (0.83, 1.95)	0.27
	No	11	40		
Treatment	Low-dose 13cRA	9	26	1.00 (0.46, 2.18)[a]	0.99
	β-carotene	11	33	1.06 (0.77, 1.46)[a]	0.74
	Induction only	2	11		
3-month response	Response	11	39	0.85 (0.55, 1.30)	0.45
	No response	10	26		
12-month response	Response	5	24	0.58 (0.35, 0.98)	0.04
	No response	13	27		

[a] Compared to the induction-only group.

To assess the treatment effect, 9 of 26 (34.6%), 11 of 33 (33.3%), and 2 of 11 (18.2%) patients developed cancer in the low-dose 13cRA, β-carotene, and induction-only treatment groups, respectively. The relatively smaller percentage of cancer development in the induction-only group was associated with a shorter follow-up period due to loss of follow-up or progression. Figure 20.3C shows the cancer-free survival of the low-dose 13cRA and β-carotene treatment groups. Although the overall time-to-cancer development was not statistically significant between the two groups ($P=0.70$), it appeared that cancer onset was somewhat delayed in the low-dose 13cRA treatment group between years 1 and 3. With a smoothing bandwidth of 0.8 year, the estimated cancer rates at years 1, 2, and 3 were 2.5, 5.7 and 9.6% for the low-dose 13cRA treatment group and 6.2, 7.1, and 10.4% for the β-carotene treatment group, respectively. As a reminder, patients in this study received only up to 1 year of treatment. The agents may not have a prolonged effect after the completion of treatment.

Clinical response at 12 months, but not at 3 months, was statistically significantly predictive of cancer. Subjects showing a continued response to maintenance therapy developed fewer cancers than nonresponders (risk ratio=0.58, $P=0.04$). Older patients had a higher risk of developing cancer (risk ratio=1.61, $P=0.05$). In our sample, smoking status, alcohol use, and chewing tobacco did not predict cancer risk, however.

B. Combined Predictive Effect of Histology and Prior Cancer History on Cancer Development

Applying the interval event chart [35], Fig. 20.4 provides a graphic assessment of the combined effect of histology and prior cancer history on time-to-cancer development. Event charts are very useful in revealing the association of multiple time-to-event data at the individual level. Starting from the bottom of Fig. 20.4, 6 of 9 patients with moderate or severe dysplasia developed cancer. One of these 6 patients had a prior cancer (shown in a solid circle), which occurred approximately 10 years before trial registration, and then developed a new cancer approximately 4 years after registration. The events of 59 patients with hyperplasia or mild dysplasia are plotted in the upper part of Fig. 20.4. Among them, 6 of 10 patients with prior cancer history developed new cancer. Although patients with a prior history of cancer were more likely to develop new cancers, it appears that time since the prior cancer did not correlate with time to new cancer development. Furthermore, the relatively long time from prior cancer to new cancer diagnosis (range=6.0–26.6 years, median=9.2 years, $N=7$) suggests that the new cancer is unlikely the result of locoregional recurrence.

Figure 20.4 shows that six patients developed cancer within 5 years but had no prior history of cancer or moderate/severe dysplasia. However, one patient with moderate/severe dysplasia and three patients with a history of

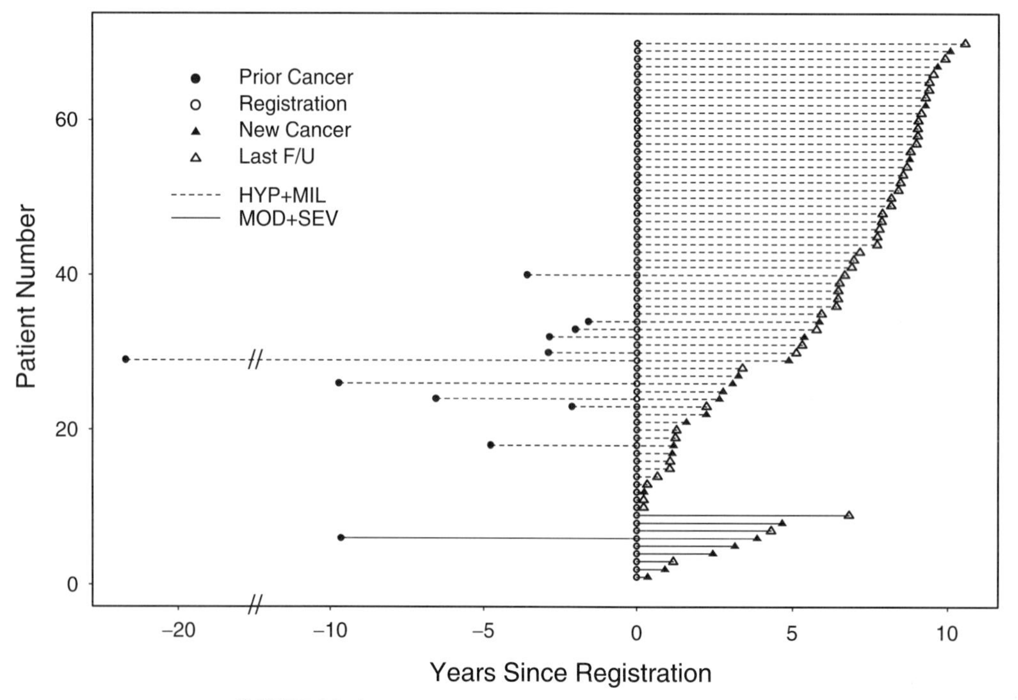

FIGURE 20.4 Interval event chart in years since registration.

prior cancer did not develop cancer after more than 5 years of follow-up. These findings indicate that prior cancer history and histology combined cannot completely explain the cancer risk. Results call for an investigation of the predictive effect of the molecular and cellular biomarkers to enhance the accuracy of cancer risk assessment.

C. Biomarkers as Predictors for Cancer: One Covariate Case

Although several biomarkers were analyzed in the OPL study, we chose five main biomarkers for analysis.

1. Chromosome Polysomy

Baseline tissue samples were available from 40 patients for the analysis (Fig. 20.2). Due to the skewed distribution (range: 0.2–39.5; mean: 6.6; median: 3.3), we chose 3 as the cutoff for dichotomizing the chromosomal polysomy into the low and high groups. The choice of the cut point is close to the median of data and was a standard cutoff used in the past from the same laboratory. It was not the optimal cut point resulting in the smallest *P* value described previously. Table 20.11 shows that 13 out of 20 (65%) patients with high polysomy developed cancer, whereas only 5 of 20 (25%) patients had cancer in the low polysomy group. The risk ratio was 1.85 with a 95% CI of 1.05–3.25 ($P = 0.03$).

2. p53 Protein

The p53 protein in OPL was measured at baseline and at 3 months after trial registration. At each time point, the percentage of cells staining positive was counted in the basal layer, parabasal layer, and whole layer. Therefore, we have six measures for p53. How do we choose which marker or what combination of markers to use?

One standard statistical method for dimension reduction is principal component analysis (PCA). In the presence of a large number of variables, the goal of PCA is to find a small number of transformed variables to explain most of the variability introduced by the original variables. It can be achieved by forming orthogonal linear combinations of the original variables. Fig. 20.5 plots the first and second principal components, which explain 80.5 and 11.5% of the total variability contained in the data. The accession number of each patient is shown in Fig. 20.5 with the principal component direction of each of the original six variables marked by a pointed line. As illustrated, the three baseline measurements and three measurements at an interval of 3 months cluster with themselves, respectively. Also, not much difference is seen among basal, parabasal, and whole layers within each time point. The PCA suggests that instead of using all six variables, it is sufficient to use one or at most two principal components for data analysis. We decided to use the baseline parabasal layer p53 measure as the marker

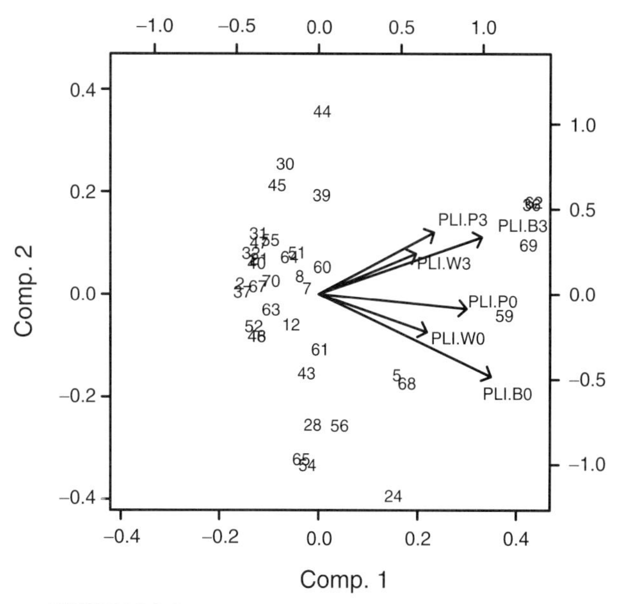

FIGURE 20.5 Principal component analysis on p53 status.

TABLE 20.11 Analysis of Cancer Risk by Cellular and Molecular Biomarkers Using the Cox Proportional Hazards Model: One Covariate

Variable	Category	Cancer	*N*	Risk ratio (95% confidence interval)	*P* value
Chromosomal	> 3%	13	20	1.85 (1.05, 3.25)	0.03
polysomy	≤ 3%	5	20		
p53 in	> 0.2	6	10	1.53 (0.91, 2.59)	0.11
parabasal layer	≤ 0.2	10	42		
LOH in	Yes	8	19	1.94 (0.89, 4.23)	0.09
3p or 9p	No	2	18		
Last measured	Loss	3	10	0.91 (0.26, 3.21)	0.88
RAR-β	Present	15	46		
Last measured	>2	6	18	0.91 (0.56, 1.47)	0.70
micronuclei	≤2	16	45		

for predicting cancer risk. The decision was based on our prior published report showing no p53 modulation during treatment and worsening histology associated with higher p53 accumulation in the parabasal layer [14]. The usage of original variables rather than the principal components will also help enhance the interpretability of the model. In addition, consistent with our prior papers, we chose 0.2 as a cutoff for dichotomizing the p53 labeling index [14,17]. Among the 52 evaluable patients, 6 of 10 (60%) in the high p53 group developed cancer, whereas only 10 of 42 patients (24%) developed cancer in the low p53 group ($P=0.11$, Table 20.11).

3. LOH at 3p or 9p

In 37 evaluable patients, 8 of 19 (42%) patients with 3p or 9p LOH and 2 of 18 (11%) patients without LOH developed cancer. Our initial report showed that 3p or 9p LOH was associated with a higher cancer risk [7]. However, with a longer follow-up, this significant association was somewhat weakened ($P=0.09$, Table 20.11).

4. RAR-β Expression

RAR-β expression was measured at baseline, 3 months, and 12 months after trial registration. Due to the change of RAR-β status (upregulated to 90% expression at 3 months) [12] in our retinoid-based trial, we used the last measured RAR-β status as the covariate for modeling cancer development. We found that RAR-β expression measured at the last follow-up in the trial was not a predictor for long-term cancer risk ($P=0.88$, Table 20.11).

5. Micronuclei

Similar to the RAR-β expression, micronuclei were also measured at baseline, 3 months, and 12 months after trial registration. Because baseline micronuclei were reduced after the 13cRA induction treatment [13], we chose the last measured micronuclei as the predictor for cancer development. Table 20.11 shows that there was no difference in cancer risk between the low and high micronuclei groups ($P=0.70$).

D. Biomarkers as Predictors for Cancer: Multiple Covariates Case

Figure 20.6 (see also color insert) shows the scatter plot matrix of the follow-up time (time to cancer development or lost to follow-up), cancer status, chromosomal polysomy, LOH, and p53 status. Each panel is one scatter plot of a pair of variables corresponding to the label on the x axis and y axis. Small random variations were added to the values of the discrete variables (cancer status and LOH: 0=No, 1=Yes) to break the ties for a better visualization of data. Selected points were marked with colors.

The scatter plot matrix revealed that there was a weak to moderate correlation between chromosomal polysomy and p53 expression (Pearson's $r=0.61$, Spearman's $r=0.23$) but no correlation between polysomy and LOH (Spearman's $r=-0.07$) or p53 and LOH (Spearman's $r=0.07$). The second row of plots indicates that patients with cancer were inclined to have higher chromosomal polysomy (CP), higher p53 expression, and LOH, consistent with the results presented in Table 20.11. Figure 20.6 also shows that patients having two or three of the high-risk factors were more likely to develop cancer. Better visualization can be achieved by "brushing" or highlighting points with certain features interactively on a computer monitor. For example, the magenta circle represents a patient with high CP, high p53, and LOH. Sure enough, the patient developed cancer 3 years after registration. The patient marked by a red circle also had high CP, high p53, and LOH. The patient developed cancer at year 10. Interestingly, one patient marked with orange had high CP and high p53, but no LOH. The patient was cancer free after about 5 years of follow-up. Brushing, in conjunction with a scatter plot matrix, provides a powerful tool for visualizing the complex interactions among multiple variables.

Because of the limited sample size and missing biomarker values, only 24 patients had complete data on these three variables, which restricts the use of conventional regression analysis with multiple covariates. To overcome this limitation, we devised a combined score, denoted as CP.p53.LOH, to capture the collective information contained in these three markers. CP.p53.LOH was computed as the sum of the three indicators, one for each of the three biomarkers. Each indicator was assigned a value of either 0 or 1, denoting either a low- or a high-risk marker value, respectively. If a biomarker value was missing, the indicator was set as 0. Nine patients who were missing all CP, p53, and LOH information were removed from the analysis. With this combined score, CP.p53.LOH has a risk ratio of 2.27 (95% CI: 1.41–3.66) and was highly significant in predicting cancer development ($P=0.0008$, Table 20.12).

E. Combined Biomarker Score and Patient Characteristics as Predictors for Cancer: Multiple Covariates Case

When the combined biomarker score and cancer history were entered in the Cox regression analysis, model 2 in Table 20.12 shows that CP.p53.LOH remained highly significant in predicting cancer. Cancer history remained significant ($P=0.046$) for cancer risk, but was not as significant in combination as it was when used as a single predictor ($P=0.009$, Table 20.10). The result indicated that, after incorporating the biomarker information, the predictive value of cancer history was less important than we had seen

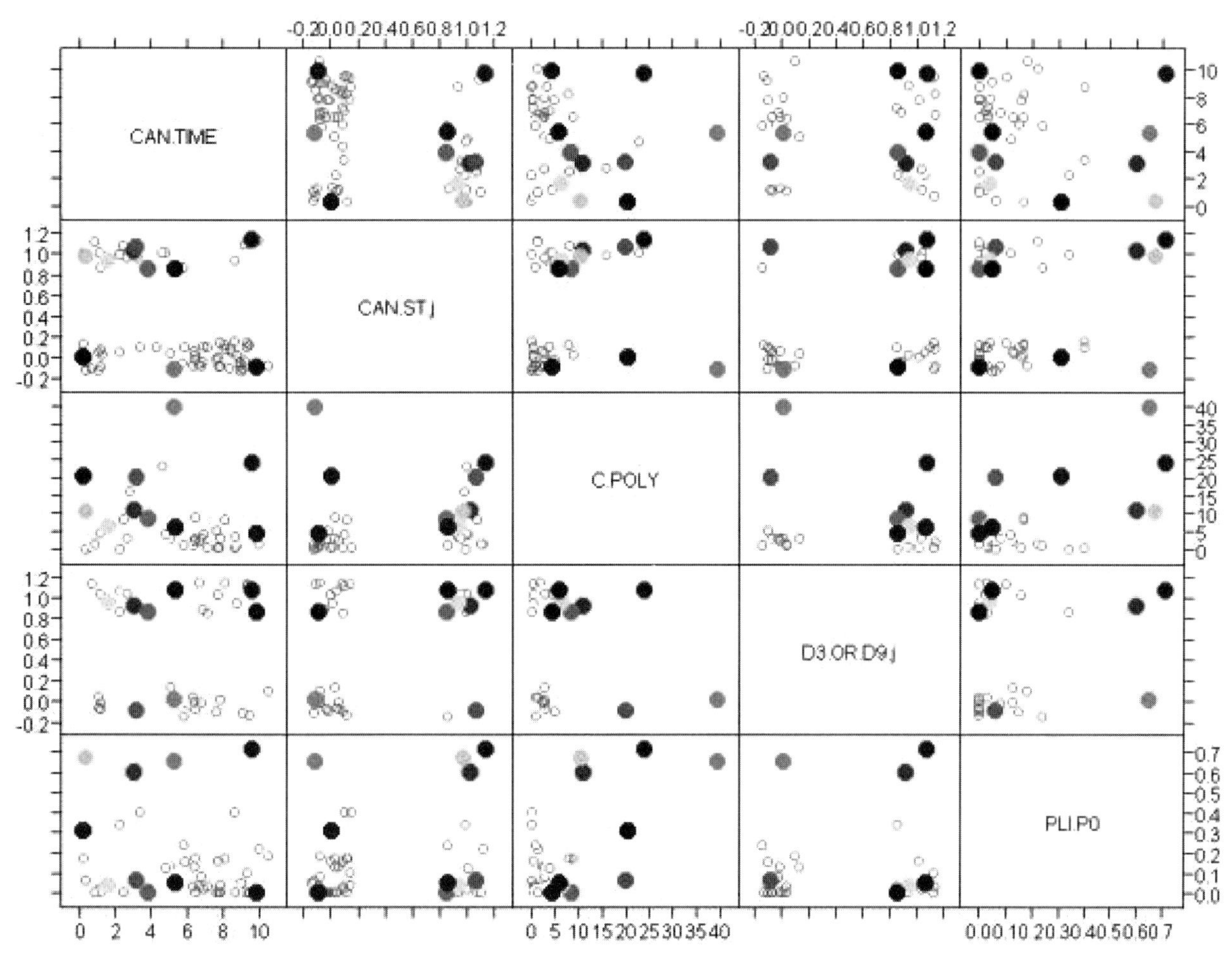

FIGURE 20.6 Scatter plot matrix of time (CAN.TIME), cancer status (CAN.ST), chromosomal polysomy (C.POLY), LOH at 3p or 9p, and p53. (See also color insert.)

previously. In other words, a major part of the information contained in cancer history for predicting new cancer was associated with high chromosomal polysomy, high p53 expression, and LOH. Model 3 in Table 20.12 shows the joint predictive effect of CP.p53.LOH and histology. Both variables remained highly significant (CP.p53.LOH, $P=0.0008$; histology, $P=0.0003$), suggesting that the histology and the combined biomarker score jointly were important in predicting cancer development. Figure 20.7 plots the probability of cancer incidence by histology and CP.p53.LOH, showing that among patients with hyperplasia/mild dysplasia lesions, the cancer incidence rates were 1/22 (4.5%), 6/22 (27.3%), 4/7 (57.1%), and 2/2 (100%) for the combined biomarker scores of 0, 1, 2, and 3, respectively. Cancer incidence rates for patients with moderate/severe dysplasia were 1/3 (33.3%) if the combined biomarker score was 0 and 5/5 (100%) if patients had at least one biomarker in the high-risk range. The same trend held in 24 patients with complete biomarker information (data not shown). Although the sample size in several subgroups was small, the trend shown in Fig. 20.7 illustrates the value of using both histology and biomarker information in modeling cancer risk.

TABLE 20.12 Analysis of Cancer Risk by Combined Biomarker Score and Patient Characteristics Using the Cox Proportional Hazards Model: Multiple Covariates

Model	*N*	Variable	Risk ratio (95% confidence interval)	*P* value
1	61	CP.p53.LOH	2.27 (1.41, 3.66)	0.0008
2	61	CP.p53.LOH	2.02 (1.21, 3.35)	0.007
		Cancer history	2.90 (1.02, 8.22)	0.046
3	61	CP.p53.LOH	2.41 (1.44, 4.03)	0.0008
		Histology	2.68 (1.57, 4.59)	0.0003
4	61	CP.p53.LOH	2.07 (1.17, 3.63)	0.012
		Histology	2.49 (1.45, 4.28)	0.001
		Cancer history	2.38 (0.78, 7.22)	0.13

When CP.p53.LOH, histology, and cancer history were entered in the Cox regression analysis, model 4 indicated that cancer history was no longer significant ($P=0.13$)

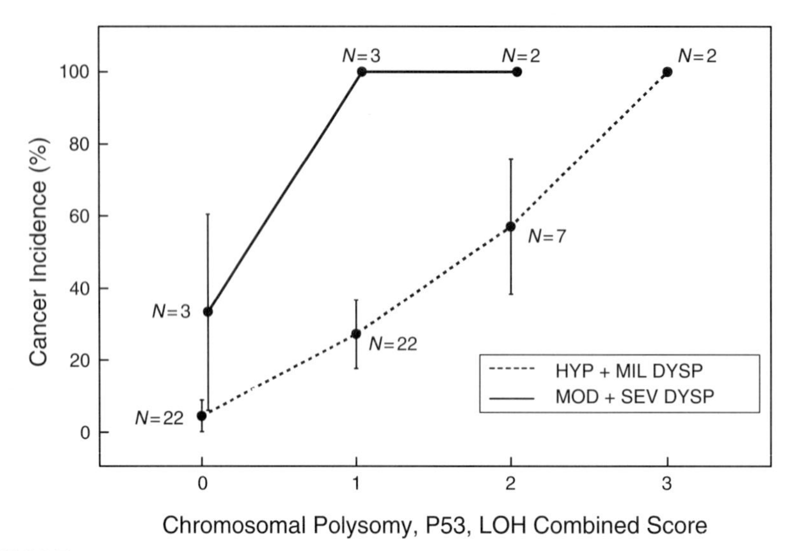

FIGURE 20.7 Cancer incidence rate (± 1 standard error) by the composite score of chormosomal polysomy, p53, and LOH at 3p or 9p. (HYP, hyperplasia; MIL, mild; MOD, moderate; SEV, severe; DYSP, dysplasia.)

but CP.p53.LOH and histology remained significant (Table 20.12). The likelihood ratio test showed that model 4 did not provide significant improvement over model 3 by adding prior cancer history ($P=0.13$). Therefore, the best model for predicting cancer risk in our data contains the combined biomarker score and histology presented in model 3.

A test of proportional hazards assumption for model 3 was assessed by the cox.zph function in S-Plus [36,37]. Plots of regression coefficients versus time are all flat (figures not shown), indicating that the proportional hazards model provides a reasonable fit to data. The P values for testing nonproportionality was 0.74, suggesting that there is no significant deviation from the proportional hazards assumption.

F. Sensitivity Analysis on Patients with Missing Biomarker Information

We also performed sensitivity analysis by assigning missing biomarkers a value of 0.5. Results are consistent with models 1 through 4 in Table 20.12. Specifically, if 0.5 was assigned to missing biomarkers, the combined biomarker score was still significant in predicting cancer development as a single covariate ($P=0.002$), and in combination with histology (combined biomarker score, $P=0.002$ and histology, $P=0.001$).

G. Recursive Partitioning for the Classification of Cancer Risk: Case of Multiple Covariates

In the aforementioned analyses, we determined that histology, history of cancer, chromosomal polysomy, p53, and LOH are important predictors for cancer development. We have also applied recursive partitioning (RP) to construct an alternative classification model for cancer risk using these five covariates (Fig. 20.8). Except for p53, four of the five covariates were chosen in the model. The classification tree started with node 1 at the top where 22 out of 70 total patients developed cancer. The standardized event rate (a special case of the event rate in the Poisson model for censored data with exponential scaling) was set to 1 for the entire sample in node 1. The RP model chose histology for the first split. Patients with hyperplasia or mild dysplasia were placed in node 2, where 16 of 61 patients had cancer. The remaining patients with moderate or severe dysplasia formed node 3, where 6 of 9 patients had cancer. The standardized event rates were 0.8 and 2.6 in nodes 2 and 3, respectively. Patients in node 2 were further split into two groups according to their prior cancer history, to nodes 4 and 5. For patients with hyperplasia or mild dysplasia and no history of cancer (node 4), LOH was the next variable chosen for classification. Finally, chromosomal polysomy with a cutoff value of 4.25 was selected to split patients in node 7 to nodes 8 and 9. Nodes 6, 8, 9, 5, and 3 are terminal nodes with increasing standardized event rates. The observed cancer incidence rates in these groups were 5.3, 13.3, 33.3, 60, and 66.7%, respectively. Note that 8 patients were not shown in the terminal nodes due to missing biomarker information. The RP algorithm uses all available information at each split. Therefore, variables with higher predictive power and less missing values are more likely to be chosen in earlier steps. Results show that patients with either moderate/severe dysplasia or a history of cancer had a high risk for new cancer. In the remaining patients, i.e., patients with hyperplasia/mild dysplasia

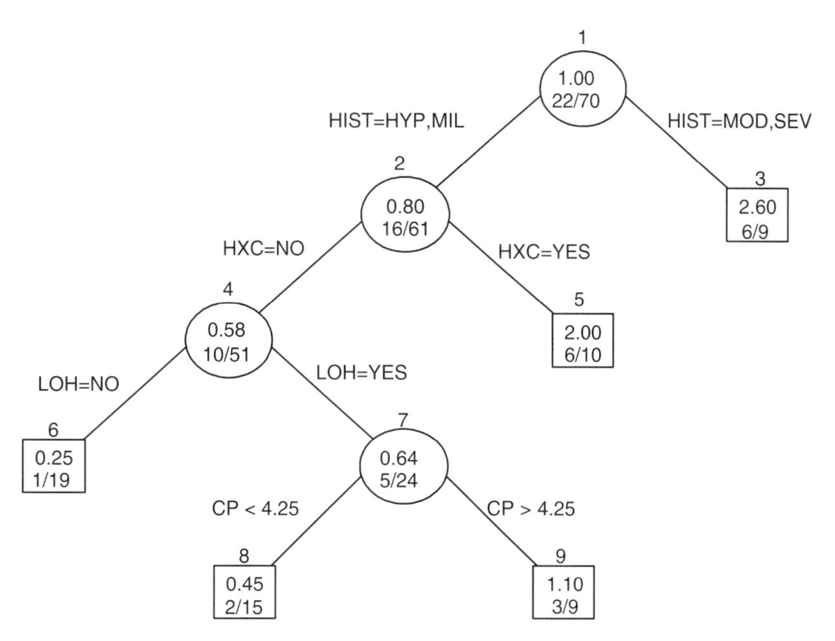

FIGURE 20.8 Recursive partitioning tree for time to cancer development. (HIST, histology; HYP, hyperplasia; MIL, MOD, or SEV, mild, moderate, or severe dysplasia, respectively; HXC, history of cancer; LOH, LOH at 3p or 9p; CP, chormosomal polysomy.)

and no history of cancer, LOH and CP can provide additional information to classify patients according to their cancer risk.

VIII. SUMMARY AND DESIGN CONSIDERATIONS FOR BIOMARKER-INTEGRATED TRANSLATIONAL STUDIES

Many key questions are often raised in cancer prevention trials. Typically, the questions include:

1. Are the treatments effective in preventing or delaying cancer?
2. How can biomarkers be used as prognostic factors for treatment effect?
3. How can biomarkers be used in predicting cancer development?
4. How can biomarkers be used in identifying high-risk individuals?
5. How can we identify suitable biomarkers to be used as surrogate end points?
6. How can the information of multiple biomarkers be combined and utilized in data analysis?
7. How do we perform a comprehensive cancer risk assessment using medical/demographic variables, epidemiologic measures, and multiple biomarker information in predicting cancer development?
8. Can the surrogate end point biomarkers and/or the risk assessment models be validated?

All these questions are important questions and demand answers. However, no single prevention trial can provide answers for all the questions. The major challenges are (a) limited number of patients and/or follow-up; (b) only a small fraction of patients develop cancer; (c) cancer development is a long and complicated process with many factors and pathways; (d) loss of information due to loss of follow-up, refusal, death, or other competing risks with some data being censored before cancer development; (e) noncompliance; (f) large number of correlated variables; (g) biomarkers are often measured with errors; and (h) missing values due to patients dropping out or due to insufficient amount of tissue for biomarker analysis, etc. With the help of applying relevant statistical methods, we illustrate a practical data analysis scheme to address some of the aforementioned questions using the example of the OPL trial stated in earlier sections.

To ensure the success of translational chemoprevention trials, the following important design issues must be prospectively planned and considered:

1. Identify the biomarkers to be measured in the study, including clearly specified rationale and hypothesis.
2. Collect sufficient tissue specimens for complete biomarker analysis over time to minimize the amount of missing data.
3. Keep a uniform follow-up schedule.
4. Prioritize the biomarkers to be analyzed.
5. Reduce interpatient variability by selecting subjects with comparable characteristics by carefully chosen eligibility criteria.

6. Reduce the measurement variability by increasing the process reliability.
7. Store tissues for both premalignant and cancer lesions to facilitate comparison.
8. Increase sample size to have enough power for testing multiple end points, interactions, and subset analysis.

More design and statistical considerations for the biomarker development can be found elsewhere [38,39].

Acknowledgments

This work was supported in part by Grants CA52051, CA16672, and DE11906 from the National Institutes of Health and the Tobacco Settlement Funds as appropriated by the Texas State Legislature.

References

1. Lippman, S. M., Batsakis, J. G., Toth, B. B., Weber, R. S., Lee, J. J., Martin, J. W., Hays, G. L., Goepfert, H., and Hong, W. K. (1993). Comparison of low-dose 13cRA with β-carotene to prevent oral carcinogenesis. *N. Engl. J. Med.* **328**, 15–20.
2. Shin, D. M., Mao, L., Papadimitrakopoulou, V. M., Clayman, G., El-Naggar, A., Shin, H. J., Lee, J. J., Lee, J. S., Gillenwater, A., Myers, J., Lippman, S. M., Hittelman, W. N., and Hong, W. K. (2000). Biochemopreventive therapy for patients with premalignant lesions of the head and neck and p53 gene expression. *J. Natl. Cancer Inst.* **92**, 69–73.
3. Torrisi, R., and Decensi, A. (2000). Fenretinide and cancer prevention. *Curr. Oncol. Rep.* **2**, 263–270.
4. Hong, W. K., Spitz, M. R., and Lippman, S. M. (2000). Cancer chemoprevention in the 21st century: Genetics, risk modeling, and molecular targets. *J. Clin. Oncol.* **18**, 9S–18S.
5. Lippman, S. M., and Hong, W. K. (2001). Molecular markers of the risk of oral cancer. *N. Engl. J. Med.* **344**, 1323–1326.
6. Lippman, S. M., Lee, J. J., and Sabichi, A. L. (1998). Cancer chemoprevention: Progress and promise. *J. Natl. Cancer Inst.* **90**, 1514–1528.
7. Mao, L., Lee, J. S., Fan, Y. H., Ro, J. Y., Batsakis, J. G., Lippman, S., Hittelman, W., and Hong, W. K. (1996). Frequent microsatellite alterations at chromosomes 9p21 and 3p14 in oral premalignant lesions and their value in cancer risk assessment. *Nature Med.* **2**(6), 682–685.
8. Mao, L., El-Naggar, A. K., Papadimitrakopoulou, V., Shin, D. M., Shin, H. C., Fan, Y., Zhou, X., Clayman, G., Lee, J. J., Lee, J. S., Hittelman, W. N., Lippman, S. M., and Hong, W. K. (1998). Phenotype and genotype of advanced premalignant head and neck lesions after chemopreventive therapy. *J. Natl. Cancer Inst.* **90**, 1545–1551.
9. Beenken, S. W., Sellers, M. T., Huang, P., Peters, G., Krontiras, H., Dixon, P., Stockard, C., Listinsky, C., and Grizzle, W. E. (1999). Transforming growth factor alpha (TGF-alpha) expression in dysplastic oral leukoplakia: Modulation by 13-cis retinoic acid. *Head Neck* **21**, 566–573.
10. Kim, J., Shin, D. M., El-Naggar, A., Lee, J. S, Corrales, C., Lippman, S. M., Hong, W. K., and Hittelman, W. N. (2001). Chromosome polysomy and histological characteristics in oral premalignant lesions. *Cancer Epidemiol. Biomark. Prev.* **10**, 319–325.
11. Sudbo, J., Kildal, W., Risberg, B., Koppang, H. S., Danielsen, H. E., and Reith, A. (2001). DNA content as a prognostic marker in patients with oral leukoplakia. *N. Engl. J. Med.* **344**, 1270–1278.
12. Lotan, R. L., Xu, X.-C., Lippman, S. M., Ro, J. J., Lee, J. S., Lee, J. J., and Hong, W. K. (1995). Suppression of retinoic acid receptor-β in premalignant oral lesions and its up-regulation by 13cRA. *N. Engl. J. Med.* **332**, 1405–1410.
13. Benner, S. E., Lippman, S. M., Wargovich, M. J., Lee, J. J., Velasco, M., Martin, J. W., Toth, B. B., and Hong, W. K. (1994). Micronuclei, a biomarker for chemoprevention trials: Results of a randomized study in oral pre-malignancy. *Int. J. Cancer* **59**, 457–459.
14. Lippman, S. M., Shin, D. M., Lee, J. J., Batsakis, J. G., Lotan, R., Tainsky, M. A., Hittelman, W. N., and Hong, W. K. (1995). *p53* and retinoid chemoprevention of oral carcinogenesis. *Cancer Res.* **55**, 16–19.
15. Xu, X.-C., Zile, M. H., Lippman, S. M., Lee, J. S., Lee, J. J., Hong, W. K., and Lotan, R. (1995). Anti-retinoic acid (RA) antibody binding to human premalignant oral lesions, which occurs less frequently than binding to normal tissue, increases after 13-cis-RA treatment *in vivo* and is related to RA receptor ß expression. *Cancer Res.* **55**, 5507–5511.
16. Shin, D. M., Lee, J. S., Lippman, S. M., Lee, J. J., Tu, Z. N., Choi, G., Heyne, K., Shin, H. J. C., Ro, J. Y., Goepfert, H., Hong, W. K., and Hittelman, W. N. (1996). *p53* expression: Predicting recurrence and second primary tumors in head and neck squamous cell carcinoma. *J. Natl. Cancer Inst.* **88**, 519–529.
17. Shin, D. M., Xu, X. C., Lippman, S. M., Lee, J. J., Lee, J. S., Batsakis, J. G., Ro, J. Y., Martin, J. W., Hittelman, W. N., Lotan, R., and Hong, W. K. (1997). Accumulation of *p53* protein and retinoic acid receptor-β in retinoid chemoprevention. *Clin. Cancer Res.* **3**, 875–880.
18. Spitz, M. R., McPherson, S., Jiang, H., Hsu, T. C., Trizna, Z., Lee, J. J., Lippman, S. M., Khuri, F. R., Steffen-Batey. L., Chamberlain, R. M., Schantz, S. P., and Hong, W. K. (1997). Correlates of mutagen sensitivity in patients with upper aerodigestive tract cancer. *Cancer Epidemiol. Biomark. Prev.* **6**, 687–692.
19. Spitz, M. R., Lippman, S. M., Jiang, H., Lee, J. J., Khuri, F., Hsu, T. C., Trizna, Z., Schantz, S. P., Benner, S., and Hong, W. K. (1998). Mutagen sensitivity as a predictor of tumor recurrence in patients with cancer of the upper aerodigestive tract. *J. Natl. Cancer Inst.* **90**, 243–245.
20. El-Naggar, A. K., Lai, S., Clayman, G., Lee, J. J., Luna, M. A., Goepfert, H., and Batsakis, J. G. (1997). Methylation: A major mechanism of p16/CDKN2 gene inactivation in head and neck squamous carcinoma. *Am. J. Pathol.* **151**, 1767–1774.
21. Lee, J. J., and Tu, Z. N. (1997). A versatile one-dimensional distribution plot: The *BLiP* plot. *Am. Stat.* **51**, 353–358.
22. Klein, J. P., and Moeschberger, M. L. (1997). "Survival Analysis." Springer-Verlag, New York.
23. Therneau, T. M., and Grambsch, P. M. (2000). "Modeling Survival Data: Extending the Cox Model." Springer-Verlag, New York.
24. Becker, R. A., Cleveland, W. S., and Shyu, M. (1996). The visual design and control of trellis display. *J. Computat. Graph. Stat.* **5**, 123–155.
25. Cook, R. D., and Weisberg, S. (1999). "Applied Regressing Including Computing and Graphics." Wiley, New York.
26. Altman, D. G., Lausen, B., Sauerbrei, W., and Schumacher, M. (1994). Dangers of using optimal cutpoints in the evaluation of prognostic factors. *J. Natl. Cancer Inst.* **86**, 829–835.
27. Lee, J. J., Hong, W. K., Hittelman, W. N., Mao, L., Lotan, R., Shin, D. M., Benner, S. E., Xu, X.-C., Lee, J. S., Papadimitrakopoulou, V. M., Geyer, C., Perez, C., Martin, J. W., El-Naggar, A. K., and Lippman, S. M. (2000). Predicting cancer development in oral leukoplakia:Ten years of translational research. *Clin. Cancer Res.* **6**, 1702–1710.
28. Papadimitrakopoulou, V. A., Hong, W. K., Lee, J. S., Martin, J. W., Lee, J. J., Batsakis, J. G., and Lippman, S. M. (1997). Low-dose 13cRA versus beta-carotene to prevent oral carcino genesis: Long-term follow-up. *J. Natl. Cancer Inst.* **89**, 257–258.
29. Prentice, R. L. (1989). Surrogate end points in clinical trials: Definition and operational criteria. *Stat. Med.* **8**, 431–440.
30. Freedman, L. S., Graubard, B. I., and Schatzkin, A. (1992). Statistical validation of intermediate end points for chronic disease. *Stat. Med.* **11**, 167–178.
31. Lin, D. Y., Fleming, T. R., and De Gruttola, V. (1997). Estimating the proportion of treatment effect explained by a surrogate marker. *Stat. Med.* **16**, 1515–1527.

32. Buyse, M., and Molenberghs, G. (1998). Criteria for the validation of surrogate end points in randomized experiments. *Biometrics* **54**, 1014–1029.
33. Xu, J., and Zeger, S. L. (2001). The joint analysis of longitudinal data comprising repeated measures and time-to-events. *Appl. Stat.* **50**, 375–387.
34. De Gruttola, V. G., Clax, P., DeMets, D. L., Downing, G. J., Ellenberg, S. S., Friedman, L., Gail, M. H., Prentice, R., Wittes, J., and Zeger, S. L. (2001). Considerations in the evaluation of surrogate endpoints in clinical trials: Summary of a National Institutes of Health workshop. *Control Clin. Trials* **22**, 485–502.
35. Lee, J. J., Hess, K. R., and Dubin, J. A. (2000). Extensions and applications of event charts. *Am. Stat.* **54**, 63–70.
36. Grambsch, P., and Therneau, T. (1994). Proportional hazards tests and diagnostics based on weighted residuals. *Biometrika* **81**, 515–526.
37. S-PLUS 2000 Guide to Statistics (2000). Seattle, Washington MathSoft, Inc., Data Analysis Products Division, Seattle, WA.
38. Lee, J. J., Lieberman, R., Sloan, J. A., Piantadosi, S., and Lippman, S. M. (2001). Design considerations for efficient prostate cancer chemoprevention trials. *Urology* **57**(4 Suppl. 1), 206–212.
39. Sullivan Pepe, M., Etzioni, R., Feng, Z., Potter, J. D., Thompson, M. L., Thornquist, M., Winget, M., and Yasui, Y. (2001). Phases of biomarker development for early detection of cancer. *J. Natl. Cancer Inst.* **93**(14), 1054–1061.

CHAPTER

21

Molecular Detection of Head and Neck Cancer

WAYNE M. KOCH and DAVID SIDRANSKY

Department of Otolaryngology—Head and Neck Surgery
Johns Hopkins University School of Medicine
Baltimore, Maryland 21286

I. Introduction 305
II. Detection of Head and Neck Squamous Cell Carcinoma in Exfoliated Cell Samples 306
A. Gene Mutations 306
B. Microsatellite Panels 307
C. Hypermethylation 309
D. Viral Markers 309
III. Detection of Circulating Tumor Markers 310
A. Tumor DNA 310
B. Circulating Tumor-Related Antibodies 310
IV. Microarrays 311
V. Spectroscopic Analysis 311
VI. Diagnostic or Prognostic Biomarkers in Biopsy Material 311
VII. Conclusion 312
References 312

I. INTRODUCTION

The potential benefit of an efficient and powerful screening tool for early detection of the upper aerodigestive tract [head and neck squamous cell carcinoma (HNSCC)] can hardly be overstated. When discovered early in its clinical course, HNSCC is controlled readily using a single treatment modality with a high rate of long-term cure. For example, a T1 true vocal cord cancer should be cured by endoscopic excision or narrow field radiation alone in nearly 90% of cases. In contrast, the rate of cure for a stage IV larynx cancer even when treated with combined therapy, with concomitant severe morbidity, is just over 50% [1]. The contrast in success is equally great for early vs late cancers of the oral cavity and pharynx. Even without the development of new treatment strategies and agents, HNSCC would be eliminated as a major health concern if it could be detected at its earliest stages in the population at large.

At the same time that the benefits of early detection of HNSCC are compelling, so are the challenges that combine to cause most cases to elude diagnosis until the tumor reaches an advanced stage. Alcohol and tobacco use put people at a risk to develop the disease. The habitual use of these products characterizes a life style in which individuals are not predisposed to give careful attention to health issues. Furthermore, the social classes most at risk for HNSCC are not those with the most ready access to quality health care. Add to these facts the technical challenges of a thorough evaluation of the mucosa of the upper aerodigestive tract, which requires special expertise to perform, and it is easy to see why delays in a patient reaching a head and neck oncologist are the rule rather than the exception. Even when a qualified expert has been consulted, the hidden nature of many early lesions amidst the crypts and crevices of the oral cavity, pharynx, and larynx accentuates the need for new detection strategies.

The explosion of information about the molecular basis of neoplastic transformation that occurred in the 1990s has provided a myriad of opportunities for progress in understanding and managing HNSCC. It is now accepted that cancers, including HNSCC, develop from a clonal population of cells that has become increasingly altered through the accumulation of genetic damage. Each additional mutation or other genetic alteration provides either a growth advantage for the cell, permitting it to overgrow its neighbors as it divides more rapidly, or conveys some other phenotypic characteristic of malignancy such as invasion, stimulation of new blood vessel growth, and metastasis. With the advent of the polymerase chain reaction (PCR), small samples of DNA could be amplified to provide material sufficient for study by direct sequencing of the genetic code. This technique, together with numerous sophisticated methods for manipulating DNA in order to gain insight into the function of individual genes, yielded evidence for some of the seminal events that produce malignancy. By comparing the sequence of DNA isolated and amplified from tumor cells with that of normal DNA from the same individual, highly specific markers for the clonal cancer cells became available. For example, the *p53* gene is a tumor suppressor gene frequently involved in tumor progression and is commonly inactivated by point mutations that can serve as detection markers for HNSCC. However, the identification of other genes involved directly in malignant transformation that contain point mutations in HNSCC and other tumors stalled in the late 1990s. Molecular detection schemes using other types of alteration have been explored, using DNA, mRNA, protein, and antibody as the markers.

An ideal tumor marker for molecular detection must have distinct attributes. It must be highly sensitive, able to provide a signal for a very small number of cancer cells amidst a background of normal cells. It must also be highly specific, able to distinguish between cancer and noncancerous tissue accurately. The event that produces the marker should occur relatively early in the malignant transformation process so that the appearance of the marker heralds the presence of disease at its initial stages. The method of detection must be made readily available for widespread use. Thus, it must be inexpensive, amenable to automation in order to facilitate the screening of large numbers of individuals quickly, and simple and reliable enough to be performed reproducibly in laboratories throughout the world. Ideally, the use of a single marker would suffice to screen every individual, i.e., the marker would be universal. Of course, no marker possessing all of these attributes for HNSCC detection has been identified to date.

Molecular detection strategies can be applied to a variety of clinical scenarios that have slightly different requirements. *De novo* identification of cancer is perhaps the most challenging and the most urgent application. *De novo* screening requires universal markers that can be used to test individuals with no previous tumor available to identify a specific marker. Although all cancers may ultimately possess similar biologic capacities, no single known oncogene is altered in all HNSCC. All cancers express proteins not commonly seen in normal tissue, or at levels that are abnormal, however, no protein highly specific for HNSCC has yet been identified.

Once a cancer is discovered, the malignant cells can be analyzed, and specific alterations that are present only in the malignant cells can be identified. Those alterations can then serve as highly specific markers for the detection of cancer in that individual under a variety of circumstances. One example is the use of tumor-specific markers for the surveillance of cancer recurrence after primary therapy. Early detection of recurrent cancer may have less impact on outcome than early *de novo* detection, but it may permit salvage treatment with increased success if minimal residual disease can be identified promptly. Molecular detection may also provide information about the extent and potential behavior of disease, directing adjuvant therapy to those individuals that are in greatest need.

Before a cancer cell can be detected, the cells or their products (DNA, protein, etc.) must be accessible for testing. HNSCC is an ideal candidate for molecular detection because of the ready access to the mucosa of the upper aerodigestive tract. Epithelial cells can be inspected *in situ* or harvested by biopsy, brushing, or simply rinsing the surface with a fluid. Lymph node metastases are available through fine needle aspiration biopsy, and circulating metastatic cells may be identified in lymphatic fluid, peripheral blood, or bone marrow.

Patients with HNSCC cannot wait for the identification of ideal markers to help manage their disease. Technological advancements gleaned from other research endeavors can be applied to tumor-screening efforts and new and better markers can be added as they become available. Methods for high throughput analysis at low cost are on the horizon, including silicon chip microarray technologies and fluorescent tags permitting automated detection.

II. DETECTION OF HEAD AND NECK SQUAMOUS CELL CARCINOMA IN EXFOLIATED CELL SAMPLES

A. Gene Mutations

Because cancer is caused by alterations in the genetic code, DNA provides one of the most direct sources for potential markers. Tumor-specific alterations have been described in several target genes, including *p53* and *p16*. Factors exist with both of these genes that limit their use as ideal markers. Less than half of all HNSCC have any identifiable alteration in the *p53* gene [2]. The spectrum of mutations of the *p53* gene present in a population of HNSCC tumors includes numerous base pair substitutions and deletions that can occur at many positions throughout the gene. The use of mutations at *p53* hot spots for cancer detection could permit the implementation of this molecular approach for *de novo* detection. In contrast, the *p16* gene is altered in the majority of HNSCC cases, but the alteration is caused most commonly by homozygous deletion or promoter hypermethylation rather than by point mutation [3]. Other oncogenes harboring specific mutations (PTEN, etc.) are rarely detected in HNSCC.

B. Microsatellite Panels

Tumor-specific microsatellite alterations provide a fertile source of potential tumor markers. Microsatellites are composed of short sequences of DNA that are repeated with variable copy numbers and are scattered throughout the genome. They are not necessarily contained within the encoded portion of genes, and their purpose is uncertain. To be useful for cancer detection, a microsatellite must be highly polymorphic. That is, the copy number must vary among most individuals so that it is likely that the repeat number in a given person's maternal and paternal allele is different. If that is the case, the microsatellite is said to be "informative." Tumor DNA can display an alteration at an informative microsatellite in two ways. If one allele is lost by all members of the clonal population comprising the cancer, the alteration is called loss of heterozygosity (LOH). If, instead, the tumor cells develop an altered allele with a number of repeats that is different from either the maternal or paternal copy, the alteration is called a shift or microsatellite instability (MSI).

In order to detect an alteration in a microsatellite, the DNA including and surrounding it is amplified using specific PCR primers. The product is then separated electrophoretically by size using a variety of different matrices and platforms. Radiolabeling or fluorescent labeling of the PCR product permits detection of the different sized fragments. LOH in tumor cells produces a decrease in the intensity of the band for the lost allele. Because there are always some normal cells mixed with the tumor in a clinical specimen, a complete absence of the band is not the rule. Convention has it that a specimen is scored as showing LOH if there is a diminution of one band of 30–50% when the tumor specimen is compared with normal DNA from the same individual. This normal DNA can be conveniently isolated from peripheral blood leukocytes (PBL). In the case of tumor-specific MSI, a new additional band will appear in the tumor sample together with the normal bands (Fig. 21.1, see also color insert).

Several early studies yielded important information providing the basis for early detection of *de novo* lesions in the general population. A survey of alterations in a population of 35 HNSCC samples showed several key regions within the genome that frequently display alteration. Chief among these are portions of chromosomal arm 9p, 3p, and 17p. In fact, there are many regions that display alterations in over a third of all HNSCC [4]. Later, a study comparing the rate of alteration at various chromosomal sites found in invasive cancer with that in premalignant

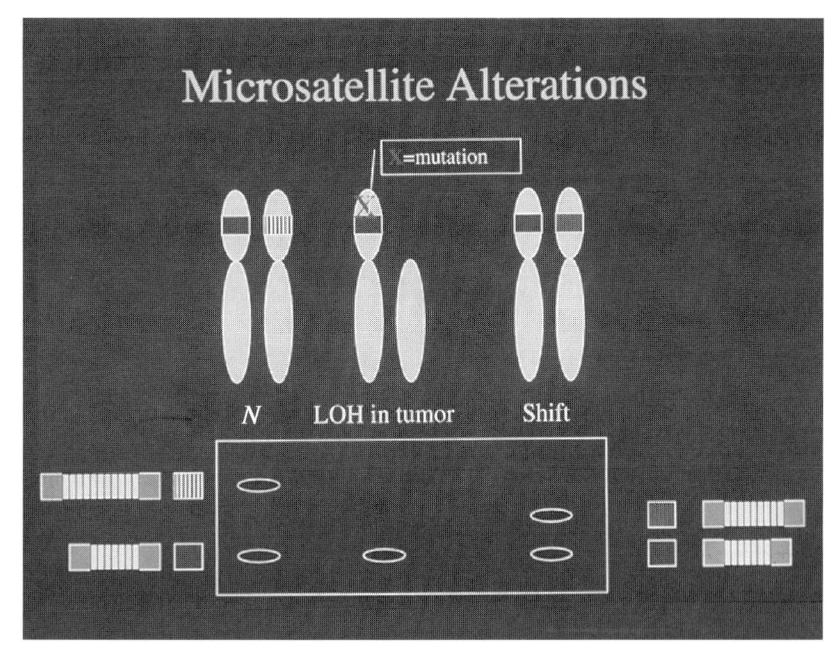

FIGURE 21.1 Schema for altered band pattern on polyacrylamide gel electropheresis (PAGE) for tumor-associated loss of heterozygosity (LOH) and microsatellite shift. *N* = normal (control) DNA displaying polymorphism of the hypothetical microsatellite. With LOH, a deletion of one chromosomal arm results in absence of the band for the microsatellite allele located on the deleted segment. A shift pattern occurs when the tumor cells display a new form of the microsatellite containing a novel number of repeats. (See also color insert.)

disease demonstrated that microsatellite alterations located on 3p and 9p are particularly valuable for the detection of cancer in its earliest state [5]. Each of these sites displays alterations in a large percentage of cases of premalignant lesions, and the rate of alteration does not increase substantially in invasive lesions. From this evidence, it is postulated that events involving genes located on 9p and 3p are among the earliest changes in the development of HNSCC. The target on 9p may well be *p16* and its close neighbor, $p14^{ARF}$, whereas the target(s) on 3p has yet to be determined. Regardless of the target gene at these sites, alterations in microsatellites on 3p and 9p are particularly valuable as potential markers for molecular detection.

Early studies reported successful molecular detection using microsatellite markers when visible lesions were scraped or brushed [6–8]. In one of these studies, microsatellite alterations were detected in each of 32 laryngeal lesions, both benign and malignant, sampled by a cytologic brush [8]. Scraping or brushing an obvious tumor is closely akin to taking a biopsy and is not equal to the rigor of primary detection strategies.

Microsatellite shifts result in the presence of a new visible band or peak derived from tumor cells. Hence, MSI is easier to detect in a clinical sample than LOH. The loss of an allele by rare cancer cells may be masked by the overwhelming majority of normal cells within a sample. Combining the information about early changes with the favorable nature of MSI, a panel of microsatellites was compiled in an effort to include markers that could detect the vast majority of HNSCC cancers in the population at large. Twenty-three microsatellites were selected using cues derived from similar efforts in other tumor types, especially bladder cancer [9]. In order to test the efficacy of the panel, it was applied to 44 cases of known HNSCC for which samples of tumor were available. In each case, the patient had provided a rinse and swab sample of the oral cavity prior to tumor treatment. Swabs were taken with a cotton-tipped applicator from prescribed surfaces of the oral cavity, avoiding the obvious visible cancer. The exfoliated cells were collected, DNA extracted, and amplified by PCR for all 23 microsatellites. Tumor DNA from biopsy material and normal PBL DNA were also extracted and tested and results compared with DNA from oral rinses of 43 healthy volunteers.

At least one of the microsatellite markers displayed an alteration in 86% of the tumors. When a marker was present in the tumor, it could be detected in the oral rinse specimen, indicating the presence of cancer cells in 92% of cases. MSI was more likely to be detected than LOH, as predicted. Most encouraging was the fact that 12 of 13 early lesions were detected, including three of three cases of metastatic cancer from clinically occult primary tumors (Tx) and 9 of 10 T1 tumors. In addition, all four cases in which tumor had recurred after prior radiation were detected successfully [10]. This approach is promising for the successful application of a microsatellite panel for early detection and posttreatment surveillance screening. The results of this and other early detection studies are summarized in Table 21.1.

One flaw in the microsatellite screening panel is the failure to detect 14% of cases with known tumor. In order to be useful for actual screening efforts, the panel must be augmented with a goal of capturing over 95% of all cases. Another problem is the massive amount of work involved in amplifying and separating 23 microsatellite assays for test

TABLE 21.1 Early Detection Studies

Marker	Population	Tissue	Sensitivity/specificity	References
23 microsatellites	44 HNSCC 43 healthy controls	Oral rinse	92%/100%	Spafford *et al.* [10]
3 genes	30 HNSCC 30 healthy controls	Oral rinse	65%/97%	Rosas *et al.* [13]
EBV DNA	21 NPC 157 controls	NP brushing	90%/99%	Lin *et al.* [15]
12 microsatellites	21 HNSCC	Serum	29% positive	Nawroz *et al.* [19]
4 genes	50 HNSCC	Serum	42%/?	Sanchez-Cespedes *et al.* [22]
EBV DNA	167 NPC 77 controls	Serum	59%/87%	Shotelersuk *et al.* [23]
HPV DNA	70 HNSCC	Serum	40%/?	Capone *et al.* [26]
Anti-p53 ab	70 HNSCC 50 premalignant 63 controls	Serum	34% HNSCC 30% premalignant 6% controls	Ralhan *et al.* [27]
Anti-EBV TK ab	87 NPC 52 control	Serum	93%/94%	Connolly *et al.* [28]

and normal DNA for each subject screened. Finally, reading of data by visual inspection introduces unwanted labor and subjectivity that could adversely affect the successful implementation of the test on a widespread basis.

Other centers have attempted a similar approach using microsatellite alterations to detect HNSCC in saliva with mixed results. A cohort of 37 cancer patients at the M.D. Anderson Cancer Center was assessed using a panel of 25 microsatellites. The markers chosen were mostly LOH markers and few displayed instability. Most tumors had a marker altered among the panel [32 of 37 (86%)]. However, LOH of a marker was detected in only 18 saliva specimens (49%), and most of the alterations in the saliva did not correlate with those in tumor. It is noteworthy that 3 subjects had a total of 11 shift markers detected in saliva, suggesting again that microsatellite instability is easier to detect than LOH [11]. These investigators did not include a healthy control population in the study. They concluded that (1) certain combinations of markers improved the rate of detection over that of the entire panel and (2) clonal heterogeneity may explain the discordant results between tumor and saliva.

The method of sample procurement in early detection schemes may have a profound impact on results. The feasibility of self-collection of useful oral rinse samples by subjects at home has been piloted. Successful PCR amplification of DNA from the home collection samples was possible in over 85% of cases [12].

C. Hypermethylation

Methylation of cytosine moieties in DNA is a normal occurrence that plays a role in X inactivation and tissue specialization during embryogenesis. CpG islands in promoter regions are subject to hypermethylation, turning off transcription of downstream genes. The hypermethylation of CpG islands in promoter regions in tumor cells is part of an of inactivation pattern for putative target suppressor genes. This is the mechanism responsible for the inactivation of *p16* in a substantial proportion of cases of HNSCC.

Promoter hypermethylation provides an opportunity to improve on the results of molecular detection of HNSCC with microsatellite panels in two ways. First, methylated genes constitute additional markers to improve the yield of detected tumors in the general population. Assay methods using fluorescent probes released in methylation-specific PCR (MSP) offer the potential for automated detection. Hypermethylation has been demonstrated to successfully identify cancer cells in exfoliative oral rinse samples using three target genes. Aberrant methylation was detected in at least one of the three genes in 17 of 30 primary tumors (56%). When the tumor displayed a hypermethylation marker, the same aberration could be identified in 11 of 17 samples of exfoliated cells in saliva (65%). The cohort of successfully detected tumors included six T1 lesions, five from the oral cavity and one from an oropharyngeal site. There were no cases in which hypermethylated DNA was detected in saliva of patients whose tumors did not display that marker, although 1 of 30 healthy control subjects provided a saliva sample in which hypermethylation was detected. This subject was a smoker and had hypermethylation of two of the three target genes. The individual was not followed longterm in the study in order to determine whether the appearance of abnormal DNA in the saliva was a false-positive signal or a case of early detection of mucosal neoplasia [13]. Hypermethylation is detectable using automated semiquantitative fluorescence methods and can be used to estimate tumor burden. These advantages make hypermethylation an attractive category of tumor marker for the further development of molecular detection schemes.

D. Viral Markers

Nasopharyngeal carcinoma (NPC) must be considered as a distinctive entity in light of its unique etiologic association with Epstein–Barr virus (EBV). That association provides a powerful approach to molecular detection, which constitutes a line of investigation that has several advantages over early detection of HNSCC using gene mutations, hypermethylation, and microsatellite alterations. Viral DNA is distinct from that of the host and thus is potentially a highly sensitive marker. The specificity of cancer detection using viral markers may be more problematic in that some individuals may display viral DNA as a result of infection or merely a carrier state rather than indicating the presence of virally induced cancer.

The nasopharynx is accessible via an endoscopic approach with either biopsy or brush cytologic sampling performed easily. In one study, brush samples were analyzed using PCR to identify viral DNA in a group of 21 NPC patients and 157 controls from high-risk populations. Nineteen tumor patients tested positive for EBV DNA (90%) compared to 2 of 149 informative control samples (1.3%). One of the positive control samples was due to the presence of an EBV-positive inverted papilloma [14]. A similar study with a smaller control arm provided nearly identical results with 36 of 38 cancer patients testing positive for EBV by nasal swab [15]. Furthermore, EBV DNA detection methods may be applied to fine needle aspiration biopsy (FNAB) samples for the assessment of neck metastases. This may help indicate the site of origin when the primary disease has an occult presentation. Specimens from 10 patients with NPC, 19 with other HNSCC, and 1 with lymphoma were analyzed using *in situ* hybridization for EBV. All FNAB samples from the NPC patients demonstrated the presence of EBV, while none of the others tested positively [16].

III. DETECTION OF CIRCULATING TUMOR MARKERS

A. Tumor DNA

It is well recognized that naked tumor DNA circulates in a relatively concentrated state in the blood of patients with solid malignancy. This phenomenon may be due to tumor necrosis or apoptosis in the primary site or to destruction of circulating tumor cells (metastatic or premetastatic). In the latter case, the presence of circulating tumor DNA may be a marker for increased risk of metastatic disease. However, if tumor DNA is released from the primary site early in the development of disease, it may lend itself to early detection of cancer via analysis of a blood sample, a very appealing concept. Indeed, numerous efforts to detect HNSCC using tumor antigens have appeared in the literature over recent decades [17,18] but have suffered from a low specificity with results for cancer patients overlapping with those of healthy control subjects.

Tumor-specific genetic alterations may be detected in plasma or serum DNA of HNSCC patients. This approach should eliminate concern over test specificity; however, the sensitivity under various circumstances must be evaluated. When a panel of 12 microsatellite alterations were analyzed in a small series of 21 HNSCC patients who provided peripheral blood serum samples, 6 (29%) were found to contain DNA with one or more tumor-specific markers. All 6 patients with positive test results had advanced disease (stage III or IV), 5 had regional lymph node metastases, and 3 developed distant metastases [19]. An attempt to streamline this approach using only one microsatellite and *p53* gene mutation as markers in 117 cancer patients was largely unsuccessful in detecting circulating tumor-specific DNA [20]. Microsatellite analysis on plasma is difficult due to the isolation of minute quantities of DNA and is easily misinterpreted. The discordant results achieved in these reports may also be due to the use of more numerous microsatellite markers in the successful study. A third report confirmed the ability to detect tumor DNA in a subset of HNSCC patients (17 of 58) using a panel of 8 microsatellite markers, but found no correlation with disease behavior [21].

Gene promoter hypermethylation, which can be detected using high-throughput, automated methods may prove more useful for detection of circulating tumor DNA. In a pilot study of the serum of 50 HNSCC patients, four genes were tested using methylation-specific PCR (MSP). Over half of the tumors displayed hypermethylation in at least one of the four marker genes. Twenty-one paired tumor and plasma DNA samples had the same methylation pattern. Five of these subjects went on to develop distant metastatic disease compared to only 1 of the 29 individuals without evidence of circulating tumor DNA [22].

Circulating viral DNA is also a promising marker for the presence of tumor in NPC patients. In a study of 167 NPC patients and 77 healthy blood donors, EBV DNA could be detected in 98 of 167 patients' serum prior to treatment, but was also found in 10 of the 77 control samples. This study went on to pilot the use of EBV detection in serial posttreatment samples for tumor surveillance. They found a high degree of specificity for the continued presence or return of EBV DNA in that setting with no false-positive results among 37 patients [23]. A smaller study of 17 NPC patients corroborated these findings. Circulating EBV DNA titers remained low in the 11 patients who remained in remission, while rising substantially in 6 with recurrent disease. In some cases, significant elevations in serum EBV DNA predated the clinical documentation of recurrence by up to 6 months [24,24a].

Nearly one-quarter of all nonnasopharyngeal HNSCC and over one-half of all tonsil and tongue base HNSCC contain DNA of the human papillomavirus (HPV) [25]. Thus, HPV DNA may be a useful marker for a subset of HNSCC cases, especially in surveillance paradigms similar to those just discussed using EBV. In a pilot study of 70 HNSCC patients, serum was analyzed for the presence of HPV DNA using PCR followed by dot-blot analysis. HPV-16 DNA was present in 15 of the 70 tumors (21%). When real-time quantitative PCR was used, six serum samples were found to contain HPV DNA. Four of these individuals went on to develop distant metastatic disease [26]. It appears that HPV DNA may be a useful marker for tumor burden for the subset of oropharyngeal HNSCC for prognostic and surveillance applications.

B. Circulating Tumor-Related Antibodies

The immune response to tumor antigens in the form of circulating tumor-specific antibodies may serve as a marker for the presence of HNSCC. In theory, the production of antibodies to unique tumor antigens may be highly specific and could occur after stimulation with a very small tumor innoculum. The search for tumor-specific antigens has been ongoing for several decades. Most proteins displayed by tumor cells are also present to some degree in normal cell populations. Furthermore, tumor cells seem to enjoy protection from an effective immune response. This may be because of a generalized immunodeficiency in tumor patients, absence of unique antigens, or development of cancer antigens outside of an effective immune-stimulating context.

New information about the genetic alterations that are unique and responsible for tumorigenesis has provided clues as to possible antibody markers for tumor burden. Once again, mutant *p53* gene sequences and viral DNA have served as prototypes for this approach to tumor marker development. In one such study, sera of 120 patients with premalignant or invasive oral cancer lesions were analyzed for the presence of antibodies to p53 protein (wild type) by an Ab-captured enzyme immunoassay. Circulating antibodies were detected in 24 of 70 serum samples from patients with invasive cancer (34%) compared to those of 15 of 50 patients with preinvasive lesions (30%). Four of 63 healthy normal control subjects also had circulating p53 antibodies, although the quantitative level of antibodies was lower, on average, in the control group. All cancer patients that had circulating anti-p53 antibodies had p53 protein expression in their tumors as detected by immunohistochemistry. However, 13 patients whose tumors displayed p53 protein failed to develop detectable levels of anti-p53 antibodies [27]. From these results, it seems that p53 antibodies are not a particularly sensitive or specific marker for HNSCC.

Epstein–Barr virus DNA encodes several proteins that may serve as antigens for the stimulation of antibody production. One of these antigens, thymidine kinase (TK), is highly immunogenic in its full-length form. This antigen was used to develop a simple ELISA test for the serum IgA response to EBV-related NPC. Serum from 52 healthy subjects who were seropositive for EBV was studied, along with that of 87 NPC patients. Three control subjects and 81 of the cancer patients had circulating anti-TK antibodies, for a sensitivity of 93% and a specificity of 94% [28]. Once again, the highly unique nature of EBV-related NPC makes this subset of HNSCC stand apart from other forms of the disease. However, the results are promising for serum detection of disease by the measurement of circulating antibodies if a highly immunogenic tumor-specific antigen has been identified.

IV. MICROARRAYS

The adaptation of silicon chip technology to molecular biology has introduced a powerful and potentially fertile new approach to cancer detection. Using microarrays, a survey of thousands of proteins or mRNA transcripts can be performed in a single assay for a given clinical sample. While the methods involved are not without difficulty and need rigorous adherence to laboratory standards, what is truly challenging is the interpretation of the plethora of data that is produced from each chip. The fluorescent signal reading of the chip is automated, producing individual values for each transcript or protein in the array compared with that of a normal control sample. However, the shear number of molecules for which data are produced renders impractical any approach that considers them individually. Instead, the moieties are clustered or grouped using sophisticated mathematical approaches in an effort to separate samples into categories in which molecular data mirror the clinical parameter(s) (e.g., malignant vs premalignant vs benign).

Another challenge to the application of mRNA expression arrays for the molecular detection of HNSCC is the need for undegenerated, high-quality RNA. In a pilot project to collect useful RNA from exfoliated cells, only the highly abundant transcripts were detectable. At the transcript level of most genes, accurate measure was not possible. This would be expected to be even more problematic in the case of a tumor-related gene present in a small subset of the cells found in a clinical specimen [29].

The results of microarray studies have not matured at present to the point for use in early detection schemes, although it may be expected that such applications will be more commonplace in the near future. Already, profiles of cancer specimen collections are being compiled and reported [30].

V. SPECTROSCOPIC ANALYSIS

In some ways, microarray assessment of thousands of proteins in a tissue specimen at a molecular level conceptually parallels spectroscopic evaluations of gross tissue. Both gather broad-sweeping data that are related to the spectrum of proteins or ribonucleic acid present within the tissue. Spectroscopy in various forms has been piloted for detection of malignancy of the mouth and throat. Midinfrared fiberoptic attenuated total reflectance spectroscopy reportedly can distinguish between benign and malignant oral mucosa due to a weakening or absence of the band at 1745 cm, which is assigned to the vibration of the ester group on triglycerides [31]. Similarly, Raman spectroscopy has been suggested as a means of detecting early malignancy. This approach analyzes scattered photons after monochromatic laser excitation of the tissue. The detailed spectral array can be compared peak by peak or using mathematical manipulations such as linear discriminant analysis to separate the results of diseased from healthy mucosa. Scottish investigators have documented predictive sensitivities of 80–90% for normal, dysplastic, and invasive laryngeal lesions analyzed in this fashion [32]. Similar results have been reported using fluorescence spectroscopy to distinguish between malignant and normal oral and nasopharyngeal mucosal lesions. Optimal excitation wavelengths for the detection of cancer have been proposed [33–35]. Spectroscopic analysis has yet to be field tested for actual *de novo* detection of malignant mucosal disease.

VI. DIAGNOSTIC OR PROGNOSTIC BIOMARKERS IN BIOPSY MATERIAL

Genetic alterations may enhance the information available from tissue biopsies to further refine diagnosis under certain circumstances. For example, directed biopsies from the upper aerodigestive tract taken during the workup of

HNSCC metastatic from an occult primary may show dysplasia or other premalignant change. The question then arises as to whether the metastatic focus and the premalignant cells are related clonally, i.e., is the dysplastic area accurately indicative of a nearby primary invasive cancer site? Another clinical dilemma surrounds the proper management of individuals with multiple or recurring premalignant lesions. Areas destined to progress to invasive cancer would be best treated aggressively with wide local excision or even radiation therapy, whereas those that may remain indolent could be watched expectantly.

A retrospective molecular analysis of directed biopsy tissue taken from patients with metastatic HNSCC from occult primaries was performed. Tumor cells from the metastatic focus were analyzed using a panel of microsatellites. Available markers were then used to screen the histologically normal tissue from directed biopsies that had been preserved in formalin. In 6 of 10 cases, concordant alterations in at least one marker were identified in tumor and at least one biopsy site. In two of these cases, the location of the biopsy identified as containing cells related clonally to the tumor (bearing identical microsatellite alterations) corresponded closely to the location of the subsequent development of clinically apparent disease presumed to be the primary [36].

Loss of heterozygosity of microsatellite markers mapped to 3p, 9p, and 17p has been shown by several investigators to accurately predict which premalignant lesions will progress to invasive cancer. LOH at 3p and 9p confers a relative risk of 3.8 for tumor progression, whereas the additional presence of LOH on 4q, 8p, 11q, or 17p is associated with a risk of 33-fold [37–39]. Combined with enhanced means of inspection of the mucosa using a supravital stain such as Toluidine blue, microsatellite analysis may prove of value in not only detecting early lesions, but in distinguishing clinically relevant changes [40].

VII. CONCLUSION

Molecular detection of HNSCC using high-throughput, low-cost, and accurate methods will be integrated into clinical practice in the near future. The marriage of molecular biology to clinical practice in this manner promises to be a fruitful one, providing oncologists with an opportunity to identify patients during the earliest and most amenable stages for treatment and cure with currently available treatment modalities.

References

1. Adam, G. L. (1998). Malignant tumors of the larynx and hypopharynx. *In* "Otolaryngology-Head and Neck Surgery," 3rd Ed. (C. W. Cummings *et al.*, eds.), pp. 2130–2175. Mosby-Year Book, St. Louis, MO.
2. Boyle, J. O., Hakim, J., Koch, W., *et al.* (1993). The incidence of p53 mutations increases with progression of head and neck cancer. *Cancer Res.* **53**, 4477–4480.
3. Reed, A. L., Califano, J., Cairns, P., *et al.* (1996). High frequency of p16 (CDKN2/MTS-1/INK4A) inactivation in head and neck squamous cell carcinoma. *Cancer Res.* **56**, 3630–3633.
4. Nawroz, H., van der Riet, P., and Hruban, R., *et al.* (1994). Allelotype of head and neck squamous cell carcinoma. *Cancer Res.* **54**, 1152–1155.
5. Califano, J., van der Riet, P., Clayman, G., *et al.* (1996). A genetic progression model for head and neck cancer: Implications for field cancerization. *Cancer Res.* **56**, 2488–2492.
6. Rosin, M. P., Epstein, J. B., Berean, K., *et al.* (1997). The use of exfoliative cell samples to map clonal genetic alterations in the oral epithelium of high-risk patients. *Cancer Res.* **57**, 5258–5260.
7. Nunes, D. N., Kowalski, L. P., and Simpson, A. J. (2000). Detection of oral and oropharyngeal cancer by microsatellite analysis in mouth washes and lesion brushings. *Oral Oncol.* **36**, 525–528.
8. Rizos, E., Sourvinos, G., Spandidos, D. A. (1998). Loss of heterozygosity at 8p, 9p, and 17q in laryngeal cytological specimens. *Oral Oncol.* **34**, 519–523.
9. Mao, L., Schoenberg, M. P., Scicchitano, M., *et al.* (1996). Molecular detection of primary bladder cancer by microsatellite analysis. *Science* **271**, 659–662.
10. Spafford, M. F., Koch, W. M., Reed, A. L., *et al.* (2001). Detection of head and neck squamous cell carcinoma among exfoliated oral mucosal cells by microsatellite analysis. *Clin. Ca. Res.* **7**, 607–612.
11. El-Naggar, A. K., Mao, L., and Staerkel, G., *et al.* (2001). Genetic heterogeneity in saliva from patients with oral squamous carcinomas: Implications in molecular diagnosis and screening. *J. Mol. Diagn.* **3**, 164–170.
12. Harty, L. C., Shields, P. G., Winn, D. M., *et al.* (2000). Self-collection of oral epithelial cell DNA under instruction from epidemiologic interviewers. *Am. J. Epidemiol.* **151**, 199–205.
13. Rosas, S. L. B., Koch, W., Carvalho, M. G. C., *et al.* (2001). Promoter hypermethylation patterns of p16, O^6-methylguanine-DNA-methyltransferase, and death-associated protein kinase in tumors and saliva of head and neck cancer patients. *Ca. Res.* **61**, 939–942.
14. Tune, C. E., Liavaag, P. G., Freeman, J. L., *et al.* (1999). Nasopharyngeal brush biopsies and detection of nasopharyngeal carcinoma in a high risk population. *J. Natl. Cancer Inst.* **91**, 796–800.
15. Lin, S., Tsang, N., Kao, S., *et al.* (2001). Presence of Epstein-Barr virus *latent membrane protein-1* gene in the nasopharyngeal swabs from patients with nasopharyngeal carcinoma. *Head Neck* **23**, 194–200.
16. Lee, W. Y., Hsiao, J. R., Jin, Y. T., *et al.* (2000). Epstein-Barr virus detection in neck metastases by in-situ hybridization in fine-needle aspiration cytologic studies: An aid for differentiating the primary site. *Head Neck* **22**, 336–340.
17. Eibling, D. E., Johnson, J. T., and Wagner, R. L. (1989). SCC-RIA in the diagnosis of squamous cell carcinoma of the head and neck. *Laryngoscope* **99**, 117–124.
18. Yen, T. C., Lin, W. Y., Kao, C. H., *et al.* (1998). A study of a new tumour marker, CYFRA 21-1 in squamous cell carcinoma of the head and neck, and comparison with squamous cell carcinoma antigen. *Clin. Otolaryngol.* **23**, 82–86.
19. Nawroz, H., Koch, W., Anker, P., *et al.* (1996). Microsatellite alterations in serum DNA of head and neck cancer patients. *Nature Med.* **2**, 1035–1037.
20. Coulet, F., Blons, H., Cabelguenne, A., *et al.* (2000). Detection of plasma tumor DNA in head and neck squamous cell carcinoma by microsatellite typing and p3 mutation analysis. *Cancer Res.* **60**, 707–711.
21. Nunes, D. N., Kowalski, L. P., and Simpson, A. J. (2001). Circulating tumor-derived DNA may permit the early diagnosis of head and neck squamous cell carcinomas. *Int. J. Cancer* **92**, 214–219.

22. Sanchez-Cespedes, M., Esteller, M., Wu, L., *et al.* (2000). Gene promoter hypermethylation in tumors and serum of head and neck cancer patients. *Cancer Res.* **60**, 892–895.
23. Shotelersuk, K., Khorprasert, C., Sakdikul, S., *et al.* (2000). Epstein-Barr virus DNA in serum/plasma as a tumor marker for nasopharyngeal cancer. *Clin. Cancer Res.* **6**, 1046–1051.
24. Lo, Y. M. D., Chan, Y. S. L., Chan, A. T. C., *et al.* (1999). Quantitative and temporal correlation between circulating cell-free Epstein Barr virus DNA and tumor recurrence in nasopharyngeal carcinoma. *Cancer Res.* **59**, 5452–5455.
24a. Lo, Y. M. D. (2001). Quantitative analysis of Epstein-Barr virus DNA in plasma and serum. *Ann. N. Y. Acad. Sci.* **945**, 68–72.
25. Gillison, M. L., Koch, W. M., Capone, R. B., *et al.* (2000). Evidence for a causal association between human papillomavirus and a subset of head and neck cancers. *J. Natl. Cancer Inst.* **92**, 709–720.
26. Capone, R. B., Pai, S. I., Koch, W. M., *et al.* (2000). Detection and quantitation of human papillomavirus (HPV) DNA in the sera of patients with HPV-associated head and neck squamous cell carcinoma. *Clin. Cancer Res.* **6**, 4171–4175.
27. Ralhan, R., Nath, N., Agarwal, S., *et al.* (1998). Circulating p53 antibodies as early markers of oral cancer: Correlation with p53 alterations. *Clin. Cancer Res.* **4**, 2147–2152.
28. Connolly, Y., Littler, E., Sun, N., *et al.* (2001). Antibodies to Epstein-Barr virus thymidine kinase: A characteristic marker for the serological detection of nasopharyngeal carcinoma. *Int. J. Cancer* **91**, 692–697.
29. Klaassen, I., Copper, M. P., Brakenhoff, R. H., *et al.* (1998). Exfoliated oral cell messenger RNA: Suitability for biomarker studies. *Cancer Epid. Biomark. Prev.* **7**, 469–472.
30. Alevizos, I., Mahadevappa, M., Zhang, X., *et al.* (2001). Oral cancer in vivo expression profiling assisted by laser capture microdissection and microarray analysis. *Oncogene* **20**, 6196–6204.
31. Wu, J. G., Xu, Y. Z., Sun, C. W., *et al.* (2001). Distinguishing malignant from normal oral tissues using FTIR fiber-optic techniques. *Biopolymers* **62**, 185–192.
32. Stone, N., Stavroulaki, P., Kendall, C., *et al.* (2000). Raman spectroscopy for early detection of laryngeal malignancy: Preliminary results. *Laryngoscope* **110**, 1756–1763.
33. Heintzelmann, D. L., Utzinger, U., Fuchs, H., *et al.* (2000). Optimal excitation wavelength for in vivo detection of oral neoplasia using fluorescence spectroscopy. *Photochem. Photobiol.* **72**, 103–113.
34. Qu, J. Y., Wing, P., Huang, Z., *et al.* (2000). Preliminary study of in vivo autofluorescence of nasopharyngeal carcinoma and normal tissue. *Lasers Surg. Med.* **26**, 432–440.
35. Leunig, A., Betz, C. S., Mehlmann, M., *et al.* (2000). Detection of squamous cell carcinoma of the oral cavity by imaging 5-aminolevulinic acid-induced protoporphyrin IX fluorescence. *Laryngoscope* **110**, 78–83.
36. Califano, J., Westra, W., Koch, W., *et al.* (1999). Unknown primary head and neck squamous cell carcinoma: Molecular identification of site of origin. *J. Natl. Cancer Inst.* **91**, 599–604.
37. Mao, L., Lee, J. S., Fan, Y. H., *et al.* (1996). Frequent microsatellite alterations at chromosomes 9p21 and 3p14 in oral premalignant lesions and their value in cancer risk assessment. *Nature Med.* **2**, 682–685.
38. Partridge, M., Pateromichelakis, S., Phillips, E., *et al.* (2000). A case-control study confirms that microsatellite assay can identify patients at risk of developing oral squamous cell carcinoma within a field of cancerization. *Cancer Res.* **60**, 3893–3898.
39. Rosin, M. D., Cheng, X., Poh, C., *et al.* (2000). Use of allelic loss to predict malignant risk for low-grade oral epithelial dysplasia. *Clin. Cancer Res.* **6**, 357–362.
40. Guo, Z., Yamaguchi, K., Sanchez-Cespedes, M., *et al.* (2001). Allelic losses in Orascan directed biopsies of patients with prior upper aerodigestive tract malignancy. *Clin. Cancer Res.* **7**, 1963–1968.

PART IV

CURRENT APPROACHES

Surgical Considerations (Chapters 22–29)

Nonsurgical Considerations (Chapters 30–33)

CHAPTER

22

Management of the Neck in Squamous Cell Carcinomas of the Head and Neck

JESUS E. MEDINA

Department of Otorhinolaryngology
The University of Oklahoma Health Sciences Center
Oklahoma City, Oklahoma 43190

I. Introduction 317
II. Incidence of Cervical Metastases 317
A. Cancer of the Pharynx 317
B. Cancer of the Larynx 318
C. Cancer of the Oral Cavity 318
III. Patterns of Lymphotic Spread 319
IV. Prognostic Implications of Neck Node Metastases 320
V. Current Management of Lymph Nodes of the Neck 321
A. Oropharyngeal Cancers 321
B. Cancer of the Larynx 321
C. Cancer of the Oral Cavity 323
VI. Types of Neck Dissection 324
References 324

I. INTRODUCTION

Appropriate management of the regional lymph nodes is important in the treatment of patients with squamous cell carcinoma (SqCC) of the upper aerodigestive tract. Significant changes have occurred in the management of the lymph nodes of the neck in the past two decades. Today, in addition to radical neck dissection, described by Crile [1] in 1906 and popularized by Martin and colleagues [2] during the 1950s, surgical treatment of the neck includes modifications of this operation and other types of neck dissection. Also, cervical lymph node metastases are frequently treated with combinations of surgery and radiation therapy and sometimes with radiation therapy alone. Thus, appropriate treatment planning requires the clinician to have a working understanding of the incidence, patterns, and prognostic implications of lymph node metastases for the different head and neck tumor sites and of the current treatment strategies.

II. INCIDENCE OF CERVICAL METASTASES

Although the ability to detect occult cervical lymph node metastases has improved with the advent of modern imaging techniques, particularly when combined with fine needle aspiration cytology, a considerable number of such metastases remain undetectable without histopathological examination of the lymph nodes. Currently, therapeutic decisions in patients with cancer of the upper aerodigestive tract and stage N0 neck are, for the most part, based on the probability of lymph node metastases for a given tumor. For many years, clinicians have used the stage of the primary tumor and its site of origin to estimate the probability of lymph node metastases. Lindberg's [3] classical incidence figures, which were based on the presence of palpablely phadenopathy in over 2000 patients with SqCC of the head and neck, have been refined by the studies of Byers *et al.* [4] and Shah [5] of a large number of neck dissection specimens and by numerous studies trying to identify histological, cytological, and molecular features of the primary tumor that can predict its propensity to metastasize to the lymph nodes.

A. Cancer of the Pharynx

Pharyngeal carcinomas have a higher propensity to metastasize to the lymph nodes than carcinomas of the larynx and oral cavity. In fact, this propensity is so high for carcinomas of the nasopharynx, tonsillar fossa, base of the tongue, and hypopharynx that the rate of occurrence of lymph node

TABLE 22.1 Incidence of Lymph Node Metastases by Site and Stage (Based on Physical Examination)

Site of primary tumor	Patients with lymph node metastases (%)			
	T1	T2	T3	T4
Tonsil	70.5	67.5	70	89.5
Base of tongue	70	71	74.5	84.5
Pharyngeal walls	25	30	67	76

metastases in patients with small (T1 and T2) and large (T3 and T4) tumors is between 70 and 90% (Table 22.1).

B. Cancer of the Larynx

The frequency of nodal metastases according to site of origin within the larynx and T stage is shown in Tables 22.2 and 22.3.

In addition to the T stage, McGavran *et al.* [11] have studied the relationship between various characteristics of the primary tumor and the probability of lymph node metastases in a homogeneous series of 96 patients who underwent *en bloc* laryngectomy and radical neck dissection. They found that the incidence of lymph node metastases was significantly higher in patients whose tumors measured more than 2 cm in diameter, were poorly differentiated, or exhibited an infiltrating rather than a pushing margin or perineural invasion. Unfortunately, these authors did not perform a multivariate analysis. More recently, Kowalski *et al.* [12] performed a similar study of 103 cases of laryngeal carcinoma who underwent either unilateral or bilateral comprehensive neck dissection. Interestingly, a logistic regression analysis demonstrated that tumor site (supraglottic origin) and poor histological differentiation were the only predictors of nodal metastases. When they consider only cases staged N0, the probability of occult lymph node metastases was influenced significantly only by a supraglottic origin of the primary tumor [12]. In studies of advanced laryngeal cancer, the highest incidence of cervical lymph node metastases is associated with supraglottic tumors (65%), whereas metastases are found in only 20–22% of advanced glottic tumors [13,14].

A number of other putative predictive factors have been studied. For instance, some investigators have found a correlation between DNA index and lymph node status in laryngeal carcinomas, with a higher incidence of nodal metastases in patients with aneuploid tumors [15,16].

Expression of epidermal growth factor (EGFR) is another potentially useful biological marker. A significant correlation between expression of EGFR and the risk of lymph node metastases was observed by Maurizi *et al.* [17] in a study of 140 cases of laryngeal carcinoma.

TABLE 22.2 Frequency of Nodal Metastases in Glottic Cancer

Stage	Yuen *et al.* [6] (%)	Daly and Strong [7] (%)	DeSanto *et al.* [8] (%)
T1–T2	—	5–8	1–7
T3	11	15	<20
T4	16	—	—

TABLE 22.3 Frequency of Nodal Metastases in Supraglottic Cancer

"T" stage	Shah and Tollefsen [9] (%)	DeSanto [10] (%)
T1	40	17
T2	42	39
T3	55	52
T4	65	90

C. Cancer of the Oral Cavity

The frequency of nodal metastases according to site of origin within the oral cavity and T stage is shown in Table 22.4.

Cancers of the oral cavity have been investigated more extensively than other tumors of the upper aerodigestive tract regarding clinical, histological, biochemical, and genetic factors that may be useful predictors of the propensity of a tumor to metastasize to the lymph nodes, and thus may be useful in treatment planning. In a study of 126 patients with squamous cell carcinoma of the oral cavity who underwent neck dissection as part of their treatment, Martinez-Gimeno and colleagues [18] found a statistically significant association between lymph node metastasis and the presence of microvascular invasion, grade of differentiation, tumor thickness, inflammatory infiltration, tumoral interphase, and the presence of perineural spread. These investigators designed a scoring system based on their results with scores ranging from (a) 7 to 12, (b) 13 to 16, (c) 17 to 20, and (d) 21 to 30 points. The respective risk of metastasis was (a) 0%, (b) 20%, (c) 63.6%, and (d) 86.3%. Prognostic score systems like this have not been embraced

TABLE 22.4 Incidence of Lymph Node Metastases in Oral Cavity[a]

Site of primary tumor	Necks with nodal Metastases (%)			
	T1	T2	T3	T4
Oral tongue	14	30	47.5	76.5
Floor of mouth	11	29	43.5	53.5
Retromolar trigone	11.5	37.5	54	67.5

[a]Modified from Lindberg [3].

widely, possibly because they are not simple and easily reproducible and they require full histopathological evaluation of the primary tumor on permanent sections, which are the characteristics of an ideal "predictive factor." Thus, the search continues for such a factor or parameter.

1. Tumor Thickness

The incidence of nodal metastases in patients with squamous cell carcinoma of the floor of the mouth and oral tongue appears to increase as a function of the thickness of the primary tumor. In 1986, Mohit-Tabatabai *et al.* [19] found a significant correlation between tumor thickness greater than 1.5 mm and subsequent development of neck metastases in a series of patients with stage I and II carcinomas of the floor of mouth. In the same year, Spiro *et al.* [20] found, in a study of 105 patients with oral and oropharyngeal carcinoma with N0 necks, that lymph node metastases occurred more frequently in patients whose tumors measured 2 mm in thickness or more.

Using mathematical modeling, the estimated risk of cervical node metastasis, relative to the thickness of squamous cell carcinomas of the oral cavity, was found to be approximately 3.9% for 1-mm-thick cancers, 17% for 2-mm-thick cancers, and 25% for 3-mm-thick cancers [21]. Similar findings by Frierson and Cooper [22] in patients with carcinoma of the lower lip and by Rasgon *et al.* [23] in patients with oral and oropharyngeal carcinoma support the importance of thickness as a predictor of nodal metastases. However, its usefulness in planning treatment of the neck is limited by the difficulty in determining tumor thickness by inspection or in biopsy specimens.

2. Histologic Grade

Although there is no general agreement about the prognostic impact of histologic differentiation, several studies have demonstrated a higher incidence of neck node metastases in patients with poorly differentiated squamous cell carcinoma [22,24–26]. In oral cavity tumors, a higher grade of malignancy, in terms of degree of differentiation and character of the borders, may increase the risk of metastases beyond the first and second echelon of regional lymph nodes [27] and may increase the risk of occult metastases in lower levels of the neck. In a multivariate analysis of a cohort of patients with T1 and T2 squamous cell carcinoma of the oral tongue, Beenken *et al.* [28] found that the worse prognosis in this population of patients with early tongue cancer was associated with the presence of a poorly differentiated tumor.

3. Invasive Margin of the Tumor

More recently, attention has focused on the "malignancy grading" and character of the invasive margin of the tumor (pushing vs infiltrating). Bryne *et al.* [29] have shown that oral cavity carcinomas that exhibit an infiltrating margin with abundant mitoses and nuclear polymorphism are associated with a dismal survival. Others have shown an increased risk of occult lymph node metastases for such tumors [22,26,27,30,31].

4. Vascular Invasion

The role of vascular invasion as a risk factor for lymph node metastases is unclear. In a study of 43 patients, Close *et al.* [32] found lymph node metastases in 77% of the cases in which vascular invasion was present, but only in 25% of the cases in which it was not present. While Crissman and co-workers [30] and Poleksic and colleagues [33] have found a similar correlation, others have not [22,24].

5. Perineural Invasion

Several studies have suggested a strong correlation between the presence of perineural invasion at the primary site and cervical lymph node metastases [22,26].

6. Inflammatory Infiltrate

A marked inflammatory infiltrate in the stroma surrounding the tumor has been found to correlate with a lower incidence of lymph node metastases in some studies [19,22,26]. Unfortunately, other studies have not shown such a correlation [24,30,32].

7. DNA Content

A recent prospective study correlated various tumor and patient factors with the presence or absence of pathologically positive nodes in a group of 91 patients with cancer of the oral tongue who underwent glossectomy and elective neck dissection. The best predictors of nodal metastases were depth of muscle invasion, double DNA aneuploidy, and poor histologic differentiation [34].

III. PATTERNS OF LYMPHATIC SPREAD

Anatomical and radiographical studies of the lymphatics of the head and neck have demonstrated that lymphatic drainage of the different areas of the upper aerodigestive tract occurs along predictable pathways [35,36]. Furthermore, multiple clinical studies have demonstrated that tumors from these areas metastasize to the lymph nodes following the same pathways, at least as long as the neck has not been treated previously with surgery or radiation [5,37]. It is now commonly accepted that the lymph node groups that harbor metastases most frequently in patients with carcinomas of the oral cavity are the submental (Ia), submandibular (Ib), upper jugular (II), and midjugular (III), whereas the lymph nodes along the jugular vein (II, III, IV) are involved more frequently in patients with tumors of the oropharynx, larynx, and hypopharynx. It has been shown convincingly that these patterns of distribution occur both in patients staged clinically N0 who are found to have occult metastases and in patients with palpable, histologically proven lymph node metastases [5,37]. It must be kept in mind, however, that skip metastasis—meaning direct metastases beyond the first or second echelon of expected lymphatic drainage—does occur [38].

In addition to these lymphatic pathways, the retropharyngeal nodes are a common site for metastases in tumors of the hypopharynx, tonsillar fossa, soft palate, and posterior and lateral oropharyngeal walls [39,40]. The paratracheal nodes are a common site of metastases for laryngeal carcinomas that involve the subglottic region and for carcinomas of the cervical esophagus. In a study that included 91 patients with carcinoma of the larynx who underwent paratracheal lymph node dissection, Weber and colleagues found that metastases to these lymph nodes occurred more often in patients with subglottic tumors (40%) and transglottic tumors (21.4%), but they also occurred in patients with glottic (13%) and supraglottic tumors (15.7%). The presence of paratracheal lymph node (PTLN) metastases had a significant negative impact on survival. While the survival at 48 months for the entire group of 141 patients was 60%, none of the 29 patients with PTLN metastases survived beyond 42 months ($p < 0.001$) [41–43].

IV. PROGNOSTIC IMPLICATIONS OF NECK NODE METASTASES

The presence of clinically obvious, histologically proven lymph node metastases is the single most important prognostic factor in patients with SqCC of the head and neck. In general, it decreases the overall survival by at least one-half [44]. However, the unfavorable impact on survival varies depending on the following factors.

a. Presence of extracapsular spread (ECS) of tumor. Numerous studies have demonstrated that tumor extension beyond the capsule of a lymph node worsens the prognosis. Johnson *et al.* [45] have reported that less than 40% of patients with histologic evidence of ECS were free of disease 24 months after therapy. Furthermore, the survival of these patients was significantly lower than comparable patients whose metastases were confined to the lymph nodes [46]. Similarly, Steinhart *et al.* [47] found that the rate of extracapsular spread was especially high (70%) in patients with carcinomas of the hypopharynx and that the 5-year survival rate was greatly different for patients with extracapsular spread of tumor (28%) and patients with no metastases (77%). Interestingly, a correlation between the degree of ECS and prognosis has not yet been established clearly, although Carter *et al.* [48] have reported that macroscopically recognizable ECS carries a worse prognosis than microscopic spread.

b. Desmoplastic stromal pattern. In a study done at the Mayo Clinic, a desmoplastic stromal pattern in the lymph nodes involved by tumor was associated with almost a sevenfold increase in the risk for recurrence in the neck. The study included 284 patients with pathologically confirmed metastatic squamous cell carcinoma who underwent neck dissection and received no adjuvant therapy. This finding has not been reported by others [49].

c. The number of lymph nodes involved. The survival of patients with histologically positive nodes is significantly lower when multiple nodes are involved [44]. In a study by Leemans and colleagues [96], the overall incidence of distant metastases in a group of 281 head and neck cancer patients who underwent a neck dissection was 10.7%, whereas it was 46.8% in the group of patients with three or more positive nodes.

d. The level of the neck metastases. Several studies have suggested that survival decreases as lymph nodes in lower levels of the neck become involved [25,50–52]. This tendency has been best demonstrated by Ho [53] and Teo *et al.* [54] for nasopharyngeal carcinoma: the lower the level, the worse the prognosis. Grandi *et al.* [55] have defined three levels in the neck (upper, middle, and lower) divided by two imaginary lines that pass through the hyoid bone and through the lower border of the thyroid cartilage. They found that the worse prognosis was associated with the presence of nodal metastases at the lower level. Based on these observations, a new staging system for the neck in patients with nasopharyngeal carcinoma has been implemented [55]. Likewise, the most recent American Joint Committee on Cancer Staging System includes a notation on the level location of the nodes for the N1–N3 staging categories [97].

The prognostic significance of nonpalpable (occult) metastases in the N0 neck is less clearly defined. It is more difficult to study because the T stage also affects prognosis and may overshadow the effect of occult neck nodes [56]. However, it has been shown that ECS does occur in nonpalpable lymph nodes [4]. Also, as in patients with palpable nodal metastases, the number and location of involved nodes in the N0 neck appear to affect prognosis [52,57–59]. For instance, Kalnins *et al.* [60] found that patients with SqCC of the oral cavity and uninvolved neck nodes had a 75% 5-year survival. Survival decreased to 49% when one node was involved histopathologically, 30% when two nodes were involved, and 13% when three or more nodes were involved by tumor.

V. CURRENT MANAGEMENT OF LYMPH NODES OF THE NECK

A. Oropharyngeal Cancers

Tumors of the oropharynx have a high propensity to metastasize to the regional lymph nodes even in early stages. Therefore, the regional lymph nodes should be treated electively, regardless of the stage of the primary tumor. With the exception of early, well-lateralized lesions, carcinomas of the oropharynx have a tendency to metastasize to both sides of the neck, and therefore treatment of both sides of the neck is often indicated. Treatment of the retropharyngeal nodes should be considered for tumors extending onto the pharyngeal walls, as historically, retropharyngeal nodal metastasis occurred in up to 44% of cases [61].

1. N0

In general, the treatment modality used for neck nodes depends on the treatment modality used for the primary tumor. When the primary is treated with radiation, elective irradiation of the neck nodes can result in regional control in over 95% of the cases [62]. If the intent is to treat the primary tumor with surgery alone, elective treatment of the neck should consist of bilateral neck dissections. Postoperative irradiation is indicated when metastases are found in multiple lymph nodes or when there is extracapsular extension of the tumor. This usually consists of a radiation dose of 50 to 63 Gy and is delivered as soon as healing is adequate, preferably before 6 weeks after surgery.

2. N1

Elective neck irradiation or neck dissection results in a high rate of regional control, and the treatment choice is often based on the preferences of the institution. Ideally though, treatment of the neck should again be tailored by treatment of the primary tumor.

A number of head and neck oncologists believe that when patients present with clinically palpable disease, they should undergo combined surgery and radiation. This preference must be identified prior to the initiation of radiation therapy if that is the selected treatment for the primary. In these cases, while the upper jugular and midjugular lymph node groups are treated with the same high dose of radiation as the primary tumor, the remainder of the neck can be treated with a minimal dose of 50 Gy. The surgeon then performs neck dissection 5 to 6 weeks after the completion of radiation.

Those who argue for surgical management of all N1 necks point out the additional staging information gained through histologic evaluation of the lymph nodes. Assessment for extracapsular spread or microscopic involvement of additional subclinical lymph nodes would again indicate the need for postoperative radiation therapy.

3. N2 or N3

Patients with oropharyngeal cancer and advanced neck metastases can be managed in different ways, depending on the size of the primary tumor and the choice of treatment for it. If the primary tumor is advanced and requires surgery, an appropriate ipsilateral or bilateral neck dissection, including the retropharyngeal nodes, is performed. Postoperatively, radiation is delivered to both the primary and the neck. If the primary tumor is amenable to radiotherapy, but the tumor in the neck warrants combined therapy, the patient can be treated with radiotherapy to the primary tumor and the neck, followed by neck dissection 6–8 weeks later. When organ preservation protocols combining chemotherapy and radiation are utilized, even in the case of a clinical complete response in the neck, dissection is indicated for these patients, as 25% will have an incomplete histologic response and are at risk for higher rates of recurrence [63].

Interestingly, patients with neck metastases from oropharyngeal cancer who respond completely to treatment with the "concomitant boost" regimen do not appear to need a planned neck dissection. In a recent analysis of 75 patients treated in this manner who did not undergo neck dissection, the recurrence rate in the neck at 2 years was 13% among patients whose nodal disease was 3 cm or larger and 15% among those whose nodes were less than 3 cm in diameter [64].

B. Cancer of the Larynx

1. N0

The efficacy of elective treatment of the neck in patients with larynx cancer has been compared to that of therapeutic neck dissection at the time metastases become clinically apparent in an interesting retrospective study by Gallo *et al.* [65]. They found no significant difference in the 5-year survival of a group of 76 patients who underwent elective neck dissection and were found to have histologically positive nodes and a group of 96 patients who were initially staged N0, but subsequently developed lymph node metastases and underwent therapeutic neck dissection. However, other retrospective studies have found that elective neck dissection decreases the neck recurrence rates significantly in patients treated for N0 supraglottic carcinoma [66].

Self-examination by the patient and reliable follow-up are essential for watchful waiting to succeed in the management of the N0 neck. Unfortunately, a significant number of the patients who do not undergo elective neck dissection cannot be salvaged later, when they present with palpable metastases, because the disease is too far advanced [67].

Consequently, most head and neck surgeons prefer to treat the neck electively, even though the impact of this decision on patient survival remains controversial.

Another alternative consists of performing intraoperative frozen sections of the jugulodigastric lymph nodes. If the presence of metastasis is demonstrated, neck dissection is performed [68].

Indications for elective treatment of the neck vary depending on the site of origin and extent of the primary tumor. The probability of occult lymph node metastases in glottic cancers is low; accordingly, elective treatment of the regional lymph nodes is not indicated in patients with T1 and T2 cancers. While some surgeons believe that elective treatment of the lymph nodes is not indicated in patients with T3 glottic tumors, others have observed lymph node metastases in 17–22% of the cases and recommend elective treatment of the neck [14,69]. The incidence of lymph node metastases in patients with T4 tumors is sufficiently high to warrant elective treatment.

The probability of occult metastases when the primary tumor is located in the supraglottic larynx is sufficiently high, regardless of the tumor stage, that elective treatment is warranted. A possible exception to this may be T1 tumors of the suprahyoid epiglottis. Because "as many as 75% of cervical failures occur in the undissected" contralateral side of the neck, elective treatment of the neck in supraglottic cancers should include the lymph nodes at risk on both sides of the neck [70].

It is difficult to make recommendations about elective treatment of the cervical lymph nodes in subglottic cancers. On the one hand, because of our knowledge about the distribution of lymph node metastases in these cases, it seems prudent to treat the paratracheal lymph nodes on both sides. On the other hand, elective treatment of the lateral compartments of the neck does not seem necessary, as the reported incidence of metastases to these nodes is relatively low (10%) [71,72].

In general, when surgery is selected as the treatment modality for the primary tumor and elective treatment of the lymph nodes is indicated, a neck dissection is performed. Elective surgical treatment of the neck in patients with larynx cancer should consist of a lateral neck dissection [73] or a modified radical neck dissection type III ("functional neck dissection"), which removes lymph node levels II, III, IV, and V and preserves the sternocleidomastoid muscle, the internal jugular vein, and the spinal accessory nerve. In addition, ipsilateral paratracheal node dissection may be beneficial for "transglottic" tumors and for tumors that exhibit subglottic extension.

The rationale to include level V in elective neck dissections for patients with cancer of the larynx has to be questioned in light of the results of studies of functional neck dissection specimens from patients with larynx cancer. These studies have shown that the lymph nodes of the posterior triangle of the neck are seldom involved in these patients [15,65]. Interestingly, Ambrosch *et al.* [74] advocate dissecting levels II and III only. They have reported very low recurrence rates in level IV or anywhere in the neck using this approach. Finally, the efficacy of elective lateral neck dissection and modified radical neck dissection has been compared in a prospective randomized study of patients with supraglottic and transglottic cancers [75]. This study found that the rates of 5-year overall survival and neck recurrence were similar for the two operations. Consequently, it is not necessary to dissect level V nodes routinely when treating the neck electively in patients with larynx cancer.

Bilateral neck dissection is indicated in the surgical management of cancer of the supraglottic larynx [76]. With the use of selective neck dissection, the issue of postoperative morbidity has been reduced significantly. In a review of the clinical course of 76 patients undergoing excision of supraglottic squamous carcinoma combined with bilateral neck dissection, a decrease in neck recurrences from 20 to 9% was attributed to the use of bilateral neck dissection [76].

However, if a laryngeal cancer is treated with radiation, the lymph nodes at high risk are treated with radiation as well. Following elective neck irradiation, the risk of developing clinically positive nodes in an N0 neck is about 5% [77,78]. In another study, isolated neck failure was observed in 1% of 413 patients with larynx cancer treated by elective neck irradiation at the Institute Gustave Roussi [79]. This compares favorably with a 2.9% rate of recurrence in the neck in 328 cases treated at the same institution with selective neck dissection extended to a modified radical neck dissection if positive at frozen section.

2. N1–N3

When palpable cervical lymph node metastases are present in patients with cancer of the larynx, a neck dissection is the mainstay of treatment. The extent of the dissection to be performed depends on the extent of the disease in the neck.

Numerous studies suggest that the rate of tumor recurrence in the neck is decreased by the addition of radiation when multiple nodes are involved at multiple levels of the neck and when ECS of tumor is found [80–83]. The dose of postoperative radiation therapy is essential to achieve optimal results. Daily fractions of 1.8 Gy to a total dose of 57.6 Gy to the entire operative bed is currently recommended. Sites of increased risk for recurrence, such as areas of the neck where ECS of tumor was found, should be boosted to 63 Gy [84]. Timing of the initiation of radiotherapy is also important; delays beyond 6 weeks may compromise tumor control [80].

In cases with advanced neck metastases (>3 cm in diameter) with a small primary tumor in the larynx in which the neck has to be treated with surgery while the primary can be treated successfully with radiation alone, Verschuur *et al.*

[85] have performed a neck dissection first, followed within 4 weeks by radiotherapy to the primary and to the neck. In a group of 15 patients treated in this manner, no recurrence in the neck was observed at 5 years [85]. Using a similar treatment approach in 65 patients with tumors of the larynx and hypopharynx, the French Head and Study Group observed a rate of recurrence in the neck of 4.6% [86].

C. Cancer of the Oral Cavity

1. N0

The value of elective treatment of the neck is not universally accepted. The notion of watching the neck and treating it only when metastases become clinically apparent is allegedly supported by two prospective randomized studies involving cancers of the oral cavity [67,87]. In both studies, the survival of patients who underwent "elective" neck dissection was not significantly better than the survival of patients who underwent a delayed therapeutic neck dissection. Unfortunately, these studies did not resolve the controversy; in fact, they have been criticized because the number of patients studied was insufficient to be conclusive. More recently, a retrospective study done by Yuen *et al.* [88] and a prospective, controlled study reported by Kligerman *et al.* [89] have shown a statistically significant survival advantage for patients with cancer of the tongue in whom the neck nodes were treated electively.

Surgical treatment of the N0 neck is preferred when surgery is selected for the treatment of the primary tumor, particularly when the expectations of controlling the primary with surgery alone are reasonably good. In such cases (i.e., T2 and selected T3 tumors), appropriate dissection of the regional lymph nodes alone is, in most cases, sufficient to control the disease in the neck. Surgical dissection of the cervical lymph nodes is also desirable when the neck must be entered, and certain structures, such as the hypoglossal nerve or the carotid artery, need to be exposed to facilitate adequate resection of the primary tumor.

Postoperative radiation therapy may also be beneficial when the following features are found on histopathological examination of the node dissection specimen(s).

Presence of tumor in more than two lymph nodes.
Presence of tumor in multiple node groups.
Presence of extracapsular spread of tumor.

When a single metastasis is found in the lymph nodes (pN_1), surgery alone has been considered adequate treatment. Regional recurrence rates from 16 to 25%, however, have been reported with surgery alone, and it has been suggested that postoperative radiation may be beneficial [93]. Unfortunately, the role of postoperative radiation has not been clearly defined for this group of patients, as a review from multiple institutions showed no statistically significant difference in the rate of recurrence in the neck in patients with tongue cancer, staged N0 clinically and staged N1 pathologically, that were treated with either surgery alone or surgery and postoperative radiation therapy [93,94].

Surgical dissection of the cervical lymph nodes is also desirable when the neck must be entered and certain structures, such as the hypoglossal nerve or the carotid artery, need to be exposed to facilitate the resection of the primary tumor.

The elective neck dissection most commonly done in patients with oral cancer is the supraomohyoid neck dissection, which consists of the *enbloc* removal of nodal regions I, II, and III. The operation is performed on both sides of the neck in patients with cancers of the anterior tongue and floor of the mouth. A bilateral dissection is performed when the lesion is located at or near the midline.

Using the supraomohyoid dissection in this manner, the rate of recurrence in the neck, ranges from zero when the nodes removed were histologically negative to 12.5% when there were multiple positive nodes or ECS. Incorrect clinical staging of the N0 occurs in approximately 20% of the patients, even when imaging studies are used. Intraoperative palpation and inspection does not improve significantly the surgeon's ability to predict nodal stage. Thus, upstaging the neck without frozen-section examination of suspected lymph nodes is not reliable. While some authors recommend converting the selective dissection to a radical or modified radical neck dissection on the basis of the results of frozen sections, this is not necessary [98]. Removal of the steruocleidomastoid muscle (SCM), the internal jugular vein (IJV) or of the posterior triangle of the neck is not done unless they are obviously involved by the tumor. The decision to extend a selective neck dissection, to include the jugular vein, the spinal accessory nerve or occasionally the hypoglossal or the vagus nerve, is based on the findings at the time of surgery and on an objective assessment of the extent of nodal disease by the surgeon. In patients with carcinoma of the oral tongue, Byers *et al.* [91] reported finding "skip metastases" to lymph nodes in level III or IV in 15.8% of the patients. As a result, they recommend performing a dissection of level IV, whenever an elective neck dissection is performed in a patient with cancer of the oral tongue. Since then, other have found that the incidence of metastases in level IV lymph nodes in patients with T1–3 N0 tumors of the tongue is low (4%) and that dissecting level IV only when suspicious nodes (enlarged or hardened) are found in level III or in multiple levels is not associated with increased rates of recurrence in the neck [92].

2. N1–N3 Neck

Surgery continues to be the mainstay in the treatment of patients with palpable cervical lymph node metastases. The surgical treatment of the N+ neck is, in our practice, a matter of surgical judgment, as we accept the premise that it is surgically feasible and oncologically sound to remove lymph nodes obviously involved by tumor, with the surrounding fibrofatty tissue of the neck and without removing important uninvolved structures, such as the spinal accessory nerve. In addition, with the judicious combination of surgery and radiation therapy, excellent tumor control in the neck can be obtained while preserving function and cosmesis. It cannot be overemphasized, however, that the main goal of neck dissection is to *adequately* remove the tumor in the neck and that radiation therapy does not compensate for poor surgical judgment and technique. Preservation of adjacent structures should be pursued *only* when a clearly identifiable plane exists between the tumor and such a structure. Creation of a plane by cutting into the tumor and tumor spillage must be avoided.

VI. TYPES OF NECK DISSECTION

The classification of neck dissections presented here has been sanctioned by the American Academy of Otolaryngology–Head and Neck Surgery, Inc., and by the American Society for Head and Neck Surgery in 1989 [66]. This classification standardized the names used for the various modifications of the radical neck dissection, thereby reducing the confusion that existed for many years when surgeons used diverse names to refer to the same operation. It is based primarily on the lymph node groups of the neck that are removed and secondarily on the anatomic structures that may be preserved, such as the spinal accessory nerve and the internal jugular vein. It also relies on the acceptance of a uniform nomenclature for the lymph node groups of the neck, in groups or levels I through VI.

This classification groups neck dissections into four categories: radical, modified radical, selective, and extended.

Radical neck dissection consists of the removal of all five lymph node groups of one side of the neck, including the sternocleidomastoid muscle, the internal jugular vein, and the spinal accessory nerve.

Modified radical neck dissections include modifications of the radical neck dissection that were developed with the intention of reducing the morbidity of this operation by preserving one or more of the following structures: the spinal accessory nerve, the internal jugular vein, or the sternocleidomastoid muscle. Like radical neck dissection, modified radical neck dissections remove all five nodal groups in one side of the neck. The three neck dissections that can be included in this category differ from each other only in the number of neural, vascular, and muscular structures that are preserved. Therefore, Medina [95] suggested subclassifying these neck dissections into a type I in which only "one" structure, the spinal accessory nerve, is preserved, type II in which "two" structures, the spinal accessory nerve and the internal jugular vein, are preserved, and type III in which all "three" structures, the spinal accessory nerve, the internal jugular vein, and the sternocleidomastoid muscle, are preserved.

Selective neck dissections (SND) consist of the removal of only the lymph node groups that are at the highest risk of containing metastases according to the location of the primary tumor. The spinal accessory nerve, the internal jugular vein, and the sternocleidomastoid muscle are preserved. Four different neck dissections can be included in this category. The updated classification calls for abandoning the terms lateral or supraomohyoid and instead using the terms selective neck dissection (SND) with the lymph node levels removed [99].

SND of lymph node levels II, III, and IV are removed (lateral).
SND of lymph node levels I, II, and III are removed (supraomohyoid).
Posterolateral neck dissection: suboccipital and retroauricular nodes are removed in addition to lymph node levels II, III, IV, and V.
SND of level VI (anterior).

The term *extended neck dissection* is used, in addition to any of the aforementioned designations, when a given neck dissection is "extended" to include either lymph node groups or structures of the neck that are not removed routinely, such as the retropharyngeal nodes or the carotid artery.

References

1. Crile, G. (1906). Excision of cancer of the head and neck. *JAMA* **47**, 1780–1786.
2. Martin, H., DelValle, B., Erhlich, H., *et al.* (1951). Neck dissection. *Cancer* **4**, 441–449.
3. Lindberg, R. (1972). Distrbution of cervical lymph node metastases from squamous cell carcinoma of the upper respiratory and digestive tracts. *Cancer* **29**, 1446–1449.
4. Byers, R. M., Wolf, P. F., and Ballantyne, A. J. (1988). Rationale for elective modified neck dissection. *Head Neck Surg.* **10**, 160–167.
5. Shah, J. P. (1990). Patterns of cervical lymph node metastasis from squamous carcinomas of the upper aerodigestive tract. *Am. J. Surg.* **160**, 405–409.
6. Yuen, A., Medina, J., Goepfert, H., *et al.* (1984). Management of stage T3 and T4 glottic carcinomas. *Am. J. Surg.* **148**, 467–472.
7. Daly, C. J., and Strong, E. (1975). Carcinoma of the glottic larynx. *Am. J. Surg.* **130**, 489–493.
8. DeSanto, L., Devine, K. D., and Lillie, J. C. (1977). Cancers of the larynx: Glottic cancers. *Surg. Clin. North Am.* **57**, 611–620.
9. Shah, J., and Tollefsen, H. R. (1974). Epidermoid carcinoma of the supraglottic larynx: Role of neck dissection in initial surgical treatment. *Am. J. Surg.* **128**, 494–499.
10. DeSanto, L. (1985). Cancer of the supraglottic larynx: A review of 260 patients. *Otolaryngol Head Neck Surg.* **93**(6), 705–711.
11. McGavran, M. H., Bauer, W. C., and Ogura, J. H. (1961). The incidence of cervical lymph node metastases from epidermoid carcinoma of the larynx and their relationship to certain characteristics of the primary tumor. *Cancer* **14**, 55–66.

12. Kowalski, L. P., Franco, E. L., and Sobrinho, J. (1994). Factors influencing regional lymph node metastasis from laryngeal carcinoma. *Ann. Otol. Rhinol. Laryngol.* **104**, 442–447.
13. Moe, K., Wolf, G., Fisher, S., *et al.* (1996). Regional metastases in patients with advanced laryngeal cancer. *Arch. Otolaryngol. Head Neck Surg.* **122**, 644–648.
14. Terhaard, C. H., Hordijk, G. J., van den Broek, P., de Jong, P. C., Snow, G. B., Hilgers, F. J., Annyas, B. A., Tjho-Heslinga, R. E., de Jong, J. M. (1992). T3 laryngeal cancer: A retrospective study of the Dutch Head and Neck Oncology Cooperative Group: Study design and general results. *Clin. Otolaryngol. Allied Sci.* **17**(5), 393–402.
15. Redaelli de Zinis, L., Nicolai, P., Barezzani, M., *et al.* (1994). Incidence and distribution of lymph node metastases in supraglottic squamous cell carcinoma: Therapeutic implications. *Acta Otorhinol. Ital.* **14**(1), 19–27.
16. Wolf, G., Fisher, S., Truelson, J., *et al.* (1994). DNA content and regional metastases in patients with advanced laryngeal squamous carcinoma. Department of Veterans Affairs Laryngeal Study Group. *Laryngoscope* **104**(4), 479–483.
17. Maurizi, M., Almadori, G., Cadoni, J., *et al.* (1997). EGFR expression in primary laryngeal cancer patients: An independent prognostic factor for lymph node metastases. *Br. J. Cancer* **7**(Suppl. 1), 1.37a.[Abstract]
18. Martinez-Gimeno, C., Rodriguez, R. M., Navarro, C., *et al.* (1995). Squamous cell carcinoma of the oral cavity: A clinicopathologic scoring system for evaluating risk of cervical lymph node metastasis. *Laryngoscope* **105**, 728–733.
19. Mohit-Tabatabai, M. A., Sobel, H. J., Rush, B. F., *et al.* (1986). Relation of thickness of floor of mouth stage I and II cancers to regional metastasis. *Am. J. Surg.* **152**, 351–353.
20. Spiro, R. H., Huvos, A. G., Wong, G. Y., *et al.* (1986). Predictive value of tumor thickness in squamous cell carcinoma confined to the tongue and floor of the mouth. *Am. J. Surg.* **152**, 345–350.
21. Long, J. P., Schechter, G. L., Nettleton, J. M., *et al.* (1996). Correlating primary head and neck cancer thickness, growth and treatment delay with the risk of cervical nodal metastasis. *In* "4th International Conference on Head and Neck Cancer Final Program and Abstract Book," 80. [Abstract]
22. Frierson, H. F., and Cooper, P. H. (1986). Prognostic factors in squamous cell carcinoma of the lower lip. *Hum. Pathol.* **17**, 346–354.
23. Rasgon, B. M., Cruz, R. M., Hilsinger, R. L. *et al.* (1989). Relation of lymph node metastasis to histopathologic appearance in oral cavity and oropharyngeal carcinoma: A case series and literature review. *Laryngoscope* **99**, 1103–1110.
24. McGavran, M. H., Bauer, W. C., and Ogura, J. H. (1991). The incidence of cervical lymph node metastases from epidermoid carcinoma of the larynx and their relationship to certain characteristics of the primary tumor: A study based on the clinical and pathological findings for 96 patients treated by primary en bloc laryngectomy and radical neck dissection. *Cancer* **14**, 55–65.
25. Mendelson, B. C., Woods, J. E., and Beahrs, O. H. (1976). Neck dissection in the treatment of carcinoma of the anterior two thirds of the tongue. *Surg. Gynecol. Obstet.* **143**, 75–80.
26. Willen, R., Nathanson, A., Moberger, C., *et al.* (1975). Squamous cell carcinoma of the gingiva: Histological classification and grading of malignancy. *Acta Otolaryngol.* **79**, 146–154.
27. Umeda, M., Yokoo, S., Take, Y., *et al.* (1992). Lymph node metastasis in squamous cell carcinoma of the oral cavity: Correlation between histologic features and the prevalence of metastasis. *Head Neck Surg.* **14**, 263–272.
28. Beenken, S. W., Krontiras, H., Maddox, W. A., *et al.* (1999). T1 and T2 squamous cell carcinoma of the oral tongue: prognostic factors and the role of elective lymph node dissection. *Head Neck* **21**, 124–130.
29. Bryne, M., Koppang, H. S., Lilleng, R., *et al.* (1992). Malignancy grading of the deep invasive margins of oral squamous cell carcinomas has high prognostic value. *J. Pathol.* **166**, 375–381.
30. Crissman, J. D., Liu, W. Y., Gluckman, J. L., *et al.* (1984). Prognostic value of histopathologic parameters in squamous cell carcinoma of the oropharynx. *Cancer* **54**, 2995–3001.
31. Yamamoto, E., Miyakawa, A., and Kohama, G. (1984). Mode of invasion and lymph node metastasis in squamous cell carcinoma of the oral cavity. *Head Neck Surg.* **6**, 938–947.
32. Close, L. G., Burns, D. K., Reisch, J., *et al.* (1987). Microvascular invasion in cancer of the oral cavity and oropharynx. *Arch. Otolaryngol. Head Neck Surg.* **113**, 1191–1195.
33. Poleksic, S., and Kalwaic, H. J. (1978). Prognostic value of vascular invasion in squamous cell carcinoma of the head and neck. *Plast. Reconstr. Surg.* **61**, 234–240.
34. Byers, R. M., El-Naggar, A. K., Lee, Y. Y., *et al.* (1998). Can we detect or predict the presence of occult nodal metastases in patients with squamous carcinoma of the oral tongue? *Head Neck* **20**(2), 138–144.
35. Fisch, U. P., and Sigel, M. E. (1964). Cervical lymphatic system as visualized by lymphography. *Ann. Otol. Rhinol. Laryngol.* **73**, 869–882.
36. Tobias, M. J. (1938). Lymphatics of the larynx, the trachea, and the oesophagus. *In* "Anatomy of the Human Lymphatic System: A Compendium," (H. Rouviere, ed.) pp. 57–62. Edwards Brothers, Ann. Arbor,MI.
37. Skolnik, E. M. (1976). The posterior triangle in radical neck surgery. *Arch. Otolaryngol. Head Neck Surg.* **102**, 1–4.
38. Toker, C. (1963). Some observations on the distribution of metastatic squamous carcinoma within cervical lymph nodes. *Ann. Surg.* **157**(3), 419–426.
39. Hasegawa, Y., and Matsuura, H. (1994). Retropharyngeal node dissection in cancer of the oropharynx and hypopharynx. *Head Neck* **16**, 173–180.
40. Amatsu, M., Mohri, M., and Kinishi, M. (2001). Significance of retropharyngeal node dissection at radical surgery for carcinoma of the hypopharynx and cervical esophagus. *Laryngoscope* **111**, 1099–1103.
41. McLaughlin, M. P., Mendenhall, W. M., Mancuso, A., *et al.* (1995). Retropharyngeal adenopathy as a predictor of outcome in squamous cell carcinoma of the head and neck. *Head Neck* **17**, 190–198.
42. Ballantyne, A. J. (1964). Significance of retropharyngeal nodes in cancer of the head and neck. *Am. J. Surg.* **108**, 500–503.
43. Weber, R. S., Marvel, J., Smith, P., *et al.* (1993). Paratracheal lymph node dissection for carcinoma of the larynx, hypopharynx and cervical esophagus. *Otolaryngol. Head Neck Surg.* **108**, 11–17.
44. O'Brien, C. J., Smith, J. W., Soong, S. J., *et al.* (1986). Neck dissection with and without radiotherapy; Prognostic factors, patterns of recurrence and survival. *Am. J. Surg.* **152**, 456–463.
45. Johnson, J. T., Barnes, E. L., and Myers, E. N. (1981). The extracapsular spread of tumors in cervical node metastasis. *Arch. Otolaryngol. Head Neck Surg.* **107**, 725–728.
46. Johnson, J. T., Myers, E. N., Bedetti, C. D., *et al.* (1985). Cervical lymph node metastases. *Arch. Otolaryngol. Head Neck Surg.* **111**, 534–537.
47. Steinhart, H., Schroeder, H. G., Buchta, B., *et al.* (1994). Prognostic significance of extra-capsular invasion in cervical lymph node metastases of squamous epithelial carcinoma. *Laryngo-Rhino-Otologie* **73**(12), 620–625.
48. Carter, R. L., Barr, L. C., and O'Brien, C. J. (1985). Transcapsular spread of metastatic squamous cell carcinoma. *Am. J. Surg.* **150**, 495–499.
49. Olsen, K. D., Caruso, M., Foote, R. L., *et al.* (1994). Primary head and neck cancer: Histopathologic predictors of recurrence after neck dissection in patients with lymph node involvement. *Arch. Otolaryngol. Head Neck Surg.* **120**(12), 1370–1374.
50. Spiro, R. H., Alfonso, A. E., Farr, H. W., *et al.* (1974). Cervical node metastasis from epidermoid carcinoma of the oral cavity and oropharynx. A critical assessment of current staging. *Am. J. Surg.* **128**, 562–567.

51. Shah, J. P., and Tollefsen, H. R. (1974). Epidermoid carcinoma of the supraglottic larynx. *Am. J. Surg.* **128**, 494–500.
52. Tulenko, J., Priore, R. L., and Hoffmeister, F. S. (1966). Cancer of the tongue. *Am. J. Surg.* **112**, 562–568.
53. Ho, J. H. C. (1978). An epidemiological and clinical study of nasopharyngeal carcinoma. *Int. J. Radiat. Oncol. Biol. Phys.* **4**, 183–198.
54. Teo, P. M. L., Leung, S. F., Yu, P., *et al.* (1991). A comparison of Ho's, International Union Against Cancer, and American Joint Committee Stage Classifications for nasopharyngeal carcinoma. *Cancer* **67**, 434–439.
55. Grandi, C., Boracchi, P., Mezzanotte, G., *et al.* (1990). Analysis of prognostic factors and proposal of a new classification for nasopharyngeal cancer. *Head Neck Surg.* **12**, 31–40.
56. Baatenburg de Jong, R. J., Gullane, P. G., and Freeman, J. L. (1993). The significance of false positive nodes in patients with oral cancer. *J. Otolaryngol.* **22**, 154–159.
57. Fall, H. W., Goldfarb, P. M., and Farr, C. W. (1980). Epidermoid carcinoma of the mouth and pharynx at Memorial Sloan Kettering Cancer Center. *Am. J. Surg.* **140**, 563–567.
58. DeSanto, L. W., Holt, J. J., and Beahrs, O. H. (1982). Neck dissection: Is it worthwhile? *Laryngoscope* **92**, 502–509.
59. Shaha, A. R., Spiro, R. H., and Shah, J. P. (1984). Squamous carcinoma of the floor of the mouth. *Am. J. Surg.* **148**, 455–459.
60. Kalnins, I. K., Leonard, A. G., and Sako, K. (1977). Correlation between prognosis and degree of lymph node involvement in carcinoma of the oral cavity. *Am. J. Surg.* **134**, 450–454.
61. Ballantyne, A. J. (1964). Significance of retropharyngeal nodes in cancer of the head and neck. *Am. J. Surg.* **108**, 500.
62. Weber, R. S., Peters, L. J., Wolf, P., *et al.* (1988). Squamous cell carcinoma of the soft palate, uvula, and anterior faucial pillar. *Otolaryngol. Head Neck Surg.* **99**(1), 16–23.
63. Lavertu, P., Adelstein, D. J., Saxton, J. P., *et al.* (1997). Management of the neck in a randomized trial comparing concurrent chemotherapy and radiotherapy with radiotherapy alone in resectable stage III and IV squamous cell head and neck cancer. *Head Neck* **19**, 559–566.
64. Peters, L. J., Weber, R. S., Morrison, W. H., *et al.* (1996). Neck surgery in patients with primary oropharyngeal cancer treated by radiotherapy. *Head Neck* **18**, 552–559.
65. Gallo, O., Boddi, V., Parrella, F., *et al.* (1996). Treatment of the clinically negative neck in laryngeal cancer patients. *Head Neck* **18**(6), 566–572.
66. Robbins, K. T., Medina, J. E., Wolfe, G. T., Levine, P., Sessions, R., and Pruet, C. (1991). Standardizing neck dissection terminology. *Arch. Otolaryngol.* **117**, 601–605.
67. Fakih, A., Rao, R., Borges, A., *et al.* (1989). Elective versus therapeutic neck dissection in early carcinoma of the oral tongue. *Am. J. Surg.* **158**, 309.
68. Sun, X., Guo, Z., and Lu, C. (1994). Clinical significance of intraoperative frozen biopsy of cervical lymph node for laryngeal cancer. *Chin. J. Otorhinolaryngol.* **29**(2), 104–106.
69. Hao, S. P., Myers, E., and Johnson, J. (1995). T3 glottic carcinoma revisited. Transglottic vs pure glottic carcinoma. *Arch. Otolaryngol. Head Neck Surg.* **121**(2), 166–170.
70. Pillsbury, H. (1997). A rational for therapy of the N0 neck. *Laryngoscope* **107**, 1294–1314.
71. Lederman, M. (1971). Cancer of the larynx. Natural history in relation to treatment. *Br. J. Radiol.* **44**, 569–578.
72. Martensson, B., Fluur, E., and Jacobsson, F. (1967). Aspects on treatment of cancer of the larynx. *Ann. Otol. Rhinol. Laryngol.* **76**, 313.
73. Johnson, J. (1994). Carcinoma of the larynx: Selective approach to the management of cervical lymphatics. *ENT J.* **73**(5), 303–305.
74. Ambrosch, P., Freudenberg, L., Kron, M., *et al.* (1996). Selective neck dissection in the management of squamous cell carcinoma of the upper digestive tract. *Euro. Arch. Oto-Rhino-Laryngol.* **253**(6), 329–335.
75. Brazilian Head and Neck Cancer Study Group. (1999). End results of a prospective trial on elective lateral neck dissection versus type III modified radical neck dissection in the management of supraglottic and transglottic carcinomas. *Head Neck* **21**, 694–702.
76. Weber, P., Johnson, J., and Myers, E. (1994). The impact of bilateral neck dissection on pattern of recurrence and survival in supraglottic carcinoma. *Arch. Otolaryngol. Head Neck Surg.* **120**(7), 703–706.
77. Fletcher, G. (1972). Elective irradiation of subclinical disease in cancers of the head and neck. *Cancer* **29**, 1450.
78. Mendenhall, W., Million, R., and Cassisi, N. (1980). Elective neck irradiation in squamous cell carcinoma of the head and neck. *Head Neck Surg.* **3**, 15.
79. Schwaab, G., Luboinski, B., Julieron, M., *et al.* (1994). Management and results of no neck in laryngela cancers: Experience of the Gustave Roussy Institute 1975–1984. Second World Congress on Laryngeal Cancer. [Abstract]
80. Vikram, B., Strong, E., Shah, J., *et al.* (1984). Failure in the neck following multimodality treatment for advanced head and neck cancer. *Head Neck Surg.* **6**, 724.
81. O'Brien, C., Smith, J., Soong, S., *et al.* (1986). Neck dissection with and without radiotherapy - prognostic factors, patterns of recurrence and survival. *Am. J. Surg.* **152**, 456–463.
82. Johnson, J., Myers, E., Bedetti, C., *et al.* (1985). Cervical lymph node metastases. *Arch. Otolaryngol. Head Neck Surg.* **111**, 534–537.
83. Carter, R., Barr, L., O'Brien, C., *et al.* (1985). Transcapsular spread of metastatic squamous cell carcinoma from cervical lymph nodes. *Am. J. Surg.* **150**, 495.
84. Peters, L., Goepfert, H., Kiang, A., *et al.* (1993). Evaluation of the dose for postoperative radiation therapy of head and neck cancer: First report of a prospective randomized trial. *Int. J. Radial. Oncol. Biol. Phys.* **26**, 3–11.
85. Verschuur, H. P., Keus, R. B., Hilgers, F., *et al.* (1996). Preservation of function by radiotherapy of small primary carcinomas preceded by neck dissection for extensive nodal metastases of the head and neck. *Head Neck* **18**, 277–282.
86. French Head and Neck Study Group. (1991). Early pharyngolaryngeal carcinomas with palpable nodes. *Am. J. Surg.* **162**, 377–380.
87. Vandenbrouck, C., Sancho-Garnier, H., and Chassagne, D. (1980). Elective versus therapeutic radical neck dissection in epidermoid carcinoma of the oral cavity: Results of a randomized clinical trial. *Cancer* **46**, 386.
88. Yuen, A. P., Wei, W. I., Wong, Y. M., *et al.* (1997). Elective neck dissection versus observation in the treatment of early oral tongue carcinoma. *Head Neck* **19**, 583–588.
89. Kligerman, J., Lima, R. A., Soares, J. R., *et al.* (1994). Supraomohyoid neck dissection in the treatment of T1/T2 squamous cell carcinoma of oral cavity. *Am. J. Surg.* **168**, 391–394.
90. Houck, J. R., and Medina, J. E. (1995). Management of cervical lymph nodes in squamous carcinomas of the head and neck. *Semin. Surg. Oncol.* **11**, 228–239.
91. Byers, R., Weber, R. S., Andrews, T., *et al.* (1997). Frequency and therapeutic implications of "skip metastases" in the neck from squamous cell carcinoma of the oral tongue. *Head Neck* **19**, 14–19.
92. Khafif, R. A., Lopez-Garza, J. R., and Medina, J. E. (2001). Is dissection of level IV necessary in patients with early (T1–T3, N0) tongue cancer. *Laryngoscope* **111**, 1088–1090.
93. Krempl, G. A., Kowalski, L., Yuen, A., *et al.* (2001). Is postoperative radiation beneficial in the pathologically N1 (pN1) neck? Presented at AHNS Meeting, Palm Desert, CA.

94. Byers, R. M., Clayman, G. L., and McGill, D. (1999). Selective neck dissections for squamous carcinoma of the upper aerodigestive tract: Patterns of regional failure. *Head Neck* **21**(6), 499–505.
95. Medina, J. E. (1989). A rational classification of neck dissections. *Otolaryngol. Head Neck Surg.* **100**, 169–176.
96. Leemans, C. R., Tiwari, R., Nauta, J. J., van der Waal, I., and Snow, G. B. (1993). Regional lymph node involvement and its significance in the development of distant metastases in head and neck carcinoma. *Cancer*, **71**(2), 452–456.
97 American Joint Committee on Cancer (2002). "Cancer Staging Manual," Sixth ed. Springer-Verlag, New York.
98 Wein Richard, O., Winkle Mark, R., Norante John, D., and Coniglio John, U. (2002). Evaluation of selective lymph node sampling in the node-negative neck. *Laryngoscope* **112**(6), 1006–1009.
99. Robbins, K T., Clayman, G., Levine, P. A., Medina, J. E., *et al.* (2002). Neck dissection classification update: Revisions proposed by the American head and neck society and the American academy of otolaryngology—head and neck surgery. *Arch. Otolaryngol. Head Neck Surg.* **128**, 751–758.

CHAPTER

23

Management of the Carotid Artery in Advanced Head and Neck Cancer

JOSE E. OTERO-GARCIA, GEORGE H. YOO, and JOHN R. JACOBS

Department of Otolaryngology—Head and Neck Surgery
Karmanos Cancer Institute
Wayne State University
Detroit, Michigan 48201

I. Management of the Carotid Artery in Advanced Head and Neck Cancer 329
II. Patient Presentation: Clinical and Radiological Diagnosis 329
III. Clinical Outcome 330
IV. Nonsurgical Treatment Options 331
V. Surgical Treatment Options 332
VI. Preoperative Evaluation 332
VII. Surgical Procedure 333
VIII. Postoperative Complications 335
IX. Conclusion 335
References 336

Invasion of the carotid artery by squamous cell carcinoma is associated with a very poor prognosis. Understanding of the clinical history, diagnostic testing, and various therapeutic options for management of the carotid artery in advanced head and neck carcinoma is crucial for optimal patient management. Surgical options that have been utilized in the past have included ligation, peeling off the tumor from the artery, and resection followed by immediate reconstruction with a venous graft. These options all have various degrees of success associated with their usage. We feel that carotid artery resection with immediate arterial graft reconstruction in select patients produces local control and reduced morbidity and should be the preferred technique for this difficult problem.

I. MANAGEMENT OF THE CAROTID ARTERY IN ADVANCED HEAD AND NECK CANCER

The first ligation of the common carotid for hemorrhage was attributed to Ambroise Pare in 1552 [1]. In the late 1700s, John Abernathy of St. Bartholemew's Hospital in London made the first complete report of ligation of the carotid artery for hemorrhage [2]. In the 19th century, carotid artery ligation was performed to treat various disease entities, such as tumors, aneurysms, epilepsy, trigeminal neuralgia, psychosis, hemiplegia, and headache [3]. As expected, the morbidity and mortality of this procedure eventually limited enthusiasm [3]. In 1879, Wyeth [4] reported a 41% mortality rate with the procedure and went on to condemn the procedure as "wrong in principle, unsafe in practice and (one that) should cease to be a surgical procedure."

Martin [5] was the first to recognize that by maintaining a normal blood pressure and intravascular volume, the mortality and morbidity of carotid ligation could be improved significantly. Until that observation the mortality of ligation approached 50%. Later in 1953 [6] and in 1957 [7], Conley reported the use of autogenous vein grafts for reconstruction after carotid artery resection. This was followed by other reports from Keirle and Altemeier [8], Rella *et al.* [9] and Lore [10,11] in the 1960s describing the use of autogenous and prosthetic grafts. Current treatment of advanced neck carcinoma could involve the resection of the carotid artery. Controversies exist in defining the role this procedure may have in the treatment of these patients. Clinical presentation, radiological diagnosis, treatment options, surgical procedures, and complications are discussed in this chapter.

II. PATIENT PRESENTATION: CLINICAL AND RADIOLOGICAL DIAGNOSIS

Patients with carcinoma invading the carotid artery usually present with a fixed neck mass. This means that the mass either has no motion upon palpation or movement of the mass can be achieved in the horizontal axis only, but not in the vertical axis. The size and location of the mass in the neck are important to note in order to establish surgical options. Patients with very high cervical masses that invaded the skull base or alternatively whose removal will not leave enough distal carotid artery to sew to do not have a surgical option. Patients who present with skin invasion and ulceration associated with episodes of bleeding may have an impending carotid artery rupture and clinical management should be instituted expeditiously. Clinical impression alone has a rate of inaccuracy for predicting carotid artery invasion of 32 to 63% [12].

A Computed Tomography (CT) scan with contrast of the neck helps evaluate the size and extension of the mass and its relation to the carotid artery wall (Fig. 23.1). Loss of tissue planes and arterial wall definition between the artery and the tumor may suggest the presence of carotid artery invasion [13]. CT can exclude patients without carotid artery involvement, but has a high false-positive rate of 94% [14]. The combination of clinical assessment with CT scan results appears to be the most predictive of tumor invasion [12]. Yoo *et al.* [12] have demonstrated that more than 180° of carotid artery encasement as seen by a CT scan predicts a poor clinical outcome, but not the extent of tumor invasion into the carotid artery. Magnetic resonance imaging (MRI) criterion for invasion is evidence of loss of the fascial plane around the internal or common carotid artery [13]. MRI is helpful in distinguishing between fibrosis and edema from tumor in postradiotherapy patients [13].

Using MRI, CT, or ultrasound, carotid involvement of 270° or more predicts arterial wall invasion with a sensitivity of 92 to 100% and a specificity of 88 to 93% [17,18]. More than 80% of patients with carotid encasement greater than 270° had direct tumor invasion [12]. Ultrasound criteria for invasion of the carotid artery are a step off in the echo-dense signal in both planes (transverse and longitudinal) and luminal invasion by tumor [13]. Adequate sonographic imaging of the carotid artery in bulky neck cancers can be troublesome. In view of a significant incidence of distant metastasis reported in the literature (21–67%) [15,16] in this group of patients, a complete workup for distant metastasis should be performed. This includes a chest X-ray and liver enzyme levels. If abnormalities are found in these tests, then a CT scan of the chest and abdomen and bone scans are indicated.

Sessa *et al.* [19] presented a series of 30 patients who underwent carotid artery resection. During surgery the tumor either encased the artery or it was intimately adherent to the vessel wall. Twenty-nine patients underwent preoperative carotid angiography. Arteriograms showed carotid invasion in 2 cases, nonatherosclerotic lumenal narrowing, suggesting carotid invasion in 6 patients, and normal arteriogram in 15 cases. These results demonstrate that carotid arteriography is not reliable in predicting which patients will need carotid artery resection. Reilly *et al.* [20] found that arteriography correctly identified tumor invasion of the carotid artery in only 1 of 12 patients.

III. CLINICAL OUTCOME

Prognosis of patients with advanced carcinoma in the neck with carotid artery invasion is very poor. It is known that the 5-year survival of patients with N+ disease is reduced by half in patients with squamous cell carcinoma of the head and neck (SCCHN)[21]. In patients with advanced

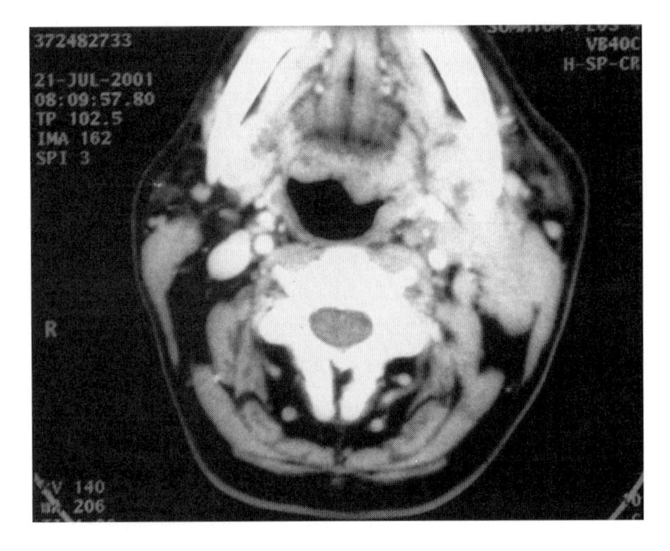

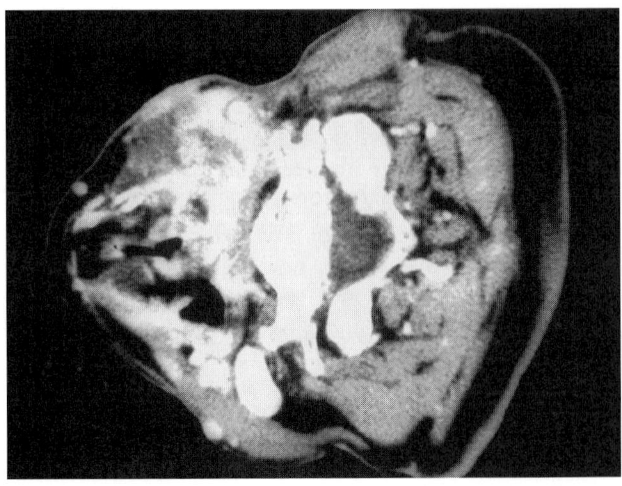

FIGURE 23.1 Axial CT scan of the neck showing (a) less than 180° and (b) more than 180° of carotid artery encasement.

neck carcinoma, the incidence of distant metastasis is four times that associated with N0 or N1 disease [21]. Massive fixed neck disease is usually associated with macroscopic extracapsular lymph node spread with tumoral invasion of the soft tissues of the neck, which significantly increases the rates of recurrence and distant metastasis and decreases survival [21]. Snyderman *et al.* [15] made a retrospective review of 22 series of patients ($n = 158$) undergoing carotid artery resection for SCCHN and found a 2-year disease-free survival rate of 22%. They also reported an incidence of recurrence in 74% of patients in a period of 24 months (32% local, 47% regional, 21% distant) [15]. More than 180° of artery encasement predicts a poor clinical outcome [12]. Although the carotid artery can be resected out, complete excision of cancer is difficult, as the adjacent soft tissue is diffusely infiltrated by tumor. Incomplete resection of tumor is associated with high locoregional recurrence rates, which is an important factor for the development of distant metastasis [22].

Direct tumor invasion of the carotid artery wall has been found in 37 to 42% of specimens after carotid resection [23,24]. The adventitia of the carotid artery serves as a barrier to tumor invasion [23]. High proportions of patients who have carotid artery resections do not have microscopic arterial wall invasion. This fact can explain why there is not a significant difference in survival rates between patients with N3 disease with extracapsular spread and patients undergoing carotid artery resection [15,25]. However, patients with direct tumor invasion of the carotid artery wall in resection specimens have a dismal prognosis [25]. One hundred percent recurrence and mortality rates have been reported in patients with carcinoma microscopically invading the fascia and wall of the carotid artery [25]. We demonstrated previously that overall survival was worse in patients who had tumor within 1.8 mm from the elastic lamina of the vessel wall [12].

The outcome is very poor in patients with recurrence in a previously operated neck [26]. Initial therapy might destroy natural tissue barriers to tumor spread and reduce the efficacy of salvage surgery. The use of preoperative radiotherapy in this patient population can be as high as 70 to 85% [3,24,27]. Salvage surgery in previously irradiated patients with fixed neck nodes is associated with a poor survival (7% survival at 2 years) [28].

IV. NONSURGICAL TREATMENT OPTIONS

Nonsurgical modalities for the treatment of advanced neck carcinoma include radiotherapy (conventional external beam, altered fractionation regimens, brachytherapy, or reirradiation), chemotherapy, or chemoradiotherapy. Conventional external beam radiotherapy consists of once-a-day fractionation treatments (150–200 cGy/fraction) for total doses of 7000 to 8000 cGy [29,30].

Altered fractionation radiotherapy schedules have been compared to conventional radiotherapy for the treatment of advanced head and neck carcinoma [29]. Hyperfractionation (twice-a-day, lower dose per fraction) treatment permits an increase in the total amount of radiation dose delivered to the tumor [29]. Accelerated fractionation (treating once per day on 6 to 7 days of the week or using a split-course treatment or a concomitant boost) prevents treatment failure due to tumor cell regeneration during treatment [29]. Two phase III randomized studies done by Radiation Therapy Oncology Group (RTOG) (9003) and European Organization for Research and Treatment of Cancer (EORTC) demonstrated that accelerated fractionation was significantly better in terms of local-regional control than conventional fractionation in patients with locally advanced head and neck cancer [31,32]. However, overall and disease-free survival was not improved.

The use of intraoperative radiation therapy (IORT) as a supplement to surgical resection in advanced neck carcinoma is controversial. One study of 16 patients undergoing carotid artery resection for advanced neck carcinoma and IORT (1500–2000 cGY) showed a 1-year local control rate of 60% [23]. In a retrospective study of 75 patients with advanced neck carcinoma, Freeman *et al.* [33] found a local control rate of 68% using a combination of extended neck dissection, including carotid artery resection if necessary, and IORT (2000-cGy electron beam dose) [33]. Nag *et al.* [34] presented a series of 38 patients with previously irradiated recurrent advanced head and neck malignancies that were treated with radical surgery and IORT. Locoregional control rates at 6 months, 1 year, and 2 years were 33, 11, and 4%, respectively. It was concluded that intraoperative radiotherapy is not sufficient for the control of recurrent, previously irradiated head and neck cancer. A limited external beam radiation therapy and/or brachytherapy should be added to increase local control [34].

Additional local radiation can be delivered using interstitial implants (i.e., ^{192}Ir, ^{198}Au, ^{125}I) to improve local control. Werber *et al.* [35] demonstrated that carotid-sheath contents in rabbits could tolerate high doses (13,000 cGy) of low-dose interstitial brachytherapy (^{192}Ir implants) without complications. Ultrasound examination of the carotid artery done in 24 patients with advanced recurrent head and cancer treated with salvage surgery and intraoperative iodine (^{125} I) implants showed that after 1 year of treatment there were minimal or no changes in the carotid artery wall compared to the contralateral side [36]. Local control rates of 58 to 64% have been reported with the use of ^{125}I implants in advanced head and neck carcinoma [37,38]. However, other institutions have abandoned the use of these implants because of high recurrence rates in areas outside the implant site [26]. The patients also develop extreme fibrosis of the soft tissue of the neck and experience chronic pain [26].

Initial disappointment with chemotherapy came about because no trial demonstrated a survival benefit in postoperative adjuvant [39,40] or neoadjuvant trials [41]. However, platinum-based chemotherapy has shown to improve survival when used in conjunction with radiation for nasopharyngeal carcinoma [42] and unresectable head and neck squamous carcinoma (HNSCC) [43–46]. When cisplatinum was used together with 5-fluorouracil, organ preservation of the larynx [47] and hypopharynx [48] has been achieved. Chemoradiation using cisplatinum after surgical resection has demonstrated a benefit in local control for high-risk patients in a phase II setting [49] and overall survival (phase III, France) [46] and is currently being tested in a phase III trial (expected release of information spring 2002). Organ preservation approaches have now been converted into nonsurgical approaches because of the benefit of chemoradiation [43].

V. SURGICAL TREATMENT OPTIONS

The axiom of surgical oncology is to resect a margin of normal tissue around a tumor to assure that no cancer cells are left behind and to assure that resection will improve overall survival and not cause undue morbidity. The definition of resectability is variable among surgeons and institutions and is subject to opinion and bias. Some proposed criteria of an unresectable cancer are tumor invasion into the floor of neck or prevertebral fascia, direct extension into mediastinum, skull base, nasopharynx, or cranial cavity along with carotid artery encasement or invasion. An indirect definition of unresectability is the presence of distant metastasis, medically ineligible for surgery, no improvement in survival, or if the overall 2-year survival is less than 20%. If a group of patients have a less than 20% overall survival, this cohort should be placed in clinical trials to find regimens that can improve outcomes.

Carotid artery resection is recommended when the goal is complete tumor removal. Complete tumor removal with negative histological margins is the most important prognostic factor in head and neck surgery [25]. In our institution, 63% of patients ($n = 30$) who underwent carotid artery resection had local control at their time of their death or at their most recent follow-up (mean follow up of 20 months). Disease-free survival rates were 60, 31, and 10% at 1, 2, and 5 years, respectively [19]. The final decision to resect the carotid artery *en bloc* with the tumor mass is made by the surgeon in the operating suite. Complete tumor encasement of the carotid artery and an inability to find a tumor-free plane are important factors to consider when deciding to resect the vessel. Others perform frozen sections to confirm the presence of cancer cells in the adventitia [20].

In a small group of patients, palliation of pain and prevention of carotid blowout and carotid hemorrhage can be achieved with carotid artery resection and revascularization. In the presence of imminent carotid artery rupture with evidence of gross infection, carotid artery ligation without revascularization or placement of a detachable balloon is often necessary, yet is associated with high morbidity and mortality.

Peeling off the tumor from the carotid artery and removing all gross disease is a surgical option that needs to be considered carefully. Leaving microscopic tumor cells behind can increase the risk of locoregional recurrence in more than 60% of patients and increase the risk of late carotid artery blowout from continuous erosion from malignant cells and the loss of supporting adventitia [19]. It has been demonstrated that 50% of patients with tumor fixation to the carotid artery with less than 180° involvement of the carotid artery by CT have tumor in a distance greater than 1.8 mm from the elastic lamina. Peeling off tumor in these patients should be considered [12].

VI. PREOPERATIVE EVALUATION

Various techniques have been described that are designed to assess collateral circulation in the circle of Willis. These tests are supposed to help predict which patients will tolerate carotid sacrifice without neurologic sequelae. The Matas test consists of compressing the carotid artery against a vertebral transverse process for 10 min while observing deterioration of neurologic function [50]. Oculoplethysmography is performed with ipsilateral carotid artery occlusion and assumes that the pressure in the ophtalmic artery is proportional to the efficiency of the collateral cerebral flow [25]. Direct intraoperative carotid stump pressures can estimate collateral circulation [51]. If the stump pressure is greater than 70 mm Hg, there is little risk of neurologic impairment [25]. If the stump pressure is less than 50 mm Hg, there is a high risk of cerebrovascular accident [25]. Stump pressure between 50 and 70 mmHg indicates an increased risk for stroke [25]. Carotid arteriography can demonstrate nonstenotic patent carotid arteries, spontaneous "crossover" and an intact circle of Willis [52]. Temporary balloon occlusion of the carotid artery is performed while the patient is awake. The balloon is inflated for 30 min and the patient is monitored for any neurologic sequelae [52]. Then a xenon-133 (Xe^{133}) CT scan is performed in order to make quantitative assessments of cerebral blood flow [51]. Inhaled xenon acts as a contrast agent whose uptake and diffusion from cerebral tissue are proportional to blood flow [51]. With this technique, a cerebral blood flow map is obtained after balloon occlusion. Positron emission tomography (PET) can also measure cerebral blood flow in addition to metabolic parameters, simultaneously [51]. PET provides the most accurate knowledge regarding cerebral hemodynamics [51].

Despite the aforementioned techniques, stroke risk continues to be high because there are other factors that cause postcarotid ligation brain infarction. Urken *et al.* [53] cited the unreliable nature of preoperative or intraoperative testing in predicting the risk of stroke. Stroke rates after a negative balloon occlusion test are approximately 30%, which is not significantly different from the stroke rate that would be expected after unselected ligation alone [54]. Two causes of stroke after carotid artery ligation without reconstruction are hemispheric hypotension from poor collateral flow and thrombus extension from the internal carotid artery to the middle or anterior cerebral arteries [54].

Vascular reconstruction is important even in patients with adequate collateral cerebral blood flow. Patients that have a high risk of developing a stroke may still undergo carotid artery resection with revascularization. In our institution, a complete physical examination describing the tumor staging and the patient's performance status is followed by a thorough evaluation of the extension of the neck metastasis and the condition of the carotid artery wall using CT scans (and MRI scans as needed). Then preoperative carotid arteriography is performed to define the intracranial vascular anatomy, including the collateral pathways, and to evaluate atherosclerotic disease and search for other vascular abnormality [19]. The vascular surgeon is consulted preoperatively. Carotid artery resection is performed followed by immediate reconstruction with an autogenous vascular graft.

VII. SURGICAL PROCEDURE

Normal systemic blood pressure is maintained during the procedure. The patient is anticoagulated with intravenous heparin to avoid propagation of thrombus in the carotid system. Lore [55] recommended avoiding entering the respiratory or digestive tract with the use of any reconstructive graft because of the risk of graft infection. He also suggested not performing vessel reconstruction at the same stage of the primary tumor surgery. In our experience, all of the neck surgical fields were exposed intraoperatively to pharyngeal or tracheal contamination when the aerodigestive tract was sectioned and thus were considered contaminated [19].

The entire extracranial (distal) and extrathoracic (proximal) portion of the carotid artery is exposed. Additional exposure superiorly can be achieved by performing a mandibulotomy posterior to the third molar tooth [56]. Proximal and distal vessel control is necessary for graft reconstruction. We do not perform carotid artery resection if the base of the skull is involved by disease and if less than 2 cm of distal internal carotid artery from the base of skull is available [55]. However, Urken *et al.* [57] has reported anastomosis to the carotid artery in the distal temporal bone when tumor involves the base of the skull. The proximal portion of the common carotid artery should be exposed at least 3 cm above the clavicle [54]. The proximal side-to-end anastomosis of the vascular graft to carotid is performed using an angulated vascular clamp that tangentially occludes the common carotid artery while still allowing blood to pass through the vessel [55]. The internal carotid artery distal to the tumor is cross clamped and is then transected. Prior to carotid artery cross clamping, Decadron and Etomidate are administered to the patient. An end to end anastomosis is performed at the distal internal carotid artery. Ligation and transection of the common carotid artery distal to the proximal anastomosis are performed [55]. The carotid segment with the vagus nerve is removed *en bloc* with the tumor (Fig. 23.2). The average time of interruption of

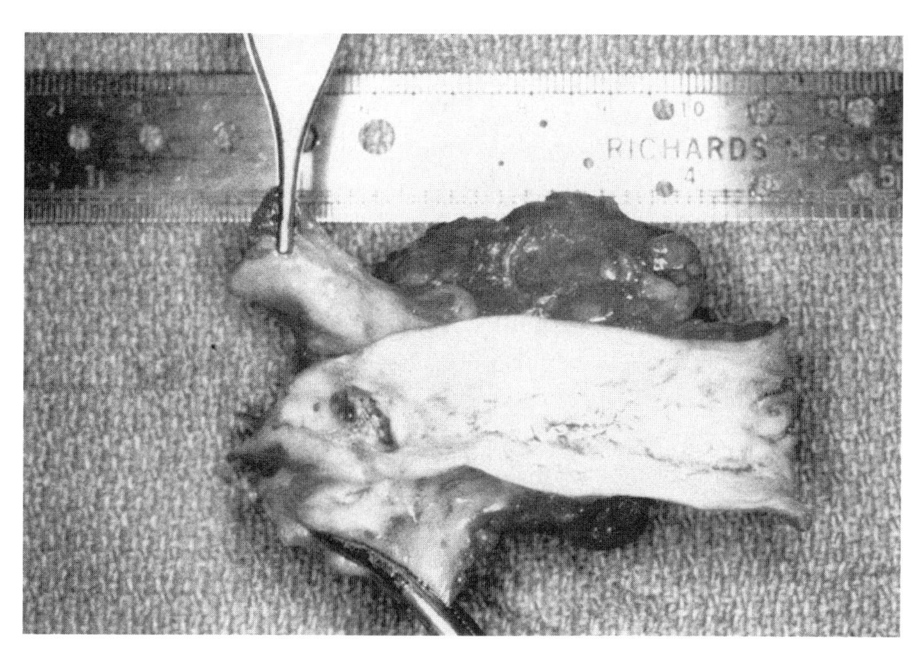

FIGURE 23.2 Resected specimen demonstrating tumor and intraluminal carotid artery.

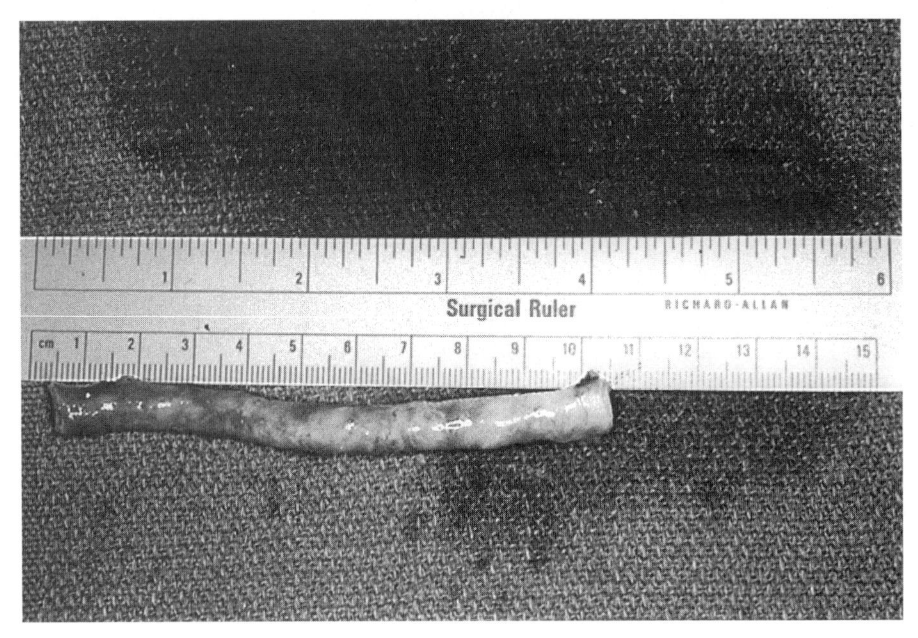

FIGURE 23.3 Superficial femoral artery graft.

cerebral blood flow can be as low as 4 min, which negates the need for intraoperative vascular shunts for cerebral protection [27].

In our institution, superficial femoral artery grafts (Fig. 23.3) are preferred for revascularization of the carotid artery (superficial femoral artery are screened preoperatively to assess its quality and diameter using either selective femoral arteriography or duplex imaging). Grafting between the common carotid to internal carotid artery is performed more frequently (Fig. 23.4). Biller *et al.* [27] prefer to perform proximal anastomosis in the subclavian artery for various reasons: it is located at a distance from the pharynx and from the area of recurrence and it is outside the field of irradiation. Superficial femoral artery discontinuity in the

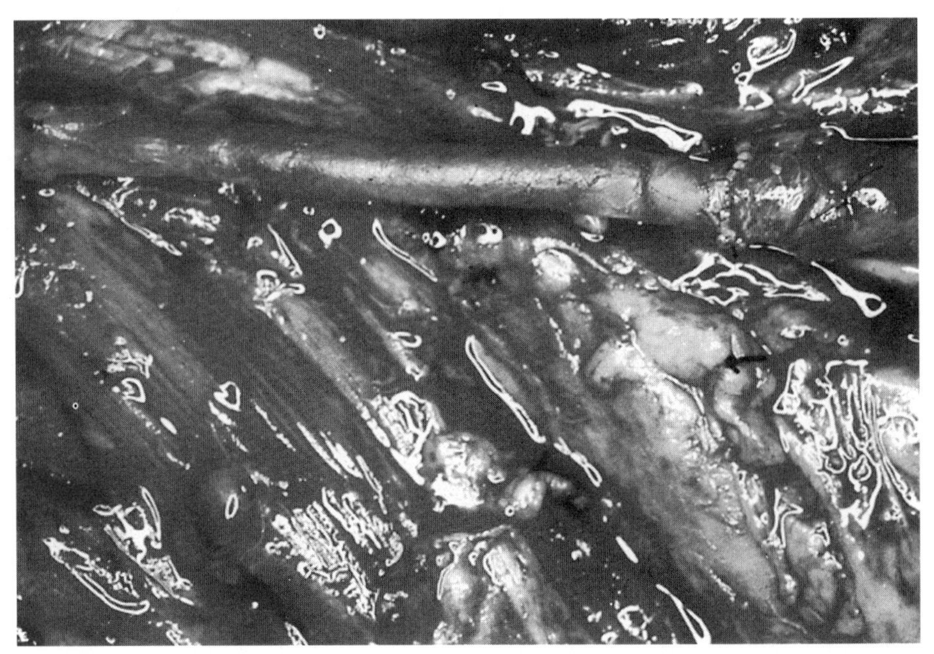

FIGURE 23.4 Superficial femoral artery graft in place after anastomosis was performed to proximal and distal carotid artery.

thigh is replaced with a segment of expanded polytetrafluoroethylene by a separate surgical team.

Intraoperative arteriography is carried out to assess the technical result. Vascular anastomotic suture lines in the neck are covered with soft tissue (i.e., local soft tissue and skin or pedicled myocutaneous flaps). Closed-system suction catheters are used to drain the dead space. Perioperative intravenous antibiotics are administered, and postoperative anticoagulation is not required. Prior to discharge, patients should have their neck and thigh imaged using duplex ultrasound to determine carotid and superficial femoral artery patency [19].

VIII. POSTOPERATIVE COMPLICATIONS

When carotid artery resection and ligation have been performed, high rates of neurological complications (45%) and mortality (31–41%) have been reported [2,7]. Major neurological complications include coma and hemiplegia. Transient ischemic attacks with mild paresis are considered minor complications. Many patients may have a delayed development of neurologic deficits (up to 6 days after ligation), which could be caused by a late propagation of thrombus into the low-flow intracranial circulation [54]. Late stroke rates have been reported to be 5 and 6% of the patients at 6 and 12 months, respectively [23,54]. Perioperative death mainly results from cerebral infarction. Maintaining a normal blood pressure and normal volume status are imperative in order to reduce cerebral complications and mortality rates [5].

The complication rates have been reduced with the use of carotid artery reconstruction with synthetic, venous (greater saphenous vein), or arterial (superficial femoral artery) grafts [58]. The stroke rate and perioperative mortality (less than 5%) have been lowered significantly with carotid artery reconstruction [19,54]. In 1981, Lore and Boulous [56] presented a series of 10 patients who underwent carotid artery resection. Revascularization was done with either autogenous saphenous vein or synthetic grafts. The main complication they reported was wound infection (sometimes associated with salivary fistula), which can lead to graft exposure and occlusion and breakdown of the vascular anastomosis [56]. Although they advocate the use of synthetic grafts for carotid replacement, we feel that if the upper aerodigestive tract is violated and the wound is contaminated, the complications associated with infection make the use of synthetic material inadvisable in most circumstances [19].

Saphenous vein grafts may have difficulty tolerating an infection or fistula. Graft coverage with local soft tissue and myocutaneous flaps help significantly in the protection of the graft from a hostile environment, i.e., infected wounds, salivary leaks, and irradiated fields [19]. The long-term deleterious effect that scarring and fibrosis may have on this type of graft has to be investigated. Wright *et al.* [54] reported only one saphenous graft blowout in a series of 20 patients (90% cases had intraoperative contamination and 50% had intraoperative radiotherapy). This carotid reconstruction that failed was only covered with a skin graft. None of the remaining 19 patients who had their graft covered with local or myocutaneous flaps had a graft blowout. The rate of a major stroke was 5% [54].

Arterial grafts may be more resistant to infection and may have superior patency rates [19]. Stoney and Wiley [59] were the first to describe the use of autogenous arterial bypass grafts for use in septic and irradiated wounds. Our experience with the superficial femoral artery graft in carotid reconstruction has been reported [19]. In our series of 30 patients, no case of suture line disruption or graft blowout was observed despite the high incidence of wound infection (43%) and fistula (30%) [19]. The size and mechanical strength, along with durable resistance to infection and thrombosis, make arterial grafts an excellent alternative in carotid artery reconstruction. Thirteen percent of the patients had complications at the superficial femoral artery harvest site, which include asymptomatic occlusion of the graft, superficial thigh wound infection, persistent lymph leak, and an infected pseudoaneurysm. There were no long-term complications. Only one stroke-related death (3%) occurred during the postoperative period [19].

Vascular complications associated with radiation therapy are accelerated atherosclerosis, radiation arteritis, and poor wound healing [54]. Salvage surgery after radiation therapy can be very difficult and is associated with a 60 to 70% rate of major postoperative complications, often resulting in the need for multiple reconstructive operations and prolonged hospitalizations [60].

Resection of the vagus nerve is associated with motor and sensory impairment of the palate, pharynx, and larynx, and with visceral deficits (i.e., decreased gastric emptying) [61]. Patients may need swallowing therapy, vocal cord medialization techniques, and prokinetic agents to improve their symptoms. Bilateral-staged carotid artery resections have been performed. These patients may experience a decreased response to hypoxia and deregulation of the blood pressure secondary to removal of both the carotid body and the carotid sinus [27].

IX. CONCLUSION

The management of patients with advanced neck carcinoma invading the carotid artery should be discussed in a multidisciplinary head and neck tumor conference, as it is subject of debate. Proponents of carotid artery resection contend that the surgical morbidity has been decreased significantly with the use of autogenous vascular grafts for reconstruction done by experienced surgeons. Although survival rates are low, improved local-regional control can be achieved with complete tumor removal and perhaps adjuvant radiotherapy and chemotherapy. Surgical resection can alleviate pain and massive carotid bleeding. If the patient chooses surgery, the head and neck surgeon should accomplish a complete tumor resection with adequate margins and the vascular surgeon should reconstruct the carotid artery. Nonsurgical modalities are a second option in the treatment of patients with advanced neck carcinoma in order to avoid complications of carotid artery resection. However, radiotherapy and chemotherapy are associated with significant morbidity, a decrease in the quality of life, and loss of function (affecting mastication, swallowing, and speech) [62]. A prospective randomized trial comparing both treatment modalities is difficult to perform, as the number of patients needed to reach statistical power is too large.

References

1. Watson, W., and Silverstone, S. (1939). Ligature of the common carotid artery in cancer of the head and neck. *Ann. Surg.* **109**,1–27.
2. Moore, O., and Baker, H. W. (1955). Carotid artery ligation in surgery of head and neck. *Cancer* **8**, 712–726.
3. Jacobs, J. R., Arden, R. L., Marks, S. C., Kline, R., and Berguer, R. (1994). Carotid artery reconstruction using superficial femoral arterial grafts. *Laryngoscope* **104**, 689–693.
4. Wyeth, J. A. (1879). "Essays in Surgical Anatomy and Surgery." William Wood and Company, New York.
5. Martin, H. (1957). "Surgery of the Head and Neck Tumors," pp. 80–81. Paul B. Hoeber, New York.
6. Conley, J. J. (1953). Free autogenous vein graft to the internal and external carotid arteries in the treatment of tumors of the neck. *Ann. Surg.* **137**, 205.
7. Conley, J. J. (1957). Carotid artery surgery in the treatment of tumors of the neck. *Arch. Otolaryngol.* **65**, 437–446.
8. Keirle, A. M., and Altemeier, W. A. (1960). Resection of carotid arteries for neoplastic invasion with maintenance of circulation. *Am. Surg.* **26**, 588.
9. Rella, A. J., Rongetti, J. R., and Bisi, R. (1962). Replacement of carotid arteries with prosthetic graft. *Arch. Otolaryngol.* **76**, 550–554.
10. Lore, J. M. (1973). Vascular procedures. *In* "Atlas of Head and Neck Surgery," Vol. II, pp. 844–848. Saunders, Philadelphia.
11. Lore, J. M. (1966). Surgery in head and neck tumors. *Arch. Otolaryngol.* **83**, 51.
12. Yoo, G. H., Hocwald, E., Korkmaz, H., *et al.* (2000). Assessment of carotid artery invasion in patients with head and neck cancer. *Laryngoscope* **110**, 386–390.
13. Langman, A. W., Kaplan, M. J., Dillon, W. P., *et al.* (1989). Radiologic assessment of tumor and the carotid artery. *Head Neck* **11**, 443–449.
14. Rothstein, S. G., Persky, M. S., and Horii, S. (1988). Evaluation of malignant invasion of the carotid artery by CT scan and ultrasound. *Laryngoscope* **88**, 321–324.
15. Snyderman, C. H., and D'Amico, F. (1992). Outcome of carotid artery resection for neoplastic disease: A meta-analysis. *Am. J. Otolaryngol.* **13**, 373–380.
16. Kennedy, J. T., Krause, C. J., and Loevy, S. (1977). The importance of tumor attachment to the carotid artery. *Arch. Otolaryngol.* **103**, 70–73.
17. Yousem, D. M., Hatabu, H., Hurst, R. W., *et al.* (1995). Carotid artery invasion by head and neck masses: Prediction by MR imaging. *Radiology* **195**, 715–720.
18. Gritzmann, N., Grasl, M. C., Helmer, M., and Steiner, E. (1990). Invasion of the carotid artery and jugular vein by lymph node metastases: Detection with sonography. *AJR Am. J. Roentgenol.* **154**, 411–414.
19. Sessa, C. N., Morasch, M. D., Berguer, R., Kline, R. A., Jacobs, J. R., and Arden, R. L. (1998). Carotid resection and replacement with autogenous arterial graft during operation for neck malignancy. *Ann. Vasc. Surg.* **12**, 229–235.
20. Reilly, M. K., Perry, M. O., Netterville, J. L., and Meacham, P. W. (1992). Carotid artery replacement in conjunction with resection of squamous cell carcinoma of the neck: Preliminary results. *J. Vasc. Surg.* **15**, 324–329.
21. Collins, S. L. (1999). Controversies in management of cancer of the neck. *In* "Comprehensive Management of Head and Neck Tumors" (S. E. Thawley, ed.), 2nd Ed., Vol. II, pp. 1504–1508. Saunders, Philadelphia.
22. Leibel, S. A., Scott, C. B., and Mohiuddin, M. (1991). The effect of local-regional control on distant metastasic dissemination in carcinoma of the head and neck: Results of an analysis from the RTOG head and neck database. *Int. J. Radiat. Oncol. Biol. Phys.* **21**, 549–556.
23. McCready, R. A., Miller, S. K., Hamaker, R. C., *et al.* (1989). What is the role of carotid arterial resection in the management of advanced cervical cancer? *J. Vasc. Surg.* **10**, 274–280.
24. Huvos, A. G., Leaming, R. H., and Moore, O. S. (1973). Clinicopathologic study of the resected carotid artery: Analysis of sixty-four cases. *Am. J. Surg.* **126**, 570–574.
25. Brennan, J. A., and Jafek, B. W. (1994). Elective carotid artery resection for advanced squamous cell carcinoma of the neck. *Laryngoscope* **104**, 259–263.
26. Collins, S. L. (1999). Management of the carotid artery and residual neck disease. *In* "Comprehensive Management of Head and Neck Tumors" (S. E. Thawley, ed.), 2nd Ed., Vol. II, pp. 1522–1525. Saunders, Philadelphia.
27. Biller, H. F., Urken, M., Lawson, W., and Haimov, M. (1988). Carotid artery resection and bypass for neck carcinoma. *Laryngoscope* **98**, 181–183.
28. Stell, P. M., Dalby, J. E., Singh, S. D., *et al.* (1984). The fixed cervical lymph node. *Cancer* **53**, 336.
29. Saunders, M. I. (1999). Head and neck cancer: Altered fractionation schedules. *Oncologist* **4**, 11–16.
30. Lindbergh, R. D., Paris, K. J., and Fletcher, G. H. (1999). Radiation only for clinically positive neck nodes. *In* "Comprehensive Management of Head and Neck Tumors" (S. E. Thawley, ed.), 2nd Ed., Vol. II, p. 1463. Saunders, Philadelphia.
31. Fu, K. K, Pajak, T. F., Trotti, A., *et al.* (2000). A Radiation Therapy Oncology Group (RTOG) phase III randomized study to compare hyperfractionation and two variants of accelerated fractionation to standard fractionation radiotherapy for head and neck squamous cell carcinomas: First report of RTOG 9003. *Int. J. Radiat. Oncol. Biol. Phys.* **48**, 7–16.
32. Horiot, J. C, Bontemps, P., van den Bogaert, W., *et al.* (1997). Accelerated fractionation (AF) compared to conventional fractionation (CF) improves loco-regional control in the radiotherapy of advanced head and neck cancers: results of the EORTC 22851 randomized trial. *Radiother.Oncol.* **44**, 111–121.
33. Freeman, S. B., Hamaker, R. C., Rate, W. R., *et al.* (1995). Management of advanced cervical metastasis using intraoperative radiotherapy. *Laryngoscope* **105**, 575–578.
34. Nag, S., Schuller, D. E., Martinez-Monge, R., *et al.* (1998). Intraoperative electron beam radiotherapy for previously irradiated advanced head and neck malignancies. *Int. J. Radiat. Oncol. Biol. Phys.* **42,** 1085–1089.
35. Werber, J. L., Sood, B., Alfieri, A., McCormick, S. A., and Vikram, B. (1991). Tolerance of the carotid-sheath contents to brachytherapy: An experimental study. *Laryngoscope* **101**, 587–591.

36. Chen, K. Y., Mohr, R. M., and Silverman, C. L. (1996). Interstitial iodine 125 in advanced recurrent squamous cell carcinoma of the head and neck with follow-up evaluation of carotid artery by ultrasound. *Ann. Otol. Rhinol. Laryngol.* **105**, 955–961.
37. Martinez, A., *et al.* (1983). ^{125}I implants as an adjuvant to surgery and external beam radiotherapy in the management of locally advanced head and neck cancer. *Cancer* **51**, 973.
38. Vikram, B., *et al.* (1983). ^{125}I implants in head and neck cancer. *Cancer* **51**, 1310.
39. Anonymous (1987). Adjuvant chemotherapy for advanced head and neck squamous carcinoma: Final report of the Head and Neck Contracts Program. *Cancer* **60**, 301–311.
40. Laramore, G. E., Scott, C. B., al-Sarraf, M., *et al.* (1992). Adjuvant chemotherapy for resectable squamous cell carcinomas of the head and neck: Report on Intergroup Study 0034. *Int. J. Radiat. Oncol. Biol. Phys.* **23**, 705–713.
41. Schuller, D. E., Metch, B., Stein, D. W., Mattox, D., and McCracken, J. D. (1988). Preoperative chemotherapy in advanced resectable head and neck cancer: Final report of the Southwest Oncology Group. *Laryngoscope* **98**, 1205–1211.
42. al-Sarraf, M., LeBlanc, M., Giri, P. G., *et al.* (1998). Chemoradiotherapy versus radiotherapy in patients with advanced nasopharyngeal cancer: Phase III randomized Intergroup study 0099. *J. Clin. Oncol.* **16**, 1310–1317.
43. Adelstein, D. J., Adams, G., Li, Y., *et al.* (2000). A phase III comparison of standard radiation therapy (RT) versus RT plus concurrent cisplatin (CDDP) versus split-course RT plus concurrent CDDP and 5-fluorouracil in patients with unresectable squamous cell head and neck cancer (SCHNC): An intergroup study. *In* "Proceedings, Annual Meeting American Society of Clinical Oncology," 1624. [Abstract]
44. Wendt, T. G., Grabenbauer, G. G., Rodel, C. M., *et al.* (1998). Simultaneous radiochemotherapy versus radiotherapy alone in advanced head and neck cancer: A randomized multicenter study. *J. Clin. Oncol.* **16**, 1318–1324.
45. Calais, G., Alfonsi, M., Bardet, E., *et al.* (1999). Randomized trial of radiation therapy versus concomitant chemotherapy and radiation therapy for advanced-stage oropharynx carcinoma. *J. Natl. Cancer Inst.* **91**, 2081–2086.
46. Bachaud, J. M., Cohen-Jonathan, E., Alzieu, C., David, J. M., Serrano, E., and Daly-Schveitzer, N. (1996). Combined postoperative radiotherapy and weekly cisplatin infusion for locally advanced head and neck carcinoma: Final report of a randomized trial. *Int. J. Radiat. Oncol. Biol. Phys.* **36**, 999–1004.
47. Anonymous (1991). Induction chemotherapy plus radiation compared with surgery plus radiation in patients with advanced laryngeal cancer. The Department of Veterans Affairs Laryngeal Cancer Study Group. *N. Engl. J. Med.* **324**, 1685–1690.
48. Lefebvre, J. L., Chevalier, D., Luboinski, B., Kirkpatrick, A., Collette, L., and Sahmoud, T. (1996). Larynx preservation in pyriform sinus cancer: Preliminary results of a European Organization for Research and Treatment of Cancer phase III trial. EORTC Head and Neck Cancer Cooperative Group. *J. Natl. Cancer Inst.* **88**, 890–899.
49. al-Sarraf, M., Pajak, T. F., Byhardt, R. W., Beitler, J. J., Salter, M. M., and Cooper, J. S. (1997). Postoperative radiotherapy with concurrent cisplatin appears to improve locoregional control of advanced, resectable head and neck cancers: RTOG 88–24. *Int. J. Radiat. Oncol. Biol. Phys.* **37**, 777–782.
50. Matas, R. (1940). Personal experiences in vascular surgery. *Ann. Surg.* **112**, 802.
51. March, M. A., Jenkins, H. A., and Coker, N. J. (1993). Preoperative carotid artery assessment. *In* "Otolaryngology-Head and Neck Surgery" (C.W. Cummings, ed.), 2nd Ed., Vol. IV, p.3373. Mosby-Year Book, St. Louis.
52. Adams, G. L., Madison, M., Remley, K., and Gapany, M. (1999). Preoperative permanent balloon occlusion of internal carotid artery in patients with advanced head and neck squamous cell carcinoma. *Laryngoscope* **109**, 460–466.
53. Urken, M., Biller, H. F., and Lawson, W. (1986). Salvage surgery for recurrent neck carcinoma after multimodality therapy. *Head Neck Surg.* **8**, 332–342.
54. Wright, J. G., Nicholson, R., Schuller, D. E., and Smead, W. L. (1996). Resection of the internal carotid artery and replacement with greater saphenous vein: A safe procedure for en bloc cancer resections with carotid involvement. *J. Vasc. Surg.* **23**, 775–782.
55. Lore, J. M. (1988). Resection of portion of common and internal carotid arteries involved by cancer. *In* "An Atlas of Head and Neck Surgery," 3rd Ed., pp. 1074–1081. Saunders, Philadelphia.
56. Lore, J. M. J., and Boulos, E. J. (1981). Resection and reconstruction of the carotid artery in metastatic squamous cell carcinoma. *Am. J. Surg.* **142**, 437–442.
57. Urken, M., Biller, H. F., and Haimov, M. (1985). Intratemporal carotid artery bypass in resection of a base of skull tumor. *Laryngoscope* **95**, 1472–1477.
58. Melecca, R. J., and Marks, S. C. (1994). Carotid artery resection for cancer of the head and neck. *Arch. Otolaryngol. Head Neck Surg.* **120**, 974–978.
59. Stoney, R. J., and Wiley, E. (1970). Arterial autografts. *Surgery* **67**, 18–25.
60. Collins, S. L. (1999). Salvage surgery after chemoradiation: A dangerous exercise in futility. *In* "Comprehensive Management of Head and Neck Tumors" (S. E. Thawley, ed.), 2nd Ed., Vol. I, pp. 239–241. Saunders, Philadelphia.
61. Jackson, C. G., and Bohrer, P. S. (1999). Cranial nerve rehabilitation. *In* "Comprehensive Management of Head and Neck Tumors" (S. E. Thawley, ed.), 2nd Ed., Vol. I, pp. 457–462. Saunders, Philadelphia.
62. Collins, S. L. (1999). Controversies in quality-of-life issues: Our impressions often differ from those of the patient. *In* "Comprehensive Management of Head and Neck Tumors" (S. E. Thawley, ed.), 2nd Ed., Vol. I, pp. 251–255. Saunders, Philadelphia.

CHAPTER

24

Skull Base Surgery

TERRY Y. SHIBUYA, WILLIAM B. ARMSTRONG, and JACK SHOHET

Department of Otolaryngology—Head and Neck Surgery
University of California, Irvine, College of Medicine and
Craniofacial/Skull Base Surgery Center
University of California, Irvine, Medical Center
Orange, California 92868

I. Introduction 339
II. Evaluation of Patient 340
A. History and Physical Examination 340
B. Radiographic Imaging 340
C. Carotid Artery Assessment 340
D. Specialized Diagnostic Tests 341
III. Anesthetic Considerations 341
A. Preoperative Consideration 341
B. Anesthetic Technique 341
IV. Physiological Monitoring 342
V. Pathology 342
A. Malignant Lesions 343
B. Benign Lesions 344
VI. Surgery of Skull Base 346
A. Anterior Skull Base 346
B. Anterolateral Skull Base 349
C. Middle and Posterior Skull Base 353
VII. Complications 355
A. Intraoperative Complications 355
B. Postoperative Complications 356
VIII. Conclusion 356
References 356

Surgery of the skull base has made tremendous strides over the past several decades. Areas of the base of skull considered surgically unresectable are now commonly operated within. Advances in intraoperative monitoring, radiographic imaging, anesthetic technique, and the concept of a skull base surgery team composed of multiple specialists have all improved surgical treatment and outcome. The following chapter provides an introduction to the diagnostic evaluation of the skull base tumor patient, the anesthetic and intraoperative monitoring options available, common pathologies encountered, surgical approaches for tumors involving the cranial base, and a discussion of potential complications. The skull base has been divided into anatomical regions, anterior, anterior–lateral, middle, and posterior, with each section reviewed individually.

I. INTRODUCTION

Surgery of the skull base for the treatment of benign tumors and malignancies has made considerable advancements since the 1960s. Ketcham *et al.* [1] published their series of craniofacial resections and demonstrated the feasibility of safely operating in the sinonasal/anterior cranial base region. Although they reported a complication rate of 80% and a mortality of 7%, their outcomes were far better than results reported previously up to that time. Follow-up of their patients revealed 65% were alive and without evidence of disease at 3 years [1]. Another study by Terz *et al.* [2] found that 72 and 50% of their patients were without disease at 3 and 5 years, respectively. Additional advances in posterior skull base surgery by House [3] and his development of the translabyrinthine approach for acoustic neuroma removal reduced surgical morbidity and mortality significantly. These early developments provided a foundation for future surgeons to build upon.

Technological advancements have made a significant impact on skull base surgery. Advances in radiological imaging using magnetic resonance imaging (MRI), computerized tomography (CT) scan, angiography, computer guidance systems, and positron emission tomography (PET) scan have all enhanced the diagnostics, surgical planning, treatment, and postoperative follow-up of patients. Advances in anesthetic techniques have permitted longer and safer procedures, whereas the electrophysiological monitoring of cranial nerves during surgery has enhanced the ability of surgeons to preserve function.

Advances in otolaryngology/head and neck surgery, plastics and reconstructive surgery, ophthalmology, oral/maxillofacial surgery, and neurosurgery have all resulted in improved treatment for skull base tumors. The development of a skull base surgery team has improved patient outcomes by blending the expertise of each specialist and combining it into one team focused on an optimal outcome.

II. EVALUATION OF PATIENT

A. History and Physical Examination

The evaluation of the patient with a cranial base lesion begins with a thorough history and physical examination. A complete head and neck examination, along with a neurological assessment, is performed. Particular emphasis in the examination is placed on the region of tumor involvement. Lesions of the anterior skull base frequently present with symptoms localized to the orbit, nose, or sinus area. Specific symptoms, such as nasal obstruction, epistaxis, anosmia, reduced visual acuity, proptosis, diplopia, epiphora, sinusitis, cerebral spinal fluid leak, or facial pain, may be present. Lesions of the anterolateral/middle skull base frequently present with symptoms localized to the posterior orbit, nasopharynx, or infratemporal fossa region. Specific symptoms, such as facial numbness, trismus, serous otitis media, endocrinopathy, or ocular gaze palsy, may be present. Lesions of the posterior skull base frequently present with symptoms localized to the temporal bone, cerebellum, or brain stem region. Specific symptoms, such as facial palsy, hearing loss, tinnitus, vertigo, vocal cord paralysis, cough, aspiration, shoulder weakness, lingual atrophy, ataxia, paralysis of conjugate gaze, or glossopharyngeal neuralgia, may be present. At the completion of the history and physical examination, the examiner should have an idea of the extent and location of the skull base lesion.

B. Radiographic Imaging

After completing the history and physical examination, it is important to obtain a high-quality imaging study, which is used for defining tumor location, extension, and position relative to vital structures. MRI provides excellent visualization of the intracranial contents, including the dura, meninges, and brain, with the addition of gadolinium (gadopentetate meglumine) providing enhancement of conspicuous pathology. We request axial, coronal, and sagittal images with 3-mm sections. All signal intensities, such as T1, T2, and proton-density images, are ordered. T1 images provide excellent normal anatomical detail, whereas T2 images enhance abnormal tissue pathology. Gadolinium infusion and fat suppression are used to define the intracranial extension of tumor and differentiate cystic from solid lesions. The addition of gradient-echo imaging and magnetic resonance angiography helps define the vascular anatomy. Cortical bone is visualized as a signal void, whereas the medullary cavity, if filled with fat or marrow, shows signal enhancement. Manipulation of the proton excitation time (TE) and relaxation time (TR) is used to help differentiate tumor from muscle and fat. Gadolinium with fat suppression helps in demonstrating perineural tumor spread and in visualizing the cavernous sinus.

A CT scan is preferred for visualizing bony structures, foramina, fissures, and canal (carotid). We request 3- to 4-mm overlapping axial and coronal cuts with soft tissue and bone windows. For temporal bone visualization, 1- to 2-mm overlapping fine cuts are requested. Two major disadvantages of CT are the inadequate display of intracranial structures and an inability to obtain direct sagittal images. Direct sagittal images are available with MRI and improve visualization of the sella turcica, clivus, foramen magnum, cribriform plate, and nasopharynx. Obtaining both MRI and CT scans will frequently complement each other and enhance the surgeon's preoperative knowledge concerning tumor extension into soft tissue, bone, and foramen.

C. Carotid Artery Assessment

Skull base imaging will occasionally reveal tumor invading the carotid artery. In this situation it is necessary to determine preoperatively whether the patient can tolerate carotid artery occlusion. There is a stroke rate of 26% for individuals undergoing abrupt carotid artery occlusion and a mortality rate of 12% [4]. In cases necessitating carotid artery occlusion, we obtain a balloon occlusion test preoperatively. For this test, the interventional radiologist will transfemorally cannulate the carotid artery and occlude the lumen for 15 min while monitoring the patient's neurologic function. If any neurologic deficit develops, the balloon is deflated immediately and the test is terminated. Those who fail balloon test occlusion should be considered for nonoperative therapies. Immediate neurologic deficits occur in 5–10% of individuals tested [4,5]. If the patient tolerates the balloon occlusion, a second phase of testing is performed with a xenon-CT scan, which quantifies the cerebral blood flow of the patient. If the cerebral blood flow rate is between 15 and 35 ml/100 g/min (25% of patients), there is a small-to-moderate risk of stroke with carotid artery ligation.

In this situation, reconstruction with an interposition graft at the time of surgery is recommended. If the cerebral blood flow rate is >35 ml/100 g/min (70% of patients), there is minimal risk of stroke with permanent artery occlusion. Even with minimal risk, it is still best to provide an interposition graft and maintain carotid artery perfusion [5]. Even though balloon occlusion testing and cerebral blood flow studies are very helpful, they can never fully predict the neurological outcome after carotid artery occlusion, even in low-risk patients [6].

D. Specialized Diagnostic Tests

Additional tests may be requested based on tumor location. For anterior lesions involving the orbit or periorbital region, neuroophthalmologic consultation and testing should be considered. Testing of structures traversing the superior orbital fissure can be performed. Visual acuity, visual fields, ocular mobility, optic nerve, and lacrimal apparatus function can also be assessed. Tumors arising from the posterior fossa will frequently need audiologic testing, which includes an audiogram, brain stem auditory-evoked responses, otoacoustic emissions, and electronystagmography. Tests are selected at the discretion of the evaluating specialist.

III. ANESTHETIC CONSIDERATIONS

A. Preoperative Consideration

Preparation of the patient undergoing skull base surgery can be complicated and requires effective communication among the team of surgeons, anesthesiologists, nurses, and ancillary staff. Prior to surgery a complete patient assessment is performed. Preoperative counseling by the surgeon(s) and anesthesiologist(s) regarding the operative and anesthetic risks is performed. The anesthesiologist must understand the surgical exposure needed, physiological monitoring planned, and patient positioning required. Because of these issues, the patient's head and extremities may not be readily accessible intraoperatively to the anesthesiologist. The development of a team with individuals familiar with the complexities involving skull base surgery is extremely valuable.

Because of the long duration of the procedure, proper patient positioning and padding are vital in the prevention of pressure sores and/or neuropathies. The patient is positioned on an egg crate or soft mattress. The arms are protected in a neutral position with additional soft padding. If an oblique, park bench, or sitting position is required, a beanbag mattress with an auxiliary roll is used. Heel cushions are also utilized. Frequently, a Mayfield head holder is used for head positioning (horseshoe or pins). The horseshoe head holder is used frequently to allow turning of the patient's head intraoperatively. It is important to further cushion the holder with foam and gauze padding to prevent pressure necrosis of the face.

The endotracheal tube must also be secured properly to the patient. Orotracheal tubes can be sutured to the mandible to prevent extrusion should frequent head turning be necessary. Nasotracheal tubes must be secured carefully with a stitch and pressure necrosis of the alar cartilage prevented. For extensive skull base procedures where the development of a tension pneumocephalus is a significant risk, performing a tracheotomy at the start of the procedure should be considered.

Additional use of warming blankets, compression stockings, and sequential compression devices are used to decrease the risk of deep vein thrombosis and pulmonary embolism. Two large-bore intravenous catheters provide adequate venous access. An arterial line allows for continuous blood pressure monitoring and access for blood sampling. A central venous pressure line or Swan–Ganz catheter is utilized frequently. In preparation for surgery, the selected skull base region is widely exposed and cleansed. If access to the eye is necessary, a corneal shield or tarsorraphy stitch may be placed for protection.

B. Anesthetic Technique

The induction, maintenance, and emergence from anesthesia can be a challenge for the anesthesiologist. A smooth induction without hypertension, hypotension, hypoxia, or hypercarbia is important. Many patients with skull base tumors will have difficult airways and a decreased tolerance of fluctuations in blood pressure or oxygenation. The anesthesiologist must also understand the neurophysiologic monitoring requirements of the case and plan accordingly. If electromyography (EMG) monitoring is planned, paralytic agents should not be used after the induction of anesthesia. In this instance, a narcotic-based anesthetic technique is commonly used. Narcotic agents are favorable for skull base surgery because they decrease the cerebral oxygen consumption rate, have fewer hypotensive side effects, avoid elevation of the intracranial pressure, and are reversible, allowing for rapid emergence from anesthesia. Intraoperatively, proper fluid management is critical. Diuretics are often administered to shrink the brain during the intracranial portion of the surgery to enhance surgical exposure. Simultaneously, hypovolemia and hypotension are to be avoided, whereas excessive fluid administration can cause cerebral edema. Fluids are generally replaced with isotonic crystalloid solutions (normal saline, lactated Ringer's, or plasmalyte) and colloid solutions (albumin, hetastarch). Extensive blood loss requires transfusions. The replacement of clotting factors (fresh frozen plasma, platelets) should be administered after four to six units of packed red blood cells have been transfused to prevent coagulopathy.

Proper preoperative preparation and anesthetic technique are cornerstones in a successful skull base operation. The coordination of the skull base team and preparation of the patient may be as complex as the surgical procedure and should not be underestimated.

IV. PHYSIOLOGICAL MONITORING

A number of neurophysiological monitoring options are available during skull base surgery. For comprehensive coverage of intraoperative monitoring, see reviews by Sclabassi and others [7]. The principal reason for monitoring is that the cerebral cortex, brain stem, and cranial nerves are sensitive to physiologic change, physical manipulation, or minor trauma. Therefore, the intraoperative monitoring of these structures can be used to minimize or prevent neurological damage. For difficult or revision cases, neurophysiological monitoring can aid in the identification of distorted nervous structures. It is important to emphasize that monitoring is not a substitute for having a detailed knowledge of the anatomy and precise operative technique.

Neurophysiological monitoring is carried out at three levels: the cerebral cortex, the brain stem, and the cranial nerves. Electroencephalography (EEG) can provide basic information about the cerebral cortex function and warn of decreased cerebral perfusion or oxygenation. More detailed information can be provided by somatosensory-evoked potentials. A peripheral nerve (most commonly median nerve) is stimulated and the cerebral cortex response is recorded. Visual-evoked potentials record the effects of light stimulation on the visual axis, measuring response from the retina to the occipital cortex. Brain stem function can be monitored by recording the somatosensory brain stem-evoked potentials or the brain stem auditory-evoked responses (BAER). Peripheral cranial nerve monitoring by EMG provides information about the cranial nerve and the corresponding brain stem nucleus. All motor cranial nerves (III, IV, V_3, VI, VII, IX, X, XI, XII) can be evaluated by placing an electrode into the muscle they individually innervate. Muscle action potentials are monitored, and activity caused by trauma or irritation to the nerve can be detected. Additionally, the location of intact nerves can be confirmed by stimulating each with specialized probes and observing the action potential generated. BAER is used to monitor hearing intraoperatively. EMG monitoring requires the patient to be nonparalyzed during surgery and this must be communicated to the anesthesiologist at the outset of the procedure. Today, the facial nerve is the most commonly monitored nerve during skull base surgery, other cranial nerves can be selectively monitored at the discretion of the surgeon.

V. PATHOLOGY

There are a diverse number of pathologic lesions, which may develop in or involve the skull base. Intracranial extension of extracranial tumors is much more common than primary intracranial tumors with extradural extension. This section describes briefly the common benign and malignant tumors involving the skull base.

A variety of tumors can involve the anterior skull base. The ethmoid sinus is the most common extradural site from which tumor extends into the anterior cranial base. A benign tumor originating from this location, such as inverting papilloma, has the potential to develop into a malignant lesion in 10% of cases. Other benign lesions include neurofibroma, nasal glioma, and anterior meningocele. Sinonasal malignancies, which may extend into the anterior skull base, include esthesioneuroblastoma, squamous cell carcinoma, adenocarcinoma, adenoid cystic carcinoma, and various sarcomas. Benign and malignant lesions originating from bone or cartilage involving the anterior skull base include fibrous dysplasia, ossifying fibroma, chordoma, chondrosarcoma, and osteogenic sarcoma. Other tumors include malignant schwannoma, plasmacytoma, melanoma, lymphoma, and metastatic tumors. In children, rhabdomyosarcoma is the most common malignant tumor of this region. Juvenile nasopharyngeal angiofibroma presents in young male adolescents. Olfactory groove meningiomas arise as intracranial tumors with extradural extension.

A variety of lesions arise from or extend into the anterolateral skull base or infratemporal fossa region. Juvenile nasopharyngeal angiofibroma can extend laterally through the pterygomaxillary fissure into the infratemporal fossa. Often, this lateral extension cannot be accessed adequately via a transpalatal or transmaxillary approach. Benign lesions involving this region include schwannomas, glomus tumors, chondromas, meningiomas, and fibrous dysplasia. An unusual condition seen occasionally in the infratemporal fossa is inflammatory pseudotumor, usually originating from the orbit. This condition is characterized by inflammatory tissue on biopsy and is managed by steroid treatment.

Malignant tumors can involve the infratemporal fossa. Squamous cell carcinoma extending from the maxillary sinus, nasopharynx, or temporal bone can invade this region. Malignant salivary gland tumors arising from the deep lobe of the parotid gland or minor salivary gland may also extend into the area. Lymphomas occasionally present as an infratemporal fossa mass. Rhabdomyosarcoma should also be considered in differential diagnosis, especially in children and young adults. Chordomas, chondrosacromas, and other metastatic tumors such as breast, prostate, melanoma, and kidney cancers, also present in this region.

Neoplasms arising from the middle or posterior cranial base often spread and involve adjacent skull base regions. Both benign and malignant lesions occur in this area.

Malignancies involving the middle or posterior skull base include squamous cell carcinomas, chordomas, or chondrosarcomas. Benign lesions commonly involving these regions include meningiomas, epidermal cysts, paragangliomas, acoustic neuromas, and schwannomas. Treatment options are limited primarily to surgical resection and radiation therapy. Both are quite challenging given the risk of injury to normal adjacent structures with either modality.

A. Malignant Lesions

1. Squamous Cell Carcinoma

Squamous cell carcinoma (SCCA) is a malignancy derived from the squamous epithelia of the upper respiratory tract. It is the most common malignancy of the head and neck region and may extend into the anterior or anterolateral skull base from the sinonasal cavity, oropharynx, hypopharynx oral cavity, or nasopharynx. Extension into the middle or posterior skull base may come from the external auditory canal. SCCA can directly invade soft tissue or bone and spread intracranially via perineural spread through skull base foramina. Histologically, SCCA is divided into well, moderate, and poorly differentiated tumors based on the degree of intercellular bridges, amount of keratinization, nuclear pleomorphism, mitotic rate, and degree of cell cohesion. Immunohistochemistry show this tumor to be cytokeratin positive. Treatment options of surgery, radiation therapy, and/or chemotherapy are based on tumor location and extension within the skull base.

2. Adenoid Cystic Carcinoma

Adenoid cystic carcinoma (ACC) is a tumor of salivary gland origin. ACC is composed of small basal duct lining cells and myoepithelia cells arranged in a characteristic cribriform pattern. It is the most common malignant tumor of minor salivary gland origin occurring in the sinuses, oral cavity, or oropharynx. ACC is characterized by a slow continuous growth over a long time period. Adenoid cystic carcinoma is neurotropic and extends intracranially by perineural spread. The incidence of lymphatic metastasis is 13–16% [8], whereas hematogenous metastasis is 40%. Three histological patterns of ACC have been described, which include tubular, solid, and cribriform. The cribriform pattern consists of the "classic" pattern with nests of cells demonstrating multiple circular spaces filled with bluish or pink mucinous material. The tubular pattern consists of cells arranged in individual ducts or tubules, and the solid pattern has sheets of basaloid cells with few cyst formations. Surgical resection with postoperative radiation therapy is the primary therapy for this cancer. Chemotherapy is reserved for residual, recurrent, metastatic, or unresectable disease.

3. Esthesioneuroblastoma

Esthesioneuroblastoma or olfactory neuroblastoma is a neuroectodermal tumor arising from the olfactory epithelium. It is characteristically identified high in the nasal vault and is attached to the cribriform plate. Symptoms of nasal obstruction, epistaxis, or change in smell are common at the time of diagnosis. Cervical lymph node metastasis occurs in 10–30% of cases and systemic metastasis occurs in 8–46% of cases [9–12]. The incidence of local recurrence is 30–40% [9–11]. The 5- and 10-year survival rates are 70–80 and 60–70%, respectively [9,10]. Histologically, this tumor is characterized by small, round cells slightly larger than lymphocytes growing in a lobular or diffuse pattern or a combination of both. There are hyperchromatic nuclei with uniform chromatin distribution. The cytoplasm is sparse, whereas the stroma is highly vascular and often shows a neurofibrillary appearance. Homer Wright and Flexner–Wintersteiner rosettes are seen frequently. The immunohistochemistry stain is positive for synaptophysin and neuron-specific enolase, whereas the S-100 protein may be focally positive [11,13]. Wide excision with postoperative radiation therapy is frequently recommended for this cancer. Chemotherapy is reserved for recurrent, residual, metastatic, or unresectable disease.

4. Chordoma

Chordomas are slow-growing, locally aggressive malignant neoplasms derived from the vestigial remnants of the notochord. They are relatively rare and comprise only 1% of intracranial tumors [14]. Most large series indicate that 35% of chordomas originate from the base of skull/clivus. They typically present in the third and fourth decades of life. The most common presenting symptoms include diplopia from cranial nerve VI involvement and headache [15]. The tumor affects males more frequently. MRI is important to delineate the extent of the tumor. Findings include posterior extension to the pontine cistern and a lobulated, "honeycomb" appearance after gadolinium enhancement. A CT scan demonstrates osteolysis [16]. The treatment is wide surgical excision and radiation therapy. Complete surgical resection is difficult and local recurrence is common but systemic metastases are rare. The average survival of patients with a cranial chordomas is 4–8 years.

Histologically, chordomas are classified into three types: conventional or classic, chondroid, and dedifferentiated. Chordomas are characterized by epithelioid cells with vesicular nuclei and copious granular to vacuolated-appearing cytoplasm creating the characteristic physaliferous cells. Chondroid chordomas, a variant of chordoma, have a better prognosis than nonchondroid chordomas. Chondroid chordomas are noted for their predominance in females, younger patients, and presence of a cartilaginous component [17].

5. Chondrosarcoma

Chondrosarcomas are often confused with chordomas but should be differentiated, as their prognosis is significantly better. Females are affected more commonly and the average age is 39 years old. They originate form the petro-occipital junction in 66% and the clivus in 28% of cases [18]. MRI findings in both chondrosarcomas and sarcomas are a low signal intensity on T1-weighted images and a medium-to-high signal intensity on T2-weighted scans. There is a heterogeneous but usually intense enhancement with contrast [19].

Sarcomas are generally radio resistant, and surgical excision is the primary treatment modality. Both anterior and lateral approaches to the tumor are effective and should be tailored to the individual tumor location. A lateral approach offers the advantages of better control of major vascular structures and avoidance of cerebral spinal fluid (CSF) contamination with pharyngeal secretions. Radiation has a variable role with some advocating postoperative radiation only for incomplete tumor excision [15] or in all patients with chordoma postoperatively [18]. Morbidity is quite high in terms of cranial nerve deficits and CSF leakage [15]. One series reported a 5-year recurrence-free survival rate for chordoma and chondrosarcoma as 65 and 90%, respectively [15]. In another series of 200 skull base chondrosarcomas, the 5- and 10-year local control rates were 99 and 98%, respectively.

B. Benign Lesions

1. Fibrous Dysplasia

Fibrous dysplasia is a benign fibroosseous process in which normal bone is replaced by fibrous connective tissue and structurally weak fibrillar bone. It may be monostotic or polyostotic and frequently involves the craniofacial complex, most commonly occurring in the calvarium and maxilla. It is usually a self-limiting process, which starts in childhood and may not give symptoms until adulthood. Albright's syndrome consists of polyostotic fibrous dysplasia, pigmented skin lesion, and endocrine abnormalities such as precocious puberty. Histologically, fibrous dysplasia is characterized by irregularly woven bony trabeculae surrounded by loose, uniform fibroblastic spindle cells. The trabeculae are often described as having a "Chinese character" shape. The treatment is surgical resection or curettage.

2. Juvenile Nasopharyngeal Angiofibroma

Juvenile nasopharyngeal angiofibromas (JNA) typically occur in young adolescent male and frequently present as recurrent epistaxis. JNA is a vascular tumor, which may be locally aggressive and originates from the posterolateral nasal wall in the region of the sphenopalatine foramen. It comprises only 0.5% of all head and neck lesions, but up to 20% will have intracranial extension at the time of diagnosis. There is a 10% recurrence rate after surgical resection. Histologically, the tumor is nonencapsulated with slit-like vascular channels (stag-horn shaped) interspersed within a fibrous stroma. Immunohistology is positive for vimentin and occasionally the androgen receptor. Primary treatment is surgical resection with radiation therapy used for advance unresectable lesions, which invade vital structures.

3. Meningioma

Meningiomas are usually benign, slow-growing, extra-axial brain tumors arising from the meningocytes of arachnoid granulations [20]. They present in middle-to-late adult life and show a female predominance. Meningiomas comprise 15% of all intracranial tumors and occur frequently in the parasagittal region along the dural sinuses, lateral cerebral convexities, sphenoid ridge, and areas of dural penetration by cranial nerves. In 20% of cases, extracranial extension occurs into the calvarium, skull base, orbit, middle ear, sinonasal cavity, or parapharyngeal space. Meningiomas are the second most common tumor originating in the cerebellopontine angle (CPA) and may also be found ectopically in the middle ear, sinonasal cavities, orbit, oral cavity, and parotid gland. Symptoms usually include progressive hearing loss, dysequilibrium, headache, and cranial neuropathies. Overall, 90% of meningiomas are benign, 6% are atypical, and 2% are malignant.

Meningiomas demonstrate a pathognomonic speckled calcification within the substance of the tumor on CT. Lesions are usually solitary, well demarcated, firm, smooth or lobulated. Histologically, there are four basic growth patterns: syncytial, fibroblastic, transitional, or angioblastic. Only the angioblastic variant has any clinical relevance, which displays a more aggressive behavior. Neurofibromatosis II predisposes patients to multiple meningiomas and acoustic neuromas.

Complete surgical resection is usually curative and is recommended whenever possible. Stereotactic radiation should be considered for lesions in vital areas [21], as adjuvant therapy after subtotal resections [21], or in poor surgical risk patients. Symptomatic postradiation perilesional edema has occurred in a number of patients and does not appear to be related to tumor volume, radiation dosage, or amount of brain in the radiation field [22]. Chemotherapy with hydroxyurea, interferon α, tomoxifen, and mifepristone (RU-486) has been somewhat successful in unresectable tumors or treatment failures [23].

4. Epidermoid Cyst (EC)

Epidermoid cysts of the skull base are congenital lesions believed to originate from epithelial rests. They consist of desquamated keratin debris within a stratified squamous epithelial lining. EC are identified most commonly within the petrous portion of the temporal bone (40%). Because of their slow growth, symptoms do not typically present until the fourth decade of life. EC are identical to middle ear cholesteatomas and expand along the path of least resistance. They can displace cranial nerves, brain, and vascular structures causing neurological symptoms, which may include sensorineural hearing loss, imbalance, ataxia, facial tic, facial paresis, or trigeminal neuralgia.

CT and MRI aid in making the diagnosis. By CT, EC appear hypointense compared to brain and do not enhance with contrast. The capsule can demonstrate calcification. With MRI, EC are homogeneous and isointense to brain on T2 and heterogeneous and hypointense to brain on T1 images.

The treatment of EC is surgical resection as they are radioresistant. The primary objective is to decompress the mass and remove the nonadherent portions of the capsule. Dissection of the tumor capsule on vital structures is not recommended and therefore subtotal resection may be necessary in a significant number of patients. Long-term follow-up with serial imaging studies is necessary. The 13-year recurrence-free survival rate for totally resected tumors is 95% compared to 65% for subtotally resected tumors [24].

5. Paragangliomas

The so-called "glomus" tumors involving the skull base are of neural crest origin and arise in close association with the autonomic ganglion cells surrounding the jugular bulb or vagus nerve. Jugulotympanic paragangliomas spread along the path of least resistance through the air cell tracts within the temporal bone [25,26]. Extratemporal spread occurs through the lumen of the internal jugular vein, the eustachian tube, internal auditory canal, jugular foramen, and other neuromuscular foramina. The posterior cranial fossa is involved much more commonly than the middle cranial fossa.

Approximately 5% of paragangliomas are malignant [27] and are typically slow growing and asymptomatic until advanced size. Initial symptoms include pulsatile tinnitus, hearing loss, and aural fullness followed by facial paralysis and lower cranial nerve symptoms. Intravagal tumors usually present as a vague, painless mass high in the neck behind the angle of the mandible [28].

Jugulotympanic paragangliomas are the most common tumor of the middle ear [29] and the second most common tumor of the temporal bone [30]. Females are affected more commonly than males, and the typical age at presentation is between the fifth and sixth decades of life. There appears to be a familial tendency in some cases [31]. Multicentricity occurs in 10% of nonfamilial cases, with other primary ipsi- or contralateral paragangliomas of the carotid body, intravagal, or jugulotympanic region present [30,32,33].

One to 3% of tumors secrete catecholamines, and preoperative screening for urinary catecholamines is generally recommented [34]. CT is useful in determining the amount of bony destruction, whereas MRI delineates the limits of the tumor. Both studies should be part of the preoperative workup. Diagnostic angiography is usually not necessary; however, preoperative embolization is performed frequently to reduce intraoperative bleeding.

Surgery is the mainstay of therapy, although stereotactic radiation therapy can be used in cases of residual or recurrent tumor or in poor surgical candidates [35]. Watchful waiting may be more appropriate for patients with small tumors or advanced age.

6. Acoustic Neuromas

Acoustic neuromas are benign tumors that arise from the supporting Schwann cells of the vestibular portion of the vestibulocochlear nerve. These tumors account for 10% of all intracranial neoplasms and up to 90% of all CPA tumors. There are approximately 1:100,000 cases annually in the general population. Up to 96% of individuals with neurofibromatosis type 2 will have bilateral acoustic tumors [36].

Symptoms include progressive sensorineural hearing loss, tinnitus, and dysequilibrium. Facial paresthesias and numbness develop as the tumor extends superiorly. Although the facial nerve is adjacent to the cranial nerve VIII in the brain stem and internal auditory canal, facial paralysis or paresis is typically a late finding. The tumor growth rate is quite variable. The average growth rate is between 2.5 and 4 mm/year, although a significant proportion will not demonstrate any growth. Tumors typically expand in a lateral-to-medial direction, resulting in brain stem compression, hydrocephalus, elevated intracranial pressure, and eventual death if left untreated.

To identify tumors, MRI with gadolinium is the study of choice for diagnosis. Acoustic neuromas are isointense or mildly hypointense compared to brain on T1 images and mildly hyperintense on T2 images. The addition of gadolinium contrast increases the sensitivity of the scan.

Therapeutic options for acoustic neuromas include watchful waiting, stereotactic radiotherapy, or surgical resection. Studies, which have performed serial observation of tumors by MRI, have reported that only 30–50% of tumors increase in size after initial detection [37–39]. These findings have significant implications on the management of acoustic neuromas. Age at diagnosis, tumor size, contralateral hearing status, current health status, and symptoms must all be taken into consideration when selecting a course of therapy. The majority of patients over the age of 65 with small tumors are best managed by observation using serial imaging [18]. A significant number will have regression of tumor size [18].

Stereotactic radiotherapy can be attractive, as a craniotomy and hospitalization are avoided. However, long-term studies have shown that there are still complications of hearing loss and facial nerve paresis associated with this treatment. As an alternative, surgery offers a much better long-term tumor control rate of 98% [40], whereas the long-term results of low-dose radiation are still not available. When observation is not indicated, surgical resection is the treatment of choice unless the patient's health is poor or refuses surgery [40,41].

7. Facial and Lower Cranial Nerve Schwannomas

Facial nerve schwannomas can arise from any location along the course of the nerve. Hemifacial spasm is a common symptom and can help differentiate facial schwannomas from vestibular schwannomas. Facial weakness does not present until the tumor becomes very large. MRI characteristics are identical to those of acoustic neuromas.

Schwannomas of cranial nerves IX, X, or XI typically form within the jugular foramen and produce deficits of the nerve involved. Larger tumors can cause deficiencies of all nerves within the foramen. Treatment options include surgical resection or stereotactic radiation therapy. Surgical resection can be accomplished usually through a combined infratemporal fossa and transjugular approach.

VI. SURGERY OF SKULL BASE

A number of neoplasms can involve the cranial base. Each is unique and is approached on an individual basis with the goal of obtain optimal surgical resection while preserving vital structures and neurologic function. The ideal surgical exposure avoids prolonged retract of the skull base maintains separation of the cranial contents from the sinues, nasal cavity, or middle ear space, which is important for the prevention of infection.

For purposes of discussion, the skull base has been divided anatomically into anterior, anterolateral, and middle/posterior regions. For each region, the anatomy, surgical approaches, and reconstructive techniques are discussed.

A. Anterior Skull Base

Anterior cranial base surgery addresses tumors arising from the central craniofacial region. Anatomically, this includes the olfactory region, paranasal sinus, orbit, nasal cavity, nasopharynx, and clivus. Tumors are characterized as arising from the extracranial or intracranial region. Extracranial tumors arising from the sinonasal region extending into the brain or orbit are more common than intracranial tumors extending externally.

1. Surgical Approaches

There are a number of approaches available for accessing the anterior skull base, each with their own advantages and disadvantages. The traditional *transfrontal approach* (Fig. 24.1) was described by Derome [42,43] to access inferiorly down to the level of the sphenoid sinus and clivus. Lateral access to the optic canal, orbit, superior orbital fissure, and cavernous sinus is possible through this exposure. The disadvantage of this approach is the frontal lobe retraction required to expose the region. An improvement to this approach has been the *subfrontal approach* (Fig. 24.1) pioneered by Raveh *et al.* [44,45], which reduces retraction on the brain and improves posterior and inferior exposure of the skull base. The *transethmoidal approach* (Fig. 24.2) is another option, which provides direct access to the anterocentral skull base. It is not used as frequently due to incisions placed directly over the face. This exposure can be extended to access the frontal sinus, medial orbit, maxillary sinus, nasal cavity, sphenoid, pituitary, clivus, pterygopalatine fossa, and petrous apex [46]. The *transeptal/transnasal approach* (Fig. 24.3) is the simplest method to approach the pituitary and upper clivus. Surgery in this region may be performed with a microscope or endoscope. Instrumenting laterally within this exposure can be a limitation. To visualize the entire clivus inferiorly, a *transmaxillary approach* (Fig. 24.4) may be added to the transeptal/transnasal approach by mobilizing the lower maxilla via a LeFort I osteotomy with or without palatal split. The *transmaxillary approach* may also be used to access the pterygoid region by degloving the face and removing the medial and posterior maxillary sinus walls. The *transoral approach* (Fig. 24.5) is used to access midline lesions, which extend from the lower clivus down to C2–C3. Further superior extension may be

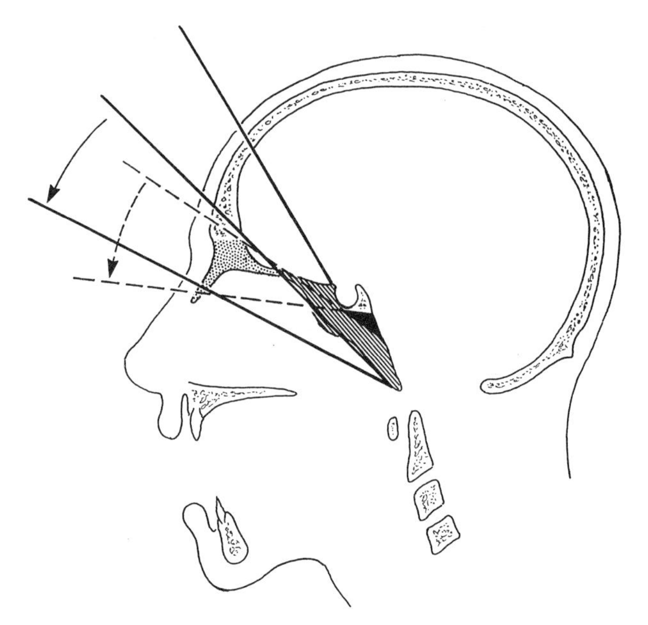

FIGURE 24.1 The transfrontal and subfrontal to approach the skull base. The arrow indicates the improved view with subfrontal extension.

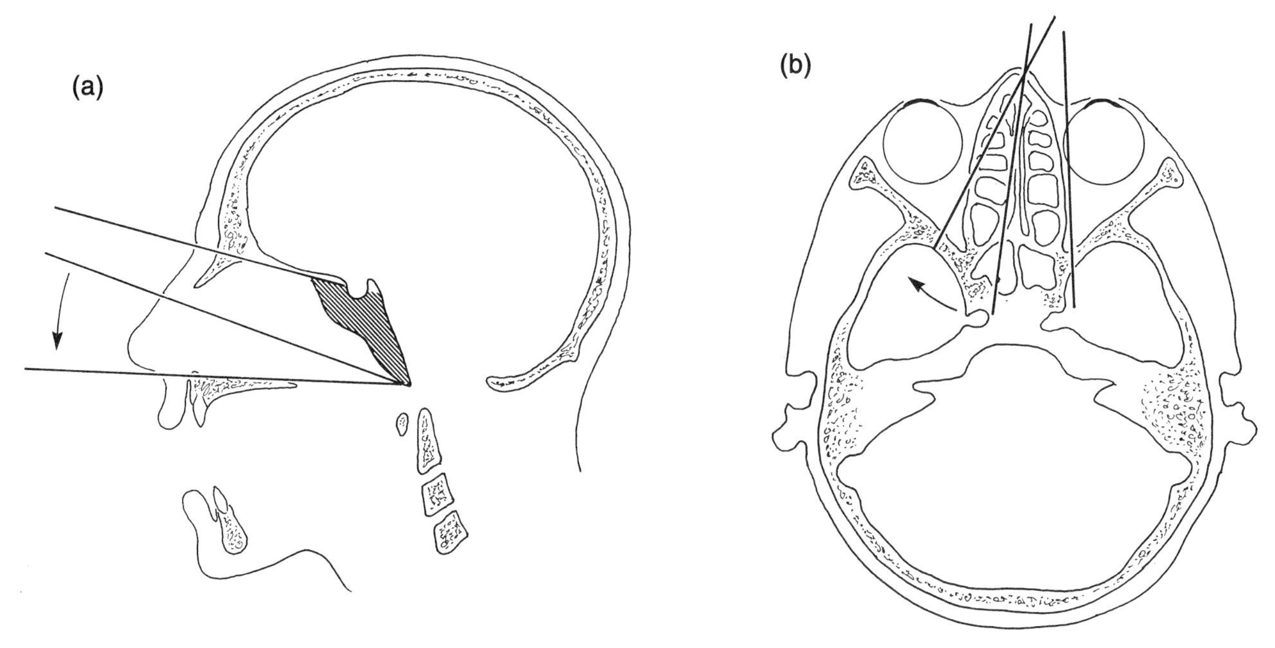

FIGURE 24.2 The transethmoidal approach to the skull base. (a) The arrow indicates addition of lateral rhinotomy improving the ability to instrument in this area. (b) Lines show the amount of lateral extension available through this exposure.

obtained by splitting the soft palate and additional inferior extension by depressing the tongue.

2. Subfrontal Approach

Raveh, in 1978, pioneered the subfrontal approach to the anterior skull base. Initially, this approach was use to repair high-velocity skull base trauma and congential anomalies. It provides vertical access from the anterior ethmoid roof down to the clivus and horizontal access across both orbital roofs, extending toward the temporal bone. Visualization of the nasal cavity and maxillary sinuses is obtained easily. The exposure allows for intra- and extradural tumor resection with minimal or no frontal lobe retraction.

The subfrontal approach begins with a coronal incision performed from preauricular crease to preauricular crease. A pericranial flap is preserved and based on the supratrochlear and supraorbital arteries for later use in reconstruction of the

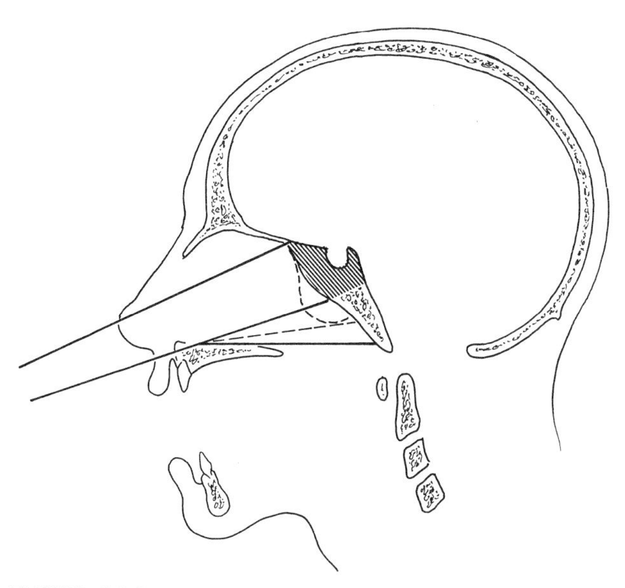

FIGURE 24.3 The transseptal–transsphenoidal approach to the skull base. Lines indicates the extent of exposure.

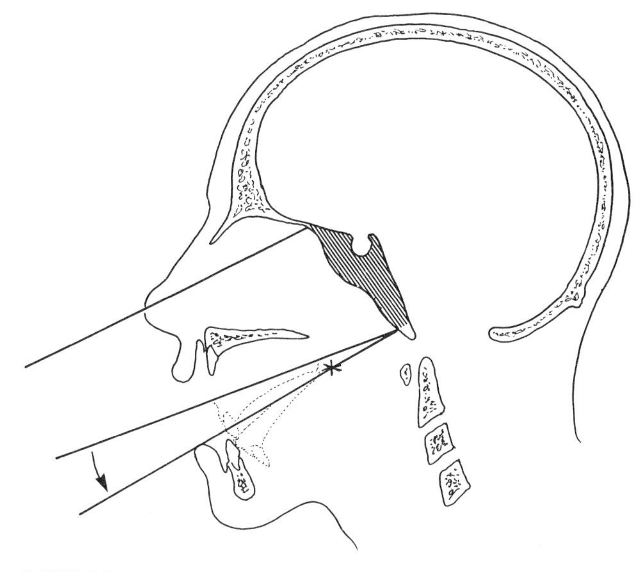

FIGURE 24.4 The transmaxillary with the Le Fort I osteotomy approach to the skull base. Lines indicate the extent of exposure.

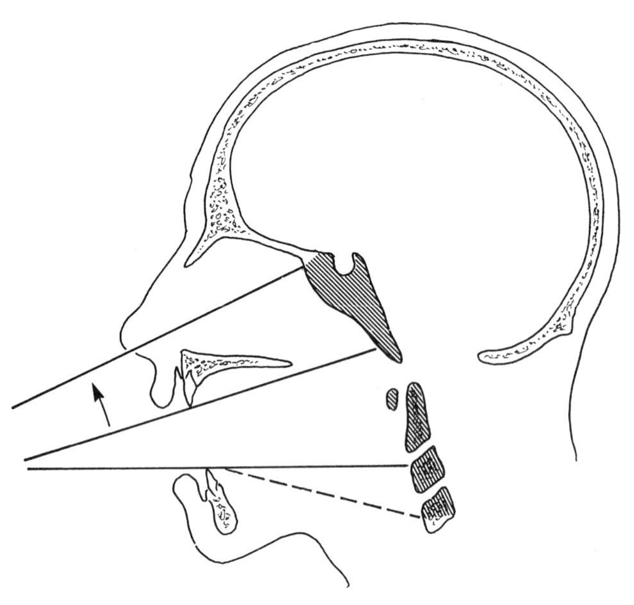

FIGURE 24.5 The transoral approach to the clivus and cervical spine. Lines indicate the amount of superior and inferior exposure available through this approach.

anterior skull base defect. The flap is carried down to the frontozygomatic suture line laterally and to the rhinion and piriform aperture medially. The orbit is accessed and the periorbita is elevated off the medial, superior, and lateral orbital walls. The anterior ethmoidal artery is ligated. Osteotomies are marked out, and the size of the frontal bone flap removal is dependent on the size of tumor being removed. Osteotomies are usually performed as follows: superior osteotomy—placed across the frontal bone in a horizontal plane, lateral osteotomies—cut from the superior osteotomy ends inferiorly down to the superior orbital rims bilaterally, orbital osteotomies—placed from the superior orbital rim cut 1 cm posterior into the superior orbital roof, then cut 90° medially to the medial orbital wall, then cut inferiorly down the medial orbital wall to the level of the nasolacrimal duct, and then cut anteriorly and out the medial orbital wall, anterior osteotomy—placed along the nasomaxillary groove horizontally just anterior to the lacrimal duct and connected with the opposite side (Fig. 24.6). A final vertical osteotomy is performed anterior to the crista galli detaching the frontonasal segment. The orbit and dura are protected at all times with ribbon retractors. There are variations in the size of bone flap removed. A Raveh type I approach removes the frontonasal segment while preserving the posterior wall of the frontal sinus. The posterior wall is removed in a second step and is indicated when tumor abuts this region. A Raveh type II approach removes the frontonasal segment, which includes the posterior wall of the frontal sinus. This is performed when tumor does involve the posterior wall or broader intracranial exposure is needed to access the tumor. Visualization and removal of tumor extending to the sphenoid sinus and clivus are achieved easily (Fig. 24.7). Craniofacial resections may also be performed through this exposure. For sinonasal tumors extending through the olfactory groove, the olfactory cleft may be keyholed easily and dropped inferiorly into the sinus cavity for an en bloc resection. If tumor only involves one side of the olfactory groove, the involved side may be visualized and resected easily while preserving the opposite side.

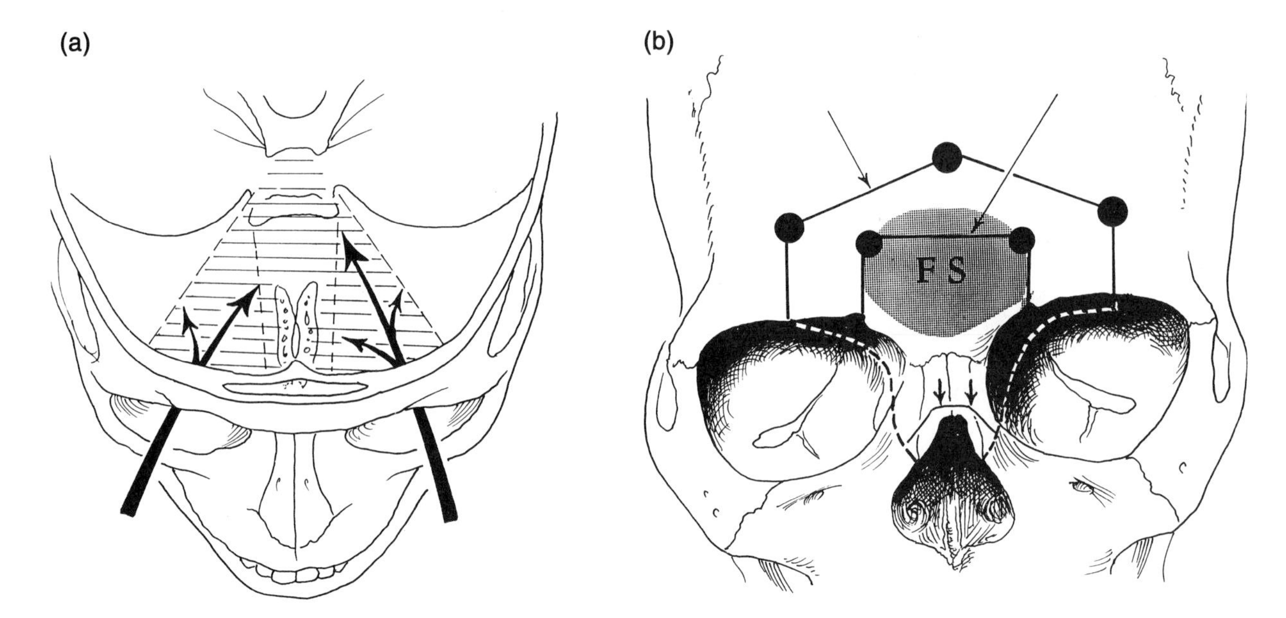

FIGURE 24.6 The subfrontal approach. (a) Subcranial access viewed from above. (b) Osteotomies performed, FS, frontal sinus.

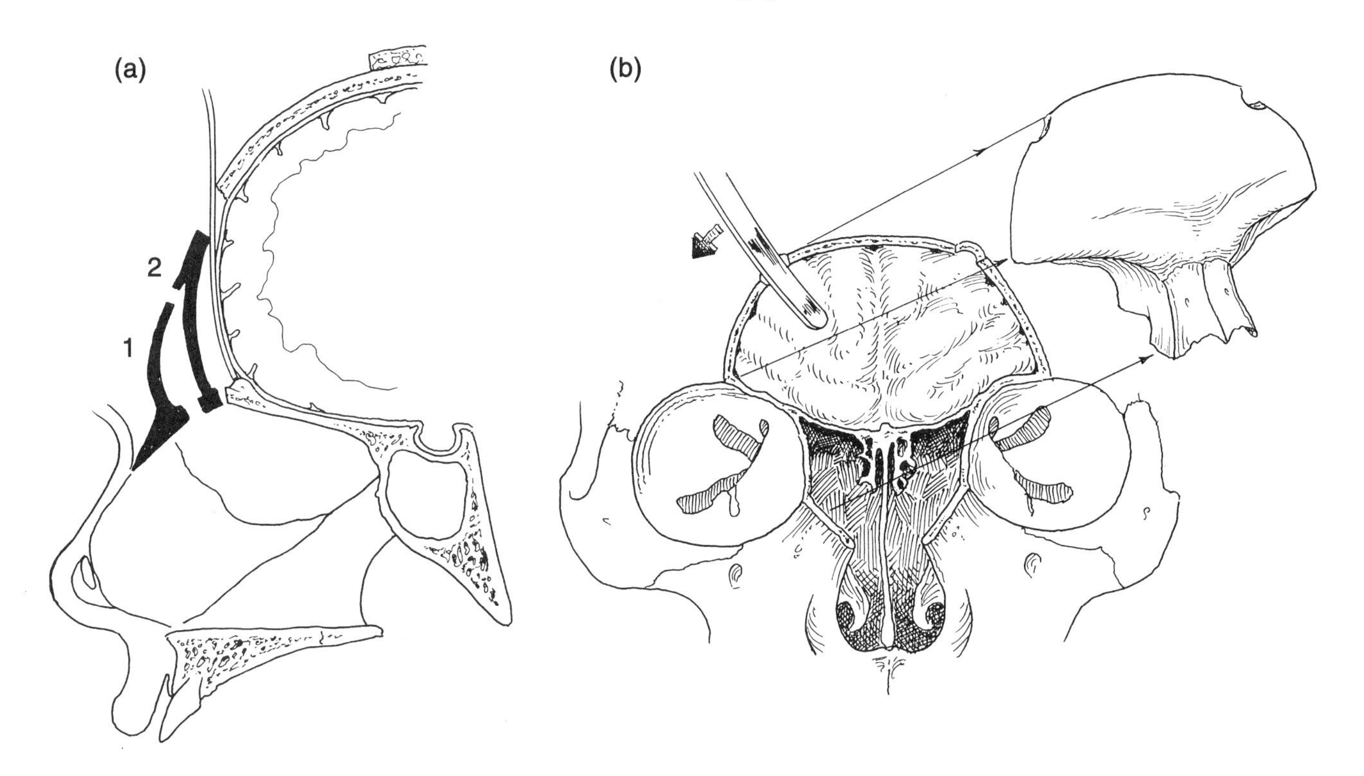

FIGURE 24.7 Raveh type 1 and 2 approaches. (a) The frontal bone flap created with the type 1 and 2 approaches. (b) Removal of the frontal bone flap and view obtained.

4. Reconstruction

Meticulous dural reconstruction is performed to avoid CSF leak and infection. Small dural defects are sutured shut while larger defects may be repaired using an anteriorly based pericranial flap or a laterally based temporalis-pericranial flap, which is supplied by the superficial and deep temporal arteries. Either flap is rotated into the defect and used to separate the dura from the sinonasal cavity. Other options include using tensor fascia lata, temporalis fascia, lyophilized dura, or bovine pericardium to separate the regions. For large defects a free or pedicled myocutaneous flap may be used. To prevent herniation of the medial orbital contents, temporalis fascia or tensor fascia lata may be used to line the medial wall. Gel foam (Pharmacia & Upjohn Comp., Kalamazoo, MI) is then place on top of the fascia, and Xeroform petrolatum gauze (Sherwood Medical, St. Louis, MO) is then used to line the cavity [47]. Bacitracin-impregnated packing is placed to hold the grafts in position for 1 week and is then removed. Free bone grafts to reconstruct the medial orbital wall are rarely performed. If used, they must be completely surrounded by vascularized tissue or have a high risk of failure. This is especially true in a previously irradiated tissue bed. The frontal bone flap is replaced and the frontal sinus is cranialized prior to reinsertion, while the nasofrontal duct region is also obliterated.

B. Anterolateral Skull Base

Surgery of the anterolateral cranial base requires a detailed anatomical knowledge of the infratemporal fossa, parapharyngeal space, pterygopalatine fossa, orbit, floor of the middle cranial fossa, and cavernous sinus. Several anatomic barriers are encountered during the surgical approach to the infratemporal fossa. Superficially located are the facial nerve, parotid gland, temporalis muscle, zygomatic arch, and mandible. The internal maxillary artery passes just deep to the mandible and provides deep branches to the temporalis muscle. The lateral wall of the orbit and squamous portion of the temporal bone lie under the temporalis muscle. The inferior orbital fissure is encountered on the deep aspect of the lateral orbital wall, and further dissection along the greater wing of the sphenoid leads to the lateral pterygoid process of the sphenoid bone. At the posterior-lateral edge of the pterygoid process lies the third division of the trigeminal nerve (V_3) emanating from foramen ovale. Directly behind foramen ovale, the middle meningeal artery passes through foramen spinosum. Posterior and lateral to foramen spinosum lies the spine of the sphenoid bone, which serves as a landmark for identifying the internal carotid artery (ICA) and its bony canal (Fig. 24.8). Anterior to the pterygoid plate lies a gap between the plate and the posterior wall of the maxillary sinus called the pterygomaxillary fissure. The interorbital

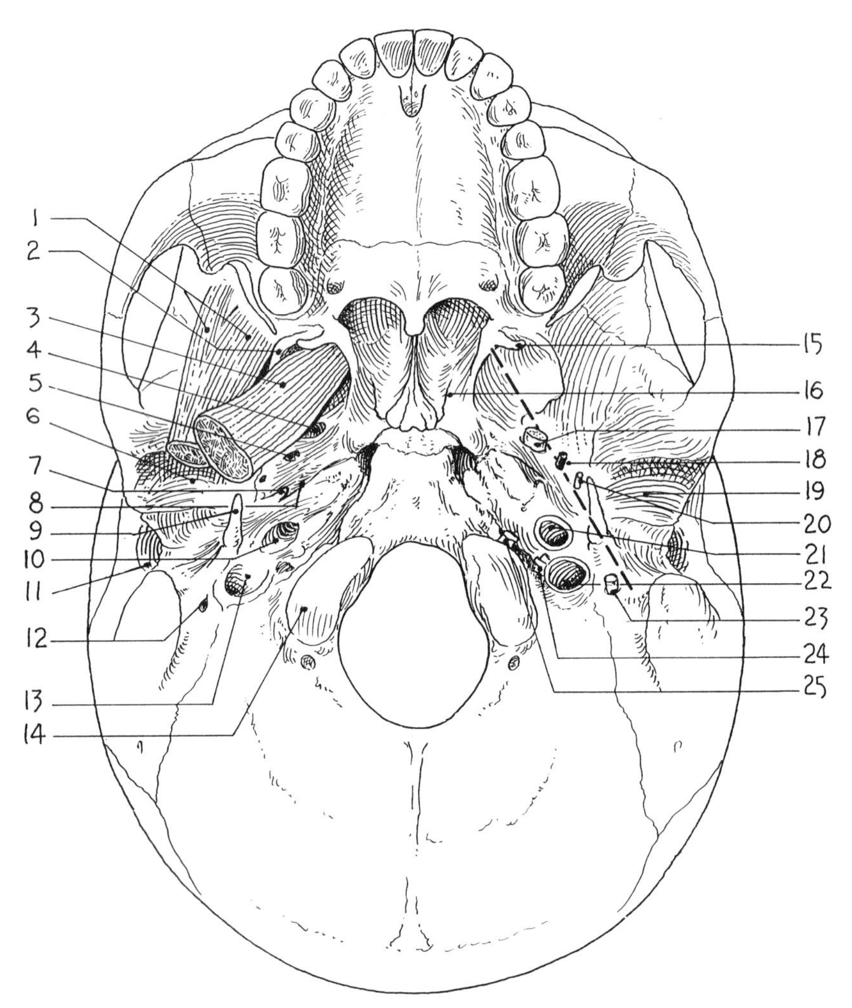

FIGURE 24.8 Anatomy of the skull base and infratemporal fossa. The most important structures of the infratemporal fossa are located along the dashed line. 1, lateral pterygoid muscle; 2, lateral pterygoid plate; 3, medial pterygoid muscle; 4, foramen ovale; 5, foramen spinosum; 6, glenoid fossa; 7, spine of the sphenoid; 8, groove for Eustachian tube; 9, styloid process; 10, carotid atery canal; 11, external auditory canal; 12, stylomastoid foramen; 13, jugular foramen; 14, occipital condyle; 15, pterygoid hamulus; 16, Eustachian tube; 17, trigeminal nerve, third division; 18, middle meningeal artery; 19, temporal mandibular joint; 20, chorda typmani nerve; 21, internal carotid artery; 22, jugular vein; 23, cranial nerve IX, X and XI; 24, cranial nerve XII.

fissure lies anterior to the pterygomaxillary fissure. The ICA ascends superiorly into the skull base directly medial to the neck of the condyle. Access to the superior portion of the ICA requires mobilisation or removal of the condyle and temporal mandibular joint (TMJ). Once the ICA enters the temporal bone below the cochlea, it turns horizontally and medially. It passes medially to the eustachian tube and travels toward the trigeminal nerve as it enters the cavernous sinus. The cavernous sinus is a dural ensheathed venous plexus that contains the ICA and cranial nerves III, IV, V_1, V_2, and VI.

1. Surgical Approach

Several surgical approaches have been described for accessing tumors of the anterolateral skull base. They can be broadly classified into two types; those traversing through the temporal bone and those anterior approaches in front of the temporal bone. The transtemporal approach commonly utilized for accessing the infratemporal fossa was popularized by Fisch and Pillsbury [48] and modified by others, whereas the particular transzygomatic approach was described by Obwegeser [49], Sehkar *et al.* [50], and others [51]. This section concentrates on the preauricular transzygomatic approach and the facial translocation approach described by Janecka and co-workers [46] for accessing this region.

2. Preauricular Infratemporal Fossa Dissection

The surgical procedure starts with a coronal incision extended inferiorly into the preauricular skin crease. If access to the carotid artery or jugular vein is needed, the incision is extended into the neck. The coronal flap is elevated in a subgaleal plane, and the pericranium is preserved for possible use during reconstruction. A subcutaneous flap is elevated over the parotid gland for increased exposure. Care is taken to elevate directly on top of the temporal fat pad, just deep to the superficial layer of the deep temporalis muscle. It is important to stay deep to the fascia and error deep into the fat when elevating this flap to avoid injury to the temporal branch of the facial nerve. Careful dissection is then performed down to the zygomatic arch. The arch and lateral orbital rim are exposed by subperiosteal dissection (Fig. 24.9a). The zygomatic arch is then transected and retracted inferiorly (Fig. 24.9b). Titanium microplates are precontoured and predrilled prior

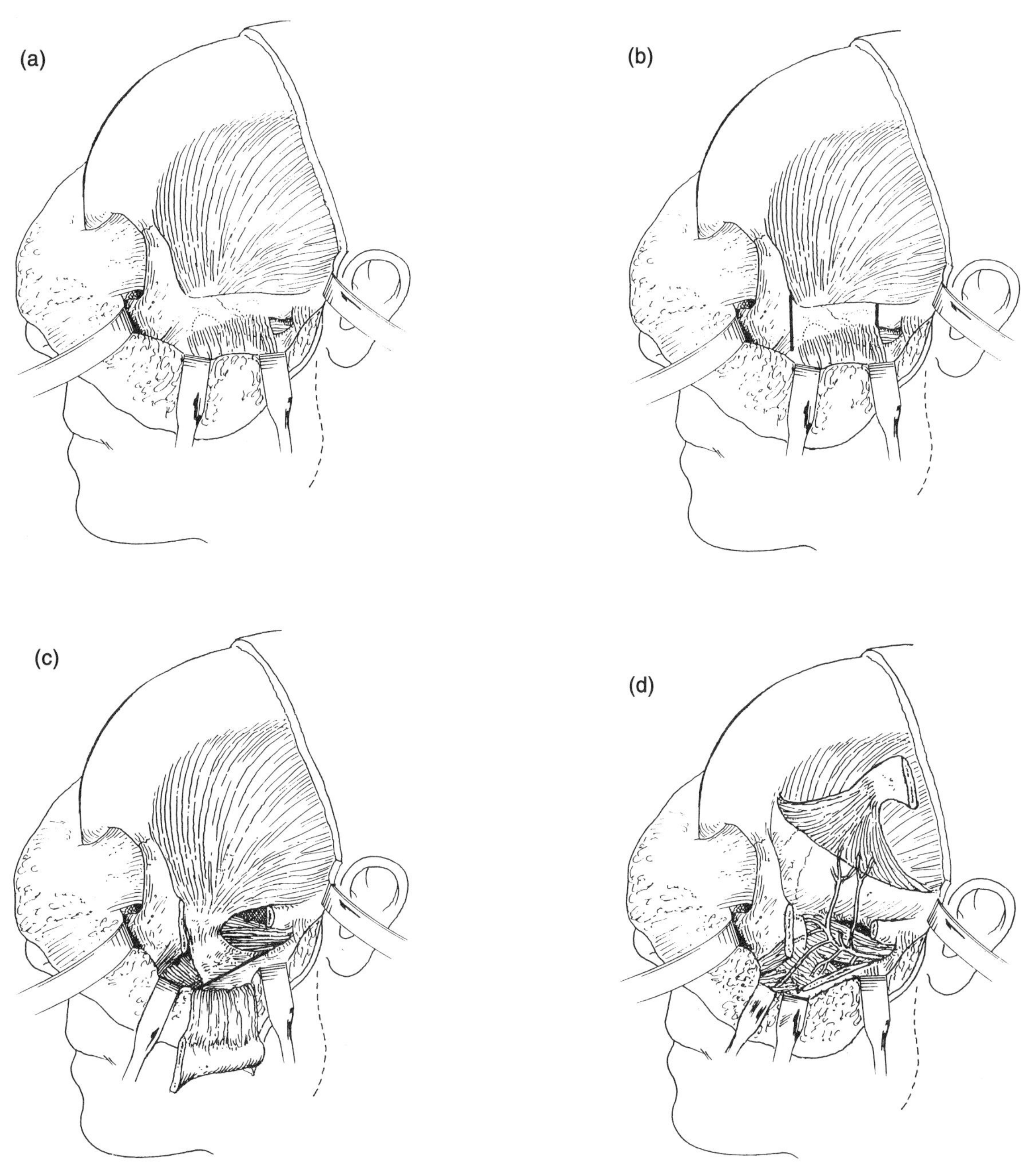

FIGURE 24.9 (a) The preauricular retromaxillary-infratemporal fossa approach is performed over the lateral orbital rim and zygomatic arch. (b) Osteotomy of the zygomatic arch is performed, with the anterior osteotomy as far as needed and the posterior osteotomy art the root of the zygomatic arch. (c) The zygomatic arch with attached masseter muscle is the retracted inferiorly. (d) The carotid process of the mandible with attached temporalis muscle is the retracted superiorly, exposing the intratemporal fossa.

to performing osteotomies for precise anatomical reconstruction later. The anterior osteotomy can be made either behind the lateral orbital rim or include a portion of the rim for extended exposure. The posterior osteotomy is performed at the root of the zygoma. The masseter muscle is usually left attached to the inferiorly retracted zygomatic arch or may be detached from the arch if necessary (Fig. 24.9c). The temporalis muscle can be handled one of two ways. The muscle can be transected from its insertion onto the coronoid process of the mandible and reflected superiorly (Fig. 24.9d). The superior and posterior portions of the muscle are left attached to the squamous portion of the temporal bone to preserve the blood supply [49,52]. Alternatively, the muscle can be detached from the squamous portion of the temporal bone and reflected inferiorly while preserving the internal maxillary artery and deep temporal branches to the muscle [50]. The inferiorly pedicled temporalis muscle can be used later as a vascularized flap for soft tissue reconstruction. The mandibular condyle can be handled in several ways. For anterior lesions where exposure of the internal carotid artery is not necessary, the condyle may be left undisturbed. For extensive lesions that require exposure of the intratemporal portion of the internal carotid artery, the TMJ and condyle can be either removed or reflected inferiorly. Unfortunately, condylar resection can result in mandibular drift and malocclusion, whereas mobilization of the TMJ can result in trismus postoperatively. Mobilization of the mandible is aided by carefully dissecting the parotid gland off the underlying masseter muscle, thereby freeing the mandible for inferior displacement. Cervical exposure provides access and control of the carotid sheath structures, as well as exposure to cranial nerves IX–XII.

The lateral orbital wall, squamous portion of the temporal bone, and greater wing of the sphenoid bone are exposed after displacement of the temporalis muscle. Inferomedially the greater wing of the sphenoid bone and pterygoid process are encountered as the lateral pterygoid muscle is elevated. The lateral pterygoid plate is identified and is followed posteriorly to locate V_3 as it exits through foramen ovale. Just posterior to V_3 lies the middle meningeal artery entering foramen spinosum.

At this point in the procedure, further dissection is tailored to the lesion being resected (Fig. 24.10). Access to the cavernous sinus is obtained by performing a pterional craniotomy extending inferomedially to foramen ovale. Removing the cranial bone flap provides additional exposure to the lateral orbital wall. Removal of the pterygoid plates and posterior wall of the maxillary sinus provides access to the nasopharynx and sphenoid sinus. Bone remaining inferomedially is rongeured away to expose V_3, whereas V_2 can be visualized extradurally by drilling away the base of the lateral pteryoid process. The internal carotid artery may be followed superiorly in the skull base

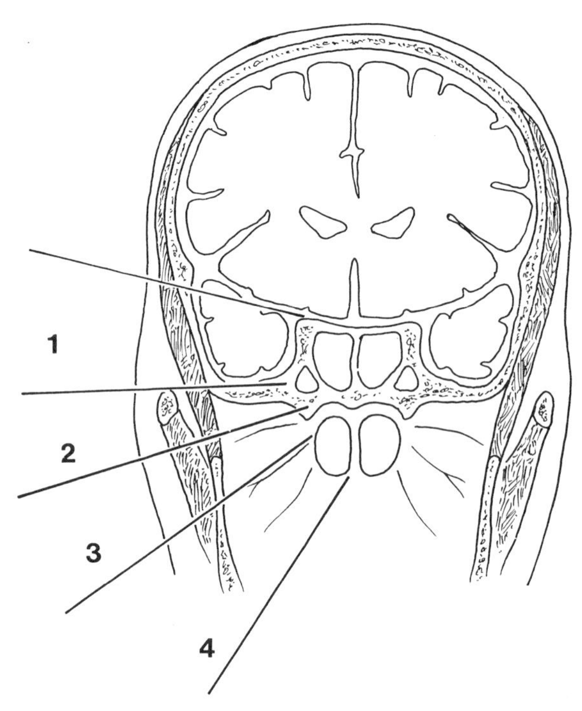

FIGURE 24.10 Anterior–lateral approach to the skull base showing exposure via (1) requires significant retraction of the temporal lobe to reach cavernous sinus and clivus, (2) addition of osteotomy of the zygomatic arch improves the exposure but still requires some retraction, (3) coronoid resection improves exposure even more, and (4) division of the mandible extends exposure to the retropharyngeal craniocervical region.

up to its entrance into the carotid canal. The eustachian tube may be resected to further expose the carotid within the skull base. The cavernous sinus is encountered by removing the bony ring surrounding V_3 and dissecting in a posterior and medial direction. Here the cavernous sinus and contents are encountered. The extent of dissection varies depending on tumor size and location. This lateral approach is often combined with an anterior craniofacial or transmaxillary approach, depending on the extent anterior exposure required.

3. Facial Translocation

The facial translocation approach [46] provides panoramic access to the infratemporal fossa, lateral orbit, and maxilla. The approach is oriented in a more anterior direction than the preauricular transzygomatic approach, and the anatomical barriers are the same as described previously. The primary difference between the facial translocation approach and the transzygomatic preauricular approach is the choice of incisions. The facial incision starts with a lateral rhinotomy incision that extends laterally through the lacrimal duct into a transconjunctival incision to completely mobilize the lower eyelid. The lateral canthus is detached and the incision is extended across the face posteriorly to the helical root. The transected lacrimal duct is repaired and stented at the end of the procedure. The temporal branch of the facial nerve is identified, tagged, and transected. It is later repaired at the conclusion of the case. From the helical root, the incision is extended superiorly into a coronal incision. The preauricular incision may also be extended into the cervical region if necessary. The anterior cheek flap is elevated subperiosteally and the infraorbital nerve is transected, helping to expose the maxilla, lateral orbital rim, and zygomatic arch. The coronal incision is dissected anteriorly and the pericranium is preserved for use later in reconstruction. Osteotomies are performed to remove the lateral orbital rim, maxilla, and zygoma *en bloc*. From here, the dissection is very similar to the infratemporal fossa dissection described previously. The authors who pioneered this approach emphasize the preservation of the blood supply to the temporalis muscle for later soft tissue reconstruction [50].

4. Reconstruction

After the tumor has been resected, reconstruction of the defect requires a watertight dural repair, soft tissue reconstruction, and skeletal frame restoration. In certain instances, replacement of resected skin is also necessary. For dural repair, small openings can be closed primarily, whereas larger defects require grafting. Tensor fascia lata, temporalis fascia, lyophilized dura, bovine pericardium, or other graft materials can be used. Vascularized soft tissue for reinforcing dual repair can be provided by local flaps, such as temporalis muscle, pericranial, or temporoparietal fascia flaps. Regional myocutaneous flaps are occasionally utilized for large defects. Extensive defects may require a vascularized free tissue flap [52], which provides abundant vascularized tissue without the constraints inherent in myocutaneous flaps. Fibrin glue is also helpful in providing a watertight seal. Lumbar drainage can reduce the CSF pressure and help facilitate the healing process.

The zygomatic arch and orbital rim are replaced and plated into proper position using the precontoured and predrilled plates and screws. If an open cavity into the maxillary sinus or nasal cavity is present, this may be skin grafted and packed with gauze for 7–10 days to allow for healing. In cases where nonvascularized bone grafts are used for reconstruction, it is critical that they be surrounded by vascularized tissue. This is especially important if preoperative radiation therapy has been administered or postoperative radiation treatment is planned. Failure to cover all surfaces with vascularized tissue can result in bone necrosis. If there is a significant cutaneous defect after tumor resection, local advancement flaps may be used for skin closure. If this is not adequate, a myocutaneous or vascularized free tissue transfer can provide coverage.

C. Middle and Posterior Skull Base

There are several approaches to the middle and posterior skull base and fossa. Tumors of the middle skull base are approached most often through an infratemporal or anterolateral approach discussed earlier in this chapter. The middle cranial base can also be accessed by extending any of the posterior cranial base approaches medially through the cochlea and IAC. This often involves transposing the intratemporal course of the facial nerve. The choice of approach depends in large part on the location of the tumor and hearing status. Occasionally, these approaches are combined to provide access for larger tumors.

1. Middle Fossa Approach

The middle fossa approach is primarily indicated for small intracanalicular tumors or tumors with minimal cerebellopontive angle (CPA) involvement and serviceable preoperative hearing (Fig. 24.11). A temporal craniotomy is performed and the IAC is outlined after identifying the superior semicircular canal and geniculate ganglion (Fig. 24.12). The primary advantage of this approach is the excellent exposure of the IAC fundus with hearing preservation. The disadvantages are limited access to the CPA and inferior IAC.

The middle fossa approach can be extended to include removal of the petrous ridge and posterior aspects of the temporal bone up to the labyrinth. Anterior bone removal is extended up to foramen lacerum and the internal carotid

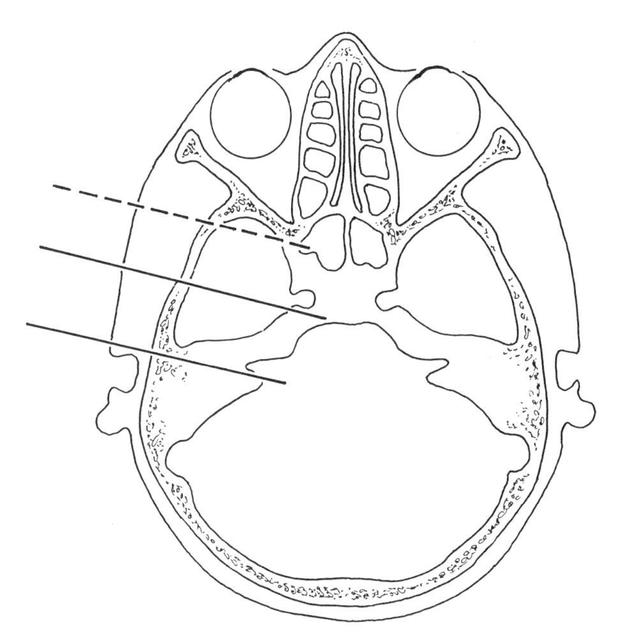

FIGURE 24.11 Middle fossa transpetrous apex approach demonstrating the transverse access through this exposure.

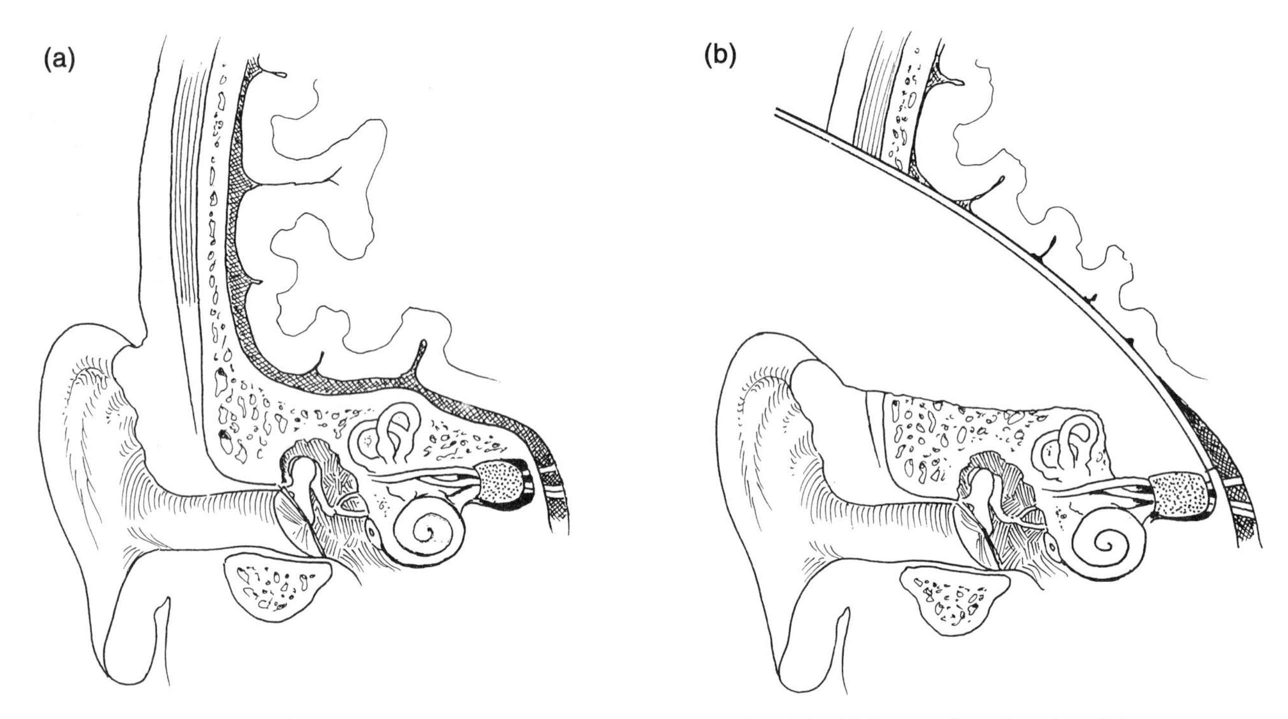

FIGURE 24.12 (a) Coronal through the temporal bone illustrating the relationship between the undersurface of the temporal lobe and the roof of the petrous pyramid. (b) Extradural elevation of the temporal lobe demonstrating exposure to the internal auditory canal.

artery. This extended approach is indicated for patients with serviceable hearing for the removal of extensive petrous ridge and petroclival lesions involving the temporal bone. This includes lesions extending through the tentorium into the middle fossa.

2. Translabyrinthine Approach

The translabyrinthine approach is indicated for tumors of all sizes involving the CPA. It is accomplished through a transmastoid labyrinthectomy in which the sigmoid sinus, middle and posterior fossa dura, and the external genu and vertical segment of the facial nerve are skeletonized. The IAC is encountered medial to the semicircular canals and can be opened to expose the contents. Dissection can be carried more medially to access petroclival lesions as well (Fig. 24.13).

Primary advantages of the translabyrinthine approach include wide exposure of the CPA, avoidance of retraction of the cerebellum, and easy identification of the facial nerve at the fundus of the IAC. The primary disadvantage is hearing loss.

The translabyrinthine approach can be extended anteriorly to include removal of the external auditory canal, tympanic membrane, ossicles, and cochlea. This transcochlear approach requires transposition of the facial nerve posteriorly from its intratemporal course and removal of the posterior external auditory canal (EAC) with oversewing of the external auditory meatus. The primary benefit of this approach is access to the petrous apex and clivus. Petroclival meningiomas, epidermoid cysts, glomus jungulare tumors, temporal bone carcinomas, and extensive schwannomas are accessed readily with this technique.

3. Retrosigmoid (Suboccipital) Approach

The retrosigmoid approach provides wide exposure of the posterior fossa for the excision of CPA lesions. It is performed (initiated) by making a craniotomy bounded by the transverse sinus superiorly and the sigmoid sinus anteriorly. The posterior fossa dura is opened and the cerebellum is reacted posteriorly. The sigmoid can be retracted to provide further anterior exposure. It is often necessary to drill off the posterior lip of the IAC to expose its contents fully (Fig. 24.13).

The advantages of this approach are that hearing preservation is possible and wide exposure of the posterior fossa is available. The disadvantages are the risk of permanent hearing loss during IAC exposure and hydrocephalus from prolonged retraction of the cerebellum. Furthermore, the intradural drilling of bone has been associated with severe postoperative headaches in a significant proportion of patients [54].

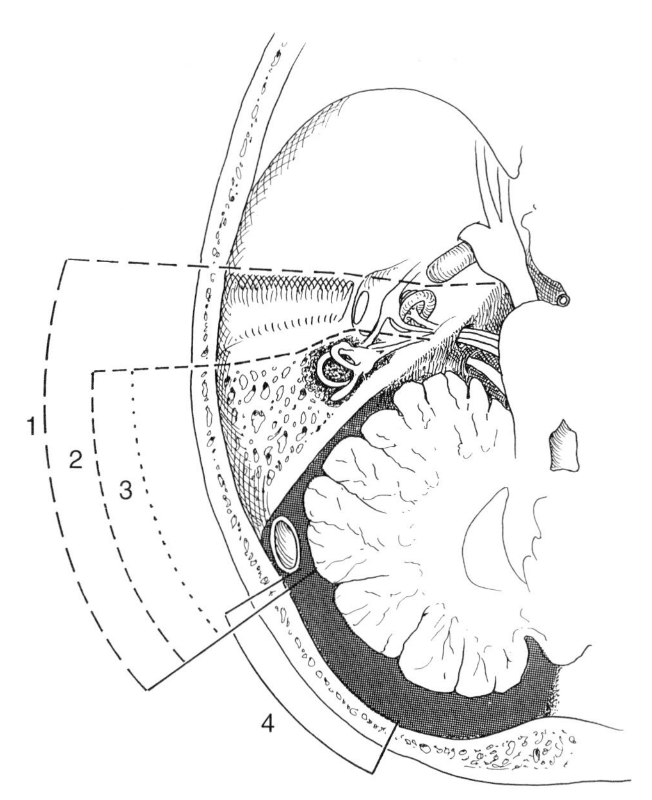

FIGURE 24.13 An axial view of the common approaches to posterior and middle fossa. Demonstrated are (1) transcochlear, (2) translabyrinthine, (3) retrolabyrinthine, and (4) retrosigmoid accesses.

4. Retrolabyrinthine Approach

Primary indications for retrolabyrinthine approach include vestibular nerve section and trigeminal neurotomy for tic douloureux. This approach involves a simple mastoidectomy with posterior fossa dura exposure from 1 to 2 cm posterior to the sigmoid sinus to the level of the semicircular canals anteriorly. The posterior fossa dura anterior to the sigmoid is opened and the sigmoid sinus is retracted laterally to provide exposure into the CPA (Fig. 24.13). This approach allows preservation of hearing and requires less cerebellar retraction than the retrosigmoid procedure, which is used for the similar indications. There is no problem with the postoperative headaches that are associated with the retrosigmoid approach. The primary disadvantage of the retrolabyrinthine approach is the limited exposure into the posterior fossa, which may be critical in the case of a intraoperative hemorrhage.

VII. COMPLICATIONS

The complex anatomy and aggressive behavior of tumors in the skull base make perioperative complications a valid concern for both the patient and the surgeon. Appropriate patient selection, preoperative planning, and optimization of the medical condition can help reduce the rate of complications. However, despite the most exhaustive preparations, complications will continue to occur both intraoperatively and postoperatively.

A. Intraoperative Complications

Given the long duration of skull base procedures, proper positioning of the patient on the operating room table is imperative. Padding must be placed to minimize risks of pressure necrosis or peripheral nerve injury. Compression stockings or sequential compressive devices are placed to minimize risks of thromboembolism from deep venous structures in the lower extremities.

With advances in available anesthetics and intraoperative monitoring, anesthesia-related complications have been reduced. Intraoperative hemorrhage can be minimized by the preoperative embolization of prominent feeder vessels in selected tumors. With large vascular structures in the surgical field, transfusions of blood products are often necessary. It is important to provide the patient an opportunity for autologous blood donation, as well as typing and cross matching sufficient units of blood. Preoperative balloon test occlusion can help predict risk of stroke should carotid artery ligation be necessary. Preparation for arterial grafting should be made preoperatively should the patient be in a high-risk group for stroke. Venous bleeding is usually controlled by vessel ligation, bipolar electrocauterization, or packing with absorbable hemostatic agents.

Along with venous bleeding comes the risk of air embolism. This occurs when the head is positioned higher than the heart to minimize venous bleeding. Large amounts of air can enter through a venotomy and travel to the right heart where it can get trapped and eliminate cardiac output or enter the pulmonary circulation where it will produce pulmonary arterial vasoconstriction, leading to cor pulmonale and pulmonary edema. The management of this complication requires immediate occlusion of the open vein, compressing both jugular veins, discontinuing any nitrous oxide inhalants, and placing the patients in a left lateral Trendelenburg position (right side up) in an attempt to trap the air in the right side of the heart. At this point, the air may be aspirated by a right atrial catheter or a needle inserted subxiphoid into the right ventricle.

One of the most common complications in skull base surgery is a cerebrospinal fluid leak. The incidence of a CSF leak varies proportionally to the size of the dural defect created. Recognized leaks are closed primarily, in a watertight fashion. Larger defects can be repaired with pericranium, temporalis fascia, cadveric dura, fascia lata, or a microvascular myofascial flap. Reinforcement with muscle or fibrin glue may also be helpful.

B. Postoperative Complications

Cerebrospinal fluid leaks detected in the postoperative period occur typically through the surgical wound, ear canal (otorrhea), or nose (rhinorrhea). Early identification and measures to correct this are important to minimize the risk of CSF contamination and the development of meningitis. Depending on the side of origin and the severity of the leak, treatment measures include bed rest, oversewing the skin incision, lumbar drainage, dural grafting, soft tissue obliteration, and eustachian tube or canal obliteration.

Neurological complication can include an isolated cranial nerve deficit or a severe life threatening cerebrovascular accident. Early detection of neurological deficiency and rapid treatment by the interventional neurovascular service or thrombolytic therapy can potentially avert a disaster. Cranial nerve deficits can be addressed postoperatively with appropriate consultation and rehabilitation. Elevated intracranial pressure causing hydrocephalus needs early identification and an appropriate shunting procedure.

Wound infections are usually delayed several days postoperatively and many require incision, drainage, debridement, and possibly flap reconstruction, as well as antimicrobial therapy. Meningitis must be detected early and empiric antibiotic therapy initiated immediately after the CSF is cultured. Antibiotics can be adjusted based on culture sensitivity results. As with all surgery, anticipation and preparation for complications before they happen will minimize adverse events.

VIII. CONCLUSION

Advances in otolaryngology/head and neck surgery, plastics and reconstruction surgery ophthalmology, oral/maxillofacial surgery, and neurosurgery have all resulted in improved treatment for skull base tumors. The development of a skull base surgery team has improved results greatly by combining the expertise of each specialists and forming one team focused on an optimal outcomes. Future advances in diagnostic imaging, anesthetic techniques, intraoperative monitoring, and surgical techniques should continue to improve the surgical outcome of tumors involving the cranial base.

References

1. Ketcham, A. S., Wilkins, R. H., Van Buren, J. M., and Smith, R. R. (1963). Combined intracranial facial approach to the paranasal sinuses. *Am. J. Surg.* **106**, 698–703.
2. Terz, J. J., Young, H. F., and Lawrence, W., Jr. (1980). Combined craniofacial resection for locally advanced carcinoma of the head and neck. *Am. J. Surg.* **140**, 618–624.
3. House, W. F. (1979). A history of acoustic tumor surgery. *In* "Acoustic Tumors" (W. F. House, and C. M. Luetje, eds.), Vol. I, pp. 3–41. University Park Press, Baltimore.
4. Linkskey, M. E., Jungreis, C. A., Yonas, H., Hirsch, W. L., Sekhar, L. N., Horton, J. A., and Janosky, J. E. (1994). Stroke risk after abrupt internal carotid artery sacrifice: Accuracy of preoperative assessment with balloon test occlusion and stable xenon-enhanced CT. *AJNR Am. J. Neuroradiol.* **15**, 829–843.
5. Steed, D. L., Webster, M. W., DeVries, E. J., Jungreis, C. A., Horton, J. A., Sehkar, L., and Yonas, H. (1990). Clinical observations on the effect of carotid artery occlusion on cerebral blood flow mapped by xenon computed tomography and its correlation with carotid artery back pressure. *J. Vasc. Surg.* **11**, 38–44.
6. Nemzek, W. R. (1998). Carotid artery assessment and interventional radiologic procedures before skull base surgery. *In* "Surgery of the Skull Base" (P. J. Donald, ed.), pp. 105–118. Lippincott-Raven, Philadelphia.
7. Sclabassi, R. J., Balzer, J. R., and Krieger, D. N. (1998). Intraoperative neurophysiologica monitoring. In "Surgery of the Skull Base" (P. J. Donald, ed.), pp. 137–161. Lippincott-Raven, Philadelphia.
8. Matsuba, H. M., Spector, G. J., Thawley, S. E., Simpson, J. R., Mauney, M., and Pikul, F. J. (1986). Adenoid cystic salivary gland carcinoma: A histologic review of treatment failure patterns. *Cancer* **57**, 519–524.
9. Dulgerov, P., and Calcaterra, T. (1992). Esthesioneuroblastoma: The UCL A experience 1970–1990. *Laryngoscope* **102**, 843–849.
10. Eden, B. V., Debo, R. F., Larner, J. M., Kelly, M. D., Levine, P. A., Stewart, F. M., Cantrell, R. W., and Constabel, W. C. (1994). Esthesioneuroblastoma: Long term outcome and patterns of failure. *Cancer* **73**, 2556–2562.
11. Hirose, T., Scheitauer, B. W., Lopes, M. B., Gerber, H. A., Altermatt, H. J., Harner, S. G., and VandenBerg, S. R. (1994). Olfactory neuroblastoma: An immunohistochemical, ultrastructural and flow cytometric study. *Cancer* **73**, 4–19.
12. Schwaab, G., Micheau, C., LeGuillou, C., Pacheco, L., Marandas, P., Comenge, C., Richard, J. M., and Wibault, P. (1988). Olfactory esthesioneuroma: A report of 40 cases. *Laryngoscope* **98**, 872–876.
13. Taxy, J. B., Hharani, N. K., Mills, S. E., Frierson, H. F., Jr., and Bould, V. E. (1986). The spectrum of olfactory neural tumors: A light-microscopic, immunohistochemical and ultrastructural analysis. *Am. J. Surg. Pathol.* **10**, 687–695.
14. Barnes, L. (1991). Pathobiology of selected tumors of the base of the skull. *Skull Base Surg.* **1**, 207–213.
15. Gay, E., Sekhar, L. N., Rubinstein, Wright, D. C., Sen, C., and Janecka, I. P. (1995). Chordomas and chondrosarcomas of the cranial base: Results and follow-up of 60 patients. *Neurosurgery* **36**, 887–896.
16. Doucet, V., Peretti-Viton, P., Figarella-Branger, D., Manera, L., and Salamon, G. (1997). MRI of intracranial chordomas: Extent of tumour and contrast enhancement. *Neuroradiology* **39**, 571–576.
17. Wenig, B. M. (1993). "Atlas of Head and Neck Pathology," p. 188. Saunders, Philadelphia.
18. Rosenberg, A. E., Nielsen, G. P., Keel, S. B., Renard, L. G., Fitzek, M. M., Munzenrider, J. E., and Liebsch, N. J. (1999). Chondrosarcoma of the base of the skull: A clinicopathologic study of 200 cases with emphasis on its distinction from chordoma. *Am. J. Surg. Pathol.* **23**, 1370–1378.
19. Weber, A. L., Brown, E. W., Hug, E. B., and Liebsch, N. J. (1995). Cartilaginous tumors and chordomas of the cranial base. *Otolaryngol. Clin. North Am.* **28**, 453–471.
20. Barnes, L., and Kapadia, S. B. (1994). The biology and pathology of selected skull base tumors. *J. Neurooncol.* **20**, 213–240.
21. Roche, P. H., Regis, J., Dufour, H., Fournier, H.D., Delsanti, C., Pellet, W., Grisoli, F., and Peragut, J. C. (2000). Gamma knife radiosurgery in the management of cavernous sinus meningiomas. *J. Neurosurg.* **93**(Suppl. 3), 68–73.

22. Singh, V. P., Kansai, S., Vaishya, S., Julka, P. K., and Mehta, V. S. (2000). Early complications following gamma knife radiosurgery for intracranial meningiomas. *J. Neurosurg.* **93**(Suppl. 3), 57–61.
23. Chamberlain, M. C. (2001). Meningiomas. *Curr. Treat. Options Neurol.* **3**, 67–76.
24. Talacchi, A., Sala, F., Alessandrini, F., Turazzi, S., and Bricolo, A. (1998). Assessment and surgical management of posterior fossa epidermoid tumors: Report of 28 cases. *Neurosurgery* **42**, 242–251.
25. Rosenwasser, H. (1968). Monograph on glomus jugulare tumors. *Arch. Otolaryngol.* **88**, 3–40.
26. Myers, E. N., Newman, J., Kaseff, L., and Black, F. O. (1971). Glomus jugulare tumor: A radiographic-histologic correlation. *Laryngoscope* **81**, 1838–1851.
27. Manolidis, S., Shohet, J.A., Jackson, C. G., and Glasscock, M. E. (1999). Malignant glomus tumors. *Laryngoscope* **109**, 30–34.
28. Conley, J. J, and Clairmont, A. A. (1977). Glomus intravagale. *Laryngoscope* **87**, 2096–2100.
29. Spector, G. J., Sobol, S., Thawley, S. E., Maisel, R. H., and Ogura, J. H. (1979). Panel discussion: Glomus jagulare tumors of the temporal bone. *Laryngoscope* **89**, 1628–1639.
30. Spector, G. J., Gado, M., Ciralsky, R., Ogura, J. H., and Maisel, R. H. (1975). Neurologic implications of glomus tumors in the head and neck. *Laryngoscope* **85**, 1387–1395.
31. van Baars, F. M., Cremers, C. W., van den Broek, P., and Veldman, J. E. (1981). Familiar non-chromaffinic paragangliomas (glomus tumors): Clinical and genetic aspects. *Acta Otolaryngol.* **91**, 589–593.
32. Irons, G. B., Weiland, L. H., and Brown, W. L. (1977). Paragangliomas of the neck: Clinical and pathologic analysis of 116 cases. *Surg. Clin. North Am.* **57**, 575–583.
33. Bickerstaff, E. R., and Howell, J. S. (1953). Neurological importance of tumors of the glomus jugulare. *Brain* **76**, 576–593.
34. Jackson, C. G. (1993). Diagnosis for treatment planning and treatment options. *Laryngoscope* **103**(Suppl. 60), 7–15.
35. Jordan, J. A., Roland, P. S., McManus, C., Weiner, R. L., and Giller, C. A. (2000). Stereotactic radiosurgery for glomus jugulare tumors. *Laryngoscope* **110**, 35–38.
36. Martuza, R. L., and Eldridge, R. (1988). Neurofibromatosis 2 (bilateral acoustic neurofibromatosis). *N. Engl. J. Med.* **318**, 684–688.
37. Fucci, M. J., Buchman, C. A., Brackmann, D. E., Berliner, K. I. (1999). Acoustic tumor growth: Implications for treatment choices. *Am. J. Otol.* **20**, 495–499.
38. Tschudi, D. C., Linder, T. E., and Fisch, U. (2000). Conservative management of unilateral acoustic neuromas. *Am. J. Otol.* **21**, 722–728.
39. Shin, Y. J., Fraysse, B., Cognard, C., Gafsi, I., Charlet, J. P., Berges, C., Deguine, O., and Tremoulet, M. (2000). Effectiveness of conservative management of acoustic neuromas. *Am. J. Otol.* **21**, 857–862.
40. Kaylie, D. M., Horgan, M. J., Delashaw, J. B., and McMenomey, S. O. (2000). A meta-analysis comparing outcomes of microsurgery and gamma knife radiosurgery. *Laryngoscope* **110**, 1850–1856.
41. Prasad, D., Steiner, M., and Steiner, L. (2000). Gamma surgery for vestibular schwannoma. *J. Neurosurg.* **92**, 745–759.
42. Derome, P. (1983). Chirurgie des tumeurs osseuses de la base du crane. *Rev. Laryngol.* **104**, 283–286.
43. Deromone, P. (1988). The transbasal approach to tumors invading the skull base. *In* "Operative Neurosurgical Techniques" (H. H. Schmidek and W. H. Sweet, eds.), pp. 619–633. Grune & Stratton, Orlando, F. L.
44. Raveh, J., Laedrach, K., Vuillemin, T., and Zingg, M. (1992). Management of combined frontonaso-orbital/skull base fractures and telecanthus in 355 cases. *Arch. Otolaryngol. Head Neck Surg.* **118**, 605–614.
45. Reveh, J., Reidli, M., and Markwalder, T. M. (1984). Operative management of 194 cases of combined maxillofacial-frontobasal fractures: Principles and modification. *J. Oral Maxillofac. Surg.* **42**, 555–564.
46. Janecka, I. P., Sen, C. N., Sekhar, L.N., and Arriaga, M. (1990). Facial translocation: A new approach to the cranial base. *Otolaryngol. Head Neck Surg.* **103**, 413–419.
47. Mathog, R. H., Shibuya, T. Y., Leider, J. S., and Marunick, M. T. (1997). Rehabilitation for extended facial and craniofacial resection. *Laryngoscope* **107**, 30–39.
48. Fisch, U., and Pillsbury, H. C. (1979). Infratemporal fossa approach to lesions in the temporal bone and base of the skull. *Arch. Otolaryngol.* **105**, 99–107.
49. Obwegeser, H. L. (1985). Temporal approach to the TMJ, the orbit, and the retromaxillary-infracranial region. *Head Neck* **7**, 185–199.
50. Sekhar, L. N., Schram, F. L., and Jones, N. F. (1987). Subtemporal-preauricular infratemporal fossa approach to large lateral and posterior cranial base neoplasms. *J. Neurosurg.* **67**, 488–499.
51. Al-Mefty, O., and Anand, V. K. (1990). Zygomatic approach to skull-base lesions. *J. Neurosurg.* **73**, 668–673.
52. Shibuya, T. Y., Doerr, T. D., Mathog, R. H., Burgio, D. L., Meleca, R. J., Yoo, G. H., and Guthikonda, M. (2000). Functional outcomes of the retromaxillary-infratemporal fossa dissection for advanced head and neck skull base lesions. *Skull Base Surg.* **10**, 109–117.
53. Urken, M. L., Catalano, P. J., Sen., C., Post, K., Futran, N., and Biller, H. F. (1993). Free tissue transfer for skull base reconstruction: Analysis of complications and a classification scheme for defining skull base defects. *Arch. Otolaryngol. Head Neck Surg.* **119**, 1318–1325.
54. Jackson, C. G., McGrew, B. M., Forest, J. A., Hampf, C. R., Glasscock, M. E., III, Brandes, J. L., and Hanson, M. B. (2000). Comparison of postoperative headache after retrosigmoid approach: Vestibular nerve section versus vestibular schwannoma resection. *Am. J. Otol.* **21**, 412–416.

C H A P T E R

25

Management of Laryngeal Cancer

R. KIM DAVIS
Division of Otolaryngology—Head and Neck Surgery
University of Utah School of Medicine
Salt Lake City, Utah 84132

I. Introduction 359
II. Treatment of Carcinoma *in Situ* and Minimally Invasive T_1 Glottic Cancer 360
III. Treatment of T_1 Glottic Cancer 361
IV. Treatment of T_2 Glottic Cancer 363
V. Treatment of T_3 and T_4 Glottic Cancer 366
VI. Treatment of Supraglottic Cancer 369
VII. Summary 371
References 372

I. INTRODUCTION

Today is a fascinating time in the treatment of cancers of the larynx. Historical paradigms of management are either being challenged or actually changed. Early cancers of the larynx were treated by full-course irradiation therapy given as daily fractionations 5 days a week for a period of 6 to 7 weeks. Advanced cancer was typically treated by total laryngectomy with or without pre- or postoperative irradiation. Currently, radiation therapy for early cancer of the larynx is yet often employed in the country, but increasingly more altered fractionation or hyperfractionation schemes are used. More significantly, the primary role of radiation therapy in these lesions has been altered by the introduction of the CO_2 laser, utilized transorally with sophisticated microlaryngeal surgical techniques. Even the role of the laser is currently undergoing significant change, as carcinoma *in situ* and very early T_1 lesions of the true vocal cord, the very lesions first treated by the laser, are now best recognized to be treated by cold instrument excision.

The major historical change in surgery has been movement away from transcutaneous procedures, especially total laryngectomy, to a variety of procedures that conserve laryngeal function. External approaches to conservation surgery of the larynx have evolved from open supraglottic laryngectomy and a variety of partial laryngectomy procedures for glottic cancer to the various modifications of supracricoid laryngectomy and Pearson's near total laryngectomy. Endoscopic supraglottic laryngectomy is becoming increasingly advocated for T_1 through selected T_3 supraglottic cancers, whereas the endoscopic resection of glottic cancer is extending well beyond treatment of the early cancers already mentioned.

Another significant change in treatment paradigms has been the introduction of chemotherapy as an additional modality to treat laryngeal cancer. As is well known to all head and neck oncologists, this role of chemotherapy has already evolved, both by the schemes of application and by the chemotherapeutic agents employed. Examples of these changes include the introduction of induction chemotherapy before definitive surgery followed by postoperative irradiation to the more current thoughts of concomitant chemotherapy and irradiation.

In the face of the many changes and treatment options available to surgeons today, it is especially important to understand the principles governing the different treatment modalities and to develop as clearly as possible guidelines to the appropriate applications of these various approaches. The intent of this chapter is to describe principles of cancer treatment and then to list the application of these principles as it relates to surgery, irradiation therapy, and chemothe-apy. Woven into the discussion of these techniques will be the author's personal preferences for treatment in each of the different stages of cancer and the rationale developed to justify these choices.

II. TREATMENT OF CARCINOMA *IN SITU* AND MINIMALLY INVASIVE T_1 GLOTTIC CANCER

Patients presenting with dysplastic-appearing true vocal cords typically undergo surgical microlaryngoscopy with, at a minimum, selected biopsies from the dysplastic-appearing areas. Historically supravital dyes such as toluidine blue have been used to best direct these biopsies. Areas of greatest staining have been biopsied or very limited areas of dysplasia simply removed by microlaryngeal surgical techniques. When such biopsy techniques have shown carcinoma *in situ* two treatment approaches were historically taken. In some centers, patients with extensive carcinoma *in situ* have received primary irradiation therapy, usually with single daily fractions of 180 to 200 cGy given 5 days a week for 6 or 7 weeks. Success rates for this approach are reported in the 80–95% range.

The primary surgical alternative to this has been microsurgical resection of the areas of dysplasia, initially described as a cord stripping type approach. With the precision of the surgical microscope, incisions were placed posterior to the area of dysplasia and carried through the mucosa and into the lamina propria. A microsurgical cup was then used to hold the area posteriorly while gentle traction was applied to further dissect or literally strip the abnormal epithelium off the vocal cord from posterior to anterior. Attention was directed to preservation of the anterior commissure mucosa. When dysplastic changes involved both vocal cords, two separate microlaryngeal resection sessions were used wherein one cord at a time was operated on. These procedures typically did not take any special precautions to preserve the lamina propria.

With introduction of the carbon dioxide laser, microsurgical cold instrument approaches were often replaced by transoral resection of these dysplastic areas using the laser. The advantage of the laser was the near total hemostasis obtained along the laser incisional lines. The laser was first used to incise the vocal cord posteriorly, after which microsurgical cups were applied and the cord was again stripped by applying tension with selective further laser resection in areas not elevated by tension alone. Where the epithelium did not cleanly lift off of the underlying lamina propria, some surgeons advocated starting the laser incision posteriorly and using the laser to separate the epithelium along the full length of the cord. As was true in earlier cold surgical techniques, special attention was given at the anterior commissure where it was quickly recognized that overvigorous laser excision would often lead to scar tissue formation and webbing.

After introduction of the laser, it became readily apparent that resection along the full true vocal cord resulted in significant scarring in the underlying lamina propria. This problem of overvigorous laser excision led to the innovation of "microspot" laser technology wherein the initial 1- to 2-mm spot sizes of the first lasers were replaced with microspots of 0.3 to 0.64 mm. This microspot technique allowed more precise incision of the epithelium with less thermal damage occurring to underlying structures. Judicious application of microspot cutting techniques did eliminate some of the notable thermal damage seen using the first lasers.

For a brief period of time, cryotherapy was also used to treat carcinoma *in situ* and limited T_1 glottic cancer. Success rates with this approach were reported in the 80–90% range, but few surgeons embraced this technique, as there were no definitive pathology margins possible to observe. Additionally, the very few people who advocated laser vaporization of dysplastic areas versus excision also fell into rapid disfavor for the same reason. In addition to the lack of specific tumor margins, thermal damage to the underlying lamina propria was extensive.

Today, transoral resection of carcinoma *in situ* or early glottic cancer has developed into a phonomicrosurgical approach pioneered by Zeitels and other laryngologists [1–4]. This approach resulted from the convergence of a microlaryngoscopy surgical technique with the body cover mucosal wave theory of voice production. Whereas the previously described procedures resulted in cure rates of carcinoma *in situ* and early glottic cancer in the 90% range, the vocal results were not adequate. Vocal outcome using the phonomicrosurgical approach was improved by minimizing the deep resection margin, thereby preserving the vocal folds' normal layered microstructure of epithelium and the lamina propria.

Zeitels described four basic procedures, which varied with the depth of resection, to accomplish a narrow field deep cancer margin. In an early study of 20 patients, 13 with T_1 cancers and 7 with carcinoma *in situ*, promising results were seen [3]. No patient who underwent cancer resection developed recurrence with a minimum follow-up of 2 years and a mean follow-up of 42 months. In patients with carcinoma *in situ*, two patients later developed microinvasive carcinoma, despite block resection, but were both retreated successfully.

Zeitels' retrospective review of 307 microsurgical procedures included 263 procedures for glottic cancer. Cold instruments alone were used in 203 surgeries, whereas 60 procedures utilized both cold instruments and the CO_2 laser [2]. The governing principle in this study was that voice would best be preserved by maximally preserving the lamina propria and epithelium. Zeitels advocated precise tangential dissection in order to accomplish this goal. In limited lesions, this dissection was best accomplished with cold instruments alone. The CO_2 laser was utilized for selected larger glottic lesions, in which bleeding would obscure visualization of the microanatomy of the musculo-membranous vocal fold. Both cure rates and voice preservation were enhanced by this approach.

One of the especially helpful phonomicrosurgical techniques is to inject saline with epinephrine immediately below the vocal cord epithelium in patients with apparent dysplasia or early cancer [4]. Careful injection in this manner will literally balloon up or lift off the dysplastic epithelium or very limited early cancer. Where these lesions lift up, cold instrument dissection can be used readily to resect the lesion and preserve maximally the underlying lamina propria. When the injection technique is used, any area of the vocal cord that is tethered down to the underlying vocalis muscle suggests a deeper invasion of cancer. These areas of deeper tethering can then be selectively resected with the laser to gain a deep tumor margin. Using this technique, the lamina propria of the true vocal cord is certainly better preserved with resultant voice improvement.

III. TREATMENT OF T_1 GLOTTIC CANCER

T_1 squamous cell carcinoma of the true vocal cord has historically been treated by full course irradiation therapy. This has generally been accomplished using opposed lateral wedged portals with daily fractionation doses of 180 to 220 cGy. Historically, it became apparent that daily fractionation doses of at least 200 cGy per day produced higher cure rates than doses less than 200 cGy. Refinements in this approach have included using oblique portals to limit overirradiation of the arytenoid and interarytenoid areas, as well as hyperfractionation or altered fractionation schemes. This is discussed in Section IV.

Success rates in treating stage I glottic cancers have depended on the location and extent of these T_1 lesions. Cancer confined to one true vocal cord without extension to the vocal process of the arytenoid cartilage or to the anterior commissure is reported to be cured from 80 to 95% of the time [5–9]. Large teaching institutions with excellent radiation therapy departments have typically reported cure rates over 90%. The actual cure rate in a countrywide setting, including all radiation therapy facilities, probably falls below the 90% rate. Very few studies have actually assessed countrywide cure rates, but most head and neck oncologists would accept cure rates of 85±5% for these limited T_1 lesions (T_1a). Certainly the initial paradigm and standard of care in stage I laryngeal cancer have been to use irradiation therapy.

More extensive T_1 glottic cancers, especially those involving the anterior commissure area or involving the posterior true vocal cord, are generally considered more difficult to cure by radiation therapy alone. Cure rates for these anterior commissure and posterior glottic T_1 cancers (T_1b) are clearly less than T_1a cancers. This poorer rate of cure using conventional radiation therapy for these T_1b lesions has been one of the impetuses to current hyperfractionation schemes, or potentially even the use of concomitant chemotherapy and irradiation in these more threatening early cancers.

The poorer cure rates with T_1b lesions or with bulky T_1 glottic cancers in general are consistent with the principle that irradiation therapy in squamous cell carcinomas is more effective when treating smaller volumes of cancer. The full rationale of this statement is discussed later in the chapter. What is clear is that conventional daily fractionation schemes alone do not cure a significant number of patients with these T_1b lesions. An extension of this same thought is that radiation clearly does not cure all squamous cell cancers, even T_1a lesions.

The actual number of squamous cell carcinomas able to be cured by irradiation is not as apparent as studies reporting success rates using irradiation would imply. In an interesting study in 1973 by Lillie and DeSanto [10], patients were treated by endoscopic transoral cordectomy using nonlaser techniques. The main purpose of this paper was to present the excellent surgical results obtained by this approach. A very important issue raised by this paper was that more than 20% of the patients they took to surgery had no residual cancer. In these patients the initial biopsies of their early glottic cancer had in fact removed all demonstrable cancer. In the second operation where more tissue was taken, there was no residual cancer.

Lillie and DeSanto's study suggests that a certain number of patients are actually cured by their biopsy procedures and do not need subsequent surgery or irradiation. If this group of patients is as high as 20%, and this certainly seems plausible to the author as his own personal experience has been that approximately 30% of his patients have had no residual cancer after staging biopsies, then these patients, if irradiated, would certainly have been reported as cured by irradiation. The implication of this thought is that the number of patients actually cured by irradiation is overstated. The actual cure rate of T_1 glottic cancer treated by irradiation is less than the reported series of 85±5%. The point of this discussion is not to state that irradiation is unimportant, which is clearly not the case, but to make a simple point that not all patients are curable by irradiation and that even early cancers are sometimes radioresistant.

Very early in the evolution of treating stage I glottic cancer, surgery developed a role parallel to that of irradiation. When the only alternative to radiation therapy was extensive oblative laryngeal surgeries, then it was apparent that radiation was in fact the treatment of choice. When external conservation surgery of the larynx was developed, then this paradigm shifted.

Through the pioneering work of Ogura and many other surgeons, the principles of partial laryngectomy were carefully elucidated [11–15]. Partial laryngectomy for glottic cancer was a very significant advance and offered a clear alternative to irradiation. Whereas these procedures were an especial advantage in stage II glottic cancer, application of partial laryngectomy to stage I glottic cancer was highly successful.

Over time, a large variety of external conservation operations were introduced ranging from hemilaryngectomy to a variety of anterior vertical partial laryngectomies. The success rate in treating both T_1a and T_1b glottic cancers by these external approaches has historically been greater than 90%. These procedures still remain an important part of the armamentarium of head and neck surgeons, but even here the paradigm has shifted.

Lillie and DeSanto [10] reported excellent results in patients with T_1 glottic cancer who were treated by endoscopic transoral cordectomy using the nonlaser technique. In the 1970s, transoral CO_2 laser excision of early glottic cancer was introduced in America by Strong, Jako, and Vaughan [16,17]. The principle of this microsurgical approach has already been discussed for carcinoma *in situ* and minimally invasive T_1 glottic cancers [3]. Transoral CO_2 laser excision proved to be an excellent modality for treating T_1 glottic cancers.

Initial transoral laser excision of stage I glottic cancer involved partial cordectomies, which were limited to the anterior true vocal cord. Resection of the posterior true vocal cord to include the arytenoid cartilage was felt to be absolutely contraindicated due to the high likelihood of postoperative aspiration. This philosophy had been developed by Vaughan in animal studies where full cordectomy, including arytenoid cartilage resection, led to intolerable aspiration in almost all of the study animals. This principle was then extrapolated to use of the CO_2 laser in human patients.

With this premise of the inability to endoscopically resect the arytenoid cartilage (without immediate reconstruction), initial guidelines to the resection of glottic cancer were investigated by Davis and colleagues [18] in a cadaver study in the early 1980s. In this study, it was determined that excision could be accomplished at a right angle across the vocal process of the arytenoid cartilage through the paraglottic space to the thyroid cartilage. This incision line was determined to be the posterior lateral extent to which safe laser excision could be accomplished. Anteriorly, it was found that the thyroid cartilage could be approached readily in almost all patients were the anterior commissure could be exposed by the large bore tubed laryngoscopes of that era.

In keeping with earlier well-established guidelines related to external conservation surgery of the larynx, Davis and Jako suggested limitations to endoscopic resection. The limitation at the anterior commissure of cancer extension inferiorly was set at 5 mm, well within the 10-mm guideline for open partial laryngectomy. The rationale for this decision was that as endoscopic surgery was in fact a new modality, and as cancer can escape through the cricoid thyroid membrane, which is normally 10 mm below the inferior edge of the anterior true vocal cord, then laser excisions were felt to necessarily be limited for the safety of these patients.

A very important consideration in limiting the posterior resection line to the already described tangential incision across the vocal process of the arytenoid cartilage came from the remarkable work of John Kirchner [19]. In elegant studies of whole mount laryngeal specimens, Dr. Kirchner had clearly shown that a main avenue of cancer escape from the glottis both inferiorly and superiorly was in this posterior paraglottic space. Cancers that approached this space were clearly not able to be resected by the transoral laser resection techniques then advocated. Additionally, using endoscopic laser techniques of that period, the posterior paraglottic space was very poorly visualized deep to the vocal process. Further, large blood vessels were known to traverse this space, which were difficult to control endoscopically.

With the anatomical guidelines suggested by Davis and Jako, most surgeons confined CO_2 laser transoral excision to T_1a lesions that could be encompassed readily by the laser under clear vision. The CO_2 laser was used to outline the intended area of resection. The cancer was grasped with a small forceps and pulled anteromedially. The outlined area was then incised more deeply with the laser, and the cancer was resected from posterior to anterior. The cancer was then removed on block and oriented so that definitive margins could be determined.

As the CO_2 laser caused some charring at the incision line, this technique had the potential of somewhat obscuring margins, which were narrow in the first place. True margins after laser excision needed to be taken from the excisional bed using the standard cold instrument technique.

Using the aforementioned technique, the Boston University experience in 100 patients was reported by Blakesley *et al.* [20]. Patients who underwent laser cordectomy and had negative cold instrument biopsies after the laser excision were not further treated. Patients who had positive biopsies after the laser excision were treated either by open vertical partial laryngectomy or, more typically, by postoperative irradiation. Cure rates using this approach were over 90% for patients with negative margins and approximately 85% for those patients with positive margins who then underwent full course irradiation after laser excision. The limited number of patients retreated by open vertical partial laryngectomy were all cured.

Data reported by the Boston group have been duplicated by other investigators [21–23]. When T_1a lesions have been the focus of these reports, the cure rates have remained uniformly over 90% and are basically similar to the radiation therapy reports showing the highest success rates. CO_2 laser excision of T_1a lesions was felt to be at least comparable to radiation therapy results and represented an appropriate alternative in selected patients.

One advantage of laser excision for these early cancers was that this was a very cost-efficient procedure. Patients who resided at a distance from a major therapy center were subjected to one limited surgery and then could return to their homes without the need of the protracted time away from home and employment or the expense of radiation therapy techniques. Conversely, patients with greater availability to radiation therapy facilities could yet undergo radiation therapy.

Whereas cure rates were felt to be similar between these two modalities, voice preservation was felt to be best with irradiation. While this concept is being challenged today, radiation therapy nonetheless was and is employed more commonly to treat T_1a glottic cancer.

Refinements in transoral laser microsurgery of T_1 glottic cancer have largely come from Germany where Steiner, Rudert, and other surgeons have both refined these techniques and notably extended endoscopic resection beyond T_1 glottic cancer [24]. Within the realm of purely T_1 glottic cancer, one significant refinement introduced by Steiner was to initially transect the cancer at a right angle to the lateral surface of the true vocal cord. This literal transection and incision of cancer was carried to a point in the paraglottic space beyond the lateral edge of cancer. Once this point had been determined, then the lesion was removed in several surgical fragments. Typically the posterior part of the lesion was first removed and then under very clear vision with a known lateral and inferior extent the anterior-most part of the cancer was removed. This approach allowed better visualization of the cancer and ultimately better preservation of the true vocal cord as the cancer was not simply excised on block by a full cordectomy. Margins were determined by direct visualization and pathological review.

The main criticism with laser excision of T_1 cancers has been the poor cure rates reported by some investigators in cancers involving the anterior commissure. This criticism has almost certainly been justified as the series reporting such poor results have used simple laser cordectomy approaches without the appropriate dissection at the anterior commissure, which is now advocated. The principles of approaching the anterior commissure are discussed in the following section.

IV. TREATMENT OF T_2 GLOTTIC CANCER

In the United States, T_2 glottic cancer has historically been treated most frequently by either full-course irradiation therapy or by the open partial laryngectomy procedures initially developed by Ogura and associates and refined by many other surgeons over time. The AJCC staging system defines T_2 glottic cancer as tumors extending to the subglottis and/or the supraglottis with normal true vocal cord mobility, or glottic lesions with impaired true vocal cord mobility. This section presents the principles of radiation therapy and surgery for these T_2 glottic cancers.

Radiation therapy for T_2 glottic cancer is generally given using opposed lateral-wedged portals with field sizes of 4×4 to 5×5 cm^2. The Co-60 unit or a 4-MV linear accelerator is used in most institutions. For bulky T_2 lesions, a field size of 6×6 cm^2 is often advocated. The most common fractionation schemes include 66 Gy in 33 daily fractions of 200 cGy or 63 Gy in 28 daily fractions of 225 cGy. Most institutions do not support the prophylactic treatment of regional lymph nodes in patients with T_2 glottic cancer [9,25,26].

The success rate in unselected patients treated by radiation therapy averages approximately 70%. The range in several large series is from 69 to 75% [5,9,27,28]. Radiation therapy series often include data of surgical salvage in patients treated unsuccessfully by irradiation. Surgical salvage of radiation failure can be accomplished 60 to 75% of time, leading to ultimate cure rates when salvage surgery is included of approximately 90%. However, the country norm for the salvage surgery of T_2 glottic cancer is total laryngectomy.

As at least 60% of patients with T_2 glottic cancer can be cured by irradiation, there is certainly a need to define prognostic indicators to preselect to the maximum extent possible those patients who will be cured by radiation. Prognostic factors identified and defined in the literature include gender, age, hemoglobin (Hb) level prior to and during radiotherapy, anterior commissure involvement, posterior glottic involvement, and the continuation or not of smoking after therapy [29–33]. Most series show that gender and age have little value as prognostic indicators. Anemia before or after radiation therapy has been shown to be an indicator, but it is unclear whether this association is due to tumor hypoxia or tumor volume.

Tumor volume has clearly been shown to be a major prognostic indicator in supraglottic cancer, but it is certainly logical that tumor volume would have prognostic significance in more bulky T_2 glottic lesions. As cigarette smoking is one of the major contributing factors to the development of glottic cancer in the first place, it is logical to assume that patients who continue to smoke after being treated will have a higher cancer recurrence. This has been shown in several series where up to a sixfold increase in local recurrence from cancer has been reported.

When treatment-related factors are analyzed in patients with T_2 glottic cancer treated by irradiation, the variables include the overall duration of radiation therapy, the dose per fraction, the total dose beam energy, and the radio-therapeutic technique. As is true in most cancers, prolonging the overall treatment time in patients undergoing continuous-course irradiation has been shown to affect survival adversely [34–37]. Investigations have shown decreased survivals of up to 10 to 15% when treatment time has exceeded 35 days in one series or 45 days in another study. The dose per fraction also has bearing on the ultimate survival time. It is well known that better local control rates have been obtained using ≥200-cGy daily fractions versus the original standard 180 cGy per fraction despite similar total dose [38].

The most significant tumor-related variables involve tumor bulk and extension of T_2 glottic cancer to either the anterior commissure or the posterior glottis [26,32,33]. When these two sites have been involved, some reported series show survival rates as low as 50 to 60%. Critical analysis of these series by radiation therapists usually attributes the poor local control rates to understaging of the initial lesion or inadequate radiation doses. Inadequate dosing can be due to inappropriately small field sizes or difficulties with dose distribution in either the anterior or the posterior commissure [7,39]. In several critical reviews of anterior commissure cancer involvement where these problems were felt to be addressed appropriately, the only negative prognostic indicator was felt to be subglottic extension of cancer.

Impaired true vocal cord mobility is uniformly regarded as a poor prognostic indicator in T_2 lesions [5,8,9,27,28,40]. Related to this, several authors have advocated the subclassification of T_2 glottic cancer to T_2a lesions, meaning those with normal true cord mobility, and T_2b lesions being those of impaired cord mobility. Most radiation therapists currently advocate higher total doses or altered fractionation schemes in patients with T_2b cancers or, more recently, treatment of these lesions with concomitant chemotherapy and irradiation [41].

While increasing fractionation size appears to improve ultimate tumor response, it also leads to more significant acute and long-term complications. As the dose per fraction is increased, the potential for late tissue damage to connective tissue, muscle, bone, and nervous tissue increases. As these late tissue effects are less influenced by overall treatment time than by fraction size, and as ultimate cure rates depend on a shorter treatment period, the principle of smaller treatment doses and a shorter time frame should be oncologically sound. Several altered fractionation schemes using this principle have been developed.

The most commonly used altered fractionation scheme is hyperfractionation. A typical scheme of hyperfractionation involves giving twice-daily fractions of 120 cGy to a total dose of 74 to 80 Gy. In actual reported series it is yet unclear whether such fractionation does significantly improve the effectiveness of treating T_2 glottic cancer.

The rationale for concomitant chemotherapy and irradiation has developed over time after initial studies using induction chemotherapy. The literature of induction chemotherapy is very expansive, involves a number of chemotherapeutic agents, and is not fully discussed here. Principles of induction chemotherapy did emerge in these studies and have led to the current thoughts regarding concomitant therapy with chemotherapy and irradiation.

It was clearly shown in initial chemotherapeutic studies that multiple agents had significantly higher response rates than single agents [42]. Certainly the most studied and most effective regiments have used cisplatinum and 5-fluorouracil. Patients who achieve a complete clinical and radiological response to induction chemotherapy have been shown to have significantly higher chances of ultimate tumor cure than patients with partial or no response [43,44]. Complete response rates have averaged in the literature between 15 and 40%, with most series showing a complete response in and around 25 ± 5% [44–46]. In general, patients who have achieved a complete response to chemotherapy for solid cancers without boney involvement have then shown cure rates with definitive radiation given after chemotherapy in the 70 to 90% range. The implication derived from these results is that full-course irradiation has certainly been able to sterilize microscopic residual cancer in a high percentage of patients who achieve a complete response to induction chemotherapy.

When all head and neck cancer patients given induction chemotherapy have been looked at, the overall cure rates have not improved. Patients with only a partial response to chemotherapy at best follow expected survival patterns as seen in historical controls. Nonresponse to chemotherapy is a very poor prognostic indicator with most patients dying in less than 1 year from the onset of treatment time [47].

Most oncologists agree that induction chemotherapy is at a minimum a prognostic indicator selecting out patients who can be cured by radiation. Proponents of radiation therapy without chemotherapy would argue that the complete responders to chemotherapy are only those patients who would have been cured by radiation therapy alone.

The movement to concomitant chemotherapy and irradiation has been based on several principles. Clearly some patients derive benefit from chemotherapy when used in an inductive mode, and certainly from radiation, which is known to be curative in many patients. The idea of using both modalities together is therefore logical. Additionally, chemotherapeutic agents, beyond their ability to kill cancer cells, are also known to be sensitizers for definitive irradiation. Using these modalities together is an extension of these thoughts.

The yet unresolved problem with concomitant therapy has been the significant morbidity seen in patients using this treatment technique. Whether the expected severe acute effects or, in some cases, the profound long-term effects will justify concomitant therapy in stage II glottic cancer remains very much to be seen. Certainly the judicious approach of concomitant therapy in appropriate treatment protocols for higher risk T_2 cancers should be investigated further.

The argument for primary irradiation to include hyperfractionation schemes or concomitant chemotherapy and radiotherapy is that cure rates are comparable to those achieved by partial laryngectomy surgical techniques (especially when salvage surgery is added) and, equally importantly, that functional preservation is better. Vocal quality following irradiation is accepted as a country norm to be superior to voice quality following open partial laryngectomy. Glottic sphincteric function is also clearly better in patients irradiated successfully than in those undergoing any partial surgical procedure.

Despite the purported advantages of irradiation, partial laryngectomy procedures are advocated by head and neck surgeons as an alternative to radiation therapy. The most significant single rationale for this bias is that ultimate control by radiation is in fact dependent on subsequent surgical salvage procedures, which often has meant total laryngectomy. In other words, ultimate local control by irradiation will mean that a certain number of patients will undergo total laryngectomy. Proponents of partial laryngectomy would suggest that ultimate cure rates are at least as high as radiation therapy techniques when salvage is included. When salvage is excluded, partial laryngectomy techniques are felt to provide higher survival rates with ultimate better functional preservation.

All initial partial laryngectomy techniques for T_2 glottic cancer involved external approaches. These approaches have been named hemilaryngectomy and a variety of vertical laryngectomy approaches to include anterior vertical, lateral vertical, anterior lateral vertical laryngectomy, and so on. The surgical principle has been that the area of cancer involvement is approached by a laryngofissure technique wherein the thyroid cartilage is divided in the midline for unilateral lesions or across the midline in lesions involving the anterior commissure. The involved area of the glottic larynx is removed with adequate margins and the preservation of at least one functional true vocal cord based on superior and recurrent laryngeal nerve preservation, and preservation of at least one arytenoid cartilage and cricoarytenoid joint.

In these techniques, when the cancer has been removed and the larynx preserved as just stated, restoration of voice and sphincteric function has depended on the introduction into the surgical defect of enough adynamic bulky tissue to allow glottic closure against this buttress for the preservation of speech and laryngeal sphincteric function. Clearly partial laryngectomy surgeries, which preserve more normal tissue, will have better functional results. Patients undergoing external approaches almost always need tracheotomy placement at surgery and continue until such time as more normal laryngeal function has returned. Patients as well need feeding tubes while glottic sphincteric function is returning.

The indications and contraindications to partial laryngectomy were initially introduced by Ogura and have by and large remained intact over time [12–14]. Some conventional vertical partial laryngectomy procedures to include resection of part of the cricoid have been proposed, but generally have not been used countywide due to the functional problems of aspiration, poor voice, or the need for long-term tracheotomy. More recently supracricoid laryngectomy approaches developed in Europe have been introduced to address these issues. These techniques are discussed in depth in Section V.

As familiarity with transoral laser microsurgery has increased, and more significantly due to the pioneering efforts of Dr. Wolfgang Steiner and other German physicians, transoral laser excision of stage II glottic cancer has developed into an exciting alternative to conventional external partial laryngectomy procedures. Steiner, in particular, has extended the bounds of transoral laser microsurgery significantly beyond the anatomical limitations originally proposed by Davis and Jako, which were described previously [18].

The principles involved in the extension of endoscopic surgery from selected T_1 cancer to T_2 glottic cancer do not differ from the principles espoused for open vertical partial laryngectomy. The difference involves the approach to these cancers.

As is true in any cancer surgery, exposure of the cancer and surrounding normal structures is critical. As was described originally by Davis and Jako when the anterior commissure can be well seen with a tubed laryngoscope, cancer excision can extend all the way to the thyroid cartilage. Visualization of this area has been improved by both the introduction of adjustable laryngoscopes and with newer, better contoured tubed scopes. Additionally, the ability to endoscopically expose the paraglottis and anterior commissure has been enhanced greatly by techniques of endoscopic resection of supraglottic tissue, which open the full view of the underlying glottic larynx.

Resection of the false vocal cord ipsilateral to a glottic cancer literally unroofs the ventricle and anterior aspect of the true vocal cord. When the resection of the false vocal cord is taken all the way to the superior aspect of the thyroid cartilage, then the anterior lateral paraglottic space is also fully opened. Additionally, when the ipsilateral hemiepiglottis and aryepiglottic fold are resected endoscopically, the full posterior paraglottic space can be visualized. These refinements of exposure of the glottic larynx were developed in the evolution of endoscopic supraglottic laryngectomy, which is discussed later. Importantly though, exposure of the glottis and the full paraglottic space is possible endoscopically in most patients. In contrast to external partial laryngectomy for glottic cancer, in endoscopic partial laryngectomy the thyroid cartilage need not be divided nor the skin and subcutaneous tissues to gain this full exposure. When either the hemiepiglottis and false vocal cord are resected, or false vocal cord alone is moved, the upper edge of the thyroid cartilage can be identified. With this exposure, endoscopic surgery can comfortably remove all cancer anterior to the body of the arytenoid cartilage. Very importantly, the internal perichondrium of the thyroid lamina can be exposed from above and literally dissected from the thyroid cartilage to clearly define the extent of a T_2 glottic cancer. If cancer in fact is involving the thyroid cartilage, then this is definitively known and the surgery appropriately adjusted.

In contrast to open vertical partial laryngectomy where the arytenoid cartilage can be removed, most transoral laser microsurgeons do not feel the arytenoid cartilage can be removed completely without significant aspiration occurring. The obvious reason for this is that endoscopic approaches to date do not reconstruct the posterior glottis where the arytenoid cartilage is removed in a single stage. Open vertical partial laryngectomy techniques in contrast do reconstruct this area at the time of cancer resection.

In the evolution of CO_2 microsurgery for T_1 laryngeal cancer, it became very evident that when all of the arytenoid cartilage, with the exception of part of the vocal process, was preserved, a neocord would develop from granulation tissue that was adequate to prevent aspiration. While the vocal qualities of this neocord were obviously not the same as the uninvolved true vocal cord on the opposite side, these neocords were in fact mobile, provided good voice, and certainly allowed preservation of sphincteric function. When laser resection techniques were expanded, it became apparent that further resection of the arytenoid cartilage could be accomplished but without the same functional normalcy. If the main aspect of the body of the arytenoid was preserved and the vocal process and part of the arytenoid resected, then the resultant neocord formation still resulted in a fixed structure against which the other true vocal cord could buttress for the return of speech and sphincteric function. This was much more akin to the functional results seen in full hemilaryngectomy where the total arytenoid cartilage was removed and replaced by soft tissue. In contrast to the open techniques, though, the full extent to which the neocord would buttress the true cord and provide posterior glottic competence was more limited. It became apparent in extended endoscopic resections where part of the arytenoid cartilage was preserved that subsequent medialization procedures could be performed to further improve sphincteric function and voice.

When these observations were fully appreciated, it became apparent that in patients who can be adequately visualized endoscopically, most T_2a glottic cancers could in fact be resected with the same cancer margin and functional result as those obtained by open vertical partial laryngectomy. The limitation of the endoscopic approach related to tumor involvement in and around the arytenoid cartilage. If the arytenoid body could be preserved, then the endoscopic resection of T_2 lesions could in fact be accomplished.

Endoscopic laser resection in T_1 glottic cancer has been reported to result in poor results when there was anterior commissure involvement. These initial published reports of CO_2 laser resection involved using endoscopic techniques of simple cordectomy without the elaborate exposure and control of the anterior commissure earlier stated in this chapter.

When cancer involves the anterior commissure area, exposure must be first obtained superiorly as presented. As the perichondrium of the thyroid cartilage is microdissected from above if cancer is found to involve the internal thyroid cartilage perichondrium, then the endoscopic approach either needs to be converted to an open approach or be coupled to a limited external approach wherein a selected part of the thyroid cartilage is resected after being exposed through a small transcutaneous incision. This simple refinement has allowed the anterior commissure to be controlled well in a predominantly endoscopic procedure.

As mentioned earlier in the radiation therapy section, both radiation therapists and surgeons agree that significant infraglottic cancer extension is a main precipitating cause of failure, both by radiation therapy and by partial laryngectomy. Advocates of transoral laser microsurgery would suggest that given the magnification and ability to see the anterior subglottis when this area is appropriately approached endoscopically, actual cancer mapping can be accomplished surgically. Anterior, inferior cancer extension is mapped endoscopically by carefully resecting cancer down to the area of the cricothyroid membrane or along the anterior tracheal wall. Obviously if cancer is escaping the cricoid thyroid membrane, then endoscopic resection alone is inadequate. It has been noted, however, that a number of cancer extensions in this area tend to be superficial and can in fact be resected with margins controlled by appropriate pathological analysis.

The treatment options for T_2 glottic cancer include irradiation with recently developed hyperfractionation techniques or concomitant radiation therapy with chemotherapy. These techniques show promise but must yet be better elucidated in controlled studies. Partial laryngectomy techniques yet remain very viable and certainly are preferred when the arytenoid cartilage must be resected completely. However, in patients where adequate exposure can be obtained, and where newer techniques of endoscopic resection are employed, almost all T_2a glottic cancers can be resected by transoral laser microsurgical techniques without violation of oncological principles and with expected excellent functional result.

V. TREATMENT OF T_3 AND T_4 GLOTTIC CANCER

For several decades the gold standard in the treatment of T_3 and T_4 glottic carcinoma has been total laryngectomy.

Cure rates for T_3 glottic cancer treated in this way are predictably higher than treatment of T_4 cancers. As very few conservation surgical options are available for T_4 glottic cancer, the focus of this section is on T_3 glottic cancer. The challenge for the head and neck oncologist is to maintain the cure rate able to be obtained by total laryngectomy for these advanced glottic cancers while introducing surgical innervations that better preserve laryngeal function.

As total laryngectomy is remarkably morbid, primary radiation therapy to treat T_3 glottic cancer has been historically used and is strongly advocated by some. Radical irradiation is the primary modality followed by salvage total laryngectomy in cases of radiation failure [48]. More recently, induction chemotherapy has been used as a prognosticator of radiotherapeutic response in laryngeal preservation regimens. In these series, patients obtaining at least partial responses (meaning greater than 50% reduction in tumor volume) to induction chemotherapy have been treated by subsequent irradiation therapy only. Patients who achieve less than a partial response to chemotherapy have been treated by total laryngectomy with or without postoperative irradiation. The rationale for each of these different approaches is now presented.

Treatment of T_3 glottic cancer by primary irradiation has yielded local control rates in the 50–60% range [49,50]. Of the patients who fail radical irradiation, total laryngectomy is reported to salvage approximately half of this group, yielding an overall cure rate of 70±5% with a laryngeal preservation rate of the original 50% [48]. Centers that follow this particular regimen generally are proponents of radial radiotherapy for favorable lesions, typically those that involve only one side of the larynx, do not impair the airway, and are in reliable patients who are easy to examine. In 1984, DeSanto [51] presented a strong rebuttal to this approach in a series involving T_3 glottic cancers in patients who were treated by total laryngectomy. Ultimate survival in this group was in the 90% range versus the 75% figure reported in the radial irradiation and salvage surgery group. Many patients were spared two-modality therapy and with the development of tracheal esophageal speech did significantly well.

When total laryngectomy with postoperative irradiation is compared to radical radiotherapy and surgical salvage, actuarial 4-year disease rates have been seen by some investigators to be significantly better in the combined treatment group. Thakar *et al*, [52] showed a 4-year disease-free survival rate of 79.3% in the combined therapy group versus a 65.3% control rate in the radical radiotherapy and surgical salvage group. Failure in the latter group was almost always at the primary site. The probability of surviving with an intact larynx in this series was only 30%. These authors strongly felt that radical irradiation with surgical salvage was not indicated.

As an alternative to total laryngectomy, Pearson and associates [53] at the Mayo Clinic published their experience with near-total laryngectomy. The basic principle of this procedure involved the preservation of the posterior half of an uninvolved hemilarynx. By bringing a superiorly based pharyngeal flap to this preserved tissue, an internal shunt could be created that provided voice by using the innervated arytenoid to allow air passage into the pharynx and, at the same time, prevented aspiration of food and saliva into the trachea. Patients undergoing this procedure therefore had voice preservation, but nonetheless had a stoma for respiration.

As an alternative to total laryngectomy or Pearson's near-total laryngectomy, Brasnu and Laccourreye [54,55] of France advocated the supracricoid partial laryngectomy-cricohyoidopexy for selected T_3 glottic laryngeal patients. In patients amenable to this approach, a stoma can be avoided. This procedure is also possible in T_4 glottic cancers with limited thyroid cartilage invasion.

As discussed in Section IV, supracricoid partial laryngectomy is even better amenable to T_2 glottic cancers with extension to the ventricle, false vocal cord, petiole of the epiglottis, anterior aspect of the arytenoid cartilage, and/or impaired vocal cord mobility [56].

Transoral laser excision of T_3 glottic cancers has been advocated by Steiner, but is being done infrequently in America [57]. Most American endoscopic advocates do not feel T_3 cancer is amenable for endoscopic resection if the arytenoid cartilage area is involved by cancer, whereas cancers that fix the true vocal cord due to direct extension into the vocalis muscle anteriorly could be treated endoscopically if appropriate visualization is possible. If endoscopic resection would require arytenoid cartilage removal, then American surgeons have not advocated this approach.

Significant interest has been developed in the last decade in the use of induction chemotherapy followed by irradiation or potentially followed by conservation surgical procedures. It has long been established in patients treated initially by radical irradiation who recur that the only safe and viable surgical salvage procedure must encompass the area where the original cancer had been. Brandenburg's work emphasized this principle by showing that limited clinical recurrence following irradiation meant that the probability of viable cancer cells in other areas of previous cancer was still very high [58]. Any surgical procedure that removes less than the full original area of cancer with appropriate margins would be fraught with a high likelihood of later recurrence.

Induction chemotherapy has introduced a new concept to the original thought of surgical salvage after radical irradiation. When a patient with laryngeal cancer undergoes induction chemotherapy and achieves a complete response, then almost all oncologists would agree that subsequent irradiation therapy alone is a very reasonable and likely successful therapeutic approach to organ preservation. An interesting question is raised when induction chemotherapy results in a significant partial response in glottic cancer not originally able to be treated by partial laryngectomy. Could partial laryngectomy be used to resection all residual clinical disease as a second adjunctive therapy before full-course irradiation? If the surgical procedure removed all known remaining cancer following partial response to chemotherapy, then it would seem logical that subsequent full-course irradiation given to the full origin area of cancer with appropriate margins could be given with a high expectation of therapeutic success. The basic issue is whether partial response further treated by surgery to gain complete response and then irradiated would be similar to complete response by chemotherapy followed by irradiation alone. This idea seems plausible and may be an appropriate alternative if the conservation surgery can be accomplished and followed by irradiation without undue morbidity.

The issue of whether induction chemotherapy could be followed by conservation surgery alone to less than the full area of the original cancer is a more tenuous idea. If the work of Brandenburg and others related to recurrence after irradiation would also pertain to chemotherapeutic response, then it would seem unlikely that partial laryngectomy procedures could be done as a final therapeutic modality. At this point in time there are no definitive series in the literature to either justify or refute this concept, but experience from the past would suggest that partial laryngectomy alone after initial chemotherapy likely would not be adequate if the partial laryngectomy would not have been possible in the first place.

There has been significant interest in the laryngeal preservation study conducted in the Veterans Administration Hospital system in the United States [58]. The principles of this approach have already been stated. Responders to chemotherapy are treated by radiation alone, whereas nonresponders received total laryngectomy. Very important in this approach is the determination of whether there are specific tumor or biologic factors that are predictive of chemotherapeutic response, organ preservation, and subsequent survival. When the large VA study was analyzed, the potential prognostic variables included clinical and histological factors, immunohistochemical expression of proliferating cell nuclear antigen and p53, and adjusted DNA index measurements [60]. Multivariate analysis revealed that the best predictor of a complete response to induction chemotherapy was low t class. Additionally, p53 overexpression and elevated proliferating cell nuclear antigen index were independent predictors of successful organ preservation.

In summary, there are a number of different therapeutic approaches to T_3 glottic cancer. Total laryngectomy still remains the most proven treatment modality with the highest ultimate survival. Pearson's near total laryngectomy is a very good alternative in appropriately trained hands [61]. Radical irradiation followed by surgical salvage will save 30 to 50% of larynges in these patients but with a likely poorer overall survival (10 to 20% less) than total laryngectomy alone. Patients who are complete responders to induction chemotherapy have a high chance of cure by subsequent full-course irradiation alone. Partial responders to induction chemotherapy *may* be salvaged adequately with laryngeal preservation by conservation surgeries if subsequent full-course irradiation is later given to the full original area of cancer. Supracricoid laryngectomy does benefit selected patients with T_3 glottic lesions. Endoscopic resection of T_3 glottic cancer may be possible if vocal cord fixation is due to anterior true vocal cord invasion.

Current trials of hyperfractionated irradiation or concomitant chemotherapy and irradiation for T_3 glottic cancer offer some hope that these approaches may yield better results than radiation alone. Whether these approaches ultimately gain widespread use will be dependent not only on the cure rates obtainable by these measures, but also by both short- and long-term morbidity, which accompany these approaches.

Patients with T_3 and T_4 glottic cancers also have a risk of regional spread of cancer [62,63]. Related to this elective treatment of the neck is recommended in T_3 transglottic cancers, most T_4 glottic cancers, and in patients with subglottic cancers. The N_0 neck may be treated electively by either surgery or irradiation, but irradiation is best reserved for cases where that modality is employed for the primary tumor. Elective neck dissection can provide important information for prognostic purposes and therapeutic decisions. This is accomplished by establishing the presence, number and location, and nature of occult lymph node metastases. Elective neck dissection involving removal of lymph node areas 2, 3, and 4 on the side ipsilateral to the cancer is the procedure of choice. Paratracheal nodes (lymph node level 6) should be dissected in stage IV glottic and subglottic cancers. Dissection of lymph nodes levels 1 and 5 are almost never indicated in these glottic cancers, as tumor very rarely involves these regions.

The role of sentinel lymph node biopsy is not fully elucidated in head and neck cancers in general to include cancers of the larynx. The initially expressed concern is that sentinel lymph node biopsies may fail to detect tumor on frozen section analysis or may not reveal skipped metastases. As functional neck dissections can be done with minimal morbidity, sentinel node procedures seem less likely to be of help than in other cancer models.

T_3N_1 or occasional T_3N_2 patients may be treated by selective neck dissection without postoperative irradiation. Almost all oncologists agree that any patient with an extracapsular extension is best treated by postoperative irradiation. Patients with N_3 neck disease almost always require either modified radical or full radical neck dissections where again lymph node levels 2, 3, and 4 must be dissected but with jugular vein and sternocleidomastoid muscle resection.

The survival in patients with T_4 glottic cancer still remains poor. Total laryngectomy followed by full-course irradiation remains the treatment of choice. Patients who undergo induction chemotherapy with complete response have a much better prognosis than partial or nonresponders. The theoretical area of concern in T_4 glottic cancer patients with cartilage invasion is whether subsequent irradiation is adequate to eliminate cancer in cartilage.

Current regimens of concomitant chemotherapy and irradiation may offer better potential cure rates with an increase in laryngeal preservation than irradiation alone. For this to be true, complete response rates to this modality will need to be shown to be higher than those obtained by irradiation alone. Additionally, salvage surgery when indicated will need to be able to be done without morbidity beyond that seen with radiation and salvage surgery.

VI. TREATMENT OF SUPRAGLOTTIC CANCER

Supraglottic carcinomas have historically been managed by radiation alone, surgery alone, or by combined therapy in the most advanced stage lesions. In contrast to early glottic cancer, supraglottic cancers more extensive than T_1 suprahyoid supraglottic cancers require treatment of the neck due to the high propensity for regional metastatic cancer spread. This propensity for neck spread derives from a significantly richer field of lymphatics in the supraglottis which develop due to the significantly different embryology of the supraglottic versus the glottic larynx.

Stage I suprahyoid supraglottic cancers are very rare and are treated either by primary irradiation therapy or by epiglottectomy. As there is little risk of spread to the neck, radiation portals can be reduced significantly from fields used to treat other supraglottic primary cancers.

Surgical approaches to T_1 supraglottic cancer originally were transcutaneous resection of the epiglottis, which often necessitated tracheotomy and postoperative feeding tubes. In the late 1970s and early 1980s the concept of transoral laser resection of the epiglottis was well developed and found to be a very viable alternative to open supraglottic resection [64]. Patients with transoral epiglottectomy faired extremely well with minimal postoperative airway or other functional problems to include aspiration.

T_2 supraglottic cancer has historically been treated most often by full-course irradiation therapy. In patients with adequate pulmonary function, open supraglottic laryngectomy was developed as a very viable alternative to irradiation. When patients with T_2 supraglottic cancers are treated by irradiation, the radiation portals include both sides of the neck due to the high propensity of regional metastatic spread. Patients treated by surgery only have undergone supraglottic laryngectomy with a functional neck dissection(s). As is true in glottic carcinomas, supraglottic cancers rarely metastasis to lymph node levels 1 or 5. Selective neck dissections have therefore involved lymph node levels 2, 3, and 4.

As many supraglottic cancers are either midline or near midline structures, the concept of bilateral functional neck dissection was advocated when surgery alone was used. In America, these neck dissections were almost always done at the time of open supraglottic laryngectomy. Some European surgeons have advocated supraglottic laryngectomy in one setting followed by neck dissection(s) approximately 1 week after primary surgery. The principle espoused in this approach has been that some cancer cells may be in transit from the primary site to the neck and would not necessarily be included in selective neck dissection done at the same time as supraglottic laryngectomy.

When T_2 supraglottic cancers are lateralized to one side of the supraglottis, some surgeons have advocated supraglottic laryngectomy coupled with ipsilateral functional neck dissection. If pathological review shows cancer in the neck dissection specimen, then the contralateral neck would need to be treated either by simultaneous dissection, as a later dissection, or by irradiation therapy when the contralateral neck was N_0 and irradiation was otherwise indicated.

In midline T_2 supraglottic lesions treated by surgery only, many patients have undergone simultaneous bilateral functional neck dissections. DeSanto [65] has advocated an interesting alternative approach. In T_2 (or T_3) N_0M_0 supraglottic cancer located to any degree more on one side of the supraglottis, then neck dissection is first done on that side. Frozen section analysis of the neck dissection specimen is done immediately at the time of surgery while the supraglottic laryngectomy is being performed. When the frozen sections are completely negative for cancer, then the contralateral neck is not dissected simultaneously. If any lymph node spread is found in the ipsilateral neck dissection specimen, then the contralateral neck is also operated on in that same setting.

In DeSanto's series, only 1 of 98 N_0 patients who were node negative on frozen section analysis had subsequent contralateral neck cancer and underwent later neck dissection. Conversely, when the first side of the neck dissected was positive for lymph node spread, 31 of these 90 patients required contralateral neck dissection for cancer at some point in time. Nineteen of these 32 (59%) patients died of their disease. In contrast, only 25 of 156 (16%) of patients with unilateral neck dissections died of their cancers.

DeSanto further compared the results of delayed dissection on the second side of the neck to simultaneous bilateral neck dissection when the first dissected side was positive for cancer. His data showed a higher death rate from disease as well as a higher frequency of distant metastasis when the contralateral neck dissection was delayed.

In series that present results of radiation therapy alone for stage II supraglottic cancers, the 3-year disease-free survival rates are reported from 60 to 80% [66,67]. The probability of response to radiotherapy is definitely influenced by the actual volume of tumor present in the supraglottis. Work done at the University of Florida has shown that tumors less than 6 cm^3 had response rates in the 80% range, whereas tumors larger than 6 cm^3 had response rates less than 50% [67,68].

One of the problems in evaluating success with radiation therapy for supraglottic lesions that are clinically and radiologically T_2 relates to the probability of understaging these cancers. Zeitels [69] has reported that nearly 30% of apparent T_2 infrahyoid epiglottic cancers are actually T_3 lesions based on invasion of the preepiglottic space, which is not apparent clinically, but is proven pathologically. These results were seen in the analysis of supraglottic cancer resection specimens obtained by transoral laser microsurgery. Obviously, if irradiation only is given it is not possible to sort out which of the apparent T_2 lesions are in fact more aggressive and are in reality T_3 cancers.

Success rates in treating T_3 supraglottic cancer by radiation therapy alone are as expected less than results seen in T_2 lesions. These results range from 40 to 70% and in all likelihood are somewhat site dependent. T_3 cancers that have developed by extending into the paraglottic space to produce immobility of the true vocal cord are likely more aggressive and less likely to be cured than cancers staged T_3 due to extension through the epiglottis into the preepiglottic space.

Surgical approaches to supraglottic cancers started with total laryngectomy and then advanced to the conservation procedures described by Ogura and associates [70–73] of horizontal partial laryngectomy or supraglottic laryngectomy. As the resection of supraglottic tissue removes part of the sphincteric function of the larynx, patients so treated are more susceptible to postoperative aspiration and subsequent pulmonary compromise. Related to this pulmonary function test standards have been developed to help select which patients are amenable to open supraglottic resection.

The limitations for supraglottic laryngectomy initially defined by Ogura and others have largely remained in place up to this time. The original external supraglottic laryngectomies entailed removal of the full supraglottis above the level of the true vocal cords and included the preepiglottic space and hyoid bone. The first modification to this approach came from Italian surgeons, who found that resection of the hyoid bone was in fact not necessary in many patients [73]. Leaving the hyoid bone in place allowed the resected residual larynx to be suspended to the hyoid, which helped diminish postoperative aspiration.

In patients staged T_3 based on paraglottic space invasion and vocal cord fixation, some extended supraglottic laryngectomy procedures have been proposed that entail resection of the full supraglottis with the resection carried down into the glottic larynx. Pearson's near-total laryngectomy is one example of such a procedure that has been shown to successfully remove T_3 and even T_4 supraglottic cancers with limited thyroid cartilage invasion. This procedure, as already discussed, leaves the patient stoma dependent but does leave a shunt for the production of speech.

More recently, French physicians have introduced supracricoid partial laryngectomies for both glottic and supraglottic squamous cell carcinomas, as discussed earlier. Supracricoid partial laryngectomy with cricohyoidopexy is a very viable alternative to extended supraglottic laryngectomy or near-total laryngectomy in lesions that involve the anterior aspect of the larynx. In this procedure, after resection of the cancer, the cricoid cartilage is pexed to the remaining hyoid bone, helping to prevent aspiration, leaving a viable voice, and avoiding a tracheostoma.

As discussed in the section on advanced glottic cancer, investigations have been done and are ongoing to enhance the effect of primary irradiation for these more extensive supraglottic cancers. Nakfoor, and colleagues [66] at the Massachusetts General Hospital and the Massachusetts Eye and Ear Infirmary have used accelerated radiotherapy in the treatment of these lesions. Their experience from 1981 to 1992 included 164 evaluable patients for stages I–IV supraglottic cancer. Their approach was to use accelerated hyperfractionated radiotherapy to 67.2–72 Gy in 1.6-Gy fractions given twice daily for 6 weeks. Their success rates in treating T_3 and T_4 supraglottic cancers are reported to be 76 and 43% 5-year actuarial local and regional control. When surgical salvage is included, the ultimate local control rates for T_3 and T_4 lesions were 88 and 51%. Voice preservation for T_3 and T_4 patients was 72 and 43%, respectively.

Current studies are ongoing in institutions using other hyperfractionation regimes, as well as studies using concomitant chemotherapy and irradiation. Advanced stage supraglottic cancer was also included in the large VA cooperative study, which has been discussed previously. Success rates in stage IV supraglottic cancer are comparably poor, similar to stage IV glottic cancer. The protocols of larynx preservation are the same for both.

Transoral laser resection of supraglottic cancer is now becoming used more prevalently. As stated earlier in this section, transoral resection of T_1 suprahyoid epiglottic cancer has been performed for almost 20 years and is an accepted alternative for this rare and early supraglottic cancer [74]. Endoscopic supraglottic resection for stage II and more recently stage III lesions has developed very slowly in America. This procedure developed in America as basically an adjunctive procedure done to stabilize the airway of patients with extensive supraglottic cancers before definitive therapy usually total laryngectomy and postoperative irradiation [75]. In some cases, patients had extended epiglottic, aryepiglottic fold, or false vocal cord resections prior to definitive irradiation, again with the intent of stabilizing the airway and preventing obstruction during the course of irradiation. As it became apparent that many of these patients tolerated the endoscopic resections extremely well with minimal postoperative morbidity to include airway loss or bleeding, the concept of resection of all visible cancer followed by definitive radiation was advocated.

Zeitels and Davis [75–77] have reported their preliminary experiences in resecting supraglottic cancers endoscopically and either coupling these resections with postoperative irradiation or relying on surgery alone. This determination was based on the size of the primary lesion and the degree of confidence with which the cancer was fully resected endoscopically.

Initial endoscopic resections involved piecemeal resection of cancers through large-bore laryngoscopes. As the full surgical field could rarely be seen in one laryngoscope placement, these scopes were often adjusted during the surgery in a mosaic-type way. As results of this approach were reported, there was justifiable concern that these resections may not have been oncologically sound due to the challenges of exposure. It was certainly recognized that these endoscopic supraglottic resections were more difficult to perform than open supraglottic laryngectomy and were fraught with some chance of inadequate cancer resection. As adjustable laryngoscopes were introduced, and as evolutions in surgical technique developed, transoral laser supraglottic resection has become better understood and accepted in America.

In Germany, Steiner and Rudert, among others, have long advocated transoral laser supraglottic resection of T_2 and selected T_3 or even T_4 supraglottic cancer [78–80]. Primary endoscopic resection in the hands of these surgeons has often been done in one setting with neck dissection done within a week of the primary procedure. Success rates reported by these German physicians are similar to success rates reported by other surgeons worldwide using open supraglottic laryngectomy.

While it is beyond the intent of this chapter to be a surgical atlas, there are key surgical steps that allow endoscopic resection to be safe and effective. Use of the adjustable laryngoscope has already been addressed. Excellent visualization must be obtained for any endoscopic (or other) surgery to be successful. The single surgical innovation that has most helped endoscopic surgery has been the initial transection of the epiglottis from pharyngoepiglottic fold to pharyngoepiglottic fold as the first step of endoscopic resection. This epiglottectomy often transects cancer, but totally opens a view of the residual supraglottis and preepiglottic space superiorly and anteriorly, which allows supraglottic laryngectomy to be done to the same extent that would be done in an open procedure. The obvious difference is that normal tissues are not divided to gain tumor access, thereby notably lessening postoperative morbidity in the endoscopic approach.

In the initial work by Davis endoscopic supraglottic laryngectomy was coupled with irradiation as stated previously. Importantly, patients undergoing endoscopic resection and postoperative irradiation did not have the same magnitude of postoperative problems to include inability to swallow that had been seen in open supraglottic laryngectomy followed by irradiation. Additionally, while the same guidelines of assuring adequate pulmonary reserve were followed in the endoscopic resections, it certainly appeared anecdotally that patients had significantly less aspiration, suggesting that patients with poorer pulmonary function could even be treated this way.

As the work of the pioneering German surgeons became well known in America, the limitations to supraglottic laryngectomy done endoscopically were certainly broadened. More notably, the need to give postoperative irradiation for control of the primary site in the supraglottis was appropriately called into question. At the current time, endoscopic supraglottic laryngectomy in patients in whom the supraglottis can adequately be exposed endoscopically is felt to be an alternative to open supraglottic laryngectomy. The same ultimate tumor margins can be obtained to include hyoid bone resection and resection of the upper part of the thyroid cartilage where needed.

Treatment of T_4 supraglottic cancer has been alluded to in the discussion of T_2 and T_3 cancers. T_4 supraglottic cancer still remains highly threatening with survival rates from primary irradiation less than 50%, even with the more innovative approaches. Many T_4 supraglottic cancers still require a total laryngectomy with postoperative irradiation with an expected ultimate survival near 50%. Laryngeal-sparing regimes are used appropriately with T_4 supraglottic cancer and will continue to be investigated.

VII. SUMMARY

Treatment options in the management of laryngeal cancer have been presented from a principal basis. The focus has been on why different modalities of therapy could be selected versus any in-depth discussion of how these approaches are done. The author's own preference for treatment of laryngeal cancer is as follows.

1. Carcinoma *in situ* of the true vocal cords is best treated by phonomicrosurgical techniques as advocated by Zeitels and others.

2. T_1a glottic cancers in patients who can be exposed endoscopically are best treated by transoral laser microsurgery. T_1b lesions involving the anterior larynx are also treated preferentially by laser resection utilizing the principles listed beyond standard cordectomy techniques. T_1b lesions involving the posterior glottis are treated preferentially by radiotherapy with the hope that altered

fractionation or chemo-irradiation schemes may produce better results.

3. T_2a glottic lesions are best treated by transoral laser resection in patients adequately able to be exposed or by open vertical partial laryngectomy or supracricoid partial laryngectomy in patients where exposure is more difficult. T_2b cancers are best treated by these open surgical approaches. Radiotherapy is reserved for patients who are not surgical candidates.

4. The gold standard for cure in T_3 glottic cancer still remains total laryngectomy. Where cancer involves the anterior glottis more significantly, open partial laryngectomy or supracricoid partial laryngectomy is advocated. The use of laryngeal preservation schemes to include induction chemotherapy are an important alternative therapy. Additionally, concurrent chemotherapy and irradiation will likely become a viable alternative.

5. Stage IV glottic and supraglottic cancers are highly threatening and are best treated by total laryngectomy and postoperative irradiation in most cases. Complete response to either induction chemotherapy followed by irradiation or to concurrent chemotherapy and radiation may save the need for total laryngectomy.

References

1. Zeitels, S. M. (1995). Premalignant epithelium and microinvasive cancer of the vocal fold; the evolution of phonomicrosurgical management. *Laryngoscope* **105**(3 Pt 2), 1–51.
2. Zeitels, S. M. (1996). Laser versus cold instruments for microlaryngoscopic surgery. *Laryngoscope* (5 Pt 1), 545–552.
3. Zeitels, S. M. (1996). Phonomicrosurgical treatment of early glottic cancer and carcinoma-in-situ. *Am. J. Surg.* **172**(6), 704-709.
4. Kass, E. S., Hillman, R. E., and Zeitels, S. M. (1996). Vocal fold submucosal infusion technique in phonomicrosurgery. *Ann. Otol. Rhinol. Laryngol.* **105**(5), 341–347.
5. Wang, C. C. (1990). Carcinomas of the larynx. *In* "Radiation Therapy for Head and Neck Neoplasms: Indications, Techniques, and Results." Year Book Medical, Chicago.
6. Rudoltz, M. S., Benammar, A., and Mohiuddin, M. (1993). Prognostic factors for local control and survival in T_1 squamous cell car cinoma of the glottis. *Int. J. Radiat. Oncol. Biol. Phys.* **26,** 767.
7. Olszewski, S. J., Vaeth, J. M., Green J. P., *et al.* (1985). The influence of field size, treatment modality, commissure involvement and histology in the treatment of early vocal cord cancer with irradiation. *Int. J. Radiat. Oncol. Biol. Phys.* **11**, 1333.
8. Kelly, M. D., Hahn, S. S., Spaulding, C. A., *et al.* (1989). Definitive radiotherapy in the management of stage I and II carcinoma of the glottis. *Ann. Otol. Rhinol. Laryngol.* **98**, 235.
9. Mendenhall, W. M., Parsons, J. T., Million, R. R., *et al.* (1988). T_1-T_2 squamous cell carcinoma of the glottic larynx treated with radiation therapy: Relationship of dose-fractionation factors to local control and complications. *Int. J. Radiat. Oncol. Biol. Phys.* **15**, 1267.
10. Lillie, J. C., and DeSanto, L. W. (1973) Transoral surgery of early cordal carcinoma. Trans. *Am. Acad. Ophthalmol. Otolaryngol.* **77**, 92–96.
11. Som, M. L. (1951). Hemilaryngectomy. *Arch. Otolaryngol.* **54**, 524.
12. Ogura, J., and Biller, H. (1969). Conservation surgery in cancer of the head and neck. *Otolaryngol. Clin. North. Am.* **12**, 641.
13. Biller, H., Ogura, J., and Pratt, L. (1971). Hemilaryngectomy for T_2 glottic cancers. *Arch. Otolaryngol.* **93**, 238.
14. Biller, J., and Lawson, W. (1984). Partial laryngectomy for transglottic cancers. *Ann. Otol. Rhinol. Laryngol.* **93**, 297.
15. Kirchner, J. (1984). Pathways and pitfalls in partial laryngectomy. *Ann. Otol. Rhinol. Laryngol.* **93**, 301.
16. Strong, M. S. (1975). Laser excision of carcinoma of the larynx. *Laryngoscope* **85**, 1286–1289.
17. Vaughan, D. W., Strong, M. S., and Jako, G. J. (1978). Laryngeal carcinoma: Transoral treatment utilizing the CO_2 laser. *Am. J. Surg.* **136**, 490–493.
18. Davis, R. K., Jako, G. J., Hyams, V. J., and Shapshay, S. M. (1982). The anatomical limitations of CO_2 laser cordectomy. *Laryngoscope* **92**, 980–984.
19. Kirchner, J. A., Cornog, J. L., and Holmes, R. E. (1974). Transglottic cancer: Its growth and spread within the larynx. *Arch. Otolaryngol.* **99**, 247–251.
20. Blakeseley, D., Vaughan, C. W., Shapshay, S. M., *et al.* (1984). Excisional biopsy in the selected management of T_1 glottic cancer: A three-year follow-up study. *Laryngoscope* **94**, 488–494.
21. Westmore, S. J., Key, M., and Suen, J. Y. (1986). Laser therapy for T_1 glottic carcinoma of the larynx. *Arch. Otolaryngol. Head Neck Surg.* **112**, 853–855.
22. McGuirt, W. F., and Koufman, J. A. (1987). Endoscopic laser surgery. *Arch. Otolaryngol. Head Neck Surg.* **113**, 501–505.
23. Davis, R. K., Kelly, S. M., Parkin, J. L., *et al.* (1990). Selective management of early glottic cancer. *Laryngoscope* **100**, 1306–1309.
24. Steiner, W. (1993). Results of curative laser microsurgery of laryngeal carcinoma. *Am. J. Otolaryngol.* **14**, 116–121.
25. Mendenhall, W. M., Parsons, J. T., Brant, T. A., *et al.* (1989). Is elective neck treatment indicated for T_2N_0 squamous cell carcinoma of the glottic larynx? *Radiother. Oncol.* **14**, 199.
26. Howell-Burke, D., Peters, L. J., Goepfert, H., *et al.* (1990). T_2 glottic cancer: Recurrence, salvage and survival after definitive radiotherapy. *Arch. Otolaryngol. Head Neck Surg.* **116**, 830.
27. Fein, D. A., Mendenhall, W. M., Parsons, J. T., *et al.* (1993). T_1-T_2 squamous cell carcinoma of the glottic larynx treated with radiotherapy: A multivariate analysis of variables potentially influencing local control. *Int. J. Radiat. Oncol. Biol. Phys.* **25**, 605.
28. Pellitteri, P. K., Kennedy. T. L., Vrabec, D. P., *et al.* (1991). Radiotherapy: The mainstay in the treatment of early glottic carcinoma. *Arch. Otolaryngol. Head Neck Surg.* **117**, 297.
29. Fein, D. A., Lee, W. R., and Hanlon, A. L. (1995). Pretreatment hemoglobin level influences local control and survival of T_1-T_2 squamous cell carcinomas of the glottic larynx. *J. Clin. Oncol.* **13**, 2077.
30. Van Acht, M. J. J., Hermans, J., Boks, D. E. S., *et al.* (1992). The prognostic value of hemoglobin and a decrease in hemoglobin during radiotherapy in laryngeal carcinoma. *Radiother. Oncol.* **23**, 229.
31. Benninger, M. S., Gillen, J., Thieme, P., *et al.* (1994). Factors associated with recurrence and voice quality following radiation therapy for T_1 and T_2 glottic carcinomas. *Laryngoscope* **104**, 294.
32. Rucci, L., Gallo, O., and Fini-Storchi, O. (1991). Glottic cancer involving the anterior commissure: Surgery vs. radiotherapy. *Head Neck* **13**, 403.
33. Ton-Van, J., Lefebvre, J. L., Stern, J. C., *et al.* (1991). Comparison of surgery and radiotherapy in T_1 and T_2 glottic carcinomas. *Am. J. Surg.* **162**, 337.
34. Barton, M. B., Keane, T. J., and Gadalla, T. (1992). The effect of treatment time and treatment interruption on tumour control following radical radiotherapy of laryngeal cancer. *Radiother. Oncol.* **23**, 137.
35. Parsons, J. T., Bova, F. J., and Million, R. R. (1980). A re-evaluation of split-course technique for squamous cell carcinoma of the head and neck. *Int. J. Radiat. Oncol. Biol. Phys.* **6**, 1645.
36. Robertson, A. G., Robertson, C., Boyle, R. P., *et al.* (1993). The effect of differing radiotherapeutic schedules on the response of glottic carcinomas of the larynx. *Eur. J. Cancer* **29A**, 501.

37. Slevin, N. J., Hendry, J. H., Roberts, S. A., *et al.* (1992). The effect of increasing the treatment time beyond three weeks on the control of T_2 and T_3 laryngeal cancer using radiotherapy. *Radiother. Oncol.* **24**, 215.
38. Ricciardelli, E. J., Weymuller, E. A., Koh, W. J., *et al.* (1994). Effect of radiation fraction size on local control rates for early glottic carcinoma. *Arch. Otolaryngol. Head Neck Surg.* **120**, 737.
39. Small, W., Mittal, B. B., Brand, W. N., *et al.* (1992). Results of radiation therapy in early glottic carcinoma: Multivariate analysis of prognostic and radiation therapy variables. *Radiology* **183**, 789.
40. Terhaard, C. H. J., Snippe, K., Ravasz, L. A., *et al.* (1991). Radiotherapy in T_1 laryngeal cancer: Prognostic factors for locoregional control and survival, uni- and multivariate analysis. *Int. J. Radiat. Oncol. Biol. Phys.* **21**, 1179.
41. Parsons, J. T., Mendenhall, W. M., Stringer, S. P., *et al.* (1993). Twice-a-day radiotherapy for squamous cell carcinoma of the head and neck: The University of Florida experience. *Head Neck* **15**, 87.
42. DeVita, V. T. (1993). Principles of chemotherapy. *In* "Cancer: Principles and Practice of Oncology" (V. T. DeVita, S. Hellman, and S. A. Rosenberg, eds.), 4th Ed., p. 276. Lippincott, Philadelphia.
43. Al-Kourainy, K., Kish, J., Ensley, J., *et al.* (1987). Achievement of superior survival for histologically negative versus histologically positive clinically complete responders to cisplatin combination chemotherapy in patients with locally advanced head and neck cancer. *Cancer* **59**, 233–238.
44. Al-Sarraf, M., Drelichman, A., Peppard, S., *et al.* (1981). Adjuvant cis-platinum and 5-fluorouracil 96 hour infusion in previously untreated epidermoid cancers of the head and neck. *Proc. Am. Soc. Clin. Oncol.* **22**, 428.
45. Weaver, A., Fleming, S., Ensley, J., *et al.* (1984). Superior complete clinical response and survival rates with initial bolus cis-platinum and 120 hour 5-FU infusion before definitive therapy in patients with locally advanced head and neck cancer. *Am. J. Surg.* **148**, 525–530.
46. Jacobs, J. R., Pajak, T. F., Kinzie, J., *et al.* (1987). Induction chemotherapy in advanced head and neck cancer. *Arch. Otolaryngol. Head Neck Surg.* **113**, 193–197.
47. Davis, R. K., Stoker, K., Harker, G., Davis, K., Gibbs, F. A. Jr., Harnsberger, H. R., Stevens, M. H., Parkin, J. L., and Johnson, L. P. (1989). Prognostic indicators in head and neck cancer patients receiving combined therapy. *Arch. Otolaryngol. Head Neck Surg.* **115**(12), 1443–1446.
48. MacKenzie, R. G., Franssen, E., Balogh, J. M., Gilbert, R. W., Birt, D., and Davidson, J. (2000). Comparing treatment outcomes of radiotherapy and surgery in locally advanced carcinoma of the larynx: A comparison limited to patients eligible for surgery. *Int. J. Radiat. Oncol. Biol. Phys.* **47**(1), 65–71.
49. Harwood, A.R., Beal, F. A., Cummings, B. J., *et al.* (1980). T_3 glottic cancer: An analysis of dose-time-volume factors. *Int. J. Radiat. Oncol. Biol. Phys.* **6**, 675.
50. Mendenhall, W. M., Parsons, J. T., Stringer, S. P., *et al.* (1992). Stage T_3 squamous cell carcinoma of the glottic larynx: A comparison of laryngectomy and irradiation. *Int. J. Radiat. Oncol. Biol. Phys.* **23**, 725.
51. DeSanto, L. W. (1984). T_3 glottic cancer: Options and consequences of the options. *Laryngoscope* **94**, 1311.
52. Thaker, A., Bahadun, S., Mohanti, B. K., and Nivsarker, S. (2000). Clinically staged $T_3N_0M_0$ laryngeal cancer: How is it best treated? *J. Laryngol. Otol.* **114**(2), 108–112.
53. Pearson, B. W., Woods, R. D., and Hartman, D. E. (1980). Extended hemilaryngectomy for T_3 glottic carcinoma with preservation of speech and swallowing. *Laryngoscope* **90**, 1950.
54. Laccourreye, H., Laccourreye, O., Weinstein, G., *et al.* (1990). Supracricoid laryngectomy with cricohyoidoepiglottopexy: A partial laryngeal procedure for glottic carcinoma. *Ann. Otol. Rhinol. Laryngol.* **99**, 421.
55. Bron, L., Brossard, Monnier, P., and Pasche, P. (2000). Supracricoid partial laryngectomy with cricohyoidoepiglottopexy and crico-hyoidopexy for glottic and supraglottic carcinomas. *Laryngoscope* **110**(4), 627–634.
56. Laccourreye, O., Laccourreye, L., Garcia, D., Gutierrez-Fonseca, R., and Brasnu, D. (2000). Vertical partial laryngectomy versus supracricoid partial laryngectomy selected carcinomas of the true vocal cord classified as T_2N_0. *Ann. Otol. Rhinol. Laryngol.* **109**(10, Pt. 1), 965–971.
57. Steiner, W., and Ambrosch, P. (2000). Laser microsurgery for laryngeal carcinoma. *In* "Endoscopic Laser Surgery of the Upper Aerodigestive Tract" (W. Steiner ed.), pp. 47–82. Thieme Stuttgart, New York.
58. Brandenberg, J. H., Condon, K. G., and Frank, T. W. (1986). Coronal sections of larynges from radiation-therapy failures: A clinical- pathologic study. *Otolaryngol. Head Neck Surg.* **95**(2), 213–218.
59. Department of Veterans Affairs Laryngeal Cancer Study Group. (1991). Induction chemotherapy plus radiation compared with surgery plus radiation in patients with advanced laryngeal cancer. *N. Engl. J. Med.* **324**, 1685–1690.
60. Bradford, C. R., Wolf, G. T., Carey, T. E., Zhu, S., Beals, T. F., Truelson, J. M., McClatchey, K. D., and Fisher, S. G. (1999). Predictive markers for responsse to chemotherapy, organ preservation, and survival in patients with advanced laryngeal carcinoma. *Otolaryngol. Head Neck Surg.* **121**(5), 534–538.
61. Pearson, B. W., DeSanto, L. W., Olsen, K. D., and Salassa, J. R. (1988). Results of near-total laryngectomy. *Ann. Otol. Rhinol. Laryngol.* **107**(10, Pt. 1), 820–825.
62. Schuller, D. E, and Bier-Laning, C. M. (1997). Laryngeal carcinoma nodal metastases and their management. *Otolaryngol. Clin. North Am.* **30**(2), 167–177.
63. Ferlito, A., Silver, C. E., Rinaldo, A., and Smith, R. V. (2000). Surgical treatment of the neck in cancer of the larynx. *ORL J. Otorhinolaryngol. Relat. Spec.* **62**(4), 217–225.
64. Davis, R. K., Kelly, S. M., and Hayes, J. (1991). Supraglottic resection with the CO_2 laser. *Laryngoscope* **101**, 680–683.
65. DeSanto, L. W., Magrina, C., and O'Fallon, Wm. (1990). The 'second' side of the neck in supraglottic cancer. *Otol.-HN. Surgery* **102**(4), 351–361.
66. Nakfoor, B. M., Spiro. I. J., Wang, C. C., Martins, P., Montgomery, W., and Fabian, R. (1988). Results of accelerated radiotherapy for supra glottic carcinoma: A Massachusetts General Hospital and Massachusetts Eye and Ear Infirmary experience. *Head Neck* **20**(5),379–384.
67. Mendenhall, W. M, Parsons, J. T., Mancuso, A. A., *et al.* (1996). Radiotherapy for squamous cell carcinoma of the supraglottic larynx: An alternative to surgery. *Head Neck* **18**, 24–35.
68. Freeman, D. E., Mancuso, A. A., Parsons, J. T., *et al.* (1990). Irradiation alone for supraglottic larynx carcinoma: Can CT findings predict treatment results? *Int. J. Radiat. Oncol. Biol. Phys.* **19**, 485–490.
69. Zeitels, S. M., and Vaughan, C. W. (1991). Preepiglottic space invasion in "early" epiglottic cancer. *Ann. Otol. Rhinol. Laryngol.* **100**, 789–792.
70. Ogura, J. H., Marks, J. E., and Freeman, R. B. (1980). Results of conservation surgery for cancer of the supraglottic and pyriform sinus. *Laryngoscope* **90**, 591.
71. Ogura, J. H., Sessions, D. G., and Ciralsky, R. H. (1975). Supraglottic carcinoma with extension to the arytenoid. *Laryngoscope* **85**, 1327.
72. Ogura, J. H., Sessions, D. G., and Spector, G. J. (1975). Conservation surgery for epidermoid carcinoma of the supraglottic larynx. *Laryngoscope* **85**, 1808.
73. Bocca, E., Pignataro, O., and Oldini, C. (1983). Supraglotttic laryngectomy; 30 years of experience. *Ann. Otol. Rhinol. Laryngol.* **92**, 14.
74. Davis, R. K., Kelly, S. M., and Hayes, J. (1991). Endoscopic CO_2 laser excisional biopsy of early supraglottic cancer. *Laryngoscope* **100**, 680–683.
75. Davis, R. K., Shapshay, S. M., Strong, S. M., and Hyams, V. (1983). Transoral partial supraglottic resection using the CO_2 laser. *Laryngoscope* **93**, 429–432.

76. Zeitels, S. M., Koufman, J. A., Davis, R. K., and Vaughan, C. W. (1994). Endoscopic treatment of supraglottic and hypopharynx cancer. *Laryngoscope* **104**, 71–78.
77. Zeitels, S. M., Vaughan, C. W., and Domanowski, G. F. (1990). Endoscopic management of early supraglottic cancer. *Am. Otol. Rhinol. Laryngol.* **99**, 951–956.
78. Ambrosch, P., Kron, M., and Steiner, W. (1998). Carbon dioxide laser microsurgery for early supraglottic carcinoma. *Ann. Otol. Rhinol. Laryngol.* **107**(8), 680–688.
79. Iro, H., Waldfahrer, F., Altendorf-Hoffman, A., Weidenbecher, M., Sauer, R., and Steiner, W. (1998). Transoral laser surgery of supra glottic cancer: follow-up of 141 patients. *Arch. Otolaryngol. Head Neck Surg.* **124**(11), 1245–1250.
80. Rudert, H. H., Werner, J. A., and Hoft, S. (1999). Transoral carbon dioxide laser resection of supraglottic carcinoma. *Ann. Otol. Rhinol. Laryngol.* **108**(9), 819–827.

CHAPTER

26

Partial Laryngeal Procedures

GREGORY S. WEINSTEIN

Department of Otorhinolaryngology—Head and Neck Surgery, The Center for Head and Neck Cancer
The University of Pennsylvania Medical Center, Philadelphia, Pennsylvania 19104

I. Introduction 375
II. Basic Concepts: Beyond Conservation Laryngeal Surgery to Organ Preservation Surgery of the Larynx 376
III. Principle One: Nonsurgical Organ Preservation Strategies versus Organ Preservation Surgery 377
A. T Staging in Laryngeal Carcinoma 380
B. Pretreatment Endoscopy for Laryngeal Carcinoma 380
C. Laryngeal Mobilities in Laryngeal Carcinoma 380
D. Radiology prior to Treatment of Laryngeal Carcinoma 380
E. Confirmation of Findings Following Treatment of Laryngeal Carcinoma 380
IV. Principle Two: The Cricoarytenoid Unit 380
V. Principle Three: The Spectrum Concept of Laryngeal Carcinoma and Laryngeal Cancer Surgery 381
VI. Principle Four: The Resection of Normal Tissue in OPS to Maintain Postoperative Function 382
VII. Principle Five: The Importance of Local Control 383
VIII. Organ Preservation Surgical Techniques 383
A. Preoperative Clinical Assessment 383
IX. Organ Preservation Surgery for Carcinomas Arising at the Glottic Level 384
A. Endoscopic Approaches 384
B. Vertical Partial Laryngectomy 384
C. Supracricoid Partial Laryngectomy with Cricohyoidoepiglottopexy 385
X. Organ Preservation Surgery for Carcinomas Arising at the Supraglottic Level 386
A. Endoscopic Approaches 386
B. Supraglottic Partial Laryngectomy 386
C. Supracricoid Partial Laryngectomy with Cricohyoidopexy 387
D. Role of Organ Preservation Surgery Following Radiation Failure 388
XI. Summary 388
References 388

I. INTRODUCTION

When we think of "organ preservation" in the treatment of laryngeal cancer, many of us think of either radiation alone or chemotherapy and radiation in combination. The reason for this is likely to be that the clinicians who have advocated the development of nonsurgical approaches such as chemotherapy/radiation therapy protocols have often used the words "organ preservation" in response to the common use of total laryngectomy for a high percentage of laryngeal cancers. What role does surgery play for the clinician who limits their organ preservation approach to "nonsurgical" modalities such as radiation or chemotherapy and radiation therapy? In the nonsurgical "organ preservation" view, surgery most often means total laryngectomy and essentially constitutes a "back-up" approach for patients who are either not candidates for aggressive nonsurgical regimens or who have failed locally or regionally after nonsurgical therapy. Although it was recognized that surgical therapies existed to save the voice box, in the nonsurgical "organ preservation" viewpoint, these surgical techniques were considered dubious because either very few surgeons performed them with any regularity or the functional results, when applied for the intermediate and larger lesions, were suspect. Nonetheless, a renaissance is evident in the literature for the partial surgical approaches for laryngeal preservation. So although larynx sparing techniques have been available for over a century to treat laryngeal cancer, the spectrum of surgical organ preservation techniques has been broadened during the past decade. The results of the new organ preservation surgery approaches are not only excellent in terms of oncologic outcomes, but also in terms of speech and swallowing outcomes. The purpose of this chapter is to present the role of organ preservation surgery for laryngeal cancer.

A situation has arisen where a dichotomy has developed among many clinicians and even medical centers, in which essentially one of two approaches for laryngeal organ preservation dominates, either surgical or nonsurgical. This pattern developed during an era when the conservation or partial surgical approach was dominated by the vertical hemilaryngectomy and the supraglottic laryngectomy. The standard form of these two procedures has only limited indications for the management of mostly early laryngeal carcinoma. Surgeons that limit themselves to utilizing only vertical hemilaryngectomy and the supraglottic laryngectomy as their "conservation" approaches can only treat a very select group of patients and therefore must offer the remaining patients total laryngectomy or perhaps near total laryngectomy, both of which require a permanent stoma. The result has been that some surgical residents who are being trained in centers where nonsurgical organ preservation approaches dominate may not understand the surgical organ preservation techniques. The reality is, however, that both surgical and nonsurgical approaches for laryngeal preservation are important and valuable.

In a chapter of this size it is not possible to provide a comprehensive analysis of all aspects of organ preservation surgery. The goal of this chapter is to provide an introduction that allows clinicians to assess which organ preservation surgical options may be of benefit for their patients and counsel their patients concerning those options. It is critical for any clinician counseling a patient with laryngeal cancer to discuss both nonsurgical and surgical approaches for organ preservation. In the "internet" era, patients now have access to information about all approaches, surgical and nonsurgical, so it behooves the clinician to be prepared to answer educated questions. In most cases, however, the depth of understanding that is required for a surgeon to actually perform these procedures can be found in the literature that is referenced throughout.

II. BASIC CONCEPTS: BEYOND CONSERVATION LARYNGEAL SURGERY TO ORGAN PRESERVATION SURGERY OF THE LARYNX

Conservation laryngeal surgery is a type of surgery in which the skin is incised and the larynx is opened to remove the cancer. In the late 19th century, the vertical partial laryngectomy was developed and remained the dominant conservation laryngeal technique for most of the 20th century [1]. Alonso [2] made a significant contribution in 1947 by introducing the supraglottic partial laryngectomy for the treatment of selected supraglottic carcinomas. Both vertical partial laryngectomies and supraglottic partial laryngectomies are excellent in terms of oncologic control for early laryngeal carcinomas were utilized for very selected, mostly early carcinomas [3,4]. The problem with the conservation surgery paradigm, which was dependent on the vertical partial laryngectomy and the supraglottic partial laryngectomy, was that there was still a large number of patients that required either near-total or total laryngectomy, both of which require a permanent tracheostoma [5]. The strategy that was developed to deal with this dilemma, by some surgeons, was to extend the indications of vertical partial laryngectomy and supraglottic partial laryngectomy. These extended procedures posed problems because the original procedures were designed for small lesions. The pundits who published the results following extended vertical partial laryngectomies and extended supraglottic partial laryngectomies tended to report their techniques in small numbers, with variability in terms of both functional outcome and local control [6–9]. In addition, it was not realistic to think that many of these techniques could ever be adopted by surgeons, other than by the original authors and their teams, as they were reported with a plethora of difficult to reproduce laryngeal reconstructions frequently with very little in the way of technical detail. So in trying to salvage the conservation surgery approach by extending the standard techniques, the pundits actually pushed the field into the hands of a few experts and out of the hands of the front-line clinicians and their daily practices. Why perform a vertical partial laryngectomy for T1 cancer when the voice outcome was so much worse than radiation therapy and the cure rates are the same? Furthermore, why perform an extended vertical hemilaryngectomy for a T2 glottic carcinoma when the control rates were clearly variable and the functional outcome of the difficult to reproduce reconstruction was unsure [10]? Similar statements can be made for higher staged glottic lesions, and parallel arguments can be made for the role of supraglottic laryngectomy. So for the otolaryngologist–head and neck surgeon in the community, who most commonly is the first to diagnosis these carcinomas, the conservation surgery approaches played little role in their own practices, the result of which was turning to an approach of radiation for small and intermediate lesions followed by total laryngectomy for failure and total laryngectomy for very advanced cancers as primary treatment [11].

Nonsurgical organ preservation protocols utilizing novel combinations of radiation and chemotherapy were developed in response to the limitations of the conservation surgical approach to avoid a permanent tracheostoma in higher stage tumors. In 1985 the VA Cooperative Studies Program prospectively randomized patients between induction chemotherapy and radiation versus the standard therapy of total laryngectomy with postoperative radiation therapy [12]. While survival was not compromised in the induction chemotherapy and radiation arm, over one-third of these patients still underwent total laryngectomy [12]. Shortly after publication of the VA Cooperative Studies program, Intergroup 91-11 was instituted comparing radiation alone, induction chemotherapy followed by radiation therapy (VA protocol), and concomitant chemotherapy and radiation therapy in an effort to define the optimal nonsurgical organ preservation approach. The results of Intergroup 91-11 indicated that the superior arm was concomitant chemotherapy and radiation therapy, yielding an actuarial 85% laryngeal preservation rate at 5 years. In the United States, at the inception of the VA trial and Intergroup 91-11, the only standard surgical approaches for laryngeal preservation were supraglottic partial laryngectomy and vertical hemilaryngectomy. At the time it was reasonable to search for nonsurgical alternatives to total laryngectomy because these two surgical approaches had limited application for patients with early lesions, leaving the surgical option for the majority of patients with intermediate and advanced carcinomas being total laryngectomy. In European chemotherapy radiation trials, patients who are candidates for organ preservation surgery were explicitly excluded from nonsurgical organ preservation trials; this was not the case in either the VA trial or Intergroup 91-11 [12–15]. Why did such a different approach evolve in Europe?

By the late 1950s in Europe, a new strategy was developed to avoid total laryngectomy in intermediate and advanced laryngeal cancer [16]. This involved the introduction of new surgical techniques known as supracricoid partial laryngectomies [17,18]. There are actually two version of the supracricoid partial laryngectomy. The supracricoid partial laryngectomy used for selected glottic carcinomas results in the resection of both true cords, both false cords, the bilateral paraglottic spaces, and the entire thyroid cartilage. Reconstruction is performed by suturing the cricoid to the epiglottis and hyoid, and hence the name cricohyoidoepiglottopexy (CHEP). When supracricoid partial laryngectomy is used for selected supraglottic carcinomas, all of the tissue that is resected in the supracricoid partial laryngectomy with CHEP but also resects the whole preepiglottic space and epiglottis. Reconstruction in this case is done by suturing the cricoid to the hyoid and hence the name cricohyoidopexy (CHP). Supracricoid partial laryngectomies have been reported from centers in numerous European countries to have reproducibly excellent local control and functional outcome for selected cases of laryngeal carcinoma [17–20]. Supracricoid partial laryngectomies have been noted to be an excellent addition to the surgical armamentarium for the management of laryngeal carcinoma; in addition, the quality of life following supracricoid partial laryngectomy has been shown to be superior to total laryngectomy with tracheoesophageal puncture [21,22].

During the two decades since the inception of the VA trial and the decade since the inception of Intergroup 91-11, supracricoid partial laryngectomies, which have been utilized for over 40 years in Europe [16], have become increasingly popular surgical therapy in the United States. Supracricoid partial laryngectomies allow for a surgical larynx preservation rate of approximately 95% when utilized for selected cases of intermediate and advanced laryngeal carcinomas [23,24]. At present, in the United States, the choice of organ preservation surgery versus nonsurgical organ preservation tends to vary based on the experience of the clinicians involved and the "culture" of the institution where the patient seeks treatment. The only oncologic criteria that preclude performing organ preservation surgery, when supracricoid partial laryngectomies are included among the options, are (1) arytenoid cartilage fixation, (2) greater than 1 cm of cancer extension into the subglottis at the midcord level, and (3) interarytenoid involvement. Therefore, while advancements have been made over the past two decades in nonsurgical approaches to avoid total laryngectomy, synchronous developments in the area of organ preservation surgery have decreased the percentage of patients requiring total laryngectomy when surgery is the primary modality.

The second major development in surgical approaches for organ preservation over the past two decades was the introduction of endoscopic resection of laryngeal carcinoma. Endoscopic approaches, utilizing both laser and standard techniques, have been shown to have high local control rates for early lesions. In addition, contrary to popular belief among patients and clinicians, the voice outcome following endoscopic resection of selected glottic and supraglottic lesions has been shown to be equivalent to radiation therapy [25].

The addition of supracricoid partial laryngectomies on one end of the surgical spectrum and endoscopic approaches on the other has fueled a renaissance in the surgical management of laryngeal carcinoma. Integration of endoscopic techniques, conservation surgical techniques, and supracricoid partial laryngectomies can be done by uniting them under a new paradigm known as "organ preservation surgery" for laryngeal cancer. By necessity, to understand the organ preservation surgery approach, we must move beyond the older conservation surgery paradigm. Adopting the organ preservation surgery approach requires not only a change in perspective, but also an understanding of five critical principles.

III. PRINCIPLE ONE: NONSURGICAL ORGAN PRESERVATION STRATEGIES VERSUS ORGAN PRESERVATION SURGERY

During the past few years publications and protocols focusing on chemotherapy radiation schemes to avoid surgery at the primary site have utilized the term organ preservation as a description of their purpose. The implication is that there is nonsurgical organ preservation and then there is surgery. In some institutions, surgery most often means total laryngectomy or some other partial laryngeal procedure that "we do not do here" or "might work but at the expense of terrible swallowing problems." This type of perspective may have been reasonable at a time when the conservation surgery paradigm depended only on vertical partial laryngectomy and supraglottic laryngectomy and their difficult to reproduce extended variations. However, with the addition of endoscopic approaches and supracricoid partial laryngectomies, the myopic perspective that only nonsurgical approaches are the "true" organ preservation approaches is both untrue and unacceptable.

So the first concept of the organ preservation surgery paradigm is that there are nonsurgical as well as surgical approaches for organ preservation and that both have a place in the management of laryngeal cancer. The first question that may come to mind is "If we are resecting a large portion of abnormal and even normal larynx, how can this possibly be considered organ preservation?" Doesn't organ preservation mean that we should end up with a "normal larynx" at the end of treatment? In the first place, a larynx that has cancer in it is no longer a "normal larynx." Even after successful nonsurgical organ preservation therapy, such as radiation and chemoradiation, the presence of a cancer itself, particularly large cancers, can result in the destruction of normal tissue and permanent scarring, affecting all aspects of laryngeal function [26]. In the second place, all treatments of larynx cancer, both surgical and nonsurgical, have both acute and chronic permanent side effects, both in the larynx and in the surrounding tissues.

A common misconception among both surgeons and patients is that radiation therapy is the preferred modality of treatment for early laryngeal carcinoma because the voice quality is superior to surgery. This fallacy most likely arose during a time when the only surgical options were procedures such as the vertical partial laryngectomy for glottic carcinoma rather than on the contemporary approach of partial endoscopic cordectomy. The reality is that with the advent of glottic and supraglottic endoscopic approaches, and open approaches such as the supraglottic laryngectomy, enough of the vocal cords can be saved so that the postoperative voice can either be normal after surgical resections or equivalent to radiation therapy. The classic example of this is when a patient has a very superficial, small or medium sized, midcord, T1 glottic carcinoma, when the reasons given to a patient for choosing radiation is that the cure rate is the same as for surgery and the voice after radiation is superior. The present literature does not support this concept; at present, simply put, this concept is inaccurate. When radiation is utilized, even for early glottic carcinoma, there is an impact on the tissues of the larynx, resulting in abnormal vocal quality [27]. Additionally, data now clearly show that the key issue for voice quality after endoscopic resections is the amount of tissue that is removed, with less of course being better. Studies demonstrate that voice after partial endoscopic cordectomy for early glottic cancer is equivalent to voice after radiation therapy for similar lesions [25,28]. Voice after chemoradiation for advanced laryngeal cancer significantly deviates from normal [29].

Of course, it would be ludicrous to say that all larynx sparing surgical techniques leave patients with voice quality that is equivalent to radiation or chemotherapy radiation therapy. For intermediate or advanced laryngeal cancer, the surgical organ preservation techniques may result in worse voice quality than the nonsurgical approaches (although this point has not been studied as yet). Nonetheless, the major determinant of quality of life is the stoma and not post-treatment voice quality [30]. Given this finding, the most important factor in recommending either a surgical or a nonsurgical technique is not voice quality, but the overall local control, which of course has a major impact on the ultimate number of patients who will need a permanent tracheostomas after salvage laryngectomy. This point is particularly important when we compare oncologic and function results following radiation therapy alone versus supracricoid partial laryngectomy. Although it has not been studied, it is our experience that voice quality is superior following radiation therapy for a T2 glottic carcinoma when compared to voice results following supracricoid partial laryngectomy. The key issue is that the local control for supracricoid partial laryngectomy is approximately 95% [18], whereas the local control following radiation therapy for the same lesion is approximately 30% [31]. The much higher local failure rate for radiation results in almost a third of patients ultimately requiring total laryngectomy and therefore a permanent stoma; it is the permanent stoma that patients dislike. Therefore, on the basis of vocal quality, the nonsurgical organ preservation approach, in this case radiation, is a better choice than supracricoid partial laryngectomy for T2 glottic carcinoma. Conversely, if the primary goal is avoiding total laryngectomy for a larger proportion of the patients, and hence lower the incidence of permanent tracheostoma for the group, then supracricoid partial laryngectomy is superior to radiation therapy. Almost all patients who fail radiation therapy for a T2 glottic carcinoma require total laryngectomy for salvage [32]. Because the presence of a permanent stoma has a greater impact on quality of life than hoarseness [30], it is reasonable to offer the patient supracricoid laryngectomy as the surgical alternative to radiation therapy for T2 glottic carcinoma.

In light of the 85% organ preservation with concurrent chemotherapy and radiation therapy, as reported in the Intergroup 91-11 study [33], new questions arise concerning the role of organ preservation surgery. If this type of excellent result can be achieved, why should a surgeon learn these complex techniques or offer these techniques as an option? First, retrospective surgical literature indicates that if the appropriate organ preservation surgery technique is utilized, than the expected local control is on the order of 90–95%, which is superior to the 85% achieved by concurrent chemoradiation therapy [33]. Some would argue that a weakness of the organ preservation approach is that since there is always some possibility of intraoperative conversion to total laryngectomy, all patients must be consented for a total laryngectomy at the time of organ preservation surgery. This should be a rare occurrence if patient selection has been done appropriately. In any case, there is a percentage of patients that immediately following concurrent chemotherapy and radiation therapy will require total laryngectomy for persistent disease, which is the nonsurgical equivalent to intraoperative conversion to total laryngectomy. One interesting finding in Intergroup 91-11 is that although the 5-year actuarial organ preservation rate for chemotherapy and radiation (85%) was superior to either induction chemotherapy followed radiation (71%) or radiation alone (64%). Nonetheless, the overall survival was not statistically different in any of the three groups (54, 53, and 59%, respectively) [33]. Of note, in a study reported by Laccourreye *et al.* for 60 patients that underwent supracricoid partial laryngectomy with CHP with neoadjuvant chemotherapy for advanced endolaryngeal carcinoma classified as T3–T4, the overall laryngeal preservation rate was 91.7% and the 5-year actuarial survival estimate was 72.7%. The 5-year actuarial survival was almost 20% higher in the organ preservation surgery group compared to the concurrent chemotherapy radiation therapy group (72.7% versus 54%) [34]. One might say that this is an unfair comparison, as Laccourreye's group utilized induction chemotherapy; however, numerous other studies have yielded similar results without the use of induction chemotherapy [19,35,36]. It is possible that differences in nodal status, underlying medical illness, and other factors may account for the differences in survival between Intergroup 91-11 data and organ preservation surgery data. However, it is also possible that if patients are dying of other causes in the Intergroup 91-11 study, that if they had lived longer as a group that the local failure rate would have been higher. At present, literature indicates that both the local control and the survival are superior following organ preservation surgery regimens that include supracricoid partial laryngectomy when compared to concurrent chemotherapy radiation therapy.

These differences in local control and survival between surgical and nonsurgical organ preservation are particularly important based on the inclusion and exclusion criteria of Intergroup 91-11. Inclusion criteria for Intergroup 91-11 were (1) glottic or supraglottic carcinoma, (2) T2–T4, and (3) T4 tumors had to meet the following criteria: (a) questionable cartilage invasion by CT scan only (clinical T3) without penetration beyond the larynx and (b) supraglottic primary with invasion of the base of tongue up to 1 cm clinically and by imaging studies. The only oncologic criteria that preclude performing organ preservation surgery when supracricoid partial laryngectomies are included among the options are (1) fixation of the arytenoid cartilage, (2) greater than 1 cm of cancer extension into the subglottis at the midcord level, or (3) interarytenoid involvement. Therefore, a large number of patients that were included in this nonsurgical organ preservation trial would have been candidates for an organ preservation surgery technique. In addition, Intergroup 91-11 only included patients in whom the nutritional, pulmonary, and cardiac status must be considered adequate to tolerate the proposed chemotherapy, radiation therapy, and surgical treatment. In our experience, patients that cannot tolerate the postoperative course of a supraglottic or supracricoid partial laryngectomy for medical reasons would also not be a candidate for the rigorous concurrent chemotherapy radiation therapy regimen of Intergroup 91-11. Given the improvements in both local control and survival following organ preservation surgery, a reasonable approach is to offer all patients with stage III and IV glottic and supraglottic carcinoma, who are organ preservation surgery candidates, the organ preservation surgery approach, with concurrent chemotherapy radiation therapy offered to those patients with cancers that are not suitable for organ preservation surgery. Total laryngectomy is then reserved for those patients who are not candidates for organ preservation surgery and who do not fit inclusion criteria for Intergroup 91-11.

While it would be ideal for all surgeons to perform all of the available organ preservation surgical techniques, this is unlikely to ever be the case because (1) there are many surgeons who do not see a high enough volume of laryngeal cancer to gain expertise in these techniques, (2) some surgeons do not have the desire to perform these techniques themselves, and (3) just as in some institutions it is clear that surgical approaches dominate, there are institutions where nonsurgical approaches are the tradition. Nonetheless, given that the literature now shows that there is safety and efficacy of both surgical and nonsurgical organ preservation approaches, insular institutional traditions are not acceptable alternatives to appropriate informed consent based on that literature. The fact that a clinician does not choose to perform a given surgical technique him/herself is not an excuse for a lack of full understanding of the indications and posttreatment impact of the technique in terms of both function and cancer control. The surgeon counseling patients with laryngeal cancer must understand both nonsurgical and surgical organ preservation approaches, even if they never actually perform these procedures themselves. Either a surgeon or a nonsurgeon should, from an ethical standpoint, also be prepared to refer a patient to a head and neck surgeon who has expertise with organ preservation surgical techniques if, after appropriate counseling, the patient prefers to undergo or requires additional information concerning the surgical approach.

A. T Staging in Laryngeal Carcinoma

The organ preservation surgery approach and nonsurgical organ preservation approaches differ in their pretreatment workups. The role of T staging in nonsurgical approaches such as radiation and chemoradiation is to predict prognosis based on a large series of patients reported in the literature. The T staging system, however, plays no role in assessing which organ preservation surgery to perform. This is because the T staging system is an artificial division of a large spectrum of cancers into four stages based on a gross extension of tumor and vocal cord mobility [37]. The choice of which organ preservation surgery is the best option in a particular case requires a very precise assessment of the two- and three-dimensional extent of the cancer; a much more precise assessment than is possible with four T stages. While T staging is not useful for choosing a particular organ preservation procedure, once the procedure is chosen, the T staging system is useful for comparing the prognosis to nonsurgical organ preservation approaches.

B. Pretreatment Endoscopy for Laryngeal Carcinoma

The nonsurgeon assesses the gross findings in the larynx, such as whether the tumor is exophytic or endophytic, the sites of involvement of the cancer, i.e., true cord and epiglottis, and the vocal cord mobility when they perform indirect laryngoscopy. In other words, indirect laryngoscopy is performed to T stage the cancer. Because the organ preservation surgeon must precisely map the surface extent of the carcinoma with greater accuracy than the T staging system, they should use pretreatment endoscopy indirectly in the office and under general anesthesia directly in the operating room to accomplish this goal [38].

C. Laryngeal Mobilities in Laryngeal Carcinoma

While the nonsurgeon assesses vocal cord mobility to T stage the patient, the organ preservation surgeon assesses laryngeal mobilities for other reasons as well. An excellent approach used to study the clinical pathologic nature of laryngeal cancer is to section the whole pathologic specimen and then correlate malignant extension seen on the sections with clinical findings. These whole organ section studies have correlated changes in laryngeal mobilities with the depth and extent of the cancer. Kirchner and Som [39] have shown that fixation of the vocal cord correlates with deep invasion into the thyroarytenoid muscle. Brasnu *et al.* [40] have shown that arytenoid fixation is due either to cricoarytenoid muscle invasion or to cricoarytenoid joint invasion. The organ preservation surgeon uses both vocal cord and arytenoid mobility to assess depth of invasion of the cancer, which in turn enables them to make the appropriate choice of surgical technique.

D. Radiology prior to Treatment of Laryngeal Carcinoma

As in the remainder of the pretreatment workup, the nonsurgeon is more focused on the gross characteristics of the tumor, such as tumor volume, and radiographs are utilized to asses these findings. The greater the tumor volume of the tumor, the worse the prognosis following radiation therapy [41]. The tumor volume itself is not a relevant issue for the organ preservation surgeon. There are areas of cancer invasion that are hard to evaluate clinically, such as early cartilaginous invasion, moderate invasion of the preepiglottic space or subglottic extension, and extension of the carcinoma out of the larynx; assessment of these areas is most valuable aspect of radiographs to the organ preservation surgeon [42].

E. Confirmation of Findings Following Treatment of Laryngeal Carcinoma

A key difference in nonsurgical versus surgical organ preservation approaches is in the confirmation of the clinical examination findings. The surgeon confirms their pretreatment clinical assessment at the time of surgery, very shortly after the workup. The nonsurgeon, however, never removes the tumor and therefore at no time actually confirms the pretreatment assessment. This difference has a major impact on the treatment. Critical reliance on the pretreatment clinical examination means that if the assessment is inadequate, the wrong procedure may be recommended, the outcome of which may be total laryngectomy. Therefore, only surgeons who have attained expertise in the organ preservation surgery workup (not only the techniques themselves) can apply these techniques successfully.

IV. PRINCIPLE TWO: THE CRICOARYTENOID UNIT

In the old conservation surgery paradigm the vocal cord was the central focus, largely ignoring the role of the mobility of the arytenoid cartilage. In the organ preservation surgery paradigm the focus shifts posteriorly to the cricoarytenoid unit. The cricoarytenoid unit includes one arytenoid cartilage, the cricoid, the ipsilateral superior and recurrent laryngeal nerves, and the ipsilateral cricoarytenoid musculature. The cricoarytenoid units are in fact the engine of the larynx. They are responsible for opening and closing the larynx in speech, respiration, and swallowing. So speech, swallowing, and respiration without a permanent tracheostomy are possible if the organ preservation surgery approach spares at least one functional cricoarytenoid unit. So sparing at least one vocal cord (or most of one), which is so central to the conservation surgery approach, has changed to sparing at least one cricoarytenoid unit in the organ preservation surgery approach. Sparing a vocal cord does not preserve voice, rather it preserves vocal quality.

V. PRINCIPLE THREE: THE SPECTRUM CONCEPT OF LARYNGEAL CARCINOMA AND LARYNGEAL CANCER SURGERY

We have already discussed the inadequacies of the T staging system in the workup of a patient for organ preservation surgery. This is evident to any head and neck surgery resident who reads the literature and discovers that numerous surgical techniques have been recommended for a given T stage, but when they get to the specifics of the indications, there is tremendous overlap in the literature. The literature in some cases seems almost contradictory, with one surgical expert strongly recommending one procedure for a particular lesion and another recommending an entirely different procedure. With such a large number of organ preservation techniques and such a large diversity in the superficial and deep extent of laryngeal carcinoma at the time of presentation, it can be quite challenging to plan treatment for these patients. In the face of the challenges of organ preservation surgery workup, some surgeons might rather simply leave these options out of the discussion and only recommend nonsurgical organ preservation surgery to their patients. The reality is that the literature supports the feasibility of the organ preservation surgical techniques so to ignore them is both ethically unsound and medicolegally unwise. Even if a surgeon would rather not perform these procedures, they are obliged to know when the techniques are best utilized. The organ preservation surgery spectrum may be utilized to decrease some of the complexity of the pretreatment workup, which can be particularly challenging for those residents in training. There is a spectrum of surgical procedures that may be utilized to preserve the larynx, and a corresponding spectrum of lesions, from small to large (Figs. 26.1 and 26.2).

The upper portion of the spectrum shows numerous schematic larynges, with the lesions being the smallest to

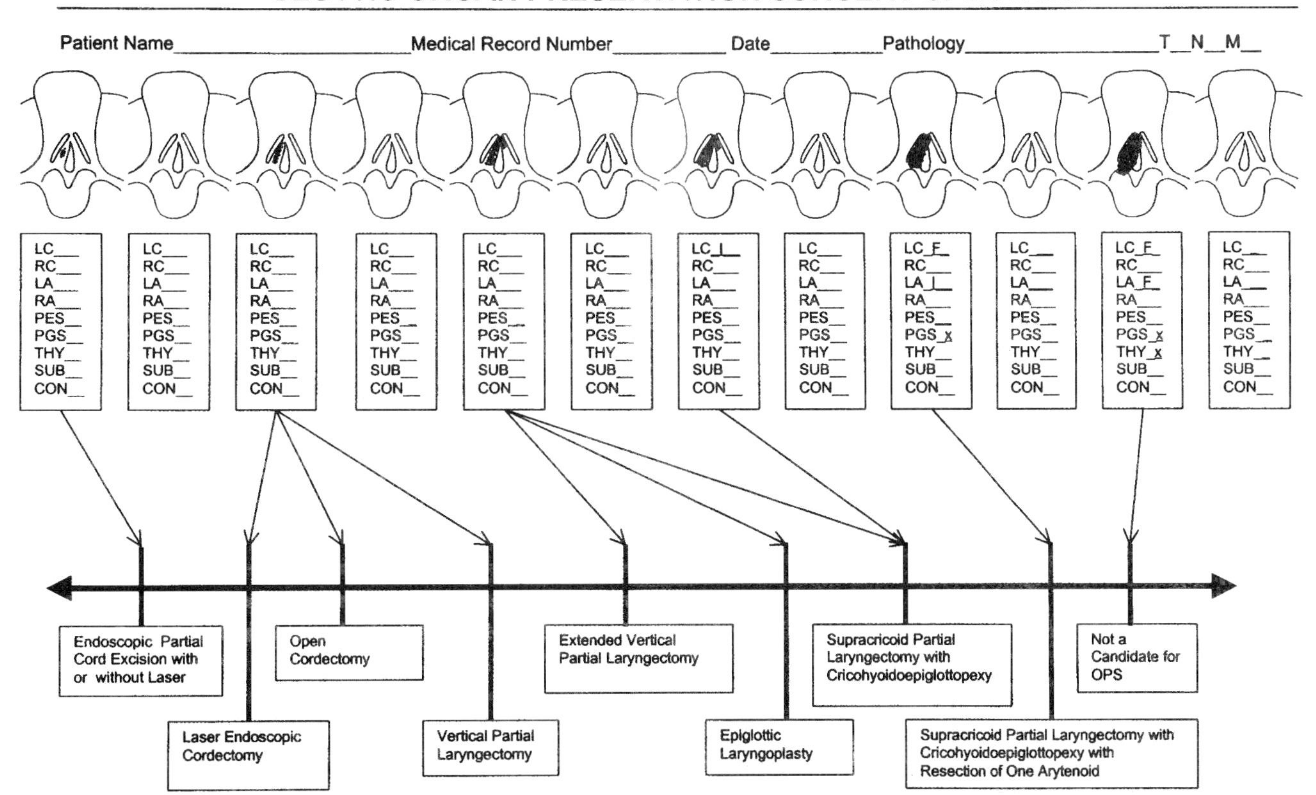

LC: left vocal cord mobility; RC: right vocal cord mobility; LA: left arytenoid mobility; RA: right arytenoid mobility; PES: preepiglottic space involvement; PGS: paraglottic space invasion; THY; early radiologic evidence of thyroid cartilage invasion; SUB: subglottic extension > 1m at midcord, or > 0.5 cm posterior cord level; CON: tumor and non-tumor contraindications. At the top of the form is a series of schematic diagrams of the larynx with additional descriptors in the text box below. At the bottom of the form is the spectrum of organ preservation surgeries that are applicable. Representative cancers are shown in a spectrum from least extensive on the left through most extensive on the right. The mucosal extent is documented on the schematic. Mobility is documented with either a "blank" for normal mobility, "I" for impaired mobility, and "F" for fixation. An "x" documents involvement of the preepiglottic space, paraglottic space, thyroid cartilage, and a tumor- or non-tumor-related contraindication. This form may be used in surgical planning by drawing the lesion into a blank laryngeal schematic, in the appropriate place on the spectrum.

FIGURE 26.1 The organ preservation surgery spectrum for glottic carcinoma published with permission from "Organ Preservation Surgery for Laryngeal Carcinoma," G. S. Weinstein, O. Laccourreye, D. Brasnu, and H. Laccourrye, Singular Publishing, San Diego, 1999.

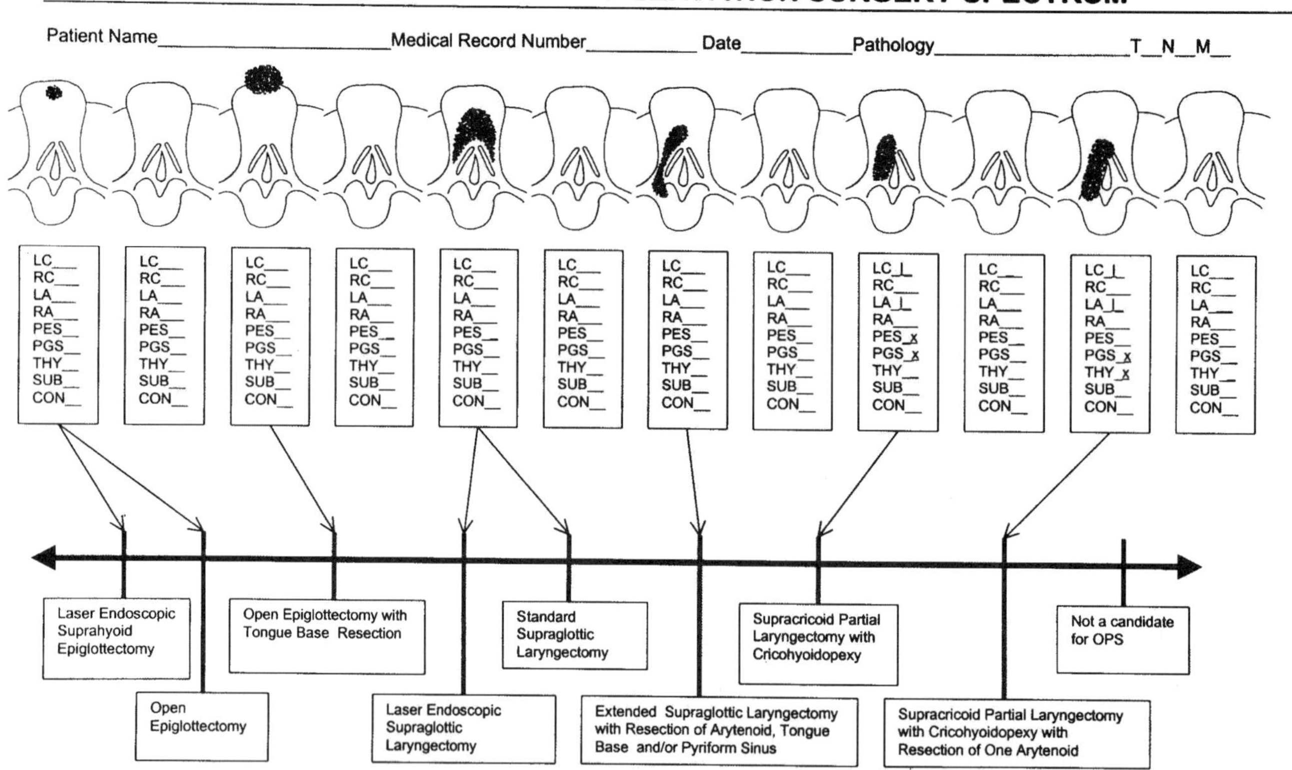

LC: left vocal cord mobility; RC: right vocal cord mobility; LA: left arytenoid mobility; RA: right arytenoid mobility; PES: preepiglottic space involvement; PGS: paraglottic space invasion; THY; early radiologic evidence of thyroid cartilage invasion; SUB: subglottic extension > 1m at midcord, or > 0.5 cm posterior cord level; CON: tumor and non-tumor contraindications. At the top of the form is a series of schematic diagrams of the larynx with additional descriptors in the text box below. At the bottom of the form is the spectrum of organ preservation surgeries that are applicable. Representative cancers are shown in a spectrum from least extensive on the left through most extensive on the right. The mucosal extent is documented on the schematic. Mobility is documented with either a "blank" for normal mobility, "I" for impaired mobility, and "F" for fixation. An "x" documents involvement of the preepiglottic space, paraglottic space, thyroid cartilage, and a tumor- or non-tumor-related contraindication. This form may be used in surgical planning by drawing the lesion into a blank laryngeal schematic, in the appropriate place on the spectrum.

FIGURE 26.2 The organ preservation surgery spectrum for supraglottic carcinoma. Published with permission from "Organ Preservation Surgery for Laryngeal Carcinoma," G. S. Weinstein, O. Laccourreye, D. Brasnu, and H. Laccourrye, Singular Publishing, San Diego, 1999.

the largest, from left to right with a blank schematic in between. The box below the schematic is a box used to document the mobility of the vocal cords, the arytenoids, and other details, including cartilage invasion or subglottic extension. On the bottom portion of Fig. 26.1 is the spectrum of organ preservation surgical techniques, from least to most extensive. There is a different spectra for glottic and supraglottic lesions (Figs. 26.1 and 26.2). The techniques included in the spectra have been well documented in the literature of having not only consistent oncologic control, but also a consistent functional outcome in terms of speech and swallowing without a tracheostomy. The clinician can use these spectra by drawing the lesion that they are evaluating either in a blank laryngeal schematic or one of the examples if they match. The surgeon then matches their lesions with the procedures. Note that in some cases there is more than one surgical option for a given lesion. This is because there are some controversies in the literature. The surgeon can then use the technique that in their experience has the best outcomes.

VI. PRINCIPLE FOUR: RESECTION OF NORMAL TISSUE TO MAINTAIN POSTOPERATIVE FUNCTION

In the conservation laryngeal surgery paradigm, the goal of the surgeon is to resect only the amount of tissue necessary to remove the cancer, preserving as much normal tissue as possible. The reconstruction is then performed based on the remaining tissue. This approach led to a plethora of difficult to reproduce complex reconstructions. Because the surgeon is constantly attempting to create a novel approach for reconstruction, depending on the defect, the functional outcome by necessity is unpredictable in many cases.

The quantum leap beyond the conservation surgery paradigm made by the surgeons who developed the supracricoid partial laryngectomys was accepting that it was reasonable to remove "normal" laryngeal tissues, beyond that which is considered an oncologic margin, so that the postoperative functional outcome would be improved. So when performing a supracricoid partial laryngectomy for a unilateral T3 carcinoma, the surgeon resects both true vocal cords, not just the ipsilateral cord. The opposite cord and the entire thyroid cartilage are resected in this case, not for oncologic reasons but to allow for a predictable functional reconstruction. The principle is that there are certain postoperative anatomical configurations of the larynx, which, by the experience of numerous surgeons, yields consistent functional outcomes. The organ preservation surgeon must be willing to resect uninvolved laryngeal tissue to allow for a reconstruction that will result in consistent functional outcome. The supracricoid partial laryngectomy is just one example of how this concept is applied to organ preservation surgery. Hirano *et al.* [43] have shown that in the supraglottic partial laryngectomy that the surgeon should always remove both false vocal cords, even when it is not indicated from an oncologic perspective, or the patient will have increased dysphagia. Another example is imbrication laryngoplasty following vertical partial laryngectomy [3]. In that procedure, a strip of thyroid cartilage is resected with the glottic level and then reconstruction is performed by imbricating the upper portion of the remaining thyroid cartilage medial to the lower portion (Fig. 26.3). The false cord is then advanced in place of the missing vocal cord. In this case the thyroid cartilage is not removed for oncologic reasons, but is resected to allow the false cord to be medialized by the imbrication of the cartilage and a better cord reconstruction with an improved voice outcome [44].

VII. PRINCIPLE FIVE: THE IMPORTANCE OF LOCAL CONTROL

The final fundamental principle of the organ preservation surgery approach is that local control is critical. Numerous studies have shown that local failure following either surgical or nonsurgical organ preservation approaches has been associated with a decrease in survival [32,45–47]. In addition, as mentioned earlier, the reason local control is important is that in the vast majority of local failures, the patient must undergo total laryngectomy. The importance of avoiding total laryngectomy is that studies have shown that the major determinant of quality of life after the treatment of laryngeal cancer is the need for a permanent tracheostoma [30].

VIII. ORGAN PRESERVATION SURGICAL TECHNIQUES

A. Preoperative Clinical Assessment

Both indirect and direct laryngoscopy are necessary in the pretreatment workup for organ preservation surgery. The indirect laryngoscopy typically done in the office is particularly important to assess both vocal cord and arytenoid mobilities, which cannot be done at the time of direct laryngoscopy when the patient is under general anesthesia. During direct laryngoscopy in the operating room, some assessments can be made more easily than in the office, including palpation of the vallecula to assess the extent of preepiglottic space invasion, precise measurement of the extension into the subglottis, or the extent of involvement of the ventricle, among others.

The role of bronchoscopy and esophagoscopy remains controversial [38]. With the exception of carcinoma *in situ* or T1 glottic carcinoma, the neck and primary site should be evaluated with a magnetic resonance imaging scan or a computerized tomographic scan, basing the choice of scan on the desires of the radiologist. In the absence of other symptoms, the only metastatic workup necessary is a chest X ray.

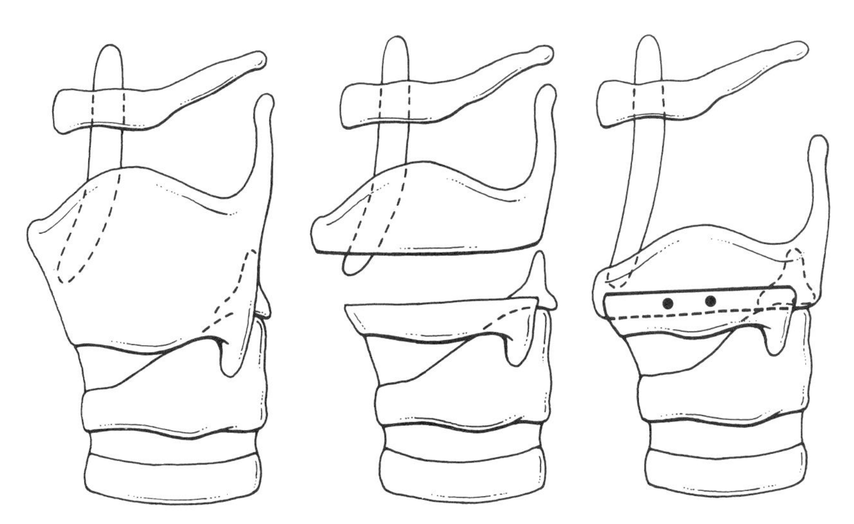

FIGURE 26.3 Imbrication laryngoplasty. Published with permission from "Organ Preservation Surgery for Laryngeal Carcinoma," G. S. Weinstein, O. Laccourreye, D. Brasnu, and H. Laccourrye, Singular Publishing, San Diego, 1999.

IX. ORGAN PRESERVATION SURGERY FOR CARCINOMAS ARISING AT THE GLOTTIC LEVEL

A. Endoscopic Approaches

A common scenario is for the surgeon to encounter a white lesion of the larynx that is discrete on the midlevel of the mobile true vocal cord. It is not uncommon for surgeons to do a small incisional biopsy of this type of lesions, while leaving most of the lesion on the vocal cord. There are a number of weaknesses to this incisional biopsy approach of the discrete laryngeal lesion. First, only a small portion of the lesion is biopsied so the surgeon never knows if this small sample actually represents the whole lesion. Second, the pathologist typically receives a minute specimen, and in many cases can only give a "rule out diagnosis" such as "at least carcinoma *in situ*, possible invasion." This then leads to a therapeutic dilemma: (1) do you get more tissue, (2) remove the whole lesion, or (3) radiate the lesion? When the cord is mobile and the lesion is superficial, the possible pathologic diagnoses include (1) a benign process, (2) some degree of dysplasia, (3) carcinoma *in situ*, (4) microinvasive carcinoma, and (5) invasive malignancy. The reality is that if the surgeon removes the whole lesion, then diagnoses one through four need no further treatment. The last diagnosis, invasive malignancy, may be treated with either radiation or further partial cordectomy depending on the voice outcome expectations. The concern for most surgeons in excising the entire lesion is that there will be a poor voice outcome, and they do not want to risk this for a potentially benign lesion. Zeitels [48,49] has overcome this problem by popularizing a technique in which the mucosa is injected with an epinephrine solution, causing a sort of submucosal bleb or blister [48,49]. This process raises the mucosa from the vocal ligament so that when the entire cord mucosa is excised, the ligament may be spared, providing a scaffold upon which the cord mucosa may regenerate. If after injection, a portion of the mucosa is adherent to the underlying muscle, then this correlates with invasion by the carcinoma. In this case, Zeitels advocated excising the adherent area with a cuff of normal tissue. Because this technique is for relatively small white lesions, only a minimal resection of healthy tissue is needed and therefore the voice results can still be expected to be excellent. We use a 25-gauge butterfly needle with the "wings" cut off and a tuberculin syringe to inject 1% lidocaine with 1:100,000 epinephrine to do the injection. We use a microalligator forceps to control the needle. Zeitels reports excellent oncologic and functional outcomes following this technique. Surgeons interested in utilizing this technique should review the technique described by Zeitels and then do this technique [48–50]. If it works in your hands, it will likely become a standard part of your armamentarium.

In general for discrete resectable carcinoma *in situ* surgical resection is superior to radiation because it takes less time, it is less expensive, the cure rates are the same, and the functional outcome is comparable [48,51–53]. Radiation does have a definite role in carcinoma *in situ* of the larynx. Radiation is useful for patients who have multiple recurrences following repeated surgical resections, contraindications to general anesthesia, or extensive carcinoma *in situ* throughout the larynx precluding adequate surgical excision [54].

Because T1 glottic carcinomas may range in size from minute midcord lesions to extensive lesions involving both true vocal cords, the anterior commissure, and the subglottis bilaterally, clinical judgment is required on the part of the clinician evaluating the patient to assess the role of endoscopic resection. In one study, the local control rates for early glottic carcinoma were equivalent for radiation versus surgery, the voice quality after laser resection was as good as after radiotherapy, and the cost of laser cordectomy was much lower [28]. The degree of hoarseness after endoscopic excision correlates directly with the extent of the endoscopic resection [55]. When the clinician's judgement, based on literature and clinical experience, is that the voice outcome will be comparable with radiation therapy, then the approach of choice is endoscopic excision. This would commonly be the case for discrete white lesions that do not involve the anterior commissure, microinvasive carcinomas, or when a primary invasive carcinoma would permit a partial rather than a total endoscopic cordectomy. The role of endoscopic excision of T2 glottic carcinoma continues to be controversial. While the most aggressive laser endoscopic resections are presently being published from German centers, Eckel and Thumfart [56] have recommended that endoscopic resections not be attempted for T3 glottic carcinomas. The description of these techniques by Steiner and Ambrosch [57] is the most comprehensive at present.

B. Vertical Partial Laryngectomy

The commonality among all the procedures in this category is a vertical transection through the thyroid cartilage to gain access to the endolarynx for the resection of tumors on the glottic level. In these procedures, a temporary tracheostomy is indicated. Vertical partial laryngectomies are really a family of procedures including open laryngofissure and cordectomy, vertical hemilaryngectomy, and epiglottic laryngoplasty. A range of local control of 89–100% can be expected from vertical hemilaryngectomy selected T1 glottic carcinomas [3,10,58–61]. For larger cancers, the role of vertical partial laryngectomy is less clear. For instance, the local control for T2 glottic carcinoma with vertical partial laryngectomy is much more variable, with a local failure rate ranging from 4 to 26% [3,10,59,60,62–65].

It is because of this wide range of local failure rates that this author prefers supracricoid partial laryngectomy, which has been shown in the literature to have a consistently higher local control rate for T2 glottic carcinoma [22]. Although the oncologic outcome following the vertical partial laryngectomy is excellent when there is strict adherence to the indications, the weakness of the procedures is the need for a temporary tracheostomy, as well as some degree of postoperative hoarseness. The latter weakness has decreased the use of these techniques, in many cases in favor of radiation therapy. On the surgical side, the role of vertical partial laryngectomy has been limited by the introduction of endoscopic approaches for very early lesions and by supracricoid partial laryngectomy with cricohyoidoepiglottopexy for selected intermediate and advanced glottic lesions [66].

Although hoarseness can be expected following vertical partial laryngectomy, Brasnu *et al.* [44] have demonstrated that reconstruction with a false cord advancement flap yields improved voice results. Imbrication laryngoplasty goes one step further than Brasnu's work by medializing the false vocal cord advancement flap with a portion of the thyroid cartilage [3,67].

C. Supracricoid Partial Laryngectomy with Cricohyoidoepiglottopexy

When supracricoid partial laryngectomy is utilized for selected glottic carcinoma, resection includes the entire thyroid cartilage, both true and false cords, sparing at least one cricoarytenoid unit. The cricoid is sutured to the hyoid and epiglottis to accomplish the reconstruction and hence the name cricohyoidoepiglottopexy (CHEP). (Fig. 26.4, see also color insert). T2 and T3 glottic carcinomas are the primary lesions that are amenable to this procedure. Laccourreye *et al.* [68] reported a 5-year local control rate for 62 patients of 98.2%(61/62) for T1b and T2 lesions with invasion of the anterior commissure. In another report, Laccourreye *et al.* [69] noted a 5-year local control rate of 95.5% among 67 patients with T2 lesions (31 with impaired motion of the true vocal cord and 36 without impaired motion). The 5-year actuarial local control estimate was 90% in 20 patients with T3 glottic carcinoma with vocal cord fixation [70]. Chevalier *et al.* [71] reported a local control rate of 94.6% among a group of 112 previously untreated patients with impaired motion (T2=90) or fixation (T3=22) of the true vocal cord.

The typical outcome in terms of function is temporary difficulty swallowing, a temporary tracheostomy, and permanent hoarseness. Although the voice is hoarse, quality of life data indicate that the patients perception of the voice following supracricoid partial laryngectomy is superior to total laryngectomy with tracheoesophageal puncture [22]. Voice analysis following supracricoid partial laryngectomy

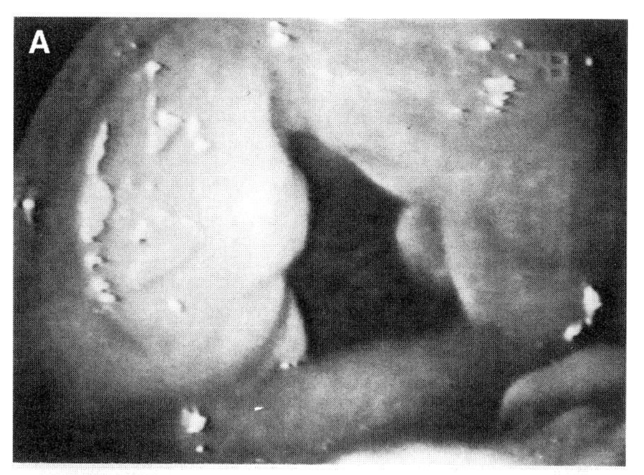

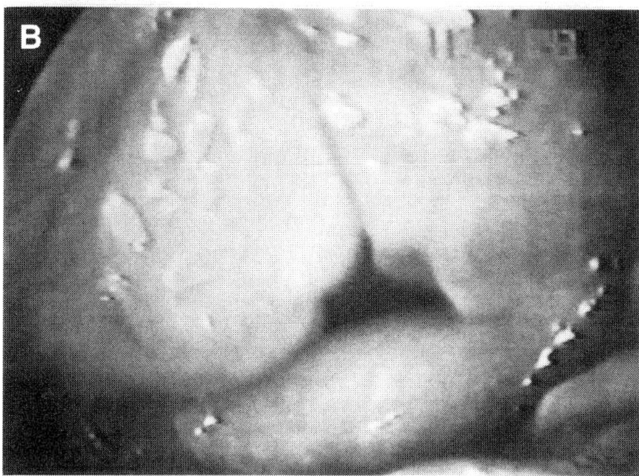

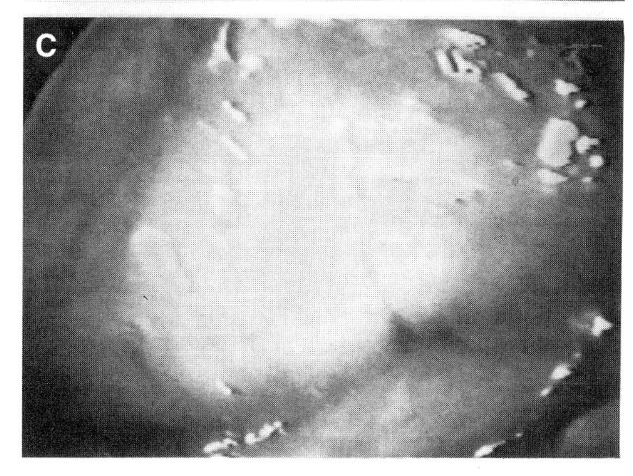

FIGURE 26.4 Postoperative appearance of the larynx after supracricoid partial laryngectomy with cricohyoidoepiglottopexy in (A) respiration (arytenoids open), (B) phonation beginning (arytenoids partially closed), and (C) phonation (arytenoids closed). Published with permission from G. S. Weinstein and Laccourreye O. (1994). Supracricoid laryngectomy with cricohyoidoepiglottopexy. *Otolaryngol. Head Neck Surg.* **111**, 684–685. (See also color insert.)

reveals that the phrase grouping and number of words per minute are similar to normal speakers, whereas the fundamental frequency is lower and wider than normal, suggesting voice instability [72]. Nonetheless, qualitatively the voice is hoarse. The technique is well described in a number of publications [18,73].

X. ORGAN PRESERVATION SURGERY FOR CARCINOMAS ARISING AT THE SUPRAGLOTTIC LEVEL

A. Endoscopic Approaches

The value of endoscopic approaches for supraglottic carcinoma is that the procedures can be done transorally and typically they may be performed without a tracheostomy. Carcinoma *in situ* of the supraglottis should be managed by endoscopic excision unless the lesion is so widespread that this is not possible. Davis and Hayes [74] reported a 4.2% local failure rate among 24 patients treated with transoral laser supraglottic laryngectomy and postoperative radiation therapy for T2 or microscopic T3 tumors. Although these authors used postoperative radiation therapy, the preponderance of data now indicates that this is not indicated, except in the typical postoperative cases such as positive margins, perineural invasion, multiple lymph node metastases, or extracapsular spread from lymph node metastases. The key point in endoscopic excision is not just a "debulking procedure," it is as a true a resection as can be accomplished by any open approach. Eckel and Thumfart [56] reported on 15 patients undergoing laser supraglottic laryngectomy with no local recurrences. Zeitels *et al.* [75] published the results of a series of patients undergoing laser resections without postoperative radiation therapy, including 22 patients with mostly T1 cancers of the supraglottis and hypopharynx, without local failure. Another German report noted that among 48 patients with early supraglottic carcinoma managed by laser supraglottic laryngectomy, there was a 5-year local control rate of 100% for T1 disease and 89% for T2 [76]. The necks should be dissected bilaterally. Therapeutic neck dissections can be done 4–8 days after the primary resection, whereas elective neck dissections can be done 4 to 6 weeks following the initial surgery [57].

When laser supraglottic laryngectomy has been utilized for more advanced supraglottic carcinomas, the results are more variable. Eckel and Thumfart [56] recommended that laser supraglottic carcinoma should not be used for T3 supraglottic carcinomas. Steneir *et al.* [77], nonetheless, advocated the use of laser resection for T3 supraglottic carcinoma. The timing of neck dissection is done as discussed earlier [57]. Rudert, from Keil, Germany, advocates laser for T3 supraglottic lesions in which the tumor invades only the central portion of the preepiglottic space and does not extend laterally and submucosally into the aryeipglottic region (personal communication).

Excellent functional and oncologic findings are dependent on performing the techniques appropriately. The technique as decribed by either Rudert or Steiner is recommended [57,78]. The functional outcome following laser supraglottic laryngectomy has the advantage of not requiring a tracheostomy in most cases and temporary dysphagia in the vast majority of patients. Voice quality is excellent, as the vocal cords are not resected (Fig. 26.5, see also color insert).

B. Supraglottic Partial Laryngectomy

Open supraglottic partial laryngectomy results in excision of the epiglottis, both false cords, and the upper half of the thyroid cartilage. The three types of extended resections include resection of (1) the tongue base, (2) the arytenoid, or (3) the pyriform sinus. A similar trend of decreased utilization has been seen with supraglottic partial

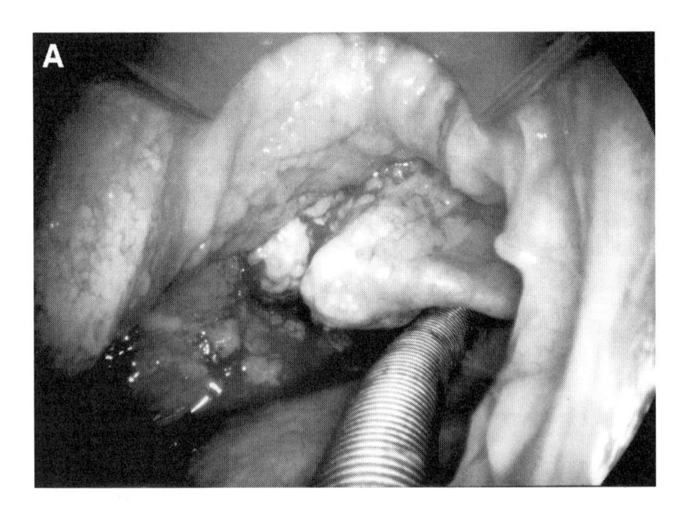

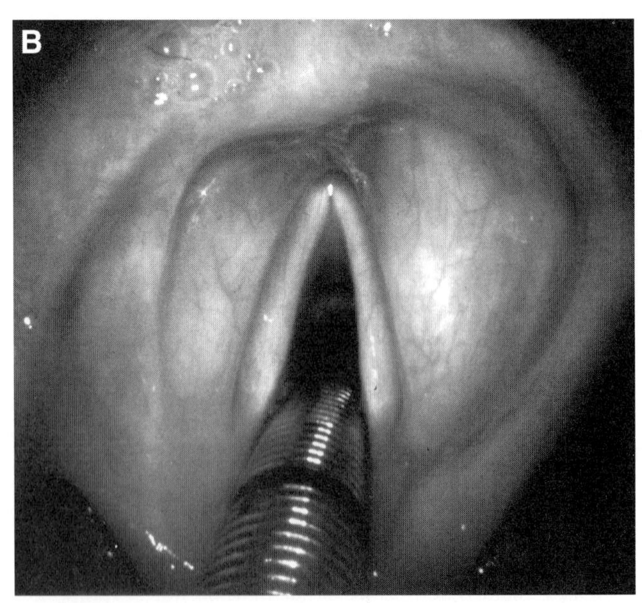

FIGURE 26.5 (A) A carcinoma of the epiglottis and the left aryepiglottic fold in a 89-year-old patient before laser resection (initially palliatively indicated). No tracheostomy was performed. (B) The larynx 1 year later without evidence of disease. Published with permission from "Organ Preservation Surgery for Laryngeal Carcinoma," G. S. Weinstein, O. Laccourreye, D. Brasnu, and H. Laccourrye, Singular Publishing, San Diego, 1999. (See also color insert.)

laryngectomy as we have seen with vertical partial laryngectomy. The reasons for this are (1) the use of nonsurgical procedures such as radiation or chemotherapy radiation for selected supraglottic carcinomas, (2) the introduction of endoscopic supraglottic laryngectomy as a replacement for early lesions, and (3) the increased use of supracricoid partial laryngectomy with CHP for more advanced supraglottic lesions. Local failure following open supraglottic partial laryngectomy for T1 carcinoma ranged from 0 to 10% and ranged from 0 to 15% for T2 lesions [4,79–82]. As in endoscopic supraglottic laryngectomy, the results when open supraglottic laryngectomy is utilized for T3 and T4 supraglottic carcinomas are much more variable then when it is used for T1 or T2 lesions. The range of local failure rates for T3 supraglottic carcinomas is from 0 to 75% and for T4 carcinomas is from 0 to 67% [4,79–82]. Given the variability in oncologic outcomes when supraglottic laryngectomy is utilized for T3 and T4 carcinomas, caution is recommended when considering this option. If the clinical examination reveals an extension of carcinoma to the glottic level, ventricle, or impaired or fixed mobility of the vocal cord, supraglottic partial laryngectomy is contraindicated and supracricoid partial laryngectomy with CHP becomes the organ preservation surgery option.

While functional outcome has been reported to be consistently good after standard supraglottic laryngectomy [43], the incidence of impaired swallowing has been noted to be higher following extended supraglottic laryngectomy [83]. Again, as in other procedures, attention to technical details will improve the chance for successful oncologic and functional outcome [84].

C. Supracricoid Partial Laryngectomy with Cricohyoidopexy

When supracricoid partial laryngectomy is utilized for supraglottic carcinoma, resection includes the entire thyroid cartilage, both true and false cords, and the entire epiglottis and preepiglottic space, sparing at least one cricoarytenoid unit. The cricoid is sutured to the hyoid to accomplish the reconstruction and hence the name cricohyoidopexy (Fig. 26.6, see also color insert). Indications for supracricoid partial laryngectomy with CHP include supraglottic carcinomas with glottic level invasion either via the ventricle or anterior commissure, impaired vocal cord mobility, preepiglottic space invasion, or limited thyroid cartilage invasion. Laccourreye *et al.* [17] reported on 68 patients who underwent supracricoid partial laryngectomy with CHP (T1 = 1, T2 = 40, T3 = 26, T4 = 1) and found no local recurrences. In another study, Laccourreye *et al.* [85] reported on 19 patients who underwent supracricoid partial laryngectomy with cricohyoidopexy for supraglottic carcinomas involving the preepiglottic space and found that the local failure rate was 5.6% (1/19). Chevalier and Piquet [35] reported on a 3.3% local failure rate among 61 patients with supraglottic carcinoma managed by supracricoid partial laryngectomy with cricohyoidopexy. An Italian series reported a local failure rate of 6% among 98 supraglottic and transglottic carcinomas following supracricoid partial laryngectomy with CHP [20].

This procedure is the most extensive organ preservation surgical technique, particularly when one arytenoid is

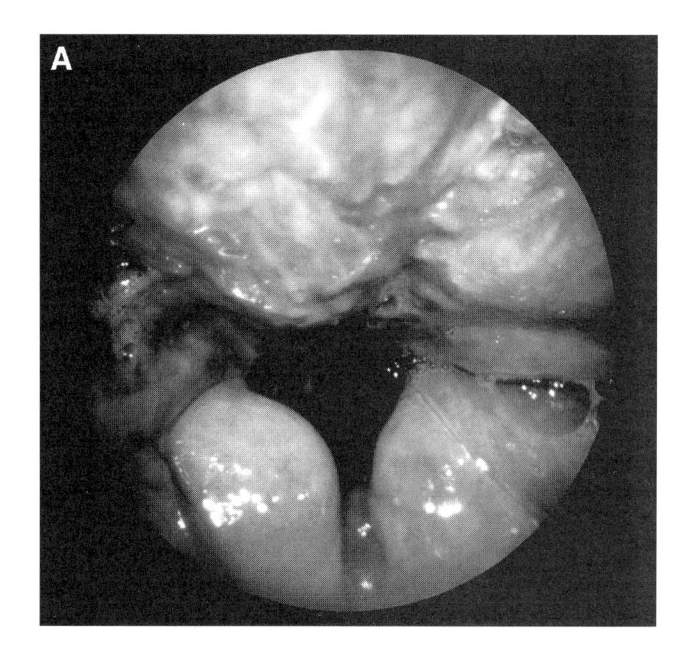

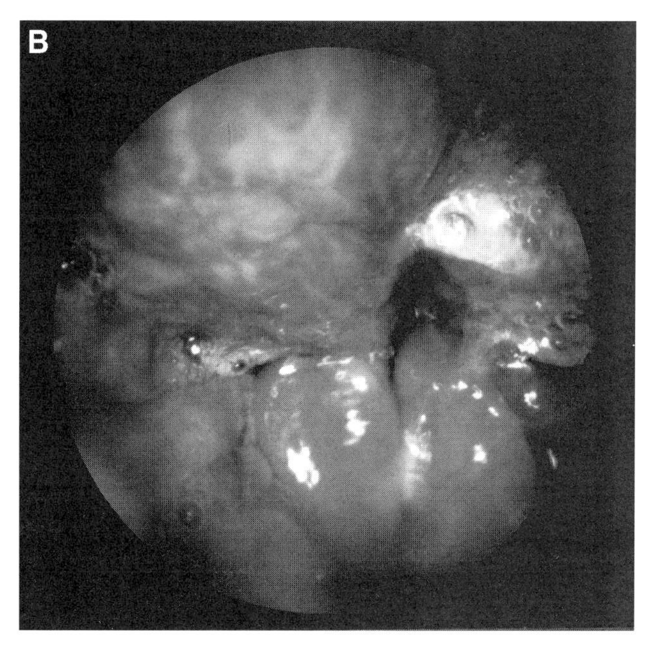

FIGURE 26.6 Postoperative appearance of the neolarynx after supracricoid partial laryngectomy with cricohyoidopexy sparing both arytenoids and (A) arytenoids open and (B) arytenoids closed. Published with permission from "Organ Preservation Surgery for Laryngeal Carcinoma," G. S. Weinstein, O. Laccourreye, D. Brasnu, and H. Laccourrye, Singular Publishing, San Diego, 1999. (See also color insert.)

resected, and it is critical to follow the technique precisely [17,86]. Overall, the reported functional outcome in terms of speech and swallowing and tracheal decannulation has been consistently excellent. One study noted significantly less efficient jitter, shimmer, and maximum phonation time, which is consistent with the long-term qualitative finding of hoarseness [87].

D. Role of Organ Preservation Surgery Following Radiation Failure

A common misconception is that a patient may undergo nonsurgical organ preservation such as radiation or chemotherapy radiation and then if they have local failure they surely will be able to undergo some organ preservation surgical technique. The simple fact is that patients who fail nonsurgical therapy are highly unlikely to remain candidates for surgical laryngeal preservation [88,89]. This even applies to T1 lesions [32]. The issue may be late recognition of recurrence because of postradiation changes, including edema, erythema, or changes in arytenoid or vocal cord mobility secondary to treatment or scar tissue [90]. It is important for the clinician who is recommending a nonsurgical organ preservation approach that he/she makes it clear that if the patient fails therapy then it is extremely likely that they will need a total laryngectomy for salvage. However, there are some patients who fail either radiation or chemotherapy and radiation and are still candidates for organ preservation surgery. Biller and colleagues [91] noted a local control rate of 80% following postradiation vertical hemilaryngectomy. Sorenson *et al.* [92] cautioned that both functional and oncologic outcome was poor when supraglottic partial laryngectomy was utilized in the postradiation setting. Laccourreye and co-workers [93] have shown that supracricoid partial laryngectomy with either CHEP or CHP is a viable alternative for recurrent lesions that are too large or otherwise not amenable to either vertical partial laryngectomy or suparglottic laryngectomy.

XI. SUMMARY

There has been a revolution in the laryngeal organ preservation arena. The catalyst for the change has been (1) the refinement and successful utilization of transoral endoscopic resections for earlier cancers (2) and the introduction of supracricoid partial laryngectomies for intermediate to advanced cancers. The result of this revolution has been to change the role of surgery both in perspective and in management for laryngeal cancer. The impact has been to broaden our perspective so that the limited view that organ preservation means only nonsurgical approaches have been abandoned for the broader view that there are both nonsurgical and surgical organ preservation approaches, which each having a useful role in the management of laryngeal carcinomas.

Utilizing the organ preservation surgery approach requires a change in perspective. The adoption of a number of principles allows the surgeon to apply these techniques with success: (1) accepting that there are both surgical and nonsurgical organ preservation approaches, both of which have a role in the management of laryngeal carcinoma, (2) the concept that in the organ preservation surgery paradigm the focus is no longer on the vocal cord but on the cricoarytenoid unit, (3) understanding that it is not possible to use the T staging system in the pretreatment evaluation of candidates for organ preservation surgery and that a more useful approach is to understand that there is a spectrum of cancers from smallest to largest and a spectrum of procedures, also from smallest to largest, and the goal of the surgeon is to match the appropriate procedure with the appropriate lesion, (4) although the goal in the conservation surgery approach was to spare as much normal tissue as possible, in the organ preservation surgery approach, it is acceptable to resect normal tissue if the outcome is an anatomical configuration that consistently yields good functional results, and (5) the goal of the organ preservation surgeon is to achieve a high local control rate to decrease the need for salvage laryngectomy and the concomitant stoma that would be required. Data also show that local failure impacts negatively on survival.

This chapter provides an introduction to the concepts of organ preservation surgery of the larynx. The literature is replete with detailed descriptions of the technical aspects of these techniques. In the internet age, the patient, their friends, and family frequently have access to the latest information available. It is no longer possible for the clinician to ignore either surgical or nonsurgical approaches when counseling patients. Even if the patient does not seem to be aware of all of the options available to them prior to treatment, it is becoming more and more likely that at the time of a recurrence they will come across what their actual options might have been prior to treatment. Clinicians who are counseling patients concerning the alternatives for laryngeal organ preservation should understand surgical as well as nonsurgical options presently available and be prepared to provide an honest appraisal of all approaches.

References

1. Thompson, S. (1912). Intrinsic cancer of the larynx: Operation by laryngo-fissure: Lasting cure in 80% of cases. *Br. Med. J.* 355–359.
2. Alonso, J. M. (1947). Conservative surgery of the larynx. *Trans. Am. Acad. Opthalmol. Otolaryngol.* **51**, 633–642.
3. Liu, C., Ward, P. H., and Pleet, L. (1986). Imbrication reconstruction following partial laryngectomy. *Ann. Otol. Rhinol. Laryngol.* **95**(6 Pt 1), 567–571.
4. Burstein, F. D., and Calcaterra, T. C. (1985). Supraglottic laryngectomy: Series report and analysis of results. *Laryngoscope* **95**(7 Pt 1), 833–836.

5. Pearson, B. W., Woods, R. D. D., and Hartman, D. E. (1980). Extended hemilaryngectomy for T3 glottic carcinoma with preservation of speech and swallowing. *Laryngoscope* **90**(12), 1950–1961.
6. Biller, H. F., and Som, M. L. (1977). Vertical partial laryngectomy for glottic carcinoma with posterior subglottic extension. *Ann. Otol. Rhinol. Laryngol.* **86**(6 Pt 1), 715–718.
7. Calcaterra, T. C. (1987). Bilateral omohyoid muscle flap reconstruction for anterior commissure cancer. *Laryngoscope* **97**(7 Pt 1), 810–813.
8. Friedman, W. H., Katsantonis, G. P., Siddoway, J. R., and Cooper, M. H. (1981). Contralateral laryngoplasty after supraglottic laryngectomy with vertical extension. *Arch. Otolaryngol.* **107**(12), 742–745.
9. Nagahara, K., Hirose, A., and Iwai, H. (1976). Laryngeal reconstruction by free flap transfer. *Plast. Reconstr. Surg.* **57**(5), 604–610.
10. Laccourreye, O., Weinstein, G., Brasnu, D., *et al.* (1991). Vertical partial laryngectomy: A critical analysis of local recurrence. *Ann. Otol. Rhinol. Laryngol.* **100**(1), 68–71.
11. Singer, M. I. (1995). A clinical trial of continuous cisplatin-fluorouracil induction chemotherapy and supracricoid partial laryngectomy for glottic carcinoma classified as T2. *Cancer* **76**(1), 149–151.
12. The Department of Veterans Affairs Laryngeal Cancer Study Group (1991). Induction chemotherapy plus radiation compared with surgery plus radiation in patients with advanced laryngeal cancer. *N. Engl. J. Med.* **324**(24), 1685–1690.
13. Richard, J. M., Sancho-Garnier, H., Pessey, J. J., *et al.* (1998). Randomized trial of induction chemotherapy in larynx carcinoma. *Oral. Oncol.* **34**(3), 224–228.
14. Lefebvre, J. L., Chevalier, D., Luboinski, B., *et al.* (1996). Larynx preservation in pyriform sinus cancer: Preliminary results of a European Organization for Research and Treatment of Cancer phase III trial. *J. Natl. Cancer Inst.* **88**(13), 890–899.
15. Lefebvre, J. L. (1998). Larynx preservation: The discussion is not closed. *Otolaryngol. Head Neck Surg.* **118**(3 Pt 1), 389–393.
16. Majer, H., and Reifer, W. (1959). Technique de laryngecomie permetant de conserver la permeabilite' respiratoire la cricohyoido-pexie. *Ann. Otolaryngol. Chir. Cervicofac.* **76**, 677–683.
17. Laccourreye, H., Laccourreye, O., Weinstein, G., *et al.* (1990). Supracricoid laryngectomy with cricohyoidopexy: A partial laryngeal procedure for selected supraglottic and transglottic carcinomas. *Laryngoscope* **100**(7), 735–741.
18. Laccourreye, H., Laccourreye, O., Weinstein, G., *et al.* (1990). Supracricoid laryngectomy with cricohyoidoepiglottopexy: A partial laryngeal procedure for glottic carcinoma. *Ann. Otol. Rhinol. Laryngol.* **99**(6 Pt 1), 421–426.
19. Bron, L., Brossard, E., Monnier, P., and Pasche, P. (2000). Supracricoid partial laryngectomy with cricohyoidoepiglottopexy and cricohyoidopexy for glottic and supraglottic carcinomas. *Laryngoscope* **110**, 627–634.
20. de Vincentiis, M., Minni, A., Gallo, A., and DiNardo, A. (1998). Supracricoid partial laryngectomies: Oncologic and functional results. *Head Neck* **20**, 504–509.
21. Coman, W. B., Grigg, R. G., Tomkinson, A., and Gallagher, R. M. (1998). Supracricoid laryngectomy: A significant advance in the management of laryngeal cancer. *Aust. N. Z. J. Surg.* **68**(9), 630–634.
22. Weinstein, G. S., Ruiz, C. R., Dooley, P., *et al.* (2001). Larynx preservation with supracricoid partial laryngectomy results in improved quality of life when compared to total laryngectomy. *Laryngoscope* **111**(2), 119–129.
23. Weinstein, G. S., El-Sawy, M. M., Ruiz, C., *et al.* (2001). Laryngeal preservation with supracricoid partial laryngectomy results in improved quality of life when compared with total laryngectomy. *Laryngoscope* **111**, 191–199.
24. Weinstein, G. S., Laccourreye, O., and Rassekh, C. (1998). Conservation laryngeal surgery. *In* "Otolaryngology—Head and Neck Surgery," (C. Cummings, J. M. Frederickson, L. A. Harker, *et al.*, eds.), Vol. 3, pp. 2220–2228. Saunders, Philadelphia.
25. Delsupehe, K. G., Zink, I., Lejaegere, M., and Bastian, R. W. (1999). Voice quality after narrow-margin laser cordectomy compared with laryngeal irradiation. *Otolaryngol. Head Neck Surg.* **121**, 528–533.
26. Lazarus, C. L., Logemann, J. A., Pauloski, B. R., *et al.* (1996). Swallowing disorders in head and neck cancer patients treated with radiotherapy and adjuvant chemotherapy. *Laryngoscope* **1996**(106), 1157–1166.
27. Honocodeevar-Boltezar, I., and Zargi, M. (2000). Voice quality after radiation therapy for early glottic cancer. *Arch. Otolaryngol. Head Neck Surg.* **126**, 1097–1100.
28. Cragle, S. P., and Brandenburg, J. H. (1993). Laser cordectomy or radiotherapy, cure rates, communication and cost. *Otolaryngol. Head Neck Surg.* **108**, 648–654.
29. Woodson, G. E., Rosen, C. A., Murry, T., *et al.* (1996). Assessing vocal function after chemoradiation for advanced laryngeal carcinoma. *Arch. Otolaryngol. Head Neck Surg.* **122**, 858–864.
30. DeSanto, L. W., Olsen, K. D., and Perry, W. C. (1995). Quality of life after surgical treatment of cancer of the larynx. *Ann. Otol. Rhinol. Laryngol.* **104**, 763–769.
31. Howell-Burke, D., Peters, L. J., Goepfert, H., and Oswald, M. J. (1990). T2 glottic cancer: Recurrence, salvage, and survival after definitive radiotherapy. *Arch. Otolaryngol. Head Neck Surg.* **116**(7), 830–835.
32. Viani, L., Stell, P. M., and Dalby, J. E. (1991). Recurrence after radiotherapy for glottic carcinoma. *Cancer* **67**(3), 577–584.
33. Forastiere, A. A., Berkey, B., Maor, M., *et al.* (2001). Phase III trial to preserve the larynx: Induction chemotherapy and radiotherapy versus concommitant chemoradiotherapy versus radiotherapy alone, Intergroup Trial RTOG91-11. *In* "37th Annual Meeting of the American Society of Clinical Oncology," San Fransisco.
34. Laccourreye, O., Brasnu, D., Biacabe, B., *et al.* (1998). Neo-adjuvant chemotherapy and supracricoid partial laryngectomy with cricohyoidopexy for advanced endolaryngeal carcinoma classified as T3- T4: 5-year oncologic results. *Head Neck* **20**(7), 595–599.
35. Chevalier, D., and Piquet, J. J. (1994). Subtotal laryngectomy with cricohyoidopexy for supraglottic carcinoma: Review of 61 cases. *Am. J. Surg.* **168**(5), 472–473.
36. Lima, R. A., Freitas, E. Q., Kligerman, J., *et al.* (2001). Supracricoid laryngectomy with CHEP: Functional results and outcome. *Otolaryngol. Head Neck Surg.* **124**, 258–260.
37. Johns, M. E., Farrior, E., Boyd, J. C., and Cantrell, R. W. (1982). Staging of supraglottic cancer. *Arch. Otolaryngol.* **108**(11), 700–702.
38. Hartig, G. H., Truelson, J. T., and Weinstein, G. S. (2000). Supraglottic cancer. *Head Neck* **22**, 426–434.
39. Kirchner, J. A., and Som, M. L. (1971). Clinical significance of fixed vocal cord. *Laryngoscope* **81**, 1029–1044.
40. Brasnu, D., Laccourreye, H., Dulmet, E., and Jaubert, F. (1990). Mobility of the vocal cord and arytenoid in squamous cell carcinoma of the larynx and hypopharynx: An anatomical and clinical comparative study. *Ear Nose Throat J.* **69**(5), 324–330.
41. Pameijer, F. A., Mancuso, A. A., Mendenhall, W. M., *et al.* (1997). Can pretreatment computed tomography predict local control in T3 squamous cell carcinoma of the glottic larynx treated with definitive radiotherapy? *Int. J. Radiat. Oncol. Biol. Phys.* **37**(5), 1011–1021.
42. Weinstein, G. S., Laccourreye, O., Brasnu, D., and Yousem, D. (1996). The role of CT and MR in planning conservation laryngeal surgery. The Neuroimaging Clinics of North America.
43. Hirano, M., Kurita, S., Tateishi, M., and Matsuoka, H. (1987). Deglutition following supraglottic horizontal laryngectomy. *Ann. Otol. Rhinol. Laryngol.* **96**(1), 7–11.
44. Brasnu, D., Laccourreye, O., Weinstein, G., *et al.* (1992). False vocal cord reconstruction of the glottis following vertical partial laryngectomy: A preliminary analysis. *Laryngoscope* **102**(6), 717–719.
45. Parsons, J. T., Mendenhall, W. M., Stringer, S. P., *et al.* (1995). Salvage surgery following radiation failure in squamous cell carcinoma of the supraglottic larynx. *Int. J. Radiat. Oncol. Biol. Phys.* **32**(3), 605–609.

46. Laccourreye, O., Guiterrez-Fonseca, R., Barcia, D., *et al.* (1999). Local recurrence after vertical partial laryngectomy, a conservative modality of treatment for patients with stage I-II squamous cell carcinoma of the glottis. *Cancer* **85**, 2549–2556.
47. Stell, P. M., and Dalby, J. E. (1985). The treatment of early (T1) glottic and supra-glottic carcinoma: Does partial laryngectomy have a place? *Eur. J. Surg. Oncol.* **11**(3), 263–266.
48. Zeitels, S. M. (1996). Phonomicrosurgical treatment of early glottic cancer and carcinoma in situ. *Am. J. Surg.* **172**(6), 704–709.
49. Zeitels, S. M. (1995). Premalignant epithelium and microinvasive cancer of the vocal fold: The evolution of phonomicrosurgical management. *Laryngoscope* **105**(3 Pt 2), 1–51.
50. Zeitels, S. M. (1999). Endoscopic management of vocal cord atypia and carcinoma in situ. *In* "Organ Preservation Surgery for Laryngeal Cancer," (G. S. Weinstein, O. Laccourreye, D. Brasnu, and H. Laccourreye, eds.), pp. 73–94. Singular Publishing Group, San Diego.
51. Pene, F., and Fletcher, G. H. (1976). Results in irradiation of the in situ carcinomas of the vocal cords. *Cancer* **37**(6), 2586–2590.
52. MacLeod, P. M., and Daniel, F. (1990). The role of radiotherapy in n-situ carcinoma of the larynx. *Int. J. Radiat. Oncol. Biol. Phys.* **18**(1), 113–117.
53. Miller, A. H., and Fisher, H. R. (1971). Clues to the life history of carcinoma in situ of the larynx. *Laryngoscope* **81**(9), 1475–1480.
54. Fein, D. A., Mendenhall, W. M., Parsons, J. T., *et al.* (1993). Carcinoma in situ of the glottic larynx: The role of radiotherapy. *Int. J. Radiat. Oncol. Biol. Phys.* **27**(2), 379–384.
55. McGuirt, W. F., Blalock, D., Koufman, J. A., *et al.* (1994). Comparative voice results after laser resection or irradiation of T1 vocal cord carcinoma. *Arch. Otolaryngol. Head Neck Surg.* **120**(9), 951–955.
56. Eckel, H. E., and Thumfart, W. F. (1992). Laser surgery for the treatment of larynx carcinomas: indications, techniques, and preliminary results. *Ann. Otol. Rhinol. Laryngol.* **101**(2 Pt 1), 113–118.
57. Steiner, W., and Ambrosch, P. (2000). Laser microsurgery for laryngeal carcinoma. *In* "Endoscopic Laser Surgery of the Upper Aerodigestive Tract with Special Emphasis on Cancer Surgery," (W. Steiner, and P. Ambrosch, eds.), pp. 47–83. Thieme, New York.
58. Thomas, J. V., Olsen, K. D., Neel, H. B., III, *et al.* (1994). Early glottic carcinoma treated with open laryngeal procedures. *Arch. Otolaryngol. Head Neck Surg.* **120**(3), 264–268.
59. Bailey, B. J. (1971). Conservation surgery in carcinoma of the laryngeal anterior commissure. *South. Med. J.* **64**(3), 305–310.
60. Mohr, R. M., Quenelle, D. J., and Shumrick, D. A. (1983). Verticofrontolateral laryngectomy (hemilaryngectomy). Indications, technique, and results. *Arch. Otolaryngol.* **109**(6), 384–395.
61. Rothfield, R. E., Johnson, J. T., Myers, E. N., and Wagner, R. L. (1989). The role of hemilaryngectomy in the management of T1 vocal cord cancer. *Arch. Otolaryngol. Head Neck Surg.* **115**(6), 677–680.
62. Kirchner, J. A., and Som, M. L. (1975). The anterior commissure technique of partial laryngectomy: Clinical and laboratory observations. *Laryngoscope* **85**, 1308–1317.
63. Som, M. L. (1975). Cordal cancer with extension to the vocal process. *Laryngoscope* **85**, 1298–1307.
64. Biller, H. F., Ogura, J. H., and Pratt, L. L. (1971). Hemilaryngectomy for T2 glottic cancers. *Arch. Otolaryngol.* **93**, 238–243.
65. Johnson, J. T., Myers, E. N., Hao, S. P., and Wagner, R. L. (1993). Outcome of open surgical therapy for glottic carcinoma. *Ann. Otol. Rhinol. Laryngol.* **102**(10), 752–755.
66. Ferlito, A., Silver, C. E., Howard, D. J., *et al.* (2000). The role of partial laryngeal resection in current management of laryngeal cancer: A collective review. *Acta. Otolaryngol.* **120**, 456–465.
67. Pleet, L., Ward, P. H., DeJager, H. J., and Berci, G. (1977). Partial laryngectomy with imbrication reconstruction. *Trans. Am. Acad. Ophthalmol. Otolaryngol.* **84**(5),ORL882–ORL889.
68. Laccourreye, O., Muscatello, L., Laccourreye, L., *et al.* (1997). Supracricoid partial laryngectomy with cricohyoidoepiglottopexy for "early" glottic carcinoma classified as T1-T2N0 invading the anterior commissure. *Am. J. Otolaryngol.* **18**(6), 385–390.
69. Laccourreye, O., Weinstein, G., Brasnu, D., *et al.* (1994). A clinical trial of continuous cisplatin-fluorouracil induction chemotherapy and supracricoid partial laryngectomy for glottic carcinoma classified as T2. *Cancer* **74**(10), 2781–2790.
70. Laccourreye, O., Salzer, S. J., Brasnu, D., *et al.* (1996). Glottic carcinoma with a fixed true vocal cord: Outcomes after neoadjuvant chemotherapy and supracricoid partial laryngectomy with cricohyoidoepiglottopexy. *Otolaryngol. Head Neck Surg.* **114**(3), 400–406.
71. Chevalier, D., Laccourreye, O., Brasnu, D., *et al.* (1997). Cricohyoidoepiglottopexy for glottic carcinoma with fixation or impaired motion of the true vocal cord: 5-year oncologic results with 112 patients. *Ann. Otol. Rhinol. Laryngol.* **106**(5), 364–369.
72. Crevier-Buchman, L., Laccourreye, O., Weinstein, G., *et al.* (1995). Evolution of speech and voice following supracricoid partial laryngectomy. *J. Laryngol. Otol.* **109**(5), 410–413.
73. Laccourreye, O., Laccourreye, H., El-Sawy, M., and Weinstein, G. S. (1999). Supracricoid partial laryngectomy with cricohyoidoepiglottopexy. *In* "Organ Preservation Surgery for Laryngeal Cancer," (G. S. Weinstein, O. Laccourreye, D. Brasnu, and H. Laccourreye, eds.), pp. 73–94. Singular Publishing Group, San Diego.
74. Davis, R. K., and Hayes, J. K. (1995). Management of supraglottic cancer: Selected endoscopic laser resection and postoperative irradiation. *Adv. Otorhinolaryngol.* **49**, 231–236.
75. Zeitels, S. M., Koufman, J. A., and Vaughn, C. W. (1994). Endoscopic treatment of supraglottic and hypopharynx cancer. *Laryngoscope* **104**, 71–78.
76. Ambrosch, P., Kron, M., and Steiner, W. (1998). Carbon dioxide laser microsurgery for early supraglottic carcinoma. *Ann. Otol. Rhinol. Laryngol.* **107**(8), 680–688.
77. Steiner, W. (1993). Results of curative laser microsurgery of laryngeal carcinomas. *Am. J. Otolaryngol.* **14**(2), 116–121.
78. Rudert, H. H. (1999). Endoscopic management of supraglottic carcinoma. *In* "Organ Preservation Surgery for Laryngeal Carcinoma,"(G. S. Weinstein, O. Laccourreye, D. Brasnu, and H. Laccourreye, eds.), pp. 97–107. Singular Publishing Group, San Diego.
79. Bocca, E., Pignataro, O., Oldini, C., *et al.* (1987). Extended supraglottic laryngectomy: Review of 84 cases. *Ann. Otol. Rhinol. Laryngol.* **96**, 384–386.
80. Lee, N. K., Goepfert, H., and Wendt, C. D. (1990). Supraglottic laryngectomy for intermediate-stage cancer: U. T. M. D. Anderson Cancer Center experience with combined therapy. *Laryngoscope* **100**(8), 831–836.
81. Spaulding, C. A., Constable, W. C., Levine, P. A., and Cantrell, R. W. (1988). Partial Laryngectomy and radiotherapy for supraglottic cancer: A conservative approach. *Ann. Otol. Rhinol. Laryngol.* **98**, 125–129.
82. Herranz-Gonzalez, J., Gavilan, J., Martinez-Vidal, J., and Gavilan, C. (1996). Supraglottic laryngectomy: Functional and oncologic results. *Ann. Otol. Rhinol. Laryngol.* **105**(1), 18–22.
83. Rademaker, A. W., Logemann, J. A., Pauloski, B. R., *et al.* (1993). Recovery of postoperative swallowing in patients undergoing partial laryngectomy. *Head Neck* **15**(4), 325–334.
84. Laccourreye, O., and Weinstein, G. S. (1999). Supraglottic laryngectomy. *In* "Organ Preservation Surgery for Laryngeal Cancer," (G. S. Weinstein, O. Laccourreye, D. Brasnu, and H. Laccourreye, eds.), pp. 107–126. Singular Publishing Group, San Diego.
85. Laccourreye, O., Brasnu, D., Merite-Drancy, A., *et al.* (1993). Cricohyoidopexy in selected infrahyoid epiglottic carcinomas presenting with pathological preepiglottic space invasion. *Arch. Otolaryngol. Head Neck Surg.* **119**(8), 881–886.
86. Brasnu, D., Hartl, D. M., and Laccourreye, H. (1999). Supracricoid partial laryngectomy with cricohyoidopexy. *In* "Organ Preservation Surgery for Laryngeal Cancer," (G.S. Weinstein, O. Laccourreye,

D. Brasnu, and H. Laccourreye, eds.), pp. 127–148. Singular Publishing Group, San Diego.

87. Laccourreye, O., Crevier-Buchmann, L., Weinstein, G., *et al.* (1995). Duration and frequency characteristics of speech and voice following supracricoid partial laryngectomy. *Ann. Otol. Rhinol. Laryngol.* **104**, 516–521.
88. DeSanto, L. W., Lillie, J. C., and Devine, K. D. (1976). Surgical salvage after radiation for laryngeal cancer. *Laryngoscope* **86**(5), 649–657.
89. Skolnik, E. M., Martin, L., Yee, K. F., and Wheatley, M. A. (1975). Radiation failures in cancer of the larynx. *Ann. Otol. Rhinol. Laryngol.* **84**(6), 804–811.
90. Ward, P. H., Calcaterra, T. C., and Kagan, A. R. (1975). The enigma of post-radiation edema and recurrent or residual carcinoma of the larynx. *Laryngoscope* **85**(3), 522–529.
91. Biller, H. F., Barnhill, F. R., Jr., Ogura, J. H., and Perez, C. A. (1970). Hemilaryngectomy following radiation failure for carcinoma of the vocal cords. *Laryngoscope* **80**(2), 249–253.
92. Sorensen, H., Hansen, H. S., and Thomsen, K. A. (1980). Partial laryngectomy following irradiation. *Laryngoscope* **90**(8 Pt 1), 1344–1349.
93. Laccourreye, O., Weinstein, G., Naudo, P., *et al.* (1996). Supracricoid partial laryngectomy after failed laryngeal radiation therapy. *Laryngoscope* **106**(4), 495–498.

CHAPTER

27

High-Dose Intraarterial Chemotherapy/ Radiation for Advanced Head and Neck Cancer

K. THOMAS ROBBINS

Department of Otolaryngology—Head and Neck Surgery
University of Florida, Gainesville, Florida

I. Introduction 393
II. RADPLAT Drug Delivery Technique 394
III. RADPLAT Results 395
A. Temporal Bone Carcinoma 397
B. Massive Neck Metastases (N3 Disease) 397
C. Advanced Paranasal Sinus Cancer 397
D. Bone and Cartilage Invasion 398
IV. RADPLAT Experience at Other Centers 398
V. New Directions Using the RADPLAT Concept 399
A. PENTORADPLAT 399
B. NEORADPLAT 400
C. Multi-RADPLAT 400
VI. Conclusions 402
References 403

I. INTRODUCTION

RADPLAT refers to a high-dose cisplatin intraarterial (IA) chemoradiation protocol developed to treat patients with advanced head and neck cancer. The treatment program incorporates a novel technique for infusing cisplatin directly into the tumor bed while minimizing the effects of the drug systemically. This is achieved by using microcatheters placed angiographically to permit superselective rapid infusions while sodium thiosulfate, a neutralizing agent for cisplatin, is simultaneously infused systemically. Because of this, it is feasible to increase the dose intensity of cisplatin by a magnitude that is at least five times higher relative to standard chemotherapy protocols, thereby enabling the delivery of an enormous amount of drug over a relatively short time interval.

The theoretical advantage the high-dose intensity chemotherapy regimen employed in the RADPLAT program relates to the phenomenon of acquired drug resistance. Head and neck tumors have a high rate of response to combination chemotherapy, particularly cisplatin based, when used in the neoadjuvant setting. However, despite this initial sensitivity by the tumor, a survival advantage for patients treated in this manner has yet to be demonstrated. One explanation for such treatment failure is drug resistance that is known to develop rapidly following exposure to chemotherapy agents [1–3]. Using a sponge gel-supported histoculture, 43 tumor specimens from patients with squamous cell carcinoma (SCC) of the upper aerodigestive tract (UADT) were grown and exposed to cisplatin [4]. Growth inhibition by the drug, in concentrations equivalent to peak therapeutic doses (1.5 μg/ml) and concentrations 10- and 25-fold greater (15 and 37.5 μg/ml), were measured in specimens from patients with previously untreated and recurrent lesions. *In vitro*, the rate of sensitivity of the tumor samples to cisplatin concentrations of 1.5, 15, and 37.5 μg/ml was 25.9, 63.3, and 79.3%, respectively, among patients with previously untreated disease compared to 10.0, 55.6, and 85.6% for patients with recurrent disease. These results indicate that resistance to standard doses of cisplatin by squamous cell carcinoma (SCC) of the UADT can be substantially overcome with a decadose (10×) increase and is more pronounced in tumors from patients with recurrent disease.

The pharmacokinetics of IA therapy were described by Eckman *et al.* [5]. The relative advantage of IA infusion is directly proportional to the plasma clearance of the drug and is inversely proportional to the plasma flow to the tumor. Thus, the faster the drug is excreted, and the slower the tumor plasma flow, the larger the relative advantage for infusing by the IA route. In the case of head and neck lesions, Wheeler *et al.* [6] reported that the mean tumor blood flow estimated by radionucleotide washout techniques was 13.6±6.7 ml 100 g/ml compared to 4.2±2.1 ml/100 g/min for normal tissue (scalp). This provided a mean tumor: normal tissue blood flow ratio of 3.9±2.7 ml/100 g/min, thus favoring drug delivery to the tumor contained within the infused region.

Results from the IA trials are variable, but several studies, particularly those using cisplatin-based regimens, indicate that a high response rate can be achieved. Others argue that the best results of IA chemotherapy do not surpass the best response rates from IV chemotherapy and when combined with the increased risk of local toxicity, there is no advantage to this approach. Most of the technical complications have been related to catheter placements, including thrombosis, dislodgement, and hemorrhage. However, newer and safer angiographic techniques as used in our study now permit highly selective placements of microcatheters into small arteries under direct vision using fluoroscopy [7]. These advances in interventional vascular radiology make it possible to selectively and repeatedly infuse chemotherapeutic agents into head and neck tumors with minimal side effects [8].

II. RADPLAT DRUG DELIVERY TECHNIQUE

Cisplatin is infused over 3–5 min through a microcatheter placed intraarterially using angiographic techniques to selectively encompass only the dominant blood supply of the targeted tumor. Simultaneous with starting the IA infusion of cisplatin, sodium thiosulfate (9 g/m^2 over 30 min, followed by 12 g/m^2 over 2 h) is given intravenously. This allows the tumor bed to initially receive the full dose of cisplatin prior to the neutralizing agent and for the systemic organs to receive the neutralizing agent prior to the cisplatin (Fig. 27.1). Patients receive pretreatment IV hydration over 2 h consisting of 2 liters of normal saline containing 20 mEq KCl and 2 g magnesium sulfate. Cisplatin is dissolved in 400 ml normal saline. Posttreatment hydration consists of 1 liter of normal saline containing 20 mEq KCl and 2 g magnesium sulfate over 6 h. Decadron is also administered intravenously (IV) or orally (PO), 4 mg every 6 h until the following morning.

Thiosulfate reacts covalently with cisplatin to produce a complex that is still soluble but is totally devoid of either toxicity or antitumor activity [9]. When this neutralization occurs in the plasma, it effectively increases the plasma "clearance" of cisplatin. The extent of reaction is a function of the concentration of both agents. Thiosulfate is not a very potent neutralization agent, and molar thiosulfate/cisplatin ratios in excess of 10 are required before the reaction is fast enough to contribute significantly to the clearance of cisplatin [10]. Thiosulfate itself is very nontoxic and doses in excess of 72 g can be given acutely, which is well above that needed to provide effective cisplatin neutralization. Pharmacokinetic studies have demonstrated an important additional feature of thiosulfate: it is concentrated

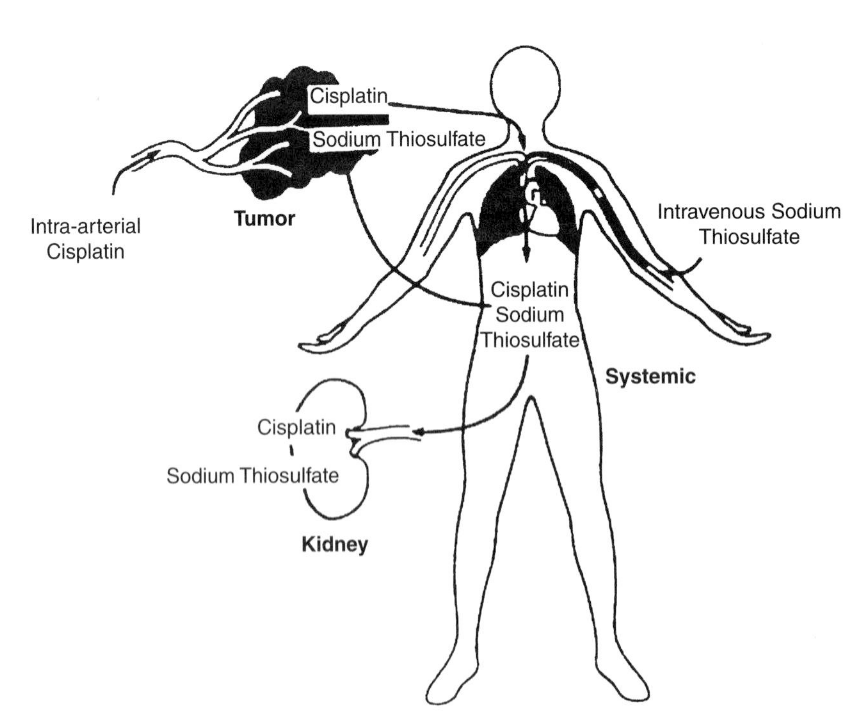

FIGURE 27.1 Schematic representation of RADPLAT infusion technique with systemic sodium thiosulfate neutralization.

extensively (>25-fold) in the urine, which provides excellent protection against cisplatin-induced nephrotoxicity.

Patients are admitted to the hospital and hydrated overnight. Catheterization is done by interventional radiologists under local anesthesia in the angiography suite. Transfemoral carotid arteriography is first done to assess the vascular anatomy and any vessel pathology [7]. The appropriate arteries supplying the region of primary disease will then be infused with cisplatin. This is usually achieved by placing a microcatheter, introduced transaxially through a tracker catheter, into the external carotid artery at the level of the orifice of the dominant branching artery to the tumor. Thus, cisplatin can be infused rapidly to selectively encompass on its initial exposure only the territory of the targeted tumor. In selected patients with bulky disease crossing the midline, bilateral transfemoral catheterizations are performed to permit simultaneous infusions of the contralateral disease. In each patient, the goal is to infuse the component of the disease considered to be bulky and/or infiltrative and likely to fail radiotherapy alone. Surgery and radiation therapy are used to treat the regional lymphatics, although in selected patients who have unresectable neck disease, it is often possible to infuse its blood supply through the superior thyroid artery or the thyrocervical trunk.

III. RADPLAT RESULTS

The original clinical trial was a phase I study designed to determine the maximum tolerated dose of cisplatin that could be administered [11]. Cohorts of patients received dose intensity schedules of 50, 100, 150, and 200 mg/m^2/week. The dose-limiting toxicity was a severe leakage of electrolytes from the kidneys that could be alleviated by intravenous replacement therapy until recovery was established. The maximum tolerated dose was 150 mg/m^2/week × 4. None of the patients had irreversible nephrotoxicity. Other grade III–IV toxicities noted during this trial were gastrointestinal symptoms (four events) and neutropenia (one event). Among 22 patients with previously untreated disease, 9 (41%) had a complete response (CR) and 10 (45%) had a partial response (PR). Thus, the major response rate was 86%. Among the 16 patients with recurrent disease, 4 had a CR (25%), and 6 had a PR (38%), for a major response rate of 63%. Because this was a dose-escalating study in which the dose intensity schedule ranged between 32.5 and 200 mg/m^2/week, a more detailed analysis of the response effects of cisplatin at specific dose intensities was done [12]. The overall response rate (CR and PR) to cisplatin therapy at dose intensity intervals of 0 to 74, 75, to 149 and 150 to 200 mg/m^2/week was 45.5, 72.7, and 100%, respectively. The average received dose intensity for nonresponders (NR) versus responders (CR and PR) was 57.8 and 120.7 mg/m^2/week, respectively ($P=0.031$). These data indicate that high-dose cisplatin exposure increases the rate of tumor response in SCC of the UADT and support the hypothesis that acquired cisplatin resistance by these tumors is usually mild to moderate and can be circumvented by 10-fold concentrations of the drug. The decadose effect was achieved by increasing the dose and condensing the overall treatment time. Calculated as the dose intensity value, this increased cisplatin exposure reaches an equivalent that is 10-fold greater relative to standard cisplatin protocols.

When radiotherapy was given concomitantly with targeted cisplatin chemotherapy, i.e., RADPLAT, preliminary observations indicated an extremely high complete pathologic response rate, sustained disease control above the clavicles, and a relatively low rate of toxicity [13–15]. Conventional external beam irradiation was used in daily fractions (180 to 200 cGy/fraction) to a total dose of 68.5 to 74.0 Gy given over 7 to 8 weeks. All patients received IA cisplatin and IV sodium thiosulfate infusions concurrently on days 1, 8, 15, and 22 of radiotherapy. Planned neck dissection was done 2 months after treatment in patients whose original nodal disease was considered to be N2 or N3. Salvage surgery was done in patients who developed recurrent disease that was considered to be resectable.

The most recent analysis included 213 patients treated between 1993 and 1998 at the University of Tennessee, Memphis, for whom the follow-up ranged between 16 and 74 months (median, 30 months) [16]. Organized by a specific subsite, the distribution of patients is as follows: oral cavity, 22 (10.3%); oropharyngeal wall, 89 (41.8%); hypopharynx, 44 (20.7%); larynx, 44 (20.7%); nasopharynx, 7 (3.3%); and other sites, 7 (3.3%). One-third of the lesions were massive and anatomically unresectable, whereas two-thirds were potentially resectable by technical (anatomic) criteria, but removal would have caused the loss of one or more organs necessary for speech and swallowing. Ninety-four patients (44%) had T4 disease; 102 (48%) had T3 disease; 15 (7%) had T2 disease; and two (1%) had T1 disease (both had N3 nodal disease). Distributed by N stage, the patient breakdown is as follows: N0 disease, 61 (29%); N1, 36 (17%); N2, 90 (42%); and N3, 26 (1%). Thus, only 28.6% of patients had stage III disease, whereas the remaining 71.4% had stage IV disease.

Of the 213 patients entered into the treatment program, a complete response in the primary site was obtained in 171 (80%). However, the complete response rate in the primary site for those patients who completed the treatment and were available for restaging at 2 months following radiotherapy was 171 of 189 (90.5%). Of the 152 patients with clinically node-positive disease, a complete response was obtained in 92 (61%). Among the 17 patients who had a partial response to treatment in the primary site, 12 subsequently underwent salvage surgery, of whom 9 had complete eradication of disease. However, 6 of these patients subsequently developed recurrent disease, 4 within the primary site and 2 at distant sites. Thus, only 3 of 17 patients who had a partial response in the primary site remained disease free despite attempts at surgical salvage. The single patient who had no response had massive unresectable disease extending from the nasopharynx to the piriform sinus and had no further active treatment after failing intraarterial cisplatin and radiotherapy.

There were 95 grade III–IV events of toxicity among the 213 patients undergoing a cumulative total of 717 infusions. This total included 56 events of mucositis, 2 incidents of gastrointestinal toxicity, 17 hematologic events, 9 neurologic events, and 8 cardiovascular events. There were 6 grade V toxicities (treatment-related deaths). No events of grade III–IV toxicity were noted in 130 of 213 patients (61%), whereas 73 patients experienced 1 event, 8 patients experienced 2 events, and 2 patients had 3 events. One hundred and seventy patients (80%) received all four planned infusions of cisplatin; 22 (10%) received three infusions, and 21 (10%) received less than three infusions. The total number of central nervous system events was 7 (1% of all infusions), 5 of which were cerebrovascular accidents and 2 transient ischemic attacks. Although each of the cerebrovascular patients had residual motor deficits, these were mild, and all of these patients were able to remain ambulatory and physically active.

With a median follow-up interval of 30 months (range, 16 to 69 months), the Kaplan–Meier plot projected disease-specific survival at 5 years to be 53.6% (SD, 3.9%) and the overall survival to be 38.8% (SD, 3.7%) (Fig. 27.2). Comparison of survival rates between patients with T3 versus T4 lesions did not show any significant difference between the two subsets ($P=0.095$). However, there was a significant difference between patients with N0–N1 versus N2–N3 disease ($P=0.014$). The Kaplan–Meier plot for the rate of disease control above the clavicle for all patients was 74.3% (SD, 3.6%) at 5 years. Eighteen patients (8%) had persistent disease after treatment, most of whom did not complete the protocol. Among the remaining 195 patients rendered disease free following treatment, 51 patients (26%) developed recurrences: 11 (6%) within the primary site, 5 (3%) within the regional lymphatics, and 35 (18%) in distant sites (lung, 12; bone, 10; multiple sites, 13).

Death from persistent or recurrent disease within the primary site or the neck is often associated with catastrophic suffering, including alterations in important bodily functions, severe pain, and disfigurement. Very few patients in the study succumbed to persistent disease following initial therapy, and only 16 patients died of recurrent disease above the clavicle. The majority of patients who died of disease did so because of recurrent tumor at distant sites, most commonly the lungs. This pattern of cancer death is different from that seen in

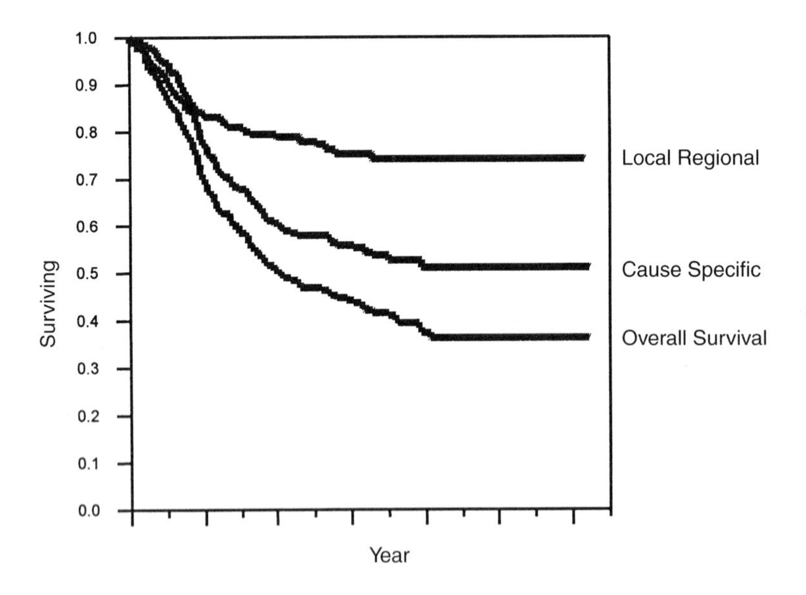

Number of Patients at Risk

213	139	109	72	40	23	3	Local Regional
213	151	112	76	41	24	3	Cause Specific
213	151	112	76	41	24	3	Overall Survival

FIGURE 27.2 Targeted cisplatin chemoradiation survival.

most head and neck cancer trials in which death from locoregional disease is by far more common. It is likely that distant metastatic disease in patients with head and neck cancer is usually masked by locoregional disease. With improved methods to control disease above the clavicle, one can expect an unmasking of clinical distant disease among patients who had occult metastatic disease prior to therapy. The emerging problem of death from distant disease will require designing of subsequent studies to include a systemic treatment component, particularly for patients who are at greatest risk.

A. Temporal Bone Carcinoma

Supradose cisplatin infusions delivered selectively to the skull base combined with concurrent systemic cisplatin neutralization were given to 14 patients with temporal bone cancer [17]. In 5 patients, the epithelial site of origin was the external auditory canal, in another 5 the parotid gland, and in the remaining 4 the mucosa of the pharynx. Ten patients had squamous cell carcinoma, 2 had mucoepidermoid carcinoma, and 2 had adenoid cystic carcinoma. Six patients had previously untreated tumors, whereas 8 presented with recurrent disease. The extent of temporal bone invasion, as determined by physical examination and radiographs, was as follows: external bony canal only (4); external bony canal/mastoid (1); external bony canal/mastoid/petrous apex (4); mastoid/petrous apex (2); mastoid/hypotympanum (1); and petrous apex only (2 patients). Four patients received chemotherapy alone, 4 had concomitant radiotherapy, and 6 had subsequent radiotherapy and/or temporal bone surgery. All of the patients tolerated the chemotherapy without any significant complications or toxicity. All 3 of the patients with previously untreated disease responded to chemotherapy (2 CR, 1 PR); 3 of the seven patients with recurrent disease responded to chemotherapy and all 4 patients treated with chemoradiation had a complete response (including 1 patient with recurrent disease). At a median follow-up of 19 months (range, from 5 to 63 months), 9 of 14 patients were alive, including the 4 who were treated with targeted chemoradiation. Data support further investigations using targeted high-dose chemotherapy, particularly when it is given simultaneously with radiotherapy, for patients with malignant skull base lesions.

B. Massive Neck Metastases (N3 Disease)

Patients who present with bulky nodal disease are at high risk for failure both in the regional site and distantly. Results with targeted chemoradiation for this unfavorable group of patients have been quite encouraging with regard to achieving effective regional control, but the rate of distant metastases continues to be excessive [18]. Thirty-one patients with N3 nodal disease were treated at the University of Tennessee Health Science Center between June 1993 and June 1997 [19]. All patients received the RADPLAT protocol as described previously, including a high total dose of radiotherapy (68 to 74 Gy/7 week) to the nodal disease. Five patients did not complete the protocol and were thus unevaluable. Of the 26 evaluable patients, disease response in the neck at 2 months following the completion of radiotherapy based on clinical criteria [physical examination and computerized tomography (CT) scans] was as follows: CR, 4 patients; PR, 21 patients; and NR, 1 patient. Nineteen patients subsequently underwent a salvage neck dissection, 5 of whom had histologic evidence of residual disease. The 7 evaluable patients who did not have a neck dissection following RADPLAT included 4 with a complete response and 3 who died of intercurrent disease prior to restaging. There were no recurrences in the neck among the 23 patients who were rendered disease free following treatment, whereas 1 patient had a recurrence at the primary site and 11 patients had recurrences at distant sites. With a median follow-up of 15 months (range, 4 to 41 months), the 3-year overall and disease-free survival rate is 41 and 43%, respectively. Thus, targeted chemoradiation followed by surgical salvage is a highly effective approach for regional control of patients with N3 nodal disease, whereas additional strategies are required to address the problem of distant metastases [19].

C. Advanced Paranasal Sinus Cancer

RADPLAT has also been used to treat patients with advanced paranasal sinus cancer. Eleven patients (10 T4; 1 T3) were treated between June 1994 and June 1998 at the University of Tennessee Health Science Center [20]. Patients received three ($n = 2$) or four ($n = 9$) intraarterial infusions of cisplatin (150 mg/m^2), simultaneous systemic neutralization with intravenous sodium thiosulfate (9 g/m^2), and concomitant radiotherapy (median dose, 50 Gy; range, 48 to 70 Gy). The protocol included a planned surgical resection 2 months after the completion of radiotherapy. Four patients did not have surgery (2 refused and 2 died of intercurrent disease). Seven patients were resected (6 through a bifrontal craniotomy approach and 1 through a transfacial approach). Three specimens were found to contain microscopic foci of residual tumor, whereas the remaining four specimens were negative. At a median follow-up of 13 months (range, 6 to 23 months), 8 of 11 patients are alive and free of disease. No recurrences have yet been observed. Orbital exenteration and/or palatectomy was not required for any patient primarily because the surgical approach was almost always from above (transfrontal) without the use of *en bloc* resection of tissue. The results, although preliminary, support the use of preoperative targeted supradose intraarterial chemotherapy and concomitant radiotherapy as a neoadjuvant treatment of advanced paranasal sinus carcinomas, facilitating surgical resections with organ preservation.

D. Bone and Cartilage Invasion

It is particularly enlightening to analyze this subset of patients with T4 lesions of the head and neck who have been treated with chemoradiation because common philosophy dictates surgical management for resectable disease. At the University of Tennessee, Memphis, 135 of 293 patients treated with targeted chemoradiation between 1993 and 1998 had T4 primary disease. Within this group, 45 patients had lesions with evidence of bone and/or cartilage invasion. A retrospective analysis was done to compare the efficacy of RADPLAT in patients with invasion of bony or cartilaginous structures (group I, $n=45$) versus other patients having T4 disease without bone or cartilage involvement (group II, $n=90$) [21]. The presence of bone/cartilage invasion was established by a review of tumor diagrams of clinical findings and CT or magnetic resonance imaging. Thirty patients had evidence of bone invasion, with the following breakdown: mandible, 12; maxilla, 9; sphenoid, 3; and hyoid, 6. Thirty-eight patients had evidence of cartilage invasion: epiglottic, 18; thyroid, 16; and cricoid, 4. The rate of complete response obtained in group I (69%) was not significantly different from that in group II (71%, $P=0.79$). The 2-year overall actuarial survival for group I [46.3%; 95% confidence interval (CI)=30.3% to 62.3%] was not significantly different (generalized Wilcoxon test, $P=0.36$) from that of group II (36.9%; 95% CI=25.5% to 48.4%). A marked trend was noted for higher response rates in cases of cartilage invasion (83%) than in those with bone invasion (62%, $P=0.15$). This may be reflective of patients with laryngeal cancer who typically have smaller volume disease. Equivalence of the efficacy of treatment in the two groups suggests that targeted chemoradiation can be a definitive therapeutic option in patients with advanced head and neck cancer invading bony or cartilaginous structures.

IV. RADPLAT EXPERIENCE AT OTHER CENTERS

At the Netherlands Cancer Institute in Amsterdam, Tan and associates have been using the RADPLAT protocol since 1997. The initial phase II study using the exact protocol, as developed by Robbins and colleagues, included patients primarily with T4 lesions of the upper aerodigestive tract who were considered to be inoperable. Eighty-five patients were treated between April 1997 and December 1999 with radiotherapy (70 Gy, 7 weeks, 35 fractions) and concomitant superselective intraarterial cisplatin (150 mg/m^2, days 1, 8, 15, and 22) and systemic sodium thiosulfate neutralization. There were 73 patients with T4 lesions and 12 with T3, whereas 61 patients had N+ nodal disease. Complete remission was achieved in 88%. At 2 years, the overall and disease-free survival was 65 and 45%, respectively. The rates of loco-regional and local control were 57 and 61%, respectively. No treatment interruptions or dose limitations resulted from acute toxicity. There was one treatment-related death, whereas 17% of patients had a grade IV hematologic toxicity. There were no grade IV nonhematologic toxicities. Grade III mucositis was seen in 43% of patients, and grade III GI toxicity was seen in 60%. Ten percent of the patients had evidence of hearing loss. These authors concluded that the RADPLAT treatment schedule is feasible to transport to other centers and provides an excellent response and organ preservation rate. This study has now led to the initiation of a multicenter phase III trial comparing radiotherapy and concomitant systemic cisplatin vs RADPLAT [22].

A concurrent analysis of the Amsterdam patients was done for quality of life. Fifty patients were interviewed by means of the Functional Assessment of Cancer Therapy (FACT) questionnaire, including a special head and neck module and the University of Washington questionnaire, pretreatment, and 3, 6, 9, and 12 months after starting therapy. Results indicated that most subscales of the facts showed a significant decline between the pretreatment and 3 months after initial treatment. Thirty-seven patients could be analyzed at the 6-month measuring point. Analysis showed that some stabilization occurred between 3 and 6 months. Only the functional well-being scale already showed a slightly positive trend. Feelings of nausea did occur sporadically during the total period. After 6 months, 13 patients (26%) still had a feeding tube. Xerostomia and feelings of pain were reported by 18 (36%) and 11 (22%) patients, respectively. Of the 23 patients who had a job pretreatment, 6 were back at work. The authors concluded that the overall quality of life aspects declined during the treatment period and then seemed to stabilize after 3 months. These findings are similar to a study performed at the University of Tennessee by Murray and colleagues. They noted decline of quality of life and swallowing function during therapy but a recovery to pretreatment levels at 6 months follow-up. However, swallowing recovery was not complete compared to pretreatment levels [23].

At the George Washington University Medical Center in Washington, DC, Wilson and co-workers [24] have used the RADPLAT approach for patients with advanced head and neck cancer since 1994. The treatment regimen consisted of four weekly intraarterial infusions of cisplatin (150 mg/m^2) targeting the tumor bed, followed by 6 weeks of radiation therapy. Thus the program is of a sequential nature rather than the simultaneous type used by the group at the University of Tennessee. Wilson *et al.* [24] reported their experience with 58 patients with a median follow-up of 25 months. Forty-two had previously untreated disease, whereas 16 had recurrent lesions. Of the 42 previously untreated patients, 27 were alive and disease free, corresponding to a sustained complete response rate of 64.3% at a median follow-up of 30 months. Among the 16 patients with recurrent lesions, there were 4 survivors, corresponding to a sustained complete response rate of 25% with a median follow-up of 15.5 months. This group reported no deaths or serious complications related to the treatment in either group. Only 1 patient required resection of the tumor site because of relapse. The authors concluded that the combination of high-dose, intraarterial cisplatin and radiation therapy is effective in improving both survival and organ preservation rates in previously untreated advanced squamous cell carcinoma of the head and neck [24].

At the University of Kentucky, Valentino and colleagues applied the RADPLAT principle in a different manner to patients with advanced head and neck cancer. Rather than delivering the chemotherapy at the beginning of radiation therapy, these authors tested the concept of a concomitant chemotherapy boost toward the end of radiation therapy when there is a known biologic risk of resistant tumor cell repopulation. Referred to as HYPERRADPLAT, 20 patients with locally advanced disease were treated between December 1995 and November 1997 with hyperfractionated radiation therapy (76.8–79.2 Gy at 1.2 Gy twice daily over 6–7 weeks) and high-dose intraarterial cisplatin (150 mg/m^2) given at the start of radiation therapy boost treatment, i.e., start of week 6. Seventeen patients (85%) had T4 disease and 14 (70%) had N2–N3 disease. Results indicated that grade III–IV acute toxicity was limited to 1 grade IV (5%) and 14 grade III (70%) mucosal events. No grade III-IV hematologic toxicity was observed. Median weight loss during therapy was 9% (2–16%). Eighteen patients had a complete response (90%) at the primary site; 14 were confirmed pathologically. Among 17 patients with positive neck disease, 16 (94%) achieved complete response in the neck, including 12 of 13 patients with N2–N3 disease who underwent planned neck dissection. At a follow-up of 12 to 32 months and a median follow-up of 20 months, 11 patients were alive without disease, 5 died of disease, and 4 died of intercurrent disease. Eighteen patients (90%) remain disease free at the primary site, and the locoregional control rate was 80%. These authors concluded that high-dose intraarterial cisplatin and concurrent hyperfractionated radiation therapy is very feasible and warranted further study. The high complete response rate and low-grade IV toxicity rate in this highly unfavorable subset of patients studied appeared better than previously reported chemoradiation regimens for more favorable patients [25].

V. NEW DIRECTIONS USING THE RADPLAT CONCEPT

Work to date using the technique of rapid intraarterial infusions of supradose cisplatin, combined with systemic sodium thiosulfate neutralization and concomitant radiotherapy, indicates that this method is very effective at eradicating disease in the treatment field. Its weaknesses relate to the chronic soft tissue side effects seen locally, i.e., fibrosis and xerostomia and its low therapeutic effect systemically. We have initiated two new strategies (PENTORADPLAT and NEORADPLAT) to address the former concern and are currently planning to add a systemic treatment regimen using an antiangiogenesis agent for the latter. We have also conducted a multicenter phase II trial (multi-RADPLAT) to determine whether targeted chemoradiotherapy can be done safely and effectively by other institutions.

A. PENTORADPLAT

Patients eligible for the PENTORADPLAT protocol included those treated on the RADPLAT protocol who would have undergone a planned limited neck dissection because of initial bulky lymphadenopathy, i.e., N2 or N3 disease. In an attempt to reduce this soft tissue neck toxicity, treatment with pentoxiphyline, 400 mg orally, four times a day, was initiated on day 1 of the RADPLAT protocol and was continued daily for a total of at least 6 months, or 2 months following the neck dissection, whichever was longer.

Seventeen patients initiated and/or completed the PENTORADPLAT protocol and also underwent assessment of the neck. Two of the 17 patients experienced neck relapses and were excluded from this toxicity analysis. In the primary site, 15 of 15 patients had complete responses. In neck disease, 12 of 15 had complete responses, and 3 had partial responses. Thirteen of 15 patients are alive without any evidence of recurrent disease, whereas the other 2 died of distant metastases. The median interval between the neck dissection (or at the end of radiotherapy) and the neck assessment was 12 months (range, 2 to 26 months). The rate of "worst" neck fibrosis among the 15 patients was grade 0 to 2, 12 patients; and grade III–IV, 3 patients. Twelve of the 15 patients showed good, i.e., 60 to 89° to full, i.e., 90°, lateral range of neck motion. None of the 15 patients have experienced a wound complication. These results were compared with those in a similar group of stage IV, N2–N3 patients ($n=15$) who were relatively matched for clinical and disease characteristics but were unable to take pentoxiphyline for various reasons, including percutaneous endoscopic gastrostomy placement prior to therapy or cost of the drug. Comparison between the two groups of patients revealed the rate of neck fibrosis to be lower in the PENTORADPLAT group compared with the control group: grade 0–II = 80% ($n=12/15$) vs 60% ($n=9/15$), respectively, and grade III–IV = 20% ($n=3/15$) vs 40% ($n=6/15$), respectively.

Our preliminary results indicate that the prophylactic use of pentoxiphyline reduces the rate and severity of late neck fibrosis following treatment of stage IV, N2–N3 squamous cell carcinoma of the head and neck with intraarterial cisplatin and radical doses of radiotherapy followed by a limited neck dissection.

B. NEORADPLAT

Intermediate (T3 or larger T2) cancers of the oral cavity and oropharynx are currently treated on chemoradiation protocols incorporating high-dose radiotherapy or with extensive surgery with or without adjuvant radiation. To minimize toxic effects of radiation, i.e., fibrosis, xerostomia, and loss of taste, and those of surgery (functional and cosmetic impairment), we initiated a new treatment regimen of reduced radiotherapy (50 Gy) and concomitant IA cisplatin followed by limited surgery ("tumorectomy" or biopsy only). The objective was to study the efficacy of this regimen in patients with T2/T3 squamous cancers of the oral cavity and oropharynx. Treatment consisted of IA cisplatin (150 mg/m^2/week × 4) with sodium thiosulfate protection (9 mg/m^2) and concurrent radiotherapy to a dose of 50 Gy at 2.0 Gy/fraction to the primary site/overt nodal disease. Surgery consisted of a conservative excision of any residual tumor or scar tissue and was performed if frozen-section biopsy at 8 weeks following treatment restaging revealed residual cancer or if there remained some suspicion of residual disease despite a negative biopsy. Twenty-one patients with a T2/T3 primary (T2=3; T3=18) and N0/N2 nodal disease (N0=12; N1=5; N2=4) are available for analysis after a median follow-up of 36 months (range, 15 to 57 months). Four cisplatin infusions could be delivered in 19 of 21 patients. One patient died 3 months after completion of radiotherapy from aspiration pneumonia. Otherwise, the toxicity profile was favorable (hematologic: grade 3, 3; grade 4, 1; neurologic: cranial neuropathy, 1). Complete histologic response at the primary site was obtained after chemoradiation in 15 of 20 evaluable patients (75%), with 19 of 21 patients rendered disease free after surgery. No surgery was necessary in 8 patients, whereas 10 patients required limited surgery: transoral (retromolar trigone excision, 2; tonsillectomy, 1; partial glossectomy, 5) or lip splitting (retromolar trigone excision, 2). One patient required a composite resection and another a transoral hemiglossectomy. Six of the 9 patients with node-positive disease underwent a selective neck dissection; no histologic evidence of cancer was found in the neck contents in 4 patients. Local recurrence developed in 3 patients (14%). The actuarial overall and disease-specific survival rates at 2 years are 73 and 83%, respectively. We concluded that this regimen of IA cisplatin and reduced-dose radiation with or without limited surgery is highly effective in the treatment of intermediate squamous oral and oropharyngeal cancers [26]. Further analysis is necessary to determine whether the reduced dose of radiation therapy had any ameliorative effects on soft tissue function, including salivary flow.

C. Multi-RADPLAT

Although we have proven conclusively that the RADPLAT protocol is feasible with very promising results in a single institutional setting, questions still remain about the "exportability" and feasibility of this highly technical concept to the "community at large." In an attempt to answer this question, an 11 center National Cancer Institute/National Institutes of Health–funded trial was conducted under the auspices of the Radiation Therapy Oncology Group (RTOG) to determine the feasibility of using intraarterial cisplatin (P) and radiotherapy (RT) for advanced head and neck squamous cell carcinoma in a multi-institutional setting, i.e., multi-RADPLAT [27].

Major eligibility requirements for the multi-RADPLAT protocol included stage IV, T4, N0/N3 SCC of the oral cavity, oropharynx, hypopharynx, or larynx. Treatment consisted of IA-P (150 mg/m^2 on days 1, 8, 15, and 22) and concurrent RT at a total dose of 70 Gy at 2.0 Gy/fraction/7 week. Protocol feasibility was defined as the ability to deliver three to four cycles of IA-P and the full dose of RT. Between May 1997 and December 1999, 62 of the 67 patients enrolled in the trial were eligible for analysis; 2 were determined to be ineligible (1 due to a previous prostate cancer and 1 due to below normal creatinine clearance). Data on the other 3 patients are still pending. Fourteen patients were enrolled by the experienced institutions and the remaining 48 came from inexperienced centers. Of the 48 patients entered by the inexperienced institutions, a subtotal of 16 patients represented the "first 2 patients" from each center and thus part of the "learning curve." The other 32 patients compromised the population of patients from the inexperienced institutions, which was to be considered for the primary analysis of feasibility of the multi-RADPLAT protocol.

Clinical characteristics of the 62 patients included a median age at diagnosis of 52 years (range, 21–75), 56 (90%) males and 6 (10%) females, and 49 (79%) Caucasians, 11 (18%) African-Americans, and 2 (3%) Native American/Alaskan. Sites of disease origin included oropharynx in 32 (52%), 20 of which originated from the base of tongue, hypopharynx in 8 (13%), larynx in 7 (11%), and oral cavity in 14 (23%). (Table 27.1). The Karnofsky performance status was as follows: "100" in 7 (12%) patients, "90" in 30 (49%), "80" in 16 (26%), "70" in 6 (10%), and "60" in 2 (3%). All patients were diagnosed with AJC stage IV disease (Table 27.2). Fifty-seven (95%) patients presented with T4 lesions, and 3 had T3 disease. Lymph nodal staging was as follows: N0 in 20 (33%), N1 in 5 (8%), N2 in 30 (50%), and N3 in 5 (8%).

1. Intraarterial Cisplatin Infusions

Overall, among all 62 patients from all centers, cisplatin infusions were delivered as follows: three or four infusions in 12 (19%) and 46 (74%) patients respectively; two infusions in 2 (3%); one infusion in 1 (2%) patient, and data pending in the other 1 (2%) patient. For the 14 patients

TABLE 27.1 Feasibility of Intraarterial Cisplatin and Concurrent Radiation Therapy on RTOG 9615

Intraarterial cisplatin and concurrent RT	All patients $n=62$ (%)	Experienced institutions $n=14$ (%)	Inexperienced institutions: Learning curve (first two patients) $n=16$ (%)	Inexperienced institutions: 3^+ patients $n=32$ (%)
Feasible (three or four IA Infusions, full-dose RT)	52 (84%)	12 (86%)	12 (75%)	28 (88%)
Three or four IA infusions, RT < full dose	4 (7%)	1 (7%)	2 (12.5%)	1 (3%)
Three or four IA infusions, RT data pending	2 (4%)	0	2 (12.5%)	0
One or two IA infusions, RT per protocol	3 (5%)	0	0	3 (9%)
Chemotherapy data pending, RT per protocol	1 (2%)	1 (7%)	0	0

enrolled from the experienced centers, 10 (71%) underwent four infusions, 3 (21%) had three infusions, and data are still pending in the other 1 patient. Among the 16 patients who formed the learning curve at inexperienced institutions (i.e., first 2 patients from each center), the delivery of IA cisplatin infusions was as follows: three or four infusions in 2 (13%) and 14 (88%) patients, respectively. Excluding the "learning curve" patients, the delivery of IA infusions among the remaining 32 patients was as follows: three or four infusions in 7 (22%) and 22 (69%) patients, respectively; two infusions in 2 (6%); and one infusions in 1 (3%).

Overall, multi-RADPLAT was feasible (i.e., three or four infusions of IA cisplatin and full dose of RT) in 52 (84%) patients. In 4 (7%) and 2 (4%) patients, three or four infusions of IA cisplatin were delivered, but the radiation was either less than the protocol dose or RT data are still pending, respectively. Three patients underwent only one or two infusions of the IA cisplatin, although the RT was per protocol. In 1 patient, the RT was per protocol but chemotherapy data are still pending. For the 14 patients enrolled from the experienced centers, multi-RADPLAT was feasible in 12 (86%) patients; in 1 patient, four IA infusions were delivered but the RT was less than the protocol dose. In another 1 patient, the RT was per protocol but the chemotherapy administration is still pending. Analyzing just the first 2 "learning curve" patients (i.e., $n=16$ patients) from the inexperienced centers, the protocol therapy was feasible in 12 (75%) patients; four IA infusions were delivered but the RT was not per protocol in 2 (12.5%) patients; and in another 2 (12.5%) patients, three ($n=1$) or four ($n=1$) cycles of cisplatin were infused but RT data are still pending. Excluding the "learning curve" patients, multi-RADPLAT was feasible in 28 (88%) patients; in 1 patient, three IA infusions were given but the RT was not per protocol; and in 3 patients, only one ($n=1$) or two ($n=2$) cycles of cisplatin were infused, although the RT was per protocol. Overall, at the inexperienced centers, the protocol therapy was feasible in 40 (83%) of 48 patients, with radiotherapy data still pending in 2 patients. Response rates were available in 61 of the 62 patients. At the primary site, a complete response was obtained in 51 (84%) of 61 patients. At the nodal regions, a complete response was attained in 34 (87%) of 39 patients; 4 of the 34 patients achieved a CR by salvage neck dissection. Overall, a complete response was

TABLE 27.2 RTOG 9615L Grade III–V Toxicities

Institution type	Hematologic, grade III	Hematologic, grade IV	Hematologic, grade V	Nonhematologic, grade III	Nonhematologic, grade IV	Nonhematologic, grade V	Overall III	Overall IV	Overall V
Experienced, $n=14$	2	0	0	7	2	0	7	2	0
Inexperienced									
First two patients, $n=16$	2	3	1[a]	7	4	2[a]	5	7	2
All subsequent patients, $n=32$	13	7	0	21	6	0	16	13	

[a]DVT = 1, leukopenia/pneumonia = 1.

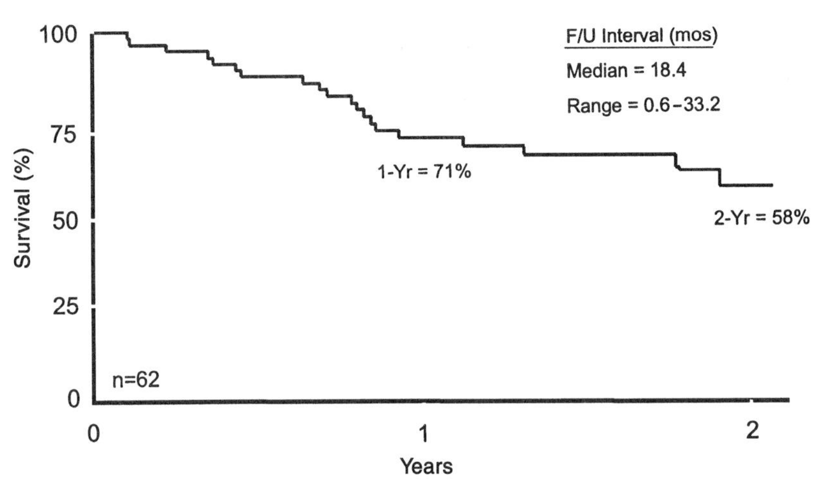

FIGURE 27.3 RTOG 9615: Overall survival.

attained both at the primary site and at the nodal region in 49 (80%) of 61 patients.

Survival data were available for all 62 patients. At a median follow-up of 18.4 months (range, 0.6–33.2 months), the 1- and 2-year Kaplan–Meier survival rates are 71% (95% CI: 59.1 and 83.1%) and 58% (95% CI: 42.5 and 73.8%), respectively (Fig. 27.3).

Overall, among all 62 patients, the rate of grade III, IV, and V toxicities was 45, 35, and 3%, respectively. Of significance, the rate of grade III and IV mucositis was 47 and 10%, respectively. Grade III and IV neurological toxicity was seen in 3 (5%) and 1 (2%) patients, respectively. Additionally, 2 patients experienced transient ischemic attacks that fully resolved without residual neurological deficits. Only 2 of 14 patients from the experienced institutions had grade III hematologic toxicity; no grade IV or V hematologic toxicity was observed at the experienced institution (Tables 27.1 and 27.2). However, at the inexperienced institutions, the rate of grade III, IV, and V hematologic toxicity by the number of patients was 15 (31%), 10 (21%), and 1 (2%), respectively. The single patient who died due to leukopenia at an inexperienced institution was the first patient enrolled in the multi-RADPLAT protocol at this center. In terms of nonhematologic toxicity, the rate of nonhematologic grade III and IV toxicity was 50% ($n=7$ patients) and 14% ($n=2$ patients), respectively at the experienced centers. At the inexperienced institutions, the rate of nonhematologic grade III, IV, and V toxicities was 45% ($n=28$ patients), 16% ($n=10$ patients), and 3% ($n=2$ patients), respectively. The 2 patients who experienced nonhematologic lethal toxicity at the inexperienced institution included the other previously reported patient who had developed leukopenia and pneumonia, as well as another patient who developed a deep vein thrombosis.

The pattern of recurrence for patients who had initially attained a complete remission was as follows: primary site only, $n=5$ patients; primary site and nodal region, $n=3$ patients; primary site, nodal region and distant metastases (bone), $n=1$ patient; distant metastases only, $n=3$ patients, all of whom failed in the lung. Second primaries have been noted in 5 patients. Two patients have experienced another second primary in the head and neck region, one patient experienced another primary in the head and neck and skin region, while one patient developed a lung primary and another developed a carcinoma arising from the liver. The 1- and 2-year local-regional failure rates are 28% (95% CI: 16.6 and 39.4%) and 37.3% (95% CI : 24.0 and 50.5%), respectively (Fig. 27.4).

Hence, we were able to prove that RADPLAT is indeed feasible in a multi-institutional setting. The feasibility rate of multi-RADPLAT (85%) is very similar to that of the initial phase II single-institution experience (87%), as reported previously [15]. The difference in the rate of severe toxicities is indicative of a learning curve at inexperienced institutions.

VI. CONCLUSIONS

Despite the earlier failures to demonstrate an effective role for chemotherapy in head and neck cancer, the promise remains for successfully incorporating this approach into the multimodality therapy for this disease. The targeted supradose cisplatin program is one strategy that appears to provide a lasting state of disease control for patients with T3–T4 resectable lesions without having to sacrifice the function of major organs and for patients with massive unresectable cancers. This approach is proving to be particularly effective for treating lesions that are otherwise difficult to manage, such as those in patients with massive lymphadenopathy (N3 disease), advanced paranasal sinus cancer, and temporal bone malignancies. We have also demonstrated that this technique can be used safely in

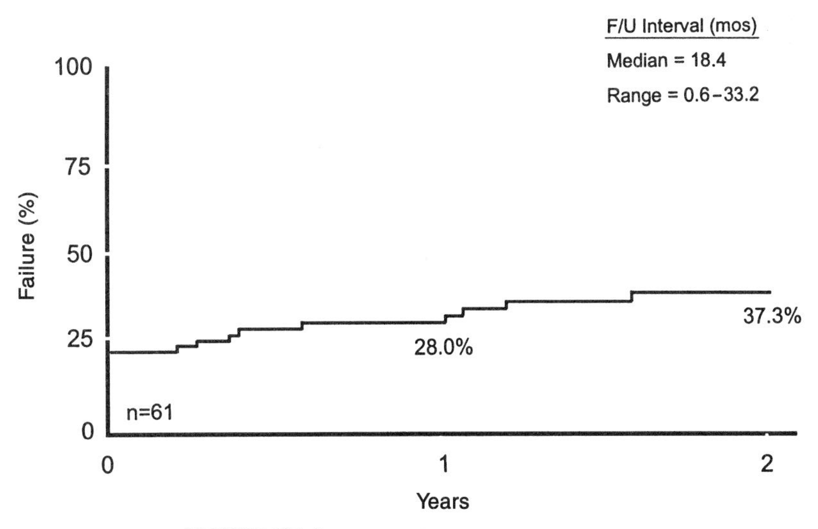

FIGURE 27.4 RTOG 9615: Local-regional failure.

multiple centers, where its effectiveness has been duplicated. Ultimately, randomized trials may be indicated to determine whether this approach can increase survival, maintain organ function, and improve upon the quality of life relative to other chemoradiotherapy protocols.

References

1. Teicher, B. A., Holden, S. A., Kelley, M. J., Shea, T. C., Cucchi, C. A., Rosowsky, A., Henner, W. D., and Frei, E, III (1987). Characterization of a human squamous carcinoma cell line resistant to cis-diamminedichloroplatinum [II]. *Cancer Res.* **47**, 388–393.
2. Waud, W. (1987). Differential uptake of cis-diamminedichloroplatinum [II] by sensitive and resistant murine L1210 Leukemia cells. *Cancer Res.* **47**, 6549–6555.
3. Van Hoff, D. D., Clark, G. M., Weiss, G. R., Marshall, M. H., Buchok, J. B., Knight, W. A., and LeMaistre, C. F. (1986). Use of in vitro dose response effects to select antineoplastics for high-dose or regional administration regimens. *J. Clin. Oncol.* **4**, 1827–1834.
4. Robbins, K. T., and Hoffman, R. (1996). Decadose effects of cisplatin on squamous cell carcinoma of the upper aerodigestive tract: I. Histoculture experiments. *Laryngoscope* **106**, 37–42.
5. Eckman, W. W., Patlak, C. S., and Fenstermacher, J. D. (1974). A critical evaluation of the principles governing the advantages of intra-arterial infusions. *J. Pharmacokinet. Biopharm.* **2**, 257–285.
6. Wheeler, R. H., Ziessman, H. A., Medvec, B. R., Juni, J. E., Thrall, J. H., Keyes, J. W., Pitt, S. R., and Baker, S. R. (1986). Tumor blood flow and systemic shunting in patients receiving intraarterial chemotherapy for head and neck cancer. *Cancer Res.* **46**, 4200–4204.
7. Kerber, C. W., Wong, W. H. M., and Robbins, K. T. (1998). A treatment for head and neck cancer by direct arterial infusion. *In* "Interventional Neuroradiology" (J. J. Conners and J. Womack, eds.). Saunders, Philadelphia.
8. Robbins, K. T., Storniolo, A. M. S., Kerber, C., Vicario, D., Seagren, S., Shea, M., Hanchett, C., Los, G., and Howell, S. B. (1994). Phase I study of highly supradose cisplatin infusions for advanced head and neck cancer. *J. Clin. Oncol.* **12**, 2113-2120.
9. Elferink, W. J. F., van der Vijah, I. K., and Pinedo, H. M. (1986). Interaction of cisplatin and carboplatin with sodium thiosulfate: Reaction rates and protein binding. *Clin. Chem.* **32**, 642–645.
10. Shea, M., Koziol, J. A., and Howell, S. B. (1984). Kinetics of sodium thiosulfate, a cisplatin neutralizer. *Clin. Pharmacol. Ther.* **35**, 419–425.
11. Robbins, K. T., Storniolo, A. M. S., Kerber, C., Vicario, D., Seagren, S., Shea, M., Hanchett, C., Los, G., and Howell, S. B. (1994). Phase I study of highly selective supradose cisplatin infusions for advanced head and neck cancer. *J. Clin. Oncol.* **12**, 2113–2120.
12. Robbins, K. T., Storniolo, A. M. S., Hryniuk, W. H., and Howell, S. B. (1996). Decadose effects of cisplatin on squamous cell carcinoma of the upper aerodigestive tract. II, Clinical studies. *Laryngoscope* **106**, 37–42.
13. Robbins, K. T., Vicario, D., Seagren, S., Weisman, R., Pellitteri, P., Kerber, C., Orloff, L., Los, G., and Howell, S. B. (1994). A targeted supradose cisplatin chemoradiation protocol for advanced head and neck cancer. *Am. J. Surg.* **168**, 419–421.
14. Robbins, K. T., Fontanesi, J., Wong, F. S. H., Vicario, D., Seagren, S., Kumar, P., Weisman, R., Pellitteri, P., Thomas, J. R., Flick, P., Palmer, R., Weir, A., Kerber, C., Murry, T., Ferguson, R., Los, G., Orloff, L., and Howell, S. B. (1996). A novel organ preservation protocol for advanced carcinoma of the larynx and pharynx. *Arch. Otolaryngol.* **122**, 853–857.
15. Robbins, K. T., Kumar, P., Regine, W. F., Wong, F. S., Weir, A. B., Flick, P., Kun, L. E., Palmer, R., Murry, T., Fontanesi, J., Ferguson, R., Thomas, R., Hartsell, W., Paig, C. U., Salazar, G., Norfleet, L., Hanchett, C. B., Harrington, V., and Niell, H. B. (1997). Efficacy of supradose intra-arterial targeted (SIT) cisplatin (P) and concurrent radiation therapy (RT) in the treatment of unresectable stage III-IV head and neck carcinoma: The Memphis experience. *Int. J. Radiat. Oncol. Biol. Phys.* **38**, 263–271.
16. Robbins, K. T., Kumar, P., Weisman, R. A., *et al.* (1996). Phase II trial of targeted supradose cisplatin (DDP) and concomitant radiation ther apy (RT) for patients with stage III-IV head and neck cancer. *Proc ASCO* **15**, 323.[Abstract]
17. Robbins, K. T., Pellitteri, P., Vicario, D., *et al.* (1996). Targeted infusions of supradose cisplatin with systemic neutralization for carcinomas invading the temporal bone. *Skull Base Surg.* **6**, 53–60.
18. Robbins, K. T., Wong, F. S., Kumar, P., Hartsell, W. F., Vieira, F., Mullins, B., and Niell, H. B. (1999). Efficacy of targeted chemoradiation and planned selective neck dissection to control bulky nodal disease in advanced head and neck cancer. *Arch. Otolaryngol. Head Neck Surg.* **125**, 670–675.
19. Ahmed, K. A., Robbins, K. T., Wong, F., and Salazar, J. E. (2000). Efficacy of concomitant chemoradiation and surgical salvage for N3 nodal disease associated with upper aerodigestive tract carcinoma. *Laryngoscope* **110**, 1789–1793.

20. Shannon, K. F., Robertson, J. T., Kumar, P., and Robbins, K. T. (1999). Intra-arterial cisplatin and concurrent radiotherapy in the treatment of paranasal sinus cancer. *Skull Base Surg*. **9**, 28. [Abstract]
21. Samant, S., Vieira, F., Hanchett, C., and Robbins, K. T. (2000). Bone or cartilage invasion by advanced head and neck cancer: Intra-arterial supradose cisplatin chemotherapy and concomitant radiotherapy for organ preservation. Proc. Int. Head Neck Cancer 5. [Abstract]
22. Ackerstaff, A. H., Tan, I. B., Rasch, C. R. N., Hilgers, F. J. M., Balm, A. J. M., Keus, R. B., and Schornagel, J. H. (2000). Quality of life assessment after chemoradiation (RADPLAT) in locally advanced, inoperable stage IV head and neck cancer patients. Proc. 5th Int. Conf. Head Neck Cancer 79. [Abstract]
23. Schornagel, J. H., and Ackerstaff, A. H. (2000). Intra-arterial cisplatin and concomitant radiation for inoperable head and neck cancer. Proc. 5th Int. Conf. Head and Neck Cancer 265. [Abstract]
24. Wilson, W. R., Siegel, R. S., Harisiadas, L., Davis, D. O., Bank, W. O., and Nguyen, H. H. (2000). High-dose intra-arterial cisplatin followed by radiation therapy is effective in treating advanced squamous cell carcinoma of the head and neck. Proc. 5th Int. Conf. Head and Neck Cancer 79. [Abstract]
25. Regine, W. F., Valentino, J., John, W., Storey, G., Sloan, D., Kenady, D., Patel, P., Pulmano, C., Arnold, S. M., and Mohiuddin, M. (2000). High-dose intra-arterial cisplatin and concurrent hyperfractionated radiation therapy in patients with locally advanced primary squamous cell carcinoma of the head and neck: Report of a phase II study. *Head Neck* **22**, 543–549.
26. Robbins, K. T., and Samant, S. (2000). Concomitant intra-arterial (IA) cisplatin, abbreviated radiotherapy and limited surgery for intermediate oral and oropharyngeal cancers. Proc. Int. Head Neck Cancer 5. [Abstract]
27. Robbins, K. T., Harris, J., Kumar, P., *et al.* (2000). Targeted chemoradiation (RADPLAT) for T4 carcinoma of the upper aerodigestive tract (RTOG 96-15): Interim analysis of a multi-institutional trial. Proc. Int. Head Neck Cancer 5. [Abstract]

C H A P T E R

28

Thyroid Cancer

STEPHEN Y. LAI and RANDAL S. WEBER

Department of Otorhinolaryngology–Head and Neck Surgery
University of Pennsylvania Medical Center
Philadelphia, Pennsylvania 19104

I. Introduction 405
II. Molecular Basis for Thyroid Cancer 406
III. Risk Factors and Etiology 406
IV. Tumor Staging/Classification 407
A. Tumor Node Metastasis Classification 407
B. AMES 408
C. AGES and MACIS 408
V. Evaluation of a Thyroid Nodule 408
A. Clinical Assessment: History and Physical Examination 408
B. Diagnostic Studies 409
VI. Review of Thyroid Cancers 412
A. Papillary Carcinoma 412
B. Follicular Carcinoma 414
C. Hürthle Cell Tumor 416
D. Medullary Carcinoma 416
E. Anaplastic Carcinoma 417
F. Other Forms of Thyroid Cancer 418
VII. Surgical Management and Technique 419
A. Surgical Anatomy and Embryology 419
B. Surgical Technique 421
C. Special Surgical Situations 422
D. Extent of Surgery 423
E. Complications 425
VIII. Postoperative Management and Special Considerations 426
A. Thyroid Hormone Replacement 426
B. Radiation Treatment 426
C. External Beam Radiotherapy and Chemotherapy 427
D. Follow-Up Management 427
IX. Conclusions 427
References 427

I. INTRODUCTION

Thyroid cancer is somewhat uncommon, accounting for approximately 1% of all malignancies [1]. The incidence of thyroid cancer in the United States is 40 per 1,000,000 people each year, resulting in 6 deaths per 1,000,000 persons annually. The approximately 18,000 annual cases of thyroid cancer represent greater than 90% of all endocrine tumors and lead to approximately 1100 deaths per year.

Although thyroid cancer is relatively rare, the incidence of thyroid nodules is significantly higher, affecting approximately 4–7% of the U.S. population [2]. While the overwhelming majority of these nodules are benign, the challenge is to identify the 5% or so of those patients with a malignant lesion. Furthermore, a subset of thyroid cancers are particularly aggressive with a potential for devastating morbidity. No reliable indicators are currently available to determine which patients will develop aggressive or recurrent disease, although risk categories based on clinical and pathologic criteria do yield important prognostic information.

The great majority (85–90%) of thyroid carcinomas are well-differentiated tumors of follicular cell origin [3]. These lesions are defined histologically as papillary, follicular, and Hürthle cell carcinoma. A small proportion (6%) of patients with these lesions have a family history of thyroid cancer. Medullary thyroid cancer, which arises from parafollicular C cells, accounts for about 6% of thyroid carcinomas. Approximately 30% of patients with these lesions have a strong genetic contribution. Anaplastic carcinomas, lymphoma, and metastatic disease comprise a small portion of thyroid malignancies.

The most common presentation of a thyroid cancer is the development of a thyroid mass or nodule. Assessment of the lesion requires a careful history, physical examination, fine-needle aspiration cytology (FNAC), and perhaps imaging studies. With correct diagnosis and management, most patients with well-differentiated thyroid carcinomas have an excellent prognosis. Controversy regarding the treatment of thyroid carcinomas and the extent of thyroidectomy to be performed arises because of the indolent course of the majority of thyroid cancers. Interventions for thyroid cancer have been difficult to evaluate because of the long follow-up and large number of patients needed to determine differences in survival. Furthermore, the morbidity that may accompany any aggressive intervention needs to be balanced with the generally good prognosis of thyroid cancer patients.

This chapter, begins with a review of the present understanding of pathogenetic mechanisms leading to thyroid cancer. After a brief review of risk factors and staging of thyroid carcinomas, we describe an algorithm for the evaluation of a thyroid nodule and the available diagnostic tools. A review of the different forms of thyroid cancer, ranging from well-differentiated carcinomas to anaplastic and other less common malignancies, is followed by a discussion of surgical management and postoperative adjuvant treatment.

II. MOLECULAR BASIS FOR THYROID CANCER

A number of genetic and molecular abnormalities have been described in thyroid carcinomas. As with other head and neck cancers, an accumulation of genetic alterations appears to be required for progression to a thyroid carcinoma. However, the events and their order have not been well delineated.

Alterations noted in the development of thyroid carcinomas include changes in total cellular DNA content. The loss of chromosomes, or aneuploidy, has been noted in 10% of all papillary carcinomas, but is present in 25–50% of all patients who die from these lesions [4]. Similarly, development of follicular adenomas is associated with a loss of the short arm of chromosome 11 (11q) and transition to a follicular adenocarcinoma appears to involve a deletion of 3p [5].

Several oncogenes, altered genes that contribute to tumor development, have been identified in early thyroid tumor progression. Mutations in the TSH receptor and G-protein mutations are found in hyperfunctioning thyroid nodules [6]. These changes can lead to the constitutive activation of cell signaling pathways, such as the adenylate cyclase–protein kinase A system, leading to a well-differentiated tumor. Point mutations of the G-protein *ras* found in thyroid adenomas, multinodular goiters, and follicular carcinomas are believed to be an early mutation in tumor progression [7]. Mutations in TRK-A, a receptor for nerve growth factor, are associated with papillary carcinomas. Mutations in *met*/hepatic growth factor have been linked to papillary and poorly differentiated thyroid carcinomas. Furthermore, mutations in the tumor suppressor gene *p53*, a transcriptional regulator, appears to be involved in the progression from papillary to anaplastic carcinoma.

The role of mutations of the *ret* oncogene in the development of papillary and medullary thyroid carcinomas has been studied extensively [8]. Located on chromosome 10, *ret* codes for a transmembrane tyrosine kinase receptor that binds glial cell line-derived neurotrophic factor (GDNF). During embryogenesis, RET is normally expressed in the nervous and excretory systems. Abnormalities in RET expression result in developmental defects, including the disruption of the enteric nervous system (Hirschsprung's disease). Medullary thyroid cancer and pheochromocytoma arise from neural crest cells containing *ret* point mutations. These point mutations have been well documented in patients with familial medullary thyroid cancer, multiple endocrine neoplasia (MEN) IIA and MEN IIB.

Rearrangements of the *ret* gene by fusion with other genes also create transforming oncogenes. These oncogene proteins—RET/PTC1, RET/PTC2, and RET/PTC3—are found in papillary thyroid cancers and are associated more frequently with childhood thyroid carcinomas. However, not all patients with papillary carcinomas express a RET/PTC gene [9]. There are marked geographical differences and the gene rearrangement is strongly associated with radiation exposure. Following the Chernobyl nuclear disaster, 66% of the papillary thyroid cancers removed from affected patients had RET/PTC1 and/or RET/PTC3 rearrangements [10].

III. RISK FACTORS AND ETIOLOGY

While the specific molecular events related to the development of thyroid carcinomas remain to be completely defined, several patient and environmental factors have been examined closely. Epidemiologic studies have not demonstrated a clear association between dietary iodine with thyroid carcinomas [11]. Additionally, there does not appear to be a simple relationship between benign goiter and well-differentiated thyroid carcinomas. Although papillary thyroid carcinomas are not associated with goiter, follicular and anaplastic thyroid carcinomas occur more commonly in areas of endemic goiter. Additionally, two particularly important risk factors, exposure to radiation and a family history of thyroid cancer, have been studied extensively.

Exposure to ionizing radiation increases patient risk for the development of thyroid carcinoma [12,13]. Low-dose ionizing radiation treatments (<2000 cGy) were used in the treatment of "enlarged thymus" to prevent "sudden crib death," enlarged tonsils and adenoids, acne vulgaris, hemangioma, ringworm, scrofula, and other conditions. The risk increases linearly from 6.5 to 2000 cGy and typically has a latent period between 10 and 30 years. Although higher doses of ionizing radiation typically lead to the destruction of thyroid tissue, Hodgkin's disease patients who receive 4000 cGy also have a higher incidence of thyroid cancer. Palpable thyroid nodularity may be present in 17–30% of patients exposed to ionizing radiation [14]. A patient with a history of radiation exposure who presents with a thyroid nodule has up to a 50% chance of having a malignancy [15]. Of these patients with thyroid cancer, 60% have cancer within the nodule, while the remaining 40% have cancer located in another area of the thyroid. The thyroid carcinoma tends to be papillary and in frequently multifocal. Additionally, there is a higher risk of cervical metastases.

Similarly, patients exposed to radiation from nuclear weapons and accidents have a higher incidence of thyroid cancer. Children near the Chernobyl nuclear power facility had a 60-fold increase in thyroid carcinoma following the nuclear accident in 1986 [16]. Most of these children were infants at the time of the accident and a great number of these cases developed without the typical latency period.

Finally, familial and genetic contributions need to be fully evaluated. The patient with a family history of thyroid carcinoma may require specific diagnostic testing. Approximately 6% of patients with papillary thyroid cancer have familial disease. Papillary thyroid cancer occurs with increased frequency in certain families with breast, ovarian, renal, or central nervous system malignancies [17]. Gardner's syndrome (familial colonic polyposis) and Cowden's disease (familial goiter and skin hamartomas) are associated with well-differentiated thyroid carcinomas. Furthermore, patients with a family history of medullary thyroid cancer, MEN IIA, or IIB warrant evaluation for the RET point mutation.

IV. TUMOR STAGING/CLASSIFICATION

A number of staging and classification systems have been devised to stratify patients with thyroid carcinomas. These classifications have identified key patient- and tumor-specific characteristics that predict patient outcome. Risk grouping has been employed to focus aggressive treatment for high-risk patients and to avoid excessive treatment and its potential complications in patients with a lower risk for tumor recurrence and/or tumor-related death.

A. Tumor Node Metastasis Classification

The American Joint Commission on Cancer (AJCC) and the Union International Contre le Cancer (UICC) adopted a tumor node metastasis (TNM) classification system (Table 28.1). In this system, patient age at presentation influences the clinical staging of a thyroid carcinoma. Eighty-two percent (82%) of patients with stage I disease had a 20-year survival of nearly 100%, whereas the 5% of patients with stage IV disease experienced a 5-year survival of only 25% [18].

TABLE 28.1 TNM Staging for Thyroid Cancer[a]

Primary tumor (T)	
TX	Primary tumor cannot be assessed
T0	No evidence of primary tumor
T1	Tumor ≥ 2 cm in greatest dimension, limited to thyroid
T2	Tumor > 2 and ≤ 4 cm in greatest dimension, limited to thyroid
T3	Tumor > 4 cm in greatest dimension, limited to the thyroid or any Tumor with minimal extrathyroid extension (e.g., extension to sternothyroid muscle or perithyroid soft tissues)
T4a	Tumor of any size extending beyond the thyroid capsule to invade subcutaneous soft tissues, larynx, trachea, esophagus or recurrent laryngeal nerve
T4b	Tumor invades prevertebral fascia or encases carotid artery or mediastinal vessels
All anaplastic carcinomas are considered T4 tumors.	
T4a	Intrathyroidal anaplastic carcinoma—surgically resectable
T4b	Extrathyroidal anaplastic carcinoma—surgically unresectable
Regional lymph nodes (N)	
NX	Regional lymph nodes cannot be assessed
N0	No regional lymph node metastasis
N1	Regional lymph node metastasis
N1a	Metastasis to level VI (pretracheal, paratracheal and prelaryngeal/Delphian lymph nodes)
N1b	Metastasis to unilateral, bilateral or contralateral cervical or superior mediastinal lymph nodes
Distant metastasis (M)	
MX	Distant metastasis cannot be assessed
M0	No distant metastasis
M1	Distant metastasis

Stage grouping	<45 years	≥45 years
Papillary/follicular		
Stage I	Any T, any N, M0	T1, N0, M0
Stage II	Any T, any N, M1	T2, N0, M0
Stage III		T3, N0, M0
		Any T, N1a, M0
Stage IVA		T4a, N0, M0
		T4a, N1a, M0
		T1–4a, N1b, M0
Stage IVB		T4b, any N, M0
Stage IVC		Any T, any N, M1
Medullary		
Stage I	T1, N0, M0	
Stage II	T2, N0, M0	
Stage III	T3, N0, M0	
	T1–3, N1a, M0	
Stage IVA	T4a, N0, M0	
	T4a, N1a, M0	
	T1–4a, N1b, M0	
Stage IVB	T4b, any N, M0	
Stage IVC	Any T, any N, M1	
Anaplastic		
Stage IVA	T4a, any N, M0	
Stage IVB	T4b, any N, M0	
Stage IVC	Any T, any N, M1	

[a]From: American Joint Committee on Cancer (2002). "AJCC Cancer Staging Manual" 6th ed. Springer, New York.

B. AMES

In this system, patient age, tumor size, extent of tumor invasion, and the presence of metastases were used to stratify patients into low-risk and high-risk groups (Table 28.2). Low-risk patients were young (men <41 years, women <51) without distant metastases and all older patients without extrathyroidal papillary carcinoma, major invasion of the tumor capsule by follicular carcinoma, or a primary tumor less than 5 cm in diameter. In a review of 310 patients from 1961 to 1980, low-risk patients (89%) had a mortality of 1.8% compared to a mortality rate of 46% in high-risk patients (11%). Recurrence in low-risk patients was 5% and in high-risk patients was 55% [19].

TABLE 28.2 Factors Employed in Prognostic Classification Systems

	AMES	AGES	MACIS
Patient factors			
Age	X	X	X
Sex	X		
Tumor factors			
Size	X	X	X
Histologic grade		X	
Histologic type	X	[a]	[a]
Extrathyroidal spread	X	X	X
Distant metastasis	X	X	X
Incomplete resection			X

[a]AGES/MACIS classifications for papillary carcinomas only.

C. AGES and MACIS

In the original AGES system, (a)ge at diagnosis, histologic tumor (g)rade, (e)xtent of disease at presentation, and tumor (s)ize were used to calculate a prognostic score [20]. Given the infrequent practice of tumor grading, a more recent modification of the system eliminated histologic tumor grade and incorporated metastasis and extent of resection. The MACIS system accounts for (m)etastasis, (a)ge at diagnosis, (c)ompleteness of surgical resection, extrathyroidal (i)nvasion, and (s)ize [21]. The MACIS score is calculated as 3.1 (patient age<40 years) or 0.08×age (patient age≥40 years)+0.3×tumor size (in cm)+1 (if extrathyroidal extension)+1 (if incomplete resection)+3(if distant metastases). Patients were stratified by their prognostic scores into four groups with statistically significant differences in 20-year disease-specific mortality.

A number of multivariable prognostic scoring systems have been developed, but none are universally accepted. Application of these classifications to a single population has demonstrated incompatible findings when compared to the original studies [22]. Furthermore, these systems do not necessarily apply to patients with poorly differentiated and more aggressive thyroid carcinomas. Nevertheless, some general conclusions can be drawn from these studies regarding the prognosis of patients with well-differentiated thyroid carcinomas. A low risk for tumor recurrence and disease-specific mortality is noted in patients who are younger at diagnosis, have smaller primary tumors that lack extrathyroidal extension or regional/distant metastases, and have complete gross resection of disease at the initial surgery. Delay in treatment will impact prognosis negatively. However, the single most significant indicator overall of a poor prognosis is distant metastases, especially to bone [23].

V. EVALUATION OF A THYROID NODULE

The incidence of thyroid nodule(s) is quite high, occurring spontaneously at a rate of 0.08% per year starting in early life and extending into the eighth decade [14]. Although thyroid nodules represent a wide spectrum of disease, the great majority are colloid nodules, adenomas, cysts, focal thyroiditis, and carcinoma. With a lifetime incidence of 4 to 7%, approximately 225,000 nodules are identified in the United States each year [24]. The vast majority of these nodules are benign and do not require removal. However, with 13,000 to 14,000 new thyroid cancers each year, about 1 in 20 new thyroid nodules will contain carcinoma and approximately 1 in 200 nodules will be lethal. Thus, the challenge in treating patients with a thyroid nodule(s) is to identify those with malignant lesions and to balance the potential morbidity of treatment with the aggressiveness of their disease.

A. Clinical Assessment: History and Physical Examination

A number of findings should raise the physician's suspicion of malignancy in a patient presenting with a thyroid nodule(s). Both younger and older patients are more likely to have a malignant thyroid nodule. Patients less than 20 years of age have an approximately 20–50% incidence of malignancy when presenting with a solitary thyroid nodule [25]. Nodular disease is more common in older patients, usually men over 40 and women over 50 years of age. Even though children may present with more advanced disease and even cervical metastases, malignancy in older patients has a considerably worse prognosis. Men often have more aggressive malignancies than women, but the overall incidence of both thyroid nodules and malignancy is higher in women.

A family history of thyroid carcinoma should be evaluated carefully. Similarly, any history of medullary carcinoma, pheochromocytoma, or hyperparathyroidism should raise suspicion for the MEN syndromes. Additionally, Gardner's syndrome (polyposis coli) or Cowden's disease has been associated with well-differentiated thyroid carcinomas. As described previously, a history of previous head and neck radiation exposure significantly increases the risk of malignancy in those patients with a thyroid nodule.

In evaluating the patient, rapid growth of a preexisting or new thyroid nodule is concerning, although the change may represent hemorrhage into a cyst. Throat or neck pain is rarely associated with carcinoma, but occurs frequently with hemorrhage into a benign nodule. Patients should be questioned carefully regarding any compressive or invasive symptoms, such as voice change, hoarseness, dysphagia, or dyspnea. However, the clinician should not rely on these findings alone, as unilateral vocal cord paralysis can be present without voice change or swallowing difficulties. Although most patients with thyroid cancer are euthyroid at presentation, symptoms of hyperthyroidism and hypothyroidism should be explored. Patients with large carcinomas that have replaced a significant portion of the normal thyroid gland may be hypothyroid, and patients with Hashimoto's thyroiditis may develop lymphoma. Although the history alone cannot determine the presence of thyroid cancer, important historical features are associated with thyroid carcinoma and should not be discounted even if diagnostic tests indicated a benign lesion.

The physical examination of a patient with a thyroid nodule begins with careful palpation of the thyroid to assess the lesion. One should determine whether the lesion is solitary or the dominant nodule in a multinodular gland, although the risk of carcinoma in either setting is the same [2,15]. Having the patient swallow will assist in the examination, as nonthyroid pathology does not typically elevate with the thyroid during swallowing. Palpable nodules are typically 1 cm or larger. Smaller nodules may be found incidentally on radiographic studies for other reasons and may be monitored. Lesions greater than 1 cm in size warrant a complete workup. Firmness of the nodule may be associated with an increased risk of carcinoma by two- to threefold [26].

Larger lesions are of more concern, as nodules greater than 2 cm in diameter have an increased incidence of harboring carcinoma. The evaluation of larger lesions also requires more caution, as the rate of false-negative results during fine-needle aspiration also increases [27]. Potential substernal extension can be estimated by the relationship of the inferior aspect of the mass to the clavicle. Potential thoracic inlet obstruction due to a substernal goiter can be assessed with Pemburton's maneuver. The patient raises his/her arms over the head and positive findings of obstruction include subjective respiratory discomfort or venous engorgement. Radiographic studies are more definitive in determining substernal involvement.

Further assessment of the patient may reveal the extent of involvement of a thyroid lesion. Palpable cervical nodes adjacent to the thyroid nodule certainly increases the suspicion for malignancy and may even be the only presenting sign of a thyroid carcinoma. However, adenopathy may be present in a patient affected by Hashimoto's thyroiditis, Grave's disease, or infection [28,29]. Large lesions may potentially shift the larynx and trachea within the neck. The mobility of the nodule relative to the laryngotracheal complex and adjacent neck structures should be evaluated. Malignant lesions are more likely to be fixed to the trachea, esophagus, or strap muscles.

All patients with a thyroid lesion should have a complete vocal cord examination. Extension into the thyroid cartilage and larynx may result in a complete vocal cord paralysis that is clinically silent. Laryngoscopy should be performed to assess vocal cord motion.

Despite the importance of the initial clinical assessment, the history and physical examination are unreliable in the prediction of carcinoma. Many of the clinical signs of malignancy are manifest late in the course of disease. Additionally, many of these same findings may be caused by events associated with benign disease (e.g., hemorrhage into a benign nodule). Thus, clinical assessment should provide a justification and a context for the interpretation of diagnostic studies, such as fine-needle aspiration. Of particular note would be any patient and thyroid nodule features that might be concerning for aggressive carcinoma behavior (Table 28.3).

B. Diagnostic Studies

1. Laboratory Studies

The majority of patients who present with a thyroid nodule are euthyroid. The finding of hypothyroidism or hyperthyroidism tends to shift the workup away from thyroid carcinoma to a functional disorder of the thyroid gland, such as Hasimoto's thyroiditis or a toxic nodule [30]. While many thyroid hormone tests are available, few are needed in the initial patient evaluation. Thyroid-stimulating

TABLE 28.3 Risk Factors for Aggressive Behavior of Well-Differentiated Thyroid Carcinomas

Risk factors
Patient factors
Age
Male >40 years
Female >50 years
Gender
Male > female
Histopathologic factors (at initial presentation)
Size (>4 cm)
Extrathyroidal spread
Vascular invasion
Lymph node metastasis
Distant metastasis
Histologic type
Tall cell variant of papillary carcinoma
Follicular carcinoma
Hürthle cell carcinoma

hormone (TSH) measurement serves as an excellent screening test. Full thyroid function tests can be performed if the TSH level is abnormal.

Measurement of thyroglobulin is generally not performed initially, as thyroglobulin is secreted by both normal and malignant thyroid tissue. Levels of thyroglobulin cannot differentiate between benign and malignant processes, unless levels are extremely high, as in metastatic thyroid cancer. Furthermore, antithyroglobulin antibodies may also interfere with the assay. Thyroglobulin levels may be useful in following patients who have undergone total thyroidectomy for well-differentiated thyroid cancer.

Serum calcitonin levels are not a typical initial test for patients with a thyroid nodule unless the patient has a family history of medullary thyroid cancer or MEN II. However, if fine-needle aspiration cytology demonstrates or is suspicious for medullary thyroid carcinoma, calcitonin levels should be obtained. Additionally, if the patient has RET oncogene mutations, the existence of a coexisting pheochromocytoma should be evaluated through a 24-h urine collection to measure vanillylmandelic acid (VMA), metanephrine, and catecholamine. The serum calcium level should be measured to exclude hyperparathyroidism.

2. Imaging

Ultrasonographic imaging is tremendously useful and sensitive. These studies detect nonpalpable nodules and differentiate between cystic and solid nodules. Identifying solid nodules is clearly important, as these lesions have an increased likelihood of harboring carcinoma [28,29]. In patients with a difficult neck to examine (e.g., a patient with previous history of head and neck irradiation), sonography can also clarify findings. Sonography can identify hemiagenesis and contralateral lobe hypertrophy, which may be misdiagnosed as a thyroid nodule. Additionally, these studies can identify cervical nodes, which may contain metastatic disease. Sonography is also useful in the evaluation of cervical lymph nodes in patients with a history of thyroid cancer who present with adenopathy or rising thyroglobulin levels.

These studies provide key baseline information regarding nodule size and architecture. Thus, sonography is also a noninvasive and inexpensive method for following changes in the size of benign nodules. However, there is no role for sonography in screening asymptomatic patients for thyroid nodules. Additionally, these studies are not useful in the evaluation of substernal extent or the involvement of adjacent structures.

Computed tomography (CT) and magnetic resonance imaging (MRI) scans are usually unnecessary in the evaluation of thyroid tumors, except for large or retrosternal lesions. Although these studies are not as effective as sonography in the evaluation of thyroid nodules, they are more reliable in evaluating the relationship of the thyroid lesion to adjacent neck structures, such as the trachea and esophagus. These studies are useful in determining substernal extension, identifying cervical and mediastinal adenopathy, and evaluating possible tracheal invasion [31]. Caution must be exercised in the use of iodine-containing contrast material in patients with multinodular goiter if a hyperthyroid state is suspected and in patients with well-differentiated thyroid cancer. In the latter group, iodinated contrast media will preclude the use of postoperative radioactive iodine therapy for 2 to 3 months. Finally, the MRI scan is more accurate than a CT scan in distinguishing recurrent or persistent thyroid tumor from postoperative fibrosis.

3. Thyroid Isotope Scanning

Radionuclide scanning with iodine-123 (^{123}I) or technetium-sestamibi (^{99m}Tc) assesses the functional activity of a thyroid nodule and the thyroid gland. Nodules that retain less radioactivity than the surrounding thyroid tissue are termed "cold," nonfunctioning, or hypofunctional. These cold nodules are thought to have lost functions of fully differentiated thyroid tissue and to be at increased risk of containing carcinoma. In a meta-analysis of patients with scanned nodules that were removed surgically, 95% of all modules were cold [28,29]. The incidence of malignancy in cold nodules was 10–15%, but only 4% in hot nodules.

^{99m}Tc scanning only tests iodine transport, but can be performed in 1 day and involves less radiation exposure than ^{123}I.. Cold nodules identified with this test will also be cold with iodine scanning. However, any "hot" nodules require ^{123}I scanning for confirmation. ^{123}I scanning tests both transport and organification of iodine. The test is more expensive and requires 2 days to complete. Cold lesions can be more difficult to visualize because of overlying thyroid tissue and glandular asymmetry, although oblique views during scanning can improve detection. Additionally, ^{99m}Tc does not penetrate the sternum and is not useful in confirming substernal extension.

With the evolution of fine-needle aspiration cytology, radionuclide scanning is not performed routinely in the evaluation of a thyroid nodule. More frequently, "cold" nodules are detected in patients during evaluation for thyroid metabolic disorders and these patients are subsequently referred to a surgeon for further assessment. However, patients who present initially with a thyroid nodule and are found to be hyperthyroid on preliminary thyroid function testing should have radionuclide scanning to differentiate toxic nodule versus Grave's disease. Additionally, patients suspected of having Hashimoto's thyroiditis may also be scanned. A small firm thyroid lobe in a patient with this condition can be mistaken for a nodule. In a hypothyroid patient with positive thyroid peroxidase (TPO) antibodies, radionuclide scanning can clarify this situation. Thus, the diagnosis of

Hashimoto's thyroiditis can be confirmed and fine-needle aspiration, which has a high false-positive rate in this condition, can be avoided.

4. Fine-Needle Aspiration Cytology

Although other studies may be helpful in the diagnosis of a thyroid nodule, fine-needle aspiration cytology has largely replaced radionuclide scanning and ultrasonographic imaging as the central diagnostic test in the initial evaluation of thyroid nodules. The procedure is minimally invasive and may be performed quickly with little patient discomfort. Unlike large-bore needle biopsies, such as the Tru-cut or Vim-Silverman needle, there are fewer complications. The findings are highly sensitive and specific, although the accuracy of FNAC is related to the skill of the aspirator and the experience of the cytopathologist [32]. With the advent of this technique, the number of patients requiring surgery has decreased by 35–75% and the cost in managing patients with thyroid nodules has been reduced substantially [33,34]. Additionally, the yield of malignancies has almost tripled in those patients who have had thyroid surgery following FNAC [34,35]. The accuracy of an FNA diagnosis of papillary carcinoma is 99% with a false-positive rate of less than 1% [36].

FNAC should be one of the initial steps in the surgical evaluation of a thyroid nodule. Approximately 15% of all aspirates are inadequate or nondiagnostic, largely because of the sampling from cystic, hemorrhagic, hypervascular, or hypocellular colloid nodules. Reaspiration of the nodule is critical, as a nondiagnostic finding should never be interpreted as a negative finding for carcinoma. In fact, surgical diagnoses following repeated nondiagnostic aspirations revealed malignant nodules in 4% of women and 29% of men [37]. Nodules that are difficult to localize and those that have yielded nondiagnostic aspirates on previous attempts may benefit from ultrasound-guided aspiration.

Cytopathologic evaluation of a successful FNA will categorize a nodule into the following groups: benign, malignant, and suspicious. In 60–90% of nodules, FNAC will reveal a benign or "negative" diagnosis. The likelihood of malignancy (false-negative rate) is 1–6% [38]. The diagnosis of malignancy, particularly papillary (including follicular variant), medullary, and anaplastic carcinomas and lymphomas, can be determined in about 5% of nodules. The likelihood of a false-positive finding is less than 5% [38]. Frequently, false-positives result from difficulties in interpreting cytology in patients with Hashimoto's thyroiditis, Grave's disease, or toxic nodules. The remaining "suspicious" samples are composed of lesions that contain abnormal follicular epithelium with varying degrees of atypia. This finding needs to be evaluated in the context of patient history and physical findings that may be suggestive of malignancy.

Follicular neoplasms cannot be classified by FNAC alone. The least worrisome finding is a macrofollicular lesion or colloid adenomatous nodule, which has a very low malignant potential. The presence of hypercellular, microfollicular arrays with minimal colloid increases the concern for carcinoma. However, the differentiation between follicular adenoma and follicular carcinoma depends on the histologic finding of capsular or vascular invasion, which requires evaluation of the entire thyroid nodule. Occasionally, patients with a diagnosis of follicular neoplasm on FNAC will have a ^{123}I-thyroid scan. If the suspicious nodule is "cold," surgery is indicated. Overall, 20% of nodules diagnosed as follicular neoplasms by FNAC will contain thyroid carcinomas [39].

Similarly, Hürthle cells (oxyphilic) neoplasms can be difficult to evaluate. The presence of Hürthle cells in an aspirate may indicate an underlying Hürthle cell adenoma or carcinoma, but may also be present in thyroid disorders, such as multinodular goiter and Hashimoto's thyroiditis. Carcinomas can be found in up to 15% of nodules identified as follicular and oxyphilic neoplasms [40]. Because of the risk of underlying carcinoma in these cases, surgery is recommended.

5. Management of a Thyroid Cyst

Approximately 15–25% of all thyroid nodules are cystic or have a cystic component [14]. The presence of a cyst does not signify a benign lesion, as papillary carcinomas and parathyroid tumors may present with cystic masses. When encountered during fine-needle aspiration, the cyst should be drained completely. This may prove curative in the majority of simple cysts, although one or two additional drainage procedures may be required. However, if a cyst persists after three drainage attempts or reaccumulates quickly, the suspicion for carcinoma should increase. Brown fluid withdrawn from a cyst may represent old hemorrhage into an adenoma, but red fluid is more suspicious for carcinoma [26]. Clear, colorless fluid may be withdrawn from a parathyroid cyst and can be assessed for parathyroid hormone [41]. Overall, the incidence of carcinoma in all thyroid cysts is less than 9%, although the incidence is much higher in cysts greater than 4 cm in diameter [28]. In suspicious cases, the surgeon and patient should consider an ultrasound-guided FNA to sample a solid component of the lesion or a unilateral thyroid lobectomy to obtain a definitive diagnosis.

6. A Rational Approach to Management of a Thyroid Nodule (Fig. 28.1)

A number of diagnostic algorithms have been proposed for the evaluation of a thyroid nodule [32,42]. In general, evaluation begins with a thorough history and physical

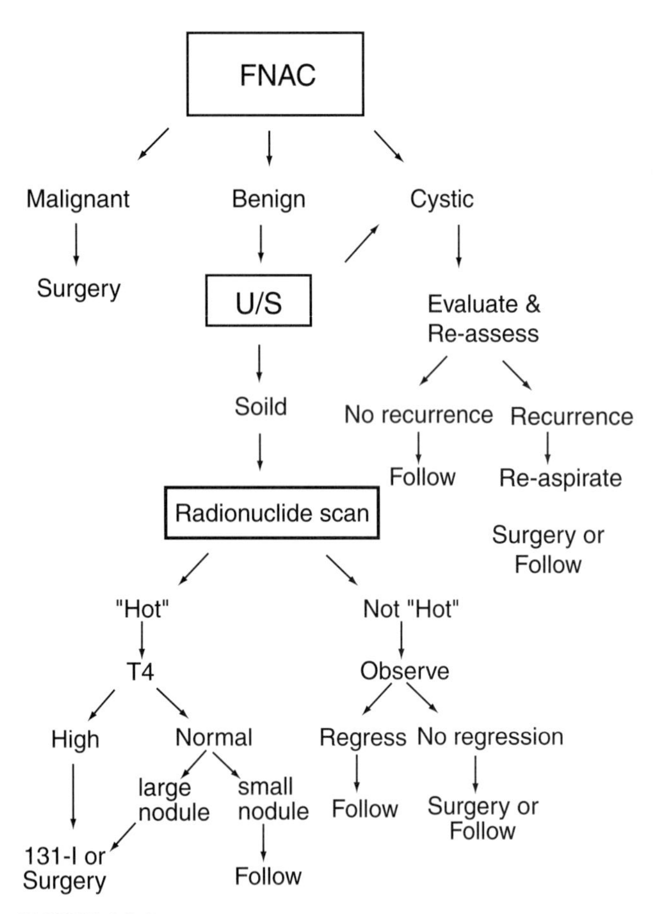

FIGURE 28.1 An algorithm for the evaluation and management of a thyroid nodule.

examination to identify significant risk factors. Surgery may be deemed appropriate based solely on high-risk factors, such as age, sex, history of radiation exposure, rapid nodule growth, upper aerodigestive tract symptoms, and/or fixation.

Baseline TSH screening then determines the diagnostic course. Patients with hyperthyroidism (suppressed serum TSH level) should receive radionuclide scanning to determine the presence of a toxic "hot" nodule or Marine–Lenhart syndrome or Graves' disease with a concomitant "cold" nodule [43]. The majority of patients will be euthyroid (normal serum TSH level) and an FNA should be performed. Cytologic findings that are diagnostic or strongly suggestive of malignancy should direct the surgeon to removal of the lesion.

A diagnosis of follicular neoplasm by FNAC requires surgery to determine the presence of follicular adenoma or follicular carcinoma. An FNAC sample suspicious for medullary carcinoma may be subject to immunohistochemical techniques to detect calcitonin. Prior to surgical intervention, a patient with an FNAC suggestive of medullary carcinoma will require genetic studies and additional testing that will be discussed later (see Section VI,D). Suspicious findings on FNAC must be assessed in the context of patient risk factors in determining the need for surgery. If a nonsurgical approach is taken, the nodule must be monitored closely, usually with ultrasonography. Benign lesions are usually observed and require surgical removal only in cases of cosmetic or symptomatic concerns. These nodules may be reaspirated after 1 year to confirm the diagnosis or sooner if growth is detected.

VI. REVIEW OF THYROID CANCERS

A. Papillary Carcinoma

1. Clinical Presentation

Papillary carcinoma is the most common form of thyroid malignancy, accounting for 60–70% of all thyroid cancer [44,45]. This lesion typically presents in patients 30 to 40 years of age and is more common in women, with a female-to-male ratio of 2:1. Interestingly, this ratio has decreased steadily since the 1960s as the incidence in men has risen) [46]. Papillary carcinomas are the predominant thyroid malignancy in children (75%). Although children present more commonly with advanced disease, including cervical and distant metastases, their prognosis remains quite favorable.

The majority of cases of papillary carcinoma occur spontaneously. Patients with a history of low-dose radiation exposure tend to develop papillary carcinomas (85–90%) [47]. Additionally, these lesions are more common in patients with Cowden's syndrome and familial polyposis. Only 6% of papillary carcinomas are associated with familial disease.

Papillary carcinoma may be classified into three categories based on size and extent of the primary lesion [48,49]. Minimal or occult/microcarcinoma tumors are up to 1.5 cm in size and demonstrate no evidence of invasiveness through the thyroid capsule or to cervical lymph nodes. These lesions are typically non palpable and are usually incidental findings during operative or autopsy examination. Intrathyroid tumors are greater than 1.5 cm in diameter, but are confined to the thyroid gland with no evidence of extrathyroid invasion. Extrathyroid tumors extend through the thyroid capsule to involve the surrounding viscera. This latter form of papillary carcinoma is associated with a substantial morbidity and decreased survival [48,50].

Most patients present with a slow-growing, painless mass in the neck and are often euthyroid. Often, the primary lesion is confined to the thyroid gland, although up to 30% of patients may have clinically evident cervical nodal disease [51,52]. Histologic studies have demonstrated the strong lymphotropic nature of papillary carcinoma, leading to multifocal disease within the thyroid and regional lymphatics. Microscopic disease has been identified in the cervical nodes of 50–80% of patients and in the contralateral lobe in up to 80% of patients with papillary carcinoma [53]. However, the significance of this microscopic disease is unclear, as clinical recurrence in the neck and in the contralateral lobe occurs less than 10% of patients [54]. More likely, the prevalence of microscopic disease suggests that the majority of papillary carcinomas have an indolent course that only occasionally becomes clinically evident. However, definite predictors of the clinical course for papillary carcinoma are not well defined.

Advanced disease may be associated with symptoms of local invasion, including dysphagia, dyspnea, and hoarseness. Occasionally, cervical nodal involvement may be more apparent than the thyroid nodule. Distant metastases, especially to the lungs, are encountered more commonly in children, although up to 10% of all patients may ultimately develop distant disease [40].

Thyroid cancer is often suspected in these patients following a thorough history and physical examination. The diagnosis is usually established by FNAC. Thyroid function tests are done routinely in the preoperative assessment. Radiographic imaging (CT scan or MRI) is performed selectively to define extensive local or substernal disease and to evaluate possible lymph node involvement.

2. Pathology

On gross examination, papillary carcinoma is firm, white, and not encapsulated. The lesion tends to remain flat on sectioning, rather than bulging like normal thyroid tissue or benign nodular lesions. Macroscopic calcifications, necrosis, or cystic changes may be readily apparent [55].

Histologically, these lesions arise from thyroid follicular cells and contain papillary structures, which consist of a neoplastic epithelium overlying a true fibrovascular stalk [56]. Cells are cuboidal with a pale, abundant cytoplasm. Large, crowded nuclei with folded and grooved nuclear margins may have intranuclear cytoplasmic inclusions. Prominent nucleoli account for the "orphan Annie eye" appearance. Laminated calcium densities, psammoma bodies, are likely the remnants of necrotic calcified neoplastic cells and are present in 40% of cases.

Lesions with any papillary features, even though a follicular component may predominate, behave clinically as papillary carcinomas. Thus, the designation of papillary carcinoma includes mixed papillary follicular carcinoma and the follicular variant of papillary carcinoma. A more unfavorable prognosis is associated with the certain histologic forms of papillary carcinoma, including diffuse sclerosing and tall cell variants [56,57].

Papillary carcinomas have a strong tendency for lymphatic spread within the thyroid and to local lymph nodes in the paratracheal and cervical regions. The tendency for intraglandular spread may lead to the multifocal disease often present in patients. However, discrete lesions may be due to *de novo* formation, especially in patients exposed previously to ionizing radiation [58].

Local invasion occurs in 10–20% of these tumors, leading to involvement of the overlying strap muscles, laryngeal and tracheal framework, recurrent laryngeal nerves, pharynx, and esophagus. This extension may evolve from the primary lesion or from the extracapsular extension of metastatic nodes. Angioinvasion is a clear harbinger of increased risk for recurrence and worse prognosis [45]. Interestingly, a coexisting lymphocytic thyroiditis has been correlated with decreased recurrence and better overall prognosis.

3. Management and Prognosis

The majority of patients with papillary carcinoma do well regardless of treatment. Prolonged survival, even with recurrent disease, has led to controversy regarding the extent of thyroidectomy for patients with well-differentiated thyroid carcinomas (see Section VII,D). A balance must be achieved between an effective surgical treatment for these malignancies and the potential morbidity of this surgery. A number of studies have attempted to categorize patients by their risk factors and to justify more aggressive surgical intervention for high-risk patients (see Section IV).

Minimal papillary thyroid carcinoma is usually identified in a thyroid specimen removed for other reasons. Unilateral thyroid lobectomy and isthmusectomy is usually sufficient surgical treatment unless there is angioinvasion or tumor at the margins of resection. These patients may then be treated with thyroid hormone to suppress TSH and closely followed. In patients with a small, encapsulated papillary thyroid carcinoma (< 1.5 cm in diameter), a total lobectomy is sufficient.

When patients present with more extensive disease or indications of disease in both lobes, total or near-total thyroidectomy is the procedure of choice. Additionally, patients stratified into high-risk categories in any of the classification schemes described previously (see Section IV) would probably benefit from a more extensive surgical procedure. This would permit possible thyroid hormone suppression therapy and radioiodine ablation of remaining disease.

Multifocal disease is present in as many as 80% of patients in some reports [52,57]. This may represent *de novo* multicentric tumor formation or intraglandular metastasis. The prevalence of multifocal disease lends credence to the argument for more complete surgical removal of the thyroid gland in patients with papillary thyroid cancer. Patients with partial thyroidectomy had higher local recurrence rates and increased pulmonary and cervical metastases [59,60]. Controversy remains, however, because this increased local recurrence did not compromise disease survival in some studies [19,61].

Generally, invasive tumors are associated with a compromise in survival. Woolner *et al.* [55] reviewed 1181 thyroid cancer patients and found that no patient died of papillary cancer when the lesion was less than 1.5 cm in size. Only 3% of patients died when the lesion was larger, but remained intrathyroid. Mortality rose to 16% of patients when extrathyroid disease was present.

Following total thyroidectomy, patients may be monitored by following thyroglobulin levels (should remain below 3 ng/ml). Any rise in thyroglobulin levels is suspicious for disease recurrence and will require appropriate screening with 131-iodine. Approximately 12% of patients with papillary carcinoma are not cured by initial treatment, leading to a prolonged clinical course [62]. Recurrent disease may occur after many years, involving the thyroid bed (5–6%), regional lymphatics (8–9%), and/or distant sites (4–11%) [63]. Successful treatment of recurrence varies by site of involvement and the patients initial risk classification.

Local recurrence is a serious complication and is associated with a disease-related mortality of 33–50% [63]. Typically, patients with nodal recurrence fare better than those with tumor recurrence in the thyroid bed or distant sites. Studies have been inconsistent regarding the impact of cervical metastases on survival. Patients older than age 40 may have clinically evident nodal disease in 36–75% of cases and overall increased mortality [23,46]. Surprisingly, some studies suggest prognosis is better with more cervical node involvement [21,64]. Although the role of cervical metastasis in survival may be controversial, there is an association with an increased recurrence rate, especially in elderly patient [62,64,65].

Given these findings and the overall prevalence of microscopic cervical disease with uncertain prognostic implications, management of cervical metastasis tends to be conservative. There is no role for elective neck dissection in the clinically disease-free neck, especially given the effectiveness of radioiodine therapy in ablating microscopic disease [54]. In patients with a primary tumor larger than 2 cm in diameter or involving extrathyroidal structures, nodal dissection in the central compartment from hyoid bone to mediastinum between the internal jugular veins should be performed [66]. In patients with palpable neck disease, a selective neck dissection should be performed. Dissection of the posterior triangle and suprahyoid regions is typically unnecessary, but should be performed if disease is present.

The presence of distant metastasis is clearly associated with a worse prognosis. Approximately 10% of patients with papillary thyroid carcinoma develop distant metastasis at some point during their disease course [40]. Most commonly, the lungs are involved, although bone sites and the central nervous system may also be affected. In patients with macroscopic metastasis, death from disease occurs within 1 year of detection in almost 50% of patients [67].

In nearly every study, patient age at the time of diagnosis is an important prognostic variable [23,64,68]. Older patients, especially above the age of 40, with papillary carcinoma have a worse prognosis. Furthermore, extrathyroidal invasion appears to be more common in older patients. The prognosis for men younger than age 40 is comparable to women of the same age. However, overall survival is worse for men and the risk of death from papillary thyroid carcinoma may be twice as great [19,64]. Furthermore, since the 1960s, the increase in incidence of papillary thyroid carcinoma in men has decreased the gender ratio of men to women with the disease from 1:4 to 1:2 [46].

Children clearly fare better with this disease. In patients younger than 15, 90% demonstrate cervical metastasis at some time during their disease course [69]. Furthermore, up to 20% of children may present with pulmonary metastases [70]. However, neither factor seems to have any impact on survival. Perhaps these differences may be related to biological differences in the disease process or between age groups.

Finally, the tall cell variant of papillary thyroid carcinoma is clearly different from other forms of this disease. Review of patients with tall cell variant papillary carcinoma demonstrates a more aggressive natural history in all age groups and a worse prognosis [71].

B. Follicular Carcinoma

1. Clinical Presentation

Follicular carcinomas represent 10% of thyroid malignancies. The mean age of presentation is 50 years compared to the younger mean age of patients with papillary carcinoma (35 years). Women have this lesion more commonly, with a female-to-male ratio of 3:1 [72]. These lesions occur more frequently in iodine-deficient areas, especially areas of endemic goiter. Follicular carcinomas have been correlated with pregnancy and with certain HLA subtypes (DR1, DRw, and DR7). Also, a rare form of familial follicular carcinoma is reported in patients with dyshormonogenesis. Interestingly, the overall incidence of follicular carcinoma is decreasing in the United States.

Patients usually present with a solitary thyroid nodule, although some patients may have a history of long-standing goiter and recent rapid size increase. These lesions are typically painless, although hemorrhage into the nodule may cause pain. Cervical lymphadenopathy is uncommon at initial presentation, although distant metastases are encountered more frequently than with papillary carcinomas. In rare cases (1%), the follicular carcinoma may be hyperfunctioning and the patient will present with signs and symptoms of thyrotoxicosis.

Other than characterization of a follicular neoplasm, a definitive preoperative diagnosis is usually not possible by FNAC. Differentiation between follicular adenoma and follicular carcinoma requires an evaluation of the thyroid capsule for invasion or identification of vascular invasion. Typically, about 20% of thyroid nodules demonstrating follicular cells will contain carcinoma.

Unlike papillary carcinoma, follicular thyroid carcinomas are less likely to metastasize via lymphatic pathways (found in less than 10% of patients) [31]. More commonly, follicular carcinoma spread through local extension and hematogenous spread. Often, the presence of cervical lymph node disease indicates significant local disease and visceral invasion [46,73]. Distant metastasis is also more common in follicular cancers than papillary cancers, especially at presentation [74,75]. A pathologic bone fracture may be the initial presentation of follicular carcinoma. Other common sites include the liver, lung, and brain.

2. Pathology

Follicular thyroid carcinoma tends to present as solitary, encapsulated lesions. Cytologic analysis of follicular neoplasms reveals small follicular arrays or solid sheets of cells [56]. The follicular structures have lumen that do not contain colloid, and the overall architectural pattern depends on the degree of tumor differentiation. Increased cellularity may increase the suspicion for carcinoma, but cytology alone is not sufficient to distinguish between a follicular adenoma and a carcinoma.

Histologic findings are necessary to distinguish benign and malignant lesions. Malignant lesions are differentiated by the identification of capsular invasion and potential microvascular invasion of vessels along the tumor capsule [76,77]. Complete capsular evaluation must be performed. Thus, frozen-section analysis is often inadequate and definitive diagnosis requires complete assessment of permanent sections.

The degree of capsular invasion is important for patient prognosis. Follicular carcinomas can be divided into two broad categories. Minimally invasive tumors demonstrate evidence of invasion into, but not through, the tumor capsule at one or more sites. These lesions do not exhibit small vessel invasion. Frankly invasive tumors demonstrate invasion through the tumor and often exhibit vascular invasion [78]. Tumor infiltration and invasion may be apparent at surgery with tumor present in the middle thyroid or jugular veins.

Many other factors have been investigated as means to differentiate between adenomas and carcinomas. To date, no molecular markers have been clinically useful. DNA ploidy varies in both adenomas and carcinomas with considerable overlap [79]. Aneuploid follicular carcinomas though are noted to behave in a more aggressive manner.

3. Management and Prognosis

Patients diagnosed with a follicular lesion by FNAC should have a thyroid lobectomy with isthmusectomy performed. The pyramidal lobe, if present, should be included in the resection. As described previously, cytologic findings alone cannot determine the presence of adenoma or carcinoma. Intraoperative frozen-section analysis is not helpful given the incomplete assessment of the tumor capsule. However, frozen sections should be analyzed to confirm gross evidence of adjacent cervical lymphadenopathy. A total thyroidectomy is recommended if carcinoma is identified. In patients with a clinical suspicion for follicular carcinoma, the tendency is to perform a more complete resection. Total thyroidectomy is performed in older patients with a nodule greater than 4 cm in size diagnosed by FNAC as follicular neoplasm. In these patients, the risk of carcinoma is approximately 50% [80].

A diagnosis of follicular carcinoma following a thyroid lobectomy usually necessitates a completion thyroidectomy. Patients with minimally invasive follicular cancer have a very good prognosis and the initial thyroid lobectomy may be sufficient treatment. However, invasiveness of follicular carcinoma correlates directly with decreased survival. In patients with invasive follicular carcinomas, many surgeons tend toward completion and total thyroidectomy to permit radioiodine scanning for detection and ablation of metastatic disease. More aggressive surgical intervention does not clearly improve survival given that invasiveness already indicates the increased likelihood of distant metastasis. Neck dissection is performed if cervical lymphadenopathy is present. Elective neck dissections are unwarranted because nodal involvement is relatively infrequent [66].

The recurrence rate following initial management is approximately 30% [81]. Recurrence is related to the degree of invasiveness of the initial lesion, not the extent of initial thyroid surgery. Minimally invasive disease behaves similarly to follicular adenoma and is typically cured with conservative surgical procedures (thyroid lobectomy) [82]. Recurrence of minimally invasive follicular carcinoma is approximately 1%. About 15% of those with recurrent or metastatic disease can be cured. The prognosis of these patients relates to the site of recurrence as well as the patient's initial risk stratification. Survival outcomes are significantly poorer in those with capsular invasion and angioinvasion [83]. Those with cervical node recurrence have a 50% cure rate, whereas those with distant metastases have a cure rate of about 9% [63,81].

The prognosis of patients with follicular carcinoma has typically been reported to be worse than for those with papillary carcinoma. Five-year survival is 70% and decreases to 40% for 10 years. The presence of distant metastasis diminishes 5-year survival to 20% [84]. Factors that worsen prognosis include age greater than 50 years old at presentation, tumors greater than 4 cm in size, higher tumor grade, marked vascular invasion, extrathyroidal invasion, and distant metastasis at the time of diagnosis [66]. Extrathyroidal invasion beyond the capsule and into the thyroid parenchyma and local structures is the key factor decreasing patient survival. Some reports that match age, sex, and stage at time of diagnosis suggest that patients with papillary and follicular carcinomas have similar survival patterns and that the poor prognosis of those with follicular carcinoma is related to the increased number of patients who present at an older age and a more advanced disease stage [23,85]. Additionally, unlike papillary carcinoma, mortality is directly related to recurrence in patients with follicular carcinoma [46].

C. Hürthle Cell Tumor

1. Clinical Presentation

According to World Health Organization classification, Hürthle cell tumor (HCT) is a subtype of follicular cell neoplasm. HCT nodules may be found in patients with Hashimoto's thyroiditis, Graves' disease, or within a nodular goiter. These tumors are derived from oxyphilic cells of the thyroid gland. While the precise function of these cells is unknown, Hürthle cells express thyroid-stimulating hormone receptors and produce thyroglobulin.

Hürthle cell neoplasms are typically diagnosed by FNAC. Approximately 20% of these lesions are malignant. As with follicular lesions, histologic criteria are required to diagnose carcinomas. Hürthle cell carcinomas represent approximately 3% of all thyroid malignancies. Hürthle cell carcinomas are more aggressive than follicular carcinomas. They are often multifocal and bilateral at presentation. Additionally, these malignancies are more likely to metastasize to cervical nodes and distant sites [86].

2. Pathology

FNAC of HTC typically demonstrates hypercellularity and the presence of eosinophilic cells. These neoplasm are characterized by sheets of eosinophilic cells packed with mitochondria. Cytologic differentiation between adenoma and malignant tumor by FNAC is extremely difficult. f capsular or vascular invasion confirm le cell carcinoma.

3. Management and Prognosis

The clinical approach to a Hürthle cell tumor is similar to that for follicular neoplasms. With a Hürthle cell adenoma, resection of the affected lobe and isthmus appears to be sufficient for cure. Invasive findings for a Hürthle neoplasm by frozen section examination intraoperatively or on formal pathology warrants a total or completion thyroidectomy. Hürthle cell carcinomas tend to be more aggressive than other follicular carcinomas and are less amenable to radioiodine therapy given their decreased tendency to uptake radiolabeled iodine. A careful examination for local disease extension or adjacent cervical lymphadenopathy should be performed. These patients should routinely have removal of paratracheal nodes, as in patients with medullary thyroid cancer. Also, a neck dissection should be undertaken if lateral neck nodes are palpable. Overall, survival rates for Hürthle cell tumors are significantly worse than for follicular thyroid cancer.

Postoperative management should include TSH suppression and thyroglobulin monitoring. A technicium-sestamibi scan may be useful for detecting persistent local or metastatic disease. A radioiodine scan and ablation may be performed to remove any residual normal thyroid tissue to allow for better surveillance. However, this therapy is unlikely to be effective in tumor ablation, as few (approximately 10%) Hürthle cell carcinomas uptake radioiodine [87].

D. Medullary Carcinoma

1. Clinical Presentation

Medullary thyroid carcinomas (MTC) are a distinct category of disease and represent approximately 5% of all thyroid carcinomas. These malignancies arise from parafollicular C cells and may secrete calcitonin, carcinoembryonic antigen (CEA), histaminadases, prostaglandins, and serotonin. Measurement of secreted calcitonin is useful for the diagnosis of medullary carcinoma and for postsurgical surveillance for residual and recurrent disease.

MTC demonstrate an intermediate behavior between well-differentiated thyroid cancers and anaplastic carcinomas. Women and men are equally affected by medullary carcinomas [88]. Patients usually present with a neck mass associated with palpable cervical lymphadenopathy (up to 20%) [89]. Local pain is more common in these patients, indicating the presence of local invasion, and may be associated with dysphagia, dyspnea, and/or dysphonia. MTC may present along with papillary thyroid carcinoma, as related mutations in *ret* are present in both diseases. While MTC spreads initially to cervical nodes, distant metastases may be found in the mediastinum, liver, lung, and bone and are present in up to 50% of patients at diagnosis [90].

The majority (70%) of medullary carcinomas are spontaneous unifocal lesions in patients aged 50–60 years old without an associated endocrinopathy [89]. The remaining 30% of cases affecting younger patients are familial. These hereditary medullary carcinomas are inherited as autosomal-dominant traits with nearly 100% penetrance. Medullary carcinoma in these patients is preceded by multifocal C-cell hyperplasia and leads to disease that is multicentric and bilateral in 90% of cases [91,92]. Familial medullary thyroid carcinoma (FMTC) is not associated with any other endocrine pathology. Two forms of multiple endocrine neoplasm (MEN) syndrome are associated with MTC. Patients with MEN IIA exhibit MTC, pheochromocytoma, and hyperparathyroidis [92,93]. Patients with MEN IIB have a marfanoid body habitus and may suffer MTC, pheochromocytoma, and mucosal neuromas. Although penetrance for MTC approaches 100% in these patients, expression of other features is variable [91,94].

FNAC diagnosis of MTC is confirmed by elevated serum calcitonin. Additionally, these patients should have testing for mutation of the *ret* protoncogene. Genetic screening has largely replaced provocative pentagastrin-stimulation testing. Careful screening for hereditary diseases is also necessary when a patient is diagnosed with MTC. Hyperparathyroidism can be assessed by serum calcium levels and appropriate imaging studies. Patients should also be screened for the presence of a pheochromocytoma with 24-h urinary levels for VMA, catecholamine, and metanephrine. An undiagnosed pheochromocytoma could lead to an intraoperative hypertensive crisis and death. Additionally, the detection of any hereditary form of MTC in a patient should lead to family screening. Affected family members can often be identified and treated at earlier stages of disease with improved survival [95,96].

2. Pathology

MTC originates from parafollicular C cells of neuroectomdermal origin [97]. They descend to join the thyroid gland proper and are concentrated mainly in the lateral portions of the superior poles. Thus, most MTC lesions are located in the middle and upper thyroid poles. In patients with hereditary forms of MTC, the disease is often multifocal.

These lesions are composed of sheets of infiltrating neoplastic cells that are heterogeneous in shape and size. These cells are separated by collagen, amyloid, and dense irregular calcification. The amyloid deposits are likely polymerized calcitonin and are virtually pathognomonic for MTC, although not all MTC contain amyloid [98]. More aggressive tumors typically have increased mitotic figures, nuclear pleomorphism, and areas of necrosis. Immunohistochemistry for calcitonin and CEA are useful diagnostic studies.

3. Management and Prognosis

Total thyroidectomy is the treatment of choice in patients with MTC, as the lesions have a high incidence of multicentricity and an aggressive disease course. Patients with FMTC or MEN II should have the entire gland removed, even in the absence of a palpable mass.

Given the frequent involvement of cervical nodes, initial surgical management should include a central compartment neck dissection. When central compartment nodes are involved or when palpable lateral cervical nodes are present, treatment including an ipsilateral or bilateral modified radical neck dissection (MRND) should be considered. Additionally, when the primary lesion is greater than 2 cm, the patient should undergo an elective ipsilateral MRND, as nodal metastases may be present in more than 60% of these patients [99,100]. If superior mediastinal lymph node disease is noted intraoperatively, involved areas should be removed. This rarely requires a median sternotomy.

Potentially associated conditions such as hyperparathyroidism and pheochromocytoma must be evaluated carefully and, if necessary, treated prior to thyroidectomy. A pheochromocytoma may need to be removed prior to treatment of the thyroid lesion. In the presence of hypercalcemia, the parathyroid glands need to be identified during thyroidectomy. If the parathyroid glands are abnormal, they should be removed. Otherwise, they should be adequately marked to facilitate future identification, especially in patients with MEN IIA.

Children with any of the genetic disorders leading to MTC need to be treated aggressively. Typically a total thyroidectomy should be performed by 5 to 6 years of age or before C-cell hyperplasia occurs. Pentagastrin stimulation of calcitonin secretion may be useful for monitoring [101]. Removal of the thyroid gland should prevent development of MTC in these patients and improve survival. However, MTC has been diagnosed in MEN 2B patients as young as 7 months old [102].

Following surgery, patients require close follow-up and monitoring of serum calcitonin and CEA levels. Calcitonin is more sensitive for detecting persistent or recurrent disease, but CEA levels appear to be predictive for survival [99]. Rising or persistent calcitonin levels should increase suspicion for residual or recurrent disease. Localization studies should be performed to identify potential sites of disease involvement. Tumor debulking for metastatic disease or local recurrence can decrease symptoms of flushing and diarrhea and may reduce the risk of death resulting from recurrent central neck disease [90,103]. Unfortunately, though, MTC do not respond to radioiodine therapy or TSH suppression therapy given their parafollicular C-cell origin [95]. External beam radiation therapy has been controversial for patients with positive tumor margins or unresectable tumor and there are no effective chemotherapy regimen.

Prognosis for patients with MTC is directly related to disease stage. The overall 10-year survival rate is between 61 and 75%, but decreases to 45% if cervical nodes are involved [95,104]. The best outcome is for patients with FMTC, then MEN IIA, sporadic disease, and MEN IIB.

E. Anaplastic Carcinoma

1. Clinical Presentation

Anaplastic thyroid carcinomas are one of the most aggressive malignancies with few patients surviving 6 months beyond initial presentation [24,105]. These lesions represent less than 5% of all thyroid carcinomas [106]. These tumors affect patients 60–70 years of age and presentation prior to the age of 50 years is extremely rare. Women are affected more commonly than men with a ratio of 3:2. Eighty percent may occur with a coexisting carcinoma and may represent transformation of a well-differentiated thyroid cancer [105,107].

Typically, patients have a long-standing neck mass that enlarges rapidly. This sudden change is often accompanied by pain, dysphonia, dysphagia, and dyspnea. Often the mass is quite large and fixed to the tracheolaryngeal framework, resulting in vocal cord paralysis and tracheal compression. Over 80% have jugular lymph node involvement at the time of presentation and greater than 50% have systemic metastases [108]. Most patients will succumb to superior vena cava syndrome, asphyxiation, and/or exsanguinations.

2. Pathology

The gross specimen will demonstrate areas of necrosis and macroscopic invasion of surrounding tissues, often with lymph node involvement. Microscopically, sheets of cells with marked heterogeneity are present. Spindle, polygonal, and giant multinucleated cells are present with occasional foci of differentiated cells. These cells do not produce thyroglobulin, do not transport iodine, and do not express thyroid hormone receptors [105]. These findings can often be established on FNAC, although a formal biopsy is occasionally necessary to exclude a diagnosis of lymphoma.

3. Management and Prognosis

Management of anaplastic carcinoma is extremely difficult, requiring a multidisciplinary approach and close consultation with the patient and family. All treatment forms are disappointing and median survival is 2 to 6 months [106,109]. Surgical debulking may be performed for palliation and a palliative airway may need to be established. One current treatment protocol involves doxorubicin, hyperfractionated radiation therapy, and potentially surgical debulking [110,111]. Although survival beyond 2 years is only 12%, this is the one of the only regimens available currently for these patients.

F. Other Forms of Thyroid Cancer

1. Insular Thyroid Carcinoma

Insular carcinoma was named for the clusters of cells that contain small follicles resembling pancreatic islet cells [112]. These tumors are very rare and present as an independent lesion or concomitantly with papillary or follicular thyroid carcinomas. These cells stain with thyroglobulin antibodies, but not for calcitonin. Typically, capsular and vascular invasion is present at the time of diagnosis.

These lesions are very aggressive when compared to follicular and papillary carcinoma and appear to have an increased recurrence and mortality rate when present as an independent process [113]. However, insular carcinoma located within follicular and/or papillary thyroid cancer does not seem to affect the clinical course adversely. Fortunately, many insular thyroid carcinomas are able to concentrate radioiodine.

2. Lymphoma

Primary thyroid lymphoma is unusual and represents less than 5% of all thyroid malignancies [114]. Women are affected more commonly at a ratio of 3:1 and typically present above 50 years of age. Patients may present with symptoms similar to anaplastic carcinoma, although the rapidly enlarging mass is often painless. Symptoms may also include regional adenopathy, dysphagia, and vocal cord paralysis secondary to recurrent laryngeal nerve invasion. Many affected patients are clinically hypothyroid or already receiving thyroid replacement therapy for conditions such as Hashimoto's disease [115]. Non-Hodgkin's B-cell type lymphoma is most common, although Hodgkin's disease and plasmacytomas do occur more rarely [116]. Thyroid lymphoma may arise as part of a generalized lymphomatous condition, as many of these patients have Hashimoto's disease. Current hypotheses include chronic antigenic lymphocyte stimulation that results in lymphocyte transformation.

A definite diagnosis needs to be made and frequently can be established by FNAC. Occasionally, a needle-core or open cervical lymph node biopsy may be necessary. Given the rare incidence of primary thyroid lymphoma, a comprehensive survey must be performed to exclude the presence of lymphoma at other sites.

Patients typically respond rapidly to chemotherapy, especially cyclophosphamide, doxorubicin, vincristine, and prednisone (CHOP) [117,118]. Combined radiation and chemotherapy regimens have also been developed and have been promising. Thyroidectomy and nodal resection may be considered to alleviate symptoms of airway obstruction in patients who do not respond rapidly to treatment, but surgical options are not primary treatment modalities.

The prognosis of patients depends on the histologic grade of the tumor and the presence of extrathyroidal disease. Overall, the 5-year survival is about 50%. However, intrathyroid disease survival is 85% and decreases to 40% for patients with extrathyroid disease.

3. Metastatic Carcinoma

The thyroid is a rare site for metastases from other cancers. However, metastasis may occur from primary lesions in the kidney, breast, lung, and skin (melanoma). The most common metastatic tumor to the thyroid is from a hypernephroma. Also, approximately 3% of bronchogenic carcinomas metastasize to the thyroid, but account for 20% of all metastases to the thyroid [119].

Typically, the history and physical examination identify the source of the metastasis. FNAC is performed for definitive diagnosis. Thyroidectomy may be considered for palliation, especially when the primary lesion is very slow growing (e.g., renal cell carcinoma) [120].

4. Squamous Cell Carcinoma

Squamous cell carcinoma (SCC) of the thyroid is very rare, representing less than 1% of thyroid cancers [121]. Older patients are affected most commonly and the disease can progress rapidly with local invasion and metastasis. During the workup, metastasis from another site within the upper aerodigestive tract needs to be excluded. Early detection and aggressive surgical treatment appear to represent the best option for palliation and cure. As with other SCC of the head and neck, radiation therapy is probably important, although not well characterized [122].

VII. SURGICAL MANAGEMENT AND TECHNIQUE

A. Surgical Anatomy and Embryology

The thyroid medial anlage derives from the ventral diverticulum from the endoderm of the first and second pharyngeal pouches at the foramen cecum [66,123]. The diverticulum descends from the base of tongue to its adult pretracheal position through a midline anterior path during weeks 4 to 7 of gestation. Lateral thyroid primordial arise from the fourth and fifth pharyngeal pouches and descend to join the central component. Parafollicular C cells arise from the neural crest of the fourth pharyngeal pouch as ultimobranchial bodies and infiltrate the upper portion of the thyroid lobes [124].

The thyroid gland is composed of two lateral lobes connected by a central isthmus, weighing 15 to 25 g in the adult. A thyroid lobe measures about 4 cm in height, 1.5 cm in width and 2 cm in depth. The superior pole lies posterior to the sternothyroid muscle and lateral to the inferior constrictor muscle and posterior thyroid lamina. The inferior pole can extend to the level of the sixth tracheal ring. Approximately 40% of patients have a pyramidal lobe that arises from either lobe or the midline isthmus and extends superiorly (Fig. 28.2).

The thyroid is enclosed between layers of the deep cervical fascia in the anterior neck. The true thyroid capsule is tightly adherent to the thyroid gland and continues into the parenchyma to form fibrous septa separating the parenchyma into lobules. The surgical capsule is a thin, film-like layer of tissue lying on the true thyroid capsule. Posteriorly, the middle layer of the deep cervical fascia condenses to form the posterior suspensory ligament, or

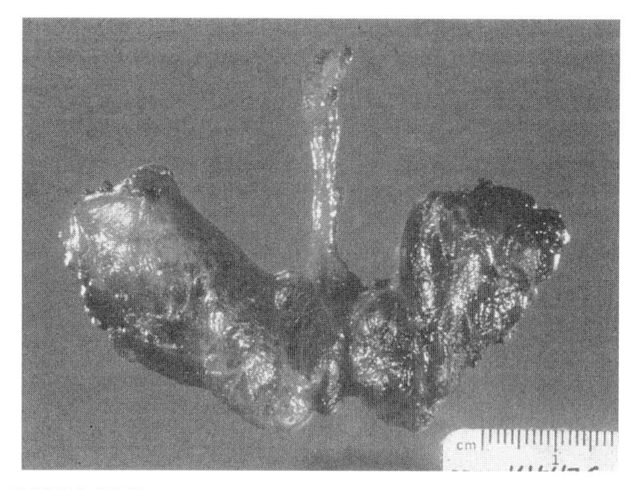

FIGURE 28.2 A pyramidal lobe of the thyroid gland may occasionally arise from the isthmus. This portion of the gland should be identified and removed carefully with the surgical specimen.

Berry's ligament, connecting the lobes of the thyroid to the cricoid cartilage and the first two tracheal rings.

Blood supply to and from the thyroid gland involves two pairs of arteries, three pairs of veins, and a dense system of connecting vessels within the thyroid capsule. The inferior thyroid artery arises as a branch of the thyrocervical trunk. This vessel extends along the anterior scalene muscle, crossing beneath the long axis of the common carotid artery to enter the importation of the thyroid lobe. Although variable in its relationship, the inferior thyroid artery lies anterior to the recurrent laryngeal nerve in approximately 70% of patients [125]. The inferior thyroid artery is also the primary blood supply for the parathyroid glands.

The superior thyroid artery is a branch of the external carotid artery and courses along the inferior constrictor muscle with the superior thyroid vein to supply the superior pole of the thyroid. This vessel lies posterolateral to the external branch of the superior laryngeal nerve as the nerve courses through the fascia overlying the cricothyroid muscle. Care should be taken to ligate this vessel without damaging the superior laryngeal nerve. Occasionally, a thyroid ima artery may arise from the innominate artery, carotid artery, or aortic arch and supply the thyroid gland near the midline [125]. Many veins within the thyroid capsule drain into the superior, middle, and inferior thyroid veins, leading to the internal jugular or innominate veins. The middle thyroid vein travels without an arterial complement, and division of this vessel permits adequate rotation of the thyroid lobe to identify the recurrent laryngeal nerve and parathyroid glands.

Careful management of thyroid carcinomas requires a thorough knowledge of the course of the recurrent laryngeal nerve (RLN) (Fig. 28.3). The RLN provides motor supply to the larynx and some sensory function to the upper trachea and subglottic area. The right RLN leaves the vagus nerve at the base of the neck, loops around the right subclavian

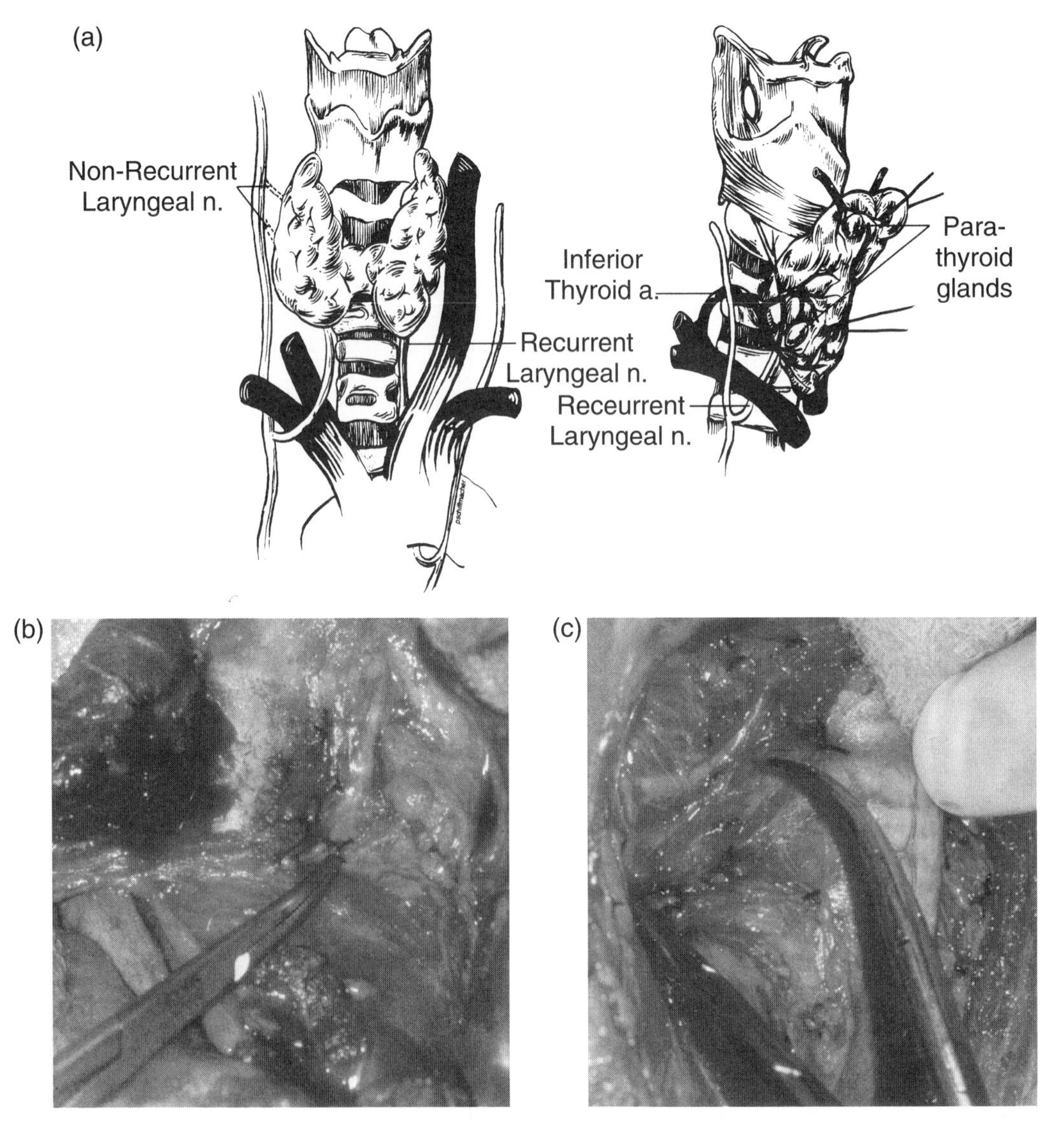

FIGURE 28.3 (a) Schematic of thyroid gland and important adjacent structures. In the lateral view, the gland has been mobilized medially to demonstrate the recurrent laryngeal nerve and its close relationship to the inferior thyroid artery. This relationship can vary between sides within a patient. Please refer to the text for details. In the frontal view, the potential course(s) of the nonrecurrent laryngeal nerve has been indicated (dashed lines). (b) The course of the recurrent laryngeal nerve along the tracheoesophageal groove is demonstrated intraoperatively. (c) The lateral course of the nonrecurrent laryngeal nerve has been revealed intraoperatively.

artery, and extends back into the thyroid bed approximately 2 cm lateral to the trachea. The nerve enters the larynx between the arch of the cricoid cartilage and the inferior cornu of the thyroid cartilage. A "nonrecurrent" laryngeal nerve may rarely occur on the right side and enters from a more lateral course. Typically, a retroesophageal subclavian artery is present. The left RLN leaves the vagus at the level of the aortic arch and loops around the arch lateral to the obliterated ductus arteriosus. The nerve returns to the neck posterior to the carotid sheath and travels near the tracheoesophageal groove along a more medial course than the right RLN. The nerve will cross deep to the inferior thyroid artery approximately 70% of the time and often branches above the level of the inferior thyroid artery prior to entry into the larynx [56].

Identification of the RLN is best achieved through an inferior approach in a space defined by Lore *et al.* [127] as the recurrent laryngeal nerve triangle. The triangle is bounded by the trachea medially, the carotid sheath laterally, and the undersurface of the retracted inferior thyroid pole superiorly. Careful dissection in this area parallel to the course of the RLN should safely identify the nerve. A thyroid goiter or unusually large thyroid mass will potentially displace the nerve. The RLN, in these cases, can become fixed to and splay across the undersurface of the enlarged thyroid lobe. Great care needs to be taken in these situations and identification of the nerve may require a superior approach, identifying the RLN at its entry into the larynx. Finally, the RLN may be closely associated with or penetrate through the ligament of Berry before entering the larynx. Retraction along this area during removal of the gland can lead to RLN injury. Thus, caution must be taken to carefully dissect out the course of the RLN and subject the nerve to as little traction injury as possible.

The superior laryngeal nerve (SLN) descends medially to the carotid sheath and divides into an internal and external branch about 2 cm above the superior pole of the thyroid. The internal branch travels medially and enters through the posterior thyroid membrane to supply sensation to the supraglottis. The external branch extends medially along the inferior constrictor muscle to enter the cricothyroid muscle. Along its course, the nerve travels with the superior thyroid artery and vein. The nerve typically diverges from the superior thyroid vascular pedicle about 1 cm from the thyroid superior pole [128]. Because of the close association of these structures, superior pole dissection can be performed as the final step in the thyroid lobectomy to improve mobilization and exposure in this area. (Fig. 28.3)

Proper management of the parathyroid glands during thyroid surgery is critical to avoid potential calcium metabolism alterations. The parathyroid glands are caramel-colored glands, weighing 30 to 70 mg. The superior parathyroid glands are derived from the fourth pharyngeal pouch, whereas the inferior counterparts originate from the third pharyngeal pouch. The subtle distinction of tan and yellow coloration permits differentiation from adjacent fatty tissue, although with trauma, the glands can become mahogany in color. Eighty percent of patients have four glands and at least 10% have more than four glands [129]. The glands are situated on the undersurface of the thyroid gland in fairly predictable locations. The superior glands are located at the level of the cricoid cartilage, usually medial to the intersection of the RLN and the inferior thyroid artery [129]. The inferior glands are more variable in location than their superior counterparts. These glands may be on the lateral or posterior surface of the lower pole. In many patients, the position of the parathyroid glands on one side is similar to the other side and should be a useful guide.

B. Surgical Technique

Prior to any thyroid surgery, any voice changes or previous neck surgery should prompt assessment of vocal cord mobility by indirect laryngoscopy. Although many patients with thyroid carcinomas are euthyroid, necessary medical therapy should be instituted for patients demonstrating thyrotoxicosis or hypothyroidism to avoid intraoperative metabolic derangements, such as hypertensive crisis. Details of this management are beyond the scope of this chapter, but should include consultation with an endocrinologist.

The patient should be positioned supine on the operating table with an inflatable pillow or shoulder roll and adequate head support to permit full neck extension for optimal exposure. A symmetrical, transverse incision along a skin crease approximately 1 cm below the cricoid cartilage is made through the platysma. The length of the incision will depend on the size of the thyroid gland. Larger incisions will be necessary for patients with short, thick necks, difficulty with neck extension, or a low-lying thyroid gland. Subplatysmal skin flaps are raised superiorly to the level of the thyroid cartilage notch and inferiorly to within 1 cm of the clavicle.

Exposure of the thyroid gland is obtained through a midline, vertical incision through the superficial layer of the deep cervical fascia between the sternohyoid and the sternothyroid muscles. The strap muscles are separated by blunt dissection and then proceed laterally along the thyroid capsule until the ansa cervicalis is noted at the lateral edge of the sternohyoid muscle/medial aspect of the internal jugular vein. Rarely, the strap muscles must be divided to gain access to a large thyroid tumor. This division should be done high on the muscle to preserve innervation from the ansa hypoglossal nerve. The strap muscles should be reapproximated prior to skin closure. Any evidence of frank invasion of thyroid carcinoma into the strap muscles should result in the *en bloc* resection of section of the affected muscle with the thyroid lobe.

Through blunt dissection, the thyroid lobe is swept anteromedially to the tracheolaryngeal framework (Fig.28.4). The middle thyroid vein(s) should be identified, and division of this vessel should improve lateral exposure. The cricoid and trachea should be identified in the midline, and continued mobilization is achieved by sweeping dorsally all tissue along the posterolateral border of the thyroid lobe. Meticulous hemostasis should be maintained to facilitate identification of the SLN, RLN, and parathyroid glands.

The RLNs should be identified relatively early (Fig. 28.3). As the dissection is accomplished along the inferior pole and the RLN triangle is revealed, attention should focus on finding these nerves. Especially during repeat surgery, nerve stimulation can be used to facilitate and confirm nerve identification. Once identified, the RLN should be followed to its laryngeal entry at the level of the cricoid cartilage, passing under or through Berry's ligament, and entering the larynx deep to the inferior constrictor muscle. The most difficult portion of the operation is typically during the dissection where the recurrent nerve passes through Berry's ligament. The RLN is in close proximity to the thyroid, tethered down by the ligament. Bleeding may occur at this site and should be controlled by gentle pressure prior to identification of the nerve to avoid injury. Use of electrocautery in this region should be strictly avoided. A small portion of thyroid tissue may be embedded with the

(a)

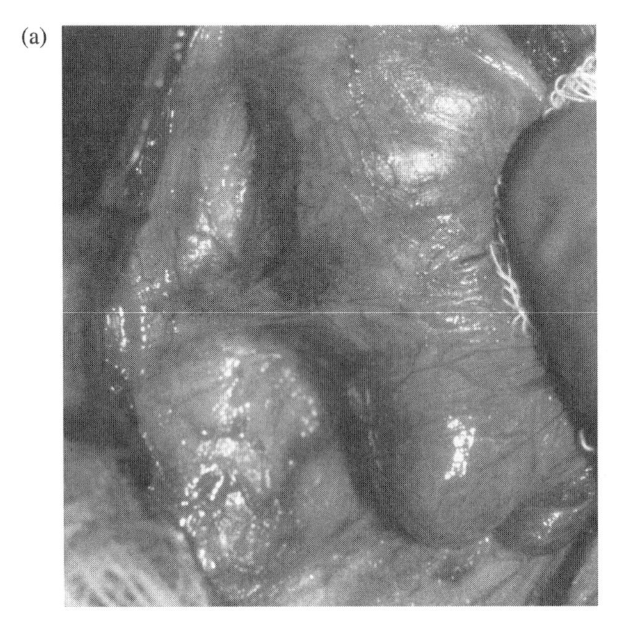

(b)

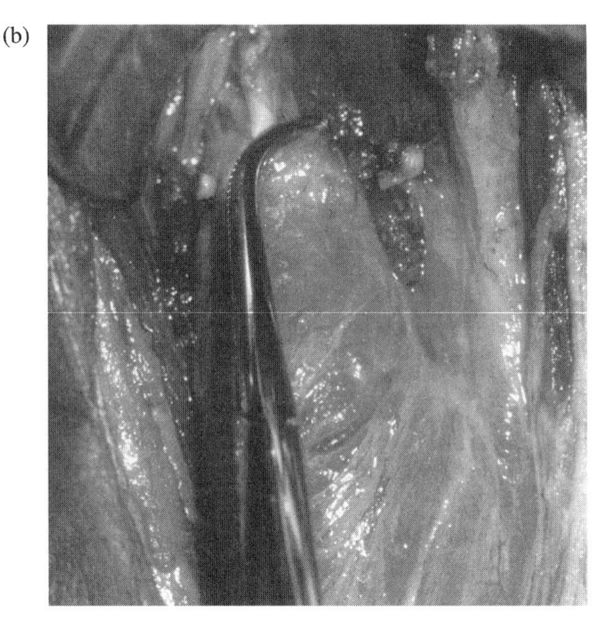

FIGURE 28.4 (a) Careful dissection along the lateral portion of the thyroid lobe permits mobilization of the gland medially. (b) The superior pole of the thyroid gland can be mobilized and the superior thyroid vessels ligated. Care should be taken to ligate the vessels near the thyroid capsule to prevent injury to the superior laryngeal nerve.

ligament, accounting for a remnant of thyroid tissue left following total thyroidectomy.

All vessels are ligated and divided on the capsule to reduce the risk of parathyroid devascularization. Ischemic parathyroids and those situated anteriorly on the thyroid gland or removed with the thyroid lobe should be examined. They should be biopsied and confirmed by frozen-section examination. The parathyroid glands can be minced into 1-mm^3 cubed pieces and reimplanted in the ipsilateral sternohyoid muscle or sternocleidomastoid muscle with a silk suture/clip to mark the reimplantation site.

Once the lobe is mobilized and key structures are identified, the isthmus can be transected close to the contralateral side. The edge of the isthmus is oversewn and a careful search is made for a pyramidal lob, which should be removed in continuity with the thyroid lobe, and isthmus when present.

The remaining pedicle along the superior thyroid pole is identified by retracting the thyroid inferiomedially. Dissection should be carried out close to the thyroid capsule to avoid possible injury to the external branch of the SLN. Occasionally, the external branch of the SLN that supplies the cricothyroid muscle can be visualized. The superior pole vessels should be identified individually and isolated for ligation close to the thyroid lobe. At this point, tissues poterolateral to the superior pole can be swept away from the lobe in a posteromedial direction.

For a total thyroidectomy, the same procedure is then repeated on the contralateral side. However, the decision to proceed to excision of the opposite thyroid lobe should depend on the course of the initial thyroid lobectomy. Following removal of the thyroid specimen, hemostasis is verified and a Jackson–Pratt suction drain is placed in the thyroid lobe bed. Divided strap muscles are reapproximated with absorbable sutures and closed along the midline to prevent tracheal adhesion to the skin. The platysma is reapproximated with absorbable sutures and the skin is closed with a running subcuticular suture.

C. Special Surgical Situations

1. Cervical Lymph Nodes

Management of the neck in patients with well-differentiated thyroid carcinomas is typically conservative. Cervical node disease is rare in follicular carcinomas, but more common in papillary carcinomas. There is no role currently for elective neck dissection in cases of follicular carcinoma. More controversy exists regarding central compartment dissection in patients with papillary carcinoma [130,131]. Typically, patients with a primary papillary carcinoma greater than 2 cm will have a concurrent central compartment lymph node dissection with their total thyroidectomy or the completion thyroid lobectomy.

Given the higher frequency of microscopic tumor spread in Hürthle cell and medullary carcinomas and the inability of these tumors to uptake radioiodine, elective central compartment lymph node dissection is commonly performed. When these patients present with palpable anterior lymph node disease, a central compartment dissection is performed with the total thyroidectomy and the lateral cervical nodes are inspected carefully. When palpable lateral cervical nodes are present, a neck dissection, including levels II–IV, should be performed bilaterally. Retrospective analysis revealed a 10-year survival rate of 67% for those patients with medullary carcinoma treated with neck dissection versus 43% for those who were not [100]. Extension of the collar incision will provide exposure of the lateral neck and level V should also be removed if palpable disease is present. Careful inspection and dissection should be performed along the spinal accessory nerve, as this is a frequent site of metastatic thyroid disease [66]. A more extensive modified radical neck dissection or radical neck dissection may be necessary to remove gross disease. Finally, removal of upper mediastinal nodal groups may also be reasonably performed through the cervical approach.

2. Intrathoracic Goiter

Less than 1% of patients may have a thyroid gland that is partially or completely intrathoracic [132,133]. In the great majority of these patients, the intrathoracic goiter can be removed via a collar incision in the neck without resorting to a sternotomy. The vascular supply, typically originating in the neck, is identified, ligated, and divided. Division of the isthmus may facilitate mobilization of the substernal goiter from beneath the sternum. Large sutures may be placed deeply into the goiter to facilitate traction and blunt dissection to permit delivery of the thyroid through the neck.

Patients with substernal goiters who have had previous thyroid operations, some with invasive malignant tumors, and patients with no thyroid tissue in the neck may require a median sternotomy. Occasionally, the substernal goiter may simply be too large to deliver through a cervical incision. In these cases, the thyroidectomy is usually performed in collaboration with a thoracic surgeon.

3. Recurrent Laryngeal Nerve Invasion

Surgical situations necessitating the sacrifice of the RLN are quite uncommon. If preoperative vocal cord paralysis is present and carcinoma invasion is seen intraoperatively, the nerve may be sacrificed. More commonly, the RLN should be dissected free of gross disease. In patients with well- differentiated thyroid carcinomas, there is no survival difference between those with RLN sacrifice and those treated postoperatively with radioiodine for gross disease left on the nerve [118]. Given the lack of response by medullary carcinoma to postoperative radioiodine treatment, RLN sacrifice must be considered to achieve complete removal of gross disease.

Sacrifice of the RLN requires the exclusion of nerve infiltration by benign disease processes. Graves' disease, Hashimoto's thyroiditis, and Reidel's thyroiditis can involve the RLN with or without vocal cord paralysis. Benign processes can cause stretch injuries to the RLN that resolve with surgical removal of the mass. Finally, lymphomas can involve the RLN, but treatment is rarely surgical and should not involve excision of the nerve.

4. Extended Surgical Resection

Surgical treatment of thyroid carcinomas should remove all gross disease, especially in patients with medullary carcinoma. Fixation to the thyroid cartilage or trachea may require partial- to full-thickness removal of those structures. A thyroid cartilage lamina can be removed without major morbidity if the internal thyroid perichondrium is left intact. The trachea can be partially resected and repaired to permit *en bloc* tumor removal. Primary anastamosis can be performed for resections involving up to four tracheal rings [134]. Additionally, tracheal shaving can be performed, leaving the internal mucosa intact. Isolated full-thickness defects can be repaired with composite mucosa–cartilage grafts from the nasal septum. In patients with more extensive skeletal involvement, a partial laryngectomy may be required and has demonstrated improved survival [135,136]. Total laryngectomy should be performed in only the most extreme cases of extensive intraluminal invasion. Typically, this would be done following the failure of radioiodine treatment and/or external beam radiation therapy. Pharyngeal and esophageal local invasion typically requires resection of the immediate area and primary closure (Fig. 28.5).

D. Extent of Surgery

Surgery is the primary modality for the treatment of thyroid carcinomas. Although well-differentiated thyroid cancers can be extremely aggressive and lethal, the vast majority of patients have prolonged survival even with residual or recurrent disease. Thus, controversy exists regarding the extent of surgery to be performed in patients with well-differentiated carcinomas. The primary goals of surgical treatment should be to eradicate primary disease, to reduce the incidence of local/distant recurrence, and to facilitate the treatment of metastases. These oncologic goals should be achieved with minimal morbidity.

The large body of literature regarding thyroid carcinoma and its treatment has created a wide variety of terms for describing the extent of thyroid tissue removal. Most surgeons would agree that a subtotal excision of a thyroid nodule is not acceptable and that the minimum amount of thyroid removed should be a lobectomy and isthmusectomy. Some surgeons advocate a near-total thyroidectomy that preserves the posterior portion of the gland on one side to

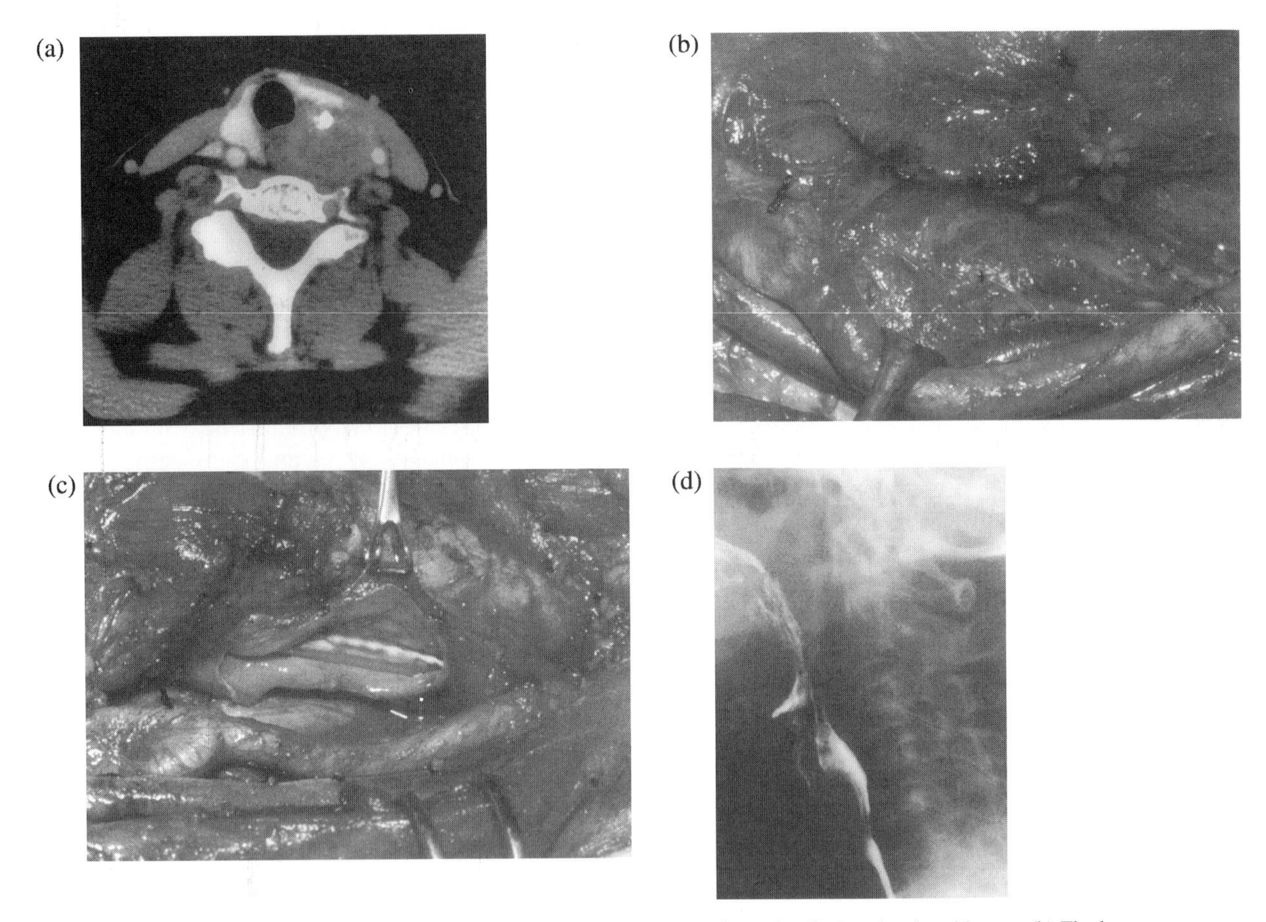

FIGURE 28.5 (a) An axial view from a CT scan demonstrates a large, locally invasive thyroid mass. (b) The large thyroid mass has been dissected free from adjacent structures, but involves the esophagus. (c) Resection of the thyroid mass required excision of a portion of the esophagus. The esophagus was closed primarily. (d) A postoperative barium swallow study demonstrated that the esophagus was intact and the patient was able to tolerate a regular diet.

avoid injury to the recurrent laryngeal nerve and at least one parathyroid gland. A total thyroidectomy involves the complete removal of both thyroid lobes, although radioiodine scans may reveal 2–5% residual tissue [137].

Proponents of a more conservative surgical approach suggest that the thyroid lobectomy and isthmusectomy is sufficient treatment for the great majority of patients (>80%). These patients would be categorized as low risk in the different classification schemes (see Section IV). The thyroid lobectomy and isthmusectomy is simpler to perform and less time-consuming than a total thyroidectomy. The overall risk of morbidity from recurrent laryngeal nerve, superior laryngeal nerve, and parathyroid gland injury is less. Finally, compared with total thyroidectomy, this conservative approach does not adversely affect prognosis and survival.

A large number of studies support the conservative approach. In the AGES classification, low-risk patients with papillary carcinoma had the same 2% 25-year mortality rate for those treated with thyroid lobectomy or total thyroidectomy [20]. A study by Shah *et al.* [45] based risk classifications on the AJC staging system. There was no difference in 20-year survival based on surgical treatment for patients with intrathyroidal tumors of less than 4 cm. However, patients treated with total thyroidectomy had an increased risk of complications.

Although survival differences are not found, several studies reported an increased incidence of local recurrence in patients treated with only a thyroid lobectomy [138,139]. Some controversy remains regarding the relationship of local recurrence to disease survival, especially with papillary carcinomas [68,140]. Finally, patients categorized as high risk in these classification systems are treated with near-total or total thyroidectomy and radioiodine ablation therapy.

A more aggressive surgical approach favors total thyroidectomy in most cases and a lobectomy for small, single lesions. Proponents of this approach favor total thyroidectomy as a better oncologic operation. Removal of the entire thyroid gland encompasses potential extracapsular extension and multicentric lesions. Morbidity of this procedure is relatively low with a good technique and an experienced surgeon [141]. Several studies demonstrate improved survival and decreased local/distant recurrence [48,62,142]. Cautionary studies also demonstrate the risk of aggressive disease even in patients determined to be low risk in the various classification schemes [143,144].

Postoperative thyroglobulin levels are more valid with removal of all normal thyroid tissue. Total thyroidectomy also facilitates the use of postoperative diagnostic scans to evaluate for metastases and recurrence. Radioiodine does not need to be used to ablate excess normal tissue and can be concentrated upon the removal of residual carcinoma and distant metastases. Finally, although rare, removal of the entire thyroid reduces the risk of transformation from a well-differentiated thyroid carcinoma to an anaplastic carcinoma.

Clearly, controversy surrounding the extent of surgery will continue to persist, as no uniform set of characteristics can be used to classify the aggressiveness of well-differentiated thyroid carcinomas. Results of treatment will also continue to differ among institutions and surgeons. Nevertheless, the primary goal should remain the effective eradication of thyroid cancer with minimal morbidity.

E. Complications

Among head and neck surgical procedures, thyroid surgery is very safe. Mortality rates are extremely low and morbidity is relatively low. In general, serious complications occur in less than 2% of all thyroid cases [66]. Complications can typically be divided into nonmetabolic and metabolic complications. Of particular concern are injuries to the recurrent laryngeal nerve and the parathyroid glands.

Perioperative complications are fairly unusual in thyroid surgery. Postoperative infections are very unusual given the abundant blood supply in the thyroid bed. The prevention of scar widening or hypertrophy depends on proper placement of the incision. The incision can often be hidden within existing skin creases. Additionally, the incision should not be placed too low in the neck to avoid the increased skin tension over the sternal notch. Pneumothorax is very rare and is often associated with extended procedures that involve subclavicular dissection. Chylous fistula may occur more often on the left side, but are usually self-limiting when wound drainage is adequate.

Hemorrhage is uncommon when surgery is performed with meticulous hemostasis. However, bleeding may occur due to an undetected coagulopathy or a technical mishap. Significant hemorrhage in the immediate postoperative period can lead to life-threatening airway compression. A rapidly expanding hematoma requires immediate opening of the surgical incision and evacuation of blood. Airway control can then be established and the patient can be returned to the operating room for complete exploration to identify bleeding sites. A wound seroma may occur, especially after the removal of a large goiter. Should a fluid collection present, simple needle aspiration should manage the problem and prevent the risk of infection.

Injury to the external branch of the superior laryngeal nerve is thought to be rare, but the exact frequency is unknown. Often, disturbance of SLN function is temporary and frequently unrecognized by the patient and the surgeon [145]. Injury to the SLN will alter function of the cricothyroid muscle. Patients may have difficulty shouting and singers will find difficulty with pitch variation, especially in the higher frequencies. The external branch of the SLN is not often visualized and lies near the superior pole vessels. Adequate exposure of the superior thyroid pole and close ligation of the individual vessels on the thyroid capsule may prevent SLN injury. Voice therapy may help patients to compensate in cases of SLN injury.

Injury to the RLN has a much greater impact and is more noticeable than SLN injury. The incidence of permanent RLN paralysis is approximately 1 to 1.5% for total thyroidectomy and less for near-total procedures [48,146,147]. Temporary dysfunction due to nerve traction occurs in 2.5 to 5% of patients [127]. Incidence increases with second and third procedures. RLN injury is also more common in thyroidectomy with neck dissection, although this may reflect more advanced disease states [148]. Disease-specific risk factors for permanent nerve damage include recurrent thyroid carcinoma, substernal goiter, and various thyroiditis conditions. Vocal cord function should be evaluated and documented by indirect laryngoscopy, especially in patients who have had previous surgery.

Unilateral RLN injury leads to a vocal cord in the paramedian position and the voice may be breathy and lack volume. Concurrent injury of the SLN will result in a more laterally positioned vocal cord and worsen voice quality and glottic competence [149]. Occasionally, patients may have difficulty with aspiration and pneumonia [148].

Bilateral RLN injury may present very dramatically. Immediate postoperative stridor and dyspnea may require immediate reintubation and a possible tracheostomy. Occasionally, bilateral RLN injury may not be immediately noticeable and patients may adapt to the reduced airway. Over time, though, the vocal cords will move to the midline, compromising the airway.

Identification and careful dissection along the course of the RLN decrease the incidence of permanent injury. The surgeon should also be aware of the possibility of a nonrecurrent nerve, most commonly on the right side. If the nerve is transected during surgery, microsurgical repair of the nerve is recommended. While the repair is unlikely to restore normal function, reanastamosis of the RLN may decrease the extent of vocal cord atrophy [150]. Some surgeons advocate anastamosis of the ansa hypoglossal nerve to the distal end of the severed RLN to prevent laryngeal synkinesis and possible vocal cord hyperadduction [151,152].

Comprehensive management of vocal cord injury is beyond the scope of this chapter. In the majority of cases, RLN injury will be detected postoperatively. Management is supportive, although some surgeons favor reexploration when vocal cord paralysis is noted in the immediate postoperative period [66]. Return of normal vocal cord function occurs as late as 6 to 12 months when temporary RLN injury has occurred. Speech therapy can be valuable. Serial examinations should document potential return of function or compensation by the contralateral vocal cord. In patients with continued vocal incompetence or aspiration, treatment directed toward vocal cord medialization may consist of vocal cord injection, thyroplasty, and/or arytenoids medialization. In cases of bilateral RLN injury, management is directed at improving the airway while not completely sacrificing airway quality and may involve arytenoidectomy or transverse cordotomy.

Transient symptomatic hypocalcemia after total thyroidectomy occurs in approximately 7 to 25% of cases, but permanent hypocalcemia is less common (0.4–13.8%) [40,153]. The risk of hypoparathyroidism is related to the size and degree of invasion of the tumor, pathology, and the extent of procedure and surgeon experience [20,154]. Changes in serum calcium levels are often transient and may not always be related to parathyroid gland trauma or vascular compromise.

Transient hypocalcemia is often related to variations in serum protein binding due to perioperative alterations in acid–base status, hemodilution, and albumin concentration. These changes do not produce hypocalcemic symptoms. However, sudden changes in levels of ionized serum calcium can result in perioral and distal extremity paresthesias. As calcium levels continue to decline, patients may experience tetany, bronchospasm, mental status changes, seizures, laryngospasm, and cardiac arrhythmias. Chvostek's sign and Trousseau's sign may develop with increased neuromuscular irritability as serum calcium levels drop below 8.0 mg/dl.

Typically, serum calcium levels are measured in the immediate postoperative period and the next morning for patients with a total or completion thyroidectomy. Patients should demonstrate a stable or rising serum calcium level. Patients undergoing a thyroid lobectomy do not usually require serum calcium monitoring. Findings that should be worrisome for hypoparathyroidism include hypocalcemia, hyperphospatemia, and metabolic alkalosis.

Treatment for hypocalcemia is typically initiated if the patient is symptomatic or serum calcium levels fall below 7.0 mg/dl. In these patients, cardiac monitoring is warranted. Patients should receive 10 ml of 10% calcium gluconate and 5% dextrose in water intravenously, titrated to symptom resolution and subsequent serum calcium levels. Oral calcium supplementation should begin with 2–3 g of calcium carbonate per day. Additionally, calcitriol (1,25-dihydroxyvitamin D_3) should be initiated. Adjustments in supplemental calcium and vitamin D should be done in consultation with an endocrinologist.

VIII. POSTOPERATIVE MANAGEMENT AND SPECIAL CONSIDERATIONS

A. Thyroid Hormone Replacement

Following total or completion thyroidectomy, exogenous supplementation of thyroid hormone is necessary [23,155]. Long-term supplementation with levothyroxine (T_4) is monitored to suppress TSH to below-normal levels. Patients receiving suppressive therapy have a lower recurrence rate and improved survival [74,156].

In the immediate postoperative period, patients are frequently given cytomel (T_3). Cytomel has a shorter half-life than thyroxine, decreasing the waiting period before radioiodine body scanning and possible ablative therapy can be performed.

B. Radiation Treatment

Radiolabeled iodine has been used since the early 1960s to ablate normal thyroid tissue and to treat residual tumor and metastases [157]. The ^{131} I isotope emits β particles that penetrate and destroy tissue within a 2-mm zone. Patients classified as high risk with papillary carcinoma and most patients with follicular carcinoma are considered for treatment [75]. Despite poor radioiodine uptake in Hürthle cell and medullary carcinomas, these patients will often be treated to provide any possible benefit.

Whole-body scans stage the patient and determine the need and potential benefit of radioiodine therapy. Elevated TSH levels are necessary to enhance the uptake of iodine by thyroid cancer cells. Thus, patients are taken off thyroid hormone suppression therapy for 4 to 6 weeks prior to scanning and placed on a low-iodine diet. Additionally, some work has been done to provide exogenous TSH [158].

At our institution, ^{123}I has replaced ^{131}I for the initial diagnostic radioiodine scan. The physical properties of ^{123}I permit better image quality and decrease possible stunning of functioning thyroid cells by the β particle emission of ^{131}I [159]. This permits the maximal benefit of ^{131}I in ablation following the diagnostic procedure [160].

Following the diagnostic scan, therapeutic ablative doses of ^{131}I can be given. Typically, 100 mCi of ^{131}I is given for uncomplicated cases with only thyroid bed uptake. Although lower doses (30–60 mCi) have been commonly employed, studies suggest that they are less effective in disease treatment (A. Alavi, personal communication). Patients with uptake in cervical nodes or distant metastases will receive

125–200 mCi [161,162]. Doses higher than 200 mCi have not been shown to be more effective in most cases [66].

Diverse alternatives and protocols exist regarding the use of radioiodine therapy. Surgeons who favor total thyroidectomy for most thyroid carcinomas argue that removal of normal tissue enhances radioiodine ablation therapy. However, proponents of more conservative treatment suggest that radioiodine therapy can be used to remove even a remaining thyroid lobe prior to whole-body scanning for residual tumor or metastases. Additionally, the use of ^{123}I for diagnostic scanning has only recently gained favor. Continued work in this area should improve patient outcomes and decrease disease recurrence.

C. External Beam Radiotherapy and Chemotherapy

Given the effectiveness of surgery and radioiodine treatment for the majority of thyroid carcinomas, experience with external beam radiation therapy (EBRT) and chemotherapy is more limited. EBRT appears to improve local control of well-differentiated carcinomas, especially when employed in combination with doxorubicin [163–165]. However, the effect of EBRT on survival is uncertain [165,166]. Palliation of patients with distant metastasis of well-differentiated carcinomas, especially to the bone, appears to be improved with EBRT [163].

Limited success with EBRT in combination with doxorubicin/cisplatin has been noted with anaplastic carcinomas [167,168]. However, this disease remains uniformly fatal, and palliation through local control and airway protection are the only realistic goals. Patients with metastatic medullary carcinoma may benefit from EBRT and chemotherapy to decrease local recurrence [169,170]. Typically, patients with regional nodal metastases recognized during surgery will receive postoperative EBRT. In general, the poorer outcomes for patients with medullary and anaplastic carcinomas have not been altered significantly by EBRT and chemotherapy.

D. Follow-Up Management

Patients treated for thyroid carcinomas will require long-term follow-up and monitoring. In addition to regular physical examination, thyroid hormone and TSH levels are monitored to ensure adequate suppression. Thyroglobulin levels should be monitored closely and diagnostic radioiodine scanning should be performed. These tests should be performed annually for the first 2 years and then every 5 years for 20 years [171]. Typically, thyroglobulin levels should be less than 2 ng/ml following total thyroidectomy and radioiodine ablation therapy (less than 3 ng/ml if patient is off thyroid replacement therapy). Rising serum thyroglobulin levels are highly sensitive (97%) and specific (100%) for thyroid cancer recurrence [172]. Elevation of thyroglobulin levels warrants repeat radioiodine scanning and therapy.

Patients with medullary carcinomas require serial measurements of calcitonin and CEA. Suspected recurrences may also be detected with the pentagastrin-stimulation test.

IX. CONCLUSIONS

A great deal of literature addresses the treatment of thyroid carcinomas. Questions remain unresolved regarding the extent of thyroidectomy to perform in patients with well-differentiated carcinomas and the effect on patient survival. Because the great majority of patients with well-differentiated thyroid carcinomas do well, failure to resolve issues like this are tolerated. Additionally, studies to address these issues require very large cohorts with long-term follow-up that needs to extend 20 to 30 years.

Nevertheless, surgeons need to remain wary of the devastating progression of disease in medullary and anaplastic thyroid carcinomas. Despite clinical classification schemes, even well-differentiated carcinomas can have unpredictably aggressive clinical manifestations. As our understanding of the molecular and genetic mechanisms of thyroid carcinomas improves better diagnostic tests should improve our ability to treat these cancers.

References

1. Landis, S. H., Murray, T., Bolden, S., *et al.* (1999). Cancer statistics. *CA Cancer J Clin.* **49**, 8–31.
2. Mazzaferri, E. L. (1992). Thyroid cancer in thyroid nodules: Finding a needle in a haystack. *Am. J. Med.* **93**, 359–362.
3. Hay, I. D., and Klee, G. G. (1993). Thyroid cancer diagnosis and management. *Clin. Lab. Med.* **13**, 725–734.
4. Sozzi, G., Bongarzone, I., Miozzo, M., *et al.* (1992). Cytogenetic and molecular genetic characterization of papillary thyroid carcinomas. *Genes Chromosomes Cancer* **5**, 212–218.
5. Fagin, J. A. (1994). Molecular pathogenesis of human thyroid neoplasms. *Thyroid Today* **18**, 1–6.
6. Williams, E. D. (1995). Mechanisms and pathogenesis of thyroid cancer in animals and man. *Mutat Res.* **333**, 123–129.
7. Namba, H., Rubin, S. A., and Fagin, J. A. (1990). Point mutations of ras oncogenes are an early event in thyroid tumorigenesis. *Mol. Endocrinol.* **4**, 1474–1479.
8. Takahashi, M. (1995). Oncogenic activation of the ret proto-oncogene in thyroid cancer. *Crit. Rev. Oncog.* **6**, 35–46.
9. Grieco, M., Santoro, M., Berlingieri, M. T., *et al.* (1990). PTC is a novel rearranged form of the ret proto-oncogene and is frequently detected in vivo in human thyroid papillary carcinomas. *Cell* **60**, 557–563.
10. Santoro, M., Grieco, M., Melillo, R. M., *et al.* (1995). Molecular defects in thyroid carcinomas: Role of the ret oncogene in thyroid neoplastic transformation. *Eur. J. Endocrinol.* **133**, 513–522.
11. Williams, E. D., Doniach, I., Bjarnason, O., *et al.* (1977). Thyroid cancer in an iodine rich area. *Cancer* **39**, 215–222.
12. Duffy, B. J., and Fitzgerald, P. J. (1950). Cancer of the thyroid in children: A report of 28 cases. *J. Clin. Endocrinol.* **10**, 1296–1308.

13. Becker, F. O., Economou, S. G., Wouthwick, H. W., *et al.* (1975). Adult thyroid cancer after head and neck irradiation in infancy and childhood. *Ann. Intern. Med.* **83**, 347–351.
14. Rojeski, M., and Gharib, H. (1985). Nodular thyroid disease. *N. Engl. J. Med.* **313**, 418–436.
15. Daniels, G. H. (1996). Thyroid nodules and nodular thyroids: A clinical overview. *Comprehen. Ther.* **22**, 239–250.
16. Malone, J., Unger, J., Delange, F., *et al.* (1991). Thyroid consequences of Chernobyl accident in the countries of the European Community. *J. Endocrinol. Invest.* **14**, 701–717.
17. McTiernan, A., Weiss, N. S., and Daling, J. R. (1987). Incidence of thyroid cancer in women in relation to known or suspected risk factors for breast cancer. *Cancer Res.* **47**, 292–295.
18. DeGroot, L. J., Kaplan, E. L., Straus, F. H., *et al.* (1994). Does the method of management of papillary thyroid carcinoma make a difference in outcome? *World J. Surg.* **18**, 123–130.
19. Cady, B., and Rossi, R. (1988). An expanded view of risk-group definition in differentiated thyroid carcinoma. *Surgery* **104**, 947–953.
20. Hay, I. D., Grant, C. S., Taylor, W. F., *et al.* (1987). Ipsilateral lobectomy versus bilateral lobar resection in papillary thyroid carcinoma: A retrospective analysis of surgical outcome using a novel prognostic scoring system. *Surgery* **102**, 1088–1095.
21. Hay, I. D., Bergstralh, E. J., Goellner, J. R., *et al.* (1993). Predicting outcome in papillary thyroid carcinoma: Development of a reliable prognostic scoring system in a cohort of 1779 patients surgically treated at one institution during 1940 through 1989. *Surgery* **114**, 1050–1058.
22. Hannequin, P., Liehn, J. C., and Delisle, M. J. (1986). Multifactorial analysis of survival in thyroid cancer: Pitfalls of applying the results of published studies to another population. *Cancer* **58**, 1749–1755.
23. Mazzaferri, E. L., and Jhiang, S. M. (1994). Long-term impact of initial surgical therapy and medical therapy on papillary and follicular thyroid cancer. *Am. J. Med.* **97**, 418–428.
24. Leeper, R. (1985). Thyroid carcinoma. *Med. Clin. North Am.* **69**, 1079–1096.
25. McHenry, C., Smith, M., Lawrence, A., *et al.* (1988). Nodular thyroid disease in children and adolescents. *Ann. Surg.* **54**, 444–447.
26. Sadler, G. P., Clark, O. H., van Heerden, J. A., *et al.* (1999). Thyroid and parathyroid. *In* "Principles of Surgery" (S. I. Schwartz, ed.), 7th Ed., Vol. 2, pp. 1661–1713. McGraw-Hill, New York.
27. Miller, J., Kinsi, S. R., and Hamburger, J. I. (1985). Diagnosis of malignant follicular neoplasm of the thyroid by needle biopsy. *Cancer* **55**, 2812–2817.
28. Ashcraft, M., and van Herle, A. (1981). Management of thyroid nodules I. *Head Neck Surg.* **3**, 216–227.
29. Ashcraft, M., and van Herle, A. (1981). Management of thyroid nodules II. *Head Neck Surg.* **3**, 297–322.
30. Ahuja, S., and Ernst, H. (1991). Hyperthyroidism and thyroid carcinoma. *Acta Endocrinol.* **142**, 146–151.
31. Clark, O. H. (1980). Thyroid nodules and thyroid cancer: Surgical aspects. *West. J. Med.* **133**, 1–8.
32. Mazzaferri, E. L. (1993). Management of a solitary thyroid nodule. *N. Engl. J. Med.* **320**, 553–559.
33. Bisi, H., Camargo, R., and Filho, A. (1991). Role of fine needle aspiration cytology in the management of thyroid nodules: Review of experience with 1925 cases. *Diagn. Cytopathol.* **8**, 504–510.
34. Hamburger, J. (1982). Consistency of sequential needle biopsy findings for thyroid nodules: Management implications. *Arch. Intern. Med.* **147**, 97–99.
35. Pepper, G., Zwicker, D., and Rosen, Y. (1989). Fine needle aspiration of the thyroid nodule: Results of a start-up project in a general teaching hospital setting. *Arch. Intern. Med.* **149**, 594–596.
36. Caruso, D., and Mazzaferri, E. (1991). Fine needle aspiration in the management of thyroid nodules. *Endocrinologist* **1**, 194–202.
37. McHenry, C. R., Walfish, P. G., and Rosen, I. B. (1993). Non-diagnostic fine needle aspiration biopsy: A dilemma in management of nodular thyroid disease. *Am. J. Surg.* **59**, 415–419.
38. Gharib, H., and Goellner, J. R. (1993). Fine needle aspiration biopsy of the thyroid: An appraisal. *Ann. Intern. Med.* **118**, 282–289.
39. Grant, C. (1995). Operative and post-operative management of the patient with follicular and Hürthle cell carcinoma. *Surg. Clin. North Am.* **75**, 395–403.
40. Callender, D., Sherman, S., Gagel, R., *et al.* (1996). Cancer of the thyroid. *In* "Cancer of the Head and Neck" (J. Suen and E. Myers, eds.), 3rd Ed. Saunders, Philadelphia.
41. Hathaway, H. (1990). Diagnosis and management of the thyroid nodule. *Otolaryngol. Clin. North Am.* **23**, 303–337.
42. Tyler, D. S., Winchester, D. J., Caraway, N. P., *et al.* (1994). Indeterminate fine-needle aspiration biopsy of the thyroid: Identification of subgroups at high risk for invasive carcinoma. *Surgery* **116**, 1054–1060.
43. Chandramouly, B., Mann, D., Cunningham, R. P., *et al.* (1992). Marine-Lenhart syndrome: Graves' disease with poorly functioning nodules. *Clin. Nuclear Med.* **17**, 905–906.
44. Jossart, G. H., and Clark, O. H. (1994). Well-differentiated thyroid cancer. *Curr. Probl. Surg.* **21**, 933–1012.
45. Shah, J. P., Loree, T. R., Dharker, D., *et al.* (1992). Prognostic factors in differentiated carcinoma of the thyroid gland. *Am. J. Surg.* **164**, 658–661.
46. Cady, B., Sedgwick, C. E., Meissner, W. A., *et al.* (1976). Changing clinical, pathologic, therapeutic and survival pattern in differentiated thyroid carcinoma. *Ann. Surg.* **184**, 541–553.
47. Scheider, A. B., Favus, M. J., Srachura, M. E., *et al.* (1978). Incidence, prevalence and characteristics of radiation-induced thyroid tumors. *Am. J. Surg.* **64**, 243–252.
48. Mazzaferri, E., and Young, R. (1981). Papillary thyroid carcinoma: A 10 year follow-up report of the impact of therapy in 576 patients. *Am. J. Med.* **70**, 511–518.
49. Woolner, L. B., Lemmon, M. L., Beahrs, O. H., *et al.* (1960). Occult papillary carcinoma of the thyroid gland: A study of 140 cases observed over a thirty year period. *J. Clin. Endocrinol.* **20**, 89–105.
50. Frazell, E., and Foote, F. (1958). Papillary cancer of the thyroid. *Cancer* **11**, 895.
51. Breaux, E., and Guillamondegui, O. (1980). Treatment of locally invasive carcinomas of the thyroid: How radical? *Am. J. Surg.* **140**, 514–517.
52. McConahey, W. M., Hay, I., Woolner, C., *et al.* (1986). Papillary thyroid carcinoma treatment at Mayo Clinic 1946 through 1970: Initial manifestations, pathologic findings, treatment and outcome. *Mayo Clin. Proc.* **61**, 978–996.
53. Noguchi, S., and Muracami, N. (1987). The value of lymph node dissection inpatients with differentiated thyroid carcinoma. *Surg. Clin. North Am.* **67**, 251–261.
54. Tollefsen, H. R., Shah, J. P., and Huvos, A. G. (1972). Papillary carcinoma of the thyroid: Recurrence in the thyroid gland after initial surgical treatment. *Am. J. Surg.* **124**, 468–472.
55. Woolner, L. B., Beahrs, O., Block, B., *et al.* (1961). Classification and prognosis of thyroid carcinoma: A study of 885 cases observed in a 30 year period. *Am. J. Surg.* **102**, 354–387.
56. Rosai, J., Carcangui, M. L., and DeLellis, R. A. (1992). Tumors of the thyroid gland. *In* "Atlas of Thyroid Patholog." AFIP, Washington, DC.
57. LiVolsi, V. A. (1992). Papillary neoplasms of the thyroid, pathologic and prognostic features. *Am. J. Clin. Pathol.* **3**, 426–434.
58. Favas, M., Schneider, A., and Stachvra, M. (1976). Thyroid cancer occurring as a consequence of head and neck irradiation. *N. Engl. J. Med.* **294**, 1019.
59. Carcangiu, M., Zampi, G., and Pupi, A. (1985). Papillary carcinoma of the thyroid: A clinicopathologic study of 241 cases treated at the University of Florence, Italy. *Cancer* **55**, 805–828.

60. Massin, J., Savoie, J., Garnier, H., *et al.* (1984). Plmonary metastases in differentiated thyroid carcinoma: Study of 58 cases with implications for the primary tumor treatment. *Cancer* **53**, 982–992.
61. Crile, G., Jr., Antunez, A., and Esselstyn, C. (1985). The advantages of subtotal thyroidectomy and suppression of TSH in the primary treatment of papillary carcinoma of the thyroid. *Cancer* **55**, 2691–2697.
62. DeGroot, L. J., Kaplan, E. L., McCormick, M., *et al.* (1990). Natural history, treatment and course of papillary thyroid carcinoma. *JCEM* **71**, 414–424.
63. Rossi, R., Nieroda, C., Cady, B., *et al.* (1985). Malignancy of the thyroid gland: The Lahey Clinic experience. *Surg. Clin. North Am.* **65**, 211–230.
64. Mazzaferri, E., Young, R., Oertel, J., *et al.* (1977). Papillary thyroid carcinoma: The impact of therapy in 576 patients. *Medicine* **56**, 171–196.
65. Sellers, J. (1992). Prognostic significance of cervical lymph node metastasis in differentiated thyroid carcinoma. *Am. J. Surg.* **164**, 578–581.
66. Sessions, R. B., Taylor, T., Roller, C. A., *et al.* (1999). Cancer of the thyroid gland. *In* "Head and Neck Cancer: A Multidisciplinary Approach" (L. B. Harrison, R. B. Sessions, and W. K. Hong, eds.). Lippincott-Raven, Philadelphia.
67. Hoie, J., Stennig, H., Kullman, G., *et al.* (1988). Distant metastases in papillary thyroid cancer: A review of 91 patients. *Cancer* **61**, 1–6.
68. Cady, B., Rossi, R., Silverman, M., *et al.* (1985). Further evidence of the validity of risk group definition. *Surgery* **98**, 1171–1178.
69. Hayles, A., Kennedy, R., Beahrs, O., *et al.* (1960). Management of the child with thyroid cancer. *JAMA* **173**, 21.
70. Exelby, P., and Frazell, E. (1969). Carcinoma of the thyroid in children. *Surg. Clin. North Am.* **49**, 249–259.
71. Prendiville, S., Burman, K., Ringeil, M., *et al.* (2000). Prognostic Implications of the tall cell variant of papillary thyroid carcinoma. *Otolaryngol. Head Neck Surg.* **122**, 352–357.
72. Brennan, M. D., Bergstralh, E. J., van Heerden, J. A., *et al.* (1991). Follicular thyroid carcinoma treatment at Mayo Clinic 1946 to 1970: Initial manifestation, pathologic findings, therapy and outcome. *Mayo Clin. Proc.* **66**, 11–22.
73. Kahn, N., and Perzin, K. (1983). Follicular carcinoma of the thyroid. *Pathol. Annu.* **18**, 221–253.
74. Young, R. L., Mazzaferri, E. L., Rahe, A. J., *et al.* (1980). Pure follicular carcinoma: Impact of treatment in 214 patients. *J. Nuclear Med.* **21**, 733–737.
75. Harness, J., Thompson, N. W., McLeod, M. K., *et al.* (1984). Follicular carcinoma of the thyroid gland: Trends and treatment. Surgery **96**, 972–980.
76. Lange, W., Georgii, A., Stauch, G., *et al.* (1980). The differentiation of atypical adenomas and encapsulated follicular carcinoma in the thyroid gland. *Virch. Arch(A).* **385**, 125–141.
77. Yamashina, M. (1992). Follicular neoplasms of the thyroid. *Am. J. Surg. Pathol.* **16**, 392–400.
78. Woolner, L. (1971). Thyroid carcinoma: Pathologic classification with data on prognosis. *Semin. Nuclear Med.* **1**, 481–502.
79. Johannessen, J. V., Sobrinho-Simmoes, M., Lindmot, T., *et al.* (1982). The diagnostic value of flow cytometric DNA measurements in selected disorders of the human thyroid. *Am. J. Clin. Pathol.* **77**, 20–25.
80. Lange, W., Choritz, H., and Hundeshagen, H. (1986). Risk factors in follicular thyroid carcinoma: A retrospective follow-up study covering a fourteen year period with emphasis on morphologic findings. *Am. J. Surg. Pathol.* **10**, 246–255.
81. Maxon, H. R., and Smith, H. S. (1990). Radioactive ^{131}I in the diagnosis and treatment of metastatis well-differentiated thyroid carcinoma. *Endocr. Metab. Clin. North Am.* **19**, 695–717.
82. Silverman, M. (1991). Pathology of thyroid and parathyroid glands. *In* "Surgery of the Thyroid and Parathyroid Glands" (B. Cady, and R. Rossi, eds.). Saunders, Philadelphia.
83. van Heerden, J., Hay, I., and Goellner, J. (1992). Follicular thyroid carcinoma with capsular invasion alone: A non-threatening malignancy. *Surgery* **112**, 1136–1138.
84. Crile, G., Pontius, K., and Hawk, W. (1985). Factors influencing the survival of patients with follicular carcinoma of the thyroid gland. *Surg. Gynecol. Obstet.* **160**, 409–413.
85. Donohue, J., Goldfien, S., and Miller, T. (1984). Do the prognosis of papillary and follicular thyroid cancer differ? *Am. J. Surg.* **148**, 168–173.
86. Thompson, M., Dunn, E., Batsakis, J., *et al.* (1973). Hürthle cell lesions of the thyroid gland. *Surg. Gynecol. Obstet.* **139**, 555–560.
87. El-Naggar, A., Batsakis, J., and Luna, M. (1988). Hürthle cell tumors of the thyroid: A flow cytometric DNA analysis. *Arch. Otolaryngol. Head Neck Surg.* **114**, 520–521.
88. Hazard, J. B. (1977). The C-cell of the th thyroid carcinoma: A review. *Am. J. Pathol*
89. Chong, F., Beahrs, O., Sizemore, G., *et al.* (1975). Medullary carcinoma of the thyroid gland. *Cancer* **35**, 695–704.
90. Ellenhorn, J., Shah, J., and Brennan, M. F. (1993). Impact of therapeutic regional lymph node dissection for medullary carcinoma of thyroid gland. *Surgery* **114**, 1078–1082.
91. Raue, F., Frank-Raue, K., and Grauer, A. (1994). Multiple endocrine neoplasia type 2: Clinical features and screening. *Endocrinol. Metab. Clin. North Am.* **23**, 137–156.
92. DeLellis, R. (1995). Biology of disease: Multiple endocrine neoplasia syndromes revisited. *Lab. Invest.* **72**, 494–505.
93. Moley, J. (1995). Medullary thyroid cancer. *Surg. Clin. North Am.* **75**, 405–420.
94. Goodfellow, P., and Wells, S. (1995). RET gene and its implications for cancer. *JNCI* **87**, 1515–1523.
95. Saad, M. F., Ordonez, N. G., Rashid, R. K., *et al.* (1984). Medullary thyroid carcinoma: A study of the clinical features and prognostic factors in 161 patients. *Medicine* **63**, 319–342.
96. Bergholm, U., Adami, H. O., and Bergstrom, R. (1989). Clinical characteristics in sporadic and familial medullary thyroid carcinoma. *Cancer* **63**, 1196–1204.
97. Sessions, R. B., Harrison, L. B., and Forastiere, A. A. (1996). Tumors of the salivary glands and paragangliomas. *In* "Cancer: Principles and Practice of Oncology" (V. T. DeVita, S. Hellman, and S. A. Rosenberg, eds.), 5th Ed. Lippincott-Raven, Philadelphia.
98. Stepanas, A., Samaan, N., Hill, C., *et al.* (1979). Medullary thyroid carcinoma: importance of serial calcitonin measurements. *Cancer* **43**, 825–837.
99. Donovan, D. T., and Gagel, R. F. (1997). Medullary thyroid carcinoma and the multiple endocrine neoplasia syndromes. *In* "Thyroid Disease: Endocrinology, Surgery, Nuclear Medicine and Radiotherapy" (S. A. Falk, ed.), 2nd Ed. Lippincott-Raven, Philadelphia.
100. Block, M. A. (1990). Surgical treatment of medullary carcinoma of the thyroid. *Otolaryngol. Clin. North Am.* **23**, 453–473.
101. Lips, C. J., Landsvater, R. M., Höppener, J. W., *et al.* (1994). Clinical screening as compared with DNA analysis in families with multiple endocrine neoplasia type 2A. *N. Engl. J. Med.* **331**, 828–835.
102. O'Riordain, D., O'Brien, T., Crotty, T., *et al.* (1995). Multiple endocrine neoplasia type 2B: More than an endocrine disorder. *Surgery* **118**, 936–942.
103. Wells, S., Baylin, S., and Gann, P. (1978). Medullary thyroid carcinoma: Relationship of method of diagnosis to pathologic staging. *Ann. Surg.* **188**, 377–383.
104. Kakudo, K., Carney, J. R., and Sizemore, G. W. (1985). Medullary carcinoma of the thyroid: Biological behavior of the sporadic and familial neoplasm. *Cancer* **55**, 2818–2821.
105. LiVolsi, V., and Merino, M. (1987). Pathology of thyroid tumors. *In* "Comprehensive Management of Head and Neck Tumors" (S. Thawley, W. Panje, J. Batsakis, *et. al.*, eds.). Saunders, Philadelphia.

106. Nel, C., Van Heerden, J. L., and Goellner, J. (1985). Anaplastic carcinoma of thyroid: A clinicopathologic study of 82 cases. *Mayo Clin. Proc.* **60**, 51–58.
107. Kapp, D., LiVolsi, V., and Sanders, M. (1982). Anaplastic carcinoma following well-differentiated thyroid cancer: Etiological considerations. *Yale J. Biol. Med.* **55**, 521–528.
108. Demeter, J., DeJong, S., Lawrence, A., *et al.* (1991). Anaplastic thyroid carcinoma: Risk factors and outcome. *Surgery* **110**, 956–963.
109. Venkatesh, Y., Ordonez, N., and Schultz, P. (1990). Anaplastic carcinoma of thyroid: A clinicopathologic study of 121 cases. *Cancer* **66**, 321–330.
110. Kim, J., and Leeper, R. (1987). Treatment of locally advanced thyroid carcinoma with combination of doxorubicin and radiation therapy. *Cancer* **60**, 2372–2375.
111. Tennvall, J., Lundell, G., Hallquist, A., *et al.* (1994). Combined doxorubicin, hyperfractionated radiotherapy and surgery in anaplastic thyroid carcinoma: Report on two protocols. The Swedish Anaplastic Thyroid Cancer Group. *Cancer* **74**, 1348–1354.
112. Carcangiu, M. L., Zampi, G., and Rosai, J. (1984). Poorly-differentiated ("insular") thyroid carcinoma: A reinterpretation of Langham's "Wuchemde Struma." *Am. J. Surg. Pathol.* **8**, 655–668.
113. Burman, K., Ringel, M., and Wartofsky, L. (1996). Unusual types of thyroid neoplasms. *Endocrinol. Metab. Clin. North Am.* **25**, 49–68.
114. Sirota, D. K., and Segal, R. L. (1979). Primary lymphomas of the thyroid gland. *JAMA* **242**, 1743–1746.
115. Tsang, R. W., Gospodarowicz, M. K., Sutcliffe, S. B., *et al.* (1993). Non-Hodgkin's lymphoma of the thyroid gland: prognostic factors and treatment outcome. The Princess Margaret Hospital Lymphoma Group. *Int. J. Radiat. Oncol. Biol. Phys.* **27**, 559–604.
116. Salhany, K. E., and Pietra, G. G. (1993). Extranodal lymphoid disorders. *Am. J. Clin. Pathol.* **99**, 472–485.
117. Divine, R. M., Edis, A., and Banks, P. (1981). Primary lymphoma of the thyroid: A review of the Mayo Clinic experience through 1978. *World J. Surg.* **5**, 33–38.
118. Rasbach, D. A., Mondschein, M. S., Harris, N. L., *et al.* (1985). Malignant lymphoma of the thyroid gland: A clinical and pathologic study of 20 cases. *Surgery* **6**, 1166–1170.
119. Ivy, H. K. (1984). Cancer metastatic to the thyroid: A diagnostic problem. *Mayo Clin. Proc.* **59**, 856–859.
120. Green, L., Ro, J. Y., Mackay, B., *et al.* (1989). Renal cell carcinoma metastatic to the thyroid. *Cancer* **63**, 1810–1815.
121. Meissner, W., and Adler, A. (1958). Papillary carcinoma of thyroid. *Arch. Pathol.* **656**, 518.
122. Tubiana, M., Haddad, E., and Schlumberger, M. (1985). External radiotherapy in thyroid cancers. *Cancer* **55**, 2062–2071.
123. Hayes, B., Anthony, A., and Kersham, R. (1985). Anatomy and development of the thyroid gland. *Ear Nose Throat J.* **64**, 10.
124. Copp, D., Cockcroft, D., and Kuch, Y. (1967). Calcitonin from ultimobranchial glands of dogfish and chickens. *Science* **158**, 924–925.
125. Hollingshead, W. H. (1958). Anatomy of the endocrine glands. *Surg. Clin. North Am.* **39**, 1115–1140.
126. Nemiroff, P. M., and Katz, A. D. (1982). Extralaryngeal divisions of the recurrent laryngeal nerve: Surgical and clinical significance. *Am. J. Surg.* **144**, 466–469.
127. Lore, J. M., Kim, D. J., and Elias, S. (1977). Preservation of the laryngeal nerves during total thyroid lobectomy. *Ann. Otol. Rhinol. Laryngol.* **86**, 777–788.
128. Lennquist, S., Kahlin, C., and Smeds, S. (1987). The superior laryngeal nerve in thyroid surgery. *Surgery* **102**, 1000–1008.
129. Wang, C. (1976). The anatomic basis of parathyroid surgery. *Ann. Surg.* **183**, 271–275.
130. Hutter, R., Frazell, E., and Foote, F. (1970). Elective radical neck dissection: An assessment of its use in the management of papillary thyroid cancer. *Cancer* **20**, 87–93.
131. Siperstein, A. E., and Clark, O. H. (1996). Surgical therapy. *In* "Wener and Ingbar's the Thyroid: A Fundamental and Clinical Text" (L.E. Braverman, and R. D. Utiger, eds.), 7th Ed. Lippincott-Raven, Philadelphia.
132. Mack, E. (1995). Management of patients with substernal goiters. *Surg. Clin. North Am.* **75**, 377–394.
133. Shah, A., Alfonso, A. E., and Jaffe, B. M. (1989). Operative treatment of substernal goiters. *Head Neck* **11**, 325–330.
134. Grillo, H. C., and Zannini, P. (1986). Resectional management of airway invasion by thyroid carcinoma. *Ann. Thorac. Surg.* **42**, 287–298.
135. Ballantyne, A. J. (1994). Resections of the upper aerodigestive tract for locally invasive thyroid cancer. *Am. J. Surg.* **168**, 636–639.
136. Shelton, V. K., Skolnick, G., and Berlinger, F. G. (1982). Laryngotracheal invasion by thyroid carcinoma. *Ann. Otol. Rhinol. Laryngol.* **91**, 363–369.
137. Park, H. M., Park, Y. H., and Zhou, X. H. (1997). Detection of thyroid remnant/metastasis without stunning: An ongoing dilemma. *Thyroid* **7**, 277–280.
138. Segal, K., Fridental, R., Lubin, E., *et al.* (1995). Papillary carcinoma of the thyroid. *Otolaryngol. Head Neck Surg.* **113**, 356–363.
139. Grant, C. S., Hay, I. D., Gough, I. R., *et al.* (1988). Local recurrence in papillary thyroid carcinoma: Is the extent of surgical resection important? *Surgery* **104**, 954–962.
140. Pasieka, J. L., Thompson, N. W., McLeod, M. K., *et al.* (1992). The incidence of bilateral well-differentiated thyroid cancer found at completion thyroidectomy. *World J. Surg.* **16**, 711–717.
141. Kupferman, M. E., Mandel, S. J., DiDonato, L., *et al.* (2002). Safety of completion thyroidectomy following unilateral lobectomy for well-differentiated thyroid cancer. *Laryngoscope* **112**, 1209–1212.
142. Schlumberger, M., Tubiana, M., and DeVathaire, F. (1986). Long term results of treatment of 283 patients with lung and bone metastases from differentiated thyroid carcinoma. *J. Clin. Endocrinol. Metab.* **63**, 960–967.
143. Chonkich, G. D., and Petti, G. H. (1992). Treatment of thyroid carcinoma. *Laryngoscope* **102**, 486–491.
144. Attie, J. N., Bock, G., and Moskowitz, G. W. (1992). Postoperative radioactive evaluation of total thyroidectomy for thyroid carcinoma: Reappraisal and therapeutic implications. *Head Neck* **14**, 297–302.
145. Ward, P. H., Berci, G., and Calcaterra, T. C. (1977). Superior laryngeal nerve paralysis, an often overlooked entity. *Trans. Am. Acad. Ophthalmol. Otolaryngol.* **84**, 78–89.
146. Beahrs, O. (1977). Complications of surgery of the head and neck. *Surg. Clin. North Am.* **57**, 823–829.
147. Flynn, M. B., Lyons, K. J., and Tartar, J. W. (1994). Local complications after surgical resection for thyroid cancer. *Am. J. Surg.* **168**, 404–407.
148. Netterville, J. L., Aly, A., and Ossoff, R. H. (1990). Evaluation and treatment of complications of thyroid and parathyroid surgery. *Otolaryngol. Clin. North Am.* **23**, 529–552.
149. Dedo, H. (1970). The paralyzed larynx: An electromyographic study in dogs and humans. *Laryngoscope* **80**, 1455–1517.
150. Boles, R., and Fritzell, B. (1969). Injury and repair of the recurrent laryngeal nerves in dogs. *Laryngoscope* **70**, 1405–1418.
151. Crumley, R., and Izdensk, K. (1986). Voice quality following laryngeal reinnervation by ansa hypoglossal transfer. *Laryngoscope* **96**, 611–616.
152. Tucker, H. M. (1977). Reinnervation of unilateral paralyzed larynx. *Ann. Otol. Rhinol., Laryngol.* **86**, 789–794.
153. Beahrs, O. (1979). Complications in thyroid and parathyroid surgery. *In* "Complications in Head and Neck Surgery" (J. Conley, ed.). Saunders, Philadelphia.
154. Harasch, H. R., Franssila, K. O., and Wasenus, V. M. (1985). Occult papillary carcinoma of the thyroid: A "normal" finding in Finland. A systematic autopsy study. *Cancer* **56**, 531–538.
155. Gharib, H., and Goellner, J. R. (1993). Fine-needle aspiration biopsy of the thyroid: An appraisal. *Ann. Intern. Med.* **118**, 282–289.
156. Cunningham, M. P., Duda, R. B., Recant, W., *et al.* (1990). Survival discriminants for differentiated thyroid cancer. *Am. J. Surg.* **160**, 344–347.

157. Samaan, N., Schultz, P., Hickey, R., *et al.* (1992). The results of various modalities of treatment of well-differentiated thyroid carcinoma: A retrospective review of 1599 patients. *J. Clin. Endocrinol. Metab.* **75**, 714.
158. Meier, C. A., Braverman, L. E., and Ebner, S. A. (1994). Diagnostic use of recombinant human thyrotropin in patients with thyroid carcinoma (phase I/II study). *J. Clin. Endocrinol. Metab.* **78**, 188–196.
159. Mandel, S. J., Shankar, L. K., Benard, F., and Alavi, A. (2001). Superiority of iodine-123 compared with iodine-131 scanning for thyroid remnants in patients with differentiated thyroid cancer. *Clin. Nuclear Med.* **26**, 6–9.
160. Shankar, L. K., Yamamoto, A. J., Alavi, A., *et al.* (2002). Comparison of ^{123}I scintigraphy at 5 and 24 hours in patients with differentiated thyroid cancer. *J. Nuclear Med.* **43**, 72–76.
161. Beirwaltes, W. (1978). The treatment of thyroid cancer with radioactive I. *Semin. Nuclear Med.* **8**, 79–94.
162. Beierwaltes, W. H., Nishiyama, R. H., Thompson, N. W., *et al.* (1982). Survival time and "cure" in papillary and follicular thyroid carcinoma with distant metastases: Statistics following University of Michigan therapy. *J. Nuclear Med.* **23**, 561–568.
163. Tubiana, M., Schlumberger, M., Rougier, P., *et al.* (1985). Long-term results and prognostic factors in patients with differentiated thyroid carcinoma. *Cancer* **55**, 794–804.
164. Simpson, W., Panzarella, T., Carruthers, J., *et al.* (1988). Papillary and follicular thyroid cancer: Impact of treatment in 1578 patients. *Int. J. Radiat. Oncol. Biol. Phys.* **14**, 1063–1075.
165. Brunt, L. M., and Wells, S. H. (1987). Advances in the diagnosis and treatment of medullary carcinoma. *Surg. Clin. North Am.* **67**, 263–279.
166. Samaan, N. A., Schultz, P. N., and Hickey, R. C. (1988). Medullary thyroid carcinoma: Prognosis of familial versus sporadic disease and the role of radiotherapy. *J. Clin. Endocrinol. Metab.* **67**, 801–805.
167. Tan, R., Finley, R., Driscoli, D., *et al.* (1995). Anaplastic carcinoma of the thyroid: A 24 year experience. *Head Neck* **17**, 41–48.
168. deBesi, P., Busnardo, B., Toso, S., *et al.* (1991). Combined chemo-therapy with bleomycin, adriamycin and platinum in advanced thyroid cancer. *J. Endocrinol. Metab.* **14**, 475–480.
169. Rougier, P., Parmentier, C., Laplanche, A., *et al.* (1983). Medullary thyroid carcinoma: Prognostic factors and treatment. *Int. J. Radiat. Oncol. Biol. Phys.* **9**, 161–169.
170. Samaan, N., Yang, K., and Schultz, P. (1989). Diagnosis, management and pathogenetic studies in medullary thyroid carcinoma syndrome. *Henry Ford Hosp. Med. J.* **37**, 132.
171. Szanto, J., Vincze, B., and Sinkovics, I. (1989). Postoperative thyroglobulin level determination to follow-up patients with highly differentiated thyroid cancer. *Oncology* **46**, 99.
172. Ozata, M., Suzuki, S., and Miyamoto, T. (1994). Serum thyroglobulin in the follow-up of patients with treated differentiated thyroid cancer. *J. Clin. Endocrinol. Metab.* **79**, 98.

CHAPTER

29

Assessment of Outcomes in Head and Neck Cancer

URJEET PATEL and JAY PICCIRILLO

Department of Otolaryngology
Washington University
St. Louis, Missouri 63110

I. Background 433
II. Survival 434
III. Determinate Survival 434
IV. Actuarial Survival 435
V. Kaplan–Meier Survival 436
VI. Morbidity 436
VII. Health Status 436
VIII. Quality of Life 438
IX. Cost of Care 441
X. Summary 441
References 442

Outcomes research is the scientific study of outcomes derived from diverse therapies used to treat a particular disease, illness, or condition [1–3]. The primary goal of outcomes research is to assess treatment effectiveness, or the positive impact that a given therapy has on aspects of care that are of interest to both patient and clinician [4]. Identifying the most effective treatment through outcomes research is a growing field due, in part, to the cost-conscious environment of health care in the United States [5]. Key elements of outcomes research are the nonrandomized study of diverse therapies for a particular illness, the expanded definition of outcome, and the central role of the patient in treatment selection.

The management of head and neck cancer is an ideal field for the application of outcomes research [6]. The criterion standard for assessing the merits of a given treatment is the prospective, blinded randomized clinical trial. Accordingly, such trials have long been advocated; however, such trials in head and neck cancer treatment are inherently problematic and rare [7–9]. Death is the end result of untreated cancer; subsequently, there are ethical considerations when constructing control groups for a study. Another problem is the heterogeneity of the study population in terms of tumor stage, primary site, histologic grade, and age. Given the relative rarity of head and neck cancer, it is difficult to generate a sufficient sample size of similar patients for a given research trial. In addition, the nature of surgical and radiation therapy makes it virtually impossible to blind a researcher regarding treatment received in order to eliminate bias. One other factor that further discourages such study is the lack of standardization of surgical resection and reporting. Weymuller *et al.* [10] noted a substantial heterogeneity in surgical resections and a large spectrum of reporting formats when examining the head and neck cancer literature. They offered a solution to this lack of standardization through an outcome-reporting instrument designed to provide consistency of information for study purposes [11]. Until such standard reporting mechanisms are widely adopted, the heterogeneity of data further discourages well-performed clinical trials. Finally, it is often difficult to randomize patients to treatments that are so markedly different, although the Veterans Affairs (VA) cooperative study of chemotherapy and radiation versus surgery for laryngeal cancer was a notable exception [12]. Patients and physicians are hesitant to leave such grossly dissimilar options to chance alone. After examining these difficulties associated with prospective randomized studies, it is not surprising that the majority of research to assess the treatment effectiveness in head and neck cancer comes from observational studies.

I. BACKGROUND

Head and neck oncology has evolved into a highly complicated, multimodality specialty. Historically, patients were treated with radical surgery that included wide local resection of important structures to rid them of their oncologic disease. Over time, new therapeutic options have sprung forth in the area of radiation therapy and chemotherapeutics that together offer a variety of effective therapies.

Central to the task of outcome assessment is the definition of a clinically relevant outcome. In contrast to non-neoplastic processes, the usual outcome of unsuccessful treatment in head and neck cancer is death from uncontrolled disease. Subsequently, the primary outcome measure used to assess treatment effectiveness has been survival. Historically, the benefit of a given therapy for head and neck cancer was gauged solely by its ability to improve survival or reduce mortality. In modern outcomes research, however, the emphasis is broadened to encompass measures of outcome other than survival alone [13]. This broader approach evaluates not only traditional end points, such as mortality and local control, but also other parameters, such as patient satisfaction, quality of life, functional status, and cost of care. Factors related to a patient's overall health status receive equal if not greater consideration in modern outcomes assessment as compared to survival alone. This becomes especially true when different therapeutic approaches yield similar survival rates, which is frequently observed to be the case with treatment for head and neck cancer.

Expanded patient outcomes have been described in a hierarchical fashion by Fries and Spitz [14]. In their formulation, the first two tiers of patient outcome description are formed by mortality followed by morbidity. The third level is then described broadly by the "health status" of the patient. This broad term can be further described as the physical, functional, and emotional limitations experienced by the patient in relation to the disease process [15,16]. The next level of outcome is related to quality of life. Finally, cost of medical care comprises the final tier in the evaluation of treatment outcome. In the current environment of cost-conscious health care, cost effectiveness is a crucial factor in the evaluation of treatment effectiveness. By combining information regarding cost, quality of life, health status, and symptoms with the traditional end points of morbidity and mortality, the clinician can better define the effectiveness and efficiency of treatment and thus improve treatment recommendations by physician and choice by patient [13].

II. SURVIVAL

Historically, survival has been the critical outcome measure for patients with head and neck cancer. In an editorial addressing the uniformity of results, Weymuller and Goepfert [17] focused on such objective measures as mortality and survival to evaluate treatments in head and neck cancer. As described by Fries and Spitz [14], survival continues to form the basic level of outcome measure in cancer studies. Some contend that a properly calculated survival rate is the best single statistical index available for measuring the efficacy of one cancer therapy compared with another [18].

Survival data from studies of cancer patients are generally given in terms of survival at a particular timepoint following treatment (e.g., 3- or 5-year survival). Various reference dates are used as starting points for evaluating the effects of a given treatment on survival. Such dates include the date of first symptom, date of first visit to physician, date of diagnosis, and date of initiation of treatment. The specific type of reference date used may vary with the purpose of a given study; however, it should be clearly stated and remain constant for all members within a study. For evaluation of therapy, for example, the date of initiation of treatment is the ideal reference point. For patients not receiving treatment, the most comparable date is the time at which the decision was made that no treatment would be given. In both cases, the aforementioned times from which survival rates are calculated will generally coincide with the date of initial cancer staging [18].

The results of treatment of head and neck cancer as they relate to survival are then given in relation to a particular time interval following the reference point. The interval that is used most frequently is the 5-year overall survival. It should be noted that this term does not imply cure of the patient, but rather that the patient has survived for 5 years since the reference time. Accordingly, quoting 5-year statistics for success of therapy must be stated as survival rates as opposed to cure rates. This illustrates the need for strict definitions and calculations regarding survival data. Depending on particular definitions and formulations, many types of survival data can be generated.

III. DETERMINATE SURVIVAL

The *determinate survival* method for arriving at 5-year survival, also known as the direct method, is perhaps the simplest procedure for summarizing patient survival. It was first proposed by Martin and colleagues in 1935 and later modified in 1948 by McDonald [19]. This method is used to characterize patient survival after each member has been followed for a given time interval. The determinate method stratifies patients into two basic groups. The *indeterminate* group is composed of patients lost to follow-up, patients who refused treatment, those who sought treatment elsewhere, and those dead of intercurrent disease. This group is subtracted from the initial total population to arrive at the determinate group. This *determinate* group is composed of treatment successes, treatment failures, and those dead of disease, all of whom have been followed for at least 5 years since the reference point. A 5-year survival rate can then be calculated by the ratio of the patients alive in the determinate to the total determinate group.

An assumption must be made to assume the reliability of the resulting survival rate. The survival rate for the indeterminate group must be equal to the rate of the determinate group. If patients in the indeterminate group have a strikingly different survival, then the accuracy of the determinate group's survival should be questioned. As a rule, if the number of patients in the indeterminate group is small, the variance in survival rates will be small and the assumption of similarity in survival will likely hold.

An additional problem with the determinate method deals with the efficiency of data use. If patients are being collected longitudinally, one may frequently have to wait several years after the given interval period to collect an adequate number of patients who have all been followed for the requisite amount of time. It may require 10 years to collect 200 patients, all of whom have been followed for a full 5 years. There may be many recently treated patients who cannot be included in the study because they have not been exposed to the entire 5-year risk of death. Consequently, the analysis of data must be significantly delayed to a time rather remote from the initial study. Despite these potential pitfalls of the determinate method, it still remains the most simple means for calculating a valid survival rate for a study population.

IV. ACTUARIAL SURVIVAL

The *actuarial survival* method provides a means of using all patient data accumulated right up to the closing date of a study. It was first proposed by Berkson and Gage [20] and may be considered a more scientific method of survival calculation. This method gives approximately similar values for survival as the determinate method when evaluating a large series of patients; however, in a smaller series or one where many are lost to follow-up, the actuarial method may be more useful. Data are analyzed at periodic intervals with a survival percentage calculated at the end of each interval. Accordingly, the percentage survival for each interval can be displayed graphically to generate a survival curve of a given treatment group. This survival curve can then be compared to the curve for a control group and the mortality hazard can be viewed graphically.

The critical feature of the actuarial method is the use of censored patient data. In the determinate method, a patient who has survived for 3 years after treatment but who has not had a complete 5-year follow-up must be excluded from analysis. With the actuarial method, such potentially valuable data are included in the survival calculation. At the end of the first year, one can examine all patients regardless of whether they completed 1 year of follow-up. At the end of the second year, all patients regardless of whether the completed 2 years of follow-up are included. By this system, most cases can be analyzed in less time needed than with the determinate method. Patients in actuarial analysis considered censored are those that are lost to follow-up or those who did not complete the full duration of follow-up. With the actuarial method, they still contribute valuable information in the overall analysis.

To better understand survival calculation by the actuarial method, an example is given in Table 29.1. In this case, 50 patients are included in the study. The time interval is set at 1 year. The number of patients during each year of follow-up is shown. Also, the number of patients who are lost to follow-up is also displayed. These patients are considered to be followed for one-half of a year on average and are thus able to contribute to survival data. Each such patient counts as one-half of a year of survival, which is added to the denominator when calculating the proportion surviving for a given time interval. The survival rate can then be calculated for each time interval. A composite survival rate can be derived from the product of survival rates for each time interval. This information can then be displayed graphically as seen in Fig. 29.1 to generate a survival curve for this population. Such survival

TABLE 29.1 Calculation of Survival Method[a]

Interval of last observation	No. alive at beginning of year	No. dying during year	No. last seen alive during year	Effective no. exposed to risk of dying	Proportion dying during year	Proportion surviving year	Proportion surviving from first Rx to end of interval
0–<1	50	5	0	50	0.100	0.900	0.900
1–<2	45	4	0	45	0.089	0.911	0.820
2–<3	41	4	4	39	0.102	0.898	0.736
3–<4	33	3	3	31.5	0.095	0.904	0.665
4–<5	27	2	2	26	0.077	0.923	0.614
>5	23	3	20				
Total		21	29				

[a]Fifty patients are presented in this population. Survival rates are given for each time interval as is cumulative survival.

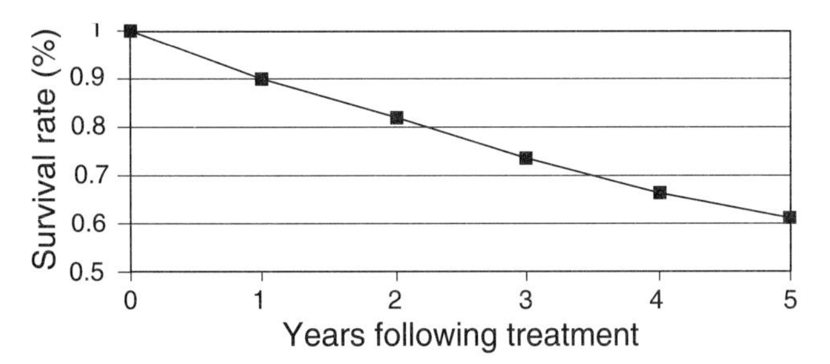

FIGURE 29.1 Survival rate following treatment.

curves can then be compared for subgroups receiving different treatments to assess clinical outcomes.

When calculating actuarial survival, one can adjust for disease-specific survival by deriving an adjusted survival rate. In this situation, attention is given to the cause of death and tumor status at that time. The adjusted rate is that proportion of patients who escaped death due to cancer when all other causes of death were not operating. The calculation of these adjusted rates relies heavily on accurate information regarding cause of death. Such an adjusted survival rate is of particular importance when comparing groups that may differ significantly with respect to such factors as age, gender, and socioeconomic status.

V. KAPLAN–MEIER SURVIVAL

A third method of survival calculation is the Kaplan–Meier method [21]. This method was first described in 1958 and has become widely used since then. It is similar to the actuarial method in that censored data are included. This method measures the probability that a patient survives past a given time point. The patients are organized in such a fashion that those surviving the shortest periods are placed first and then followed by those surviving longer. A survival distribution can then be constructed and a survival curve generated. Survival percentage is calculated at intervals each ending with the death of a given patient. Censored patients are then eliminated at the end of each interval. Of note is that the time intervals are not evenly spaced and predetermined as with the actuarial method but instead vary in length based on the time to each subsequent event. Survival curves can then be generated for patients receiving alternate treatment and standard treatment in order to compare treatment effect.

VI. MORBIDITY

The second tier of outcome assessment is composed of patient morbidity. Morbidity can be described as the presence of a diseased state or condition and can be measured in a given patient. It is generally considered sequelae to either the primary disease or the subsequent treatment of disease. A patient's mortality is a binary variable (alive or deceased) and thus very simple to measure. In contrast, morbidity is as varied as disease itself and may present with varying degrees of severity. The list of different morbidities encountered in the treatment of head and neck cancer is extensive and beyond the scope of this chapter. The presence of a specific disease process in a given patient is calculated easily; however, it generally makes little sense to lump the wide variety of encountered morbidity together for the purpose of outcome assessment. Each specific morbidity (wound infection, stomatitis, myocardial infarction) is usually best analyzed and reported by itself in any given study.

Certain schemes of classification can be used to stratify morbidity into categories that may be useful for analysis. Morbidity can be considered by its location and divided into local versus systemic. Stomatitis and pharyngocutaneous fistula are examples of local morbidity, whereas anemia and malnutrition are systemic. They can also be categorized according to time frame and classified as transient or permanent. Chyle leak after neck dissection may frequently be transisent without lasting effects. While mucositis after radiation therapy may be temporary, xerostomia is permanent. Perioperative cerebrovascular accident and myocardial infarction are examples of long-term morbidity that often have permanent sequelae. Finally, morbidity can be considered iatrogenic or idiopathic. Pneumonia in a cancer patient may be idiopathic, whereas recurrent laryngeal nerve injury during thyroidectomy would be iatrogenic. Because outcomes studies are concerned with the impact of treatment, it is frequently such iatrogenic morbidity that is being evaluated in a given study. Thus, classification of morbidity along the afore-mentioned parameters may be useful to the clinician to establish which morbidity needs to be evaluated and how it will be measured in overall outcomes assessment.

VII. HEALTH STATUS

Health status is a description of the physical, functional, and emotional impairments experienced by the patient. As one might imagine, these attributes may be more difficult to measure than the other outcome end points of survival and morbidity. Subsequently, detailed data in these areas have not been provided routinely in reports of treatment results. Instruments to measure health status have been developed and validated that do measure these outcomes and frequently take the form of patient-based questionnaires [13]. Unlike other end points that can be derived from routine patient care with a minimal investment of additional time or resources, such questionnaires are sometimes considered cumbersome to administer. Perhaps a larger problem, however, is the perception by many investigators that such data are "soft" and inferior to "hard" traditional clinical outcome end points [22]. Feinstein [23] argued that consistency and reproducibility are the key components of "hard" data, and many health status instruments have demonstrated both of these attributes.

Numerous instruments have been developed to measure the health status of patients. Different instruments may focus on a particular aspect of health status that includes physical, functional, and overall performance parameters. Some indices assess overall functional status, such as the Karnofsky performance status scale (KPS) [24] and the sickness impact profile [25]. The KPS, for example, is a rating scale used to quantify a patient's functional capacity for work and daily activities of self-care. Factors that influence ratings are the degree to which symptoms hinder that ability to work of the amount of assistance required in self-care. KPS scores can be assigned to patients prior to treatment and repeatedly throughout treatment as well as afterward for two purposes. First, KPS scores have been shown to be of prognostic value with regard to outcome. Pretreatment KPS scores were shown to have an impact on survival in a study of laryngeal cancer patients where decreasing KPS scores were associated with decreased survival [26]. In this study, the strength of this association with survival was noted to be greater for KPS than T stage. Second, KPS ratings can be performed following treatment in order to serve as an independent outcome measure of health status. The impact of a given treatment on health status can be assessed and compared with one another.

While such global scales are useful for comparing treatment effects and outcomes across a wide range of cancer conditions, there are other instruments more specific to head and neck cancer patients that may assess outcomes more relevant to this patient population. Mobility, for example, is a key parameter in the KPS equation; however, it is not usually a critical descriptor for head and neck cancer patients. Many of these patients regain mobility soon after treatment and are thus given high KPS scores. These patients may still have severe impairments of health status that are not reflected by this global score.

Over the past decade, considerable work has been done to develop and validate cancer-specific instruments [27]. Several of them were designed specifically for patients with head and neck cancer. While some of these tools assess global performance, others focus on a symptom or aspect of health status that is of particular relevance to the head and neck cancer patient. Examples of these instruments and their relevance to assessment of health status outcomes are presented later.

List *et al.* [28] developed the performance status scale of head and neck cancer patients (PSSHNCP). This scale is composed of three subscales that each assess unique areas of dysfunction that are commonly experienced by this patient population. The three components assess speech, diet, and ability to eat in public. Each subscale is rated by the clinician on a scale from zero to 100. Descriptions of the scoring system are displayed in Table 29.2. The three subscales assess three different areas of functioning with relatively low correlation between them. In addition, each subscale was found to correlate poorly with the KPS score. It follows that this scale does indeed measure independent information that is not captured in global performance scales. This scaling system was shown to be reliable across raters and is able to demonstrate relevant differences in health status across a spectrum of head and neck cancer patients. This instrument has since been used in subsequent studies to assess outcome following treatment of head and neck cancer to capture the unique problems of head and neck cancer patients [29].

Baker and Schuller [30] developed a patient-based measure of functional status called the functional status in head and neck cancer–self-report (FSH&N-SR). It is composed of a patient-based questionnaire that inquires into problems specific to head and neck cancer patients. Numerous parameters of function are assessed, including chewing, swallowing, taste, speech, and breathing. Each variable is graded on a five-point scale with descriptors to

TABLE 29.2 Performance Status Scale of Head and Neck Cancer Patients

Normalcy of diet	
100	Full diet (no restrictions)
90	Peanuts
80	All Meat
70	Carrot, celery
60	Dry bread and crackers
50	Soft chewable foods (e.g., macaroni, canned fruits)
40	Soft food requiring no chewing (e.g., mashed potatoes, pudding)
30	Pureed foods (in blender)
20	warm liquids
10	Cold liquids
0	Nonoral feeding (tubefed)
Eating in public	
100	No restriction of place, food, or companion (eats out at any opportunity)
75	No restriction of place, but restricts diet when in public (eats any where, but may limit intake to less "messy" foods)
50	Eats only in presence of selected persons in select places
25	Eats only at home in presence of selected persons
0	Always eats alone
Understandability of speech	
100	Always understandable
75	Understandable most of the time, occasional repetition necessary
50	Usually understandable, face-to-face contact needed
25	Difficult to understand
0	Never understandable; may use written communication

aid in the assignment of ratings. This measure differs from the PSSHNCP in that it is administered to the patient with no ratings preformed by the clinician. This instrument was found to be a valid measure of dysfunction in head and neck cancer patients. It was found to have a moderate correlation with KPS and a stronger correlation with the PSSHNCP. This measure may serve a similar function in the assessment of head and neck cancer-specific outcomes as the PSSH-NCP; however, it differs fundamentally in that data are obtained directly from patient polling.

While voice quality and characteristics are partly assessed in the aforementioned instruments, several strategies have been developed to exclusively evaluate voice quality for the purposes of outcomes studies. Clary *et al.* [31] performed a study to evaluate voice function in children following surgical procedures. While this study did not examine head and neck cancer patients specifically, the methods of measuring this aspect of function could be applied to such a population. A multidimensional approach was taken to assess voice quality. This was composed of an analysis of taped voice samples that were then evaluated by speech therapists using 16 parameters described in the voice profile analysis protocol developed by Laver [32]. This involved assessment of such qualities as harshness, pitch, and whisper and were rated according to a six-point system. In addition, video laryngoscopy was performed by otolaryngologists to assess the appearance of the larynx and its function during vocalization. Finally, a questionnaire was administered to the parents of the study patients to evaluate speech. Questions were directed toward subjective quality of speech and volume, as well as to change in voice following treatment. Through a combination of endoscopic evaluation, voice profile analysis, and patient questionnaires, the authors proposed a means of voice assessment for the purpose of outcome study.

This approach has since been refined further by Dejonckere *et al.* [33] through development of a protocol for voice assessment. Analysis of the literature regarding voice quality following phonosurgery, for example, reveals a large diversity in methods to assess voice quality. This raises the question of development of universal standards of voice evaluation that can be used by clinicians to assess outcomes accurately. Dejonckere notes five aspects of voice assessment that provide useful information for voice characterization. These are perception, videostroboscopy, acoustics, aerodynamics/efficiency, and subjective rating by the patient. For each of these aspects, certain parameters are measured and then graded according to the protocol. Looking at videostroboscopy, for example, characteristics that are examined and then graded are glottic closure, regularity, mucosal wave, and symmetry. For each stroboscopic parameter, a four-point grading scale is used ranging from zero (no deviance) to three (severe deviance). Following grading across all parameters by the clinician, a quantitative profile of the patient's voice is generated and used to compare patients. In addition, such a voice quality instrument can be used to evaluate head and neck cancer patients after therapy to assess treatment outcomes with respect to functional status.

Salassa [34] investigated the issue of dysphagia and developed the functional outcome swallowing scale (FOSS). The purpose of this scale is to determine the severity of oropharyngeal dysphagia in order to assess the effectiveness of therapy or outcome. Salassa compiled a list of relevant symptoms and findings that are used as criteria for swallowing morbidity. This list consists of 20 signs and symptoms found commonly in patients with complaints of dysphagia, including throat clearing, cough, weight loss, gagging, and reflux. These were ranked by severity and then stratified into six categories or grades of dysphagia. Symptoms were noted to be either present or absent in a patient that ultimately guided placement of patients into appropriate FOSS grades. The six grades range from 0 (normal physiologic function with no symptoms) to V (nonoral feeding for all nutrition). Grades I–IV represent progressively increasing severity of dysphagia and are described by the number and severity of dysphagia symptoms. Studies in the past have demonstrated that many patients with dysphagia may have normal physiologic function and yet have significant subjective disability [35,36]. By using patient symptoms as well as physiologic function to grade dysphagia, the FOSS is able to include such patient-based aspects of impairment in the swallowing scale.

The instruments mentioned earlier serve as an example of previously described instruments. Usefulness of a given system may vary with the nature of a given study and the exact outcome being evaluated. Such instruments provide critical information regarding physical and functional performance that allows health status to be included with survival and morbidity in outcomes analysis.

VIII. QUALITY OF LIFE

The symptoms and treatments associated with head and neck cancer often have a significant impact on a patient's quality of life [37]. Critical basic daily functions of breathing, eating, and speaking, in addition to physical appearance, may be affected and impaired [38]. In light of these impairments, the need to address patients' health-related quality of life (HRQOL) has been recognized by clinicians treating head and neck cancer [39,40]. Over the past several decades, significant efforts have been made to measure the health-related quality of life for the purpose of outcomes study. In contrast to tools for health status, quality of life instruments are designed to assess a broader range of physical, psychological, and social functioning of the patient.

Many definitions of quality of life have been proposed; however, they generally all share certain commonly accepted elements. Quality of life is best measured from the perspective of the patient and is thus a subjective measure. It is dynamic in that quality of life changes over the course of time and varying situations. It is often characterized as multidimensional in that it integrates information over a range of areas relating to a patient's life, such as physical, emotional, and social well-being [41,42]. Some authors have alluded that quality of life delineates the gap between health status and the ideal standard of health for a given patient [43,44]. In a broad sense, the health-related quality of life can be considered a description of a patient's health status and the value placed on that condition by the patient [45,46].

It follows that instruments designed to measure quality of life should incorporate these features in their measurement. In addition, it is generally agreed that such instruments should be self-administered by the patient to minimize the potential for health care worker bias and to maximize the reflection of patient perspective. They should be concise and easy to understand to facilitate patient comprehension. It then follows that administering such instruments will require minimal expense of institutional time or expense, as clinician involvement in this step will be minimal [47–49].

There is considerable choice of validated and reliable HRQOL instruments that may be used by the head and neck oncologist [40,50,51]. Although there is often overlap in their fields of inquiry, they can be divided into two groups for present purposes: general cancer and head and neck cancer specific. Evidence shows that disease-specific instruments are more sensitive to handicaps imparted by a particular disease, as well as the longitudinal changes in HRQOL seen during treatment of that disease [52]. General instruments, however, have the advantage of being comparable across disease states and may offer a more global picture of HRQOL [53]. Because both types of tools offer distinct advantages, it may be useful to combine two instruments when assessing clinical outcomes [54]. Accordingly, HRQOL assessment has been moving toward a more modular approach, which permits a more multidimensional HRQOL measurement as it relates to a patient's specific disease [55–57]. In this case, one may use a general instrument that can then be supplemented with a head and neck cancer-specific questionnaire. A description of commonly used instruments and their applications to the study of head and neck cancer is presented.

The functional assessment of cancer therapy scale (FACT) was developed by Cella *et al.* [58]. It is a comprehensive, self-administered questionnaire that takes 5 to 10 min to complete. It consists of 33 general questions that apply to all cancer patients and addresses five areas: physical well-being, emotional well-being, relationship with the physician, social and family well-being, and functional well-being. Reliability, validity, and sensitivity to changes over time were found to be high. In addition, a separate module designed specifically for the head and neck cancer population can be added. This incorporates questions that inquire into fields particularly relevant to head and neck cancer in order to customize the FACT for these patients.

The European organization for the research and treatment of cancer (EORTC) quality of life questionnaire is a general cancer instrument [57,59]. It is a 30-item instrument (EORTC QLQ-C30) that is designed to be self-administered and measures HRQOL. The EORTC QLQ-C30 is composed of five functional scales (physical condition, cognitive function, emotional well-being, role and social functioning, and global QOL). In addition, there are three symptom scales that address fatigue, nausea/vomiting, and pain. Six single items (dyspnea, insomnia, appetite, diarrhea, constipation, and financial difficulties) are also included. Patients are instructed to provide answers based on a 1-week time frame. Scores are then transformed into a 0 to 100 scale. High scores for functional scales represent high levels of function, whereas high scores of symptom scales correspond to increased symptoms or problems. Global HRQOL scores are then calculated according to the guideline delineated in the EORTC QLQ-C30 scoring manual [60].

In order to increase the usefulness of the EORTC QLQ-C30 for head and neck patients, a separate module has been created specifically to address complaints particular to this patient population [61]. This is a 35-item questionnaire (EORTC QLQ-H&N35) that addresses symptoms pertaining to tumor location and treatment. Seven multiple item scales (e.g., pain, swallowing, speech) are included in addition to 11 single items, such as mouth opening, thick saliva, and dry mouth. Most items are scored on a one to four scale. This module has been reviewed and revised; however, the final version has been validated and found to have good psychometric quality [62]. The EORTC QLQ-H&N35 has been designed for use across a broad range of patients with head and neck cancer, varying in disease stage, location, and treatment modality [50].

The EORTC QLQ-H&N35 has also demonstrated useful application in the study of head and neck cancer outcomes. Sherman *et al.* [55] sought to verify the validity and reliability of the QLQ-H&N35 as a supplemental module. The questionnaire was administered to 120 patients with advanced head and neck disease at different points in the course of their treatment. Results of the study showed good reliability and validity. Correlation with the QLQ-C30 was noted to be low to moderate, further establishing that the head and neck module provides unique information not captured by the general instrument. Epstein *et al.* [63] examined this further in a study of patients receiving radiation therapy. The QLQ-H&N35 was given to patients before, during, and after treatment. Changes in HRQOL were measured and correlated with time point during treatment. They found the instrument to be sensitive to the HRQOL changes that occurred after treatment and useful in determining which aspect of patients' dissatisfaction with oral performance was transient as opposed to permanent. They further considered that use of the QLQ-H&N35 may provide a useful measure of outcome to assess possible oral care prevention and management strategies for such patients. Hammerlid *et al.* [64] performed a similar study using the QLQ-H&N35 in a longitudinal study of patients extending 3 years after diagnosis of head and neck cancer. To follow changes in HRQOL, the instrument was administered at predetermined time points in the patients' posttreatment course. The dependence of HRQOL on tumor characteristic and choice of treatment was calculated. After separating patients by treatment group, differences in HRQOL were noted and were attributed in part to treatment given. Accordingly, the QLQ-H&N35 performed as a valid measure of clinical outcome and captured changes in the HRQOL that varied by treatment choice.

The University of Washington Head and Neck quality of life questionnaire (UW QOL) is a head and neck-specific instrument designed to measure QOL [37]. It is also self-administered with data derived exclusively from the patient. The scale consists of nine categories, which each describes aspects of daily living that are frequent complaints for the head and neck cancer patient. Table 29.3 displays the nine areas: pain, disfigurement, activity, recreation, employment, eating, swallowing, speech, and shoulder function. Each category is scored from 0 to 100 for a total possible score of 900 with high scores reflecting improved QOL. In their initial study, Hassan and Weymuller [37] compared the UW QOL to the Karnofsky performance scale and the sickness impact scale [25]. They found their instrument to have comparable validity and reliability as the other two scales. They also found that UW QOL was the preferred test format for 97% of patients studied, as it was concise and easier to complete than the other questionnaires. Since that time, several other authors have used the UW QOL as clinical outcomes measure. Netscher *et al.* [53] administered this questionnaire to assess QOL in patients undergoing microvascular reconstruction of the oropharynx. Their analysis suggested that this is a reliable, valid, and easy-to-use measure of QOL specific to patients with head and neck cancer.

After development and validation of this instrument, Weymuller *et al.* [65] commented on their experience using it in a prospectively designed study. Pretreatment QOL data were collected on 549 patients and were subsequently collected at predetermined time intervals. They noted that achieving statistically significant results in such QOL studies was difficult in a single institution setting after stratifying patients according to cancer site, stage, and treatment. In addition, the composite QOL score was subject to "internal cancellation," where scores improved for certain domains while worsening for others. This suggests that QOL domains may be best analyzed separately in order to be most sensitive.

TABLE 29.3 UW Head and Neck Questionnaire (UW QOL)

Pain
I have no pain
There is mild pain not needing medication
I have moderate pain—requires regular medication
I have severe pain controlled only by narcotics
I have severe pain not controlled by medication
Disfigurement
Peanuts
There is no change in my appearance
The change in my appearance in minor
My appearance bothers me but I remain active
I feel significantly disfigured and limit my activities due to my appearance
I cannot be with people due to my appearance
Activity
I am as active as I have ever been
There are time when I cannot keep up my old pace, but not often
I am often tired and I have slowed down my activities, although I still get out
I do not go out because I do not have the strength
I am usually in a bed or chair and do not leave home
Recreation/environment
There are no limitations to recreation home and away from home
There are a few things I cannot do but I still get out and enjoy life
There are many times when I wish I could get out more but I am not up to it
There are severe limitations to what I can do, mostly I stay home and watch TV
I cannot do anything enjoyable
Employment
I work full time
I have a part-time but permanent job
I only have occasional employment
I am unemployed
I am retired (circle one below)
Not related to cancer treatment
Due to cancer treatment
Eating
Chewing
I can chew as well as ever
I can eat soft solids but cannot chew some foods
I cannot even chew soft solids
Swallowing
I can swallow as well as ever
I cannot swallow certain solid foods
I can only swallow liquid food
I cannot swallow because it "goes down the wrong way" and chokes me
Speech
My speech is always the same
I have difficulty with saying some words but I can be understood over the phone
Only my friends can understand me
I cannot be understood
Shoulder disability
I have no problem with my shoulder
My shoulder is stiff but is has not affected my activity or strength
Pain or weakness in my shoulder has caused me to change my work
I cannot work due to problems with my shoulder

Finally, their analysis suggests that incremental changes in QOL may be most representative of what we seek to measure in QOL studies of head and neck cancer treatment.

Regardless of choice of instrument, HRQOL is an important variable that is relevant to the treatment of head and neck cancer. Rogers *et al.* [50] compared the UW QOL to the EORTC QLQ-C30 and the QLQ-H&N35. His findings suggest that all three instruments are useful measures of HRQOL. Through use of such tools either alone or in combination with one another, authors are able to measure HRQOL in a multidimensional fashion that is sensitive to changes in patients' well-being over time. These instruments will be used with likely increasing frequency in future clinical studies to assess outcomes of diverse treatment options.

IX. COST OF CARE

The final tier to be considered in outcomes assessment is the cost of care. The monetary cost of an illness and associated indirect costs can be used as an outcome measure. In the current health care environment, health care cost continues to rise [66]. Given the incidence and morbidity associated with malignant disease, the cost of cancer care is a major contributor to this increase. Medicare expenditures for the provision of oncology services increased by 17% per year during the 1980s [67]. Because approximately 60% of all incident cancers and 70% of all cancer mortality occur in 12.8% of the U.S. population age segment of 65 years and older [68,69], an aging population will result in additional increases in Medicare expenditures for cancer. Therefore, there is growing interest in economic studies of cancer care [13,70–72].

Economic evaluation of health care is composed of several different types of analyses. These include cost identification, cost effectiveness, and cost benefit [73]. Cost identification analyses simply report on the cost of a given aspect of health care. Frequently, these analyses rely upon institution or provider charges as their cost measure. Cost effectiveness studies analyze cost per unit of outcome and report results in noneconomic units. Such analyses do not report directly on economic gain, but instead on the cost to attain nonmonetary gains (cancers detected, life-years saved, etc.) [13]. Cost benefit analyses report both cost and benefits in monetary terms. They reflect what monetary amount is being attained or recouped through expenditure of a given sum. From this, a cost–benefit ratio can be calculated to see whether costs exceed benefit or vice versa.

Kezirian and Yueh [74] determined the extent to which recent published economic analyses in the otolaryngology literature adhered to established cost-effectiveness method guidelines [75–77]. They performed a MEDLINE search and identified 71 articles published from 1990 to 1999 in six peer-reviewed otolaryngology journals with terms such as "cost effective" in their title or representing economic analyses. Over half of the terms, such as "cost effectiveness," were used incorrectly and 60% of articles confused "charge" with "cost." About half of the articles reported a summary measure such as a cost-effectiveness ratio. Only one-third performed sensitivity analyses. The authors concluded by saying that adherence to accepted definitions and research methods of economic analyses in the otolaryngology literature is inconsistent and is worse than the general medical literature. They also conclude that there are generally few economic analyses in the published literature specifically looking at head and neck malignancies.

Pfister and Ruchlin [13] reviewed the literature searching for articles on the economics of head and neck malignancies. They identified six studies [78–84] that were primarily cost identification analyses with only secondary consideration to cost effectiveness. An example of a cost-effectiveness article was by Myers *et al.* [85], who examined the cost of different treatment options for patients with T1N0 glottic cancer. They found radiation therapy to be less expensive than hemilaryngectomy; however, microlaryngoscopic surgery was associated with the least cost. Mittal *et al.* [86] conducted a similar study where hemilaryngectomy was found to be associated with double the cost as radiation therapy in the treatment of T1N0 epidermoid glottic cancer. Similar analysis has been performed for treatment options of the retromolar trigone [87]. Tsue *et al.* [88] performed a cost analysis comparing the functional results of free flap reconstruction in the oral cavity versus pedicled soft tissue reconstruction. They found free flap reconstruction to be associated with increased functional benefits and only a slightly greater cost that is then perhaps justified in this select population. Each of the aforementioned studies used cost of care as the outcome measure of treatment options.

Rigorous cost analysis clearly adds an additional dimension to overall outcome assessment. Outcome measures, such as survival and quality of life, help answer questions regarding benefit of treatment. It is clearly useful to know what advantage in survival or functional status is gained by one form of treatment over another. In the cost-conscious environment of health care, one must address the other side of the equation, which is the cost of these treatments. By calculating cost of care, the benefits of treatment as measured by the aforementioned methods of outcomes assessment can be weighed against cost. Such cost analyses are necessary for effective decision making regarding choice of treatment.

X. SUMMARY

Outcomes research is a research methodology that is finding increasing use in the current environment of cancer health care. It is designed to identify the most effective form of treatment for cancer among the variety of treatment options. Treatment of cancer is an evolving field with an increasing number of treatment options often brought about through advances in technology and cancer research. Relatively innovative approaches in the treatment of head and neck cancer, such as microvascular tissue transfer, endoscopic partial laryngectomy, and new chemotherapy regimens, make the choice of treatment increasingly complex. Subsequently, there is a need to clearly identify the benefits of medical treatment on outcome.

It follows that an expanded definition of outcome is required to fully address the diversity of treatment options and their impact on the patient. The five tiers of outcome assessment as described earlier demonstrate the multifaceted nature of this expanded definition. While traditional end points continue to play a prominent role, measures of functional status and quality of life as described by the patient now receive greater attention than had been given previously. The patients' perspective regarding the beneficial outcomes of treatment is equally relevant as the perspective of the physician when it comes to making choices regarding treatment.

The synthesis of the five tiers forms a model for outcomes study. By combining cost analysis with a thorough calculation of treatment outcome according to the expanded definition, the physician and the patient obtain the most complete information regarding treatment for head and neck cancer. We have presented numerous examples of outcomes analysis as they relate to head and neck cancer. These serve primarily as a model for further study that will more completely address the vast array of treatment options currently available. Further research in this area will play an instrumental role in the future care of patients with head and neck cancer.

References

1. Roper, W. L., Winkenwerder, W., Hackbarth, G. M., and Krakauer, H. (1988). Effectiveness in health care: An initiative to evaluate and improve medical practice. *N. Engl. J. Med.* **319**, 1197–1202.
2. Piccirillo, J. F. (1994). Outcomes research and otolaryngology. *Otolaryngol. Head Neck Surg.* **111**, 764–769.
3. Piccirillo, J. F., Stewart, M. G., Gliklich, R. E., and Yueh, B. (1997). Outcomes research primer. *Otolaryngol. Head Neck Surg.* **117**, 380–387.
4. Marwick, C. (1993). Federal agency focuses on outcomes research. *JAMA* **270**, 164–165.
5. Wennberg, J. E., Gittelsohn, A. (1982). Variations in medical care among small areas. *Sci. Am.* **246**, 120–134.
6. Deleyiannis, F. W. B, and Weymuller, E. A. (1996). Outcomes research in head and neck oncology. *Curr. Opin. Otolaryngol.* **4**, 73–77.
7. Byar, D. P., Simon, R. M., Friedewald, W. T., *et al.* (1976). Randomized clinical trials: Perspectives on some recent ideas. *New Engl. J. Med.* **295**, 74–80.
8. Chalmers, T. C., Smith, H. J., Blackburn, B., *et al.* (1981). A method for assessing the quality of a randomized control trial. *Controlled Clin. Trials* **2**, 31–49.
9. Chalmers, T. C., Block, J. B., and Lee, S. (1972). Controlled studies in clinical cancer research. *New Engl. J. Med.* **287**, 75–78.
10. Weymuller, E. A., (1997). Clinical staging and operative reporting for multi-institutional trials in head and neck squamous cell carcinoma. *Head and Neck Surg.* **19**(8), 650–658.
11. Weymuller, E. A., Jr., Ahmad, K., Casiano, R. R., Schuller, D., Scott, C. B., Laramore, G., *et al.* (1994). Surgical reporting instrument designed to improve outcome data in head and neck cancer trials. *Ann. Oto. Rhino. Laryngol.* **103**, 499–509.
12. Veterans Affairs Laryngeal Cancer Study Group. (1991). Induction chemotherapy plus radiation compared with surgery plus radiation in patients with advanced laryngeal cancer. *New Engl. J. Med.* **324**, 1685–1690.
13. Pfister, D. G., Ruchlin, H. S. (1998). Outcome and economic issues in head and neck cancer. *In* "Comprehensive Management of Head and Neck Tumors" (S. E. Thawley and W. R. Panje, eds.), p. 296. Saunders, Orlando.
14. Fries, J. F., and Spitz, P. W. (1990). The hierarchy of patient outcome. *In* "Quality of Life Assessments in Clincal Trials" (B. Spilker, ed.), pp. 25–35. Raven Press, New York.
15. (1991). "Healthy People 2000," Public Health Service, U.S. Department of Health and Human Services Publication DHHS 91-50212. Washington, DC.
16. Institute of Medicine (1991). "Disability in America: Toward a National Agenda for Prevention." National Academy Press, Washington, DC.
17. Weymuller, E. A. and Goepfert, H. (1991). Uniformity of results reporting in head and neck cancer. *Head Neck* **13**, 275–276.
18. American Joint Committee on Cancer. (1997). "AJCC Cancer Staging Manual," 5th Ed. Lippincott-Raven, Philadelphia.
19. Davis, W. E., and Zitsch, R. P. (1998). Statistics of head and neck cancer. *In* "Comprehensive Management of Head and Neck Tumors." (S. E. Thawley and W. R. Panje, eds.), p. 283. Saunders.
20. Berkson, J., and Gage, R. P. (1950). Calculation of survival rates for cancer. *Proc. Staff Meetings Mayo Clin.* 270.
21. Kaplan, E. L., and Meier, P. (1958). Nonparametric estimation from incomplete observations. *JASA* **53**, 457–481.
22. Feinstein, A. R. (1977). Clinical biostatistics. XLI. Hard science, soft data, and the challenges of choosing clinical variables in research. *Clin. Pharm. Thera.* **22**(4), 485–497.
23. Feinstein, A. R. (1983). An additional basic science for clinical medicine. IV. The development of clinimetrics. *Ann. Intern. Med.* **99**, 843–848.
24. Karnofsky, D. A., Abelmann, W. H., Craver, L. F., and Burchenal, J. H. (1948). The use of the nitrogen mustards in the palliative treatment of carcinoma. *Cancer* **1**, 634–656.
25. Gilson, B. S., Gilson, J. S., Bergner, M. *et al.* (1975). The sickness impact profile: Development of an outcome measure of health care. *Am. J. Public Health* **65**, 1304–1310.
26. Stell, P. M. (1990). Prognosis in laryngeal carcinoma: host factors. *Clin. Otolaryngol.* **15**, 111–119.
27. Schipper, H., Clinch, J., McMurray, A., and Levitt, M. (1984). Measuring the quality of life of cancer patients: The Functional Living Index-Cancer: Development and validation. *J. Clin. Oncol.* **2**, 472–483.
28. List, M. A., Ritter-Sterr, C. and Lansky, S. B. (1990). A performance status scale for head and neck cancer patients. *Cancer* **66**, 564–569.
29. Harrison, L. B., Zelefsky, M. J., Armstrong, J. G., Carper, E., Gaynor, J. J., and Sessions, R. B. (1994). Performance status after treatment for squamous cell cancer of the base of tongue: A comparison of primary radiation therapy versus primary surgery. *Int. J. Radiat. Oncol. Biol. Phys.* **30**, 953–957.
30. Baker, C., and Schuller, D. E. (1995). A functional status scale for measuring quality of life outcomes in head and neck cancer patients [published erratum appears in *Cancer Nurs*, **19**(1), 79 (1996)]. *Cancer Nurs.* **18**, 452–457.
31. Clary, R. A., Pengilly, A., Bailey, M., *et al.* (1996). Analysis of voice outcomes in pediatric patients following surgical procedures for laryngotracheal stenosis. *Arch. Otolaryngol. Head Neck Surg.* **122**, 1189–1194.
32. Laver, J. (1991). The gift of speech. *In* "The Gift of Speech." Edinburgh University Press, Edinburgh.

33. Dejonckere, P. H., Bradley, P., Clemente, P., *et al.* (2001). A basic protocol for functional assessment of voice pathology, especially for investigating the efficacy of (phonosurgical) treatments and evaluating new assessment techniques. Guideline elaborated by the Committee on Phoniatrics of the European Laryngological Society (ELS). *Eur. Arch. Oto-Rhino-Laryngol.* **258**, 77–82.
34. Salassa, J. R. (1999). A functional outcome swallowing scale for staging oropharyngeal dysphagia. *Digest. Dis.* **17**, 230–234.
35. Buchholz, D. W. (1996). What is dysphagia? *Dysphagia* **11**, 23–24.
36. Barofsky, I., and Fontaine, K. R. (1998). Do psychogenic dysphagia patients have an eating disorder? *Dysphagia* **13**, 24–27.
37. Hassan, S. J., and Weymuller, E. A., Jr. (1993). Assessment of quality of life in head and neck cancer patients. *Head Neck* **15**, 485–496.
38. Long, S. A., D'Antonio, L. L., Robinson, E. B., Zimmerman, G., Petti, G., and Chonkich. G. (1996). Factors related to quality of life and functional status in 50 patients with head and neck cancer. *Laryngoscope* **106**, 1084–1088.
39. Gotay, C. C., and Moore, T. D. (1992). Assessing quality of life in head and neck cancer. *Qual. Life Res.* **1**, 5–17.
40. Morton, R. P. (1995). Evolution of quality of life assessment in head and neck cancer. *J. Laryngol. Otol.* **109**, 1029–1035.
41. Cella, D. F. (1992). Quality of life: The concept. *J. Palliat. Care* **8**, 8–13.
42. Cella, D. F. (1994). Quality of life: Concepts and definition. *J. Pain Sympt. Manage* **9**, 186–192.
43. D'Antonio, L. L., Zimmerman, G. J., Cella, D. F., and Long, S. A. (1996). Quality of life and functional status measures in patients with head and neck cancer. *Arch. Otolaryngol. Head Neck Surg.* **122**, 482–487.
44. Calman, K. C. (1984). Quality of life in cancer patients: An hypothesis. *J. Med. Eth.* **10**, 124–127.
45. de Haes, J. C. J. M., and van Knippenberg, F. C. E. (1987). Quality of life of cancer patients: Review of the literature. *In* "The Quality of Life of Cancer Patients" (N. K. Aaronson, and J. Beckmann, eds.), pp. 167–182. Raven Press, New York.
46. Gill, T. M., and Feinstein, A. R. (1994). A critical appraisal of the quality of quality-of-life measurements. *JAMA* **272**, 619–626.
47. Moinpour, C. M., Feigl, P., Metch, B., Hayden, K. A., Meyskens, F. L., and Crowley, J. (1989). Quality of life endpoints in cancer clinical trials: Review and recommendations. *J. Natl. Cancer Inst.* **81**, 485–495.
48. Fayers, P. M. and Jones, D. R. (1983). Measuring and analysing quality of life in cancer clinical trials: A review. *Stat. Med.* **2**, 429–446.
49. Selby, P., and Robertson, B. (1987). Measurement of quality of life in patients with cancer. *Cancer Surv.* **6**, 521–543.
50. Rogers, S. N., Lowe, D., Brown, J. S., and Vaughan, E. D. (1998). A comparison between the University of Washington Head and Neck Disease-Specific measure and the Medical Short Form 36, EORTC QOQ-C33 and EORTC Head and Neck 35. *Oral Oncol.* **34**, 361–372.
51. de Boer, M. F., McCormick, L. K., Pruyn, J. F., Ryckman, R. M., and van den Borne, B. W. (1999). Physical and psychosocial correlates of head and neck cancer: A review of the literature. *Otolaryngol. Head Neck Surg.* **120**, 427–436.
52. Patrick, D. L., and Deyo, R. A. (1989). Generic and disease-specific measures in assessing health status and quality of life. *Med. Care* **27**, S217–S232.
53. Netscher, D. T., Meade, R. A., Goodman, C. M., Alford, E. L., and Stewart, M. G. (2000). Quality of life and disease-specific functional status following microvascular reconstruction for advanced (T3 and T4) oropharyngeal cancers. *Plast. Reconstr. Surg.* **105**, 1628–1634.
54. Gliklich, R. E., Goldsmith, T. A., and Funk, G. F. (1997). Are head and neck specific quality of life measures necessary? *Head Neck* **19**, 474–480.
55. Sherman, A. C., Simonton, S., Adams, D. C., Vural, E., Owens, B., and Hanna, E. (2000). Assessing quality of life in patients with head and neck cancer: Cross-validation of the European Organization for Research and Treatment of Cancer (EORTC) Quality of Life Head and Neck module. *Arch. Otolaryngol. Head Neck Surg.* **126**, 459–467.
56. Cella, D. F., and Tulsky, D. S. (1990). Measuring quality of life today: Methodologic aspects. *Oncol.* **4**, 29–38.
57. Aaronson, N. K., Ahmedzai, S., Bergman, B., *et al.* (1993). The European Organization for Research and Treatment of Cancer QLQ-C30: A quality-of-life instrument for use in international clinical trials in oncology. *J. Nat. Cancer Inst.* **85**, 365–376.
58. Cella, D. F., Tulsky, D. S., Gray, G., *et al.* (1993). The Functional Assessment of Cancer Therapy Scale:Development and validation of the general measure. *J. Clin. Oncol.* **11**, 570–579.
59. Bjordal, K., and Kaasa, S. (1992). Psychometric validation of the EORTC Core Quality of Life Questionnaire, 30-item version and a diagnosis-specific module for head and neck cancer patients. *Acta Oncol.* **31**, 311–321.
60. Fayers, P., Aaronson, N. K., Bjordal, K., and Sullivan, M. (1995). EORTC QLQ-C30, Scoring Manual. EORTC Data Center, Belgium.
61. Bjordal, K., Ahlner-Elmqvist, M., Tollesson, E., Jensen, A. B., Razavi, D., Maher, E. J., *et al.* (1994). Development of a European Organization for Research and Treatment of Cancer (EORTC) questionnaire module to be used in quality of life assessments in head and neck cancer patients. EORTC Quality of Life Study Group. *Acta Oncologica* **33**(8), 879–885.
62. Bjordal, K., Hammerlid, E., Ahlner-Elmqvist, M., *et al.* (1999). Quality of life in head and neck cancer patients: Validation of the European Organization for Research and Treatment of Cancer Quality of Life Questionnaire. *J. Clin. Oncol.* **17**, 1008–1019.
63. Epstein, J. B., Robertson, M., Emerton, S., Phillips, N., and Stevenson-Moore, P. (2001). Quality of life and oral function in patients treated with radiation therapy for head and neck cancer. *Head Neck* **23**, 389–398.
64. Hammerlid, E., Silander, E., Hornestam, L., and Sullivan, M. (2001). Health-related quality of life three years after diagnosis of head and neck cancer: A longitudinal study. *Head Neck* **23**, 113–125.
65. Weymuller, E. A., Yueh, B., Deleyiannis, F. W., Kuntz, A. L., Alsarraf, R., and Coltrera, M. D. (2000). Quality of life in patients with head and neck cancer: Lessons learned from 549 prospectively evaluated patients. *Arch. Otolaryngol. Head Neck Surg.* **126**, 329.
66. Freudenheim, M. (2001). Medical costs surge as hospitals force insurers to raise payments. New York Times 5-25-2001.
67. Berenson, R., and Holahan, J. (1992). Sources of the growth in Medicare physician expenditures. *JAMA* **267**, 687–691.
68. Ries, L. A. G., Eisner, M. P., Kosary, C. L., Hankey, B. F., Miller, B. A., Clegg, L., and Edwards, B. K. (2000). SEER Cancer Statistics Review, 1973–1997. National Cancer Institute, Bethesda, MD.
69. Yancik, R., Ganz, P. A., Varricchio, C. G., and Conley, B. (2001). Perspectives on comorbidity and cancer in older patients: Approaches to expand the knowledge base. *J. Clin. Oncol.* **19**, 1147–1151.
70. Scheffler, R. M., and Andrews, N. C. (1989). "Cancer Care and Cost: DRGs and Beyond." Health Administration Press, Ann Arbor.
71. Bennett, C. L., Hillner, B. E., and Smith, T. J. (1995). Economic analysis of cancer treatment. *In* "Clinical Oncology" (A. M. Armitage, A. S. Lichter, and J. E. Niederhuber, eds.), pp. 187–200. Churchill L. Livingstone, New York.
72. Sherman, E. J., Ruchlin, H. S., Holden, J. S., and Pfister, D. G. (1999). Clinical economics of head and neck malignancies. *Hemato. Oncol. Clin. North Am.* **13**, 867.
73. Eisenberg, J. M. (1989). Clinical economics: A guide to the economic analysis of clinical practices. *JAMA* **262**, 2879–2886.
74. Kezirian, E. J., and Yueh, B. (2001). Accuracy of terminology and methodology in economic analyses in otolaryngology. *Otolaryngol. Head Neck Surg.* **124**, 496–502.
75. Russell, L. B., Gold, M. R., Siegel, J. E., Daniels, N., and Weinstein, M. C. (1996). The role of cost-effectiveness analysis in health and medicine. Panel on Cost-Effectiveness in Health and Medicine. *JAMA* **276**, 1172–1177.

76. Siegel, J. E., Weinstein, M. C., Russell, L. B., and Gold, M. R. (1996). Recommendations for reporting cost-effectiveness analyses. Panel on Cost-Effectiveness in Health and Medicine. *JAMA* **276**, 1339–1341.
77. Weinstein, M. C., Siegel, J. E., Gold, M. R., Kamlet, M. S., and Russell, L. B. (1996). Recommendations of the Panel on Cost-effectiveness in Health and Medicine. *JAMA* **276**, 1253–1258.
78. Piccirillo, J. F., and Pugliano, F. A. (1996). Evaluation, Classification, and staging of the patient with head and neck cancer. *In* "Cancer of the Head and Neck" (E. N. Myers, and J. Y. Suen, eds.), pp. 33–49. Saunders, Philadelphia.
79. Cragle, S. P., and Brandenburg, J. H. (1993). Laser cordectomy or radiotherapy: Cure rates, communication, and cost. *Otolaryngol. Head Neck Surg.* **108**, 648–654.
80. Blair, E. A., Johnson, J. T., Wagner, R. L., Carrau, R. L., and Bizakis, J. G. (1995). Cost analysis of antibiotic prophylaxis in clean head and neck surgery. *Arch. Otolaryngol. Head Neck Surg.* **121**, 269–271.
81. Chen, H., Nicol, T. L., and Udelsman, R. (1995). Follicular lesions of the thyroid: Does frozen section evaluation alter operative management? *Ann. Surg.* **222**, 101–106.
82. Valk, P. E., Pounds, T. R., Tesar, R. D., Hopkins, D. M., Haseman, M. K. (1996). Cost-effectiveness of PET imaging in clinical oncology. *Nuclear Med. Biol.* **23**, 737–743.
83. Foote, R. L., Buskirk, S. J., Grado, G. L., and Bonner, J. A. (1997). Has radiotherapy become too expensive to be considered a treatment option for early glottic cancer? *Head Neck* **19**, 692–700.
84. Coyle, D., and Drummond, M. F. (1997). Costs of conventional radical radiotherapy versus continuous hyperfractionated accelerated radiotherapy (CHART) in the treatment of patients with head and neck cancer or carcinoma of the bronchus. *Clin. Oncol.* 313–211.
85. Myers, E. N., Wagner, R. L., and Johnson, J. T. (1994). Microlaryngoscopic surgery for T1 glottic lesions: A cost-effective option. *Ann. Otol. Rhino. Laryngol.* **103**, 28–30.
86. Mittal, B., Rao, D. V., Marks, J. E. and Ogura, J. H. (1983). Comparative cost analysis of hemilaryngectomy and irradiation for early glottic carcinoma. *Int. J. Radia. Oncol. Biol. Phys.* **9**, 407–408.
87. Glenn, M. G., Komisar, A., and Laramore, G. E. (1995). Cost-benefit management decisions for carcinoma of the retromolar trigone. *Head Neck* **17**, 419–424.
88. Tsue, T. T., Desyatnikova, S. S., Deleyiannis, F. W., *et al.* (1997). Comparison of cost and function in reconstruction of the posterior oral cavity and oropharynx: Free vs pedicled soft tissue transfer. *Arch. Otolaryngol. Head Neck Surg.* **123**, 731–737.

CHAPTER

30

Combination of Chemotherapy with Radiation for Head and Neck Cancer

NAOMI R. SCHECHTER and K. KIAN ANG

Department of Radiation Oncology
The University of Texas M. D. Anderson Cancer Center
Houston, Texas 77030

I. Introduction 445
II. Radiotherapy Fractionation Regimens 445
III. Goals of Combining Chemotherapy and Radiation 447
IV. Randomized Studies of Induction Chemotherapy Followed by Locoregional Therapy 447
V. Randomized Trials Addressing Concurrent Chemotherapy with Radiation 449
A. Conventional Radiation Therapy Plus Concurrent Chemotherapy 449
B. Altered Fractionation Radiotherapy Plus Concurrent Chemotherapy 451
C. Alternating Chemotherapy and Radiation Therapy 454
VI. Meta-Analyses Addressing the Role of Chemotherapy 454
VII. Summary of Available Trial Results 456
VIII. Future Research Directions 457
A. Toxicity Reduction 457
B. Mechanism- or Target-Driven Combined Therapy Strategies 458
References 459

I. INTRODUCTION

Combining chemotherapy with radiation to improve tumor control and organ preservation rates has been the subject of intensive investigation in various cancers during the last several decades. Cytotoxic agents have been given before (induction or neoadjuvant chemotherapy), after (adjuvant chemotherapy), or concurrently with radiation. This chapter summarizes the rationale and data of these different modes of chemotherapy in combination with radiation treatment for the management of advanced head and neck squamous cell carcinoma (HNSCC). Although a very large number of phase I–II studies have been conducted, to date only a fraction of regimens investigated have undergone proper testing in randomized clinical trials. The results of completed trials in aggregate have begun to change the standard of care for patients with advanced HNSCC, predominantly those with carcinomas of the oropharynx, larynx–hypopharynx, and nasopharynx. In this review, we elected to discuss the results of representative phase III trials and the outcome of comprehensive meta-analyses instead of an exhaustive listing of all studies ever undertaken. Because chemotherapy has been combined with different types of radiotherapy regimens, we begin with a brief description of the terminology, definition, and rationale of radiation dose-fractionation schedules to facilitate further discussion.

II. RADIOTHERAPY FRACTIONATION REGIMENS

The standard or conventional fractionation regimen commonly used in the United States has been delivered in 1.8–2 Gy per fraction, five times a week, to total doses ranging from 66 to 72 Gy in 7–8 weeks depending on the tumor stage. Advances in radiobiological concepts stimulated the development of two classes of new, altered fractionation regimens referred to as hyperfractionation and accelerated fractionation schedules [1,2]. Briefly, hyperfractionation exploits the difference in the sensitivity between tumors and normal tissues manifesting as late morbidity to changes in the fraction size (dose per fraction). Because of this differential fractionation sensitivity, it is thought to be possible to administer higher biological doses to the tumor using small dose fractions (i.e., 1.15–1.2 Gy instead of the standard 1.8–2 Gy) without increasing the late toxicity. In contrast, accelerated fractionations attempt to reduce tumor proliferation during the course of fractionated radiotherapy as a major cause of failure by shortening the radiotherapy duration. The results of representative large-scale clinical trials testing altered fractionation regimens are summarized in Table 30.1.

With hyperfractionation, a total dose of 76.8–81.6 Gy is given in about 7 weeks by delivering 1.15–1.2 Gy per fractions, 2 fractions a day with an interval (i.e., 10 fractions a week). This regimen has been shown to improve the local-regional control rate over standard fractionation in patients with intermediate to advanced HNSCC. The European Organization for Research on Treatment of Cancer (EORTC) group [3], for example, compared the conventional 70 Gy in 7 weeks to a hyperfractionated schedule of 80.5 Gy in 7 weeks. Only patients with T2 or T3, N0 or N1 (<3 cm), M0 oropharyngeal carcinoma were eligible for this phase III trial ($n=356$). No difference was reported in late normal tissue damage, although acute mucositis and treatment interruptions were increased in the hyperfractionated accelerated radiation therapy arm (66.5% versus 49%, and 7.5% versus 4.5%, respectively). Nevertheless, local control was improved significantly with hyperfractionated radiation therapy compared with conventional radiation therapy (5-year local control of 59% versus 40%; $p=0.02$). Results of a larger randomized trial of the Radiation Therapy Oncology Group (RTOG) enrolling a total of 1113 patients with more advanced HNSCC of various sites (mainly oropharynx, larynx, and hypopharynx) to test the relative efficacy of hyperfractionation and two types of accelerated fractionations were reported more recently [4]. This trial also revealed the benefit of hyperfractionation over standard fractionation in achieving local-regional tumor control (54% versus 46% at 2 years, $p=0.05$).

TABLE 30.1 Selected Randomized Trials Comparing Altered Fractionation with Conventional Fractionation[a]

Tumor site and stage	No. of patients	d (Gy)	*n* and Ti (hr)	D (Gy)	T (W)	Tumor response	Complications	Reference
Oropharynx T2-3 N0-1	356	1.15	2 (6–8)	80.5	7	5-year LRC: 59% vs 40% ($p=0.02$).	More acute mucositis with HFX	Horiot *et al.* [3]
		2.0	1	70.0	7	No significant difference in survival	No significant difference in late toxicity	
Supraglottic larynx and pharynx	977	2.0	1	68	6	Improved LRC and survival by odds ratio of 1.3 (1.1–1.7) and 1.3	Higher acute toxicity with AFX	Overgaard *et al.* [6]
		2.0	1	68	7	(0.9–1.8), respectively	Similar late toxicity	
Various sites Stage IV	268	2.0	2	~63	3	2-year LRC: 58% vs 34%, $p < 0.01$	Grade 3–4 mucositis: 83% vs 28%, $p < 0.01$	Bourhis *et al.* [7]
		2.0	1	70	7	No difference in survival	Similar late toxicity	
Various sites Stages III–IV (stage II BOT and HP)	1073	1.2 1.6 1.8–1.5 2	2 (≥ 6) 2 (≥ 6) 1–2 (≥ 6) 1	81.6 67.2 72 70	7 6 6 7	2-year LRC was higher with HFX and AFX (CB), $p=0.05$ for both arms Trend for improved DFS but no significant difference in survival	More grade 3–4 mucositis in all altered fractionation arms No significant difference in late toxicity	Fu *et al.* [4]

[a]d, dose per fraction; *n*; number of fractions per day; Ti, time interval between fractions; D, total dose; T, overall treatment time; w, weeks; LRC, local-regional control; HFX, hyperfractionation. AFX (CB), accelerated fractionation (concomitant boost); BOT, base of tongue; HP, hypopharynx; DFS, disease-free survival.

There are many types of accelerated fractionation schedules, some of which have combined features of accelerated and hyperfractionated regimens [5]. The two regimens, studied by the Danish and RTOG investigators, respectively, yielded local-regional control benefit over standard fractionation in patients with intermediate to advanced HNSCC. A Danish trial (DAHANCA 7) randomized patients eligible for primary radiotherapy alone to receive 66–68 Gy in 33–34 fractions given either in 5 or 6 fractions per week [6]. An analysis of data on 977 patients revealed a higher incidence of severe acute mucositis and dysphagia in patients receiving 6 fractions per week but no difference in the incidence of late edema or fibrosis. The accelerated arm yielded a significantly higher tumor control rate with an odd ratio of 1.3 [95% confidence interval (CI): 1.1–1.7].

A French GORTEC 94-02 trial tested a regimen using a similar fractionation scheme as the Vancouver trial showed promising results by introducing a slight reduction (~ 10%) in the total radiation dose in the accelerated arm [7]. This study randomized 268 patients with locally advanced head and neck squamous cell carcinoma to treatment with 70 Gy over 7 weeks or 62–64 Gy over 3 weeks (2 Gy twice a day). Seventy-five percent of the patients had oropharynx primaries and 70% had T4 disease. Again, acute toxi-city was higher in the accelerated group (83% vs 28% RTOG grade 3–4), with feeding tube placement required for 90% of patients receiving accelerated fractionation compared to 41% of patients receiving conventional fractionation. Although the follow-up period was short (median 28 months), the late toxicity was reportedly similar between the two arms. Accelerated treatment yielded a higher 2-year actuarial local-regional control rate (58% vs 34%, $p<0.01$). No significant difference in overall survival was seen yet.

A large randomized trial of the Radiation Therapy Oncology Group (RTOG 90-03) as discussed earlier [4] showed that one of the altered fractionation schedules, i.e., the concomitant boost regimen that administered 72 Gy in 42 fractions over 6 weeks (i.e, 1.8 Gy daily for 3.6 weeks and 1.8 Gy + 1.5 Gy, 6 h apart, for 2.4 weeks), yielded a significantly higher 2-year locoregional control rate (54% versus 46%, $p=0.049$) than standard fractionation.

III. GOALS OF COMBINING CHEMOTHERAPY AND RADIATION

The goals of induction chemotherapy are to reduce the primary tumor and, when present, nodal size (down staging), thereby increasing the chance of cure with subsequent local therapy and also to eradicate systemic microscopic metastases [8]. Unfortunately, induction chemotherapy may induce accelerated repopulation of tumor clonogens, making the tumor more difficult to control locally by means of radiation therapy [9]. The primary goal of concurrent chemotherapy is mainly to enhance the cytotoxicity of radiation therapy against macroscopic disease [8]. It may also eradicate systemic microscopic disease, although to avoid severe side effects, the dose of chemotherapy used for concurrent chemoradiation may be too low to yield a demonstrable effect on micrometastases. Adjuvant chemotherapy is used to eradicate microscopic loci presumed to remain after local therapy and to destroy microscopic metastatic deposits [8].

IV. RANDOMIZED STUDIES OF INDUCTION CHEMOTHERAPY FOLLOWED BY LOCOREGIONAL THERAPY

Table 30.2 summarizes the results of phase III trials addressing the role of induction chemotherapy in combination with local-regional therapy. The Southwest Oncology Group (SWOG) [10] compared surgery plus postoperative radiotherapy with or without chemotherapy in patients with stage III or IV HNSCC ($n=158$). The induction chemotherapy consisted of three courses (given at a 3-week interval) of intravenous (iv) administration of 50 mg/m^2 cisplatin, 40 mg/m^2 methotrexate, and 2 mg vincristine on day 1 and iv or intramuscular (im) injection of bleomycin 15 U/m^2 on days 1 and 8. Surgery was done 3 weeks after chemotherapy. In the induction chemotherapy arm, the extent of surgical resection was determined at the time of randomization and was unaltered by the response to chemotherapy. This study showed no significant ($p=0.27$) difference in the median survival time with a trend for patients in the induction chemotherapy arm to develop wound problems (34.4% versus 24.2%). There was also one death, related to radiation-induced myelitis, in the induction chemotherapy arm.

The Department of Veterans Affairs (VA) Laryngeal Cancer Study Group [11] performed a widely quoted trial on patients with previously untreated stage III or IV laryngeal carcinomas. This trial enrolled 332 patients to compare the outcome of induction chemotherapy followed by radiation therapy in responders or by surgery and radiotherapy in nonresponders with that of surgery and postoperative radiation therapy. Chemotherapy consisted of three cycles of cisplatin (100 mg/m^2 given in rapid iv infusion) and 5-fluorouracil (FU) (1 g/m^2/day in continuous infusion for 4 days following each dose of cisplatin). The clinical tumor response was assessed after two cycles of chemotherapy, and patients with a response received another cycle of chemotherapy followed by primary radiation therapy (66 to 76 Gy). Patients having no response and those experiencing local disease recurrence after chemotherapy and radiation therapy underwent salvage laryngectomy. This VA trial showed no difference in overall survival between the arms (68% at 2 years in both groups, $p=0.98$) but a significant difference in the patterns of recurrence. Relative to surgery plus postoperative radiotherapy,

TABLE 30.2 Selected Randomized Trials Comparing Locoregional Therapy with or without Induction Chemotherapy[a]

Tumor site and stage	No. of patients	Therapy regimens	Tumor response	Complications	Reference
Various sites Stages III–IV	158	CDDP, MTX, BLM, VCR followed by S + PoRT	No significant difference in LC: (52% vs 60%) or OS	No significant difference in mucositis (14% vs 19%) or wound healing (34% vs 24%)	Schuller *et al.* [10]
		S + PoRT	DM: 28% vs 49% (*p* value not reported)	No significant difference in late toxicity	
Larynx Stages III–IV	332	CDDP and 5-FU followed by RT (responders) or S + PoRT (nonresponders)	2-year LC: 88% vs 98%, $p=0.001$ 2-year DM: 11% vs 17%, $p=0.001$ Larynx preservation: 64%	No significant difference in acute morbidity or iatrogenic mortality (2%)	Department of Veterans Affairs [11]
		S + PoRT	No significant difference in NC and OS	Second tumors: 2% vs 6%, $p=0.048$	
Various sites T2-T4 N0-3	325	Carboplatin and 5-FU followed by S + PoRT	2-year LRC: 65% vs 75%, $p=0.04$ (LC: 71% vs 85%; NC: 82% vs 91%)	Chemotherapy induced-death in 5 patients	Depondt *et al.* [12]
		S + PoRT	No significant difference in OS, DFS, or DM	No significant difference in late toxicity	
Pyriform sinus T2-T4 N0-N2b	202	CDDP and 5-FU followed by RT (responders) or S + PoRT (nonresponders)	No significant difference in LRC or OS	No significant difference in acute, surgical, or late morbidity	Lefebvre *et al.* [13]
		S + PoRT	5-year larynx preservation: 35% DM: 25% vs 36%, $p=0.041$	Chemotherapy induced-death in 1 Patient	
Nasopharynx Mainly T3-4 or N2–3	339	BLM, epirubicin, CDDP followed by conventional RT	LC: 61% vs 45.3%, $p<0.01$	Therapy-induced mortality: 8.2% vs 1.2%, $p<0.01$	INCSG [14]
		Conventional RT	No significant difference in OS or DM	Drug-induced N/V, renal toxicity: common No significant difference in acute RT toxicity (no details on late toxicity)	

[a]BLM, bleomycin; CDDP, cisplatin; DFS, disease-free survival; DM, distant metastasis; LC, local control; LRC, local-regional control; MTX, methotrexate; NC, nodal control; N/V, nausea–vomiting; OS, overall survival; PoRT, post-operative radiotherapy; S, surgery; RT, radiotherapy; VCR, vincristine.

there were more local recurrences ($p=0.0005$) and fewer distant metastases ($p=0.016$) in the induction chemotherapy group. In the induction chemotherapy arm, the larynx was preserved in 64% of the patients overall and 64% of the patients who were alive and free of disease. Because patients in the control arm underwent laryngectomy, it is not possible to resolve whether induction chemotherapy plus radiotherapy results in higher larynx preservation rate than radiation alone.

Depondt *et al.* [12] reported a multicenter trial performed in France that enrolled 325 patients. This study also evaluated surgery plus postoperative radiotherapy with or without neoadjuvant chemotherapy consisting of iv administration of 400 mg/m^2 carboplatin on day 1 and 1 g/m^2/day 5-FU on days 1–5 for three cycles at 3-week intervals. No difference was found in overall or event-free survival after a median follow-up of 25 months. There seemed to be a reduction in the need for ablative surgery, as 21 of the 73 (29%) patients with primary indications for surgery (total laryngectomy, total glossectomy, or transmaxillary buccopharyngectomy) received radiation therapy alone. A statistically significant worsening in locoregional control rate (65% versus 75%, $p=0.04$) was attributed to primary and nodal disease progression in the induction chemotherapy arm. In addition, patients in the induction chemotherapy arm suffered hematologic side effects, such as thrombocytopenia (19%) and neutropenia (24%), which contributed to toxic deaths secondary to septicemia in 2 patients. The distant metastasis rate was not significantly different between the arms.

The EORTC Head and Neck Cancer Cooperative Group [13] investigated the role of induction chemotherapy, identical to that used in the VA study, in patients with pyriform sinus cancer. A total of 202 patients were randomized prospectively to surgery and postoperative radiation therapy or induction chemotherapy followed by local-regional treatment. An endoscopic evaluation was performed after each cycle of chemotherapy, and only partial and complete responders received a third cycle. Patients with a complete response after two or three cycles of chemotherapy were subsequently treated with radiation therapy (70 Gy) and those with poor response underwent conventional surgery with postoperative radiation (50–70 Gy). Although no significant difference in survival was noted, there were fewer distant failures in the induction chemotherapy arm (25% versus 36%, $p = 0.041$). The 3- and 5-year functional larynx retention rates were 42 and 35%, respectively.

The International Nasopharyngeal Cancer Study Group [14] compared the effect of a combination of induction chemotherapy consisting of a combination of bleomycin, epirubicin, and cisplatin (BEC), given in 3-week cycles, with radiation to that of radiation alone in patients with relatively advanced nasopharyngeal carcinoma. Each BEC cycle consisted of 15 mg bleomycin iv on day 1 followed by 12 mg/m^2/day continuous iv infusion for 5 days, 70 mg/m^2 epirubicin slow iv bolus on day 1, and 100 mg/m^2 cisplatin over 1 h on day 1 with pre- and posthydration and mannitol diuresis. In both arms, radiotherapy was delivered in 2-Gy fractions, five fractions of per week, to a total dose of 65–70 Gy in 6.5–7.5 weeks to the primary tumor. Data showed that compared with radiation therapy alone induction chemotherapy increased the 5-year disease-free survival rate from 30 to 39% ($p < 0.01$) but not the overall survival rate.

A number of investigators addressed predictive biomarkers for the HNSCC response to therapy. Two French groups reported p53 status as an independent predictor of HNSCC response to cisplatin and 5-FU neoadjuvant chemotherapy. Temam *et al.* [15] studied 105 patients with locally advanced HNSCC treated with three cycles of cisplatin (20 mg/m^2/day, 24 h continuous infusion) and 5-FU (1 g/m^2/day, 4-day continuous infusion). The p53 status was determined by immunohistochemistry and sequencing analysis (for mutation). These parameters were included in multivariate analysis along with tumor stage and node stage. It was found that the p53 mutation was the only variable to significantly predict for an objective response to neoadjuvant chemotherapy and the strongest predictor for a major response to neoadjuvant chemotherapy defined as an $\geq 80\%$ reduction in tumor size. Specifically, compared to tumors with wild-type p53, those with a p53 gene mutation had a 77% reduction in the chance of an objective response (odds ratio 0.23; 95% confidence interval, 0.1 to 0.6, $p = 0.002$) and a 70% reduction in chance of a major response (odds ratio 0.3; 95% confidence interval, 0.1 to 0.7, $p = 0.01$). Cabelguenne *et al.* [16] studied a group of patients treated with three cycles of a 4-day continuous infusion of cisplatin (25 mg/m^2/day) and 5-FU (1 mg/m^2/day) using $\geq 50\%$ tumor regression as the criterion for response. Similarly, they found that the prevalence of p53 mutations was significantly lower in 37 responding than that in 69 nonresponding tumors (81% versus 61%, $p < 0.04$).

V. RANDOMIZED TRIALS ADDRESSING CONCURRENT CHEMOTHERAPY WITH RADIATION

A. Conventional Radiation Therapy Plus Concurrent Chemotherapy

Lo *et al.* [17] randomized 136 patients to receive 60–70 Gy in 30–35 fractions with or without 5-FU (Table 30.3). Fluorouracil was given in a dose of 10 mg/kg in iv infusion on days 1–3 and then 5 mg/kg on day 4 and the subsequent Mondays, Wednesdays, and Fridays until the end of the radiation course or occurrence of severe reactions. The 2-year relapse-free survival rate was significantly higher with chemoradiation therapy than with conventional radiation therapy alone (49% versus 18%, $p < 0.05$). The presence or absence of lymph node metastases had less impact on patient outcome in the chemoradiation therapy arm than in the conventional radiation therapy arm. The 2-year disease-free survival rates for N+ and N0 oropharynx cancer patients were 41 and 50%, respectively, for the chemoradiation arm. The corresponding rates were 21% and 50%, respectively, for the radiotherapy alone arm. The concurrent chemoradiation regimen induced, however, severe late toxicity in 5 patients who had no primary relapse. These late effects included bone necrosis and fistula, which caused death in 1 patient.

Fu *et al.* [18] randomized 104 patients with stage III or IV inoperable HNSCC to receive 70 Gy given in 1.8 Gy per fraction with or without bleomycin (5 U given in iv infusion twice a week) (Table 30.3). Patients in the chemoradiation therapy arm subsequently received 16 weekly iv infusions of bleomycin (15 U) and methotrexate (25 mg/m^2). The locoregional complete response rate was higher in the chemoradiation arm (67% versus 45%, $p = 0.001$), as was the 2-year locoregional control rate, including salvage surgery (64% versus 26%, $p = 0.001$) and the 3-year relapse-free survival rate (31% versus 15%, $p = 0.041$). There was a trend for a higher distant metastasis rate in the chemoradiation therapy arm (38% versus 24%). Again, chemoradiation induced a significantly higher incidence of mucositis (grade 3–4: 76% versus 25%, $p < 0.001$).

Similarly, Eschwege *et al.* [19] enrolled 199 patients with T2-4 oropharyngeal carcinoma into a phase III EORTC trial comparing 70 Gy in 7 weeks with or without twice weekly iv infusion of bleomycin (15 mg) (Table 30.3). This trial, however, revealed that chemoradiation did not improve the complete response and overall survival rates but increased the incidence of acute reactions significantly (71% versus 21%, $p < 0.01$), which led to an increase in

TABLE 30.3 Selected Randomized Trials Comparing Conventional Radiotherapy with or without Concurrent Chemotherapy[a]

Tumor site and stage	No. of patients	Therapy regimens	Tumor response	Complications	Reference
Oral cavity and oropharynx Mainly stages III–IV	136	60–70 Gy + 5-FU (10 mg/kg iv days 1–3, 5 mg/kg iv day 4, then 3×/W) 60–70 Gy	2-year LC: 49% vs 18%, $p<0.05$ OS: C-RT > RT, $p<0.05$ (benefits mainly oral cavity patients) DM: no significant difference	No significant difference in acute side effects Fistula and necrosis occurred mainly in C-RT arm	Lo *et al.* [17]
Various sites Stages III–IV Inoperable	104	70 Gy + bleomycin (15 U/W × 16) and MTX (25 mg/m^2/W × 16) 70 Gy/35–39 Fxs	2-year LRC (with salvage surgery): 64% vs 26%, $p=0.001$ DFS: 31% vs 15%, $p=0.041$ No significant difference in OS (3-year) or DM	Grade 3–4 mucositis: 76% vs 25%, $p<0.001$ No significant difference in skin toxicity or speech and swallowing	Fu *et al.* [18]
Oropharynx T2-4 N0-3	199	70 Gy + bleomycin (15 mg, 2×/W) 70 Gy/35 Fxs	No significant difference in LC, OS, DM	Mucositis with epidermatitis: 71% vs 21%, $p<0.01$ Completed RT as scheduled: 70% vs 95%, $p<0.01$ No significant difference in late toxicity	Eschwege *et al.* [19]
Various sites stage III–IV Resectable	100	66–72 Gy + cisplatin (20 mg/m^2/day) and 5FU (1 g/m2/day) ×4 days on day 1 and 22 66–72 Gy/33–35 Fxs	5-year LC: 62% vs 51%, $p=0.04$ 5-year LC without S: 77% vs 45%, $p<0.001$ No significant difference in OS, DM 5-year survival with preservation of organ: 42% vs 34%, $p=0.004$	Higher incidence of acute cytopenia, mucosal and skin reactions, weight loss, tube feeding in C-RT arm: $p<0.01$–0.001 Minimal details on late toxicity More second tumors in C-RT arm ($p=0.03$)	Adelstein *et al.* [20]
Nasopharynx Stages III–IV	193	70 Gy + cisplatin (100 mg/m^2, days1, 22, and 43), then cisplatin (80 mg/m^2, day1) and 5-FU (1 g/m^2/day, days1–4) × 3-4 70 Gy/35 Fxs	3-year OS: 78% vs 47%, $p=0.005$ 3-year DFS: 69% vs 24%, $p<0.001$ C-RT arm had lower local failure and DM rates (p values not reported)	Grade 3–4 leukopenia: 29% vs 1%, $p<0.05$ Grade 3–4 emesis: 14% vs 3%, $p<0.05$ Late toxicity not reported in detail	Al-Sarraf *et al.* [21]
Oropharynx Stages III–IV	226	70 Gy + carboplatin and 5 FU (70 mg/m^2/day and 600 mg^{2}day ×4) 70 Gy/35 Fxs	3-year OS: 51% vs 31%, $p=0.02$ 3-year DFS: 42% vs 20%, $p=0.04$ 3-year LRC: 66% vs 42%, $p=0.03$ No significant difference in DM	Grade 3–4 mucositis: 71% vs 39%, $p=0.005$ C-RT had more severe skin reactions, weight loss, tube feeding, and myelosuppression; $p<0.02$-0.05 No significant difference in late toxicity	Calais *et al.* [22]
Various sites Unresectable		70 Gy + cisplatin (100 mg/m^2, days 1, 22, and 43) 30 Gy and 40 Gy + cisplatin (75 mg/m^2 day 1) and 5-FU (1000 mg/m^2 day 1–4) q4W 70 Gy/35 Fxs	3-year OS: arm 1 vs 3: 37% vs 20%, $p = 0.016$ arm 2 vs 3: 29% vs 20%; $p = 0.13$	Grade ≥3 toxicity: arm 1 vs 3: 86% vs 53%, $p=0.0001$ arm 2 vs 3: 77% vs 53%; $p=0.001$ Details of late toxicity have not been reported	Adelstein *et al.* [23]

[a]C-RT, chemoradiotherapy; DFS, disease-free survival; DM, distant metastasis; Fxs, fractions; LC, local control, LRC, local-regional control; OS, overall survival.

weight loss (19% versus 5%, $p < 0.05$) and profound weakness (18% versus 1%, $p < 0.01$). More importantly, the increase in acute sideeffects led to a significant delay in radiation therapy in the chemoradiation arm as only 70% of patients treated with chemoradiation completed their radiation as planned as opposed to 95% of patients treated with radiotherapy alone ($p < 0.01$).

Adelstein *et al.* [20] randomized 100 patients to receive conventional radiation therapy (1.8–2 Gy per fraction to 66–72 Gy total dose) with or without concurrent cisplatin and 5-FU (Table 30.3). Both cisplatin and 5-FU were given in 4-day continuous infusions starting on days 1 and 22 of radiation, and the doses were 20 mg/m^2/day and 1 g/m^2/day, respectively. With a median follow-up of 5 years (range 3–8 years), the projected 5-year recurrence-free rate was significantly higher with chemoradiation (62% versus 51%, $p = 0.04$). In addition, chemoradiation conferred an advantage in 5-year local control without surgical resection (77% versus 45%, $p < 0.001$) and overall survival with organ preservation (42% versus 34%, $p = 0.004$). However, the distant metastasis-free interval was unaffected. Chemoradiation therapy patients had a significantly higher rate of grade 3–4 neutropenia (38% versus 0%, $p < 0.001$), thrombocytopenia (16% versus 0%, $p < 0.01$), cutaneous reaction (44% versus 10%, $p < 0.001$), and mucositis (84% versus 26%, $p < 0.001$). A greater percentage of patients treated with chemoradiation therapy lost ≥10% of body weight (12.5% versus 6.3%, $p < 0.001$) and required tube feeding (58% versus 32%, $p < 0.01$). In addition, 36% of the patients undergoing chemoradiation therapy required hospitalization for care of neutropenic fever but no toxic deaths occurred in either arm. In terms of late effects, significantly more second malignancies occurred in the chemoradiation group ($p = 0.03$). Nine patients in the chemotherapy arm developed second cancers (including four aerodigestive tract tumors), as opposed to two in the radiotherapy arm (one in the aerodigestive tract). Of these 11 patients, 8 died of second malignancies.

The intergroup study on advanced nasopharyngeal carcinoma (0099) [21] compared the efficacy of radiation (70 Gy in 35–39 fractions) with or without concurrent and subsequent adjuvant chemotherapy (Table 30.3). The stratification variables were tumor stage, nodal stage, performance status, and histology. Chemotherapy consisted of cisplatin in a dose of 100 mg/m^2 given on days 1, 22, and 43 of radiation followed 4 weeks later by three cycles of cisplatin (80 mg/m^2 on day 1) and 5-FU (1 g/m^2/day on days 1–4) given 4 weeks apart. Of the 193 patients registered, 147 were eligible for primary analysis, which showed a signi-ficantly better 3-year progression-free survival rate (69% versus 24%, $p < 0.001$) and overall survival rate (78% versus 47%, $p = 0.005$) in favor of combined therapy. A secondary analysis on 185 patients revealed 3-year survival rates of 76 and 46% for combined therapy and radiation alone, respectively ($p < 0.001$). In terms of toxicity, the incidence of grade 3–4 leukopenia and vomiting was higher in the combined therapy arm ($p < 0.05$). Overall, 63% of patients received three courses of concurrent chemotherapy and 55% received all three cycles of adjuvant chemotherapy. The late treatment toxicities have not been reported in detail.

Calais *et al.* [22] enrolled 226 patients with stage III or IV oropharyngeal carcinoma with Karnofsky performance status ≥60 in a phase III trial (Table 30.3). These patients were randomized to receive 70 Gy in 35 fractions with or without three cycles of a concurrent 4-day continuous infusion of carboplatin (70 mg/m^2/day) and 5-FU (600 mg/m^2/day) given 3 weeks apart. Compared with conventional radiation therapy, concurrent chemoradiation yielded significantly higher 3-year actuarial rates of locoregional control (66% versus 42%, $p = 0.03$), disease-free survival (42% versus 20%, $p = 0.04$), and overall survival (51% versus 31%, $p = 0.02$). Chemoradiation therapy did not, however, change the risk of distant metastasis (11%) or alter the pattern of relapse. Of the 65 patients who experienced recurrence after radiation therapy alone, 58 (89%) had local recurrence and 35 (54%) had nodal relapse. The corresponding numbers in the chemoradiation therapy arms were 36/40 (90%) and 21/40 (52%), respectively. No difference was found in cutaneous reaction. Chemoradiation, however, induced increased mucositis (grade 3–4: 71% versus 39%, $p = 0.005$) and hematologic side effects, such as neutropenia (<0.9 cells/m^2: 4% versus 0%) and thrombocytopenia (≤50 cells/mm^3: 6% versus 1%, $p = 0.04$). In addition, combined therapy also induced significantly more weight loss (>10% of body weight: 14% versus 6%, $p = 0.04$) and the need for tube feeding (36% versus 15%, $p = 0.02$).

Data of a recently completed intergroup study enrolling patients with unresectable HNSCC were presented at the 2000 Annual Meeting of the American Society of Clinical Oncology (ASCO) [23]. This trial randomized patients to radiation alone (70 Gy in 7 weeks), radiation (70 Gy) plus cisplatin (100 mg/m^2 iv every 3 weeks×3), or split course radiation (30 Gy + 30–40 Gy) given with the first and third cycles of cisplatin (75 mg/m^2 on day 1) plus fluorouracil (1000 mg/m^2 on days 1–4) given every 4 weeks. The 2- and 3-year actuarial survival rates were 30 and 20% for radiation alone, 43 and 37% for radiation + cisplatin ($p = 0.016$), and 40 and 29% for split-course radiation + cisplatin–fluorouracil ($p = 0.13$) respectively. Grade 3 or worse toxicity, however, occurred in 53, 86 ($p < 0.0001$), and 77% ($p < 0.001$) of the patients, respectively.

B. Altered Fractionation Radiotherapy Plus Concurrent Chemotherapy

Weissler *et al.* [24] randomized 58 patients with advanced HNSCC to an altered fractionation regimen with or without two cycles of cisplatin and 5-FU. Radiotherapy consisted of two courses of 1.5 Gy given twice a day, 5 days a week, for 2 weeks separated by a 2-week break. In the combined therapy arm, cisplatin (100 mg/m^2) was given on day 1 and 5-FU (1 g/m^2/day) on days 1–4 of each radiation course. Chemoradiation was found to improve outcome measured by rather unconventional end points, i.e., the mean time to disease progression (284.0 versus 123.8 person-months, $p=0.013$) and the mean time to death (246.6 versus 85.9 person-months, $p=0.002$). The major side effect was leukopenia (grade 3–4: 33.3% versus 0%, $p=0.001$). However, mucositis was not found to be increased by the addition of chemotherapy ($p=0.59$). Of note is that patients randomized to the radiotherapy alone arm only received a suboptimal dose of 60 Gy in 40 fractions over 6 weeks.

Sailer *et al.* [25] randomized 58 patients with advanced HNSCC to receive a predominantly accelerated fractionation (1.5 Gy twice a day to approximately 68 Gy over a total of 6.5 to 8.5 weeks, including treatment breaks) with or without two cycles of chemotherapy consisting of 100 mg/m^2 cisplatin given on day 1 and 1 g/m^2/day 5-FU on days 1–4, repeated after a 4-week interval. After a dose of 30 Gy, a 2-week break was given, and after a dose of 54 Gy, a second break was allowed. With a mean follow-up of 22.9 months (range 1.9 to 42.1 months), chemoradiation therapy appeared to yield a better complete response, locoregional control, overall survival rate, and disease-specific survival rates, but, unfortunately, statistical tests for differences were not performed. Again, the major side effects were leukopenia, mucositis, nausea, and vomiting.

Magno *et al.* [26] randomized 97 patients with stage II–IV HNSCC to a predominantly accelerated fractionation regimen (1.5 Gy twice a day, 5 days a week to 60–66 Gy over 4–5 weeks) with or without lonidamine. Lonidamine, an indazole carboxylic acid derivative, was administered in a dose of 150 mg, three times a day, beginning 3 days before the start of radiation and continuing for 3 months. The rate of tumor clearance was not significantly different between the groups (65–66%), as was the overall survival (median 18.1 versus 21.8 months, $p=0.311$). Nevertheless, the median duration of locoregional control in adequately treated patients tended to be longer in lonidamine-treated patients (28.3 months versus 9.3 months, $p=0.06$), as did the median time to disease progression (14.5 versus 7.3 months, $p=0.027$). The incidences of acute and late side effects were found to be similar in both groups. The most frequent side effects of lonidamine were myalgia (8.5%) and testicular pain (4.2%).

Wendt *et al.* [27] randomized 298 patients with stage III–IV unresectable HNSCC to receive an altered fractionation regimen with or without three cycles of concurrent chemotherapy given during weeks 1, 4, and 7. Radiotherapy consisted of three courses of 13 fractions of 1.8 Gy given twice a day to a total dose of 70.2 Gy in an overall duration of 8 weeks. Chemotherapy regimens were 60 mg/m^2 of cisplatin (30 min iv infusion), 350 mg/m^2 of 5-FU (iv bolus), and 50 mg/m^2 of leucovorin (iv bolus) on day 2 followed by 350 mg/m^2/day of 5-FU and 100 mg/m^2/day of leucovorin in continuous iv infusion over days 2–5. Chemoradiation yielded improvement in the 3-year locoregional control rate (36% versus 17%, $p<0.004$) and survival rate (48% versus 24%, $p<0.0003$). Distant metastasis rate was equivalent in both arms (9%). The chemoradiation schedule induced grade 3–4 leukopenia in 15% and nausea–emesis in 10.8% of patients and a significantly worse cutaneous reaction (17% versus 6.4%, $p<0.05$) and grade 3–4 mucositis (38% versus 16%, $p<0.001$). Although not statistically significant, the incidence of severe late side effects was slightly higher in the chemoradiation therapy group (10% versus 6.4%).

The Vienna variation of a continuous hyperfractionated, accelerated radiation therapy (V-CHART) regimen with or without concurrent mitomycin-C was tested against conventional fractionation by Dobrowsky and colleagues [28] (Table 30.4). This trial randomized 188 patients with HNSCC to receive one of three treatment regimens: conventional fractionation to 70 Gy over 7 weeks, CHART alone to 55.3 Gy over 17 days (2.5 Gy on day 1 followed by 1.65 BID on days 2 to 17), or CHART with 20 mg/m^2 mitomycin-C on day 5 of radiation. With a median follow-up of 48 months, local control and actuarial survival were improved significantly in the group receiving combined treatment. Concurrent CHART/mitomycin-C was found to improve local control and actuarial survival. Local control for conventional fractionation, CHART, and CHART/mitomycin-C was 28, 32, and 56%, respectively ($p<0.05$). Actuarial rates were 29, 31, and 51%, respectively ($p<0.05$). One-third of the patients treated with conventional fractionation and nearly all patients treated with accelerated fractionation developed grade 3 mucositis, but the major part of the acute reaction started at the end or after the completion of accelerated therapy, therefore not causing interruption of the radiation treatment. The addition of mitomycin did not increase mucositis. Grade 3–4 hematologic toxicity was seen in 12/61 patients who received mitomycin-C, with the majority consisting of thrombocytopenia. Late toxicity was not reported in this study.

Brizel *et al.* [29] randomized 122 patients with advanced HNSCC to receive a hyperfractionated radiation with or without concurrent cisplatin (12 mg/m^2/day × 5 days) and 5-FU (600 mg/m^2/day × 5 days) given during weeks 1 and 6 of radiotherapy. The fractionation regimen was the same in both arms (i.e., 1.25 Gy twice daily, 5 days a week), but the total radiation dose was slightly lower (70 Gy versus 75 Gy) and a planned 1-week interruption was introduced in the chemotherapy arm. With a median follow-up of 41 months, the 3-year actuarial locoregional control rate of the chemoradiation arm was significantly higher than the radiotherapy arm (70% versus 44%, $p=0.01$). There was also a trend toward improved relapse-free survival (55% versus

TABLE 30.4 Selected Randomized Trials Comparing Altered Fractionation with or without Concurrent Chemotherapy[a]

Tumor site and stage	No. of patients	Therapy regimens	Tumor response	Complications	Reference
Various sites Stages III–IV	270	Three courses of 23.4 Gy (1.8 Gy, bid) + cisplatin (60 mg/m^2), 5-FU (350 mg/m^2), and leucovorin (50 mg/m^2) with 1 week break after weeks 2 and 4 Same RT regimen	3-year OS: 48% vs 24%, $p<0.0003$ 3-year LRC: 36% vs 17%, $p<0.004$ No significant difference in DM	Grade 3–4 mucositis: 38% vs 16%, $p=0.001$ Grade 3–4 dermatitis: 17% vs 7.1%, $p<0.05$ Details of late toxicity not reported	Wendt *et al.* [27]
Various sites T1-4 N0-3	188	V-CHART: 55.3 Gy/17 days (2.5 Gy day1 then 1.65 Gy bid days 2–17) V-CHART: 55.3 Gy/17 days plus mitomycin C (20 mg/m^2 day5) 70 Gy/7 weeks	LRC and OS: V-CHART + mitomycin C significantly better then conventional RT ($p<0.05$ for both end points) No significant difference in DM	More mucositis with V-CHART not increased further by mitomycin C Details of late toxicity not reported	Dobrowsky *et al.* [28]
Various sites T2-4 N0-3	122	RT: 70 Gy/47 days in 1.25 Gy bid (7–10-day break after 40 Gy) + cisplatin (12 mg/m^2/day) and 5-FU (600 mg/m^{2}day) × 5 days at weeks 1 and 6 RT alone: 75 Gy/42 days in 1.25 Gy bid alone: 75 Gy/42 day in 1.25 Gy bid	3-year LRC: 70% vs 44%, $p=0.01$ 3-year OS: 55% vs 34%, $p=0.07$	No significant difference in mucositis. increased need for enteral feeding and sepsis in C-RT arm No significant difference in late complications	Brizel *et al.* [29]
Various sites Stages III–IV	130	77 Gy/7 weeks + cisplatin (6 mg/m^2/day) 77 Gy/7 weeks (1.1 By, bid)	5-year OS: 46% vs 25%. $p=0.008$ 5-year PFS: 46% vs 25%, $p=0.007$ 5-year LRPFS: 50% vs 36%, $p=0.04$ 5-year DMFS: 86% vs 57%, $p=0.001$	No significant difference in acute morbidity (except for leukopenia, $p=0.006$) or late toxicity	Jeremic *et al.* [30]
Various sites Stages III–IV	240	69.9 Gy/5.5 weeks + carboplatin (70 mg/m^2/day) and 5-FU (600 mg/m^2/day) for 5 days × 2 69.9 Gy/5.5 weeks (1.8 Gy QD for 3.5 weeks and then (1.8 Gy + 1.5 Gy) for 2 weeks	No significant difference in OS ($p=0.11$, 2 years: 48% vs 39%) No significant difference in LC ($p=0.14$, 2 years: 51% vs 45%) Patients receiving G-CSF had worse LRC ($p=0.007$)	Grade 3–4 mucositis: 68% vs 52%, $p=0.01$ Grade 3–4 vomiting: 8.2% vs 1.6%, $p=0.02$ Late swallowing problems and feeding tube dependency: 51% vs 25%, $p=0.02$	Staar *et al.* [31]

[a]bid, twice a day; C-RT, chemoradiotherapy; DM, distant metastasis; DMPF, distant metastasis-free survival; G-CSF, granulocyte–colony-stimulating factor; LC, local control; LRC, local-regional control; LRPFS, local-regional progression-free survival; OS, overall survival; PFS, disease-free survival.

34%, $p=0.07$) and overall survival (61% versus 41%, $p=0.08$). However, sepsis occurred in 14 of 56 patients (25%) in the chemoradiation arm compared with 4 of 60 (7%) in the radiation arm. Chemoradiation-induced mucositis also took longer to resolve (mean duration: 6 week versus 4 weeks) and more patients in the chemoradiation arm required a feeding tube (79% versus 48%).

Jeremic *et al.* [30] randomized 130 patients with stage III–IV HNSCC to receive predominantly hyperfractionated radiation therapy (1.1 Gy twice a day to 77 Gy in 35 treatment days over 7 weeks) with or without concurrent low dose cisplatin (6 mg/m^2/day iv bolus 3 to 4 h after the first daily fraction, i.e., 1–2 h before the second fraction). This trial revealed that chemoradiation yielded significantly higher 5-year rates in locoregional progression-free survival (50% versus 36%, $p=0.041$), progression-free survival (46% versus 25%, $p=0.0068$), distant-metastasis-free survival (86% versus 57%), and overall survival (46% versus 25%, $p=0.0075$ at 5 years; 68% versus 49% at 2 years). Chemoradiation, however, induced more severe grade 3–4 thrombocytopenia ($p=0.058$) and leukopenia ($p=0.006$). These investigators did not observe an increase in the incidence of either acute or late high-grade toxicity in the chemoradiation arm (Table 30.4).

A trial supported by the German Cancer Society [31] randomized 240 patients to receive 69.9 Gy (one fraction of 1.8 Gy per day for 3.5 weeks and then 1.8 Gy plus 1.5 Gy per day for 2 weeks) with or without two cycles of carboplatin and 5-FU (70 mg/m^2 short iv infusion and 600 mg/m^2/day continuous iv infusion, respectively, on days 1–5 and 29–33)(Table 30.4). No significant differences were detected between the treatment arms in the overall survival rates at 1 and 2 years (66% vs 60% and 48% vs 39%, respectively, $p=0.11$) or in the local-regional control rates at 1 and 2 years (69% vs 58% and 51% vs 45%, respectively, $p=0.14$). Subset analysis revealed that patients with oropharyngeal carcinomas have benefited from combined therapy ($p=0.046$ for 1-year overall survival) but not those with hypopharyngeal cancer ($p>0.5$). Relative to the radiation only group, patients receiving combined therapy experienced significantly more grade 3–4 mucositis (68% vs 52%, $p=0.01$), vomiting (8.2% vs 1.6%, $p=0.02$), and long-term swallowing problems and continuous use of feeding tube (51% vs 25%, $p=0.02$).

C. Alternating Chemotherapy and Radiation Therapy

Merlano *et al.* [32,33] randomized 157 patients to receive either 70 Gy in 2-Gy fractions over 7 weeks or four cycles of cisplatin (20 mg/m^2 intravenous infusion during a 2-h period of forced hydration at days 1–5) and 5-FU (200 mg/m^2 intravenous bolus at the end of hydration days 1–5) given on weeks 1, 4, 7, and 10, alternating with three courses of 20 Gy in 2-Gy fractions on weeks 2–3, 5–6, and 8–9. These investigators reported a significant improvement in the complete response rate from 22 to 43% ($p=0.037$), a locoregional relapse-free survival rate from 32 to 64% ($p=0.03$), a progression-free survival rate from 9 to 21% ($p=0.008$), and overall survival improved from 10 to 24% ($p=0.01$). The incidence of grade 3–4 mucositis was reported to be equivalent in both arms (18 and 19%, respectively). It should be noted that the outcome of the control arm of this trial was quite poor compared to other series.

VI. META-ANALYSES ADDRESSING THE ROLE OF CHEMOTHERAPY

Several meta-analyses have been conducted during the last decade to assess the role of combination of induction or concurrent chemotherapy with radiation. Although there are many pitfalls associated with such analysis, their results have been remarkably consistent [34–38]. Details of these analyses are summarized Table 30.5.

In the early 1990s, Stell *et al.* [34,35] reviewed 28 trials testing combination of chemotherapy and radiation for HNSCC and found that concurrent and concurrent plus maintenance (adjuvant) chemoradiation therapy reduced mortality significantly, i.e., by 6 and 23%, respectively. In contrast, induction chemotherapy and induction plus maintenance chemotherapy in combination with radiation did not significantly reduce cancer-related mortality. In the 11 studies that gave full details of the outcome of all patients at a specified interval, the cancer-related mortality rate was 1% less, but the total mortality rate was 7% greater for the chemotherapy arm. This study showed that an analysis restricted to cancer mortality would have underestimated the death rate by 8%, which would likely include deaths due to treatment toxicity.

Two other groups of investigators undertook a similar type of analysis. Munro [36] reviewed 54 randomized trials and found that compared to radiotherapy alone, induction chemotherapy plus radiation yielded 3.7% (95% confidence interval: 0.9–6.5%, $p=0.01$) and concurrent chemoradiation resulted in 12.1% (95% CI: 5–19%, $p=10^{-13}$) survival advantage. El-Sayed and Nelson [37] reviewed 42 randomized trials and divided them in three categories, i.e., induction chemotherapy, induction plus maintenance chemotherapy, and concurrent chemotherapy in combination with locoregional treatment. Overall, the addition of chemotherapy to local treatment reduced the mortality rate by 11% (i.e., mean relative hazard of 0.89, $p<0.05$). The mortality rate of induction chemotherapy was 5% lower (mean relative hazard of 0.95) and that of induction plus maintenance chemotherapy was 2% higher (mean relative hazard of 1.02) than locoregional treatment alone. These differences did not reach a 0.05 significant level.

TABLE 30.5 Meta-analysis on the Role of Chemotherapy in the Management of Head and Neck Carcinomas[a]

No. trials	No. patients	Findings	Reference
28	4,292	Overall C-RT was associated with 1% lower cancer mortality rate but 7% higher total mortality rate Neoadjuvant and/or adjuvant chemotherapy had no significant effect on LRC or total mortality Concurrent C-RT reduced total mortality from 61 to 55% ($p<0.025$). Concurrent + adjuvant C-RT reduced total mortality from 72 to 49% ($p<0.005$) C-RT-related mortality occurred in 6% (43/729) of patients (11 trials) but late toxicity was not addressed	Stell and Rawson [34]; Stell [35]
54	4,536	Neoadjuvant and concurrent C-RT reduced mortality by 3.7% ($p=0.011$) and 12.1% ($p=10^{-13}$), respectively C-RT increased LC by 7.9% ($p=10^{-8}$) but was associated with a 1.9% higher DM rate ($p=0.02$). Treatment toxicity was not addressed	Munro [36]
42	5,583	Neoadjuvant and concurrent C-RT reduced mortality by 5% ($p=0.05$) and 22% ($p<0.005$), respectively Neoadjuvant + adjuvant C-RT increased mortality by 2%, $p<0.01$ C-RT improved LC by a factor of 1.27 ($p<0.001$) C-RT increased mucositis by 2.97 ($p<0.001$), skin reaction by 1.43 ($p=0.01$), nausea by 76.9 ($p<0.001$), myelosuppression by 15.9 ($p<0.001$), RT delay by 2.87 ($p<0.001$), and therapy-related death by 2.40 ($p<0.001$)	El Sayed and Nelson [37]
63	10,741	Overall, C-RT yielded a hazard ratio of death of 0.9 (95% CI: 0.85–0.94, $p<0.0001$) corresponding to an absolute survival benefit of 4 percentage points at both 2 and 5 years Neoadjuvant and adjuvant chemotherapy had no significant effect on survival Concurrent C-RT increased 5-year absolute survival by 8% ($p<0.0001$), and the effect was greater with multiagent than with single agent ($p<0.01$), but a considerable heterogeneity was found between the trials Treatment toxicity was not addressed in detail	Pignon *et al.* [38]

[a]C-RT, combination of chemotherapy and radiotherapy; DM, distant metastasis; LRC, local-regional control; OS, overall survival.

However, concurrent chemoradiation reduced the mortality rate by 22% (mean relative hazard of 0.78, $p<0.005$), which corresponds to an absolute survival benefit of 8 percentage points. This means that when 50% of patients in the locoregional treatment alone group were alive, 58% of the patients who received chemoradiation would be expected to be alive. El-Sayed and Nelson [37] also assessed treatment toxicity and found that combined therapy increased the side effects significantly (odds ratio 2.17 with a 95% CI of 1.84–2.56, $p<0.001$), such as skin reactions, mucositis, delay in radiotherapy, and treatment-induced deaths.

The Meta-analysis of Chemotherapy on Head and Neck Cancer Collaborative Group [38] undertook the most extensive meta-analysis of 63 randomized trials, including five unpublished series and six series reported in abstracts only, comparing locoregional treatment with or without chemotherapy in a total of 10,741 patients. Follow-up data of all patients in randomized trials between 1965 and 1993 were updated. The investigators noted a marked heterogeneity among the trials regarding to tumor and patient characteristics, therapy regimen, and follow-up, which complicates analysis and data interpretation. Overall, chemotherapy yielded a hazard ratio of death of 0.9 (95% CI: 0.85–0.94, $p<0.0001$) corresponding to an absolute survival benefit of 4 percentage points at both 2 years (from 50 to 54%) and 5 years (from 32 to 36%).

On analysis by the mode of combined therapy, it was found that no significant benefit resulted from neoadjuvant (induction) and adjuvant chemotherapy, which were associated with a 2% ($p=0.10$) and 1% ($p=0.75$) higher absolute survival at 5 years, respectively. In contrast, concurrent chemoradiation produced an 8% overall increase in 5-year absolute survival ($p<0.0001$), but considerable heterogeneity was found between the trials. It was also noted that the effect of concurrent chemoradiation was significantly ($p<0.01$) greater with multiagent chemotherapy than with single-agent chemotherapy (hazard ratio 0.69 versus 0.87).

Examination of data of six randomized trials that compared neoadjuvant with or without adjuvant chemotherapy plus radiotherapy versus concurrent or alternating radiochemotherapy revealed a pooled hazard ratio of death of 0.91 for alternating or concurrent radiochemotherapy (95% CI: 0.79–1.06, $p=0.23$). Analysis of the three larynx-preservation trials comparing radical surgery plus radiation (control arm) with neoadjuvant chemotherapy plus radiotherapy in responders or radical surgery and radiotherapy in nonresponders showed a trend toward an excess in death associated with chemotherapy. The hazard ratio of death was 1.19 (95% CI: 0.97–1.46, $p=0.1$). However, larynx preservation was successful in close to two-thirds of the patients randomized to the chemotherapy arm.

VII. SUMMARY OF AVAILABLE TRIAL RESULTS

More than 10,000 patients with relatively advanced head and neck carcinomas have been enrolled onto randomized clinical trials, mainly during the last two decades, to address the value of the combination of chemotherapy with locoregional treatment, predominantly radiation. Critical evaluation of data of representative randomized trials and details of the meta-analyses reveal rather consistent findings, which have begun to transform the standard of care for patients with relatively advanced HNSCC.

It is interesting to note that despite a lack of convincing data, induction or neoadjuvant chemotherapy in combination with radiotherapy has been adopted in many centers as the standard of care of patients with stage III–IV HNSCC. The enthusiasm and popularity of this combined modality therapy stemmed from the relatively high initial response rate and the reported "high" organ preservation rate deduced from trials using ablative surgery as the control arm. Sufficient evidence now concludes that induction chemotherapy using the available cytotoxic agents has no or minimal impact on the natural history of advanced HNSCC, despite a reduction in the systemic relapse rate detected in a few trials. A recently completed RTOG phase III trial (RTOG 91-11) comparing the efficacy of induction chemotherapy plus radiation, as tested by the VA group, directly with that of radiation alone in patients with relatively advanced laryngeal carcinomas failed to demonstrate a significant improvement in the larynx preservation rate in patients with T2-4N0-3 larynx carcinoma (79% had T3 and 48% had N1-2 disease) [39]. The lack of benefit along with its toxicity and the cost to the patient and society strongly advised against routine prescription of induction chemotherapy for patients with advanced HNSCC. A rare exception may be when the disease extent is at the border between selection of aggressive locoregional therapy and consideration for palliative treatment, a good response to one to two cycles of chemotherapy would justify taking a more aggressive therapy approach.

In contrast to induction chemotherapy, an increasing body of evidence indicates that concurrent chemoradiation yields a better outcome than radiotherapy alone in patients with advanced HNSCC. In addition, data reveal that the survival benefit of concurrent chemoradiation results from an improvement in locoregional disease control rather than from a decrease in systemic relapse. The recently completed phase III RTOG larynx preservation trial showed that concurrent chemoradiation yielded a significantly higher larynx preservation rate ($p=0.0047$) than induction chemotherapy plus radiation [39]. Unfortunately, the results also show that the improvement in locoregional control and survival comes at the expense of increased acute toxicity and late morbidity. Pignon *et al.* [38] concluded after a detailed meta-analysis of updated data bases of randomized trials that concurrent chemoradiation should remain experimental, particularly when toxicity and the cost–benefit ratio are taken into account in addition to survival.

In the absence of better therapeutic options, however, many centers have adopted concurrent chemoradiation as the new, organ-preserving standard of care for patients with advanced HNSCC. A more detailed examination of data of positive randomized trials testing various combined schedules could not identify a clearly superior regimen. In addition, none of the trials properly addressed the value of altered fractionation in the concurrent chemoradiation setting. Consequently, it would be logical to choose a logistically simple and least expensive regimen having the best track record for routine use at the present time. The combination of conventional fractionation (70 Gy in 2-Gy fractions over 7 weeks) and cisplatin (100 mg/m^2 given iv every 3 weeks) as tested in intergroup trials for unresectable HNSCC [23], advanced nasopharyngeal carcinoma [21], and the RTOG trial for advanced laryngeal cancer [39], for example, seem to meet these criteria best. Some investigators are now testing whether induction chemotherapy followed by concurrent chemoradiation can further improve the results of concurrent chemoradiation.

A therapeutic issue that has not been addressed sufficiently in randomized trials is the role of neck dissection in the combined chemoradiation setting. Because the therapeutic benefit from concurrent chemoradiation results mainly from improvement in locoregional disease eradication, optimizing nodal control is an important element in determining the therapeutic ratio. In a retrospective review, Clayman *et al.* [40] found that patients who had a complete response at the primary site but a partial response in the neck benefited significantly from a selective neck dissection performed 6–10 weeks after completion of a combination of chemotherapy and radiation. With a median follow-up time of 27 months (range 3–108.5 months), the disease-free survival was 100% (10/10) with neck dissection versus 33% (2/6) without neck dissection ($p=0.02$). This study also revealed that none of the 29 patients who had a radiologically documented complete response in the neck experienced a neck recurrence. This finding suggests that patients with a bulky nodal disease that has regressed completely after chemoradiation can be spared the morbidity of neck surgery. It also suggests that selection of patients with residual nodal disease but complete response of the primary tumor for selective nodal dissection may improve locoregional control. It will be interesting to reproduce these findings in a prospective trial.

VIII. FUTURE RESEARCH DIRECTIONS

Long-term investment on laboratory and clinical research has finally begun to turn the tide against a number of neoplasms [41]. After rising for decades, for the first time the mortality rate from all cancers in the United States fell 2.6% between 1991 and 1995. Most of this decrease in the mortality rate was observed in patients below the age of 65. Of interest is that the highest decline in the mortality rate (9.6%) occurred in patients with head and neck cancers, irrespective of age. It is also obvious, however, that there is still a relatively long way to go in optimizing treatment outcome in terms of quality of life and tumor control in patient with advanced HNSCC.

A. Toxicity Reduction

Data presented earlier clearly show that the addition of chemotherapy to radiation, particularly when given concurrently, increases treatment-induced toxicity, hence compromising the therapeutic index. In addition, acute treatment toxicity also prevents completion of chemotherapy and/or radiation as planned, partially offsetting its therapeutic efficacy. Therefore, in addition to improving tumor control, further research should also address strategies to reduce normal tissue injury. Refinement of radiotherapy technology can reduce the volume of normal tissues exposed to a high radiation dose, thereby reducing morbidity or increasing the compliance to the combined modality therapy. Chemical compounds having the potential to protect normal tissues from radiation and/or cytotoxic agent induced damage are being developed and tested.

1. Conformal Radiotherapy

Advances in computerized radiotherapy planning and delivery technology open the possibility to conform irradiation to an irregular tumor target volume (conformal radiation therapy) [42]. Consequently, more of the critical normal tissues surrounding the tumor can be spared from high radiation doses, resulting in a reduction in morbidity. Reduced toxicity would in turn permit escalation of the radiation dose or combining radiotherapy with intensive chemotherapy, each of which has the prospect of improving HNSCC control.

Such precision radiotherapy can be accomplished by the use of an array of X-ray beams individually shaped to conform to the projection of the target, which is referred to as three-dimensional conformal radiation therapy (3-D CRT). In addition, technology is also available to modify the intensity of the beams across the irradiation field as an added degree of freedom to enhance the capability of conforming dose distributions in three dimensions. This radiotherapy technique is called intensity-modulated radiation therapy (IMRT).

The role of 3-D CRT, particularly IMRT in reducing morbidity and perhaps improving the control of HNSCC through radiation dose escalation, is being tested in a number of centers. Results so far already reveal that it is effective in sparing parotid glands from receiving a high radiation dose, thereby preventing radiation-induced permanent xerostomia in selected patients [43].

The study undertaken at the University of California–San Francisco to test the role of IMRT in combination with chemotherapy (cisplatin during and cisplatin plus 5-FU after radiotherapy) in the management of patients with nasopharyngeal carcinoma yielded very encouraging results [44]. Of the 35 patients treated, the worst acute toxicity was grade 2 in 16 (46%), grade 3 in 18 (51%), and grade 4 in 1 (3%) patients. The worst late morbidity was grade 1 in 15 (43%), grade 2 in 13 (37%), and grade 3 in 5 (14%) patients. Only 1 patient had a transient grade 4 soft tissue necrosis and, at 2 years after treatment, 50% of the evaluated patients had grade 0, 50% had grade 1, and none had grade 2 xerostomia. In terms of tumor control, with a median follow-up of 21.8 months, the local-regional progression-free rate was 100%. The 4-year overall survival and distant metastasis-free rates were 94 and 57%, respectively. More data on critical normal tissue protection will be forthcoming from many institutions in the near future.

2. Amifostine

Amifostine, WR-2721, has been investigated as a radiation protectant. Brizel *et al.* [45] reported the results of a multicenter phase III trial randomizing 303 previously untreated patients with HNSCC to conventional radiation therapy to 50–70 Gy with or without daily administration of amifostine (200 mg/m^2 iv 15–30 min before each radiation fraction). Whole saliva production was quantified before radiation therapy and regularly during follow-up. Patients evaluated their symptoms through a questionnaire during and after treatment.

With a median follow-up of 26 months, no significant difference was detected in local control, disease-free survival, and overall survival rates (58, 53, and 71%, respectively, with amifostine versus 63, 57, and 66%, respectively, without amifostine. The most common side effects of amifostine treatment were nausea (44% versus 16%, $p<0.0001$), vomiting (37% versus 7%, $p<0.0001$), hypotension (15% versus <2%, $p<0.0001$), and allergic reaction (5% versus 0%, $p=0.003$). Overall, 53% of patients had at least one episode of nausea or vomiting, but it only occurred in 5% of the doses (233/4314). This study showed that amifostine reduced the incidence of grade 3 xerostomia from 78 to 51% ($p<0.001$) and the incidence of grade 2 or greater xerostomia declined from 57 to 34% ($p=0.002$). The median saliva production improved from 0.10 to 0.26 g ($p=0.04$), and the percentage of patients with >0.1 g of basal saliva production improved from 49 to 72% ($p=0.003$). The median dose to onset of xerostomia was 60 Gy with as opposed to 42 Gy without amifostine ($p=0.0001$). The clinical benefit of amifostine treatment was evaluated on an eight item validated patient benefit questionnaire filled out during and up to 11 months after radiation therapy [46]. Amifostine-treated patients had significantly better scores, particularly on scores relating to chronic xerostomia. Specifically, patients receiving amifostine suffered 31% less chronic xerostomia consequences than untreated patients. The authors suggested that this should lead to improved dental and oral health and improved diet, nutrition, and sleep. As a test of validity, scores on the patient benefit questionnaire at first follow-up were correlated with the quantity of saliva produced, and the correlation was found to be highly significant ($p<0.0001$).

3. Growth Factors

The *Granulocyte–colony–stimulating factor (G-CSF)* used to reduce the incidence and severity of chemotherapy-induced acute and cumulative myelosuppression has been assessed for its protective effect on therapy-induced severity and duration of mucositis, yielding conflicting data. Abitbol *et al.* [47] reported that G-CSF reduced the incidence of grade 3 or 4 mucositis from 69 to 31% ($p=0.001$). In contrast, Vokes *et al.* [48] found no improvement in the incidence of grade 3–4 mucositis with the use of G-CSF in a rather small series of patients (5/11 versus 14/29). Similarly, Mascarin *et al.* [49] did not observe a significant reduction in objective mucositis. A report of a randomized trial conducted in Germany showed a trend for less mucosal toxicity ($p=0.07$) in patients receiving G-CSF. Alarmingly, this study also revealed that patients receiving G-CSF had a significantly worse local-regional control ($p=0.007$) [31].

Similar to G-CSF, results of studies testing *granulocyte–macrophage colony-stimulating factor (GM-CSF)* are also conflicting. Several small-scale studies suggest that GM-CSF may reduce therapy-induced mucositis. In a series of 10 patients, Kannan *et al.* [50] found that daily subcutaneous (sc) administration GM-CSF (1 μg/kg) starting from 20 Gy until the completion of radiotherapy to 60–66 Gy resulted in minimal mucosal reactions. Similar impressions were reported by Rosso *et al.* [51] in 29 patients receiving alternating chemoradiotherapy and by Wagner *et al.* [52] in 16 patients receiving postoperative radiotherapy using different GM-SCF doses. In contrast, the randomized study of Makkonen *et al.* [53] comparing GM-CSF (150–300 μg) plus sucralfate with sucralfate alone did not show evidence of activity. A larger scale randomized study is going (RTOG 99-01) to test the efficacy of GM-CSF properly.

Keratinocyte growth factor (KGF), a member of the heparin-binding fibroblast growth factor family, has potent mitogenic activity on epithelial cell types and may be a specific paracrine mediator for normal epithelial growth and differentiation [54–61]. It has no detectable effects on fibroblasts, endothelial cells, or other nonepithelial cells that respond to other fibroblast growth factor family members. Ning *et al.* [54] have found that recombinant human KGF results in little or no stimulation of the proliferation of human head and neck squamous cell carcinoma cell lines *in vitro* and has no effect on the radiosensitivity of these cell lines *in vitro* or *in vivo*. They, therefore, recommend that KGF be investigated clinically for the prevention of radiation-induced mucositis. A multicenter prospective study testing its value in reducing mucositis has completed patient accrual but results are not yet available.

B. Mechanism- or Target-Driven Combined Therapy Strategies

Most combined radiation–chemotherapy regimens tested so far have evolved empirically by administering drugs found to have some activity against tumors of interest in a dose and time sequence known to be tolerated in a single modality therapy setting. This strategy has yielded some improvement in outcome but at a cost of increased morbidity. Further progress should, therefore, come from more rational integration of therapy modalities. In this effort, the primary treatment objective should guide the design of a combination of radiation with cytotoxic and biologic agents to maximize the likelihood of yielding a therapeutic benefit. For example, systemic therapy aiming for the reduction of distant metastatic relapse rate is best comprised of least toxic agents proven to have high antitumor activity in the treatment of patients with metastatic disease. It is also logical to administer systemic therapy sequentially with radiation (i.e., in an adjuvant or neoadjuvant setting) to prevent the occurrence of toxicity resulting from a drug–radiation interaction. In contrast, for improving locoregional tumor control, it is more rational to choose agents based on their mechanisms of action to offset known causes of radioresistance rather than on their independent antitumor activity. It is also prudent to select the proper timing of administration such as to yield the maximal drug–radiation interaction or to prevent specific normal tissue toxicity.

Improved insights into the molecular action mechanisms of various agents and molecular biology of HNSCC, along with well-orchestrated preclinical investigations, facilitate the design of biologically sound clinical combined therapy regimens, many of which are undergoing clinical testing. Examples include a combination of radiation with tirapazamine (a hypoxic toxin), the C-225 monoclonal antibody against epidermal growth factor receptor, and receptor tyrosine kinase inhibitors. It is hoped that such integrated translational research will accelerate progress in the management of advanced HNSCC.

References

1. Thames, H. D., Withers, H. R., and Peters, L. J. (1982). Changes in early and late radiation responses with altered dose fractionation: Implications for dose-survival relationships. *Int. J. Radiat. Oncol. Biol. Phys.* **8**, 219–226.
2. Thames, H. D., Bentzen, S. M., and Turesson, I. (1990). Time-dose factors in radiotherapy: A review of the human data. *Radiother. Oncol.* **19**, 219–235.
3. Horiot, J. C., LeFur, R. N., and Guyen, T. (1992). Hyperfractionation versus conventional fractionation in oropharyngeal carcinoma: Final analysis of a randomized trial of the EORTC cooperative group of radiotherapy. *Radiother. Oncol.* **25**, 231–241.
4. Fu, K. K., Pajak, T. F., and Trotti, A. (2000). A radiation therapy oncology group (RTOG) phase III randomized study to compare hyperfractionation and two variants of accelerated fractionation to standard fractionation radiotherapy for head and neck squamous cell carcinomas: First report of RTOG 9003. *Int. J. Radiat. Oncol. Biol. Phys.* **48**, 7–16.
5. Ang, K. (1998). Altered fractionation in head and neck cancer. *Semin. Radiat. Oncol.* **8**, 230–236.
6. Overgaard, J., Sand Hansen, H., and Sapru, W. (1996). Conventional radiotherapy as the primary treatment of squamous cell carcinoma (SCC) of the head and neck: A randomized multicenter study of 5 versus 6 fractions per week — Preliminary report from DAHANCA 6 and 7 trial. *Radiother. Oncol.* **40**, S31.
7. Bourhis, J., Lapeyre, M., and Tortochaux, J. (2000). Very accelerated versus conventional radiotherapy in HNSCC: Results of the GORTEC 94–02 randomized trial. In Proceedings of the American Society for Therapeutic Radiology and Oncology 42nd Annual Meeting, October 22–26, Boston, MA, [Abstract 2]
8. Vokes, E. E., Weichselbaum, R. R., and Lippman, S. M. (1993). Head and neck cancer. *N. Engl. J. Med.* **328**, 184–194.
9. Withers, H. R., Taylor, J. M. G., and Maciejewski, B. (1988). The hazard of accelerated tumor clonogen repopulation during radiotherapy. *Acta Oncol.* **27**, 131–146.
10. Schuller, D. E., Metch, B., and Stein, D. W. (1988). Preoperative chemotherapy in advanced resectable head and neck cancer: Final report of the southwest oncology group. *Laryngoscope* **98**, 1205–1211.
11. The Department of Veterans Affairs Laryngeal Cancer Study Group. (1991). Induction chemotherapy plus radiation compared with surgery plus radiation in patients with advanced laryngeal cancer. *N. Engl. J. Med.* **324**, 1685–1690.
12. Depondt, J., Gehanno, P., and Martin, M. (1993). Neoadjuvant chemotherapy with carboplatin/5-fluorouracil in head and neck cancer. *Oncology* **50**, 23–27.
13. Lefebvre, J. L., Chevalier, D., and Luboinski, B. (1996). Larynx preservation in pyriform sinus cancer: Preliminary results of a European Organization for Research and Treatment of Cancer phase III trial: EORTC Head and Neck Cancer Cooperative Group. *J. Natl. Cancer Inst.* **88**, 890–899.
14. INCSG (1996). Preliminary results of a randomized trial comparing neoadjuvant chemotherapy (cisplatin, epirubicin, bleomycin) plus radiotherapy vs. radiotherapy alone in stage IV (>or = N2, M0) undifferentiated nasopharyngeal carcinoma: A positive effect on progression-free survival. VUMCA I trial. *Int. J. Radiat. Oncol. Biol. Phys.* **35**, 463–469.
15. Temam, S., Flahault, A., and Perie, S. (2000). p53 gene status as a predictor of tumor response to induction chemotherapy of patients with locoregionally advanced squamous cell carcinomas of the head and neck. *J. Clin. Oncol.* **18**, 385–394.
16. Cabelguenne, A., Blons, H., and de Waziers, I. (2000). p53 alterations predict tumor response to neoadjuvant chemotherapy in head and neck squamous cell carcinoma: A prospective series. *J. Clin. Oncol.* **18**, 1465–1473.
17. Lo, T. C., Wiley, A. L., and Ansfield, F. J. (1976). Combined radiation therapy and 5-fluorouracil for advanced squamous cell carcinoma of the oral cavity and oropharynx: A randomized study. *Am. J. Roentgenol.* **126**, 229–235.
18. Fu, K. K., Phillips, T. L., and Silverberg, I. J. (1987). Combined radiotherapy and chemotherapy with bleomycin and methotrexate for advanced inoperable head and neck cancer: Update of Northern California Oncology Group randomized trial. *J. Clin. Oncol.* **5**, 1410–1418.
19. Eschwege, F., Sancho-Garnier, H., and Gerard, J. P. (1988). Ten-year results of randomized trial comparing radiotherapy and concomitant bleomycin to radiotherapy alone in epidermoid carcinomas of the oropharynx: Experience of the European Organization for Research and Treatment of Cancer. *NCI Monogr.* **6**, 275–278.
20. Adelstein, D. J., Saxton, J. P., and Lavertu, P. (1997). A phase III randomized trial comparing concurrent chemotherapy and radiotherapy with radiotherapy alone in resectable stage III and IV squamous cell head and neck cancer: Preliminary results. *Head Neck* **19**, 567–575.
21. Al-Sarraf, M., LeBlanc, M., and Shanker, G. (1998). Chemo-radiotherapy versus radiotherapy in patients with locally advanced nasopharyngeal cancer: Phase III randomized intergroup study (0099) (SWOG 8892, RTOG 8817, ECOG 2388). *J. Clin. Oncol.* **16**, 1310–1317.
22. Calais, G., Alfonsi, M., and Bardet, E. (1999). Randomized trial of radiation therapy versus concomitant chemotherapy and radiation therapy for advanced-stage oropharynx carcinoma. *J. Natl. Cancer Inst.* **91**, 2081–2086.
23. Adelstein, D. J., Adams, G. L., and Li, Y. (2000). A phase III comparison of standard radiation therapy (RT) versus RT plus concurrent cisplatin (DDP) versus split-course RT plus concurrent DDP and 5-fluorouracil (5FU) in patients with unresectable squamous cell head and neck cancer (SCHNC): An intergroup study. *Proc. ASCO* **19**, 411a. [Abstract]
24. Weissler, M. C., Melin, S., and Sailer, S. L. (1992). Simultaneous chemoradiation in the treatment of advanced head and neck cancer. *Arch. Otolaryngol. Head Neck Surg.* **118**, 806–810.
25. Sailer, S. L., Weissler, M. C., and Merlin, S. A. (1992). Toxicity and preliminary results from a trial of hyperfractionated radiation with or without simultaneous 5-fluorouracil-cisplatin in advanced head and neck squamous cell carcinomas. *Semin. Radiat. Oncol.* **2**, 38–40.
26. Magno, L., Terraneo, F., and Bertoni, F. (1994). Double-blind randomized study of lonidamine and radiotherapy in head and neck cancer. *Int. J. Radiat. Oncol. Biol. Phys.* **29**, 45–55.
27. Wendt, T. G., Grabenbauer, G. G., and Rodel, C. M. (1998). Simultaneous radiochemotherapy versus radiotherapy alone in advanced head and neck cancer: A randomized multicenter study. *J. Clin. Oncol.* **16**, 1318–1324.
28. Dobrowsky, W., Naude, J., and Widder, J. (1998). Continuous hyperfractionated accelerated radiotherapy with/without mitomycin C in head and neck cancer. *Int. J. Radiat. Oncol. Biol. Phys.* **42**, 803–806.
29. Brizel, D. M., Albers, M. E., and Fisher, S. R. (1998). Hyperfractionated irradiation with or without concurrent chemotherapy for

locally advanced head and neck cancer. *N. Engl. J. Med.* **338**, 1798–1804.
30. Jeremic, B., Shibamoto, Y., and Milicic, B. (2000). Hyperfractionated radiation therapy with or without concurrent low-dose daily cisplatin in locally advanced squamous cell carcinoma of the head and neck: A prospective randomized trial. *J. Clin. Oncol.* **18**, 1458–1464.
31. Staar, S., Rudat, V., and Stuetzer, H. (2001) . Intensified hyperfractionated accelerated radiotherapy limits the additional benefit of simultaneous chemotherapy: Results of a multicentric randomized german trial in advanced head-and-neck cancer. *Int. J. Radiat. Oncol. Biol. Phys.* **50**, 1161–1171.
32. Merlano, M., Vitale, V., and Rosso, R. (1992). Treatment of advanced squamous-cell carcinoma of the head and neck with alternating chemotherapy and radiotherapy. *N. Engl. J. Med.* **327**, 1115–1121.
33. Merlano, M., Benasso, M., and Corvo, R. (1996). Five-year update of a randomized trial of alternating radiotherapy and chemotherapy compared with radiotherapy alone in treatment of unresectable squamous cell carcinoma of the head and neck. *J. Natl. Cancer Inst.* **88**, 583–589.
34. Stell, P. M., and Rawson, N. S. B. (1990). Adjuvant chemotherapy in head and neck cancer. *Br. J. Cancer* **61**, 779–787.
35. Stell, P. M. (1992). Adjuvant chemotherapy for head and neck cancer. *Semin. Radiat. Oncol.* **2**, 195–205.
36. Munro, A. J. (1995). An overview of randomised controlled trials of adjuvant chemotherapy in head and neck cancer. *Br. J. Cancer* **71**, 83–91.
37. El-Sayed, S., and Nelson, N. (1996). Adjuvant and adjunctive chemotherapy in the management of squamous cell carcinoma of the head and neck region: A meta-analysis of prospective and randomised trials. *J. Clin. Oncol.* **14**, 838–847.
38. Pignon, J. P., Bourhis, J., and Domenge, C. (2000). Chemotherapy added to locoregional treatment for head and neck squamous-cell carcinoma: Three meta-analyses of updated individual data. *Lancet* **355**, 949–955.
39. Forastiere, A. A., Berkey, B., and Maor, M. (2001). Phase III trial to preserve the larynx: induction chemotherapy and radiotherapy versus concomitant chemoradiotherapy versus radiotherapy alone, Intergroup Trial R91-11. *Am. Soc. Clin. Oncol.* **1**, 2a.
40. Clayman, G. L., Johnson, C. J., II, and Morrison, W. (2001). The role of neck dissection after chemoradiotherapy for oropharyngeal cancer with advanced nodal disease. *Arch. Otolaryngol. Head Neck Surg.* **127**, 135–139.
41. The National Cancer Institute. (1998). The nation's investment in cancer research.
42. Verhey, L. J. (1999). Comparison of three-dimensional conformal radiation therapy and intensity-modulated radiation therapy systems. *Semin. Radiat. Oncol.* **9**, 78–98.
43. Eisbruch, A., Ten Haken, R. K., and Kim, H. M. (1999). Dose, volume, and function relationships in parotid salivary glands following conformal and intensity-modulated irradiation of head and neck cancer. *Int. J. Radiat. Oncol. Biol. Phys.* **45**, 577–587.
44. Sultanem, K., Shu, H. K., and Xia, P. (2000). Three-dimensional intensity-modulated radiotherapy in the treatment of nasopharyngeal carcinoma: The university of California-San Francisco experience. *Int. J. Radiat. Oncol. Biol. Phys.* **48**, 711–722.
45. Brizel, D. M., Wasserman, T. H., and Henke, M. (2000). Phase III randomized trial of amifostine as a radioprotector in head and neck cancer. *J. Clin. Oncol.* **18**, 3339–3345.
46. Wasserman, T. H. (2000). Effect of amifostine on patient assessed clinical benefit in irradiated head and neck cancer. *Int. J. Radiat. Oncol. Biol. Phys.* **48**, 1035–1039.
47. Abitbol, A. A., Sridhar, K. S., Lewin, A. A., *et al.* (1997). Hyperfractionated radiation therapy and 5-fluorouracil, cisplatin, and mitomycin-c (± granulocyte-colony stimulating factor) in the treatment of patients with locally advanced head and neck cancer. *Cancer* **80**(2), 266–276.
48. Vokes, E. E., Haraf, D., and Mick, R. (1994). Intensified concomitant chemoradiotherapy with and without filgrastim for poor-prognosis head and neck cancer. *J. Clin. Oncol.* **12**, 2351–2359.
49. Mascarin, M., Franchin, G., and Minatel, E. (1999). The effect of granulocyte colony-stimulation factor on oral mucositis in head and neck cancer patients treated with hyperfractionated radiotherapy. *Oral Oncol.* **35**, 203–208.
50. Kannan, V., Bapsy, P. P., and Anantha, N. (1997). Efficacy and safety of granulocyte macrophage-colony stimulating factor (GM-CSF) on the frequency and severity of radiation mucositis in patients with head and neck carcinoma. *Int. J. Radiat. Oncol. Biol. Phys.* **37**, 1005–1010.
51. Rosso, M., Blasi, G., and Gherlone, E. (1997). Effect of granulocyte-macrophage colony-stimulating factor on prevention of mucositis in head and neck cancer patients treated with chemo-radiotherapy. *J. Chemother.* **9**, 382–385.
52. Wagner, W., Alfrink, M., and Haus, U. (1999). Treatment of irradiation-induced mucositis with growth factors (rhGM-CSF) in patients with head and neck cancer. *Anticancer Res.* **19**, 799–803.
53. Makkonen, T. A., Minn, H., and Jekunen, A. (2000). Granulocyte macrophage-colony stimulating factor (GM-CSF) and sucralfate in prevention of radiation-induced mucositis: A prospective randomized study. *Int. J. Radiat. Oncol. Biol. Phys.* **46**, 525–534.
54. Ning, S., Shui, C., and Khan, W. B. (1998). Effects of keratinocyte growth factor on the proliferation and radiation survival of human squamous cell carcinoma cell lines *in vitro* and *in vivo*. *Int. J. Radiat. Oncol. Biol. Phys.* **40**, 177–187.
55. Alarid, E. T., Rubin, J. S., and Young, P. (1994). Keratinocyte growth factor functions in epithelial induction during seminal vesicle development. *Proc. Natl. Acad. Sci. USA* **91**, 1074–1078.
56. Housley, R. M., Morris, C. F., and Boyle, W. (1994). Keratinocyte growth factor induces proliferation of hepatocytes and epithelial cells throughout the rat gastrointestinal tract. *J. Clin. Invest.* **94**, 1767–1777.
57. Pierce, G. F., Yanagihara, D., and Klopchin, K. (1994). Sitmulation of all epithelial elements during skin regeneration by keratinocyte growth factor. *J. Exp. Med.* **179**, 831–840.
58. Ulich, T. R., Yi, E. S., and Cardiff, R. (1994). Keratinocyte growth factor is a growth factor for mammary epithelium *in vivo*: The mammary epitherlium of lactating rats is resistant to the proliferative action of keratinocyte growth factor. *Am. J. Pathol.* **144**, 862–868.
59. Ulich, T. R., Yi, E. S., and Longmuir, K. (1994). Keratinocyte growth factor is a growth factor for type II pneumocytes *in vivo*. *J. Clin. Invest.* **93**, 1298–2128.
60. Yi, E. S., Yin, S., and Harclerode, D. L. (1994). Keratinocyte growth factor induces pancreatic ductal epithelial proliferation. *Am. J. Pathol.* **145**, 85–88.
61. Finch, P. W., Rubin, J. S., and Miki, T. (1989). Human KGF is FGF-related with properties of a paracrine effector of epithelial cell growth. *Science* **245**, 752–755.

CHAPTER

31

Clinical Trials in Advanced Unresectable Head and Neck Cancer

DAVID J. ADELSTEIN

Department of Hematology and Medical Oncology
Cleveland Clinic Foundation
Cleveland, Ohio 44195

I. Altered Fractionation Radiation 462
II. Induction Chemotherapy 463
III. Alternating Chemotherapy and Radiation 466
IV. Concurrent Chemotherapy and Radiation 466
V. Future Directions 469
References 470

The designation of a squamous cell head and neck cancer as being unresectable implies extensive disease and a poor prognosis. Historically, radiation therapy has been the only treatment modality with demonstrated benefit. Despite aggressive radiotherapeutic schedules and doses, the prognosis of these patients remains poor. The definition of unresectability, however, has been quite variable and the literature remains confusing and inconsistent. This is compounded by the fact that radiation therapy rather than surgery may be used preferentially for some head and neck cancer subsites and disease stages, thus making the definition of unresectable disease less important.

Efforts to improve upon the disappointing results after radiation therapy have included altered radiation therapy fractionation schedules and the use of systemic chemotherapy in conjunction with radiation. Altered radiation fractionation schedules can be characterized as either accelerated fractionation, when the dose of radiation is unchanged but the overall treatment time is reduced, or hyperfractionation, when the overall radiation dose is increased but the treatment time is left unchanged. Hyperfractionated schedules have been associated with some increase in acute toxicity, but have also demonstrated promising survival benefits. Accelerated fractionation schedules, however, have been more toxic, and a survival benefit has not been demonstrated consistently. The large Radiation Therapy Oncology Group four-arm clinical trial, which explored altered radiation fractionation in advanced head and neck cancer, demonstrated an improvement in locoregional control with a hyperfractionated regimen and with an accelerated fractionation/concomitant boost treatment schedule.

Systemic chemotherapy is an active treatment modality in patients with advanced disease and, in the previously untreated patient, results in a very high response rate. Induction chemotherapeutic schedules, although theoretically promising, have not produced any survival benefit. There has been limited exploration of alternating radiation and chemotherapy treatment schedules, and encouraging results have been reported for this approach. Most exciting, however, have been the concomitant chemotherapy and radiation therapy schedules. A clear survival benefit for patients with unresectable disease has been demonstrated for concurrent radiation and single-agent cisplatin in the recently completed North American Intergroup study. Other more aggressive multiagent chemotherapy regimens have also produced significant survival advantages. Careful meta-analysis data have confirmed these results.

Future efforts in the management of patients with advanced head and neck cancer will include toxicity modulation and an exploration of newer treatment modalities, including new chemotherapeutic agents and new ways of chemotherapy and radiation administration, as well as genetic and immunologic approaches.

The designation of a head and neck cancer as advanced and unresectable identifies disease with a particularly poor prognosis. A major obstacle in understanding the results of clinical trials in patients with advanced unresectable head and neck cancer, however, has been just this definition of unresectability [1]. While the appellation "unresectable" does indicate the limitations of initial disease management, anatomic criteria for unresectability vary from institution to institution and from surgeon to surgeon. Patient factors such as age, performance status, and willingness to accept surgical morbidity further impact on the technical decision as to what represents "resectable" disease. Even when formal definitions are proposed, the ultimate decision about resectability still remains that of the responsible surgeon.

For patients who are deemed unresectable, definitive management has been radiation therapy based. Radiation therapy, however, has also proven successful in the definitive treatment of patients with less disease; disease that might be considered technically resectable. The inherent assumption in attempting to define a population with "advanced unresectable" cancer is that surgical resection represents the preferred therapeutic option; an unproven and debatable hypothesis for many stages and subsites of head and neck cancer [2,3]. Clearly, if nonsurgical treatment is preferable, precise definitions of unresectability become irrelevant.

This has produced considerable confusion in the literature. While some clinical trials have taken great pains to define patient eligibility based on disease extent, other studies have rather loosely included anybody deemed "unresectable" or "inoperable" without clear criteria given. Some studies consider unresectability to include patients with a low likelihood of surgical cure or patients deemed medially unfit for surgical resection, again without careful definition. It is also not uncommon to find radiation therapy-based clinical trials in patients with "advanced disease" in which the question of resectability is never raised. Comparison of clinical experiences in these widely varying patient populations is quite difficult. Furthermore, because the patient resources available for clinical trial are relatively limited in this disease, many studies have included patients with multiple primary tumor sites and widely varying clinical stages. This produces even further heterogeneity and further confusion.

This chapter reviews the clinical trials of definitive, not palliative, treatment for patients with advanced squamous cell head and neck cancer. The studies of interest will be those in which surgery was felt to be a poor therapeutic choice based on disease extent rather than those studies in which radiation therapy was deemed the preferred curative treatment approach. This is obviously an imprecise and arbitrary distinction. In general, however, this patient population includes those with advanced locoregional but nonmetastatic disease, usually American Joint Committee on Cancer (AJCC) stage III or IV in extent, for which the prognosis has historically been quite poor [4]. Such patients, irrespective of treatment, have an expected overall survival below 25% [5–8]. This chapter will also be restricted to a discussion of squamous cell cancers of the oral cavity, oropharynx, larynx, and hypopharynx and will not address primary nasopharyngeal cancers or primary malignancies of the paranasal sinuses or of the salivary glands.

Clinical trials for these patients have focused on either altered radiation therapy fractionation schedules or on combinations of radiation therapy and chemotherapy. While many phase II, single-arm clinical experiences have been reported, the heterogeneity of patients studied precludes anything but a feasibility or toxicity analysis of this work. Meaningful conclusions can only be drawn from randomized phase III trials.

It is important to pay attention to end points in these clinical trials. Survival is the "gold standard" measure of any treatment success; however, survival can be affected by many factors. While locoregional disease remains the most common cause for treatment failure in head and neck cancer [5,7], distant metastases represent an increasing concern, particularly as the success of our locoregional treatments improves [9,10]. Even in those patients achieving disease control, second primary neoplasms and multiple medical comorbidities also impact on overall survival and may obscure the benefit derived from a therapeutic intervention [5,7,11–13]. It thus becomes critical to look at other end points, such as locoregional control, distant control, and cause-specific survival in addition to the overall survival.

It is also important to recognize that with each increase in treatment intensity, there is an associated increase in toxicity [14,15]. Much of the success of recent multimodality treatments has come from an improvement in our ability to manage these toxicities. Perhaps of even greater importance are the so-called "consequential late effects" of treatment, including impairment of swallowing, xerostomia, and osteonecrosis. Clearly, each additional layer of either early or late toxicity alters the risk/benefit ratio of our interventions.

I. ALTERED FRACTIONATION RADIATION

The expectations after definitive radiation therapy for advanced head and neck cancer have been established from the Radiation Therapy Oncology Group database [8]. This is a large prospectively gathered database, which confirms the poor prognosis of patients with stage III and IV disease. Indeed, most subcategories of patients with stage IV disease have a 4-year survival of 20% or less. This database, however, is not defined in terms of resectability and one can only surmise that those patients deemed unresectable have a significantly poorer chance of success. Data from the control arms of the two North American Intergroup unresectable studies are equally dismal [16,17]. Both studies used conventional daily radiation therapy for the control arm and both reported median survivals of approximately 13 months, with long-term disease-free survival below 20%.

One approach taken in attempting to improve on these results has been the exploration of altered fractionation treatment schedules [18–20]. The goal of such schedule manipulation has been to optimize locoregional control without undue or excessive locoregional toxicity.

In the United States, conventional fractionation has been considered to be once daily treatment at a rate of 1.8 to 2.5 Gy per fraction, five fractions per week, in a continuous course over a 4- to 8-week overall treatment time. Typical treatment courses include 50 Gy in 20 fractions over 4 weeks or 70 Gy in 35 fractions over 7 weeks. It has been established that, if tolerable, a shorter overall time to completion of treatment, a greater radiation therapy dose, and an avoidance of any treatment interruptions are all important in overall outcome [18,20–22].

Altered fractionation schedules can be characterized as either (1) accelerated fractionation, where the overall time to treatment completion is reduced, or (2) hyperfractionation, where the overall total dose is increased and the overall time remains unchanged. Many permutations on these themes have been tested, often representing hybrid mixtures of both hyperfractionation and accelerated fractionation schedules.

The rationale behind accelerated fractionation is that the reduction in overall treatment time reduces the opportunity for tumor cell regrowth during treatment. This increases the probability of tumor control for any given dose, without increasing the probability of late normal tissue injury. Pure accelerated fractionation regimens do not alter individual fraction size and might include twice-daily fractionation or radiation administered 6 of 7 or 7 of 8 days of the week. Because of the expected increase in acute toxicity experienced with this kind of approach, some accelerated fractionation regimens have employed a split-course technique, an intervention that is suboptimal but may be compensated for by the accelerated schedule. Smaller fraction sizes are also often utilized, thus merging this schedule alteration with the hyperfractionation regimens.

The rationale for hyperfractionation is based on the differential sensitivities of tumors and late-responding normal tissues to radiation therapy. If given in smaller fractions, a greater total dose can, in theory, be administered, resulting in a greater tumor kill, but an equivalent effect on late-responding normal tissues.

Several large randomized trials of altered fractionation radiation therapy have been conducted. Although these studies have been conducted in patients with advanced head and neck tumors, unresectability was rarely a criterion for patient entry. In general, the hyperfractionated treatment schedules have produced an improvement in survival with some increase in acute toxicity [23–28]. Consequential late effects, however, did not appear to be worse when compared to conventional treatment schedules.

The accelerated fractionation regimens and the hybrid accelerated fractionation regimens proved more difficult to tolerate. Toxicity was significantly worse and survival not favorably impacted [29–34]. The continuous hyperfractionated accelerated radiation therapy (CHART) regimen is of particular note [33]. This study randomized patients to either 1.5 Gy, three times daily for 12 consecutive days to a total dose of 54 Gy, or to 2 Gy once daily to a total dose of 66 Gy over 6.5 weeks (five treatment days per week). Locoregional control and survival were equivalent between the two treatment arms. Acute toxicity was significantly worse with the CHART regimen, although late morbidities were not.

The recent publication of the results of the large Radiation Therapy Oncology Group Study 9003 are of particular import [35]. This trial randomized 1113 patients among four treatment arms (Fig. 31.1): (1) standard fractionation given at 2 Gy per day to 70 Gy in 35 fractions over 7 weeks; (2) hyperfractionation given at 1.2 Gy per fraction, twice daily, 5 days per week to 81.6 Gy in 68 fractions over 7 weeks; (3) accelerated fractionation with a split given at 1.6 Gy per fraction, twice daily, 5 days per week to 67.2 Gy, in 42 fractions over 6 weeks, including a 2-week rest after 38.4 Gy; or (4) accelerated fractionation with a concomitant boost given at 1.8 Gy per fraction, once daily 5 days per week to the larger field, with a boost field being administered at 1.5 Gy per fraction as a second daily treatment for the last 12 treatment days to a total dose of 72 Gy in 42 fractions over 6 weeks.

This study demonstrated significantly better locoregional control for those patients treated with hyperfractionation (arm 2) or accelerated fractionation with a concomitant boost (arm 4) when compared to the standard fractionation schedule. Overall survival was not improved significantly, although there was a trend toward improved disease-free survival in both of the more successful treatment arms. Those treated on arm 3, with the split course of accelerated fractionation, did not fare better than the standard fractionation treatment arm. Acute side effects were worse in all three of the altered fractionation arms. Late effects did not appear to be different. It should be noted, however, that this study was not limited to those with unresectable disease. Patients with previously untreated stage III or IV tumors (and patients with stage II tumors of base of tongue or hypopharynx) were eligible. Nonetheless, the observations derived from this study should still be valid in a study limited to those with more advanced tumors.

Altered fractionation radiation schedules have also been employed in conjunction with concurrent chemotherapy, as discussed later. Although benefit has been seen when compared to conventional radiation alone, it remains unclear whether the altered fractionation radiation or the concurrent chemotherapy is of greater importance.

II. INDUCTION CHEMOTHERAPY

It has been recognized for some time that recurrent and/or metastatic head and neck cancer is surprisingly

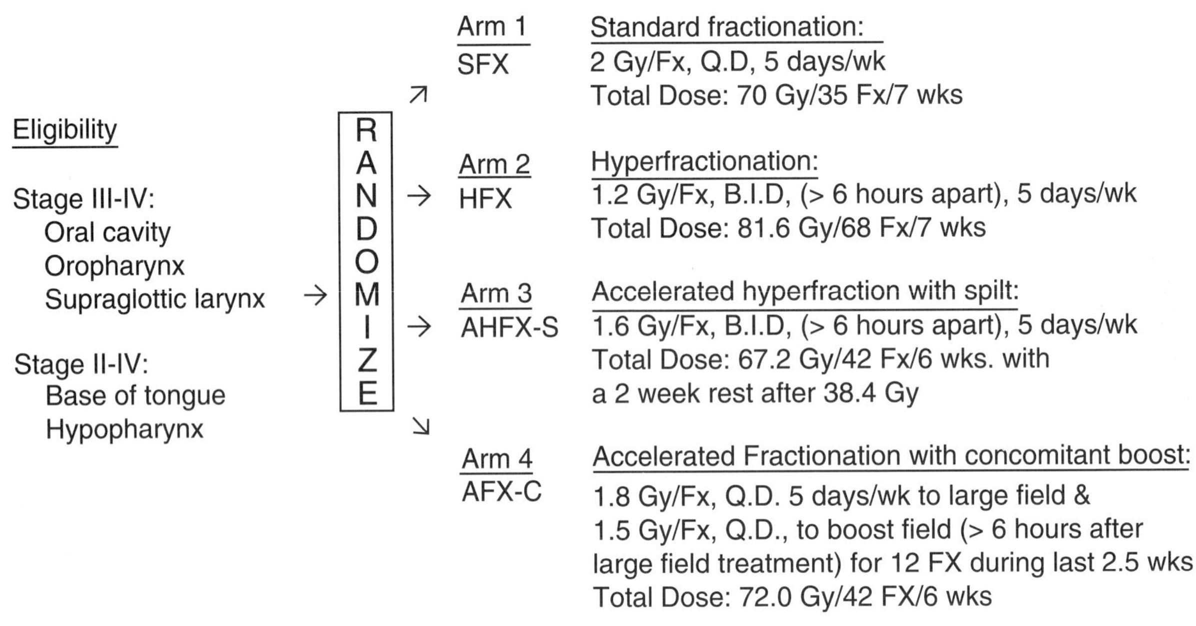

FIGURE 31.1 Radiation Therapy Oncology Group 9003: Treatment schema.

sensitive to systemic chemotherapy [5–7]. Response rates of 30 to 35% have been reported reproducibly after aggressive chemotherapy regimens using combinations such as 5-fluorouracil and cisplatin (Table 31.1) [36–39]. Recent explorations of chemotherapy using the taxanes (paclitaxel, docetaxel) have been similarly encouraging [40–42]. This chemosensitivity led to the exploration of this treatment modality in patients with advanced but previously untreated disease. Similar to observations made in other solid tumors, it was noted that the previously untreated patient responded to chemotherapy almost twice as often as the patient who had undergone surgery and/or radiation [43,44]. This improved responsiveness could be attributed both to the better performance status found in the previously untreated patient and to the presence of an intact blood supply undisturbed by prior intervention [45,46].

For example, when the 5-fluorouracil and cisplatin combination regimen is used in newly diagnosed patients, an overall response rate between 68 and 93% is seen, with complete response rates as high as 54% (Table 31.1) [47–52]. Chemotherapy sensitivity of this magnitude is remarkable for an epithelial tumor and approaches the success seen in the highly chemosensitive tumors, such as the lymphomas, small cell carcinoma of the lung, and testicular cancer. These results produced considerable optimism about the potential for incorporation of chemotherapy into multimodality treatment schedules and for an improvement in overall treatment success. This potential benefit from chemotherapy would be particularly important for those patients with advanced unresectable tumors, in whom radiation alone had only limited success.

Induction or neoadjuvant chemotherapy is the multimodality treatment schedule explored most extensively [53]. The recognition that chemotherapy was most effective if used prior to radiation therapy suggested that the best utilization of chemotherapy would be as the initial treatment approach, to downstage the size of the tumor and maximize its response to locoregional intervention [5,7,45,54]. The impact of chemotherapy on the low but finite incidence of distant metastases was unknown, but also a theoretical advantage.

Many small phase II trials were initially conducted using induction chemotherapy prior to definitive surgical or nonsurgical management in patients with both resectable and unresectable cancers [55,56]. Most of these trials employed cisplatin-based combination regimens [57–60], although an experience using noncisplatin-containing combinations also emerged [56]. The most widely used combination remained the same 5-fluorouracil and cisplatin regimen that had proven successful in the palliative treatment of patients with recurrent disease. Although somewhat cumbersome to administer, it is a well-tolerated combination with a predictable toxicity profile. Attempts have been made to improve on this drug combination by modulating the 5-fluorouracil with agents such as leucovorin; however, response rates have increased only marginally at the cost of a significant increase in toxicity [61–63].

This phase II experience resulted in several major observations and conclusions. Tumor regression was found in between 60 and 90% of previously untreated patients and

TABLE 31.1 Selected Trials Using 5-Fluorouracil and Cisplatin

	Year	No. of patients	Response (%)	Complete response (%)	Median survival (months)	Reference
Recurrent or metastatic disease	1987	20	35	5	6	Mercier *et al.* [36]
	1992	79	32	6	5.5	Jacobs *et al.* [37]
	1992	87	32	6	6.6	Forastiere *et al.* [38]
	1994	116	33	2	Not stated	Clavel *et al.* [39]
Previously untreated disease	1985	61	93	54		Rooney *et al.* [47]
	1987	27	85	19		Toohill *et al.* [48]
	1990	37	68	46		Martin *et al.* [49]
	1991	166	85	31		Wolf *et al.* [50]
	1994	118	80	31		Paccagnella *et al.* [51]
	1997	71	83	32		Athanasiadis *et al.* [52]

a complete response was possible in between 20 and 50% [5,6,45–47,54–60,64,65]. Even in these complete responders, however, relapse could be expected without definitive treatment. Head and neck cancer does not yet appear to be a "chemocurable" disease such as lymphoma or testicular cancer, and single modality treatment with chemotherapy cannot be recommended [66]. Chemotherapy responses were observed to continue for at least the first three courses of treatment. Whether more than three courses are of value is unclear [47,64,65,67]. Furthermore, the use of induction chemotherapy did not appear to adversely affect a patient's tolerance for subsequent definitive management [47,52,54,58,59,68].

An observation felt important in organ preservation strategies was the recognition that a response to chemotherapy was predictive for a subsequent response to radiation therapy [69]. Furthermore, those patients with aneuploid or poorly differentiated tumors appeared to respond better to chemotherapy than those with diploid cancers [70].

Unfortunately, the success of initial chemotherapeutic intervention also interfered with patient compliance [45,46,71]. Complex multimodality treatment regimens and protracted treatment schedules often proved difficult in this poorly compliant patient population. A distressing and not infrequent occurrence was the patient who achieved an excellent response to induction therapy, but then chose to delay or decline the necessary definitive disease management.

This extensive phase II experience convincingly demonstrated the feasibility of induction chemotherapy prior to definitive management. Results were rapid, dramatic, and gratifying. Convincing demonstration of a survival benefit, however, was necessary if the cost and toxicity of this approach were to be justified. Phase III randomized trials were required and, to the credit of the oncologists caring for this group of patients, a large number of high-quality studies were completed [53,65,72]. Results have been reported for both resectable patients treated with definitive surgery after induction chemotherapy and patients considered either inoperable or more appropriate for definitive radiation rather than for surgery. Table 31.2 details the results of these trials in patients for whom definitive management consisted of radiation therapy alone [51,73–78]. Although several of these studies characterized their patients as "inoperable," others may have included potentially resectable patients treated preferentially with radiation.

Unfortunately, these studies have been quite convincingly negative. Trials using both single agent and combination chemotherapy followed by definitive radiation therapy have failed to demonstrate a reproducible survival advantage with the neoadjuvant treatment schedule. Only Paccagnella *et. al.* [51] demonstrated any survival advantage in the unresectable subset of a larger patient population. Many of the other studies used a 5-fluorouracil and platinum-based chemotherapy regimen and were quite large with significant statistical power.

Failure of the induction chemotherapeutic approach was counterintuitive and multiple explanations were proposed [5–7,45,46,64,72]. Although some of the earlier studies were methodologically flawed or were flawed by the use of suboptimal chemotherapeutic combinations, more recent studies used 5-fluorouracil and platinum-based regimens [51,76,78]. Perhaps a better argument that can be made to explain the failure of induction chemotherapy is the fact that chemotherapy is the least effective of the treatment modalities available for this disease. Oncologically, it makes little sense to use a suboptimal treatment as the first intervention, particularly when that intervention requires several months to administer [53,64].

Despite this failure of induction chemotherapy, widespread adoption of neoadjuvant approaches has been the rule. In 1997, Harrari [79] reported the results of a mail survey of 300 community cancer specialists, which attempted to identify the most frequent treatments for patients with locoregionally, advanced nonmetastatic squamous cell head and neck cancer. The single most

TABLE 31.2 Randomized Trials of Induction Chemotherapy Followed by Definitive Radiation versus Radiation Therapy Alone

Year	No. of patients	Chemotherapy[a]	Survival benefit	Reference
1980	638	M	No	Fazekas *et al.* [73]
1983	86	VMFBHuMpC	No	Stell *et al.* [74]
1987	116	VBPD	No	Szpirglas *et al.* [75]
1992	100	PBVdMi	No	Jaulerry *et al.* [76]
1992	108	FPVd	No	Jaulerry *et al.* [76]
1992	42	CpFt	No	Tejedor *et al.* [77]
1994	171[b]	FP	Advantage: chemotherapy[b]	Paccagnella *et al.* [51]
1996	166	FP	No	Domenge *et al.* [78]

[a]F, 5-fluorouracil; P, cisplatin; V, vincristine; B, bleomycin; M, methotrexate; Vd, vindesine; Hu, hydroxyurea; Mi, mitomycin C; Cp, carboplatin; Ft, ftorafur; Mp, mercaptopurine; C, cyclophosphamide; D, doxorubicin.

[b]Unresectable subset.

common treatment approach proved to be induction chemotherapy with 5-fluorouracil and cisplatin followed by radiation therapy. Physicians cited the desire to improve both locoregional tumor control and overall survival using this treatment schedule, despite the fact that improvement in neither of these end points has been demonstrated successfully by randomized clinical trials. The persistence of the use of this therapeutic approach in the community, despite data that do not support it, remains a difficult problem. At present, one can only justify induction chemotherapy within the context of a clinical trial.

Whether the recent incorporation of several newer chemotherapeutic agents, such as the taxanes, into more aggressive drug combinations might improve on the results of neoadjuvant therapy is unknown. It is difficult to imagine, however, that any drug or drug combination can significantly better the reproducible 90% response rates and up to 50% complete response rates achieved with the 5-fluorouracil and cisplatin combination. As such it appears unlikely that any new regimen will impact survival when an induction schedule is utilized.

III. ALTERNATING CHEMOTHERAPY AND RADIATION

Rapidly alternating schedules of chemotherapy and radiation have also been tested in this disease. The rationale for this approach is that by interdigitating different treatment modalities rapidly, the development of treatment resistant tumor clones can be minimized. Furthermore, an interdigitating treatment schedule can allow for full therapeutic doses of chemotherapy and radiation to be given in as short a time, with as little toxicity, as possible. Merlano *et al.* [80, 81] have reported two randomized trials testing this approach in patients deemed unresectable at presentation. The first compared a sequential treatment schedule of four cycles of chemotherapy followed by definitive radiation with a schedule alternating four courses of chemotherapy with three partial courses of radiation given between the chemotherapy cycles. The chemotherapy was identical on both treatment arms and consisted of vinblastine, bleomycin, and moderate dose methotrexate with leucovorin rescue. The complete response rate, progression-free survival, and overall survival were significantly better for patients on the alternating treatment arm. Even for this arm, however, the 4-year overall survival was only 22% [80].

The second randomized trial compared a similar alternating treatment schedule of chemotherapy and radiation to radiation therapy alone [81]. The chemotherapy in this trial consisted of bolus injections of 5-fluorouracil and cisplatin on 5 consecutive days every 3 weeks. Once again, the complete response rate, progression-free survival, and overall survival were significantly improved by the interdigitation of chemotherapy into the radiation therapy schedule. Once again, the long-term (5-year) overall survival, even on the multimodality arm, was only 24%. Further investigation of this concept is nonetheless clearly justified.

IV. CONCURRENT CHEMOTHERAPY AND RADIATION

Another approach taken in multimodality therapy has been an attempt to maximize the benefits of chemotherapy and radiation by giving these two treatments concurrently. The rationale for this recognizes that both chemotherapy and radiation therapy are independently active treatment modalities and that when used together there is the further potential for synergism (i.e., chemotherapeutic radiosensitization). Furthermore, the concurrent use of these two modalities shortens the total treatment duration, improves patient compliance, and allows for a potential effect on micrometastases [6,14,46].

There are several disadvantages to concurrent treatment, however. The simultaneous use of two independently toxic treatment modalities will produce greater toxicity than either treatment modality alone. This increase in toxicity frequently results in compromise of either the radiotherapy or the chemotherapy delivery; a therapeutic strategy that is clearly flawed [5,14,46]. These compromises have included the choice of single agent rather than combination chemotherapy, chemotherapy dose reductions, or suboptimal radiotherapy administration such as the use of split-course treatment schedules. Even with these compromises, however, phase II studies were sufficiently encouraging to prompt the performance of phase III clinical trials.

The first concurrent chemoradiotherapy schedules studied used full-course, uninterrupted radiation therapy along with single-agent chemotherapy. Radiation fractionation was, in general, conventional, although several of the recent trials used altered fractionation regimens. Although the chemotherapy by itself was recognized as suboptimal, in conjunction with the radiation it was, in theory, radiosensitizing [46]. Randomized trials have been conducted using many of the active single agents in this disease, including 5-fluorouracil [25,82,83], methotrexate [84], bleomycin [85–88], mitomycin C [89,90], cisplatin [16,17,91,92], and carboplatin [91,93] (Table 31.3). This phase III experience since the mid-1970s has demonstrated a relatively consistent survival benefit using 5-fluorouracil, cisplatin, and carboplatin and an inconsistent benefit with bleomycin. It should be pointed out that the one negative randomized trial with single-agent cisplatin and simultaneous radiation was the first North American Intergroup unresectable trial, a study that used a weekly low-dose cisplatin schedule [16]. Subsequent studies using either a low-dose daily cisplatin regimen [91,92] or a high-dose regimen given every 3 weeks [17] demonstrated a clear survival advantage over radiation therapy alone.

It should also be pointed out that these studies in general were performed in patients with locally advanced unresectable tumors. In particular the first intergroup study [16], which tested the weekly concurrent low-dose cisplatin regimen, had carefully defined entry criteria to ensure that only unresectable patients were enrolled. Although a negative study, the same entry criteria were again used in INT 0126, the second-generation intergroup randomized trial [17] (Fig. 31.2). This study compared radiation therapy alone, given in a conventional fractionation schedule (arm A), to a high-dose cisplatin regimen given every 3 weeks with radiation therapy (arm B), to a third arm using an unconventional split-course radiation and concurrent combination chemotherapy schedule with 5-fluorouracil and cisplatin (arm C). This third treatment schedule was designed with the hope that patients initially deemed unresectable might be rendered surgically resectable after limited preoperative chemotherapy and radiation.

The results of this study clearly demonstrated an improved survival for patients treated with the concurrent chemoradiotherapy regimen using high-dose cisplatin given every three weeks when compared to radiation therapy alone. The split-course radiation therapy and combination chemotherapy arm, however, was not statistically different than either of the other two treatment regimens. The lack of improvement seen from

TABLE 31.3 Randomized Trials of Definitive Radiotherapy versus Concurrent Single-Agent Chemoradiotherapy

Year	No. of patients	Chemotherapy[a]	Radiation (Gy)	Survival benefit	Reference
1976	136	F	60–70	Yes	Lo *et al.* [82]
1990	859	F	60	Yes	Sanchiz *et al.* [25]
1994	175	F	66	Marginal	Browman *et al.* [83]
1987	313	M	45–55	Marginal	Gupta *et al.* [84]
1980	157	B	55–60	Yes	Shanta and Krishnamurthi [85]
1985	222	B	65	No	Vermund *et al.* [86]
1987	104	B	70	Yes[c]	Fu *et al.* [87]
1988	199	B	70	No	Eschwege *et al.* [88]
1989	39[b]	Mi	68	No	Weissberg *et al.* [89]
1998	188	Mi	70	Yes	Dobrowsky *et al.* [90]
1990	319	P	68–78	No	Haselow *et al.* [16]
1997	159	P	70	Yes	Jeremic *et al.* [91]
		Cp	70	Yes	
2000	130	P	77	Yes	Jeremic *et al.* [92]
2000	295	P	70	Yes	Adelstein *et al.* [17]
1996	130	Cp	70	Yes	Gabriele *et al.* [93]

[a]F, 5-fluorouracil; M, methotrexate; B, bleomycin; Mi, mitomycin C; P, cisplatin; Cp, carboplatin.
[b]Nonsurgical patients only.
[c]Relapse-free survival.

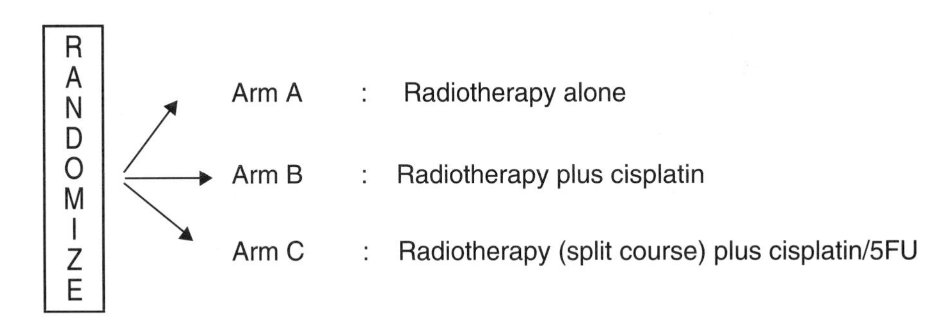

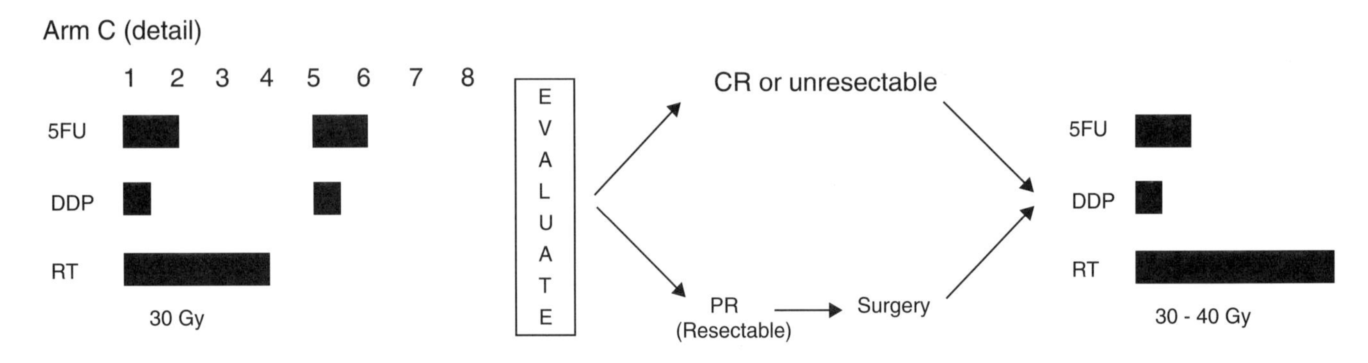

5FU : 5-Fluorouracil 1000 mg/m^2/day IV continuous infusion × 4 days
DDP : Cisplatin 75 mg/m^2 IV bolus
RT : Radiation therapy 2 Gy/day (single daily fractions)

FIGURE 31.2 Second-generation Intergroup Unresectable Trial INT 0126: Treatment schema.

the use of a multiagent chemotherapy regimen was felt to represent the detrimental impact of the split-course radiation therapy, which was not compensated for by the number of surgical resections ultimately performed. This intergroup study, as well as its predecessor, allowed for salvage surgical procedures to be performed for patients with less than complete responses or with a locoregional recurrence after definitive nonoperative treatment. This proved possible on both studies. Oftentimes, however, the surgical procedure performed was a neck dissection after achievement of a complete response at the primary site.

Concurrent multiagent chemotherapy and radiation regimens have also been pursued by a number of other investigators, often quite tentatively, however, due to the expected toxicity. As a result, as in INT 0126, suboptimal radiotherapy treatment schedules (and/or drug combinations) have often been employed. Studies using aggressive and uncompromised radiation therapy along with concurrent multiagent chemotherapy have been much more consistent in demonstrating both a survival and a locoregional control benefit (Table 31.4) [94–99]. Although unresectable disease was not always an entry criterion for all of these studies, these results are highly encouraging. Toxicity with these schedules has been, as expected, significant, but manageable, and further study of these approaches is clearly indicated.

Several of the single-agent trials [90,92] and several of the multiagent trials [94,97,98] also employed altered fractionation radiation schedules. All demonstrated further benefit by the addition of concurrent chemotherapy.

Taylor *et al.* [100] conducted a randomized comparison in unresectable patients between concurrent 5-fluorouracil and cisplatin chemoradiotherapy and a sequential induction treatment schedule using the same drugs followed by definitive radiation. An improved progression-free survival but not overall survival was seen for the concurrent treatment approach. Pinnaro *et al.* [101] tested a similar question, but used single-agent cisplatin concurrently with radiation compared to induction 5-fluorouracil and cisplatin followed by radiation. No differences in survival were identified.

Meta-analysis data have been confirmatory. Four separate meta-analyses, performed to assess the value of chemotherapy in patients with head and neck cancer, failed to demonstrate a survival advantage for the neoadjuvant approach [102–105]. The most recent and largest of these meta-analyses, from the MACH-NC group based in France, reviewed 63 randomized trials, including over 10,000 patients, using updated individual patient data [105]. No survival advantage was identified for either a neoadjuvant or an adjuvant treatment schedule. Patients treated with concomitant chemoradiotherapy, however,

TABLE 31.4 Randomized Trials of Definitive Radiotherapy versus Concurrent Multiagent Chemoradiotherapy

Year	No. of patients	Chemotherapy[a]	Radiation (Gy)	Survival benefit	Locoregional control benefit	Reference
1992	32[b]	FP	72 (bid, split)	Yes[c]	-	Weissler *et al.* [94]
1993	209	FMi	50 (split)	No	No	Keane *et al.* [95]
1995	49	MiB	66–70	Yes[d]	-	Smid *et al.* [96]
1998	270	FPLV	70.2 (bid, split)	Yes	Yes	Wendt *et al.* [97]
1998	116[e]	FP	70–75 (bid)	Yes	Yes	Brizel *et al.* [98]
1999	226	FCp	70	Yes	Yes	Calais *et al.* [99]

[a]F, 5-fluorouracil; P, cisplatin; Mi, mitomycin C; B, bleomycin; LV, leucovorin; Cp, carboplatin.
[b]Unresectable subset.
[c]Disease-specific survival.
[d]Disease-free survival.
[e]Only 62 patients considered unresectable.

predominantly an unresectable group, had an absolute 5-year survival benefit of 8% ($p<0.0001$). It is of particular note that this meta-analysis did not include those trials in Table 31.4, which were reported since 1993, many of which are even more convincing.

It must be stressed that these kinds of concurrent treatment schedules are associated with significant toxicity, far in excess of that commonly seen after single modality radiation alone [14,15]. The success of such concurrent treatment is possible only with intensive supportive measures. Close nursing and physician follow-up, active dental prophylaxis, early and appropriate antibiotic usage, and aggressive supportive alimentation are necessary. Feeding tubes are now almost routinely employed in an effort to avoid the treatment breaks that would previously have been required due to mucositis and dysphagia [98,106]. Clearly it does very little good to initiate an aggressive concurrent treatment plan only to find that it must be delayed due to treatment toxicity. The importance of maintaining dose intensity cannot be overemphasized.

The use of feeding tubes during aggressive treatment with concurrent chemotherapy and radiation, however, has been associated with the emergence of newer consequential late toxicities [107]. One significant concern has been the persistence of feeding tube dependence in patients treated with aggressive chemoradiotherapy [108]. While some of this may reflect the locally advanced nature of the tumor at presentation, which produced a baseline compromise in swallowing, much also appears to reflect the protracted period of time during which the chemoradiotherapy-induced mucosal inflammation precludes any adequate oral intact. This, coupled with baseline tumor-induced functional changes, allows stricturing and scarring of the oropharyngeal structures. The resultant dysphagia may be irreversible and certainly requires aggressive diagnostic and therapeutic intervention.

V. FUTURE DIRECTIONS

Building on recent success, a number of future therapeutic possibilities have emerged. Since the early 1990s, several new chemotherapeutic agents have been made available for clinical study and use. Many of these have radiosensitizing properties and are currently undergoing intensive investigation in the head and neck cancer population. In particular the taxanes (paclitaxel, docetaxel) have demonstrated significant activity when used in the metastatic disease setting and as first-line therapy in previously untreated patients [40–42,109–111]. Whether they will add significantly to the results of definitive management in this disease or whether they have an improved toxicity profile remains to be seen. Other agents, including gemcitabine, vinorelbine, and the topoisomerase I inhibitors, require further testing [112–114].

The University of Tennessee Group has revived and updated an older treatment approach of intraarterial chemotherapy. Robbins *et al.* [115] have explored the use of high-dose intraarterial cisplatin and concomitant radiation therapy for locoregionally advanced head and neck cancers. Promising results have been reported both from a single institutional trial and from a multi-institutional Radiation Therapy Oncology Group pilot study [116]. Locoregional control appears excellent, although concern has been raised about the development of distant metastases. The chemotherapy administration procedure is technically challenging and requires considerable experience, but has proven feasible.

Gene therapy, particularly that directed at the p53 tumor suppressor gene, has been studied in several centers [117, 118]. Restoration of wild-type p53, however, is dependent on viral vectors for delivery. These delivery methods are cumbersome and at present are confined to locoregional tumors [119]. Efficacy has, nonetheless, been demonstrated, and the integration of this treatment modality into more definitive management remains a theoretical and attractive possibility.

Monoclonal antibodies and other agents directed against an epidermal growth factor have also emerged as a potentially powerful intervention. Clinical trials are underway and preliminary results are encouraging [120,121]. These agents are particularly well tolerated from a toxicity viewpoint and may represent an exciting new treatment modality for this disease.

Along with these newer therapeutic initiatives has come a renewed interest in the end result of our interventions. It has long been recognized that disease eradication, while a valid and appropriate end point for antineoplatic therapy, must also consider the quality of the life preserved. As discussed previously, aggressive chemoradiotherapeutic interventions produce significant toxicities, some of which may prove permanent. The implications, for example, of long-term feeding tube dependence after aggressive treatment of a head and neck cancer must be addressed and quality of life issues formally studied. A number of validated tools have been developed over the past decade and are currently being tested in a randomized fashion in an effort to answer these kinds of questions [122].

Toxicity modification thus becomes an important direction for further study. The notion of chemotherapy and radiotherapy protectants is an attractive one and has been explored with the hope that we can better support our patients through these intensive therapeutic interventions [123]. Well-established protectants include such drugs as mesna for Ifosfamide-induced urothelial toxicity and the hematopoietic colony-stimulating factors (filgrastim and sargramostim) used in the prevention of neutropenia after myelosuppressive chemotherapy [124].

Several agents have been suggested as protectants to obviate the mucosal toxicity of aggressive chemoradiotherapy. Amifostine, in particular, has been studied in this regard. Although initial studies were promising, the single randomized trial performed thus far failed to demonstrate a significant mucosal protective effect from this agent [125]. The drug did appear to decrease the incidence of acute and late xerostomia in patients undergoing radiation therapy to the head and neck region, however. Other potential agents of interest include hematopoietic colony-stimulating factors [126]. Data in support of their use, however, are limited. Glutamine has also been suggested as a mucosal protectant and is currently being studied in patients undergoing radiation therapy.

The potential role of surgery in the management of the unresectable patient continues to be raised. In general, the radiation therapist and the medical oncologist have managed patients deemed unresectable at presentation, with no further thought given to surgical intervention. Several studies, however, have considered the role of salvage surgical resection after initial nonoperative management for those with locoregionally advanced previously "unresectable" tumors. The first intergroup unresectable trial allowed for such surgical salvage, which was successful in a small percentage of individuals [16].

The second intergroup unresectable study (INT 0126) allowed for both midcourse surgery on the multiagent chemoradiotherapy arm (Fig. 31.2) and for salvage surgery on all three arms if deemed appropriate [17]. Although primary site surgery was performed in only 8% of the patients, an additional 12% of these patients had achieved local control of their primary tumor after radiation therapy alone or after chemoradiotherapy, and subsequently underwent a neck dissection in an effort to achieve disease control in the neck. Thus 20% of these well-defined "unresectable" patients underwent some kind of surgical procedure after initial nonoperative management. The role of neck dissection in patients initially treated nonoperatively has been reviewed extensively [127,128]. It remains a valuable and important role for surgery, even in these patients with advanced "unresectable" disease.

The increasing ability to achieve locoregional disease control has been accompanied by an increasing incidence of distant metastases, an alteration in natural history resulting from treatment success [9,10]. Therapeutic interventions, such as the use of more aggressive or alternative chemotherapeutic agents, may be appropriate, even in induction or adjuvant treatment schedules, with the specific goal of decreasing distant recurrences.

What emerges is a progressive blurring of the distinction between resectable and unresectable patients. Nonoperative interventions, particularly concomitant chemotherapy and radiation, have been successful in achieving locoregional disease control and in allowing surgery to be reserved for salvage therapy, neck disease control, or for specific worrisome indications such as bone involvement. Randomized clinical trials have already suggested equivalent survival when surgery and radiation are compared to chemotherapy and radiation for advanced larynx cancer [50] and advanced cancer of the hypopharynx [129]. Study is underway extending these observations to other primary sites. Allocation of patients at diagnosis to either a "resectable" or an "unresectable" group becomes less critical than a decision about the best initial oncologic intervention and whether that should be surgical or nonsurgical.

References

1. Forastiere, A., Geopfert, H., Goffinet, D., *et al.* (1998). NCCN practice guidelines for head and neck cancer. *Oncology* **12**, 39–145.
2. Mendenhall, W. M., Stringer, S. P., Amdur, R. J., Hinerman, R. W., Moore-Higgs, G. J., and Cassisi, N. J. (2000). Is radiation therapy a preferred alternative to surgery for squamous cell carcinoma of the base of tongue? *J. Clin. Oncol.* **18**, 35–42.
3. Mendenhall, W. M., Amdur, R. J., Stringer, S. P., Villaret, D. B., and Cassisi, N. J. (2000). Radiation therapy for squamous cell carcinoma

of the tonsillar region: A preferred alternative to surgery? *J. Clin. Oncol.* **18**, 2219–2225.
4. Fleming, I. D., Cooper, J. S., Henson, D. E., *et al.* (1997). "AJCC Cancer Staging Manual," 5th Ed. Lippincott-Raven, Philadelphia.
5. Vokes, E. E., Weichselbaum, R. R., Lippman, S. M., and Hong, W. K. (1993). Head and neck cancer. *N. Engl. J. Med.* **328**, 184–194.
6. Dimery, I. W., and Hong W. K. (1993). Overview of combined modality therapies for head and neck cancer. *J. Natl. Cancer Inst.* **85**, 95–111.
7. Adelstein, D. J., Tan, E. H., and Lavertu, P. (1996). Treatment of head and neck cancer: The role of chemotherapy. *Crit. Rev. Oncol. Hematol.* **24**, 97–116.
8. Cooper, J. S., Farnan, N. C., Asbell, S. O., Rotman, M., Marcial, V., Fu, K. K., McKenna, W. G., and Emami, B. (1996). Recursive partitioning analysis of 2105 patients treated in Radiation Therapy Oncology Group studies of head and neck cancer. *Cancer* **77**, 1905–1911.
9. Vokes, E. E., Kies, M. S., Haraf, D.J., Stenson, K., List, M., Humerickhouse, R., Dolan, M. E., Pelzer, H., Sulzen, L., Witt, M. E., Hsieh, Y. C., Mittal, B. B., and Weichselbaum, R. R. (2000). Concomitant chemoradiotherapy as primary therapy for locoregionally advanced head and neck cancer. *J. Clin. Oncol.* **18**, 1652–1661.
10. Adelstein, D. J., Saxton, J. P., Lavertu, P., Rybicki, L. A., Esclamado, R. M., Wood, B. G., Strome, M., and Larto, M. A. (2000). A phase II trial of concurrent chemotherapy (CT) and hyper-fractionated radiation (HRT) in patients with stage IV squamous cell head and neck cancer (SCHNC). *In* "Proceedings of the Fifth International Conference on Head and Neck Cancer," p. 165. San Francisco, CA. [Abstract].
11. Cooper, J. S., Pajak, T. F., Rubin, P., Tupchong, L., Brady, L. W., Leibel, S. A., Laramore, G. E., Marcial V. A., Davis, L. W., and Cox, J. D. (1989). Second malignancies in patients who have head and neck cancer: Incidence, effect on survival and implications based on the RTOG experience. *Int. J. Radiat. Oncol. Biol. Phys.* **17**, 449–456.
12. Day, G. L., Blot, W. J., Shore, R. E., McLaughlin, J. K., Austin, D. F., Greenberg, R. S., Liff, J. M., Preston-Martin, S., Sarkar, S., Schoenberg, J. B., and Fraumeni, J. F., Jr. (1994). Second cancers following oral and pharyngeal cancers: Role of tobacco and alcohol. *J. Natl. Cancer Inst.* **86**, 131–137.
13. Schwartz, L. H., Ozsahin, M., Zhang, G. N., Touboul, E., De Vataire, F., Andolenko, P., Lacau-Saint-Guily, J., Laugier, A., and Schlienger, M. (1994). Synchronous and metachronous head and neck carcinomas. *Cancer* **74**,1933–1938.
14. Adelstein, D. J. (1998). Recent randomized trials of chemoradiation in the management of locally advanced head and neck cancer. *Curr. Opin. Oncol.* **10**, 213–218.
15. Trotti, A. (2000). Toxicity in head and neck cancer: A review of trends and issues. *Int. J. Radiat. Oncol. Biol. Phys.* **47**, 1–12.
16. Haselow, R. E., Warshaw, M.G., Oken, M. M., *et al.* (1990). Radiation alone versus radiation with weekly low dose cis-platinum in unresectable cancer of the head and neck. *In* "Head and Neck Cancer" (W. E. Fee, Jr., H. Goepfert, M. E. Johns, *et al.,* eds.), Vol. II, pp. 279–281. Lippincott, Philadelphia.
17. Adelstein, D. J., Adams, G. L., Li, Y., Wagner, H., Jr., Kish, J. A., Ensley, J. F., Schuller, D. E., and Forastiere, A. A. (2000). INT 0126: An Intergroup phase III comparison of standard radiation therapy (RT) and two schedules of concurrent chemoradiotherapy in patients with unresectable squamous cell head and neck cancer (SCHNC). *In* "Proceedings of the 5th International Conference on Head and Neck Cancer," p. 107. San Francisco, CA. [Abstract]
18. Ang, K. K. (1998). Altered fractionation trials in head and neck cancer. *Semin. Radiat. Oncol.* **8**, 230–236.
19. Mendenhall, W. M., and Parsons, J. T. (1998). Altered fractionation in radiation therapy for squamous-cell carcinoma of the head and neck. *Cancer Invest.* **16**, 594–603.
20. Hu, K. S., and Harrison, L. B. (1999). Altered fractionation in the treatment of head and neck cancer. *Curr. Oncol. Rep.* **1**, 110–123.
21. Pajak, T. F., Laramore, G. E., Marcial, V. A., *et al.* (1991). Elapsed treatment days-a critical item for radiotherapy quality control review in head and neck trials: RTOG report. *Int. J. Radiat. Oncol. Biol. Phys.* **20**, 13–20.
22. Cox, J. D., Pajak, T. F., Marcial, V. A., Coia, L., Mohiuddin, M., Fu, K. K., Selim, H. M., Byhardt, R. W., Rubin, P., Ortiz, H. G., and Martin, L. (1992). Interruptions adversely affect local control and survivalwith hyperfractionated radiation therapy of carcinomas of the upper respiratory and digestive tracts. *Cancer* **69**, 2744–2748.
23. Marcial, V. A., Pajak, T. F., Chang, C., Tupchong, L., and Stetz, J. (1987). Hyperfractionated photon radiation therapy in the treatment of advanced squamous cell carcinoma of the oral cavity, pharynx, larynx, and sinuses, using radiation therapy as the only planned modality: Preliminary report by the Radiation Therapy Oncology Group (RTOG). *Int. J. Radiat. Oncol. Biol. Phys.* **13**, 41–47.
24. Datta, N. R., Choudhry, A. D., and Gupta, S. (1989). Twice a day versus once a day radiation therapy in head and neck cancer. *Int. J. Radiat. Oncol. Biol. Phys.* **17**, 132.
25. Sanchiz, F., Millá, A., Torner, J., *et. al.* (1990). Single fraction per day versus two fractions per day versus radiochemotherapy in the treatment of head and neck cancer. *Int. J. Radiat. Oncol. Biol. Phys.* **19**, 1347–1350.
26. Pinto, L. H. J., Canary, P. C. V., Araújo, C. M. M., Bacelar, S. C., and Souhami, L. (1991). Prospective randomized trial comparing hyperfractionated versus conventional radiotherapy in stages III and IV oropharyngeal carcinoma. *Int. J. Radiat. Oncol. Biol. Phys.* **21**, 557–562.
27. Horiot, J. C., Le Fur, R., N'Guyen, T., Chenal, C., Schraub, S., Alfonsi, S., Gardani, G., Van Den Bogaert, W., Danczak, S., Bolla, M., Van Glabbeke, M., and De Pauw, M. (1992). Hyperfractionation versus conventional fraction in oropharyngeal carcinoma: Final analysis of a randomized trial of the EORTC cooperative group of radiotherapy. *Radiother. Oncol.* **25**, 231–241.
28. Cummings, B., O'Sullivan, B., Keane, T., Pintilie, M., Liu, F., McLean, M., Payne, D., Waldron, P., Warde, P., Gullane, P., and the ENT Group (2000). A prospective randomized trial of hyperfractionated versus conventional once daily radiation for advanced squamous cell carcinomas of the larynx and pharynx: 5 year results. *Int. J. Radiat. Oncol. Biol. Phys.* **48** (ASTRO 2000), 151. [Abstract]
29. Maciejewski, B., Skladowski, K., Pilecki, B., Taylor, J. M. G., Withers, R. H., Miszczyk, L., Zajusz, A., and Suwinski, R. (1996). Randomized clinical trial on accelerated 7 days per week fractionation radiotherapy for head and neck cancer: Preliminary report on acute toxicity. *Radiother. Oncol.* **40**, 137–145.
30. Overgaard, J., Hansen, H. S., Sapru, W., *et al.* (1996). Conventional radiotherapy as the primary treatment of squamous cell carcinoma (SCC) of the head and neck: A randomized multicenter study of 5 versus 6 fractions per week. *Radiother. Oncol.* **40**, S31.
31. Jackson, S. M., Weir, L. M., Hay, J. H., *et al.* (1997). A randomised trial of accelerated versus conventional radiotherapy in head and neck cancer. *Radiother. Oncol.* **43**, 39–46.
32. Horiot, J. C., Bontemps, P., van den Bogaert, W., *et al.* (1997). Accelerated fractionation (AF) compared to conventional fractionation (CF) improves locoregional control in the radiotherapy of advanced head and neck cancers: Results of the EORTC 22851 randomized trial. *Radiother. Oncol.* **44**, 111–121.
33. Dische, S., Saunders, M., Barrett, A., *et al.* (1997). A randomised multicentre trial of CHART versus conventional radiotherapy in head and neck cancer. *Radiother. Oncol.* **44**, 123–136.
34. Bourhis, J., Lapeyre, M., Tortochaux, J., *et al.* (2000). Very accelerated versus conventional radiotherapy in HNSCC: Results of the GORTEC 94–02 randomized trial. *Int. J. Radiat. Oncol. Biol. Phys.* **48** (ASTRO 2000), 111. [Abstract]
35. Fu, K. K., Pajak, T. F., Trotti, A., *et al.* (2000). A Radiation Therapy Oncology Group (RTOG) phase III randomized study to compare hyperfractionation and two variants of accelerated fractionation to standard

fractionation radiotherapy for head and neck squamous cell carcinomas: First report of RTOG 9003. *Int. J. Radiat. Oncol. Biol. Phys.* **48**, 7–16.
36. Mercier, R. J., Neal, G. D., Mattox, D. E., *et al.* (1987). Cisplatin and 5-fluorouracil chemotherapy in advanced or recurrent squamous cell carcinoma of the head and neck. *Cancer* **60**, 2609–2612.
37. Jacobs, C., Lyman, G., Velez-Garcia, E., *et al.* (1992). A phase III randomized study comparing cisplatin and fluorouracil as single agents and in combination for advanced squamous cell carcinoma of the head and neck. *J. Clin. Oncol.* **10**, 257–263.
38. Forastiere, A. A., Metch, B., Schuller, D. E., *et al.* (1992). Randomized comparison of cisplatin plus fluorouracil and carboplatin plus fluorouracil versus methotrexate in advanced squamous-cell carcinoma of the head and neck: A Southwest Oncology Group Study. *J. Clin. Oncol.* **10**, 1245–1251.
39. Clavel, M., Vermorken, J. B., Cognetti, F., *et al.* (1994). Randomized comparison of cisplatin, methotrexate, bleomycin and vincristine (CABO) versus cisplatin and 5-fluorouracil (CF) versus cisplatin (C) in recurrent or metastatic squamous cell carcinoma of the head and neck. *Ann. Oncol.* **5**, 521–526.
40. Hussain, M., Gadgeel, S., Kucuk, O., *et al.* (1999). Paclitaxel, cisplatin, and 5-fluorouracil for patients with advanced or recurrent squamous cell carcinoma of the head and neck. *Cancer* **86**, 2364–2369.
41. Shin, D. M., Glisson, B. S., Khuri, F. R., *et al.* (1998). Phase II trial of paclitaxel, ifosfamide, and cisplatin in patients with recurrent head and neck squamous cell carcinoma. *J. Clin. Oncol.* **16**, 1325–1330.
42. Schrijvers, D., Van Herpen, C., Kerger, J., *et al.* (1999). Phase I–II study with docetaxel (D), cisplatin (C) and 5-fluorouracil (5-FU) in patients (PTS) with locally advanced inoperable squamous cell carcinoma of the head and neck (SCCHN). *Proc. Am. Soci. Clin. Oncol.* **18**, 396a. [Abstract]
43. Randolph, V. L., Vallejo, A., Spiro, R. H., *et al.* (1978). Combination therapy of advanced head and neck cancer: Induction of remissions with diamminedichloroplatinum(II), bleomycin and radiation therapy. *Cancer* **41**, 460–467.
44. Brown, A. W., Blom, J., Butler, W. M., *et al.* (1980). Combination chemotherapy with vinblastine, bleomycin, and cis-diamminedichloroplatinum (II) in squamous cell carcinoma of the head and neck. *Cancer* **45**, 2830–2835.
45. Jacobs, C. (1991). Adjuvant and neoadjuvant treatment of head and neck cancers. *Semin. Oncol.* **18**, 504–514.
46. Stupp, R., Weichselbaum, R. R., and Vokes, E. E. (1994). Combined modality therapy of head and neck cancer. *Semin. Oncol.* **21,** 349–358.
47. Rooney, M., Kish, J., Jacobs, J., *et al.* (1985). Improved complete response rate and survival in advanced head and neck cancer after three-course induction therapy with 120-hour 5-FU infusion and cisplatin. *Cancer* **55**, 1123–1128.
48. Toohill, R. J., Anderson, T., Byhardt, R. W., *et al.* (1987). Cisplatin and fluorouracil as neoadjuvant therapy in head and neck cancer. *Arch. Otolaryngol. Head Neck Surg.* **113**, 758–761.
49. Martin, M., Hazan, A., Vergnes, L., *et al.* (1990). Randomized study of 5-fluorouracil and cisplatin as neoadjuvant therapy in head and neck cancer: A preliminary report. *Int. J. Radiat. Oncol. Biol. Phys* . **19**, 973–975.
50. Wolf, G. T., Hong, W. K., Fisher, S. G., *et al.* (1991). Induction chemotherapy plus radiation compared with surgery plus radiation in patients with advanced laryngeal cancer. *N. Engl. J. Med.* **324**, 1685–1690.
51. Paccagnella, A., Orlando, A., Marchiori, C., *et al.* (1994). Phase III trial of initial chemotherapy in stage III or IV head and neck cancers: A study by the Gruppo di Studio sui Tumori della Testa e del Collo. *J. Natl. Cancer Inst.* **86**, 265–272.
52. Athanasiadis, I., Taylor, S., Vokes, E. E., *et al.* (1997). Phase II study of induction and adjuvant chemotherapy for squamous cell carcinoma of the head and neck. *Cancer* **79**, 588–594.
53. Adelstein, D. J. (1999). Induction chemotherapy in head and neck cancer. *Hematol./Oncol. Clin. North Am.* **13**, 689–698.
54. Clark, J. R., and Frei, E. (1989). Chemotherapy for head and neck cancer: Progress and controversy in the management of patients with M_0 disease. *Semin. Oncol.* **16**(Suppl. 6), 44–57.
55. Urba, S. G., and Forastiere, A. A. (1989). Systemic therapy of head and neck cancer: Most effective agents, areas of promise. *Oncology* **3**, 79–88.
56. Hill, B. T., and Price, L. A. (1990). The role of adjuvant chemotherapy in the treatment of advanced head and neck cancer. *Acta. Oncol.* **29**, 695–703.
57. Cognetti, F., Pinnaro, P., Carlini, P., *et al.* (1988). Neoadjuvant chemotherapy in previously untreated patients with advanced head and neck squamous cell cancer. *Cancer* **62**, 251–261.
58. Vogl, S. E., Lerner, H., Kaplan, B. H., *et al.* (1982). Failure of effective initial chemotherapy to modify the course of stage IV (M_0) squamous cancer of the head and neck. *Cancer* **50**, 840–844.
59. Ervin, T. J., Clark, J. R., Weichselbaum, R. R., *et al.* (1987). An analysis of induction and adjuvant chemotherapy in the multidisciplinary treatment of squamous cell carcinoma of the head and neck. *J. Clin. Oncol.* **5**, 10–20.
60. Decker, D. A., Drelichman, A., Jacobs, J., *et al.* (1983). Adjuvant chemotherapy with cis-diamminodichloroplatinum II and 120-hour infusion 5-fluorouracil in stage III and IV squamous cell carcinoma of the head and neck. *Cancer* **51**, 1353–1355.
61. Vokes, E. E., Schilsky, R. L., Weichselbaum, R. R., *et al.* (1990). Induction chemotherapy with cisplatin, fluorouracil, and high-dose leucovorin for locally advanced head and neck cancer: A clinical and pharmacologic analysis. *J. Clin. Oncol.* **8**, 241–247.
62. Pfister, D. G., Bajorin, D., Motzer, R., *et al.* (1994). Cisplatin, fluorouracil, and leucovorin: Increased toxicity without improved response in squamous cell head and neck cancer. *Arch. Otolaryngol. Head Neck Surg.* **120**, 89–95.
63. Clark, J. R., Busse, P. M., Norris, C. M., *et al.* (1997). Induction chemotherapy with cisplatin, fluorouracil, and high-dose leucovorin for squamous cell carcinoma of the head and neck: Long-term results. *J. Clin. Oncol.* **15**, 3100–3110.
64. Al-Sarraf, M., Kish, J. A., and Ensley, J. F. (1991). The Wayne State University experience with adjuvant chemotherapy of head and neck cancer. *Hematol./Oncol. Clin. North Am.* **5**, 687–700.
65. Brockstein, B. E., and Vokes, E. E. (1996). Chemoradiotherapy for head and neck cancer. *PPO Updates/Principles Pract. Oncol.* **10(9)**.
66. Tannock, I. F., and Browman, G. (1986). Lack of evidence for a role of chemotherapy in the routine management of locally advanced head and neck cancer. *J. Clin. Oncol.* **4**, 1121–1126.
67. Clark, J. R., Fallon, B. G., Dreyfuss, A. L., *et al.* (1988). Chemotherapeutic strategies in the multidisciplinary treatment of head and neck cancer. *Semin. Oncol.* **15**(Suppl. 3), 35–44.
68. Posner, M. R., Weichselbaum, R. R., Fitzgerald, T. J., *et al.* (1985). Treatment complications after sequential combination chemotherapy and radiotherapy with or without surgery in previously untreated squamous cell carcinoma of the head and neck. *Int. J. Radiat. Oncol. Biol. Phys.* **11**, 1887–1893.
69. Ensley, J. F., Jacobs, J. R., Weaver, A., *et al.* (1984). Correlation between response to cisplatinum-combination chemotherapy and subsequent radiotherapy in previously untreated patients with advanced squamous cell cancers of the head and neck. *Cancer* **54**, 811–814.
70. Ensley, J.F., and Maciorowski, Z. (1994). Clinical applications of DNA content parameters in patients with squamous cell carcinomas of the head and neck. *Semin. Oncol.* **21**, 330–339.
71. Al-Sarraf, M. (1988). Head and neck cancer: Chemotherapy concepts. *Semin. Oncol.* **15**, 70–85.
72. Fu, K. K. (1997). Combined-modality therapy for head and neck cancer. *Oncology* **11**, 1781–1796.
73. Fazekas, J. T., Sommer, C., and Kramer, S. (1980). Adjuvant intravenous methotrexate or definitive radiotherapy alone for advanced squamous cancers of the oral cavity oropharynx, supraglottic larynx or hypopharynx. *Int. J. Radiat. Oncol. Biol. Phys.* **6**, 533–541.

74. Stell, P. M., Dalby, J. E., Strickland, P., *et al.* (1983). Sequential chemotherapy and radiotherapy in advanced head and neck cancer. *Clin. Radiol.* **34**, 463–467.
75. Szpirglas, H., Nizri, D., Marneur, M., *et al.* (1987). Neo-adjuvant chemotherapy: A randomized trial before radiotherapy in oral and oropharyngeal carcinomas: End results. *In* "Proceedings of the Second International Head and Neck Oncology Research Conference," pp. 261–264. Kugler Publications, Amsterdam.
76. Jaulerry, C., Rodriguez, J., Brunin, F., *et al.* (1992). Induction chemotherapy in advanced head and neck tumors: Results of two randomized trials. *Int. J. Radiat. Oncol. Biol. Phys.* **23**, 483–489.
77. Tejedor, M., Murias, A., Soria, P., *et al.* (1992). Induction chemotherapy with carboplatin and ftorafur in advanced head and neck cancer. *Am. J. Clin. Oncol.* **15**, 417–421.
78. Domenge, C., Coche-Dequeant, B., Wibault, P., *et al.* (1996). Randomized trial of neoadjuvant chemotherapy before radiotherapy in oropharyngeal carcinoma. *In* "Proceedings of the Fourth International Conference on Head and Neck Cancer," p. 99. Toronto, Canada.[Abstract]
79. Harari, P. M. (1997). Why has induction chemotherapy for advanced head and neck cancer become a United States community standard of practice? *J. Clin. Oncol.* **15**, 2050–2055.
80. Merlano, M., Corvo, R., Margarino, G., *et al.* (1991). Combined chemotherapy and radiation therapy in advanced inoperable squamous cell carcinoma of the head and neck. *Cancer* **67**, 915–921.
81. Merlano, M., Benasso, M., Corvó, R., *et al.* (1996). Five-year update of a randomized trial of alternating radiotherapy and chemotherapy compared with radiotherapy alone in treatment of unresectable squamous cell carcinoma of the head and neck. *J. Natl. Cancer Inst.* **88**, 583–589.
82. Lo, T. C. M., Wiley, A. L., Ansfield, F. J., *et al.* (1976). Combined radiation therapy and 5-fluorouracil for advanced squamous cell carcinoma of the oral cavity and oropharynx: A randomized study. *Am. J. Roentgenol.* **126**, 229–235.
83. Browman, G. P., Cripps, C., Hodson, D. I., *et al.* (1994). Placebo-controlled randomized trial of infusional fluorouracil during standard radiotherapy in locally advanced head and neck cancer. *J. Clin. Oncol.* **12**, 2648–2653.
84. Gupta, N. K., Pointon, R. C. S., and Wilkinson, P. M. (1987). A randomized clinical trial to contrast radiotherapy with radiotherapy and methotrexate given synchronously in head and neck cancer. *Clin. Radiol.* **38**, 575–581.
85. Shanta, V., and Krishnamurthi, S. (1980). Combined bleomycin and radiotherapy in oral cancer. *Clin. Radiol.* **31**, 617–620.
86. Vermund, H., Kaalhus, O., Winther, F., *et al.* (1985). Bleomycin and radiation therapy in squamous cell carcinoma of the upper aero-digestive tract: A phase III clinical trial. *Int. J. Radiat. Oncol. Biol. Phys.* **11**, 1877–1886.
87. Fu, K. K., Phillips, T. L., Silverberg, I. J., *et al.* (1987). Combined radiotherapy and chemotherapy with bleomycin and methotrexate for advanced inoperable head and neck cancer: Update of a Northern California Oncology Group randomized trial. *J. Clin. Oncol.* **5**, 1410–1418.
88. Eschwege, F., Sancho-Garnier, H., Gerard, J. P., *et al.* (1988). Ten-year results of randomized trial comparing radiotherapy and concomitant bleomycin to radiotherapy alone in epidermoid carcinomas of the oropharynx: Experience of the European Organization for Research and Treatment of Cancer. *N.C.I. Monogr.* **6**, 275–278.
89. Weissberg, J. B., Son, Y. H., Papac, R. J., *et al.* (1989). Randomized clinical trial of mitomycin C as an adjunct to radiotherapy in head and neck cancer. *Int. J. Radiat. Oncol. Biol. Phys.* **17**, 3–9.
90. Dombrowsky, W., Naud, J. J., Widder, J., *et al.* (1998). Continuous hyperfractionated accelerated radiotherapy with/without mitomycin C in head and neck cancer. *Int. J. Radiat. Oncol. Biol. Phys.* **42**, 803–806.
91. Jeremic, B., Shibamato, Y., Stanisavljevic, B., *et al.* (1997). Radiation therapy alone or with concurrent low-dose daily either cisplatin or carboplatin in locally advanced unresectable squamous cell carcinoma of the head and neck: A prospective randomized trial. *Radiat. Oncol.* **43**, 29–37.
92. Jeremic, B., Shibamoto, Y., Milicic, B., *et al.* (2000). Hyperfractionated radiation therapy with or without concurrent low-dose daily cisplatin in locally advanced squamous cell carcinoma of the head and neck: A prospective randomized trial. *J. Clin. Oncol.* **18**, 1458–1464.
93. Gabriele, P., Tessa, M., Ragona, R., *et al.* (1996). An interim analysis of phase III study on radiotherapy (RT) versus RT plus carboplatin (CBDCA) in inoperable stage III-IV head and neck (H&N) carcinoma. *In* "Proceedings of the Fourth International Conference on Head and Neck Cancer," p. 116. Toronto, Canada. [Abstract]
94. Weissler, M. C., Melin, S., Sailer, S. L., *et al.* (1992). Simultaneous chemoradiation in the treatment of advanced head and neck cancer. *Arch. Otolaryngol. Head Neck Surg.* **118**, 806–810.
95. Keane, T. J., Cummings, B. J., O'Sullivan, B., *et al.* (1993). A randomized trial of radiation therapy compared to split course radiation therapy combined with mitomycin C and 5-fluorouracil as initial treatment of advanced laryngeal and hypopharyngeal squamous carcinoma. *Int. J. Radiat. Oncol. Biol. Phys.* **25**, 613–618.
96. Smid, L., Lesnicar, H., Zakotnik, B., *et al.* (1995). Radiotherapy, combined with simultaneous chemotherapy with mitomycin C and bleomycin for inoperable head and neck cancer: Preliminary report. *Int. J. Radiat. Oncol. Biol. Phys.* **32**, 769–775.
97. Wendt, T. G., Grabenbauer, G. G., Rödel, C. M., *et al.* (1998). Simultaneous radiochemotherapy versus radiotherapy alone in advanced head and neck cancer: A randomized multicenter study. *J. Clin. Oncol.* **16**, 1318–1324.
98. Brizel, D. M., Albers, M. E., Fisher, S. R., *et al.* (1998). Hyperfractionated irradiation with or without concurrent chemotherapy for locally advanced head and neck cancer. *N. Engl. J. Med.* **338**, 1798–1804.
99. Calais, G., Alfonsi, M., Bardet, E., *et al.* (1999). Randomized trial of radiation therapy versus concomitant chemotherapy and radiation therapy for advanced-stage oropharynx carcinoma. *J. Natl. Cancer Inst.* **91**, 2081–2086.
100. Taylor, S. G., Murthy, A. K., Vannetzel, J. M., *et al.* (1994). Randomized comparison of neoadjuvant cisplatin and fluorouracil infusion followed by radiation versus concomitant treatment in advanced head and neck cancer. *J. Clin. Oncol.* **12**, 385–395.
101. Pinnaro, P., Cercato, M. C., Giannarelli, D., *et al.* (1994). A randomized phase II study comparing sequential versus simultaneous chemoradiotherapy in patients with unresectable locally advanced squamous cell cancer of the head and neck. *Ann. Oncol.* **5**, 513–519.
102. Stell, P. M. (1992). Adjuvant chemotherapy in head and neck cancer. *Semin. Radiat. Oncol.* **2**, 195–202.
103. Browman, G. P. (1994). Evidence-based recommendations against neoadjuvant chemotherapy for routine management of patients with squamous cell head and neck cancer. *Cancer Invest.* **12**, 662–671.
104. El-Sayed, S., and Nelson, N. (1996). Adjuvant and adjunctive chemotherapy in the management of squamous cell carcinoma of the head and neck region: A meta-analysis of prospective and randomized trials. *J. Clin. Oncol.* **14**, 838–847.
105. Pignon, J. P., Bourhis, J., Domenge, C., *et al.* (2000). Chemotherapy added to locoregional treatment for head and neck squamous-cell carcinoma: Three meta-analyses of updated individual data. *Lancet* **355**, 949–955.
106. Lavertu, P., Adelstein, D. J., Saxton, J. P., *et al.* (1999). Aggressive concurrent chemoradiotherapy for squamous cell head and neck cancer. *Arch. Otolaryngol. Head Neck Surg.* **125**, 142–148.
107. Buentzel, J., Glatzel, M., Kuettner, K., *et al.* (2000). Late toxicities due to multimodal treatment of head and neck cancer (HNC). *Proc. Am. Soc. Clin. Oncol.* **19**, 413a. [Abstract]
108. Mekhail, T. M., Adelstein, D. J., Rybicki, L. A., *et al.* (1998). Enteral nutrition during head and neck cancer treatment: Is percutaneous

endoscopic gastrostomy (PEG) preferable to a nasogastric (NG) tube? *Proc. Am. Soc. Clin. Oncol.* **17**, 44a. [Abstract]
109. Forastiere, A. A., Shank, D., Neuberg, D., *et al.* (1998). Final report of a phase II evaluation of paclitaxel in patients with advanced squamous cell carcinoma of the head and neck. *Cancer* **82**, 2270–2274.
110. Colevas, A. D., and Posner, M. R. (1998). Docetaxel in head and neck cancer: A review. *Am. J. Clin. Oncol.* **21,** 482–486.
111. Colevas, A. D., Tishler, R., Fried, M., *et al.* (2000). A phase I/II study of outpatient docetaxel (Taxotere), cisplatin, 5-FU, and leucovorin as induction chemotherapy for patients with squamous cell carcinoma of the head and neck. *Proc. Am. Soc. Clin. Oncol.* **19**, 420a. [Abstract]
112. Humerickhouse, R., Haraf, D., Brockstein, B., *et al.* (1999). Phase I study of gemcitabine, 5-FU, and paclitaxel with RT in recurrent or advanced head and neck cancer. *Proc. Am. Soc. Clin. Oncol.* **18**, 396a. [Abstract]
113. Airoldi, M., Marchionatti, S., Pedani, F., *et al.* (1999). Docetaxel (TXT) + vinorelbine (VNB) in recurrent (R) heavily pre-treated head and neck cancer (HNC) patients (pts). *Proc. Am. Soc. Clin. Oncol.* **18**, 402a. [Abstract]
114. Humerickhouse, R. A., Haraf, D., Stenson, K., *et al.* (2000). Phase I study of irinotecan (CPT-11), 5-FU, and hydroxyurea with radiation in recurrent or advanced head and neck cancer. *Proc. Am. Soc. Clin. Oncol.* **19**, 418a. [Abstract]
115. Robbins, K. T., Kumar, P., Regine, W. F., *et al.* (1997). Efficacy of targeted supradose cisplatin and concomitant radiation therapy for advanced head and neck cancer: The Memphis experience. *Int. J. Radiat. Oncol. Biol. Phys.* **38**, 263–271.
116. Kumar, P., Harris, J., Robbins, K. T., *et al.* (2000). The feasibility of using intraarterial cisplatin and radiation therapy for stage IV-T4 head/neck (H/N) squamous cell carcinoma in a multi-institutional setting: Preliminary results of Radiation Therapy Oncology Group (RTOG) Trial 9615. *Int. J. Radiat. Oncol. Biol. Phys.* **48** (ASTRO 2000), 152.[Abstract]
117. Clayman, G. L., El-Naggar, A. K., Lippman, S. M., *et al.* (1998). Adenovirus-mediated p53 gene transfer in patients with advanced recurrent head and neck squamous cell carcinoma. *J. Clin. Oncol.* **16**, 2221–2232.
118. Kirn, D., Nemunaitis, J., Ganly, I., *et al.* (1998). A phase II trial of intratumoral injection with an E1B-deleted adenovirus, ONYX-015, in patients with recurrent, refractory head and neck cancer. *Proc. Am. Soc. Clin. Oncol.* **17**, 391a.[Abstract]
119. Clayman, G. L., and Dreiling, L. (1999). Local delivery for gene therapy. *Curr. Oncol. Rep.* **1**, 138–143.
120. Ezekiel, M. P., Bonner, J. A., Robert, F., *et al.* (1999). Phase I trial of chimerized anti-epidermal growth factor receptor (anti-EGFr) antibody in combination with either once-daily or twice daily irradiation for locally advanced head and neck malignancies. *Proc. Am. Soc. Clin. Oncol.* **18**, 388a. [Abstract]
121. Mendelsohn, J., Shin, D. M., Donato, N., *et al.* (1999). A phase I study of chimerized anti-epidermal growth factor receptor (EGFr) monoclonal antibody, C225, in combination with cisplatin (CDDP) in patients (PTS) with recurrent head and neck squamous cell carcinoma (SCC). *Proc. Am. Soc. Clin. Oncol.* **18**, 389a. [Abstract]
122. Hanna, E., and Sherman, A. C. (1999). Quality-of-life issues in head and neck cancer. (1999). *Curr. Oncol. Rep.* **1**, 124–128.
123. Hensley, M. L., Schuchter, L. M., Lindley, C., *et al.* (1999). American Society of Clinical Oncology Clinical Practice Guidelines for the use of chemotherapy and radiotherapy protectants. *J. Clin. Oncol.* **17**, 3333–3355.
124. Ozer, H., Armitage, J. O., Bennett, C. L., *et al.* (2000). 2000 update of recommendations for the use of hematopoietic colony-stimulating factors: Evidence-based, clinical practice guidelines. *J. Clin. Oncol.* **18**, 3558–3585.
125. Brizel, D. M., Wasserman, T. H., Henke, M., *et al.* (2000). Phase III randomized trial of amifostine as a radioprotector in head and neck cancer. *J. Clin. Oncol.* **18**, 3339–3345.
126. Chi, K. H., Chen, C. H., Chan, W. K., *et al.* (1995). Effect of granulocyte-macrophage colony-stimulating factor on oral mucositis in head and neck cancer patients after cisplatin, fluorouracil, and leucovorin chemotherapy. *J. Clin. Oncol.* **13,** 2620–2628.
127. Mendenhall, W. M., Parsons, J. T., Stringer, S.P., *et al.* (1992). Squamous cell carcinoma of the head and neck treated with irradiation: Management of the neck. *Semin. Radiat. Oncol.* **2**, 163–170.
128. Lavertu, P., Adelstein, D. J., Saxton, J. P., *et al.* (1997). Management of the neck in a randomized trial comparing concurrent chemotherapy and radiotherapy with radiotherapy alone in resectable stage III and IV squamous cell head and neck cancer. *Head Neck* **19**, 559–566.
129. Lefebvre, J. L., Chevalier, D., Luboinski, B., *et al.* (1996). Larynx preservation in pyriform sinus cancer: Preliminary results of a European Organization for Research and Treatment of Cancer phase III trial. *J. Natl. Cancer Inst.* **88**, 890–899.

CHAPTER

32

Organ Preservation in Head and Neck Cancer

ARLENE A. FORASTIERE and MAURA GILLISON

Department of Oncology
Johns Hopkins University School of Medicine
Baltimore, Maryland 21287

I. Introduction 475
II. Laryngeal Preservation: Cancers of the Larynx and Hypopharynx 475
III. Role of Chemotherapy in Preservation of the Oropharynx 478
A. Clinical Trials Demonstrating Benefit of Chemoradiation in Patients with Advanced Oropharyngeal Cancer 479
B. Clinical Trials of Conventional Fractionation Radiotherapy with or without Concurrent, Single-Agent Chemotherapy 480
C. Randomized Clinical Trials of Conventional Fractionation Radiation Therapy with or without Concomitant, Multiagent Chemotherapy 481
D. Randomized Clinical Trials of Conventional Fractionation and Split-Course Concurrent Chemoradiation or Alternating Chemotherapy and Radiation 484
E. Clinical Trials of Altered Fractionation Schedules of Radiation Therapy with or without Chemotherapy 484
F. Summary 488
References 489

I. INTRODUCTION

The integration of chemotherapy into the initial management of locally advanced cancers of the head and neck has been under study since the 1970s. Over the course of nearly three decades, the design of these trials has shifted from simply adding chemotherapy to radical surgery and radiotherapy with the goal of improved survival to focus on organ preservation or, more precisely, organ function conservation. This can be achieved by combining chemotherapy and radiotherapy and reserving surgery to manage the neck, when indicated. Two nonsurgical strategies that have been tested extensively are (1) induction chemotherapy followed by radiotherapy in responding patients and (2) concurrent radiotherapy and chemotherapy. Data from uncontrolled feasibility studies and prospective randomized trials are available to guide the appropriate use of nonoperative therapies for squamous cell cancers originating in the larynx, hypopharynx, and oropharynx. Although the distinction between resectable and unresectable disease has become somewhat blurred, organ function preservation, by definition, applies to those situations in which the alternative treatment strategy is surgery that results in substantial impairment of speech or swallowing function.

II. LARYNGEAL PRESERVATION: CANCERS OF THE LARYNX AND HYPOPHARYNX

Induction chemotherapy trials utilizing cisplatin-based combination chemotherapy for several cycles followed by surgery and radiotherapy were designed in the late 1970s and early 1980s to improve local-regional control and survival. These trials were stimulated by the observation that patients with newly diagnosed, untreated disease demonstrated rapid reduction of tumor in response to cisplatin-based chemotherapy. It was noted that not only were the highest rates of clinical complete and partial response occurring with larynx cancer, but larynges were being removed that had no evidence of residual squamous cell cancer [1].

Cisplatin (100 mg/m^2 day 1) and infusional 5-fluorouracil (5-FU) (1000 mg/m^2/day, days 1–5) is the standard treatment regimen and results in an overall response rate of 85% and clinical complete response ranging from 30 to 50% [2–4]. Pathologic complete response confirmed by biopsy or resection occurs in two-thirds of clinical complete responders. Furthermore, when sequenced with radiotherapy, the response to chemotherapy is predictive of radiosensitivity. These observations led investigators to conduct feasibility trials of induction chemotherapy followed by radiotherapy to preserve the larynx. Surgery was reserved for salvage of chemotherapy nonresponders or for management of persistent or recurrent disease. The results of these small, uncontrolled trials suggested that the larynx could be preserved without compromise in survival [5–7].

The second strategy for organ preservation is the administration of chemotherapy concurrent with radiotherapy to take advantage of radiation-enhancing properties of cisplatin, 5-FU, and other cytotoxics active against head and neck cancer. This strategy has been investigated in the treatment of unresectable head and neck cancer for decades. Recently completed trials have demonstrated significant improvement in local-regional control, disease-free survival, or overall survival when compared to radiotherapy alone in unresectable cancers and in site-specific trials of advanced cancers of the nasopharynx and the oropharynx [8–12].

Published data with long follow-up are available using the induction chemotherapy strategy to preserve the larynx for resectable stage III and IV cancers of the larynx and stages II, III, and IV cancers of the hypopharynx [13,14] (Table 32.1). A large randomized trial evaluating concurrent chemotherapy and radiotherapy in advanced larynx cancer has been completed by the U.S. Intergroup and the results reported in preliminary abstract form [15]; a similar trial in hypopharynx cancer is in progress in Europe. The results of these trials form the basis for current treatment guidelines and will be reviewed.

Two multicenter trials for resectable stage III and IV larynx cancers have been completed in the United States and both serve as landmark studies [13,15]. The first, conducted by the Department of Veterans Affairs Laryngeal Study Group, directly compared the standard of care, total laryngectomy followed by radiotherapy, to an organ preservation approach consisting of induction cisplatin/5-FU for a maximum of three courses followed by 70 Gy of radiotherapy [16]. This trial showed that preservation of the larynx was successful in 62% of surviving patients treated with induction chemotherapy and there was no significant difference in survival of this group compared to those in the surgery treatment group. The details of the trial and results have been analyzed extensively and published [13,16–19]. A total of 332 patients with stage III or IV squamous cell

TABLE 32.1 Laryngeal Preservation Trials

Group	No. of patients	Site T stage		Treatment[a]	Survival		Laryngeal preservation in survivors
VACSP	332	Supraglotic	63%		3 years	10 years	5 years
		Glottic	37%	PF→RT	53%	25%	62%
		T1–2	9%	S→RT	56%	30%	
		T3	65%				
		T4	26%				
INT R91–11	547	Supraglotic	69%		2 years		2 years[b]
		Glottic	31%	PF→RT	76%		58% (74%)
		T2	11%	RT+P	76%		68% (88%)[c]
		T3	79%	RT	77%		53% (69%)
		T4	10%				
EORTC	202	Epilarynx	22%		3 years		3 years
		Hypopharynx	78%	PF→RT	57%		48%
		T2	19%	S→RT	43%		
		T3	75%				
		T4	6%				
MACH-NC meta-analysis	602	Larynx	73%		5 years		5 years
		Hypopharynx	26%	PF→RT	39%		58%
		T1–2	12%	S→RT	45%		
		T3–4	77%				

[a]PF, cisplatin +5-FU × 2–3 cycles; RT, radiotherapy, S, surgery.
[b]Larynx preserved at last follow-up or time of death.
[c]Statistically significant difference.

cancer of the larynx were enrolled. One-third had glottic cancer and two-thirds had supraglottic cancer; 57% of patients had stage III disease and 43% stage IV. The study was designed to utilize laryngectomy as salvage treatment for patients in the organ preservation arm whose response after two courses of cisplatin/5-FU was less than partial or for persistent or recurrent disease following completion of radiotherapy.

A total of 166 patients were randomized to the organ preservation treatment arm. Of these patients, 78% proceeded to radiotherapy whereas 16% underwent laryngectomy because of poor response to chemotherapy and 6% died or refused further therapy. Biopsies taken in 101 patients after the third course of chemotherapy documented a 63% pathologic complete response to induction chemotherapy. Five- and ten-year survival rates were 46 and 30% for the surgery group and 42 and 25% for the chemotherapy organ preservation group. An important finding in this trial was a difference in the pattern of first failure for the two treatments. More local recurrences (12% vs 2%, $p=0.001$) and fewer distant metastases (11% vs 17%, $p=0.001$) were observed in the organ preservation treatment group as a site of first failure. Factors predicting response to chemotherapy and successful organ preservation were a lower T stage (T1–3 vs T4), p53 overexpression, and elevated proliferating cell nuclear antigen [17].

The Veterans Administration trial included assessments of functional outcomes related to communication, swallowing, and eating [20]. These assessments showed very clearly that patients who had a preserved larynx had significantly better speech intelligibility, reading rates, and communication profiles across 6-, 12-, and 24-month time points following randomization. Only a small percentage of patients undergoing total laryngectomy developed usable esophageal speech (6%) or remained nonvocal (8%), whereas the majority communicated with the aid of an artificial electrolarynx (55%) or tracheo-esophageal speech (31%). No significant differences were found in swallowing function. These favorable results for preservation of speech without decrement in survival led to the acceptance of induction chemotherapy with cisplatin and 5-fluorouracil followed by radiation therapy as a standard of care alternative to total laryngectomy for the management of locally advanced larynx cancer.

The precise contribution of chemotherapy, however, was not addressed by this trial and remained a contentious issue. Outside of the United States, Canada and Great Britain in particular, radiotherapy as a single modality is the treatment of choice for larynx preservation [21], but this approach is generally reserved for selected patients with T3–4, N0–1 larynx cancers, a less advanced patient population than those enrolled in the Veterans Administration trial. To formally address this question of the contribution of chemotherapy and also to address the question of optimal sequencing of chemotherapy and radiotherapy, the U.S. Intergroup embarked on a three-arm follow-up trial, which has recently undergone a first analysis [15].

The Intergroup trial, designated R91-11, had three treatment groups into which 547 patients were randomized. The control arm consisted of induction cisplatin and 5-fluorouracil followed by radiotherapy and was identical to the organ preservation arm of the Veterans Administration trial. The two experimental arms were (1) radiotherapy and concurrent cisplatin (100 mg/m^2) administered on days 1, 22, and 43 and (2) radiotherapy alone. The radiotherapy was the same in all treatment groups, a total of 70 Gy delivered using once-daily factions of 200 cGy. Laryngectomy was reserved for salvage of treatment failures in all groups. Because R91-11 was a study of nonoperative treatment of larynx cancer, eligible patients had less advanced disease than those in the Veterans Administration trial. Patients with high-volume T4 disease (tumor penetrating through cartilage or greater than 1 cm into the base of tongue) were not eligible. The characteristics of the enrolled patients (80% T3 and 72% N0–1 stage) showed that this trial was mainly an evaluation of T3, N0–1 disease. The key findings of the trial were the following: (1) a 2-year survival rate of 76%, which did not differ by treatment; (2) no significant difference in larynx preservation rates for patients in the induction treatment group compared to treatment with radiotherapy alone; (3) significantly higher rates of larynx preservation after treatment with radiotherapy and concurrent chemotherapy compared to either induction chemotherapy or radiotherapy alone; and (4) a significant reduction in distant metastases for patients receiving chemotherapy compared to those in the radiotherapy alone treatment group.

These results will influence treatment guidelines [22] for advanced cancer of the larynx in the following manner. First, it is now clear that for the purpose of achieving larynx preservation for T3 and low-volume T4 cancers, radiotherapy alone is inferior to radiotherapy combined with chemotherapy and that sequential (induction) chemotherapy and radiotherapy is inferior to concurrent chemotherapy and radiotherapy. Therefore, the standard of care for larynx preservation for patients with T3, N0–3 glottic or supraglottic cancers and low-volume T4 supraglottic cancers should be radiotherapy with concurrent cisplatin, plus management of the neck as dictated by the initial N stage and response to treatment [21,22]. Most patients with T4 glottic cancers and high-volume T4 supraglottic cancers are unlikely to achieve successful preservation of a functioning larynx and generally will be managed with primary surgery.

It is important to note that chemotherapy added to radiotherapy is associated with substantially more toxicity than radiotherapy alone. Therefore, in view of the failure of chemotherapy to improve survival, patients who desire preservation of their voice but have poor performance status or weak psychosocial support systems should be offered radiotherapy alone.

A similar approach to study organ preservation has been followed for cancer of the hypopharynx. The European Organization for Research and Treatment of Cancer (EORTC) conducted a multicenter trial in which 202 patients were randomly assigned treatment with the standard of care, total laryngectomy with partial pharyngectomy and neck dissection, or induction cisplatin and 5-fluorouracil for up to three courses followed by radiotherapy [14]. Eligible patients had stage T2–4, N0–2b resectable, squamous cell cancer of the aryepiglottic fold (lateral epilarynx) or pyriform sinus. This trial demonstrated survival equivalence with an overall survival rate of 47% at 4 years and 18% at 6 years. The rates of survival with a functioning larynx at 4 and 6 years were 23 and 10%, respectively [14,23]. There were no differences in rates of local and regional failure between the two treatment groups; however, the development of distant metastases was delayed in the chemotherapy group. These results closely parallel those of the Veterans Administration Laryngeal Study Group trial and form the basis for treatment guidelines for advanced cancer of the hypopharynx, which has a notably poor prognosis.

In progress in Europe is a follow-up EORTC trial addressing the question of whether organ preservation and survival may be improved with the administration of chemotherapy concurrent with radiotherapy. This trial directly compares (1) induction cisplatin and 5-fluorouracil (standard dosing for up to three courses) followed by radiotherapy and (2) radiotherapy and concurrent cisplatin (100 mg/m^2) on days 1, 22, and 43. Institutions have the option of using conventional or twice-a-day fractionated radiotherapy. A survival benefit and/or improved local-regional control has emerged from a number of recently completed trials of concurrent chemoradiation compared to radiotherapy alone in mixed site trials of resectable and unresectable patients [10–12,24]. These data are encouraging and suggest that the concurrent treatment approach for organ preservation may also emerge as a superior treatment strategy for cancer of the hypopharynx. At present, the evidence-based treatment guidelines recommend induction chemotherapy with cisplatin and 5-fluorouracil followed by radiotherapy in complete responders (per the EORTC trial) for patients with resectable T2–4, N0–3 cancers or enrollment in a multimodality clinical trial [22]. The alternative is laryngopharyngectomy and neck dissection. Radiation alone is limited to selected patients with early T stage disease to preserve function and has no role in the management of more advanced resectable disease. Notably poor survival results (2.4% at 5 years) have been reported using radiotherapy alone to treat more advanced stage disease [21,22] compared with 30% survival rates from surgical series [25,26].

Four meta-analyses of chemotherapy added to local-regional treatment covering the published literature up to 1993 have been reported [27–30]. All four studies demonstrated a small survival advantage from the addition of chemotherapy and this was mainly attributable to those patients receiving chemotherapy concurrent with radiotherapy. In the analysis reported by Pignon and colleagues [27] using individual patient data, a separate analysis of 601 patients enrolled in trials to preserve the larynx showed a larynx preservation rate of 58% of surviving patients at 5 years and a nonsignificant difference in survival (45% for surgery vs 39% for induction chemotherapy and radiotherapy) after a median follow-up of 5.8 years.

It is apparent that there are few well-designed large site-specific trials that can be used to develop new standards of care. The theoretical rationale for concurrent chemotherapy and radiotherapy is well described [31,32]. However, successfully mounting a trial that compares radical surgery with nonoperative management is unlikely to be feasible due to physician biases and patient preferences. Therefore, as new cytotoxic combinations and novel therapeutics come into clinical trials in head and neck cancer, the evaluation of nonoperative therapies must include rigorous assessments of speech and swallowing function and indices of quality of life.

III. ROLE OF CHEMOTHERAPY IN PRESERVATION OF THE OROPHARYNX

The success of laryngeal preservation trials has stimulated interest in extending radiation therapy-based organ preservation strategies to other head and neck sites, particularly the oropharynx. Until recently, the only alternative to surgical resection and its associated morbidities was conventional fractionation external beam radiation therapy. Improvements in the local control of oropharyngeal cancers with primary radiotherapy have been achieved recently with the use of altered fractionation radiotherapy protocols (hyperfractionated and accelerated) [33,34]. In addition, single institutional case series have demonstrated impressive local control rates for cancers of the tonsil and base of tongue with conventional fractionation radiotherapy and iridium brachytherapy boost [35–37]. Local control is further enhanced by concomitant administration of platinum-based chemotherapy and either conventional fractionation [9,38] or hyperfractionated [12,39] radiotherapy. To date, no randomized trials have compared local control or survival achieved with surgical resection of oropharyngeal cancer to that achieved with primary radiation therapy with or without induction or concomitantly administered chemotherapy. It remains the standard in many institutions for patients with resectable, oropharyngeal cancer to be treated with surgery and postoperative radiation therapy. When a nonsurgical approach is considered for a patient with resectable, oropharyngeal cancer to preserve speech and swallowing function, therapeutic options now include conventional fractionation radiation therapy with brachytherapy boost, hyperfractionated radiotherapy, accelerated radiotherapy, or combined modality therapy with conventional or altered fractionation radiotherapy and cisplatin-based chemotherapy.

Local control of tongue base or tonsillar tumors with conventional fractionation radiotherapy has never formally been compared to that achieved with conventional fractionation plus brachytherapy boost. Single institutional studies that have included predominantly T1, T2, and T3 tumors have reported local control rates at 5 years of over 80% [36,37,40,41] and demonstrated improved speech and swallowing function in patients treated with conventional fractionation plus brachytherapy boost compared to surgically treated patients [35,36]. However, similar control rates have been reported in case series of patients treated with hyperfractionated radiotherapy [42], and therefore both of these treatment approaches are acceptable for T1, T2, and most T3 tumors. Factors reportedly associated with poor local control with primary radiation therapy include bone invasion [42], deeply invasive as compared to exophytic tumors [43], and tonsillar tumors that extend into the base of tongue or mobile tongue [37]. Factors associated with poor outcomes with brachytherapy treatment of base of tongue tumors include extension into the hyoid, posterior pharyngeal wall, or preepiglottic space [40]. Surgical resection is therefore preferred in patients with these features. However, surgical morbidities are considerable in this patient population, given that total glossectomy and/or laryngectomy may be required in approximately 20% to achieve negative surgical margins [44]. Local control and survival at 5 years is 25–30% in this patient population in surgical series [42,44].

Concurrent chemoradiation therapy is the standard of care for patients with unresectable, oropharyngeal cancer and a good performance status. This combined modality therapy demonstrated significant improvement in survival in a meta-analysis of individual data from over 10,000 patients participating in trials comparing primary radiation therapy plus chemotherapy to a control arm of standard locoregional therapy alone [27]. After a median follow-up of 5.9 years, chemotherapy reduced the hazard of death by 10% [hazard ratio (HR) 0.90, 95% confidence interval (CI) 0.85–0.94)], which corresponded to an absolute survival benefit of 4% at 2 and 5 years. Both adjuvant (HR 0.98, 95% CI 0.85–1.19) and neoadjuvant (HR 0.95, 95% CI 0.88–1.01) chemotherapy had no effect on overall survival. However, concomitantly administered chemotherapy resulted in a 19% reduction in risk of death (HR 0.81, 95% CI 0.76–0.88), which corresponded to a 7 and 8% absolute benefit at 2 and 5 years, respectively. Chemotherapy regimens that included more than one drug had a greater effect on overall survival than single-agent chemotherapy (HR 0.69 versus 0.87). The benefit of chemotherapy on survival declined as a function of age: patients under the age of 60 had the greatest benefit.

Several additional meta-analyses have reported similar improvements in overall survival in patients receiving concomitantly administered chemotherapy and radiation [30,45] and reported additional outcomes of interest. In a literature-based meta-analysis of 42 randomized clinical trials, patients who received chemotherapy were significantly more likely to experience delays in the administration of radiotherapy, mucositis, nausea, bone marrow toxicity, and treatment-related deaths. However, response rates and local-regional control at 2 years were improved significantly with the addition of chemotherapy [30]. Cisplatin and 5-fluorouracil-based combination chemotherapy regimens were found to have the greatest impact on mortality. In contrast, mitomycin, 5-fluorouracil, and bleomycin-based regimens were not found to reduce mortality significantly [45]. The role of chemotherapy in oropharyngeal preservation with primary radiation therapy is reviewed.

A. Clinical Trials Demonstrating Benefit of Chemoradiation in Patients with Advanced Oropharyngeal Cancer

Early clinical trials evaluating the addition of chemotherapy to conventional fractionation radiotherapy suggested that the benefits were most marked for patients with oropharyngeal cancers. Local control was improved with the addition of mitomycin and bleomycin chemotherapy (24 versus 63%, $p=0.015$) to conventional fractionated radiotherapy in a small trial in which the majority of patients had oropharyngeal cancers (33 of 49) [46]. Improvements in local control (18 versus 81%, $p<0.001$) and disease-free survival at 18 months (0 versus 66%, $p<0.001$) were most marked for patients with oropharyngeal tumors. Single-agent methotrexate did not improve local control or overall survival when administered with radiation therapy in a randomized trial of 313 patients with stage I to IV (M0) head and neck cancer [47]. However, patients with oropharyngeal cancer who comprised 33% of all enrolled patients had a significant improvement in local control and overall survival when treated with chemotherapy [47].

Combination chemotherapy with a platinum agent and 5-fluorouracil produces overall response rates of 68 to 93% in previously untreated head and neck cancer patients, with complete response rates ranging from 19 to 54% [48]. Improvements in local control and survival were achieved when multiagent chemotherapy was administered concomitant with accelerated hyperfractionation in oropharyngeal cancer patients [49]. Two hundred and forty patients with oropharyngeal or hypopharyngeal cancer were randomized to receive a total dose of 69.9 Gy accelerated radiation alone or together with two cycles of 5-fluorouracil and carboplatin. Seventy four percent of patients had oropharyngeal tumors. Patients in the chemotherapy arm were more likely to experience grade three or four mucositis and vomiting and were more likely to remain dependent on a feeding tube. After a median follow-up of 22 months, locoregional control was not significantly different in the two arms of the trial when the hypopharynx patients were included in the analysis (40 versus 34%, $p=0.34$). However, when analysis was limited to patients with oropharyngeal cancers, patients in the chemotherapy arm had both improved local control (60% vs 40%, $p=0.010$) and estimated overall survival at 1 year (57% vs 68%, $p=0.04$).

As the heterogeneity of treatment response and prognosis in cancers originating from different primary sites in the head and neck is appreciated, site-specific trials are being performed. Two trials have been completed to evaluate the role of concurrent chemoradiation in a uniform population of patients with oropharyngeal cancer. These studies have included patients with both resectable and unresectable disease. An early study completed by the EORTC compared conventional fractionation radiation therapy to radiation therapy administered concomitantly with biweekly bleomycin chemotherapy in patients with T2–T4 oropharyngeal cancer [50]. The toxicity of therapy was intensified with the addition of chemotherapy, and no improvements in response rates (67% vs 68%) or survival at 6 years (24% vs 22%) were observed [50]. This lack of benefit may be attributed to the poor single-agent activity of bleomycin in head and neck cancers (~21%).

Improved locoregional control, disease-free, and overall survival were demonstrated in patients with locally advanced oropharyngeal cancer who received carboplatin and 5-fluorouracil administered concomitantly with conventional fractionation radiation therapy when compared to patients treated with radiation alone [9]. In this study conducted by the Groupe d'Oncologie Radiotherapie Tete et Cou (GORTEC), 226 patients with stage III or IV (M0) squamous cell carcinoma of the oropharynx were randomized to receive 70 Gy in 35 fractions to the primary and involved lymph nodes with or without three cycles of carboplatin and 5-fluorouracil. Seventy four percent of the enrolled patients had tonsillar carcinomas. Ninety-seven, 94, and 65% of patients in the chemotherapy arm of the trial received one, two, and three cycles of chemotherapy, respectively. The duration of radiation therapy treatment breaks was significantly longer in the chemotherapy arm of the trial; however, the frequency of treatment breaks and the mean total dose of radiation therapy received were not different in the two arms of the trial. Chemotherapy increased the frequency of toxicity, including cytopenias, radiation dermatitis, mucositis, and associated weight loss. Local-regional control at 3 years was improved significantly by the addition of chemotherapy (66% vs 42%, $p=0.04$). After a median follow-up of 35 months, patients in the chemotherapy arm of the trial had a median survival of 29.2 months as compared to 15.4 months in the control arm. Overall survival (51% vs 31%, $p=0.02$) and disease-free survival (42% versus 20%, $p=0.04$) at 3 years were improved significantly in the chemotherapy arm of the trial. This trial, together with results supporting improved survival when a similar regimen was added to accelerated, hyperfractionated radiotherapy of oropharyngeal cancers [49], has established the current standard for oropharyngeal preservation with concurrent chemoradiation.

The potential benefit of induction chemotherapy followed by concurrent chemoradiation therapy for oropharyngeal cancer is currently being evaluated in phase II clinical trials in the cooperative groups. A single randomized trial has demonstrated an improvement in overall survival with administration of neoadjuvant (induction) chemotherapy to patients with oropharyngeal cancer [51]. Three hundred and eighteen patients with stage T2 to T4, N0 to N2b squamous cell carcinoma of the oropharynx were randomized to receive three cycles of induction chemotherapy followed by local-regional therapy or local-regional therapy alone. Chemotherapy consisted of cisplatin (100 mg/m^2) on day 1 followed by a continuous infusion of 5-fluorouracil (1000 mg/m^2/day for 5 days). Patient enrollment was stratified by local-regional therapy: surgery followed by postoperative radiation therapy or radiation therapy alone. Seventy-eight percent of the 157 patients in the chemotherapy arm received three cycles of chemotherapy. A complete response to induction chemotherapy was observed in 20%, with an additional 36% achieving a partial response. After a median survival of 5 years, an improvement in overall survival was observed in the chemotherapy arm, after adjustment for local-regional therapy received (HR 0.71, 95% CI 0.40–1.02). Median survival was 5.1 years in the chemotherapy arm versus 3.3 years in the control group.

Although there are only two published randomized controlled trials of primary radiation therapy with and without chemotherapy exclusively for patients with oropharyngeal primary tumors, the majority of patients participating in randomized trials of primary radiation therapy with or without concomitant chemotherapy have oropharyngeal primaries. Therefore, these trials are reviewed here and summarized in Tables 32.2–32.5.

B. Clinical Trials of Conventional Fractionation Radiotherapy with or without Concurrent, Single-Agent Chemotherapy

Several clinical trials have evaluated the ability of single-agent chemotherapy (hydroxyurea, bleomycin, mitomycin, methotrexate, 5-fluorouracil, carboplatin, and cisplatin) to improve local control when administered concurrently with conventional radiation therapy. Mitomycin, 5-fluorouracil, carboplatin, and cisplatin were associated with improvements in local control, but single-agent hydroxyurea and bleomycin were not. In a small trial of 40 patients randomized to receive palliative conventional fractionation radiation therapy alone with or without hydroxyurea, no difference between the two treatment arms in response or local control at 24 months was reported [52] (Table 32.2). Administration of low-dose daily bleomycin chemotherapy prior to radiation therapy also did not improve response rates, progression-free, disease specific, or overall survival in a randomized clinical trial of primary conventional fractionation radiotherapy in patients with stage II to IV squamous cell carcinoma [53] (Table 32.2).

A randomized controlled trial to evaluate the addition of single-agent, infusional 5-FU has been completed in patients with stage III and IV squamous cell carcinoma of the oral cavity, oropharynx, larynx, or hypopharynx [54]. One hundred seventy-five patients were randomized to receive radiation therapy and two cycles of infusional 5-FU or a placebo saline infusion (Table 32.2). Despite stratification by primary site and tumor stage, the chemotherapy arm of the trial had a greater number of patients with T4 N2–3 disease (13 versus 4). Compete response occurred more frequently in the chemotherapy arm (68% vs 56%, $p=0.04$), and there was a trend toward improved progression-free and overall survival in the chemotherapy arm after a median of 3.5 years. The overall survival at 24 months was 63% versus 50% ($p=0.08$).

Investigators at Yale University performed two sequential randomized trials to evaluate whether concurrent chemoradiation would improve clinical outcomes in patients treated with radiation therapy [55]. The trials included patients who received radiation therapy administered in the neoadjuvant ($n=2$) and adjuvant ($n=119$) settings as well as primary therapy ($n=74$). Eligible patients included all stage I to IV squamous cell carcinomas of the upper aeodigestive tract. Two hundred and three patients were randomized, after statification by treatment type, primary site, and stage, to receive radiation therapy alone or with two doses of mitomycin C (15 mg/m^2) with or without dicumarol during the first and sixth weeks of radiation therapy. After a median follow-up of 138 months, patients randomized to the chemotherapy arm had improved local (85% versus 66%, $p=0.002$), local-regional (76% versus 54%, $p=0.003$), and cause-specific survival (74% versus 52%, $p=0.005$) at 5 years when compared to the radiation alone arm.

The radiation-enhancing effects of cisplatin and carboplatin are augmented if the chemotherapy is present in the target cells at the time of radiation therapy [56]. Therefore, Jeremic and colleagues have conducted two randomized clinical trials to evaluate the benefits of concomitant chemoradiation therapy with low-dose daily cisplatin or carboplatin in patients with unresectable disease [38]. In a three-arm trial, 159 patients with stage III or IV (M0) unresectable head and neck cancer were randomized to receive conventional fractionation radiation therapy alone or together with low-dose daily cisplatin or carboplatin (Table 32.2). The addition of chemotherapy significantly increased hematologic toxicity (6 and 10% versus 0%) when compared to the radiation alone arm, but there was no difference in the incidence of nonhematologic toxicity, including mucositis/stomatitis. Locoregional control was improved significantly with the addition of either cisplatin or carboplatin (clinical complete response rates of 72 and 68% versus 38%, $p<0.01$). Median survival was significantly longer in both chemotherapy arms of the trial (32 and 30 months versus 16 months, $p=0.02$). Five-year survival was 32 and 29% in the cisplatin and carboplatin arms of the trial, respectively, versus 15% in the radiation therapy alone arm ($p=0.02$). There was no difference in distant metastasis-free survival observed in the three arms of the trial. A subsequent trial confirmed that the benefits of low-dose daily cisplatin extend to hyperfractionated therapy in patients with resectable and unresectable disease (Table 32.3) [39]. This efficacy of this dose administration schedule of cisplatin has not yet been compared to the more familiar dosing schedule of cisplatin (100 mg/m^2) with or without 5-fluorouracil. Given the low incidence of toxicity and the efficacy associated with this regimen, low-dose daily cisplatin or carboplatin should be used on this schedule in patients who may not tolerate multiagent, concurrent chemoradiation therapy with a platinum agent and 5-fluorouracil. It would be of interest to evaluate this daily low-dose platinum regimen in resectable patients seeking organ preservation.

C. Randomized Clinical Trials of Conventional Fractionation Radiation Therapy with or without Concomitant, Multiagent Chemotherapy

A single trial has evaluated primary site preservation with concurrent, multiagent chemotherapy and radiation in patients with resectable stage III and IV squamous cell carcinomas of the oral cavity, oropharynx, larynx, or hypopharynx [57]. One hundred patients received radiation therapy alone or together with two cycles of 5-fluorouracil and cisplatin (Table 32.3). Locoregional control was improved in the chemotherapy arm (94% versus 66%, $p<0.01$). Patients with persistent disease or local recurrence were salvaged with surgical resection. Local failure was observed in 54% of patients in the radiation therapy alone arm as compared to 22% in the chemotherapy arm of the trial ($p<0.001$). The 5-year survival with primary site intact was greater in the chemotherapy arm of the trial (42% versus 34%, $p=0.004$). In subset analysis, primary site preservation was improved with chemotherapy in patients with laryngeal and hypopharyngeal tumors but not in patients with oropharyngeal primaries. Overall survival was not improved by the addition of chemotherapy [57].

TABLE 32.2 Randomized Trials of Standard Fractionation Radiotherapy with or without Concurrent Single-Agent Chemotherapy[a]

Patient no. and primary tumor site (%)	Tumor stage	XRT	Chemotherapy	CR rates	Progression-free survival	Disease-specific survival	Overall survival	Disease-free survival	Reference
40 OP(40) OC(3) L(5) HP(15) NP(3) UP(5) SG(5) RD(25)	Squamous cell, mucoepidermoid, and adenoid cystic histologies included Newly diagnosed and recurrent	Arms A and B: Standard fractionation variable 45–75 Gy	Arm B only: Hydroxyurea 60 mg/kg QMWF	Arm A 50% Arm B 72% (NS)	Arm A 27% at 2 years Arm B 35% (NS)	NR	NR	NR	Hussey and Abrams [52]
222 OP(10) OC(38) L(29) HP(9) NP(5) S(8)	II, III, and IV M0 Resectable and Unresectable newly diagnosed	Arms A and B: Standard fractionation 65–70 Gy ± Iridium-192 interstitial implants	Arm B only: Bleomycin 5 mg IM QD 5 days/week	Arm A 58% Arm B 63% (NS)	Local regional Arm A 58% at 5 years Arm B 53% ($p = 0.97$)	Arm A 61% at 5 years Arm B 50% ($p = 0.33$)	Arm A 42% at 5 years Arm B 38% ($p = 0.70$)	NR	Vermund *et al.* [53]
175 OP(42) OC(12) L(27) HP(14) OL(5)	III and IV M0 resectable or Unresectable newly diagnosed	Arms A and B: Standard fractionation 66 Gy/2 Gy per fraction	Arm B only: 5-FU 1.2 g/m²/day as CI×3D Weeks 1 and 3	Arm A 56% Arm B 68% ($p = 0.04$)	Improved in Arm B ($p = 0.057$)	NR	Arm A 50% at 2 years Arm B 63% ($p = 0.08$)	NR	Browman *et al.* [54]
159 OP(38) OC(16) L(18) HP(18) NP(10)	III and IV M0 Unresectable newly diagnosed	Arms A, B, and C: Standard fractionation 70 Gy/2 Gy per fraction	Arm B: Cisplatin 6 mg/m²/day of XRT Arm C Carboplatin 25 mg/m²/day of XRT	Arm A 38% Arm B 72% Arm C 68% ($p = 0.002$)	Local regional Arm A 27% at 5 years Arm B 51% Arm C 48% ($p = 0.04$)	NR	Arm A 15% at 5 years Arm B 32% Arm C 29% ($p = 0.02$)	NR	Jeremic *et al.* [38]
195 OP(24) OC(22) L(27) HP(16) NP(6) UP(5)	I, II, III, and IV M0 Newly diagnosed Preoperative, postoperative, and untreated patients Included	Postoperative patients Standard fractionation 60 Gy/1.8 to 2.0 per fraction Primary XRT Standard fractionation 68 Gy/1.8 to 2.0 per fraction	Arm B: Mitomycin 15 mg/m²/day Weeks 1 and 6± dicumarol	NR	Local regional Arm A 54% at 5 years Arm B 76% ($p = 0.003$)	Arm A 51% at 5 years Arm B 74% ($p = 0.005$)	Arm A 42% at 5 years Arm B 48% (NS)	NR	Haffty *et al.* [55]
224 OP(100)	T2 T3 and T4 N0–N3 Newly diagnosed Resectable and unresectable	Arms A and B Standard fractionation radiation 70 Gy/7–8.5 weeks	Arm B only: Bleomycin 15 mgs im or iv twice weekly for 5 weeks	Arm A 69% Arm B 65% (NS)	NR	NR	Arm A 22% at 6 years Arm B 24% (NS)	Arm A 22% at 5 years Arm B 22% (NS)	Eschwege *et al.* [50]
313 OP (33) OC(22) L(18) HP(16) NP+S(10)	I, II, III, and IV M0 Newly diagnosed Resection status not reported	Arms A and B Variable (<40–55) Majority 50 Gy in 15 or 16 fractions over 3 weeks	Arm B only: Methotrexate 100 mg/m² days 0–14	NR	NR	Improved in OP subset	Arm A 42% at 15 years Arm B 38% (NS) Significant improvement in OP subset	NR	Gupta and Swindell [47]

[a]OP, oropharynx; OC, oral cavity; L, larynx; HP, hypopharynx; NP, nasopharynx; UP, unknown primary; S, sinus; SG, salivary gland; RD, recurrent disease; NR, not reported; NS, not significant.

TABLE 32.3 Randomized Trials of Concurrent Multiagent Chemotherapy[a,b]

Patient no. and primary tumor site (%)	Tumor stage	XRT	Chemotherapy	CR rates	Progression-free survival	Disease-specific survival	Overall survival	Disease-free survival	Reference
104 OP(34) OC(13) L(5) HP(13) NP(33)	III and IV M0 Newly diagnosed inoperable	Arms A and B: Standard fractionation 70 Gy/1.8 Gy per fraction 5 fractions per week	Arm B only: bleomycin 5U 2x/week and adjuvant chemotherapy bleomycin 15U weekly×16 weeks methotrexate 25 mg/m^2 weekly ×16 weeks	Arm A 45% Arm B 67% ($p = 0.056$)	Local regional Arm A 26% at 2 years Arm B 64% ($p = 0.001$)	Arm A 15% at 3 years Arm B 31% ($p = 0.024$)	Arm A 24% at 3 years Arm B 43% ($p = 0.112$)	NR	Fu 1987
90 OP(30) OC(47) HP(23)	IV unresectable Newly diagnosed	Standard fractionation Arms A, B, and C: 70 Gy/2 Gy per fraction	Arm B only: 2 cycles neoadjuvant vinblastine: 4 mg/m^2 day 1 mitomycin C: 8 mg/m^2 on day1 cisplatin: 30 mg/m^2 days 2 and 4 bleomycin: 10 mg/m^2 days 2 and 4 repeated on day 21 Arm C only: Two cycles concurrent with radiation bleomycin: 5 mg days 1 and 5 cisplatin: 20 mg/m^2 days 2 and 3	Arm A 10% Arm B 30% Arm C 23% $p = 0.099$	NR	NR	No significant difference	NR	Salvajoli 1992
49 OC(16) OP(67) HP(8) S(8)	Inoperable Newly diagnosed	Arms A and B: Standard fractionation 66–70 Gy/2 Gy per fraction	Arm B only: Bleomycin: 5U twice a week Mitomycin C: 15 mg/m^2 week 1 10 mg/m^2 week 7	Arm A 24% Arm B 63% ($p = 0.005$) Oropharynx only: Arm A 18% Arm B 81% ($p = 0.0003$)	Arm A 9% at 18 months Arm B 48% ($p = 0.001$) Oropharynx only: Arm A 0% Arm B 66% ($p = 0.00001$)	NR	NR	NR	Smid *et al.* [46]
226 OP(100)	III and IV M0 Resectable and unresectable Newly diagnosed	Arms A and B: Standard fractionation 70 Gy /35 fractions 2 Gy per fraction	Arm B only: 3 cycles carboplatin 70 mg/m^2/day 5 fluorouracil 600 mg/m^2/day CI on days 1–4, 22–25, 43–46	NR	Local regional Arm A 42% at 3 years Arm B 66% ($p = 0.03$)	NR	Arm A 31% at 3 years Arm B 51% ($p = 0.02$)	Arm A 20% Arm B 42% ($p = 0.04$)	Calais *et al.* [9]
100 OP(44) OC(4) L(36) HP(16)	III and IV M0 Resectable only Newly diagnosed	Arms A and B: Standard fractionation 66–72/1.8–2 Gy per fraction	Arm B only: 2 cycles cisplatin: 20 mg/m^2/day CI×4 days, days 1 + 22 5 flourouracil: 1000 mg/m^2/day × CI 4 days, days 1 + 22	Arm A 66% Arm B 94% ($p = 0.001$)	Recurrence-free survival Arm A 51% at 5 years Arm B 62% ($p = 0.04$) Distant metastasis-free survival Arm A 75% at 5 years Arm B 84% ($p = 0.09$)	NR	Arm A 48% at 5 years Arm B 50% ($p = 0.55$)	NR	Adelstein *et al.* [24]

[a]Standard fractionation; [b]OP, oropharynx; OC, oral cavity; L, larynx; HP, hypopharynx; NP, nasopharynx; UP, unknown primary; S, sinus; NR, not reported.

D. Randomized Clinical Trials of Conventional Fractionation and Split-Course Concurrent Chemoradiation or Alternating Chemotherapy and Radiation

Split-course and rapidly alternating chemotherapy and radiation protocols were designed to evaluate whether the improved local control observed with combined modality therapy could be sustained while minimizing local toxicities. Many of these clinical trials were limited to unresectable patients but are included here as potential regimens for future site-specific trials designed to evaluate organ function preservation in resectable patients. A randomized trial compared sequential chemotherapy and conventional fractionation radiotherapy to concurrent chemoradiotherapy in patients with newly diagnosed, unresectable stage III or IV head and neck cancer or recurrent local disease after primary surgical resection [58] (Table 32.4). Patients in the induction chemotherapy arm were treated with three cycles of cisplatin and fluorouracil followed by 70 Gy of radiation therapy. Patients in the concurrent therapy arm of the trial received seven weekly cycles of cisplatin, 5-fluorouracil, and daily radiation therapy alternating with a week-long break, to a total dose of 70 Gy. One hundred and seven patients were randomized to each arm of the trial. Local control was not different in the two arms of the trial (complete response rates of 50 and 52%). Significant differences with regard to primary tumor site and stage occurred in the two arms of the trial: there were more patients in the concomitant arm with oropharyngeal primaries and T4 N2–3 disease and fewer hypopharyngeal patients. As a result, proportional hazards regression analysis was performed to adjust for the differences in prognostic factors in the two arms of the trial. Concomitant therapy was found to be significant in an analysis of disease-free survival.

Two sequential trials performed by the same investigators have compared conventional fractionation radiation therapy with or without induction chemotherapy to rapidly alternating chemotherapy and radiation with two different multiagent chemotherapy regimens [59,60]. One hundred sixteen patients with unresectable head and neck cancer were enrolled in the first trial, 47% of whom had oropharynx primaries [59] (Table 32.4). Patients were randomized to receive four cycles of vinblastine, bleomycin, methotrexate, and leucovorin every 2 weeks as neoadjuvant chemotherapy followed by 70 Gy total of conventional fractionation radiation therapy or alternating with three courses of 20 Gy radiation therapy. Complete response (31% versus 13%, $p<0.03$), progression-free survival (12% versus 4%, $p<0.02$), and improved overall survival at 4 years (22% versus 10%, $p<0.02$) were improved in the alternating therapy arm of the trial. Patients with stage IV disease had improved complete response rates (51% versus 21%), progression-free survival (15% versus 4%, $p<0.009$), and overall survival (23% versus 4%, $p<0.02$) in the alternating chemo-radiation arm.

In a subsequent trial [60], 157 patients with stage III or IV unresectable, metastatic pharyngeal, oral cavity, or laryngeal carcinoma were randomized to receive chemotherapy with cisplatin and fluorouracil alternating with radiotherapy or to conventional fractionation radiotherapy to 70 Gy (Table 32.4). Locoregional control was improved significantly in the chemotherapy arm (43% versus 22%, $p<0.037$), and the difference was more marked after surgical salvage (53% versus 26%, $p=0.001$). Five-year locoregional relapse-free survival was 64% in the chemotherapy arm versus 32% in the radiation therapy alone arm. This improvement in locoregional control was associated with an improvement in 5-year progression free (21% versus 9%, $p=0.008$) and overall survival (24% versus 10%, $p=0.01$). The risk of death was reduced by 41% in the chemotherapy arm (0.59, 95% CI: 0.41–0.85). An organ preservation strategy with rapidly alternating chemotherapy remains to be compared to concurrent chemoradiation without planned treatment breaks, which may have compromised outcome.

E. Clinical Trials of Altered Fractionation Schedules of Radiation Therapy with or without Chemotherapy

The poor survival for patients with unresectable head and neck cancer treated with conventional fractionation radiation therapy is largely attributed to poor locoregional control of disease. Several randomized clinical trials have suggested that local control of head and neck cancer can be improved with altered fractionation schedules of radiation therapy [42].

Hyperfractionated radiotherapy schedules involve the delivery of smaller fractions (1.1 to 1.2 Gy twice per day) than conventional radiation (2 Gy per fraction) therapy and allow a greater total dose of radiation to be administered in the same time period. Two randomized clinical trials have demonstrated improved local control of oropharyngeal cancers with hyperfractionated versus conventional radiotherapy [33,34]. Several trials have investigated whether the improved local control achieved with the addition of chemotherapy to conventional fractionation schedules extends to hyperfractionated schedules of radiotherapy. A small, randomized trial compared split-course hyperfractionated radiation therapy with or without two cycles of cisplatin and 5-fluorouracil in patients with either unresectable head and neck cancer or postoperative patients at high risk for relapse [61]. Thirty-two patients had unresectable disease and 26 had high-risk disease. Patients in the unresectable subset were found to have improved progression-free and overall survival with the addition of chemotherapy.

Accelerated radiation therapy protocols in which the overall treatment time is reduced by the use of multiple daily

TABLE 32.4 Randomized Trials of Concurrent Multiagent Chemotherapy[a,b]

Patient no. and primary tumor site (%)	Tumor stage	XRT	Chemotherapy	CR rates	Progression-free survival	Disease-specific survival	Overall survival	Disease-free survival	Reference
116 OP(47) OC(25) L(4) HP(16) NP(8)	III and IV M0 Inoperable Newly diagnosed	Arm A: Standard fractionation 70 Gy/2 Gy per fraction Arm B: Split-course alternating with chemotherapy 60 Gy in 3 courses of 20 Gy/2 Gy per fraction separated by 2 weeks ± 10 Gy boost	Arm A: 4 cycles neoadjuvant Arm B 4 cycles alternating with XRT Vinblastine: 6 mg/m^2 day 1 Bleomycin: 30 Im day 1 Methotrexate: 200 mg day 2 Leucourin:45 mg day 3 Q 2 weeks × 4	Arm A 13% Arm B 31% ($p \leq 0.03$)	Arm A 4% at 4 years Arm B 12% ($p < 0.02$)	NR	Arm A 10% at 4 years Arm B 22% ($p < 0.02$)	Arm A 14% at 4 years Arm B 22% (NS)	Merlano *et al.* [59]
215 OP(23) OC(32) L(11) HP(23) NP(6) S(1)	III and IV M0 Newly diagnosed Unresectable or recurrent after surgical resection (9 of 215)	Arm A: Standard fractionation 70 Gy/2 Gy per fraction ± boost Arm B: 70 Gy/2 Gy per fraction on days 1–5 of chemotherapy alternating with 1 week break for 7 cycles	Arm A 3 cycles neoadjuvant cisplatin 100 mg/m^2 day 1 5-fluorouracil 1000 mg/m^2/day CI × day 1–5 Arm B: 7 cycles concomitant chemotherapy with cisplatin 60 mg/m^2 on day 1 5-FU 800 mg/m^2/day CI × 5 days	Arm A 50% Arm B 52% (NS)	Significant improvement in arm B	NR	No significant difference	Significantly improved in concurrent arm B after adjustment for significant prognostic differences in two arms	Taylor *et al.* [58]
157 OP(34) OC(29) L(10) HP(19) NP(8)	III or IV Newly diagnosed Unresectable	Arm A: Standard fractionation 70 Gy/2 Gy per fraction for 7 weeks Arm B: 60 Gy/2 Gy per fraction weeks 2–3, 5–6, 8–9	Arm B only: Cisplatin 20 mg/m^2/day CI days, 1–5 5-Fluorouracil 200 mg/m^2/day CI days 1–5 weeks 1,4,7, and 10	Arm A 22% Arm B 43% ($p = 0.037$)	Arm A 9% at 5 years Arm B 21% ($p = 0.008$) Local Regional Arm A 32% at 5 years	NR Arm B 64% ($p = 0.038$)	Arm A 10% at 5 years Arm B 24% ($p = 0.01$)	NR	Merlano *et al.* [60]

[a]Standard fractionation versus split-course chemoradiation therapy or alternating chemotherapy.
[b]OP, oropharynx; OC, oral cavity; L, larynx; HP, hypopharynx; NP, nasopharynx; S, sinus; NR, not reported; NS, not significant.

TABLE 32.5 Randomized Trials of Hyperfractionated Radiotherapy with or without Concurrent Chemotherapya

Patient no. and site (%)	Tumor stage	XRT	Chemotherapy	CR rates	Progression-free survival	Disease-specific survival	Overall survival	Disease-free survival	Reference
58 OP(38) OC(16) L(10) HP(22) UP(9) OTH(3) CE(2)	III or IV Newly diagnosed Inoperable or high-risk postoperative patients	Inoperable patients: 69 Gy/1.5 Gy BID × 10 days followed by 2-week break, then 1.5 Gy BID × 8–13 days High-risk patients: Positive margins 60 Gy Negative margins 54 Gy	Arm B: cisplatin 100 mg/m^2 day 1 5-Fluorouracil 1000 mg/m^2/day CI days 1–4 Repeated on day 29	NR	NR	NR	NR	Improved in chemotherapy Arm	Weissler *et al.* [61]
122 OP(45) OC(5) L(16) HP(20) NP(6) S(5) OTH(3)	T3, T4, N0–N3 T2N0 BOT Resectable or unresectable Newly diagnosed	Arm A: Hyperfractionated 75 Gy/1.25 Gy per fraction BID Arm B: Hyperfractionated 70 Gy/1.25 per fraction BID	Arm B only: Cisplatin 12 mg/m^2 × 5d 5-Fluoruoracil 600 mg/m^2/day CI × 5d weeks 1 + 6 +2 cycles adjuvant with cisplatin dose increased to 16 + 20 mg/m^2	Arm A 73% Arm B 88% ($p = 0.52$)	Arm A 41% at 3 years Arm B 61% ($p = 0.08$)	NR	Arm A 34% at 3 years Arm B 55% ($p = 0.07$)	NR	Brizel *et al.* [10]
270 OP(42) OC(22) HP+L(36)	III and IV M0 Unresectable	Arms A and B: Hyperfractionated 70 Gy/1.8 Gy fractions BID to total of 23.4 Gy× 3 separated by two 11-day breaks	Arm B only: 3 cycles, days 1, 22, and 44 Cisplatin: 60 mg/m^2 days 1 5-Fluorouarocil: 350 mg/m^2 day 1 350 mg/m^2/day days 2–5 as CI Leucovorin: 50 mg/m^2 day 2 100 mg/m^2/day days 2–5 as CI	NR	Arm A 17% at 3 years Arm B 35% ($p < 0.004$) (local regional)	NR	Arm A 24% at 3 years Arm B 49% ($p < 0.0003$)	NR	Wendt *et al.* [12]

130 OP(37) OC(21) L(17) HP(16) NP(9)	III and IV M0 Resectable or unresectable	Arms A and B: Hyperfractionated 77 Gy/1.1 Gy fractions BID	Arm B only: Cisplatin 6 mg/m^2/day of XRT to total dose 210 mg/m^2	Arm A 48% Arm B 75% ($p = 0.002$)	Arm A 25% at 5 years Arm B 46% ($p = 0.0068$)	NR	Arm A 25% at 5 years Arm B 46% ($p = 0.0075$)	NR	Jeremic *et al.* [39]
239 OP(41) OC(30) L(12) HP(17)	III and IV Majority inoperable Newly diagnosed	Arm A: Standard fractionation 70 Gy 7 weeks /35 fractions 7 weeks Arms B and C: Hyperfractionated accelerated 2.5 Gy day 1 55.3 Gy/1.65 Gy BID/17 days	Arm C only: Mitomycin C 20 mg/m^2 day 5	Arm A 44% Arm B 55% Arm C 61%	Local regional (NS)	NR	Arm A 24% Arm B 31% Arm C 41% ($p = 0.03$)	NR	Dobrowsky and Naude [62]
240 OP(74) HP(26)	III and IV Unresectable	Arms A and B: Hyperfractionated accelerated 69.9 Gy/1.8 and 1.5 Gy fractions QD/38 days	Arm B only: 2 cycles 5-FU 600 mg/m^2/day CI Carboplatin 70 mg/m^2 IVP days 1–5 and 29–33	Arm A 34% Arm B 40% ($p = 0.34$)	Local regional Arm A 45% at 2 years Arm B 51% ($p = 0.14$)	NR	Arm A 48% at 2 years Arm B 39% Oropharynx: Arm A 57% at 1 year Arm B 68% ($p = 0.04$)	NR	Staar *et al.* [49]

[a] OP, oropharynx; OC, oral cavity; L, larynx; HP, hypopharynx; NP, nasopharynx; UP, unknown primary; S, sinus; NR, not reported; NS, not significant; CE, cervical esophagus; OTH, other.

fractions were designed to reduce the accelerated repopulation of tumor cells and thereby hopefully reduce local recurrence rates [42]. The benefit of administering multiagent chemotherapy with cisplatin and 5-fluorouracil concurrently with accelerated radiation therapy was investigated by Brizel [10] in a randomized controlled trial (Table 32.5). Eligible patients included those with newly diagnosed resectable or unresectable T3–T4N0–N3 M0 (and T2N0 BOT primaries) squamous cell carcinoma of the head and neck, including nasopharyngeal and paranasal sinus primaries. Patients were stratified according to the resectablility of the tumor and hemoglobin concentration. Accelerated hyperfractionated radiotherapy was administered twice daily in 1.2-Gy fractions to a total dose of 7500 cGy to patients in the radiation therapy alone arm. Two cycles of cisplatin and 5-fluorouracil were administered during weeks 1 and 6 of radiation therapy in the chemotherapy arm of the trial. Two cycles of adjuvant cisplatin/5-fluorouracil chemotherapy were administered after the completion of radiation therapy. The total dose of radiation therapy in the chemotherapy arm was reduced to 7000 cGy, and a 1-week treatment break was also incorporated after 4000 cGy. A complete response to therapy was observed in 88% of the patients in the chemotherapy arm versus 73% in the radiation therapy arm ($p = 0.52$). After a median follow-up of 41 months, patients in the chemotherapy arm of the trial had improved local regional control of disease (70% versus 44%, $p = 0.01$), as well as a trend toward improved relapse-free (61% versus 41%, $p = 0.08$) and overall survival (55% versus 34%, $p = 0.07$) at 3 years.

Wendt and colleagues [12] also evaluated the contribution of multiagent chemotherapy to local control and overall survival when administered concurrently with accelerated, fractionated split-course radiation therapy. Two hundred and seventy patients with unresectable stage III and IV HNSCC were randomized to receive a total dose of 70.2 Gy with or without chemotherapy (Table 32.5). Radiation therapy was administered in 39 fractions of 1.8 Gy twice daily in three cycles of 23.4 Gy separated by two 11-day treatment breaks. Patients in the chemotherapy arm received three 21-day cycles of cisplatin, 5-fluorouracil, and leucovorin. Patients in the chemotherapy arm had improved locoregional control (35% versus 17%, $p < 0.004$) and overall survival (49% versus 24%, $p < 0.0003$) at 3 years. These trials provide compelling data that the benefits of concurrent cisplatin-based chemotherapy extend to altered fractionation radiotherapy.

Dobrowsky and Naude [62] have completed a clinical trial to evaluate the effect of mitomycin C on survival rates in patients receiving hyperfractionated, accelerated radiotherapy. Two hundred thirty-nine patients with resectable or unresectable oral cavity, oropharyngeal, laryngeal, or hypopharyngeal cancer were randomized to three treatment arms: conventional fractionation radiation therapy to 70 Gy or continuous hyperfractionated accelerated radiotherapy of 55.3 Gy in 33 fractions over 17 days with or without a single dose of mitomycin C (20 mg/m^2) (Table 32.5). Complete response occurred in 44% of the patients in the standard arm, whereas 55 and 61% of patients in the accelerated fractionation arm without and with mitomycin C, respectively, achieved a complete response. After a median survival of 48 months, overall survival was improved in the concurrent chemoradiation arm of the study (41% versus 31 and 24%, $p < 0.05$).

F. Summary

The current standard of care for patients with unresectable, oropharyngeal cancer and a good performance status is concurrently administered platinum-based chemotherapy and primary radiation therapy. Combination chemotherapy with a platinum agent and 5-fluorouracil has been utilized in the majority of clinical trials demonstrating benefits in local control and survival. However, this combination of chemotherapy is frequently associated with significant local-regional toxicity as well as hematologic toxicity and should be administered by physicians familiar with managing the toxicities of combined modality therapy. Concomitant, single-agent, low-dose daily cisplatin is an alternative to combination chemotherapy, as this regimen has demonstrated improved local control, progression-free survival, distant metastasis-free survival, and overall survival in two well-controlled trials when administered with conventional fractionation or hyperfractionated radiotherapy. Although survival is improved by the addition of chemotherapy in this population, survival at 5 years is very poor (25%). Therefore, patients should be encouraged to participate in clinical trials investigating the radiosensitizing potential of newer chemotherapeutic agents, rationally targeted small molecule and antibody-based therapies.

Patients with early stage oropharyngeal cancer (T1 and T2) tumors may be treated with surgical resection or primary radiation therapy. Nonrandomized, single institutional case series have suggested that patients treated with radiation therapy may have improved speech and swallowing function when compared to patients treated with surgical resection. Local control of early stage oropharyngeal cancers with conventional fractionated radiotherapy with a brachytherapy boost and altered fractionation schedules both achieve rates of local control of greater than 80%, but have yet to be compared in a formal clinical trial. Patients with resectable, locally advanced oropharyngeal cancer (T3 and T4) with a good performace status desiring an organ preservation strategy should be encouraged to participate in a clinical trial incorporating new chemotherapeutic agents as radiosensitizers. Current clinical trials of organ preservation for oropharyngeal cancer should incorporate formal assessments to evaluate functional preservation as well.

Patients not participating in clinical trials should receive cisplatin-based chemotherapy administered concurrently with radiation therapy.

References

1. Hong, W., O'Donoghue, G., Sheets, S., *et al.* (1985). Sequential response patterns to chemotherapy and radiotherapy in head and neck cancer: Potential impact of treatment in advanced laryngeal cancer. *Prog. Clin. Biolog. Res.* **201**, 191–197.
2. Al-Kourainy, K., Kish, J., Ensley, J., *et al.* (1987). Achievement of superior survival for histologically negative versus histologically positive clinically complete responders to cisplatin combination in patients with locally advanced head and neck cancer. *Cancer* **59**, 233–238.
3. Weaver, A., Fleming, S., Ensley, J., *et al.* (1984). Superior clinical response and survival rates with initial bolus of cisplatin and 120 hour infusion of 5 fluorouracil before definitive therapy for locally advanced head and neck cancer. *Am. J. Surg.* **148**, 525–529.
4. Ensley, J., Jacobs, J., Weaver, A., *et al.* (1984). Correlation between response to cisplatinum combination chemotherapy and subsequent radiotherapy in previously untreated patients with advanced squamous cell cancers of the head and neck. *Cancer* **54**, 811–814.
5. Pfister, D., Strong, E., Harrison, L., *et al.* (1991). Larynx preservation with combined chemotherapy and radiation therapy in advanced but resectable head and neck cancer. *J. Clin. Oncol.* **9**, 850–859.
6. Demard, F., Chauvel, P., Santini, J., *et al.* (1990). Response to chemotherapy as justification for modification of the therapeutic strategy for pharyngolaryngeal carcinomas. *Cancer* **12**, 225–231.
7. Clayman, G., Weber, R., Guillamondegui, O., *et al.* (1995). Laryngeal preservation for advanced laryngeal and hypopharyngeal cancers. *Arch. Otolaryngol. Head Neck Surg.* **121**, 219–223.
8. Al-Saraf, M., Leblanc, M., Giri, P., *et al.* (1998). Chemoradiotherapy versus radiotherapy in patients with advance nasopharyngeal cancer: Phase III randomized intergroup study. *J. Clin. Oncol.* **16**, 1310–1317.
9. Calais, G., Alfonso, M., Bardet, E., Sier, C., Germain, T., Bergerot, P., Rhein, B., Tortochaux, J., Oudinot, P. and Bertrand, P. (1999). Radiation therapy versus concomitant chemoradiotherapy for stages II/IV oropharynx carcinoma: A GORTEC randomized trial. *J. Natl. Cancer Inst.* **91**, 2081–2086.
10. Brizel, D., Albers, M., Fisher, S., Scher, R., Richtsmeier, W., Hars, V., George, S., Huang, A. and Prosnitz, L. (1998). Hyperfractionated irradiation with or without concurrent chemotherapy for locally advanced head and neck cancer. *N. Engl. J. Med.* **338**, 1798–1804.
11. Merlano, M., Vitale, V., Rosso, R., Benasso, M., Corvo, R., Cavallari, M., *et al.* (1992). Treatment of advanced squamous cell carcinoma of the head and neck with alternating chemotherapy and radiotherapy. *N. Engl. J. Med.* **327**, 1115–1121.
12. Wendt, T., Grabenbauer, G., Rodel, C., Thiel, H., Aydin, H., Rohloff, R., Sustrow, T., Iro, H., Popella, C., and Schalhorn, A. (1998). Simultaneous radiochemotherapy versus radiotherapy alone in advanced head and neck cancer: A randomized multicenter study. *J. Clin. Oncol.* **16**, 1318–1324.
13. Wolf, G., Hong, W., and Fisher, S. (1996). Neoadjuvant chemotherapy for organ preservation: Current status. *In* "Fourth International Conference on Head and Neck Cancer," pp. 89–96.
14. Lefebvre, J., Chevalier, D., Loboniski, B., *et al.* (1996). Larynx preservation in pyriform sinus cancer: Prelimary results of a European Organization for Research and Treatment of Cancer Phase III Trial. EORTC Head and Neck Cancer Cooperative Group. *J. Natl. Cancer Inst.* **88,** 890–899.
15. Forastiere, A., Berkey, B., Maor, M., Weber, R., Goepfert, H., Morrisson, W., Glisson, B., Trotti, A., Ridge, J., Chaio, C., Peters, G., Lee, D., Leaf, A., and Ensley, J. (2001). Phase III trial to preserve the larynx: Induction chemotherapy and radiotherapy versus concomitant chemoradiotherapy versus radiotherpy alone. *Proc. Am. Soc. Clin. Oncol.* **20**, 2a.
16. Department of Veterans Affairs Laryngeal Study Group (1991). Induction chemotherapy plus radiation compred with surgery plus radiation in patients with advanced laryngeal cancer. *N. Engl. J. Med.* **324**, 1685–1690.
17. Bradford, C., Wolf, G., Carey, T., *et al.* (1999). Predictive markers for response to chemotherapy, organ preservation and survival in patients with advanced laryngeal carcinoma. *Otolaryngol. Head Neck Surg.* **121**, 534–538.
18. Spaulding, M., Fisher, S., Wolf, G., *et al.* (1994). Tumor response, toxicity and survival after neoadjuvant organ preseving chemotherapy in advanced laryngeal cancer. *J. Clin. Oncol.* **12**, 1592.
19. Wolf, G., and Fisher S. (1992). Laryngeal Cancer Study Group: Effectiveness of salvage neck dissection for advanced regional metastases when induction chemotherapy is used for organ preservation in laryngeal cancer. *Laryngoscope* **102**, 934.
20. Hillman, R., Walsh, M., Wolf, G., Fisher, S., and Hong, W. (1998). functional outcomes following treatment for advanced laryngeal cancer. I. Voice preservation in advanced laryngeeal cancer. II. Laryngectomy rehabilitation: The state of the art In the VA system. *Ann. Otolaryngol. Suppl.* **172**, 1–27.
21. Gillison, M., and Forastiere, A. (1999). Larynx preservation in head and neck cancers. *Hematol. Oncol. Clin. North Am.* **13**, 699–700.
22. Forastiere, A. (2000). NCCN practice guidelines for head and neck cancers. *Oncology* **14**, 163.
23. Lefebvre, J. Personal communication.
24. Adelstein, D., Adams, G., Li, Y., Wagner, H. J., Kish, J., Ensley, J., Schuller, S., and Forastiere, A. (2000). A phase II comparison of standard radiation therapy (RT) versus RT plus concurrent DDP and 5-fluorouracil (5U) in patients with unresectable squamous cell head and neck cancer: An intergroup study. *Proc. Am. Soc. Clin. Oncol.* **19**, 411a.
25. Dubois, J., Guerrier, B., Ruggiero, J. D., *et al.* (1984). Correlation between response to cisplatinum-combination chemotherapy and subsequent radiotherapy in previously untreated patients with advanced squamous cell cancers of the head and neck. *Cancer* **548**, 811–814.
26. Mendenhall, W., Parsons, J., Devine, J., *et al.* (1987). Squamous cell carcinoma of the pyriform sinus treated with surgery and/or radiotherapy. *Head Neck Surg.* **10**, 88–92.
27. Pignon, J., Bourhis, J., Domenge, C., and Designe, L. (2000). Chemotherapy added to locoregional treatment for head and neck squamous cell carcinoma: Three meta-analyses of updated individual data. *Lancet* **9208**, 949–955.
28. Stell, P. (1992). Adjuvant chemotherapy for head and neck cancer. *Semin. Radiat. Oncol.* **2**, 195–205.
29. Munro, A. (1995). An overview of randomised controlled trials of adjuvant chemotherapy in head and neck cancer. *Br. J. Cancer* **71**, 83–91.
30. El-Sayed, S., and Nelson, N. (1996). Adjuvant and adjunctive chemotherapy in the management of squamous cell carcinoma of the head and neck region: A meta-analysis of prospective and randomized trials. *J. Clin. Oncol.* **14**, 838–847.
31. Vokes, E., and Weichselbaum, R. (1990). Concomitant chemoradiotherapy: Rational and clinical experience in patients with solid tumors. *J. Clin. Oncol.* 911–934.
32. Fu, K., and Phillips, T. (1991). Biologic rationale of combined radiotherapy and chemotherapy. *Hematol. Oncol. Clin. North Am.* **5**, 737–751.
33. Pinto, L., Canary, P., Araujo, C., Bacelar, S., and Souhami, L. (1991). Prospective randomized trial comparing hyperfractionated versus conventional radiotherapy in stages III and IV oropharyngeal carcinoma. *Int. J. Radiat. Oncol. Biol. Phys.* **21**, 557–562.
34. Horiot, J., LeFur, R., N'Guyen, T., Chenal, C., Schraub, S., Alfonsi, S., Gardani, G., VanDenBogaert, W., Danczak, S., Bolla, M.,

VanGlabbeke, M., and DePauw, M. (1992). Hyperfractionation versus conventional fractionation in oropharyngeal carcinoma: Final analysis of a randomized trial of the EORTC cooperative group of radiotherapy. *Radiother. Oncol.* **25**, 231–241.

35. Harrison, L., Zelefsky, M., Armstrong, J., Carper, E., Gaynor, J., and Sessions, R. (1994). Performance status after treatment for squamous cell cancer of the base of tongue: A comparison of primary radiation therapy versus primary surgery. *Int. J. Radiat. Oncol. Biol. Phys.* **30**, 953–957.
36. Harrison, L., Zelefsky, M., Sessions, R., Fass, D., Armstrong, J., Pfister, D., and Strong, E. (1992). Base-of-tongue cancer treated with external beam irradiation plus brachyherapy: Oncologic and functional outcome. *Radiology* **184**, 267–270.
37. Pernot, M., Malissard, L., Hoffstetter, S., Luporsi, E., Aletti, P., Peiffeert, D., Allavena, C., Kozminski, P., and Bey, P. (1994). Influence of tumoral, radiobiological, and general factors on local control and survival of a series of 361 tumors of the velotonsillar area treat by exclusive irradiaiton (external beam irradiation + brachytherapy or brachytherapy alone). *Int. J. Radiat. Oncol. Biol. Phys.* **30**, 1051–1057.
38. Jeremic, B., Shibamoto, Y., Stanisavljevic, B., Milojevic, L., Milicic, B., and Nikolic, N. (1997). Radiation therapy alone or with concurrent low-dose daily either cisplatin or carboplatin in locally advanced unresectable squamous cell carcinoma of the head and neck: A prospective randomized trial. *Radiat. Oncol.* **43**, 29–37.
39. Jeremic, B., Shibamoto, Y., Milicic, B., Nikolic, N., Dagovic, A., Aleksandrovic, J., Vaskovic, Z., and Tadic, L. (2000). Hyperfractionated radiation therapy with or without concurrent low-dose daily cisplatin in locally advanced squamous cell carcinoma of the head and neck: A prospective randomized trial. *J. Clin. Oncol.* **18**, 1458–1464.
40. Harrison, L., Lee, H., Pfister, D., Kraus, D., White, C., Raben, A., Zelefsky, M., Strong, E., and Shah, J. (1998). Long term results of primary radiotherapy with/without neck dissection for squamous cell cancer of the base of tongue. *Head Neck* 668–673.
41. Horwitz, E., Frazier, A., Martinez, A., Keidan, R., Clarke, D., Lacerna, M., Gustafson, G., Heil, E., Dmuchowski, C., and Vicini, F. (1996). Excellent functional outcome in patients with squamous cell carcinoma of the base of tongue treated with external irradiation and interstitial iodine 125 boost. *Cancer* **78**, 948–957.
42. Mendenhall, W., Amdur, R., Siemann, D., and Parsons, J. (2000). Altered fractionation in definitive irradiaiton of squamous cell carcinoma of the head and neck. *Curr. Opin. Oncol.* **12**, 207–214.
43. Weber, R., Gidley, P., Morrison, W., Peters, L., Hankins, P., Wolf, P., and Guillamondegui, O. (1990). Treatment selection for carcinoma of the base of the tongue. *Am. J. Surg.* **160**, 415–419.
44. Gourin, C., and Johnson, J. (2001). Surgical treatment of squamous cell carcinoma of the base of the tongue. *Head Neck* 653–660.
45. Browman, G., Hodson, D., Mackenzie, R., Bestic, N., Zuraw, L., and Cancer Care Ontario Practice Guideline Initiative Head and Neck Cancer Disease Site Group (2001). Choosing a concomitant chemotherapy and radiotherapy regimen for squamous cell head and neck cancer: A systematic review of the published literature with subgroup analysis. *Head Neck* 579–589.
46. Smid, L., Lesnicar, H., Zakotnik, B., Soba, E., Budihna, M., Furlan, L., Zargi, M., and Rudolf, Z. (1995). Radiotherapy, combined with simultaneous chemotherapy with mytomycin C and bleomcyin for inoperable head and neck cancer: Preliminary report. *Int. J. Radiat. Oncol. Biol. Phys.* **32**, 769–775.
47. Gupta, N., and Swindell, R. (2001). Concomitant methotrexate and radiotherapy in advanced head and neck cancer: 15-year follow-up of a randomized clinical trial. *Clin. Oncol.* **13**, 339–344.
48. Adelstein, D. (1999). Induction chemotherapy in head and neck cancer. *Head Neck* **13**, 689–698.
49. Staar, S., Volker, R., Stuetzer, H., Dietz, A., Volling, P., Schroeder, M., Flentje, M., Eckel, H., and Mueller, R. (2001). Intensified hyperfractionated accelerated radiotherpy limits the additional benefit of simultaneous chemotherapy: Results of a multicentric randomized German trial in advanced head-and-neck cancer. *J. Radiat. Oncol. Biol. Phys.* **50**, 1161–1171.
50. Eschwege, F., Sancho-Garnier, H., Gerard, J., Madelain, M., DeSaulty, A., Jortay, A., and Cachin, Y. (1988). Ten-year results of randomized trial comparing radiotherapy and concomitant bleomycin to radiotherapy alone in epidermoid carcinomas of the oropharynx: Experience of the European Organization for Research and Treatment of Cancer. *NCI Monogr.* **6**, 275–278.
51. Domenge, C., Hill, C., Lefebvre, J., DeRaucourt, D., Rhein, B., Wibault, P., Marandas, P., Coche-Dequeant, B., Stomboni-Luboinski, M., Sancho-Garnier, H., and Luboinski, B. (2000). Randomized trial of neoadjuvant chemotherapy in oropharyngeal carcinoma. *Br. J. Cancer* **83**, 1594–1598.
52. Hussey, D., and Abrams, J. (1975). Combined therapy in advanced head and neck cancer: Hdroxyurea and radiotherapy. *Prog. Clin. Cancer* **6**, 79–86.
53. Vermund, H., Kaalhus, O., Winthier, F., Trausjo, J., Thorud, E., and Harang, R. (1985). Bleomycin and radiation therapy in squamous cell carcinoma of the upper aero-digestive tract: A phase III clinical trial. *J. Radiat. Oncol. Biol. Phys.* **11**, 1877–1886.
54. Browman, G., Cripps, C., Hodson, I., Eapen, L., Sathya, J., and Levine, M. (1994). Placebo-controlled randomized trial of infusional flourouracil during standard radiotherapy in locally advanced head and neck cancer. *J. Clin. Oncol.* **12**, 2648–2263.
55. Haffty, B., Son, Y., Papac, R., Sasaki, C. T., Weissberg, J., Fischer, D., Rockwell, S., Sartorelli, A., and Fischer, J. (1997). Chemotherapy as an adjunct to radiation in the treatment of squamous cell carcinoma of the head and neck: Results of the Yale mitomycin randomized trials. *J. Clin. Oncol.* **15**, 268–276.
56. Bartelink, H., Kallman, R., Rapacchietta, D., and Hart, G. (1986). Therapeutic enhancement in mice by clinically relevant dose and fractionation schedules of cis-diamminedichloroplatinum(II) and irradiation. *Radiother. Oncol.* **6**, 61–74.
57. Adelstein, D., Lavertu, P., Saxton, J., Secic, M., Wood, B., Wanamaker, J., Eliachar, I., Strome, M., and Larto, M. (2000). Mature results of a phase III randomized trial comparing concurrent chemoradiotherapy with radiation therapy alone in patients with stage III and IV squamous cell carcinoma of the head and neck. *Cancer* **88**, 876–883.
58. Taylor, S., Murthy, A., Vannetzel, J., Colin, P., Dray, M., Caldarellli, D., Sholl, S., Vokes, E., Showel, J., Hutchinson, J., Witt, T., Griem, K., Hartsell, W., Mittal, B., Rebishung, J., Coupez, D., Desphieux, J., Bobin, S., and LePajole, C. (1994). Randomized comparison of neoadjuvant cisplatin and flourouracil infusion followed by radiation versus concomitant treatment in advanced head and neck cancer. *J. Clin. Oncol.* **12**, 385–395.
59. Merlano, M., Corvo, R., Margarino, G., Benasso, M., Rosso, R., Sertoli, M., Cavallari, M., Scala, M., Guenzi, M., Siragusa, A., Brema, F., Luzi, G., Bottero, G., Biondi, G., Scasso, F., Garaventa, G., Accomando, E., Santelli, A., Cordone, G., Comella, G., Vitriolo, S., and Santi, L. (1991). Combined chemotherapy and radiation therapy in advanced inoperable squamous cell carcinoma of the head and neck. *Cancer* **67**, 915–921.
60. Merlano, M., Benasso, M., Corvo, R., Rosso, R., Vitale, V., Blengio, F., Numico, G., Margarino, G., Bonelli, L., and Santi, L. (1996). Five-year update of a randomized trial of alternating radiotherapy and chemotherapy compared with radiotherapy alone in treatment of unresectable squamous cell carcinoma of head and neck. *J. Natl. Cancer Inst.* **88**, 583–589.
61. Weissler, M., Meilin, S., Sailer, S., Qaqish, B., Rosenman, J., and Pilsbury, H. (1992). Simultaneous chemoradiation in the treatment of advanced head and neck cancer. *Arch. Otolayngol. Head Neck Surg.* **118**, 806–810.
62. Dobrowsky, W., and Naude, J. (1999). Continuous hyperfractionated accelerated radiotherapy with/without mytomycin C in head and neck cancers. *Radiother. Oncol.* **57**, 119–124.

C H A P T E R

33

Nasopharyngeal Cancer

P.G. SHANKIR GIRI and MUHYI AL SARRAF

Department of Radiation Oncology
Eastern Virginia Medical School
Norfolk, Virginia 23507

I. Introduction 491
II. Epstein–Barr Virus 491
III. Genetics 492
IV. Environmental Factors 492
A. Dietary Factors 492
B. Cigarette Smoking 492
C. Occupational Exposures 492
V. Anatomy 492
VI. Pathology 492
VII. Natural History 493
VIII. Workup 493
IX. Pretreatment Evaluation 493
X. Staging 493
XI. Radiation Therapy 494
XII. Volume Treated 494
XIII. Boost Volume 494
XIV. Dose Response 494
A. T3 T4 495
XV. Reirradiation 495
XVI. Survival 496
XVII. Chemotherapy 496
XVIII. Chemotherapy for Recurrent/Metastatic Disease 497
XIX. Chemotherapy for Locally Advanced and Previously Untreated Nasopharyngeal Cancer 498
XX. Induction Chemotherapy 498
XXI. Concurrent Chemoradiotherapy 499
XXII. Adjuvant Chemotherapy 499
XXIII. Randomized Trials 500
XXIV. Conclusion 501
References 501

I. INTRODUCTION

Nasopharyngeal cancer (NPC) is an epithelial tumor with worldwide distribution. Its incidence, however, varies with geographical location. It is rare in North American and European Caucasians [1]. In the United States the incidence is about 1 per 100,000. It is 0.8 per 100,000 in males and 0.3 per 100,000 in females with a male to female ratio of 2.4 to 1. The peak age is between 50 and 59 years [2]. However, the incidence is high in China and southeast Asia where it is 25–30 per 100,000 population. It is also high in Eskimos. The risk is intermediate in the Middle East and north African Arab populations [3,4].

II. EPSTEIN–BARR VIRUS

Epstein–Barr virus (EBV) is the primary cause of infectious mononucleosis. It has been linked to other malignancies, including nasopharyngeal carcinoma and Burkitts lymphoma. EB virus is distributed ubiquitously in the human population, resulting from an inapparent infection [5].

The virus is found in tumor cells in the nasopharynx but not in the surrounding lymphocytes. The presence of the virus has been demonstrated by serum antibodies to the virus capsid antigen (VCA) and early antigen (EA) [6]. Elevated antibody titers are present in nasopharyngeal cancer and not in other cancers or normal individuals [7,8].

The exact mechanism by which EBV induces cancer is not well understood. It is most commonly associated with nonkeratinizing and undifferentiated carcinomas [9–12]. Of particular interest is the expression of a latent membrane protein gene whose protein has been shown to transform epithelial cells *in vitro*. Some investigators have found this gene in 65% of nasopharyngeal carcinoma [13].

The significance of these findings is not well understood, but preliminary data suggest that the latent membrane protein expression in EB virus-infected cells might protect the cells from programmed cell death (apoptosis) [14].

Finally, there are at least two recognized strains of EBV (A and B). Studies have demonstrated the A strain to transform B lymphocytes, indicating a possible role in tumor induction. However, other studies do not support such a hypothesis [15,16].

III. GENETICS

The geographical distribution, familial clustering, and increased risk of developing NPC in members of the same family suggest the role of genetics in NPC [17–19]. In addition, the incidence of NPC is high in Chinese living outside the United States [20,21]. Numerous studies have suggested an association with human leukocyte antigen [22,23] and p53 mutations [24,25] in the development of NPC. Current evidence suggests that these are associated factors and in themselves are not causative factors in the development of nasopharyngeal carcinoma.

IV. ENVIRONMENTAL FACTORS

The possibility that environmental factors may play a role is suggested by numerous studies. Among Chinese living in the United States, American-born second-generation Chinese have a lower risk of NPC as compared to Asian-born first-generation Chinese [26,27].

A. Dietary Factors

Dietary factors have been implicated in the development of NPC. The consumption of salted fish has been associated with an increased risk of developing NPC [28–30]. The risk appears to increase when the fish is consumed at a young ages. Analysis of food samples demonstrated the presence of nitrosoamine precursors in salted fish [31]. The nitrosoamines are not carcinogenic by themselves, but their DNA adducts have been shown to be carcinogenic and implicated in the development of cancer of the upper respiratory tract [32–34].

B. Cigarette Smoking

Cigarette smoking has been linked to the development of cancer in the lung and upper respiratory tract. However, only a few studies show an association between smoking and NPC. In a study from Taiwan, Lin *et al.* [35] reported that heavy smokers (>30 cigarettes per day) had a 3.1-fold increase in risk of developing NPC when compared to nonsmokers. If the association of tobacco to NPC is real, the possible mechanism of tumor induction would be the nitrosamines and its precursors in tobacco.

C. Occupational Exposures

Numerous studies have suggested an association of NPC with occupational exposures. Wood and wood dust have been implicated with NPC [36,37], but this finding has not been consistent. Exposure to formaldehyde, especially to high levels and long durations, has been implicated in the development of NPC [38,39]. Exposure to formaldehyde vapor induces cancers of the nasal passages in laboratory animals. It is therefore possible it could also cause NPC.

V. ANATOMY

The nasopharynx is a cuboidal structure. Its anterior extent is defined by the posterior choanae of the nasal cavity. It has a posteriosuperior roof that gradually slopes to become the posterior wall of the nasopharynx. The roof is formed by the basisphenoid and basiocciput, whereas the posterior wall is formed by the first cervical vertebra. Its lateral walls contain cartilagenous openings of the eustachian tubes. The eustachian tubes traverse through an opening in the pharyngobasilar fascia (sinus of Morgagni) as they enter into the nasopharynx. In the nasopharynx the opening is thrown into a fold that forms a tubercle in its superior aspect (Torus Tubarus). A recess, the fossa of Rosenmuller, is located above the torus. This is often the most common site of origin for nasopharyngeal cancers. The floor is at the level of the soft palate.

The nasopharynx is supplied extensively with lymphatics. They drain into the deep retropharyngeal lymph nodes located medial to the carotid artery at around the first cervical vertebral body. The highest of these nodes is referred to as the node of Rouvier. These nodes are seen only by computerised tomography (CT) or magnetic resonance imaging (MRI) examinations. Furthermore, lateral spread of the tumor often encases these nodes so that a distinct node is not seen. The nasopharynx also drains along the jugular vein to all levels of the neck. These nodes are palpated easily.

VI. PATHOLOGY

The nasopharynx is covered by a stratified columnar epithelium. It also has lymphatic tissue as part of the Waldeyeres ring. The most common types of tumors that arise in the nasopharynx are squamous cell carcinomas. The WHO has classified these carcinomas as WHO type I or keratinizing carcinoma, WHO type II or nonkeratinizing carcinoma, and WHO type III or undifferentiated carcinoma.

Undifferentiated carcinomas are often characterized by the presence of a lymphocytic infiltrate. This has led these tumors to be called lymphoepitheliomas. Type I are found more commonly in North America, whereas types II and III are found more commonly in China, Hong Kong, Taiwan, and other far eastern countries.

Other tumors that occur in the nasopharynx include lymphomas, juvenile angiofibromas, and adenocarcinomas from the minor salivary glands and plasmacytomas.

VII. NATURAL HISTORY

Nasopharyngeal carcinomas spread by expansion or infiltration. The primary tumor could be small and the patient presents with enlarged neck nodes typically in the postauricular area. In most cases the tumor can be seen either clinically as a mucosal abnormality or by CT/MRI scanning submucosally.

The most common presenting complaint of NPC is the presence of enlarged neck nodes, which occurs in over 90% of patients at diagnosis. Anterior extension of the tumor toward the nose results in nasal stuffiness and epistaxis. Lateral extension around the eustachian tube causes decreased hearing, pain, and otitis media. Tumors can extend into the parapharyngeal space by passing laterally through the sinus of morgagni, an opening in the lateral wall of the nasopharynx through which the eustachian tubes traverse. In the parapharyngeal space the tumor can involve the pterygoid muscles, resulting in trismus. The tumors then extend posteriorolaterally into the poststyloid compartment. They can also extend inferiorly into the neck. NPC can also extend inferiorly into the oropharynx and hypopharnx along the lateral and posterior pharyngeal walls. Superiorly NPC can destroy the base of the skull and extend into the cavernous sinus where they can cause cranial nerve abnormalities of III, IV, and VI cranial nerves.

Neck node metastasis at diagnosis is common in nasopharyngeal cancer. The incidence varies from 50 to 75%. Despite having advanced disease, they respond well to radiation therapy. The overall neck control rate with radiation therapy ranges from 70 to 85%. Therefore, routine neck dissections are not warranted. Neck dissections should be performed in those with residual neck disease 6–8 weeks following the completion of radiation therapy. Patients with large bulky nodes (N3) are more likely to fail in the neck and may require a neck dissection. Lymph node metastasis into the retroparotidean space results in palsies of cranial nerves IX, X, and XI due to enlarged lymph nodes (retroparotidean syndrome).

VIII. WORKUP

Clinical workup prior to treatment should include a detailed history and physical examination and complete blood chemistries, including liver and renal functions. A chest X-ray and a CT of the chest, including the liver, should be performed to identify metastasis to the lung and/or liver. An endoscopic examination of the nasopharynx and upper airway should be done in all patients. A multidisciplinary team that consists of the surgeon, radiation oncologist, and medical oncologist should see all patients.

Extension into the parapharyngeal space, bone erosion, or intracranial involvement is seen only by CT/MRI examinations. Yu and colleagues [40] reported a 55% upstaging of T2 and 56% of T3 patients when comparing staging done by CT scans versus plain X-rays. Similar results have been reported by Alni *et al.* [41] and Kraiphebul *et al.* [42].

Whether MRI or CT should be the imaging modality of choice depends on the stage. MRI is superior to CT in detecting infiltration along the pharyngobasilar fascia and retropharyngeal lymph nodes. CT, however, may be useful in detecting early bone invasion [43–45].

MRI may be more useful when compared to CT in the evaluation of NPC after treatment. Its ability to differentiate between radiation effect and tumor in soft tissues may make it the preferred method for follow-up. However, Chong and Ian [46] compared CT and MRI to detect recurrent NPC. They found CT had a sensitivity of 45% versus 56% for MRI. The specificity was 70 and 83%, respectively, for CT and MRI. They concluded that both have low sensitivity and either could be used for follow-up evaluation postradiation.

Thus CT/MRI scans form an important part of the initial staging system because they detect (a) parapharyngeal space extension/infiltration, (b) enlarged retropharyngeal lymph nodes, (c) erosion of bone at base of skull, (d) intracranial extension of tumor, and (e) enlarged neck nodes. Because these two techniques are complementary, it is common to employ both.

IX. PRETREATMENT EVALUATION

Pretreatment evaluation should include a complete physical examination, including a head and neck examination. Fiberoptic nasopharyngoscopy should be performed on all patients. A dental evaluation must be performed prior to initiating radiation therapy. Dental prophylaxis with fluoride carriers should be done, as most of the patients will have some degree of xerostomia following radiation therapy treatment. A dietary consult for maintaining adequate nutrition during treatment is equally important.

X. STAGING

NPC are staged according to the American Joint Commission on Cancer (AJCC) staging system (Table 33.1). A new system, which standardizes staging of NPC

TABLE 33.1 Staging of Nasopharyngeal Cancer[a]

T1 Tumor confined to the nasopharynx

T2 Tumor extends to soft tissues of oropharynx and/or nasal cavity

T2a without parapharyngeal extension

T2b with parapharyngeal extension

T3 Tumor invades bony structures and/or paranasal sinuses

T4 Tumor with intracranial extension and/or involvement of cranial nerves, hypopharynx, infratemporal fossa or orbit

N0 No regional lymph node metastasis

N1 Unilateral metastasis in nodes(s) 6 cm or less, above the supraclavicular fossa

N2 Bilateral metastasis in node(s) 6 cm or less, above the supraclavicular fossa

N3 Metastasis in a node(s)

3a >6 cm

3b Extension to the supraclavicular fossa

M0 No distant metastasis

M1 Distant metastasis

Stage:	I	II	III	IV
Grouping:	T1	T2	T3	T4
AJCC 1977				
N0	I	II	III	IV
N1	II	II	III	IV
N2	III	III	III	IV
N3	IV	IV	IV	IV

[a]AJCC 1997 classification for nasopharyngeal carcinoma.

across continents, was adopted by the AJCC [47]. This system is a blend of the old AJCC and HO staging systems and reflects the biology of NPC and differs from AJCC staging in other head and neck sites.

XI. RADIATION THERAPY

Radiation therapy has been the mainstay in the management of nasopharyngeal cancer. Simulation of all patients is required. CT/MRI scans should be used as many of these tumors spread outside the nasopharynx into the adjacent sites. CT-based three-dimensional conformal/intensity-modulated radiation therapy (IMRT) plans are ideal as they enhance the ability to treat all of the tumor and deliver a high dose to the tumor and decrease the volume of normal tissue in the irradiated field.

XII. VOLUME TREATED

Nasopharyngeal carcinomas are usually treated by an upper neck field that encompasses the nasopharynx and the upper neck nodes. The nasopharyngeal field should include the following.

Anteriorly: At least post one-third of the maxillary sinus.
Superiorly: Post one-third of orbit, posterior ethmoid sinus, entire sphenoid sinus, clivus, and lower part of the occipital lobe.
Posteriorly: Include foramen magnum and entire bodies of C1 and C2 vertebrae. The field extends to the tip of the spinous process. The field extends further back in case of large post auricular lymph nodes. There should be a 1.5-cm margin beyond the enlarged neck node.
Inferiorly: Entire soft palate.

This field arrangement has to be adjusted depending on the extent of the tumor spread. The upper neck field is generally stopped above the larynx. A matching lower neck field is added starting from the level of the larynx to the suprasternal notch. Laterally the field includes the supraclavicular fossa. This usually corresponds to the field being 1–2 cm inside of the coracoid process of the scapula. A midline block is introduced from the start at the upper border of this lower neck field. This will shield the larynx and spinal cord.

After the upper neck has been treated to 39.6 Gy, a spinal cord block is introduced into the upper neck field and the posterior neck is treated with appropriate energy electrons to bring the dose of the posterior neck up to 50 Gy. The upper neck and lower neck are treated to 50 Gy and then a boost volume is treated, which encompasses the nasopharynx and enlarged lymph nodes. The dose is given at 1.8 to 2.0 cGy per fraction once a day 5 days a week for 7–8 weeks. The total doses to the various sites are:

a. Areas at risk for microscopic disease 50 Gy.
b. Nasopharynx T–T2, 70 Gy and T3–T4, 72–75 Gy.
c. Neck nodes: microscopic disease 50 Gy
   ≤2 cm 66 Gy
   ≥2 cm 70–75 Gy
d. Spinal cord: The maximum dose should not exceed 45 Gy.

XIII. BOOST VOLUME

The nasopharynx itself should be treated using a CT-based plan to ensure adequate coverage of any tumor extension, especially in the parapharyngeal space. Opposed lateral fields rarely cover these extensions and the retropharyngeal lymph nodes adequately except in very early stages. Care should be taken to minimize the dose to the temporal lobe of the brain, spinal cord, parotid, and temporomandibular joints.

XIV. DOSE RESPONSE

Most NPC are treated at 180–200 cGy per fraction. Other fractionation schemes using a higher dose per fraction and

TABLE 33.2 T1,T2

No.	Number of patients	Dose	Local control (%)	Reference
1[a]	8	65	100	Marks *et al.* [49]
2	196	65	91	Lee *et al.* [52]
3[b]	226	65–68	91	Tang *et al.* [51]
4	173	65	86	Sanguineti *et al.* [53]
5	555	75	88	Teo *et al.* [55]
6	73	65	96	Mesic *et al.* [48]
7	71	65–70	71	Bailet *et al.* [54]
8	21	66	81	Marcial *et al.* [58]

[a]Includes T3–T4 who received >65 Gy.
[b]Patient also received a 16-Gy boost by intracavitary cobalt-60.

shorter periods of time have been reported mainly from the Far East. Early (T1, T2) stage tumors are usually treated to a dose of 65–70 Gy (Table 33.2). Mesic *et al.* [48] reported on the experience at M.D. Anderson Hospital on 73 patients with T1 and T2 cancers. They found that by increasing the dose from 60 to 67.5 Gy, the local failure of T1, T2 NPC decreased from 28 to 7%. Similarly, Marks *et al.* [49] showed that patients properly simulated and treated with doses of 65–70 Gy had a local control of 100% with no difference between the dose range. Hoppe and co-workers [50] reported a local failure rate of 11% in patients treated with 65–70 Gy and field sizes of 92 cm^2. When this was increased to 128 cm^2, they had no failures. Thus patients with T1, T2 NPC treated to adequate fields and given doses of 65–70 Gy can be controlled adequately with radiation therapy (XRT) alone.

A. T3 T4

The results of treatment of T3 and T4 patients are shown in Tables 33.3 and 33.4 [48–60]. The usual doses have ranged from 60 to 75 Gy. Doses of 70 Gy or more are required to control this stage of tumor. T3 tumors have a better control than T4 tumors. Tumors invading into the skull and those with cranial nerve palsies have a poor prognosis. The local control is also poor in this group of patients. Bailet *et al.* [54] reported an overall local control of 60% in patients with T3, T4 tumors. Tang *et al.* [51] treated patients to 70–72 Gy and then added a boost to the nasopharynx by means of an intracavitary cobalt-60 source. A dose of 10–16.5 Gy was added. They reported a local control of 75% in this group of patients [51]. A similar improvement in local control has been reported by others, who added the boost by intracavitary sources [59], additional external beam radiation through multiple fields, or IMRT [60,61].

Thus all patients with advanced stage disease should be treated to a dose of at least 70 Gy and then evaluated for residual disease. If residual disease is present, then an additional boost of 10–15 Gy should be delivered either by an intracavitary source or external beams using IMRT or by similar three-dimensional plans to minimize the volume of normal tissues irradiated. Tumors with erosion of the base of the skull can still be controlled with radiation therapy. However, if there is gross disease in the middle cranial fossa or posterior fossa, local control and survival are poor even with radiation therapy [62].

TABLE 33.3 T3

No.	Number of patients	Dose	Local control (%)	Reference
1[a]	20	65–70	75	Marks *et al.* [49]
2	18	70	68	Sanguineti *et al.* [53]
3[b]	61	70–72	75	Tang *et al.* [51]
4		60–70	67	Perez *et al.* [56]
5	18	40–70	63	Chu *et al.* [57]
6	27	65–70	70	Hoppe *et al.* [50]
7	71	65–70	71	Bailet *et al.* [54]
8	21	66	81	Marcial *et al.* [58]

[a]Includes T3–T4 who received >65 Gy.
[b]Patient also received a 16-Gy boost by intracavitary cobalt-60.

TABLE 33.4 T4

No.	Number of patients	Dose	Local control (%)	Reference
1[a]	136	70–72	41	Tang *et al.* [51]
2[b]	179	70–74	68	Lin and Jan [59]
3[c]	71	65–70	61	Bailet *et al.* [54]
4	24	60–70	54	Cooper *et al.* [60]
5	31	70	68	Mesic *et al.* [48]
6	22	60–70	75	Chu *et al.* [57]
7	30	70–72	53	Sanguineti *et al.* [53]
8	20	65–68	75	Marks *et al.* [49]
9	50	66	74	Marcial *et al.* [58]

[a]All received a 10- to 16-GY IC boost.
[b]Included T3–T4 patients.
[c]Included T3–T4 patients.

XV. REIRRADIATION

Patients who fail radiation therapy can be retreated successfully with additional radiation therapy. These patients must be selected carefully and their treatment planned using CT/MRI. The time to recurrence appears to have an impact on local control and survival. Recurrences occurring in 12 months or less do worse than those in whom the recurrence occurs at 24 months or longer [63,64]. Patients whose tumors are confined to the NP have a better chance of achieving control then those where the tumor has extended outside the nasopharynx. Pryzant *et al.* [65] reported a 5-year survival of 32% when the tumor was confined to the NP as opposed to

9% with extension outside the NP. Lee *et al.* [66] have reported similar results that intracranial extension has a worse survival.

Treatment technique varies from external beam radiation therapy, brachytherapy, or a combination of both. If the entire treatment is given by EBRT, there is a greater risk of complications. Chua *et al.* [67] reported a 30% incidence of treatment and a 34% rate of neurological sequelae using EBRT as compared to 3–7% with brachytherapy alone. The complication rate also appears to be dose dependent. Hwang *et al.* [63] reported a 5% incidence of severe complications when the total cumulative dose was ≤ 120 Gy. This increased to 32% when the dose was greater than 120 Gy. Similarly, Pryzant *et al.* [65] reported a 4% incidence of severe complications with cumulative doses of ≤ 100 Gy and 39% when the doses exceeded > 100 Gy.

The dose of irradiation for retreatment has to be greater than 50 Gy. Wang [68] reported a local control rate of 38% with doses greater than 60 Gy. Similar results have been reported by other authors [63,64]. Because normal tissue tolerance limits the dose that can be delivered, retreatment has been carried out by brachytherapy [69,70], a combination of EBRT and brachytherapy, stereotactic three-dimensional treatment [71,72], or charged particles [73,74]. Lee *et al.* [74] reported that a combination of EBRT and brachytherapy resulted in a 5-year local control rate of 45% as compared to 29% by brachytherapy alone or 32% with external beam radiation. Retreatment is associated with the possibility of significant complication, such as soft tissue necrosis, brain necrosis, cranial nerve palsies, bone necrosis, and swallowing dysfunction. The incidence of severe complication ranges from 10 to 50% with higher rates being reported in those treated to large volumes and doses with external beam radiation therapy. The results of retreatment as reported by several sources are shown in Table 33.5 [75–86].

XVI. SURVIVAL

The survival of early stage T1, T2 tumors is excellent when treated with EBRT (Table 33.2). The survival rate of those with early T1, T2 tumors ranges from 85 to 100%. T3 tumors with minimal base of skull erosion have a good prognosis (Table 33.3). However, T4 tumors, especially those extending into the middle and posterior cranial fossa, have poor survivals (Table 33.4). A majority of the patients will present with neck nodes metastasis. The nodes respond well to radiation therapy. The neck control with external beam radiation therapy varies from 60 to 80% for N1, N2 neck disease. Neck node involvement signifies a greater risk of development of subsequent distant disease usually to lung, liver, and bone. The development of distant metastasis is correlated with neck disease rather than the T stage of the disease. The survival of patients treated with radiation therapy is shown in Table 33.5.

TABLE 33.5 Overall 5-Year Survival with Radiotherapy Alone in Locally Advanced Nasopharyngeal Carcinoma

Number of patients in trial	Percentage of patients alive at 5 years	References
170	39	Meyer and Wang [75]
146	37	Moench and Phillips [76]
82	62	Hoppe *et al.* [50]
99	24	Baker [77]
1605	32	Shu-Chen [78]
109	40	Marcial *et al.* [58]
251	52	Mesic *et al.* [48]
89	37	Applebaum *et al.* [79]
1578	48	Hsu *et al.* [80]
107	54	Vikram *et al.* [81]
91	62	Rahima *et al.* [82]
1379	41	Qin *et al.* [83]
464	35	Wang *et al.* [84]
166	36	Laramore *et al.* [85]
103	58	Bailet *et al.* [54]
126	54	Sutton *et al.* [86]
378	48	Sanquineti *et al.* [53]

XVII. CHEMOTHERAPY

Until 1985, locally advanced nasopharyngeal cancer was treated almost exclusively with radiotherapy, but in recent years chemotherapy has assumed an important role in its management. Nasopharyngeal carcinomas are highly sensitive and responsive to chemotherapy (Tables 33.1 and 33.2). Nasopharyngeal cancer has a high rate of distant metastases, more than 30% as compared to the incidence of systemic involvement from other head and neck tumor sites, which may mandate the use of chemotherapy up front in patients with locally advanced disease as part of the combined modality treatment. Single agents identified as being active include methotrexate, bleomycin, adriamycin, cisplatin, carboplatin, and, more recently, taxans and, to a lesser extent, 5-fluorouracil (5-FU) and vinca alkaloids. In general, for patients with head and neck cancers, especially in patients with NPC, combination agents seem more active than single drugs, and cisplatin-based combinations are more active than nonplatinol-based combination chemotherapy [87–89] (Table 33.6). The most widely used cisplatin combination in these patients is the combination of cisplatin and 5-FU 96 to 120-h infusion. The regimen is easy to administer and is well tolerated by most patients. 5-FU is given in the form of a 4- to 5-day continuous infusion at a dose of 1000 mg/m^2/day and cisplatin is given at a dose of 80–100 mg/m^2 on the first day of therapy [90,91]. Courses are repeated every 3 to 4 weeks.

TABLE 33.6 Response Rate to Combination Chemotherapy in Patients with Locally Advanced Nasopharyngeal Carcinoma

Chemotherapy	No.	RR%	CR%	Reference
Non-P based	12	81	33	Galligioni *et al.* [105]
P based	49	75	22	Tannock *et al.* [106]
P based	65	86	11	Garden *et al.* [107]
P based	24	75	29	Clark *et al.* [108]
PEB	39	98	66	Bachouchi *et al.* [109]
PEB	67	98	66	Azli *et al.* [110]
CCMBA	21	86	5	Siu *et al.* [93]
PF	8	75	50	Al Kourainy *et al.* [88]
PF	47	93	21	Dimery *et al.* [111]
PF				Geara *et al.* [112]
Local	61	79	18	
Nodal	50	76	32	
PF	27	93	37	Zidan *et al.* [113]
PF				Teo *et al.* [114]
Local	209	62	6	
Nodal	209	83	20	
PF	21	94	33	Onat *et al.* [115]
PFL	14	86	14	Chi *et al.* [95]
PFB	30	83	10	Azli *et al.* [116]
PFB	11	100	45	Zubizarreta *et al.* [118]

[a]B, bleomycin; E, epirubicin; F, flurouracil; P, cisplatin; CCMBA, cyclophosphamide, cisplatin, methotrexate, bleomycin, doxorubicin.

TABLE 33.7 Response Rate to Combination Chemotherapy in Patients with Recurrent/ Metastatic Nasopharyngeal Carcinoma

Chemotherapy[a]	No.	RR%	CR%	Reference
P and non-P based	17	53	22	Decker *et al.* [87]
P based	12	75	25	Al Kourainy *et al.* [88]
Non-P based	28	39	11	Choo and Tannok *et al.* [89]
P based	22	63	18	
PEB	44	50	20	Mahjoubi *et al.* [92]
CCMBA	61	52	16	Siu *et al.* [93]
PF	26	73	23	Marchini *et al.* [94]
PLF	21	86	14	Chi *et al.* [95]
CP/F	42	38	17	Yeo *et al.* [96]
PFB	25	40	4	Su *et al.* [97]
PFB	49	79	19	Boussen *et al.* [98]
PFE-M	46	61	9	Cvitkovic *et al.* [99]
CP/T	32	75	3	Tan *et al.* [100]
CP/T	14	57	14	Fountzilas *et al.* [101]
CP/T	27	59	11	Yeo *et al.* [102]
PI	18	59	15	Stein *et al.* [103]

[a]B, bleomycin; CP, carboplatin; E, epirubicin; F, flurouracil; M, mitomycin; P, cisplatin; T, paclitaxel: I, ifosfamide; CCMBA, cyclophosphamide, cisplatin, methotrexate, bleomycin, doxorubicin.

XVIII. CHEMOTHERAPY FOR RECURRENT/METASTATIC DISEASE

Patients with recurrent or metastatic cancer of the nasopharynx should be considered distinct from those patients with cancer originating elsewhere in the head and neck. Patients with NPC presenting with metastatic disease and those patients who recur after initial curative treatments, and are beyond further local treatment(s) of surgery and/or reirradiation, are treated with systemic chemotherapy for palliative purpose. Because of the rarity of this disease, especially in Western countries, all reported trials of chemotherapy for recurrent/metastatic NPC are phase II studies (Table 33.7). Before 1980, nonplatinol based combinations have been used [87–89]. Since then, these combinations have been most widely investigated and reported in the world literature in patients with NPC (Table 33.8). Before discussing the results of these trials, it is important to emphasize the important prognostic factors that may influence the overall response rate (RR) and especially the complete response (CR) to chemotherapy in these patients [90,91]. The most important of these prognostic factors are performance status, histopathology (WHO I, II, III), local vs systemic disease, site of the systemic metastasis, the bulk of recurrent or metastatic cancers being treated, types of chemotherapy combinations used, and adequacy of treatment given. Because some patients with recurrent/metastatic NPC treated with systemic chemotherapy may achieve a prolonged remission and possible cure [88,104], it is also important to mention some of the prognostic factors that may effect survival in these patients. These are performance status, the quality of response to the systemic chemotherapy, number of courses, and adequacy of chemotherapy administered.

As shown in Table 33.7 the majority of cisplatin-based combinations reported are cisplatin and 5-FU infusion, with or without other agents, such as leucovorin or bleomycin. The overall RR is between 50 and 65%, but, more important, the CR rate is about 15–20% of the patients treated. It does not seem from these phase II trials that the addition of other agents to cisplatin and 5-FU infusions has produced higher overall RR or CR, but may have added to the toxicities of the treatment. More recently, other active combinations have been investigated utilizing newer active agents such as Taxol [100–102] or Ifosfamide [103]. The majority of these patients who achieve CR to chemotherapy are alive more than 2 years, or even considered cured [88,89,104]. This is especially true in patients with histological cancers classified as WHO II or III. Fandi *et al.* [104] reported on 20 long-term unmaintained complete responders to chemotherapy; 14 of 20 were still alive with no evidence of disease after treatment 82+ to 190+ months. No standard duration of further treatment after achieving CR has been reported. It is our opinion that after a CR has been confirmed by biopsy (for local recurrence) or X-rays (CT scan, bone scan) in patients with nodal or systemic recurrence, an additional

TABLE 33.8 Results of Prospective, Randomized Phase III Trials of Chemoradiotherapy [CT-RT vs Radiotherapy (RT)] in Locally Advanced Disease[a]

Year	Timing	Agent[a]	Survival	*p* Value	Comments[b]	Reference
1988	Adjuvant	CAV	67.3% RT 58.5% CT-RT	NS	4 year	Rossi *et al.* [134]
1995	Induction + adjuvant	PF	80.5% RT 80% CT-RT	NS	2 year	Chan *et al.* [141]
1996	Induction	PEB	45% RT 40% CT-RT	NS	5 year	Cvitkovic *et al.* [142], International Nasopharyngeal Cancer Study Group [143], El-Guedari [144]
1998	Induction	PE	71% RT 78% CT-RT	NS	3 year	Chua *et al.* [145]
1998/2001	Concurrent + adjuvant	P,PF	37% RT 67% CT-RT	< 0.001	5 year, DRS 46% vs 74%	Al-Sarraf *et al.* [146–148]
2001	Induction	PFB	56% RT 63% CT-RT	 NS	RFS 0.05 FLR 0.04	Ma *et al.* [151]

[a]A, adriamycin; B, bleomycin; E, epirubicin; F, 5-fluorouracil; P, cisplatin; V, vincristine.
[b]DRS, disease-related survival; RFS, relapse-free survival; FLR, freedom from local recurrence.

four to six courses of chemotherapy may need to be administered. These patients with CR need to be followed closely, especially the first 2 years after the end of chemotherapy. In addition, biological agents, such as inhibitors of the epidermal growth factor receptor (EGFR), and angiogenesis given simultaneously with chemotherapy, or after significant cytoreduction postchemotherapy, may play an important role in the future of this disease.

XIX. CHEMOTHERAPY FOR LOCALLY ADVANCED AND PREVIOUSLY UNTREATED NASOPHARYNGEAL CANCER

Because of the high incidence of locoregional failure to radiotherapy, despite the initial high clearance rate, and because of the high incidence of distant metastasis in patients with nasopharyngeal cancers, combined chemotherapy with standard radiotherapy is a very attractive concept. The different treatments may be complementary and may even be synergistic. There are several ways of combining chemotherapy with radiotherapy: it can be given as induction chemotherapy followed by irradiation treatment, concomitantly, or adjuvant (radiotherapy followed by chemotherapy). Also, a combination of these has been investigated.

XX. INDUCTION CHEMOTHERAPY

Induction chemotherapy was explored in the mid-1970s in patients with locally advanced stage III and IV head and neck cancers of all sites. Most of these patients were inoperable or unresectable and included patients with nasopharyngeal carcinomas. High overall and clinical complete response rates were obtained. The feasibility of sequential induction chemotherapy followed by radiotherapy was established. This led to many phase II studies using the same treatments in patients with locally advanced and previously untreated nasopharyngeal cancers (Table 33.7). Again it is important to mention the factors that may influence the response rate, complete response to chemotherapy, and overall survival. The factors that might affect the response rate are stage of the NPC, especially the N stage, performance status, histopathology, type of chemotherapy combinations, number of courses, the adequacy of the chemotherapy doses given, and evaluation of response at the primary site, whether it was evaluated clinically or by CT scan or MRI. Because the majority of patients present initially with a T stage of 3 or 4, some changes may continue to persist locally when CT or MRI is done, even though the patients may have repeated negative biopsies of these areas and are considered cured without further treatment. Factors that may affect survival are stage of the disease, type of response to chemotherapy (partial vs complete), histopathology, and performance status of these patients [118]. The majority of the trials are with cisplatin-based combination, especially the combination of cisplatin and 5-FU infusion. In a few trials, leucovorin or bleomycin is added to this combination. These studies demonstrated an overall response rate to chemotherapy of 80–90%. In some trials, up to 66% of the responses were complete. Because of the various prognostic factors that may have influenced the response rate, no statement could be made of the best possible chemotherapy combination. Geara *et al.* [112] and

Teo *et al.* [114] reported a much higher CR to chemotherapy for the nodal disease than at the primary tumor. Both these investigators utilized CT scans to evaluate the response rate. In our opinion, both in patients with head and neck cancers and in patients with NPC treated with chemotherapy, the complete response rate is usually higher at the primary site than at regional lymph nodes. Many investigators have reported an improved survival when chemotherapy given before radiation therapy is compared to historic controls treated with radiation therapy alone. This is especially true with the administration of adequate doses of two to three courses of cisplatin and 5-FU infusion. Teo *et al.* [114] reported on the results of two courses of cisplatin and 5-FU infusion followed by radiotherapy in patients with node-positive locally advanced nasopharyngeal cancers (209) compared to similar patients treated with radiotherapy alone during the same period. The chemotherapy group had significantly more bulky nodes, lower cervical/supraclavicular nodes, and more advanced overall stages than the nonchemotherapy patients. Unfortunately, the duration of the 5-FU infusion was for 3 days, instead of the standard 5 days, and only two courses of chemotherapy were given instead of the usual three courses. Despite this, the addition of chemotherapy to radiation treatment significantly enhanced the local control in node-positive nasopharyngeal cancer patients in general and node-positive T3 and T4 stage IV in particular. The improvement of 5-year survival to sequential chemoradiotherapy in patients with stage IV cancers ranged between 50 and 55%, as compared to the 5-year survival of less than 30% to radiotherapy only. It is important when evaluating survival in patients with NPC to have a minimum follow-up of 3 years for all patients before such an evaluation can be considered adequate. Induction chemotherapy before radiotherapy was administered in children and adolescence patients with stage IV NPC, with an achievement of 100% overall response rate and 45% CR to chemotherapy [117].

XXI. CONCURRENT CHEMORADIOTHERAPY

The concurrent approach of chemoradiotherapy is attractive, with the possible advantages of synergy and enhancement between chemotherapy and radiotherapy, in addition to the additive effects. Combination chemoradiotherapy has been investigated in many other solid malignancies, which report an improved disease-free survival and overall survival of the combined approach over radiotherapy alone. Many phase II trials of concurrent chemoradiotherapy in locally advanced patients with NPC, especially those with stage IV disease, have been reported [119–130]. Many chemotherapeutic agents have such a synergistic and/or enhancing effect when used concomitantly with radiotherapy, especially cisplatin [113–133].

One possible advantage of the use of concomitant cisplatin and radiotherapy in patients with head and neck cancers, especially in patients with nasopharyngeal carcinomas, is the lack of increased local side effects, especially mucositis, as compared with other agents such as methotrexate, bleomycin, or 5-FU. Most of the investigators gave standard one fraction per day irradiation. The radiation dose was the same without chemotherapy (>6400 cGy) in most of the cases. When the results of concomitant chemoradiotherapy were compared to historical matched control patients, most of the investigators reported improvement in local control, disease-free survival, and overall survival.

Other agents, such as taxanes, and gemzar, have shown to be radiation sensitizers and may need to be investigated in the future with concomitant radiotherapy in patients with NPC. In an early study, the RTOG [128–130] reported the results of treating 124 patients with locally advanced, inoperable, or unresectable stages III (24) and IV (100) cancers. It included 28 patients (27 stage IV) with nasopharyngeal cancers with concomitant single-agent cisplatin and radiotherapy. The dose of cisplatin was 100 mg/m^2 given intravenously with hydration and mannitol diuresis once every 3 weeks, concurrent with standard radiation therapy to a total dose of up to 7380 cGy. The complete response rate for all patients was 70%, and for patients with nasopharyngeal cancers, the complete response rate was 89%. When the survival of patients with nasopharyngeal cancers was compared with those with tumors of other sites, significant improvement with concurrent chemoradiotherapy was observed. What was more interesting was the fact that about 55% of the patients with cancers of the nasopharynx were still alive after 5 years. No additive local toxicities were observed to the chemoradiotherapy when compared to matched historical patients with head and neck cancers treated with the same radiotherapy alone. Because of the good results achieved with concomitant chemoradiotherapy in locally advanced NPC, the same treatment was administered to patients with early stage (stage II) cancer with good results.

Wolden *et al.* [150] reported on the results of accelerated concurrent boost radiotherapy and cisplatin for two courses in 50 patients with advanced (stages II–IVb) nasopharyngeal cancer. Adjuvant chemotherapy of cisplatin and 5-FU infusion for three courses was given to 37 patients. These patients were compared to 51 patients treated with standard fractionation radiotherapy without chemotherapy. They reported significant improvement with 3-year actuarial local control (89% vs 74%), progression-free survival (66% vs 54%), and survival (84% vs 71%).

XXII. ADJUVANT CHEMOTHERAPY

The benefits of giving active systemic chemotherapy after the curative intent of radiotherapy in patients with locally advanced NPC are to consolidate the local control achieved by radiation treatment and to reduce the incidence of microscopic systemic metastasis. This may result in the improvement of disease-free survival, and eventually improvement of overall survival of these patients. One prospective randomized trial [134] utilizing a nonplatinol combination did not show a benefit of adjuvant chemotherapy and is discussed later. In another phase II trial [135], the combination chemotherapy used was too toxic, resulting in high treatment-related mortality. However, other reports [136–140,149] of single arm studies do show the possible benefit of adjuvant chemotherapy in patients with NPC. The type of chemotherapeutic agents given, doses, and length of treatment may determine the incidence of possible local side effects and overall morbidity of adjuvant chemotherapy after curative radiation treatment in these patients. Many newer agents, and combinations, may have less local mucosal side effects and may need to be investigated in the future. Eventually, the question of the possible benefit of adjuvant chemotherapy in locally advanced NPC will need to be answered with prospective randomized phase III trials.

XXIII. RANDOMIZED TRIALS

Six prospective randomized phase III trials (Table 33.8) comparing chemoradiotherapy to the same radiotherapy alone have been reported so far in patients with locally advanced nasopharyngeal cancers [134,142–148]. Three studies gave only sequential chemotherapy followed by radiotherapy. In two of these trials, other cisplatin combinations than with 5-FU infusion were used. Results of phase II studies using sequential initial chemotherapy followed by radiotherapy have been reported in recent years with great interest. A study was conducted by the International Nasopharyngeal Cancer Study Group [142,143] where 339 patients were randomized to receive three induction courses of chemotherapy using the combination of bleomycin, cisplatin, and epirubicin followed by radiation therapy or radiation therapy alone. An initial report of this study showed that induction chemotherapy prolonged disease-free survival but there was no difference in overall survival between the two groups. A recent update of this trial after a median follow-up of 74 months confirmed these results [144]. Unfortunately, the mortality rate on the chemotherapy arm was 9%. This study also demonstrated a high salvage rate in chemotherapy naive patients treated initially with radiotherapy only. The high mortality rate with this study may be due to the possible use of other cisplatin combinations, than 5-FU infusion, and the high salvage rate of the chemotherapy naive patients (standard arm) may contribute to the results of this trial.

Chan *et al.* [141] reported a smaller study of 82 patients that did not demonstrate an advantage to induction chemotherapy with two courses of cisplatin and 5-FU infusion followed by radiotherapy vs radiation treatment alone. In this study, two courses of chemotherapy were given and a smaller dose of 5-FU infusion (60% of standard dose) was used, which may have affected the results of the study. The number of patients randomized was also small, and the median follow-up of these patients was short (28 months).

Chua *et al.* [145] compared cisplatin and epirubicin in two to three cycles followed by radiotherapy versus radiation therapy alone and found no difference in the 3-year relapse-free and 3-year overall survival rates between the two groups. When only the evaluable patients (286/334) were analyzed, a trend of improved 3-year relapse-free survival was observed (58% vs 46%, $p=0.053$). In the subgroup of 49 patients with bulky neck nodes >6 cm, a significantly improved 3-year relapse-free survival rate was observed in favor of the combined group (63% vs 28%, $p=0.026$), and significant improvement in 3-year overall survival (73% vs 37%, $p=0.057$) was observed. We do feel that the follow-up of these patients was short, as most of the difference in survival was usually observed after 3 years.

A randomized adjuvant chemotherapy trial was reported by Rossi *et al.* [134]. In that study, 229 patients were randomized to radiotherapy alone or radiation treatment followed by six courses of chemotherapy of noncisplatin combination. No significant overall survival or relapse-free survival rates were observed. In our opinion, although based on nonrandomized trials, noncisplatin combinations are inferior to cisplatin combination chemotherapy in this disease and other sites of head and neck cancers, which may have contributed to the negative results in this study.

The encouraging results obtained with concurrent cisplatin and radiotherapy led to a prospective randomized phase III Intergroup trial conducted in North America by the Southwest Oncology Group with the participation of the RTOG and ECOG [146–148]. In this study, 193 patients were registered and 185 patients were randomized and stratified to receive radiation therapy alone (92) to a total dose of 70 Gy versus a combination of chemotherapy and radiotherapy (93). Chemotherapy consisted of three cycles of cisplatin given at a dose of 100 mg/m^2 concurrent with the same radiation therapy every 3 weeks. This was followed by three cycles every 4 weeks of cisplatin at a dose of 80 mg/m^2 on day 1 and 1000 mg/m^2/day of 5-FU days 1 through 4 given as a continuous infusion. Adjuvant chemotherapy was given to enhance the locoregional control and to decrease systemic metastasis. Patients were stratified according to the T and N stages, performance status, and histopathology. All slides were reviewed centrally by independent pathologists. This important trial demonstrated a highly significant difference in the overall survival at 5 years (37% in RT vs 67% in CT-RT, $p<0.001$), as well as a significant difference in progression-free survival (29% for RT group vs 58% for CT-RT patients, $p<0.001$). Significantly less local, nodal, and distant metastasis was observed with the combined treatment arm. The possible reasons for the success of this trial were that patients were stratified to important and accepted prognostic factors and total chemoradiotherapy was used in the investigational arm. This consists of initial concurrent cisplatin and radiotherapy to produce the best locoregional control, followed by adjuvant chemotherapy of cisplatin and 5-FU infusion for three courses to consolidate this control and to reduce the incidence of systemic metastasis. This study demonstrates clearly that chemoradiotherapy is highly effective in the treatment of locally advanced nasopharyngeal carcinoma and is now considered as the standard of care in the treatment of patients with nasopharyngeal cancers.

A sixth randomized phase III study was recently reported by Ma *et al.* [151]. They compared induction chemotherapy with cisplatin, 5-FU infusion, and belomycin followed by radiotherapy to patients who received radiation treatment only. They reported significant improvement in relapse-free survival (59% vs 49%) and a significant improvement in the 5-year freedom from the local recurrence rate (82% vs 74%) in favor of the combined group. The 5-year overall survival rates were 63% for the combined group vs 56% for radiotherapy, and this difference was not significant. Although numerous patients were included in this randomized trial, the majority of patients were stage III disease. Also, 20% less 5-FU dose was given because of the addition of bleomycin to the chemotherapy, and patients were given two or three courses of induction chemotherapy. The authors did not specify the number of patients that received two cycles of the chemotherapy. The difference in overall survival was not significant. This may be due to the number of patients randomized in this study that prevented from detecting the less than 10% difference.

In our experience with locally advanced NPC previously untreated, we gave induction chemotherapy of cisplatin (100 mg/m^2 IV) on day 1 and 5-FU (1000 mg/m^2/day) on days 1 through 5 by continuous infusion every 3 weeks, repeated for three courses. This was followed by standard radiotherapy and concurrent cisplatin (80 mg/m^2 IV) every 4 weeks for three courses. The early results of this combined approach, about 90% of patients with stage IV, are alive more than 3 years and the majority are disease free.

XXIV. CONCLUSION

Chemotherapy is a very important part in the treatment and improved results in patients with nasopharyngeal carcinoma. In patients with recurrent/metastatic disease, effective and adequate chemotherapy alone and other palliative treatments, are highly recommended, with possibly achieving a 15–20% long-term survival and possibly a cure. Combination platinol-based chemotherapy is the most active and recommended in these patients. In patients with locally advanced NPC (stages III and IV), combination chemotherapy given neoadjuvantly (induction), concomitant with, or after curative total radiotherapy did improve local control, decreased the systemic metastasis, and produced improvement of disease-free survival and overall survival in the majority of the studies reported. Concurrent chemoradiotherapy, followed by adjuvant chemotherapy, significantly improved local control systemic control, progression-free survival, and overall survival in a phase III prospective randomized trial. Different sequencing of induction chemotherapy followed by concomitant chemoradiotherapy, new agents, and combinations need to be investigated in patients with locally advanced NPC.

References

1. Chu, A. M., Cutler, S. J., and Young, J. L. (1975). Third National Cancer Survey: Incidence data. *Natl. Cancer Inst. Monogr.* 411–454.
2. Buell P. (1965). Nasopharyngeal cancer in Chinese of California. *Bx J. Cancer* **19**, 459–470.
3. Quisenberry, W., Zieman, and Jasinki, D. (1967). Ethnic differences of nasopharyngeal cancer in hawaii. *In* "Cancer of Nasopharynx," UICC Monograph Services No. 1, pp. 77–86. Munkogaard, Copenhagen.
4. King, H., and Haenszelk. (1972). Cancer mortality among foreign and native born Chinese in the United States. *J. Chron.* **26**, 623–646.
5. Epstein, M. A., and Actiong, B. G. (1979). "Introduction Discovery and General Biology of the Virus in the Epstein-Barr Virus" (M.A. Epstein and B.A. Achong, eds.) pp. 1–22. Springer, New York.
6. Henle, W., and Henle, A. (1976). Epstein-Barr virus-specific IgA serum antibodies as an outstanding feature of nasopharyngeal carcinoma. *Int. J. Cancer* **17**, 1–7.
7. Epstein, M. A., and Actiong, B. G. (eds.) (1979). "Introduction Discovery and General Biology of the Virus in the Epstein-Barr Virus," p. 21. Springer, New York.
8. Epstein, M. A., and Actiong, B. G. (eds.) (1979). "Introduction Discovery and General Biology of the Virus in the Epstein-Barr Virus," p. 22. Springer, New York.
9. Epstein, M. A., and Actiong, B. G. (eds.) (1979). "Introduction Discovery and General Biology of the Virus in the Epstein-Barr Virus," p. 17. Springer, New York.
10. Epstein, M. A., and Actiong, B. G. (eds.) (1979). "Introduction Discovery and General Biology of the Virus in the Epstein-Barr Virus," p. 18. Springer, New York.
11. Epstein, M. A., and Actiong, B. G. (eds.) (1979). "Introduction Discovery and General Biology of the Virus in the Epstein-Barr Virus," p. 19. Springer, New York.
12. Epstein, M. A., and Actiong, B. G. (eds.) (1979). "Introduction Discovery and General Biology of the Virus in the Epstein-Barr Virus," p. 20. Springer, New York.
13. Fahraeus, R., Rymo, L., Rhim, J. S., and Klein, G. (1990). Morphological transformation of human keratinocytes expressing the LMP gene of Epstein-Barr virus. *Nature* **345**, 447–449.
14. Gregory, C. D., Dive, C., Henderson, S., Smith, C. A., Williams, G. T., Gordon, J., and Rickinson, A. B. (1991). Activation of Epstein-Barr virus latent genes protects human B cells from death by apoptosis. *Nature* **349**, 612–614.
15. Sixbey, J. W., Shirley, P., Chesney, P. J., Buntin, D. M., and Resnick, L. (1989). Detection of a second widespread strain of Epstein-Barr virus. *Lancet.* **2**, 761–765.
16. Chen, X. Y., Pepper, S. D., and Arrand, J. R. (1992). Prevalence of the A and B types of Epstein-Barr virus DNA in nasopharyngeal carcinoma biopsies from southern China. *J. Gen. Virol.* **73**, 463–466.

17. Levine, P. H., Pocinki, A. G., Madigan, P., and Bale, S. (1992). Familial nasopharyngeal carcinoma in patients who are not Chinese. *Cancer* **70**, 1024–1029.
18. Henderson, B. E., Louse, E., and Sooltoo, J. J. (1976). Risk factors associated with nasopharyngeal carcinoma. *New Engl. J. med.* **295**, 1101–1106.
19. Yan, L., Xi, Z., and Dreitner, B. (1989). Epidemiological studies of nasopharyngeal cancer in the Guangzhou area. *China Acta. Otolaryngol.* **107**, 424–427.
20. Epstein, M. A., and Actiong, B. G. (eds.) (1979). "Introduction Discovery and General Biology of the Virus in the Epstein-Barr Virus," p. 5. Springer, New York.
21. Epstein, M. A., and Actiong, B. G. (eds.) (1979). "Introduction Discovery and General Biology of the Virus in the Epstein-Barr Virus," p. 6. Springer, New York.
22. Simons, M. J., Wee, G. B., Day, N. E., Morris, P. J., Shanmugaratnam, K., and De-The, G. B. (1974). Immunogenetic aspects of nasopharyngeal carcinoma: I. Differences in HL-A antigen profiles between patients and control groups. *Int. J. Cancer* **13**, 122–134.
23. Moore, S. B., Pearson, G. R., Neel, H. B., III and Weiland, L. H. (1983). HLA and nasopharyngeal carcinoma in North American Caucasoids. *Tissue Antigens* **22**, 72–75.
24. Effert, P., McCoy, R., Abdel-Hamid, M., Flynn, K., Zhang, Q., Busson, P., Tursz, T., Liu, E., and Raab-Traub, N. (1992). Alterations of p53 gene in nasopharyngeal carcinoma. *J. Virol.* **66**, 3768–3775.
25. Spruck, C. H., III, Tsai, Y. C., Huang, D. P., Yang, A. S., Rideout, W. M., III, Gonzalez-Zulueta, M., Choi, P., Lo, K. W., Yu, M. C., and Jones, P. A. (1992). Absence of p53 gene mutations in primary nasopharyngeal carcinomas. *Cancer Res.* **52**, 4787–4790.
26. Epstein, M. A., and Actiong, B. G. (eds.) (1979). "Introduction Discovery and General Biology of the Virus in the Epstein-Barr Virus," p. 7. Springer, New York.
27. Epstein, M. A., and Actiong, B. G. (eds.) (1979). "Introduction Discovery and General Biology of the Virus in the Epstein-Barr Virus," p. 8. Springer, New York.
28. Epstein, M. A., and Actiong, B. G. (eds.) (1979). "Introduction Discovery and General Biology of the Virus in the Epstein-Barr Virus," p. 9. Springer, New York.
29. Epstein, M. A., and Actiong, B. G. (eds.) (1979). "Introduction Discovery and General Biology of the Virus in the Epstein-Barr Virus," p. 10. Springer, New York.
30. Epstein, M. A., and Actiong, B. G. (eds.) (1979). "Introduction Discovery and General Biology of the Virus in the Epstein-Barr Virus," p. 11. Springer, New York.
31. Fong, Y. Y., and Walsh, E. O. (1971). Carcinogenic nitrosamines in Catonese salt-dried fish. *Lancet* **2**, 1032.
32. Bartsch, H., and Montesano, R. (1984). Relevance of nitrosamines to human cancer. *Carcinogenesis* **5**, 1381–1393.
33. O'Neill, I. K., Chen, J., and Bartsch, H. (ed.) Relevance to human cancer of N-nitroso compounds, tobacco smoke and mycotoxines. *Int. Agency Res. Cancer Sci. Publ.* **105**
34. Epstein, M. A., and Actiong, B. G. (eds.) (1979). "Introduction Discovery and General Biology of the Virus in the Epstein-Barr Virus," p. 14. Springer, New York.
35. Lin, T. M., Chen, K. P., and Lin, C. C. (1973). Retrospective study on nasopharyngeal carcinoma. *J. Natl. Cancer Instit.* **51**, 1403–1408.
36. Vaughan, T. L. (1989). Occupation and squamous cell cancers of the pharynx and sinonasal cavity. *Am. J. Ind. Med.* **16**, 493–510.
37. Ng, T. P. (1986). A case-referent study of cancer of the nasal cavity and sinuses in Hong Kong.*Int. J. Epidemiol.* **15**, 171–175.
38. Epstein, M. A., and Actiong, B. G. (eds.) (1979). "Introduction Discovery and General Biology of the Virus in the Epstein-Barr Virus," p. 13. Springer, New York.
39. Roush, G. C., Walrath, J., Stayner, L. T., Kaplan, S. A., Flannery, J. T., and Blair, A. (1987). Nasopharyngeal cancer, sinonasal cancer and occupations related to formaldehyde: A case-control study. *J. Natl. Cancer Inst.* **79**, 1221–1224.
40. Yu, Z. H., Xu, G. Z., Huang, Y. R., Hu, Y. H., Su, X. G., and Gu, X. Z. (1985). Value of computed tomography in staging the primary lesion (T-staging) of nasopharyngeal carcinoma (NPC): An analysis of 54 patients with special reference to the parapharyngeal space. *Int. J. Radiat. Oncol. Biol. Phys.* **11**(12), 2143–2147.
41. Alni, P., Cellai, E., Chiavacci, A., Fallai, C., Giannardi, G., Fargnoli, R., and Villari, N. (1990). Computed tomography in nasopharyngeal carcinoma. Part I. T-stage conversion with CT staging. *Int. J. Radiat. Oncol. Biol. Phys.* **19**(5), 1171–1175.
42. Kraiphibul, P., Atichartakarn, V., Clongsusuek, P., Kulapaditharom, B., Ratanatharathorn, V., and Jenjitranant, J. (1989). Changes in T-staging of nasopharyngeal carcinoma by CT scan. *J. Med. Assoc. Thai* **72**(12), 661–665.
43. Olmi, P., Fallai, C., Colagrande, S., and Giannardi, G. (1995). Staging and follow-up of nasopharyngeal carcinoma: Magnetic resonance imaging versus computerized tomography. *Int. J. Radiat. Oncol. Biol. Phys.* **32**(3), 795–800.
44. Sakata, K., Hareyama, M., Tamakawa, M., Oouchi, A., Sido, M., Nagakura, H., Akiba, H., Koito, K., Himi, T., and Asakura, K. (1999). Prognostic factors of nasopharynx tumors investigated by MR imaging and the value of MR imaging in the newly published TNM staging. *Int. J. Radiat. Oncol. Biol. Phys.* **43**(2), 273–278.
45. Ng, S. H., Chang, T. C., Ko, S. F., Yen, P. S., Wan, Y. L., Tang, L. M., and Tsai, M. H. (1997). Nasopharyngeal carcinoma: MRI and CT assessment. *Neuroradiology* **10**, 741–746.
46. Chong, V. F., and Ian, F. F. (1997). Detection of recurrent nasopharyngeal carcinoma: MRI versus CT. *Radiology* **202**(2), 463–470.
47. Fleming, I. D., Cooper, J. S., Henson, D. E., *et al*, (eds.) (1997). American Joint Committee on Cancer (AJCC) Cancer Staging manual, 5th ed. Lippincott-Raven, Philadelphia.
48. Mesic, J. B., Fletcher, G. H., and Goepfert, H. (1981). Megavoltage irradiation of epithelial tumors of the nasopharynx. *Int. J. Radiat. Oncol. Biol. Phys.* **7**, 447–453.
49. Marks, J. E., Bedwinek, J. M., Lee, F., Purdy, J. A., and Perez, C. (1982). Dose response analysis for nasopharyngeal carcinoma. *Cancer* **50**, 1042–1050.
50. Hoppe, R. T., Goffinet, O. R., and Bagshaw, M. A. (1976). Carcinoma of the nasopharynx: Eighteen years experience with megavoltage. *Cancer* **37**, 2605–2612.
51. Tang, S. G., See, L. C., Chen, W. C., Tsang, S., Chang, J. T., and Hong, J. H. (2000). The effect of nodal status on determinants of initial treatment response and patterns of relapse-free survival in nasopharyngeal carcinoma. *Int. J. Radiat. Oncol. Biol. Phys.* **47**(4), 867–873.
52. Lee, A. W., Sham, J. S., Poon, Y. F., and Ho, J. H. (1989). Treatment of stage I nasopharyngeal carcinoma: Analysis of the patterns of relapse and the results of withholding elective neck irradiation. *Int. J. Radiat. Oncol. Biol. Phys.* **17**(6), 1183–1190.
53. Sanguineti, G., Geara, F. B., Garden, A. S., *et al*. (1997). Carcinoma of the nasopharynx treated by radiotherapy alone: Determinants of local and regional control. *Int. J. Radiat. Oncol. Biol. Phys.* **37**, 985–996.
54. Bailet, J. W., Mark, R. J., Abemayor, E., *et al*. (1992). Nasopharyngeal carcinoma: Treatment results with primary radiation therapy. *Laryngoscope* **102**, 965–972.
55. Teo, P. M., Leung, S. F., Fowler, J., Leung, T. W., Tung, Y., O, S. K., Lee, W. Y., and Zee, B. (2000). Improved local control for early T-stage nasopharyngeal carcinoma: A tale of two hospitals. *Radiother. Oncol.* **57**(2), 155–166.
56. Perez, C. A., Devineni, V. R., Marcial, V. V., Marks, J. E., Simpson, J. R., and Kucik, N. (1992). Carcinoma of the nasopharynx: Factors affecting prognosis. *Int. J. Radiat. Oncol. Biol. Phys.* **23**(2), 271–280.
57. Chu, A. M., Flyn, M. B., Achino, E., Mendoza, E. F.,Scott, R. M., and Jose, B. (1984). Irradiation of the nasopharygeal carcinoma: Correlations

with treatment factors and stage. *Int. J. Radiat. Oncol. Biol. Phys.* **10**(12), 2241–2249.

58. Marcial, V. A., Hanley, J. A., Chang, C., *et al.* (1980). Split-course radiation therapy of carcinoma of the nasopharynx: Results of a national collaborative clinical trial of the Radiation Therapy Oncology Group. *Int. J. Radiat. Oncol. Biol. Phys.* **6**, 409–414.
59. Lin, J. C., and Jan, J. S. (1999). Locally advanced nasopharyngeal cancer: Long term outcomes of radiation therapy. *Radiology* **211**(2), 513–518.
60. Cooper, J. S., DelRowe, J., and Newell, J. (1983). Regional stag IV carcinoma of the nasopharynx treated by aggressive radiotherapy. *Int. J. Radiat. Oncol. Biol. Phys.* **9**(11), 1737–1745.
61. Leibel, S. A., Kutcher, G. H., Harrison, L. B., Fass, D. E., Bwimann, C. M., Hunt, M. A., Mohan, R., Brewster, L. J., Ling, C. C., and Juks, Z. Y. (1991). Improved dose distribution for 3D conformal boost treatments in carcinoma of the nasopharynx. *Int. J. Radiat. Oncol. Biol. Phys.* **29**(4), 823–833.
62. Yan, J. H., Qin, D. X., Hu, Y. H., Cai, W. M., Xu, G. Z., Wu, X. L., Li, S. Y., and Gu, X. Z. (1989). Management of local residual primary lesion of nasopharyngeal carcinoma (NPC): Are higher doses beneficial? *Int. J. Radiat. Oncol. Biol. Phys.* **16**, 1465–1469.
63. Hwang, J. M., Fu, K. K., and Phillips, T. L. (1998). Results and prognostic factors in the retreatment of locally recurrent nasopharyngeal carcinoma. *Int. J. Radiat. Oncol. Biol. Phys.* **5**, 1099–1111.
64. Chang, J. T. C., See, L. C., Liao, C. T., Ng, S. H., Wang, C. H., Chen, I. H., Tsang, N. M., Tseng, C. K., Tang, S. G., and Hong, J. H. (2000). Locally recurrent nasopharyngeal carcinoma. *Radiother. Oncol.* **54**, 135–142.
65. Pryzant, R. M., Wendt, C. D., Delclos, L., and Peters, L. J. (1992). Retreatment of nasopharyngeal carcinoma in 53 patients. *Int. J. Radiat. Oncol. Bio. Phys.* **22**, 941–947.
66. Lee, A. W., Foo, W., Law, S. C. K., Poon, Y. F., Sze, W. M., Tung, S. T., and Lau, W. H. (1997) Reirradiation for recurrent nasopharyngeal carcinoma: Factors affecting the therapeutic ratio and ways for improvement. *Int. J. Radiat. Oncol. Biol. Phys.* **38**, 43–52.
67. Chua, D. T., Sham, J. S., Kwong, D. L., Wei, W. I., Au, G. K., and Choy, D. (1998). Locally recurrent nasopharyngeal carcinoma: Treatment results for patients with computed tomography assessment. *Int. J. Radiat. Oncol. Biol. Phys.* **41**(2), 379–386.
68. Wang, C. C. (1987). Re-irradiation of recurrent nasopharyngeal carcinoma—treatment techniques and results. *Int. J. Radiation Oncology Biol. Phys.* **13**, 953–956.
69. Syed, A. M. N., Puthawala, A. A., Damore, S. J., Cherlow, J. M., Austin, P. A., Sposto, R., and Ramsinghani, N. S. (2000). Brachytherapy for primary and recurrent nasopharyngeal carcinoma: 20 years' experience at Long Beach Memorial. *Int. J. Radiat. Oncol. Biol. Phys.* **47**, 1311–1321.
70. Nishioka, T., Shirato, H., Kagei, K., Fukuda, S., Hashimoto, S., and Ohmori, K. (2000). Three-dimensional small-volume irradiation for residual or recurrent nasopharyngeal carcinoma. *Int. J. Radiat. Oncol. Biol. Phys.* **48**, 495–500.
71. Mitsuhashi, N., Sakurai, H., Katano, S., Kurosaki, H., Hasegawa, M., Akimoto, T., Nozaki, M., Hayakawa, K., and Niibe, H. (1999). Stereotactic radiotherapy for locally recurrent nasopharyngeal carcinoma. *Laryngoscope* **109**, 805–807.
72. Chang, J. T. C., See, L. C., Liao, C. T., N, S. H., Wang, C. H., Chen, I. H., Tsang, N. M., Tseng, C. K., Tang, S. G., and Hong, J. H. (2000).Locally recurrent nasopharyngeal carcinoma. *Radiother. Oncol.* **54**, 135–142.
73. Feehan, P. E., Castro, J. R., Phillips, T. L., Petti, P., Collier, M., Daftari, I., and Fu, K. (1992). Recurrent locally advanced nasopharyngeal carcinoma treated with heavy charged particle irradiation. *Int. J. Radiat. Oncol. Biol. Phys.* **23**, 881–884.
74. Lee, A. W. M., Foo, W., Law, S. C. K., Poon, Y. F., Sze, W. M., Tung, W. Y., and Lau, W. H. (1997). Reirradiation for recurrent nasopharyngeal carcinoma: Factors affecting the therapeutic ratio and ways for improvement. *Int. J. Radiat. Oncol. Biol. Phys.* **38**, 43–52.
75. Meyer, J. E., and Wang, C. C. (1971). Carcinoma of the nasopharynx: Factors influencing results of therapy. *Radiology* **100**, 385–388.
76. Moench, H. C., and Phillips, T. L. (1972). Carcinoma of the nasopharynx: Review of 146 patients with emphasis on radiation dose and time factors. *Am. J. Surg.* **124**, 515–518.
77. Baker, S. (1980). Nasopharyngeal carcinoma: Clinical course and results of therapy. *Head Neck Surg.* **3**, 8–14.
78. Shu-Chen, H. (1980). Nasopharyngeal cancer: A review of 1605 patients treated radically with cobalt 60. *Int. J. Radiat. Oncol. Biol. Phys.* **6**, 401–407.
79. Applebaum, E. L., Mantravadi, P., and Haas, R. (1982). Lymphoepithelioma of the nasopharynx. *Laryngoscope* **92**, 510–514.
80. Hsu, M. M., Huang, S. C., Lynn, T. C., *et al.* (1982). The survival of patients with nasopharyngeal carcinomas. *Otolaryngol. Head Neck Surg.* **90**, 289–295.
81. Vikram, B., Strong, E. W., Manolatoss, S., *et al.* (1984). Improved survival in carcinoma of the nasopharynx. *Head Neck Surg.* **7**, 123–128.
82. Rahima, M., Rakowsky, E., Barzilay, J., *et al.* (1986). Carcinoma of the nasopharynx: An analysis of 91 cases and a comparison of different treatment approaches. *Cancer* **58**, 843–849.
83. Qin, D., Hu, Y., Yan, J., *et al.* (1988). Analysis of 1379 patients with nasopharyngeal carcinoma treated by radiation. *Cancer* **61**, 1117–1124.
84. Wang, C. C., Cai, W. M., Hu, V. C., *et al.* (1988). Long term survival of 1035 cases of nasopharyngeal carcinoma. *Cancer* **61**, 2338–2341.
85. Laramore, G. E., Clubb, B., and Quick, C. (1988). Nasopharyngeal carcinoma in Saudi Arabia: A retrospective study of 166 cases treated with curative intent. *Int. J. Radiat. Oncol. Biol. Phys.* **15**, 1119–1127.
86. Sutton, J. B., Green, J. P., Meyer, J. L., *et al.* (1995). Nasopharyngeal carcinoma: A study examining Asian patients treated in the United States. *Am. J. Clin. Oncol.* **18**, 337–342.
87. Decker, D. A., Drelichman, A., Al-Sarraf, M., *et al.* (1983). Chemotherapy for nasopharyngeal cancer, a ten year experience. *Cancer* **52**, 602–605.
88. Al Kourainy, K., Crissman, J., Ensley, J., *et al.* (1988). Excellent response to cisplatin based chemotherapy in patients with recurrent or previously untreated advanced nasopharyngeal carcinoma. *Am. J. Clin. Oncol.* **11**, 427–430.
89. Choo, R., and Tannok, I. (1991). Chemotherapy for recurrent and metastatic carcinoma of the nasopharynx: The Princess Margaret Hospital Experience. *Cancer* **68**, 2120–2124.
90. Al Sarraf, M. (1987). Chemotheraputic management of head and neck cancer. *Cancer Metast. Rev.* **6**, 181–198.
91. Al Sarraf, M. (1988). Head and neck cancer: Chemotherapy concepts. *Semin. Oncol.* **15**, 70–85.
92. Mahjoubi, R., Azli, N., Bachouchi, M., *et al.* (1992). Metastatic undifferentiated carcinoma of the nasopharynx treated with bleomycin,epirubicin cisplatin final report. *Proc. Am. Soc. Clin. Oncol.* **11**. [Abstract 772.]
93. Siu, L. L., Czaykowski, P. M., and Tannok, I. (1998). Phase I/II study of the CAPABLE regimen for patients with poorly differentiated carcinoma of the nasopharynx. *J. Clin. Oncol.* **16**, 2514–2521.
94. Marchini, S., Licitra, L., Grandi, L., *et al.* (1991). Cisplatin and flurouracil in recurrent and or disseminated nasopharyngeal cancer. *Proc. Am. Soc. Clin. Oncol.* **10**, 202.
95. Chi, K. H., Chan, W. K., Cooper, D. L., *et al.* (1994). A phase II study of outpatient chemotherapy with cisplatin, 5-fluorouracil, and leucovorin in nasopharyngeal carcinoma. *Cancer* **73**, 247–252.
96. Yeo, W., Leung, T. W. T., Leung, S. F., *et al.* (1996). Phase II study of combination carboplatin and 5-fluorouracil in metastatic nasopharyngeal carcinoma. *Cancer Chemother. Pharmacol.* **38**, 466–470.
97. Su, W. C., Chen, T. Y., Kao, R. H., and Tsao, C. J. (1993). Chemotherapy with cisplatin and continuous infusion of 5-fluorouracil and

belomycin for recurrent and metastatic nasopharyngeal carcinoma in Taiwan. *Oncology* **50**, 205–208.

98. Boussen, H., Cvitrovic, E., and Wendling, J. L. (1991). Chemotherapy of metastatic and/or recurrent undifferentiated nasopharyngeal carcinoma with cisplatin, bleomycin and flurouracil. *J. Clin. Oncol.* **9**, 1675–1681.
99. Cvitkovic, E., Mahjoubi, R., and Lianes, P. (1991). Flurouracil, mitomycin, epirubicin cisplatin in recurrent and/or metastatic undifferentiated nasopharyngeal carcinoma (UCNT). *Proc. Am. Soc. Clin. Oncol.* **10**, 200.
100. Tan, E. H., Khoo, K. S., Wee, J., *et al.* (1999). Phase II trial of paclitaxel and carboplatin combination in Asian patients with metastatic nasopharyngeal carcinoma. *Ann. Oncol.* **10**, 235–237.
101. Fountzilas, G., Skarlos, D., Athanassiades, A., *et al.* (1997). Paclitaxel by a three hour infusion and carboplatin in advanced carcinoma of the nasopharynx and other head and neck cancers. *Ann. Oncol.* **8**, 451–455.
102. Yeo, W., Leung, T. W., and Chan, A. T. (1998). A phase II study of combination paclitaxel and carboplatin in advanced nasopharyngeal carcinoma. Eu*r. J. Cancer* **34**, 2027–2031.
103. Stein, M. E., Ruff, P., Weaving, A., *et al.* (1996). A phase II study of cisplatin/ifosfamide in recurrent/metastatic undifferentiated nasopharyngeal carcinoma among young blacks in southern Africa. *Am. J. Clin. Oncol.* **19**, 386–388.
104. Fandi, A., Bachouchi, M., Taamma, A., *et al.* (2000). Long-term disease-free survivors in metastatic undifferentiated carcinoma of nasopharyngeal type. *JCO* **18**, 1324–1330.
105. Galligioni, E., Carbone, A., Tirelli, U., *et al.* (1982). Combined chemotherapy with doxorubicin, bleomycin, vinblastine, dacarbazin, and radiotherapy for advanced lymphoepithelioma. *Cancer Treat. Rep.* **66**, 1207–1210.
106. Tannock, I., Payne, D., Cummings, B., *et al.* (1987). Sequential chemotherapy and radiation for nasopharyngeal cancer: Absence of long-term benefit despite a high rate of tumor response to chemotherapy. *J. Clin. Oncol.* **5**, 629–634.
107. Garden, A. S., Lippman, S. M., Morrison, W. H., *et al.* (1996). Does induction chemotherapy have a role in the management of nasopharyngeal carcinoma? Results of treatment in the era of computerized tomography. *Int. J. Radiat. Oncol. Biol. Phys.* **5**, 1005–1012.
108. Clark, J. R., Norris, C. M., Dreyfuss, A. I., *et al.* (1987). Nasopharyngeal carcinoma: The Dana-Farber Cancer Institute experience with 24 patients treated with induction chemotherapy and radiotherapy. *Ann. Otol. Rhinol. Laryngol.* **96**, 608–614.
109. Bachouchi, M., Cvitkovic, E., Azli, N., *et al.* (1990). High complete response in advanced nasopharyngeal carcinoma with bleomycin, epirubicin, and cisplatin before radiotherapy. *J. Int. Cancer Inst.* **82**, 616–620.
110. Azli, N., Bachouchi, M., Chadjaa, M., *et al.* (1992). Update on treatment of locally advanced undifferentiated carcinoma of nasopharyngeal type (UCNT). *In* "Proc. IV Int. Symposium, Interaction of Chemotherapy and Radiotherapy," p. 28.
111. Dimery, I. W., Peters, L. J., Goepfert, H., *et al.* (1993). Effectiveness of combined induction chemotherapy in advanced nasopharyngeal carcinoma. *J. Clin. Oncol.* **11**, 1919–1928.
112. Geara, F., Glisson, B. S., Sanguineti, G., *et al.* (1997). Induction chemotherapy followed by radiotherapy versus radiotherapy alone in patients with advanced nasopharyngeal carcinoma. *Cancer* **79**, 1279–1286.
113. Zidan, J., Kuten, A., and Robinson, E. (1996). Intensive short course chemotherapy followed by radiotherapy of locally advanced nasopharyngeal carcinoma. *Cancer* **77**, 1973–1977.
114. Teo, P. M. L., Chan, A. T. C., Lee, W. Y., *et al.* (1999). Enhancement of local control in locally advanced node-positive nasopharyngeal carcinoma by adjunctive chemotherapy. *Int. J. Radiat. Oncol. Biol. Phys.* **43**, 261–271.
115. Onat, H., Altun, M., Bilge, N., *et al.* (1992). Chemotherapy and radiotherapy for locally advanced undifferentiated nasopharyngeal carcinoma. *In* "Proc. IV Int. Symposium, Interaction of Chemotherapy and Radiotherapy," p. 31.
116. Azli, N., Armand, J. P., Rahal, M., *et al.* (1992). Alternating chemoradiotherapy with cisplatin and 5-fluorouracil plus bleomycin by continuous infusion for locally advanced undifferentiated carcinoma nasopharyngeal type. *Eur. J. Cancer* **28A**, 1792–1797.
117. Heng, D. M. K., Wee, J., Fong, K.-W., *et al.* (1999). Prognostic factors in 677 patients in Singapore with non-disseminated nasopharyngeal carcinoma. *Cancer* **86**, 1912–1920.
118. Zubizarreta, P. A., D'Antonio, G., Raslawski, E., *et al.* (2000). Nasopharyngeal carcinoma in children and adolescence: A single-institution experience with combined therapy. **89**, 690–695.
119. Richards, G. J., and Chambers, R. G. (1973). Hydroxyurea in the treatment of neoplasm of the head and neck. *Am. J. Surg.* **126**, 513–518.
120. Van Andel, J.G., and Hop, W. J. C. (1982). Carcinoma of the nasopharynx: Review of 86 cases. *Clin. Radiol.* **33**, 95–99.
121. Huang, S. C., Tak Lui, L., and Lynn, T. S. (1985). Nasopharyngeal cancer. Study III. A review of 1206 patients treated with combined modalities. *Int. J. Radiat. Oncol. Biol. Phys.* **1**, 1789–1793.
122. Flores, A., Dickson, R. I., Riding, K., *et al.* (1986). Cancer of the nasopharynx in British Columbia. *Am. J. Clin. Oncol.* **9**, 281–291.
123. Souhami, L., and Babinowits, M. (1998). Combined treatment in carcinoma of the nasopharynx. *Laryngoscope* **98**, 881–883.
124. Choi, K. N., Rotman, M., Aziz, H., *et al.* (1997). Concomitant infusion cisplatin and hyperfractionated radiotherapy for locally advanced nasopharyngeal and paranasal sinus tumors. *Int. J. Radiat. Oncol. Biol. Phys.* **39**, 823–829.
125. Pendjer, I., Krejovic, B., and Vucicevic, S. (1997). A comparative study of undifferentiated nasopharyngeal carcinoma treated with radiotherapy or combined treatment with zorubicin-cisplatin and radiotherapy. *Eur. Aech. Oro-Rhino-Laryngol. Suppl.* **1**, 127–129.
126. Maoleekoonpairoj, S., Pharomratanapongse, P., and Puttanuparp, S. (1997). Phase II study: Concurrent chemo-radiotherapy in advanced nasopharyngeal carcinoma. *J. Med. Assoc. Thailand* **80**, 778–784.
127. Lin, J. C., Chen, K. Y., Jan, J. S., and Hsu, C. Y. (1996). Partially hyperfractionated accelerated radiotherapy and concurrent chemotherapy for advanced nasopharyngeal carcinoma. *Int. J. Radiat. Oncol. Biol. Phys.* **36**, 1127–1136.
128. Al-Sarraf, M., Zundmanis, M., Marcial, V., *et al.* (1986). Concurrent cisplatin and radiotherapy in patients with locally advanced nasopharyngeal carcinoma. RTOG study. *Proc. Am. Soc. Clin. Oncol.* **5**, 142.
129. Al-Sarraf, M., Pajak, T. F., Marcial, V. A., *et al.* (1987). Concurrent radiotherapy and chemotherapy with cisplatin in in-operable squamous cell carcinoma of the head and neck: RTOG study. *Cancer* **59**, 259–265.
130. Al-Sarraf, M., Pajak, T. F., Cooper, J. S., *et al.* (1990). Chemo-radiotherapy in patients with locally advanced nasopharyngeal carcinoma: A Radiation Therapy Oncology Group study. *J. Clin. Oncol.* **8**, 1342–1351.
131. Douple, E. B., Richmond, R. C., and Logan, M. E. (1977). Therapeutic potentiation in a mouse mammary tumor and an intracerebral rat brain tumor by combined treatment with cis-dichlorodiamineplatinum (II) and radiation. *J. Clin. Hematol. Oncol.* **7**, 585–603.
132. Soloway, M. S., Morris, C. R., and Sudderth, B. (1979). Radiation therapy and cis-dichlorodiamineplatinum (II) in transplantable and primary urinary bladder cancer. *Int. J. Radiat. Oncol. Biol. Phys.* **5**, 1355–1360.
133. Szumiel, I., and Nias, A. H. W. (1976). The effect of combined treatment with a platinum complex and ionizing radiation on Chinese hamster ovary cells in vitro. *Br. J. Cancer* **3**, 450–458.
134. Rossi, A., Molinari, R., Boracchi, P., *et al.* (1988). Adjuvant chemotherapy with vincristine, cyclophosphamide, and doxorubicin

after radiotherapy in local-regional nasopharyngeal cancer: Results of a 4-year multicenter randomized study. *J. Clin. Oncol.* **6**, 1401–1410.

135. Lin, J. C., Jan, J. S., and Hsu, C. Y. (1996). Preliminary report of outpatient weekly adjuvant chemotherapy for high-risk nasopharyngeal carcinoma. *Am. J. Clin. Oncol.* **19**, 624–627.
136. Roper, H. P., Essex-Carter, A., Marsden, H. B., *et al.* (1986). Nasopharyngeal carcinoma in children. *Pediatr. Hematol. Oncol* **3**, 143–152.
137. Rahima, M., Rakowsky, E., Barzilay J and Sidi, J. (1986). Carcinoma of the nasopharynx: An analysis of 91 cases and a comparison of different treatment approaches. *Cancer* **58**, 843–849.
138. Tsuji, H., Kamada, T., Tsuji, H., *et al.* (1989). Improved results in the treatment of nasopharyngeal carcinoma using combined radiotherapy and chemotherapy. *Cancer* **63**, 1668–1672.
139. Droz, J. P., Domenge, C., Marin, J. L., *et al.* (1987). Adjuvant chemotherapy of regionally advanced UCNT with monthly vindesine, adriamycin, bleomycin, cyclophosphamide and cisplatin: Increasing disease free survival. *Proc. ASCO* **6**, 545.
140. Kim, T. H., McLaren, J., Alvarado, C. S., *et al.* (1989). Adjuvant chemotherapy for advanced nasopharyngeal carcinoma in childhood. *Cancer* **63**, 1922–1926.
141. Chan, T. C., Teo, P. M. L., Leung, T. W. T., *et al.* (1995). Prospective randomized trial conducted to compare chemo-radiotherapy against radiotherapy alone in the treatment of loco-regionally advance nasopharyngeal carcinoma. *Int. J. Radiat. Oncol. Biol. Phys.* **33**, 560–577.
142. Cvitkovic, E., Grange, G. R. L., and Temple, S. (1994). Neoadjuvant chemotherapy (NACT) with epirubicin (EPI), cisplatin (CDDP), bleomycin (BLEO) (BEC) in undifferentiated nasopharyngeal cancer (UCNT): Preliminary results of an international phase III trial. *Proc. Am. Soc. Clin. Oncol.* **13**, 283.
143. International Nasopharyngeal Cancer Study Group. (1996). Preliminary results of the randomized trial comparing neoadjuvent chemotherapy (cisplatinum, epirubicin, bleomycin) plus radiotherapy vs radiotherapy alone in stage IV (≥N2, M0) undifferentiated nasopharyngeal cancer. *Int. J. Radiat. Oncol. Biol. Phys.* **35**, 463–469.
144. El-Guedari, B. (1998). Final results of the VUMCA 1 randomized trial comparing neoadjuvant chemotherapy (CT) (BEC) plus radiotherapy (RT) to RT alone in undifferentiated nasopharyngeal carcinoma (UCNT). *Proc. Am. Soc. Clin. Oncol.* **17**, 1482.
145. Chua, D. T. T., Sham, J. S. T., Choy, D., *et al.* (1998). Preliminary report of the Asian-Oceanian clinical oncology association randomized trial comparing cisplatin and epirubicin followed by radiotherapy versus radiotherapy alone in the treatment of patients with locoregionally advanced nasopharyngeal carcinoma. *Cancer* 2270–2283.
146. Al-Sarraf, M., LeBlanc, M., Giri, P. G. S., *et al.* (1998). Chemo-radiotherapy vs radiotherapy in patients with advanced nasopharyngeal cancer: Phase III randomized Intergroup study 0099. *J. Clin. Oncol.* **16**, 1310–1317.
147. Al-Sarraf, M., LeBlanc, M., Giri, P. G. S., *et al.* (1998). Chemo-radiotherapy (CT-RT) vs radiotherapy (RT) in patients with advanced nasopharyngeal cancer. Intergroup (0099) SWOG 8892, RTOG 8817, ECOG 2388, Progress Report. *Proc. ASCO* **17**, 385a.
148. Al-Sarraf, M., LeBlanc, M., Giri, P.G.S., *et al.* (2001). Superiority of five year survival with chemoradiotherapy (CT-RT) vs radiotherapy in patients with locally advanced nasopharyngeal cancer (NPC). Intergroup (0099), (SWOG 8892, RTOG 8817, ECOG 2388) Phase III study: Final report. *Proc. ASCO* 20.
149. Cheng, S. H., Tsai, S. Y. C., Yen, K. L., *et al.* (2000). Concomitant radiotherapy and chemotherapy for early stage nasopharyngeal carcinoma. *JCO* **18**, 2040–2045.
150. Wolden, S. L., Zelefsky, M. J., Kraus, D. H., *et al.* (2001). Accelerated concomitant boost radiotherapy and chemotherapy for advanced nasopharyngael carcinoma. *J. Clinc. Oncol.* **19**, 1105–1110.
151. Ma, J., Mai, H. Q., Hong, M. H., *et al.* (2001). Results of a prospective randomized trial comparing neoadjuvant chemotherapy plus radiotherapy with radiotherapy alone in patients with locoregionally advanced nasopharyngeal carcinoma. *J. Clin. Oncol.* **19**, 1350–1357.

PART V

NOVEL APPROACHES

C H A P T E R

34

Squamous Carcinomas of the Head and Neck: Novel Genomic Approaches

VYOMESH PATEL, CHIDCHANOK LEETHANAKUL, PANOMWAT AMORNPHIMOLTHAM, and J. SILVIO GUTKIND

Oral and Pharyngeal Cancer Branch
National Institute of Dental and Craniofacial Research
National Institutes of Health
Bethesda, Maryland 20892

I. Introduction 509
II. Genetic Alterations in Head and Neck Squamous Cell Carcinomas (HNSCC) 510
III. Techniques used for Molecular Profiling of HNSCC 510
A. Microdissection 510
B. Laser Capture Microdissection 511
C. Array Technologies 513
IV. Use of Gene Arrays to Evaluate Gene Expression Profiles in HNSCC 514
V. Gene Discovery Efforts in HNSCC: The Cancer Genome Anatomy Project 514
VI. Proteomics 518
VII. Conclusion 519
References 519

Cancers of the oral cavity, salivary glands, larynx, and pharynx, referred to collectively as squamous cell carcinomas of the head and neck (HNSCC), represent the sixth most common cancer among men in the developed world and are often associated with unfavorable survival and morbidity rates. Although risks factors for HNSCC are well recognized, very little is known about the molecular mechanisms that characterize these lesions. A major challenge, therefore, has been the assessment of genetic alterations in HNSCC as they exist *in vivo*. This chapter describes some of the recent developments of sensitive molecular techniques and approaches that are likely to identify altered gene products that are causal of HNSCC development and progression. These include the use of laser capture microdissection (LCM), driven primarily by the need to isolate pure cell population from their native tissue, and the use of DNA array technology for the comprehensive molecular characterization of normal, precancerous, and malignant cells, which enable the analysis of expression patterns of thousands of genes simultaneously. Furthermore, we describe the use of the LCM platform for the Head and Neck Cancer Genome Anatomy Project (HN-CGAP), whose primary goal has been to systematically identify and catalogue known and novel genes expressed during HNSCC development, and subsequently has become the leading effort in gene discovery in squamous epithelium. Similarly, we describe the use of LCM for proteomics in order to identify those proteins that may characterize different stages of HNSCC development and could potentially predict patient outcomes. These revolutionary approaches are likely to have an unprecedented impact in cancer biology and provide exciting opportunities to unravel the still unknown mechanisms involved in squamous cell carcinogenesis. They are also expected to provide a molecular blueprint of HNSCC, thus helping to identify suitable markers for the early detection of preneoplastic lesions, as well as novel targets for pharmacological intervention in this disease.

I. INTRODUCTION

Cancers of the oral cavity, salivary glands, larynx, and pharynx are the sixth most common cancer among men in the developed world [1,2]. As more than 90% of these neoplastic lesions are of squamous cell origin, they are usually referred to, collectively, as squamous cell carcinomas of the head and neck (HNSCC). The 5-year survival rate after diagnosis for this cancer type is relatively low, approximately 53% [1,2], and this high morbidity rate can be attributed to many factors, which include failure to respond to available therapy, late presentation of the lesions, and the lack of suitable markers for early detection [3–5]. As described in other chapters, current efforts aimed to elucidate fully the genetic changes leading to the development of HNSCC are expected to have important implications for the early diagnosis, therapy, and prognosis of HNSCC patients.

II. GENETIC ALTERATIONS IN HEAD AND NECK SQUAMOUS CELL CARCINOMAS (HNSCC)

Cancer arises in a multistep process resulting from the sequential accumulation of genetic defects and the clonal expansion of selected cell populations [6]. In the case of HNSCC, the precise nature of the genetic alterations occurring at each step is still unclear, but a recent report has described a preliminary HNSCC molecular progression model, thus providing a framework for the understanding of the molecular pathogenesis of this cancer type [7]. In this model, the sequential loss of chromosomal material is thought to result in changes leading to dysplasia (9p21, 3p21, 17p13), carcinoma *in situ* (11q13, 13q21, 14q31), and invasive tumors (4q26-28, 6p, 8p, 8q). These initial observations have been expanded by the use of a variety of highly sophisticated techniques, such as comparative genomic hybridization (CGH), fluorescence *in situ* hybridization, and the use of polymorphic microsatellite markers. The latter has helped identify a number of areas of loss of heterozygosity (LOH), including 3p, 4q, 5q21-22, 8p21-23, 9p21-22, 11q13, 11q23, 13q, 14q, 17p, 18q, and 22q [8–21], thus suggesting the contribution of several known tumor suppressor genes in HNSCC, such as *p16* (9p21), *APC* (5q21-22), and *p53* (17p13), as well as the existence of many novel putative tumor suppressor genes affected in HNSCC [22,23]. In this regard, the near completion of the human genome project and recent technical developments, including the use of laser-assisted microdissection and gene arrays, are expected to help decipher the nature of the genetic alterations responsible for this devastating disease [24,25].

III. TECHNIQUES USED FOR MOLECULAR PROFILING OF HNSCC

A major scientific challenge in HNSCC is our understanding of the molecular events that drive tumor progression *in vivo* [26]. In this regard, determining the identity of those genes that are expressed in any given cell type is the first step toward elucidating the nature of the molecules that may mediate normal cellular physiology or disease processes [27]. For example, even normal cellular functions characterizing squamous epithelium, such as proliferation, differentiation, apoptosis, and DNA repair, require the coordinated expression of a diverse set of genes [28–31]. Indeed, maintaining the complex integrity of the epithelium and ensuring that these intrinsic processes occur correctly require a large assortment of genes being turned on and off at specific times [32]. Thus far, only a limited set of genes have been described that are specific for squamous epithelium and include those that encode for proteins involved in cornified envelope assembly and several cytokeratins [33, 34]. Also included are regulatory proteins, e.g., the activator proteins (AP-1 and AP-2), which are now known to have major roles in modulating gene expression in these cells [35]. However, perturbation of the normal balance of some of these gene products that drive these cellular processes forms the basis of many human diseases, including malignancies, and information such as normal or aberrant function of genes in this tissue is expected to provide an opportunity to understand and control the complex processes that ultimately result in tumor development and progression [36]. Furthermore, if molecular markers representing early and late events could be isolated, it would then be possible to identify persons at high risk for HNSCC, namely those whose lesions are progressing through the premalignant state. Identifying the gene products that are altered in HNSCC may also help identify targets for new treatments modalities [37]. Clearly, full elucidation of the genetic changes leading to the development of HNSCC will lead to improved molecular assays with important implications for the early diagnosis, therapy, and prognosis of HNSCC patients [38,39]. This chapter focuses on describing the development of sensitive molecular techniques and approaches, with the expectation that these efforts may soon help identify altered gene products, which might be causal for the conversion of normal epithelium to a malignant phenotype, and that this information may yield novel biomarkers of tumor development and progression, as well as candidate drug targets for pharmacological intervention [40].

A. Microdissection

Technological advances and the development of analytical tools and kits now enable the comparative analysis of macromolecules that may be causal in tumor development and progression [41]. Although studies may have identified molecules that may be involved in the pathogenesis of certain types of neoplasia, tissue heterogeneity may have limited the value of this body of information. Indeed, only a fraction of the total tissue volume (<5%) is suitable for this type of analysis, thus the use of bulk tissue may include abundant contamination, such as cells of lymphatic or stromal origin, and may not be representational to those cells of interest [42]. In this regard, the development of microdissection techniques now allows the molecular analysis of subpopulations of normal and pathologically altered cells from heterogeneous specimens. Some of these techniques in current use are summarized.

Manual tissue microdissection holds certain advantages, which include low cost and the ease and quickness to procure large homogeneous areas with a sterile needle or a scalpel [43]. However, this approach is usually associated with a very high risk of cell contamination if the tissue area of interest is heterogeneous and thus not suitable for procuring a small cell population. More recently, the ultraviolet (UV) laser has been adapted for use with microdissection, which allows for more precise and rapid procurement of homogeneous cell populations from both frozen and formalin-fixed, paraffin-embedded tissue sections, thus making this approach highly suitable for cell-specific analysis of DNA, RNA, and proteins [44]. A major limitation with this approach, excluding its high and sometimes prohibitive cost, is that the quality of data obtained is highly dependent on the precision with which target cells can be identified. Because no coverslip or mounting medium is used during laser-assisted microdissection, cellular detail is often poor, making it extremely difficult to distinguish different cell types reliably by ordinary morphology in routinely stained sections. Examples using this technology include laser microbeam microdissection (LMM), which uses a pulsed UV narrow beam focus laser to cut out target cells and to photoablate unwanted adjacent tissue [45]. In this way, it is possible to procure with or without a computer-assisted micromanipulator only those cells of interest, leaving behind the unwanted cells. This procedure is of very high precision, but very time-consuming. LMM can also be used for microbeam microdissection of membrane-mounted native tissue (MOMeNT) [46], whereby tissue sections mounted on a thin supporting polyethylene membrane, which is attached to a slide by nail polish, can be cut together with the target tissue by laser and collected in a tube with a single laser pulse (LPC, see later). While this particular technique facilitates the dissection and transfer of large intact tissue fragments with a minimal risk of contamination, the process does involve elaborate tissue section preparation and it is only suitable for membrane-mounted tissue sections. Similarly for laser pressure catapulting (LPC), which relies on a laser to cut the target area of cells and their collection by noncontact laser pressure catapulting into sample tubes [47]. While this has a minimal risk of including contaminating cells, the actual procedure of procurement may result in damaged cells and hence a loss in integrity of the macromolecules. More recently, the laser microdissection system (LMD) allows small areas, including single cells within tissue sections mounted on a thin plastic film, to be cut by a computer-assisted use of laser along a predetermined contour [48]. Avoiding direct contact, areas of tissue microdissected by laser can be collected directly in tubes by gravity. In our laboratory, we have accumulated ample experience on the use of another popular system known as laser capture microdissection (LCM) [49]. Thus, although we strongly encourage the readers to explore which one of these recently developed systems is more suitable for the microdissection of their particular clinical samples under investigation, we will describe in more detail the use of LCM as an example of the use of microdissection techniques for the study of the molecular basis of HNSCC.

B. Laser Capture Microdissection

The method of laser capture microdissection, developed at the Laboratory of Pathology (National Cancer Institute), was driven by the need to isolate pure premalignant cells from their native tissue and facilitate the study of molecular events leading to invasive cancer. It has subsequently found widespread interest as an attractive addition to the repertoire of microdissection techniques. LCM allows a precise and more accurate assessment of molecular alterations in cancers and the potential to identify those genes that have never been described previously. This procedure enables the procurement of pure cell populations from frozen or archival human tissue sections in one step, under direct visualization [50–53], and thus provides a platform for current efforts in defining some of the molecular basis of neoplasias as they exist *in vivo*. In this regard, the accurate procurement of specific cell types for RNA isolation using LCM remains a critical step in the analysis of genes expressed in HNSCC and the potential of addressing their possible contribution to neoplasia, if comparison can be made with normal tissue.

LCM, as shown in Fig. 34.1 (see also color insert), primarily involves the PixCell II instrument (Arcturus Engineering, Mountain View, CA). The system consists of an inverted microscope, a solid-state near-infrared laser diode, a laser control unit, a joystick-controlled microscope stage with a vacuum chuck for slide immobilization, a CCD camera, and a color monitor. The LCM microscope is connected to a personal computer for additional laser control and image archiving. Accessories include optically clear caps that are coated at one end with a film of ethylene vinyl acetate (EVA) of an approximately diameter of 6 mm, which are in direct contact with the tissue and are used for the transfer of selected cells. The caps then fit onto standard 0.5-ml microcentrifuge tubes for further tissue processing.

The actual procedure of LCM, as described by the manufacture (www.arctur.com), involves using the transporter arm to maneuver the cap onto the tissue section. Cells of interest, based on exhibiting a specific morphology under

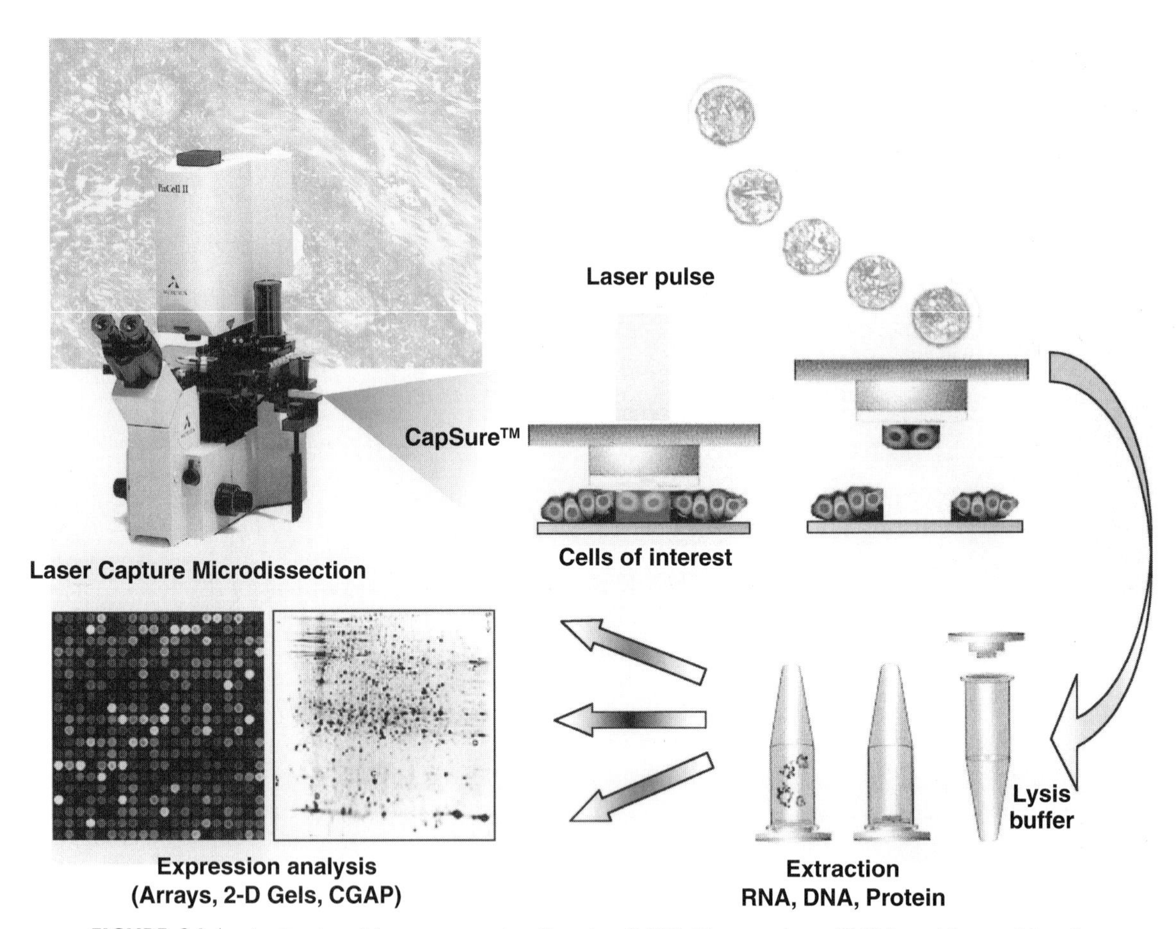

FIGURE 34.1 Application of laser capture microdissection (LCM). The procedure of LCM provides a quick and reliable method for the procurement of a pure population of cells from their native tissues. The process involves using the handle of the transport arm to manipulate the platform holding the vacuum-secured glass slide until an area of interest is determined with the microscope. A pulse of laser beam is then used to capture areas of tissues of interest onto ethylene vinyl acetate (EVA)-coated caps (CapSure). These caps are then transferred after LCM to Eppendorf tubes containing appropriate lysis buffer for extraction of macromolecules (RNA, DNA, and proteins). Using this method, frozen tissue sections can be microdissected immediately after staining, and thus a pure population of cells of interest (>95% purity) can be concentrated rapidly with maximum preservation of macromolecules. Adapted from Leethanakul *et al.* [67]. (See also color insert.)

light microscopy after staining with heamotoxylin and eosin, are visualized and the image is subsequently transferred to a computer screen. After visual selection of the desired cells using the joystick and guided by a positioning beam, laser activation leads to focal melting of the EVA membrane, which has its absorption maximum near the wavelength of the laser. The melted polymer expands into the section and fills the extremely small hollow spaces present in the tissue. The polymer resolidifies within milliseconds and forms a composite with the tissue. The adherence of the tissue to the activated membrane exceeds the adhesion to the glass slide and allows selective removal of the desired cells. Laser impulses, usually between 0.5 and 5 ms in duration, can be repeated multiple times across the whole cap surface, which allows the rapid isolation of large numbers of cells. Of importance, as most of the energy is absorbed by the membrane, biological macromolecules of interest are thus left intact. The selected tissue fragments are harvested by simple lifting of the cap, which is then transferred to a microcentrifuge tube containing the buffer solutions required for the isolation of the molecules of interest, e.g., RNA. The homogeneity of the captured cells is then confirmed under a light microscope prior to proceeding with RNA extraction. With this method, a pure population of cells of interest (>95% purity) can be concentrated rapidly, and with maximum preservation of RNA. The isolated RNA from these cells can then be used with the currently available technology for the detailed analysis of gene expression. The remaining tissue on the slide is intact and can be subjected to further dissection. In addition, a video camera on top of the microscope enables the user to monitor the progress of the microdissection, thus providing a convenient reference point for subsequent serial sections.

Certain advantages of LCM include speed, precision, and versatility, which together allow for the procurement of thousands of cells from several tissue components from the same slide, e.g., normal and neoplastic cells, within a relatively short period of time. Additionally, the PixCell II instrument allows for the capture of single cell(s) within a population of cells of interest. Its limitations include the restrictive optical resolution of routinely stained, dehydrated tissue sections without a coverslip, thus making the precise procurement of cells from certain complex tissues, which may lack architectural features, such as lymphoid tissues or diffusely infiltrating carcinomas, almost impossible. This problem can be circumvented by special stains, particularly immunohistochemistry, which help highlight the cell population to be isolated or avoided. However, when compared with conventional stains, immunostaining always results in a measurable decrease in RNA recovery, probably because of the longer exposure to aqueous media. In contrast, the use of precipitating fixatives such as acetone or ethanol and the rapid dehydration process of frozen sections by inactivating endogenous RNase may help preserve RNA integrity, and the use of hematoxylin and eosin for cell visualization has been demonstrated to have a minimum effect on the recovery of these nucleic acids [54].

C. Array Technologies

The development of several high-throughput, hybridization-based methods utilizing gene–specific polynucleotides derived from the 3′ end of RNA transcripts, arrayed individually on a single matrix (nylon membranes or glass slides), now allows the analysis of thousands of genes simultaneously and thus provides a unique opportunity to examine the identity of genes expressed in normal and tumor tissues, as well as to obtain valuable information about their biological function [55,56].

The principle behind any array experiment is to match through base-pairing rules (A-T and G-C for DNA; A-U and G-C for RNA) known immobilized cDNA probes (500–5000 bp) with unknown labeled nucleic acid targets by hybridization [57]. However, central to the success of such experiments is the choice of "probes" to be printed on the array, which is usually based on available information on clones, e.g., from databases (GenBank, dbEST, UniGene) and collection of full–length cDNAs, partially sequenced cDNAs (or ESTs), or randomly chosen cDNAs from any library of interest. Once an array format has been designed with a choice of clones, these are then polymerase chain reaction (PCR) amplified, purified, resuspended, and prepared for printing. For both glass and membrane matrices, each element is generated by the deposition of a few nanoliters of purified PCR product using workstations; after fixation of DNA, e.g., by UV cross-linking, the arrays are ready for analysis [58,59]. Currently, high-density arrays can consist of approximately 40,000 elements. The standard approach to an array experiment involves the fluorescent labeling of control (Cye3-dUTP) and test (Cye5-dUTP) total RNA by reverse transcription and subsequently these targets are then pooled and allowed to hybridize to cDNAs arrayed onto glass slides and the signal is detected and measured by laser detectors. Using appropriate software, these images can then be analyzed, data from a single experiment can be viewed as a normalized ratio, and the relative abundance of transcripts present in the samples can be assessed [60]. The scheme is similar when using a radiolabeled probe, but it is not possible to carry out a simultaneous hybridization of test and reference samples, and detection involves the use of PhosphorImagers. In such cases, serial or parallel hybridization is required, introducing the possibility of higher variability in comparisons of expression level. In this case, [^{33}P]dCTP as a source of radioactive label is preferred as it avoids spot saturation and interference of surrounding weak hybridization signal.

Another genomic approach uses high-density oligonucleotide microarrays, which offer certain advantages over glass slides arrayed with cDNAs. Using photolithography, up to 400,000 oligonucleotides, representing nearly 25,000 genes, can be synthesized directly onto a chip from sequence information currently available, thus avoiding the need of maintaining large numbers of cDNA clones. Furthermore, 20 pairs of oligonucleotides are usually included for each gene represented and among these are controls that ensure the specificity and reliability of the gene expression data obtained. Although this approach avoids repetitive or homologous gene regions and chip-to-chip variations, expensive and specialized equipment is required [61].

A major limitation for this type of approach includes tissue heterogeneity, which may make it very difficult to assign expressed genes to specific cell populations if gross tissue extracts are used as a source of mRNA. Thus, an accurate procurement of specific cell types for RNA isolation is a critical step influencing the validity of this type of analysis [62]. LCM, as described earlier, fulfills this criterion. However, the relative amounts of RNA obtainable from LCM material are usually low and not suitable for representative gene expression analysis, as those genes of moderate or low abundance may remain undetected. For example, new micro RNA isolation kits, involving guanidium isothiocyanate (GITC), phenol/chloroform, isopropanol precipitation, and treatment with DNase I in the presence of RNase inhibitors, have been used successfully to obtain RNA (~20 ng) from approximately 5000 cells [63,64]. RNA recovered from LCM-derived cells is usually fragmented; therefore, prior to gene expression studies, it is essential to monitor the quality and integrity of the nucleic acid, e.g., by reverse transcribing and amplifying a "housekeeping gene." Additionally, as the total amount of RNA required for expression analysis (10–100 μg) is usually much greater than the amount that can be extracted from microdissected tissues, elaborate protocols are now available that allow the synthesis of first-strand cDNAs from small amounts of total RNA, such as those extracted from microdissected tissues [65]. More recently, methods using T7-oligo(dT) primers, reverse transcriptase and T7 transcription, allow for the linear amplification of total RNA (>1000-fold), including transcripts of low abundance, resulting in gene expression profiles that are comparable to those observed when using larger amounts of total RNA [66].

IV. USE OF GENE ARRAYS TO EVALUATE GENE EXPRESSION PROFILES IN HNSCC

Many studies have examined the differences in the expression of one or a few genes in HNSCC, but no comprehensive and systematic study of gene expression profiles in these neoplastic cells had been undertaken. Therefore, available molecular models are not able to explain fully the complexity of the genetic changes that occur during the development and progression of HNSCC. Thus, to address some of these shortfalls, the complementary use of LCM with gene array approaches has been applied to gene expression of HNSCC. In particular, these studies involved lesions from different anatomical sites (tongue, larynx, pharynx), contrasting histopathology (poorly, moderate to well differentiated and invasive) and clinical stage (T1–T4) and comparing differences with matching normal tissue, either adjacent or distant [67]. In these cases, total RNA extracted from specific cell populations (squamous epithelium) procured by LCM was reverse transcribed and amplified, and the resulting cDNAs were labeled and used as targets to hybridize to filter membranes arrayed with known cancer genes. Using this strategy, it was possible to ascertain reliable and specific gene expression information from these tissues (Fig. 34.2), thus enabling direct comparisons to be made with normal cells [67]. A distinct pattern of expression of differentiation and growth-related genes was clearly demonstrated in this tumor type. For example, most of the cytokeratins were underrepresented and, in contrast, key signaling molecules, growth factors, and angiogenic cytokines were overexpressed in this cancer type when compared directly with the adjacent normal tissue (Table 34.1). Additionally, these initial efforts demonstrated the feasibility of recovering complex transcriptomes from microdissected cell populations More recently, this approach has been extended to the use of high-density arrays and T7-amplified mRNA procured by LCM for the gene expression profiling of oral lesions [68]. This study revealed that many known genes with reported roles in neoplasia are expressed differentially in HNSCC. These included transcription factors, oncogenes, tumor suppressors, and many genes often associated with differentiation, proliferative, and invasive pathways. However, a subset of genes that have not been previously implicated in oral neoplasias, e.g., neuromedin U, were also reported.

In addition to gene expression profiling of neoplasias, the array approach has been applied to improve our understanding of the extent of molecular heterogeneity that exists within most types of cancers and which may reflect the overall clinical outcome. Therefore, it is likely that the molecular classification of tumors will identify previously undetected lesions and thus lead to more specific and effective treatment options. For instance, in tumors from patients with a type of non-Hodgkins lymphoma, gene expression analysis using microarrays was able to identify two molecularly distinct subgroups of lesions that contrasted in their responses to standard chemotherapy regimes and survival [69]. Similar expression profiles on human breast carcinomas identified at least five subgroups related to different molecular features of this cancer type [70]. It is likely that similar studies to include more patients and different lesions may be able to classify all human tumors in clinically homogeneous groups and thus identify molecular determinants that will reflect overall survival.

V. GENE DISCOVERY EFFORTS IN HNSCC: THE CANCER GENOME ANATOMY PROJECT

Laser-assisted microdissection techniques are well suited for expression analysis of known and unknown genes in a tissue and cell-specific pattern, which may help elucidate the etiology and pathogenesis of certain cancers, including colon, lung, breast, prostate, adrenal, and ovary [24]. In this regard, the development of these techniques was expedited by their integration into the "cancer genome anatomy project" (CGAP) sponsored by the National Cancer Institute. The purpose of this effort was the development of technology, information, and material infrastructure toward the goal of the comprehensive molecular analysis of cells in human tissues as they transition from the normal state to the precancerous state and, ultimately, to cancer [71,72]. The CGAP efforts initially focused on the five most prominent tumor types (breast, colon, lung, prostate, ovarian), but as the project developed and its potential was realized, other lesions, including HNSCC, were included (http://cgap.nci.nih.gov/). In this regard, the Head and Neck CGAP (HN-CGAP) was established as a cooperative effort between the National Institute of Dental and Craniofacial Research (NIDCR) and the NCI CGAP. The rationale was to generate sequence information initially from representational cDNA libraries constructed from cell

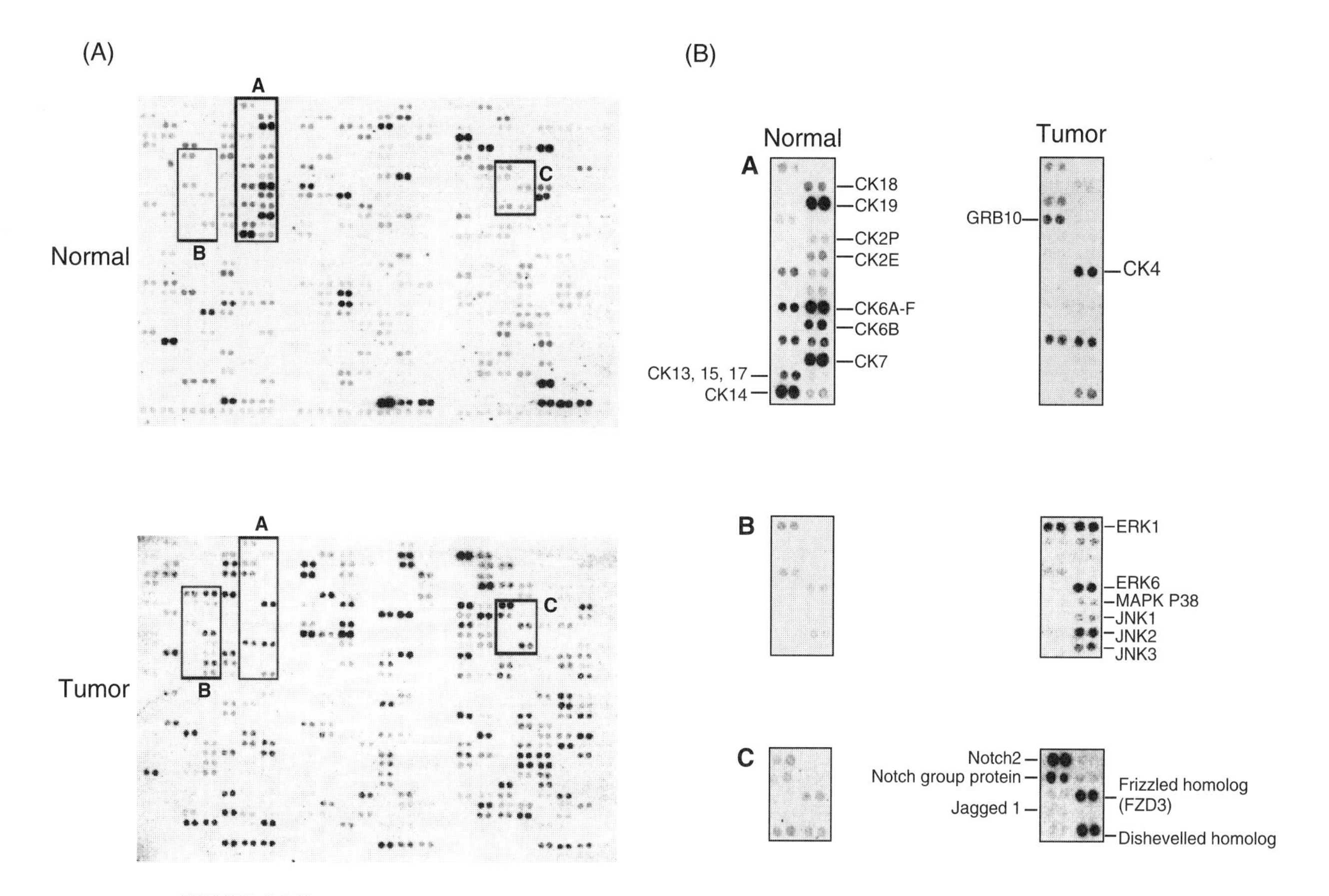

FIGURE 34.2 Analysis of gene expression in HNSCC using cDNA arrays. For each HNSCC tissue set (normal and tumor), amplified cDNA (AcDNA) probes were prepared and used simultaneously for the hybridization of nylon membranes arrayed in duplicate with human cancer and housekeeping genes (BD Biosciences). Conditions used for hybridization were essentially as described in the manufacturer's protocol and were analyzed by PhosphorImaging and autoradiography. Pattern of gene expression for a representative tissue set from the same HNSCC patient is shown (A). Differentially expressed genes in three or more HNSCC tissue sets were considered of likely biological significance, and examples of those are indicated (B). Data from Leethanakul *et al.* [67].

lines derived from primary and secondary HNSCC lesions of contrasting clinical staging (T2 to T4) and primary oral keratinocytes and E6/E7 of HPV-immortalized oral keratinocytes, annotated HN1-4 and HN5-6, respectively, in the CGAP database to gain information on the nature of those genes expressed in this particular cell type [73]. Overall, from the 1160 clones sequenced, 38 of these were identified as novel genes. Furthermore, a distinct pattern of gene expression can already be observed when comparing normal and cancerous cells using analytical tools available at the CGAP site [73].

However, cells in culture often exhibit genetic mutations similar to those of the tumor of origin, and the molecules expressed are likely to be quite different from those present in their natural setting and highly dependent on the particular culturing conditions [74,75]. Thus, sequence information from highly complex cDNA libraries derived from LCM-procured HNSCC tissue (normal and tumor tissue resected from HNSCC patients) may be more representational and help to begin defining more accurately those genes that are expressed in this cancer type and their molecular changes as they occur during tumor progression *in vivo* [76]. The choice of tissue samples for constructing cDNA libraries was based primarily on clinical features such as anatomical sites, which included the retromolar trigone region, floor of the mouth, and the tongue, thus representing the most frequently detected HNSCC sites [77]. Furthermore, the actual lesions were assessed to be either carcinoma *in situ* (CIS), well-differentiated invasive carcinoma or moderated to poorly differentiated carcinoma, and the matching normal epithelium, either adjacent or distant, to be of squamous origin. Seven high-quality, LCM-derived cDNA libraries have been thus far generated from microdissected oral epithelium; however, six of these are from patient sets, which include carcinoma and adjacent matching normal epithelium (annotated HN8, 10, 12, and HN7, 9, 11, respectively, in the CGAP database; http://cgap.nci.nih.gov [63]). As there is very limited

TABLE 34.1 Genes Highly Represented in HNSCC[a]

Accession	Gene
Cell cycle	
L33264	cdc-2-related kinase
X59768	Cyclin D1
U11791	Cyclin H
U40345	p19INK4D
U18422	DP-2
Angiogenesis	
L12350	Thrombospondin 2 precursor
U43142	VEGF-C
X07819	MMP-7
X07820	MMP-10
D50477	MMP-16
Z30183	TIMP-3
U76456	TIMP-4
Growth factors and oncogenes	
M11730	HER2
M37722	bFGFR 1
Z12020	MDM2
M31213	c-RET
D17517	SKY
X87838	β-Catenin
X14445	FGF 3
M37825	FGF 5
X63454	FGF 6
M65062	IGFBP5
M60718	HGF precursor
K03222	TGF-α
X02812	TGF-β
M96956	EGF Cripto protein CR1 and 2
X06374	PGDF-a
X02811	c-SIS
M74088	APC
Signaling molecules	
X60811	ERK1
X79483	ERK6
U82532	GDI dissociation protein
L35253	MAP kinase p38
L26318	JNK1
L31951	JNK2
U34819	JNK3
U39657	MKK6
U78876	MEKK3
M31470	Ras-like protein (TC10)
L25080	Rho A
Apoptosis	
U45878	Inhibitor of apoptosis protein
U37448	Caspase 7 precursor
U60520	Caspase 8 precursor
U28014	Caspase 5 precursor
M77198	AKT2
U59747	BCL-W
U78798	TRAF2
S83171	BAG-1
L22474	BAX
WNT/Notch signaling system	
M73980	Notch 1
U77493	Notch 2
M99437	Notch group protein (N)
AF028593	Jagged 1
L37882	Frizzled
U82169	Frizzled homologue (FZD3)
U46461	Disheveled homologue
U43148	Patched homologue
U94352	Manic fringe
U94354	Lunatic fringe

[a]Gene expression pattern for each HNSCC tissue set ($n = 5$) was analyzed by PhosphorImaging, and the relative amount of expression was compared with those of housekeeping genes. Those genes judged to be differentially expressed at >2-fold in at least three of the cancer tissue sets were considered of likely biological significance and are listed in their corresponding functional groups. The GeneBank access number for each gene is included. Data are from Leethanakul *et al.* [67].

information on genes expressed in normal oral epithelium and HNSCC, we chose to use nonnormalized cDNA libraries. The advantage of non normalized, nonamplified libraries is that the transcript abundance of the original cell or tissue is reflected accurately in the frequency of clones in the libraries, thus they can be used for both gene discovery and to compare the expression of highly expressed genes in different cells or tissue samples. To this end, nucleotide sequence information from clones that underwent 3′ end single-pass high-through put sequencing was analyzed further using the BLAST search algorithm to identify homology to sequences already available in databases [78]. Unexpectedly, this approach revealed that the gene discovery rate in these libraries was relatively high, approximately 4–7%, and of particular interest was the presence of a very high number of unknown unique genes in these libraries (168 genes), which is probably because these represent the first group of cDNA libraries from microdissected HNSCC constructed thus far [63]. In addition, 69% of the unknown unique genes were discovered from normal cDNA libraries, which might represent cell-type-specific transcripts in normal squamous epithelium.

A very important contribution made by NCBI is a weekly report posted on the website http://www.ncbi.nlm.nih.gov/UniGene/Hs_DATA/lib_report.html, with details such as EST library diversity and future gene discovery prospects, which is used by CGAP to guide decisions about future library production and sequencing efforts. In this regard, it is likely that the number of unknown unique genes detected initially in microdissected-derived HNSCC cDNA libraries may change over time as more sequence information is deposited in GenBank. For example, at the time of the first round of sequencing, the total number of unique genes was 182 genes. At the time of this analysis, the total number of unique genes was 168, and this information is updated regularly. Thus far, a total of 19 cDNA libraries, including those described earlier and others derived from bulk HNSCC, have been constructed, and detailed information regarding the libraries and gene discovery is available in the CGAP database (Fig. 34.3).

However, a parallel effort sponsored by the Brazilian funding agency FAPESP and the U.S.-based Ludwig Institute for Cancer Research has focused on providing sequences from genes expressed in tumors that are important within the context of public health in the state of São Paulo, Brazil [79]. This included the production of 847 libraries of bulk HNSCC. Sequence information has been released recently to the public and now constitutes part of the CGAP database. A total of 122, 167 short sequences were generated, which upon analysis using the CGAP Gene Library Summarizer (GLS) tool, revealed the existence of 60 unknown unique genes and 8 known unique genes. Although this monumental effort did not result in a high rate of gene discovery, it nonetheless facilitated gene expression analysis, as many known (8068) and unknown (3181) previously described genes were also found in this extensive collection of HNSCC cDNA libraries. Because library preparation did not include the isolation of squamous epithelium, the actual cell type of origin for each transcript is uncertain, but will surely be analyzed in the foreseeable future using gene array techniques.

A major challenge of CGAP currently facing researchers is the organization, analysis, and interpretation of sequence

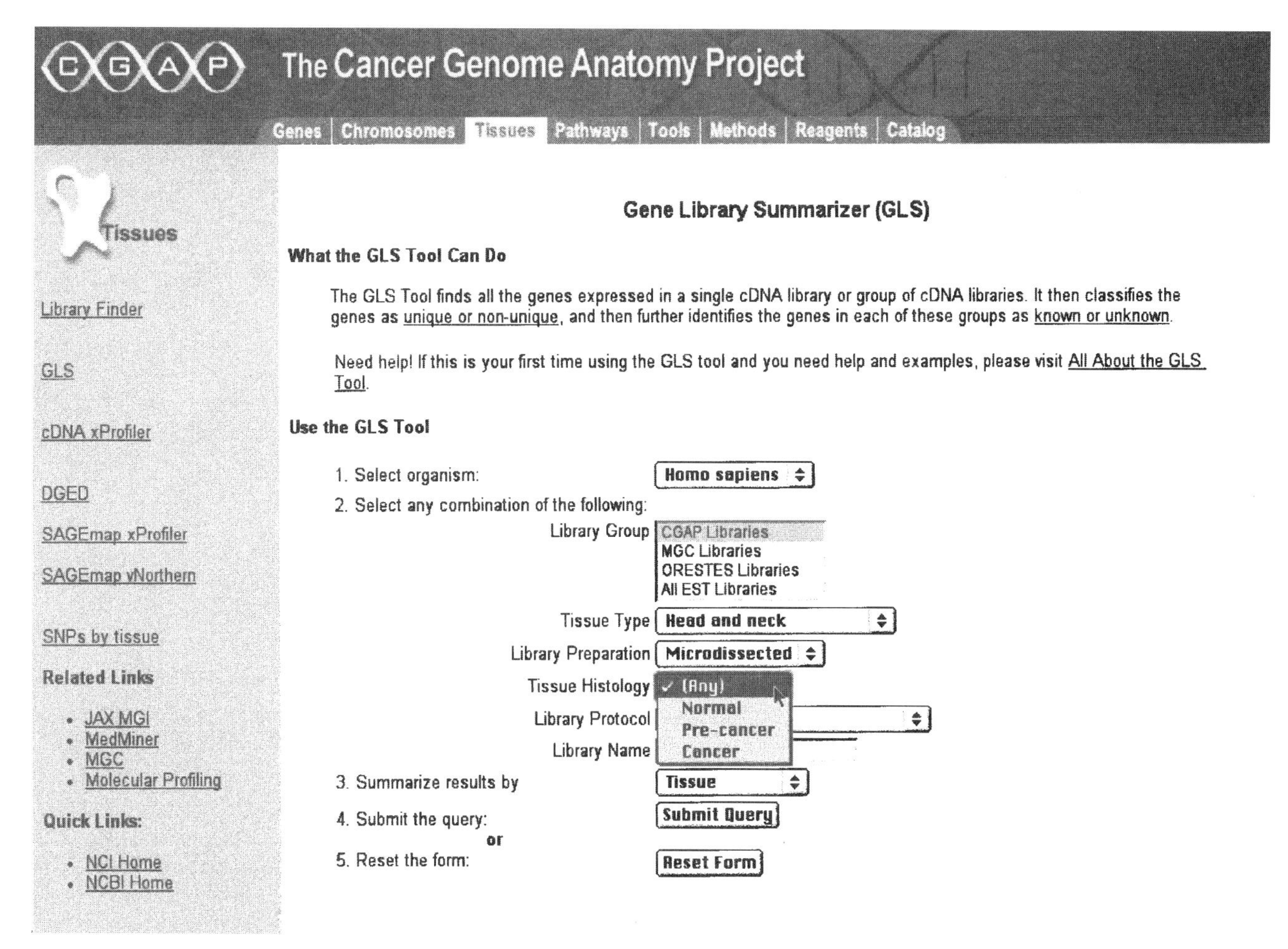

FIGURE 34.3 The Head and Neck Cancer Genome Anatomy Project (HN-CGAP). Total RNA extracted from LCM material is used to construct cDNA libraries, which are then used as templates for gene discovery and gene annotation. This procedure involves a single-pass high-throughput sequencing of the cDNA clones from the 3′ end, which represents a unique sequence tag for a particular transcript. Stringent bioinformatics used by CGAP partitions the information into a nonredundant set of gene-oriented clusters, with each cluster indicative of a unique gene. All information thus far generated from all HNSCC cDNA libraries is accessible from the database http://cgap.nci.nih.gov/Tissues and the web page is shown for this.

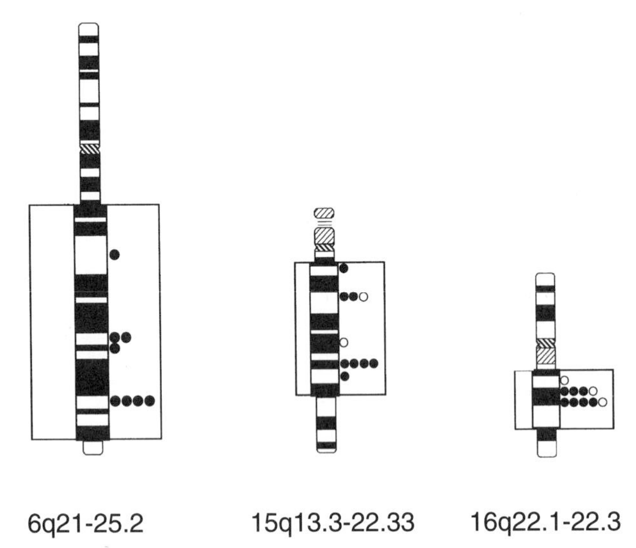

FIGURE 34.4. Localization of unique HNSCC clones to the human genome. The first draft of the human sequencing project can now be used to carry out chromosome-specific BLAST searches of unknown sequences. This approach can identify the affected chromosomal loci and genes located at these sites. All unique (●) and nonunique (○) unknown sequences from cDNA clones from LCM-derived HNSCC libraries were analyzed and were found to localize to different chromosomal regions. Clusters of sequences that localize to the same chromosomal region, e.g., chromosomes 6, 15, and 16, as indicated, may help define chromosomal loci that may be causal to HNSCC development, which were not described previously.

data, which may provide new insights in our understanding of the molecular basis that give rises to HNSCC development and whether this information can be used for treatment and prevention. In this regard, the first draft of the human sequencing project, by providing chromosome-specific BLAST searches, all unique and nonunique unknown sequences can be localized to chromosomal regions and then comparisons can be made with those genes often affected in HNSCC. As shown in Fig. 34.4, some clusters of sequences, by locating to the same chromosomal region, e.g., chromosomes 6q, 15q, and 16q, may identify genes that may be causal to HNSCC development not described previously. Furthermore, analysis of the pattern of expression of these genes and functional assays will surely help identify gene products that may determine the transformed and/or the metastatic phenotype, as well as additional molecules that, without playing an obvious role in the neoplastic process, can nevertheless be used as clinically useful markers of HNSCC. To this end, an application of technologies that can expedite the simultaneous analysis of multiple genes and suited for use with tissues is mandatory.

VI. PROTEOMICS

The human genome project has thus far identified approximately 19,000 genes, with more likely to be discovered from the estimated total number. However, the transcriptional activity of these molecules, as assessed by cDNA arrays, may not fully coincide with those of the protein products, as expression levels and forms cannot be usually predicted from mRNA analysis [80]. Furthermore, protein expression and function are subject to modulation through transcription as well as through posttranscriptional and translational events. For instance, a number of RNA molecules can result from one gene through a process of differential splicing. Equally, the many posttranslation modifications that proteins undergo, such as glycosylation, phosphorylation, acetylation, sumoylation, and sulfation, affect function, which includes protein–protein and nucleotide–protein interaction, stability, and half-life, all contributing to a potentially large number of protein products from one gene [81–85].

It follows that protein analysis provides a suitable platform for understanding the interaction between the functional pathways of a cell and its environmental milieu, irrespective of RNA level, and helping unravel those critical changes that may occur, e.g., during cancer pathogenesis. In this regard, altered protein expression or those that may be modified differentially in normal and tumor cells may affect cellular function, and identification of these changes, using available technologies, is likely to provide useful information for diagnosis and/or prognosis and, equally, improve our understanding of molecular mechanisms and thus the biology underlying this disease process [86,87]. However, as cellular heterogeneity characterizing many cancers, including HNSCC, may restrict this body of information, the use of LCM for this type of analysis would be favorable over whole tissue sections [88,89]. Information obtained using this approach may aid drug discovery, as those that are currently available for treating various pathological conditions are based on approximately 500 protein targets [90]. Therefore, by molecular profiling human cancers, additional targets are likely to identify those that could be used potentially for drug development, specifically for malignancies [91].

Thus far, only a small proportion of the proteome has been documented using conventional approaches, e.g., immuno-based assays. More recently, two-dimensional gel electrophoresis (2-DE) is now being used extensively for protein characterization and discovery and by integating the data to analytical instrumentation, to improve reproducibility and sensitivity, the resulting information will help to increase our knowledge on the proteome [92]. In this approach, protein samples are usually denatured and separated on the basis of their charge through isoelectric focusing. In this regard, immobilized pH gradients have greatly enhanced reproducibility in resolving almost the complete spectrum of basic to acidic proteins and have allowed both analytical and preparative amounts of proteins to be resolved. Furthermore, narrow-range pH gradients may help increase protein detection, e.g., a range of 1 pH unit was demonstrated to resolve approximately 1000 protein spots. The proteins are separated further by migration in a polyacrylamide gel on the basis of their molecular weights [93]. Using silver staining techniques, many proteins can be visualized on a single gel, and specialized software can then allow means to compare 2-DE gel patterns with one another, and quantitative and qualitative differences in protein profiles can then be detected for biomarker identification and discovery.

Fluorescent dyes have been developed to overcome some of the drawbacks of silver staining in making the protein samples more amenable to mass spectrometry [94]. Mass spectrometry can now provide a powerful means for obtaining peptide mass fingerprints for proteins resolved by 2-DE. Protein databases and reference maps can then be established to catalog proteins resolved by 2-DE from various cell types in both healthy and diseased states. This approach, for molecular profiling of human cancers is likely to identify proteins that may be used to distinguish between different types and sub-types of malignancies, and predict outcome among cancer patients. Ultimately, by creating a database of proteins, the information can be potentially used to identify targets for drug development [95].

VII. CONCLUSION

Recent discoveries have dramatically increased our understanding of the most basic mechanisms controlling normal cell growth, and have also greatly enhanced our ability to investigate the nature of the biological processes that lead to cancer. Nevertheless, these studies were carried out by concentrating all efforts on one or a few genes at a time, based largely on prior knowledge of function in cellular pathways such as proliferation, differentiation, apoptosis, and DNA repair. However, with the use of new technologies and the wealth of readily available sequence information, it is now possible to identify novel areas for hypothesis-driven investigation into the molecular basis of neoplasias.

DNA sequencing and microarray technologies, among others, have provided a unique opportunity to monitor, simultaneously, thousands of genes, which allows the systematic scanning of expression patterns of molecules and the possibility of identifying those correlating with a particular disease state. In addition, gene expression profiles can now be investigated within a histologically defined, homogeneous populations of cells, thus affording the possibility of applying these newly available techniques to investigate expression patterns in normal as well as neoplastic tissue. Cancer represents a complex family of genetic diseases resulting from genetic changes that involve a multiplicity of both inherited and acquired alterations in the DNA sequence, ranging from point mutations to deletions, amplifications, and translocations. Invariably, these changes in genetic information result in alterations in expression patterns at both transcriptional and protein levels. Thus, availability of an emerging body of information on gene expression and function is expected to help elucidate the molecular mechanisms responsible for human malignancies.

Furthermore, information from various human and nonhuman sequencing efforts and chip-based assays has resulted in the establishment of large public databases. Thus, with online resources for the retrieval of gene expression data from any organism, it is now possible to translate all DNA sequences into amino acids and ascribe functions to most of the newly discovered genes, thus addressing biological questions relevant to cancer development and progression. For instance, information encoded by the genome can be studied at the level of the proteome to further yield an understanding of cancer, as well as identify new targets for therapeutic intervention and markers for early detection.

These revolutionary approaches are likely to have an unprecedented impact in cancer biology, particularly in the search for the still unknown mechanisms involved in squamous cell carcinogenesis. However, this will also require the development of effective methods to validate the biological relevance of the newly identified candidate genes. Indeed, the use of tissue culture systems and animal models to recapitulate this complex disease will be a central component of these efforts. Furthermore, the effective use of DNA- and RNA-labeling techniques and the development of immunological tools would be expected to allow the direct examination of expression profiles in clinical specimens, including potential premalignant lesions. We can conclude that exciting opportunities are ahead to understand the molecular basis of oral and pharyngeal cancers. However, it is becoming increasingly clear that it will take a concerted effort from the entire scientific and health professional community to battle the ravaging consequences of this disease.

References

1. Landis, S. H., Murray, T., Bolden, S., and Wingo, P. A. (1999). Cancer statistics, 1999. *CA Cancer J. Clin.* **49**, 8–31.
2. Parkin, D. M., Pisani, P., and Ferlay, J. (1999). Global cancer statistics. *CA Cancer J. Clin.* **49**, 33–64.
3. Lewin, F., Damber, L., Jonsson, H., Andersson, T., Berthelsen, A., Biorklund, A., Blomqvist, E., Evensen, J. F., Hansen, H. S., Hansen, O., Jetlund, O., Mercke, C., Modig, H., Overgaard, M., Rosengren, B., Tausjo, J., and Ringborg, U. (1997). Neoadjuvant chemotherapy with cisplatin and 5-fluorouracil in advanced squamous cell carcinoma of the head and neck: A randomized phase III study. *Radiother. Oncol.* **43**, 23–28.
4. Brunin, F., Mosseri, V., Jaulerry, C., Point, D., Cosset, J. M., and Rodriguez, J. (1999). Cancer of the base of the tongue: Past and future. *Head Neck* **21**, 751–759.
5. Veneroni, S., Silvestrini, R., Costa, A., Salvatori, P., Faranda, A., and Molinari, R. (1997). Biological indicators of survival in patients treated by surgery for squamous cell carcinoma of the oral cavity and oropharynx. *Oral Oncol.* **33**, 408–413.

6. Dewanji, A., Goddard, M. J., Krewski, D., and Moolgavkar, S. H. (1999). Two stage model for carcinogenesis: Number and size distributions of premalignant clones in longitudinal studies. *Math Biosci.* **155**, 1–12.
7. Califano, J., van der Riet, P., Westra, W., Nawroz, H., Clayman, G., Piantadosi, S., *et al.* (1996). Genetic progression model for head and neck cancer: Implications for field cancerization. *Cancer Res.* **56**, 2488–2492.
8. Ransom, D. T., Barnett, T. C., Bot, J., de Boer, B., Metcalf, C., Davidson, J. A., *et al.* (1998). Loss of heterozygosity on chromosome 2q: Possibly a poor prognostic factor in head and neck cancer. *Head Neck* **20**, 404–410.
9. Gonzalez, M. V., Pello, M. F., Ablanedo, P., Suarez, C., Alvarez, V., and Coto, E. (1998). Chromosome 3p loss of heterozygosity and mutation analysis of the *FHIT* and *beta-cat* genes in squamous cell carcinoma of the head and neck. *J. Clin. Pathol.* **51**, 520–524.
10. Wang, X. L., Uzawa, K., Imai, F. L., and Tanzawa, H. (1999). Localization of a novel tumor suppressor gene associated with human oral cancer on chromosome 4q25. *Oncogene* **18**, 823–825.
11. Mao, E. J., Schwartz, S. M., Daling, J. R., and Beckmann, A. M. (1998). Loss of heterozygosity at 5q21–22 (adenomatous polyposis coli gene region) in oral squamous cell carcinoma is common and correlated with advanced disease. *J. Oral. Pathol. Med.* **27**, 297–302.
12. Pai, S. I., Wu, G. S., Ozoren, N., Wu, L., Jen, J., Sidransky, D., *et al.* (1998). Rare loss-of-function mutation of a death receptor gene in head and neck cancer. *Cancer Res.* **58**, 3513–3518.
13. El-Naggar, A. K., Coombes, M. M., Batsakis, J. G., Hong, W. K., Goepfert, H., and Kagan, J. (1998). Localization of chromosome 8p regions involved in early tumorigenesis of oral and laryngeal squamous carcinoma. *Oncogene* **16**, 2983–2987.
14. Ishwad, C. S., Shuster, M., Bockmuhl, U., Thakker, N., Shah, P., Toomes, C., *et al.* (1999). Frequent allelic loss and homozygous deletion in chromosome band 8p23 in oral cancer. *Int. J. Cancer* **80**, 25–31.
15. Waber, P., Dlugosz, S., Cheng, Q. C., Truelson, J., and Nisen, P. D. (1997). Genetic alterations of chromosome band 9p21–22 in head and neck cancer are not restricted to *p16INK4a*. *Oncogene* **15**, 1699–1704.
16. Venugopalan, M., Wood, T. F., Wilczynski, S. P., Sen, S., Peters, J., Ma, G. C., *et al.* (1998). Loss of heterozygosity in squamous cell carcinomas of the head and neck defines a tumor suppressor gene region on 11q13. *Cancer Genet. Cytogenet.* **104**, 124–132.
17. Lazar, A. D., Winter, M. R., Nogueira, C. P., Larson, P. S., Finnemore, E. M., Dolan, R. W., *et al.* (1998). Loss of heterozygosity at 11q23 in squamous cell carcinoma of the head and neck is associated with recurrent disease. *Clin. Cancer Res.* **4**, 2787–2793.
18. Ogawara, K., Miyakawa, A., Shiba, M., Uzawa, K., Watanabe, T., Wang, X. L., *et al.* (1998). Allelic loss of chromosome 13q14.3 in human oral cancer: Correlation with lymph node metastasis. *Int. J. Cancer* **79**, 312–317.
19. Lee, D. J., Koch, W. M., Yoo, G., Lango, M., Reed, A., Califano, J., *et al.* (1997). Impact of chromosome 14q loss on survival in primary head and neck squamous cell carcinoma. *Clin. Cancer Res.* **3**, 501–505.
20. Pearlstein, R. P., Benninger, M. S., Carey, T. E., Zarbo, R. J., Torres, F. X., Rybicki, B. A., *et al.* (1998). Loss of 18q predicts poor survival of patients with squamous cell carcinoma of the head and neck. *Genes Chromosomes Cancer* **21**, 333–339.
21. Miyakawa, A., Wang, X. L., Nakanishi, H., Imai, F. L., Shiiba, M., Miya, T., *et al.* (1998). Allelic loss on chromosome 22 in oral cancer: Possibility of the existence of a tumor suppressor gene on 22q13. *Int. J. Oncol.* **13**, 705–709.
22. Scully, C., Field, J. K., and Tanzawa, H. (2000). Genetic aberrations in oral or head and neck squamous cell carcinoma 2: Chromosomal aberrations. *Oral Oncol.* **36**, 311–327.
23. Miyashita, H., Mori, S., Tanda, N., Nakayama, K., Kanzaki, A., Sato, A., Morikawa, H., Motegi, K., Takebayashi, Y., and Fukumoto, M. (2001). Loss of heterozygosity of nucleotide excision repair factors in sporadic oral squamous cell carcinoma using microdissected tissue. *Oncol. Rep.* **8**, 1133–1138.
24. Gillespie, J. W., Ahram, M., Best, C. J., Swalwell, J. I., Krizman, D. B., Petricoin, E. F., Liotta, L. A., and Emmert-Buck, M. R. (2001). The role of tissue microdissection in cancer research. *Cancer J.* **7**, 32–39.
25. Zhang, W., Laborde, P. M., Coombes, K. R., Berry, D. A., and Hamilton, S. R. (2001). Cancer genomics: Promises and complexities. *Clin. Cance. Res.* **7**, 2159–2167.
26. Laconi, S., Pani, P., Pillai, S., Pasciu, D., Sarma, D. S., and Laconi, E. (2001). A growth-constrained environment drives tumor progression invivo. *Proc. Natl. Acad. Sci. USA* **98**, 7806–7811.
27. Eckert, R. L., Crish, J. F., and Robinson, N. A. (1997). The epidermal keratinocyte as a model for the study of gene regulation and cell differentiation. *Physiol. Rev.* **77**, 397–424.
28. Compagni, A., and Christofori, G. (2000). Recent advances in research on multistage tumorigenesis. *Br. J. Cancer* **83**, 1–5.
29. Patterson, T., Vuong, H., Liaw, Y. S., Wu, R., Kalvakolanu, D. V., and Reddy, S. P. (2001). Mechanism of repression of squamous differentiation marker, SPRR1B, in malignant bronchial epithelial cells: Role of critical TRE-sites and its transacting factors. *Oncogene* **20**, 634–644.
30. Bloor, B. K., Seddon, S. V., and Morgan, P. R. (2001). Gene expression of differentiation-specific keratins in oral epithelial dysplasia and squamous cell carcinoma. *Oral Oncol.* **37**, 251–261.
31. Bradley, G., Irish, J., MacMillan, C., Mancer, K., Witterick, I., Hartwick, W., Gullane, P., Kamel-Reid, S., and Benchimol, S. (2001). Abnormalities of the ARF-p53 pathway in oral squamous cell carcinoma. *Oncogene* **20**, 654–658.
32. Lemon, B., and Tjian, R. (2000). Orchestrated response: A symphony of transcription factors for gene control. *Genes Dev.* **14**, 2551–2569.
33. Zhu, S., Oh, H. S., Shim, M., Sterneck, E., Johnson, P. F., and Smart, R. C. (1999). C/EBPbeta modulates the early events of keratinocyte differentiation involving growth arrest and keratin 1 and keratin 10 expression. *Mol. Cell. Biol.* **19**, 7181–7190.
34. Jarnik, M., Kartasova, T., Steinert, P. M., Lichti, U., and Steven, A. C. (1996). Differential expression and cell envelope incorporation of small proline-rich protein 1 in different cornified epithelia. *J. Cell Sci.* **109**, 1381–1391.
35. Eckert, R. L., Crish, J. F., Banks, E. B., and Welter, J. F. (1997). The epidermis: Genes on – genes off. *J. Invest. Dermatol.* **109**, 501–509.
36. Hu, Y. C., Lam, K. Y., Law, S., Wong, J., and Srivastava, G. (2001). Identification of differentially expressed genes in esophageal squamous cell carcinoma (ESCC) by cDNA expression array: Overexpression of Fra-1, Neogenin, Id-1, and CDC25B genes in ESCC. *Clin. Cancer Res.* **7**, 2213–2221.
37. Workman, P., and Clarke, P. A. (2001). Innovative cancer drug targets: Genomics, transcriptomics and clinomics. *Expert Opin. Pharmacother.* **2**, 911–915.
38. Sidransky, D. (1995). Molecular genetics of head and neck cancer. *Curr. Opin. Oncol.* **7**, 229–233.
39. Rodrigo, J. P., Suarez, C., Gonzalez, M. V., Lazo, P. S., Ramos, S., Coto, E., Alvarez, I., Garcia, L. A., and Martinez, J. A. (2001). Variability of genetic alterations in different sites of head and neck cancer. *Laryngoscope* **111**, 1297–1301.
40. Bange, J., Zwick, E., and Ullrich, A. (2001). Molecular targets for breast cancer therapy and prevention. *Nature Med.* **7**, 548–552.
41. Luk, C., Tsao, M. S., Bayani, J., Shepherd, F., and Squire, J. A. (2001). Molecular cytogenetic analysis of non-small cell lung carcinoma by spectral karyotyping and comparative genomic hybridization. *Cancer Genet. Cytogenet.* **125**, 87–99.
42. Crnogorac-Jurcevic, T., Efthimiou, E., Capelli, P., Blaveri, E., Baron, A., Terris, B., Jones, M., Tyson, K., Bassi, C., Scarpa, A., and Lemoine, N. R. (2001). Gene expression profiles of pancreatic cancer and stromal desmoplasia. *Oncogene* **20**, 7437–7446.
43. Gupta, S. K., Douglas-Jones, A. G., and Morgan, J. M. (1997). Microdissection of stained archival tissue. *Mol. Pathol.* **50**, 218–220.

44. Walch, A., Specht, K., Smida, J., Aubele, M., Zitzelsberger, H., Hofler, H., and Werner, M. (2001). Tissue microdissection techniques in quantitative genome and gene expression analyses. *Histochem. Cell Biol.* **115**, 269–276.
45. Gjerdrum, L. M., Lielpetere, I., Rasmussen, L. M., Bendix, K., and Hamilton-Dutoit, S. (2001). Laser-assisted microdissection of membrane-mounted paraffin sections for polymerase chain reaction analysis: Identification of cell populations using immunohistochemistry and in situ hybridization. *J. Mol. Diagn.* **3**, 105–110.
46. Bohm, M., Wieland, I., Schutze, K., and Rubben, H. (1997). Microbeam MOMeNT: Non-contact laser microdissection of membrane-mounted native tissue. *Am. J. Pathol.* **151**, 63–67.
47. Nagasawa, Y., Takenaka, M., Matsuoka, Y., Imai, E., and Hori, M. (2000). Quantitation of mRNA expression in glomeruli using laser-manipulated microdissection and laser pressure catapulting. *Kidney Int.* **57**, 717–723.
48. Kolble, K. (2000). The LEICA microdissection system: Design and applications. *J. Mol. Med.* **78**, B24–25.
49. Rubin, M. A. (2001). Use of laser capture microdissection, cDNA microarrays, and tissue microarrays in advancing our understanding of prostate cancer. *J. Pathol.* **195**, 80–86.
50. Dean-Clower, E., Vortmeyer, A. O., Bonner, R. F., Emmert-Buck, M., Zhuang, Z., and Liotta L. A. (1997). Microdissection-based genetic discovery and analysis applied to cancer progression. *Cancer J. Sci. Am.* **3**, 259–265.
51. Emmert-Buck, M. R., Bonner, R. F., Smith, P. D., Chuaqui, R. F., Zhuang, Z., Goldstein, S. R., Weiss, R. A., and Liotta, L. A. (1996). Laser capture microdissection. *Science* **274**, 998–1001.
52. Pappalardo, P. A., Bonner, R., Krizman, D. B., Emmert-Buck, M. R., and Liotta, L. A. (1998). Microdissection, microchip arrays, and molecular analysis of tumor cells (primary and metastases). *Semin. Radiat. Oncol.* **8**, 217–223.
53. Simone, N. L., Bonner, R. F., Gillespie, J. W., Emmert-Buck, M. R., and Liotta, L. A. (1998). Laser-capture microdissection: Opening the microscopic frontier to molecular analysis. *Trends Genet.* **14**, 272–276.
54. Ehrig, T., Abdulkadir, S. A., Dintzis, S. M., Milbrandt, J., and Watson, M. A. (2001). Quantitative amplification of genomic DNA from histological tissue sections after staining with nuclear dyes and laser capture microdissection. *J. Mol. Diagn.* **3**, 22–25.
55. Weinstein, J. N. (2001). Searching for pharmacogenomic markers: The synergy between omic and hypothesis-driven research. *Dis. Mark.* **17**, 77–88.
56. Martin, K. J., Kritzman, B. M., Price, L. M., Koh, B., Kwan, C. P., Zhang, X., Mackay, A., O'Hare, M. J., Kaelin, C. M., Mutter, G. L., Pardee, A. B., and Sager, R. (2000). Linking gene expression patterns to therapeutic groups in breast cancer. *Cancer Res.* **60**, 2232–2238.
57. Southern, E. M. (2001). DNA microarrays. History and overview. *Methods Mol. Biol.* **170**, 1–15.
58. Schena, M., Shalon, D., Davis, R. W., and Brown, P. O. (1995). Quantitative monitoring of gene expression patterns with a complementary DNA microarray. *Science* **270**, 467–470.
59. Galbraith, D. W., Macas, J., Pierson, E. A., Xu, W., and Nouzova, M. (2001). Printing DNA microarrays using the Biomek 2000 laboratory automation workstation. *Methods Mol. Biol.* **170**, 131–140.
60. Duggan, D. J., Bittner, M., Chen, Y., Meltzer, P., and Trent, J. M. (1999). Expression profiling using cDNA microarrays. *Nature Genet.* **21**(1 Suppl.), 10–14.
61. Lipshutz, R. J., Fodor, S. P., Gingeras, T. R., and Lockhart, D. J. (1999). High density synthetic oligonucleotide arrays. *Nature Genet.* **21**(1 Suppl.), 20–24.
62. Goldsworthy, S. M., Stockton, P. S., Trempus, C. S., Foley, J. F., and Maronpot, R. R. (1999). Effects of fixation on RNA extraction and amplification from laser capture microdissected tissue. *Mol. Carcinog.* **25**, 86–91.
63. Leethanakul, C., Patel, V., Gillespie, J., Shillitoe, E., Kellman, R. M., Ensley, J. F., Limwongse, V., Emmert-Buck, M. R., Krizman, D. B., and Gutkind, J. S. (2000). Gene expression profiles in squamous cell carcinomas of the oral cavity: Use of laser capture microdissection for the construction and analysis of stage-specific cDNA libraries. *Oral Oncol.* **36**, 474–483.
64. Dolter, K. E., and Braman J. C. (2001). Small-sample total RNA purification: Laser capture microdissection and cultured cell applications. *Biotechniques* **30**, 1358–1361.
65. Bertucci, F., Van, Hulst. S., Bernard, K., Loriod, B., Granjeaud, S., Tagett, R., Starkey, M., Nguyen, C., Jordan, B., and Birnbaum, D. (1999). Expression scanning of an array of growth control genes in human tumor cell lines. *Oncogene* **18**, 3905–3912.
66. Wang, E., Miller, L. D., Ohnmacht, G. A., Liu, E. T., and Marincola, F. M. (2000). High-fidelity mRNA amplification for gene profiling. *Nature Biotechnol.* **18**, 457–459.
67. Leethanakul, C., Patel, V., Gillespie, J., Pallente, M., Ensley, J. F., Koontongkaew, S., Liotta, L. A., Emmert-Buck, M., and Gutkind, J. S. (2000). Distinct pattern of expression of differentiation and growth-related genes in squamous cell carcinomas of the head and neck revealed by the use of laser capture microdissection and cDNA arrays. *Oncogene* **19**, 3220–3224.
68. Alevizos, I., Mahadevappa, M., Zhang, X., Ohyama, H., Kohno, Y., Posner, M., Gallagher, G. T., Varvares, M., Cohen, D., Kim, D., Kent, R., Donoff, R. B., Todd. R., Yung, C. M., Warrington, J. A., and Wong, D. T. (2001). Oral cancer in vivo gene expression profiling assisted by laser capture microdissection and microarray analysis. *Oncogene* **20**, 6196–6204.
69. Alizadeh, A. A., Eisen, M. B., Davis, R. E., Ma, C., Lossos, I. S., Rosenwald, A., Boldrick, J. C., Sabet, H., Tran, T., Yu, X., Powell, J. I., Yang, L., Marti, G. E., Moore, T., Hudson, J. Jr., Lu, L., Lewis, D. B., Tibshirani, R., Sherlock, G., Chan, W. C., Greiner, T. C., Weisenburger, D. D., Armitage, J. O., Warnke, R., Staudt, L. M., *et al.* (2000). Distinct types of diffuse large B-cell lymphoma identified by gene expression profiling. *Nature* **403**, 503–511.
70. Sorlie, T., Perou, C. M., Tibshirani, R., Aas, T., Geisler, S., Johnsen, H., Hastie, T., Eisen, M. B., van de Rijn, M., Jeffrey, S. S., Thorsen, T., Quist, H., Matese, J. C., Brown, P. O., Botstein, D., Eystein, Lonning, P., and Borresen-Dale, A. L. (2001). Gene expression patterns of breast carcinomas distinguish tumor subclasses with clinical implications. *Proc. Natl. Acad. Sci. USA* **98**, 10869–10874.
71. Riggins, G. J., and Strausberg, R. L. (2001). Genome and genetic resources from the Cancer Genome Anatomy Project. *Hum. Mol. Genet.* **10**, 663–667.
72. Schaefer, C., Grouse, L., Buetow, K., and Strausberg, R. L. (2001). A new cancer genome anatomy project web resource for the community. *Cancer J.* **7**, 52–60.
73. Shillitoe, E. J., May, M., Patel, V., Lethanakul, C., Ensley, J. F., Strausberg, R. L., and Gutkind, J. S. (2000). Genome-wide analysis of oral cancer; early results from the Cancer Genome Anatomy Project. *Oral Oncol.* **36**, 8–16
74. Pivarcsi, A., Szell, M., Kemeny, L., Dobozy, A., and Bata-Csorgo, Z. (2001). Serum factors regulate the expression of the proliferation-related genes alpha5 integrin and keratin 1, but not keratin 10, in HaCaT keratinocytes. *Arch. Dermatol. Res.* **293**, 206–213.
75. Poumay, Y., and Pittelkow, M. R. (1995). Cell density and culture factors regulate keratinocyte commitment to differentiation and expression of suprabasal K1/K10 keratins. *J. Invest. Dermatol.* **104**, 271–276.
76. Emmert-Buck, M. R., Strausberg, R. L., Krizman, D. B., Bonaldo, M. F., Bonner, R. F., Bostwick, D. G., Brown, M. R., Buetow, K. H., Chuaqui, R. F., Cole, K. A., Duray, P. H., Englert, C. R., Gillespie, J. W., Greenhut, S., Grouse, L., Hillier, L. W., Katz, K. S., Klausner, R. D., Kuznetzov, V., Lash, A. E., Lennon, G., Linehan, W. M., Liotta, L. A., Marra, M. A., Munson, P. J., Ornstein, D. K., Prabhu, V. V., Prang, C., Schuler, G. D., Soares, M. B., Tolstoshev, C. M., Vocke, C. D., and Waterston, R. H.

(2000) Molecular profiling of clinical tissues specimens: Feasibility and applications. *J. Mol. Diagn.* **2**, 60–66.
77. Heng, C., and Rossi, E. P. (1995). A report on 222 cases of oral squamous cell carcinoma. *Mil. Med.* **160**, 319–323.
78. Krizman, D. B. (1996). Gene identification by 3′ terminal exo trapping. *Genet. Eng. (N. Y.)* **18**, 49–56.
79. Bonalume, and Neto, R. (1999). Brazilian scientists team up for cancer genome project. *Nature* **398**, 450.
80. Humphery-Smith, I., Cordwell, S. J., and Blackstock, W. P. (1997). Proteome research: Complementarity and limitations with respect to the RNA and DNA worlds. *Electrophoresis* **18**, 1217–1242.
81. Dwek, M. V., Ross, H. A., and Leathem, A. J. (2001). Proteome and glycosylation mapping identifies post-translational modifications associated with aggressive breast cancer. *Proteomics.* **1**, 756–62
82. Post, S., Weng, Y. C., Cimprich, K., Chen, L. B., Xu, Y., and Lee, E. Y. (2001). Phosphorylation of serines 635 and 645 of human Rad17 is cell cycle regulated and is required for G1/S checkpoint activation in response to DNA damage. *Proc. Natl. Acad. Sci. USA* **98**, 13102–13107.
83. Chen, L. F., Fischle, W., Verdin, E., and Greene, W. C. (2001). Duration of nuclear NF-kappaB action regulated by reversible acetylation. *Science* **293**, 1653–1657.
84. Kahyo, T., Nishida, T., and Yasuda, H. (2001). Involvement of PIAS1 in the sumoylation of tumor suppressor p53. *Mol. Cell* **8**, 713–718.
85. Lauder, R. M., Huckerby, T. N., and Nieduszynski, I. A. (2000). Increased incidence of unsulphated and 4-sulphated residues in the chondroitin sulphate linkage region observed by high-pH anion-exchange chromatography. *Biochem. J.* **347**(Pt. 2), 339–348.
86. Mills, G. B., Bast, R. C., Jr., and Srivastava, S. (2001). Future for ovarian cancer screening: Novel markers from emerging technologies of transcriptional profiling and proteomics. *J. Natl. Cancer Inst.* **93**, 1437–1439.
87. Srinivas, P. R., Srivastava, S., Hanash, S., and Wright, G. L., Jr. (2001). Proteomics in early detection of cancer. *Clin. Chem.* **47**, 1901–1911.
88. Lawrie, L. C., Curran, S., McLeod, H. L., Fothergill, J. E., and Murray, G. I. (2001). Application of laser capture microdissection and proteomics in colon cancer. *Mol. Pathol.* **54**, 253–258.
89. Banks, R. E., Dunn, M. J., Forbes, M. A., Stanley, A., Pappin, D., Naven, T., Gough, M., Harnden, P., and Selby, P. J. (1999). The potential use of laser capture microdissection to selectively obtain distinct populations of cells for proteomic analysis: Preliminary findings. *Electrophoresis* **20**, 689–700.
90. Ricci, M. S., and el-Deiry, W. S. (2000). Novel strategies for therapeutic design in molecular oncology using gene expression profiles. *Curr. Opin. Mol. Ther.* **2**, 682–690.
91. Bichsel, V. E., Liotta, L. A., and Petricoin, E. F., III. (2001). Cancer proteomics: From biomarker discovery to signal pathway profiling. *Cancer J.* **7**, 69–78.
92. Emmert-Buck, M. R., Gillespie, J. W., Paweletz, C. P., Ornstein, D. K., Basrur, V., Appella, E., Wang, Q. H., Huang, J., Hu, N., Taylor, P., and Petricoin, E. F., III. (2000). An approach to proteomic analysis of human tumors. *Mol. Carcinog.* **27**, 158–165.
93. Gygi, S. P., Corthals, G. L., Zhang, Y., Rochon, Y., and Aebersold, R. (2000). Evaluation of two-dimensional gel electrophoresis-based proteome analysis technology. *Proc. Natl. Acad. Sci. USA* **97**, 9390–9395.
94. Westbrook, J. A., Yan, J. X., Wait, R., and Dunn, M. J. (2001). A combined radiolabelling and silver staining technique for improved visualisation, localisation, and identification of proteins separated by two-dimensional gel electrophoresis. *Proteomics* **1**, 370–376.
95. Appel, R. D., Hoogland, C., Bairoch, A., and Hochstrasser, D. F. (1999). Constructing a 2-D database for the World Wide Web. *Methods Mol. Biol.* **112**, 411–416.

CHAPTER

35

Molecular Assessment of Surgical Margins

WAYNE M. KOCH

Department of Otolaryngology—Head and Neck Surgery
Johns Hopkins University School of Medicine
Baltimore, Maryland 21286

I. Introduction 523
- A. Current Use of Margins 523
- B. Problems with Current Practice 524
- C. Patterns of Cancer Growth 524

II. Insight from the Molecular Biology Revolution 524
- A. Cancer Is a Clonal Genetic Disease 524
- B. Strategies for Molecular Detection 524

III. The Margin Dilemma 524
- A. Clinical Utility of Margins 524
- B. Failure of Margins to Predict Outcome 525

IV. Technical Factors Influencing Margin Assessment 525
- A. Some Definitions 525
- B. Why Do Margins Fail to Predict Outcome? 526

V. Is Margin Status Reflective of Biologic Aggressiveness? 527
- A. Growth Pattern Is a Prognostic Factor 527
- B. Field Cancerization and Margin Status 527
- C. Tumor–Host Interaction at the Margin 528

VI. Theoretical Basis for Molecular Margin Analysis 528

VII. Published Molecular Margin Studies 528
- A. Mutant p53 Oligomer Probing 528
- B. Overexpression of p53 Protein Detected with Immunohistochemistry 529
- C. Generalized Markers for Head and Neck Squamous Cell Carcinomas 530

VIII. Clinical Response to Positive Margins 530
- A. Further Resection 530
- B. Radiation 531
- C. Chemotherapy 531
- D. Innovative Therapy 531

IX. Concluding Remarks 531

References 532

I. INTRODUCTION

Surgical resection with curative intent is required at some point in the management of many cases of solid malignancies involving the tissues of the upper aerodigestive tract, head, and neck. Effective surgical treatment dictates that the cancer be removed in its entirety. If even a tiny amount of tumor is left in place, it is likely to grow and again pose a threat to the health and life of the patient within a short period of time.

A. Current Use of Margins

In order to achieve the goal of complete cancer extirpation, the surgeon must judge the extent of disease using physical observations that combine tactile assessments of the thickness and turgor of the lesion with its visual qualities, such as white or red color, ulceration, or exophytic growth. Experience and awareness of specific patterns of tumor extension are combined with these observations to guide resection with an appropriate "margin" of "normal" tissue. The width of this clinically normal rim around the obvious tumor may range from several millimeters, as in the case of minute laryngeal cancers, to several centimeters or more, as when perineural extension is suspected in adenoid cystic carcinoma.

The surgeon may then select areas around the rim of resection for intraoperative frozen section histopathological analysis. Tissue with functional importance, at the limits of safe resection, or of questionable physical characteristics are prime candidates for this immediate margin assessment. When a definite answer can be provided within the time constraints of the anesthetic, further surgical resection may render an individual "free of disease."

Final assessment of fixed tissue margin samples is completed within days of the surgical resection. The results are useful in several ways. In some cases, further resection may be undertaken during a second procedure directed on the basis of the margin analysis. When this is undesirable or impossible, the addition of adjuvant therapy, usually radiation, may be selected, at least in part, on the basis of the margin status. Finally, margin data can be used as a prognostic factor for the risk of both local recurrence and death. Indeed, the status of the surgical resection margins is often judged to be among the strongest of prognostic indicators.

B. Problems with Current Practice

Histologic analysis of surgical margins is far from flawless, despite its widespread acceptance in standard practice. False-negative margin evaluation must be assumed whenever there is a local recurrence of tumor after a report of all margins of resection being clear of disease. Less commonly, false-positive evaluations are encountered and a cancer does not recur despite an involved final margin. Both situations do occur in any busy surgical oncology practice with unnerving frequency.

C. Patterns of Cancer Growth

Efforts have been made for decades to classify individual tumors according to their patterns of growth. Some squamous cell carcinomas of the head and neck (HNSCC) demonstrate pushing borders in which the leading edge of the tumor–stroma interface is broad and the cancer cells remain attached to one another in large groups. Other HNSCC spreads in narrow finger projections or as separate clusters or even individual cells that may extend well beyond the clinically detectable tumor boundary. Some tumors have a propensity for perineural or lymphangitic spread and may even have skip regions that lack identifiable neoplastic cells. The less organized the leading edges of the tumor, the more difficult the task of the pathologist attempting to assess the completion of resection through sampling of the margin tissue and light microscopy.

II. INSIGHT FROM THE MOLECULAR BIOLOGY REVOLUTION

A. Cancer Is a Clonal Genetic Disease

The advent of molecular biology has provided a new dimension in understanding the process of malignant transformation to initiate cancer. In addition, research results have provided a host of powerful tools for the detection of rare cancer cells amidst a background of normal cells. It is commonly accepted that cancers develop as a result of changes in the DNA (mutation, deletion, insertion, methylation), leading to alterations in protein function that cause the characteristic cancer phenotype. The first genetic alterations produce a loss of growth control that leads to the overgrowth of related daughter cells, a phenomenon known as clonal expansion. Members of the clone share tumor-specific genetic alterations that are distinct from all of the individual's normal cells.

B. Strategies for Molecular Detection

Alterations indicative of malignant transformation can be detected in a clinical sample in a variety of ways. Immunohistochemical staining for altered protein expression and fluorescent *in situ* hybridization detecting changes in the copy number of specific gene fragments can be performed on tissue sections parallel to those used for standard light microscopy. More powerful detection of rare tumor cells is possible through amplification of DNA using the polymerase chain reaction, followed by separation and detection of altered tumor-specific fragments.

Since the mid-1990s, several of these methods have been applied to the problem of surgical margin analysis for tumors of the head and neck. Initial results are promising, indicating that molecular assessment of surgical margins provides improved prediction of local-regional tumor recurrence and patient survival. Technical problems remain. Chief among these is the wherewithal to analyze samples within the time limits of the surgical procedure so as to direct further resection as needed. Molecular assessment of margins has not yet been validated in a large, multi-institutional trial confirming the clinical accuracy of the information and the potential for widespread application of the techniques.

III. THE MARGIN DILEMMA

A. Clinical Utility of Margins

The concept that complete removal of every cancer cell is necessary for curative surgical therapy of cancer is intuitively compelling. Because cancer begins with a single, fully genetically altered cell that spawns a clonal population eventuating in a malignant tumor, even a single remaining cancer cell should be able to repopulate a tumor bed and appear some time later as a clincal recurrence. The utility of standard light microscopic margin evaluation is borne out in the results of numerous studies showing a clear survival advantage among individuals whose tumors were removed with clear margins compared to those with compromised margins. For example, a study of 270 surgical patients treated at Roswell Park Memorial Institute for HNSCC demonstrated a 39% estimated 5-year disease-free survival rate for patients treated with clear margins compared to 7% for those with not-free margins [1]. Three hundred and ninety-eight patients undergoing surgical resection at Memorial Sloan Kettering Cancer Center were found to be twice as likely to recur in the primary site if margins of resection were positive (36% vs 18%). The difference in 5-year survival was less dramatic, but still statistically significant (60% vs 52%) [2]. Another study of 478 HNSCC patients from the United Kingdom demonstrated a significant decrease in 5-year survival if resection margins were positive. Fifty-two percent of the margin negative group survived compared with 46% of the margin-positive group ($p<0.025$) [3]. The same institution later published data for 352 patients undergoing salvage surgery after failed primary radiation. In this study, margin status was significantly associated with the risk of local recurrence (47% of margin negative group recurred compared to 66% of margin positive group) [4]. A report from the Head and Neck Intergroup considered the outcome of 696 patients enrolled in a clinical trial. The rate of local recurrence was much lower when the resection margins were clear (9%) than when at least one margin was compromised (21%). The median survival of the positive-margin group was 18.4 months compared to 38 months for the negative-margin group. It was asserted that survival of patients with tumor resection with positive surgical margins is not much better than that of inoperable cases, despite a possible benefit in local control when the tumor is debulked [5].

B. Failure of Margins to Predict Outcome

Despite the strong argument in favor of the clinical utility of surgical margin analysis, it is widely recognized that the status of margins does not accurately predict outcome in many cases. Often, the difference in survival between margin-positive and margin-negative groups is modest and may be eliminated when other parameters are taken into account through multivariate analysis. In virtually every published report, sizable minorities of HNSCC patients with clear surgical margins experience local recurrence of disease, whereas at least some patients with cancer cells at the margins of resection do not go on to develop recurrent cancer (see later). The phenomenon of "falsely" positive margins is particularly noted to pertain to certain anatomic sites such as the larynx. The much-cited study from Washington University involving a series of 111 hemilaryngectomies illustrates the point. Thirty-nine patients had positive margins, but were not treated further unless they developed recurrent disease. Only 7 (18%) went on to recur locally compared with 4 of 72 (6%) patients with clear margins [6]. In summarizing a retrospective review of 769 HNSCC patients resected at Washington University, the assertion was made that "as involvement of the resection margins increased, the survival rates declines." And yet, "the meaning of positive resection margins remains somewhat unclear and is site dependent" [7].

Margin status must be assessed in multivariate statistical models to demonstrate its prognostic significance independent of factors such as stage, tumor grade, site, and treatment. In the United Kingdom study of 352 patients undergoing salvage surgery after failed primary radiation, the apparent poor effect on survival of positive margins was not confirmed in multivariate analysis. Controlling for site and T stage eliminated the impact of margin status. Again, the authors note with surprise that over a third of the patients with failed attempts to resect persistent disease remained alive and disease free at 5 years having received no additional treatment [4]. Among 478 patients undergoing primary surgery at the same institution, there was only a 6% difference in survival between those with clear and compromised margins. Again, the status of margins was not independent of site [3].

IV. TECHNICAL FACTORS INFLUENCING MARGIN ASSESSMENT

A. Some Definitions

Several studies include a variety of histologic findings within the broad category of "involved" or "positive" margins. When evaluating these reports, it is critical to consider first what techniques were used to evaluate margins and what was considered a positive margin. Were margins taken by the surgeon from the patient after removal of the main specimen or harvested by the pathologist from that specimen? Were margins cut perpendicular or parallel to the tumor–host boundary? How many margins were submitted? Were they analyzed by frozen section and later as fixed tissue to confirm findings? How many levels were evaluated within the block?

John Batsakis appealed for uniformity of definition as to what constitutes a positive or negative margin [8]. He stated that the effect of closeness of disease to a margin has not been studied systematically, but that his experience indicates that *in situ* carcinoma, severe dysplasia, and microinvasive carcinoma at the margin have equal (poor) prognostic significance. Several studies, in fact, assert that tumor within 5 mm of the actual resection margin ("close margin") has the same clinical impact as a frankly positive margin. Dysplasia or carcinoma *in situ* at the cut edge also has a negative prognostic impact, although perhaps not as profound as the close or frankly positive margin. Loree and Strong [2] reviewed 129 cases of HNSCC with compromised margins (83 close, 9 *in situ*, 9 dysplasia, 28 invasive carcinoma). Overall, the incidence of local recurrence among this group was twice as great as among 269 others with clear margins (36% vs 18%). It should be pointed out that 64% of patients with compromised margins did not recur, and 18% of those with clear margins had local recurrence. The rate of local recurrence was nearly the same between each subset of compromised margins (close = dysplasia = *in situ* = invasive); however, 5-year survival decreased with greater histologic neoplastic grade at the margins [2]. Other studies support the contention that any subcategory of margin compromise portends an approximately equal risk of local recurrence [1]. In the hemilaryngectomy series from Washington University, local recurrences occurred with equal frequency in cases with gross tumor at the margins, intraepithelial tumor in margins, or margins of less than 5 mm. Intriguingly, those with premalignant disease at the margin recurred much later (4–8 years after surgery) than those with gross disease or close margins (1 year), consistent with the time needed for further tumor progression of residual premalignant subclones [6].

However, a series of 80 Dutch patients indicates that the local recurrence rate in the face of dysplasia or carcinoma *in situ* (CIS) at the margin is no greater than when margins are clear. Tumor at the deep margins and at multiple margins conveyed a much greater risk of recurrence [9]. Tumor involvement of the deep margins may have a more profound prognostic impact than mucosal surface involvement. In a study of 51 oral cancer resections, the closeness of the deep margins alone was significantly different among cases that went on to recur locally compared with those that remained disease free [10].

B. Why Do Margins Fail to Predict Outcome?

Several rationales can be proposed to explain situations in which the histologic margin status fails to accurately predict clinical outcome. The bases of this theoretical consideration include the acknowledged fact that all margin analysis involves sampling and the recognition that the status of surgical margins may reflect a more global biologic behavior potential of an individual neoplasm. Many of these factors have been examined since Bauer's seminal musings over margins in hemilaryngectomies [6]. The problems that confound standard margin analysis offer an opportunity to improve upon the clinical accuracy of margin analysis using molecular tools and markers.

1. Sampling

There are numerous reasons for inaccurate histological evaluation of surgical margins. Most of these reasons fall under the broad category of sampling error. The evaluation of surgical margins is an exacting, tedious, time-consuming, and therefore expensive process. There is a practical limit to the number of margins that can be evaluated. Even when many samples are submitted, the precise identification of the location from which an involved margin sample was taken becomes tenuous when bulky, three-dimensional, complex lesions are involved. There is also a limit to the amount of tissue volume visualized under the microscope. Viewing all of a larger sample in two dimensions may be difficult in itself. The third dimension can only be assessed by cutting through the specimen to some arbitrary depth, discarding intervening sections of tissue.

Mohs chemosurgery as applied to cutaneous malignancies represents one method that has been used to address the problem of sampling in assessing the completeness of cancer extirpation [11]. The dermatologic Mohs chemosurgeon carefully maps the resection bed and takes multiple small margin samples encompassing the entire periphery in three dimensions. Each section is processed and assessed under the microscope while the patient waits. Further resection followed by repeat mapping and margin analysis is then performed wherever malignant cells remained.

Using the general approach of Mohs, a report of 70 HNSCC patients undergoing resection with multiple parallel frozen section control has been published [12]. As in chemosurgery, maps were constructed, an effort was made to harvest tissue from the entire tumor bed, and frozen section results were used to guide additional resection until negative margins were achieved. First, the tumor was resected with a 1-cm cuff of apparently normal margin. The vast majority (51/68 or 75%) of cases had positive margins on frozen section after this first level of resection. Cords of cancer 10–20 cells across were found 2–3 cm away from the visible tumor edge. Most of the microscopic extensions were seen in the submucosa or tracking along muscle bundles.

While the Mohs approach is widely accepted for dermatologic lesions, several practical problems preclude its use for tumors of the upper aerodigestive tract. Patients cannot be maintained safely, economically, and easily under general anesthesia for the hours (and sometimes days) needed to evaluate the entire tumor bed in this thorough and meticulous fashion. The orientation of tissue specimens taken from deep soft tissue margins that may fold or collapse against each other is a much more complex task than mapping of a skin malignancy. Moreover, in the end, the Mohs approach still depends on light microscopy to identify cancer cells in a block of tissue.

2. Technical Confounding Factors

Factors such as cautery artifact, fibrosis or inflammatory infiltrates, tissue shrinkage, and precise orientation of specimens affect the interpretation of margins as well. In all cases, the pathologist is called upon to identify tiny amounts of residual cancer amidst a background of nonmalignant tissue. This task requires the recognition of clusters of cells exhibiting a histologic pattern and/or cytologic features that are recognized as abnormal. In general, the nature of a suspicious cell in a background of normal cells cannot be definitely judged as residual cancer. It is not practical to analyze every cell individually along the tumor–host interface using current technology. Rare tumor cells at the edge of resection may elude detection, particularly when the pattern of tumor growth at the perimeter involves thin strands or detached individual cells or when the background is obscured by cautery, inflammatory cells or fibrosis.

The accuracy of frozen section evaluation of tumor margins is less reliable than that using fixed tissue sections. Inflammatory infiltrates, fibrosis, and other tissue factors make the assessment of frozen tissue sections difficult. Disagreement between frozen section report and fixed tissue confirmation occurred in 2.1% of 1947 frozen sections in one study [13]. A recent series from Memorial Sloan Kettering judged the accuracy of frozen sections to be somewhat lower, with 11% of samples read differently when confirmed after fixation [14].

Different arguments must be entertained to explain why a positive margin does not always correspond to tumor recurrence. The actual physical measurement of tissue in the resected specimen may underestimate the distance of tumor cells from the cut edge for a variety of reasons. The extirpation of tumor is often accomplished using some heat-generating instrument such as bovie or bipolar cautery or Shaw scalpel. These instruments destroy some cells at the site of their use so that light microscopic analysis considers cells several micrometers away from the actual tumor–host interface. Indeed, some of the cells left in the tumor bed may have been rendered nonviable by the heat. Shrinkage of the tissue during the fixation process is another cause of underestimation of the extent of disease-free tissue margins [8]. Also, the surgeon often stretches tissues in order to facilitate tumor resection. Both epithelium and muscle have elasticity that is released upon incision, resulting in a springing back of the cut edges. It is typical for a margin judged to be 2 cm by the surgeon to appear as 5 mm or less on the histologic slide. That difference may be critical if the contention that a margin of less than 5 mm is effectively the same as a frankly involved margin is true. Finally, complex three-dimensional specimens present a substantial challenge to the pathologist who must orient the tissue and select the actual resection margins for sampling. If separation of the tissues has occurred during the surgical manipulation of the tumor, regions may be judged as true margins when they are not.

V. IS MARGIN STATUS REFLECTIVE OF BIOLOGIC AGGRESSIVENESS?

A. Growth Pattern Is a Prognostic Factor

Tumors that grow with thin strands or individually advancing cells are more likely to be resected with compromised margins. This detached growth pattern portends a poor prognosis independent of the margin status as evidenced by several reports. Growth pattern was more important than margin status for a group of 150 HNSCC patients treated surgically at Memorial Sloan Kettering Cancer Center. While the status of the margins was associated with the rate of local disease control, it did not impact survival. However, a growth pattern of diffuse infiltration and cellular dissociation did portend a significant decrease in survival with higher rates of nodal and distant metastases [14]. Among another series of 66 patients with oral tongue tumors resected with clear margins, the subset with widespread invasion by varying sized groups of tumor cells had a higher rate of locoregional failure and nodal metastasis [15]. A group of Dutch investigators have attempted to adapt the size of margins taken for a given tumor to its particular growth pattern. They counted margins as clear only if the distance from the closest tumor nest to the cut edge was greater than the distance between tumor nests when the growth pattern was "spidery." Using this stringent definition, they report only a 4% local recurrence rate among tumors resected with clear margins [16].

The molecular basis for variant growth patterns within HNSCC has not been determined rigorously. The loss of extracellular matrix and the overexpression of metalloproteinases have been shown to correspond to the invasive tendency of HNSCC [17]. The poor prognostic implication of an aggressive pattern of growth at the tumor–host interface can be joined with other clinical observations to support the contention that tumors with positive margins are inherently more biologically aggressive. For example, several studies have shown that tumors resected with positive margins have a poor prognosis despite intervening postoperative radiation therapy. Relative radiation resistance is implied. Tumors resected with positive margins may also be more apt to display both regional and distant metastatic potential [3,5]. Thus, a tumor with a gross phenotype that eludes the physical judgement of an experienced surgeon is more likely to have other clinical behaviors that are apparently unrelated. In support of this observation, it has been noted that tumors resected with initially positive margins continue to behave badly even after additional resection renders all margins free of disease [18,19].

B. Field Cancerization and Margin Status

Another phenotypic observation that may impact molecular margin analysis is that of field cancerization. Attributed to Slaughter, this concept is based on the phenomena of multiple upper aerodigestive tract primary lesions found in a subset of HNSCC patients and the presence of "condemned mucosa" with diffuse premalignant change seen in some others [20]. An ongoing debate exists as to whether all the malignant or premalignant cells in different sites in a patient with field cancerization are clonally related to a single progenitor transformed cell or have arisen after separate independent transforming events [21]. This theoretical debate may have profound significance when pursuing a molecular margin analysis strategy. While a tumor-specific genetic alteration such as a particular point mutation of a target gene (p53) offers a very specific marker for a given clonal population, it would fail to detect cells from a clone arising independently and lacking that mutation. In that circumstance, other molecular approaches employing more general markers for malignant transformation might prove superior. Their lack of specificity for a single clone and an ability to detect transformed cells from independent sources would be more successful.

For multiple primary cancers to be clonally related, one must postulate that partially transformed daughter cells can move along the mucosal surface producing skip lesions when they stop and begin to infiltrate in a new location. The work of Bedi *et al.* [22] provides molecular evidence supporting such a clonal relationship among two primary cancers in a small group of HNSCC patients. Regardless of how the phenomenon occurs, field cancerization must involve individual partially or fully transformed cells spread across a large surface of the epithelium. These cells may be difficult to detect by light microscopy amidst the background of the normal epithelium. They may be detected using any of a variety of new molecular techniques and may account for a greater level of sensitivity in margin assessment using those approaches. Indeed, molecular detection of isolated, widespread, clonally related cells in normal-appearing epithelium at a distance from an index lesion has been demonstrated successfully [23]. Other studies indicate the presence of clones with discordant p53 mutations spread throughout the mucosa [24,25]. These observations may come together if one assumes that p53 mutations come later in the tumor progression pathway, explaining the presence of subclones sharing the earliest alterations (such as 9p loss of heterozygosity) while containing different later events (p53 mutation).

C. Tumor–Host Interaction at the Margin

The failure of tumor to grow after incomplete resection may reflect an inhospitable environment. A very small tumor burden may be managed adequately by an effective immune response. Immune function has been linked to outcomes among surgically treated HNSCC patients [26]. Alternatively, a scarred tumor bed with a disrupted blood supply may be an inhospitable environment for a few residual cells.

VI. THEORETICAL BASIS FOR MOLECULAR MARGIN ANALYSIS

Molecular margin analysis hinges on the presence of proteins or DNA sequences that are highly tumor specific and can be detected in the background of normal cells at a threshold far below that of light microscopy. The widespread acceptance of a molecular approach to margin assessment will depend on clear demonstration of a clinical utility (better sensitivity and specificity) and technical feasibility involving reasonable time and expense. Finally, the usefulness of more accurate margin analysis is dependent on the availability of an effective therapeutic response to the information provided. An ideal molecular margin analysis strategy must address these issues. The molecular marker must identify only viable, dangerous cells with a high likelihood for clinical virulence. Detection limits should surpass that of light microscopy by orders of magnitude while distinguishing clinically significant from insignificant tumor burden (if such exists). Ideally, the approach would be uniform for all patients and be performed using a robust, inexpensive, and highly reproducible assay. It should improve upon the limitations imposed by sampling such that an adequate volume of the surrounding tissue is analyzed, yet maintain or improve upon the ability to pinpoint the location of residual disease in the tumor bed. Most daunting of all, it should be completed within the time limitations of the extirpative procedure so that the opportunity for further resection or other intraoperative intervention is available. To date, no molecular approach put forward accomplishes all of these ideals.

VII. PUBLISHED MOLECULAR MARGIN STUDIES

A. Mutant p53 Oligomer Probing

The first attempt to use molecular markers in the evaluation of tumor resection margins was published by Brennan and co-workers [27] at Johns Hopkins in 1995. The p53 gene of a selected group of 30 HNSCC tumors was found by direct gene sequencing to display a point mutation. Those mutated sequences provided highly specific markers for the tumor cells. An oligomer (a short single strand of DNA matching the gene sequence surrounding and including the mutation) for each case was manufactured and radiolabeled. DNA from tumor cells (as a positive control) and from histologically free margins was amplified using the polymerase chain reaction. The DNA was then packaged into bacteriophage that were diluted and plated onto a lawn of bacteria. The resultant lytic plaques each contained multiple copies of the packaged DNA from one cell, either wild type from normal cells or mutated from cells bearing the same specific alteration as found in the infiltrating cancer. The proportion of plaques with normal and mutant DNA was ascertained by hybridizing the radiolabeled probes to a membrane to which the DNA from the petri dish had been transferred. Results were interpreted to show the presence and relative frequency of cancer cells in the margin tissue.

Of 30 subjects in the Brennan series, 5 had final pathology margins read as containing tumor and were eliminated from further consideration. Thirteen of the remaining 25 tumors showed molecular evidence of histologically occult tumor cells in at least one margin sample using the p53 probe technique. In the follow-up period of 8–27 months, none of the 12 with clear molecular margins recurred locoregionally. Five of the 13 with molecularly positive margins recurred either locally or regionally. In each case, the location of the margin that was positive by p53 analysis corresponded to the site of local tumor regrowth. An unpublished review of the Brennan patients after 4 years of follow-up revealed no additional local recurrences in any of the 25 subjects, although 2 each in the molecular-positive and negative-margin group experienced distant tumor recurrence. Of the 5 cases that recurred, 4 subjects had received postoperative radiation therapy. That postoperative radiation did not prevent recurrence in the margin-positive cases is consistent with reports in the literature already discussed. In those cases, when the histologic margins were positive, radiation did not result in an improved outcome, in general.

Because DNA is extracted from a three-dimensional block of margin tissue, the Brennan approach goes well beyond the sampling capability of any light microscopy-based analysis in which only a limited number of thin sections can be viewed. The sensitivity of the oligomer probe technique was judged to be approximately 1 cancer cell per 10,000 normal cells as estimated by counting positive plaques in a dilutional assay. Among cases with molecular evidence of mutation-containing cells in the margins, the concentration of those cancer cells ranged from 0.05 to 28%. Even retrospectively, the light microscopist could only detect cancer in the margin tissue if there were 5 cancer cells per 100 normal cells present. Thus, the oligomer probe approach appears to have at least a 100-fold higher sensitivity for tumor cells compared to light microscopy.

Disadvantages of the Brennan approach included the fact that it is applicable to less than half of all HNSCC cases, those with a point mutation of the p53 gene. Furthermore, the methods involved are time-consuming, expensive, and technically demanding. It took approximately 3 days to complete the assay for a given case.

Several questions were raised by the Brennan study that have not yet been answered. Did the high rate of tumor recurrence in the molecularly positive group reflect the biological aggressiveness of tumors with p53 mutations? In particular, are these tumors relatively radioresistant, as that group has suggested [28]? Is there a threshold concentration of tumor cells that is clinically significant (indicative of recurrence)?

The oligomer probe approach has been reproduced by Partridge *et al.* [29]. The study involved 18 patients with clear traditional margins of which 12 had tumors displaying p53 point mutation and were thus amenable to oligomer probe evaluation. Six patients had molecular evidence of cancer in at least one margin and recurred locally (5) or regionally (1), despite receiving postoperative radiation. One case that recurred locally had cells in mucosal samples some distance from the tumor displaying a p53 mutation other than the one identified in the tumor. The investigators interpreted this finding as indicative of field cancerization. The recurrent tumor harbored both mutations, that found in the normal distant mucosa and that found in the original tumor. Unlike Brennan, there were no false-positive cases seen (molecular positivity without clinical recurrence). The percentage of mutant cells in molecularly positive margins ranged from 0.1 to 10%. These results largely mirror those of Brennan and serve to strengthen the case for the oligomer probe approach [29].

B. Overexpression of p53 Protein Detected with Immunohistochemistry

Investigators at the University of Cincinnati have published a report analyzing tumor margins for occult tumor cells using immunohistochemistry (IHC) to detect overexpression of p53 protein [30]. The tumor specificity of p53 overexpression is a subject of debate. In normal cells, the half-life of the p53 protein product is less than 8 min. For this reason, when tissue is processed for IHC analysis using antibodies to p53 protein, most normal tissue will not stain. Many, but not all, mutations of the p53 gene result in a protein product that is altered conformationally to have a longer half-life and will therefore stain with a p53-directed antibody. Mutations that encode a stop or truncation message cause tumors that do not stain. Wild-type p53 protein will stain if it is stabilized by cellular proteins or overexpressed temporarily in response to tissue damage. Other technical factors, such as the threshold of staining selected to indicate overexpression, methods of tissue fixation, antigen unmasking, and antibody selection, account for variability in results using IHC to detect p53 [31,32]. Despite these concerns, if IHC detection of p53-overexpressing cells in tumor margins increases the sensitivity and specificity of margin analysis, there may be clinical utility for the approach. Certainly, IHC is a less expensive, less technically demanding, and therefore a more convenient method to analyze the status of p53 compared with the approach of Brennan.

The Cincinnati study involved 24 subjects of which 10 experienced local recurrence within a 2-year follow-up period. Fourteen subjects had at least one margin that stained positively for the p53 protein. Eight of these 14 eventually had recurrent disease, whereas only 2 of 10 without IHC evidence of p53 protein overexpression in the margin recurred. Thus, the odds ratio for local recurrence in the face of IHC evidence of p53 overexpression in a margin was

5.33. One of the cases with recurrence had carcinoma *in situ* at the margin that stained positively for p53. The study does not state explicitly that all tumors stained positively or how intensely they overexpressed p53 protein. Reports in the literature indicating the rate of p53 overexpression in HNSCC vary widely in their estimation of this phenomenon. Still, this small, retrospectively assembled series is provocative, suggesting that p53 IHC is essentially as useful as the more cumbersome oligomer probe mutation analysis for the evaluation of minimal residual disease in surgical margins.

C. Generalized Markers for Head and Neck Squamous Cell Carcinomas

While the mutation spectrum of p53 renders it a highly specific marker for a particular tumor, individual probes must be constructed for each mutation in the Brennan schema. If, instead, a marker were available that was specific to all HNSCC, but present in no normal cell, such a marker would have an obvious advantage of generalized use for all clinical cases. It has been proposed that the eIF4E protein is such a marker [33]. The eIF4E protein is a protein synthesis initiation factor that binds to mRNA in the first step of mRNA recruitment for protein production. Expression of eIF4E is identified easily using IHC. The expression of eIF4E is elevated in breast and HNSCC, but not in normal mucosa. It is overexpressed in some histologically normal margins of HNSCC. A prospective study of 65 patients, all with HNSCC demonstrating elevated eIF4E, has been reported. The majority of patients (36/65) also had elevated eIF4E in apparently normal margin tissue, and 20 of these went on to develop local or regional recurrence of disease. Only 2 of the 29 individuals with margins free of eIF4E overexpression developed recurrent cancer. Over half of the margins that stained positively for eIF4E had epithelial dysplasia present, however, and would be considered histologically positive under more stringent criteria. The size of the study did not permit assessment of the independent prognostic implication of eIF4E overexpression and the presence of dysplasia at the margins, although eIF4E at the margins was associated with a sevenfold increased risk of local recurrence. Thus, eIF4E is a promising candidate for molecular margin analysis, despite the facts that the cause of its elevation in malignant cells has not been determined rigorously and that the approach continues to depend on IHC with light microscopic evaluation of sampled thin sections of tumor margins.

Another general marker for malignancy that has been piloted as a marker in margin analysis is hyperamplification of centrosomes [34]. Centrosomes are involved in chromosomal separation during mitosis, such that abnormal numbers of centrosomes are associated with aneuploidy and other chromosomal aberrations seen in malignant cells. Moreover, centrosome hyperamplification corresponds to the presence of p53 mutation or mdm2 protein overexpression. Because mdm2 binds and inactivates p53, centrosome abnormalities may reflect an array of alterations of p53 function [35]. Centrosomes can be detected using indirect immunofluorescence with antibody to tubulin, a component of the centrosome. Surgical margins from 18 HNSCC patients were analyzed using this technique. Centrosome hyperamplification was seen in nearly all tumors. It was present in a large number of cells in histologically normal margins in 8 of 10 cases that went on to recur, compared with 3 of 8 that remained disease free. While this difference was not statistically significant, the concept closely resembles that of eIF4E detection and seems worthy of further investigation [34].

VIII. CLINICAL RESPONSE TO POSITIVE MARGINS

Selection of a therapeutic response to a positive molecular or traditional margin depends on pragmatic and technical factors, on one's assessment of the biological phenomenon that led to the positive margin, and on the tools available for further therapy.

A. Further Resection

Removing an area of minimal residual tumor is an obvious response to positive tumor margins. If a positive margin reflects only the failure of a surgeon to remove enough tissue, then further resection alone might render the patient free of risk for recurrence. This might occur if the surgeon is inexperienced or if the tumor closely approaches some structure judged to be worthy of sparing. However, if the positive margin is indicative of an aggressive growth pattern, the addition of further local or systemic therapy may be necessary.

A report from M.D. Anderson Cancer Center reflects the benefit of further resection when initial margins are compromised. Using frozen section sampling to guide resection, 50 of 216 patients had at least one frozen section margin containing neoplastic cells, which was cleared with immediate further resection. Of these 50 individuals, 10 (20%) went on to recur locally. This compares favorably with an 80% local failure rate seen among 20 patients who could not be cleared of positive margins. Indeed the experience of those who were cleared with second resection was nearly as good as that of the group with initially negative margins. They had a 14% local recurrence rate [36]. The minor negative impact on survival associated with initially positive frozen section margins that were cleared with further surgery was also demonstrated in another report from M.D. Anderson two decades later [19]. Once again, further surgery appeared to be a worthwhile response to suspected minimal residual disease. Finding and clearing the positive margin may be difficult, however. In another series, 73% of subsequent resection specimens removed to address a positive frozen section did not have any cancer present. Either the tumor in the frozen section was the true edge of tumor or the surgeon did not harvest the subsequent tissue from the correct site [18].

B. Radiation

Standard radiation therapy can cover the entire surgical field, addressing the likelihood that if one margin is close or positive, others will be as well. The effect of postoperative radiation in rendering compromised margins harmless is uncertain and controversial. Data from M.D. Anderson indicate that it was helpful to reduce local recurrence rates in cases when initial frozen section margins were positive [18]. A series of 102 oral cancer patients from Memorial Sloan Kettering Cancer Center achieved equivalent local control regardless of margin status when those with compromised margins were treated with 6000 cGy or more postoperative radiation [37]. Loree and Strong [2] reported a trend toward lower rates of local recurrence and regional metastasis among 49 HNSCC patients treated with postoperative radiation in the face of compromised margins compared with 80 who did not receive adjuvant therapy. There was no impact of postoperative radiation on length of survival, however. Furthermore, the local recurrence rate was higher among those patients with positive margins who received radiation than among those with clear margins who did not have postoperative radiation [2].

However, a review of the results of the VA Laryngeal Cancer Study Group population of 144 laryngectomies showed that higher doses of postoperative radiation provided to individuals with involved margins did not eliminate the greater risk of death for this group [38]. Data from Pittsburgh more starkly suggests the inadequacy of radiation to manage minimal residual disease in the postoperative setting. Of 31 individuals with positive margins, only 2 remained disease free after 36 months. Among the 25 who received postoperative radiation, 60% failed within the radiation field [39].

C. Chemotherapy

If tumors with aggressive growth, likely to evade the surgeon and pathologist, are also radioresistant and/or widely metastatic, new treatment responses would be necessary. Initially positive margins indicating generalized tumor aggression are suggested by careful analysis of the M.D. Anderson data. Among the subset with initially positive margins, later rendered clear by further resection, local control was not significantly worse than among those with initially clear margins. However, the rate of 5-year survival in this hard to resect group was significantly worse, despite postoperative radiation [18]. One obvious option is the addition of chemotherapy postoperatively for individuals at high risk for recurrence. Review of a subset of 109 patients enrolled in a head and neck intergroup trial indicated that the addition of chemotherapy to postoperative radiation in the face of positive surgical margins failed to significantly enhance the median survival. This was a small group, the treatment for which was not driven by protocol, and results were compared to a historical control [5]. The Radiation Therapy Oncology Group completed accrual for a prospective, randomized clinical trial in which patients at high risk for recurrent disease after surgery received either radiation alone or radiation with systemic chemotherapy. Preliminary reports indicate no significant benefit of chemotherapy.

D. Innovative Therapy

Other treatment options for minimal residual disease include local injected gene therapy agents, chemotherapeutic agents impregnated into gels or sheets, intraoperative radiation, and photodynamic therapy. Systemic options may soon include a variety of vaccine approaches enhancing the immune response to targeted tumor antigens.

IX. CONCLUDING REMARKS

The application of molecular biologic technology to the evaluation of surgical margins is an appealing concept, intuitively logical, technically feasible, and already tested in pilot studies. The promised benefit of this marriage of new technology and standard clinical practice is an increased sensitivity and specificity of evaluation. Traditional margins are predictive of both locoregional control and survival. They are associated most closely with the former, while it is survival that is ultimately most meaningful for the patient. Molecular markers are potentially valuable prognostic indicators. Perhaps the combination of the strength of these markers as predictors of biologic behavior and as a means to detect minimal residual disease will provide a substantial improvement in prognostic capability.

Still, the work done to date raises more questions than it answers. What is the best marker for margin assessment or are several markers equally useful? How specific or generalized must the marker be? Is there a threshold of tumor burden below which the risk of recurrence is not clinically significant? When will a real-time molecular test of margins be available to replace the frozen section? Is fine mapping of the tumor bed to localize residual tumor possible and worthwhile? Are other factors such as radiosensitivity or metastatic capability of tumor or host immune function of greater importance to eventual outcome than the most precise evaluation of surgical margins? Finding answers to these questions will require rigorous evaluation of current strategies that have been proposed in pilot studies. Large populations, studied prospectively at multiple institutions with very careful collection and evaluation of outcome data, will be needed. Technologic advances will also be required to find the best markers of minimal residual disease and to do so within time and cost constraints that make the broad application of this approach feasible. As this approach is studied and developed, much should be revealed regarding the disease process of HNSCC, its pattern of growth, markers of biologic aggressiveness, and potential for intervention in the setting of minimal residual local disease.

References

1. Chen, T. Y., Emrich, L. J., and Driscoll, D. L. (1987). The clinical significance of pathological findings in surgically resected margins of the primary tumor in head and neck carcinoma. *Int. J. Radiat. Oncol. Biol. Phys.* **13**, 833–837.
2. Loree, T. R., and Strong, E. W. (1990). Significance of positive margins in oral cavity squamous carcinoma. *Am. J. Surg.* **160**, 410–414.
3. Cook, J. A., Jones, A. S., Phillips, D. E., and Lluch, E. S. (1993). Implications of tumour in resection margins following surgical treatment of squamous cell carcinoma of the head and neck. *Clin. Otolaryngol.* **18**, 37–41.
4. Jones, A. S., Hanafi, Z. B., Nadapalan, V., Roland, N. J., Kinsella, A., and Helliwell, T. R. (1996). Do positive resection margins after ablative surgery for head and neck cancer adversely affect prognosis? A study of 352 patients with recurrent carcinoma following radiotherapy treated by salvage surgery. *Int. J. Cancer* **74**, 128–132.
5. Jacobs, J. R., Ahmad, K., Casiano, R., Schuller, D. E., Scott, C., Laramore, G. E., and Al-Sarraf, M. (1993). Implications of positive surgical margins. *Laryngoscope* **103**, 64–68.
6. Bauer, W. C., Lesinski, S. G., and Ogura, J. H. (1975). The significance of positive margins in hemilaryngectomy specimens. *Laryngoscope* **85**, 1–13.
7. Brennan, C. T., Sessions, D. G., Spitznagel, E. L., and Harvey, J. E. (1991). Surgical pathology of cancer of the oral cavity and oropharynx. *Laryngoscope* **101**, 1175–1197.
8. Batsakis, J. (1999). Surgical excision margins: A pathologist's perspective. *Adv. Anatom. Pathol.* **6**, 140–148.
9. Ravasz, L. A., Slootweg, P. J., Hordijk, G. J., Smit, F., and van der Tweel, I. (1991). The status of the resection margin as a prognostic factor in the treatment of head and neck carcinoma. *J. Cranio.-Max.-Fac. Surg.* **19**, 314–318.
10. Kirita, T., Okabe, S., Izumo, T., and Sugimura, M. (1994). Risk factors for the postoperative local recurrence of tongue carcinoma. *J. Oral Maxillofac. Surg.* **52**, 149–154.
11. Mohs, F. E. (1978). "Chemosurgery: Microscopically Controlled Surgery for Skin Cancer." Charles C. Thomas, Springfield, IL.
12. Davidson, T. M., Nahum, A. M., Haghighi, P., Astarita, R. W., Saltzstein, S. L., and Seagren, S. (1984). The biology of head and neck cancer: Detection and control by parallel histologic sections. *Arch. Otolaryngol.* **110**, 193–196.
13. Gandour-Edwards, R. F., Donald, P. F., and Wiese, D. A. (1993). Accuracy of intraoperative frozen section diagnosis in head and neck surgery: Experience at a university medical center. *Head Neck* **15**, 33–38.
14. Spiro, R. H., Guillamondegui, O., Paulino, A. F., and Huvos, A. G. (1999). Pattern of invasion and margin assessment in patients with oral tongue cancer. *Head Neck* **21**, 408–413.
15. Hosal, A. S., Unal, O. F., and Ayhan, A. (1998). Possible prognostic value of histopathologic parameters in patients with carcinoma of the oral tongue. *Eur. Arch. Otorhinolaryngol.* **255**, 216–219.
16. van Es, R. J. J., van Nieuw Amerongen, N., Slootweg, P. J., and Egyedi, P. (1996). Resection margin as a predictor of recurrence at the primary site for T1 and T2 oral cancers. *Arch. Otolaryngol. Head Neck Surg.* **122**, 521–525.
17. Kurahara, D., Shinohara, M., Ikebe, T., Nakamura, S., Beppu, M., Hiraki, A., Takeuchi, H., and Shirasuna, K. (1999). Expression of MMPS, MT-MMP, and TIMPs in squamous cell carcinoma of the oral cavity: Correlation with tumor invasion and metastasis. *Head Neck* **21**, 627–638.
18. Scholl, P., Byers, R. M., Batsakis, J. G., Wolf, P., and Santini, H. (1986). Microscopic cut-through of cancer in the surgical treatment of squamous carcinoma of the tongue. *Am. J. Surg.* **152**, 354–360.
19. Lydiatt, D. D., Robbins, T. K., Byers, R. M., and Wolf, P. F. (1993). Treatment of stage I and II oral tongue cancer. *Head Neck* **15**, 308–312.
20. Slaughter, D. P., Southwick, H. W., and Smejkal, W. (1953). Field cancerization in oral stratified squamous epithelium: Clinical implications of multicentric origin. *Cancer* **6**, 963–968.
21. Van Oijen, M. G. C. T., Leppers vd Straat, F. G. L., Tilanus, M. G. J., and Slootweg, P. J. (2000). The origins of multiple squamous cell carcinomas in the aerodigestive tract. *Cancer* **88**, 884–893.
22. Bedi, G. C., Westra, W. H., Gabrielson, E., Koch, W., and Sidransky, D. (1996). Multiple head and neck tumors: Evidence for a common clonal origin. *Cancer Res.* **56**, 2484–2487.
23. Califano, J., van der Riet, P., Clayman, G., Westra, W., Corio, R., Koch, W., and Sidransky, D. (1996). A genetic progression model for head and neck cancer: Implications for field cancerization. *Cancer Res.* **56**, 2488–2492.
24. Nees, M., Homann, N., Discher, H., Andl, T., Enders, C., Herod-Mende, C., Schuhmann, A., and Bosch, F. X. (1993). Expression of mutated p53 occurs in tumor-distant epithelia of head and neck cancer patients: A possible molecular basis for the development of multiple tumors. *Cancer Res.* **53**, 4189–4196.
25. Waridel, F., Estreicher A., Bron, L., Flaman, J.-M., Fontolliet, C., Monnier, P., Frebourg, T., and Iggo, R. (1997). Field cancerisation and polyclonal p53 mutation in the upper aerodigestive tract. *Oncogene* **14**, 163–169.
26. Reichert, T. E., Day, R., Wagner, E. M., and Whiteside, T. L. (1998). Absent or low expression of the zeta chain in T cells at the tumor site correlates with poor survival in patients with oral carcinoma. *Cancer Res.* **58**, 5344–5347.
27. Brennan, J. A., Mao, L., Hruban, R. H., Boyle, J. O., Eby, Y. J., Koch, W. M., Goodman, S. N., and Sidransky, D. (1995). Molecular assessment of histopathological staging in squamous-cell carcinoma of the head and neck. *N. Engl. J. Med.* **332**, 429–435.
28. Koch, W. M., Brennan, J. A., Zahurak, M., Goodman, S. N., Westra, W. H., Schwab, D., Yoo, G. H., Lee, D. J., Forastiere, A. A., Sidransky, D. (1996). p53 mutation and locoregional treatment failure in head and neck squamous cell carcinoma. *J. Natl. Cancer Inst.* **88**, 1580–1586.
29. Partridge, M., Li, S.-R., Pateromichelakis, S., Francis, R., Phillips, E., Huang X. H., Tesfa-Selase, F., and Langdon, J. D. (2000). Detection of minimal residual cancer to investigate why oral tumors recur despite seemingly adequate treatment. *Clin. Cancer Res.* **6**, 2718–2725.
30. Ball, V. A., Righi, P. D., Tejada, E., Radpour, S., Pavelic, Z. P., and Gluckman, J. L. (1997). p53 immunostaining of surgical margins as a predictor of local recurrence in squamous cell carcinoma

of the oral cavity and oropharynx. *ENT-Ear Nose Throat J.* **76**, 818–823.

31. Hall, P. A., and Lane, D. P. (1994). P53 in tumour pathology: Can we trust immunohistochemistry?-revisited! *J. Pathol.* **172**, 1–4.
32. Wynford-Thomas, D. (1992). p53 in tumour pathology: Can we trust immunocytochemistry? *J. Pathol.* **166**, 329–330.
33. Nathan, C. A., Franklin, S., Abreo, F. W., Nassar, R., De Benedetti, A., and Glass, J. (1999). Analysis of surgical margins with the molecular marker eIF4E: A prognostic factor in patients with head and neck cancer. *J. Clin. Oncol.* **17**, 2909–2914.
34. Gustafson, L. M., Gleich, L. L., Fukasawa, K., Chadwell, J., Miller, M. A., Stambrook, P. J., and Gluckman, J. L. (2000). Centrosome hyperamplification in head and neck squamous cell carcinoma: A potential phenotypic marker of tumor aggressiveness. *Laryngoscope* **110**, 1798–1801.
35. Carroll, P. E., Okuda, M., Horn, H. F., Biddinger, P., Stambrook, P. J., Gleich, L. L., Li, Y. Q., Tarapore, P., and Fukasawa, K. (1999). Centrosome hyperamplification in human cancer: Chromosome instability induced by p53 mutation and/or Mdm2 overexpression. *Oncogene* **18**, 1935–1944.
36. Byers, R. M., Bland, K. I., Borlase, B., and Luna, M. (1978). The prognostic and therapeutic value of frozen section determinations in the surgical treatment of squamous carcinoma of the head and neck. *Am. J. Surg.* **136**, 525–529.
37. Zelefsky, M. J., Harrison, L. B., Fass, D. E., Armstrong, J. G., Shah, J. P., and Strong, E. W. (1993). Postoperative radiation therapy for squamous cell carcinoma of the oral cavity and oropharynx: Impact of therapy on patients with positive surgical margins. *Int. J. Radiat. Oncol. Biol. Phys.* **25**, 17–21.
38. Bradford, C. R., Wolf, G. T., Fisher, S. G., and McClatchey, K. D. (1996). Prognostic importance of surgical margins in advanced laryngeal squamous carcinoma. *Head Neck* **18**, 11–16.
39. Zieske, L. A., Johnson, J. T., Myers, E. N., and Thearle, P. B. (1986). Squamous cell carcinoma with positive surgical margins: Surgery and postoperative irradiation. *Arch. Otolaryngol. Head Neck Surg.* **112**, 863–866.

C H A P T E R

36

Novel Agents and Modalities for the Treatment of Squamous Carcinoma of the Head and Neck

ADRIAN M. SENDEROWICZ,* CARTER VAN WAES,† JANET DANCEY,‡ and BARBARA CONLEY‡

*Molecular Therapeutics Unit, Oral and Pharyngeal Cancer Branch,
National Institute of Dental and Craniofacial Research, National Institutes of Health, and
†National Institute of Deafness and Communication Disorders,
‡Cancer Treatment Evaluation Program, National Cancer Institute
National Institutes of Health, Bethesda, Maryland 20892

I. Novel Targets and Agents for the Treatment of Advanced Head and Neck Squamous Cell Carcinoma (HNSCC) 535
A. Novel Cytotopics 535
B. Modulation of Epidermal Growth Factor Receptor as a Therapeutic Strategy in HNSCC 538
C. Modulations of Cell Cycle Control for the Treatment of HNSCC 540
D. Modulation of Hypoxia and Angiogenesis 545
E. Gene Therapy Studies 545
F. Immunological Therapy of Head and Neck Cancer 546
II. Conclusions 547
References 547

I. NOVEL TARGETS AND AGENTS FOR THE TREATMENT OF ADVANCED HEAD AND NECK SQUAMOUS CELL CARCINOMA (HNSCC)

With the understanding of the molecular underpinnings of human disease and with the novel knowledge obtained with the sequence of the human genome, we are in a better position to understand the mechanisms by which a normal cell becomes neoplastic. This novel information may allow us, in the near future, to rationally design novel targeted therapies in order to modulate those specific processes involved directly in the development of cancer [1–3]. As discussed elsewhere in this book, advanced refractory HNSCC has, still, a very poor prognosis [4]. This poor prognosis may be due, in part, to the overall complexity of the genetic alterations that characterize this disease, thus compromising the task of choosing an appropriate treatment strategy. It follows that there is a clear need to identify and develop new strategies for the treatment and management of HNSCC patients.

This chapter focuses on several novel promising therapies/targets for this disease. Novel standard cytotoxics, such as taxanes, novel tubulin disruptors, novel antimetabolites, and novel derivatives of cisplatin are considered first. Then, novel targets are discussed, focusing first on strategies to modulate the epidermal growth factor receptor (EGFR) signaling pathways. Opportunities to modulate cell cycle control are followed by modulation of angiogenesis/hypoxia processes and, finally, a discussion about biological agents with therapeutic importance, such as gene therapy and immune modulation in HNSCC (see Fig. 36.1, see also color insert).

A. Novel Cytotoxics

1. Antitubulin Agents

Taxanes have shown significant activity in squamous cancers of the head and neck. In a multi-institutional study using 250 mg/m^2 paclitaxel in a 24-h continuous infusion regimen with granulocyte colony-stimulating factor (G-CSF) support, a response rate of 40% was obtained in patients with either disease recurrent after primary treatment or presenting with metastatic disease [5]. Toxicity was significant, with severe or life-threatening granulocytopenia occurring in 91% of 30 evaluable patients [5]. However, this trial did confirm results of a single institution trial using the same

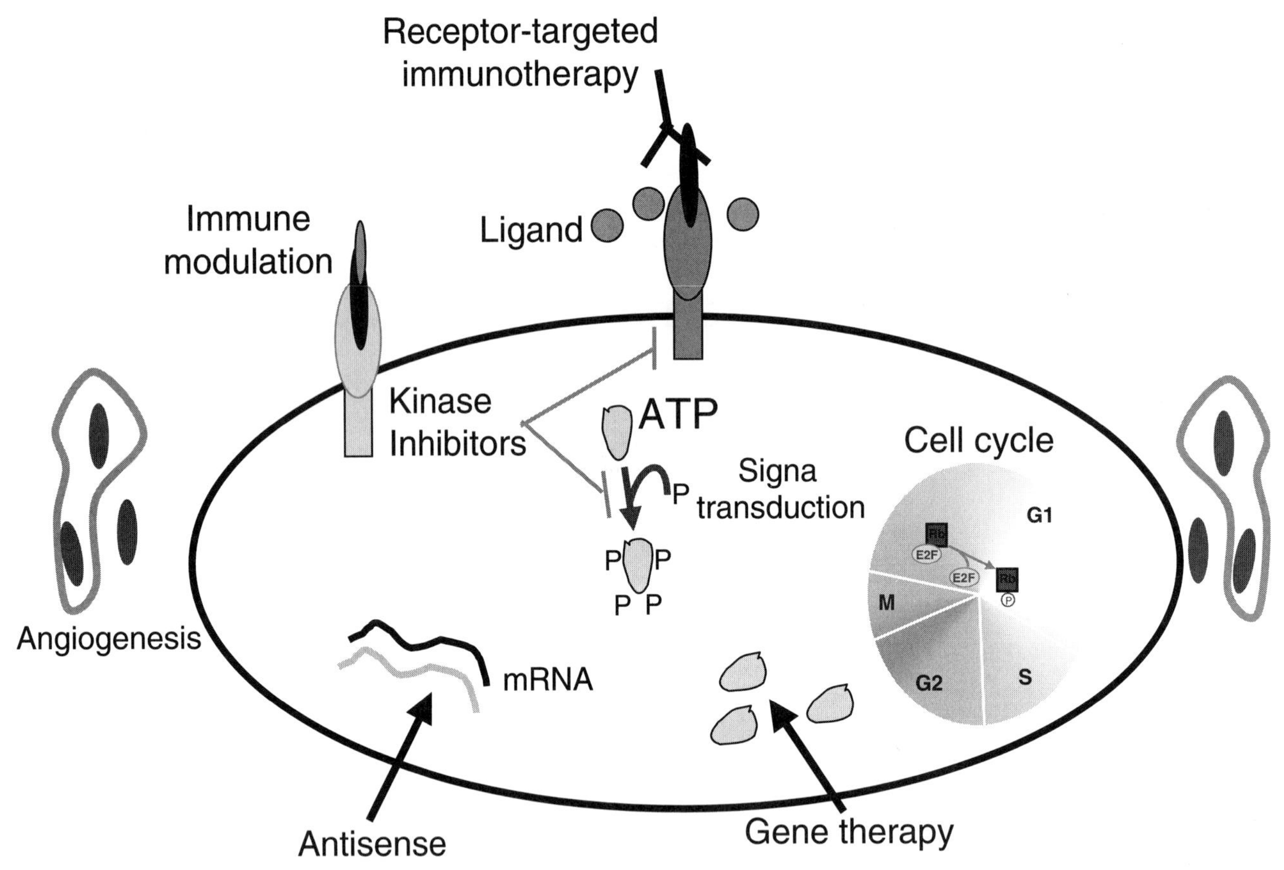

FIGURE 36.1 Gene therapy and immune modulation. (See also color insert.)

regimen [6]. In a subsequent study, the Eastern Cooperative Oncology Group studied cisplatin (75 mg/m^2) with either low-dose paclitaxel (135 mg/m^2) given over 24 h or with paclitaxel (200 mg/m^2) given over 24 h with G-CSF. There was no difference in response rate (36% and 35%, respectively) or, interestingly, in toxicity. The estimated median survival was 7.3 months (95% confidence interval 6.0–8.6 months). The 1-year survival was 29% and event-free survival was 4.0 months. Fatal toxicities occurred at a rate of approximately 10%, leading the group to discard the 24-h infusion paclitaxel regimens in subsequent studies [7]. A multi-institutional randomized phase III trial of cisplatin (100 mg/m^2) on day 1 with 5-fluorouracil (5-Fu) (1 g/m^2) by a continuous 96-h infusion days 1 to 4 every 21 days vs paclitaxel (175 mg/m^2) with cisplatin (75 mg/m^2) every 21 days was conducted in metastatic or recurrent squamous carcinoma of the head and neck patients with good performance status. There was no significant difference in 1-year survival between the two arms: 41% vs 30%, respectively. Median survival was 8 and 9 months, respectively, and response rates were 22 and 28%, respectively. However, the taxane-containing regimen appeared to be less toxic than the more standard regimen, especially with respect to hematological toxicity [8]. Combinations of paclitaxel, (175 mg/m^2) given as a 3-h infusion with carboplatin, AUC 6, have also shown activity in squamous cancer of the head and neck, with response rates of about 30% in patients with recurrent disease (40% of whom had received prior cisplatin and 5-FU) [9]. Combinations of cisplatin, paclitaxel, and ifosfamide are also under study and show response rates of up to 59% [10].

Docetaxel has proven to be about as efficacious as paclitaxel in patients with squamous cell carcinoma of the head and neck. Of 29 assessable patients with recurrent or metastatic squamous carcinoma of the head and neck who received 100 mg/m^2 docetaxel every 3 weeks in a phase II trial, 13% achieved a complete response and 29% achieved a partial response [11]. Mean response duration was 5 months. Unfortunately, 53% of patients had nadir fever, with 13 patients requiring dose reduction. Hypersensitivity reactions occurred despite premedication in 4 of 30 patients, and minimal edema occurred in 17% of patients.

Currently, several regimens using either paclitaxel or docetaxel with cisplatin, or carboplatin, and/or 5-fluorouracil are being investigated in phase II trials, with response rates of

84–93% when used as neoadjuvant therapy in previously untreated patients with locally advanced disease [12,13]. Studies using taxanes with or without platinum agents and concurrent radiation therapy report results that compare favorably with those of prior chemoradiation regimens [14–17]. In a Radiation Therapy Oncology Group randomized phase II study, 241 patients with stage III or IV, M0 squamous cancers of the oral cavity, oropharynx, or hypopharynx were randomized to (1) 70 Gy/7 weeks with cisplatin (10 mg/m^2) and 5-FU (400 mg/m^2) daily during the last 10 days of radiation (chemotherapy boost); (2) 70 Gy/13 weeks, given on alternate weeks with daily hydroxyurea (1 g twice a day) and 5-FU (800 mg/m^2) (FHx), or (3) 70 Gy/7 weeks with weekly cisplatin (20 mg/m^2) and paclitaxel (30 mg/m^2). In 227 evaluable patients, grade 4 toxicities were seen in 25% randomized to arm 1, in 32% randomized to arm 2, and in 29% randomized to arm 3. Estimated 1- and 2-year survival rates were 72 and 60% for arm 1, 87 and 65% for arm 2, and 80 and 67% for arm 3. Comparison to RTOG historical controls indicated that patients in all three arms had better survival than patients treated with radiation alone or radiation with cisplatin [18].

2. Epothilones

Epothilone analogs are currently investigational, but show promising preclinical activity in both taxane-sensitive and some taxane-resistant tumors. These agents represent a family of 16 membered ring macrolides originally isolated in 1992 from the fermentation broth of the myxobacterium *Sorangium cellulosum* [19]. Epothilones have a mode of action similar to the taxanes, i.e., microtubule stabilization, but are more potent than paclitaxel [20]. Some of these agents may be bioavailable by the oral route. Although there are not yet data on activity in squamous cancers of the head and neck, the activity of taxanes in these tumors suggest that epothilones may have equivalent or superior activity profiles. Phase I studies of weekly epothilone B demonstrated diarrhea as dose limiting [21]. Phase I evaluation of the epothilone B analog (BMS 247550) using a 1-h infusion schedule every 3 weeks observed neutropenia and sepsis as dose limiting [22].

3. Platinum Agents

Of the new platinum agents, Nedaplatin was reported to have sequence-specific, enhanced antitumor efficacy compared to cisplatin when combined with 5-fluorouracil in human squamous carcinoma xenografts [23]. This agent was selected from various platinum analogs based on preclinical antitumor activity against a variety of cancers, with less nephrotoxicity than cisplatin. Promising activity was also seen in Japanese phase II trials in head and neck cancer [23].

Oxaliplatin, a platinum agent with a 1,2-diaminocyclohexane (DACH) carrier ligand, was developed because of the observation that DACH-Pt compounds are effective against a variety of cisplatin-resistant tumors preclinically [24]. The most constant acute side effect of oxaliplatin administration is a transient peripheral neuropathy, characterized by paresthesia and dysesthesias of the hands, feet, and perioral area. In addition, some patients reported laryngo-pharyngeal dysesthesia when exposed to cold foods or drink [24]. Objective responses were observed in all phase I studies, including head and neck cancer patients [24]. Responses were observed in 6% of pretreated patients and 13% of previously untreated patients with recurrent or metastatic head and neck cancer given oxaliplatin (130 mg/m^2) every 3 weeks in a phase II study [25]. Of note, this agent was recently approved by the FDA for the treatment of metastatic colon cancer.

4. Antimetabolites

a. 5-Fluorouracil Analogs

5-Fluorouracil is a known radiation sensitizer but has a short half-life due to rapid metabolism by dihydropyrimidine dehydrogenase (DPD) [26]. In addition, its activity is dependent on its metabolism to fluorodeoxyuridine monophosphate (FdUMP) by thymidine phosphorylase and to the binding of FdUMP to thymidylate synthase, resulting in inhibition of DNA synthesis. Several compounds have now been developed that inhibit 5-FU metabolism or enhance its specificity for malignant cells. Eniluracil, an inhibitor of DPD, combined with oral 5-FU and radiation on an every other week schedule was studied in a phase I trial in patients with advanced head and neck cancer. The dose-limiting toxicity was cumulative myelosuppression. However, despite myelosuppression, no dose-limiting mucositis or dermatitis was observed, implying that radiosensitization did not occur at the maximum tolerable dose [27]. Capecitabine and UFT (Tegafur and uracil) are other agents in this class, available by the oral route. UFT produced a response rate of 38% in 43 previously treated patients with recurrent head and neck cancer [28]. In previously untreated patients, UFT, in combination with carboplatin, produced a response rate of 53% [29]. The agent also appears to be a radiation sensitizer for clinical use [30]. Use of capecitabine alone or in combination with other chemotherapy agents or radiation has not yet been widely studied. However, the requirement for the drug to be activated by thymidine phosphorylase [31] and the observation that thymidine phosphorylase is elevated in squamous cancers of the head and neck [32,33] provide a rationale to study the agent in this disease.

b. Gemcitabine

Gemcitabine is a deoxycytidine analog (2′,2′-difluoro-2′-deoxycytidine) that has greater potency and slower clearance from tumor cells compared with cytosine arabinoside. In addition, gemcitabine has activity against solid tumors, which may be schedule dependent [34]. Gemcitabine is phosphorylated intracellularly and inhibits DNA synthesis by incorporation of 2′,2′-difluoro-2′-deoxycytidine triphosphate (dFdCTP). The phosphorylated metabolite is retained in cells for a prolonged period, allowing less frequent dosing. Promising *in vitro* activity was observed against squamous head and neck cancer cells [35], and a significant increase in tumor doubling time, and even cures, were observed in human tumor xenograft-bearing mice treated with intraperitoneal gemcitabine [36].

The European Organization for Research and Treatment of Cancer Early Clinical Trials Group reported a response rate of 13% in 54 patients with advanced or recurrent squamous cancer of the head and neck, many of whom had received previous chemotherapy, treated on a weekly ×3 schedule. Toxicity was mild, with grade 3–4 neutropenia in $<10\%$ of patients [37]. Studies have also reported a response rate of 22.7% for the combination of cisplatin (50 mg/m^2) on days 1 and 8 with gemcitabine (800 mg/m^2) on days 1, 8, and 15 [38]. Combinations of paclitaxel and gemcitabine have also been studied. Fountzilas *et al.* [39] administered gemcitabine (1100 mg/m^2) on days 1 and 8 with paclitaxel (200 mg/m^2) on day 1 to patients with recurrent and/or metastatic nonnasopharyngeal head and neck cancer. Twenty-four of 44 patients completed six cycles of treatment. Of the total group (44 patients), 5 (11%) had a complete response and 13 (30%) had a partial response. The median time to progression was 4 months, but toxicity was acceptable [39].

Surprising results have been forthcoming in studies of gemcitabine combined with radiation therapy. Initial studies demonstrated radiosensitization by gemcitabine at noncytotoxic concentrations (100 nM) that were much less than plasma concentrations produced by current clinical dosing (2 μM) [40]. The lack of mucositis associated with the administration of gemcitabine was an additional factor making the drug attractive for combinations with radiation. In a phase I trial, Eisbruch *et al.* [41] explored the clinical toxicity of combining weekly gemcitabine with daily fractionated radiation to 70 Gy in patients with locally advanced, or unresectable, squamous cancers of the head and neck. In addition, tumor concentrations of dFdCTP were evaluated after the first dose of gemcitabine and prior to radiation. The initial dose level of 300 mg/m^2 was found to be intolerable, and successive dose de-escalations were necessary. Confluent acute mucositis was observed in most patients receiving doses 50 mg/m^2 and higher, which lasted about 7 weeks. Deep, persistent, mucosal ulceration or pharyngeal/esophageal obstruction that could not be relieved by dilatation was observed in at least 2 patients at each dose level of 50 mg/m^2 and above. Biopsy-proven complete remission was observed in 66–89% of patients of each cohort. Only a few patients were able to have tumor biopsies, but the difference in cellular dFdCTP concentrations at the various doses was not statistically significant. While the study bears out the preclinical observation that gemcitabine is a potent radiosensitizer at very low concentrations (at doses that are 5–30% of those tolerated when gemcitabine is administered as a single agent), the rate of late dysphagia implies that the toxicity is not selective to tumor tissue. It is possible that the mechanism of radiation sensitization involves the inhibition of ribonucleotide reductase, which occurs *in vitro* at concentrations of gemcitabine that are too low to result in DNA incorporation [41].

B. Modulation of Epidermal Growth Factor Receptor as a Therapeutic Strategy in HNSCC

Among the emerging molecular targets for therapy of HNSCC is EGFR (see Fig. 36.1). Overexpression of EGFR is present in the majority of head and neck squamous cell carcinomas, and this overexpression is associated with a poor prognosis [42,43]. Selective compounds have been developed that target either the extracellular ligand-binding region of the EGFR or the intracellular tyrosine kinase region, resulting in interference with the signaling pathways that modulate proliferation and other cancer-promoting responses, such as cell motility, cell adhesion, invasion, and angiogenesis. Potential new anticancer agents that target the extracellular ligand-binding region of the receptor include a number of monoclonal antibodies, immunotoxins, and ligand-binding cytotoxic agents. Agents that target the intracellular tyrosine kinase region include small molecule tyrosine kinase inhibitors (TKIs), which act by interfering with ATP binding to the receptor. Currently, the most advanced of the newer therapies undergoing clinical development are antireceptor monoclonal antibody C225 (cetuximab) and a number of small molecule EGFR-TKIs, principally of the quinazoline and pyrrolo-pyridopyrimidine inhibitor structural classes.

C225 (cetuximab) is an IgG antibody directed against the human EGFR, which binds to the ligand-binding domain of the receptor and competes for receptor binding with EGF. Cetuximab inhibits *in vitro* cellular proliferation in squamous cell carcinoma cell lines that overexpress the EGFR and inhibits the tyrosine kinase activity of the EGFR. Enhanced cytotoxicity was observed with a number of cytotoxic agents (e.g., doxorubicin and cisplatin) and radiation in preclinical models [44]. EGFR inhibition by C225 enhances the *in vitro* and *in vivo* radiosensitivity of HNSCC cell lines. Interestingly, collective data from xenograft experiments suggest that the enhanced antitumor activity seen when C225 is combined with radiation derives not only from inhibiting proliferation, but also from inhibiting several important processes, including DNA repair after exposure to radiation and angiogenesis [45].

In phase I trials, the most frequently reported C225-related adverse events were asthenia, fever, nausea, transaminases, and allergic reactions. Of these, skin rash and allergic reactions appear to the most clinically relevant. Eighty percent of patients developed skin rashes, and lesions appeared to be a sterile folliculitis usually affecting the face, upper chest, and/or back. The rash was usually mild to moderate in severity and resolves without treatment. Significant infusion reactions were uncommon but can be severe. The incidence of 3/4 allergic reactions is 4%, with 2% grade 4 anaphylactic reactions. These reactions usually appear within minutes of starting the initial infusion. Reactions were responsive to standard treatment and could be prevented with prophylactic administrating antihistamines and/or prolonging the infusion duration [46].

Preliminary results from phase I trials suggest that combinations of C225 with radiation and C225 with cisplatin are well tolerated. A phase 1 study reported that C225 can be given in combination with cisplatin (60 mg/m^2) every 4 weeks without increased toxicity [47]. A phase 1 trial evaluating C225 with standard irradiation (70 Gy, 2.0 Gy given once-daily) for patients with locally advanced head and neck malignancies reported that the combination is tolerable. C225 was delivered as a loading dose of 100–500 mg/m^2 intravenously followed by weekly infusions of 100–250 mg/m^2 for 7 weeks. Nine of 16 (56%) patients experienced irradiation-related grade 3 mucositis and 1 patient experienced grade 4 mucositis. Five (31%) patients had grade 3 skin toxicity, primarily skin rash. Of the 15 evaluable patients, 14 achieved complete responses [44,48]. This agent is presently in phase III trials in head and neck squamous cell cancer.

Small molecule inhibitors of the intracellular tyrosine kinase domain of EGFR are also under clinical evaluation [49]. Currently available EGFR inhibitors belong to three chemical series: 4-anilinoquinazolines, 4-[ar(alk)ylamino] pyridopyrimidines, and 4-phenylaminopyrrolo-pyrimidines. Two quinazolines that have shown promising antitumor activity in early clinical trials are ZD1839 (Iressa) and OSI-774 (Tarceva). More recently, potent, irreversible inhibitors of EGF receptor kinases such as CI-1033 and EKB-569 have been developed. CI-1033 is of interest as it is not only an irreversible inhibitor that binds covalently to a cysteine residue near the ATP-binding site of the EGFR, it also targets all four EGFR family members. Whether irreversible inhibition will result in improved therapeutic index or will remove the requirement for continuous dosing requires further clinical study.

In general, these small molecules competitively inhibit ATP binding to EGFR, hindering autophosphorylation, leading to dose-dependent tumor stasis and even tumor regression in some tumor xenograft models. However, the most promising laboratory data have been from the combination of these targeted inhibitors with standard chemotherapy or radiation therapy [50,51]. In addition to their shared mechanism of action, these small molecules are administered orally on chronic schedules and have a similar spectrum of toxicity, with diarrhea and skin rash being the most commonly reported adverse events.

ZD1839 is an anilinoquinazoline that acts as a potent and specific inhibitor of EGFR tyrosine kinase activity by competing with ATP for its binding site on the intracellular domain of the receptor [51]. The IC_{50} of ZD1839 using enzyme extracted from the A431 human squamous vulval cell line was 23–79 nM. It is approximately 100-fold less active against erbB2 kinase and has little or no enzyme inhibitory activity against several other tyrosine and serine-threonine kinases tested. ZD1839 has antitumor activity in a broad range of human tumor xenografts with both tumor stasis and regressions seen in xenograft models [51]. However, rapid regrowth of tumors was generally observed when the drug was discontinued, suggesting the need for chronic administration.

Two trials are assessing escalating doses of ZD1839 administered on a continuous daily schedule. Dose-limiting diarrhea was seen at dose levels of 1000 mg/day [52]. The most frequent adverse events were grade 1–2 skin rash, diarrhea, nausea, and vomiting. Grade 3 adverse events included diarrhea, skin rash, increased hepatic transaminases, nausea, and vomiting. Skin toxicity consisted primarily of grade 1 to 2 pustular or acne-like lesions with occasional erythema, or dry skin. The rash was usually located on the face, with involvement of the upper torso at higher doses, and was resolved rapidly after drug discontinuation. Nausea and/or emesis occurred infrequently and were mild–moderate in severity. Doses of 250 and 500 mg per day orally in combination with standard chemotherapy are currently being evaluated in phase III placebo-controlled studies in patients with advanced nonsmall cell lung cancer.

Similar to ZD 1839, OSI-774 (Tarceva) is also an anilinoquinazoline that has shown interesting antitumor activity in preclinical models and in phase 1 clinical trials. OSI-774 is a selective inhibitor of human epidermal growth factor receptor tyrosine kinase with an IC_{50} of 2 nM and selectively reduces EGFR autophosphorylation in intact tumor cells with an IC_{50} of 20 nM [53]. In preclinical studies, daily administration of the agent markedly inhibits the growth of HN5 head and neck, and A431 epidermoid carcinoma xenografts [54].

Results from phase 1 studies indicate that OSI-774 has been well tolerated and antitumor activity has been observed. Phase 1 trials in cancer patients investigated the following schedules: 3 days/week $\times 3$ weeks/q 4 weeks, daily $\times 3$ weeks q 4 weeks, daily continuous dosing, and weekly dosing [55]. With the continuous dosing schedule, a recommended phase 2 dose of 150 mg/day was established.

Among the 42 patients treated by continuous dosing in the phase 1 study, the most common toxicities were acneiform rash (55%), diarrhea (45%), nausea (29%), and headache (29%). The target average plasma concentration for biological activity of 500 ng/ml was achievable at doses >100 mg.

For the weekly dosing schedule, doses up to 1600 mg/week were evaluated without reaching dose-limiting toxicity [56]. Among 27 patients treated on this study, the following toxicities were seen: diarrhea (63%), rash (44%), and nausea (44%). There was 1 patient with grade 3 skin rash. The target plasma concentration was achieved at all dose levels; however, preliminary pharmacokinetic analysis revealed large intra- and intersubject variability [56].

Preliminary results from the phase 2 study in SCCHN evaluating the efficacy of OSI-774 administered at 150 mg daily continuously are available [57]. Of 114 head and neck patients enrolled, response data are available on 78. There were 10 patients (13%) with partial responses, 23 (29%) with stable disease, and 45 (58%) with progressive disease. Adverse events included an acneiform rash affecting the face, upper torso, and arms, diarrhea, nausea, vomiting, headache, and fatigue. The acneiform was noted in 82/114 (72%) patients. Rash was mild in 32 patients (28%), moderate in 39 (34%), and severe in 9 (8%). Diarrhea was treated with either dose reduction or oral loperamide without drug discontinuation. These data suggest that OSI-774 has antitumor activity in advanced SCCHN and that further study as an adjunct to standard therapeutic approaches is warranted.

In light of a relationship between overexpression of EGFR and clinically aggressive malignant disease, EGFR has emerged as a promising target for treatment of patients with HNSCC. Preliminary results from early clinical trials of both anti-EGFR monoclonal antibodies and EGFR small molecule tyrosine kinase inhibitors are promising. The rapid evaluation of these target-specific noncytotoxics is limited by the lack of accurate information on the relevance of target expression and its modulation to this tumor type. Early clinical trials are being designed to address these concerns. Current research goals include (1) defining the optimal dose and schedule in combinations with conventional chemotherapeutic agents and with radiation therapy and (2) determining predictive factors that identify the best patient population in which to study and administer these agents. However, the clinical impact of EGFR inhibitors in SCCHN patients must await the completion of randomized evaluations in combination with standard radiation and chemotherapeutic regimens.

C. Modulators of Cell Cycle Control for the Treatment of HNSCC

Upon activation of several growth factor/mitogenic signaling cascades, cells commit to entry into a series of regulated steps allowing traverse of the cell cycle. First, synthesis of DNA (genome duplication), also known as S phase, occurs, followed by the separation of two daughter cells (chromatid separation) or M phase. The time between S and M phases is known as G2 phase (see Fig. 1). This period is where cells can repair errors that occur during DNA duplication, preventing the propagation of these errors to daughter cells. In contrast, the G1 phase represents the period of commitment to cell cycle progression that separates M and S phases as cells prepare for DNA duplication upon mitogenic signals [58].

Over the past decade, it has been recognized that a universally utilized pathway allowing mitogenic signals to promote progression through G1 to S phase utilizes phosphorylation and inactivation of the retinoblastoma gene product (Rb), a tumor suppressor gene product important for G1 control [59,60]. Rb inactivation is the result of its phosphorylation by the serine/threonine kinases, known as cyclin-dependent kinases (cdks) [61]. These key regulators of the cell cycle are enzymes that periodically form complexes with proteins known as cyclins. There are at least 9 different cdks (cdk1–cdk9) [61–64]. Cyclin-dependent kinases clearly involved in cell cycle control are cdk1 through cdk7. In contrast, cdk8 and cdk9, although related structurally to cell cycle regulatory cdks, are important regulators of transcriptional control [62,63]. There are at least 15 different cyclins (cyclin A through T) [64–67]. These complexes are in turn regulated by a stoichiometric combination with small inhibitory proteins, called endogenous cyclin-dependent kinase inhibitors (CKIs). INK4 (inhibitor of cdk4) family members include $p16^{ink4a}$, $p15^{ink4b}$, $p18^{ink4c}$, and $p19^{ink4d}$ and specifically inhibit cyclin D-associated kinases. Members of the kinase inhibitor protein (KIP) family, $p21^{waf1}$, and $p27^{Kip1}$, and $p57^{kip2}$, bind and inhibit the activity of cyclin E/cdk2 and cyclin A/cdk2 complexes [68].

Most human neoplasms have abnormalities in some component of the Rb pathway due to hyperactivation of cdks as a result of amplification/overexpression of positive cofactors, cyclins/cdks, or downregulation of negative factors, endogenous CKIs, or mutation in the Rb gene product. These aberrations promote deregulated S-phase progression in a way that ignores growth factor signals, with loss of G1 checkpoints [58,69]. Therefore, development of pharmacologic cdk inhibitors (smCDKI), "mechanism-based therapy," would be of great interest as a treatment strategy for many neoplasms [70]. Furthermore, inappropriate or deregulated activation of cdks might have adverse consequences for cells, and indeed cdk activation/inactivation has been reported to correlate with the cellular response to apoptotic stimuli in several preclinical models [71–74] (see later).

Two cdk modulators, flavopiridol and UCN-01, have completed initial human phase I trials [70,75–82] and are described later.

Several strategies could be considered to modulate cdk activity. These strategies are divided into direct effects on the catalytic cdk subunit or indirect modulation of regulatory pathways that govern cdk activity [2,70]. The small molecular endogenous cdk inhibitors (SCDKI) are compounds that directly target the catalytic cdk subunit. Most of these compounds modulate cdk activity by interacting specifically with the ATP-binding site of cdks [2,70,83–85]. The second class of cdk inhibitors are compounds that inhibit cdk activity by targeting the regulatory "upstream pathways" that modulate the activity of cdks; by altering the expression and synthesis of the cdk/cyclin subunits or the cdk inhibitory proteins; by modulating the phosphorylation of cdks; by targeting cdk-activating kinase, cdc25, and wee1/myt1; or by manipulating the proteolytic machinery that regulates the catabolism of cdk/cyclin complexes or their regulators [2,70].

1. Modulators of Cyclin-Dependent Kinases in Clinical Trials

a. Flavopiridol

Flavopiridol (L86-8275 or HMR 1275) is a semisynthetic flavonoid derived from rohitukine, an indigenous plant from India. Initial studies with flavopiridol demonstrated modest *in vitro* inhibitory activity with respect to EGFR and protein kinase A (PKA)(IC_{50} = 21 and 122 μM, respectively) [86]. However, when this compound was tested in the National Cancer Institute (NCI) 60 cell line anticancer drug screen panel, this compound demonstrated a very potent growth inhibition (IC_{50} = 66 nM), a concentration that is about 1000 times lower than the concentration required to inhibit PKA and EGFR [86]. Initial studies with this flavonoid revealed clear evidence of G1/S or G2/M arrest due to loss in cdk1 and cdk2 (87–89). Studies using purified cdks showed that the inhibition observed is reversible and competitively blocked by ATP, with a K_i of 41 nM [87–91]. Furthermore, the crystal structure of the complex of deschloroflavopiridol and cdk2 showed that flavopiridol binds to the ATP-binding pocket, with the benzopyran occupying the same region as the purine ring of ATP [92], confirming earlier biochemical studies with flavopiridol [88]. Flavopiridol inhibits all cdks thus far examined (IC_{50} ~ 100 nM), but it inhibits cdk7 (cdk-activating kinase) less potently (IC_{50} ~ 300 nM) [88,90,91].

In addition to directly inhibiting cdks, flavopiridol promotes a decrease in the level of cyclin D1, an oncogene that is overexpressed in many human neoplasias. Of note, neoplasms that overexpress cyclin D1 have a poor prognosis [93–95]. When MCF-7 human breast carcinoma cells were incubated with flavopiridol, levels of cyclin D1 protein decreased within 3 h [96]. This effect was followed by a decline in the levels of cyclin D3 with no alteration in the levels of cyclin D2 and cyclin E, the remaining G1 cyclins, leading to loss in the activity of cdk4. Thus, depletion of cyclin D1 appears to lead to the loss of cdk activity [96]. The depletion of cyclin D1 is caused by depletion of cyclin D1 mRNA and was associated with a specific decline in the cyclin D1 promoter, measured by a luciferase reporter assay [96]. The transcriptional repression of cyclin D1 observed after treatment with flavopiridol underscores the conserved effect of flavopiridol on eukaryotic cyclin transcription [97]. In summary, flavopiridol can induce cell cycle arrest by at least three mechanisms: (1) by direct inhibition of cdk activities by binding to the ATP-binding site; (2) by prevention of the phosphorylation of cdks at threonine-160/161 by inhibition of cdk7/cyclin H [89,91]; and (3) for G1/S phase arrest, by a decrease in the amount of cyclin D1, an important cofactor for cdk4 and cdk6 activation.

Another effect of flavopiridol on transcription is attenuation on the induction of vascular endothelial growth factor (VEGF) mRNA in monocytes after hypoxia (see later). This effect is due to alterations in the stability of VEGF mRNA [98].

Chao *et al.* [99] demonstrated that flavopiridol potently inhibits P-TEFb (also known as cdk9/cyclin T), with a K_i of 3 nM, leading to inhibition of transcription by RNA polymerase II by blocking the transition into productive elongation. Interestingly, in contrast with all cdks tested so far, flavopiridol was not competitive with ATP in this reaction. P-TEFb is a required cellular cofactor for the human immunodeficiency virus (HIV-1) transactivator, Tat. Consistent with its ability to inhibit P-TEFb, flavopiridol blocked Tat transactivation of the viral promoter *in vitro*. Furthermore, flavopiridol blocked HIV-1 replication in both single round and viral spread assays with an IC_{50} of less than 10 nM [99]. With this novel knowledge, it is reasonable to test flavopiridol in clinical trials of patients with HIV-related malignancies.

An important biochemical effect involved in the antiproliferative of flavopiridol is the induction of apoptotic cell death. Hematopoietic cell lines are often quite sensitive to flavopiridol-induced apoptotic cell death [100–103], but the mechanism(s) by which flavopiridol induces apoptosis has not yet been elucidated. In certain hematopoietic cell lines, neither BCL-2/BAX nor p53 appeared to be affected [102,104], whereas in other systems, BCL-2 may be inhibited [103]. Preliminary evidence from one laboratory demonstrated that flavopiridol-induced apoptosis in leukemia cells is associated with early activation of the MAPK protein kinase family of proteins (MEK, p38, and JNK) [105]. This activation may lead to the activation of caspases [105]. As seen in this and other models, caspase inhibitors prevent flavopiridol-induced apoptosis [101,105]. It is unclear whether the putative flavopiridol-induced inhibition of cdk activity is required for the induction of apoptosis.

Clear evidence of cell cycle arrest along with apoptosis was observed in a panel of squamous head and neck cell lines, including a cell line (HN30) that is refractory to several DNA-damaging agents, such as γ-irradiation and bleomycin [106]. Again, the apoptotic effect was independent of p53 status and was associated with the depletion of cyclin D1 [106]. Moreover, when flavopiridol was administered to head and neck xenografts (HN12) as a daily 5 mg/kg ip bolus for 5 days, significant growth delay was observed [106]. Again, clear evidence of apoptosis along with cyclin D1 depletion was readily observed in tissues from xenografts treated with flavopiridol [106]. This experience has stimulated us to initiate a phase II trial of flavopiridol in patients with refractory/metastatic head and neck squamous carcinomas (see later).

Along with targeting tumor cells, flavopiridol also targets angiogenesis pathways. Brusselbach *et al.* [107] incubated primary human umbilical vein endothelial cells (HUVECs) with flavopiridol and observed apoptotic cell death even in cells that were not cycling, leading to the notion that flavopiridol may have antiangiogenic properties due to endothelial cytotoxicity. In other model systems, Kerr *et al.* [108] tested flavopiridol in an *in vivo* Matrigel model of angiogenesis and found that flavopiridol decreased blood vessel formation, a surrogate marker for the antiangiogenic effect of this compound. Furthermore, as mentioned earlier, Melillo *et al.* [98] demonstrated that, at low nanomolar concentrations, flavopiridol prevented the induction of VEGF by hypoxic conditions in human monocytes. This effect was caused by a decreased stability of VEGF mRNA, which paralleled the decline in VEGF protein. Thus, the antitumor activity of flavopiridol observed may be in part due to antiangiogenic effects. Whether the various antiangiogenic actions of flavopiridol result from its interaction with a cdk target or other targets requires further study.

Several investigators have attempted to determine if flavopiridol has synergistic effects with standard chemotherapeutic agents. For example, synergistic effects in A549 lung carcinoma cells were demonstrated when treatment with flavopiridol followed treatment with paclitaxel, cytarabine, topotecan, doxorubicin, or etoposide [109,110]. In contrast, a synergistic effect was observed with 5-fluorouracil only when cells were treated with flavopiridol for 24 h before the addition of 5-fluorouracil. Furthermore, synergistic effects with cisplatin were not schedule-dependent [110]. However, Chien *et al.* [111] failed to demonstrate a synergistic effect between flavopiridol and cisplatin and/or γ-irradiation in bladder carcinoma models. One important issue to mention is that most of these studies were performed in *in vitro* models. Thus, confirmatory studies in *in vivo* animal models are needed.

Experiments using colorectal (Colo205) and prostate (LnCaP/DU-145) carcinoma xenograft models in which flavopiridol was administered frequently over a protracted period demonstrated that flavopiridol is cytostatic [86,112]. This demonstration led to human clinical trials of flavopiridol administered as a 72-h continuous infusion every 2 weeks [113] (see later). Subsequent studies in human leukemia/lymphoma xenografts demonstrated that flavopiridol administered intravenously as a bolus rendered animals tumor free, whereas flavopiridol administered as an infusion only delayed tumor growth [100]. Moreover, as shown earlier, in head and neck (HN-12) xenografts, flavopiridol administered as an intraperitoneal bolus daily at 5 mg/kg for 5 days demonstrated a substantial antitumor effects [106]. Based on these results, a phase I trial of 1-h infusional flavopiridol for 5 consecutive days every 3 weeks has been completed at the NCI (see later).

b. Clinical Experience with Flavopiridol

Two phase I clinical trials of flavopiridol administered as a 72-h continuous infusion every 2 weeks have been completed [79,113]. In the NCI phase I trial of infusional flavopiridol, 76 patients were treated. Dose-limiting toxicity (DLT) was secretory diarrhea with a maximal tolerated dose (MTD) of 50 mg/m^2/day for 3 days. In the presence of antidiarrheal prophylaxis (a combination of cholestyramine and loperamide), patients tolerated higher doses, defining a second maximal tolerated dose, 78 mg/m^2/per/day for 3 days. The dose-limiting toxicity observed at the higher dose level was a substantial proinflammatory syndrome (fever, fatigue, local tumor pain, and modulation of acute phase reactants) and reversible hypotension [113]. Minor responses were observed in patients with non-Hodgkin's lymphoma, colon, and kidney cancer for more than 6 months. Moreover, one patient with refractory renal cancer achieved a partial response for more than 8 months [113]. Of 14 patients who received flavopiridol for more than 6 months, 5 patients received flavopiridol for more than 1 year and 1 patient received flavopiridol for more than 2 years [113]. Plasma concentrations of 300–500 nM flavopiridol, which inhibit cdk activity *in vitro*, were safely achieved during this trial [113].

In a complementary phase I trial also exploring the same schedule (72-h continuous infusion every 2 weeks), Thomas *et al.* [79] found that the dose-limiting toxicity was diarrhea, corroborating the NCI experience. Moreover, plasma concentrations of 300–500 nM flavopiridol were also observed. Interestingly, there was one patient in this trial with refractory metastatic gastric cancer that progressed after a treatment regimen containing 5-fluorouracil. When treated with flavopiridol, this patient achieved a sustained complete response without any evidence of disease for more than 2 years after treatment was completed.

The first phase I trial of a daily 1-h infusion of flavopiridol for 5 consecutive days every 3 weeks has been completed [76]. This schedule was based on antitumor results observed in leukemia/lymphoma and head and neck xenografts treated with flavopiridol [100,106]. A total of 55 patients were treated in this trial. The recommended phase II dose is 37.5 mg/m^2/day for 5 consecutive days. Dose-limiting toxicities observed at 52.5 mg/m^2/day are nausea/vomiting, neutropenia, fatigue, and diarrhea [76]. Other (nondose-limiting) side effects are "local tumor pain" and anorexia. To reach higher flavopiridol concentrations, the protocol was amended to administer flavopiridol for 3 days (initially) and then for 1 day only. These protocol modifications allowed achievement of higher flavopiridol concentrations (~4 μM) [76]. Unfortunately, the half-life observed in this trial is much shorter (~3 h) than the infusional trial (~10 h). Thus, high micromolar concentrations in the 1-h infusional trial will be maintained for very short periods of time [76]. Several phase 2 trials in patients with refractory head and neck cancer, chronic lymphocytic leukemia (CLL), and mantle cell lymphoma (MCL) are currently being tested using this schedule (see later).

A phase 1 trial testing the combination of paclitaxel and infusional (24 h) flavopiridol demonstrated good tolerability with a dose-limiting pulmonary toxicity [78].

Phase II trials of flavopiridol given as a 72-h continuous infusion with the MTD in the absence of antidiarrheal prophylaxis (50 mg/m^2/day) to patients with chronic lymphocytic leukemia, nonsmall cell lung cancer, non-Hodgkin's lymphoma, and colon, prostate, gastric, head and neck, and kidney cancer and phase I trials of flavopiridol administered on novel schedules and in combination with standard chemotherapeutic agents are being performed [114–118]. In a phase 2 trial of flavopiridol in metastatic renal cancer, two objective responses (response rate = 6%, 95% confidence interval, 1 to 20%) were observed. Most patients developed grade 1–2 diarrhea and asthenia [118]. In this trial, patients that demonstrated glucuronide flavopiridol metabolites in plasma, as measured by HPLC methodology, have less pronounced diarrhea in comparison to nonmetabolizers [119]. Thus, it may be possible that patients with higher metabolic rates may tolerate higher doses of flavopiridol.

Phase II trials of shorter (1 h) infusional flavopiridol are being conducted in MCL, CLL, and HNSCC. Of interest, several refractory CLL and MCL patients demonstrated clear evidence of responses (partial responses) in these trials being considered in lung and colon cancer (J. R. Suarez, personal communications).

Although the initial studies of flavopiridol in humans are encouraging, the best schedule of administration of flavopiridol needs to be determined. Furthermore, phase 3 studies in combination with standard chemotherapy are being considered (J. R. Suarez, personal communications).

c. UCN-01

Staurosporine is a potent nonspecific protein and tyrosine kinase inhibitor with a very low therapeutic index in animals [120]. Thus, efforts to find staruosporine analogs of staurosporine have identified compounds specific for protein kinases. One staurosporine analog, UCN-01 (7-hydroxy-staurosporine), has potent activity against several protein kinase C isoenzymes, particularly the Ca^{2+}-dependent protein kinase C with an IC_{50} ~30 nM [121–123]. In addition to its effects on protein kinase C, UCN-01 has antiproliferative activity in several human tumor cell lines [124–128]. In contrast, another highly selective potent protein kinase C inhibitor, GF 109203X, has minimal antiproliferative activity, despite a similar capacity to inhibit protein kinase C *in vitro* [125]. These results suggest that the antiproliferative activity of UCN-01 cannot be explained solely by the inhibition of protein kinase C. Although UCN-01 moderately inhibited the activity of immunoprecipitated cdk1(cdc2) and cdk2 (IC_{50}=300–600 nM), exposure of UCN-01 to intact cells leads to "inappropriate activation" of the same kinases [125]. This phenomenon correlates with the G2 abrogation checkpoint observed with this agent (see later).

Experimental evidence suggests that DNA damage leads to cell cycle arrest to allow DNA repair. In cells where the G1 phase checkpoint is not active because of p53 inactivation, irradiated cells accumulate in the G2 phase due to activation of the G2 checkpoint (inhibition of cdc2). In contrast, Wang *et al.* [129] exposed CA46 cell lines to radiation followed by UCN-01, promoting the inappropriate activation of cdc2/cyclin B and early mitosis with the onset of apoptotic cell death. These effects could be partially explained by the inactivation of Wee1, the kinase that negatively regulates the G2/M-phase transition [130]. Moreover, UCN-01 can have a direct effect on chk1, a protein kinase that regulates the G2 checkpoint [131–133]. Thus, although UCN-01 at high concentrations can directly inhibit cdks *in vitro*, UCN-01 can modulate cellular "upstream" regulators at a much lower concentration, leading to inappropriate cdc2 activation. Studies from other groups suggest that not only is UCN-01 able to abrogate the G2 checkpoint induced by DNA-damaging agents, but also, in some circumstances, UCN-01 is able to abrogate the DNA damage-induced S-phase checkpoint [134,135].

UCN-01 also arrests cells in the G1 phase of the cell cycle [124,127,128,136–140]. For example, when human epidermoid carcinoma A431 cells (mutated p53) are incubated with UCN-01, these cells were arrested in G1 phase with Rb hypophosphorylation and $p21^{waf1}/p27^{kip1}$ accumulation [128].

Another interesting pharmacological feature of UCN-01 is the observed increased cytotoxicity in cells that harbors mutated p53 [129]. In CA-46 and HT-29 tumor cell lines carrying mutated p53 genes, potent cytotoxicity results following exposure to UCN-01. To extend these observations further, the MCF-7 cell line with no endogenous p53 because of the ectopic expression of E6, a human papillomavirus type-16 protein, showed enhanced cytotoxicity when treated with a DNA-damaging agent, such as cisplatin, and UCN-01 compared with the isogenic wild-type MCF-7 cell line. Thus, a common feature observed in more than 50% of human HNSCC associated with poor outcome and refractoriness to standard chemotherapies [141–145] may render tumor cells more sensitive to UCN-01.

As mentioned earlier, synergistic effects of UCN-01 have been observed with many chemotherapeutic agents, including mitomycin C, 5-fluorouracil, carmustine, and camptothecin, among others [134,146–153]. Therefore, it is possible that combining UCN-01 with these or other agents could improve its therapeutic index. Clinical trials exploring these possibilities are being developed.

Studies of UCN-01 in HNSCC demonstrated a very potent antiproliferative effect in a battery of HNSCC cell lines with an IC_{50} ~ 50 nM (A. Senderowicz, unpublished results), even in HN30 cells lines, refractory to DNA-damaging agents [106]. Furthermore, HN12 xenografts, treated for only one cycle with UCN-01 (7.5 mg/kg/ip QDx5), demonstrated no growth over 6 weeks of observation (A. Senderowicz, unpublished results). Moreover, immunohistochemistry studies from samples obtained after UCN-01 demonstrated not only frank apoptosis, but also upregulation of $p27^{kip1}$ and depletion of cyclin D3 (A. Senderowicz, unpublished results).

UCN-01 administered by an intravenous or intraperitoneal route displayed antitumor activity in xenograft model systems with breast carcinoma (MCF-7 cells), renal carcinoma (A498 cells), leukemia (MOLT-4 and HL-60), and HN12 cells (A. Senderowicz, unpublished results). The antitumor effect was greater when UCN-01 was given over a longer period. This requirement for a longer period of treatment was also observed in *in vitro* models, with greatest antitumor activity observed when UCN-01 was present for 72 h [124]. Thus, a clinical trial using a 72-h continuous infusion every 2 weeks was conducted (described later).

d. Clinical Trials of UCN-01

The first phase I trial of UCN-01 has been completed [77,82]. UCN-01 was initially administered as a 72-h continuous infusion every 2 weeks based on data from *in vitro* and xenograft preclinical models. However, it became apparent in the first few patients that the drug had an unexpectedly long half-life (~30 days). This half-life was 100 times longer than the half-life observed in preclinical models, most likely due to the avid binding of UCN-01 to α_1-acid glycoprotein [154,155]. Thus, the protocol was modified to administer UCN-01 every 4 weeks (one half-life) and subsequent courses, the duration of infusion was decreased by half (total 36 h). Thus, it was possible to reach similar peak plasma concentrations in subsequent courses with no evidence of drug accumulation. There was no evidence of myelotoxicity or gastrointestinal toxicity (prominent side effects observed in animal models), despite very high plasma concentrations achieved (35–50 μM) [77,82,154,155]. Dose-limiting toxicities were nausea/vomiting (amenable to standard antiemetic treatments), symptomatic hyperglycemia associated with an "insulin resistance state" (increase in insulin and c-peptide levels while receiving UCN-01), and pulmonary toxicity characterized by substantial hypoxemia without obvious radiologic changes. The recommended phase II dose of UCN-01 given on a 72-h continuous infusion schedule was 42.5 mg/m^2/day [82]. One patient with refractory metastatic melanoma developed a partial response that lasted 8 months. Another patient with a refractory anaplastic large cell lymphoma that had failed multiple chemotherapeutic regimens, including high-dose chemotherapy, has no evidence of disease 4 years after the initiation of UCN-01. Moreover, a few patients with leiomyosarcoma, non-Hodgkin's lymphoma, and lung cancer demonstrated stable disease for ≥6 months [82,156]. Of note, one patient with refractory large cell lymphoma that failed prior high-dose combination chemotherapy protocol-EPOCH-2 combination chemotherapy had rapidly progressive disease after one cycle of UCN-01. He required immediate systemic chemotherapy due to hepatic and bone marrow failure (thrombocytopenia) due to the progression of disease. Based on the poor status of this patient, a dose-reduced EPOCH combination chemotherapy was administered. His liver function and thrombocytopenia resolved completely with significant improvement in performance status within 2 weeks after combination chemotherapy. Unfortunately, he developed *Candida kruzei* septicemia and expired. His postmortem examination revealed a pathological complete response after only one cycle of chemotherapy [157]. Thus, this refractory large cell lymphoma patient became "chemotherapy-sensitive" after only one dose of UCN-01. This phenomenon recapitulates the synergistic effect observed in preclinical models with several chemotherapeutic agents. Several combination trials are being developed based on this observation.

In order to estimate "free UCN-01 concentrations" in body fluids, several efforts were considered. Plasma ultracentrifugation and salivary determination of UCN-01 revealed similar results. At the recommended phase II dose (37.5 mg/m^2/day over 72 h), concentrations of "free-salivary" UCN-01 (~45 nM) that may cause G2 checkpoint abrogation can be achieved. As mentioned earlier, UCN-01 is a potent PKC inhibitor. In order to determine the putative signaling effects of UCN-01 in tissues, bone marrow aspirates and tumor cells were obtained from patients before and during the first cycle of UCN-01 administration. Western blot studies were performed in those samples against phosphorylated adducin, a cytoskeletal membrane protein, an specific substrate phosphorylated by PKC [158].

A clear loss in phospho-adducin content in the posttreatment samples was observed in all tumor and bone marrow samples tested.

Several groups are conducting shorter (3 h) infusional trials of UCN-01. Interestingly, the toxicity profile of shorter infusions is similar to the toxicities observed with the 72-h infusion trial [80,81]. However, with shorter infusions, more pronounced hypotension was observed [80,81]. Determination of free UCN-01 in these trials is of utmost importance, as higher free concentrations for shorter periods may be more or less beneficial compared with the free concentrations observed in the 72-h infusion trial.

Based on the unique pharmacological features and anecdotal clinical evidence of synergistic effects in one patient with refractory B-cell lymphoma [157], several combination trials with standard chemotherapeutic agents have been commenced. A phase I/II trial of gemcitabine followed by 72-hr infusional UCN-01 in chronic lymphocytic leukemia started at the NCI. Other studies in combination with cisplatin, 5-fluorouracil, and dexamethasone among others, also commenced.

D. Modulation of Hypoxia and Angiogenesis

1. Hypoxia-Active Agents

Tirapazimine is a hypoxic cytotoxic agent, i.e., a bioreductive compound that demonstrates differential toxicity for hypoxic cells [159]. In preclinical studies, additive antitumor effects were observed in combination with radiation and in combination with cisplatin [160,161]. Because hypoxia is thought to be one of the major reasons for resistance to radiation in head and neck cancer, this agent has been studied in combination with radiation and chemotherapy in this disease. A phase I trial in patients with locally advanced squamous cancer of the head and neck administered conventional fractionated radiation (70 Gy in 7 weeks) with cisplatin (75 mg/m^2) and tirapazamine (290 mg/m^2) (given prior to cisplatin) in weeks 1, 4, and 7 and tirapazamine alone, (160 mg/m^2) given three times a week in weeks 2, 3, 5, and 6. Febrile neutropenia occurred toward the end of radiation in 3 of the 6 patients treated at the initial dose. In addition, 2 of the 6 patients also developed grade 4 acute radiation reactions. After elimination of the tirapazamine doses on weeks 5 and 6, only one dose-limiting toxicity (febrile neutropenia) occurred among 10 patients, and 8 of the 10 patients were able to complete treatment without any dose omissions or obvious increased radiation toxicity. 18-F misonidazole scans detected hypoxia at baseline in 14 of 15 patients studied. With a median follow-up of 2.7 years, the 3-year failure-free survival rate was 69% [162]. These failure-free survival results appear to be improved over previously reported results for patients with similar advanced stage disease. A phase II multi-institutional trial is in progress, and a phase III trial, to compare a tirapazamine containing regimen with a nontirapazamine-containing regimen, is being planned.

2. Antiangiogenesis

Angiogenesis appears to be necessary for tumors to grow to a large size and/or to metastasize. Markers of angiogenesis include VEGF, microvessel density, and thymidine phosphorylase (or platelet-derived endothelial cell growth factor). VEGF is a 34- to 50-kd a dimer composed of two identical disulfide-linked subunits that are generated from differential splicing of a single gene [163]. VEGF is induced under hypoxic conditions in a variety of tissues. A high expression of VEGF has been associated with more frequent recurrence and shorter survival in head and neck squamous [163]. However, other studies found conflicting results: either increased expression or downregulation of VEGF has been correlated with progression from premalignant lesions to head and neck cancer [164,165]. Preclinical studies of antiangiogenesis agents have shown inhibition of xenograft tumor growth with TNP-470, a synthetic analog of fumagillin [166]. Antisense VEGF transfection was shown to downregulate VEGF secretion from VEGF-secreting SCCHN cells, leading to a reduction in endothelial cell migration. However, the antisense vector did not affect growth of the tumor *in vivo* [167]. Preclinical data suggest a synergistic role of antiangiogenesis agents with radiation [168], and an association of angiogenesis and resistance to radiation therapy has been suggested [32].

Currently, phase I and II studies of single agent antiangiogenesis agents, such as thalidomide [169], endostatin [170], SU 5916 [171], SU 6668 [171], and bevacizumab (an antibody to VEGF) [172], as well as combinations with radiation and chemotherapy, are ongoing.

E. Gene Therapy Studies

Gene therapy for head and neck cancer has received considerable interest due to the localized nature of the disease and to the fact that, at present, gene therapy must be delivered in proximity to the target cells for acceptable transduction. Methods of administration include direct injection into tumor or injection through an endoscope. Gene therapy strategies have included transduction of a mutated gene (p53), adenovirus lysis of cells without normal p53, transduction of cytokines, or genes dependent on radiation for activation. To date, none of these strategies has undergone a randomized phase III trial.

The E1 b-attenuated adenovirus, ONYX-015, is an agent that can infect, replicate in, and lyse cells without wild-type p53. This agent is of interest for tumors, such as head and neck cancer, in which about half the tumors lack normal p53. The agent demonstrated antitumor activity in squamous cancers of the head and neck in initial trials. In a phase II trial, patients with recurrent or relapsed squamous cell carcinoma of the head and neck received daily times 5 every 2 weeks or twice daily for 2 consecutive weeks direct intratumoral injection of ONYX-015. On the daily x 5 arm, partial or complete regression of the injected lesions was seen in 14%, whereas the twice-daily regimen produced a response rate of 10% [173]. A second phase II trial in patients with unresectable recurrent squamous cancer of the head and neck who had not had previous systemic chemotherapy for recurrence was performed with tumor injections daily for 5 consecutive days repeated every 3 weeks or until tumor progression was noted. Patients were also treated with intravenous cisplatin (80 mg/m^2) and 5-fluorouracil, (800–1000 mg/m^2/day) continuous infusion on days 1–5 every 3 weeks. Locoregional tumor control was obtained in all nine evaluable patients. Viral presence in target tumors was confirmed by *in situ* hybridization [174].

Trials have also been done with Ad5CMVp53, an agent that transduces cells with p53. In head and neck cancer, RPR/ING trials of p53 gene transfection with Ad5CMV-p53 showed an objective response rate of 6% in recurrent tumors, with little toxicity [175]. While responses in head and neck cancer have been few when used as a single agent, combinations of gene therapy with chemotherapy and/or radiation may be more promising.

Cytokines and immune modulators have also been investigated as gene therapy strategies (see below). Murine models were used to study the treatment of established squamous tumors with recombinant adenovirus constructs containing the genes for murine granulocyte–macrophage colony-stimulating factor, murine interleukin 2 (IL-2) or herpes simplex virus thymidine kinase. Either GM-CSF or IL-2 significantly enhanced the response to adenovirus constructs containing genes for herpes simplex thymidine kinase. However, immunogenicity of the HNSCC was not enhanced [176].

Radiation may also be useful to increase the replication efficiency of attenuated herpesviruses, another targeted, and possibly less toxic, therapeutic approach [177].

F. Immunological Therapy of Head and Neck Cancer

IL-12 is a immunomodulatory cytokine that has been shown to modulate delayed-type hypersensitivity, T cytotoxic, and natural killer cell responses [178], which are regulated by T helper type 1 (Th1) lymphocyte responses. IL-12 has been reported to induce antitumor immunity [178], and preclinical evidence shows that IL-12 and IL-2 may be more effective in inducing tumor rejection when given together [179,180]. The combination of IL-12 and IL-2 has been shown to induce greater cytolytic activity when cultured with peripheral blood lymphocytes from patients with head and neck cancer than either cytokine alone [181]. Another cytokine, interferon (IFN)-γ, has been reported to be an important factor in the induction of both immunologic and antiangiogenic effects of IL-12 [182]. The effectiveness of combination therapy with cytokines IL-12 and IL-2 and the role of IFN-γ, in the regression of head and neck carcinoma have not been determined previously and have been hindered by lack of a syngeneic murine model.

Thomas *et al.* [183] developed a new syngeneic murine model of oral squamous cell carcinoma in which tumorigenicity is associated with decreased expression of the immune costimulatory molecule CD80. Along with cytokine signals provided by IL-12, IL-2, and IFN-γ, effective antitumor responses appear to depend on adequate priming and stimulation of T cells by antigen and appropriate cell surface immune recognition molecules, such as the major histocompatibility complex and costimulatory molecules CD80 or CD86 [184]. Without costimulation, exposure of T cells to antigen and cytokines can cause anergy or lead to programmed cell death of T cells [185]. Studies have shown that expression of CD80 on some tumors reduces their tumorigenicity and that transfection of tumors with CD80 combined with IL-12 or IL-2 cytokine therapy induces tumor regression at a greater rate than that achieved with either CD80 or cytokine treatment alone [186,187]. Thus, a lack of sufficient CD80 costimulation could also contribute to the low tumor response rates observed in patients with SCC treated with IL-2 or IL-12 alone.

Thomas *et al.* [188] demonstrated that combined IL-12 and IL-2 therapy can induce tumor regression in a new murine model of oral SCC and determined that the antitumor response is promoted by expression of the immune costimulatory molecule CD80 and cytokine IFN-γ. In CD80 positive or negative subclones of a BALB/c oral SCC line in syngeneic mice, they showed that systemic rIL-12 alone was comparable in effectiveness to combined therapy with IL-12 and peritumoral rIL-2, inducing complete regression of the CD80$^+$ B7E11-4 SCID. However, therapy with these cytokines had no effect on growth of the CD80$^-$ subclone B7E3-4 SCID and did not induce complete regression of the CD80$^+$ subclone B7E11-4 scid in congenic BALB IFN-γ knockout mice, indicating that expression of the CD80 costimulatory molecule and IFN-γ contribute to tumor regression [188]. In cytokine-treated mice that rejected the CD80$^+$ SCC line, an increase in infiltrating CD4$^+$ lymphocytes and apoptotic bodies within the tumor specimens was observed, and resistance to rechallenge with the same tumor was detected in 50% of recipients, consistent with an immune response.

These results provide evidence that regression of oral head and neck SCC may be induced by therapy with systemic IL-12 and that expression of the CD80 costimulatory molecule by SCC and IFN-γ by the host promote IL-12-induced regression of SCC.

Improved induction of antitumor immune responses following therapy with both IL-12 and CD80 has been observed in other preclinical tumor models. This response has been generally superior to treatment with either CD80 or IL-12 alone. Studies by others have provided evidence that IL-12 and CD80 costimulation are synergistic in inducing IFN-γ, proliferation of $CD4^+$ and $CD8^+$ T cells in both murine and human models [186,187] and for induction of lasting protection [186]. Thomas *et al.* [188] found that regression was associated with an increase in the number of infiltrating $CD4^+$ cells in the cytokine-treated SCC tumor specimens. In some tumor models, $CD4^+$ cells have been shown to be most important in mediating immunity induced by CD80 and IL-12 [189]. Thomas *et al.* [188] did not observe a significant increase in $CD8^+$ cells, and were unable to detect significant T or NK cytolytic activity in splenocytes in mixed lymphocyte tumor cultures established from animals that were resistant to rechallenge.

In a study by Thomas *et al.* [188], an increase in cell death as measured by the TUNEL assay (established method to determine apoptosis) was evident within tumor specimens from animals treated with IL-12 alone or with the combination of IL-12/IL-2, and this effect was greater than the baseline level of apoptosis observed in untreated tumor specimens. These apoptotic bodies likely represent SCC cells. IL-12-induced apoptosis has been shown in tumors in several studies [190], yet the mechanism for this remains unclear. Thomas *et al.* [188] did not observe any direct toxicity of IL-12 on the oral SCC lines *in vitro*. Interestingly, apoptosis associated with IL-12 is more likely induced by an IFN-γ-dependent mechanism [191,192]. They did observe that the growth of oral SCC lines could be inhibited directly by IFN-γ. Alternatively, apoptosis could also be the result of immune-mediated cytotoxicity or hypoxia and cell death due to the antiangiogenic activity of IL-12.

Several pilot clinical studies have been conducted to explore the use of cytokines in therapy of HNSCC. IL-2 and IL-12 have been tested as adjuvants in the treatment of patients with advanced HNSCC because of their potential immune stimulatory and antiangiogenic activities. Partial or complete tumor regression has been detected in only a few patients undergoing therapy with these cytokines, and most of these responses have been of short duration [193–195]. Clinical trials in which IL-2 alone was administered to patients with recurrent, advanced, or unresectable HNSCC demonstrated only partial response rates between 6 and 10%, although these studies were conducted in patients with advanced disease who had prior chemotherapy. IL-2 alone, at therapeutically tolerated dosages, may provide an insufficient signal for induction of an effective antitumor response [196,197].

To date, clinical studies using combined IL-12 and IL-2 or IL-12 alone in patients with head and neck SCC and other neoplasms have not been completed. The preclinical studies by Thomas *et al.* [188] provide a rationale for clinical immunotherapy trials with IL-12 with or without IL-2 in combination with CD80 for the treatment of HNSCC. IL-12 cytokine therapy may be beneficial in providing the necessary signals for the activation and reversal of immunological defects in patients with HNSCC expressing CD80 in which there is preexisting T-cell and/or NK cell unresponsiveness. However, previous results appear to indicate that there is a low likelihood that many HNSCC will express CD80 [183]. Based on these studies, it will be important to determine the effect of systemic IL-12 in combination with intralesional delivery and expression of CD80 by gene therapy.

II. CONCLUSIONS

The fundamental breakthroughs achieved in cancer research since the 1990s were made possible by the conversion of multiple disciplines, including structural biology, computational chemistry, structurally-directed medicinal chemistry, molecular and cellular biology, and novel array of screening methodologies. This emerging body of information has provided a unique opportunity to identify novel targets for cancer therapy, zeroing in on those molecules necessary for tumor progression. Although patients afflicted with advanced HNSCC have a grim prognosis, there are novel promising modalities in the clinic that modulate these "molecular targets". As practicing oncologists dealing with patients with HNSCC, our main challenge now is to incorporate these novel modalities early in the treatment of this disease, to combine these modalities with standard chemotherapeutic agents in a rational and effective way, and to develop predictive assays based on these novel targets that will allow us to select *a priori* patients that will benefit the most from these novel therapies.

References

1. Senderowicz, A. M. (2000). Development of cyclin-dependent kinase modulators as novel therapeutic approaches for hematological malignancies. *Leukemia* **15**, 1–9.
2. Senderowicz, A. M. (2000). Small molecule modulators of cyclin-dependent kinases for cancer therapy. *Oncogene* **19**, 6600–6606.
3. Druker, B. J., and Lydon, N. B. (2000). Lessons learned from the development of an abl tyrosine kinase inhibitor for chronic myelogenous leukaemia. *J. Clin. Invest.* **105**, 3–7.
4. Forastiere, A. A., Metch, B., Schuller, D. E., Ensley, J. F., Hutchins, L. F., Triozzi, P., Kish, J. A., McClure, S., VonFeldt, E., Williamson, S. K., *et al.* (1992). Randomized comparison of cisplatin plus fluorouracil and carboplatin plus fluorouracil versus methotrexate in advanced

squamous-cell carcinoma of the head and neck: A Southwest Oncology Group study. *J. Clin. Oncol.* **10**, 1245–1251.
5. Forastiere, A. A., Shank, D., Neuberg, D., Taylor, S. G. T., DeConti, R. C., and Adams, G. (1998). Final report of a phase II evaluation of paclitaxel in patients with advanced squamous cell carcinoma of the head and neck: An Eastern Cooperative Oncology Group trial (PA390). *Cancer* **82**, 2270–2274.
6. Smith, R. E., Thornton, D. E., and J, A. (1995). A phase II trial of paclitaxel in squamous cell carcinoma of the head and neck with correlative laboratory studies. *Sem. Oncol.* **22**, 41–46.
7. Forastiere, A. A., Leong, T., Rowinsky, E., Murphy, B. A., Vlock, D. R., DeConti, R. C., and GL., A. (2001). Phase III comparison of high-dose paclitaxel + cisplatin + granulocyte colony stimulating factor versus low-dose paclitaxel + cisplatin in advanced head and neck cancer: Eastern Cooperative Oncology Group Study 1393. *J. Clin. Oncol.* **19**, 1088–1095.
8. Murphy B, Li, Y., Cella, D., Karnad, A., Hussain, M., and A., F. (2001) Phase III study comparing cisplatin (C) and 5-flurouracil (F) versus cisplatin and paclitaxel (T) in metastatic/recurrent head and neck cancer (MHNC). *In* "Proc ASCO." [Abstract 894]
9. Pivot, X., Cals, L., Cupissol, D., Guardiola, E., Tchiknavorian, X., Guerrier, P., Merad, L., Wendling, J. L., Barnouin, L., Savary, J., Thyss, A., and M., S. (2001). Phase II trial of paclitaxel-carboplatin combination in recurrent squamous cell carcinoma of the head and neck. *Oncology* **60**, 66–71.
10. Shin, D. M., Khuri, F. R., Glisson, B. S., Ginsberg, L., Papadimitrakopoulou, V. M., Clayman, G., Lee, J. J., Ang, K. K., Lippman, S. M., and WK, H. (2001). Phase II study of paclitaxel, ifosfamide and carboplatin in patients with recurrent or metasatic head and neck squamous carcinoma. *Cancer.* **91**, 1316–1323.
11. Dreyfuss, A. I., Clark, J. R., Norris, C. M., Rossi, R. M., Lucarini, J. W., Busse, P. M., Poulin, M. D., Thornhill, L., Costello, R., and Posner, M. R. (1996). Docetaxel: An active drug for squamous cell carcinoma of the head and neck. *J. Clin. Oncol.* **14**, 1672–1678.
12. Posner, M. R., Glisson, B., Frenette, G., Al-Sarraf, M., Colevas, A. D., Norris, C. M., Seroskie, J. D., Shin, D. M., Olivares, R., and Garay, C. A. (2001). Multicenter phase I-II trial of docetaxel, cisplatin, and fluorouracil induction chemotherapy for patients with locally advanced squamous cell cancer of the head and neck. *J. Clin. Oncol.* **19**, 1096–1104.
13. Colevas, A. D., Norris, C. M., Tishler, R. B., Fried, M. P., Gomolin, H. I., Amrein, P., Nixon, A., Lamb, C., Costello, R., Barton, J., Read, R., Adak, S., and Posner, M. R. (1999). Phase II trial of docetaxel, cisplatin, fluorouracil, and leucovorin as induction for squamous cell carcinoma of the head and neck. *J. Clin. Oncol.* **17**, 3503–3511.
14. Sunwoo, J. B., Herscher, L. L., Kroog, G. S., Thomas, G. R., Ondrey, F. G., Duffey, D. C., Solomon, B. I., Boss, C., Albert, P. S., McCullugh, L., Rudy, S., Muir, C., Zhai, S., Figg, W. D., Cook, J. A., Mitchell, J. B., and Van Waes, C. (2001). Concurrent paclitaxel and radiation in the treatment of locally advanced head and neck cancer. *J. Clin. Oncol.* **19**, 800–811.
15. Kies, M. S., Haraf, D. J., Rosen, F., Stenson, K., List, M., Brockstein, B., Chung, T., Mittal, B. B., Pelzer, H., Portugal, L., Rademaker, A., Weichselbaum, R., and Vokes, E. E. (2001). Concomitant infusional paclitaxel and fluorouracil, oral hydroxyurea, and hyperfractionated radiation for locally advanced squamous head and neck cancer. *J. Clin. Oncol.* **19**, 1961–1969.
16. Rosenthal, D. I., Lee, J. H., Sinard, R., Yardley, D. A., Machtay, M., Rosen, D. M., Egorin, M. J., Weber, R. S., Weinstein, G. S., Chalian, A. A., Miller, L. K., Frenkel, E. P., and Carbone, D. P. (2001). Phase I study of paclitaxel given by seven-week continuous infusion concurrent with radiation therapy for locally advanced squamous cell carcinoma of the head and neck. *J. Clin. Oncol.* **19**, 1363–1373.
17. Suntharalingam, M., Haas, M. L., Conley, B. A., Egorin, M. J., Levy, S., Sivasailam, S., Herman, J. M., Jacobs, M. C., Gray, W. C., Ord, R. A., Aisner, J. A., and Van Echo, D. A. (2000). The use of carboplatin and paclitaxel with daily radiotherapy in patients with locally advanced squamous cell carcinomas of the head and neck. *Int. J. Radiat. Oncol. Biol. Phys.* **47**, 49–56.
18. Garden, A., Pajak, T., Vokes, E., Forastiere, A., Ridge, J., Jones, C., Horwitz, E., Nabell, L., Glisson, B., Cooper, J. S., Demas, W., and E., G. (2001). Preliminary results of RTOG 9703: A phase II randomized trial of concurrent radiation (RT) and chemotherapy for advanced squamous cell carcinomas (SCC) of the head and neck. *In* "Proc. Am. Soc. Clin. Oncol." [Abstract 891].
19. Hofle, G., Bedorf, N., Steinmetz, H., Schemburg, D., Gerth, K., and H., R. (1996). Epothilone A and B-novel 16-membered macrolides and cytotoxic activity: Isolation, crystal structure, and conformation in solution. *Angew. Chem. Int. Ed. Engl.* **35**, 1567–1569.
20. Lee, F. Y., Borzilleri, R., Fairchild, C. R., Kim, S. H., Long, B. H., Reventos-Suarez, C., Vite, G. D., Rose, W. C., and Kramer, R. A. (2001). BMS-247550: A novel epothilone analog with a mode of action similar to paclitaxel but possessing superior antitumor efficacy. *Clin Cancer Res.* **7**, 1429–1437.
21. Rubin, E., Siu, L. L., Beers, S., Moore, M. J., Thompson, C., Becker, M., Chen, T. L., Cohen, P., Rothermel, J., and AM, O. (2001). A phase I and pharmacologic trial of epothilone B in patients with advanced malignancies. *In* "Proc. Am. Soc. Clin. Oncol." p. 68a.
22. Mani, S., McDaid, H., Shen, H., Sparano, J. A., Hamilton, A., Runowicz, C., Hochster, H., Muggia, F., Fields, A., Damle, B., Letrent, S., Lebwohl, D., and SB., H. (2001). Phase I evaluation of an epothilone B analog (BMS 247555): Clinical findings and molecular correlates. *In* "Proc. Am. Soc. Clin. Oncol." p. 68a.
23. Takeda, Y., Kasai, H., Uchida, N., Yoshida, H., Maekawa, R., Sugita, K., and Yoshioka, T. (1999). Enhanced antitumor efficacy of nedaplatin with 5-fluorouracil against human squamous carcinoma xenografts. *Anticancer Res.* **19**, 4059–4064.
24. Raymond, E., Chaney, S. G., Taamma, A., and Cvitkovic, E. (1998). Oxaliplatin: A review of preclinical and clinical studies. *Ann. Oncol.* **9**, 1053–1071.
25. Degardin, M., Cappelaere, P., Krakowski, I., Fargeot, P., Cupissol, D., and Brienza, S. (1996). Phase II trial of oxaliplatin (L-OHP) in advanced, recurrent and/or metastatic squamous cell carcinoma of the head and neck. *Eur. J. Cancer B Oral Oncol.* **32B**, 278–279.
26. Diasio, R. B., and Harris, B. E. (1989). Clinical pharmacology of 5-fluorouracil. *Clin. Pharmacokinet.* **16**, 215–237.
27. Humerickhouse, R., Dolan, M. E., Haraf, D. J., Brockstein, B., Stenson, K., Kies, M., Sulzen, L., Ratain, M. J., and EE, V. (1999). Phase I study of eniluracil, a dihyropyrimidine dehydrogenase inactivator, and oral 5-fluorouracil with radiation therapy in patients with recurrent or advanced head and neck cancer. *Clin. Cancer Res.* **5**, 291–298.
28. Vokes, E., Brockstein, B. E., Humerickhouse, R., and DJ., H. (1998). Oral 5FU alternatives for the treatment of head and neck cancer. *Oncology (Huntingdon)* **12**, 35–38.
29. Fujii, M., Ohno, Y., Tokumaru, Y., Imanishi, Y., Kanke, M., Tomita, T., and T., Y. (2000). UFT plus carboplatin for head and neck cancer. *Oncology* **14**, 72–75.
30. Mercke, C. (2000). UFT/leucovorin in advanced squamous cell carcinoma of the head and neck administered with radiotherapy. *Oncology. Suppl.* **9**, 79–81.
31. Budman, D., Meropol, N. J., Reigner, B., Creaven, P. J., Lichtman, S. M., Berghorn, E., Behr, J., Gordon, R. J., Osterwalder, B., and T., G. (1998). Preliminary studies of a novel oral fluoropyrimidine carbamate: Capecitabine. *J. Clin. Oncol.* **16**, 1795–1802.
32. Koukourakis, M. I., Giatromanolaki, A., Fountzilas, G., Sivridis, E., KC, G., and AL, H. (2000). Angiogenesis, thymidine phosphorylase,

and resistance of squamous cell head and neck cancer to cytotoxic and radiation therapy. *Clin. Cancer Res.* **6**, 381–389.

33. Fukuiwa, T., Takebayashi, Y., Akiba, S., Matsuzaki, T., Hanamure, Y., Miyadera, K., Yamada, Y., and S., A. (1999). Expression of thymidine phosphorylase and vascular endothelial growth factor in human head and neck squamous cell carcinoma and their different characteristics. *Cancer* **85**, 960–969.
34 Braakhuis, B., Ruiz van Harpen, V. W. T., Boven, E., Veerman, G., and GJ., P. (1995). Schedule-dependent antitumor effect of gemcitabine in in vivo model systems. *Semin. Oncol.* **22**, 42–46.
35. Hammond, J., Lee, S., and PJ., F. (1999). [^{3}H]Gemcitabine uptake by nucleoside transporters in a human head and neck squamous carcinoma cell line. *J. Pharmacol Exp. Therap.* **288**, 1185–1191.
36. Braakhuis, B., van Dongen, G. A., Vermorken, J. B., and GB., S. (1991). Precinical in vivo activity of 2', 2'-difluorodeoxycytidine (Gemcitabine) against human head and neck cancer. *Cancer Res.* **51**, 211–214.
37. Catimel, G., Vermorken, J. B., Clavel, M., de Mulder, P., Judson, I., Sessa, C., Piccart, M., Bruntsch, U., Verweij, J., Wanders, J., *et al.* (1994). A phase II study of Gemcitabine (LY 188011) in patients with advanced squamous cell carcinoma of the head and neck: EORTC Early Clinical Trials Group. *Ann. Oncol.* **5**, 543–547.
38. Hitt, R., Castellano, D., Hidalgo, M., Garcia-Carbonero, R., Pena, M., Brandariz, A., Millan, J. M., Alvarez Vincent, J. J., and Cortes-Funes, H. (1998). Phase II trial of cisplatin and gemcitabine in advanced squamous-cell carcinoma of the head and neck. *Ann. Oncol.* **9**, 1347–1349.
39. Fountzilas, G., Stathopoulos, G., Nicolaides, C., Kalongera-Fountzila, A., Kalofonos, H., Nikolaou, A., Bacoyiannis, C., Samantas, E., Papadimitriou, C., Kosmidis, P., Danilidis, J., and N, P. (1999). Paclitaxel and gemcitabine in advanced non-nasopharyngeal head and neck cancer: A phase II study conducted by the Hellenic Cooperative Oncology Group. *Ann. Oncol.* **10**, 475–478.
40. Lawrence, T. S., E. A., and Shewach, D. S. (1997). Gemcitabine -mediated radiosensitization. *Semin Oncol.* **24**.
41. Eisbruch, A., Shewach, D. S., Bradford, C. R., Littles, J. F., Teknos, T. N., Chepeha, D. B., Marentette, L. J., Terrell, J. E., Hogikyan, N. D., Dawson, L. A., Urba, S. A., Wolf, G. T., and TS. L. (2001). Radiation concurrent with gemcitabine for locally advanced head and neck cancer: A phase I trial and intracellular drug incorporation study. *J. Clin. Oncol.* **19**, 792–799.
42. Grandis, J. R., Melhem, M. F., Gooding, W. E., Day, R., Holst, V. A., Wagener, M. M., Drenning, S. D., and Tweardy, D. J. (1998). Levels of TGF-alpha and EGFR protein in head and neck squamous cell carcinoma and patient survival. *J. Natl. Cancer Inst.* **90**, 824–832.
43. Maurizi, M., Almadori, G., Ferrandina, G., Distefano, M., Romanini, M. E., Cadoni, G., Benedetti-Panici, P., Paludetti, G., Scambia, G., and Mancuso, S. (1996). Prognostic significance of epidermal growth factor receptor in laryngeal squamous cell carcinoma. *Br. J. Cancer* **74**, 1253–1257.
44. Bonner, J. A., Raisch, K. P., Trummell, H. Q., Robert, F., Meredith, R. F., Spencer, S. A., Buchsbaum, D. J., Saleh, M. N., Stackhouse, M. A., LoBuglio, A. F., Peters, G. E., Carroll, W. R., and Waksal, H. W. (2000). Enhanced apoptosis with combination C225/radiation treatment serves as the impetus for clinical investigation in head and neck cancers. *J. Clin. Oncol.* **18**, 47S–53S.
45. Huang, S. M., and Harari, P. M. (2000). Modulation of radiation response after epidermal growth factor receptor blockade in squamous cell carcinomas: Inhibition of damage repair, cell cycle kinetics, and tumor angiogenesis. *Clin. Cancer Res.* **6**, 2166–2174.
46. Cohen, R. B., Falcey, J. W., Paulter, V. J., Fetzer, K. M., and Waksal, H. W. (2000). Safety profile of the monoclonal antibody (MoAb) IMC-C225, an anti-epidermal growth factor receptor (EGFr) used in the treatment of EGFr-positive tumors. *Proc. Am. Soc. Clin. Oncol.* **19** [Abstract 1862].
47. Baselga, J., Pfister, D., Cooper, M. R., Cohen, R., Burtness, B., Bos, M., D'Andrea, G., Seidman, A., Norton, L., Gunnett, K., Falcey, J., Anderson, V., Waksal, H., and Mendelsohn, J. (2000). Phase I studies of anti-epidermal growth factor receptor chimeric antibody C225 alone and in combination with cisplatin. *J. Clin. Oncol.* **18**, 904–914.
48. Ezekiel, M., Robert, F., Meredith, R. F., Spencer, S. A., Newsome, J., Khazaeli, M. B., Peters, G. E., Saleh, M., LoBuglio, A. F., and Waksal, H. W. (1998). Phase I study of anti-epidermal growth factor receptor (EGFR) antibody C225 in combination with irradiation in patients with advanced squamous cell carcinoma of the head and neck (SCCHN). *Proc. Ann. Meet. Am. Soc. Clin. Oncol.* **17**, A1522.
49. Woodburn, J. R. (1999). The epidermal growth factor receptor and its inhibition in cancer therapy. *Pharmacol Ther.* **82**, 241–250.
50. Inoue, K., Slaton, J. W., Perrotte, P., Davis, D. W., Bruns, C. J., Hicklin, D. J., McConkey, D. J., Sweeney, P., Radinsky, R., and Dinney, C. P. (2000). Paclitaxel enhances the effects of the anti-epidermal growth factor receptor monoclonal antibody ImClone C225 in mice with metastatic human bladder transitional cell carcinoma. *Clin Cancer Res.* **6**, 4874–4884.
51. Sirotnak, F. M., Zakowski, M. F., Miller, V. A., Scher, H. I., and Kris, M. G. (2000). Efficacy of cytotoxic agents against human tumor xenografts is markedly enhanced by coadministration of ZD1839 (Iressa), an inhibitor of EGFR tyrosine kinase. *Clin. Cancer Res.* **6**, 4885–4892.
52. Baselga, J., Herbst, R., LoRusso, P., Rischin, D., Ranson, M., Plummer, R., Raymond, E., Maddox, A.-M., Kaye, S. B., Kieback, D. G., Harris, A., and Ochs, J. (2000). Continuous administration of ZD1839 (Iressa), a novel oral epidermal growth factor receptor tyrosine kinase inhibitor (EGFR-TKI), in patients with five selected tumor types: Evidence of activity and good tolerability. *Proc. Am. Soc. Clin. Oncol.* **19** [Abstract 686].
53. Moyer, J. D., Barbacci, E. G., Iwata, K. K., Arnold, L., Boman, B., Cunningham, A., DiOrio, C., Doty, J., Morin, M. J., Moyer, M. P., Neveu, M., Pollack, V. A., Pustilnik, L. R., Reynolds, M. M., Sloan, D., Theleman, A., and Miller, P. (1997). Induction of apoptosis and cell cycle arrest by CP-358,774, an inhibitor of epidermal growth factor receptor tyrosine kinase. *Cancer Res.* **57**, 4838–4848.
54. Pollack, V. A., Savage, D. M., Baker, D. A., Tsaparikos, K. E., Sloan, D. E., Moyer, J. D., Barbacci, E. G., Pustilnik, L. R., Smolarek, T. A., Davis, J. A., Vaidya, M. P., Arnold, L. D., Doty, J. L., Iwata, K. K., and Morin, M. J. (1999). Inhibition of epidermal growth factor receptor-associated tyrosine phosphorylation in human carcinomas with CP-358,774: Dynamics of receptor inhibition *in situ* and antitumor effects in athymic mice. *J. Pharmacol. Exp. Ther.* **291**, 739–748.
55. Siu, L. L., Hidalgo, M., Nemunaitis, J., Rizzo, J., Moczygemba, J., Eckhardt, S. G., Tolcher, A., Smith, L., Hammond, L., Blackburn, A., Tensfeldt, T., Silberman, S., Von Hoff, D. D., and Rowinsky, E. K. (1999). Dose and schedule-duration escalation of the epidermal growth factor receptor (EGFR) tyrosine kinase (TK) inhibitor CP-358,774: A phase 1 and pharmacokinetic (PK) study. *Proc. Annu. Meet. Am. Soc. Clin. Oncol.* **18**, A1498.
56. Karp, D. D., Silberman, S. L., Csudae, R., Wirth, R., Gaynes, L., Posner, M., Bubley, G., Koon, H., Bergaman, M., Huang, M., and Schnipper, L. E. (1999). Phase I dose escalation study of epidermal growth factor receptor (EGFR) tyrosine kinase (TK) inhibitor CP-358,774 in patients with advanced solid tumors. *Proc. Ann. Meet. Am. Soc. Clin. Oncol.* **18**, A1499.
57. Senzer, N., Soulieres, D., Siu, L., Agarwala, S., Vokes, E., Hidalgo, M., Silberman, S., Allen, L., Ferrante, K., Fisher, D., Marsolai, S. C., and P., N. (2001). Phase 2 evaluation of OSI-774, a potent oral antagonist of the EGFR-TK in patients with advanced squamous cell carcinoma of the head and neck. *Proc. Annu. Meet. Am. Soc. Clin. Oncol.* **20**, 6.
58. Sherr, C. J. (1996). Cancer cell cycles. *Science* **274**, 1672–1677.

59. Hatakeyama, M., and Weinberg, R. A. (1995). The role of RB in cell cycle control. *Prog. Cell Cycle Res.* **1**, 9–19.
60. Sellers, W. R., and Kaelin, W. G., Jr. (1997). Role of the retinoblastoma protein in the pathogenesis of human cancer. *J. Clin. Oncol.* **15**, 3301–3312.
61. Morgan, D. O. (1997). Cyclin-dependent kinases: Engines, clocks, and microprocessors. *Annu. Rev. Cell Dev. Biol.* **13**, 261–291.
62. Rickert, P., Seghezzi, W., Shanahan, F., Cho, H., and Lees, E. (1996). Cyclin C/CDK8 is a novel CTD kinase associated with RNA polymerase II. *Oncogene.* **12**, 2631–2640.
63. Wei, P., Garber, M. E., Fang, S. M., Fischer, W. H., and Jones, K. A. (1998). A novel CDK9-associated C-type cyclin interacts directly with HIV-1 Tat and mediates its high-affinity, loop-specific binding to TAR RNA. *Cell.* **92**, 451–462.
64. Grana, X., De Luca, A., Sang, N., Fu, Y., Claudio, P. P., Rosenblatt, J., Morgan, D. O., and Giordano, A. (1994). PITALRE, a nuclear CDC2-related protein kinase that phosphorylates the retinoblastoma protein in vitro. *Proc. Natl. Acad. Sci. USA* **91**, 3834–3838.
65. MacLachlan, T. K., Sang, N., and Giordano, A. (1995). Cyclins, cyclin-dependent kinases and cdk inhibitors: Implications in cell cycle control and cancer. *Crit. Rev. Eukaryot. Gene Expr.* **5**, 127–156.
66. Edwards, M. C., Wong, C., and Elledge, S. J. (1998). Human cyclin K, a novel RNA polymerase II-associated cyclin possessing both carboxy-terminal domain kinase and Cdk-activating kinase activity. *Mol. Cell Biol.* **18**, 4291–4300.
67. Peng, J., Zhu, Y., Milton, J. T., and Price, D. H. (1998). Identification of multiple cyclin subunits of human P-TEFb. *Genes Dev.* **12**, 755–762.
68. Sherr, C. J., and Roberts, J. M. (1999). CDK inhibitors: Positive and negative regulators of G1-phase progression. *Genes Dev.* **13**, 1501–1512.
69. Weinberg, R. A. (1995). The retinoblastoma protein and cell cycle control. *Cell* **81**, 323–330.
70. Senderowicz, A. M., and Sausville, E. A. (2000). Preclinical and clinical development of cyclin-dependent kinase modulators. *J. Natl. Cancer Inst.* **92**, 376–387.
71. Kasten, M. M., and Giordano, A. (1998). pRb and the cdks in apoptosis and the cell cycle. *Cell Death Differ.* **5**, 132–140.
72. Chiarugi, V., Magnelli, L., Cinelli, M., and Basi, G. (1994). Apoptosis and the cell cycle. *Cell Mol. Biol. Res.* **40**, 603–612.
73. Shimizu, T., O'Connor, P., Kohn, K. W., and Pommier, Y. (1995). Unscheduled activation of cyclin B1/Cdc2 kinase in human promyelocytic leukemia cell line HL60 cells undergoing apoptosis induced by DNA damage. *Cancer Res.* **55**, 228–231.
74. Meikrantz, W., and Schlegel, R. (1996). Suppression of apoptosis by dominant negative mutants of cyclin-dependent protein kinases. *J. Biol. Chem.* **271**, 10205–10209.
75. Senderowicz, A. M., Headlee, D., Stinson, S., Lush, R. M., Tompkins, A., Brawley, O., Bergan, R., Figg, W. D., Smith, A., and Sausville, E. A. (1996). Phase I trial of a novel cyclin-dependent kinase inhibitor flavopiridol in patients with refractory neoplasms. *In* "9th National Cancer Institute-European Organization for Research on Treatment of Cancer Symposium Proccedings," p. 77.
76. Senderowicz, A. M., Messmann, R., Arbuck, S., Headlee, D., Zhai, S., Murgo, A., Melillo, G., Figg, W., and Sausville, E. (2000). A Phase I trial of 1 hour infusion of flavopiridol (Fla), a novel cyclin-dependent kinase inhibitor, in patients with advanced neoplasms. *In* "Proc. Annual Meeting of the American Society of Clinical Oncology," New Orleans.
77. Senderowicz, A. M., Headlee, D., Lush, R., Bauer, K., Figg, W., Murgo, A., S., Arbuck, S., Inoue, K., Kobashi, S., Kuwabara, T., and Sausville, E. (1998). Phase I trial of infusional UCN-01, a novel protein kinase inhibitor, in patients with refractory neoplasms. *In* "10th National Cancer Institute-European Organization for Research on Treatment of Cancer Symposium Proccedings," Amsterdam, Holland.
78. Schwartz, G., Kaubisch, A., Saltz, L., Ilson, D., O'Reilly, E., Barazzuol, J., Endres, S., Soltz, M., Tong, W., Spriggs, D., and Kelsen, D. (1999). Phase I trial of sequential paclitaxel and the cyclin-dependent kinase inhibitor flavopiridol. *In* "Proceedings of the American Society of Clinical Oncology," p. 160. Atlanta, GA.
79. Thomas, J., Cleary, J., Tutsch, K., Arzoomanian, R., Alberti, D., Simon, K., Feierabend, D., Morgan, K., and Wilding, G. (1997). Phase I clinical and pharmacokinetic trial of flavopiridol. *In* "Proc. 88th Annual Meeting of the American Association of Cancer Research," San Diego, CA.
80. Dees, E., O'Reilly, S., Figg, W., Elza-Brown, K., Aylesworth, C., Carducci, M., Byrd, J., Grever, M., and Donehower, R. (2000). A Phase I and pharmacologic study of UCN-01, a protein kinase C inhibitor. *In* "Proc. American Society of Clinical Oncology," New Orleans.
81. Tamura, T., Sasaki, Y., Minami, H., Fujii, H., Ito, K., Igarashi, T., Kamiya, Y., Kurata, T., Ohtsu, T., Onozawa, Y., Yamamoto, N., Yamamoto, N., Watanabe, Y., Tanigaara, Y., Fuse, E., Kuwabara, T., Kobayahsi, S., and Shimada, Y. (1999). Phase I study of UCN-01 by 3-hour infusion. *In* "Proceedings of the American Society of Clinical Oncology," p. 159. Atlanta, GA.
82. Sausville, E. A., Arbuck, S. G., Messmann, R., Headlee, D., Bauer, K. S., Lush, R. M., Murgo, A., Figg, W. D., Lahusen, T., Jaken, S., Jing, X., Roberge, M., Fuse, E., Kuwabara, T., and Senderowicz, A. M. (2001). Phase I trial of 72-hour continuous infusion UCN-01 in patients with refractory neoplasms. *J. Clin. Oncol.* **19**, 2319–2333.
83. Meijer, L., and Kim, S. H. (1997). Chemical inhibitors of cyclin-dependent kinases. *Methods Enzymol.* **283**, 113–128.
84. Zaharevitz, D. W., Gussio, R., Leost, M., Senderowicz, A. M., Lahusen, T., Kunick, C., Meijer, L., and Sausville, E. A. (1999). Discovery and initial characterization of the paullones, a novel class of small-molecule inhibitors of cyclin-dependent kinases. *Cancer Res.* **59**, 2566–2569.
85. De Azevedo, W. F., Leclerc, S., Meijer, L., Havlicek, L., Strnad, M., and Kim, S. H. (1997). Inhibition of cyclin-dependent kinases by purine analogues: Crystal structure of human cdk2 complexed with roscovitine. *Eur. J. Biochem.* **243**, 518–526.
86. Sedlacek, H. H., Czech, J., Naik, R., Kaur, G., Worland, P., Losiewicz, M., Parker, B., Carlson, B., Smith, A., Senderowicz, A., and Sausville, E. (1996). Flavopiridol (L86-8275, NSC-649890), a new kinase inhibitor for tumor therapy. *Int. J. Oncol.* **9**, 1143–1168.
87. Kaur, G., Stetler-Stevenson, M., Sebers, S., Worland, P., Sedlacek, H., Myers, C., Czech, J., Naik, R., and Sausville, E. (1992). Growth inhibition with reversible cell cycle arrest of carcinoma cells by flavone L86-8275. *J. Natl. Cancer Inst.* **84**, 1736–1740.
88. Losiewicz, M. D., Carlson, B. A., Kaur, G., Sausville, E. A., and Worland, P. J. (1994). Potent inhibition of CDC2 kinase activity by the flavonoid L86-8275. *Biochem. Biophys. Res. Commun.* **201**, 589–595.
89. Worland, P. J., Kaur, G., Stetler-Stevenson, M., Sebers, S., Sartor, O., and Sausville, E. A. (1993). Alteration of the phosphorylation state of p34cdc2 kinase by the flavone L86-8275 in breast carcinoma cells. Correlation with decreased H1 kinase activity. *Biochem. Pharmacol.* **46**, 1831–1840.
90. Carlson, B. A., Dubay, M. M., Sausville, E. A., Brizuela, L., and Worland, P. J. (1996). Flavopiridol induces G1 arrest with inhibition of cyclin-dependent kinase (CDK) 2 and CDK4 in human breast carcinoma cells, *Cancer Res.* **56**, 2973–2978.
91. Carlson, B., Pearlstein, R., Naik, R., Sedlacek, H., Sausville, E., and Worland, P. (1996). Inhibition of CDK2, CDK4 and CDK7 by flavopiridol and structural analogs. *In* "Proceedings of the American Association for Cancer Research," p. 424.

92. De Azevedo, W. F., Jr., Mueller-Dieckmann, H. J., Schulze-Gahmen, U., Worland, P. J., Sausville, E., and Kim, S. H. (1996). Structural basis for specificity and potency of a flavonoid inhibitor of human CDK2, a cell cycle kinase. *Proc. Natl. Acad. Sci. USA* **93**, 2735–2740.
93. Michalides, R., van Veelen, N., Hart, A., Loftus, B., Wientjens, E., and Balm, A. (1995). Overexpression of cyclin D1 correlates with recurrence in a group of forty-seven operable squamous cell carcinomas of the head and neck. *Cancer Res.* **55**, 975–978.
94. Gansauge, S., Gansauge, F., Ramadani, M., Stobbe, H., Rau, B., Harada, N., and Beger, H. G. (1997). Overexpression of cyclin D1 in human pancreatic carcinoma is associated with poor prognosis. *Cancer Res.* **57**, 1634–1637.
95. Fredersdorf, S., Burns, J., Milne, A. M., Packham, G., Fallis, L., Gillett, C. E., Royds, J. A., Peston, D., Hall, P. A., Hanby, A. M., Barnes, D. M., Shousha, S., O. Hare, M. J., and Lu, X. (1997). High level expression of p27(kip1) and cyclin D1 in some human breast cancer cells: Inverse correlation between the expression of p27(kip1) and degree of malignancy in human breast and colorectal cancers. *Proc. Natl. Acad. Sci. USA* **94**, 6380–6385.
96. Carlson, B., Lahusen, T., Singh, S., Loaiza-Perez, A., Worland, P. J., Pestell, R., Albanese, C., Sausville, E. A., and Senderowicz, A. M. (1999). Downregulation of cyclin D1 by transcriptional repression in MCF-7 human breast carcinoma cells induced by flavopiridol. *Cancer Res.* **59**, 4634–4641.
97. Gray, N. S., Wodicka, L., Thunnissen, A. M., Norman, T. C., Kwon, S., Espinoza, F. H., Morgan, D. O., Barnes, G., LeClerc, S., Meijer, L., Kim, S. H., Lockhart, D. J., and Schultz, P. G. (1998). Exploiting chemical libraries, structure, and genomics in the search for kinase inhibitors. *Science* **281**, 533–538.
98. Melillo, G., Sausville, E. A., Cloud, K., Lahusen, T., Varesio, L., and Senderowicz, A. M. (1999). Flavopiridol, a protein kinase inhibitor, down-regulates hypoxic induction of vascular endothelial growth factor expression in human monocytes. *Cancer Res.* **59**, 5433–5437.
99. Chao, S. H., Fujinaga, K., Marion, J. E., Taube, R., Sausville, E. A., Senderowicz, A. M., Peterlin, B. M., and Price, D. H. (2000). Flavopiridol inhibits P-TEFb and blocks HIV-1 replication. *J. Biol. Chem.* **275**, 28345–28348.
100. Arguello, F., Alexander, M., Sterry, J., Tudor, G., Smith, E., Kalavar, N., Greene, J., Koss, W., Morgan, D., Stinson, S., Siford, T., Alvord, W., Labansky, R., and Sausville, E. (1998). Flavopiridol induces apoptosis of normal lymphoid cells, causes immunosuppresion, and has potent antitumor activity in vivo against human and leukemia xenografts. *Blood* **91**, 2482–2490.
101. Byrd, J. C., Shinn, C., Waselenko, J. K., Fuchs, E. J., Lehman, T. A., Nguyen, P. L., Flinn, I. W., Diehl, L. F., Sausville, E., and Grever, M. R. (1998). Flavopiridol induces apoptosis in chronic lymphocytic leukemia cells via activation of caspase-3 without evidence of bcl-2 modulation or dependence on functional p53. *Blood* **92**, 3804–3816.
102. Parker, B., Kaur, G., Nieves-Neira, W., Taimi, M., Kolhagen, G., Shimizu, T., Pommier, Y., Sausville, E., and Senderowicz, A. M. (1998). Early induction of apoptosis in hematopoietic cell lines after exposure to flavopiridol. *Blood* **91**, 458–465.
103. Konig, A., Schwartz, G. K., Mohammad, R. M., Al-Katib, A., and Gabrilove, J. L. (1997). The novel cyclin-dependent kinase inhibitor flavopiridol downregulates Bcl-2 and induces growth arrest and apoptosis in chronic B-cell leukemia lines. Blood. **90**, 4307–4312.
104. Shapiro, G. I., Koestner, D. A., Matranga, C. B., and Rollins, B. J. (1999). Flavopiridol induces cell cycle arrest and p53-independent apoptosis in non-small cell lung cancer cell lines. *Clin. Cancer Res.* **5**, 2925–2938.
105. Lahusen, J., Loaiza-Perez, A., Sausville, E. A., and Senderowicz, A. M. (2000). Flavopiridol-induced apoptosis is associated with p38 and MEK activation and is prevented by caspase and MAPK inhibitors. *In* "Proc. Twentieth Annual Meeting of the American Association of Cancer Research," San Francisco, CA.
106. Patel, V., Senderowicz, A. M., Pinto, D., Igishi, T., Raffeld, M., Quintanilla-Martinez, L., Ensley, J. F., Sausville, E. A., and Gutkind, J. S. (1998). Flavopiridol, a novel cyclin-dependent kinase inhibitor, suppresses the growth of head and neck squamous cell carcinomas by inducing apoptosis. *J. Clin. Invest.* **102**, 1674–1681.
107. Brusselbach, S., Nettelbeck, D. M., Sedlacek, H. H., and Muller, R. (1998). Cell cycle-independent induction of apoptosis by the anti-tumor drug Flavopiridol in endothelial cells. *Int. J. Cancer.* **77**, 146–152.
108. Kerr, J. S., Wexler, R. S., Mousa, S. A., Robinson, C. S., Wexler, E. J., Mohamed, S., Voss, M. E., Devenny, J. J., Czerniak, P. M., Gudzelak, A., Jr., and Slee, A. M. (1999). Novel small molecule alpha v integrin antagonists: Comparative anti- cancer efficacy with known angiogenesis inhibitors. *Anticancer Res.* **19**, 959–968.
109. Schwartz, G., Farsi, K., Maslak, P., Kelsen, D., and Spriggs, D. (1997). Potentiation of apoptosis by flavopiridol in mitomycin-C-treated gastric and breast cancer cells. *Clin. Cancer Res.* **3**, 1467–1472.
110. Bible, K. C., and Kaufmann, S. H. (1997). Cytotoxic synergy between flavopiridol (NSC 649890, L86-8275) and various antineoplastic agents: The importance of sequence of administration. *Cancer Res.* **57**, 3375–3380.
111. Chien, M., Astumian, M., Liebowitz, D., Rinker-Schaeffer, C., and Stadler, W. (1999). *In* vitro evaluation of flavopiridol, a novel cell cycle inhibitor, in bladder cancer. *Cancer Chemother. Pharmacol.* **44**, 81–87.
112. Drees, M., Dengler, W., Roth, T., Labonte, H., Mayo, J., Malspeis, L., Grever, M., Sausville, E., and Fiebig, H. (1997). Flavopiridol (L86-8275): Selective antitumor activity in vitro and activity in vivo for prostate carcinoma cells. *Clin. Cancer Res.* **32**, 273–279.
113. Senderowicz, A. M., Headlee, D., Stinson, S. F., Lush, R. M., Kalil, N., Villalba, L., Hill, K., Steinberg, S. M., Figg, W. D., Tompkins, A., Arbuck, S. G., and Sausville, E. A. (1998). Phase I trial of continuous infusion flavopiridol, a novel cyclin- dependent kinase inhibitor, in patients with refractory neoplasms. *J. Clin. Oncol.* **16**, 2986–2999.
114. Wright, J., Blatner, G. L., and Cheson, B. D. (1998). Clinical trials referral resource. Clinical trials of flavopiridol. *Oncology (Huntingt).* **12**, 1018, 1023–1024.
115. Werner, J., Kelsen, D., Karpeh, M., Inzeo, D., Barazzuol, J., Sugarman, A., and Schwartz, G. K. (1998). The cyclin-dependent kinase inhibitor flavopiridol is an active and unexpectedly toxic agent in advanced gastric cancer. *In* "Proceedings of the American Society of Clinical Oncology," Los Angeles, CA.
116. Shapiro, G., Patterson, A., Lynch, C., Lucca, J., Anderson, I., Boral, A., Elias, A., Lu, H., Salgia, R., Skarin, A., Panek-Clark, C., McKenna, R., Rabin, M., Vasconcelles, M., Eder, P., Supko, J., Lynch, T., and Rollins, B. (1999). A phase II trial of flavopiridol in patients with stage IV non-small cell lung cancer. *In* "Proceedings of the American Society of Clinical Oncology," Atlanta, GA. 1999.
117. Bennett, S., Mani, S., O'Reilly, S., Wright, J., Schilsky, R., Vokes, E., and Grochow, L. (1999). Phase II trial of flavopiridol in metastatic colorectal cancer: Preliminary results. *In* "Proceedings of the American Society of Clinical Oncology," Atlanta, GA.
118. Stadler, W. M., Vogelzang, N. J., Amato, R., Sosman, J., Taber, D., Liebowitz, D., and Vokes, E. E. (2000). Flavopiridol, a novel cyclin-dependent kinase inhibitor, in metastatic renal cancer: A University of Chicago Phase II Consortium study. *J. Clin. Oncol.* **18**, 371–375.
119. Innocenti, F., Stadler, W., Iyer, L., Vokes, E., and Ratain, M. (2000). Flavopiridol-induced diarrhea is related to the systemic metabolism of flavopiridol to its glucuronide. *In* "Proc. American Society of Clinical Oncology," New Orleans.

120. Tamaoki, T. (1991). Use and specificity of staurosporine, UCN-01, and calphostin C as protein kinase inhibitors. *Methods Enzymol.* **201**, 340–347.
121. Takahashi, I., Kobayashi, E., Asano, K., Yoshida, M., and Nakano, H. (1987). UCN-01, a selective inhibitor of protein kinase C from Streptomyces. *J. Antibiot. (Tokyo).* **40**, 1782–1784.
122. Takahashi, I., Saitoh, Y., Yoshida, M., Sano, H., Nakano, H., Morimoto, M., and Tamaoki, T. (1989). UCN-01 and UCN-02, new selective inhibitors of protein kinase C. II. Purification, physico-chemical properties, structural determination and biological activities. *J. Antibiot. (Tokyo).* **42**, 571–576.
123. Seynaeve, C. M., Kazanietz, M. G., Blumberg, P. M., Sausville, E. A., and Worland, P. J. (1994). Differential inhibition of protein kinase C isozymes by UCN-01, a staurosporine analogue. *Mol. Pharmacol.* **45**, 1207–1214.
124. Seynaeve, C. M., Stetler-Stevenson, M., Sebers, S., Kaur, G., Sausville, E. A., and Worland, P. J. (1993). Cell cycle arrest and growth inhibition by the protein kinase antagonist UCN-01 in human breast carcinoma cells. *Cancer Res.* **53**, 2081–2086.
125. Wang, Q., Worland, P. J., Clark, J. L., Carlson, B. A., and Sausville, E. A. (1995). Apoptosis in 7-hydroxystaurosporine-treated T lymphoblasts correlates with activation of cyclin-dependent kinases 1 and 2. *Cell Growth Differ.* **6**, 927–936.
126. Akinaga, S., Gomi, K., Morimoto, M., Tamaoki, T., and Okabe, M. (1991). Antitumor activity of UCN-01, a selective inhibitor of protein kinase C, in murine and human tumor models. *Cancer Res.* **51**, 4888–4892.
127. Akinaga, S., Nomura, K., Gomi, K., and Okabe, M. (1994). Effect of UCN-01, a selective inhibitor of protein kinase C, on the cell-cycle distribution of human epidermoid carcinoma, A431 cells. *Cancer Chemother Pharmacol.* **33**, 273–280.
128. Akiyama, T., Yoshida, T., Tsujita, T., Shimizu, M., Mizukami, T., Okabe, M., and Akinaga, S. (1997). G1 phase accumulation induced by UCN-01 is associated with dephosphorylation of Rb and CDK2 proteins as well as induction of CDK inhibitor p21/Cip1/WAF1/Sdi1 in p53-mutated human epidermoid carcinoma A431 cells. *Cancer Res.* **57**, 1495–1501.
129. Wang, Q., Fan, S., Eastman, A., Worland, P. J., Sausville, E. A., and O'Connor, P. (1996). UCN-01: A potent abrogator of G2 checkpoint function in cancer cells with disrupted p53. *J. Natl. Cancer Inst.* **88**, 956–965.
130. Yu, L., Orlandi, L., Wang, P., Orr, M., Senderowicz, A. M., Sausville, E. A., Silvestrini, R. A., and O'Connor, P. (1998). UCN-01 abrogates G2 arrest through a cdc2-dependent pathway that involves inactivation of the Wee1Hu kinase. *J. Biol. Chem.* **273**, 33455–33464.
131. Sarkaria, J. N., Busby, E. C., Tibbetts, R. S., Roos, P., Taya, Y., Karnitz, L. M., and Abraham, R. T. (1999). Inhibition of ATM and ATR kinase activities by the radiosensitizing agent, caffeine. *Cancer Res.* **59**, 4375–4382.
132. Graves, P. R., Yu, L., Schwarz, J. K., Gales, J., Sausville, E. A., O'Connor, P. M., and Piwnica-Worms, H. (2000). The Chk1 protein kinase and the Cdc25C regulatory pathways are targets of the anticancer agent UCN-01. *J. Biol. Chem.* **275**, 5600–5605.
133. Busby, E. C., Leistritz, D. F., Abraham, R. T., Karnitz, L. M., and Sarkaria, J. N. (2000). The radiosensitizing agent 7-hydroxystaurosporine (UCN-01) inhibits the DNA damage checkpoint kinase hChk1. *Cancer Res.* **60**, 2108–2112.
134. Shao, R. G., Cao, C. X., Shimizu, T., O'Connor, P. M., Kohn, K. W., and Pommier, Y. (1997). Abrogation of an S-phase checkpoint and potentiation of camptothecin cytotoxicity by 7-hydroxystaurosporine (UCN-01) in human cancer cell lines, possibly influenced by p53 function. *Cancer Res.* **57**, 4029–4035.
135. Bunch, R. T., and Eastman, A. (1997). 7-Hydroxystaurosporine (UCN-01) causes redistribution of proliferating cell nuclear antigen and abrogates cisplatin-induced S-phase arrest in Chinese hamster ovary cells. *Cell Growth Differ.* **8**, 779–788.
136. Akiyama, T., Shimizu, M., Okabe, M., Tamaoki, T., and Akinaga, S. (1999). Differential effects of UCN-01, staurosporine and CGP 41 251 on cell cycle progression and CDC2/cyclin B1 regulation in A431 cells synchronized at M phase by nocodazole. *Anticancer Drugs* **10**, 67–78.
137. Kawakami, K., Futami, H., Takahara, J., and Yamaguchi, K. (1996). UCN-01, 7-hydroxyl-staurosporine, inhibits kinase activity of cyclin-dependent kinases and reduces the phosphorylation of the retinoblastoma susceptibility gene product in A549 human lung cancer cell line. *Biochem. Biophys. Res. Commun.* **219**, 778–783.
138. Shimizu, E., Zhao, M. R., Nakanishi, H., Yamamoto, A., Yoshida, S., Takada, M., Ogura, T., and Sone, S. (1996). Differing effects of staurosporine and UCN-01 on RB protein phosphorylation and expression of lung cancer cell lines. *Oncology* **53**, 494–504.
139. Chen, X., Lowe, M., and Keyomarsi, K. (1999). UCN-01-mediated G1 arrest in normal but not tumor breast cells is pRb-dependent and p53-independent. *Oncogene* **18**, 5691–5702.
140. Usuda, J., Saijo, N., Fukuoka, K., Fukumoto, H., Kuh, H. J., Nakamura, T., Koh, Y., Suzuki, T., Koizumi, F., Tamura, T., Kato, H., and Nishio, K. (2000). Molecular determinants of UCN-01-induced growth inhibition in human lung cancer cells. *Int. J. Cancer.* **85**, 275–280.
141. Koh, J. Y., Cho, N. P., Kong, G., Lee, J. D., and Yoon, K. (1998). p53 mutations and human papillomavirus DNA in oral squamous cell carcinoma: Correlation with apoptosis. *Br. J. Cancer.* **78**, 354–359.
142. Koch, W. M., Brennan, J. A., Zahurak, M., Goodman, S. N., Westra, W. H., Schwab, D., Yoo, G. H., Lee, D. J., Forastiere, A. A., and Sidransky, D. (1996). p53 mutation and locoregional treatment failure in head and neck squamous cell carcinoma. *J. Natl. Cancer Inst.* **88**, 1580–1586.
143. Somers, K. D., Merrick, M. A., Lopez, M. E., Incognito, L. S., Schechter, G. L., and Casey, G. (1992). Frequent p53 mutations in head and neck cancer. *Cancer Res.* **52**, 5997–6000.
144. Marchetti, A., Buttitta, F., Merlo, G., Diella, F., Pellegrini, S., Pepe, S., Macchiarini, P., Chella, A., Angeletti, C. A., Callahan, R., *et al.* (1993). p53 alterations in non-small cell lung cancers correlate with metastatic involvement of hilar and mediastinal lymph nodes. *Cancer Res.* **53**, 2846–2851.
145. Lowe, S. W., Bodis, S., Bardeesy, N., McClatchey, A., Remington, L., Ruley, H. E., Fisher, D. E., Jacks, T., Pelletier, J., and Housman, D. E. (1994). Apoptosis and the prognostic significance of p53 mutation. *Cold Spring Harb. Symp. Quant. Biol.* **59**, 419–426.
146. Akinaga, S., Nomura, K., Gomi, K., and Okabe, M. (1993). Enhancement of antitumor activity of mitomycin C in vitro and in vivo by UCN-01, a selective inhibitor of protein kinase C. *Cancer Chemother Pharmacol.* **32**, 183–189.
147. Bunch, R. T., and Eastman, A. (1996). Enhancement of cisplatin-induced cytotoxicity by 7-hydroxystaurosporine (UCN-01), a new G2-checkpoint inhibitor. *Clin Cancer Res.* **2**, 791–797.
148. Hsueh, C. T., Kelsen, D., and Schwartz, G. K. (1998). UCN-01 suppresses thymidylate synthase gene expression and enhances 5-fluorouracil-induced apoptosis in a sequence-dependent manner. *Clin. Cancer Res.* **4**, 2201–2206.
149. Husain, A., Yan, X. J., Rosales, N., Aghajanian, C., Schwartz, G. K., and Spriggs, D. R. (1997). UCN-01 in ovary cancer cells: Effective as a single agent and in combination with cis-diamminedichloroplatinum(II)independent of p53 status. *Clin. Cancer Res.* **3**, 2089–2097.
150. Pollack, I. F., Kawecki, S., and Lazo, J. S. (1996). Blocking of glioma proliferation in vitro and in vivo and potentiating the effects of BCNU

and cisplatin: UCN-01, a selective protein kinase C inhibitor. *J. Neurosurg.* **84**, 1024–1032.

151. Tsuchida, E., and Urano, M. (1997). The effect of UCN-01 (7-hydroxystaurosporine), a potent inhibitor of protein kinase C, on fractionated radiotherapy or daily chemotherapy of a murine fibrosarcoma. *Int. J. Radiat. Oncol. Biol. Phys.* **39**, 1153–1161.
152. Sugiyama, K., Shimizu, M., Akiyama, T., Tamaoki, T., Yamaguchi, K., Takahashi, R., Eastman, A., and Akinaga, S. (2000). UCN-01 selectively enhances mitomycin C cytotoxicity in p53 defective cells which is mediated through S and/or G(2) checkpoint abrogation. *Int. J. Cancer* **85**, 703–709.
153. Jones, C. B., Clements, M. K., Wasi, S., and Daoud, S. S. (2000). Enhancement of camptothecin-induced cytotoxicity with UCN-01 in breast cancer cells: Abrogation of S/G(2) arrest. *Cancer Chemother Pharmacol.* **45**, 252–258.
154. Sausville, E. A., Lush, R. D., Headlee, D., Smith, A. C., Figg, W. D., Arbuck, S. G., Senderowicz, A. M., Fuse, E., Tanii, H., Kuwabara, T., and Kobayashi, S. (1998). Clinical pharmacology of UCN-01: Initial observations and comparison to preclinical models. *Cancer Chemother Pharmacol.* **42** (Suppl.), S54–S59.
155. Fuse, E., Tanii, H., Kurata, N., Kobayashi, H., Shimada, Y., Tamura, T., Sasaki, Y., Tanigawara, Y., Lush, R. D., Headlee, D., Figg, W., Arbuck, S. G., Senderowicz, A. M., Sausville, E. A., Akinaga, S., Kuwabara, T., and Kobayashi, S. (1998). Unpredicted clinical pharmacology of UCN-01 caused by specific binding to human alpha1-acid glycoprotein. *Cancer Res.* **58**, 3248–3253.
156. Senderowicz, A. M., Headlee, D., Lush, R., Bauer, K., Figg, W., Murgo, A., S., Arbuck, S., Inoue, K., Kobashi, S., Kuwabara, T., and Sausville, E. (1999). Phase I trial of infusional UCN-01, a novel protein kinase inhibitor, in patients with refractory neoplasms. *In* "Proc. 35th Annual Meeting of the American Society of Clinical Oncology," Atlanta, GA.
157. Wilson, W. H., Sorbara, L., Figg, W. D., Mont, E. K., Sausville, E., Warren, K. E., Balis, F. M., Bauer, K., Raffeld, M., Senderowicz, A. M., and Monks, A. (2000). Modulation of clinical drug resistance in a B cell lymphoma patient by the protein kinase inhibitor 7-hydroxy-staurosporine: Presentation of a novel therapeutic paradigm. *Clin. Cancer Res.* **6**, 415–421.
158. Fowler, L., Dong, L., Bowes, R. C., 3rd, van de Water, B., Stevens, J. L., and Jaken, S. (1998). Transformation-sensitive changes in expression, localization, and phosphorylation of adducins in renal proximal tubule epithelial cells. *Cell Growth Differ.* **9**, 177–184.
159. Brown, J. M. (1993). SR 4233 (tirapazamine): A new anticancer drug exploiting hypoxia in solid tumours. *Br. J. Cancer* **67**, 1163–1170.
160. Brown, J. M., and Lemmon, M. J. (1990). Potentiation by the hypoxic cytotoxin SR 4233 of cell killing produced by fractionated irradiation of mouse tumors. *Cancer Res.* **50**, 7745–7749.
161. Dorie, M. J., and Brown, J. M. (1993). Tumor-specific, schedule-dependent interaction between tirapazamine (SR 4233) and cisplatin. *Cancer Res.* **53**, 4633–4636.
162. Rischin, D., Peters, L., Hicks, R., Hughes, P., Fisher, R., Hart, R., Sexton, M., D'Costa, I., and R., v. R. (2001). Phase I trial of concurrent tirapazamine, cisplatin and radiotherapy in patients with advanced head and neck cancer. *J. Clin. Oncol.* **19**, 535–542.
163. Smith, B., Smith, G. L., Carter, D., Sasaki, C. T., and BG, H. (2000). Prognostic significance of vascular endothelial growth factor protein levels in oral and oropharyngeal squamous cell carcinoma. *J. Clin. Oncol.* **18**, 2046–2052.
164. Tae, K., El-Naggar, A. K., Yoo, E., Fend, L., Lee, J. J., Hong, W. K., Hittelman, W. N., and DM., S. (2000). Expression of vascular endothelial growth factor and microvessel density in head and neck tumorigenesis. *Clin. Cancer Res.* **6**, 2821–2828.
165. Macluskey, M., Chandrachud, L. M., Pazouki, S., Green, M., Chisholm, D. M., Ogden, G. R., Schor, L. S., and AM., S. (2000). Apoptosis, proliferation and angiogenesis in oral tissues. Possible relevance to tumour progression. *J. Pathol.* **191**, 368–375.
166. Ueda, N., Kamata, N., Hayashi, E., Yokoyama, K., Hoteiya, T., and M., N. (1999). Effects of an anti-angiogenic agent, TNP-470, on the growth of oral squamous cell carcinomas. *Oral Oncol.* **35**, 554–560.
167. Nakashima, T., Hudson, J. M., and GL, G. (2000). Antisense inhibition of endothelial growth factor in human head and neck squamous cell carcinoma. *Head Neck* **22**, 483–488.
168. Mauceri, H., Hanna, N. N., Beckett, M. A., Gorski, D. H., Staba, M. J., Stellato, K. A., Bigelow, K., Heimann, R., Gately, S., Dhanabal, M., Soff, G. A., Sukhatme, V. P., Kufe, D. W., and RR., W. (1998). Combined effects of angiostatin and ionizing radiation in antitumor therapy. *Nature* (*Lond.*) **394**, 287–291.
169. Tseng, J. E., BS, G., Khuri, F. R., Teddy, S. R., Shin, D. M., Gillenwater, A. M., Myers, J. N., Clayman, G. L., El-Naggar, A. K., Fritsche, H. A., Lawhorn, K. N., Thall, P. F., Liu, D., and RS, H. (2000). Phase II trial of thalidomide in the treatment of recurrent and/or metastatic squamous carcinoma of the head and neck. *Oncology* (*Huntingdon*) **23**, 4.
170. Eder, J. P., Clark, J. W., Supko, J. G., Shulman, L. N., Garcia-Carbonaro, R., Roper, K., Proper, J., Keogan, M., Kinchla, N., Schnipper, L. E., Conners, S., Butterfield, C., Fogler, W., Xu, G., Puchalski, T., Park, S., Janicek, M. J., Gubish, E., Saker, S., Folkman, M. J., and, D. W., K. (2001). A phase I pharmacokinetic and pharmacodynamic trial of recombinant human endostatin. *In* "Proc. Am. Soc. Clin. Oncol.," p. 70a.
171. Rosen, P., Kabbinavar, F., Figlin, R. A., Parson, M., Laxa, B., Hernandez, L., Mayers, A., Cropp, G. F., Hannath, A. L., and LS., R. (2001). A phase I/II trial and pharmacokinetic study of SU 5416 in combination with paclitaxel/carboplatin. *In* "Proc. Am. Soc. Clin. Oncol.," p. 98a.
172. Hsei, V., Novotny, W. F., Margolin, K., Gordon, M., Small, E. J., Griffing, S., Seguenza, P., and, J., G. (2001). Population pharmacokinetic analysis of bevacizumab in cancer subjects. *In* "Proc. Am. Soc. Clin. Oncol.," p. 69a.
173. Nemunaitis, J., Khuri, F., Ganly, I., Arseneau, J., Posner, M., Vokes, E., Kuhn, J., McCarty, T., Landers, S., Blackburn, A., Romel, L., Randlev, B., Kaye, S., and D., K. (2001). Phase II trial of intratumoral administration of ONYX-015, a replication-selective adenovirus, in patients with refractory head and neck cancer. *J. Clin. Oncol.* **19**, 289–298.
174. Lamont, J., Nemunaitis, J., Kuhn, J. A., Landers, S. A., and TM, M. (2000). A prospective phase II trial of ONYX-015 adenovirus and chemotherapy in recurrent squamous cell carcinoma of the head and neck (the Baylor experience). *Ann. Surg. Oncol.* **7**, 588–592.
175. Nemunaitis, J., Bier-Laning, C. M., Costenla-Figueiras, M., Yver, A., and LK., D. (1999). Three phase II trials of intratumoral injection with a replication-deficient adenovirus carrying the p53 gene (AD5CMV-P53) in patients with recurrent/refractory head and neck cancer. *In* "Proc. Am. Soc. Clin. Oncol.," p. 431a.
176. Day, K., Li, D., Liu, S., Guo, M., and O'Malley, B. W., J. (2001). Granulocyte-macrophage colony stimulating factor in a combination gene therapy strategy for head and neck cancer. *Laryngoscope* **111**, 801–806.
177. Bradley, J., Kataoka, Y., Advani, S., Chung, S. M., Arani, R. B., Gillespie, G. Y., Whitley, R. J., Markert, J. M., Roizman, B., and RR, W. (1999). Ionizing radiation improves survival in mice bearing intracranial high-grade glioma injected with genetically modified herpes simplex virus. *Clin. Cancer Res.* **5**, 517–522.
178. Soiffer, R. J., Robertson, M. J., Murray, C., Cochran, K., and Ritz, J. (1993). Interleukin-12 augments cytolytic activity of peripheral blood lymphocytes from patients with hematologic and solid malignancies. *Blood* **82**, 2790–2796.

179. Pappo, I., Tahara, H., Robbins P.D., Gately, M. K., Wolf, S. F., Barnea, A., and Lotze, M. T. (1995). Administration of systemic or local interleukin-2 enhances the anti-tumor effects of interleukin-12 gene therapy. *J. Surg. Res.* **58**, 218–226.

180. Wiggington, J. M., Komschlies, K. L., Back, T. C., Franco, J. L., Brunda, M. J., and Wiltrout, R. H. (1996). Administration of Interleukin 12 with pulse interleukin 2 and the rapid and complete eradication of murine renal carcinoma. *J. Natl. Cancer Inst.* **88**, 38–43.

181. Rashleigh, S. P., Kusher, D. I., Endicott, J. N., Rossi, A. R., and Djeu, J. Y. (1996). Interleukins 2 and 12 activate natural killer cytolytic responses of peripheral blood mononuclear cells from patients with head and neck squamous cell carcinoma *Arch. Otolaryngol. Head Neck Surg.* **122**, 541–547.

182. Brunda, M. J., Luistro L., Hendrzk, J. A., Fountoulakis, M., Garotta, G., and Gately, M. K. (1995). Role of interferon-gamma in mediating the anti-tumor efficacy of interleukin-12. *J. Immunother. Emphasis Tumor Immunol.* **17**, 71–77.

183. Thomas, G. R., Chen, Z., Oechsli, M. N., Hendler, F. J., and Van Waes, C. (1999). Decreased expression of CD80 is a marker for increased tumorigenicity in a new murine model of oral squamous-cell carcinoma. *Int. J. Cancer.* **82**, 377–384.

184. Lenschow, D. J., Walunas, T. L., and Bluestone, J. A. (1996). CD28/B7 system of T cell co-stimulation. *Annu. Rev. Immunol.* **14**, 233–258.

185. Schwartz, R. H. (1996). Models of T cell anergy: Is there a common molecular mechanism?. *J. Exp. Med.* **184**, 1–8.

186. Chen, P., Geer, D. C., Podack, E. R., and Ksander, B. (1996). Tumor cells transfected with B7-1 and interleukin-12 cDNA induce protective immunity. *Ann. N. Y. Acad. Sci.* **795**, 325–327.

187. Zitvogel, L., Robbins, P. D., Storkus, W. J., Clarke, M. R., Maeurer, M. J., Campbell, R. L., Davis, C. G., Tahara, H., Schreiber, R. D., and Lotze, M. T. (1996). Interleukin-12 and B7-1 co-stimulation cooperate in the induction of effective antitumor immunity and therapy of established tumors. *Eur. J. Immunol.* **26**, 1335–1341.

188. Thomas, G. R., Chen, Z., Enamorado, I., Bancroft, C., and Van Waes, C. (2000). IL-12- and IL-2-induced tumor regression in a new murine model of oral squamous-cell carcinoma is promoted by expression of the CD80 co-stimulatory molecule and interferon-gamma. *Int. J. Cancer.* **86**, 368–374.

189. Gaken, J. A., Hollingsworth, S. J., Hirst, W. J. R., Buggins, A. G. S., Galea-Lauri, J., Peakman, M., Kuiper, M., Patel, P., Towner, P., Patel, P. M., Collins, M. K. L., Mufti, G. J., Farzaneh, F., and Darling, D. C. (1997). Irradiated NC adenocarcinoma cells transduced with both B7-1 and interleukin-2 induce CD4+ mediated rejection of established tumors. *Hum. Gene Ther.* **8**, 477–488.

190. Clerici, M., Sarin, A., Coffman, R. L., Wynn, T. A., Blatt, S. P., Hendrix, C. W., Wolf, S. F., Shearer, G. M., and Henkart, P. A. (1994). Type 1/ Type 2 cytokine modulation of T-cell programmed cell death as a model for human immunodeficiency virus pathogenesis. *Proc. Natl. Acad. Sci.* USA **91**, 11811.

191. Burke, F., East, N., Upton, C., Patel, K., and Blakwill, F. R. (1997). Interferon-gamma induces cell cycle arrest and apoptosis in a model of ovarian cancer: Enhancement of effect by batimastat. *Eur. J. Cancer.* **33**, 1114–1121.

192. Ossina, N. K., Cannas, A., Powers, V. C., Fitzpatrick, P. A., Knight, J. D., Gilbert, J. R., Shekhtman, E. M., Tomei, L. D., Umanski, S. R., and Kiefer, M. C. (1997). Interferon-gamma modulates a p53- independent apoptotic pathway and apoptosis-related gene expression. *J. Biol. Chem.* **272**, 16351–16357.

193. Cortesina, G., Destefani, A., Giovarelli, M., Barioglio, M. G., Cavallo, G. P., Jemma, C., and Forni. (1988). Treatment of recurrent squamous cell carcinoma of the head and neck with low dose interleukin-2 injected perilymphatically. *Cancer* **62**, 2482–2485.

194. Whiteside, T. L., Letessier, E., Hirabayashi, H., Vitolo, D., Bryant, J., Barnes, L., Snyderman, C., Johnson, J. T., Myers, E., Herberman, R. B., Rubin, J., Kirkwood, J. M., and Vlock, D. R. (1993). Evidence for local and systemic activation of immune cells by peritumoral injections of Interleukin 2 in patients with advanced squamous cell carcinoma of the head and neck. *Cancer Res.* **53**, 5654–5662.

195. Lotze, M. T., Zitvogel, L., Campbell, R., Robbins, P. D., Elder, E., Haluszczak, C., Martin, D., Whiteside, T. L., Storkus, W. J., and Tahara, H. (1996). Cytokine gene therapy of cancer using interleukin-12: Murine and clinical trials. *Ann. N. Y. Acad. Sci.* **795**, 401–454.

196. Dillman, R. O. (1994). The clinical experience with interleukin 2 in cancer therapy. *Cancer Biother Radiopharm.* **9**, 183–209.

197. Rosenberg, S. A., Yang, J. C., Topalian, S. L., Schwartzentruber, D. J., Weber, J. S., Parkinson, D. R., Seipp, C. A., Einhorn, J. H., and White, D. E. (1994). Treatment of 283 consecutive patients with metastatic melanoma or renal cell cancer using high-bolus interleukin-2. *JAMA.* **271**, 907–913.

C H A P T E R

37

Gene Therapy for Patients with Head and Neck Cancer

GEORGE H. YOO

Department of Otolaryngology—Head and Neck Surgery
Wayne State University
Detroit, Michigan 48201

GARY CLAYMAN

Department of Surgery
The University of Texas M. D. Anderson Cancer Center
Houston, Texas 77030

I. Background 556
- A. Genetics of Cancer 556
- B. Tumor Immunology 556

II. Approaches to Gene Therapy for Cancer 557
- A. Corrective Gene Therapy 557
- B. Cytotoxic Therapy 557
- C. Immunotherapy 557

III. Genes Used in Combination Therapy 557

IV. Vectors 558
- A. Viral Vectors 558
- B. Nonviral Vectors 558

V. Gene Transfer Delivery Sites 559
- A. Local-Based Gene Delivery 559
- B. Regional Gene Transfer 560
- C. Systemic Delivery 560

VI. Gene Therapy Trials in Head and Neck Squamous Cell Carcinoma 560
- A. Gene Therapy Using the p53 Gene 560
- B. Allovectin (HLA-B7/β_2-Microglobulin and DMRIE/DOPE) Gene Therapy 563
- C. Gene Therapy Using the E1A Gene 563
- D. IL-12 Gene Therapy 564
- E. IL-2 Gene Therapy 564
- F. Herpes Simplex Virus–Thymidine Kinase 564
- G. Antisense EGFR/Liposome 564
- H. IFN-α Gene Transfer Using PVP 564
- I. GM-CSF-Based Gene Therapy 565

VII. Future of Gene Therapy for Cancer 565
- A. Combination with Standard Therapy 565
- B. New Vector Strategies 565
- C. Targeting Vectors 566
- D. Summary of Gene Therapy 566

References 566

Currently, 239 gene therapy trials are approved by the National Institutes of Health/Recombinant DNA Advisory Committee (NIH/RAC) for cancer. In head and neck cancer, 23 trials using nine different genes are being tested. All of these trials are using intratumoral injections in recurrent head and neck squamous cell carcinoma (HNSCC). Genes that have been approved for testing in patients with HNSCC are p53, HLA-B7, E1A, interleukin-2, interleukin-12, interferon-α, granulocyte–macrophage colony-stimulating factor, antisense epidermal growth factor receptor, and thymadine kinase. The four approaches to gene therapy for cancer are (1) corrective gene therapy, (2) cytotoxic therapy, (3) immunotherapy, and (4) combination adjuvant therapy. Gene therapy approaches have been tested *in vitro* and *in vivo* using animal models and have demonstrated tumor suppression or killing. The ideal vector would have a high efficiency (100% of cells get transfected), a high specificity (only tumor cells receive the gene), and a low toxicity. No known vector meets all of these criteria. Adenoviruses and retroviruses are commonly used viruses. These viruses are attenuated to transfect genes, but they cannot replicate or cause an infection. *Nonviral vectors*, such as liposomes, proteins, and liposome–protein combinations, are easy to manufacture and use. Initial trials have demonstrated response rates between 5 and 20%. In a larger proportion of patients, the added benefit appears to be a cytostatic or stabilization of disease. In order to enhance gene therapy, combination with standard therapy is being tested. Improving vector and tumor cell targeting is the current area of intense research.

In 1999 an estimated 40,400 new patients developed head and neck cancer. Head and neck squamous cell carcinoma (HNSCC) is the eighth most common malignancy overall and the sixth most common cancer in men [1]. One-third of these patients will present with early disease and have an excellent 5-year survival rate (70–90%). In advanced head and neck HNSCC, the 5-year survival rate is less than 40%. Current treatment options (surgery, radiotherapy, and/or chemotherapy) are toxic as well as functionally and cosmetically debilitating. Research outside the standard treatment modalities is being studied. One biological approach, gene therapy, demonstrates tumor suppression in preclinical studies and is being tested in humans. Before the details of gene therapy are presented, background information on molecular genetics and tumor immunology will be reviewed.

I. BACKGROUND

In March 1989, the first gene therapy trial in humans was approved by RAC/NIH. Later in September of 1990, approval for the first trial in cancer therapy was obtained. Nine years later gene therapy had a setback with the death of a relative healthy individual at the University of Pennsylvania in September 1999. Then, in April and June 2000, two reported gene therapy trials showed successful long-term gene expression and clinical benefit [2,3]. In two patients with severe combined immunodeficiency-X1 (SCID-X1) who received stem cell gene therapy, gene expression of inherited genetic defect was maintained. These two patients had normal T-cell counts and immune status and are healthy and growing normally [2]. In another trial, patients with testicular cancer were treated with high-dose chemotherapy and two stem cell transplants. The stem cells in the second transplant were transfected with MDR1 (multiple drug resistant) gene in order to protect stem cells from the toxicity of high-dose chemotherapy. Long-term gene expression in stem and progenitor cells, along with the selection of MDR1 expressing cells with the use of high-dose chemotherapy, was shown [3]. As of September 2000, 409 trials have been approved or are under pending approval for gene therapy in humans.

A. Genetics of Cancer

Cancer develops through the accumulation of genetic alterations [4] and gains a growth advantage over normal surrounding cells. The genetic transformation of normal cells to neoplastic cells occurs through a series of progressive steps. Genetic progression models have been studied in HNSCC [5]. Both large chromosomal alterations and single nucleotide changes are found in the development of cancers. Genetic alterations include loss, gain, or translocation of chromosomal segments, activation of protooncogenes, or inactivation of tumor suppressor genes. In general, genes can promote tumor growth (oncogenes) or genes can suppress growth (tumor suppressor genes). Protooncogenes participate in normal cellular signaling, transduction, and transcription. Tumor suppressor genes are normally present in cells and are able to suppress cellular growth, along with many other regulatory functions. Cancer cells either knock out tumor suppressor genes or activate oncogenes so that they can proliferate without being checked. Gene therapy reengineers these genetic alterations so that cancer cell growth can be suppressed. After a gene is transfected into a cell, mRNA is transcribed and then its protein product is translated.

Two tumor suppressor genes, p53 and p16, have been implicated in the development of HNSCC [6,7]. The p53 gene is mutated in approximately 33–45% of HNSCC [6–8], and the p16 gene is altered in 80% of cases by homozygous deletion, promoter methylation, or point mutation [9]. We found that both Rb and BRCA2 genes are rarely inactivated in HNSCC [10,11]. The p53 protein acts by inhibiting the cell cycle, promoting apoptosis (programmed cell death) and regulating transcription [12]. p53 upregulates a cell cycle inhibitor, p21, which further acts on cyclin-dependent kinases to cause cell cycle arrest. The p53 gene also causes apoptosis in cells that have undergone severe DNA damage as a result of radiation or chemotherapy exposure and then inhibits DNA repair and synthesis. Furthermore, p53 regulates transcription by stimulating transactivator proteins [12]. The suppressive effects of p53 genes lead one to believe that clinical behavior can be correlated to alterations in the gene.

A handful of protooncogenes have been implicated in the development of HNSCC, e.g., Her2/neu and cyclin D1. Her2/neu, a transmembrane tyrosine kinase receptor that functions to promote cellular proliferation, is overexpressed in approximately 40–50% of HNSCC [13,14]. Cyclin D1 (a promoter of cyclin-dependent kinase 4/6 and cell cycle) is overexpressed in 12 to 54% of HNSCC [15], which is due to the amplification of chromosomal area 11q13 [16,17].

B. Tumor Immunology

The basis of tumor immunology is that the immune system can kill cancer cells. Cancer cells avoid this immune surveillance by suppressing the body's immune system. A "hierarchy of immunosuppression" exists in patients with HNSCC [18]. Immune reactivity is maximally suppressed in tumor-infiltrating lymphocytes (TIL), followed by lymph node lymphocytes (LNL) and peripheral blood lymphocytes (PBL) [18]. Immune cells, such as cytotoxic T lymphocytes (CTL) and natural killer (NK) cells, attack cancer cells. Cytokines, such as interleukin (IL) and interferons (INF), activate the immune system. Tumor-specific antigens are expressed on tumor cells and help the host recognize and mount a specific immune response against these cells.

The induction of a T-cell immune response by antigen-presenting cells (APCs) occurs in three distinct stages. Initially, a nonspecific adhesion occurs between an APC and T cell, followed by an antigen major histocompatiability complex (MHC) of the APC cross-links with the T-cell receptor (TcR). The final step occurs when a second or costimulatory signal is delivered by the APC to the T cell, enhancing stimulation.

Presently, the best-characterized second signal occurs when the B7.1 or B7.2 ligand of the APC binds to the CD28 receptor on the T cell, resulting in enhanced cellular activation [19]. The goal of genetic immunotherapy is to enhance the immune response against cancer and overcome the immune suppression so that the cancer cells are killed.

II. APPROACHES TO GENE THERAPY FOR CANCER

The four approaches to gene therapy are (1) corrective gene therapy, (2) cytotoxic therapy, (3) immunotherapy, and (4) combination adjuvant therapy (Table 37.1).

A. Corrective Gene Therapy

Gene therapy can be used to correct genetic alterations in tumor cells. The simplest approach is to replace the wild-type tumor suppressor gene. This results in suppression of tumor growth and cell death when these cells produce the gene product. Alternatively, an oncogene can be inhibited by either transfecting the antisense cDNA so it binds to the mRNA of the oncogene or a gene can be added that regulates and inhibits the transcription of an oncogene.

B. Cytotoxic therapy

Gene therapy can be used to augment cytotoxic therapy by either a drug sensitization or a resistance approach. In the drug sensitization approach, a gene is transfected to convert a prodrug into its active metabolite. This allows for drug conversion and a high level of active drug only in the tumor bed. One example is the herpes simplex virus thymadine kinase (TK) gene, which converts gangciclovir into its cytotoxic triphosphate. Another way to augment cytotoxic effects of chemotherapy is to use a drug resistance approach. A drug-resistant gene, such as MDR1, is added into cells that are sensitive to chemotherapy, such as hematopoetic stem cells, so that they can resist the toxicity of chemotherapy. Therefore, higher doses of chemotherapy can be used, as the most sensitive cells are now resistant to these levels of chemotherapy.

TABLE 37.1 Classification of Gene Therapy Approaches

I. Corrective gene therapy
   A. Replace tumor suppressor gene
   B. Inhibit an oncogene
      1. Antisense cDNA
      2. Gene that regulates oncogene

II. Cytotoxic therapy
   A. Drug sensitization
   B. Drug resistance

III. Immunotherapy
   A. Cytokine gene transfer
      1. *In vivo*
      2. *Ex vivo*
   B. Vaccination
      1. Tumor-specific antigen
      2. Alloantigen
      3. Foreign antigen
   C. Costimulatory gene

IV. Combination/Adjuvant therapy
   A. Adjuvant with chemo therapy
   B. Adjuvant with radiation theraphy
   C. Adjuvant with surgery

C. Immunotherapy

The goal of tumor immunotherpay is to stimulate the immune system and overcome immunosuppression in order to kill cancer cells. Immunotherapy can be augmented by cytokine gene transfer. Cytokine gene transfer is performed *in vivo* where tumor cells or immune cells, such as TILs and CTLs, are transfected in the body, which upregulates the immune and antitumor response. *Ex vivo* cytokine gene transfer is performed after fibroblasts, immune cells, such as TILs, CTLs, or APCs, or irradiated cancer cells are removed from the body, and then these cells are placed back in the body in order to obtain high levels of a cytokine with a resulting immunological effect. Irradiated tumor cells are used not only to produce high levels of cytokine, but also to provide tumor antigens for immune cells.

Immuno-gene therapy can also be used to vaccinate the body against tumors. A tumor-specific antigen gene is injected into a cancer cell, which helps the body recognize the tumor cell and reject it. The problem with HNSCC is that there are no reliably known tumor-specific antigens. Another vaccination approach is to add a gene that can produce an alloantigen. A third approach is simply adding a gene that produces a foreign antigen. If a gene for a foreign antigen is introduced into a tumor cell, the body forms an immune response against the tumor cell. Alloantigens and foreign antigens also act as costimulatory molecules in the tumor cell so that the immune system recognizes tumor-specific antigens.

III. GENES USED IN COMBINATION THERAPY

Gene therapy is based on the fact that a gene can be transfered into a cell and transcribe mRNA and then translate into a protein for a therapeutic purpose. A gene is placed a vector DNA, such as plasmids or viruses. The expression of this gene is under control of a promoter. Many genes can be used for gene therapy; however, only genes that are currently being tested in HNSCC will be explained in detail. IL-2, IL-12, INF-α, and granulocyte–macrophage colony-stimulating factor (GM-CSF) are cytokines that enhance the immune response against tumors. The HLA-B7 gene is alloantigen injected into tumors that helps the immune system recognize antigens on the tumor cells and therefore kill the tumor cells. The herpes simplex virus–TK gene converts the antiviral agent, gangcyclovir, into its toxic triphosphate metabolite. After TK is transfected into tumor cells and gangcylclovir is given to a patient, the activated drug kills not only the tumor cells, but also allows for a killing of surrounding tumor cells, the "bystander effect," because of high levels of the activated drug that are produced locally. Two genes, p53 and E1A, have been used in corrective therapy in HNSCC. The p53 gene has been transfected into tumor cells and has been shown to suppress growth. The E1A adenovirus gene functions to inhibit tumor growth by several pathways, including downregulation of an oncogene, HER-2/neu, along with reversion to an epithelial phenotype, loss of anchorage independent growth, and decreased tumorgenicity in nude mice.

IV. VECTORS

In order for gene therapy to work, the gene must access and bind to the target cell, be transported into the cytoplasm and then in the nucleus, and subsequently produce its mRNA and protein product. A *vector* is used so that a gene can be transfected into a cell and the gene can produce its protein product. The ideal vector would have a high efficiency (100% of cells get transfected), a high specificity (only tumor cells receive the gene), and a low toxicity. No known vector meets all of these criteria. Vectors are classified as viral and nonviral (Table 37.2).

A. Viral Vectors

Although many *viral vectors* are currently being studied, adenoviruses and retroviruses are used most commonly. These viruses are attenuated to transfect genes, but they cannot replicate or cause an infection. Eliminating their ability to replicate through genetic manipulation of the wild-type virus eliminates the pathogenicity of virus. Most viruses are replication deficient and need a packaging cell line to produce the virus. Adenovirus-associated virus (AAV), lentivirus, herpes simplex virus, and many others are currently being studied extensively in the preclincal setting.

1. Adenovirus

The adenovirus (Ad) is the most commonly used virus in gene therapy. Ad is a double-stranded DNA virus that causes upper respiratory tract infections. Subgroup C, usually C2 or C5, is the most common adenovirus used. A replication-deficient Ad after genetic modification, such as E1 deletion, is used in order to prevent pathological infection in

TABLE 37.2 Vectors Used in Gene Therapy

Viral vectors	Nonviral vectors
Adenovirus	Lipid complex Liposomes
Retroviruses	
Adenovirus-associated virus	Peptide/protein Polymers
Lentivirus	
Herpes simplex virus	Mechanical Electroporation Gene gun
Others	

the host. Replication-deficient Ad is grown in 293 human embryonic kidney cells, which have the missing Ad genes needed to replicate. Although adenoviruses infect almost all cell types, including quiescent or actively dividing cells, adenoviruses have tropism for keratincytes of the upper aerodigestive tract. After release of viral DNA, a nonreplicating extrachromasomal entity (episome) transcribes into RNA. The introduced gene persists for 7 to 42 days [20]. A potential risk contamination with a replication-competent virus exists when deriving replication-incompetent viruses. Adenoviruses can be produce in large quantities and high titers. Adenoviruses have a high level of transduction and can tranfect nondividing cells. The disadvantages of Ad are that an immune response against it and transfections are transient. The immune response to infected cells results in a loss of therapeutic gene expression [21,22]. The size of the gene is limited to 7 to 8 kb.

2. Retroviruses

A retrovirus is a single-standed RNA virus that replicates through DNA intermediates (reverse transcription). Retroviral vectors can permanently integrate in a random fashion into the genome. All retroviruses, except HIV, integrate only in dividing cells. Retroviruses have a high transduction efficiency. However, high titers are not achievable which makes large-scale production difficult. The cell host range is limited because cells must be dividing in order be transfected. Because retroviruses get integrated into the genome, a potential for genetic transformation exists by insertional mutagenesis. No clinical evidence for this event has happened in any clinical trial. The size of the gene is limited to 6–10 kb.

B. Nonviral Vectors

Nonviral vectors use plasmid DNA to express a transgene and are easy to manufacture and use (Table 37.3). Furthermore, nonviral vectors avoid the biohazard risks associated with viruses. However, nonviral vectors have no cell specificity and usually lower transfection efficiency. DNA is a negatively charged molecule that is condensed by positively charged molecules (histones and polyamines). Free DNA is too large and has the wrong chemical characteristics to cross the cell membrane. Therefore, other molecules must be used to transfer DNA into the cell.

1. Lipid Complex [23]

A liposome is a microscopic vesicle of lipid surrounding an aqueous compartment. Plasmids are incorporated into the liposome to enhance transcellular delivery and protect against degradation. Cationic (positive charge) liposomes, as opposed to anionic (negative charge), are used more frequently because they can bind negatively charged DNA. The mechanism of DNA transduction by liposome in thought to occur by fluid phase endocytosis, but is not fully understood. Effectiveness of the liposome–DNA complex to transfer DNA is based on proportions of each. Colipids, such as DOPE or cholesterol, are also added to facilitate liposome-mediated transfection. Liposomes have no pathogenic or infectious potential and low immunogenicity and are inexpensive and easy to produce. Liposomes do not have the cell specificity and transfer efficiency of viruses, but have less toxicity. Macrophages ingest and inactivate liposomes and transport them to the reticuloendothelial system. Liposomes are the most common nonviral vectors used in gene therapy.

2. Peptide/Protein and Polymers [23]

DNA–protein complexes use a receptor-mediated pathway to transfer genes. Polylysine conjugation to DNA–protein avoids rapid degredation. The advantages are cell targeting, large gene size capacity, transfection of nonreplicating cells, and repeated administration. Polymers are more efficient in condensing DNA than liposomes. Polymer-based gene therapy is either a noncondensing or a cationic-based system. Noncondensing polymers, such as polyvinyl pyrrolidone (PVP), bind to DNA and protect DNA from degradation, enhance tissue dispersion, and facilitate cellular uptake. Cationic polymer gene delivery can effectively condense DNA in order to tranfect cells. PVP is currently under investigation with IL-12, IL-2, and IFN–α gene therapy trials.

Two mechanical, electroporation and particle-mediated, gene transducing techniques are also being studied. Electroporation uses short electrical pulses to induce a physical and transient permeabilization of cell membranes. Electroporation therapy in combination with chemotherapy (bleomycin) has been tested in patients with recurrent HNSCC [24] and demonstrated responses (3 PR and 5 CR). Gene transfer using electroporation (electro-gene therapy) is high efficient, simple, and cost effective [25]. Using electrogene therapy, stable gene transfer and expression occur only between the electrodes in many tissues, including tumor cells [25]. Electro-gene therapy is currently being studied in eight trials. Particle-mediated gene transfection accelerates and bombards DNA-coated heavy metal (gold) particles to a sufficient velocity to penetrate target cells. Particle-mediated gene transfer has been used to transduce transgenes in animal models and has been shown to reduce tumor growth in mice with cytokine genes, such as interleukins and interferons [26–31]. Three trials using particle-mediated gene therapy are currently being studied.

TABLE 37.3 Delivery Approaches for Gene Therapy

Local
- Injection
  - Intratumoral
- Mechanical
  - Electroporation
  - Gene gun

Regional
- Intramuscular
  - Injection or mechanical
- Nodal injection
- Nodal *ex vivo* immunotherapy
- Intraarterial

Systemic
- Intravenous

V. GENE TRANSFER DELIVERY SITES

Genes can be delivered to the tumor, muscle, nodes, intravascularly, or systemically (Table 37.3). The majority of studies have performed local intratumoral injections. Because head and neck cancer is a local regional problem and access to most lesions is a relatively simple procedure using intratumoral injections, the head and neck area is a common site studied using gene therapy. Intravascular delivery, such as hepatic artery injections, systemic intravenous delivery, and immunogenic nodal site injections, is also being studied.

A. Local-Based Gene Delivery

The majority of clinical gene delivery approaches are based in direct intratumoral injections or *ex vivo* injection of lymphocytes, fibroblast, or tumor cells. Some genes have been delivered into intracavitary spaces (peritoneal or thoracic cavity). Two mechanical delivery methods, electroporation and the gene gun, are currently under investigation in preclinical models. These local-based delivery approaches are limited, as distant disease failure is not addressed.

B. Regional Gene Transfer

Cytokine gene transfer into regional lymph nodes by direct injection has been used to overcome immune suppression. The regional draining lymph nodes can undergo *ex vivo* gene transfer after nodal lymphocytes are removed from the patient, tranfected with the cytokine, and reintroduced into the patient. Intramuscular gene transfer of tumor antigens by direct injection or mechanical techniques has been used in tumor vaccination approaches, as APC can uptake and process the antigen gene, present the antigen to T cells, and initiate an immune response against cancer cells. Intravascular administration of genetic agents allows for delivery into tissue supplied by an artery. Intrahepatic artery infusion of Ad-p53 has been examined in preclincal experiments and has been approved for phase I and II trials in patients with hepatic metastasis from colon cancer.

C. Systemic Delivery

Local delivery fails to treat distant disease and lesion that are not amenable to direct injection. Systemic delivery approaches (intravenous) allow for treatment beyond local disease. However, systemic delivery has to overcome toxicity and rapid degradation of the vector and has to target tumor cells. Three trials have been approved using intravenous delivery, which are (1) the pharmacokinetic, safety, and tolerability study of intravenous RPR/INGN 201, (2) intravenous injection of CV787, a PSA cytolytic adenovirus, and (3) intravenously administered liposome/IL-2.

VI. GENE THERAPY TRIALS IN HEAD AND NECK SQUAMOUS CELL CARCINOMA

As of September 2000, 249 of the 409 approved gene therapy trials were for cancer in the United States. A review of all gene therapy trials can be found at the Office of Biological Activities of the NIH (http://www.nih.gov/od/oba/). Twenty-three trials using nine different genes are being tested in recurrent HNSCC using intratumoral injections (Table 37.4). Many potential genes and gene therapy strategies tested *in vitro* and *in vivo* have demonstrated tumor suppression or killing; however, only approaches that are currently being tested in human trials with HNSCC will be discussed in detail.

Before FDA approval of any gene therapy agent, a clinical benefit and an acceptable toxicity must be demon strated in a randomized phase III trial. However, genetic agents must also have demonstrated biological activity and biosafety. Biological activity can be proven if gene cDNA, RNA, and protein expression, along with changes in downstream effects, such as apoptosis and effector protein expression, are found when compared to baseline. If very low rates of replication-competent vectors, lack of germ line transfection, and no horizontal transmission to health care workers and family members are detected, a gene transfer agent can be considered biologically safe.

A. Gene Therapy Using the p53 Gene

The p53 gene regulates DNA repair, cell cycle, apoptosis, senescence, and genomic stability, along with many other cellular functions, and is mutated in half of human cancers [6]. In HNSCC, p53 mutations from tumor cells have been identified in histologically normal margins and have been correlated with a higher recurrence rate [32]. Overexpression of p53 in head and neck cancer cells has demonstrated tumor growth suppression using *in vitro* and *in vivo* models [33,34]. Using both mutated or wild-type p53 human HNSCC cell lines, exogenous wild-type p53 is dominant over its mutant gene and will select against proliferation. After exogenous wild-type p53 was transduced at comparable levels into nonmalignant fibroblasts, the growth of fibroblasts was not suppressed [34].

Twenty-five p53 gene therapy trials have been approved or have pending approval (Table 37.5). Over 500 patients have been treated with Ad-p53. Most of the experience is in patients with HNSCC and lung cancer, although trials are ongoing in prostate, bladder, breast, ovarian cancers and gliomas. All trials used an adenovirus vector except one that used a retroviral system. Two adenoviral-p53 agents are currently being tested. RPR/INGN 201 (Aventis Pharmaceuticals, Introgen, and NCI) is the only agent that is tested in HNSCC. SCH 58500 (Schering Plough Pharmaceuticals) has been used in other cancer sites. RPR/INGN 201 is a constructed adenoviral vector that contains the wild-type p53 gene driven by a CMV promoter.

1. Phase I Intratumoral HNSCC Trial (Ad-p53 = RPR/INGN 201)

In a phase I trial, patients with recurrent HNSCC received multiple intratumoral injections of Ad-p53 and were monitored for adverse events, p53 expression, Ad-p53 in body fluids, antiadenoviral antibodies, and clinical responses [35]. Thirty-three patients were injected (days 1, 3, 5, 8, 10, and 12 every 4 weeks) with Ad-p53 using doses ranging between 1×10^6 and 1×10^{11} plaque-forming units (PFU). Fifteen of 33 were resected but were judged to be incurable. No dose-limiting toxicity or related serious adverse events were noted. The expression of p53 protein was detected in tumor biopsies. The anti-p53 antibody response did not effect exogenous p53 expression. Two and 6 patients showed partial responses and stable disease, respectively, whereas 9 patients had progressive disease.

TABLE 37.4 Clinical Gene Therapy Trials in Head and Neck Cancer[a]

Gene	Vector	Institution (sponsor)	Trial	Results
p53	Adenovirus	MD Anderson (Introgen)	Phase I	
			1. Unresectable	Safe
			2. Resectable	Safe, no added surgical complications, 28% survival
		Multicenter (Aventis pharm.)	Phase II (three trials)	Response rate 6% Antitumor activity: 26%
		Multicenter (SWOG/Aventis Pharm.)	Phase II (surgical adjuvant)	Approval pending
		Multicenter (Aventis Pharm.)	Phase III (5FU/CDDP vs 5FU/CDDP/Ad-p53)	Ongoing
		Multicenter (Aventis Pharm.)	Phase III (MTX vs Ad-p53)	Ongoing
BL7	Liposome	Univ. Cincinnati	Phase I	Safe
		Multicenter	Phase II	Safe, two complete responses
		Multicenter	Phase II	Ongoing
BL7 + IL2 (SQ)	Liposome	Multicenter	Phase II	Ongoing
E1A	Liposome	WSU and Rush Univ. (Targeted Genetics Corp)	Phase I	Safe
		Multicenter	Phase II	5% response rate
IL-2	Liposome	Johns Hopkins (Valentis)	Phase I	Safe
		Multicenter (Valentis)	Phase II	Ongoing
TK	Liposome	Johns Hopkins	Phase I	Approved/not initiated
EGFR (antisense)	Liposome	Univ. of Pittsburgh	Phase I	Ongoing
IL12	Fibroblast	Univ. of Pittsburgh (Valentis)	Phase I	Two trials
		Multicenter (Valentis)	Phase II	Ongoing
IL-12	PVP	Multicenter (Valentis)	Phase II	Ongoing
IL-12 + IFN-γ	PVP	Multicenter (Valentis)	Phase II	Ongoing
IFN-α	PVP	Univ. of Pennslyvania (Valentis)	Phase I	Ongoing
GMCSF (tumor cells)	adenovirus	Univ. of Kansas	Phase I	Ongoing

[a]BL7, HLA-B7; IL, interleukin; SWOG, Southwest Oncology Group; MTX, methotrexate; 5FU, 5-fluorouracil; CDDP, cisplatinum; EGFR, epidermal growth factor receptor; IFN, interferon; GMCSF, granulocyte–macrophage colony-stimulating factor.

Ad-p53 was detected in body fluid, such as blood, urine, and sputum. No horizontal transmission to health care workers was found.

2. Surgical Adjuvant p53 Gene Therapy

Using a model that simulated residual microscopic disease after gross tumor resection of squamous cell cancer, the feasibility of gene therapy as an adjuvant to surgical resection was demonstrated [34]. Nude mice were implanted subcutaneously with tumor cells and treated with Ad-p53 before gross tumor development. Ad-p53 therapy prevented tumor development, as 2 of 30 (6.7%) mice grew tumors, which were treated with Ad-p53 as opposed to 27 of 30 (90%) in the control group [34]. The mechanism of growth suppression was found to be primarily apoptosis. Additional mechanisms of actions for Ad-p53 have been demonstrated, including Fas-mediated apoptosis and antiangiogenesis effects.

In the single-center phase I trial, a cohort of 15 patients that had recurrent/refractory (failed multimodalities of therapy) cancer and were eligible for palliative surgical resection were enrolled [35]. These patients were resectable, but thought to be incurable. Preoperatively, a patient's tumor was injected six times in a 2-week period. Patients underwent a surgical resection and were given an intraoperative injection of Ad-p53 in the resected tumor bed and in the neck dissection site. Three days later, their drainage catheters were injected (retrograde) with Ad-p53. All patients had extensive surgery and required flaps for closure. The surgical complications (one vascular anastomotic

TABLE 37.5 Clinical Gene Therapy Trials Using p53 Gene Therapy ($n = 25$)

Head and neck cancer ($n = 6$)	Colon cancer ($n = 2$) SCH585000
Phase I: M.D. Anderson	Phase I: Intrahepatic artery–adenovirus
Phase II: ($n = 2$) Intratumoral, multicenter	Phase II: Intrahepatic artery–adenovirus
Phase II: Surgical adjuvant trial (perioperative injection)	Breast cancer ($n = 2$)
Phase III: ($n = 2$) with chemotherapy starting	Phase I: Intratumoral
Lung cancer ($n = 6$)	Phase I: BMT tumor cell purge (SCH585000)
Phase I: ($n = 3$) retrovirus and adenovirus (RPR/INGN and NCI)	Malignant glioma ($n = 1$)
Phase I: combined with cisplatinum	Phase I: Intratumoral
Phase II: ($n = 1$) adenovirus	Bladder cancer ($n = 1$)
Phases I–II: Radiation adjuvant trial	Phase I: Intravesicular
Prostate cancer ($n = 2$)	Hepatocellular carcinoma ($n = 1$)
Phase I: UCLA	Phase I: Intratumoral
Phases I–II: M.D. Anderson	Pharmokinetic study ($n = 1$)
Ovarian cancer ($n = 3$) SCH585000	Phase I: Intravenous
Phase I: ($n = 2$) adenovirus	
Phases II–III: SCH585000/chemo vs chemo	

failure and one delayed wound healing) were expected and unlikely caused by Ad-p53 therapy. Therefore, this perioperative approach was found to be safe and well tolerated with no significant added wound complications. Fever (6), injection pain (5), and flu-like (4) symptoms were the only complications observed in these patients. Otherwise, it was felt to be safe and well tolerated. In a follow-up report [36], 4 (27%) patients were alive and free of disease at 18 months, while 1 other patient was alive with disease. Two died from other causes. A phase II trial is planned through the Southwest Oncology Group (SWOG).

3. Phase II Intratumoral HNSCC Trials (T201, T202, and T207)

Two phase II monotherapy intratumoral injection multicenter trials [37] using two dosing schedules [low dose (days 1, 2, and 3 every 4 weeks) or high dose (days 1, 3, 5, 8, 10, and 12 every 4 weeks)] enrolled heavily pretreated, recurrent, and unresectable patients with HNSCC, respectively. The first phase II trial ($N = 97$) randomized patients to two different dosing schedules (days 1, 2, and 3 or days 1, 3, 5, 8, 10, and 12 every 28 days). The median dose of Ad-p53 was 6×10^{10} PFU. Seventy patients were evaluable for response and a 26% (23/90) antitumor activity (PR, CR, and SD) in patients. A 6% (5 out of 90) response rate (PR/CR) was observed. When each lesion (total number = 167) was separated, the response rate per lesion was 18%. In a second multicenter phase II trial, 23 patients with HNSCC who were heavily pretreated, unresectable received intratumoral injections (median dose $= 2.5 \times 10^{11}$ PFU/day) on days 1, 2, and 3 every 4 weeks. In 15 evaluated for response, 9 demonstrated antitumor activity (PR, CR, and SD) and 1 had a partial response.

The related adverse events were fever/chills (74%), injection site pain (45%), asthenia (13%), nausea (1%), and injection site bleeding (10%). Twelve related severe adverse effects were reported [fever(4), tumor hemorrhage(3), chills(1), injection site pain(1), dehydration(1), Guillain–Barre syndrome(1), and infection(1)]. No treatment-related deaths were reported.

Based on the improvement of median survival in the high-dose regimen group (197 days vs 168 days, $p < 0.05$), a dose of 2×10^{10} viral particles was chosen for subsequent phase II and III trials. The ongoing Ad-p53 trials are in (1) refractory HNSCC (phase III, Methotrexate vs Ad-p53), (2) recurrent HNSCC (phase III, 5-FU/CDDP vs 5-FU/CDDP + Ad-p53), and (3) surgical adjuvant Ad-p53 trial in advanced HNSCC (phase II). Furthermore, a phase III trial comparing chemotherapy against chemotherapy and SCH58500 after surgery for ovarian cancer is enrolling patients.

4. Lung Cancer Phase I Intratumoral Trials

In a phase I clinical trial [38], 28 patients with recurrent nonsmall cell lung cancer (NSCLC) were injected with Ad-p53 by transbronchonscopy or computed tomography (CT) needle approaches. The dose range of Ad-p53 was 10^6–10^{11} PFU. No grade 4 and one grade 3 toxicity were reported. The related adverse events were fever (27%), pain (16%), pneumothorax (7%), hemoptysis(5%), and nausea (5%). In 25 evaluable patients, partial responses (2/25, 8%) and disease stabilization (16/25, 64%) were noted, whereas 7 patients (28%) exhibited progressive disease. Using reverse transcriptase polymerase chain reaction (RT-PCR), exogenous vector p53 mRNA was detected in 46% (12/26) patients tested.

Another phase I trial with 24 patients with NSCLC lung cancer combined Ad-p53 with chemotherapy and intratumoral injection of Ad-p53 [39]. Using 28-day cycles, Ad-p53 ($10^6 - 10^{11}$ PFU) was injected on day 4 and cisplatinum (80 mg/m^2) was infused on day 1. The toxicity profile was similar to the previous reported trial in that fever (8/24, 33%) was a common related adverse event (AE). No serious related AE or treatment-related deaths were reported. The response rate was similar in both trials (PR=2, SD=17, PD=4).

B. Allovectin (HLA-B7/β_2-Microglobulin and DMRIE/DOPE) Gene Therapy

Class I MHC expression is a method of tumor-specific immunological gene therapy. Cancer cells are altered genetically to express a class I MHC. If the class I MHC used is a human antigen, but foreign to the individual, it would be an alloantigen. This alloantigen is capable of provoking an intense immune response. Then class I MHC expression can also initiate immune responses throughout the tumor as a reaction to tumor-associated antigens. This theory was tested originally in a mouse tumor model. The tumors were treated with a foreign mouse class I MHC gene. The MHC expression induced a CTL response to the MHC, as well as to other antigens present on the surrounding tumor cells that were not modified. Allovectin-7 encodes for the class I MHC HLA-B7 α chain and β_2-microglobulin. The β_2-microglobulin allows for the synthesis and expression of complete MHC on the cell surface. The plasmid DNA is complexed with a liposomal vector. A cationic lipid mixture DMRIE/DOPE (1,2-dimyristyloxypropyl-3-dimethyl-hydroxyethyl ammonium bromide/dioleoyl phosphytidal ethanolamine) was used. These results led to the development of the drug Allovectin-7 (Vical Inc., San Diego, CA) for clinical investigations.

In a phase I trial [40], nine patients with recurrent HNSCC who did not express HLA-B7 were treated with Allovectin-7 by direct intratumoral injection (10 mg) on days 0, 14, 42, and 56. Allovectin-7 contains a plasmid complementary DNA complexed with a cationic lipid, which results in expression of HLA-B7. No toxic effects of Allovectin-7 gene therapy were encountered. A partial response was found in four of nine patients. One patient has remained alive with no clinical evidence of disease but with persistent histological evidence of cancer. Analysis of tumor specimens from two of the patients who responded to therapy demonstrated HLA-B7 expression and apoptosis.

In a phase II trial [41], 20 patients received 58 treatments with Allovectin-7 (10 mg) on days 0, 14, 42, and 56. All 20 patients received the first cycle of two injections. No drug-related adverse events were reported. Tumor progression resulted in one case of airway obstruction (tracheostomy tube placement) and another case of severe dysphagia (gastrostomy tube placement). At the 3-week evaluation point, 11 patients had disease progression and all but 1 eventually died of their cancer, 4 patients had a partial response, and 5 patients had stable disease. At 16 weeks, 6 patients had either a partial response (4) or stable (2) disease of which 5 later progressed. One patient underwent surgery and remains alive and cancer free. Although two complete responses were noted, biopsies revealed persistent disease in these patients. In two tumor samples, expression of HLA-B7 and induction of apoptosis were shown. These results have lead to further phase II trials in HNSCC. A phase III trial in metastatic melanoma is ongoing comparing Dacarbazine against Dacarbazine/Allovectin-7.

C. Gene Therapy Using the E1A Gene

The adenovirus E1A gene is the first gene expressed in virus-infected cells and is a well-known transcription factor. The E1A gene has antitumor activity by downregulating oncogenes, such as HER2/neu, inducing apoptosis, inhibiting metastasis, and enhancing the immune response against tumors [42,43]. E1A gene products have been shown to inhibit HER2 expression in cancer cells through inhibition of the HER-2 promoter, resulting in the suppression of tumor development and abolishing tumorigenicity and metastatic potential HER2-transformed fibroblasts [44–46]. *In vitro* and *in vivo* experiments have demonstrated tumor growth suppression and increased survival using E1A gene therapy [42]. Furthermore, E1A has an additive effect with chemotherapy and radiotherapy [47]. Therefore, E1A gene therapy is a potential treatment modality for patients with cancer.

Nine patients with HNSCC and 9 with breast cancer were enrolled in a phase I trial [48]. One tumor nodule was injected with E1A/liposome on days 1, 2, and 3 and then weekly for 7 more weeks (10 injections total). No dose-limiting toxicity was observed in the four dose groups (15, 30, 60, and 120 μg DNA/ cm tumor). Therefore, the maximum tolerated dose (MTD) was not reached in this study. All patients tolerated the injections, although several experienced pain and bleeding at the injection site. E1A gene transfer was demonstrated in 11 of 11 tumor samples tested, and downregulation of HER-2/neu was demonstrated in 1 of the 6 patients who overexpressed HER-2/neu at baseline. HER2 could not be assessed post-treatment in 5 of 6 specimens due to severe necrosis. In 1 BC patient, no pathologic evidence of tumor was found on biopsy of the treated tumor at week 12. In 16 patients evaluated for response, 9 had stable disease, 5 had progressive disease, and 2 had minor responses. Because intratumoral E1A gene therapy was performed safely and patients tolerated the procedure well, a phase II trial was initiated.

In multicenter phase 2 trial EIAliposome therapy [49], 24 patients with recurrent HNSCC were treated with E1A/liposome (30 ug/cm^3 tumor) on days 1, 2, and 3 and then weekly for 7 more weeks. Ten of 24 patients completed therapy, whereas 14 did not complete the protocol secondary to progression of tumor (11), voluntary withdrawal (1), and death

(2). One of 21 (4.3%) patients had a complete response, whereas no partial response, 2 (8.3%) minor, and 7 (29.2%) stable diseases were reported by bidimension CT measurement. Common adverse events were asthenia (42%) and pain (33%), and no serious related adverse events were noted. E1A expression was detected in patients tested using RT-PCR and immunohistochemistry. Future trials will combine E1A gene transfer with conventional radiotherapy or chemotherapy.

D. IL-12 Gene Therapy

IL-12 is an immunostimulatory cytokine with antitumor effects. IL-12 stimulates NK cells and augments CTL maturation along with induction of IFN-γ production. In a syngeneic mouse squamous cell carcinoma model, IL-12 gene therapy using irradiated tumor cells suppressed tumor growth [50]. A phase II trial of intravenous recombinant IL-12 was stopped early, as significant toxicity was found [51]. Two phase I trials using IL-12 gene therapy using autologous fibroblasts by direct injection were approved and performed in HNSCC, breast cancer, and melanoma. In a phase I trial [51], patients with solid cancers were injected with genetically engineered autologous fibroblasts transfected with the IL-12 gene. Fibroblasts from the patients were transduced using the retroviral vector carrying the human IL-12 gene. Two patients with HNSCC, along with individuals with breast cancer (6) and melanoma (5), were treated. Fibroblast cultures were established successfully from the patients' dermis in 27 of 29 attempts (93%). In 21 of 21 attempts, IL-12 was transferred into fibroblasts, and expression of IL-12 protein was observed. No "untoward effects" were observed, and "reduction of the tumor size" was noted in one patient with HNSCC and three melanomas. A phase II trial using genetically engineered autologous fibroblasts producing IL-12 is ongoing. Two more IL-12 gene therapy trials using IL-12/PVP and IL-12/IFN-γ/PVP have been approved in HNSCC.

E. IL-2 Gene Therapy

IL-2 is a T and NK cell activation and growth factor that has stimulated an antitumor immunological response [52]. Systemic administration has led to tumor regression in some patients with significant toxicity in melanoma [52] and HNSCC [53]. High-dose localized IL-2 therapy is an attractive approach to overcome local immunosuppression and to stimulate immunogical tumor rejection along with the avoidance of systemic toxicity. Injection of IL-2 and a cationic liposome (DOTMA:cholesterol) in head and neck tumors of immunocompetent mice after subtotal surgical resection in mice resulted in tumor growth suppression, and no significant toxicity was noted [54]. Treated mice had an increased hIL-2 production, as well as induction of murine IFN-γ and IL-12 as compared to controls. Similar results were found using an adenoviralinterleukin-2 agent [55]. Although the completed phase I trial using IL-2/liposome has not yet been published, the phase II trial using IL-2/PVP is ongoing.

F. Herpes Simplex Virus–Thymidine Kinase

The herpes simplex virus–thymidine kinase gene expresses an enzyme that phosphorylates a prodrug, ganciclovir, into a toxic compound. Furthermore, a "bystander effect" through the transfer of toxic metabolites via gap junction intercellular communications has been described in which surrounding nontransduced cells are killed. The majority of studies are in glioblastoma; however, *in vivo* HNSCC models has been studied using a combination of cytotoxic (TK) and immunological (IL-2) approaches [56–58]. Mice receiving TK and IL-2 demonstrated a greater regression of tumors as compared to controls and the group treated with only TK. To date, no patients have been enrolled in an approved TK ganciclovir phase I trial in recurrent HNSCC.

G. Antisense EGFR/Liposome

HNSCC cells overexpress epidermal growth factor receptor (EGFR), which is a tyrosine kinase cell surface receptor. Ligands, such as epidermal growth factor (EGF) and transforming growth factor-α (TGF-α), binding to EGFR stimulates mitogenesis and increases tumor growth, metastasis [59]. Furthermore, overexpression of EGFR predicts poor outcomes in HNSCC [59]. Because EGFR protein is required to sustain the proliferation of SCCHN cells *in vitro*, downregulating EGFR is a potential target in HNSCC. Intratumoral cationic liposome-mediated gene transfer of the antisense EGFR gene into human head and neck tumor xenografts in nude mice resulted in inhibition of tumor growth, suppression of EGFR protein expression, and an increased rate of apoptosis [59]. Based on these preclinical data, a phase I trial using liposome-mediated antisense EGFR gene therapy was approved and opened at the University of Pittsburgh.

H. IFN-α Gene Transfer Using PVP

IFN is an immunomodulator cytokine that has antitumor activity. IFN-α is the most widely used IFN. IFN-α2b is approved for use in high-risk melanoma and many other cancers. Response rates for patients with advanced head and neck cancer treated with IFN-α alone or with chemotherapy or IL-2 range between 18 and 54% [53,60–63]. However, significant toxicity has led some authors to suggest further investigations of less aggressive regimens. Preclinical data have demonstrated antitumor activity for interferon gene therapy [64]. A phase I trial is approved and ongoing at the University of Pennsylvania using intratumoral IFN-α gene therapy.

I. GM-CSF-Based Gene Therapy

Granulocyte–macrophage colony-stimulating factor (GM-CSF) stimulates the proliferation of myeloid precursors and has a vital role in the hematopoiesis of other cell lineages. Furthermore, GM-CSF has many other biologic effects on hematopoiesis and the immune system. The myeloproliferative effects of GM-CSF have led to its use in myelosuppressed patients. Additional biological effects have led to GM-CSF use in many other disease, such as immunotherapy for malignancies. Direct injection of GM-CSF gene or *ex vivo* transduction of GM-CSF into irradiated tumor cells has been tested. An *ex vivo* transduction phase I trial in renal and prostate cancer [65] has immunological activity and limited toxicity. One patient with renal cancer responded. In HNSCC, breast and colon cancer, and sarcomas, a phase I trial using *ex vivo*-transduced, irradiated cancer cells is ongoing at the University of Kansas.

1. Safety of Gene Therapy

To date, the safety in phase I and II trials of Ad-p53, Allovectin, E1A/liposome, and others has not shown any related severe adverse events [35,36,40,48]. The largest safety database exists for Ad-p53 [66]. The global safety database now includes over 412 patients with cancer enrolled on Ad-p53 trials using intratumoral injection to date. The first reported safety data analysis had 309 patients with late stage, recurrent/refractory cancer. The majority of these patients had head and neck squamous cell carcinoma ($n=226$, 73%) or nonsmall cell lung carcinoma ($n=83$, 27%). Phase I ($n=3$) or phase II ($n=3$) trials were conducted in 47 centers. The dose range for repeated intratumoral injections was 2.5×10^7–10^{12} viral particles for a total of 702 cycles of therapy (2296 treatment days). Dosing intervals varied from 1 to 6 days, with treatment cycles repeated every 4 weeks for as many as 18+ months in some patients, as long as antitumor activity was evident. No treatment-related deaths were reported. When data were analyzed for trends, the most frequent related adverse events were fever/chills/flu-like syndrome (60%) and pain at the injection site (39%). The majority of both AEs were graded as mild to moderate. The most frequent related serious adverse events (SAEs) were fever, infection, and hemorrhage (all less than 5%). Furthermore, laboratory tests (lymphocytes, platelets, creatinine, and liver enzymes) did not change significantly. Subset analyses revealed no clinically meaningful correlation or association between AEs and SAEs and any laboratory parameters.

VII. FUTURE OF GENE THERAPY FOR CANCER

With the completion of the human genome project, many more genes will be available for transfer. However, genes that are currently used can produce all desired antitumor effects. The limitations of gene therapy can be overcome by combing with standard therapy, development of new vectors, and targeting vectors.

A. Combination with Standard Therapy

Gene therapy is also being used as an adjuvant to conventional therapies, such as chemotherapy, radiotherapy, and surgery. Gene transfer is currently being combined with chemotherapy (Ad-p53 and E1A), radiotherapy (Ad-p53, TK, and PSA-based vaccine), and surgery (Ad-p53). The best-described adjuvant effect of gene transfer is p53. Chemotherapy and radiotherapy induce DNA damage, which leads to increases in p53 expression in normal cells and cell cycle arrest. If cells cannot repair DNA damage, apoptosis will result through p53 pathways. In cancer cells that have an altered p53, cell cycle arrest and apoptosis can be avoided after exposure to chemotherapy and radiotherapy. Preclinical experiments have demonstrated synergy between chemotherapy and p53 [67] and E1A [47] overexpression. This synergy has led to the development on the current ongoing trials using p53 gene transfer therapy and chemotherapy in HNSCC.

The basis for surgical adjuvant gene therapy lies in the observation that tumor cells are present in the margin of resection even with histological normal tissue. Because squamous-derived tumor cells have a higher level of adenoviral receptors than fibroblasts, adenovirus-based therapy can tranfect tumor cells more easily in the tumor microenvironment. Favorable results of the phase I p53 gene therapy surgical adjuvant trial has led to the phase II trial in newly diagnosed HNSCC where Ad-p53 gene therapy is given perioperatively and chemoradiation is given postoperatively. This trial is being conducted through the head and neck committee in the SWOG.

B. New Vector Strategies

Although numerous vectors are in use for gene therapy and many more are under investigation, no existing vector meets the criteria of an ideal vector, which is high efficiency (100% of cells get transfected), high specificity (only tumor cells receive the gene), and low toxicity. New vector strategies are based on novel vectors, replication-competent viruses, or modifications of existing vectors. One newer vector is adenovirus-associated virus (AAV). AAV is a DNA virus that requires a helper virus in order to replicate. AAV can infect nondividing cells without causing pathological infection. Wild-type AAV integrates specifically into chromosome 19; however, replication-deficient and recombinant AAV does not integrate specifically or may be maintained stably episomal. Clinical trials using AAV are just beginning. The size of the gene is limited to 4.6 kb.

Modified replication-competent adenoviruses (RCAs) are the most commonly used for cancer. RCAs consist of wild-type adenoviruses or modified viruses with or without an added gene or specific promoter of normal viral genes. Five trials have been approved using RCAs. The best-studied RCA is ONYX-015, which is a naturally occurring adenovirus with the E1b gene deleted. Because ONYX-015 has not been modified genetically, it is not truly a genetic therapeutic agent. The deleted E1b gene in ONYX-015 allows for it to selectively replicate in p53-deficient tumor cells and not in normal p53 wild-type cells [68]. ONYX-015 has been tested in phase I [69] and II trials in HNSCC and is now being tested in phase III trials.

CN706 is an adenovirus with a prostate-specific antigen (PSA) promoter to drive the E1A gene. In summary, the clinical safety and efficacy data of CN706 in organ-confined disease are very promising (www.calydon.com/cgi/results.php). Nine patients out of 11 (82%) had a PSA response to the treatment. Finally, dose-limiting toxicity was not reached and the product showed an acceptable safety profile, with some minor-to-moderate fever and/or flu-like symptoms. A similar approach has been used with AFP (AvE1a041).

C. Targeting Vectors

Vectors can be targeted by (1) altering vector–target cell interaction or (2) targeting promoter gene transcription. Viruses infect cells through cell surface receptors on the target cells by binding to the cell and being endocytosed. Two adenovirus receptors, integrin and coxsakie-adenovirus receptor (CAR), are on target cells [70]. The methods to target cells are by altering vector coat proteins or a bifunc tional cross-linker. The fiber protein on the adenovirus can be altered genetically to bind to specific tissue. Alternatively, a bifunctional cross-linker (protein or antibody) molecule can be introduced to bind to the adenovirus fiber and specifically to receptors on target cells.

Tumor-specific targeting of transgene expression can be obtained by designing promoters of transcription. Promoters can be tissue specific (PSA), tumor selective (AFP), tumor endothelium directed (VEGF-R), cell cycle regulated (E2F), or treatment responsive (egr1-*early growth response*). Tissue-specific promoters, such as PSA, would express transgene only in certain tumor or normal cells.

D. Summary of Gene Therapy

The biological activity and biosafety of gene therapy have been established. The clinical response rates have been between 5 and 20%, which is comparable to historical rates using single-agent chemotherapy in recurrent and refractory HNSCC patients. The added benefit appears to be a cytostatic or stabilization of disease in a larger proportion of patients. Therefore, gene therapy may produce other clinical effects, such as stabilization of disease and improved quality of life. Furthermore, a combination of cytostatic gene therapy and cytotoxic standard therapy may be beneficial in the future. In most trials, the maximum tolerated dose was not achieved because no significant toxicity was observed at the highest dose tested. The reported toxicity has been limited to pain, bleeding, and fever/chills without any treatment-related deaths. The initial clinical results that were translated from laboratory work using gene therapy must now be the basis for further laboratory research and then clinical trials. It is not yet known whether gene therapy will provide the ultimate benefit, which is improved survival. Although promising data have been reported, no agent has received FDA approval to date.

References

1. Greenlee, R. T., Murray, T., Bolden, S., and Wingo, P. A. (2000). Cancer statistics. *CA Cancer J. Clin.* **50**(1), 7–33.
2. Cavazzana-Calvo, M., Hacein-Bey, S., de Saint, B., Gross, F., Yvon, E., Nusbaum, P., Selz, F., Hue, C., Certain, S., Casanova, J. L., Bousso, P., Deist, F. L., and Fischer, A. (2000). Gene therapy of human severe combined immunodeficiency (SCID)-X1 disease. *Science* **288**, 669–672.
3. Abonour, R., Williams, D. A., Einhorn, L., Hall, K. M., Chen, J., Coffman, J., Traycoff, C. M., Bank, A., Kato, I., Ward, M., Williams, S. D., Hromas, R., Robertson, M. J., Smith, F. O., Woo, D., Mills, B., Srour, E. F., and Cornetta, K. (2000). Efficient retrovirus-mediated transfer of the multidrug resistance 1 gene into autologous human long-term repopulating hematopoietic stem cells. *Nature Med.* **6**, 652–658.
4. Fearon, E. R., and Vogelstein, B. A. (1990). A genetic model for colorectal tumorigenesis. *Cell* **61**, 759–767.
5. Califano, J., van der Riet, P., Westra, W., Nawroz, H., Clayman, G., Piantadosi, S., Corio, R., Lee, D., Greenberg, B., Koch, W., and Sidransky, D. (1996). Genetic progression model for head and neck cancer: Implications for field cancerization. *Cancer Res.* **56**, 2488–2492.
6. Boyle, J. O., Hakim, J., Koch, W., van der Riet, P., Hruban, R. H., Roa, R. A., Correo, R., Eby, Y. J., Ruppert, J. M., and Sidransky, D. (1993). The incidence of p53 mutations increases with progression of head and neck cancer. *Cancer Res.* **53**, 4477–4480.
7. Koch, W. M., Brennan, J. A., Zahurak, M., Goodman, S. N., Westra, W. H., Schwab, D., Yoo, G. H., Lee, D. J., Forastiere, A. A., and Sidransky, D. (1996). p53 mutation and locoregional treatment failure in head and neck squamous cell carcinoma. *J. Natl. Cancer Inst.* **88**, 1580–1586.
8. Chomchai, J. S., Du, W., Sarkar, F. H., Li, Y. W., Jacobs, J. R., Ensley, J. F., Sakr, W., and Yoo, G. H. (1999). Prognostic significance of p53 gene mutations in laryngeal cancer. *Laryngoscope* **109**, 455–459.
9. Reed, A. L., Califano, J., Cairns, P., Westra, W. H., Jones, R. M., Koch, W., Ahrendt, S., Eby, Y., Sewell, D., Nawroz, H., Bartek, J., and Sidransky, D. (1996). High frequency of p16 (CDKN2/MTS-1/INK4A) inactivation in head and neck squamous cell carcinoma. *Cancer Res.* **56**, 3630–3633.
10. Yoo, G. H., Xu, H. J., Brennan, J. A., Westra, W., Hruban, R. H., Koch, W., Benedict, W. F., and Sidransky, D. (1994). Infrequent inactivation

of the retinoblastoma gene despite frequent loss of chromosome 13q in head and neck squamous cell carcinoma. *Cancer Res.* **54**, 4603–4606.

11. Nawroz-Danish, H. M., Koch, W. M., Westra, W. H., Yoo, G., and Sidransky, D. (1998). Lack of BRCA2 alterations in primary head and neck squamous cell carcinoma. *Otolaryngol. Head. Neck Surg.* **119**, 21–25.
12. Harris, C. C. (1996). Structure and function of the p53 tumor suppressor gene: Clues for rational cancer therapeutic strategies. *J.Natl.Cancer Inst.* **88**, 1442–1455.
13. Beckhardt, R. N., Kiyokawa, N., Xi, L., Liu, T. J., Hung, M. C., el-Naggar, A. K., Zhang, H. Z., and Clayman, G. L. (1995). HER-2/neu oncogene characterization in head and neck squamous cell carcinoma. *Arch. Otolaryngol. Head Neck Surg.* **121**, 1265–1270.
14. Ibrahim, S. O., Vasstrand, E. N., Liavaag, P. G., Johannessen, A. C., and Lillehaug, J. R. (1997). Expression of c-erbB proto-oncogene family members in squamous cell carcinoma of the head and neck. *Anticancer Res.* **17**, 4539–4546.
15. Capaccio, P., Pruneri, G., Carboni, N., Pagliari, A. V., Quatela, M., Cesana, B. M., and Pignataro, L. (2000). Cyclin D1 expression is predictive of occult metastases in head and neck cancer patients with clinically negative cervical lymph nodes. *Head Neck* **22**(3), 234–240.
16. Mineta, H., Miura, K., Takebayashi, S., Ueda, Y., Misawa, K., Harada, H., Wennerberg, J., and Dictor, M. (2000). Cyclin D1 overexpression correlates with poor prognosis in patients with tongue squamous cell carcinoma. *Oral Oncol.* **36**(2), 194–198.
17. Bova, R. J., Quinn, D. I., Nankervis, J. S., Cole, I. E., Sheridan, B. F., Jensen, M. J., Morgan, G. J., Hughes, C. J., and Sutherland, R. L. (1999). Cyclin D1 and p16INK4A expression predict reduced survival in carcinoma of the anterior tongue. *Clin. Cancer Res.* **5**, 2810–2819.
18. Myers, J. N., and Whiteside, T. (1995). Immunotherapy of squamous cell carcinoma of the head and neck. *In* "Cancer of the Head and Neck" (E. N. Myers and J. Suen eds.), pp. 805–817. Saunders, Philadelphia.
19. Guinan, E. C., Gribben, J. G., Boussiotis, V. A., Freeman, G. J., and Nadler, L. M. (1994). Pivotal role of the B7: CD28 pathway in transplantation tolerance and tumor immunity. *Blood* **84**, 3261–3282.
20. Mulligan, R. C. (1993). The basic science of gene therapy. *Science* **260**, 926–932.
21. Yang, Y., Su, Q., and Wilson, J. M. (1996). Role of viral antigens in destructive cellular immune responses to adenovirus vector-transduced cells in mouse lungs. *J.Virol.* **70**, 7209–7212.
22. Yang, Y., and Wilson, J. M. (1995). Clearance of adenovirus-infected hepatocytes by MHC class I-restricted CD4+ CTLs in vivo. *J. Immunol.* **155**, 2564–2570.
23. Mahato, R. I., Smith, L. C., and Rolland, A. (1999). Pharmaceutical perspectives of nonviral gene therapy. *Adv. Genet.* **41**, 95–156.
24. Panje, W. R., Hier, M. P., Garman, G. R., Harrell, E., Goldman, A., and Bloch, I. (1998). Electroporation therapy of head and neck cancer. *Ann. Otol. Rhinol. Laryngol.* **107**, 779–785.
25. Hofmann, G. A., Dev, S. B., Nanda, G. S., and Rabussay, D. (1999). Electroporation therapy of solid tumors. *Crit. Rev. Ther. Drug Carrier Syst.* **16**, 523–569.
26. Rakhmilevich, A. L., Timmins, J. G., Janssen, K., Pohlmann, E. L., Sheehy, M. J., and Yang, N. S. (1999). Gene gun-mediated IL-12 gene therapy induces antitumor effects in the absence of toxicity: A direct comparison with systemic IL-12 protein therapy. *J. Immunother.* **22**, 135–144.
27. Mahvi, D. M., Sheehy, M. J., and Yang, N. S. (1997). DNA cancer vaccines: A gene gun approach. *Immunol. Cell Biol.* **75**, 456–460.
28. Mahvi, D. M., Sondel, P. M., Yang, N. S., Albertini, M. R., Schiller, J. H., Hank, J., Heiner, J., Gan, J., Swain, W., and Logrono, R. (1997). Phase I/IB study of immunization with autologous tumor cells transfected with the GM-CSF gene by particle-mediated transfer in patients with melanoma or sarcoma. *Hum. Gene Ther.* **8**, 875–891.
29. Mahvi, D. M., Burkholder, J. K., Turner, J., Culp, J., Malter, J. S., Sondel, P. M., and Yang, N. S. (1996). Particle-mediated gene transfer of granulocyte-macrophage colony-stimulating factor cDNA to tumor cells: Implications for a clinically relevant tumor vaccine. *Hum. Gene Ther.* **7**, 1535–1543.
30. Rakhmilevich, A. L., Turner, J., Ford, M. J., McCabe, D., Sun, W. H., Sondel, P. M., Grota, K., and Yang, N. S. (1996). Gene gun-mediated skin transfection with interleukin 12 gene results in regression of established primary and metastatic murine tumors. *Proc. Natl. Acad. Sci. USA* **93**, 6291–6296.
31. Sun, W. H., Burkholder, J. K., Sun, J., Culp, J., Turner, J., Lu, X. G., Pugh, T. D., Ershler, W. B., and Yang, N. S. (1995). In vivo cytokine gene transfer by gene gun reduces tumor growth in mice. *Proc. Natl. Acad. Sci. USA* **92**, 2889–2893.
32. Brennan, J. A., Mao, L., Hruban, R. H., Boyle, J. O., Eby, Y. J., Koch, W. M., Goodman, S. N., and Sidransky, D. (1995). Molecular assessment of histopathological staging in squamous-cell carcinoma of the head and neck. *N. Engl. J. Med.* **332**, 429–435.
33. Liu, T. J., Zhang, W. W., Taylor, D. L., Roth, J. A., Goepfert, H., and Clayman, G. L. (1994). Growth suppression of human head and neck cancer cells by the introduction of a wild-type p53 gene via a recombinant adenovirus. *Cancer Res.* **54**, 3662–3667.
34. Clayman, G. L., el-Naggar, A. K., Roth, J. A., Zhang, W. W., Goepfert, H., Taylor, D. L., and Liu, T. J. (1995). In vivo molecular therapy with p53 adenovirus for microscopic residual head and neck squamous carcinoma. *Cancer Res.* **55**, 1–6.
35. Clayman, G. L., el-Naggar, A. K., Lippman, S. M., Henderson, Y. C., Frederick, M., Merritt, J. A., Zumstein, L. A., Timmons, T. M., Liu, T. J., Ginsberg, L., Roth, J. A., Hong, W. K., Bruso, P., and Goepfert, H. (1998). Adenovirus-mediated p53 gene transfer in patients with advanced recurrent head and neck squamous cell carcinoma. *J. Clin. Oncol.* **16**, 2221–2232.
36. Clayman, G. L., Frank, D. K., Bruso, P. A., and Goepfert, H. (1999). Adenovirus-mediated wild-type p53 gene transfer as a surgical adjuvant in advanced head and neck cancers. *Clin. Cancer Res.* **5**, 1715–1722.
37. Bier-Laning, C. M., VanEcho, D., Yver, A., and Dreiling, L. (1999). A phase II multi-center study of AdCMV-p53 administered intratumorally to patients with recurrent head and neck cancer. *Proc. Am. Soci. Clin. Oncol.* **18**, 431a. [Abstract]
38. Swisher, S. G., Roth, J. A., Nemunaitis, J., Lawrence, D. D., Kemp, B. L., Carrasco, C. H., Connors, D. G., el-Naggar, A. K., Fossella, F., Glisson, B. S., Hong, W. K., Khuri, F. R., Kurie, J. M., Lee, J. J., Lee, J. S., Mack, M., Merritt, J. A., Nguyen, D. M., Nesbitt, J. C., Perez-Soler, R., Pisters, K. M., Putnam, J. B. J., Richli, W. R., Savin, M., and Waugh, M. K. (1999). Adenovirus-mediated p53 gene transfer in advanced non-small-cell lung cancer. *J. Natl. Cancer Inst.* **91**, 763–771.
39. Nemunaitis, J., Swisher, S. G., Timmons, T., Connors, D., Mack, M., Doerksen, L., Weill, D., Wait, J., Lawrence, D. D., Kemp, B. L., Fossella, F., Glisson, B. S., Hong, W. K., Khuri, F. R., Kurie, J. M., Lee, J. J., Lee, J. S., Nguyen, D. M., Nesbitt, J. C., Perez-Soler, R., Pisters, K. M., Putnam, J. B., Richli, W. R., Shin, D. M., and Walsh, G. L. (2000). Adenovirus-mediated p53 gene transfer in sequence with cisplatin to tumors of patients with non-small-cell lung cancer. *J. Clin. Oncol.* **18**, 609–622.
40. Gleich, L. L., Gluckman, J. L., Armstrong, S., Biddinger, P. W., Miller, M. A., Balakrishnan, K., Wilson, K. M., Saavedra, H. I., and Stambrook, P. J. (1998). Alloantigen gene therapy for squamous cell carcinoma of the head and neck: Results of a phase-1 trial. *Arch. Otolaryngol. Head Neck Surg.* **124**, 1097–1104.
41. Gleich, L. L. (2000). Gene therapy for head and neck cancer. *Laryngoscope* **110**, 708–726.
42. Yu, D., Matin, A., Xia, W., Sorgi, F., Huang, L., and Hung, M. C. (1995). Liposome-mediated in vivo E1A gene transfer suppressed dissemination of ovarian cancer cells that overexpress HER-2/neu. *Oncogene* **11**, 1383–1388.
43. Yu, D., Hamada, J., Zhang, H., Nicolson, G. L., and Hung, M. C. (1992). Mechanisms of c-erbB2/neu oncogene-induced metastasis and

repression of metastatic properties by adenovirus 5 E1A gene products. *Oncogene* **7**, 2263–2270.

44. Zhang, Y., Yu, D., Xia, W., and Hung, M. C. (1995). HER-2/neu-targeting cancer therapy via adenovirus-mediated E1A delivery in an animal model. *Oncogene* **10**, 1947–1954.
45. Frisch, S. M. (1991). Antioncogenic effect of adenovirus E1A in human tumor cells. *Proc. Natl. Acad. Sci. USA* **88**, 9077–9081.
46. Frisch, S. M. (1994). E1A induces the expression of epithelial characteristics. *J. Cell Biol.* **127**, 1085–1096.
47. Ueno, N. T., Yu, D., and Hung, M. C. (1997). Chemosensitization of HER-2/neu-overexpressing human breast cancer cells to paclitaxel (Taxol) by adenovirus type 5 E1A. *Oncogene* **15**, 953–960.
48. Yoo, G. H., Ensley, J. F., Jacobs, J. R., Sakr, W. A., Johnson, R., Wei, W. Z., Carey, M., and Daifuku, R. (1998). Intratumoral E1A gene therapy for patients with unresectable and recurrent head and neck cancer. *Proc. Annu. Meet. Am. Assoc. Cancer Res.* **39**, 322–322. [Abstract]
49. Villaret, D., Gleich, L. L., Glisson, B., Hanna, E., Kenady, D., Yoo, G. H., Carey, M., and Reynold, T. A. (2000). Multicenter phase II study of E1A lipid complex for intratumoral treatment of patients with recurrent head and neck squamous cell carcinoma. *In* "Proceedings of 5th International Conference on Head and Neck Cancer," pp. 233–233. [Abstract]
50. Myers, J. N., Mank-Seymour, A., Zitvogel, L., Storkus, W., Clarke, M., Johnson, C. S., Tahara, H., and Lotze, M. T. (1998). Interleukin-12 gene therapy prevents establishment of SCC VII squamous cell carcinomas, inhibits tumor growth, and elicits long-term antitumor immunity in syngeneic C3H mice. *Laryngoscope* **108**, 261–268.
51. Lotze, M. T., Zitvogel, L., Campbell, R., Robbins, P. D., Elder, E., Haluszczak, C., Martin, D., Whiteside, T. L., Storkus, W. J., and Tahara, H. (1996). Cytokine gene therapy of cancer using interleukin-12: Murine and clinical trials. *Ann. N. Y. Acad. Sci.* **795**, 440–454.
52. Atkins, M. B., Lotze, M. T., Dutcher, J. P., Fisher, R. I., Weiss, G., Margolin, K., Abrams, J., Sznol, M., Parkinson, D., Hawkins, M., Paradise, C., Kunkel, L., and Rosenberg, S. A. (1999). High-dose recombinant interleukin 2 therapy for patients with metastatic melanoma: Analysis of 270 patients treated between 1985 and 1993. *J. Clin. Oncol.* **17**, 2105–2116.
53. Urba, S. G., Forastiere, A. A., Wolf, G. T., and Amrein, P. C. (1993). Intensive recombinant interleukin-2 and alpha-interferon therapy in patients with advanced head and neck squamous carcinoma. *Cancer* **71**, 2326–2331.
54. Li, D., Jiang, W., Bishop, J. S., Ralston, R., and O'Malley, B. W. J. (1999). Combination surgery and nonviral interleukin 2 gene therapy for head and neck cancer. *Clin. Cancer Res.* **5**, 1551–1556.
55. O'Malley, B. W. J., Li, D., Buckner, A., Duan, L., Woo, S. L., and Pardoll, D. M. (1999). Limitations of adenovirus-mediated interleukin-2 gene therapy for oral cancer. *Laryngoscope* **109**, 389–395.
56. O'Malley, B. W., Cope, K. A., Chen, S. H., Li, D., Schwarta, M. R., and Woo, S. L. (1996). Combination gene therapy for oral cancer in a murine model. *Cancer Res.* **56**, 1737–1741.
57. Sewell, D. A., Li, D., Duan, L., Schwartz, M. R., and O'Malley, B. W. J. (1997). Optimizing suicide gene therapy for head and neck cancer. *Laryngoscope* **107**, t–5.
58. O'Malley, B. W. J., Sewell, D. A., Li, D., Kosai, K., Chen, S. H., Woo, S. L., and Duan, L. (1997). The role of interleukin-2 in combination adenovirus gene therapy for head and neck cancer. *Mol. Endocrinol.* **11**, 667–673.
59. He, Y., Zeng, Q., Drenning, S. D., Melhem, M. F., Tweardy, D. J., Huang, L., and Grandis, J. R. (1998). Inhibition of human squamous cell carcinoma growth in vivo by epidermal growth factor receptor antisense RNA transcribed from the U6 promoter. *J. Natl. Cancer Inst.* **90**, 1080–1087.
60. Benasso, M., Merlano, M., Blengio, F., Cavallari, M., Rosso, R., and Toma, S. (1993). Concomitant alpha-interferon and chemotherapy in advanced squamous cell carcinoma of the head and neck. *Am. J. Clin. Oncol.* **16**, 465–468.
61. Trudeau, M., Zukiwski, A., Langleben, A., Boos, G., and Batist, G. (1995). A phase I study of recombinant human interferon alpha-2b combined with 5-fluorouracil and cisplatin in patients with advanced cancer. *Cancer Chemother. Pharmacol.* **35**, 496–500.
62. Hamasaki, V. K. and Vokes, E. E. (1995). Interferons and other cytokines in head and neck cancer. *Med. Oncol.* **12**, 23–33.
63. Vlock, D. R., Andersen, J., Kalish, L. A., Johnson, J. T., Kirkwood, J. M., Whiteside, T., Herberman, R. B., Adams, G. S., Oken, M. M., and Haselow, R. E. (1996). Phase II trial of interferon-alpha in locally recurrent or metastatic squamous cell carcinoma of the head and neck: Immunological and clinical correlates. *J. Immunother. Emphasis Tumor Immunol.* **19**, 433–442.
64. Ferrantini, M. and Belardelli, F. (2000). Gene therapy of cancer with interferon: Lessons from tumor models and perspectives for clinical applications. *Semin. Cancer Biol.* **10**, 145–157.
65. Nelson, W. G., Simons, J. W., Mikhak, B., Chang, J. F., DeMarzo, A. M., Carducci, M. A., Kim, M., Weber, C. E., Baccala, A. A., Goeman, M. A., Clift, S. M., Ando, D. G., Levitsky, H. I., Cohen, L. K., Sanda, M. G., Mulligan, R. C., Partin, A. W., Carter, H. B., Piantadosi, S., and Marshall, F. F. (2000). Cancer cells engineered to secrete granulocyte-macrophage colony-stimulating factor using ex vivo gene transfer as vaccines for the treatment of genitouri nary malignancies. *Cancer Chemother. Pharmacol.* **46**(Suppl.) S67–S72.
66. Yver, A., Dreiling, L., Mohanty, S., Merritt, J. A., Proksch, S., Shu, C. H., and Tomko, L. S. (2000). Tolerance and safety of RPR/INGN 201, an adeno-viral vector containing a p53 gene, administered intratumorally in 309 patients with advanced cancer enrolled in phase I and II studies world-wide. *Proc. Annu. Meet. Am. Soc. Clin. Oncol.* **19**, 1806–1806.[Abstract]
67. Inoue, A., Narumi, K., Matsubara, N., Sugawara, S., Saijo, Y., Satoh, K., and Nukiwa, T. (2000). Administration of wild-type p53 adenoviral vector synergistically enhances the cytotoxicity of anti-cancer drugs in human lung cancer cells irrespective of the status of p53 gene. *Cancer Lett.* **157**, 105–112.
68. Bischoff, J. R., Kirn, D. H., Williams, A., Heise, C., Horn, S., Muna, M., Ng, L., Nye, J. A., Sampson-Johannes, A., Fattaey, A., and McCormick, F. (1996). An adenovirus mutant that replicates selectively in p53-deficient human tumor cells. *Science* **274**, 373–376.
69. Khuri, F. R., Nemunaitis, J., Ganly, I., Arseneau, J., Tannock, I. F., Romel, L., Gore, M., Ironside, J., MacDougall, R. H., Heise, C., Randlev, B., Gillenwater, A. M., Bruso, P., Kaye, S. B., Hong, W. K., and Kirn, D. H. (2000). A controlled trial of intratumoral ONYX-015, a selectively-replicating adenovirus, in combination with cisplatin and 5-fluorouracil in patients with recurrent head and neck cancer. *Nature Med.* **6**, 879–885.
70. Nemerow, G. R. (2000). Cell receptors involved in adenovirus entry. *Virology* **274**, 1–4.

C H A P T E R

38

Immunology and Immunotherapy of Head and Neck Cancer

TERRY Y. SHIBUYA
Department of Otolaryngology—Head and Neck Surgery and the Chao Family Comprehensive Cancer Center
University of California, Irvine
College of Medicine
Orange, California 92868

LAWRENCE G. LUM
Roger Williams Cancer Center
Providence, Rhode Island 02908

TOMASZ PAWLOWSKI
Department of Pathology
University of California, Irvine
College of Medicine
Orange, California 92868

THERESA L. WHITESIDE
Departments of Otolaryngology—Head and Neck Surgery and Pathology, Pittsburgh Cancer Institute
Pittsburgh, Pennsylvania 15213

I. Introduction 569
II. Principles of Immunotherapy 570
  A. Types of Immune Response to Tumor 570
III. Immune Therapeutic Strategies 571
  A. Historical Perspective 571
  B. Antigen-Specific Cytotoxic T Lymphocytes 572
  C. Dendritic Cells 572
  D. Donor Lymphocyte Infusions 572
  E. Activated T Cells 573
  F. Anti-CD3/Anti-CD28 Coactivated T Cells 573
  G. Arming of T Cells with Bispecific Antibodies 573
  H. T Cells with Chimeric Receptors 574
IV. Immunology of Head and Neck Squamous Cell Carcinoma (HNSCC) 574
  A. Hierarchy of Immunosuppression 574
  B. Clinical Implications of Suppressed Immune Function 574
  C. Causes of Immunosuppression 575
V. Immunotherapy in Head and Neck Cancer 576
  A. Immunotherapies Active or Passive 576
  B. History of Immunotherapy in HNSCC 577
  C. Current Immunotherapeutic Trials in HNSCC 578
  D. Future Prospects in Immunotherapy of HNSCC 580
VI. Conclusion 581
References 581

Head and neck cancer is the sixth most commonly reported cancer. Cure rates for this disease have remain unchanged since the 1970s under present treatment strategies, which include surgery, radiation therapy, and/or chemotherapy. New treatment options must be developed. Tremendous advances have occurred over the past two decades in the areas of immunology and molecular immunology. Immunotherapy is one novel therapeutic option, which has had some success in treating this cancer. This chapter reviews the (1) principles of immunotherapy, (2) current immunotherapeutic strategies, (3) immunology of head and neck squamous cell carcinoma (HNSCC), and (4) immunotherapies used to treat HNSCC past, present, and future.

I. INTRODUCTION

Head and neck squamous cell carcinoma (HNSCC) comprises 80% of cancers involving the upper aerodigestive tract with the main etiologies being tobacco and alcohol consumption. The standard therapies for this disease have been surgery and radiation therapy with adjuvant chemotherapy showing benefit in the treatment of nasopharyngeal carcinoma [1]. The addition of combined-modality chemoradiation therapy has been used to treat unresectable tumors, organ preservation of the larynx and hypopharynx, and poor-prognosis resectable disease. Unfortunately, even with advances in surgery, radiation therapy and chemotherapy survival rates have changed very little since the 1970s. Additional forms of therapy must be developed and added to our present treatment strategies to improve survival for this disease.

Advances in molecular biology and immunology have exploded over the past two decades, and advances in these areas must be added to our present treatment strategies. Experimental clinical trials using immunological agents have shown promise in the treatment of HNSCC and support the further development of immune based therapies. This chapter focuses on our present understanding of (1) principles of immunotherapy, (2) current immunotherapeutic strategies, (3) the immunology of HNSCC, and (4) immunotherapies used to treat HNSCC past, present, and future.

II. PRINCIPLES OF IMMUNOTHERAPY

The theory of immunosurveillance postulates that the immune system, composed of effector cells and associated molecules, can recognize and eradicate neoplastic cells, thereby preventing tumor formation. This theory was supported by the work of Paul Ehrlich and other notable scientists [2,3]. Evidence supporting this theory stems from cases of spontaneous cancer regression in patients with advance disease, as well as the development of malignancies in immune-suppressed individuals. Patients with primary immunodeficiencies have up to a 100-fold increased risk of developing cancer [4,5]. Examples of such immunodeficiencies include Wiskott–Aldrich syndrome, ataxia-telangiectasia, common variable immunodeficiency, or severe combined immunodeficiency [6,7]. The types of cancers observed in these individuals include non-Hodgkin's and Hodgkin's lymphoma, leukemias, and gastric carcinoma. Patients with human immunodeficiency virus (HIV) have an increased incidence of Kaposi's sarcoma and non-Hodgkin's lymphoma as well [8–10]. Head and neck cancer has also been identified in this patient population [11]. Renal transplant recipients also have an increased risk for developing HNSCC [12,13].

The concept of harnessing the immune system to treat cancer is not a new idea. Dating back to the 18th century there is documentation of physicians in Paris using pus to treat a patient who had advanced stage breast cancer and documenting an antitumor response [14]. Others in the 19th century administered streptococci from patients with erysipelas to cancer patients and noted an antitumor effect [15,16]. Coley [17] advanced these findings and treated a number of patients with extract from streptococci and *Bacillus prodigiosus* or *Serratia marcescens* and noted an antitumor response. These extracts became known as Coley's toxins and received considerable attention [17]. Today, *Bacille Calmette–Guerin* (BCG) is one of the more widely recognized immunostimulatory agents used to treat superficial bladder carcinoma [18]. A variety of nonspecific immunotherapeutic agents have been tested for cancer treatment, including levamisole and thymic extracts.

A. Types of Immune Response to Tumor

The immune system is a complex system in which the body uses effector cells and humoral mediators to react to exogenous and endogenous foreign antigens. The system attempts to reject what it identifies as foreign. The immune system is divided into natural or innate and acquired or specific immune systems. The natural or innate immune system requires no prior sensitization and is a nonspecific defense against invading microbes or foreign antigens and also recruits other inflammatory cells to the area of invasion. Immune cells associated with this defense include macrophages, neutrophils, eosinophils, basophils, and natural killer (NK) cells. Other components include the complement system and various cytokines. The acquired or specific immune system is enhanced by previous exposure to a foreign antigen and is mediated by B cells producing antibodies (humoral immunity) and T cells (cellular immunity) secreting cytokines. The acquired immune response may be a primary (first exposure) or a secondary (previous exposure) response.

In terms of an immune response to cancer, an enhanced acquired or specific immune response is highly desirable. Unfortunately, the tumor will use a variety of immune-suppressive mechanisms to fight back and prevent an immune response. The immune-specific response mediated by T lymphocytes may be divided into type 1 (Th1, cellular immune response) or type 2 (Th2, humoral immune response). In the classic example of T-cell activation, T cells respond through a sequence of events in which an antigen-presenting cell (APC) processes a peptide/protein and presents it on its surface membrane in the form of an antigen bound to an major histocompatibility complex (MHC). If the peptide is presented to T cells in the context of a class I MHC, then CD8 or cytotoxic T lymphocytes are stimulated, which kill tumor cells by releasing granules containing cytolytic enzymes and soluble inducers of apoptosis. If the peptide is presented to T cells in the context of a class II MHC, then CD4 or helper T lymphocytes are stimulated, which proliferate and secret interleukin(IL)-2 and other soluble immunostimulatory molecules. B cells, macrophages, and NK cells are also stimulated, and a humoral or antibody immune response is generated. After this first signal, a second or costimulatory signal is needed to optimize T-cell activation. The best-characterized costimulatory molecules are the B7.1 (CD80) or B7.2 (CD86) ligands of the APC binding the CD28 receptor of the T cell (Fig. 38.1). A strong

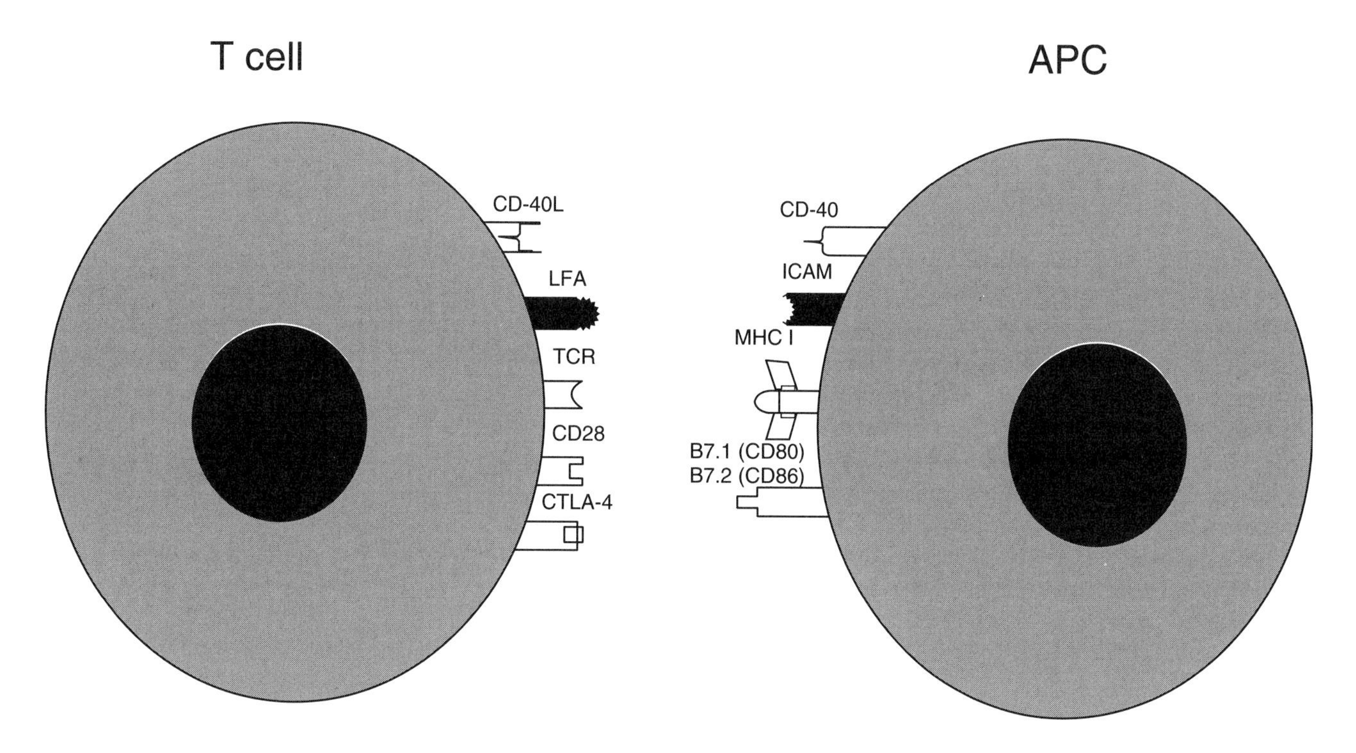

FIGURE 38.1 Activation of a T lymphocyte by an antigen-presenting cell (APC) occurs by (1) adhesion via an intercellular adhesion molecule (ICAM) to a lymphocyte function-associated antigen (LFA), (2) presentation of peptide antigen via a major histocompatability complex (MHC) to the T-cell receptor (TcR or CD3), and (3) costimulations of the B7.1 (CD80) or B7.2 (CD86) ligand to the CD28 receptor of the T cell. Alternatively, costimulation can occur via CD40:CD40L binding.

anticancer immune response is considered favorable in curing and preventing cancer. Ideally, most tumor immunologists consider a Th1 or cytolytic immune response that generates a population of cytotoxic T lymphocytes as the optimal immune response in the fight against cancer. The following section addresses the many immunotherapeutic strategies presently in use today.

III. IMMUNE THERAPEUTIC STRATEGIES

There are a vast number of immunotherapeutic strategies used to fight cancer. This section focuses primarily on approaches that use the T cell as a therapeutic platform for immunotherapy (Table 38.1). T cells serve as specific and nonspecific cytotoxic effectors and mobile cytokine factories that can be redirected by bispecific antibodies (BiAb) or engineered chimeric receptors to target specific target tumor-associated antigens (TAA). Without reviewing all of the approaches, we focus on newer strategies. Studies involving NK cells, lymphokine-activated natural killer cells (ANK), and tumor-infiltrating lymphocytes (TIL) are highlighted to provide historical perspective. Greater detail is provided on the more recent approaches using antigen (Ag)-specific cytotoxic T lymphocytes (CTL), dendritic cells (DC), activated T cells, anti-CD3/anti-CD28-coactivated T cells (COACTS), arming of T cells with BiAbs, and T cells with chimeric receptors.

A. Historical Perspective

Natural killer cells ($CD3^-$, $CD2^+$, $CD16^+$, $CD56^+$ cells) become lymphokine-activated natural killer cells in the presence of IL-2 exhibiting non-MHC-restricted cytotoxicity [19–26]. Infusions of ANK and IL-2 have been used to treat renal cell carcinoma (RCC) and malignant melanoma (MM) [24,27–32] with clinical responses up to 20%. Infusions of *ex vivo* expanded TIL along with high-dose IL-2 to treat patients with RCC or MM have shown similar responses [33–36]. TIL were reported

TABLE 38.1 Immunotherapeutic Options

Immunotherapeutic options with T-cell-related platform
Cytotoxic T lymphocytes
Dendritic cells
Donor lymphocyte infusions
Activated T cells
Anti-CD3/anti-CD28 coactivated T cells
T cells armed with bispecific antibodies
T cells armed with chimeric receptors

to be tumor specific and traffic to tumor sites. Subsequent studies showed that high-dose IL-2 was providing the antitumor effect. Preeffector T cells from tumor-draining lymph nodes can be expanded with anti-CD3 stimulation and IL-2 to mediate the regression of established metastases in a murine sarcoma model [37]. This approach has met some measure of success. In a clinical study involving 11 patients with RCC and 11 patients with MM, 6 of 11 RCC patients had clinical response and 1 of 11 MM patients had a partial response [38]. Studies using human autolymphocyte therapy (ALT) produced by stimulating peripheral blood lymphocytes (PBL) with medium conditioned by stimulating PBL with anti-CD3 (OKT3) has improved survival in a series of 90 patients with metastatic RCC randomized to receive cimetidine or cimetidine plus ALT. One billion ALT were infused monthly for 6 months without toxicity [39]. Survival in those receiving ALT was 2.5 times greater than that seen for patients in the cimetidine group ($p=0.008$). These results were confirmed in a multi-institutional study involving 355 patients [40].

B. Antigen-Specific Cytotoxic T Lymphocytes

An additional attractive approach was the development of Ag-specific CTL directed at viruses or TAA [41–44]. Cytomegalovirus (CMV)-specific CTL to prophylaxis against the development of CMV pneumonia were given to CMV seropositive bone marrow transplantation (BMT) recipients [45,46]. The production of tumor-specific CTL directed at oncogenic products such as p21 ras [47] and p53 [49–52] has been described; but this approach has not made it to the clinic. However, Epstein–Barr virus (EBV)-specific CTL produced by stimulating donor T cells with EBV-transformed recipient B lines have been used successfully to treat EBV lymphoproliferative disorders that develop after T-cell-depleted allogeneic BMT [53,54]. Because producing Ag-specific CTL is labor-intensive and time-consuming, new strategies for producing Ag-specific CTL are needed.

C. Dendritic Cells

Dendritic cells from peripheral blood or bone marrow have been used to induce tumor-specific CTL [55–61]. Because DC are the most effective Ag-presenting cells of the immune system, strategies that employ DC have become popular. DC specialize in the acquisition and transport of Ag to lymph nodes where binding to T cells with costimulatory receptors leads to activation of T cells and a cascade of events that generate Ag-specific CTL [56,62]. Although they are not a single phenotype, they can be characterized as DR^+, $CD1a^+$, $CD3^-$, $CD14^-$, $CD19^-$, $CD40^+$, $CD80^+$ (B7-1), $CD83^+$, and $CD86^+$ (B7-2) cells that exhibit perinuclear $CD68^+$ and "veils" or lamellipodia [60,63]. They express high levels of adhesion molecules such as LFA-1, LFA-2, LFA-3, ICAM-1, ICAM-3, and intercellular adhesion molecule-1 [60,63].

Adding 7-day cultured DC pulsed with proteins or peptides to PBMC can induce tumor-specific CTL [43,55,57,64]. The DC are usually prepared in cultures containing granulocyte–macrophage colony-stimulating factor (GM-CSF), IL-4, and tumor necrosis factor(TNF)-α, which are used to expand and differentiate cells into DC [58,59,63].

Peptide-pulsed DC are promising as a vaccine strategy. Exposing DC with tumor peptides and infusing the DC induce the development of tumor-specific CTL [55,59–61,65]. DC are class I restricted in their ability to present peptide Ags to T cells [55,57,63]. For example, HLA-DR1-restricted bcr-abl (b3a2)-specific $CD4^+$ cells can respond to DC pulsed with b3a2 peptide cell lysates [66]. DC produced from the PBL of patients with advanced prostate cancer, which were pulsed with prostate-specific membrane antigen, could induce a clinical anticancer response [67,68]. Infusions of DC were well tolerated and suggest that DC immunizations may be clinically useful. A vaccination trial using HLA-A2-restricted immunodominant peptides from the gp100 melanoma-associated Ag and IL-2 testing 31 patients with metastatic malignant melanoma has been performed [69]. Thirteen of 31 (42%) had objective responses and 4 patients had mixed or minor responses. Similarly, carcinoembryonic antigen (CEA) or peptides of CEA were used to load DC for the treatment of colon and breast cancer. In summary, follow-ups of vaccine studies will determine whether tumor lysates [70], tumor proteins, HLA-A2-restricted peptides [66,67,69], or RNA from tumors [71] will be optimal for inducing clinical responses.

D. Donor Lymphocyte Infusions

Donor lymphocyte infusions (DLI) have been use to treat the relapse of hematologic malignancies after allogeneic BMT. The graft-verses-leukemia (GVL) effect is most pronounced in patients who receive DLI for relapse after allogeneic BMT [72,73]. The effect of infusing donor lymphocytes is well described for relapse after allogeneic BMT for chronic myelogenous leukemia (CML), acute myelogenous leukemia (AML), acute lymphocytic leukemia (ALL), and myelodysplastic syndrome (MDS) [74]. The results are quite dramatic in patients who relapse with CML after T-cell-depleted BMT [73,74]. The development of graft-versus-host-disease (GVHD) tends to correlate with the likelihood of responses to DLI [75].

A series from 25 North American BMT programs consisting of 140 patients with CML, AML, and ALL has been published [76].

E. Activated T Cells

Binding of the T-cell receptor (TCR) with low doses of anti-CD3 mAb induces T-cell proliferation, cytokine synthesis, and cytotoxicity [77–80]. Activated T cells (ATC) mediate non-MHC-restricted cytotoxicity, secrete tumoricidal cytokines (IFN-γ, TNF-α, or GM-CSF), and serve as vehicles to deliver gene products or targeting antibodies. Infusions of ATC have been reported to reduce liver metastases in an adenocarcinoma model [81] and to prevent deaths in an established human HT29 carcinoma model [82]. ATC infused with syngeneic bone marrow increased the survival of mice preinjected with lymphoma [83], showing that ATC can provide an antilymphoma effect. Human ATC can be expanded from blood or bone marrow from normal donors and patients with malignancy and mediate non-MHC-restricted cytotoxicity [84–94]. ATC kill Daudi cells (ANK sensitive targets), K562 cells (NK sensitive targets), leukemic blasts [95,96], neuroblastomas [87], and autologous plasma cells in multiple myeloma [97].

A clinical trial using ATC in solid tumor patients revealed that PBL activated with OKT3 for 18 hr and given with IL-2 infusions to patients with RCC and MM [98] led to lymphocytosis (50,000 cells/μl) likely due to IL-2. Phase I clinical trial using OKT3-activated $CD4^+$ cells and IL-2 after cyclophosphamide showed promise with the induction of one complete responder, two partial responders, and eight minor responders in patients with advanced malignancies [99].

Because immunotherapy may be most effective in minimal disease, we evaluated the safety and efficacy of multiple infusions of ATC after peripheral blood stem cell transplantation (PBSCT) for high-risk breast cancer in combination with low-dose IL-2 and GM-CSF [100]. Twenty-three women with advanced stage breast cancer have received immunotherapy after PBSCT consisting of eight doses of 10×10^9 ATC twice/week for 3 weeks followed by six doses of 20×10^9 ATC/week, continuous infusion, or subcutaneous IL-2 for 65 days after PBSCT and GM-CSF between days 5 and 21. ATC-related toxicities were minimal and toxicities did not preclude completing the ATC infusions. Seventy percent of 23 patients survived and 50% are progression free up to 32 months after stem cell transplant [101].

F. Anti-CD3/Anti-CD28 Coactivated T Cells

Cross-linking of the TCR with anti-CD3 triggers a signaling cascade resulting in T-cell proliferation, cytokine synthesis, and immune responses [77–80]. However, optimal activation and proliferation require costimulation of CD28 receptors on T cells with anti-CD28, B7.1, or B7.2 (CD80 or CD86) molecules [102–106]. These interactions enhance proliferation and stabilization of cytokine mRNAs and GM-CSF [107]. Costimulation also leads to enhanced production of chemokines, which may augment recruitment of antitumor effector cells [108]. COACTS exhibit non-MHC-restricted cytotoxicity [109]. They produce TH1-type cytokine profiles [102,110] and may survive longer after infusions due to induction of Bcl-x_L, a gene that confers resistance to cell death [111,112]. Costimulation with anti-CD28 may overcome suppressed anti-CD3-stimulated immune responses of lymph node lymphocytes in patients with head and neck squamous cell carcinoma [113].

We completed a phase I dose study, that involved infusion of COACTS for the treatment of refractory cancers without any dose-limiting toxicities [114]. Infusions of COACTS were safe, induced detectable serum levels of IFN-γ, GM-CSF, and TNF-α, and significantly enhanced the ability of freshly isolated PBMC to secrete IFN-γ and GM-CSF upon *in vitro* anti-CD3/anti-CD28 costimulation. These data show that the host immune systems were modulated by the therapy.

G. Arming of T Cells with Bispecific Antibodies

Bispecific antibodies combine the targeting ability of monoclonal antibodies (mAbs) with the cytotoxicity mediated by T cells to lyse tumors. Arming of activated T cells with BiAb would increase the precursor frequency of CTL directed at specific tumor antigens. Treatment with armed ATC may lead to specific binding and enrichment of effector cells at the tumor site. BiAbs have been used for targeting drugs, prodrug activation, and immune recruitment strategies [115]. The anti-CD3-based bispecific antibodies reported include bispecific antibodies targeting tenascin (human glioma) [116], CD13 (acute myeloid leukemia) [117], MUC1 [118], 17-1A (EpCAM) [119], OC/TR (folate receptor on ovarian carcinoma) [120–123], kDal K29 (renal cell carcinoma) [124], G250 (renal cell carcinoma) [125], transferrin receptor [126], AMOC-31 [127], and CD19 (malignant B cells) [128–133]. T cells and chimeric mouse/human-chimeric BiAb are reactive to human carcinoembryonic Ag-expressing cells [134] and anti-CD30 (Hodgkin's lymphoma) [135-137]. A preclinical study in this laboratory shows that ATC armed with low doses of anti-CD3 x anti-HER2/*neu* BiAb effectively lyse breast cancer cell lines [138], as well as prostate cancer cell lines (L.G. Lum, unpublished results).

Clinical studies using BiAbs to arm granulocytes, monocytes, or NK cells show promise. Anti-CD16 was used to target the Fc γ receptor on white blood cells and second antibody targeted human melanoma cells [139]. In a clinical trial involving 27 patients with breast cancer, infusions of 2B1 BiAb, which binds to HER2/*neu* and Fc γ receptor III, led to two partial and three minor clinical responses [140]. Subsequently, MDX-H210 (anti-CD64×HER2/*neu*) was used to treat tumors overexpressing HER-2/*neu* in women with breast cancer, resulting in one partial response and one mixed response in 10 evaluable breast cancer patients [141].

H. T Cells with Chimeric Receptors

Although antibody and specific T cell approaches are highly specific, both approaches alone did not improve clinical results. A new approach placed the scFv that targets tumor on the surface of T cells [142–147]. Hence the term "T body" was coined for T cells transfected with a chimeric receptor containing the variable region responsible for binding to the TAA. Most T-body constructs include heavy and light chain-derived "V" regions. This type of construct redirects nonspecific cytotoxicity to TAA [145]. Clinical trials using T bodies have not been remarkable. *Ex vivo*-expanded gene-transduced T cells downregulate their expression of the transgenes [148,149]. This may be a critical barrier to the clinical application of T bodies. If the expression of T-body receptors is low when the T cell encounters Ag, the T cell may not activate. Laboratory studies suggest that the expression of the transgene in T cells may be downregulated and therefore T bodies are not effective [149].

In summary, as technology to immunize cancer patients develops, it is likely that the greatest antitumor effect will be from multidisciplinary approaches that involve surgical debulking, irradiation, chemotherapy, immunomodulation using cell products and/or biologic response modifiers, and potentially stem cell rescue.

IV. IMMUNOLOGY OF HEAD AND NECK SQUAMOUS CELL CARCINOMA (HNSCC)

Head and neck cancer patients have an abnormally functioning immune system. Studies testing the immune function of patients with HNSCC support the concept of a "hierarchy of immunosuppression" existing in these individuals. Immune reactivity is maximally suppressed in tumor-infiltrating lymphocytes, followed in descending order by proximal lymph node lymphocytes, distal LNL, and peripheral blood lymphocytes [150–155].

A. Hierarchy of Immunosuppression

Tumor-infiltrating lymphocytes are immunosuppressed; functional studies show a poor response to mitogens and reduced cytotoxicity to autologous tumor [156–166]. $CD34^+$ cells are present in TIL and secret immunosuppressive factors that block T-cell function [167]. Reduced numbers of $CD8^+$ cells and $CD4^+$ cells with an impaired ability to release IL-2 and IFN-γ have been identified [168,169]. Composition of TIL [170–175] reveals $CD3^+$ cells [175–177] with an equal distribution of $CD4^+$ and $CD8^+$ cells. $CD4^+$ cells are located primarily in the tumor stroma and $CD8^+$ cells within the tumor parenchyma [156]. NK cells [175], dendritic cells [178], and macrophages [157,179,180] comprise less than 5% of the population. Approximately 40% of the T cells present express activation marks of HLA-DR, IL-2R, or transferrin [181].

Lymph node lymphocytes are the next immunosuppressed population. Regional LNL have reduced NK and LAK cell activities, mitogen responses, and cytokine production compared to PBL [154,155,182–185]. LNL in close proximity to tumor show a weaker response to IL-2 stimulation compared to distal LNL. Lymphocytes from LN containing metastatic cancer show weak cytolytic function compared to tumor-free LN [154,184]. $CD4^+$ cells are the major T-cell population present in LN positive or negative for metastatic tumor [177,186,187].

Peripheral blood lymphocytes are the least immunosuppressed population. Early in the disease process, PBL responses to mitogens such as phytohemaglutinin (PHA), concavalin A (Con A), or other stimuli are near normal but decrease with progression of cancer [182,188–197]. Patients with early stage HNSCC have normal T-cell counts and CD4:CD8 ratios in their peripheral blood, whereas patients with advanced disease have reduced counts, decreased numbers of $CD4^+$ cells, and slight reductions in $CD8^+$ cells [187,188,198–204].

B. Clinical Implications of Suppressed Immune Function

The immune response of HNSCC patients has correlated with clinical outcomes. Strong reactivity in the regional lymph nodes (LNs) as measured by histology and flow cytometry has correlated favorably with survival [205,206]. Lymph nodes were graded as T-cell predominant, B-cell predominant, unstimulated, or lymphocyte depleted. Weak reactivity or lymphocyte depletion correlated with the poorest survival [205–208].

In prior studies from our labaratory, we identified a subpopulation of HNSCC patients who have suppressed function of their CD3 receptor. Stimulating the CD3 receptor with the anti-CD3 monoclonal antibody did not result in cellular proliferation in 25% of patients with advanced stage HNSCC who were tested. Following these patients over a 2-year period, we found 71% of individuals developed a recurrent or metastatic cancer, whereas a group of matched

HNSCC patient who responded to anti-CD3 stimulation had only a 15% rate (T.Y. Shibuya, unpublished data). These findings confirmed the findings of others that show that immune function correlates with clinical outcomes [209].

C. Causes of Immunosuppression

There are multiple causes for poor immune function in HNSCC patients. Patients frequently experience immunosuppression from tobacco, alcohol, and aging. Also, poor nutrition, tumor secretion of suppressive factors, tumor expression of suppressive factors, and alteration in T-cell function all have been identified as contributing to poor immune function (Table 38.2).

1. Nutrition

The nutritional status of HNSCC patients is frequently poor. Advanced stage cancer involving the upper aerodigestive tract disrupts the mechanical function of mastication and swallowing, resulting in reduced oral intake. In addition, patients will frequently consume large quantities of alcohol to reduce the pain while neglecting a healthy diet. Because of these problems, patients with advanced disease often experience a negative protein balance, resulting in poor wound healing and anergy. Poor nutrition has been associated with cellular immune deficiencies [210]. Moderately reduced protein levels in the presence of trace mineral and vitamin deficiencies have been associated with poor cellular-mediated immunity as well [211].

Of the trace minerals necessary for proper immune function, zinc is of particular importance. Zinc deficiencies have been associated with thymic involution and defective cellular-mediated immunity [211]. Zinc is normally incorporated into the thymus and used for the formation of zinc-thymulin, which regulates T-cell responses to interleukins [212,213] and maintains cellular immunity. Reduced levels of zinc have been identified in HNSCC patients [186,187,211,214] and zinc is an important element lacking in this patient population.

2. Tumor-Secreted Factors

HNSCC cells have been found to secrete a number of factors, which may account for immunosuppression. Prostaglandins (PGs) have been isolated from tumor extracts [169,215] as well as cancer cell lines [185]. PGs are known to downregulate T-cell responses in a number of cancers [216]. Indomethacin and other non-steroidal anti-inflammatory drugs (NSAIDs) have been used to suppress PG production in HNSCC patients [217–219].

P15-E, an immunosuppressive peptide, has also been detected in HNSCC [220,221]. P15-E is a murine retroviral peptide important in retroviral pathogenesis, and genetic material coding for this peptide has been isolate from tumors. Specifically, this peptide has been shown to impair T-cell function and inhibit monocyte chemotaxis [222].

HNSCC has also been found to secrete cytokines such as IL-4, IL-6, and GM-CSF. Extracts from homogenized tumors have also expressed these cytokines, although derivation from immune effector cells has not been ruled out [223]. Work from others has also identified GM-CSF, IL-10, and transforming growth factor-β (TGF-β) secreted from HNSCC cells [169]. Presence of these factors has been associated with a shift in immune response from a type 1 (TH1) immune response to a type 2 response (TH2), which is inhibitory in the fight against cancer.

3. Tumor-Expressed Factors

Cancers use a number of strategies to fight back against the host immune response. Interestingly, tumor expression of the Fas-ligand (Fas-L) has been identified in melanoma, hepatocellular carcinoma, gastric adenocarcinoma, esophageal carcinoma, and, more recently, HNSCC [224–227]. Tumor-expressing Fas-L binds the Fas receptor of the T cell, resulting in T-cell apoptosis and tumor escape [227]. Evidence suggests that TIL are highly susceptible to Fas-L-mediated lysis by the tumor [228]. The extent of T-cell apoptosis varies regionally within the tumor in relation to the local status of Fas-L expression. For example, in esophageal carcinoma, there was a fivefold reduction of TIL within Fas-L-positive tumor nests and a twofold increase in TIL within Fas-L-negative tumor nests [225]. These data strongly suggest that Fas-L is used as a tumor escape mechanism [229].

The IL-4 receptor has been identified on HNSCC cell lines and fresh biopsy specimens [230]. The addition of IL-4 to tumor cells expressing the IL-4 receptor has led to increased tumor growth in animal studies. Studies have also shown that

TABLE 38.2 Immunosuppression in Head and Neck Cancer Patients

Causes of immunosuppression	
Poor nutrition	Decreased protein, trace minerals, vitamins, zinc
Tumor-secreted factors	Prostaglandins, P15-E, TGF-β, cytokines (IL-4, IL-10, etc)
Tumor-expressed factors	Fas-L, IL-4 receptor
Altered immune effectors	T-cell receptor poorly functional, CD34 suppressor cells present, suppressed natural killer cell population

IL-4 levels are enhanced in the peripheral blood of smokers [231]. It is an interesting possibility that tumors use immunologic factors secreted by the host to enhance their own growth.

4. Alteration in Immune Effectors

Defective signal transduction via the TCR has been demonstrated in TIL from melanoma, ovarian, colon, and renal cell carcinomas [232–238]. Studies from our labaratory have identified a similar defect present in T cells found in HNSCC patients [239]. Specifically, we have shown impaired reactivity to anti-CD3 (αCD3) stimulation in LNL and PBL from advanced stage HNSCC patients. Others have shown that there is a decreased expression of ζ and ε chains, decreased Ca^{2+} flux, and impaired kinase activity following αCD3 stimulation [240]. A method of overcoming suppressed T-cell CD3 receptor function must be developed.

A population of immature $CD34^{+}$ suppressor cells is present in the TIL of HNSCC patients [168,240]. The maintenance of these immature $CD34^{+}$ cells in a suppressor state has been postulated to be by vascular endothelial growth factor (VEGF) [241,242]. The secretion of VEGF has been identified in HNSCC [243]. A method of maturing these immature $CD34^{+}$ suppressors into nonsuppressors must be developed.

Natural killer cell activity is suppressed in HNSCC patients. Maximal suppression occurs in TIL [150–154]. NK cell function from regional lymph nodes is also inhibited [155]. Studies have shown peripheral blood NK cell activity of normal donors to be significantly stronger than cancer patients [152]. Lymph node lymphocytes in close proximity to the tumor are more resistant to IL-2-induced stimulation and cytotoxicity compared to distally located LNL [154].

5. Testing of Immune Competence in HNSCC Patients

There are no tests considered standard for measuring the immune competence of HNSCC patients. Testing of immune competence in this patient population started in the early 1970s. Skin testing for a delayed-type hypersensitivity (DTH) response to recall antigens such as purified protein derivative (PPD) or candida has been performed [244]. Lack of a DTH response predicted a poor outcome. Skin testing for a response to dinitrochlorobenzene (DNCB) was performed on over 1000 patients with HNSCC. Approximately 46% of individuals were anergic to DNCB testing versus 5% of normal controls. Patients with late-stage inoperable disease had greater than 85% anergy. A correlation of response and recurrence was also identified with survival frequency higher in responders than in nonresponders [244]. DNCB measures the capacity of the cellular immune system to respond to a new antigen, which encompasses the following steps: antigen presentation, T-cell activation, proliferation, cytokine production, and monocyte/macrophage recruitment. Presently, this test is no longer used because of its nonspecificity and time consumption [244]. Skin testing with phytohemagglutinin (PHA) [245] and natural cytokine mixture (NCM) [246,247] has also been examined as possible methods of assessing cell-mediated immunity (CMI).

V. IMMUNOTHERAPY IN HEAD AND NECK CANCER

Present therapy of head and neck cancer includes surgery, radiotherapy, and sometimes chemotherapy. Although substantial progress has been made in the surgical resection and reconstruction of HNSCC patients and the benefits of chemotherapy and radiotherapy have been confirmed, the survival of patients with HNSCC has not improved since the 1970s [248]. Of patients with early, stage I and II disease, approximately 75% can be cured. This percentage drops to approximately 25% in patients with advanced disease. The lack of effective therapy preventing the development of a recurrence or second primary cancer is a strong motive for the development of novel therapeutic strategies in HNSCC.

Immunotherapy represents one novel approach, which has had some success in the past using biologic agents to stimulate the host immune system to fight cancer. Presently there are several active multi-institutional, phase III clinical trials testing immune-stimulating agents in patients with solid tumors. The use of immunotherapy in HNSCC has been much slower to progress compared to other cancers, and there may be a valid reason for this. Patients with HNSCC often have immunologic abnormalities, as discussed previously, which make it difficult to stimulate the compromised immune system to eradicate the cancer effectively. However, if the immune system plays an important role in controlling tumor growth, then a good rationale exists for the use of immunotherapy in HNSCC. Head and neck cancers are generally well infiltrated with activated T cells and dendritic cells [249]. Both the presence of DC and normal functioning (i.e., normally signaling) T lymphocytes in the tumor are biomarkers associated with improved prognosis and a better 5-year survival [250]. This type of evidence suggests that if immune cells in the tumor or in the peripheral circulation can be protected from the loss of antitumor function, then chances for patient survival could be improved significantly. Furthermore, it appears that the therapeutic use of immune-stimulating agents can have beneficial effects in patients with HNSCC, as reviewed briefly here.

A. Immunotherapies Active or Passive

Two types of immunotherapy have been used for the treatment of cancer: active and passive. Each of these two categories can be further subdivided into specific and nonspecific therapies. The former involves therapeutic induction of tumor-specific T cells or antibodies, whereas the latter aims at the global upregulation of host immune responses. Cancer vaccines are the best example of specific active immunotherapy, whereas cytokines or immune stimulants such as the BCG vaccine or *Corynebacterium parvum* are examples of nonspecific therapies. Adoptive cell transfers using tumor-sensitized T cells fall in the category of specific passive immunotherapy, whereas transfers of lymphokine-activated killer (LAK) or natural killer cells or genetically modified tissue cells exemplify passive nonspecific therapy. Over the years, several of these therapies have been used in patients with HNSCC.

B. History of Immunotherapy in HNSCC

The earliest attempts at immunotherapy in patients with HNSCC involve the use of nonspecific immune stimulants such as BCG, *C. parvum*, and various cytokines (Table 38.3). The BCG vaccine containing an attenuated strain of the tuberculous bacillus was used in the early 1970s in conjunction with chemotherapy, but this trial generated mixed and controversial results [251]. The controversy focused on severe toxicity experienced by a small subset of patients and only marginal therapeutic effects [251]. *C. parvum* is a potent macrophage stimulant, which was used together with methotrexate for the therapy of recurrent and metastatic HNSCC, but did not improve survival or the response rate beyond that observed with methotrexate alone [252].

Immunotherapy with cytokines, which started in the 1980s and continues to be evaluated for treatment of cancer, has met with only a limited success in HNSCC (Table 38.3). Interleukin-2 and interferon-α were the first cytokines used in therapy of HNSCC [253–266]. In a preliminary trial of nonrecombinant IFN-α in patients with HNSCC, Vlock and colleagues [253] demonstrated tolerable toxicity and potential antitumor activity of this cytokine. In the subsequent phase II trial of recombinant IFN-α2b, 71 HNSCC patients with recurrent or metastatic disease were randomized to a low (6×10^6U/m^2 daily ×3 every 4 weeks) or a high (12×10^6 U/m^2 daily ×3/week) dose of IFN-α [254]. The low, but not high, dose of IFN-α was well tolerated. Although the overall response rate was unimpressive (1 CR in low and 2 CR in high IFN groups), disease stabilization was noted in several patients [254]. Early clinical trials of IL-2 in HNSCC involved repeated perilymphatic or peritumoral injections of natural IL-2 in patients with recurrent or inoperable disease [255–258]. Local administration of low doses of IL-2 resulted in CR or PR in approximately 30% of patients; unfortunately, these responses were of short duration [257,258]. These early and somewhat promising clinical results were not reproduced in later clinical trials of locoregional-delivered IL-2 [259,260]. It was suggested that low doses of IL-2 administered to smaller tumors achieve therapeutic effects, whereas high doses of IL-2 were toxic and did not induce tumor regression [257,258]. Nevertheless, it is particularly impressive that recombinant IL-2 given perilesionally or intranodally in escalating doses to patients with HNSCC, who were clinically unresponsive, in an ECOG-sponsored phase Ib trial [260] did show a significant local antitumor responses to low doses of IL-2 [261]. Another cytokine, IFN-γ, which also upregulates cell-mediated immunity, has shown some early clinical promise but has not been evaluated extensively in clinical trials [262]. In a series of small trials, Hadden and colleagues administered natural mixtures of cytokines (IRX-2) peritumoral into patients with advanced HNSCC prior to surgery [263–265]. Clinical and pathologic findings indicated that the tumor regression observed in several cases was mediated by activated lymphocytes accumulating at the tumor site [264,265]. Combinations of cytokines with chemotherapy have also been evaluated in HNSCC [266], but there was no benefit from the addition of IL-2 or IFN-α to standard chemotherapy.

Overall, the therapeutic use of cytokines (i.e., natural or recombinant IL-2, IFN-α, IFN-γ, or natural mixtures of cytokines) in patients with HNSCC has not produced impressive clinical results, but perhaps such results should not be expected, as the trials performed were designed to evaluate toxicity and maximal tolerated doses (MTD) of cytokines and were not designed to offer protection to immune cells or restore immune cell functions. However, these immunologic end points might be achieved at optimal biologic doses (OBD) of the agent administered, which are likely to be different from the MTD and are yet to be defined for most cytokines. Also, in nearly all clinical trials testing

TABLE 38.3 Immunotherapies for Head and Neck Cancer Patients

Immunotherapies used to treat head and neck cancer patients	
Nonspecific immune stimulants	BCG, *C. parvum*, cytokines (IL-2, IFN-α, IRX-2, etc)
Retinoids and prostaglandin inhibitors	Vitamin A and analogs, retinyl palmitate, 13-*cis*-retinoic acid, indomethacin
Anticancer antibodies	Monoclonal antibodies against E48, U36, EGF receptor
Adoptive immunotherapy	Ishikawa's group
	Shu's group
Candidate target antigens for vaccination	p53, SART 1, SART 3, cyclin B

cytokine therapy, patients had advanced, inoperable, end stage disease. Locoregional delivery of cytokines, which was used largely to avoid systemic toxicities, may not be an optimal way for inducing systemic antitumor immunity, and it remains uncertain that tumor-specific systemic immune responses were generated in most of the patients receiving this form of immunotherapy in the early clinical trials.

In addition to the cytokine trials, trials testing retinoids and prostaglandin (PGE_2) inhibitors alone or in combination with surgery, cytokine therapy, or chemotherapy were performed in patients with HNSCC in the late 1980s and early 1990s (Table 38.3). Their use was based on the results of extensive preclinical investigations, which indicated that these agents can inhibit tumor growth. Vitamin A and its synthetic and natural analogs—retinoids—exert antiproliferative and differentiation effects *in vitro* and inhibit growth of HNSCC [267,268]. Experimental evidence clearly links antitumor effects of retinoids to their modulation of tumor differentiation [268]. In a multicenter European chemoprevention trial, performed by EUROSCAN, a 2-year supplementation of retinyl palmitate and/or *N*-acetylcysteine (antioxidant) resulted in no benefits in terms of survival, disease-free survival, or second primary tumors in patients with HNSCC or with lung cancer [269]. In another trial, a combination of 13-*cis*-retinoic acid and IFN-α was used for the treatment of HNSCC patients with recurrent disease without any convincing demonstration of efficacy [270]. In the case of PGE_2 inhibitors, e.g., indomethacin, it was hoped that they could prevent immunosuppression in patients with HNSCC [271,272]. However, no consistent therapeutic benefits were gained by incorporating indomethacin into therapeutic regimens [273]. Today, it is known that cyclooxygenases (COX-1 and COX-2) are the key enzymes in the conversion of arachidonic acid to prostaglandins, and novel pharmacologic strategies are being evaluated to selectively block this enzymatic pathway.

Anticancer antibodies (Abs) are the best-known immunotherapeutic agents (Table 38.3). Ab-based therapies have been utilized for imaging and delivery of radioactive isotopes to the tumor (radioimmunotherapy, RIT) or toxins, as well as for direct targeting of tumor cells [274]. In HNSCC, their therapeutic use has been limited by the paucity of Abs with tumor-restricted specificity. Monoclonal Abs E48 and U36 developed by Snow's group in Amsterdam selectively target HNSCC and have been evaluated extensively preclinically and in patients with HNSCC for pharmacokinetics, biodistribution characteristics, and the ability to bind to the tumor and mediate ADCC [275–279]. Abs labeled with rhenium or technicium are now available for adjuvant RIT in patients with HNSCC who are at risk for the development of recurrences or distant metastases [275–279].

The era of adoptive immunotherapy of cancer with TIL and LAK cells in the 1990s saw relatively little participation from oncologists treating HNSCC patients, as discussed previously. One group of investigators used LAK cells and IL-2 to treat advanced HNSCC, but with no substantial clinical benefits to the patients [280]. Aside from adoptive therapy of HNSCC with LAK cells by Ishikawa's group in Japan [281] (Table 38.3), no other clinical trials based on the adoptive transfer of immune cells were performed until quite recently. It is likely that the complexity of *ex vivo* generation of effector cells for transfers to patients and toxicities associated with a concomitant administration of cytokines (usually IL-2 at relatively high doses) dampened the enthusiasm for this form of immunotherapy. Thus, immunotherapy has a relatively modest history in head and neck cancer in comparison to its more extensive use in melanoma and renal cell carcinoma [282]. However, newer approaches and future strategies that are emerging as a result of recent progress made in basic and translational studies of the immunobiology of HNSCC promise to be more widely used and perhaps even more efficacious.

C. Current Immunotherapeutic Trials in HNSCC

Among many different immune-based approaches available today for treating HNSCC, the emphasis has clearly shifted from nonspecific to tumor-specific therapies. In the past, tumor-specific strategies were limited to the use of HNSCC-specific Abs, such as U36 or E48 [275–279] or Abs targeting the epidermal growth factor receptor (EGFR) [283]. Specific cellular therapies for HNSCC have not been possible until recently because of the lack of well-defined antigenic targets on HNSCC cells. While the repertoire of such head and neck cancer-associated proteins and peptides is still limited, antigen discovery programs ongoing worldwide have identified several potentially immunogenic candidates, including several shared antigens such as p53, particularly its HLA-A2-restricted wild-type sequence epitopes [284], and SART 1 and 3 antigens, as well as cyclin B and its peptides [285–287]. The CASP 8 epitope identified and sequenced by Mandruzzato and colleagues [288] is a mutated protein and, therefore, not a useful component of a broadly based vaccine. These proteins and peptides are being evaluated extensively for immunogenicity in preclinical studies involving autologous DC and peripheral blood lymphocytes from patients with HNSCC [284,289].

The development of therapeutic vaccines for HNSCC depends on finding the combination of epitopes that will be able to generate robust and sustained antitumor immune responses capable of eliminating the vast majority of tumor cells. In contrast, the aim of prophylactic vaccines is to generate the memory immune response capable of preventing tumor development in high-risk populations. The high-risk populations are well defined in HNSCC and, therefore, the concept of prophylactic vaccines for this cancer is particularly attractive. However, prophylactic vaccines have to await results of therapeutic vaccination trials in patients with advanced disease, which address the crucial questions regarding vaccine toxicity, safety, and immunogenicity. In general, the use of antitumor vaccines alone or in combination with the standard therapies in HNSCC is warranted in view of immune dysfunction associated with this disease [249] and the possibility that successful immunization could reverse this dysfunction. The presence of shared antigenic epitopes on HNSCC cells (e.g., p53 wild-type sequence epitopes and others being currently defined) is promising for the development of broadly applicable HNSCC vaccines [290]. The median time to tumor recurrence is relatively short in patients with HNSCC and it could provide a useful clinical end point for monitoring the effects of vaccination. The easy access to the tumor allows for locoregional immunization and facilitates monitoring of local effects of the vaccine. Finally, vaccination strategies are amenable to the incorporation of cellular, molecular, genetic, and/or cytokine interventions, providing the most comprehensive immune therapy available today.

The problem of the availability of immunogenic epitopes has been a serious obstacle in efforts to develop vaccines for HNSCC. With only few tumor-derived epitopes available to date for immunization in HNSCC, new strategies, which utilize apoptotic tumor cells (APTC) or tumor cell lysates and DC as therapeutic vaccines, have been introduced [290]. The advantage of this approach is that tumor-associated antigens (TAA) need not be either identified or purified and that DC ingesting APTC *ex vivo* can be infused into the patient and "cross-present" the processed relevant epitopes to the immune system. Convincing evidence exists in both animal tumor models [291] and human *in vitro* models using circulating mononuclear cells of patients with HNSCC that such HLA class I-restricted, tumor-specific effector cells (CTL) can indeed be generated and that they can recognize and eliminate tumor cells [292]. Based on this evidence, we are in the process of implementing two pilot feasibility trials in patients with advanced HNSCC, which are described briefly here.

In the first clinical trial, the vaccine is administered in the adjuvant setting and consists of autologous DC, which have ingested the patient's own tumor cells. These tumor cells are induced to apoptose upon exposure to UVB light [292] prior to their coincubation with DC. The vaccine is delivered to two sites: (a) intradermally on the arm, in order to observe the DTH response to the vaccine and to biopsy the site of vaccination, and (b) to the inguinal lymph node (a majority of the vaccine) using ultrasound in order to place the DC in the LN milieu, where they normally interact with T lymphocytes. We expect that the utilization of DC for processing and cross-presentation of the epitopes derived from APTC will lead to effective immunization of patients with HNSCC. The mechanism is likely to be DC-mediated "epitope spreading" in the microenvironment rich in cytokines, especially IL-12 and IFN-γ. These cytokines facilitate the Th1 type of response and might protect the generated CTL from apoptosis. While the clinical end points are toxicity and feasibility of this therapy, we also expect to demonstrate an increased frequency of tumor-specific CTL using ELISPOT assays and a decreased frequency of apoptotic T cells in the circulation of vaccinated patients.

The second clinical protocol developed is based on successful animal experiments performed by Dr. Edward Cohen at the University of Illinois College of Medicine in Chicago [293,294]. It utilizes genomic DNA obtained from the surgically removed autologous tumor and transferred by lipofection to an IL-2-secreting semiallogeneic "master" HNSCC cell line as a vaccine. This DNA-based vaccine has the potential of inducing exceptionally strong immune responses for several reasons. The DNA recipient HNSCC cell line is transduced with the IL-2 gene and is selected for high levels of IL-2 secretion [295]. This HNSCC cell line is semiallogeneic with the vaccine recipient and, thus, is expected to induce an allogeneic response, but because it also shares with the recipient a common restriction allele (HLA-A2), an opportunity for generating HLA-A2-restricted, tumor-specific responses also exists. In the "master" HNSCC cell line, tumor-derived DNA, which encodes weakly immunogenic TAA, will be integrated into its own DNA and express the transferred genes in a highly immunogenic form. In this setting, the "master" cell will present to the immune system not only its own TAA, but also antigens encoded in the transferred tumor-derived DNA. This type of vaccine combines all of the elements that are expected to overcome host tolerance to TAA and induce robust antitumor immunity, as demonstrated previously in several animal models of tumor growth [293,294]. Animals treated with this vaccine developed protective and long-lived antitumor immunity and, more importantly, survived significantly longer than nonimmunized control mice [293,294]. This clinical trial offers an opportunity to demonstrate that even in the presence of advanced head and neck cancer, it may be possible to overcome tolerance to TAA and to generate effector cells capable of eliminating tumor cells and establishing tumor-specific memory in HNSCC patients for whom no other therapeutic options are available. In the future, after toxicity studies are completed, this type of vaccination therapy might be applicable to patients with less advanced disease, who are likely to have a greater number of more responsive immune precursor cells.

Both of the described phase I clinical protocols for immune therapy of HNSCC patients are designed to be administered in the adjuvant setting, i.e., preferably only after surgery but also after surgery and radiation or chemoradiation. Aside from the fact that tumor cells are necessary for vaccine preparation, the removal of a strongly immunosuppressive tumor prior to vaccination is likely to be beneficial for the immune response. As for the effects of chemo- or radiotherapy on the host immune system, they may be a lesser evil and could even facilitate the generation of antitumor immune responses by eliminating subsets of suppressor cells [296].

Immunotherapy might involve genetic manipulations of tumor *in situ* through intratumoral injections of selected genes in order to modify the immune response. For example, delivery to the tumor of a gene for an alloantigen, human leukocyte antigen (HLA)-B7, together with the β_2-microglobulin gene, was used by Gleich [297] to restore expression of the complete MHC class I molecules on the surface of tumor cells and thus facilitate both antigen presentation and recognition of the tumor by tumor-specific immune effector cells. Twenty patients with advanced HNSCC who failed conventional therapy and whose tumors did not express HLA-B7 were treated with no adverse effects from the alloantigen gene therapy. Expression of HLA-B7 was demonstrated in the tumors, and apoptosis of tumor cells was seen in tumors presumably responding to intervention by effector cells. Two CR were observed, and the median survival of the treated patients was prolonged [297]. In another approach, Tahara and colleagues [298] used immune gene therapy to vaccinate patients with their own irradiated tumor cells admixed with the autologous fibroblasts, which were transduced with the IL-12 gene and secreted IL-12. Among 29 patients with advanced tumors participating in this phase I trial, two had HNSCC. One had a partial response to this vaccination therapy. These two examples serve to emphasize the concept of therapy aimed at inducing changes in the tumor microenvironment and thus augmenting the susceptibility of tumors to the host immune system.

Adoptive specific immunotherapy with tumor-specific T cells has also been in evidence more often in recent years. Earlier studies in murine tumor models provided convincing evidence that *ex vivo*-activated, adoptively transferred T lymphocytes can mediate antitumor activities, resulting in tumor regression [299]. In addition, several clinical studies have demonstrated the ability of adoptively transferred tumor-reactive T lymphocytes to mediate the regression of established tumors [300]. A phase I clinical trial performed by To and colleagues [301] illustrated both the strategy and the results that can be expected from adoptive immunotherapy in patients with advanced-stage, unresectable HNSCC (Table 38.3). The patients were first vaccinated with autologous-cryopreserved tumor cells, which were injected intradermally on the upper thigh together with GM-CSF. Eight to 10 days later, inguinal lymph nodes draining the vaccination site were removed surgically and the recovered LN lymphocytes were expanded in culture with staphylococcal enterotoxin A (SEA) and IL-2. Activated T lymphocytes, enriched in $CD8^+$ cells during culture, were delivered intravenously. The patients received a single treatment, except for 1/15 patients who was treated twice 5 months apart. The generation of T cells was found to be feasible, and toxicity was minimal following T-cell transfer. Ten patients had progressive disease after cell infusion with a median survival of 6 months; in 3 patients, the disease had stabilized and the median survival was 15 months; and 2 patients were alive with no evidence of disease (NED) at greater than 28 and 36 months after treatment. Thus, this type of immune therapy, although labor-intensive, appears to be feasible, nontoxic, and clinically beneficial to at least some patients with advanced HNSCC.

D. Future Prospects in Immunotherapy of HNSCC

With recent interest in anticancer vaccines and substantial progress being made in the identification of HNSCC-specific antigens, it appears that specific active immunotherapy is likely to remain in the limelight. Until prophylactic vaccines are available and proven safe, however, vaccines and other immune-based therapies have to be explored as adjuncts to conventional therapies. There is a strong rationale for using immunotherapies in this setting. For example, administration of cytokines as immune adjuvants in conjunction with vaccines or other more conventional therapies may increase the therapeutic index. It has been observed that cytokines can alter the tumor microenvironment in several ways. They can have a direct growth inhibitory effect on tumor cells, acting via cytokine receptors expressed on these cells [302]. They can upregulate expression of the TNF family receptors, as well as ligands on the tumor cell surface, and thus increase tumor sensitivity to apoptosis by a variety of mechanisms, including those mediated by immune cells. Cytokines can also upregulate antitumor functions of immune effector cells, including their ability to migrate to the tumor site and to modulate the expression of MHC molecules, as well as antigenic epitopes on tumor cells, making them more susceptible to immune intervention. Cytokines can support differentiation and maturation of antigen-presenting cells and facilitate productive interactions between APC and T cells. The therapeutic use of cytokines, which could be delivered as genetically engineered single or multiple entities, is an attractive option for HNSCC that is likely to be evaluated extensively in the future.

It can also be expected that further improvements in the quality of adoptively transferred cells to make them tumor specific rather than tumor reactive will encourage their broader utilization. The possibility of delivering tumor-specific subsets of cytotoxic and helper T cells together with cytokines, which could enhance their *in vivo* antitumor functions, may become a reality. Technical advances in cell culture and genetic engineering are likely to make this therapeutic option feasible and less labor-intensive than it has been to date.

An important future objective is to assure that the tumor-specific T cells, whether induced *in vivo* by vaccines or transferred adoptively to patients with HNSCC, can survive in the microenvironment favoring their apoptosis [303]. In addition to cytokines, many other biologic agents are capable of protecting immune cells from apoptosis and upregulate their antitumor functions. Immune therapies with apoptosis resistance genes, caspase inhibitors, and Abs blocking interactions of the TNF family of receptors and ligands, as well as vaccines aimed at expanding and maintaining clones of tumor-specific T cells, are all potentially useful for the protection of immune effector cells and restoration of the host immune system in the presence of advanced malignancy.

The use of humanized Abs for immunotherapy or RIT of HNSCC is likely to grow due to their availability and promising results of initial phase I and II trials. The best example is the effective performance of anti-EGFR mAb 425 in therapeutic as well as chemoprevention clinical trials to date [304].

An important aspect of immunotherapy for patients with HNSCC is its timing in respect to surgery as well as the subsequent chemoradiation therapy. It is reasonable to predict that surgery for primary HNSCC may be followed by a partial recovery of the host immune system. Ideally, immune therapy should be timed for the period after tumor removal, when tumor burden is minimal. Immunotherapy alone in patients with bulky or aggressive disease, even when it follows surgery and vaccination, as described earlier, is not likely to be effective because a large tumor burden and aggressive disease are often associated with particularly severe immunosuppression. However, a combination of immunotherapy with radiotherapy or chemoradiation could, in theory, be quite effective, although no adequate clinical data are available to support this assumption. The caveat to this hypothesis is that the successful combination of a conventional treatment with immunotherapy is likely to be dependent on the optimal sequence, time of administration, and dosages of the drugs, which will require extensive preclinical and clinical evaluations in the future. It is important for the oncologists treating HNSCC patients to consider this option, as immune therapy, which is skillfully selected and added to currently available conventional therapeutic regimens, might improve survival in these patients.

Finally, it is important to beep in mind that HNSCC, like most tumors, have evolved mechanisms, which enable them to avoid immune interventions. In addition to various ways of "hiding" from the host immune cells, tumors are able to disarm or kill these cells. Results from several studies suggest that *in vivo*, the immune system might play a key role in the appearance of immunoresistant tumors, which clearly have a growth advantage over tumors sensitive to immune intervention. Thus, even when the immune response to a vaccine or other biologic therapy is generated in the patient with HNSCC, the tumor, when it recurs, might become resistant to it (e.g., through the selection of "epitope loss" variants). Also, tumor-specific effector cells might be selectively eliminated by apoptosis before they have an opportunity to interfere with tumor growth. These and other possibilities need to be explored in future clinical trials with biologic agents in patients with HNSCC. New molecular technologies applied to the monitoring of patients receiving such therapies, as well as tetramer-based flow cytometry for assessing the frequencies of tumor-specific T cells in patients' samples, provide a window of opportunity for correlating immunologic with clinical end points and perhaps defining the mechanisms of tumor-effector cell interactions that might determine the therapeutic outcome.

VI. CONCLUSION

Tremendous advances have occurred over the past two decades in the areas of immunology and molecular immunology. Individuals treating the HNSCC patient should be reminded that we have made only minor advances in curing this cancer since the early 1970s with our present therapeutic strategies. The addition of immunotherapeutic-based treatment options should be developed and implemented further. We must continue to take advantage of the new scientific discoveries made daily and translate them into novel immune-based treatment strategies to improve cure of this disease.

References

1. Al-Sarraf, M., LeBlanc, M., Shanker Giri, P. G., Fu, K. K., Cooper, J., Vuong, T., Forastiere, I., Adams, G., Sakr, W. A., Schuller, D. E., and Ensley, J. F. (1998). Chemoradiotherapy versus radiotherapy in patients with advanced nasopharyngeal cancer: Phase III randomized Intergroup study 0099. *J. Clin. Oncol.* **16**, 1310–1317.
2. Himmelwiet, B. (1957). "The Collected Papers of Paul Ehrlich." Pergamon Press. Oxford.
3. Burnet, F. M. (1970). The concept of immunological surveillance. *In* "Progress in Experimental Tumor Research" (R.S. Schwartz, ed.), p. 1. Karger, Basel.
4. Penn, I. (1992). Principle of tumor immunity: Immunocompetence and cancer. *In* "Biologic Therapy of Cancer" (V.T. DeVita, S. Hellman, and S.A. Rosenburg, eds.), pp. 103–118. Lippincott, Philadelphia.
5. Robison, L. L., Stoker, V., Frizzera, G., Heinitz, K., Meadows, A. T., and Filipovich, A. (1987). Hodgkin's disease in pediatric patients with naturally occurring immune deficiency. *Am. J. Pediatr. Hematol. Oncol.* 189–192.
6. Kersy, J. H., Shapiro, R. S., and Heinitz, K. J. (1987). Lymphoid malignancy in naturally occurring and post bone marrow transplantation immunodeficiency diseases. *In* "The Nature, Cellular, and Biochemical Basis and Management of Immunodeficiencies" (R.A. Good, and E. Lindenlaub, eds.), pp. 289–294. Schattauer Verlag, Stuttgart.
7. Filipovich, A. H., Zerbe, D., and Spector, B. D. (1984). Lymphomas in persons with naturally occurring immunodeficiency disorders. *In* "Pathogenesis of Leukemias and Lymphomas: Environmental Influences" (I.T. McGrath, G.R. O'Connor, and B. Ramot, eds.), Raven Press, New York.

8. Conant, M. A., Volberding, P., Fletcher, V., Lozada, F. I., and Silverman, S., Jr. (1982). Squamous cell carcinoma in sexual partner of Kaposi's sarcoma patient. *Lancet* **1**, 286.
9. Lozada, F., Silverman, S., and Conant, M. (1982). New outbreak of oral tumors, malignancies and infectious diseases strikes young male homosexuals. *Calif. Dent. J.* **10**, 39–42.
10. Munoz, A., Gomez-Anson, B., Abad, L., and Gonzalez-Spinola, J. (1994). Squamous cell carcinoma of the larynx in a 29 year old man with AIDS. *AJR Am. J. Roentgenol.* **162**, 232.
11. Singh, B., Sabin, S., Rofim, O., Shaha, A., Har-El, G., and Lucente, F. E. (1999). Alterations in head and neck cancer occurring in HIV-infected patients. *Acta Oncol.* **38**, 1047–1050.
12. King, G. N., Healy, C. M., Glover, M. T., Kwan, J. T. C., Williams, D. M., Leigh, I. M., Worthington, H. V., and Thornhill, M. H. (1995). Increased prevalence of dysplastic and malignant lip lesions in renal-transplant recipients. *N. Engl. J. Med.* **332**, 1052–1057.
13. Bradford, C. R., Hoffman, H. T., Wolf, G. T., Carey, T. E., Baker, S. R., and McClatchey, E. (1990). Squamous cell carcinoma of the head and neck in organ transplant recipients: Possible role of oncogenic virus. *Laryngoscope* **100**, 190–194.
14. Oettgen, H. F., and Old, L. J. (1991). The history of cancer immunotherapy. *In* "Biologic Therapy of Cancer" (V.T. DeVita, S. Hellman, and S.A. Rosenberg, eds.), pp. 87–119. Lippincott, Philadelphia.
15. Feheisen, F. (1882). Uber die Zuchtung der Erysipel-Kokken auf kunstlichen Nahrboden und die Ubertragbarkeit auf den Menschen. *Dtsch. Med. Wochenschr.* **8**, 533.
16. Bruns, P. (1888). Die Heilwirkung des Erysipels auf Geschwulste. *Beitr. Klin. Chir.* **3**, 443.
17. Coley, W. B. (1893). Treatment of malignant tumors by repeated inoculations of erysipelas: With a report of ten original cases. *Am. J. Med. Sci.* **105**, 487.
18. Pinsky, C. M., Camacho, F. J., Kerr, D., Geller, N. L., Klein, F. A., Herr, H. A., and Oettgen, H. F. (1985). Intravesical administration of bacillus Calmette-Guerin in patients with recurrent superficial carcinoma of the urinary bladder. *Cancer Treat. Rep.* **69**, 47–53.
19. Lanier, L. L., Le, A. M., Phillips, J. H., Warner, N. L., and Babcock, G. F. (1983). Subpopulations of human natural killer cells defined by expression of the Leu-7 (HNK-1) and Leu-11 (NKI-15) antigens. *J. Immunol.* **131**, 1789–1796.
20. Lotzova, E., and Herberman, R. B. (1986). "Immunobiology of Natural Killer Cells." CRC Press, Boca Raton, FL.
21. Hercend, T., Griffin, J. D., Bensussan, A., Schmidt, R. E., Edson, M. A., Brennan, A., Murray, C., Daley, J. F., Schlossman, S. F., and Ritz, J. (1985). Generation of monoclonal antibodies to a human natural killer clone. Characterization of two natural killer-associated antigens, NKH1A and NKH2, expressed on subsets of large granular lymphocytes. *J. Clin. Invest.* **75**, 932–943.
22. Hersey, P., Bindon, G., Edwards, A., Murray, E., Phillips, G., and McCarthy, W. H. (1981). Induction of cytotoxic activity in human lymphocytes against autologous and allogeneic melanoma cells *in vitro* by culture with interleukin 2. *Int. J. Cancer* **28**, 685–703.
23. Lotze, M. T., Grimm, E. A., Mazumder, A., Strausser, J. L., and Rosenberg, S. A. (1981). Lysis of fresh and cultured autologous tumor by human lymphocytes cultured in T cell growth factor. *Cancer Res.* **41**, 4420–4425.
24. Grimm, E. A., Mazumder, A., Zhang, H. Z., and Rosenberg, S. A. (1982). Lymphokine-activated killer cell phenomenon: Lysis of natural killer-resistant fresh solid tumor cells by interleukin 2- activated autologous human peripheral blood lymphocytes. *J. Exp. Med.* **155**, 1823–1841.
25. Grimm, E. A., Ramsey, K. M., Mazumder, A., Wilson, D. J., Djeu, J. Y., and Rosenberg, S. A. (1983). Lymphokine-activated killer cell phenomenon. II. Precursor phenotype is serologically distinct from peripheral T lymphocytes, memory cytotoxic thymus-derived lymphocytes, and natural killer cells. *J. Exp. Med.* **157**, 884–897.
26. Lotzova, E., and McCredie, K. B. (1978). Natural killer cells and recombinant interleukin-2 *in vivo*: Direct correlation between reduction of established metastases and cytolytic activity of lymphokine-activated killer cells. *Cancer Immunol. Immunother.* **4**, 215.
27. Rosenberg, S. A., Lotze, M. T., Muul, L. M., Chang, A. E., Avis, F. P., Leitman, S., Linehan, W. M., Robertson, C. N., Lee, R. E., Rubin, J. T., Seipp, C. A., Simpson, C. G., and White, D. E. (1987). A progress report on the treatment of 157 patients with advanced cancer using lymphokine-activated killer cells and interleukin-2 or high-dose interleukin-2 alone. *N Engl. J. Med.* **316**, 889–897.
28. Mazumder, A., Eberlein, T. J., Grimm, E. A., Wilson, D. J., Keenan, A. M., Aamodt, R., and Rosenberg, S. A. (1984). Phase I study of the adoptive immunotherapy of human cancer with lectin activated autologous mononuclear cells. *Cancer* **53**, 896–905.
29. Rosenberg, S. A., Lotze, M. T., Muul, L. M., Leitman, S., Chang, A. E., Ettinghausen, S. E., Matory, Y. L., Skibber, J. M., Shiloni, E., Vetto, J. T., Seipp, C. A., Simpson, C., and Reichert, C. M. (1985). Observations on the systemic administration of autologous lymphokine-activated killer cells and recombinant interleukin-2 to patients with metastatic cancer. *N. Engl. J. Med.* **313**, 1485–1492.
30. Aebersold, P., Hyatt, C., Johnson, S., Hines, K., Korack, L., Sanders, M., Lotze, M. T., Topalian, S., Yang, J., and Rosenberg S. A. (1991). Lysis of autologous melanoma cells by tumor-infiltrating lymphocytes: Association with clinical response. *J. Natl. Cancer Inst.* **83**, 932–937.
31. Rosenberg, S. A., Lotze, M. T., Yang, J. C., Aebersold, P. M., Linehan, W. M., Seipp, C. A., and White, D. E. (1989). Experience with the use of high-dose interleukin-2 in the treatment of 652 cancer patients. *Ann. Surg.* **210**, 474–485.
32. Thompson, J. A., Shulman, K. L., Benyunes, M. C., Lindgren, C. G., Collins, C., Lange, P. H., Bush, W. H., Jr., Benz, L. A., and Fefer, A. (1992). Prolonged continuous intravenous infusion of interleukin-2 and lymphokine-activated killer-cell therapy for metastatic renal cell carcinoma. *J. Clin. Oncol.* **10**, 960–968.
33. Topalian, S. L., Solomon, D., Avis, F. P., Chang, A. E., Freerksen, D. L., Linehan, W. M., Lotze, M. T., Robertson, C. N., Seipp, C. A., Simon, P., Simpson, C. G., and Rosenberg, S. A. (1988). Immunotherapy of patients with advanced cancer using tumor-infiltrating lymphocytes and recombinant interleukin-2: A pilot study. *J. Clin. Oncol.* **6**, 839–853.
34. Rosenberg, S. A., Packard, B. S., Aebersold, P. M., Solomon, D., Topalian, S. L., Toy, S. T., Simon, P., Lotze, M. T., Yang, J. C., Seipp, C. A., Simpson, C. G., Carter, C., Bock, S., Schwartzentruber, D., Wei, J. P., and White, D. E. (1988). Use of tumor-infiltrating lymphocytes and interleukin-2 in the immunotherapy of patients with metastatic melanoma: A preliminary report. *N. Engl. J. Med.* **319**, 1676–1680.
35. Rosenberg, S. A., Aebersold, P., Cornetta, K., Kasid, A., Morgan, R. A., Moen, R., Karson, E. M., Lotze, M. T., Yang, J. C., Topalian, S. L., Merino, M. J., Culver, K., Miller, A. D., Blaese, R. M., and Anderson, W. F. (1990). Gene transfer into humans: Immunotherapy of patients with advanced melanoma, using tumor-infiltrating lymphocytes modified by retroviral gene transduction. *N. Engl. J. Med.* **323**, 570–578.
36. Goedegebuure, P. S., Douville, L. M., Li, H., Richmond, G. C., Schoof, D. D., Scavone, M., and Eberlein T. J. (1995). Adoptive immunotherapy with tumor-infiltrating lymphocytes and interleukin-2 in patients with metastatic malignant melanoma and renal cell carcinoma: A pilot study. *J. Clin. Oncol.* **13**, 1939–1949.
37. Yoshizawa, H., Chang, A. E., and Shu, S. (1991). Specific adoptive immunotherapy mediated by tumor-draining lymph node cells sequentially activated with anti-CD3 and IL-2. *J. Immunol.* **147**, 729–737.
38. Chang, A. E., and Shu, S. (1996). Current status of adoptive immunotherapy of cancer. *Crit. Rev. Oncol. Hematol.* **22**, 213–228.
39. Osband, M. E., Lavin, P. T., Babayan, R. K., Graham, S., Lamm, D. L., Parker, B., Sawczuk, I. S., Ross, S., and Krane, R. J. (1990). Effect of autolymphocyte therapy on survival and quality of life in patients with metastatic renal-cell carcinoma. *Lancet* **335**, 994–998.

40. Lavin, P. T., Maar, R., Franklin, M., Ross, S., Martin, J., and Osband, M. E. (1992). Autolymphocyte therapy for metastatic renal cell carcinoma: Initial clinical results from 335 patients treated in a multisite clinical practice. *Transplant. Proc.* **24**, 3059–3064.
41. Riddell, S. R., and Greenberg, P. D. (1990). The use of anti-CD3 and anti-CD28 monoclonal antibodies to clone and expand human antigen-specific T cells. *J. Immunol. Methods* **128**, 189–201.
42. Schultze, J. L., Michalak, S., Seamon, M. J., Dranoff, G., Jung, K., Daley, J., Delgado, J. C., Gribben, J. G., and Nadler, L. M. (1997). CD40-activated human B cells: An alternative source of highly efficient antigen presenting cells to generate autologous antigen-specific T cells for adoptive immunotherapy. *J. Clin. Invest.* **100**, 2757–2765.
43. Melief, C. J. (1993). Prospects of T-cell immunotherapy for cancer by peptide vaccination. *Semin. Hematol.* **30**, 32–33.
44. Schulz, M., Aichele, P., Schneider, R., Hansen, T. H., Zinkernagel, R. M., and Hengartner, H. (1991). Major histocompatibility complex binding and T cell recognition of a viral nonapeptide containing a minimal tetrapeptide. *Eur. J. Immunol.* **21**, 1181–1185.
45. Riddell, S. R., Watanabe, K. S., Goodrich, J. M., Li, C. R., Agha, M. E., and Greenberg, P. D. (1992). Restoration of viral immunity in immunodeficient humans by the adoptive transfer of T cell clones. *Science* **257**, 238–241.
46. Riddell, S. R., and Greenberg, P. D. (1995). Cellular adoptive immunotherapy after bone marrow transplantation. *Cancer Treat. Res.* **76**, 337–369.
47. Gedde-Dahl, T., III, Fossum, B., Eriksen, J. A., Thorsby, E., and Gaudernack, G. (1993). T cell clones specific for p21 ras-derived peptide: Characterization of their fine specificity and HLA restriction. *Eur. J. Immunol.* **23**, 754–760.
48. Hoosiers, J. G., Inman, H. W., van der Burg, S. H., Drijfhout, J. W., Kenemans, P., van de Velde, C. J., Brand, A., Momburg, F., Kast, W. M., and Melief, C. J. (1993). *In vitro* induction of human cytotoxic T lymphocyte responses against peptides of mutant and wild-type p53. *Eur. J. Immunol.* **23**, 2072–2077.
49. Schlichtholz, B., Legros, Y., Gillet, D., Gaillard, C., Marty, M., Lane, D., Calvo, F., and Soussi, T. (1992). The immune response to p53 in breast cancer patients is directed against immunodominant epitopes unrelated to the mutational hot spot. *Cancer Res.* **52**, 6380–6384.
50. Noguchi, Y., Chen, Y.-T., and Old, L. J. (1994). A mouse mutant p53 product recognized by $CD4^+$ and $CD8^+$ T cells. *Proc. Natl. Acad. Sci. USA* **91**, 3171–3175.
51. Yanuck, M., Carbone, D. P., Pendleton, C. D., Tsukui, T., Winter, S. F., Minna, J. D., and Berzofsky, J. A. (1993). A mutant p53 tumor suppressor protein is a target for peptide-induced $CD8^+$ cytotoxic T cells. *Cancer Res.* **53**, 3257–2361.
52. Nijman, H. W., Houbiers, J. G., van der Burg, S. H., Vierboom, M. P., Kenemans, P., Kast, W. M., and Melief, C. J. (1993). Characterization of cytotoxic T lymphocyte epitopes of a self-protein, p53, and a non-self-protein, influenza matrix: Relationship between major histocompatibility complex peptide binding affinity and immune responsiveness to peptides. *J. Immunother.* **14**, 121–126.
53. Papadopoulos, E. B., Ladanyi, M., Emanuel, D., Mackinnon, S., Boulad, F., Carabasi, M. H., Castro-Malaspina, H., Childs, B. H., Gillio, A. P., Small, T. N., Young, J. W., Kernan, N. A., and O'Reilly, R. J. (1994). Infusions of donor leukocytes to treat Epstein-Barr virus associated lymphoproliferative disorders after allogeneic bone marrow transplantation. *N. Engl. J. Med.* **330**, 1185–1191.
54. Smith, C. A., Ng, C. Y. C., Heslop, H. E., Holladay, M. S., Richardson, S., Turner, E. V., Loftin, S. K., Li, C., and Brenner, M. K. (1995). Production of genetically modified Epstein-Barr virus-specific cytotoxic T cells for adoptive transfer to patients at high risk of EBV-associated lymphoproliferative disease. *J. Hematother.* **4**, 73–79.
55. Celluzzi, C. M., Mayordomo, J. I., Storkus, W. J., Lotze, M. T., and Falo, L. D., Jr. (1996). Peptide-pulsed dendritic cells induced antigen-specific, CTL-mediated protective tumor immunity. *J. Exp. Med.* **183**, 283–287.
56. Steinman, R. M., Witmer-Pack, M., and Inaba, K. (1993). Dendritic cells: Antigen presentation, accessory function and clinical relevance. *Adv. Exp. Med. Biol.* **329**, 1–9.
57. Young, J. W., and Inaba, K. (1996). Dendritic cells as adjuvants for class I major histocompatibility complex-restricted antitumor immunity. *J. Exp. Med.* **183**, 7–11.
58. Sallusto, F., and Lanzavecchia, A. (1994). Efficient presentation of soluble antigen by cultured human dendritic cells is maintained by granulocyte/macrophage colony-stimulating factor plus interleukin 4 and downregulated by tumor necrosis factor a. *J. Exp. Med.* **179**, 1109–1118.
59. Romani, N., Reider, D., Heuer, M., Ebner, S., Kämpgen, E., Eibl, B., Niederwiser, D., and Schuler, G. (1996). Generation of mature dendritic cells from human blood: An improved method with special regard to clinical applicability. *J. Immunol. Methods* **196**, 137–151.
60. Young, J. W., and Steinman, R. M. (1996). The hematopoietic development of dendritic cells: A distinct pathway for myeloid differentiation. *Stem Cells* **14**, 376–387.
61. Bender, A., Sapp, M., Schuler, G., Steinman, R. M., and Bhardwaj, N. (1996). Improved method for the generation of dendritic cells from nonproliferating progenitors in human blood. *J. Immunol. Methods* **196**, 121–135.
62. Wright-Browne, V., McClain, K. L., Talpaz, M., Ordonez, N., and Estrov, Z. (1997). Physiology and pathophysiology of dendritic cells. *Hum. Pathol.* **28**, 563–579.
63. Knight, S. C., and Stagg, A. J. (1993). Antigen-presenting cell types. *Curr. Opin. Immunol.* **5**, 374–382.
64. Inaba, K., Metlay, J. P., Crowley, M. T., and Steinman, R. M. (1990). Dendritic cells pulsed with protein antigens *in vitro* can prime antigen-specific, MHC-restricted T cells *in situ*. *J. Exp. Med.* **172**, 631–640.
65. Cella, M., Sallusto, F., and Lanzavecchia, A. (1997). Origin, maturation and antigen presenting function of dendritic cells. *Curr. Opin. Immunol.* **9**, 10–16.
66. Mannering, S. I., McKenzie, J. L., Fearnley, D. B., and Hart, D. N. J. (1997). HLA-DR1-restricted bcr-abl (b3a2)-specific $CD4^+$ T lymphocytes respond to denditic cells pulsed with b3a2 peptide and antigen-presenting cells exposed to b3a2 containing cell lysates. *Blood* **90**, 290–297.
67. Tjoa, B., Erickson, S., Barren, R., III, Ragde, H., Kenny, G., Boynton, A., and Murphy G. (1995). *In vitro* propagated dendritic cells from prostate cancer patients as a component of prostate cancer immunotherapy. *Prostate* **27**, 63–69.
68. Tjoa, B., Boynton, A., Kenny, G., Ragde, H., Misrock, S. L., and Murphy, G. (1996). Presentation of prostate tumor antigens by dendritic cells stimulates T-cell proliferation and cytotoxicity. *Prostate* **28**, 65–69.
69. Rosenberg, S. A., Yang, J. C., Schwartzentruber, D. J., Hwu, P., Marincola, F. M., Topalian, S. L., Restifo, N. P., Dudley, M. E., Schwarz, S. L., Spiess, P. J., Wunderlich, J. R., Parkhurst, M. R., Kawakami, Y., Seipp, C. A., Einhorn, J. H., and White, D. E. (1998). Immunologic and therapeutic evaluation of a synthetic peptide vaccine for the treatment of patients with metastatic melanoma. *Nature Med.* **4**, 321–327.
70. Nair, S. K., Snyder, D., Rouse, B. T., and Gilboa, E. (1997). Regression of tumors in mice vaccinated with professional antigen-presenting cells pulsed with tumor extracts. *Int. J. Cancer* **70**, 706–715.
71. Boczkowski, D., Nair, S. K., Snyder, D., and Gilboa, E. (1996). Dendritic cells pulsed with RNA and potent antigen-presenting cells *in vitro* and *in vivo*. *J. Exp. Med.* **184**, 465–472.
72. Kolb, H. J., Mittermuller, J., Clemm, C., Holler, E., Ledderose, G., Brehm, G., Heim, M., and Wilmanns, W. (1990). Donor leukocyte transfusions for treatment of recurrent chronic myelogenous leukemia in marrow transplant patients. *Blood* **76**, 2462–2465.

73. Kolb, H. J., and Holler, E. (1997). Adoptive immunotherapy with donor lymphocyte transfusions. *Curr. Opin. Oncol.* **9**, 139–145.
74. Kolb, H. J., Schattenberg, A., Goldman, J. M., Hertenstein, B., Jacobsen, N., Arcese, W., Ljungman, P., and Ferrant, A. (1995). Graft-versus-leukemia effect of donor lymphocyte transfusions in marrow grafted patients. European Group for Blood and Marrow Transplantation Working Party on Chronic Leukemia. *Blood* **86**, 2041–2050.
75. Levine, J. E., Yanik, G., Hutchinson, R. J., Casper, J., Beatty, P. G., Sproles, A., and Collins, R. (1997). Donor leukocyte infusions (DLI) to treat relapses after allogeneic bone marrow transplantation in pediatric patients. *Blood* **90**, 548a.
76. Collins, R. H., Jr., Shpilberg, O., Drobyski, W. R., Porter, D. L., Giralt, S., Champlin, R., Goodman, S. A., Wolff, S. N., Hu, W., Verfaillie, C., List, A., Dalton, W., Ognoskie, N., Chetrit, A., Antin, J. H., and Nemunaitis, J. (1997). Donor leukocyte infusions in 140 patients with relapsed malignancy after allogeneic bone marrow transplantation. *J. Clin. Oncol.* **15**, 433–444.
77. Meuer, S. C., Hussey, R. E., Cantrell, D. A., Hodgdon, J. C., Schlossman, S. F., Smith, K. A., and Reinherz, E. L. (1984). Triggering of the T3-Ti anti-receptor complex results in clonal T cell proliferation through an interleukin 2 dependent autocrine pathway. *Proc. Natl. Acad. Sci. USA* **81**, 1509–1513.
78. Meuer, S. C., Hodgdon, J. C., Hussey, R. E., Protentis, J. P., Schlossman, S. F., and Reinherz, E. L. (1983). Antigen-like effects of monoclonal antibodies directed at receptors on human T cell clones. *J. Exp. Med.* **158**, 988–993.
79. Weiss, A., and Imboden, J. B. (1987). Cell surface molecules and early events involved in human T lymphocyte activation. *Adv. Immunol.* **41**, 1-38.
80. Van Wauwe, J. P., De Mey, J. R., and Gooseens, J. G. (1980). OKT3: A monoclonal anti-human T lymphocyte antibody with potent mitogenic properties. *J. Immunol.* **124**, 2708–2713.
81. Loeffler, C. M., Platt, J. L., Anderson, P. M., Katsanis, E., Ochoa, J. B., Urba, W. J., Longo, D. L., Leonard, A. S., and Ochoa, A. C. (1991). Antitumor effects of IL-2 liposomes and anti-CD3-stimulated T-cells against murine MCA-38 hepatic metastasis. *Cancer Res.* **51**, 2127–2132.
82. Murphy, W. J., Conlon, K. C., Sayers, T. J., Wiltrout, R. H., Back, T. C., Ortaldo, J. R., and Longo, D. L. (1993). Engraftment and activity of anti-CD3-activated human peripheral blood lymphocytes transferred into mice with severe combined immune deficiency. *J. Immunol.* **150**, 3634–3642.
83. Katsanis, E., Xu, Z., Anderson, P. M., Dancisak, B. B., Bausero, M. A., Weisdorf, D. J., Blazar, B. R., and Ochoa, A. C. (1994). Short-term ex vivo activation of splenocytes with anti-CD3 plus IL-2 and infusion post-BMT into mice results in vivo expansion of effector cells with potent anti-lymphoma activity. *Bone Marrow Transplant* **14**, 563–572.
84. Chen, B. P., Malkovsky, M., Hank, J. A., and Sondel, P. M. (1987). Nonrestricted cytotoxicity mediated by interleukin 2-expanded leukocytes is inhibited by anti-LFA-1 monoclonal antibodies (MoAb) but potentiated by anti-CD3 Moab. *Cell. Immunol.* **110**, 282–293.
85. Lotzova, E., Savary, C. A., Herberman, R. B., McCredie, K. B., Keating, M. J., and Freireich, E. J. (1987). Augmentation of antileukemia lytic activity by OKT3 monoclonal antibody: Synergism of OKT3 and interleukin-2. *Nature Immun. Cell Growth Regul.* **6**, 219–223.
86. Yang, S. C., Fry, K. D., Grimm, E. A., and Roth, J. A. (1990). Successful combination immunotherapy for the generation *in vivo* of antitumor activity with anti-CD3, interleukin 2, and tumor necrosis factor alpha. *Arch. Surg.* **125**, 220–225.
87. Anderson, P. M., Bach, F. H., and Ochoa, A. C. (1988). Augmentation of cell number and LAK activity in peripheral blood mononuclear cells activated with anti-CD3 and interleukin-2. Preliminary results in children with acute lymphocytic leukemia and neuroblastoma. *Cancer Immunol. Immunother.* **27**, 82–88.
88. Ochoa, A. C., Hasz, D. E., Rezonzew, R., Anderson, P. M., and Bach, F. H. (1989). Lymphokine-activated killer activity in long-term cultures with anti-CD3 plus interleukin 2: Identification and isolation of effector subsets. *Cancer Res.* **49**, 963–968.
89. Anderson, P. M., Blazar, B. R., Bach, F. H., and Ochoa, A. C. (1989). Anti-CD3+ IL-2-stimulated murine killer cells: *In vitro* generation and *in vivo* antitumor activity. *J. Immunol.* **142**, 1383–1394.
90. Ting, C.-C., Hargrove, M. E., and Yun, Y. S. (1988). Augmentation by anti-T3 antibody of the lymphokine-activated killer cell-mediated cytotoxicity. *J. Immunol.* **141**, 741–748.
91. Uberti, J. P., Joshi, I., Ueda, M., Martilotti, F., Sensenbrenner, L. L., and Lum, L. G. (1994). Preclinical studies using immobilized OKT3 to activate human T cells for adoptive immunotherapy: Optimal conditions for the proliferation and induction of non-MHC restricted cytotoxicity. *Clin. Immunol. Immunopathol.* **70**, 234–240.
92. Ueda, M., Joshi, I. D., Dan, M., Uberti, J. P., Chou, T.-H., Sensenbrenner, L. L., and Lum, L. G. (1993). Preclinical studies for adoptive immunotherapy in bone marrow transplantation. II. Generation of anti-CD3 activated cytotoxic T cells from normal donors and autologous bone marrow transplant candidates. *Transplantation* **56**, 351–356.
93. Anderson, P. M., Ochoa, A. C., Ramsay, N. K. C., Hasz, D., and Weisdorf, D. (1992). Anti-CD3+ interleukin-2 stimulation of marrow and blood: Comparison of proliferation and cytotoxicity. *Blood* **80**, 1846–8153.
94. Ochoa, A. C., Gromo, G., Alter, B. J., Sondel, P. M., and Bach, F. H. (1987). Long-term growth of lymphokine-activated killer (LAK) cell: Role of anti-CD3, beta-IL 1, interferon-gamma and -beta. *J. Immunol.* **138**, 2728–2733.
95. Sosman, J. A., Oettel, K. R., Hank, J. A., and Sondel, P. M. (1989). Isolation and characterization of human cytolytic cells (CTL) with specificity for allogeneic leukemic blasts. *FASEB J.* **3**, 506a.
96. Sosman, J. A., Oettel, K. R., Hank, J. A., Fisch, P., and Sondel, P. M. (1989). Specific recognition of human leukemic cells by allogeneic T cell lines. *Transplantation* **48**, 486.
97. Massaia, M., Attisano, C., Peola, S., Montacchini, L., Omede, P., Corradini, P., Ferrero, D., Boccadoro, M., Bianchi, A., and Pileri, A. (1993). Rapid generation of antiplasma cell activity in the bone marrow of myeloma patients by CD3-activated T cells. *Blood* **82**, 1787–1797.
98. Curti, B. C., Longo, D. L., Ochoa, A. C., Conlon, K. C., Smith, J. W., II, Alvord, W. G., Creekmore, S. P., Fenton, R. G., Gause, B. L., Holmlund, J., Janik, J. E., Ochoa, J., Rice, P. A., Sharfman, W. H., Sznol, M., and Urba W. J. (1993). Treatment of cancer patients with *ex vivo* anti-CD3-activated killer cells and interleukin-2. *J. Clin. Oncol.* **11**, 652–660.
99. Curti, B. D., Ochoa, A. C., Powers, G. C., Kopp, W. C., Alvord, W. G., Janik, J. E., Gause, B. L., Dunn, B., Kopreski, M. S., Fenton, R., Zea, A., Dansky-Ullmann, C., Strobl, S., Harvey, L., Nelson, E., Sznol, M., and Longo, D. L. (1998). Phase I trial of anti-CD3-stimulated $CD4^+$ T cells, infusional interleukin-2, and cyclophosphamide in patients with advanced cancer. *J. Clin. Oncol.* **16**, 2752–2760.
100. Lum, L. G., Treisman, J. S., Taylor, R. F., and LeFever, A. V. (1997). Phase I/II trial of activated T cells (ATC), IL-2, and GM-CSF after PBSC transplant for stage IIIb or IV breast cancer. *Blood* **90**, 381a.
101. Lum, L. G. (2000). Immunotherapy with activated T cells after high dose chemotherapy and PBSCT for breast cancer. *In* "Proc of the 10th Int'l Symposium on Autologous Blood and Marrow Transplantation," vol. 10, pp. 35–36.
102. June, C. H., Ledbetter, J. A., Linsley, P. S., and Thompson, C. B. (1990). Role of the CD28 receptor in T-cell activation. *Immunol. Today* **11**, 211–216.
103. June, C. H., Bluestone, J. A., Nadler, L. M., and Thompson, C. B. (1994). The B7 and CD28 receptor families. *Immunol. Today* **15**, 321–331.

104. Jenkins, M. K., and Johnson, J. G. (1993). Molecules involved in T-cell costimulation. *Curr. Opin. Immunol.* **5**, 361–367.
105. Schwartz, R. H. (1992). Costimulation of T lymphocytes: The role of CD28, CTLA-4, and B7/BB1 in interleukin-2 production and immunotherapy. *Cell* **71**, 1065–1068.
106. Costello, R., Cerdan, C., Pavon, C., Brailly, H., Hurpin, C., Mawas, C., and Olive, D. (1993). The CD2 and CD28 adhesion molecules induce long-term autocrine proliferation of CD4+ T cells. *Eur. J. Immunol.* **23**, 608–613.
107. Thompson, C. B., Lindsten, T., Ledbetter, J. A., Kunkel, S. L., Young, H. A., Emerson, S. G., Leiden, J. M., and June, C. H. (1989). CD activation pathway regulates the production of multiple T-cell-derived lymphokines/cytokines. *Proc. Natl. Acad. Sci. USA* **86**, 1333–1337.
108. Rosenberg, E. S., Billingsley, J. M., Caliendo, A. M., Boswell, S. L., Sax, P. E., Kalams, S. A., and Walker, B. D. (1997). Vigorous HIV-1-specific $CD4^+$ T cell responses associated with control of viremia. *Science* **278**, 1447–1450.
109. Garlie, N. K., LeFever, A. V., Siebenlist, R. E., Levine, B. L., June, C. H., and Lum, L. G. (1999). T cells co-activated with immobilized anti-CD3 and anti-CD28 as potential immunotherapy for cancer. *J. Immunother.* **4**, 335–345.
110. Levine, B. L., Ueda, Y., Craighead, N., Huang, M. L., and June, C. H. (1995). CD28 ligands CD80 (B7-1) and CD86 (B7-2) induce long-term autocrine growth of $CD4^+$ T cells and induce similar patterns of cytokine secretion *in vitro*. *Int. Immunol.* **7**, 891–904.
111. Boise, L. H., Noel, P. J., and Thompson, C. B. (1995). CD28 and apoptosis. *Curr. Opin. Immunol.* **7**, 620–625.
112. Boise, L. H., Minn, A. J., Accavitti, M. A., June, C. H., Lindsten, T., and Thompson, C. B. (1995). CD28 costimulation can promote T cell survival by inducing the expression of Bcl-X_1. *Immunity* **3**, 87–98.
113. Shibuya, T. Y., Wei, W. Z., Zormeier, M., Ensley, J., Sakr, W., Mathog, R. H., Meleca, R. J., Yoo, G., June, C.H., Levine, B., and Lum, L.G. (2000). Anti-CD3/anti-CD28 bead overcomes suppressed T cell responses in HNSCC patients. *Arch. Otolaryngol. Head Neck Surg.* **126**, 473–479.
114. Lum, L. G., LeFever, A. V., Treisman, J. S., Hanson, J. P., Jr., Garlie, N. K., Kistler, A. M., Yuille, D. L., Levine, B. L., and June, C. H. (1998). Phase I study of anti-CD3/anti-CD23 coactivated T cells (COACTS) in cancer patients: Enhanced TH_1 responses *in vivo*. *Exp. Hematol.* **26**, 772a.
115. Renner, C., and Pfreundschuh, M. (1995). Tumor therapy by immune recruitment with bispecific antibodies. *Immunol. Rev.* **145**, 179–209.
116. Bonino, L. D., De Monte, L. B., Sapnoli, G. C., Vola, R., Mariani, M., Barone, D., Moro, A. M., Riva, P., Niotra, M. R., Natali, P. G., and Malavasi, F. (1995). Bispecific monoclonal antibody anti-CD3 x anti-tenascin: An immunotherapeutic agent for human glioma. *Int. J. Cancer* **61**, 509–515.
117. Kaneko, T., Fusauchi, Y., Kakui, Y., Masuda, M., Akahoshi, M., Termura, M., Motoji, T., Okumura, K., Mizoguchi, H., and Oshimi, K. (1993). A bispecific antibody enhances cytokine-induced killer-mediated cytolysis of autologous acute myeloid leukemia cells. *Blood* **81**, 1333–1341.
118. Katayose, Y., Kudo, T., Suzuki, M., Shinoda, M., Saijyo, S., Sakurai, N., Saeki, H., Fukuhara, K., Imai, K., and Matsuno, S. (1996). MUC1-specific targeting immunotherapy with bispecific antibodies: Inhibition of xenografted human bile duct carcinoma growth. *Cancer Res.* **56**, 4205–4212.
119. Mack, M., Gruber, R., Schmidt, S., Riethmüller, G., and Kufer, P. (1997). Biologic properties of a bispecific single-chain antibody directed against 17-1A (EpCAM) and CD3: Tumor cell-dependent T cell stimulation and cytotoxic activity. *J. Immunol.* **158**, 3965–3970.
120. Bolhuis, R. L. H., Lamers, C. H. J., Goey, S. H., Eggermont, A. M. M., Trimbos, J. B. M. Z., Stoter, G., Lanzavecchia, A., di Re, E., Mioth, S., Raspagliesi, F., Rivoltini, L., and Colnaghi, M. I. (1992). Adoptive immunotherapy of ovarian carcinoma with BS-MAb-targeted lymphocytes: A multicenter study. *Int. J. Cancer* 78–81.
121. Canevari, S., Mezzanzanica, D., Mazzoni, A., Negri, D. R. M., Ramakrishna, V., Bohuis, R. L. H., Colnaghi, M. I., and Bolis, G. (1995). Bispecific antibody targeted T cell therapy of ovarian cancer: Clinical results and future directions. *J. Hematother.* **4**, 423–427.
122. Canevari, S., Stoter, G., Arienti, F., Bolis, G., Colnaghi, M. I., Di Re, E. M., Eggermont, A. M. M., Goey, S. H., Gratama, J. W., Lamers, C. H. J., Nooy, M. A., Parmiani, G., Raspagliesi, F., Ravagnani, F., Scarfone, G., Trimbos, J. B., Warnaar, S. O., and Bolhuis, R. L. H. (1995). Regression of advanced ovarian carcinoma by intraperitoneal treatment with autologous T lymphocytes retargeted by a bispecific monoclonal antibody. *J. Natl. Cancer Inst.* **87**, 1463–1469.
123. Lamers, C. H. J., van de Griend, R. J., Braakman, E., Ronteltap, C. P. M., Bénard, J., and Stoter, G. (1992). Optimization of culture conditions for activation and large-scale expansion of human T lymphocytes for bispecific antibody-directed cellular immunotherapy. *Int. J. Cancer* **51**, 973–979.
124. Zhu, Z., Ghose, T., Lee, S. H., Fernandez, L. A., Kerr, L. A., Donohue, J. H., and McKean, D. J. (1994). Tumor localization and therapeutic potential of an antitumor-CD3-heteroconjugate antibody in human renal cell carcinoma xenograft models. *Cancer Lett.* **86**, 127–134.
125. Van Dijk, J., Warnaar, S. O., van Eendenburg, J. D., Thienpont, M., Braakman, E., Boot, J. H., Fleuren, G. J., and Bolhuis, R. L. (1989). Induction of tumor-cell lysis by bi-specific monoclonal antibodies recognizing renal-cell carcinoma and CD3 antigen. *Int. J. Cancer* **43**, 344–349.
126. Tahara, H., and Lotze, M. T. (1995). Antitumor effects of interleukin -12 (IL-12): Applications for the immunotherapy and gene therapy of cancer. *Gene Ther.* **2**, 96–106.
127. Kroesen, B. J., ter Haar, A., Willemse, P., Sleijfer, D. T., de Vries, E. G. E., Mulder, N. H., Berendsen, H. H., Limburg, P. C., The, H. T., and de Leij, L. (1993). Local antitumor treatment in carcinoma patients with bispecific-monoclonal-antibody-redirected T cells. *Cancer Immunol. Immunother.* **37**, 401–407.
128. Anderson, P. M., Crist, W., Hasz, D., Carroll, A. J., Myers, D. E., and Uckun, F. M. (1992). G19.4(aCD3) x B43(aCD19) monoclonal antibody heteroconjugate triggers CD19 antigen-specific lysis of t(4;11) acute lymphoblastic leukemia cells by activated CD3 antigen-positive cytotoxic T cells. *Blood* **80**, 2826–2834.
129. Bohlen, H., Hopff, T., Manzke, O., Engert, A., Kube, D., Wickramanayake, P. D., Diehl, V., and Tesch, H. (1993). Lysis of malignant B cells from patients with B-chronic lymphocytic leukemia by autologous T cells activated with CD3 x CD19 bispecific antibodies in combination with bivalent CD28 antibodies. *Blood* **82**, 1803–1812.
130. Bohlen, H., Manzke, O., Patel, B., Moldenhauer, G., Dörken, B., von Fliedner, V., Diehl, V., and Tesch, H. (1993). Cytolysis of leukemic B-cells by T-cells activated via two bispecific antibodies. *Cancer Res.* **43**, 4310–4314.
131. Bejeck, B. E., Wang, D., Berven, E., Pennell, C. A., Peiper, S. C., Poppema, S., Uckun, F. M., and Kersey, J. H. (1995). Development and characterization of three recombinant single chain antibody fragments (scFvs) directed against the CD19 antigen. *Cancer Res.* **55**, 2346–2351.
132. de Gast, G. C., Haagen, I.-A., van Houten, A. A., Klein, S. C., Duits, A. J., de Weger, R. A., Vroom, T. M., Clark, M. R., Phillips, J., van Dijk, A. J. G., de Lau, W. B. M., and Bast, B. J. E. G. (1995). CD8 T cell activation after intravenous administration of CD3 x CD19 bispecific antibody in patients with non-Hodgkin lymphoma. *Cancer Immunol. Immunother.* **40**, 390–396.
133. Klein, S. C., Boer, L. H., de Weger, R. A., de Gast, G. C., and Bast, E. J. E. G. (1997). Release of cytokines and soluble cell surface molecules by PBMC after activation with the bispecific antibody CD3×CD19. *Scand. J. Immunol.* **46**, 452–458.
134. Kuwahara, M., Kuroki, M., Arakawa, F., Senba, T., Matsuoka, Y., Hideshima, T., Yamashita, Y., and Kanda, H. (1997). A mouse/human-chimeric bispecific antibody reactive with human carcinoembryonic

antigen-expressing cells and human T-lymphocytes. *Anticancer Res.* **16**, 2661–2668.

135. Renner, C., Jung, W., Sahin, U., Denfeld, R., Pohl, C., Trümper, L., Hartmann, F., Diehl, V., van Lier, R., and Pfreundschuh, M. (1994). Cure of xenografted human tumors by bispecific monoclonal antibodies and human T cells. *Science* **264**, 833–835.
136. Renner, C., Bauer, S., Sahin, U., Jung, W., van Lier, R., Jacobs, G., Held, G., and Pfreundschuh, M. (1996). Cure of disseminated xenografted human Hodgkin's tumors by bispecifc monoclonal antibodies and human T cells: The role of human T-cell subsets in a preclinical model. *Blood* **87**, 2930–2937.
137. Pohl, C., Denfeld, R., Renner, C., Jung, W., Bohlen, H., Sahin, U., Hombach, A., van Lier, R., Schwonzen, M., Diehl, V., and Pfreundschuh, M. (1993). CD30-antigen-specific targeting and activation of T cells via murine bispecific monoclonal antibodies against CD3 and CD28: Potential use for the treatment of Hodgkin's lymphoma. *Int. J. Cancer* **54**, 820–827.
138. Sen, M., Wankowski, D. M., Garlie, N. K., Siebenlist, R. E., Van Epps, D., LeFever, A. V., and Lum, L. G. (2001). Use of anit-CD3 x anti-HER2/neu bispecific antibody for redirecting cytotoxicity of activated T cells toward HER2/neu Tumors. *J. Hematother. Stem Cell Res.* **166**, 299–303.
139. Titus, J. A., Perez, P., Kaubisch, A., Garrido, M. A., and Segal, D. M. (1987). Human K/natural killer cells targeted with hetero-cross-linked antibodies specifically lyse tumor cells *in vitro* and prevent tumor growth *in vivo*. *J. Immunol.* **139**, 3153–3158.
140. Weiner, L. M., Clark, J. I., Davey, M., Li, W. S., de Palazzo, I. G., Ring, D. B., and Alpaugh, R. K. (1995). Phase I trial of 2B1, a bispecific monoclonal antibody targeting c-erbB-2 and Fc gamma RIII. *Cancer Res.* **55**, 4586–4593.
141. Valone, F. H., Kaufman, P. A., Guyre, P. M., Lewis, L. D., Memoli, V., Deo, Y., Graziano, R., Fisher, J. L., Meyer, L., and Mrozek-Orlowski, M. (1995). Phase Ia/Ib trial of bispecific antibody MDX-210 in patients with advanced breast or ovarian cancer that over expresses the proto-oncogene HER-2/neu. *J. Clin. Oncol.* **13**, 2281–2292.
142. Eshhar, Z., Waks, T., Gross, G., and Schindler, D. G. (1993). Specific activation and targeting of cytotoxic lymphocytes through chimeric single chains consisting of antibody-binding domains and the gamma or zeta subunits of the immunoglobulin and T-cell receptors. *Proc. Natl. Acad. Sci. USA* **90**, 720–724.
143. Hwu, P., Shafer, G. E., Treisman, J. S., Schindler, D. G., Gross, G., Cowherd, R., Rosenberg, S. A., and Eshhar, Z. (1993). Lysis of ovarian cancer cells by human lymphocytes redirected with a chimeric gene composed of an antibody variable region and the Fc receptor gamma chain. *J. Exp. Med.* **178**, 361–366.
144. Hwu, P., Yang, J. C., Cowherd, R., Treisman, J. S., Shafer, G. E., Eshhar, Z., and Rosenberg, S. A. (1995). *In vivo* antitumor activity of T cells redirected with chimeric antibody/T-cell receptor genes. *Cancer Res.* **55**, 3369–3373.
145. Eshhar, Z., Bach, N., Fitzer-Attas, C. J., Gross, G., Lustgarten, J., Waks, T., and Schindler, D. G. (1996). The T-body approach: Potential for cancer immunotherapy. *Springer Semin. Immunopathol.* **18**, 199–209.
146. Altenschmidt, U., Moritz, D., and Groner, B. (1997). Specific cytotoxic T lymphocytes in gene therapy. *J. Mol. Med.* **75**, 259–266.
147. Fitzer-Attas, C. J., and Eshhar, Z. (1998). Tyrosine kinase chimeras for antigen-selective T-body therapy. *Adv. Drug Delivery Rev.* **31**, 171–182.
148. Plavec, I., Agarwal, M., Ho, K. E., Pineda, M., Auten, J., Baker, J., Matsuzaki, H., Escaich, S., Bonyhadi, M., and Bohnlein, E. (1997). High transdominant RevM10 protein levels are required to inhibit HIV-1 replication in cell lines and primary T cells: Implication for gene therapy. *Gene Ther.* **4**, 128–139.
149. Quinn, E. R., Lum, L. G., and Trevor, K. T. (1998). T cell activation modulates retrovirus-mediated gene expression. *Hum. Gene Ther.* **9**, 1457–1467.
150. Myers, J. N., and Whiteside, T. L. (1995). Immunotherapy of squamous cell carcinoma of the head and neck. *In* "Cancer of the Head and Neck" (E. Myers and J. Suen, eds.), 3rd Ed., pp. 805–817. Saunders, Philadelphia.
151. Cortesina, G., Sacchi, M., Galeazzi, E., and De Stedfani, A. (1993). Immunology of head and neck cancer: Perspectives. *Head Neck* **15**, 74–77.
152. Mickel, R. A., Kessler, D. J., Taylor, J. M., and Lichtenstein, A. (1988). Natural killer cell cytotoxicity in the peripheral blood, cervical lymph nodes, and tumor of head and neck cancer patients. *Cancer Res.* **48**, 5017–5022.
153. Cortesina, G., Sartoris, A., DiFortunato, V., Cavallo, G. P., Morra, B., Bussi, M., Beatrice, F., Poggio, E., Marcato, P., and Rendine, S. (1984). Natural killer-mediated cytotoxicity in patients with laryngeal carcinoma. *Ann. Otol. Rhinol. Laryngol.* **93**, 189–191.
154. Wang, M. B., Lichtenstein, A., and Mickel, R. A. (1991). Hierarchical immunosuppression of regional lymph nodes in patients with head and neck squamous cell carcinoma. *Otolaryngol. Head Neck Surg.* **105**, 517–527.
155. Letessier, E. M., Sacchi, M., Johnson, J. T., Heberman, R. B., and Whiteside, T. L. (1990). The absence of lymphoid suppressor cells in tumor-involved lymph nodes of patients with head and neck cancer. *Cell. Immunol.* **130**, 446–458.
156. Kessler, D. J., Mickel, R. A., Lichtenstein, A. (1988). Depressed natural killer cell activity in cervical lymph nodes containing focal metastatic squamous cell carcinoma. *Arch. Otolaryngol. Head Neck Surg.* **114**, 313–318.
157. Heo, D. S., Whiteside, T. L., Johnson, J. T., Chen, K. N., Barnes, E. L., and Heberman, R. B. (1987). Long-term interleukin 2-dependent growth and cytotoxic activity of tumor-infiltrating lymphocytes (TIL) from human squamous cell carcinomas of the head and neck. *Cancer Res.* **47**, 6353–6362.
158. Horiuchi, K., Mishima, K., Ohsawa, M., Sugimura, M., and Aozasa, K. (1993). Prognostic factors for well-differentiated squamous cell carcinoma in the oral cavity with emphasis on immunohistochemical evaluation. *J. Surg. Oncol.* **53**, 92–96.
159. Tatake, R. J., Krishnan, N., Rao, R. S., Fakih, A. R., and Gangal, S. G. (1989). Lymphokine-activated killer-cell function of lymphocytes from peripheral blood, regional lymph nodes and tumor tissues of patients with oral cancer. *Int. J. Cancer* **43**, 560–566.
160. Murali, P. S., Somasundaram, R., Rao, R. S., Fakih, A. R., and Gangal, S. G. (1989). Interleukin-2 mediated regulation of mitogen-activated T cell reactivity from different lymphoid sources in patients with squamous cell carcinoma of the oral cavity. *J. Oral. Pathol. Med.* **18**, 327–332.
161. Leess, F. R., Bredenkamp, J. K., Lichtenstein, A., and Mickel, R. A. (1989). Lymphokine-activated killing of autologous and allogeneic short-term cultured head and neck carcinomas. *Laryngoscope* **99**, 1255–1261.
162. Ortega, I. S., Nieto, C. S., and Forelady, M. F. F. (1987). Lymph node response and its relationship to prognosis in carcinomas of the head and neck. *Clin. Otolaryngol.* **12**, 241–247.
163. Yasumura, S., Weidmann, E., Hirabayashi, H., Johnson, J. T., Herberman, R. B., and Whiteside, T. L. (1994). HLA restriction and T-cell-receptor V beta gene expression of cytotoxic T lymphocytes reactive with human squamous-cell carcinoma of the head and neck. *Int. J. Cancer* **57**, 297–305.
164. Vitolo, D., Letessier, E. M., Johnson, J. T., and Whiteside, T. L. (1992). Immunologic effector cells in head and neck cancer. *Monogr. Natl. Cancer Inst.* **12**, 203–208.
165. Vitolo, D., Kanbour, A., Johnson, J. T., Herberman, R. B., and Whiteside, T. L. (1993). In situ hybridization for cytokine gene transcripts in the solid tumor microenvironment. *Eur. J. Cancer* **29**, 371–377.

166. Tsukuda, M., Mochimatsu, I., Sakumoto, M., Furukawa, S., Yayama, S., Yanana, S., and Kubota, A. (1993). Autologous tumor cell killing activity of tumor-associated lymphocytes in patients with head and neck carcinomas. *Biotherapy* **6**, 155–161.
167. Yasamura, S., Hirabayashi, H., Schwartz, D. R., Toso, J. F., Johnson, J. T., Herberman, R. B., and Whiteside, T. L. (1993). Human cytotoxic T-cell lines with restricted specificity for squamous cell carcinoma of the head and neck. *Cancer Res.* **53**, 1461–1468.
168. Young, M. R. I., Schmidt-Pak A., Wright, M. A., Matthews, J. P., Collins, S. L., and Petruzzelli, G. (1995). Mechanisms of immune suppression in patients with head and neck cancer: Presence of immune suppressive CD34+ cells in cancers that secrete granulocyte-macrophage colony-stimulating factor. *Clin. Cancer Res.* **1**, 956–1103.
169. Young, M. R. I., Wright, M. A., and Pandit, R. (1997). Myeloid differentiation treatment to diminish the presence of immune-suppressive CD34+ cells within head and neck squamous cell carcinoma. *J. Immunol.* **159**, 990–996.
170. Young, M. R. I., Wright, M. A., Lozano, Y., Matthews, J. P., Benefield, J., and Prechel, M. M. (1996). Mechanisms of immune suppression in patients with head and neck cancer: Influence on the immune infiltrate of the cancer. *Int. J. Cancer* **67**, 333–338.
171. Hiratsuka, H., Imamura, M., Ishii, Y., Kohama, G., and Kikuchi, K. (1984). Immunohistologic detection of lymphocyte subpopulations infiltrating in human oral cancer with special reference to its clinical significance. *Cancer* **53**, 2456–2466.
172. Zeromski, J., Szmeja, A., Rewer, A., and Kruk-Zagajewska, A. (1986). Immunofluorescent assessment of tumor-infiltrating cells in laryngeal carcinoma. *Acta Otolaryngol. (Stockh.)* **102**, 325–332.
173. Wolf, G. T., Hudson, J. L., Peterson, K. A., Miller, H. L., and McClatchey, K. D. (1986). Lymphocyte subpopulations infiltrating squamous carcinomas of the head and neck: Correlations with extent of tumor and prognosis. *Otolaryngol. Head Neck Surg.* **95**, 142–152.
174. Bennest, S. H., Futrell, J. W., Roth, J. A., Hoye, R. C., and Ketcham, A. S. (1971). Prognostic significance of histological host response in cancer of the larynx or hypopharynx. *Cancer* **28**, 1255–1265.
175. Hirota, J., Vetq, E., Osaki, T., and Ogawa, Y. (1990). Immunohistologic study of mononuclear cell infiltrates in oral squamous cell carcinomas. *Head Neck* **12**, 118–125.
176. Snyderman, C. H., Heo, D. S., Chen, K., Whiteside, T. L., and Johnson, J. T. (1990). T-cell markers in tumor-infiltrating lymphocytes of head and neck cancer. *Head Neck* **12**, 118–125.
177. Whiteside, T. L., Heo, D. S., and Chen, K. (1987). Expansion of tumor-infiltrating lymphocytes from human solid tumors in interleukin-2. *In* "Progress in Clinical and Biological Research" (R.L. Truitt, R.P. Gale, and M.M. Bortin, eds), pp. 213–222. A.R. Liss, New York.
178. Snyderman, C. H., Heo, D. S., Johnson, J. T., D'Amico, F., Barnes, L., and Whiteside T. L. (1991). Functional and phenotypic analysis of lymphocytes isolated from tumors and lymph nodes of patients with head and neck cancer. *Arch. Otolaryngol. Head Neck Surg.* **117**, 899–905.
179. Nomori, H., Watanabe, S., Nakajima, T., Shimosato, Y., and Kameya, T. (1986). Histiocytes in naso-pharyngeal carcinoma in relations to prognosis. *Cancer* **57**, 100–105.
180. Neuchrist, C., Grasl, M., Scheiner, O., Lassmann, H., Ehrenberger, K., and Kraft, D. (1990). Squamous cell carcinoma: Infiltrating monocyte/macrohage subpopulations express functional mature phenotype. *Br. J. Cancer* **62**, 748–753.
181. Horst, H.-A., and Horny, H.-P. (1991). Tumor-infiltrating lymphoreticular cells. Histologic and immunohistologic investigations performed on metastasizing squamous cell carcinomas of the head and neck. *Cancer* **68**, 2397–2402.
182. Cozzolino, F., Torcia, M., Carossino, A. M., Giordani, R., Selli, C., Talini, G., Reali, E., Novelli, A., Pistoia, V., and Ferrarini, M. (1987). Characterization of cells from invaded lymph nodes in patients with solid tumors: Lymphokine requirement for tumor-specific lymphoid proliferative response. *J. Exp. Med.* **166**, 303–318.
183. Wustrow, T. P. U., and Kabelitz, D. (1987). Interleukin-2 release from lymphocytes of patients with head and neck cancer. *Ann. Otol. Rhinol. Laryngol.* **98**, 179–184.
184. Vinzenz, K., and Micksche, M. (1987). Natural cytotoxicity in draining lymph nodes of squamous cell cancer in the maxillofacial region. *J. Oral. Maxillofac. Surg.* **45**, 42–47.
185. Eura, M., Maehara, T., Ikawa, T., and Ishikawa, T. (1988). Suppressor cells in the effector phase of autologous cytotoxic reactions in cancer patients. *Cancer Immunol. Immunother.* **27**, 147–153.
186. Letessier, E. M., Heo, D. S., Okarama, T., Johnson, J. T., Herberman, R. B., and Whiteside, T. L. (1991). Enrichment in tumor-reactive CD8+ T-lymphocytes by positive selection from the blood and lymph nodes of patients with head and neck cancer. *Cancer Res.* **51**, 3891–3899.
187. Roubin, R., Bekkoucha, F., Fondaneche, M.-C., Quan, P. C., Micheau, C., Cachin, Y., and Burtin, P. (1982). Lymphoid cells in lymph nodes and peripheral blood of patients with squamous cell carcinoma of the head and neck. *J. Cancer Res. Clin. Oncol.* **102**, 277–287.
188. Schuller, D. E., Rock, R. P., Rinehart, J. J., and Koolemans-Beynen, A. R. (1986). T-lymphocytes as a prognostic indicator in head and neck cancer. *Arch. Otolaryngol. Head Neck Surg.* **112**, 938–941.
189. Eskinazi, D. P., Helman, J., Ershow, A. G., Perna, J. J., and Mihail, R. (1985). Nonspecific immunity and head and neck cancer: Blastogenesis reviewed and revisited. *Oral. Surg. Oral. Med. Oral. Pathol.* **60**, 642–647.
190. Eilber, F. R., Morton, D. L., and Ketcham, A. S. (1974). Immunologic abnormalities in head and neck cancer. *Am. J. Surg.* **128**, 534–538.
191. Wanebo, H. J., Jun, M. Y., Strong, E. W., and Oettgen, H. (1975). T-cell deficiency in patients with squamous cell cancer of the head and neck. *Am. J. Surg.* **130**, 445–451.
192. Mason, J. M., Kitchens, G. G., Eastham, R. J., and Jennings, B. R. (1977). T-lymphocytes and survival of head and neck squamous cell carcinoma. *Arch. Otolaryngol. Head Neck Surg.* **103**, 233–236.
193. Ryan, R. E., Neel, H. B., III, and Ritts, R. E. (1980). Correlation of preoperative immunologic test results with recurrence in patients with head and neck cancer. *Arch. Otolaryngol. Head Neck Surg.* **88**, 58–63.
194. Papenhausen, P. R., Kukwa, A., Croft, C. B., Borowiencki, B., Silver, C., and Emeson, E. E. (1979). Cellular immunity in patients with epidermoid cancer of the head and neck. *Laryngoscope* **89**, 538–549.
195. Eskinazi, D. P., Perna, J. J., Ershow, A. G., and Mihail, R. C. (1989). Depressed PMNC blastogenic response in patients with cancer of the head and neck: A study of IL-2 production, IL-2 consumption, and IL-2 receptor expression. *Laryngoscope* **99**, 151–157.
196. Wanebo, H. J., Jones, T., Pace, R., Cantrell, R., and Levine, P. (1989). Immune restoration with interleukin-2 in patient with squamous cell carcinomas of the head and neck. *Am. J. Surg.* **158**, 356–360.
197. Hargett, S., Wanebo, H. J., Pace, R., Katz, D., Sando, J., and Cantrell, R. (1985). Interleukin-2 production in head and neck cancer patients. *Am. J. Surg.* **150**, 456–460.
198. Wolf, G. T., Hudson, J., Peterson, K. A., Poore, J. A., and McClatchey, K. D. (1989). Interleukin-2 receptor expression in patients with head and neck squamous carcinomas: Effects of thymosin a_1 in vitro. *Arch. Otolaryngol. Head Neck Surg.* **115**, 1345–1349.
199. Wolf, G. T., Lovett, E. J., Peterson, K. A., Beauchamp, M. L., and Baker, S. R. (1984). Lymphokine production and lymphocyte subpopulations in patients with head and neck squamous carcinoma. *Arch. Otolaryngol. Head Neck Surg.* **110**, 731–735.
200. Balaram, P., and Vasudevan, D. M. (1983). Quantitation of Fc receptor-bearing T-lymphocytes (T_G and T_M) in oral cancer. *Cancer* **52**, 1837–1840.
201. Johnson, J. T., Rabin, B. S., Hirsch, B., and Thearle, P. B. (1984). T-cell subpopulations in head and neck carcinoma. *Arch. Otolaryngol. Head Neck Surg.* **92**, 381–385.

202. Dawson, D. E., Everts, E. C., Vetto, R. M., and Burger, D.R. (1985). Assessment of immunocompetent cells in patients with head and neck squamous cell carcinoma. *Ann. Otol. Rhinol. Laryngol.* **94**, 342–345.
203. Strome, M., Clark, J. R., Fried, M. P., Rodliff, S., and Blazar, B. A. (1987). T-cell subsets and natural killer cell function with squamous cell carcinoma of the head and neck. *Arch. Otolaryngol. Head Neck Surg.* **113**, 1090–1093.
204. Wolf, G. T., Schmaltz, S., Hudson, J., Robson, H., Stackhouse, T., Peterson, K. A., Poore, J. A., and McClatchey, K. D. (1987). Alterations in T-lymphocyte subpopulations in patients with head and neck cancer. *Arch. Otolaryngol. Head Neck Surg.* **113**, 1200–1206.
205. Schantz, S. P., and Liu, F. J. (1989). An immunologic profiles of young adults with head and neck cancer. *Cancer* **64**, 1232–1237.
206. Berlinger, N. T., Tsakraklides, V., Pollak, K., Adams, G. L., Yang, M., and Good, R. A. (1976). Immunologic assessment of regional lymph node histology in relation to survival in head and neck carcinoma. *Cancer* **37**, 697–705.
207. Berlinger, N. T., Tsakraklides, V., Pollak, K., Adams, G. L., Yang M., and Good, R. A. (1976). Prognostic significance of lymph node histology in patients with squamous cell carcinoma of the larynx, pharynx, or oral cavity. *Laryngoscope* **86**, 792–803.
208. Ring, A. H., Sako, K., Rao, U., Razack, M. S., and Reese P. (1985). Immunologic patterns of regional lymph nodes in squamous cell carcinoma of the floor of the mouth: Prognostic significance. *Am. J. Surg.* **150**, 461–465.
209. Kuss, I., Saito, T., Johnson, J. T., and Whiteside, T. L. (1999). Clinical significance of decreased ζ chain expression in peripheral blood lymphocytes of patients with head and neck cancer. *Clin. Cancer Res.* **5**, 329–334.
210. Brookes, G. B., and Clifford, P. (1981). Nutritional status and general immune competence in patients with head and neck cancer. *JR Soc. Med.* **74**, 132–139.
211. Good, R. A., Fernandez, G., and West, A. (1979). Nutrition, immunity and cancer: A review. I. Influence of protein or protein-calorie malnutrition and zinc deficiency on immunity. *Clin. Bull.* **9**, 3–12.
212. Hadden, J. W. (1995). Immunology of head and neck cancer. *Clin. Immunother.* **3**, 362–385.
213. Coto, J. A., Hadden, E. M., Sauro, M., Zorn, N., and Hadden, J. W. (1992). Interleukin 1 regulates secretion of zinc-thymulin by human thymic epithelial cells and its action on T-lymphocyte proliferation and nuclear kinase C. *Proc. Natl. Acad. Sci. USA* **89**, 7752–7756.
214. Abdulla, M., Biorklund, A., Mathur, A., and Walleniusk, K. (1979). Zinc and copper levels in whole blood and plasma from patients with squamous cell carcinoma of head and neck. *J. Surg. Oncol.* **12**, 107–113.
215. Johnson, J. T., Rabin, B. S., and Wagner, R. L. (1987). Prostaglandin E2 of the upper aerodigestive tract. *Ann. Otolaryngol. Rhinol. Laryngol.* **96**, 213–216.
216. Snyderman, C. H., Kaplan, I., Milanovich, M., Heo, D. S., Wagner, R., Schwartz, M., Johnson, J. T., and Whiteside, T. L. (1994). Comparison of in vitro prostaglandin E2 production by squamous cell carcinoma of the head and neck. *Otolaryngol. Head Neck Surg.* **111**, 189–196.
217. Goodwin, J. S., and Ceuppens, J. (1983). Regulation of the immune response by prostaglandins. *J. Clin. Immunol.* **3**, 295–315.
218. Panje, W. R. (1981). Regression of head and neck carcinoma with a prostaglandin-synthesis inhibitor. *Arch. Otolaryngol.* **107**, 658–663.
219. Hirsch, B., Johnson, J. T., Rabin, B. S., and Thearle, P. B. (1983). Immunostimulation of patients with head and neck cancer. *Arch. Otolaryngol.* **109**, 298–301.
220. Cross, D. S., Platt, J. L., Juhn, S. K., Bach, F. H., and Adams, G. L. (1992). Administration of a prostaglandin synthesis inhibitor associated with an increased immune cell infiltrate in squamous cell carcinoma of the head and neck. *Arch. Otolaryngol. Head Neck Surg.* **118**, 526–528.
221. Simons, P. J., Oostendorp, R. A., Tas, M. P., and Drexhage, H. A. (1994). Comparison of retroviral p15E-related factors and interferon alpha in head and neck cancer. *Cancer Immunol. Immunother.* **38**, 178–184.
222. Tan, I. B., Drexhage, H. A., Mullink, R., Hensen-Logmans, S., Snow, G. B., and Balm, A. J. (1987). Immunohistochemical detection of retro viral – P15E-related material in carcinoma of the head and neck. *Otolaryngol. Head Neck Surg.* **96**, 251–255.
223. Nelson, M., Nelson, D. S., Spradbrow, P. B., Kuchroo, V. K., Jennings P.A., and Synderman, R. (1985). Successful tumor immunotherapy: Possible role of antibodies to anti-inflammatory factors produced by neoplasms. *Clin. Exp. Immunol.* **61**, 109–117.
224. Mann, E. A., Spiro, J. D., Chen, L. L., and Kreutzer, D. L. (1992). Cytokine expression by head and neck squamous cell carcinoma. *Am. J. Surg.* **164**, 567–573.
225. Hahne, M., Rimoldi, D., Schroter, M., Romero, P., Schreier, M., French, L. E., Schneider, P., Bornand, T., Fontana, A., Lienard, D., Cerottini, J. C., and Tschopp, J. (1996). Melanoma cell expression of Fas ligand: Implications for tumor immune escape. *Science* **274**, 1363–1366.
226. Bennett, M. W., and O'Connell, J. (1998). The Fas counterattack in vivo: Apoptotic depletion of tumor-infiltrating lymphocytes associated with Fas ligand expression by human esophageal carcinoma. *J. Immunol.* **160**, 5669–5675.
227. Gratas, C., Tohma, Y., Barnas, C., Taniere, P., Hainaut, P., and Ohgaki, H. (1998). Upregulation of Fas ligand and down-regulation of Fas expression in human esophageal cancer. *Cancer Res.* **58**, 2057–2062.
228. Walker, P. R., Saas, P., and Dietrich, P. Y. (1997). Role of fas ligand (CD95L) in immune escape: The tumor cell strikes back. *J. Immunol.* **158**, 4521–4524.
229. Cardi, G., Heaney, J. A., Schned, A. R., and Ernstoff, M. S. (1998). Expression of Fas in tumor infiltrating and peripheral blood lymphocytes in patients with renal cell carcinoma. *Cancer Res.* **58**, 2078–2080.
230. O'Connell, J., Bennett, M. W., O'Sullivan, G. C., Collins, J. K., and Shanahas, F. (1999). Fas counter-attack-the best form of tumor defense? *Nature Med.* **5**, 267–268.
231. Myers, J. N., Yasamura, S., Suminiami, Y., Hirabayashi, H., Lin, W., Johnson, J. T., Lotze, M. T., and Whiteside, T. L. (1996). Growth stimulation of human head and neck squamous cell carcinoma cell lines by interleukin-4. *Clin. Cancer Res.* **2**, 127–135.
232. Byron, K. A., Varigos, G. A., and Wootton, A. M. (1994). IL-4 production is increased in cigarette smokers. *Clin. Exp. Immunol.* **95**, 333–336.
233. Finke, J. H., Zea, A. H., Stanley, J., Longo, D. L., Mizoguchi, H., Tubbs, R. R., Wiltrout, R. H., O'Shea, J. J., Kudoh, S., and Klein, E. (1993). Loss of T-cell receptor zeta chain and $p56^{lck}$ in T-cells infiltrating human renal cell carcinoma. *Cancer Res.* **53**, 5613–5616.
234. Lai, P., Rabinowich, H., Crowley-Nowich, P. A., Bell, M. C., Mantovani, G., and Whiteside, T. L. (1996). Alteration in expression and function of signal transducing proteins in tumor associated T and NK cells in patients with ovarian carcinoma. *Clin. Cancer Res.* **2**, 161–173.
235. Zea, A. H., Curti, B. D., Longo, D. L., Alvord, W. G., Strobl, S. L., Mizoguchi, H., Creekmore, S. P., O'Shea, J. J., Powers, G. C., Urba, W. J., and Ochoa, A. O. (1995). Alterations in T cell receptor and signal transduction molecules in melanoma patients. *Clin. Cancer Res.* **1**, 1327–1335.
236. Matsuda, M., Petersson, M., Lenkei, R., Taupin, J. L., Magnusson, I., Mellstedt, H., Anderson, P., and Kiessling, R. (1995). Alterations in the signal transducing molecules of T cells and NK cells in colorectal tumor infiltrating gut mucosal and peripheral lymphocytes: Correlation with the stage of the disease. *Int. J. Cancer* **61**, 765–772.
237. Rabinowich, H., Reichert, T. E., Kashii, Y., Gastman, B. R., Bell, M. C., and Whiteside, T. L. (1998). Lymphocyte apoptosis induced by Fas ligand-expressing ovarian carcinoma cells. *J. Clin. Invest.* **101**, 2579–2588.

238. Mozoguchi, H., O'Shea, J. J., Longo, D. L., Loeffler, C. M., McVicar, D. W., and Ochoa, A. C. (1992). Alterations in signal transduction molecules in T lymphocytes from tumor-bearing mice. *Science* **258**, 1795–1798.
239. Nakagomi, H., Petersson, M., Magnusson, I., Juhlin, C., Matsuda, M., Mellstedt, H., Taupin, J. L., Vivier, E., Anderson, P., and Kiessling, R. (1993). Decreased expression of the signal transducing zeta chains in tumor infiltrating T-cells and NK cells of patients with colorectal carcinoma. *Cancer Res.* **53**, 5610–5612.
240. Reichert, T. E., Rabinowich, H., Johnson, J. T., and Whiteside, T. L. (1998). Mechanisms responsible for signaling and functional defects. *J. Immunother.* **21**, 295–306.
241. Gabrilovich, D. I., Chen, H. L., Girgis, K. R., Cunningham, H. T., Meny, G. M., Nadaf, S., Kavanaugh, D., and Carbone, D. P. (1996). Production of vascular endothelial growth factor by human tumors inhibits the functional maturation of dendritic cells. *Nature Med.* **2**, 1096–1103.
242. Young, M. R. I., Schmidt-Pak, A., Wright, M. A., Matthews, J. P., Collins, S. L., and Petruzzelli, G. (1995). Mechanisms of immune suppression in patients with head and neck cancer: Presence of immune suppressive CD34+ cells in cancers that secrete granulocyte-macrophage colony-stimulating factor. *Clin. Cancer Res.* **1**, 956–10361.
243. Eisma, R. J., Spiro, J. D., and Kreutzer, D. L. (1999). Role of angiogenic factors: Coexpression of interleukin-8 and vascular endothelial growth factor in patients with head and neck carcinoma. *Laryngoscope* **109**, 687–693.
244. Hadden, J. W. (1997). The immunopharmacology of head and neck cancer: An update. *Int. J. Immunopharmacol.* **19**, 629–644.
245. Lawlor, G. J., Stiehm, E. R., Kaplan, M. S., Sengar, D. P. S., and Terasaki, P. I. (1973). Phytohemagglutinin (PHA) skin test in the diagnosis of cellular immunodeficiency. *J. Allergy Clin. Immunol.* **52**, 31–37.
246. Hadden, J. W. (1994). T-cell adjuvancy. *Int. J. Immunopharmacol.* **16**, 703–710.
247. Verastegui, E., Barrera, J. L., Zinser, J., del Rio, R., Meneses, A., de La Garza, J., and Hadden, J. W. (1997). A natural cytokine mixture (IRX-2) and interference with immune suppression induce immune mobilization and regression of head and neck cancer. *Int. J. Immunopharmacol.* **19**, 619–627.
248. Parker, S. L., Tong, T., Bolden, S., and Wingo, P. A. (1996). Cancer statistics. *CA Cancer J. Clin.* **46**, 5–27.
249. Whiteside, T. L. (1994). Tumor-infiltrating lymphocytes in head and neck cancer. *In* "Immunotherapy of Cancer With Sensitized T Lymphocytes" (A.E. Chang and S. Shu, eds.), pp. 133–154. Landes Co., Austin, TX.
250. Reichert, T. E., Day, R., Wagner, E. M., and Whiteside, T. L. (1998). Absent or low expression of the ζ chain in T cells at the tumor site correlates with poor survival in patients with oral carcinoma. *Cancer Res.* **58**, 5344–5347.
251. Medina, J. E. (1983). The controversial role of BCG in treatment of squamous cell carcinoma of the head and neck. *Arch. Otolaryngol. Head Neck Surg.* **109**, 543.
252. Vogl, S. E., Schoenfeld, D. A., Kaplan, B. H., Lerner, H. J., Horton, J., and Creech, R. H. (1982). Methotrexate alone or with regional subcutaneous *Corynebacterium parvum* in the treatment of recurrent and metastatic squamous cancer of the head and neck. *Cancer* **50**, 2295–2300.
253. Vlock, D. R., Johnson, J., Myers, E., Day, R., Gooding, W. E., Whiteside, T. L., Pelch, J., Sigler, B., Wagner, R., Colao, D., *et al.* (1991). Preliminary trial of nonrecombinant interferon alpha in recurrent squamous cell carcinoma of the head and neck. *Head Neck* **13**, 15–21.
254. Vlock, D. R., Andersen, J., Kalish, L. A., Johnson, J. T., Kirkwood, J. M., Whiteside, T. L., Herberman, R. B., Adams, G. S., Oken, M. M., and Haselow, R. E. (1996). Phase II trial of interferon-alpha in locally-recurrent or metastatic squamous cell carcinoma of the head and neck: Immunological and clinical correlates. *J. Immunother.* **19**, 433–436.
255. Forni, G., Giovarelli, M., Jemma, C., Bosco, M. C., Caretto, P., Modesti, A., Forni, M., Cortesina, G., de Stefani, A., *et al.* (1990). Perilymphatic injections of cytokines: A new tool in active cancer immunotherapy. *Ann. Ist. Super Sanita* **26**, 397–409.
256. Cortesina, G., DeStefani, A., Galeazzi, E., Cavallo, G. P., Jemma, C., Giovarelli, S., and Forni, G. (1991). Interleukin-2 injected around tumor-draining lymph nodes in head and neck cancer. *Head Neck* **13**, 125–131.
257. Cortesina, G., DeStefani, A., Galeazzi, E., Cavallo, G. P., Badellino, F., Jemma, C., and Forni, G. (1994). Temporary regression of recurrent squamous cell carcinoma of the head and neck is achieved with a low but not with a high dose of recombinant interleukin-2 injected perilymphatically. *Br. J. Cancer* **69**, 572–576.
258. DeStefani, A., Valente, G., Forni, G., Lerda W., Ragona R., and Cortesina, G. (1996). Treatment of oral cavity and oropharynx squamous cell carcinoma with perilymphatic interleukin-2: Clinical and pathologic correlations. *J. Immunother. Emphasis Tumor Immunol.* **19**, 125–133.
259. Mattijssen, V., DeMulder, P. H., Schoznagel, J. H., Verweij, J., Galazka, A., Roy, S., and Ruiter, D. J. (1991). Clinical and immunopathologic results of a phase II study of perilymphatically injected interleukin-2 in locally far advanced non-pretreated head and neck squamous cell carcinoma. *J. Immunother.* **10**, 63–68.
260. Vlock, D. R., Snyderman, C., Johnson, J. T., Myers, E. N., Eibling, D. E., Rubin, J., Kirkwood, J. M., Dutcher, J. P., and Adams, G. L. (1996). Phase Ib trial of the effect of peritumoral and intranodal injections of interleukin-2 in patients with advanced squamous cell carcinoma of the head and neck: An Eastern Cooperative Oncology Group trial. *J. Immunother.* **19**, 433–442.
261. Whiteside, T. L., Letessier, E., Hirabayashi, H., Vitolo, D., Bryant, J., Barnes, L., Snyderman, C., Johnson, J. T., Myers, E., Herberman, R. B., *et al.* (1993). Evidence for local and systemic activation of immune cells by peritumoral injections of interleukin-2 in patients with advanced squamous cell carcinoma of the head and neck. *Cancer Res.* **53**, 5654–5662.
262. Richtsmeier, W. J., Koch, W. M., McGuire, W. P., Poole, M. D., and Chang, E. H. (1990). Phase I-II study of advanced head and neck squamous cell carcinoma patients treated with recombinant interferon gamma. *Arch. Otolaryngol. Head Neck Surg.* **116**, 1271–1277.
263. Verastegui, E., Barrera, J. L., Zinser, J., Zinser J., de la Garza, J., and Hadden, J. W. (1997). A natural cytokine mixture (IRX-2) and interference with immune suppression induce immune mobilization and regression of head and neck cancer. *Int. J. Immunopharmacol.* **19**, 619–627.
264. Meneses, A., Verastegui, E., Barrera, J. L., Zinser, J., de la Garza, J., and Hadden, J. W. (1998). Histologic findings in patients with head and neck squamous cell carcinoma receiving perilymphatic natural cytokine mixture (IRX-2) prior to surgery. *Arch. Pathol. Lab. Med.* **122**, 447–454.
265. Barrera, J. L., Verastegui, E., Meneses, A., Zinser, J., de la Garza, J., and Hadden, J. W. (2000). Combination immunotherapy of squamous cell carcinoma of the head and neck: A phase 2 trial. *Arch. Otolaryngol. Head Neck Surg.* **126**, 345–351.
266. Mantovani, G., Gebbia, V., Airoldi, M., Bumma, C., Contu, P., Bianchi, A., Dessi, D., Massa, E., Curreli, L., Lampis, B., Lai, P., Mulas, C., Testa, A., Proto, I., Cadeddu, G., and Tore, G. (1998). Neoadjuvant chemo-(immuno-) therapy of advanced squamous-cell head and neck carcinoma: A multicenter, phase III, randomized study comparing cisplatin + 5-fluorouracil (5-FU) with cisplatin + 5-FU + recombinant interleukin 2. *Cancer Immunol. Immunother.* **47**, 149–156.

267. Evans, T. R., and Kaye, S. B. (1999). Retinoids: Present role and future potential. *Br. J. Cancer* **80**, 1–8.
268. Wan, H., Oridate, N., Lotan, D., Hong, W. K., and Lotan, R. (1999). Overexpression of retinoic acid receptor beta in head and neck squamous cell carcinoma cells increases their sensitivity to retinoid-induced suppression of squamous differentiation by retinoids. *Cancer Res.* **59**, 3518–3526.
269. Van Zandwijk, N., Dalesio, O., Pastorino, U., de Vries, N., and van Tinteren, H. (2000). EUROSCAN, a randomized trial of vitamin A and N-acetylcysteine in patients with head and neck cancer or lung cancer: For the European Organization for Research and Treatment of Cancer Head and Neck and Lung Cancer Cooperative Groups. *J. Natl. Cancer Inst.* **92**, 977–986.
270. Nikolaou, A. C., Fountzilas, G., and Daniilidis, I. (1996). Treatment of unresectable recurrent head and neck carcinoma with 13-cis-retinoic acid and interferon-alpha: A phase II study. *J. Laryngol. Otol.* **110**, 857–861.
271. Milanovich, M. R., Snyderman, C. H., Wagner, R., and Johnson, J. T. (1995). Prognostic significance of prostaglandin E2 production by mononuclear cells and tumor cells in squamous cell carcinomas of the head and neck. *Laryngoscope* **105**, 61–65.
272. Snyderman, C. H., Milanovich, M., Wagner, R. L., and Johnson, J. T. (1995). Prognostic significance of prostaglandin E2 production in fresh tissues of head and neck cancer patients. *Head Neck* **17**, 108–113.
273. Hirsch, B. E., Johnson, J. T., Rabin, B. S., and Thearle, P. B. (1983). Immunostimulation of patients with head and neck cancer: *In vitro* and preliminary clinical experiences. *Arch. Otolaryngol.* **109**, 298–301.
274. Cragg, M. S., French, R. R., and Glennie, M. J. (1999). Signaling antibodies in cancer therapy. *Curr. Opin. Immunol.* **11**, 541–547.
275. De Bree, R., Roos, J. C., Quak, J. J., den Hollander, W., Snow, G. B., and van Dongen, G. A. (1995). Radioimmunoscintigraphy and biodistribution of technetium-99m-labeled monoclonal antibody U36 in patients with head and neck cancer. *Clin. Cancer Res.* **1**, 591–598.
276. van Hal, N. L., van Dongen, G. A., Rood-Knippels, E. M., Van Der Valk, P., Snow, G. B., and Brankenhoff, R. H. (1996). Monoclonal antibody U36, a suitable candidate for clinical immunotherapy of squamous-cell carcinoma, recognizes a CD44 isoform. *Int. J. Cancer* **68**, 520–527.
277. van Gog, F. B., Visser, G. W., Stroomer, J. W., Roos, J. C., Snow, G. B., and van Dongen, G. A. (1997). High dose rhenium-186-labeling of monoclonal antibodies for clinical application: Pitfalls and solutions. *Cancer* **80**, 2360–2370.
278. de Bree, R., Roos, J. C., Plaizier, M. A., Quak, J. J., van Kamp, G. J., Snow, G. B., and van Dongen, G. A. (1997). Selection of monoclonal antibody E48 IgG or U36 IgG for adjuvant radioimmunotherapy in head and neck cancer patients. *Br. J. Cancer* **75**, 1049–1060.
279. van Gog, F. B., Brakenhoff, R. H., Stigter-van Walsum, M., Snow, G. B., and van Dongen, G. A. (1998). Perspectives of combined radioimmunotherapy and anti-EGFR antibody therapy for the treatment of residual head and neck cancer. *Int. J. Cancer* **77**, 13–18.
280. Squadrelli-Saraceno, M., Rivoltini, L., Cantu, G., Ravagnani, F., Parminai, G., and Molinari, R. (1991). Local adoptive immunotherapy of advanced head and neck tumors with LAK cells and interleukin-2. *Tumori* **76**, 566–571.
281. Ikawa, T., Eura, M., Fukiage, T., Murakami, H., Yamasaki, S., Fukuda, K., Arishima, S., Haehara, T., and Ishikawa, T. (1989). Adoptive immunotherapy by intraarterial infusion of ATLAK or allo-TLAK cells in patients with head and neck cancer. *Gan To Kagaku Ryoho* **16**, 1438–1447.
282. Rosenberg, S. A., Yanelli, J. R., Yang, J. C., Topalian, S. L., Schwartzentruber, D., Weber, J. S., Parkinson, D. R., Seipp, C. A., Einhorn, J. H., and White, D. E. (1994). Treatment of patients with metastatic melanoma with autologous tumor-infiltrating lymphocytes and interleukin 2. *J. Natl. Cancer Inst.* **86**, 1159–1166.
283. Mendelsohn, J. (1997). Epidermal growth factor receptor inhibition by a monoclonal antibody as anticancer therapy. *Clin. Cancer Res.* **3**, 2703–2707.
284. Hoffman, T. K., Nakano, K., Elder, E., Dworacki, G., Finkelstein, S. D., Whiteside, T. L., and DeLeo, A. B. (2000). Generation of T cells specific for the wild-type sequence $p53_{264-272}$ peptide in cancer patients: Implication for immunoselection of epitope-loss variants. *J. Immunol.* **165**, 5938–5944.
285. Shichijo, S., Nakao, M., Imai, Y., Takasu, H., Kawamoto, M., Niiya, F., Yang, D., Yamana, H., and Itoh, K. (1998). A gene encoding antigenic peptides of human squamous cell carcinoma recognized by cytotoxic T lymphocytes. *J. Exp. Med.* **187**, 277–288.
286. Ito, M., Schichijo, S., Miyagi, Y., Kobayashi, T., Tsuda, N., Yamada, A., and Saito, N. (2000). Identification of SART3-derived peptides capable of inducing HLA-A2-restricted and tumor-specific CTLs in cancer patients with different HLA-A2 subtypes. *Int. J. Cancer* **88**, 633–639.
287. Kao, H., Marto, J. A., and Hoffmann, T. K. (2001). Identification of cyclin B1 as a shared human epithelial tumor-associated antigen recognized by T cells. Submitted for publication.
288. Mandruzzato, S., Brasseur, F., Andry, G., Boon, T., and van der Bruggen, P. (1997). A CASP-8 mutation recognized by cytolytic T lymphocytes on a human head and neck carcinoma. *J. Exp. Med.* **186**, 785–793.
289. Chikamatsu, K., Nakano, K., Storkus, W. J., Appella, E., Lotze, M. T., Whiteside, T. L., and DeLeo, A. B. (1999). Generation of anti-p53 cytotoxic T lymphocytes from human peripheral blood using autologous dendritic cells. *Clin. Cancer Res.* **5**, 1281–1288.
290. DeLeo, A. B. (1998). p53-based immunotherapy of cancer. *Crit. Rev. Immunol.* **18**, 29–35.
291. Henry, F., Boisteau, O., Bretaudeau, L., Lieubeau, B., Meflah, K., and Gregoire, M. (1999). Antigen-presenting cells that phagocytose apoptotic tumor-derived cells are potent tumor vaccines. *Cancer Res.* **59**, 3329–3332.
292. Hoffmann, T. K., Meidenbauer, N., Dworacki, G., Kanaya, H., and Whiteside, T. L. (2000). Generation of tumor-specific T lymphocytes by cross-priming with human dendritic cells ingesting apoptotic tumor cells. *Cancer Res.* **60**, 3542–3549.
293. De Zoeten, E., Carr-Brendel, V., Markovic, D., and Taylor-Papdimitriou, J. (1999). Immunity to breast cancer in mice immunized with semi-allogeneic fibroblasts transfected with DNA from breast cancer cells. *J. Immunol.* **162**, 6934–6941.
294. Cohen, E. P. (2000). Cancer therapy with DNA-based vaccines. *Immunol. Lett.* **74**, 59–65.
295. Nagashima, S., Reichert, T. E., Kashii, Y., Suminami, Y., Chikamatsu, K., and Whiteside, T. L. (1997). *In vitro* and *in vivo* characteristics of human squamous cell carcinoma of the head and neck cells engineered to secrete interleukin-2. *Cancer Gene Ther.* **4**, 366–376.
296. Tzai, T. S., Lin, J. S., and Chow, N. H. (1996). Modulation of antitumor immunity of tumor-bearing mice with low dose cyclophosphamide. *J. Surg. Res.* **65**, 139–144.
297. Gleich, L. L. (2000). Gene therapy for head and neck cancer. *Laryngoscope* **110**, 708–726.
298. Tahara, H., Elder, E., and Zitvogel, L. (2001). Interleukin 12 (IL-12) gene therapy using direct injection of tumors with genetically engineered autologous fibroblasts: A phase I clinical trial. Submitted for publication.
299. Shu, S., Chou, T., and Sakai, K. (1989). Lymphocytes generated by *in vivo* priming and *in vitro* sensitization demonstrate therapeutic efficacy against a murine tumor that lacks apparent immunogenicity. *J. Immunol.* **143**, 740–748.
300. Whiteside, T. L. (1998). Cellular adoptive immunotherapy: Expectations and reality. *In* "The Biotherapy of Cancers: From immunotherapy to Gene Therapy" (S. Chouaib, ed.), pp. 239–262. Les Editions INSERM, Paris.
301. To, W. C., Wood, B. G., Krauss, J. C., Strome, M., Esclamado, R. M., Lavertu, P., Kim, J. A., Plautz, G. E., Leff, B. E., Smith, V.,

Sandstrom-Wakeling, K., and Shu, S. (2000). Systemic adoptive T cell immunotherapy in recurrent and metastatic carcinoma of the head and neck. *Arch. Otolaryngol. Head Neck Surg.* **126**, 1225–1231.

302. Whiteside, T. L., Reichert, T. E., and Dou, P. Q. (2002). Interleukin-2 and its receptors in human solid tumors: Immunobiology and clinical significance. *Cancer Metastasis Biol. Treat* (in press).

303. Saito, T., Kuss, I., Dworacki, G., Gooding, W., Johnson, J. T., and Whiteside, T. L. (1999). Spontaneous *ex vivo* apoptosis of peripheral blood mononuclear cells in patients with head and neck cancer. *Clin. Cancer Res.* **5**, 1263–1273.

304. Harari, P. M., and Huang, S. (2001). Head and neck cancer as a clinical model for molecular targeting of therapy: Combining EGFR blockade with radiation. *Int. J. Radiat. Oncol. Biol. Phys.* **49**, 427–433.

Index

A

ACC, *see* Adenoid cystic carcinoma
Acoustic neuromas, 345–346
Active immunotherapy, 576–577
Actuarial local control rate, for ACC, 20
Actuarial survival, 435–436
ADAMs proteases, related to MMPs, 138–139
Adenoid cystic carcinoma
  evaluation, 17
  histological patterns, 343
  incidence, 15
  natural history, 15–16
  pathology, 16–17
  radiation therapy, 18
  treatment, 17–20
Adenovirus vectors, 558
Adhesion receptors
  cell–matrix and cell–cell, 143
  in oral cancer invasion, 65–74
Adjuvant chemotherapy, 13
  nasopharyngeal carcinoma, 499–500
Adoptively transferred cells, 580–581
Advanced unresectable cancer
  altered fractionation radiation, 462–463
  alternating chemotherapy and radiation, 466
  chemoradiotherapeutic toxicity modification, 470
  clinical trial results, 461–462
  concurrent chemotherapy and radiation, 466–469
  gene therapy, 469–470
  induction chemotherapy, 463–466
Aerodigestive tract, upper
  biochemoprevention, 281
  chemoprevention trials, 276–281
  epidemiology, 272
  field cancerization, 275
  molecular markers of dysplasia, 44–45
  multistep carcinogenesis, 275–276
  natural history, 272–273
  normal mucosa: HPV in, 155–156
  pathological features
    histological classification, 40–44
    types, 37–40
  risk factors, 273–275
Age, micronutrients in relation to, 202
AGES staging system, thyroid carcinomas, 408
AIDS, increased risk of HNSCC with, 46
Air embolism, 355
Alcohol
  correlate of smoking behavior, 188–189
  effect on DNA damage repair, 232
  in epidemiology of head and neck cancer, 213–214
  and nutrient levels, 203
  plus tobacco, synergistic effect for HNSCC, 81
  p53 mutation and, 124–125
  risk factor for
    oral cancer, 59, 160–161
    UADT, 273–274
  role in field cancerization, 229
Algorithms, diagnostic: thyroid nodule, 411–412
Allelic imbalance, 120–121
Allovectin, gene therapy, 563
Alternative reading frame, growth control by, 107
AMES staging system, thyroid carcinomas, 407–408
Amifostine, radiation protective effect, 457–458
Anaplastic thyroid carcinoma
  clinical presentation, 417–418
  pathology, 418
Anatomy
  cancer genome anatomy project, 514–518
  nasopharynx, 492
  normal, oral cavity, 35–37
  surgical, thyroid gland, 419–421
Anesthesia, skull base surgery, 341–342
Aneuploidy
  cancer associated with, 118
  DNA, 319
  HNSCC, 169
Angiofibroma, juvenile nasopharyngeal, 344
Angiogenesis
  capillary network remodeling, 82
  in HNSCC, 83–92
  markers, 545
  tumor, mechanisms, 83
Angiogenic switch, 83
Animal models
  head and neck carcinogenesis, 230–231
  and human-based discoveries, 61
  study of oral cancer, 58–59
  using domesticated animals, 6
Anterolateral cranial base, surgery, 349–350
Antibodies
  bispecific, arming T cells with, 573–574
  circulating tumor-related, 310–311
  labeled, 578
Antimetabolites
  5-fluorouracil analogs, 537
  gemcitabine, 537–538
Antioxidants
  effect on gastrointestinal toxicities, 205
  micronutrients as, 59
Antitubulin agents
  docetaxel, 536
  paclitaxel, 536–537
  taxanes, 535–536
APC tumor suppressor protein, association with β-catenin, 71–72
Apoptotic tumor cells, as vaccine, 579
Array technologies, 513–514
Artifacts, edge, in PET, 25
Arytenoid cartilage, resection including, 362, 366
Attenuation map, 25

B

Basement membranes
  degradation, 141–142
  dysplasia-associated changes, 45
Basic fibroblast growth factor, in HNSCC, 86–88
Bcl-2, expression in carcinomas, 119
Beckwith–Wiedemann syndrome, 106
Biochemoprevention, UADT, 281
Biology, HNSCC, 81–82
Biomarkers
  in biopsy material, 311–312
  continuous
    cut points for, 290–292
    distribution, 290
  EGFR, 318–319
  integrated translational studies, 301–302
  predictors for cancer, 296–300
  qualitative measures, 288–289
  standard statistical methods of analysis, 289
  surrogate end point, 292–294
Biopsy
  directed, of potential primary sites, 10–11
  excisional, minor salivary gland tumor, 17
  material, biomarkers in, 311–312
  nasopharyngeal, LOH in, 251
  T_1 glottic cancer, 361

Bleomycin
assay, 217–218
chemoprevention trials for OPLs, 266
sensitivity to, 201–202
Blood vessels, formation in HNSCC, 83–84
Bone invasion, RADPLAT efficacy, 398
Bone morphogenetic protein-1, expression in SCC, 68
Bowman–Birk inhibitor concentrate, 266
Brachytherapy, 331

C

Cadherins
cross talk with integrin, 72–74
expression regulation, 74
and intercellular adhesion, 71–74
regulatory role in invasion, 72
Cancer, *see also* Field cancerization; *specific types*
advanced and unresectable, *see* Advanced unresectable cancer
associated with aneuploidy, 118
chemoprevention, 276
as clonal genetic disease, 524
genetics, 556
growth, patterns, 524, 527
molecular diagnosis, 125
second field, 236
Cancer development
genetic alteration in, 305–306
predicting, 295–301
Cancer genome anatomy project, 514–518
Cancer patients
adverse effects of smoking for, 186–187
carotid artery involvement
clinical outcome, 330–331
diagnosis, 329–330
postoperative complications, 335
preoperative evaluation, 332–333
surgical procedure, 333–335
treatment options, 331–332
classification of cancer risk, 300–301
histology and prior cancer history, 296
performance status scale, 437–438
quitting smoking
demographics, 188–189
psychological factors, 189–190
RADPLAT results, 395–398
recurrence of SCC, prediction, 250
skull base surgery
evaluation for, 340–341
physiological monitoring, 342
smoking behavior: future research, 195–196
smoking cessation
approaches, 192–193
intervention studies, 193–194
recommendations, 194–195
treatment program, 185–186
smoking rates among, 187–188
testing of immune competence in, 576
zinc deficiency, 207
Cancer research, translational or correlative, 6
Capecitabine, 537
Capillary growth, in angiogenesis, 82
Carcinoembryonic antigen, 572
Carcinogenesis
extracellular proteolysis in, 143
head and neck, animal models, 230–231
molecular targets for proteases in, 140–143
multistep, UADT, 275–276
tumor-associated protease systems, 137–140
Carcinogens
epithelium exposed to, 231
xenobiotic metabolism, 215–216
Carcinoma *in situ,* 39–44, 384
β-Carotene
chemoprevention trials, 264–265
plus retinol, 278–279
smoking effects, 203
Carotid artery involvement
in advanced cancer, management, 329
assessment prior to skull base surgery, 340–341
clinical outcome, 330–331
diagnosis, 329–330
nonsurgical treatment options, 331–332
postoperative complications, 335
preoperative evaluation, 332–333
surgical procedure, 333–335
surgical treatment options, 332
Cartilage invasion, RADPLAT efficacy, 398
β-Catenin, association with APC, 71–72
Cathepsins, altered expression, 139
Cdc42, activation by integrin, 70
CDK, *see* Cyclin-dependent kinase
cDNA libraries, LCM-derived, 515–517
Cell cycle genes, polymorphisms in, 220–221
Cell cycle progression
CDK activity regulation, 101–103
function of CDK complexes, 103–104
Cell cycle regulation
CDK role, 119–120
modulators, in treatment of HNSCC, 540–545
therapy and, 109
Cell lines, human cancer-derived, 5–6
Cell signaling, H-*ras,* 118–119
Cell-substrate adhesion
collagen receptors, 66–67
fibronectin/tenascin receptors, 67–68
laminin receptors, 68–70
Cerebrospinal fluid, leak, 355–356
Cervical adenopathy, clinical presentation, 9–10
Cetuximab (C225), 538–539
Chemoprevention
assessment of response to, 237–238
cancer, 276
Chemoprevention trials, 262–263
bleomycin, 266
β-carotene, 264–265
protease inhibitors, 266
retinoids, 263–264
selenium and vitamin E, 265
tea compounds, 265–266
UADT
oral premalignancy, 276–279
second primary tumors, 279–281
Chemoradiation
benefit for oropharyngeal cancer, 479–480
concurrent, HNSCC, 456–457
Chemotherapy, *see also* Induction chemotherapy
ACC, 19
adjuvant, 13
nasopharyngeal carcinoma, 499–500
alternation with radiation therapy, 454, 466, 484
antitumor effects, nutrients and, 205–206
clinical response to positive margins, 531
metabolic changes in tumor during, 30–31
multiagent, fractionation radiotherapy with or without, 481
nasopharyngeal carcinomas, 496–498
nutritional consequences, 203–204
outcome differences, 5
platinum-based, 332
plus EBRT, 427
response to, and DNA content, 178
role in preservation of oropharynx, 478–489
single-agent, fractionation radiotherapy with or without, 480–481
split-course concurrent, 484
toxicity, nutrient supplementation effect, 206–208
Chemotherapy plus irradiation
advanced unresectable cancer, 466–469
goals of, 447
intraarterial, *see* RADPLAT program
laryngeal cancers, 476–477
meta-analyses, 454–456
randomized trials, 449–451
T_2 glottic cancer, 364–365
Chk1, activation, 103
Chondrosarcoma, skull base, 344
Chordoma, skull base, 343
Chorioallantoic membrane, vessel pattern, 83–84
Chromosome 3, 3p21, allelic deletion, 121–122
Chromosome 6, LOH, 122
Chromosome 9, TSGs, 122–123
Chromosome 11, deletions at 11q13, 123
Chromosome 17, p53 TSG, 123–125
Chromosome 18, LOH at 18q21, 125
Chromosome changes, in carcinogen-exposed field, 232
Chromosome instability
in head and neck epithelium, 233–234
random, 235
Chromosome polysomy
as biomarker for predicting cancer, 296
CP.p53.LOH, 298–300
leukoplakia and, 233–234
Cip/Kip family
CKIs, 104–106
sequestration by CDK–cyclin D complexes, 103–104
Cisplatin
delivery technique in RADPLAT, 394–395
intraarterial infusions, 400–402
RADPLAT results, 395–398
Cisplatin plus 5-fluorouracil, 332, 364, 464, 476–478
CKIs, *see* Cyclin-dependent kinase inhibitors

Clinical outcome, carotid artery involvement, 330–331
Clinical patterns
correlative parameters and outcome, 5
histopathology differences, 4
TNM differences, 3–4
treatment outcome differences, 4–5
Clinical trials
flavopiridol, 542–543
immunotherapeutic, 578–580
modulators of cyclin-dependent kinases in, 541–545
oropharyngeal cancer, chemoradiation benefit, 479–480
randomized
chemotherapy plus irradiation, 449–451
nasopharyngeal cancer, 500–501
UCN-01, 544–545
Clinic-based interventions, smoking cessation, 192
Clonal outgrowths
and cancer risk, 237
during field cancerization, 234–235
molecular genetic evidence, 235–236
c-*myc,* 119
Coagulation cascade, 139
Collagen receptors, expression during SCC formation, 66–67
Comparative genome hybridization, 121
Complications
postoperative, in carotid artery involvement, 335
skull base surgery, 355–356
thyroid gland surgery, 425–426
Computed tomography
carotid artery involvement, 330
detection of recurrent disease, 31–32
evaluation of ACC, 17
imaging of larynx, 24
information from, 11
limitations, 23
thyroid nodule, 410
Concurrent chemoradiation, HNSCC, 456–457
Concurrent chemoradiotherapy
advanced unresectable cancer, 466–469
nasopharyngeal carcinomas, 499–500
Conformal radiotherapy, 457
Corrective gene therapy, 557
Cost of care, 441
Counseling, for smoking cessation, 195
Cox proportional hazards model, 295–296
Cranial nerve deficits, 356
Cribriform pattern, ACC, 17
Cricoarytenoid unit, preservation, 380–381
Cricohyoidoepiglottopexy, 377, 385
Cross talk, integrin–cadherin, 72–74
CT, *see* Computed tomography
Cut points, for continuous biomarkers, 290–293
Cyclin D1
linked to ED-L2 promoter, 60
overexpression, 109
Cyclin-dependent kinase
activity, regulation, 101–103
altered expression in tumor cells, 123
complex with cyclin D, 103–104
G1, 108–109
hyperactivation, 540
modulators, in clinical trials, 541–545
role in cell cycle regulation, 119–120
Cyclin-dependent kinase inhibitors
Cip/Kip family, 104–106
Ink4 family, 106–108
$p12^{DOC-1}$, 108
Cyst, thyroid, management, 411
Cytokeratin 5 promoter, 60–61
Cytokines
as immune adjuvants, 580
proinflammatory and proangiogenic, 90
secreted by HNSCC, 575
Cytometry, determination of DNA ploidy, 168–169
Cytotoxic T lymphocytes, antigen-specific, 572

D

Dark adaptation, abnormal, 205
DCC, putative TSG, 125
Degradation, tissue barriers, 141–142
Demographic correlates, smoking behavior, 188–189
Dendritic cells
peptide-pulsed, 572
as vaccine, 579
Design considerations, biomarker-integrated translational studies, 301–302
Determinate survival, 434–435
Diagnosis, carotid artery involvement, 329–330
Diagnostic algorithms, thyroid nodule, 411–412
Dietary factors, nasopharyngeal cancer, 492
7,12-Dimethylbenz*[a]*anthracene, powder, 58
DNA
plasmid, as nonviral vector, 558–559
ploidy, cytometric determination, 168–169
tumor, detection, 310
DNA adducts
^{32}P-postlabeling assay, 218
in tobacco-exposed head and neck tissue, 231
DNA content
ACC, 17
as correlative parameter, 5
HNSCC, 169–176
multiparameter studies, 177–178
predictor of nodal metastases, 319
salivary gland tumors, 176–177
thyroid carcinomas, 176
tissue preparation and source, 168
DNA damage, resulting from carcinogen exposure, 231–232
DNA repair
genes, polymorphisms in, 219–220
gene transcript levels, 218–219
phenotype, and cancer risk, 216–219
DNA replication, inhibition by p21, 105
Docetaxel, 536
Donor lymphocyte infusions, 572–573
Dopamine, synaptic, 215
Dopamine receptor genes, 191
Dose response, nasopharyngeal carcinomas, 494–495
Drug delivery, in RADPLAT, 394–395
Drug sensitization approach, gene therapy, 557
Dry mouth, consequence of smoking, 186–187
Dysplasia
grading, 38–39
molecular markers, 44–45

E

E1A, gene therapy with, 563–564
E-cadherin, mutational inactivation, 74
ECM, *see* Extracellular matrix
ED-L2 promoter, 60–61
Endoscopy
carcinomas arising at glottic and supraglottic level, 384–385
glottic and supraglottic, 378
pretreatment, 380
Eniluracil, 537
Epidemiology
head and neck cancer, 213–215
HNSCC, 81–82
UADT, 272
Epidermal growth factor receptor
antisense, liposome-mediated, 564
biomarker, 318–319
cell signaling, 119
modulation, in HNSCC therapy, 538–540
Epidermoid cyst, skull base, 344–345
Epithelial–mesenchymal transition, 73–74
Epithelium
carcinogen-exposed, histologic changes, 231
head and neck, chromosome instability, 233–234
lining of larynx, 36–37
Epothilones, 537
E6 protein, high-risk HPVs, 153–154
E7 protein
high-risk HPVs, 154
interaction with pRb, 153
Epstein–Barr virus
antigen expression, 45
detection in FNA specimen, 10
DNA detection, 309
in nasopharyngeal cancer, 491–492
as promoter, 59–60
risk factor for UADT, 274–275
Erythroplakia, 46
Esthesioneuroblastoma, 343
Etiology
oral leukoplakia, 46
thyroid carcinomas, 406–407
European organization for research and treatment of cancer, 439–440
Evaluation
ACC, 17
cancer patients, for skull base surgery, 340–341
of FDG PET therapy, 28–30
preoperative, carotid artery involvement, 332–333
pretreatment, nasopharyngeal carcinomas, 493

Exfoliated cell samples
detection of HNSCC, 306–309
for OPLs, 253–255
Experimental research, history of, 57–59
Extent of surgery
neck, 11–12
thyroid gland, 423–425
External beam radiotherapy, 427
Extracapsular extension, 11, 13
Extracellular matrix
bFGF solubilization from, 86–87
integrin interaction with, 66
ligation by integrin, 70–71
modification, 143

F

Facial translocation, 352–353
Fas-ligand, tumor-expressing, 575
FDG PET
accurate staging, 27–28
capitalizing on increased anaerobic metabolism, 24–25
in search for primary site, 11
therapy, evaluation of, 28–30
Femoral artery grafts, 334–335
FHIT gene, 121
Fibroblast growth factors, in HNSCC, 86
Fibronectin receptors
α5β1, 67
αvβ5, 68
Fibrous dysplasia, 344
Field cancerization
associated phenotypic changes, 236
clinical evidence, 229–230
clinical implications, 236–238
genetic evidence, 231–236
histopathologic evidence, 230–231
and margin status, 527–528
process, 228
UADT, 275
Fine needle aspiration
neck mass, 10
salivary gland tumor, 17
thyroid nodule, 411
Flavopiridol
antiproliferative effect, 541
clinical trials, 542–543
Fluoro-2-deoxy-D-glucose, *see* FDG
5-Fluorouracil
analogs, 537
plus cisplatin, 332, 364, 464, 476–478
FNA, *see* Fine needle aspiration
Focal adhesion kinase, overexpression, 70
Follicular thyroid carcinoma
clinical presentation, 414–415
management and prognosis, 415–416
pathology, 415
Follow-up management, thyroid carcinomas, 427
Fractional alleles loss, 120–121
Fractionation radiotherapy schedules, 331
accelerated, 447
altered, 462–463
concurrent chemotherapy, 451–454
with or without chemotherapy, 484, 488
hyperfractionation, 445–446
with or without single-agent chemotherapy, 480–481
T_2 glottic cancer, 364–365
Fragile histidine triad gene, 45
Functional assessment of cancer therapy scale, 439
Functional status in head and neck cancer, self-report, 437–438

G

Gastrointestinal toxicity, antioxidant effect, 205
Gemcitabine, 537–538
Gene arrays, 514
Gene therapy
advanced unresectable cancer, 469–470
augmentation of cytotoxic therapy, 557
background, 556–557
combination with standard therapy, 565
corrective, 557
genes used in combination therapy, 557–558
for HNSCC, 545–546
local-based gene delivery, 559
regional gene transfer, 560
trials, in HNSCC, 560–565
vectors, 558–559
new strategies, 565–566
targeting, 566
Genetic alterations, in HNSCC, 510
Genetic mapping, 234–235
Genetic models
with avian retroviruses, 60
cytokeratin 5 promoter, 60–61
gene targeting with promoters, 59–60
targeted ablation of genes in oral cavity, 61
Genetics
cancer, 556
nasopharyngeal cancer, 492
risk factor for UADT, 275
Genetic susceptibility, assessment, 215
Gene transfer
IFN-α, 564–565
regional, 560
Genomic structure, HPV, 152–153
Glottic cancer
organ preservation spectrum, 381–382
T_1
minimally invasive, 360–361
treatment, 361–363
T_2, treatment, 363–366
T_3 and T_4, treatment, 366–369
Glutathione, protective effect, 206
Goiter, intrathoracic, 423
Grading
dysplasia, 38–39
malignancy, 319
Grafts
femoral artery, 334
saphenous vein, 335
Granulocyte–colony-stimulating factor, 458
Granulocyte–macrophage colony-stimulating factor, 565
Grouped data, and biomarker cut points, 290–291
G1/S deregulation, 108–109
GSK3-β, 72
GTPases, Rho family, 70–71

H

Hairy leukoplakia, 46
Hamster buccal pouch model
head and neck carcinogenesis, 230
oral cancer, 58–59
Health status, measurements, 436–438
Hemidesmosomes, formation, α6β4 role, 69–70
Hepsin, overexpression, 140
HER-2/neu, 119
Herpes simplex virus-1, risk factor for UADT, 275
Herpes simplex virus–thymidine kinase, 564
Histological classification
ACC, 16–17
lesions presenting as lichen planus, 47–48
preneoplastic lesions of UADT, 40–44
Historical perspective
immunotherapy, 571–572
in HNSCC, 577–578
outcome assessment, 433–434
HNSCC
aneuploidy, 169
angiogenesis in, clinical evidence, 83–86
antiangiogenic therapies, 91–92
at-risk individuals, detection, 125
bFGF, 86–88
cell signaling in, 118–119
circulating tumor markers, detection, 310–311
concurrent chemoradiation, 456–457
cytokines, 90
detection in exfoliated cell samples, 306–309
DNA content parameters, 169–176
epidemiology, 81–82
genetic alterations, 510
histologic progression, 231
HPV detection in, 159
immunology, 574–576
molecular and genetic changes in, 214–215
molecular profiling techniques, 510–514
nitric oxide synthase, 91
novel targets and agents for treatment, 535–547
PD-ECGF, 90
point mutations, 306
prognostic factors, 125–126
progression, extracellular proteolysis in, 143
risk of, and HPV infection, 160–161
thrombospondin and p53, 90
TSGs, 120–125
vascular endothelial growth factor, 87–90
hOGG1 gene, 220
HPV, *see* Human papillomavirus
H-*ras*, mutations, 118–119
hst-1, in HNSCC, 119

hst-1 (FGF4), in HNSCC, 82
Human leukocyte antigen-B7, 580
Human papillomavirus
  and benign lesions of head and neck, 156–157
  detection in HNSCC, 159
  detection methods, 155
  high-risk
    E6 protein, 153–154
    E7 protein, 154
  life cycle, 152
  in normal mucosa of UADT, 155–156
  and premalignant lesions of head and neck, 157–158
  risk factor for UADT, 274
  and risk of HNSCC, 160–161
  survival of damaged cells enhanced by, 234
  tumors positive for, 161
  virion and genomic structure, 152–153
Hürthle cell tumor, thyroid, 416
Hybridization
  detection of HPVs, 155–161
  in situ
    evidence for clonal outgrowths, 234–235
    reversal of leukoplakia as endpoint, 233
Hyperfractionation, 445–446
Hypermethylation, promoter, 309
Hypocalcemia, transient, 426
Hypopharyngeal cancer, larynx preservation, 478
Hypoxia, modulation by tirapazamine, 545
Hypoxia-inducible factor-1, 89

I

Immune system
  responses to tumor, 570–571
  role in control of tumor development, 58–59
  suppressed function, clinical implications, 574–575
Immunohistochemistry
  detection of
    HPVs, 155–158
    p53 overexpression, 529–530
  staining for markers, 288
Immunosuppression
  causes, 575–576
  hierarchy of, 574
Immunotherapy
  augmented by cytokine gene transfer, 557
  in HNSCC, 576–581
  principles of, 570–571
  strategies, 571–574
Incidence
  ACC, 15
  cervical lymph node metastases, 317–319
  oral leukoplakia, 46
Induction chemotherapy
  advanced unresectable cancer, 463–466
  followed by locoregional therapy, 447–449
  laryngeal cancers, 475–476
  nasopharyngeal carcinomas, 498–499
  organ preservation with, 376–377
  T_3 glottic cancer, 367–369
Infection
  HPV, latent and active, 152
  viral, risk factor for UADT, 274–275
Inflammatory infiltrate, 319
Ink4 family of CKIs, p16, 106–108
Insular thyroid carcinoma, 418
Insulin-like growth factor, release, 143
Int-2, in HNSCC, 119
Integrin
  cross talk with cadherin, 72–74
  interaction with ECM, 66
  ligation of ECM by, 70–71
Intercellular adhesion, and cadherins, 71–74
Interferon-α
  antiangiogenic effect in HNSCC, 91–92
  biochemoprevention, 281
  gene transfer, 564–565
  low-dose, 577
  plus 13-*cis*-retinoic acid, 578
Interferon regulatory factor-3, HPV-16 E6 binding, 154
Interleukin-2, gene therapy, 564
Interleukin-12
  antitumor immunity induced by, 546–547
  gene therapy, 564
Interleukin-4 receptor, blood levels, 575–576
Int-2 (FGF3), in HNSCC, 82
Intraepithelial neoplasia, 42–43
Intraoperative radiation therapy, 331
Intratumoral HNSCC trial
  phase I, 560–561
  phase II, 562
Invasion
  bone and cartilage, 398
  cadherin regulatory role in, 72
  temporal bone, 397
  tumor cell, PI3K in, 71
  vascular, lymph node metastases, 319
^{123}I scanning, thyroid nodule, 410

J

Jugulotympanic paragangliomas, 345
Juvenile nasopharyngeal angiofibroma, 344

K

Kaplan–Meier survival, 436
Karnofsky performance status scale, 437
Keratinizing dysplasia
  high-grade, 39
  underreporting of, 42
Keratinocyte growth factor, 458
Keratinocytes, adhesion to laminin-5, 69
Keratins, markers of dysplasia, 44
Knockout mouse, p57, 106
K5 promoter, 60–61

L

Laboratory studies, thyroid nodule, 409–410
Laminin receptors, expression in SCC, 68–70
Laryngectomy
  partial, *see* Partial laryngectomy
  supraglottic, 371
  total, 366–369
Larynx
  cancer, 318
    spectrum concept, 381–382
    treatment of neck in, 321–323
  changes in cancer treatment paradigms, 359–360
  cricoarytenoid unit preservation, 380–381
  CT imaging, 24
  leukoplakia, 47
    malignant progression, 49–50
  normal anatomy, 36–37
  organ preservation surgery, 376–377
    and nonsurgical approaches, 377–380
  papillomatosis, and HPV, 156–157
  premalignant lesions, 158
  preservation, 475–478
Laser capture microdissection, 61, 511–513
Laser excision, transoral, 362–363, 365
Lesions
  oral premalignant, *see* Oral premalignant lesions
  premalignant, and risk of tumor development, 230
  presenting as lichen planus, 47–48
  skull base-associated
    benign, 344–346
    malignant, 343–344
  spread, assessment with scrapes, 254–255
Leukoplakia
  and chromosome polysomy, 233–234
  clinical aspects, 46–47
  definition, 45
  dysplastic, 246
  etiology, 46
  laryngeal, 47
  scrapes, 253–255
  tobacco exposure and, 214
  verrucous, 250
  vitamin A treatment, 277–278
Lichen planus, 46–48
Liposome
  mediated antisense EGFR, 564
  plasmid incorporation into, 559
Local-based gene delivery, 559
Local regional control
  ACC, 20
  advanced unresectable cancer, 470
  importance in organ preservation surgery, 383
  oropharyngeal cancer, 479–480, 488
  with surgical and nonsurgical organ preservation, 379
LOH, *see* Loss of heterozygosity
Longitudinal studies, for OPLs, 255–256
Loss of heterozygosity
  CP.p53.LOH, 298–300
  in HNSCC, 121–126
  in nasopharyngeal biopsies, 251
  pattern in OPLs, 248–250
  at 3p or 9p, 297
  preneoplastic lesions, 45
  TSGs, 120

Lung, site of distant metastasis, 16
Lung cancer, phase I clinical trials, 562–563
Lymphatic spread, patterns, 319–320
Lymph node metastases
in ACC, 15
incidence, 317–319
occult, 322, 368
from unknown primary tumors, 9
Lymph nodes
cervical, thyroid carcinoma, 422–423
considerations in staging, 25–26
neck, current management, 321–324
positive, and prognosis, 320
Lymphoma, thyroid, 418
Lysosomal cysteine proteases, 139

M

MACIS staging system, thyroid carcinomas, 408
Macrolide antibiotics, epothilones, 537
Magnetic resonance imaging
in conjunction with CT, 23
detection of recurrent disease, 31–32
for skull base surgery, 340
T1- and T2-weighted images, 24
Management
carotid artery involvement, 329
current, lymph nodes of neck, 321–324
follicular thyroid carcinoma, 415–416
OPLs
molecular markers in, 251–253
scrape assay in, 253–255
papillary thyroid carcinoma, 413–414
postsurgical, vocal cord, 426
of recurrence, 238
second primary disease, 238
thyroid carcinomas
follow-up, 427
surgical, 419–425
thyroid cyst, 411
Markers, *see also* Biomarkers
angiogenesis, 545
and DNA content, simultaneous analysis, 177–178
epithelial field carcinogenesis, 237
for HNSCC, 530
microsatellite, 120–121
molecular
dysplasia, 44–45
in management of OPLs, 251–253
viral, 309
Matrix-degrading proteases, 141–142
Matrix metalloproteinases
activation, 140–141
ADAMs and ADAMTS related to, 138–139
expression in SCC, 68–69
Medullary thyroid carcinoma
clinical presentation, 416–417
management and prognosis, 417
pathology, 417
Meningioma, skull base, 344
Meta-analyses, chemotherapy plus radiation, 454–456, 468–469
Metachronous primaries, 26–27
Metastases
cervical, incidence, 317–319
to cervical nodes, from unknown primary tumors, 9–13
considerations in staging, 26
and degradation of tissue barriers, 141–142
distant
in ACC, 15–16
in patients who recur, 31
neck node
massive, 397
prognostic implications, 320
thyroid, 418–419
Microarrays, survey of mRNA transcripts, 311
Microdissection, 510–511
laser capture, 511–513
Micronuclei
as biomarker for predicting cancer, 297
frequency, 232
Micronutrients
antioxidant, 59
cancer chemopreventive properties, 206
effects on therapy-related toxicities, 204–205
in relation to age, 202
Microsatellite analysis
identification of types 1 and 2 OPLs, 251
oral mucosal regions at risk for progression, 248–251
Microsatellite markers, 120–121
LOH, 312
Microsatellites
instability, 120
screening panels, 307–309
Microvessel density, in HNSCC, 84–85
Middle fossa approach, skull base surgery, 353–354
Migration, laminin-5-induced, 68–69
Molecular detection
biomarkers in biopsy material, 311–312
circulating tumor markers, 310–311
HNSCC, in exfoliated cell samples, 306–309
strategies for, 524
Molecular margin analysis, 528
Molecular profiling, HNSCC, 510–514
Morbidity, as tier in outcome assessment, 436
Mucositis, 187, 539
Murine models
genetic, 59–60
syngeneic animals, 5
transgenic mice, 60–61
Muscle uptake, FDG, reduction of, 25
Mutagen sensitivity, 201–202
assay, 217–218
Mutation
H-*ras,* 118–119
p53, 108, 124, 216
point, in HNSCC, 306
ret oncogene, 406

N

Nasal cavity, Schneiderian papilloma, 158
Nasopharyngeal carcinoma, 10
adjuvant chemotherapy, 499–500
anatomy, 492
biopsy: LOH in, 251
boost volume, 494
chemotherapy, 496
combined with radiotherapy, 498
for recurrent disease, 497–498
concurrent chemoradiotherapy, 499
dose response, 494–495
environmental factors, 492
Epstein–Barr virus in, 491–492
induction chemotherapy, 498–499
natural history, 493
pathology, 492–493
pretreatment evaluation, 493
radiation therapy, 494
randomized trials, 500–501
reirradiation, 495–496
staging, 493–494
survival, 496
volume treated, 494
workup, 493
Natural history
ACC, 15–16
nasopharyngeal carcinomas, 493
UADT, 272–273
Natural killer cells, activity suppression, 576
Neck dissection
contralateral, 369
elective, 322–323
indications for, 11–12
and radiation, sequence of, 12
selective, 368
types, 324–325
Nedaplatin, 537
Neocord development, 366
Nephrotoxicity, selenium effect, 205
Neuromas, acoustic, 345–346
Neutron radiation, for ACC, 19–20
Nitric oxide synthase, in HNSCC, 91
4-Nitroquinoline-*N*-oxide, 58
4-Nitroquinoline-1-oxide, 231
N-Nitroso compounds, in smokeless tobacco, 215
Nonsquamous cell carcinomas, DNA content, 176–178
Nucleotide excision repair, host-cell reactivation assay, 217
Nutrients, *see also* Micronutrients
alcohol and, 203
and antitumor effects of radiation/ chemotherapy, 205–206
plus pharmaceuticals, 206–208
tobacco and, 202–203
Nutritional consequences, radiation and chemotherapy, 203–204
Nutritional status
and cancer treatment, 203
role in immunosuppression, 575

O

Observation, clinical: importance of, 6
Occupational exposure, association with nasopharyngeal cancer, 492

Oculoplethysmography, 332
Oligomer probing, mutant p53, 528–529
Oncogenes
  overexpression, 117
  ret, mutations, 406
OPLs, *see* Oral premalignant lesions
Oral cancer
  association with HPV, 162
  candidate biomarkers, 118
  current management, 323–324
  hamster model, 58–59
  invasion, adhesion receptors in, 65–74
  micronutrients in relation to, 202
  nodal metastases, 318–319
  prevention, prediction, 266–267
  risk factors, 261–262
Oral cavity
  benign lesions, HPV and, 156
  leukoplakia, 45–47
  normal anatomy, 35–37
  targeted ablation of genes in, 61
Oral premalignant lesions
  biomarkers as predictors for cancer, 296–300
  bleomycin effect, 266
  β-carotene for, 264–265
  chemoprevention trials, 262–263, 276–279
  definitions, 246–247
    and significance, 261–262
  exfoliated cell sampling, 253–255
  HPV and, 157–158
  LOH assessment, 248
  molecular markers, 252–253
  need for longitudinal studies, 255–256
  progression risk, 247–250
  progression to invasive SCC, 48–50
  protease inhibitor effects, 266
  retinoids for, 263–264
  role in predicting efficacy for oral cancer prevention, 266–267
  selenium and vitamin E for, 265
  staging, 251–252
  tea components for, 265–266
  treatment, 247–248
  type 1 and type 2, 251
Organ preservation
  cricoarytenoid unit, 380–381
  larynx, 475–478
  oropharynx, chemotherapy role, 478–489
  spectrum, glottic cancer, 381–382
Organ preservation surgery
  carcinomas arising at glottic level, 384–385
  carcinomas arising at supraglottic level, 386–388
  following radiation failure, 388
  larynx, 376–377
  vs. nonsurgical strategies, 377–380
  techniques, 383
Oropharyngeal tumors
  lymph node management, 321
  positive for HPV, 161
  and tobacco smoke, 214–216
Oropharynx, preservation, 478–489
OSI-774 (Tarceva), 539–540
Outcome assessment
  cost of care, 441
  health status, 436–438
  historical background, 433–434
  morbidity, 436
  quality of life, 438–441
  survival, 434
    actuarial, 435–436
    determinate, 434–435
    Kaplan–Meier, 436
Overexpression
  bFGF, 87
  cyclin D1, 109
  FAK, 70
  oncogenes, 117
  p53, 124, 529–530
  serine proteases, 140
Oxaliplatin, 537
Oxidative stress, tobacco and alcohol effects, 203

P

p12^{DOC-1}, as CKI, 108
p16
  CDKN2-encoding, 122–123
  inactivation, 106–107
  interaction with high-risk HPV E6 protein, 153–154
p16/Ink4A, loss, 109
p18^{INK4c}, 108
p21
  expression independent of p53, 104
  inhibition of DNA replication, 105
p21/WAF1, 109, 122, 202
p27, role in programmed cell death, 105–106
p44/42^{MAPK}, upregulation, 87
p53
  Ad5CMVp53, 546
  as biomarker for predicting cancer, 296–297
  CP.p53.LOH, 298–300
  expression in ACC, 17
  gene therapy with, 560–563
  interaction with high-risk HPV E6 protein, 153–154
  mutant, oligomer probing, 528–529
  mutation, 108, 216
  overexpression, 237, 529–530
  role in angiogenesis in HNSCC, 89–90
  TSG on chromosome 17, 123–125
p57, *KIP2* encoding, 106
Paclitaxel, 536–538
Panendoscopy, 10–11
Papillary thyroid carcinoma
  clinical presentation, 412–413
  management and prognosis, 413–414
  pathology, 413
Papillomas, Schneiderian, of nasal cavity, 158
Paragangliomas, 345
Paranasal sinus cancer, advanced, 397
Partial laryngectomy, 365
  supracricoid, 377, 382–383
    with cricohyoidoepiglottopexy, 385
    with cricohyoidopexy, 387–388
  supraglottic, 386–387
  vertical, 384–385
Passive immunotherapy, 577
Pathology
  ACC, 16–17
  anaplastic thyroid carcinoma, 418
  follicular thyroid carcinoma, 415
  HNSCC, 81–82
  Hürthle cell tumor, 416
  medullary thyroid carcinoma, 417
  nasopharyngeal carcinomas, 492–493
  papillary thyroid carcinoma, 413
  skull base-associated, 342–346
PD-ECGF, expression in HNSCC, 90
P15-E, detection in HNSCC, 575
Performance status scale, head and neck cancer patients, 437–438
PET
  detection of recurrent disease, 31–32
  with FDG, *see* FDG PET
  staging, 27–28
Pharmacologic interventions, smoking cessation, 193–195
Pharynx, *see also* Oropharynx
  cancer, 317
  normal anatomy, 36
Phenotypic changes, associated with field carcinogenesis, 236
Phonomicrosurgical approach, T_1 glottic cancer treatment, 360
Phosphoinositide-3 OH kinase, *see* PI3K
Photon radiation, for ACC, 19–20
Physician-based intervention, smoking cessation, 192–194
Physiological monitoring, during skull base surgery, 342
Phytochemical supplementation, 207–208
PI3K, in tumor cell invasion, 71
Plakoglobin, Armadillo repeats, 71–72
Plasminogen activation system, 137–138
Platelet-derived endothelial cell growth factor, *see* PD-ECGF
Platinum agents
  chemotherapy based on, 332
  Nedaplatin, 537
Ploidy, cytometric determination, 168–169
Polymerase chain reaction, detection of HPVs, 155–161
Polymers, gene therapy based on, 559
Polymorphisms
  in cell cycle genes, 220–221
  in DNA repair genes, 219–220
  p21/WAF1, 122
Positron emission tomography, *see* PET
PRAD-1, cell signaling, 119
Preauricular infratemporal fossa, dissection, 350–352
Premature chromosome condensation, 232
Primary carcinoma
  clonal origin, 276
  unknown, metastatic to cervical nodes, 9–13
Prognosis
  follicular thyroid carcinoma, 415–416
  medullary thyroid carcinoma, 417
  papillary carcinomas, 413–414

Prognostication, 125–126
fractional allelic loss, 126
implications of neck node metastases, 320
Prognostic factors
ACC, 16
extracapsular extension as, 13
HNSCC, 125–126
Programmed cell death, p27 role, 105–106
Progression risk, of OPLs, prediction, 247–250
Proliferative verrucous leukoplakia, 49
Prostaglandins, immunosuppression and, 575
Protease inhibitors, chemoprevention trials for OPLs, 266
Protease systems, tumor-associated, 137–140
Proteolysis, extracellular, in HNSCC, 143
Proteomics, 518–519
Psychological factors, smoking cessation, 189–191
Pterygomaxillary fissure, 349–350

Q

Quality of life, 438–441
instruments of measurement, 439–441
Quinazolines, 539–540

R

Radiation therapy
ACC, 18
advanced neck carcinoma with carotid artery involvement, 331–332
alternation with chemotherapy, 454, 466, 484
antitumor effects, nutrients and, 205–206
clinical response to positive margins, 531
combined with gemcitabine, 538
conformal, 457
failure, organ preservation surgery after, 388
following thyroid surgery, 426–427
fractionation, *see* Fractionation radiotherapy schedules
locoregional, following induction chemotherapy, 447–449
nasopharyngeal carcinomas, 494
reirradiation, 495–496
neutron, for ACC, 19–20
nutritional consequences, 203–204
optimal field of, 12
outcome differences, 4–5
plus chemotherapy
goals of, 447
meta-analyses, 454–456
randomized trials, 449–451
T_2 glottic cancer, 364–365
pretreatment, for laryngeal carcinoma, 380
primary, T_3 glottic cancer, 367–369
response to, and DNA content, 174–176
supraglottic lesions, 370–371
T_2 glottic cancer, 363–365
toxicity, nutrient supplementation effect, 206–208
vitamin E effect, 204
zinc effect, 204–205
Radionuclide scanning, thyroid nodule, 410–411
RADPLAT program
drug delivery technique, 394–395
George Washington Univ. Medical Center, 398–399
high-dose chemotherapy in, 393
multi-RADPLAT, 400–402
NEORADPLAT, 400
Netherlands Cancer Institute, 398
PENTORADPLAT protocol, 399
results, 395–398
University of Kentucky, 399
Randomized trials
chemotherapy plus irradiation, 449–451
nasopharyngeal cancer, 500–501
Reconstruction
dural, 349
following anterolateral skull base surgery, 353
in partial laryngectomy, 377
Recurrence
correlation with microvessel number, 85
after initial complete remission, 402
local, β-carotene efficacy, 280–281
management, 238
nasopharyngeal carcinomas, chemotherapy for, 497–498
in neck, 322–323
PET detection, 31–32
SCC, prediction in oral cancer patients, 250
from second primary tumors, 126
tumor at distant sites, 396–397
Recurrent laryngeal nerve, 419–421, 423–426
Recursive partitioning, for classification of cancer risk, 300–301
Regional gene transfer, 560
Relapse, smoking cessation, 195
Resection
carotid artery, 332–335
craniofacial, 348
extended, thyroid carcinoma, 423
false vocal cord, 365–366
further: clinical response to positive margins, 530–531
including arytenoid cartilage, 362
normal tissue, to maintain postoperative function, 382–383
transoral laser, supraglottic cancer, 370–371
Retinoblastoma protein
inactivation, 540
interaction with HPV-16 E7, 153
phosphorylation, 122
Retinoic acid, effect on HNSCC lesions, 91–92
13-*cis*-Retinoic acid, 267, 278–280, 296
Retinoic acid receptor-β, expression, 293, 297
Retinoids, chemoprevention trials for OPLs, 263–264
Retrolabyrinthine approach, skull base surgery, 355
Retrosigmoid (suboccipital) approach, skull base surgery, 354
Retroviral vectors, 558
Risk estimation, head and neck tumor, 237
Risk factors
HNSCC, 160–161
oral cancer, studies with hamster buccal pouch, 58–59
thyroid carcinomas, 406–407
UADT
alcohol and tobacco, 273–274
genetics, 275
viral infection, 274–275

S

Safety, gene therapy, 565
Saliva, HNSCC detection in, 309
Salivary glands
ACC in, 15–20
tumors, and DNA content, 178–179
Sampling error, 526
Saphenous vein grafts, 335
SCC, *see* Squamous cell carcinoma
Schwannomas, facial and lower cranial nerve, 346
Scientific method, 6
Scrape assay, in management of OPLs, 253–255
Screening panels, microsatellite, 307–309
Second primary disease
considerations in staging, 26–27
management, 238
Second primary tumors
development, 229–230
recurrences from, 126
UADT, chemoprevention trials, 279–281
Selenium
chemoprevention trials for OPLs, 265
effect on nephrotoxicity, 205
supplementation during therapy, 206–207
Self-help, approach to smoking cessation, 192
Senescence, cellular, p16 role, 107
Sensitivity analysis, with biomarkers, 300
Serine proteases, tumor-associated, 139–140
Sexual history, and risk of oral SCC, 160–161
Simple hyperplasia, 37
Single-covariate Cox models, 295–296
Sites of occurrence, oral leukoplakia, 46–47
Skull base surgery
anesthetic considerations, 341–342
anterior skull base, 346–349
anterolateral skull base, 349–353
benign lesions, 344–346
complications, 355–356
malignant lesions, 343–344
middle and posterior skull base, 353–355
pathology, 342–346
patient evaluation for, 340–341
physiological monitoring, 342
technological advancements, 339–340
SLC6A3, genetic variation in, 191
Smoking
association with nasopharyngeal cancer, 492
behavior, genes and, 191–192
among cancer patients
adverse effects for, 186–187
rates, 187–188
demographic and medical correlates, 188–189

Smoking cessation
approaches to, 192–193
intervention studies with cancer patients, 193–194
psychological factors, 189–191
recommendations for cancer patients, 194–195
Solid pattern, ACC, 17
Spectroscopic analysis, in cancer detection, 311
sprr3 promoter, 61
Squamous cell carcinoma
collagen receptor expression, 66–67
ECM molecules in, 66
extension into skull base, 343
fibronectin/tenascin receptor expression, 67–68
head and neck, *see* HNSCC
laminin receptor expression, 68–70
oral
angiogenesis, 84
progression, 245
progression of OPLs to, 48–50
recurrence, prediction in oral cancer patients, 250
thyroid, 419
Squamous metaplasia, 37
Staging
clinical considerations, 25–27
in laryngeal carcinoma, 380
nasopharyngeal carcinomas, 493–494
OPLs, 251–252
PET, 27–28
thyroid carcinomas, 407–408
Standardized uptake ratio, 25, 28, 30
Statistical methods, analysis of biomarkers, 289
Stroke, associated with carotid artery surgery, 332–333
Subfrontal approach, skull base surgery, 347–348
Superior laryngeal nerve, 421–422, 424–425
Suppressed immune function, clinical implications, 574–575
Supraglottic cancer, treatment, 369–371
Surgery
ACC, 17–18
Ad-p53 gene therapy, 561–562
for advanced unresectable cancer, 470
carotid artery resection, 332–335
microscopically negative surgical margins, 4
neck, extent of, 11–12
organ preservation
carcinomas arising at glottic level, 384–385
carcinomas arising at supraglottic level, 386–388
following radiation failure, 388
larynx, 376–377
vs. nonsurgical approaches, 377–380
techniques, 383
outcome, and DNA content, 169, 174
palpable cervical lymph node metastases, 324
skull base, *see* Skull base surgery
thyroid carcinoma, 421–425
and timing of immunotherapy, 581
Surgical margins
assessment, definitions, 525–526
clinical utility, 524–525
current utilization of, 523–524
failure to predict outcome, 525–527
generalized markers for HNSCC, 530
mutant p53 oligomer probing, 528–529
overexpression of p53, detection, 529–530
patterns of cancer growth, 524
positive, clinical response to, 530–531
status, reflecting biologic aggressiveness, 527–528
Surrogate end point biomarkers, 292–294
Surrogates, for human cancer, 5–6
Survival
actuarial, 435–436
benefits of radiation treatment, 13
as critical outcome, 434
determinate, 434–435
following organ preservation surgery, 379
Kaplan–Meier, 436
nasopharyngeal carcinomas, 496
Swallowing, functional outcome swallowing scale, 438
Syderopenic dysphagia, 46
Synchronous lesions, 26–27
Systemic gene delivery, 560

T

Taxanes, 535–536
T cells
activated, 573
anti-CD3/anti-CD28 coactivated, 573
arming with bispecific antibodies, 573–574
with chimeric receptors, 574
tumor-specific, survival, 580–581
^{99m}Tc scanning, thyroid nodule, 410
Tea, chemoprevention trials for OPLs, 265–266
Temporal bone carcinoma, 397
Tenascin-C, expression in SCC, 67
Testisin, 140
Therapy
antiangiogenic, in HNSCC, 91–92
cell cycle regulation and, 109
complementary, 208
evaluation by FDG PET, 28–30
immunological, HNSCC, 546–547
Thrombospondin, in HNSCC, 90
Thyroid carcinomas
anaplastic, 417–418
and DNA content, 178
EBRT, 427
follicular, 414–416
follow-up monitoring, 427
Hürthle cell tumor, 416
insular, 418
lymphoma, 418
medullary, 416–417
metastatic, 418–419
molecular basis, 406
papillary, 412–414
radiation treatment, 426–427
risk factors and etiology, 406–407
squamous cell, 419
surgical complications, 425–426
surgical management and technique, 419–425
TSH replacement, 426
Thyroid hormone replacement, 426
Thyroid nodule
clinical assessment, 408–409
diagnostic studies, 409–412
Tirapazamine, 545
Tissue inhibitors of metalloproteinases, 138, 143
TNM stages
differences, 3–4
thyroid carcinomas, 407
TNP-470, effect on HNSCC progression, 92
Tobacco, *see also* Smoking
in epidemiology of head and neck cancer, 213–215
and nutrient levels, 202–203
plus alcohol, synergistic effect for HNSCC, 81
p53 mutation and, 124–125
related cancer, susceptibility to, 217
risk factor for
oral cancer, 59, 160–161, 185
UADT, 273–274
role in field cancerization, 229
smokeless products, 214–216
Tobacco smoke
DNA adducts caused by, 231
inhalation experiments, 58
interaction with β-carotene, 203
relationship to laryngeal dysplasia, 47
α-Tocopherol
effect on nephrotoxicity, 205
plus β-carotene, cancer prevention study, 278–279
Tonsil, biopsy, 11
Tonsillar carcinoma, HPV detection in, 159
Toxicity
drug–radiation interaction, 458
gemcitabine, 538
treatment-induced, reduction, 457–458
Transethmoidal approach, skull base surgery, 346
Transforming growth factor-β
ligand for αvβ5, 68
proteolytic release, 143
Translabyrinthine approach, skull base surgery, 354
Transoral laser excision
stage I glottic cancer, 362–363
T_2 glottic cancer, 365
Treatment
adenoid cystic carcinoma, 17–20
cancer, and nutrients, 203
carcinoma *in situ* and minimally invasive T_1 glottic cancer, 360–361
carotid artery involvement, 331–332
cytotoxic regimens, 167
HNSCC, modulators of cell cycle control for, 540–545
leukoplakia, with vitamin A, 277–278
OPLs, 247–248
molecular marker role, 252–253
outcome differences, 4–5

Treatment (*continued*)
radiation, survival benefits, 13
smoking cessation as, 195–196
supraglottic cancer, 369–371
T_1 glottic cancer, 361–363
T_2 glottic cancer, 363–366
toxicity induced by, reduction, 457–458
TSGs, *see* Tumor suppressor genes
Tubular pattern, ACC, 17
Tumor–host interaction, at margin, 528
Tumor immunology, 556–557
Tumor-infiltrating lymphocytes, 571–572, 575–576
Tumor markers, circulating: detection, 310–311
Tumor node metastasis, *see* TNM stages
Tumor progression
extracellular proteolysis in, 143
regulatory role of proteases, 142
Tumor suppressor genes
chromosomes 7 and 8, 122
LOH, 120
p53, 123–125
at 3p21, 121–122
at 13q14, 123
Tumor suppressors, p16 as, 107
Tumor thickness, 319

U

UADT, *see* Aerodigestive tract, upper
UCN-01
clinical trials, 544–545
pharmacological features, 543–544
Ultrasonographic imaging, thyroid nodule, 410
Urokinase plasminogen activator, 137–138, 140

V

Vaccines, new strategies for, 579–580
Validation, animal models, 6
Vascular endothelial growth factor
flavopiridol effect, 542
in HNSCC, 88–90
marker of angiogenesis, 545
mRNA, 541
Vectors
new strategies, 565–566
nonviral, 558–559
targeting, 566
viral, 558
Viral exposure, role in field cancerization, 229
Viral infection, risk factor for UADT, 274–275
Viral markers, 309
Vitamin A, treatment for oral leukoplakia, 277–278
Vitamin E
antineoplastic activity, 205
chemoprevention trials for OPLs, 265
effect on therapy-related toxicities, 204
Vocal cord
false, resection, 365–366
mobility, in laryngeal carcinoma, 380
postsurgical management, 426
Voice quality
assessment, 438
following surgery and radiation, 378

W

Workup, for nasopharyngeal carcinomas, 493

X

Xenobiotic metabolism, carcinogens, 215–216
Xenon, inhaled, 332
XPD, polymorphisms, 219

Z

ZD1839 (Iressa), 539
Zinc
deficiency, in cancer patients, 207
diltiazem plus, 206
effect on therapy-related toxicities, 204–205
Zygomatic arch, transection, 351–352
Zymogen, activation, 140–141

ISBN 0-12-239990-0

90090

9 780122 399909